THE BIRDS OF
PAKISTAN

Plate 1

MAP SHOWING ADMINISTRATIVE DISTRICTS
REFERRED TO IN THE TEXT

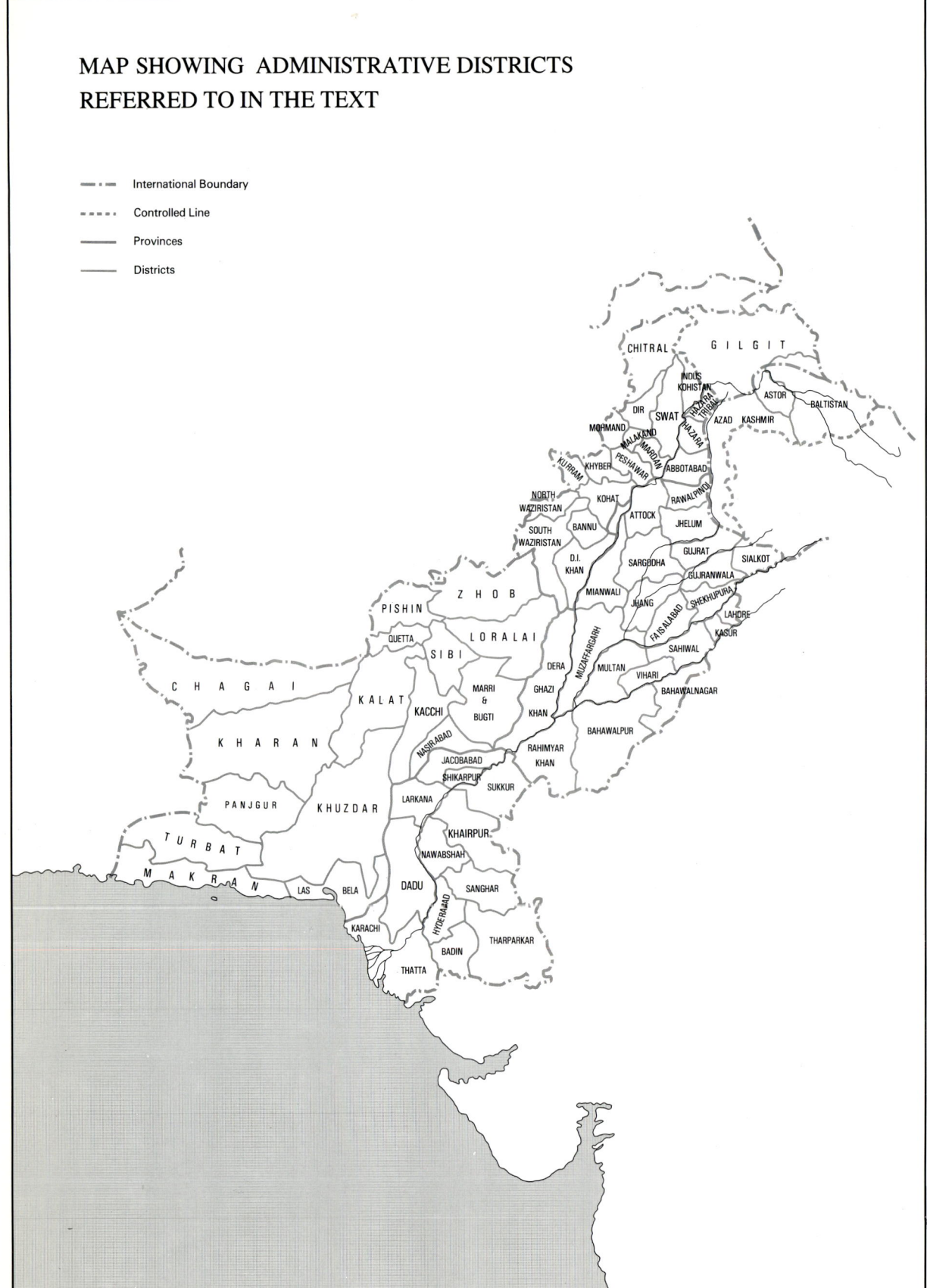

THE BIRDS OF PAKISTAN

(In 2 Volumes)

Volume 1

Regional studies and non-passeriformes

T. J. ROBERTS

M.A.(CANTAB.), M.S.A.(BRIT. COL.)

Karachi

Oxford University Press

Oxford   New York   Delhi

1991

Oxford University Press

OXFORD NEW YORK
TORONTO MELBOURNE AUCKLAND
PETALING JAYA SINGAPORE HONG KONG TOKYO
DELHI BOMBAY CALCUTTA MADRAS KARACHI
NAIROBI DAR ES SALAAM CAPE TOWN

and associates in

BERLIN IBADAN

Oxford is a trade mark of Oxford University Press

First Edition 1991

ISBN 0 19 557404 3

Printed at
Elite Publishers Limited, Karachi
Published by
Oxford University Press
5-Bangalore Town, Sharae Faisal, Karachi-75350, Pakistan.

Dedication

These volumes are dedicated to
Frances, my wife,
without whose assistance throughout,
the task could not have been accomplished.

Foreword

Tom Roberts is one of a rare species of naturalists. He has not only managed to expand his interest in wildlife from the level of a personal hobby to a life-long commitment but he has concentrated that interest in a country far from his own natural habitat.

Since the early 1950s, Tom Roberts has become a familiar figure to those vitally concerned with the study and the preservation of wildlife in Pakistan. Over the years he has made a significant contribution to the study of Pakistan's precious wildlife resources and through his various publications he has documented valuable information about the flowering plants, flora and mammalian fauna to be found in our country. This latest study focussing on the bird life of Pakistan reveals with the same sharp clarity of his previous work the colourful variety and extent of this aspect of our national heritage.

To accompany Tom Roberts in the field, as I have had the pleasure of doing during expeditions, camping together in Gilgit and the Kirthar mountain range of Sind, is to share in the deeper satisfaction that dedicated naturalists derive from stalking and observing wildlife instead of hunting it, and to be reminded of the inescapable responsibility every naturalist must feel towards the natural resources entrusted to his care.

Pakistan, in common with other rapidly developing countries, is faced with the pressures of an increasing population and the imperatives of developing its natural economic resources, often at the cost of its wildlife. Gradually, as more and more areas are forfeited to progress, beautiful birds and animals in our country are threatened with extinction and are beginning to vanish from their former haunts. Tom Roberts' present work is therefore both timely and invaluable for it serves to remind us of what we already have and what we are in serious danger of losing.

By documenting the status of each species of birds to be found in Pakistan, Tom has provided in these pages an extensive foundation of data which will hopefully stimulate further research by future naturalists. An encouraging example of this is the new members, the starling family which can now be seen populating our countryside. He has at the same time included a sobering note of warning, a checklist which includes 'recently extinct birds in Pakistan', a casualty list which every Pakistani should ensure does not lengthen.

Nowadays, with travel as easy and as speedy as the flight of birds, there is an increasing interest shown by the rest of the world and within the country itself in the wildlife of Pakistan. To cater to this the Government of Pakistan and the tourist industry are establishing better and improved facilities to accommodate everyone interested in visiting the wildlife sanctuaries and game parks established of late. I can think of no better accompanying guide, apart from Tom Roberts himself, than these two volumes. They are as informed as they are informative, helping one to identify each species of bird, its unique characteristics and plumage, and its ecological relevance. They will endure as a timely catalogue of a valuable natural and national asset.

SYED BABAR ALI
Chairman, World Wildlife International
Council Member, International Union for the
Conservation of Nature and Natural Resources

Lahore, April 1988.

Acknowledgements

Accumulating the knowledge necessary before a book of this nature can be written has to be a long-time experience and I am aware of having received much help in this process from many more individuals than I could adequately thank in the following acknowledgements.

As a nation, the people of Pakistan, by cultural tradition, are very hospitable to strangers and no region possesses a more warm-hearted or helpful populace, particularly in the more out of the way rural areas. In trying to obtain first-hand field experience, I have had to rely heavily upon the help and kindness of such people, often receiving generous hospitality from total strangers. These have ranged from local chieftains and prominent community leaders, to persons forced by circumstances to live more humbly. Many of these people who have helped me will never hear of this bird book, let alone read it, yet it is to them that I owe the greatest debt, for making it possible to explore so many corners of this fascinating region, and I gratefully acknowledge their help and contribution. Amongst their legion, too numerous to mention, I wish to record my debt to five outstanding self-trained naturalists, Natha, fisherman from Ghizri, Bachoo Ali, fisherman from Sonmiani lagoon, Nathoo, professional snake catcher from Shahyaki, Thatta district, Bashir Hussain, Forest Guard from Kuwai (Kaghan valley) and Pir Khan, retired Forest Guard from Pateka, Azad Kashmir.

Many persons who have lived and worked in the region in the past, have generously passed on to me all their unusual or interesting ornithological observations which I have been able to incorporate (with due acknowledgement in the text), into what I have written. Notable among them were the late Herbert Waite and Kenneth Eates, both high ranking Police Officers who continued to serve after Independence in the Pakistan Police Service and whose very extensive skin or egg collections and notes have been of great assistance. See also the chapter titled 'Contributions of early Ornithologists'. Besides these two, I wish to acknowledge my debt to all of the following (listed in alphabetical order) Syed Asad Ali, Mohammed Ashiq Ahmad, Major Amanullah Khan, Bruce Amstutz, Trevor H. Braham, Shahzada Prince Burhan-ud-Din, Jack Coles, David Corfield, Colonel Dastagir of the Kurram Militia, Edward Fernando, Christopher Finney, Kent Forssgren, Dr. Anthony Gaston, Paul Goriup, Richard Grimmett, J. Ainsworth Harrison, Dr. Reudi Hess, Derek A. Holmes, the late Roger Holmes, Dr. Mubashir Hasan, Kamal Islam, the late Karim Dad Junejo, Frederik Koning, Professor Afsar Mian, Mark Mallalieu, David Mallon, Dr. Donald Melnick, Zahid Beg Mirza, Rohil F. Nana, Dr. Robert Orr, Rolf Passburg, Dr. George Schaller, Mr. J. Serrao (of the Bombay Natural History staff), Dr. Michael Porter, Christopher Savage, Edward Sicely, Monsieur Jacques Vieillard, Allan Vitterey, Mrs. Ann Vollum, Miss Brenda Wheeler, Herr Walter Weitkowitz and Nicolas Van Zalinge.

I have also received much help in the identification of plants from many Forestry Department officials, but particularly wish to acknowledge the assistance from Dr. Eugene Nasir and Dr. Yasin Nasir of the National Herbarium, Islamabad and also Dr. A. Rehman Beg of the Pakistan Forest Institute. I also thank Mohammad Farooq Ahmad, Director of the Zoological Survey Department for help in identification of fishes, Dr. Samuel Telford Jr. of the

World Health Organisation for help in identifying reptiles, Nicolas Van Zalinge of FAO for identification of prawns and lobsters and Dr. Michael Turkay of the Natural History Museum in Frankfurt for identification of fresh water crabs, all of which have been relevant in describing the ecology of the birds covered by these volumes.

I also wish to record special thanks to Dr. Nikolai Formozov of Moscow State University, who has gone to great trouble to locate and send me rare recordings of Russian birds as well as otherwise unobtainable reprints.

Many serving and retired officers of the Pakistan Forest Service have not only helped me to visit normally inaccessible areas, but have freely shared their own local and specialized knowledge and encouraged me in my study of bird life. Amongst these, I would especially like to mention Mr. W.A. Kermani, former Inspector General of Forests and Secretary for Wildlife, Mr. S.K. Swathi, former Chief Conservator of Forests, Government of Baluchistan, as well as his successor, Mr. K.M. Shams (also retired), Mr. Mohammad Mumtaz Malik, Conservator Wildlife, Government of the NWFP, Mr. Khan Mohammad Khan, former Conservator Wildlife, Government of Sind, and Sheikh Abdul Qayoom, Chief Game Warden, Government of Azad Kashmir, and Ghulam Rasool, DFO Wildlife, Gilgit.

I particularly wish to acknowledge my debt to the Bombay Natural History Society and to their Curator Mr. J.C. Daniel for the many facilities extended to me over many months while preparing paintings from their bird skin collection, and also for the Society's role in stimulating and making possible my early interest in learning about the natural history of the region through the fine series of books which the Society has sponsored and published. The very numerous citations from past issues of the *Journal of the Bombay Natural History Society,* sprinkled throughout the species' accounts, will also testify to the enormous contribution which the Society has made to our knowledge of the subcontinent's birds.

I also wish to express my grateful thanks for help and facilities extended over many years by Mr. J.C. Galbraith (retired) and the Trustees of the British Museum (Natural History) both at South Kensington and now Tring for allowing me to examine their unrivalled skin collection of specimens from the subcontinent, as well as their comprehensive library.

By virtue of living in Pakistan, I have had the privilege of meeting and getting to know many eminent visiting conservationists and ornithologists and have been fortunate in obtaining their help and that of others in scrutinizing parts of the manuscript of these volumes. I am most grateful to all these persons for much helpful commentary and criticism. Foremost amongst these is the late Dr. Salim Ali who all along encouraged me in this project, who passed on to me all the field notes of Dr. E.M. Nicholson relating to his stay in southern Baluchistan, and who despite ill health at the time, read and commented upon several of the introductory chapters of Volume I. I am also privileged to have had similar advice and comments from Mr. Kermani on the chapter covering ecological zones, from Mr. Guy Mountfort, OBE (founder member and Trustee of WWF International) and Paul Goriup, Consultant to ICBP, for the chapter on migrations and to Professor Hilary Fry for looking at the chapter on the 'Problem of species'. Various species accounts have also been read and improved by persons having specialized knowledge and expertise. In alphabetical order, Dr. Euan K. Dunn (Cuculiformes), Dr. Hilary Fry (Coraciformes), Dr. Peter Garson (Galliformes), Richard Grimmett (Lairidae, Sternidae, Phylloscopi and Sylvininae), Paul Goriup (Bustards and Turdininae), Derek Goodwin (formerly of the British Museum Natural History) (Columbiformes, Corvidae and Estrildid Finches), James Hancock (Ardeidae), Carol and Tim Inskipp (Timalidae, Cardueline Finches and Buntings), Frederik Koning (Waders and Shore-birds), Professor Kai Curry Lindahl (Pipits and Wagtails), Dr. Charles Luthin (Storks, Ibises and Spoonbills), Ben King (Strigiformes), Guy Mountfort (Birds of Prey), Richard Porter (Pteroclididae), and Dr. Derek A. Scott (Anatidae).

I have throughout these volumes drawn widely upon published information contained in such important reference books as S. Dillon Ripley's *Synopsis of the Birds of India and Pakistan* (1961 and 1982), Ali and Ripley's 10 volume *Handbook of the Birds of India and Pakistan* (1968-1974), the first four volumes of *Birds of the Western Palearctic* edited by Stanley

Cramp, and the first two volumes of *Birds of Africa* compiled by Emil Urban, Hilary Fry and others. I wish to acknowledge my great debt to all the authors of these very comprehensive volumes. Building up a personal library of important reference books, especially out-of-print publications relating to this region, has been an important element in enabling me to learn about the birds of Pakistan and I would especially like to acknowledge the debt I owe to Mr. K. Swann of Wheldon and Wesley, Natural History booksellers, whose efforts over more than 25 years have enabled me to obtain many rare volumes.

Finally I wish to thank Oxford Cartographers for producing such beautiful renditions of my crude distribution maps, and to thank Mr. Zafar Ahmad of Printhouse Ltd., Lahore, for his meticulous care and skill in reproducing my bird paintings. I also acknowledge my grateful thanks to many friends who gave me duplicate copies of their best bird slides, from which some of the pen and ink sketches as well as the bird plates have been drawn.

Notwithstanding all the help received from so many people, readers will no doubt detect many errors and imperfections in these two volumes and I am solely responsible for these.

Tom Roberts

Contents

page

List of Full-page Maps and Plates

List of Illustrations

List of Figures

List of Distribution Maps

Map   page

Systematic Index of Bird Species with Checklist of the Birds included in this Volume

(Numbers in left hand column refer to Checklist)
*Species marked with an asterisk are considered as doubtfully occurring in Pakistan
+Species marked with + are considered recently extinct in Pakistan

page

ORDER COLUMBIFORMES

FAMILY COLUMBIDAE

ORDER PSITTACIFORMES

FAMILY PSITTACIDAE

ORDER CUCULIFORMES

FAMILY CUCULIDAE

ORDER STRIGIFORMES

FAMILY TYTONIDAE

Introduction

The author's approach to species accounts

t is hoped that the reader will study this relatively
hort introductory chapter before consulting other
ections of the book, since it is intended to help in the
nterpretation and understanding of the species
ccounts. But first, these explanations need to be
refaced by an apology.

The reader will soon detect some inconsistency in
ayout and presentation especially between the earlier
nd later chapters. The author is all too aware of these
hortcomings which have arisen because of the con-
iderable time span (stretching over 4 or 5 years) which
as elapsed during the actual writing of the various
ections. They are the inevitable consequence of the
ecessity of dealing with many other demands upon
ne's time (as an amateur author) and of a rather
mbitious attempt to 'do it all oneself'.

Though most of the manuscript has been re-written
nd revised several times, it was felt that further
ttempts to make the additions and alterations neces-
ary for consistent presentation would have much
elayed publication, whilst many of my friends and
cquaintances, especially in Pakistan, are already
hiding me for taking so long over the project.

These volumes are written (and illustrated) with the
ope that they will give both pleasure and helpful
nformation to anyone who is interested in wildlife and
vho likes birds. It is also hoped that these inconsisten-
ies in presentation will not detract too much from
hat pleasure.

In the introductory chapters, and throughout the
pecies accounts, the ornithological attributes of a
articular physical region are being considered. Noth-
ng in the maps, or species accounts is intended as of
olitical significance. The north-eastern boundaries
including disputed territory) in the Himalayan regions
re based upon that area which is normally accessible
o naturalists of Pakistan, and with which the author
as field experience. The term Azad Kashmir has been
n consistent use in Pakistan for over forty years and is
sed in this book to eliminate any confusion with the
olitically separate part of the pre-independence State
f Jammu and Kashmir.

1. SYSTEMATICS

Though readers principally interested in the region are
likely to be familiar with the systematic layout and
taxonomic order followed by Dr. Salim Ali and Dr. S.
Dillon Ripley, the authorities on the avifauna of the
Indian subcontinent, I have followed the sequence of
species and classification presented by Dr. K. H.
Voous in his 'List of Recent Holarctic Bird Species',
(*Ibis,* 1977), because this has been widely acclaimed as
the best modern synthesis of recent systematic researches and reflects a broad spectrum of published
opinions. The taxonomic treatment of Ali and Ripley
reflects, perhaps it would be fair to state, largely trans-
Atlantic viewpoints and especially the work of Charles
Vaurie, whose own contribution to De Voous's list is
enormous. Where species of sub-tropical Oriental
origin, are not included in Voous's 'List of Recent
Holarctic Bird Species', I have used the classification
followed in Ali and Ripley's *Handbook of the Birds of
India and Pakistan* (10 volumes, 1968-74), or where
recent studies made it desirable, the divergent taxonomic treatment of a third acknowledged expert. I
have, however, included the older scientific synonyms
which have been used in earlier standard reference
works of the region for ease of reference.

In some species' accounts the description is prefaced
by a taxonomic discussion. This is not intended in any
way as an additional contribution to the systematic
relationships of species but rather to draw the reader's
attention to the existence of such divergent views, with
occasional commentary on their validity based upon
the author's own knowledge of the species in the field.
For example, I cannot agree to the logic of splitting
Chrysomma sinense and *Moupinia altirostris* into 2
separate genera as advocated by the Oriental Bird
Club.

With regard to the choice of trivial or vernacular
names, I have also adopted those now being standard-
ized by the Oriental Bird Club, which is preparing an
up-to-date checklist of the Oriental region, in consultation with a group of experts. Much effort is now

being made by ornithological societies and institutes in different regions of the English speaking world, to standardize such names, and this seems desirable not only because scientific nomenclature is subject to constant change as new understanding emerges of phylogenetic relationships, but also in cases where trivial names are misleading, or the same species carries different regional names, or even where it is unduly long and cumbersome. Examples of the first would be dropping Black-throated Jay for *Garrulus lanceolatus,* of the second, using Northern Wren for *Troglodytes troglodytes* and the third, Grey-hooded Warbler for *Seicercus xanthoschistos.*

Generally speaking I have not given detailed descriptions or even reference to sub-species recorded from

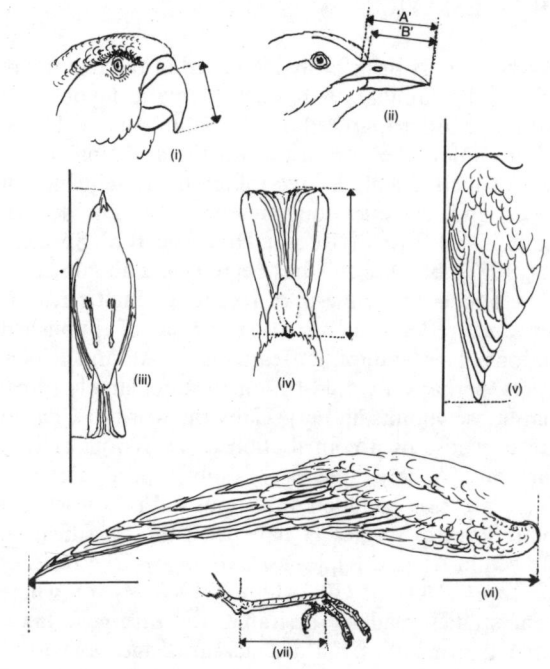

Fig. I: Showing various methods of measuring a bird's body.
- (i) Usual method for families with a fleshy cere.
- (ii) Chord 'A' from tip of bill to base of skull (used by Ali and Ripley in 'Handbook'). Chord 'B' from tip of bill to base of feathers (used by Stuart Baker in 'Fauna Series'.
- (iii) Body length—this method can be used also when measuring a live bird, provided the legs and bill tip are held between fingers of both hands (see Svensson, 'Identification Guide', 1984). The bird is laid flat on its back on the measuring ruler.
- (iv) Tail length to base of tail where central tail feather quills emerge.
- (v) Wing length of passerines and most medium sized non-passeriness—the wing is gently pressed flat onto ruler and measured from carpal joint to tip of longest primary.
- (vi) Wing chord—straight line measurement used for larger birds especially when measuring dried skins, camber is not pressed out of slightly curved wing.
- (vii) Tarsus length—measured to bottom of lowest scute on shank.

our area, largely because there is much lack of agreement as to their distinctiveness and even validity in most cases. Minute variations in plumage tone or shade have often been the basis for erecting separate sub-species in the past and in practice examples of specimens from Pakistan are often part of gradual variation or a cline, noticeable as one examines a species total population moving say from west to east across its range. However, in a few instances sub-species from the Pakistan region, are very distinctive from neighbouring sub-species and in such cases these are the ones described for the region together with their trinomial nomenclatures. For the same reasons trinomial of sub-specific rank is given to some of the species in the colour plates.

With respect to the use of botanical scientific names I have throughout both volumes followed the somewhat older nomenclature of Professor Ralph Stewart considered one of the world's authorities on the flora of the Himalayas (Stewart, 1972, *Flora of West Pakistan—An annotated catalogue of the vascular plants of West Pakistan and Kashmir*). This basic reference work is followed by the detailed *Flora of Pakistan* now under publication and so far comprising 191 Fascicles (Nasir and Ali, editors, 1972-87). These reference works also follow to some extent the nomenclature used in Parker's *Forest Flora for the Punjab* (Parker, 1956, revised ed.), as well as Sir Harry Champion's *Forest Types of Pakistan* (Champion Seth and Khattak, 1965). These are the publications more easily available in Pakistan and I have therefore not followed the more modern synonymy as used in Polunin's *Flowers of the Himalayas* (Polunin and Stainton, 1984).

2. SPECIES DESCRIPTIONS

Some measurements are given at the beginning of this section and the accompanying illustration will show how these are normally taken. Measurements are valuable in showing size comparison especially between related or allied species and wing length is the easiest one to take. Readers, however, should be cautioned that in these volumes only a selective sample of second-hand measurements (usually taken from collectors' specimen labels) have been quoted. Also that there is some inconsistency of presentation with much more detailed ranges of measurements given in the later sections (Volume 2). To be adequately informative, very detailed measurements are needed, considered beyond the scope of this book. Besides indicating the method of measuring the sample size, range and mean of such lengths all need to be stated, together with sex and geographic population of the species measured. In the case of weights where even more variation is normal, the time of year and condition of the specimens as well as sex are desirable. As indicated in the acknowledgements the author is conscious of these shortcomings and considers that if detailed measurements need to be consulted, they are becoming increasingly available in such standard reference works as have been cited at the end of the chapter titled Acknowledgements. As Paul Goriup has aptly written in commenting upon some of the measurements given in this book 'Biology is not physics and it would be misleading to imply that precise measures exist'. The reader is therefore asked to consider any measurements given, including heights of nests above ground, habitat or biotope elevation, as well as species body measurements, as being approximations only.

An attempt has been made to describe the appearance and colour pattern of plumage, whilst trying to prevent such description from becoming unduly lengthy, yet giving sufficient detail to help in field identification of a bird, perhaps seen only momentarily and probably only revealing one or two aspects or features. Colours of any bird's plumage are exceedingly complex and impossible to describe accurately in a brief summarized description. For example, many individual feathers may each have 3 contrasting colour areas. An un-grammatical sort of shorthand has become prevalent in bird book descriptions and I hope acceptable, by adding the suffix '-ish', so that the description, a 'brownish grey' may be taken to mean that such a part of the body may show distinct grey tones in certain lights but on close examination would perhaps contain white, black and brown colours combined. Similarly such terms as 'washed with grey' or 'suffused with yellow' indicate that there is an underlying pattern, perhaps brown or olive brown, which at certain angles reflects a stronger blue-grey or yellow overall tone. Intensity and tone of colour in a bird's plumage varies greatly between individuals according to age as well as to the state of plumage wear: recently moulted, or after feather abrasion.

An adequate description of all variations seen in the field would in my view have needlessly increased the length of the book. In all cases, the author has written direct descriptions from personal examination of series of museum skins and in about one-third of the accounts, from notes made of living birds in the field. After writing such preliminary descriptions, standard reference works were consulted, but the attempt has been to give evocative descriptions rather than coldly precise ones. Some authors (e.g. Ticehurst, *Review of Genus Phylloscopus,* 1938) have followed Ridgeway's colour code (Washington, D.C., 1912) in describing plumages. This code is now a rare book, and is generally only available in museums and scientific libraries and therefore of little use to the interested amateur. Thus a wider choice of comparative descriptive terms have been used, which may lack precision but should help to evoke the right impression. For example in using terms like 'rose pink', I intend to suggest a rather bluish or mauve pink colour, though

admittedly modern hybrid roses exhibit every possible shade of pink. There is a wide range of chestnut brown colours in the plumage of many birds. Rusty, tawny and orange-brown indicate closeness to the yellow side of the spectrum. Cinnamon and maroon indicate closeness to the red side of the spectrum with maroon having dark almost blackish tones, and fulvous indicates a very dilute white or buff suffusion to the tawny or chestnut colour.

In referring to the colour of bill, iris and legs, it has often been necessary to indicate that there is much variation as revealed in descriptions written by collectors on their labels from the fresh-killed specimen. Such variations are as yet not understood completely and could be related to age of the bird, its physiological state (i.e., readiness to breed) and even to diet, and sex differences. For example, a bird's iris may have to be described as yellow-brown or reddish brown, indicating that such variation typically occurs.

With respect to the bird's anatomy, most people are now familiar with the various terms used to describe particular feather tracts, or regions. Supercilium is used for the eye-brow stripe, parietal for the sides of the crown but above the region normally defined as the supercilium. Moustachial suggests a downwards ex-

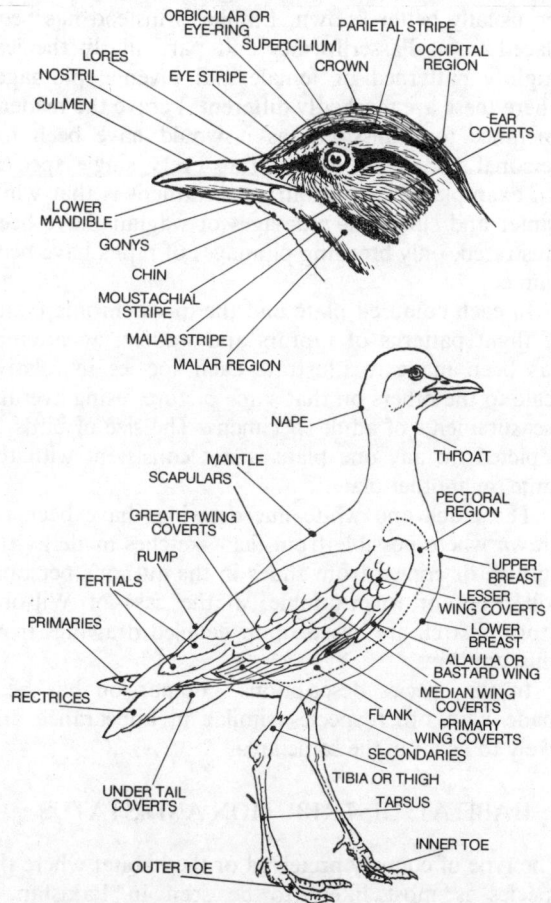

Fig. ii: Showing parts of bird's body and terms used in this book.

tending line whereas loral suggests only the immediate area between the base of the bill, gape, and the eye. Malar and moustachial have been used rather interchangeably, though malar seems more appropriate when there is a pale line or stripe between the darker ear coverts and a darker area on the sides of the throat, e.g. *Sylvia mystacea,* or for a second darker line down the sides of the throat from the base of the lower mandible, e.g. *Emberiza fucata.* Moustachial is used to indicate a darker line usually framing the ear coverts and from the base of the upper mandible. I have used the term 'tertials' rather than inner secondaries, to refer to the three usually very broad feathers that lie one on top of the other when the wing is closed and which form such a conspicuous pattern when the bird is at rest, for example in particular, the Black-billed Desert Finch (*Rhodospiza obsoleta*) or the Azure tit (*Parus cyanus*) (see colour plates).

A colour picture is worth pages of written description, but obviously publication costs have severely limited the number of species that could be illustrated. The choice, made entirely by the author who painted the plates, has had to be arbitrary, omitting most of the non-passerines, in consideration of the fact that they have been well illustrated in regional field guides and are usually better known. Emphasis instead has been placed upon Passeriformes and particularly the less brightly patterned or female and juvenile plumages where these are markedly different. I crave the readers' tolerance for omissions, as it would have been my personal preference to illustrate every single species. An example of such arbitrary treatment is that while winter and sub-adult plumages of wagtails have been illustrated, only breeding plumages of pipits have been painted.

In each coloured plate and the monochrome plates of flight patterns of raptors and waders, an attempt has been made to illustrate each species in relative scale to the others on that same picture, using average measurements of adult specimens. The size of birds as depicted on any one plate is not consistent with the same on another plate.

The black and white line drawings have been redrawn where possible from field sketches made by the author, or copied from slides in the author's personal collection, or for example, in the case of Wilson's Storm Petrel, are based upon detailed drawings from museum skins.

In the species descriptions, comparison has been made with other species similar in appearance and likely to occur in the same area.

3. HABITAT, DISTRIBUTION AND STATUS

The type of country preferred or the habitat where the species is most likely to be seen in Pakistan is indicated, not necessarily at the beginning of this section. There are distribution maps of each species, but the written accounts attempt to summarize all known records. Because the author has not lived in the NWFP or northern Punjab, but only made short (but frequent) visits to these areas, everything published by any person who has worked in these areas continuously and intensively is repeated, even though such information may seem outdated. If the author has evidence that the present day position has clearly changed, this is indicated. With personal diary notes spanning 1964 to 1986, the majority of summarized statements are based upon first hand experience, though in only a few instances for greater clarity have I indicated (in parentheses) that a statement is based upon the author's own observations.

In the case of migrating birds, regions of occurrence and preferred habitat are usually of little value, but trends are indicated.

There have been so few observers of birds in Pakistan in recent years, that more effort has actually been put into the determination of status than all the rest of the book together, though the results may not seem to show it. Only in a few instances are population estimates of particular localities available, but where any have been carried out these figures are presented. One of the major difficulties, arises from the listing of bird species as occurring in Pakistan, in such modern standard references as S.D. Ripley's 'Synopsis' (1961 and 1982, revised edition) and Ali and Ripley's 'Handbook' (10 vols. 1968-74, and first 6 vols. revised 1988). Some of these statements of a species distribution are based upon very old information, and in some known cases the source can be shown to be unreliable (see introductory chapter on the Contribution of Early Ornithologists). Where it is reasonably considered, taking all available evidence into account, that a bird has not occurred or is highly unlikely to do so, it is excluded from the **Checklist,** or implied as having become recently extinct. However, all such species excluded from the **Checklist** are referred to in the species accounts with explanations. Examples are the Spot-billed Pelican, Marshal's and the Common Iora, the Yellow-rumped Honey Guide, the Rufous-chinned Laughing Thrush and the Collared or Allied Grosbeak, to mention a few.

Birds are highly mobile and individuals can be widely unpredictable in occurrence, perhaps thereby presenting much of their charm and challenge to the enthusiast. With this in mind, and fully conscious that the information presented in this book will rapidly become outdated, it has been strongly felt that a more accurate account of the birds of this region, presented in the form of a **Checklist** and status evaluation would be a stepping stone for other workers and contribute to a better understanding of trends. Finally in this section similar or closely related species which occur sympatrically are mentioned as this is important information when the possibility of confusion in identification is considered.

At the end of each species account of distribution and status, its occurrence or status is categorized in summary fashion. COMMON means that it can invariably be seen by a careful observer, visiting those regions and specified habitats where it is described as occurring, with the proviso of course that the season is also appropriate. FREQUENT means that even when visiting appropriate habitat and regions, it will not be seen invariably, perhaps only in one visit out of three. Neither term is used to denote population density, but only as a measure of the likelihood of encounter by the human observer. ABUNDANT denotes not only that it will be invariably seen, but also in considerable numbers. FREQUENT but localized, means that it will not be seen in every area of suitable appropriate habitat even on successive visits or searches as it is erratic in occurrence with a discontinuous distribution. SCARCE indicates that it is only likely to occur in very restricted areas and that even in such areas it occurs in small numbers. RARE indicates even less likelihood of occurrence. A good example of the use of such terms as scarce and rare are the summaries of status for two Fish Eagles, viz., Pallas's Fish Eagle (*Haliaeetus leucoryphus*) which is described as SCARCE. Though it can be reliably encountered in the breeding season in the vicinity of its nest, occupied eyries are now very few and far between. The White-tailed Sea Eagle (*Haliaeetus albicilla*) described as 'Status RARE' cannot be reliably encountered even in winter time visits to previously known favoured localities, and such localities are perhaps only 2 or 3 in number. VAGRANT means that its future occurrence is highly unpredictable and that it does not occur anywhere in the region on a regular basis.

4. HABITS AND BREEDING BIOLOGY

Recent developments in the study of bird behaviour and ethology, make it easier to interpret various aspects of their activity than was possible for earlier workers and observers, and this has been exploited by the author wherever possible, especially in interpreting Kenneth Eates hand-written notes. In order to try and avoid mere repetition of information already published in Ali and Ripley's comprehensive 'Handbook Series', I have followed two approaches. When recording facts well known or published elsewhere I have tried to describe or cite observations from my own field notes. In the absence of direct observations I have cited the original author's accounts rather than Ali and Ripley's 'Handbook', in both cases with the intention of making the text fresher and more relevant. Obviously, some species are better known to the author or have been more intensively studied by other ornithologists and no attempt has been made to give consistent or equal treatment to each species; accounts being necessarily briefer for species which are only non-breeding

migrant visitors or with which the author lacks first-hand knowledge. In describing food habits, reference has been made where possible to published accounts of central Asian breeding or migrant populations, rather than western European populations, and for the Indian subcontinent, direct reference through the courtesy of the Bombay Natural History Society library, to Mason and Lefroy's detailed and exhaustive report which is otherwise out of print and unavailable (Mason and Lefroy, *Dept. of Agriculture in India*, 1912). The author's own early training in agricultural botany and entomology has been helpful in making direct gross observations of food items when watching birds foraging.

In respect of breeding biology, the literature has been searched in order to try and supplement whatever has been published about breeding habits in Ali and Ripley's 'Handbook Series'. The manuscript notes made by Kenneth Eates over 40 years of collecting birds' eggs in Sind province which he kindly loaned to me, have also been valuable in this connection.

5. VOCALIZATIONS

The recognition and identification of birds depends heavily upon the observer's ability to learn their calls and songs. By using light portable tape recording equipment the author has been able to build up an impressive collection of bird recordings (319 of the species in the **Checklist** recorded from Pakistan or India). Wherever the account is based upon play-back of such recordings, this section is marked with an asterisk so that the reader can better evaluate the description. Though sonagrams are being more and more widely used and recognized, they are difficult for the layman to comprehend, because they are two dimensional representations of only a restricted frequency range of the sound waves which make up the calls or song of a bird. Comparison of different sonagrams undoubtedly gives greater precision in examining slight variations in the repertoire or dialect of different birds, and this graphic technique will become an increasingly valuable tool. It is however, as yet a developing technique and published sonagrams do in fact differ and vary in appearance according to the apparatus and frequency bands used. Publication costs in producing sonagrams have largely influenced me to restrict my account to very imperfect verbal transliterations.

The publication of yet another book about the Indian subcontinent, a region already well documented, deserves some justification. The principal one is that it is the first attempt to describe the birds of Pakistan as a geographical entity. Secondly, due to political and foreign exchange restraints, the many books published in India are unavailable in Pakistan and this book is intended to kindle a serious interest in that country, in the study of her avifauna. In the present state of the author's knowledge **660** species are known to occur or have recently occurred in Pakistan. **27** more species are included in S. Dillon Ripley's 1982 'Revised Synopsis' or Ali and Ripley's 'Handbook Series', which are now excluded from the **Checklist.** **32** species, not included in S. Dillon Ripley's 1961 'Synopsis' as occurring in Pakistan, are now added to the **Checklist.** **6** more are described, as it seems highly probable that they do occur, but have not yet been reliably recorded. These **32** additional species are listed hereunder:

> Red-necked Grebe, Slavonian Grebe, Baer's Pochard, Red-breasted Merganser, Pied Harrier, Brown Crake, Bronze-winged Jacana, Pomarine Skua, Common Gull, White-winged Black Tern, Tengmalm's Owl, Long-tailed Nightjar, Large Pied-crested Kingfisher, Lesser Yellow-naped Woodpecker, Indian Pitta, Dusky Crag Martin, Forest Wagtail, Song Thrush, Rufous-capped Bush Warbler, Hodgson's Wren Warbler, Striated Marsh Warbler, Menetries's Warbler, Blue-throated Flycatcher, Brown Flycatcher, Yellow-bellied Fantail Flycatcher, Kashmir Red-breasted Flycatcher, Large Grey Babbler, Chestnut-bellied Nuthatch, Mrs. Gould's Sunbird, resident breeding Jackdaws, Grey-headed Myna, Pied Myna, and Red Crossbill.

Most of these records have already been published (by the author of this book) in issues of the *Journal of the Bombay Natural History Society.*

If the reader is curious to further evaluate the author's limitations in knowledge, I have seen and am familiar with 628 out of the 660 species in the **Checklist,** including slightly over 40 species in the list of 660 which have been studied by me in India, Nepal and Sri Lanka, but not seen in Pakistan. Though continuously resident in Pakistan between November 1952 and July 1984, and interested in natural history since early childhood, I was too discouraged to attempt a serious study of the birds of the region until 1964, when S. Dillon Ripley's 'Synopsis' (1961) finally came into my hands and it is to this most exhaustive and scholarly work that I would like to pay tribute in making possible this account of the birds of Pakistan.

Ecological Factors in Bird Distribution

Like all countries which form but a small part of a great land mass, Pakistan is bounded by man-made political frontiers and does not comprise an isolated entity either in zoogeographic or ecological terms. It is, however, something more than an extension of the plant and animal affinities of India due to its very considerable montane area, and attendant Palearctic characteristics. Pakistan exhibits a certain cohesion of physical geography as well as climate which has a great bearing upon the distribution and habits of its bird fauna. An attempt to sketch the overall geophysical characters of the region in broad outline is therefore helpful to an understanding of the country's plant and vertebrate fauna.

Pakistan can be pictured as a country roughly rhomboidal in shape, extending from its bottom left-hand corner at longitude 60°52′E and latitude 24 °N where the harsh arid cliffs of the Makran coast range plunge right into the Arabian Sea. At its opposite top right-hand corner, it extends to the edge of the Karakoram mountain massive on the borders of the great desert plateau of central Asia. Here it reaches longitude 75°22′E and latitude 37 °N. Pakistan covers a land area of 310,400 square miles, or 79.1 million hectares. Within these parameters to the north, are the outer mountain chains and foothills which form the watershed of the great Indus river. This river flows westwards from Ladakh up to Gilgit right across Baltistan and then bends sharply southwards and flows from north to south down to the Arabian seacoast near Karachi, covering a distance of nearly 2,900 kilometres (1,800 miles) inside Pakistan territory. Almost the entire country, therefore, can be considered as the watershed of this mighty river. The heartland of the country is the flat alluvial flood-plain which forms the drainage basin of the Indus, including its five major tributaries fanning out in a' north-easterly direction across the Punjab. The Indus basin, for it can hardly be called a valley, covers an area of about 25.9 million hectares or 100,000 square miles (*Pakistan, Past and Present,* 1977) extending from Lahore in the north-east down to the Indus delta south of Karachi, a straight line distance of about 1,126 kilometres (700 miles). It is in this region that the bulk of the human population is settled, and well over 80 per cent of the nation's agricultural wealth and food production is derived. Roughly two-thirds of Pakistan's surface area is mountainous. The Indus drainage basin is bounded not only by range upon range of high mountains to the north, but also by jagged rocky mountain ranges extending southwards and down its western flanks to the Arabian seacoast, starting from Chitral in the north, and down through the Safed Koh and other ranges in the North West Frontier and Baluchistan provinces. This perimeter of mountain ranges, we are told by geologists, forms the suture or collision front between the great land mass of Asia and the continent of Gondwanaland (Miller, *Continents in Collision,* Royal Geographic Society, 1982) which split off from the African continent over 65 million years ago (See chapter on Zoogeography). Consequently these mountain regions both in the northernmost inner ranges of the Himalayas and also down Pakistan's western boundary, are still geologically of quite recent origin and somewhat unstable, experiencing frequent earth tremors, not to speak of land slides and avalanches. This reference to the region's geological history is greatly simplified and much of the outer hill ranges of what is now Sind Kohistan, and south-western Baluchistan, comprise sedimentary cretaceous rocks which once formed the bed of the sea of Tethys, which 53 million years ago covered approximately the same region as is now occupied by the Indus plain.

Climatically, Pakistan lies mostly between sub-tropical latitudes (it lies north of the tropic of Cancer which falls at 23°27′ north, just south of the Makran coast). Consequently, the country experiences an extended hot dry summer even in the lower mountain regions. However, from early July until late September in a normal year, the western flank of the south-west monsoon sweeps up into Pakistan and is of immense significance to almost the entire plant and animal community of the country, as well as the agrarian based economy. Cool, moisture-laden winds, coming off the Indian Ocean, increase in regularity and intensity from June onwards, eventually culminating in intermittent heavy rainfall which is especially concentrated along the outer foothills of the great mountain barriers to the north (See **Fig. iii**). In the central and eastern parts of Pakistan these moisture-laden winds tend to lose their force and direction and in some years no precipitation at all may fall in the border desert areas. Similarly the direction of the monsoon is such as to largely by-pass the Baluchistan highland region and particularly the south-western corner of that province. This area, beyond the principal mountain ranges of Baluchistan, comprises a vast desert plain known as the Chagai which really forms an extension of the great Seistan or Iranian desert depression (Cloudsley-Thompson, *Desert Life,* 1965). Faunistically it has unique and fascinating affinities with the central Asian steppes, particularly with respect to less mobile herpeto and mammalian fauna (Roberts, 1977). In the far northern mountain regions also the monsoon winds cannot penetrate, and the region being mostly over 1,200 metres above sea level, experiences a cold desert climate.

The proximity of these mountain barriers to the alluvial flood plains leads to sharp contrasts in temperature as well as micro-climatic effects within the mountain regions themselves. Extremes of temperature are experienced in the more arid rocky areas, both diurnal and seasonal, due to the effects of insolation, and in winter time, temperature inversions at night time. In Skardu, the capital of Baltistan, winter temperatures drop to —22 °C and regularly rise to over 40 °C in summer. Similarly in Baluchistan even as far south as Kalat, winter temperatures have been recorded as low as —15 °C and in summer regularly rise to

38 °C. In the Sibi desert in the central part of Pakistan, the highest mean temperatures have been recorded for the whole of Asia, e.g. 52 °C from the towns of Sibi and Jacobabad. The whole of the Indus plain experiences very hot dry conditions during May and June and it is not uncommon during either of those months for the maximum shade temperature to exceed 44 °C throughout the region from Hyderabad city in the south up to Lahore in the north.

Rainfall, when it does fall, tends to be highly seasonal over most of the country and with few exceptions is not well distributed. It ranges from as low as 20-40 mm. annually in the Chagai and Sibi deserts, as well as the main valleys of Ishkoman, Yasin and Hunza in Gilgit and the Indus and Shyok valleys in Baltistan, to as high as 1,350 mm. (53 inches) in the Murree hill range (See **Plate 9c**). This hill range is unique in having a fairly evenly distributed annual rainfall with up to 6.5 metres (22 feet) of winter snow falling in the north-western part of the range, which is known as the Galis. Over the Indus plains, rain falls mainly during the monsoon season, whilst in the North West Frontier Province there is a distinctly Mediterranean character to the climate (and some Mediterranean plant affinities also) with up to 60 per cent of the rainfall occurring in the cold winter months. Further north, in Chitral and Dir, up to 90 per cent of the annual precipitation (snow and rain) falls during the winter. In the foothill regions of the Punjab and Hazara districts, rainfall is more evenly distributed with a pronounced monsoon influence, but at least 25 per cent of rain does fall in the winter and some rainfall in every month of the year.

Partly as a consequence of these geographical and climatic factors, a high percentage of Pakistan's bird fauna is migratory with a huge invasion of Palearctic winter visitors (over 30 per cent of the total bird fauna) which come to exploit the improved food availability following the monsoon season, particularly the abundance of insect life and favourable conditions associated with temporary swamplands which gradually decline from a peak in early October to the end of December. There is also a high percentage of nomadic bird species, which though residential or endemic in the sense that they do not undertake long migrations, nevertheless move in erratic patterns according to exploitable food resources. In the mountain regions there is a host of Palearctic species also which migrate northwards to breed during the Pakistan alpine summer, whilst even the truly Himalayan or endemic mountain species adopt varying strategies of altitudinal migration or latitudinal drift to avail of better food conditions. Baluchistan province has no season of above-average rainfall but the extremes of climate do offer opportunities for fluctuations in migrant bird populations, with a large number of Himalayan or sub-tropical and African wintering Palearctic species entering the province to breed

Fig. iii: Showing the strike of the south-west monsoon.

during the summer, and not a few hardy species coming in as winter migrant visitors to the upland mountain regions (See **Fig. viii**).

Finally in order to appreciate the distribution of the bird fauna in Pakistan, it is necessary to consider the effect of the human population, in a historic sense. The region has been settled by people since the earliest known civilizations but until recent times such settlements were limited and concentrated along the Indus and its tributaries. French archaeologists working at Mehrgarh since 1973, on a site where the Bolan river debouches from the Baluchistan hills, have discovered human settlements which pre-date the ancient Indus civilization of Harappa and Mohenjodaro (believed to be between 2500 and 3000 B.C.). By carbon-14 analysis they have shown that a civilization based upon grain cultivation existed in this region in 6,000 B.C. and this may well be the earliest known human civilization (Jarrige and Meadow, 'Antecedents of Civilization in the Indus Valley', *Scientific American*, 1980, Vol. 243, No. 2).

It is however only within the past century that the Indus plains have become relatively densely populated. The last official census in 1972 estimated the human population of Pakistan at 65 million, but mid-1986 official estimates now put the population at 97.67 million, making it the ninth most populous country in the world, a truly alarming increase (*Pakistan Times*, 30 May 1986). Undoubtedly within this century man has had a significant impact upon the plant vegetation. The most accessible areas and the easiest to irrigate by summer flood water, were the tropical thorn forests of the Indus flood plain. This ecological plant climax has now all but disappeared. Similarly the dry temperate semi-evergreen scrub forest which covered the outer more accessible foothill regions along the western flank of the Indus has also largely disappeared due to increasing destruction by browsing animals and cutting for human fuelwood needs. An attempt to classify Pakistan's land area into various ecological zones according to the predominant vegetation, is presented hereunder, with special reference to its effect upon bird distribution. Of necessity these zones have had to be simplified and represented diagramatically. They are delineated by lines which in reality are a gradual transition zone, and even the areas within these lines may only survive as relict pockets.

There have been several significant attempts at vegetation mapping in Pakistan and I have drawn on all of these, relying mainly upon Champion's classification of forest types, but synthesizing subsequent more detailed treatments so as to limit the number of different zones and combine them in a way which would relate to fairly distinct bird communities (See **Plate 4**). The first vegetation map covered the northern Himalayan areas and was carried out by Schweinfurth (1957). H. Champion *et al.* (*Forest types of Pakistan*). carried out a comprehensive survey in 1965, which has

been refined and made more detailed by vegetation maps in Baluchistan (Rafi, 1973) and more recently with special reference to the whole country (Beg, 1975).

Important and distinctive ecological zones can be described as under:

INDUS PLAINS REGION with characteristic plant associations. Edaphically this whole region comprises a remarkably flat level plain comprising a deep stone-free layer of alluvial silt.

i) **Littoral and Mangrove.** In particular mangroves at Sonmiani and Jiwani and the Indus delta.

   Typical plant species: Avicennia officinalis, Ceriops tagal/candolleana, Halopyrum mucronatum, Bruguiera conjugata. In most regions there is almost a pure stand of only *Avicennia.* In higher areas, not subject to daily inundation, there is a low scrub of *Salsola foetida* and *Sueda fruticosa* with scattered bunches of such grasses as *Heleochloa dura* and *Halopyrum mucronatum.*

   Typical bird species: Egretta gularis, Ardeola grayii, Gelochelidon nilotica, Ibis leucocephalus, Haliastur indus, supplemented in winter by *Pandion haliaetus, Pluvialis squatarola, Charadrius leschenaultii, Charadrius mongolus, Numenius phaeopus, Xenus cinereus* and *Larus brunnicephalus.* The only resident passerines found in this biotope are *Zosterops palpebrosa* and *Orthotomus sutorius,* and seasonally *Acrocephalus stentoreus.*

ii) **Riverain.** The Indus river from below Kalabagh in Mianwali district, and its main Punjab tributaries. Due to control of seasonal flooding through irrigation barrages and increased intensity of cultivation adjacent to the main river banks, this zone is rapidly disappearing and the riverain forest is drying out and is reduced to a few relict patches.

   Typical plant species: Climax *Acacia arabica,* and in less stable areas, *Tamarix dioica, Populus euphratica* with the grasses, *Erianthus munja, Saccharum munja* and *Saccharum spontaneum.*

   Typical bird species: Pernis ptilorhyncus, Otus bakkamoena, Remiz pendulinus, Pericrocotus cinnamomeus, and *Passer pyrrhonotus* (in riverain forest only). *Saxicola caprata, Calandrella raytal,* and *Anthus novaeseelandiae* in *Saccharum* grass areas. *Saxicola leucura, Saxicola caprata, Prinia burnesii* and *Chrysomma sinensis,* in mixed *Saccharum* and reed areas. *Calandrella raytal* prefers open sandy areas dotted with tamarisk. In winter *Jynx torquilla,*

Muscicapa parva and *Phoenicurus ochruros*, in the *Acacia* forest.

iii) **Swamps and Seasonal Inundations**

Typical plant species: Tamarix dioica, Phragmites karka, Typha angustata, Paspalum distichum, Imperata cylindrica, Nelumbium nuciferum and in water pools *Vallisneria spiralis* and *Hydrilla verticillata.*

Typical bird species: Phalacrocorax niger, Anhinga melanogaster, Egretta spp., *Ixobrychus sinensis, Ixobrychus cinnamomeus, Ardea purpurea, Turdoides earlei, Chrysomma altirostris* (very rare), *Amandava amandava, Prinia flaviventris, Prinia burnesii, Ploceus benghalensis,* and in winter *Acrocephalus* spp.

v) **Tropical Thorn Forest.** This is a major habitat originally occupying all the Indus plains from the foothills to the seacoast, but in consequence of human activity over more than one thousand years, most of this forest in its original form has disappeared. Tropical thorn forest survives in small pockets, sometimes recently regenerated in such areas as airfield peripheries, as well as around graveyards and uncultivated areas such as saline flats or 'pats'.

Typical plant species: Prosopis spicigera, Capparis aphylla, Salvadora oleoides, or *Salvadora persica, Tamarix aphylla, Ziziphus mauritiana,* (and in lower Sind *Euphorbia caducifolia*) *Calatropis procera, Suaeda fruticosa* with such grasses as *Aristida depressa, Eleusine compressa.* The first five plants form very scattered shrubby trees usually much affected by lopping.

Typical bird species: Francolinus pondicerianus, Merops orientalis, Pycnonotus leucogenys, Saxicola caprata, Prinia buchanani, Turdoides caudatus, Streptopelia decaocto, and in winter *Sylvia curruca, Oenanthe picata.*

v) **Sand dune desert.** There are five main sand dune desert areas, widely separated and one of these is located between 610 and 1,060 metres (2,000 and 3,500 feet) above sea level, the others less than 152 metres (500 feet) above sea level.

Typical plant species:
a) Thal desert: *Prosopis spicigera, Ziziphus mauritiana, Salvadora oleoides, Calligonum polygonoides, Calotropis procera, Aristida mutabilis, Saccharum bengalense.* The first three plants can grow to fair sized trees, whilst the last two are grasses and *Calligonum* is particularly adapted to colonize sand dunes.

b) Cholistan desert: *Tamarix aphylla, Prosopis spicigera, Capparis decidua, Calligonum polygonoides, Leptadenia spartium, Haloxylon griffithii,* with grasses such as *Aristida depressa, Saccharum spontaneum, Cymbopogon shoenanthus,* and *Pennisetum* spp.

c) Thar desert: *Prosopis spicigera, Tamarix aphylla, Euphorbia caducifolia, Commiphora mukul, Ziziphus nummularia, Grewia tenax, Cassia angustifolia, Calligonum polygonoides, Blepharis sindica.*

d) Sibi desert: *Capparis decidua, Suaeda fruticosa, Tamarix troupii,* with grass *Panicum antidotale.*

e) Chagai, Baluchistan— Situated between 610 and 1,060 metres (2000 and 3,500 feet) elevation: *Haloxylon ammodendron, Rhazya stricta, Astragalus sericostachys, Peganum harmala, Salsola arbuscula,* with grasses *Eleusine compressa, Pennisetum dichotomum* and *Andropogon halepensis* and *Nepeta glomerulosa.*

Typical bird species: Gyps fulvus, Aquila rapax, Francolinus pondicerianus, Chlamydotis undulata, Burhinus oedicnemus, Cursorius cursor, Pterocles senegallus, Pterocles exustus, Eremopterix nigriceps, Alaemon alaudipes, Turdoides caudatus, Lanius excubitor, supplemented in winter by *Asio flammeus, Circus macrourus, Pterocles orientalis, Anthus campestris, Oenanthe deserti, Sylvia nana* and *Sylvia curruca minula.*

FOOTHILL ZONE— HILLY TRACTS— POTOHAR AND UPLAND PLATEAUX. These are all arid tracts with ridges of sandstone and limestone escarpments and in the northern regions interspersed with loess soil deposits. Generally they are heavily overgrazed by domestic stock and pocked by severe gully erosion.

vi) **Dry Sub-tropical Semi-evergreen Scrub Forest**

Typical plant species:
a) In Sind Kohistan and southern Baluchistan, areas with pronounced humid winds during monsoon season, but hot dry and relatively frost free for the rest of the year. Clumps of cactus-like *Euphorbia* dominate the landscape of such areas. Examples, Makran range, Lakkhi and Pabb hills, Kirthar range. *Acacia jacquemontii, Acacia senegal, Commiphora mukul, Ziziphus nummularia* forming scattered stunted trees or tall bushes and *Rhazya stricta, Euphorbia caducifolia, Grewia tenax,* and *Blepharis sindica.*

b) In the Punjab Salt Range, Kalachitta hills, parts of Dera Ismail Khan and South Waziristan along the eastern flanks of the hills, forming in the more hilly areas an open canopy of stunted wild Olive and Acacia trees. *Olea cuspidata, Acacia modesta, Adhatoda vasica, Monotheca buxifolia, Dodonaea viscosa, Carissa opaca* and grasses such as *Tetrapogon villosus. Eleusine flagellifera* and *Cenchrus pennisetiformis.*

Typical bird species: *Pterocles indicus, Ammoperdix griseogularis, Pycnonotus leucogenys, Emberiza striolata, Bucanetes githagineus, Euodice malabarica, Lanius excubitor, Lanius schach, Oenanthe alboniger, Saxicoloides fulicata, Eremopterix grisea, Ammomanes deserti, Dendrocopus assimilis, Prinia gracilis* supplemented in winter by *Oenanthe xanthoprymna, Oenanthe picata capistrata* and *Coccothraustes coccothraustes.*

vii) **Dry Temperate Semi-evergreen Scrub Forest.** These areas have long cold winters often with some seasonal winter rains and show Mediterranean affinities both in plant and animal fauna. Covering a wide geographic range latitudinally, they are best sub-divided into southern and northern zones.

a) In the lower valleys and southern parts of Baluchistan, also the eastern or outer fringes of North Waziristan, Khyber and Mohmand tribal agencies, Bannu and Kohat. Hot dry summers and limited monsoon influence characterize these areas which often have some spring and winter rains. *Olea cuspidata, Acacia modesta, Artemisia maritima, Monotheca buxifolia, Rhazya stricta, Withania coagulans* with occasional trees of *Celtis eriocarpa* and in ravines *Nannorrhops ritchieana*, dwarf palm. Grasses include *Eleusine compressa, Chrysopogon aucheri, Cymbopogon jwarancusa* and *Saccharum spontaneum.* Shrubs include *Convolvulus spinosus* and *Adhatoda vasica.*

Typical bird species: *Ammomanes deserti, Anthus similis, Gyps fulvus, Ammoperdix griseogularis, Emberiza striolata.*

b) Lower slopes and main valley in the southern part of Chitral, Dir, Malakand agency, Indus Kohistan, Amb and Bunir. These areas also have hot dry summers with some winter rains. *Olea cuspidata, Acacia modesta, Monotheca buxifolia, Adhatoda vasica, Dodonaea viscosa, Mallotus philip-*

pinensis, Lannea coromandelica, with grasses such as *Cymbopogon jwarancusa* and *Saccharum spotaneum.*

Typical bird species: *Ammoperdix griseogularis, Caprimulgus affinis, Ammomanes deserti, Prinia criniger, Turdoides caudatus, Bucanetes githagineus* supplemented in winter by *Parus major, Prunella atrogularis, Phylloscopus inornatus, Coccothraustes coccothraustes* and *Emberiza cia.*

viii) **Sub-tropical Pine Forest.** This very characteristic zone is the first example listed of tall-tree forest and occurs on higher hill ranges from 910 up to 1,820 metres (3,000-6,000 feet) elevation and the dominant plant is the Chir Pine. Examples are Kahuta, parts of Mangla dam catchment, lower Kaghan valley around Kuwai, Batrasi pass, lower Swat around Murghazar and the lower Murree hills around Tret. Such forests are very subject to periodic fires. Fortunately the Chir Pine (*Pinus roxburghii*) is strongly fire resistant.

Typical plant species: *Pinus roxburghii, Quercus incana, Ficus palmata, Ficus roxburghii, Punica granatum* with understorey of *Ziziphus oxyphylla, Carissa opaca, Woodfordia fruticosa, Spiraea canescens, Buddleia paniculata, Beberis lycium, Indigofera pulchella* with grasses such as *Heteropogon contortus, Aristida cyanantha, Apluda aristata* and *Themeda anathera.*

Typical bird species include summer breeding visitors, marked with an asterisk. *Lophura leucomelana, *Streptopelia chinensis, Psittacula cyanocephala, *Otus sunia, Glaucidium cuculoides, *Caprimulgus indicus hazarae, Dendrocopos auriceps, *Hirundo daurica, Anthus sylvanus, Pericrocotus ethologus, Pycnonotus leucogenys, Prinia criniger, *Muscicapa latirostris, Seicercus xanthoschistos, Lonchura punctulata, *Melophus lathami.*

ix) **Tropical Dry Mixed Deciduous Forest.** This is perhaps the most interesting ecological zone in Pakistan, though extremely limited in extent and confined mainly to more sheltered ravines or northern facing slopes in a disjunct distribution. This ecological zone nevertheless represents an extension of the Siwalik zone further to the east with a rich association of Indo-Malayan plant genera. The best examples are the Karot valley which drains into the Jhelum river, also Kahuta, the lower Lehtrar valley and many of the side ravines in the Margalla hills to the west of Islamabad. Over 40 species of broad-leaved trees, shrubs and woody climbers have been identified in this zone and it is possible only to give a selective list of the commoner species.

Typical plant species: Acacia modesta, Bauhinia variegata, Cassia fistula, Celtis eriocarpa, Mallotus philippinensis, Morus alba, Pistacia integerrima, Punica granatum, Pyrus pashia, Salmalia malabaricum and Sterculia villosa with an understorey of Zizyphus mauritiana, Adhatoda vasica, Carissa spinarum, Clematis gouriana, Indigofera gerardiana, Porana paniculata, Elaeodendron glaucum, and Woodfordia floribunda. Amongst the grasses Heteropogon contortus and Aristida cyanantha, Apluda aristata and Themeda anathera are important sources of fodder.

Typical bird species, including summer breeding visitors: Francolinus francolinus, Lophura leucomelana, Streptopelia chinensis, Dendrocitta formosae, Clamator jacobinus, Cuculus varius, Cacomantis merulinus, Otus sunia, Taccocua leschenaultii, Glaucidium cuculoides, Caprimulgus macrurus, Megalaima asiatica, Dendrocopos macei, Pitta brachyura, Ptyonoprogne rupestris, Anthus similis, Pycnonotus leucogenys, Prinia hodgsonii, Rhipidura albicollis, Pomatorhinus erythrogenys, Strachyris pyrrhops supplemented in winter by Prunella strophiata, Phoenicurus erythrogaster, Turdus boulboul, Carpodacus rhodochlamys, Carduelis spinoides.

MONTANE REGIONS INCLUDING UPLAND VALLEYS WITHIN MAJOR MOUNTAIN RANGES

x) **Baluchistan hill ranges in southern latitudes, and lower slopes of some northern ranges.** For example Chaman, Mashlak reserve, Hazar Ganji reserve south-west of Quetta, the Harboi hills in Kalat and the Surkhab valley near Pishin. These areas have lost most of their original tree cover due to heavy lopping for goat browse and cutting for fuelwood. A few relict patches of what steppic montane scrub forest once prevailed, can only be found now in forest reserves or wildlife sanctuaries such as those listed above.

Typical plant species: Juniperus macropoda on higher ridges, Olea cuspidata, Pistacia mutica, Pistacia khinjuk, Fraxinus xanthoxyloides, with widely scattered bushes of Sophora molle, Artemisia maritima, Ephedra major, Prunus eburnea, Stocksia brahuica, and many bulbous perennials such as Tulipa sp., Ferula sp., Iris sp., and Allium sp. which make a brave show of colour in spring. Grasses include Eleusine flagellifera in southern regions, Cymbopogon parkeri and many ephemerals of Poa sp. and Bromus sp.

Typical bird species: Alectoris chukar, Ammoperdix griseogularis, Streptopelia senegalensis, Otus brucei, Caprimulgus europaeus, Merops apiaster, Ammomanes cincturus, Calandrella acutirostris Ptyonoprogne obsoleta, Hirundo daurica, Anthus similis, Scotocerca inquieta, Sylvia hortensis, Saxicola torquata, Oenanthe picata, Monticola solitarius, Sitta tephronota and Emberiza bunchanani.

xi) **Baluchistan Higher Ranges.** This is a very arid montane area with both diurnal and seasonal extremes of temperature. Winter temperatures often dropping to —14 °C and rising in summer above 35 °C. Light snowfall may carpet the higher slopes from January to March. Most of the plant fauna has Irano-Turanian affinities, though quite a few can be considered of purely Himalayan origins. The area is typified by such ranges as Takatu, Zarghun, Wam-Pilghar, Ziarat and Toba Kakar with closely similar associations in northern Malakand and the south-western part of Chitral.

Typical plant species: Juniperus macropoda (syn. polycarpus), with occasional Fraxinus xanthoxyloides, Pistacia khinjuk with scattered bushes of Prunus eburnea, Berberis baluchistanica, Caragana ambigua, Rosa moschata and Thymus serpyllum, Salvia cabulica, Sophora griffithii. On sheltered slopes the golden spires of Eremurus aurantiacus, are conspicuous in spring, with Tulipa chrysantha, and curious spiny hassock-shaped plants such as Acantholimon fasciculare, and Onobrychis cornuta. Grasses include Stipa pennata and Pennisetum orientalis.

Typical breeding birds of the juniper forest zone: Pyrrhocorax pyrrhocorax, Pica pica, Emberiza stewarti, Mycerobas carnipes, Carpodacus rhodochlamys, Carpodacus erythrinus, Serinus pusillus, Parus major, Parus rufonuchalis, Aegithalos leucogenys, Certhia himalayana, Garrulax lineatus, Muscicapa striata, Phylloscopus griseolus, Phylloscopus neglectus, Scotocerca inquieta, Turdus viscivorus, Sylvia curruca althea, Phoenicurus ochruros, Lanius isabellinus, Hieraaetus pennatus and Accipiter nisus.

xii) **Himalayan Moist Temperate Forest.** This is again a most important biotope for wildlife particularly as a breeding zone for birds with Oriental or Himalayan origins. It is predominantly a coniferous forest with glades of mixed deciduous broad-leaved species, which regrettably are gradually being eliminated due to selective lopping and felling by local villagers. This forest is typified by most of the Murree hill range (including the Galis), the lower Neelum and Kaghan valleys, extending westwards to parts of eastern Swat bordering on Indus Kohis-

Plate 2

MAP SHOWING SOME PLACES
REFERRED TO IN THE TEXT

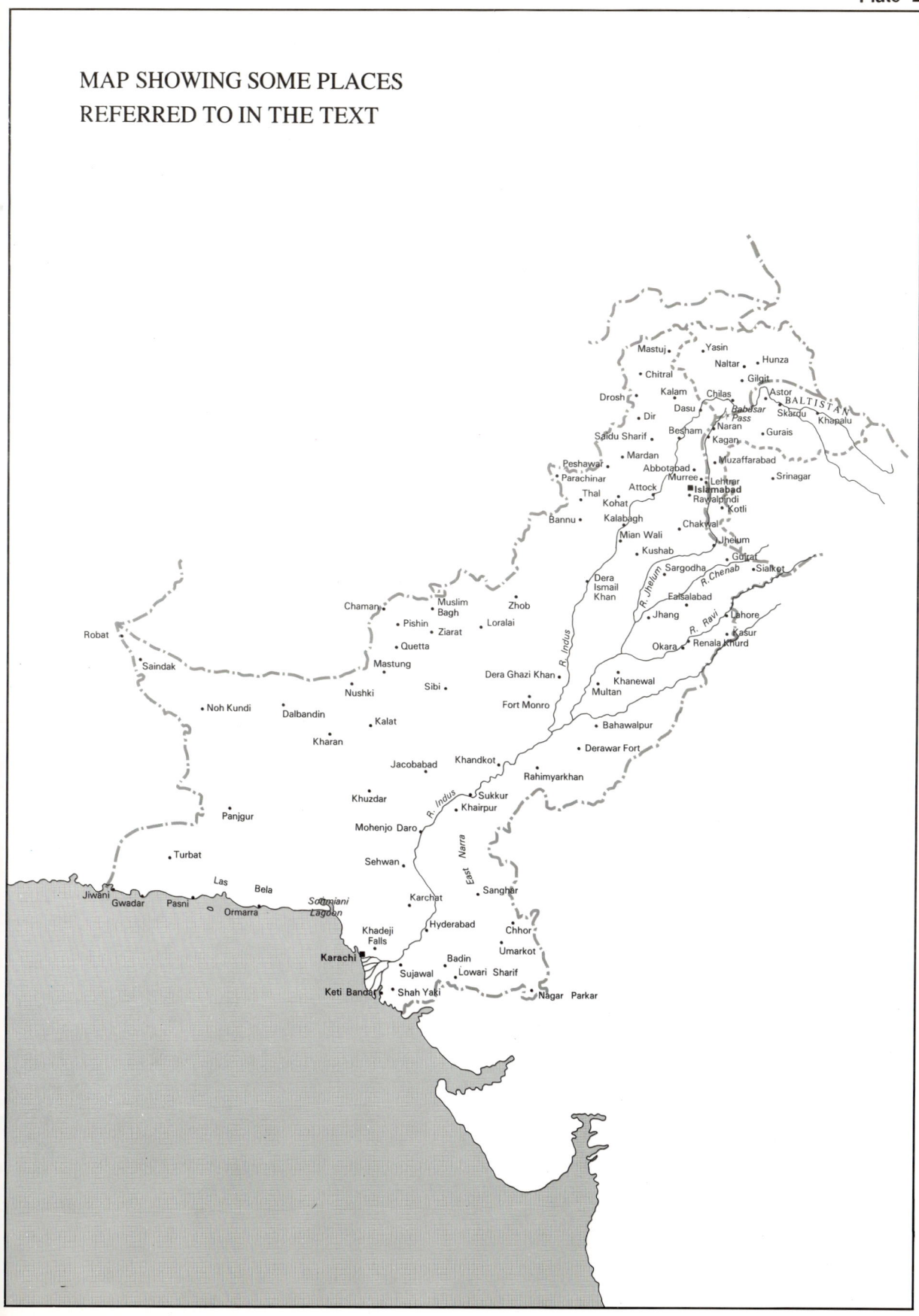

Plate 3

MAP SHOWING SOME PHYSICAL FEATURES
REFERRED TO IN THE TEXT

R.	—	River
L.	⬭	Lake
▲		Mountain Massif
⟫		Mountain Pass
⟓		Dam
⟓		River Barrage and Irrigation Headworks
✕		Parks and Forests

U.S.S.R.

CHINA

WAKHAN

Tirich Mir ▲

Mastuj R.

Shandur Pass

Naltar R.

Khunjerab Pass

Lowari Pass

Chitral

Kalam

Kargah R.

Gilgit R.

Gilgit

Shigar R.

K2

Kunhar R.

Panjkora R.

Swat R.

DIR

Indus R.

KOHISTAN

Chilas

Nanga Parbat

Skardu

Shyok R.

Kabul R.

Dubair Valley

Kaghan Valley

Burawai

Shogran

Astor R.

DROSAL PLATEAU

LADAKH

AFGHANISTAN

SAFED KOH RANGE

SAMANA HILLS

Tarbela Dam

Attock

Salkala

Muzaffarabad

Neelum R.

Burzil Pass

Kohat

KALLA CHITTA RANGE

Margalla

Margalla Hills

Jhelum R.

KASHMIR

PIR PANJAL RANGE

Baran Dam

Kurram R.

Nammal L.

SALT RANGE

Potohar

Mangla Dam

Marala Barrage

Chasma Barrage

Uchchali L.

Jhelum R.

Kushdil Khan L.

SHINGAR RANGE

Takhi-i-Suleiman ▲

THALL

Trimmu Barrage

Chenab R.

Balloki Barrage

MASHLAKH

Surkhab R.

URUK VALLEY

Ziarat

Takhatu ▲

Kaliphat

DESERT

Taunsa Barrage

SULEIMAN HILLS

Sidhnai Spill

East Channel

Changa Manga Plantation

Suleimanki Barrage

CHILTAN RANGE

Zangi Nawar L.

Bolan Pass

Pirawala Plantation

Sutlej R.

HARBOI RANGE

CHAGAI DESERT

IRAN

Panjnad Barrage

Lal Sohanran ✕

CHOLISTAN DESERT

RAJASTHAN

INDIA

SIBI DESERT

Ghauspur L.

Guddu Barrage

KIRTHAR RANGE

Manchar L.

Sukkur Barrage

Hingol R.

East Nara

MEKRAN

Akari Dail

Siranda L.

Baren R.

Jamrao Head

THAR DESERT

Sonmiani Lagoon

Hab Dam

Hab R.

Hyderabad

Astola Is.

Kinjhar L.

Cape Monze

Malir

Haleji L.

Karachi

Mehboob Shah L.

GREAT RANN OF KUTCH

ARABIAN SEA

Plate 4

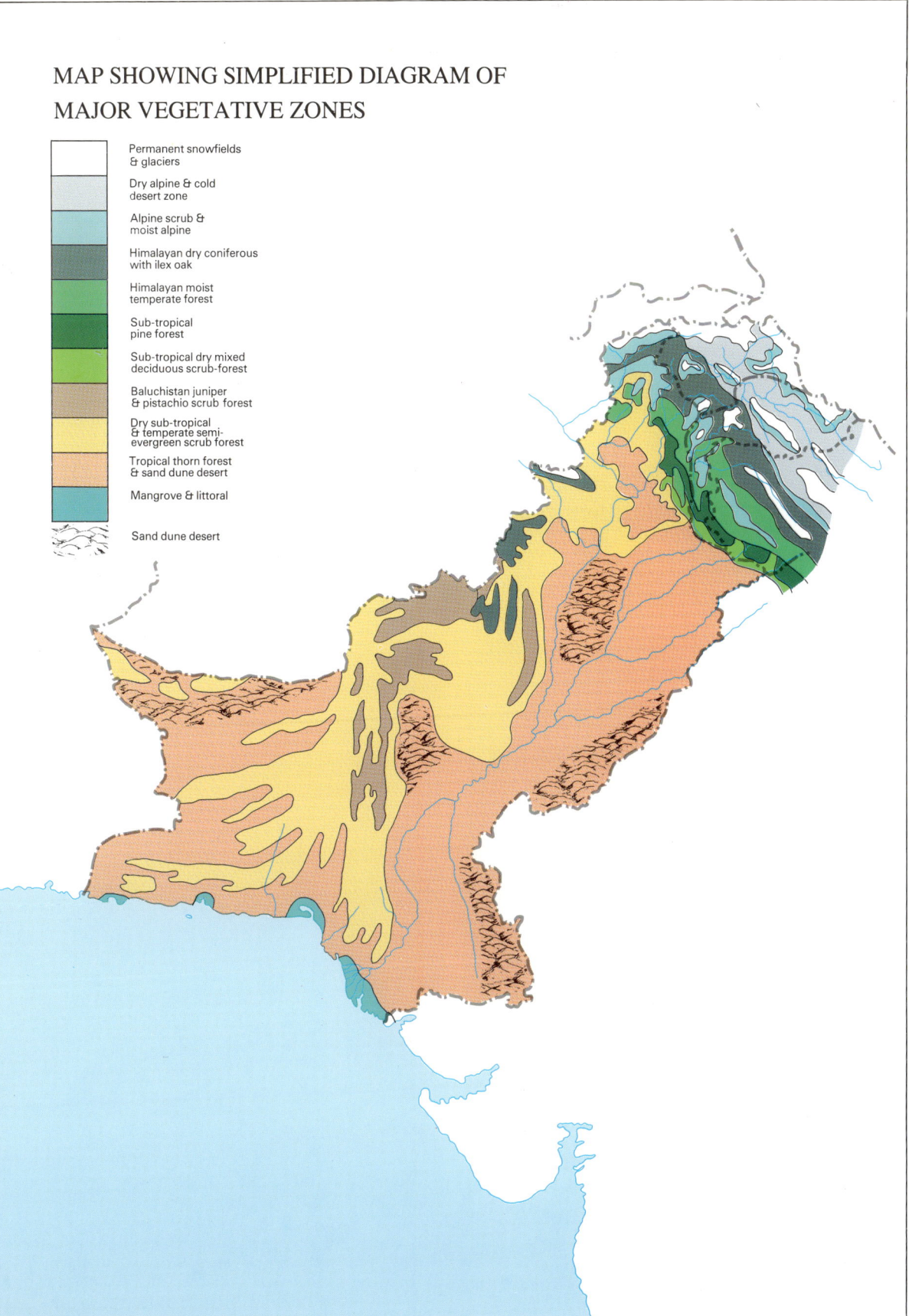

MAP SHOWING SIMPLIFIED DIAGRAM OF
MAJOR VEGETATIVE ZONES

Permanent snowfields
& glaciers

Dry alpine & cold
desert zone

Alpine scrub &
moist alpine

Himalayan dry coniferous
with ilex oak

Himalayan moist
temperate forest

Sub-tropical
pine forest

Sub-tropical dry mixed
deciduous scrub-forest

Baluchistan juniper
& pistachio scrub forest

Dry sub-tropical
& temperate semi-
evergreen scrub forest

Tropical thorn forest
& sand dune desert

Mangrove & littoral

Sand dune desert

Plate 5: Coast, Riverain and Desert Tracts

(a) Mangrove forest at low tide, Sonmiani lagoon, Lasbela, comprising *Avicennia officinalis*. Breeding habitat for *Phalacrocorax fuscicollis*, *Egretta gularis*, *Butorides striatus*, *Haliastur indus*.

(b) Riverain forest (in dry season) alongside Indus river near Sujjawal, Thatta district, with closed canopy *Acacia arabica*, and *Tamarix dioica*, in clearings. Breeding habitat of *Pernis ptilorhyncus*, *Dendrocopos assimilis*, *Pericrocotus cinnamomeus*, *Turdoides striatus*, and *Passer pyrrhonotus*.

(c) Swamps and seasonal inundation zones, Mahboob Shah, Thatta district with flowering heads of *Phragmites karka* in foreground, and *Tamarix dioica* bushes in background. Breeding habitat of *Egretta garzetta*, *Egretta intermedia*, *Porphyrio porphyrio*, and *Amaurornis phoenicurus*. Wintering habitat of migrant rallidae, Flamingoes (in picture), *Phalacrocorax niger*, and Palearctic Dabbling Ducks and Waders.

(d) Sand dune desert near Anam Bostan, Chagai with scattered bushes of *Calligonum polygonoides* and clumps of *Eleusine compressa* grass. Breeding habitat of *Cursorius cursorius*, *Alaemon alaudipes*, and *Hippolais C. rama*. Wintering ground of *Pterocles alchata*, and *Corvus ruficollis*.

Plate 6: Dry Sub-tropical Foothill Tracts

(a) Dry sub-tropical semi-evergreen in coastal tracts with Euphorbia-typical scrub.

Habb valley near Marri Manghtar hills *Euphorbia caducifolia* in foreground with scattered bushes of *Acacia jacquemontii, Capparis aphylla* and *Ziziphus nummularia.* Breeding habitat of *Burhinus oedicnemus, Pterocles lichtensteinii, Mirafra erythroptera, Eremopteryx nigriceps, Bucanetes githagineus* and *Emberiza striolata*

(b) Dry sub-tropical semi-evergreen scrub forest, interior of Sind.

Kirthar hills, Dadu district, with scattered *Acacia senegal, Ziziphus nummularia* and *Commiphora mukul,* bushes. Breeding habitat of *Ammoperdix griseogularis, Ammomanes deserti, Anthus similis, Ptyonoprogne fuligula, Pycnonotus leucogenys, Oenanthe alboniger* and *Euodice malabarica.*

(c) Dry sub-tropical semi-evergreen wild olive/Acacia scrub, Punjab.

Forest, Kala Chitta range, Attock district with open canopy *Acacia modesta* and *Olea cuspidata* and undershrubs of *Dodonea viscosa, Carissa opaca* and *Nerium odorum* (foreground) on river banks. Breeding habitat of *Columba palumbus, Caprimulgus affinis, Dendrocopos assimilis, Ammomanes deserti, Saxicoloides fulicata,* and *Saxicola caprata.*

(d) Tropical dry mixed-deciduous forest, outer Himalayan foothill zone.

Margalla hills near Islamabad with open canopy *Acacia modesta, Cassia fistula, Mallotus philippensis, Pistacia integerrima* with undershrubs of *Adhatoda vasica, Ziziphus mauritiana* and *Heteropogon contortus* grass. Breeding habitat of *Francolinus francolinus, Cacomantis merulinus, Pitta brachyura, Pomatorhinus erythrogenys, Stachyris pyrrhops, Prinia hodgsonii* and *Seicercus xanthoschistos.*

Plate 7: Dry-temperate Montane Forest

(a) Medium altitude deciduous scrub forest

Baluchistan, Hazar Ganji reserve in Chiltan hills with scattered trees of *Pistacia mutica* and *Fraxinus xanthoxyloides*, with undershrubs of *Sophora molle* and *Salvia sp;* and clumps of *Cymbopogon parkeri* grass. Breeding habitat of *Alectoris chukar, Otus brucei, Caprimulgus europaeus, Hirundo daurica, Anthus similis, Sylvia hortensis, Sitta tephronota, Lanius vittatus* and *Bucanetes githagineus*.

(b) High altitute juniper forest

Prospect Point Ziarat, with open canopy *Juniperus macropoda* and hassock shaped clumps of *Astragalus* sp. and bushes of *Berberis baluchistanica, Prunus eburnea*. Breeding habitat of *Sylvia curruca, Phylloscopus griseolus, Phoenicurus ochrurus, Parus rutoncuchalis, Serinus pusillus, Mycerobas carnipes* and *Carpodacus rhodochlamys*.

(c) High altitude 'Chilghoza' pine forest.

Shingar range, Zhob district, Baluchistan with open canopy *Pinus gerardiana* forest and *Pistacia cabulica* with shrubs of *Ephedra nebrodensis*, and *Sophora molle*. Breeding habitat of *Saxicola torquata, Phoenicurus ochrurus, Monticola solitarius, Prinia criniger, Sylvia curruca, Phylloscopus neglectus, Sitta europaea, Scotocerca inquieta* and *Garrulus lanceolatus*.

(d) Medium altitude dry oak and pine forest.

Safed Koh range above Peiwar spur with stunted *Pinus wallichiana* and *Quercus ilex* (visible middle distance), with undershrubs of *Sophora griffithi* and *Cotoneaster* sp. Breeding habitat of *Muscicapa striata, Sylvia curruca, Phylloscopus subviridis, Aegithalos leucogenys, Nucifraga caryocatactes* and *Carduelis carduelis caniceps*.

Plate 8: Himalayan or Northern Mountain Forest

(a) Himalayan moist temperate forest.

Shahran in Manshi forest, Hazara district with *Juglans regia* (mid-picture), *Acer caesium*, *Picea morinda* (extreme left) and *Abies pindrow* (behind rest house), with undershrubs of *Viburnum nervosum* (foreground). Breeding habitat of *Pucrasia macrolopha*, *Strix aluco*, *Cuculus saturatus*, *Luscinia brunnea*, *Hodgsonius phoenicuroides*, *Saxicola ferrea*, and *Monticola cinclorhyncha*.

(b) Sub-tropical pine forest.

Batrasi pass, Mansehra district with *Pinus roxburghii* canopy forest and *Pistacia integerrima* in ravines, with undershrubs of *Ziziphus oxyphylla*, *Berberis lycium* and *Ficus palmata*. Breeding habitat of *Lophura leucomelana*, *Streptopelia chinensis*, *Dendrocopos auriceps*, *Prinia criniger* and *Melophus lathami*.

(c) Himalayan alpine scrub-forest

Naltar valley, Gilgit, with *Betula utilis* birch forest and patches of *Salix himalayensis* scrub, and on cliffs, colonies of *Bergenia stracheyi*. Breeding habitat of *Lophophorus impejanus*, *Anthus trivialis*, *Prunella strophiata*, *Tarsiger cyanurus*, *Phylloscopus affinis*, *Luscinia pectoralis*, and *Phoenicurus frontalis*.

(d) Himalayan dry coniferous forest mixed with temperate deciduous trees.

Machiara sanctuary, Azad Kashmir showing *Abies pindrow* and *Picea morinda* on higher slopes with *Quercus semecarpifolia* below. Breeding habitat of *Tragopan melanocephalus*, *Glaucidium brodiei*, *Zoothera dauma*, *Ficedula tricolor*, *Phylloscopus inornatus*, *Parus rufonuchalis*, *Sitta leucopsis*, and *Mycerobas icterioides*.

Plate 9: Cold Desert and Alpine Zone in Far Northern Areas
Maps showing mean annual rainfall/temperature zones

(a) Cold dry valley bottoms in far north.

Shyok valley in eastern Baltistan near Barra with *Populus ciliata* in background and *Hippophae rhamnoides* alongside river, breeding habitat of *Upupa epops, Pica pica, Corvus corone, Motacilla alba hodgsonii, Luscinia svecica, Oenanthe picata*, and *Parus major casohmirensis*.

(b) Moist alpine zone above tree level.

Head of Jabba valley, Swat, with patches of *Juniperus squamata* on scree slopes and *Polygonum affine* by snow melt streams. Breeding biotope of *Tetraogallus himalayensis, Columba leuconota, Apus apus, Anthus roseatus, Prunella himalayana, Turdus merula maximus*, and *Leucosticte nemoricola*.

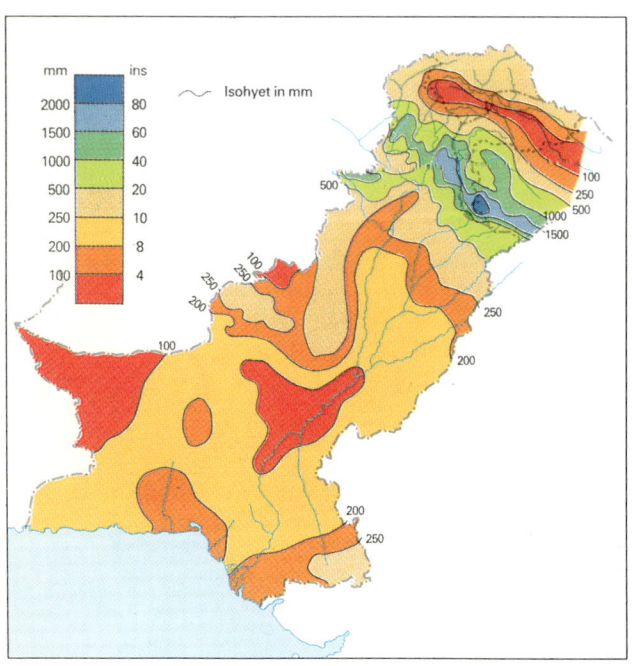

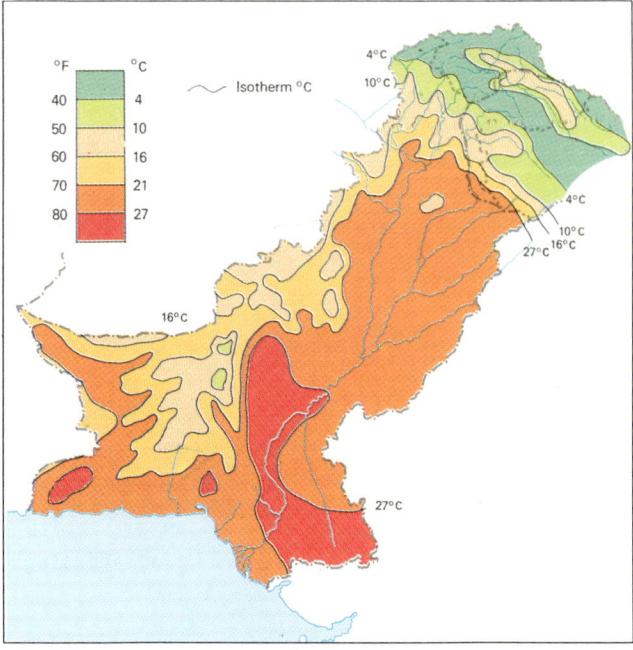

(c) Map showing mean annual rainfall zones

(d) Map showing mean annual temperature zones

tan. Again because of the great variety of associated plant species it is only possible to list a few of the more conspicuous elements.

Typical plant species: *Abies pindrow* in northern aspects or moister slopes, *Pinus wallichiana* with *Taxus baccata* as an understorey and occasional *Cedrus deodara* on dryer hotter slopes. Broad-leaved trees include *Ulmus wallichiana, Juglans regia, Quercus dilatata, Acer caesium, Acer villosum* and *Prunus cornuta.* The shrub layer comprises *Viburnum nervosum, Berberis lycium, Rosa moschata, Skimmia laureola,* and *Lonicera alpigena.* Forbes include many species of *Impatiens* and *Euphorbia* as well as *Viola, Fragaria and Gentiana* spp. Creepers include *Hedera nepalensis* and *Clematis montana.*

Typical bird species: *Dendrocopos himalayensis, Picus squamatus, Megalaima virens, Cissa flavirostris, Hypopicus hyperythrus, Streptopelia orientalis, Psittacula himalayana, Cuculus canorus, Cuculus saturatus, Otus spilocephalus, Glaucidium brodei, Strix aluco, Caprimulgus indicus hazarae, Hirundapus caudactus, Delichon urbica, Coracina melaschistos, Pericrocotus ethologus, Hypsipetes madagascariensis, Luscinia brunnea, Hodgsonius pheonicuroides, Saxicola ferrea, Zoothera dauma, Turdus rubrocanus, Turdus boulboul, Enicurus maculatus, Phylloscopus occipitalis, Phylloscopus proregulus, Niltava sundara, Muscicapa thalassina, Ficedula superciliaris, Garrulax variegatus, Pteruthius flaviscapis, Parus melanolophus* and *Parus monticolus.*

xiii) **Himalayan Dry Coniferous Forest.** This forest is associated with the inner or more northerly ranges of the Himalayas which are less subject to monsoon influence. It tends to be a gradually changing inter-face with moist temperate forest but is generally characterized by much fewer deciduous tree species and single specie stands of conifer. It occurs between 1,500 and 3,350 metres (5,000 and 11,000 feet) elevation. According to latitude it extends from Gilgit, Astor and Chitral in the north down to the Safed Koh range and Takht-i-Suleiman in the south. It can be conveniently sub-divided according to the predominant tree canopy association.

Typical plant species:
a) Dry temperate evergreen oak and deodar forest. Typified by lower Indus Kohistan, Swat Kohistan, northern Dir and parts of Chitral and inner valleys of Hazara. Higher up *Cedrus deodara* and *Pinus wallichiana,* with lower down *Quercus ilex* with a few *Juglans regia* and scattered bushes of *Artemisia maritima, Ephedra intermedia,* *Periploca aphylla, Monotheca buxifolia, Corylus corlurna, Parrotia jacquemontiana, Cotoneaster nummularia* and *Sophora mollis.*

b) Dry zone blue pine forest and spruce forest. Typified by Naltar valley and Astor in Gilgit and Takht-i-Suleiman. *Picea smithiana, Pinus wallichiana,* and a scattering of *Populus ciliata, Plectranthus rugosus, Rosa webbiana, Ribes grossularia, Prunus jacquemontii, Artemisia maritima, Berberis gambleana* and *Colutea armata.*

c) Dry zone Chilgoza and Holly-oak. Typified by parts of Chilas, Dir, Gilgit agency, Safed Koh slopes of intermediate elevation, Takht-i-Suleiman, and higher ranges of Malakand agency. *Pinus wallichiana, Pinus gerardiana, Cedrus deodara, Quercus ilex* with scattered shrubs of *Daphne oleoides, Sophora griffithii, Cotoneaster nummularia, Berberis baluchistanica, Artemisia maritima, Plectranthus rugosus* and *Berberis lycium.*

d) Higher or inner range Himalayan Dry Coniferous with *Abies pindrow, Picea smithiana, Pinus wallichiana* and *Quercus semicarpifolia*—Neelum valley, Azad Kashmir, Salkhalla and Machiara.

Typical bird species: *Buteo rufinus, Garrulus lanceolatus, Nucifraga caryocatactes, Corvus macrorhynchos, Troglodytes troglodytes, Prunella strophiata, Phoenicurus caeruleocephalus, Phylloscopus subviridis, Phylloscopus inornatus, Ficedula tricolor, Aegithalos leucogenys, Sitta leucopsis, Certhia himalayana, Serinus pusillus, Carduelis carduelis caniceps,* and *Carpodacus erythrinus.*

xiv) **Cold Desert and Dry Alpine Zone.** Most of the extreme northern regions of Chitral, Gilgit, Hunza and all of Baltistan fall into this category. Being an area of comparatively recent geologic activity, it is characterized by some of the highest and most spectacular mountain scenery in the world. But in the valley floors the landscape appears a desolate waste of huge boulders, drifting sand and sheer cliffs apparently devoid of any soil. From an ecological viewpoint there are small areas of moisture below glaciers and snowfields and on the banks of upland streams which afford a foothold for plant growth and in simplified form this can be divided into three regions.

Typical plant species:
a) Valley bottoms and stream beds.
   Hippophae rhamnoides, Myricaria elegans with occasional *Populus ciliata* especially

near villages, and *Salix viminalis*. A few widely scattered shrubs and forbes include *Capparis spinosa, Tribulus terrestris, Peganum harmala, Sophora alopecuroides* and *Lycium ruthenicum*.

b) Alpine slopes next to snow-melt or sheltered ravines. *Salix denticulata, Mertensia tibetica, Potentilla desertorum, Juniperus polycarpos, Polygonum viviparum, Berberis pachyacantha, Rosa webbiana* and *Spiraea lycioides, Bergenia stracheyi, Primula macrophylla*, grasses include *Festuca altaica* and *Poa attenuata*.

Typical bird species: *Aquila chrysaetos, Gypaetus barbatus, Gyps himalayensis, Falco tinnunculus, Alectoris chukar, Tetraogallus himalayensis, Columba rupestris, Columba leuconota, Apus apus, Eremophila alpestris, Ptyonoprogne rupestris, Delichon urbica, Anthus trivialis, Troglodytes troglodytes, Prunella collaris, Luscinia svecica, Phoenicurus erythrogaster, Oenanthe pleschanka, Tichodroma muraria, Pyrrhocorax pyrrhocorax, Pyrrhocorax graculus* and *Carpodacus puniceus*.

xv) **Himalayan Moist Alpine Zone.** This is an important zone ecologically for many breeding Himalayan migrants. Typified by continuous grass swards with scattered patches of recumbent juniper and dotted all over with screes and tumbled boulders. In spring it is criss-crossed by surface streams and runlets and often carpeted with alpine flowers. Everywhere it is subject to heavy grazing pressure from domestic flocks, aggravated in recent years by the great influx of nomadic Afghan flocks brought in by the refugees.

In ravines and sheltered alps there are often considerable patches of White Birch forest and thickets of wild rose and berberis. This zone occurs throughout the higher slopes of the Kaghan valley, Azad Kashmir and parts of Swat, Dir and Indus Kohistan.

Typical plant species: *Betula utilis, Juniperus squamata, Salix himalayensis, Rhododendron collettianum, Polygonum affine, Saxifraga sibirica, Draba trinervis*. Many grasses are annuals of the Poa genus, the beautiful purple flowered *Iris hookeriana* is conspicuous in May, also many species of *Anemone, Primula* and *Gentiana*.

Typical bird species: *Gyps himalayensis, Gypaetus barbatus, Tetraogallus himalayensis, Columba leuconota, Cuculus canorus, Motacilla citreola, Anthus roseatus* syn. *pelopus, Anthus trivialis, Pyrrhocorax graculus, Phylloscopus affinis, Luscinia pectoralis, Troglodytes troglodytes, Phoenicurus ochruros, Leucosticte nemoricola, Carpodacus thura.*

CHANGES IN THE DISTRIBUTION AND STATUS OF BIRD SPECIES AS A CONSEQUENCE OF HUMAN ACTION

As indicated earlier in this chapter, one of the most significant natural factors affecting the distribution of birds is the south-west monsoon. Though cyclical in intensity and by no means of regular occurrence, the monsoon is fortunately not affected by man's activities. However two important consequences of the monsoon are exploited by men.

The coastal waters of Pakistan have a bench extending in average for 32 kilometres (20 miles) out to sea, over which the water depth rarely exceeds 20 fathoms. This benthic region, subject to some influence from the sun's rays, experiences a major upwelling at the end of the monsoon due to the constant movement inland of the warmer surface water swept by the monsoon winds. Because the coastline lies at an angle to this movement (See **Fig. iii**) these waters move along the coast and are replaced by an upwelling of deeper cool waters which bring with them fresh supplies of mineral nutrients, which fertilize the primary production of phyto-plankton. This in turn, sets off a life cycle of zoo-plankton upon which a great variety of molluscs, crustacea and fish can subsist, which in turn are exploited by larger predatory carnivorous fish. As a consequence a large population of sea birds of considerable specific diversity, exploits this rich food source including many species which breed in very distant Arctic or Antarctic regions. There are recent very serious signs of human over-exploitation of coastal fish resources, particularly of estuarine or neritic species, such as Spiny Lobsters (*Panulirus ornatus*), shrimps (*Penaeus* and *Metapenaeus* spp.) and flatfish (*Synaptura* and *Psettodes* spp.) which spawn in these shallow waters. At present, however, there is little evidence that the sea bird population has yet been affected, largely because it often exploits food resources of no economic importance, such as Mud-skippers (*Periopthalmus* spp.) and Squids (*Sepia* spp.) Moreover as a recent sea trip around the Indus delta amply demonstrated to me, all the modern paraphernalia of radar and improved trawls employed by the fishing industry, cannot compete with the efficiency of aerial survey techniques employed by species such as terns and the Osprey.

The second and much more significant form of exploitation of the monsoon, involves the holding up of summer flood waters of the Indus river and its distribution through a network or irrigation canals, enabling crops to be grown under artificial irrigation.

This development has taken place largely over the past half century and is on a very large scale (there are an estimated 1,373 million hectares (33 million acres) in Pakistan under canal and well irrigation (*Agriculture Statistics*, 1981). Naturally there have been far-reaching ecological consequences. It has made economically worthwhile, the clearance and subsequent cultivation of huge tracts of former tropical thorn forest. It has controlled the annual flooding of the river through the succession of barrages built across the Indus and its tributaries, which have created headponds upstream of such barrages forming large storage reservoirs in summer. The negative impact upon wildlife has been the loss of habitat for specialized species due to the disappearance of tropical thorn forest as well as riverain gallery forest, which have gradually dried out or been cut down. The shrinkage or total disappearance of many oxbow lakes and seasonal swamps along the margins of former channels of the river has eliminated much suitable habitat for wetland species. But at the same time irrigation development has created compensatory effects. Paradoxically irrigation has created new swamps and wetlands, due to the major canals being unlined, particularly seepage zones around barrage headponds, as well as extensive reed beds due to waterlogging in such areas as the northern Thal, and the borders of the Thar desert, adjacent to Sanghar and Nawabshah districts. Vastly increased areas of cereal cultivation have created a new habitat equivalent to grassland steppe, but this is generally too unstable to attract many breeding bird species. Irrigation water has enabled the creation of extensive canal and roadside tree belt plantations, and especially irrigated forest plantations. This has provided a new habitat for arboreal bird species or those which formerly depended upon riverain forest. In fact the latter, though usually consisting of single tree species mono-cultures, has provided refugia for several bird species formerly heavily dependent upon riverain forest.

A summarized list is presented on page 16 of those species which were particularly associated with the Indus plains region whose status appears to have changed significantly as a result of man's actions. Whilst no precise quantitative data is available, it is fortunate that several renowned ornithologists published notes on the birds observed in Punjab and Sind provinces in the early part of the twentieth century. Referring to such comments as are available on the population status of bird species, combined with the author's own personal diary notes spanning over 30 years of observations, it is possible to present a meaningful account of this phenomenon (See **Fig. iv**).

Commenting upon **Fig. iv** the Hawk Cuckoo is a specialized feeder preferring hairy caterpillars. In irrigated forest plantations the larvae of many moths have provided an almost limitless food supply which was absent in the dryer tropical thorn-scrub biotope. Similarly an increase in the planting of *Ficus* species around villages and wells, as well as increased planting of fruit gardens has provided ample food and increased canal and roadside tree plantation has favoured the spread of the arboreal and frugivorous Koel and the Crimson Breasted Barbet. Earth embankments associated with irrigation channels have afforded widespread nesting sites for the colonial breeding Bank Myna which seems better adapted also to exploit a rice cultivation habitat than any other species. Birds highly addicted to trees growing over water, both for nesting and foraging, such as the Sind Jungle Sparrow and Sind Starling have also increased remarkably. The changed status of species adapted to *Saccharum* thickets (a sort of Pampas grass growing in tall clumps) in the riverain tracts is less clearly discernible. Some species such as the Yellow-eyed Babbler and the Yellow-bellied Wren Warbler have apparently increased and become more widespread, probably because they can adapt and exploit cultivated crops of sugarcane which has spread rapidly in agricultural importance. Two endemic riverain bird species have undoubtedly become rarer and more localized, viz., Jerdon's Babbler and the White-tailed Stonechat. The former, where it does occur, exists sympatrically in considerable densities with the closely related Yellow-eyed Babbler so that inter-specific competition does not seem to be a factor. It appears always to have been very rare. As the intensity of cultivation increases throughout Pakistan, resulting in modernization of agricultural techniques, every major irrigation barrage headwork will assume greater and greater importance as a last refuge for those specialized mammal and bird species adapted to a riverain and seasonal swampland biotope. This is because these barrages are invariably surrounded by extensive reed beds and *Saccharum* thickets, resulting from seepage from the headponds which have been created. Such areas are already subject to heavy grazing pressure and burning of the reeds to provide better grazing, but it is hoped that at least some of these barrage areas will be designated as wildlife sanctuaries before it is too late.

Figure iv

CHANGING STATUS OF BIRD SPECIES IN INDUS PLAINS AS A RESULT OF DEVELOPMENT OF CANAL IRRIGATION AND ANCILLARY TREE PLANTATION

Note: The years in brackets following early authors refer to the known periods when they actually had made field observations, not to dates of publication.

SPECIES	FORMER STATUS	1980's STATUS based upon Author's observations
Nettapus coromandelianus Cotton Teal or Cotton Pygmy goose	Very rare and local in Sind (Ticehurst, 1917). K. Eates only 2 sightings during 30 years' official touring all over Sind (MS notes 1940s)	Locally common in summer months in Punjab. Now widespread in wetlands of lower Sind, east Narra.
Hierococcyx varius Common Hawk Cuckoo	Unknown Jhang district (Whistler, 1917-1920) and Rawalpindi district (Whistler, 1910-1926).	Very common Rawalpindi/Islamabad. Common in, Punjab irrigated forest plantations.
Eudynamy scolopacea Koel	Common Rawalpindi district 1926 but rare in Jhang (Whistler, 1917-1920) uncommon Salt Range (Waite, 1918-47). In Sind confined to Karachi environs (Ticehurst, 1917-20).	Now widespread throughout Sind and Punjab particularly around towns.
Halcyon smyrnensis White-breasted Kingfisher	Scarce resident in Jhang (Whistler, 1922) but widespread Sind and western parts of Punjab (Ticehurst, 1917-20 and Waite, 1930s).	Now very common and widespread throughout canal colonies.
Megalaima haemacephala Crimson-breasted Barbet	Uncommon in northern Punjab but believed to be extending its range—not recorded in Jhang (Whistler, 1917-22).	Common resident throughout irrigated canal colonies of Punjab.
Dendrocopos assimilis Sind Pied Woodpecker	Formerly confined to semi-evergreen dry sub-tropical and dry temperate scrub-forest biotope.	Quite common in irrigated forest plantations of Punjab.
Pericrocotus cinnamomeus Wandering Minivet	More or less confined to riverain forest or semi-evergreen dry sub-tropical scrub forest.	Very common in irrigated forest plantations.
Saxicola macrorhyncha Stoliczka's Bush Chat	Very rare and local. In Sind confined to Thar desert (Ticehurst). In Punjab only recorded near Jhang (Whistler).	Apparently disappeared from all parts of Punjab. Status in Tharparkar unknown but occurs regularly further east in Rajasthan.
Saxicola leucura White-tailed Bush Chat or Stone Chat	Fairly common but local in distribution and confined to riverain swamps (Ticehurst and Whistler).	Rare and local and confined to riverain swamps and seepage around barrage headworks.
Prinia flaviventris Yellow-bellied Longtail Warbler	Not encountered Bahawalpur (Salim Ali, 1939), rare and local in Sind (Ticehurst, 1917), not recorded Rawalpindi (Whistler, 1917-26) nor in Jhang (Whistler, 1922).	Common and widespread Punjab including Bahawalpur, Sind and Rawalpindi along major canals and lakes wherever extensive reed beds occur.
Chrysomma sinense Yellow-eyed Babbler	Seen once only in Jhang (Whistler, 1917-20) nor recorded Rawalpindi (Whistler, 1925). Widespread and fairly common in Sind around seasonal swamps (K. Eates, 1940s).	Common throughout Punjab in irrigated cultivation.
Chrysomma altirostris Jerdon's or Sind Babbler	Very rare or local and confined to reed beds but absent in some equally suitable areas.	Even rarer and more localized.
Sturnus vulgaris minor Sind Starling	Rare and local in Sind (Ticehurst, 1917-20), unknown Punjab (Whistler, 1917-20).	Occurs in frequent small colonies northern Sind, Kandhkot, also Balloki headworks, and Renala Khurd also Faisalabad along major canals.
Acridotheres ginginianus Bank Myna	Bahawalpur occurring locally and sparsely (Salim Ali, 1939), unknown Salt Range (Waite, 1918-47). Fairly common Jhang (Whistler, 1917-20). locally common in Sind but confined wetter areas such as rice growing tracts (Ticehurst, 1917-20).	Extremely abundant in major rice growing tracts in Sind and Punjab, common even in Salt Range, and around Bahawalpur.
Corvus corax laurencei Punjab Raven	Widespread Punjab especially common Gujrat district, Rawalpindi (Whistler, 1917-26), also Jhang, Multan etc.	Now almost never seen. A few pairs still haunt the more hilly parts of the Salt Range plateau but they have totally disappeared from irrigated canal colonies.

Zoogeographic Aspects of Bird Distribution

Even the amateur bird watcher has today become increasingly conscious of the zoogeographic aspects of bird distribution. The science of zoogeography attempts to explain why various creatures got where they are today, yet are not found in apparently suitable areas elsewhere and why there are striking similarities between populations of animals occurring in widely separated places. Sir David Attenborough's deservedly famous BBC film and television series 'Life on Earth' (see also his book of same title, 1980 *Reader's Digest* ed.), quite apart from its magnificent aesthetic and multi-dimensional portrayal of life forms, has acquainted literally millions of persons interested in natural history, including people living in Pakistan, with the accumulating evidence of how the present diversity of bird and animal species came to inhabit different parts of this planet Earth.

Zoogeographers in studying the underlying reasons for animal distribution must draw upon many different scientific disciplines, including stratographic geological history and the fossil evidence of Palaeontology. We need to understand the degree of inter-relatedness or divergence between different kinds of birds before we can appreciate their taxonomic relationship and in turn the way in which they have spread over the planet, or conversely, may have retreated into isolated and geographically separated populations. Recent advances in the study of animal behaviour (Ethology) and biochemical systematics, particularly critical microscopic examination of chromosome numbers, egg albumen morphology and Gel electrophorensis have enabled taxonomists to determine much more closely inter-specific relatedness, often between outwardly dissimilar species (Vuilleumier in Campbell and Lack, 1985). The Palaeontologist constantly improves his ability to classify fossil remains, or recognize the probable ancestors of present day forms as our knowledge of taxonomy improves. Likewise the theory of tectonic plates has helped zoogeographers in our understanding of continental drift, and geographic changes occurring over a very extended time-scale, which have enabled birds to colonize different areas.

Our increasing understanding of genetics and evolution, has enabled us to realize that a relatively small isolated population can change quite rapidly, due to the selective pressures of the environment and the minute genetic changes which have more chance to persist in relatively small isolated populations. With this brief look at the raw materials of zoogeography, it is desirable also to consider with equal brevity what is known about the geological history of our region.

Birds are believed to have evolved from reptilian ancestors or stock about 150 million years ago and the earliest known fossil bird is *Archaeopteryx* estimated to be 130 million years old. During the early Tertiary period, 65 million years ago during the Eocene geologic epoch, there is evidence that a great land mass split off from the continent of Africa and drifted across what is now the Indian Ocean and this island was known as Gondwana. It was divided initially from the Asian land-mass by the great sea of Tethys in the north-west (where Pakistan now exists) and a much more extensive northward stretching Bay of Bengal in the north-east. Eventually Gondwana Land collided with Asia and during a great upheaval in the middle part of the Eocene epoch, 50 million years ago, the Himalyan regions were partly thrust up above the sea, creating a land bridge between south-east Asia and Gondwana, particularly in the Malaysian region. It was through this land bridge that most of the truly Oriental faunal mammals and birds are thought to have colonized or invaded Gondwana Land or the Indian subcontinent. The Oriental faunal zone is the name given by zoogeographers to one of the earth's six major faunal regions (Sclater, 1858), which are characterized by having a certain homogeneity of their bird and mammal fauna. The region has been defined as lying roughly between the longitudes 68° and 135°E, and between the latitudinal parallels of 10°S and 32°N, and is bounded on the west by the Indus river in Pakistan and on the north by the Himalayas and extending eastwards to embrace south-west China, the Philippines and most of Indonesia. It will be seen therefore that most of the dry montainous country to the west of the Indus river falls outside of the Oriental zoogeographic region and forms part of the Palearctic region so that Pakistan actually straddles two great zoogeographical zones.

A second great upheaval in the earth's crust is thought to have occurred about 13 million years ago, in the middle Pliocene epoch. This was believed to be of a more violent nature, creating the present day very high jagged Himalayan ranges as well as the European Alps. One result of this upheaval, was to create a physical barrier for the dispersal or movement of Oriental birds northwards into central Asia and likewise a climatic barrier for the warm moist monsoon winds sweeping up across India. This resulted in a gradual dessication of the Tibetan plateau region, giving Baltistan (in Pakistan), and Ladakh, their rather distinctive bird fauna, for example, *Montifringilla* and *Leucosticte* mountain finches and *Tetraogallus* Snowcocks.

During the late Pliocene, less than 4.5 to 3 million years ago, smaller upheavals or movements of tectonic plates, are thought to have created an uptilting of the Himalayan foothill zone and thus produced the region known as the Siwaliks. These areas were originally sedimentary deposits laid down 25 million years ago by great rivers such as the Indus draining into the Sea of Tethys from the Himalayas. Such sediments are the richest source of fossil remains, which is fortunate for zoogeographers, as the Siwalik beds in the Punjab Salt Range have revealed an amazing variety of fossil remains, many of which, like fossil forms of Giraffes and Hippopotamuses (Prater, 1965) support the evidence of a former land connection between Gondwana and Africa. Up to 1978, scientists from the University of Yale and Peabody Museum expeditions conducting excavations in the Salt Range have located at least 7 bird species. These dating mostly from 30 million years ago (Miocene deposits) represent families of primitive non-Passerines which still have their present day counterparts. They include a Pelican *Pelecanus* cf. *cautleyi,* a Stork *Leptoptilos siwalikensis* (an ancestor of the Oriental Adjutant Stork and the African Marabou?), A Vulture *Torgos* cf. *tracheliotus* (an ancestor to the African Lappet Faced and Indian King or Black Vulture?), a Stork of the *Xenorhynchus* genus (an ancestor of Africa's Saddle-billed Stork and the Oriental Black-necked Stork?), a Pheasant of the genus *Lophura* (ancestor of our present day Kaleej), and two Rails assigned to the genera *Urmornis* and *Prophyrio,* both of which seem to be represented more widely around the world (Pilbeam *et al., Postilla,* 1979).

During the Miocene epoch, 20 million years ago, the whole of the subcontinent was milder in climate and probably more humid than at the present time, with much more favourable conditions for life forms even extending over the Tibetan plateau and the great Indian desert of Rajasthan, than exist at the present

time. Subsequently as the Himalayas rose, they became colder and in the late Pleistocene epoch the whole planet is thought to have cooled down. There were perhaps four successive periods of glaciation when parts of the Himalayas were covered by an ice cap and sea levels sank, due to lack of water run-off. During these successive periods of climatic change, both the Himalayan region and then the great desert basins of Seistan and Rajasthan no doubt acted as refugia for plants, birds and other animals which could not tolerate the extreme conditions prevailing, either during cold dry, or during warm moist periods. It is believed that these climatic phenomena gave rise to the fascinating present day distribution of many alpine and desert adapted bird species, as well as the evolution of unique endemic faunal associations characteristic of these regions. The Himalayas in particular constitute a distinct zoogeographical region in their own right which is often referred to as the Sino-Himalayan zone.

There are many families or groups of birds which are represented in this region in greater diversity than is found elsewhere on the planet, and this would suggest that the Himalayas constituted their centre of origin. The most obvious example is in the Timalidae or Babblers and especially the Laughing Thrushes (genus *Garrulax*). There are 24 species of *Garrulax* breeding in the Himalayas. Leaf Warblers of the genus *Phylloscopus* with 15 species breeding in the Himalayas, and Rosefinches of the genus *Carpodacus* also with 15 species breeding in the Himalayas form other striking examples of typically Sino-Himalayan bird families or groups.

The Palearctic faunal region embraces approximately Europe, the Mediterranean basin countries, including north Africa, and the near East, Soviet Asia, Japan, Korea and most of north and central China.

During periods of glaciation those Palearctic bird species unable to cope with extreme cold, retreated to the Mediterranean border countries, and many of these birds have been able to spread eastwards into the dry mountainous regions of Pakistan where they are on the eastern boundary of their total range. Examples are the Cream Coloured Courser (*Cursorius cursor*), the Close-barred or Lichtenstein's Sandgrouse (*Pterocles lichtensteinii*) and many larks such as the Finch Larks of the genus *Eremopterix,* and Hoopoe Lark (*Alaemon alaudipes*), some of the Desert Larks of the genus *Ammomanes* and also the House or Striolated Bunting (*Emberiza striolata*). All these are typical birds of southern Baluchistan and the hilly tracts of Sind.

The depth of alluvial silt deposits in the Punjab and the long tortuous course of the Indus river, flowing from east to west, before it finds a break, down through the Himalayan mountain chain, both testify to the very ancient origins of this mighty river. Reaching the semi-desert regions of the wide flat plains of Pakistan, the Indus river changed from a swift mountain river into a sluggish one with wide meandering channels and belts of seasonally flooded cane grass bordered by riverain forest. There was thus created a narrow strip of comparatively humid lush plant growth with attendant insect life, which must have attracted birds and animals colonizing the region from moister and less extreme regions further east. It is not surprising therefore, that the Indus, and on the other side of the subcontinent, the Ganges and Brahmaputra rivers in India, evolved their unique endemic fauna. Some of these riverain mammals have only become extinct in recent historic times in Pakistan, such as the Great Indian Rhinoceros (*Rhinoceros unicornis*) and the Swamp Deer (*Cervus duvauceli*) (Roberts, 1977), and even the Hog Deer (*Axis porcinus*) and Gavial Fish-eating Crocodile (*Gavialis gangeticus*) are now on the verge of extinction in Pakistan.

Among endemic bird species typically associated with the Indus riverain tracts and shared with the Ganges/Brahmaputra are the White-tailed Stonechat (*Saxicola leucura*), the Longtail Grass Warbler (*Prinia burnesii*), the Yellow-bellied Wren Warbler (*Prinia flaviventris*) and the Striated Babbler (*Turdoides earlei*).

Most of the above species, including the Hog Deer and Swamp Deer, have almost disappeared from the Ganges basin but have survived often in discontinuous pockets in the immediate Himalayan foothill zone, an area typified by oxbow swamps and tall cane grass (*Saccharum*) thickets, known as the Terai and the Duars. Historically there is some evidence for believing that the original typical riverain habitat or climax vegetation disappeared from along the Ganges much earlier than has been the case with the Indus, due to a longer history of human settlement and agricultural cultivation along those rivers and this accounts for the retreat of riverain endemics up to the Duars. This supposition is not withstanding the archaeological discoveries of very ancient (up to 3000 B.C.) Harappan and Mohenjodaro civilizations on the Indus. There are only two riverain adapted species truly endemic to the Indus, which are not found on the Ganges or Brahmaputra. They are the Sind Jungle Sparrow (*Passer pyrrhonotus*) and the Sind race of the Common Starling (*Sturnus vulgaris minor*).

Because of their power of flight and unique mobility, there are no examples of desert endemics amongst the birds of the region except perhaps the Painted Sandgrouse (*Pterocles indicus*) which might better be considered an isolated sub-species of *Pterocles lichtensteinii* (see species account). There are however many examples of less mobile endemic reptiles including Gekonid and Toad Agama lizards (e.g. *Teratoscincus* and *Phrynocephalus* spp.) as well as *Eristicophis* vipers. In the Chagai desert of Baluchistan, which forms part of the great Iranian desert basin (Cloudsley-Thompson, 1965), there are several bird species which are not found anywhere else on the Indian sub-

continent, such as the Brown-necked Raven (*Corvus ruficollis*), the Egyptian Nightjar (*Caprimulgus aegypticus*) and the Afghan Scrub Sparrow (*Passer moabiticus*).

Another important aspect of zoogeography which is particularly helpful in understanding the sometimes surprising distribution of bird species in Pakistan today, relates to what could be termed the theory of invasion routes. Just as alpine regions or deserts, have acted as island refugia during post glacial warming-up or successive dry and cold periods, so a continuity of similar climatic or vegetative zones have acted as bridges for the spread or colonization of expanding bird populations. In Pakistan there appear to be four significant invasion routes, which are illustrated diagrammatically in **Fig. v.**

The first comprises two separate avenues for the colonization of Oriental species into Pakistan on either edge of the great Indian Thar desert or Rajasthan desert. One lies in the extreme south-east corner of lower Sind and the other in the north-east corner of Punjab. In Sind the mitigating effect of the sea along the coastal belt, with less extreme summer temperatures and more mesic conditions, due to persistent sea breezes during the summer months, has provided an avenue for colonization of such species as the Yellow-wattled Lapwing (*Vanellus malabaricus*) and the Watercock (*Gallicrex cinerea*), the Jungle Nightjar (*Caprimulgus asiaticus*) as well as the Pied Crested Cuckoo (*Clamator jacobinus*). In the Punjab, species such as the Common Hawk Cuckoo (*Cuculus varius*), the Crimson-breasted Barbet (*Megalaima haemacephala*), the Allied or Savannah Nightjar (*Caprimulgus affinis*)

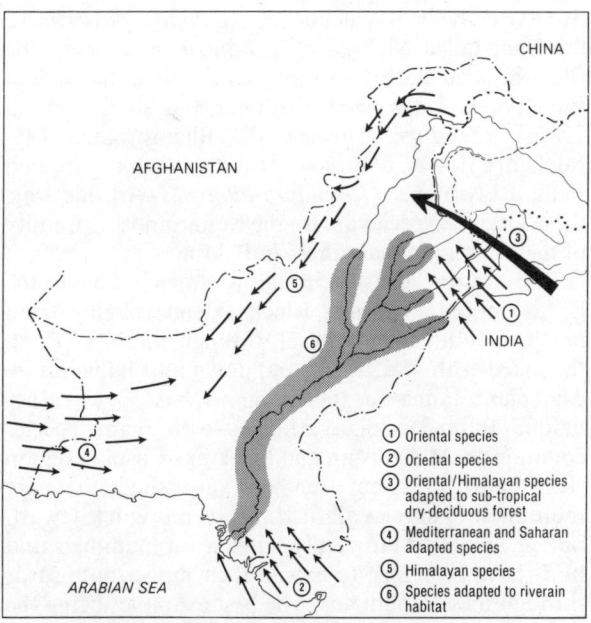

Fig. v: Invasion Routes of Birds associated with particular Faunal Regions.

and the Indian Pied Wagtail (*Motacilla maderaspaten-sis*), have been able to colonize across northern Pakistan, but do not occur in Sind and have not been able to enter via the south-eastern route. The great Rajasthan desert forms the obvious physical barrier to a wholesale invasion by a wider diversity of Oriental faunal species. A few like the Red Turtle Dove (*Streptopelia tranquebarica*) and the Yellow-throated Sparrow (*Petronia xanthocollis*) do however come each year into Pakistan to breed, travelling across the whole eastern border including the Rajasthan desert section in the middle.

The second invasion route lies in the extreme south-west corner and through the Makran from southern Iran. Mammals and birds with north African or Mediterranean affinities have colonized along this route. Most of the Sandgrouse, particularly *Pterocles alchata* and *Pterocles orientalis* have come by this route, also the Bar-tailed Desert Lark (*Ammomanes cincturus*), the Eastern Calandra Lark (*Melanocorypha bimaculata*), the Streaked Scrub Warbler (*Scotocerca inquieta*) and the Black-billed Desert Finch (*Rhodospiza obsoleta*).

The third invasion route constitutes an extremely narrow belt or zone along the outer Himalayan foothills where there is an extension of a rather stunted and degenerate tropical dry deciduous forest biotope. In sheltered ravines or north-facing slopes of these foothills will be found an astonishingly rich association of Indo-Malaysian plant species (see chapter on Ecological Factors in Bird Distribution), which provide an avenue for the invasion of many truly Oriental or Himalayan breeding birds. For example the Blossom-headed Parakeet *(Psittacula cyanocephala)*, the Plaintive (Grey-bellied) Cuckoo *(Cacomantis passerinus)*, the Long-tailed Nightjar *(Caprimulgus macrurus)*, the Blue-throated Barbet *(Megalaima asiatica)*, the Indian Pitta *(Pitta brachyura)*, the Orange-headed Ground Thrush *(Zoothera citrina)*, the Blue-throated Flycatcher *(Cyornis rubeculoides)* and the White-throated Fantail Flycatcher *(Rhipidura albicollis)*. In this zone these Oriental species are on the westernmost extremity of their distributional ranges in Pakistan.

The fourth invasion route is down through the higher mountain ranges which extend roughly from north to south from Chitral, through the NWFP via the Safed Koh, Waziristan and down into Baluchistan. The plant fauna of these regions has some rather unique Irano-Turanian affinities with many species common to Afghanistan and the Trans-Caspian region (R. Stewart, 1982) but even here, surprisingly there are more than 19 species all of Himalayan origin (Stewart, op. cit., p. 23). Many truly Himalayan mammals and birds have been able to extend their range southwards through these mountains. The best examples being the Black Bear (*Selenarctos thibetanus*), Pallas's Steppe Cat (*Felis manul*) and amongst birds, the Scaly-bellied Green Woodpecker (*Picus squamatus*), the Blue Whistl-

ing Thrush (*Myiophoneus caeruleus*), the Streaked Laughing Thrush (*Garrulax lineatus*), the Himalayan Tree Creeper (*Certhia himalayensis*) and the Simla Black Tit (*Parus rufonuchalis*). It must be kept in mind that this discussion of invasion routes does not consider normal annual migration routes but rather we are trying to focus upon those routes by which birds colonize new regions on a more permanent basis if not for the whole year, at least for the purpose of breeding.

The above summary therefore shows that Pakistan can be divided zoogeographically into the Palearctic faunal region mainly to the west of the Indus, the Oriental region mainly to the east of the river and on the Indus plains, and in the north the Himalayan region, classified usually as comprising part of the Palearctic (see K.H. Voous, 'List of Recent Holarctic Bird Species', *Ibis, 1977*). These are man-made concepts and the boundaries, as defined, are necessarily arbitrary. Invasion routes can carry a two way traffic and some Oriental species have been able to travel westwards through lower Sind and the Makran into the Middle East. Examples are the Rose Ringed Parakeet (*Psittacula krameri*), the Brown Fish Owl (*Ketupa (Bubo) zeylonensis*) and the Afghan race of the Common Babbler (*Turdoides caudatus huttoni*). The Indus river itself does not seem to be the actual physical barrier but rather it is the extensive tracts of sand dune desert or the mountainous regions to its west, which create a barrier for extension of most Oriental species. Thus the Painted Sandgrouse (*Pterocles indicus*) occurs west of the Indus in the NWFP (see species account) but in its total distribution is an endemic Oriental faunal species, and it does not appear to have spread in the southern part of its range across the sand dune desert of the Thar, where it is replaced by *Pterocles lichtensteinii*. In northern Sind, the Sibi desert extends in a north-westerly arm some 225 kilometres (140 miles) from the Indus river, but this inhospitable region has been colonized by such conspicuously Oriental species as the Indian Robin (*Saxicoloides fulicata*) the Indian House Crow (*Corvus splendens*), the Rufous-fronted Wren Warbler (*Prinia buchanani*) and the Jungle Babbler (*Turdoides striatus*), all of which can be seen at the mouth of the Bolan pass where it debouches into the Sibi plain. But none of these birds have migrated up the Bolan pass and into the central plateau regions of Baluchistan. Zoogeographic barriers are therefore associated more with edaphic features (the underlying ground topography and geology) and vegetative plant associations, rather than simple physical barriers such as rivers or temperature gradients. In fact attempts to replace Sclater's system of faunal zones with a division based upon correlation with isotherms (or temperature gradients) were made in the 1890s (Merriam, 1894), but was later recognized as having too many inconsistencies to be useful. The Sibi desert is a perfect example of the inefficacy of temperature barriers, being to my mind, the most extreme and inhospitable of Pakistan's five major desert regions (See **Plate 3**). It is at best a howling wilderness in winter with only 'dust devils' (miniature tornadoes) to adorn the horizon, and in summer a shimmering mirage dotted furnace, yet the bird fauna at the foot of the hills around Sibi town is almost entirely Oriental. I cannot help recalling an American friend and colleague, who was a seasoned traveller in hot climates and who motored by choice in early June across the Sibi plain (whilst his associates travelled by plane), taking the precaution to travel during the early morning hours and not wanting to miss an opportunity 'to see something more of the country'. Upon reaching the foot of the Baluchistan mountains where the Bolan river enters the plains, he jumped fully clothed into the first pool he encountered!

In the Sino-Himalayan region, however, it is apparent that there is a stratification or altitudinal zonation marking the preferred habitat of the resident or breeding bird species. This phenomenon could obviously be correlated with temperature gradients, but significantly there are also marked differences in associated flora, rainfall and insect fauna at different elevations. The region is so unique and important in understanding the zoogeographic distribution of birds that it deserves special consideration and does not conveniently fit into the category of Palearctic, hence this author's preference for the term Sino-Himalayan. Examples can be taken of birds which are unique to the Himalayas as summer breeders, yet they migrate each winter down into the Indian subcontinent. Such birds not occurring anywhere north of the Himalayas would seem to have closer Oriental than Palaearctic associations. The Indian Blue Chat (*Luscinia* [syn. *Erithacus*] *brunnea*) and the Large Billed Leaf Warbler (*Phylloscopus magnirostris*) being typical examples, as also those migrant birds confined in the nesting season to lower altitudes in the outer Himalayan foothills such as the Plaintive Cuckoo (*Cacomantis passerinus*) and the Orange-headed Ground Thrush (*Zoothera citrina*) and the Brown Flycatcher (*Muscicapa latirostris*).

Examples can also be cited of birds which have breeding populations in the tundra or boreal forest zones of northern Europe and Asia, but which also have breeding populations in the Himalayas and are winter time migrants down to the Indian plains or immediate foothill regions. Such birds are truly Palearctic rather than Sino-Himalayan endemic species and have presumably been able to colonize the Himalayas as providing similar suitable breeding habitat. Examples are the Blue Throat (*Luscina* [syn. *Erithacus*] *svecicus*), the Orange-flanked Bush Robin (*Tarsiger* [syn. *Erithacus*] *cyanurus*), the Kashmir Nuthatch (*Sitta europaea cashmirensis*), the White-throated Long-tailed Tit (*Aegithalos niveogularis*), the Nutcracker (*Nucifraga caryocatactes*) and the Red-mantled Rose Finch (*Carpodacus rhodochlamys*). They

may even be considered as Himalayan breeding refugia populations from post-glacial warm periods.

In concluding this chapter it remains to consider two other categories of zoogeographic distribution. A few bird species are so catholic in their food and living requirements, including tolerance of temperature extremes, that they can adapt to live in almost any conditions and occur, albeit sparsley, all over the world where suitable habitat is available. These species are either truly Cosmopolitan, occurring in both hemispheres and around the world, such as the Barn Owl (*Tyto alba*), Barn Swallow (*Hirundo rustica*) and Moorhen (*Gallinula chloropus*). Or they may be dependent upon tropical or sub-tropical conditions and yet be pan-tropical. Examples are the Cattle Egret (*Bubulcus ibis*) and the Purple Gallinule (*Porphyrio porphyrio*). Truly Holarctic species which occur all over the northern hemisphere but not in the southern hemisphere are the Golden Eagle (*Aquila chrysaetos*), the Raven (*Corvus corax*), the Eagle Owl (*Bubo bubo*) and with more limited distribution the Magpie (*Pica pica*) and the Long-eared Owl (*Asio otus*).

If the bird fauna of Pakistan is considered from the viewpoint of faunal or zoogeographic affinities, it is therefore predominantly Palearctic, especially in the winter time with an influx of migrant species. Out of the total **Checklist** of 660 species known to occur or have occurred in Pakistan, 36.6 per cent can be considered to have Oriental affinities and 63.4 per cent to be Palearctic (or Holarctic) and less than 0.5 per cent can be considered as truly Cosmopolitan or Pansub-tropical (See also **Fig. vi** in chapter on Migration).

If the Himalaya is taken as a separate and distinct zoogeographic region and only those species which do not breed widely north of the Himalayas, are excluded, then 19.5 per cent of all species are apparently endemic Sino-Himalayan species and the total of purely Palearctic, in a narrower sense, becomes reduced to about 44 per cent. It is perhaps significant here to mention that the only bird family (in a strict taxonomic sense) exclusively confined to the Oriental region, are the Leaf Birds or *Irenidae*. These arboreal birds are dependent upon sub-tropical evergreen or semi-evergreen deciduous forest and are not represented at all in Pakistan.

If the total number of individuals rather than species, in the bird community could be measured for Pakistan, it is certain that the proportion of Palearctic species would be very high in the plains regions in winter time, whereas in summer time in certain habitat types in the plains, the bird fauna would be predominantly Oriental. An analysis of the variety and approximate number of species in a typical major wetland area of Pakistan in winter time, based upon detailed counts of all species including Passerines, actually made during bird censusing operations for the IWRB, would indicate a population ratio of approximately 88 per cent Palearctic and only 12 per cent Oriental in early January. In the much more sparsely inhabited dry sub-tropical thorn forest in winter time, it is estimated that 35 per cent of the bird population comprises Oriental faunal species. During the summer monsoon season in the Indus plains, the proportion of Oriental species contributing to the total bird population rises and is estimated very roughly at just over 50 per cent of the population total, and on a species basis, irrespective of numbers, can more precisely be estimated at about 72 per cent of the total. It must be emphasized that these figures exclude hilly or mountain regions and refer to the Indus plains.

Zoogeography is, as yet, a very imperfect science based as it must be on inadequate or flimsy historical evidence, and dependent upon theorems which in large part can never be proved. This chapter presents a very incomplete discussion of the zoogeográphic distribution of birds in Pakistan and an adequate treatment would warrant a separate book in itself. But it is hoped that this account will at least reveal something of the strategic location and importance of Pakistan in contributing to our better understanding of bird distribution.

Bird Migration in
Pakistan

For a discussion of the term used to describe categories such as Palearctic, Oriental and Sino-Himalayan see chapter on Zoogeographic Aspects of Bird Distribution.

It is not too fanciful a metaphor to consider Pakistan as lying at the crossroads of Asia's major Palearctic bird migration routes. In fact the region is rather like the travel terminus of a Eurasian transport system, for here will be found at certain seasons, conspicuous numbers of birds which are merely in transit, whilst at other times there is an influx of winter visitors from northern breeding grounds, or summer breeding visitors both from the northern mountain regions and from the Indus plains, to warmer more southern latitudes. The explanation for this phenomenon appears to lie more in Pakistan's strategic geographic location rather than in any natural endowment of transport facilities. For some bird species, vast stretches of desert or ocean are deterrents to the choice of migration routes. Pakistan is bounded both on the west and east by considerable expanses of inhospitable desert, the great Iranian and Rajasthan deserts (Cloudsley-Thompson, 1965). Similarly the north of the country is bounded by a continuous pallisade of some of the highest mountain ranges in the world, stretching from the Pamirs and Hindukush in Afghanistan, eastwards to the Himalayas which continue across the northern frontier of India (See also page 31 of this chapter).

Our knowledge about bird migration is still patchy as there have been very few detailed studies carried out inside Pakistan itself. Most of the evidence from bird ringing recoveries relate to birds ringed in the Soviet Union or in India, with a few additional interesting ringing recoveries from Europe and the eastern highland regions of Africa. A mass of accumulating sight records of bird arrivals and departures, and of obvious migratory movements within Pakistan, however, gives corroboration for general conclusions about migration patterns and this has been further supported by sightings of mountaineering expeditions (bodies of migrant birds found at 6,500 metres/21,500 feet in the Karakorams, for example) and from naval and other vessels operating in the Arabian Sea. It has been demonstrated that at certain times, even small passerine species migrate at surprisingly high altitudes (many travel at an average height of 3,300 metres/11,000 feet, for example) (Baker, 1980) and that they cross right over mountain ranges rather than following along valleys. Similarly many small passerines fly across the Indian Ocean to and from east Africa and thereby sometimes by-pass the Arabian peninsula and even the Somali Horn of Africa.

The post-monsoon abundance of insect life and vegetative shelter provided by seasonal inundations, afford rich feeding conditions which attract a host of Palearctic winter visitors to the Indian subcontinent, not only from central Asia, but from as far away as western Europe and eastern Siberia. Most of the subcontinent's winter visitors come through Pakistan, many en route to India and Sri Lanka (e.g. *Anas querquedula* and *Philomachus pugnax* to the latter country). Similarly lush feeding conditions prevail in many parts of Africa during the northern hemisphere winter and a very considerable population of Siberian and Trans-Caspian breeding birds migrate each autumn through Middle Eastern countries, including Pakistan en route to highland regions both north and south of the African Equator (e.g. *Merops apiaster,* and *Aquila rapax nepalensis.* known to winter south of the Equator in Africa).

Ringing recoveries have amply demonstrated that a majority of winter visitors to the subcontinent enter via the Indus plains. Some come down the Indus river valley and its far northern tributaries such as the Kunhar, Gilgit, Hunza and Shyok rivers as well as the Chenab and Jhelum rivers further east. A very significant number enter from further west coming over the relatively low level Peiwar pass at the south-west corner of the Safed Koh range of mountains and following down the Kurram river. They thus avoid the high mountain barriers to the north altogether. Cranes, Snipe and Pelicans come by this Kurram valley route.

Some of these autumn migrants fan out eastwards into northern India and thus avoid the Rajasthan desert to the south, whilst others follow the Indus river down to its delta. Many ducks ringed in winter in Bharatpur (Rajasthan, India) have been subsequently recovered, often 1 or 2 years later, traversing Pakistan on spring or autumn passage, e.g. Shoveler (*Anas clypeata*) recovered Kabirwala 29 March and Peshawar end of March. Common Pochard (*Aythya ferina*), Bhalwal, Sargodha, March 1982 and Widgeon (*Anas penelope*) 4 April near Sargodha, Common Teal (*Anas crecca*) recovered Chashma barrage on the Indus during autumn passage and Sohan river, Jhelum district on 30 March. ('Recovery of Ringed Birds', *JBNHS*, Vol. 68, No. 1). Those reaching the Indus delta then follow the western seaboard of India until they end up in Sri Lanka (Phillips, 1965). The bulk of central Asian and western Siberian Palearctic migrants which winter in east Africa, travel in the autumn through the plateau regions of Baluchistan and the Indus basin across to the Rann of Kutch in India. During the return spring migration, it is possible that seasonal prevailing winds determine a more easterly northbound flight path, as most Palearctic migrants from Africa then pass through Iran and Iraq and are not sighted in Pakistan.

The importance of the Indus valley as a major migration route for waterfowl has already been formally acknowledged. In 1967 an internationally sponsored conference held at Ankara, Turkey to consider the wetland resources of the world, rated the Indus as the fourth major bird migration flyway in order of world importance, and a resolution to that effect was endorsed by the International Union for the Conservation of Nature, the International Wetlands Research Bureau and the International Council for Bird Preservation. Five years later at the Ramsar Wetlands Conference in Iran, it was again stressed that the Indus Wetlands were critical for a large part of the entire waterfowl population which in winter visited India, a part of Sri Lanka, and of course Pakistan (Proceedings of the International Conference of Wetlands and Waterfowl, Ramsar, Iran, 30 January-3 February 1971, IWRB, 1972).

Much of the previous discussion has focussed upon migration routes used by different summer and winter visitors, but a vast majority of Palearctic passerine migrants spend the whole winter in Pakistan. It has been amply demonstrated from ringing recoveries both in Europe as well as the Indian subcontinent that many individual birds show year to year fidelity to the same wintering territory. Examples include Orphean Warbler (*Sylvia hortensis*) (2 individuals) being recaptured 2 years later in the same place at Hingolgadh in Gujarat State, India and a Lesser Whitethroat (*Sylvia curruca*) also recovered 1 year later at the same place) (Shivraj Kumar, *JBNHS,* Vol. 59, 1962). Yet the state of ornithological knowledge in Pakistan is such that there are no overall population estimates for any species, nor bird density data though there have been a number of very valuable waterfowl, crane and House Sparrow censuses at some sites (see references under relevant species accounts).

An analysis based upon the species composition of Pakistan's bird fauna indicates the extent of these migratory patterns (See **Fig. vi**). Out of the total **Checklist** included in this book, 30 per cent of the birds are species which visit the country for a significant period of the year as long distance migrants, while 43 per cent of the total **Checklist** are either Palearctic or Oriental species which come to Pakistan only for breeding. 28 per cent of the total number of species are regular winter visitors which breed extra-limitally and mainly in trans-Himalayan northern regions.

Against the above background it is possible to classify Pakistan's birds into ten broad migration categories (See **Fig. vii** and **viii**)

Palearctic Winter Visitors

i) These are normally entirely non-breeding, (see species account of *Fulcia atra* and *Ardea cinerea* for examples of exceptions) and mainly from the Soviet Union. In terms of number and variety of species this category covers the majority of all migrants which visit Pakistan. It includes four grebes (Podicipedidae), 19 species of duck and 2 wild geese (Anseriformes), 2 storks (Cinconiiformes), 2 pelicans and 1 cormorant (Pelecaniformes), 2 herons (Ardeidae), a spoonbill

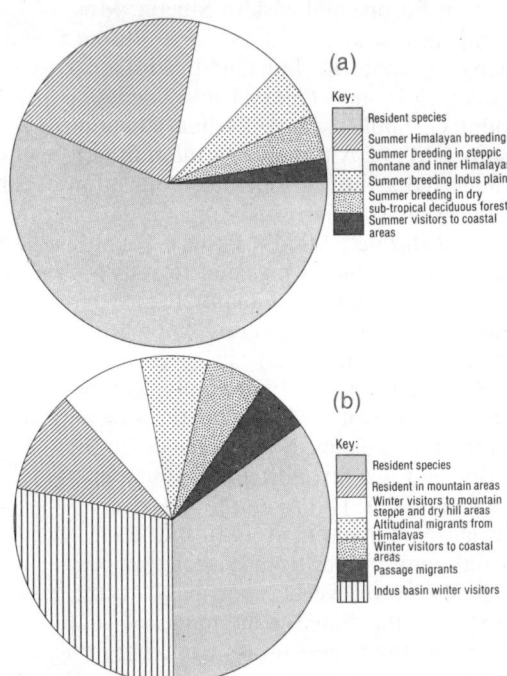

Fig. vi: Species Composition: (a) in summer and (b) in winter over whole of Pakistan.

Figure vii

SCHEMATIC PRESENTATION OF SUMMER MIGRATION PATTERNS

*Birds Breeding in Pakistan which are migratory

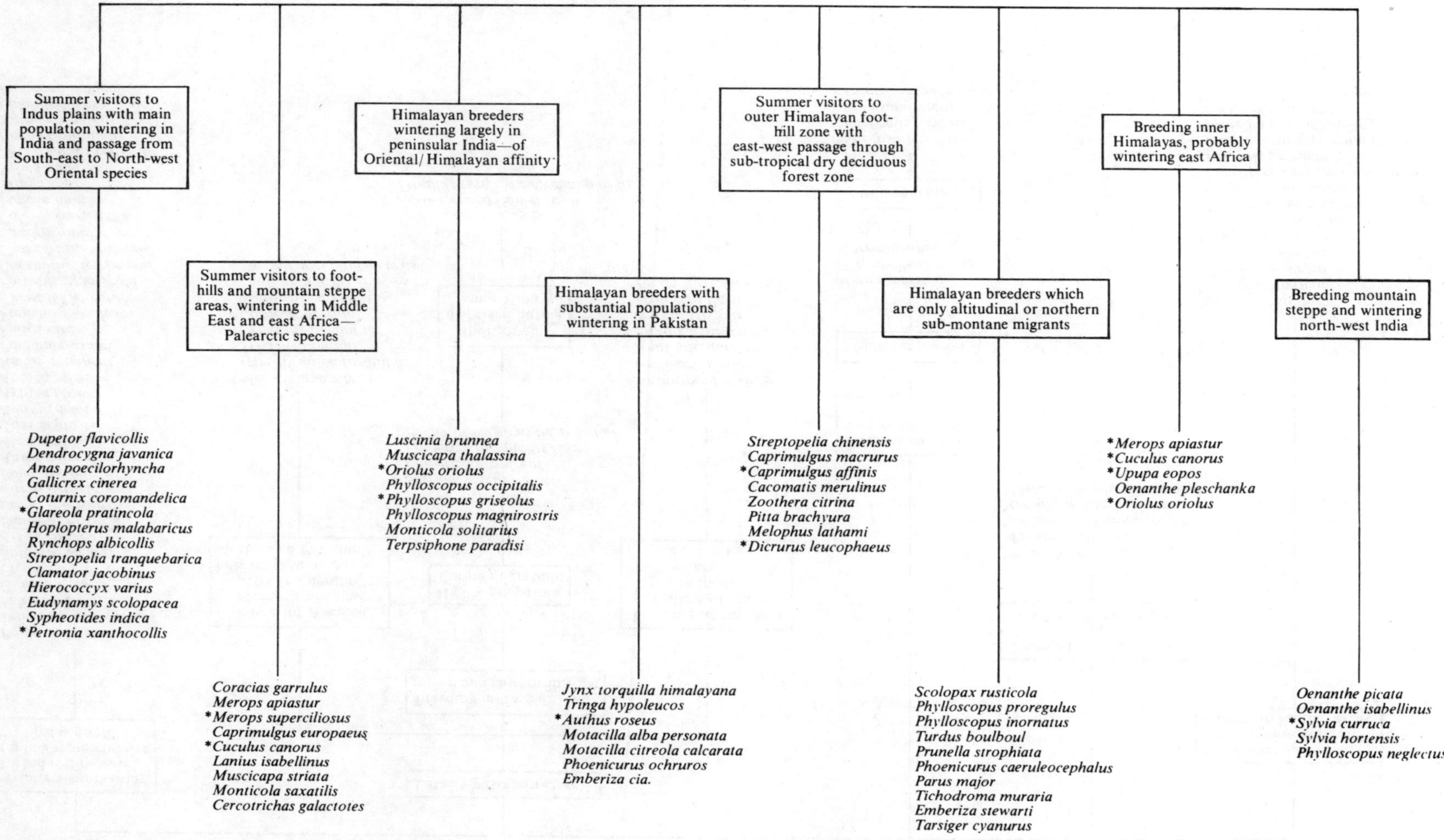

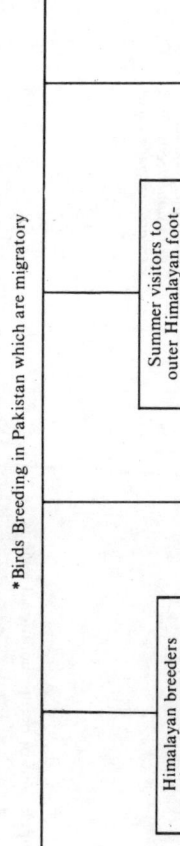

Summer visitors to Indus plains with main population wintering in India and passage from South-east to North-west Oriental species

Dupetor flavicollis
Dendrocygna javanica
Anas poecilorhyncha
Gallicrex cinerea
Coturnix coromandelica
*Glareola pratincola
Hoplopterus malabaricus
Rynchops albicollis
Streptopelia tranquebarica
Clamator jacobinus
Hierococcyx varius
Eudynamys scolopacea
Sypheotides indica
*Petronia xanthocollis

Summer visitors to foot-hills and mountain steppe areas, wintering in Middle East and east Africa—Palearctic species

Coracias garrulus
Merops apiaster
*Merops superciliosus
Caprimulgus europaeus
*Cuculus canorus
Lanius isabellinus
Muscicapa striata
Monticola saxatilis
Cercotrichas galactotes

Himalayan breeders wintering largely in peninsular India—of Oriental/Himalayan affinity

Luscinia brunnea
Muscicapa thalassina
*Oriolus oriolus
Phylloscopus occipitalis
*Phylloscopus griseolus
Phylloscopus magnirostris
Monticola solitarius
Terpsiphone paradisi

Himalayan breeders with substantial populations wintering in Pakistan

Jynx torquilla himalayana
Tringa hypoleucos
*Anthus roseus
Motacilla alba personata
Motacilla citreola calcarata
Phoenicurus ochruros
Emberiza cia.

Summer visitors to outer Himalayan foot-hill zone with east-west passage through sub-tropical dry deciduous forest zone

Streptopelia chinensis
Caprimulgus macrurus
*Caprimulgus affinis
Cacomatis merulinus
Zoothera citrina
Pitta brachyura
Melophus lathami
*Dicrurus leucophaeus

Himalayan breeders which are only altitudinal or northern sub-montane migrants

Scolopax rusticola
Phylloscopus proregulus
Phylloscopus inornatus
Turdus boulboul
Prunella strophiata
Phoenicurus caeruleocephalus
Parus major
Tichodroma muraria
Emberiza stewarti
Tarsiger cyanurus

Breeding inner Himalayas, probably wintering east Africa

*Merops apiastur
*Cuculus canorus
*Upupa eops
Oenanthe pleschanka
*Oriolus oriolus

Breeding mountain steppe and wintering north-west India

Oenanthe picata
Oenanthe isabellinus
*Sylvia curruca
Sylvia hortensis
Phylloscopus neglectus

*Species marked with an asterisk occur in other biotopes or wintering areas.

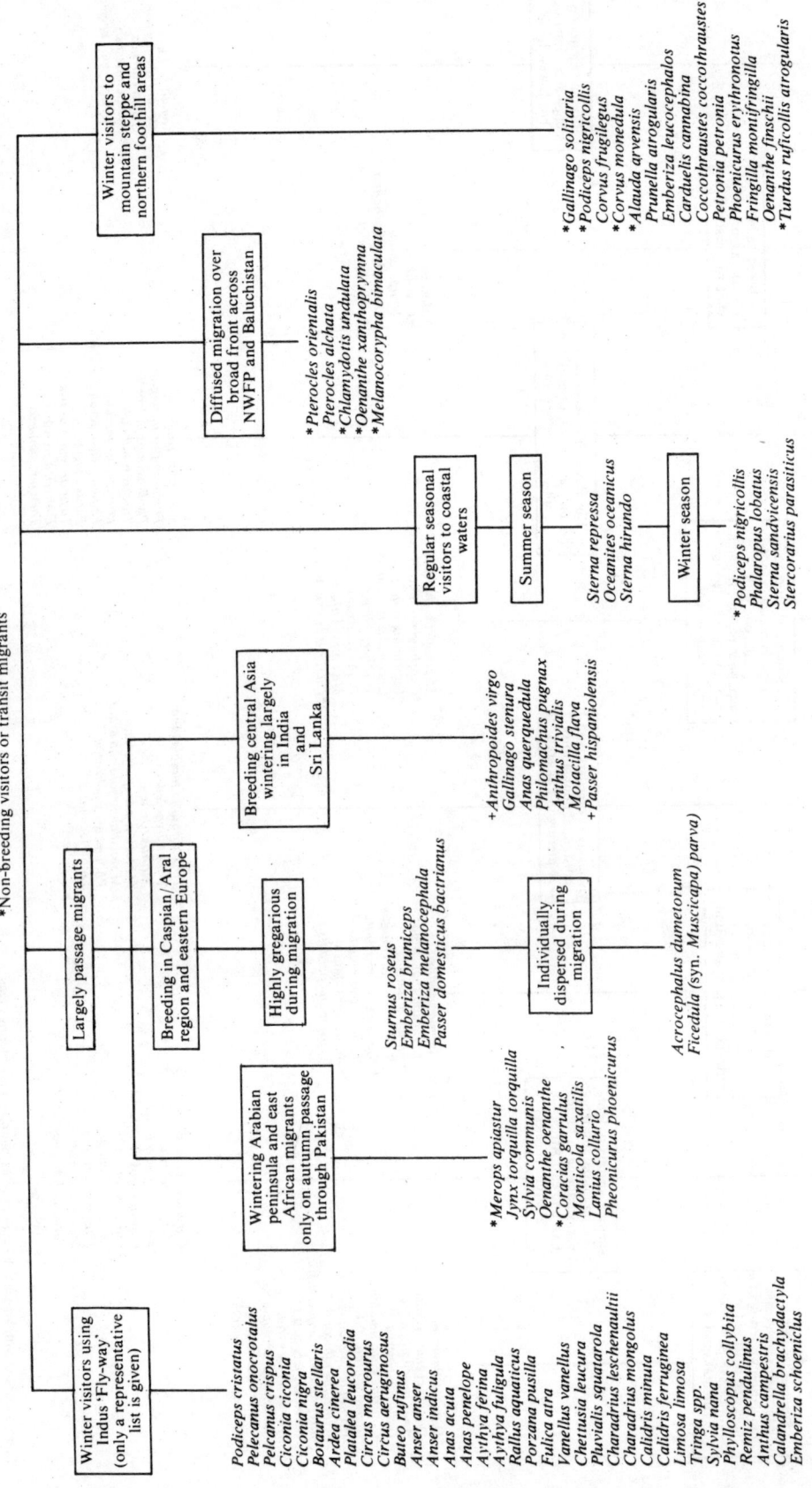

Figure viii

SCHEMATIC PRESENTATION OF WINTERING MIGRATION PATTERNS

*Non-breeding visitors or transit migrants

Winter visitors to mountain steppe and northern foothill areas

*Gallinago solitaria
*Podiceps nigricollis
Corvus frugilegus
Corvus monedula
*Alauda arvensis
Prunella atrogularis
Emberiza leucocephalos
Carduelis cannabina
Coccothraustes coccothraustes
Petronia petronia
Phoenicurus erythronotus
Fringilla montifringilla
Oenanthe finschii
*Turdus ruficollis atrogularis

Diffused migration over broad front across NWFP and Baluchistan

*Pterocles orientalis
Pterocles alchata
*Chlamydotis undulata
*Oenanthe xanthoprymna
*Melanocorypha bimaculata

Regular seasonal visitors to coastal waters

Summer season

Sterna repressa
Oceanites oceanicus
Sterna hirundo

Winter season

*Podiceps nigricollis
Phalaropus lobatus
Sterna sandvicensis
Stercorarius parasiticus

Largely passage migrants

Breeding in Caspian/Aral region and eastern Europe

Breeding central Asia wintering largely in India and Sri Lanka

+Anthropoides virgo
Gallinago stenura
Anas querquedula
Philomachus pugnax
Anthus trivialis
Motacilla flava
+Passer hispaniolensis

Highly gregarious during migration

Sturnus roseus
Emberiza bruniceps
Emberiza melanocephala
Passer domesticus bactrianus

Individually dispersed during migration

Acrocephalus dumetorum
Ficedula (syn. Muscicapa) parva)

Wintering Arabian peninsula and east African migrants only on autumn passage through Pakistan

*Merops apiastur
Jynx torquilla torquilla
Sylvia communis
Oenanthe oenanthe
*Coracias garrulus
Monticola saxatilis
Lanius collurio
Pheonicurus phoenicurus

Winter visitors using Indus 'Fly-way'. (only a representative list is given)

Podiceps cristatus
Pelecanus onocrotalus
Pelecanus crispus
Ciconia nigra
Ciconia ciconia
Botaurus stellaris
Ardea cinerea
Platalea leucorodia
Circus macrourus
Circus aeruginosus
Buteo rufinus
Anser anser
Anser indicus
Anas acuta
Anas penelope
Aythya ferina
Aythya fuligula
Rallus aquaticus
Porzana pusilla
Fulica atra
Vanellus vanellus
Chettusia leucura
Pluvialis squatarola
Charadrius leschenaultii
Charadrius mongolus
Calidris minuta
Calidris ferruginea
Limosa limosa
Tringa spp.
Sylvia nana
Phylloscopus collybita
Remiz pendulinus
Anthus campestris
Calandrella brachydactyla
Emberiza schoeniclus

* Species marked with an asterisk have resident populations or occupy additional regions.

+ Wintering in northern India.

(Threskiornithidae), probably 6 rails and crakes (Rallidae), 4 cranes and bustards (Gruiformes), 4 gulls (Laridae), at least 7 Charadridae (Plovers), 12 Tringinae, 8 Calidridinae (shore-birds and waders), 3 snipes, 2 owls (Strigiformes) and at least 12 raptors. Amongst the passerines are 2 larks, 3 pipits and a portion of all the Yellow-headed (*M. citreola*) and White Wagtail (*M. alba*) populations, plus all the *M. flava* sub-species which visit the subcontinent (ringing recoveries Kazakhstan and Kirghiz, USSR, *JBNHS*, Vol. 68, p. 79), 2 accentors, 2 Turdidae, 1 shrike (Lanidae), 2 Corvidae, 1 tit (Paridae) and at least 3 buntings (Emberizidae). Amongt the warblers, though, only 3 *Phylloscopus* and 3 *Sylvia* species breed largely extra-limitally, but the wintering populations of these last two genera certainly constitute at least one-third of the bird population in typical sub-tropical thorn forest habitat.

Palearctic Transit Migrants

These can be sub-divided into three categories.

ii) Those species wintering in east Africa, of which almost the whole population breeds outside of Pakistan and which are largely recorded on autumn passage only.

Examples include *Lanius collurio, Monticola saxatilis, Cercotrichas galactotes,* and *Sylvia communis.*

iii) Those species wintering in east Africa, a part of whose population breeds in Pakistan (though part breeds extra-limitally), and with a significant spring as well as autumn return passage through Pakistan territory. These include *Merops apiaster, Merops superciliosus, Coracias garrulax, Caprimulgus europaeus, Muscicapa striata, Oenanthe pleschanka,* and probably *Clamator jacobinus* and *Upupa epops.*

iv) Species wintering largely in India and breeding extra-limitally but with a conspicuous spring and autumn passage through Pakistan in a south-east to north-west direction. Whilst only five species come within this category, three of these are highly gregarious with very substantial migrant populations which often have an economic impact in Pakistan.

Sturnus roseus, Emberiza bruniceps, Ficedula parva, and *Emberiza melanocephala.* A specimen of *E. melanocephala* ringed in Bhuj, Gujarat state, India was recovered in Famagusta, Cyprus (*JBNHS*, Vol 68, p. 82). The two other species, *Ficedula parva* and *Acrocephalus dumetorum* winter in small numbers in Pakistan and also a portion of their population migrates in a more northerly direction, passing, for example. through the Murree hills.

v) **Summer breeding visitors to the northern mountains of Pakistan.** These can be divided into two sub-categories on the basis of zoogeographic affinities.

a) Oriental or Himalayan species, which winter in India (frequently on the Deccan peninsula) as well as Pakistan.

Monticola cinclorhyncha, Luscinia brunnea, Musicapa ruficauda, Muscicapa thalassina, Phylloscopus occipitalis, Phylloscopus magnirostris. Monticola solitarius and *Terpsiphone paradisi* also winter partly in the plains or foothills of Pakistan.

b) Palearctic species which winter in India and Pakistan, of which part of their population breeds in the Himalayan region of Pakistan and part further north in the Soviet Union.

Motacilla citreola, Motacilla alba, Phoenicurus ochruros, Luscinia svecica, Anthus trivialis and *Cuculus canorus.*

vi) **Monsoon or summer season breeding visitors from India or the Oriental region which remain in lowland areas.** A remnant population of some of these species remains during winter within Pakistan (marked*), but there is a noticeable influx in the spring or early summer months.

Ixobrychus cinnamomeus*, Dupetor flavicollis*, Pseudibis papillosa, Dendrocygna javanica*, Anas poecilorhyncha*, Coturnix coromandelica, Gallicrex cinerea, Sypheotides indica, Glareola pratincola, Hierococcyx (Culculus) varius, Petronia xanthocollis, Streptopelia tranquebarica*.

vii) **Summer breeding visitors of Oriental affinity** which perform an east-west migration. largely along the Himalayan foothill zone, the Siwaliks and Duars.

Merops philippinus, Caprimulgus macrurus, Caprimulgus affinis, Cacomantis passerinus (merulinus), Zoothera citrina, Streptopelia chinensis, Pitta brachyura, Cuculus saturatus, Melophus lathami, and *Dicrurus leucophaeus.*

viii) **Oceanic or littoral migrants which do not breed in Pakistan.** Both shoreline and pelagic species usually concentrate within 32 kilometres (20 miles) of the coast and they are assumed to have arrived by oceanic routes and not over land across Pakistan. Again they can be divided into two sub-categories:

a) Post monsoon winter visitors: *Sterna sandvicensis, Stercorarius parasiticus, Dromas ardeola, Larus argentatus* (juveniles).

b) Summer season or monsoon visitors: *Sterna repressa, Sterna hirundo, Oceanites oceanicus, Puffinus tenuirostris, Phaethon aethereus.*

ix) **Altitudinal or inter-montane local migrants.**
The great majority of species breeding in the Himalayas perform some form of limited migration, either straggling gradually down to the foothills as winter progresses (e.g. *Phoenicurus caeruleocephalus*) or crossing several high ranges to winter in the northern regions of the plains (e.g. *Parus major*). Some remain in their northern mountain breeding latitudes but descend only to immediate neighbouring valleys (e.g. *Carpodacus rubicilla*). Pakistan has such a variety of montane local migrants that it is only practicable to list a few of the more conspicuous: *Phoenicurus caeruleocephalus, Tarsiger cyanurus, Parus major, Saxicola ferrea, Phylloscopus proregulus, Phylloscopus subviridis, Myiophoneus caeruleus, Prunella strophiata, Emberiza stewarti, Pericrocotus ethologus, Certhia himalayensis, Rhyacornis fuliginosus, Enicurus scouleri, Culicicapa ceylonensis.*

x) **Palearctic winter visitors (breeding extra-limitally), which visit only montane steppe and western border foothill regions** (e.g. *Olea cuspidata* scrub forest) of Pakistan, and which apparently migrate partly in a north-west to south-east axis.

In these species, numbers fluctuate widely from year to year and some occur only as irregular stragglers. Part of the population of those species marked with an asterisk breeds within Pakistan territory.

Pterocles orientalis*, Pterocles alchata*, Melanocorypha bimaculata, Bombycilla garrulus, Corvus frugilegus, Prunella atrogularis, Phoenicurus erythronotus, Fringilla montifringilla, Fringilla coelebs, Coccothraustes coccothraustes, Oenanthe finschii*, Acanthis cannabina.

There is probably another category of migrants for which there is as yet little definite evidence. This comprises summer visitors of Palearctic or Oriental affinities, which breed in Pakistan and are known to winter in part in the countries of east Africa. Such species include: *Coturnix coturnix, Clamator jacobinus, Cuculus poliocephalus* (not recorded in Saudi Arabia), *Cuculus saturatus* (not recorded in Saudi Arabia), *Upupa epops, Oriolus oriolus* and *Ptyonoprogne obsoleta* syn. *fuligula.*

Some of these species have been observed on passage across the Arabian peninsula, others are regular winter migrants to countries of east Africa further south, (Jennings, 1981, and Moreau, 1972) and there are records of birds observed on passage on the islands of the Indus delta and in southern Baluchistan

(see account of *Otus scops*) which might indicate arrival from Africa.

A tabular representation of spring and autumn migration movements is also presented in **Figs. vii** and **viii.**

The phenomenon of migration is by no means an attribute of a particular species, applying to all its population. There are many species occurring in Pakistan, part of whose population appears to be sedentary, part extra-limital in breeding and present as long distance migrants, and part locally migratory. A case in point is the Hoopoe (*Upupa epops*) which has a resident breeding population in central Punjab, and a migrant breeding population in the far northern mountain regions which probably winters in east Africa. There is also a wintering population in Sind province which does not stay to breed but which may supplement the northern breeding population. The Rock Bunting, (*Emberiza cia*) has a distinct sub-species wintering in Baluchistan (*E. c. par*) and a different sub-species breeding in the northern dryer Himalayan regions (*E. c. stracheyi*). The Koel (*Eudynamys scolopacea*), has a small resident population in southern Sind and a visiting breeding population which comes from India to the northern regions of the Indus plains in summer. The House Sparrow race *Passer domesticus indicus* is a resident throughout the Indus plains, whilst part of the population of *Passer domesticus parkini* invades all the northern mountain valleys of Chitral, Hunza, Gilgit and Baltistan in summer, where it often nests in colonies. *P. d. bactrianus* (considerd by this author as inseparable from *parkini*), has a strong spring passage which moves across Baluchistan and the North West Frontier Province and breeds partly in our northern regions but mainly extra-limitally.

It is necessary to stress that the actual migratory movements of many bird species in Pakistan are still hardly known or understood, particularly in those species relying upon crypsis of plumage for daytime concealment which normally cannot be identified unless captured. The distribution of such species as the Scops Owls and Nightjars is known mainly from their summer breeding grounds where they are highly vocal. Winter records depend upon very small numbers of museum skins some of which could well have been birds on passage. The few museum records of *Otus brucei* seem to show a south-easterly migration into India during the winter, but it is not known where the Pakistan breeding populations of *Otus sunia* and *Otus scops turanicus* spend the winter. Similarly *Caprimulgus europaeus* and *C. mahrattensis* are sympatric in many parts of Pakistan in the breeding season, but winter sightings within the sub-tropical thorn forest regions could be of either species. There is an even greater lack of information about the winter movements of Hirundines. Out of 2,895 Common Swallows (*Hirundo rustica*) ringed all over India in the 1960s there has not been a single recovery (Mathew, *JBNHS,* Vol. 68, p. 77). *Hirundo daurica rufula* which winters in Africa (Moreau, 1972) probably contributes to the summer breeding population in Baluchistan and the North West Frontier Province whilst the sub-species *H. d. erythropygia*, which breeds around Islamabad and the foothills, disappears in October, almost certainly to winter in India. Most of the *H. rustica* which breed in Pakistan, winter in the Indus plains and there appears to be only local migration of the Pakistan population, but there is so far no information available on the winter distribution of the northern Himalayan breeding population of *Delichon dasypus* or even of *Ptyonoprogne rupestris*. Both species certainly disappear entirely from their breeding haunts in Hazara district, Gilgit and Chitral during the winter.

Regrettably, attempts to interest amateurs in bird ringing programmes as well as training programmes imparted to Government Wildlife and Forestry officials by international conservation organizations, have not been successful in getting any regular bird ringing camps started in Pakistan. Much of the above ideas about migration are therefore somewhat speculative considering the state of knowledge elsewhere in the world.

The Bombay Natural History Society did enlist the help of some Sindhi landowners in a bird ringing programme, mostly of duck, between 1926 and 1934. In 1967, Roger Holmes organized a bird ringing camp (using BNHS rings) at Balloki on the Ravi river in the Punjab when a good number of Motacillidae and 'Acrocephalines' were rung. Christopher Savage secured the help of many friends, including myself, to ring various waterfowl during the period 1965 to 1968. General Marden did a lot of ringing, Barheaded Geese (*Anser indicus*) and Common Teal (*Anas crecca*) on the Indus river at Panjnad during the 1960s, and Fred Koning carried out some bird ringing during his IWRB surveys of wetland areas between 1970 and 1974. Actual ringing recoveries of birds ringed in Pakistan and shot in the Soviet Union include Shoveler (*Anas clypeata*), Common Teal (*Anas crecca*), Garganey (*Anas querquedula*), Wigeon (*Anas penelope*), Mallard (*Anas platyrhynchos*) and Red-crested Pochard (*Netta rufina*).

Birds ringed in the Soviet Union and shot in Pakistan include *Barheaded Geese (*Anser indicus*), *Greylag Geese (*Anser anser*), Common Teal (*A. crecca*), *Mallard (*Anas platyrhynchos*), Pintail (*Anus acuta*), the *Large Cormorant (*Phalacrocorax carbo sinensis*), *Great White Egret (*Egretta alba*), *Gull-billed Tern (*Gelochelidon nilotica*), the yellow-legged race of the *Herring Gull (*Larus argentatus heuglini*) and the *Spoonbill (*Platalea leucorodia*). The *Common Teal (*A. crecca*) and *Common Pochard (*Aythya ferina*) have both been shot at Nowshera migrating up the Kabul river with rings put on them during the winter at Bharatpur, India. A Shoveler and

Gadwall ringed in Bharatpur, India were recovered in the Karakoram mountains while on migration through Hunza on 7 March 1970 (*JBNHS,* Vol. 68, p. 259 and p. 262). Indian recoveries of birds which were ringed in the Soviet Union and which most certainly have entered via Pakistan include the Rose-coloured Starling (*Sturnus roseus*) recovered in the east Punjab and the Red breasted Flycatcher (*Muscicapa (Ficedula) parva*) as well as the Great Black-headed Gull (*Larus ichthyaetus*), a *Greater Flamingo (*Phoenicpoterus roseus*) ringed in Iran was later recovered in Lasbela. Those marked with an asterisk include rings obtained, usually through purchase from local wildfowlers, by the author.

No account of bird migration in Pakistan would be complete without some consideration of the phenomenal distances known to be travelled by certain species. The White Stork (*Ciconia ciconia*), winters mainly in extreme south-eastern Pakistan or continues down into India. Ringing recoveries show that at least part of this population breeds in West Germany. Since storks avoid flying over long stretches of ocean, (which do not provide suitable updraft thermal currents), these Pakistan visitors must travel at least 7,725 kilometres (around 4,800 miles) during each migration. The Red-necked Phalarope (*Phalaropus lobatus*) breeds in the Arctic tundra and huge numbers winter off the coast of Pakistan. The wintering birds, if they followed a direct line, must fly at least 6,418 kilometres (around 3,900 miles) to reach their eastern Siberian breeding grounds. This is all the more remarkable in that there are very few land sightings of the phalarope within Pakistan during migration. Numbers can be seen each May on coastal and tidal ponds just before flying north, but subsequently, they must cover at least 3,200 kilometres (around 2,000 miles) over land nonstop. An even more spectacular migration is performed by the few specimens of the Lesser Golden Plover (*Pluvialis dominica*) which can be sighted around Karachi each winter. Known to breed only in the Kamchatka peninsula, they must travel approximately 9,650 kilometres (6,000 miles) from their breeding grounds and even this is not as far as the individuals from the same breeding population which winter in Australia. Perhaps the most dramatic example of long distance travel, is provided by the tiny starling-sized Wilson's Petrel (*Oceanites oceanicus*) to be seen in considerable numbers off the Karachi coast during the summer monsoon. They must travel, allowing for foraging activity and prevailing winds in southern latitudes, at least 12,900 to 14,500 kilometres (some 8,000 to 9,000 miles) from their breeding grounds which are on the main continent of Antarctica. Even in a straight line, the Makran coast is 11,100 kilometres (about 6,900 miles) from Antarctica.

A proper perspective of migration in Pakistan would not be obtained without some mention of the large number of Oriental and even Palearctic species which are sedentary throughout the year. As a whole the Timalinae or Babblers which are typical of Oriental fauna, are all sedentary and certainly their elaborate social patterns and cooperative breeding behaviour are connected with their year round fidelity to a fixed territory. Even the ground feeding Streaked Laughing Thrush (*Garrulax lineatus*) remains the year round in the same locality in the inner cold desert mountain regions of Pakistan, despite the sub-zero temperatures which prevail at those altitudes. Most of the prinias, mynas, bulbuls, and Oriental lark species which inhabit the plains, are sedentary. Examples of Palearctic or alpine species which are sedentary are numerous, the Alpine Chough (*Pyrrhocorax graculus*), the Lammergeier (*Gypaetus barbatus*) and the Himalayan Vulture (*Gyps himalayensis*) in Pakistan's northern areas are conspicuous examples. In Pakistan the Great Tit (*Parus major*) has a locally migratory population which spreads in winter down to the Punjab plains but it is still possible to see resident populations of this tit under snowfall conditions throughout the winter, in the montane juniper forests of Baluchistan and the far northern valleys of Gilgit and Baltistan. Palearctic nuthatches (*Sittidae*) and Himalayan Woodpeckers, such as *Dendrocopos himalayensis* and *Picus squamatus* are also sedentary in winter.

It is also possible to learn something of the varying migration techniques of birds by field observations in Pakistan. White and Black Storks (*Ciconia ciconia* and *C. nigra*), Dalmatian Pelicans (*Pelecanus crispus*), Common and Demoiselle Cranes (*Grus grus* and *Anthropoides virgo*) all do part of their travelling by night as well as by day and characteristically travel in large gregarious bands which are very conspicuous by daytime. All these species use a gliding technique as observed on migration in Pakistan. They circle round on a suitable thermal, often until they reach heights of nearly 5,000 metres (17,000 feet) and then they rapidly form into 'V' echelons or skeins and glide slowly downwards with occasional wing flapping to maintain height. They can usually travel many miles by this gliding method before it is necessary to again seek out a suitable thermal and regain cruising height. At the other extreme are Wagtails (*Motacilla* spp) and Pipits (*Anthus* spp) which stop frequently to forage on the ground during migration and otherwise seem to fly at no more than 100 metres (3-400 feet) above the ground. In late August, in Thatta district of lower Sind, I have encountered a typical passage of *Motacilla flava* during which I estimated that over 400 individuals passed by in a three hour period, usually observed as some two or three dozen birds—nearly all in sub-adult plumage, which settled within sight on cropland and wet rice fields, seeming to pick up insects, but repeatedly flew forward in the same south-southeast direction, not lingering for more than four or five minutes in any one field. Seacoasts and deltaic areas often act as

resting grounds for birds which will or have travelled long distances over ocean. Quail (*Coturnix coturnix*) in emaciated condition can regularly be encountered in the late summer and early autumn months around Karachi and the Indus delta, presumably resting before crossing the Arabian Sea.

Some migratory birds travel in loose association or small flocks though they quickly split up or occupy separate territories once they reach their breeding or wintering quarters in Pakistan. I have records of a loose flock of more than 25 Short-eared Owls (*Asio flammeus*) (E. Fernando, pers. comm., 1968) being encountered in late February in one patch of grassland in 1966 near Marala barrage on the Chenab river, a flock of 9 Pied-crested Cuckoos (*Clamator jacobinus*) in riverain forest of Thatta district on 10 June, a flock of 20 Golden Orioles (*Oriolus oriolus*) near Islamabad on 13 April (D. Corfield, pers. comm). Such concentrations certainly indicate recent arrival, as these three species are normally and regularly encountered as scattered and single idividuals in Pakistan. Captain Barnes (Barnes, *Stray Feathers*, 1880) recorded a veritable migration through Chaman on the borders of Afghanistan of Baillon's Crakes (*Porzana pusilla*) which could be caught by hand, they were so exhausted. I have seen a single migratory flock of House Sparrows (*Passer domesticus*) in northern Baluchistan in early April which numbered between 600 and 700 birds, and in Chichawatni irrigated forest plantation in late April, a migratory flock of Rosy Pastors (*Sturnus roseus*) estimated to number over 2,000 individuals. Both these are examples of highly gregarious migratory species which perform a rather leisurely passage, stopping for periods of three to four days in one locality before moving onwards in a general north north-west direction. During the daytime they are not normally seen in such large concentrations, but even on migration small flocks tend to coalesce at evening-time and share a common roosting area. The Rosy Pastor (*Sturnus roseus*) flock of 2,000 was observed at sunset with several smaller flocks flying in the vicinity, and a similar phenomenon was seen at Kalankot lake in Thatta district in Sind when an estimated 8,000 Rosy Pastors settled on bushes and even on the ground on the low hills bordering the lake, coming from all points of the compass just at dusk.

Some species can undoubtedly utilize stellar constellations as an aid to navigation and recent research has also revealed birds' sensitiveness to electro-magnetic fields so that these factors enable them to travel at night time. The Blue-cheeked Bee-eaters (*Merops superciliosus*) which cross the Arabian Sea from Africa, arrive in Pakistan with unfailing regularity starting from the last two or three days of April and up to mid-May. They tend to travel at night, in small flocks which maintain continuous contact calls, that can sometimes be heard for over 4 km. (2 miles) distance and consequently they are easily heard upon

irst arrival above Karachi. My diary records during
en years of observations in Karachi indicate that the
najority of arrivals of migratory Blue-cheeked Bee-
aters were heard calling during the night, though
often a few loose foraging parties were seen or heard in
he same area the following morning. They could first
e heard within minutes of their crossing the coastline
rom where I resided and I have only two records of
aytime arrivals compared with over 25 records of
ight time arrivals which I chanced to hear.

Bird migration is full of paradoxes and the more
nformation which scientific experiments elucidate the
nore unanswered questions arise. It is clear that birds
o not follow rigidly defined paths and may travel over
. very broad front extending in some cases to hundreds
of miles (Baker, *Mystery of Migration* 1980, and
Moreau, *Palearctic-African Bird Migration Systems,*
972). This should be borne in mind with reference to
he above details about the Indus valley as a 'fly-way'.
similarly, individuals from the breeding population of
. species, from the same locality may follow widely
lifferent migration routes and winter in quite separate
egions, and *vice versa*. Out of many Black-headed
Buntings ringed in 1962 at Bhuj in Gujarat, India one
vas recovered in the southern Ural at Krasnodar
N45°30′., E40°45′) whilst presumably on their breeding
rounds. This divergence of origin of wintering popula-
ions has been particularly demonstrated by ringing
ecoveries in India, of ducks breeding from Siberia,
nd from west Europe, as well as with wintering
opulations, at sites in Africa, of Palearctic migrants
rom all over the northern hemisphere.

RISKS OF MIGRATION

studies of many Palearctic and Nearctic bird species in
emperate latitudes have indicated that there are
nternal physiological responses in birds to changes
n length of daylight, particularly development of
ormant or regressed gonads towards active breeding
ondition, which is in turn associated with triggering
nstinctive urges to migrate (B. Lofts, and R.K. Murton,
968). This photo-periodic responsiveness is thought
o be quite sensitive even to minor changes in daylight
ength and to operate also upon many migrants which
vinter in sub-tropical latitudes such as Pakistan
Tewary *et al.,* 1983). Conversely, decreasing daylight
ength in the northern hemisphere causes regression of
reeding organs and is thought to trigger off inherent
nechanisms whereby migrant populations travel to
outhern wintering grounds. In Pakistan, as in other
ub-tropical regions which provide suitable winter
eeding conditions for Palearctic migrants, the spring
assage is frequently a much more direct and urgent
ffair. For example, birds breeding in the far northern
Arctic tundra zones such as phalaropes and many
ringinae may over-fly large distances without rest

on the ground for foraging or recuperation. Such
migrants are particularly susceptible to weather
hazards. Severe storms, persistent headwinds and
heavily cloudy weather are examples of factors which
can blow migrants off course, unduly delay their
ground travel speed and prevent them from using
navigational aids such as stellar constellations and the
position of the sun. In the case of Pakistan, the
northern mountains pose an added physical barrier for
those migrants breeding in central Asia and Siberia.

The entire northern boundaries of Pakistan, India,
Nepal, Bhutan and Bangla Desh are rimmed by lofty
mountain chains, often loosely referred to as the
Himalayas. In point of fact the Himalayas only
constitute one mountain chain, albeit the longest,
extending right across Nepal, Sikkim and Bhutan and
into Assam in the north-east, but the total width of this
chain is rather narrow in Nepal, for example, it is
about 130 kilometres (80 miles) wide in a north-south
direction. In Pakistan, by contrast the total width
stretches quite 965 kilometres (600 miles) from south
to north as it includes no less than five more or less
parallel mountain chains: the Pir Panjal range, the
Himalayas, the Karakoram range (itself 190 kilometres
(120 miles) in width), the Hindu Kush and finally the
Kun Lun mountain range (in Chinese Sinkiang). Not
only is this the greatest concentration of high mountains anywhere in the world, but it comprises most of
the world's highest peaks (seven of the ten highest
peaks in the world lie within Pakistan) and there are no
easy passes through this formidable mountain barrier,
as human explorers have found to their cost (Miller,
1982).

Some migrants undoubtedly perish en route from
exposure and starvation, or weakened by travel suc-
cumb to predators. But a more widespread hazard is
no doubt that of undue delay in arrival on the breeding
grounds of the species concerned, which results in
failure to establish favourable breeding territories or
to acquire a suitable mate. Innumerable studies have
shown that late breeders have a markedly lower
reproductive success due to these and related factors
(Brooke, 1979).

It is now an established fact that most long-distance
migrants rely upon reserves of stored fat to release the
energy necessary for prolonged and rapid spring
migration. After the physically exhausting activities of
rearing a brood of chicks, many Palearctic migrants
congregate in staging grounds before performing the
main part of their southward migration. These staging
grounds are often traditional and generally provide
the only opportunity for migratory birds to replenish
the depleted fat reserves necessary to enable them to
make prolonged flights. A few such staging areas
occur in the northern parts of Pakistan and have been
exploited by man as a predator (hunter) for several
hundred years. Such hunting pressure up to the recent
past, according to available evidence, has not seriously

affected migrating populations. Examples that come to mind are the use of duck decoys along the Kurram river as well as activities of muzzle-loading wildfowlers in the Chitral and Hunza river valleys.

However, recent economic changes (including those in their breeding grounds in the USSR) have caused severe and evidently unbearable pressure upon the populations of certain highly prized target species. For example juvenile Goshawks (*Accipiter gentilis*) which used to migrate in numbers through the Kunhar valley of Chitral into north-west Pakistan, have since ancient times been decoyed and netted by local Chitrali hunters. Interrogation of local sportsmen and hunters in Chitral state has corroborated the information that this is no longer worthwhile, as insignificant numbers of this highly prized hawk enter Pakistan. Demoiselle (*Anthropoides virgo*) and Common Cranes (*Grus grus*), following the Kurram river and certain broad valleys in northern Baluchistan use these relatively remote and unpopulated areas as staging and resting areas. Within the past couple of decades it has become increasingly fashionable amongst the local Pathan tribal populace to use captive cranes as decoys and to catch wild cranes in large numbers (Roberts and Landfried, 1983). The rapid increase in prosperity of local hunters as a result of earnings from emigration to the Persian Gulf states, together with changes in fashion, have been factors which have increased this hunting pressure. Recent detailed surveys of crane catching camps reveal that from 1,000 to 1,500 cranes are killed or captured annually and out of these about 66 per cent are Demoiselle. A recent survey of waterfowl in the wetlands of southern Rajasthan and the Rann of Kutch where the bulk of the subcontinents's population of Demoiselle Cranes are believed to winter, indicates that the total population migrating through Pakistan is about 25,000 birds (Dr. Van der Ven, pers. comm., 1983). This level of predation at one point is likely to be detrimental to the population as a whole considering the very low reproductive potential of this species, and increasing pressure on their breeding grounds in the Soviet Union as wetlands are drained for croplands.

The third example of severe human hunting pressure, of relatively recent origin, concerns the Houbara Bustard (*Chlamydotis undulata*) which normally migrates over a very broad and diffused front from central Soviet Russian deserts, mainly the Kyzyl Kum, in south-easterly direction through Baluchistan. For a short period of two or three weeks certain broad valleys or plateaux along Pakistan's border with Afghanistan, are known to harbour comparatively large concentrations of Houbara, and this fact has been exploited by local Baluchi hunters for many years. During the past fourteen years many princely families from neighbouring Arab states have been invited to Pakistan to hunt Houbara with falcons. Initially this was mainly in Punjab and Sind provinces

on the eastern desert borders. In the past three years however, these visiting hunters have concentrate their attention upon those Baluchistan staging ground Available information on the dates when such Ara parties hunt, and their daily bag reveal that the bustar population is being hunted at the time of maximum spring and autumn passage and that by carefully time and repeated visits, fresh influxes of birds are killed. I the winter of 1982-3 an estimated 4,500 Houbara wer killed by three such hunting parties in the south western part of Baluchistan alone (Prof. Afsar Miar pers. comm.). Though reliable data on the tota population of this species wintering in Pakistan is sti impossible to obtain it has been estimated that th may be in the order of 18 to 22,000 birds (Goriup 1981). Such a level of hunting again seems intolerabl when considered against the very low reproductiv potential of this species.

At the other end of the scale there are, as mentione earlier in this chapter, clearly a number of gregariou migrant species which concentrate in large flocks an by tarrying one or two months during their passag across Pakistan, use certain areas as staging posts t exploit temporarily plentiful food sources. Rosy Pas tors (*Sturnus roseus*) and Common Rosefinche (*Carpodacus erythrinus*) concentrate in large flock during spring migration, in suitable forest plantatior to feed on ripe mulberries. The mulberry tree cor cerned, *Morus alba*, is actually cultivated mainly fo its timber (valuable in the sports goods industry) an as a source of fodder for the silkworm industry Although there is some human consumption of th fruit it is by no means fully exploited in the Punja plains where these migrant flocks are mostly concentra ted. Therefore, these flocks do no significant harm During the autumn return migration, however ther are more severe problems from damage to growin grain crops by Rosy Starlings, and also by Black headed (*Emberiza melanocephala*) and Red-heade Buntings (*E. bruniceps*). These three species will fee in large flocks upon ripening crops of Sorghum an Millet (*Pennisetum typhoides*), and do inflict econom cally serious levels of loss in the hill tracts of Sind, a well as further south-east in Tharparkar district c Sind. Migrating flocks of House Sparrows (*P. c bactrianus*) also inflict severe damage upon ripenin barley fields in the Pishin and Tobakakar regions c Baluchistan. See chapter titled Birds as Pests an Beneficial Agents.

Up to the present time, the author has not been abl to find any instance of hazard to migrating birds fror crop spraying with toxic chemicals, though this ha been reported as killing numerous migratory Whit Storks (*Ciconia ciconia*) in east Africa (Ron Thomp son, Warden of Hwange National Park Zimbabwe, i paper presented at ICBP 'Birds of Prey Conference Athens, 1982). Toxic chemicals are a more seriou factor in western European countries where the use c chemical pesticides is much more widespread. The high cost of insecticides in Pakistan has up to now largely limited their use to the summer season cotton crop, which is subject to insect attack mainly from late June until late September. In the Indus plains where cotton is mainly grown, there is no very marked influx of Palearctic migrants until late September or early October.

Perhaps it is interesting to reflect that migrants can tolerate a surprisingly high level of human predation (witness the continuous annual catch of small migrant passerines from such places as Cyprus, Malta, and the Gironde region of south-west France), but that it is only when some special set of circumstances, (often the result of local cultural traditions and fashions), occur together, that a particular target species becomes unduly vulnerable. Thus in a sport-loving nation whose shooters are renowned for their marksmanship, the two migrant crane species are not being decimated by hunters with shot-guns, but by local Pathan tribesmen using ancient skills to train and decoy birds and ensnare free flying cranes with a bolero device. Similarly the Houbara Bustard population, though evidently continuously declining during the past three or four decades, is now being decimated not by local shooters, but by visiting Arab hunters practising their traditional sport of falconry. Here, however, it must be acknowledged that access to unlimited financial resources and the employment of huge motorized fleets of specially adapted vehicles, not to speak of sophisticated radio and radar communication devices, are far more important factors in obtaining the quarry than the actual use of falcons.

During the period of British imperial control in the subcontinent, there is ample evidence that these foreign rulers hunted certain rare animals to excess, though this can be excused to some extent by prevailing views and fashions of that period when the natural resources of the world, particularly wildlife, seemed unlimited. For example British army officers garrisoned at Parachinar in the Kurram valley staged snipe shoots in the rice stubbles during the autumn when the concentration of birds indicates that they must have been using this area as a staging and recuperating area. It was not uncommon for a snipe shoot comprising only four or five guns to secure 200 couple (i.e. 400 birds) in half a day's sport (Game Record Book, Officers' Mess, Parachinar, Upper Kurram tribal agency), a devastating toll on a migrant population before it has time to disperse. Today little or no snipe shooting takes place in that area, where the local people are still keen sportsmen and invariably possess locally made firearms. The reason for this is that snipe are now considerd too small a target to be worthy of the effort and cost, so the birds are benefiting from a change in human cultural attitudes.

The very incompleteness of this account indicates that a vast amount of information is still to be learned and surely many surprising discoveries will be made when opportunities finally arise for detailed studies of bird migration in Pakistan. The fascinating aspects of this phenomenon will continue to intrigue scientists and laymen alike because of the incredible distances, traverse of physical barriers, and precise return to breeding locations which are achieved, often by tiny frail birds weighing no more than a few grams.

The Problem of Species

The basis for the systematic sequence and specific nomenclature for the different species of birds described in this book is briefly explained in the introductory chapter.

There are, however, some groups of birds or populations which occur in our region, which do present special problems when considered at the species level, not only for the amateur bird watcher, but also for the most experienced taxonomists. Some commentary upon these examples which occur in Pakistan therefore will, I hope, encourage the reader to contribute further to our knowledge of the subject.

In the *New Dictionary of Birds* (Landsborough Thompson, ed., 1964) the editor stated, perhaps rather confidently, that the species, unlike other taxonomic categories, is not merely a human concept, but represents a natural reality. In the more recent *Dictionary of Birds* (Campbell and Lack, 1985) there is a more detailed discussion, with the comment that species are not constant in space or time and do change in morphology, behaviour and habitats (Endler, article on 'Species', in Campbell and Lack).

One of the processes of evolution, which has resulted in so many varied life forms on this planet and which has come to be known as speciation (following Darwin's theories), is a continuous one. The biological definitions of a species depends ultimately upon criteria which cannot always be tested in the field. There are inevitably certain bird populations or individuals which do not seem to fit very neatly into the species concept and which in a few cases are not easy to recognize as full, or 'true' species. In the latest *Dictionary of Birds* (Campbell and Lack, 1985) the definition of species is that 'they are groups of interbreeding natural populations that are reproductively isolated from other such groups'. This is based upon the definition given by Ernest Mayr (*Principles of Systematic Zoology*, 1969). Most laymen, would think of the individuals which compose a species as possessing a general morphological homogeneity or similarity; we can recognize members because they 'all look alike'. This obviously only holds good in varying degrees as we shall see when we come to consider the case of sibling species, polymorphic species, geographically isolated or allopatric species, as well as so-called Ring species and Super species.

SIBLING SPECIES

These are usually almost identical in appearance and inseparable when observed in the field. The human observer cannot tell them apart on superficial visual examination. Often they are separable only on minute morphological differences such as length of primary feathers or amount of feathering on the toes and tarsus: but they are clearly isolated reproductively, usually because of different courtship rituals and behaviour, and most importantly by different voices and calls. In Pakistan there are a number of such examples among the Sylvidae, e.g. the Paddy-field Warbler *Acrocephalus agricola* and the Blunt-winged Paddy-field or Swinhoe's Warbler *A. concinens*. Likewise amongst the *Phylloscopus* leaf warblers, the Greenish Warbler *Phyllocopus trochiloides,* and the Bright Green Leaf Warbler *P. nitidus*, or the Mountain Chiff-chaff *P. sindianus* and the Brown Chiff-chaff *P. collybita tristis*. Amongst cryptically plumaged birds, there is often an evolutionary convergence in plumage pattern and appearance and this is well illustrated by owls of the genus *Scops*, and by some groups within the genus *Caprimulgus*, of the nightjars (Marshall, J.T., 1978). Perusal of the individual species accounts given in this book will generally show that these sibling species (as defined in Campbell and Lack, 1985), are morphologically separable when examined in the hand, and easily separated by their voice. They do not in fact present any problem in our understanding of what constitutes a separate or distinct species. Nevertheless there has been disagreement in the past amongst taxonomists as to how some sibling species should be classified (see Roberts and King, 'Vocalizations of the Owls of the genus *Otus* in Pakistan', *Ornis Scandinavica,* 1986) and the varying treatment of the classification of the members of the genus *Phylloscopus*, which provides another example of continuing divergence of opinions (Ticehurst, *Systematic Review of the Genus Phylloscopus*, 1938, and K.H. Voous's 'List of Recent Holarctic Bird Species', 1977).

POLYMORPHIC SPECIES

There are two striking examples in Pakistan: the Western Reef Heron or Egret, *Egretta gularis* and the Eastern Pied Wheatear *Oenanthe picata*.

Polymorphism has been defined as the coexistence in one interbreeding population of two (dimorphism) or more, distinct and genetically determined forms, the least abundant of which is present in numbers too great to be due solely to recurrent mutation (Ford, 1945). Such variation has often in the past led to splitting into two separate species, especially if there are only two distinct colour phases. Formerly the Snow Goose *Anser hyperboreus* of north America was considered a distinct species from the Blue Goose *Anser caerulescens*, but it is nowadays recognized that they do interbreed and they are now considered as one species *A. caerulescens* (F. Cooke, 1978). In the case of the Western Reef Heron, studies by the author during 2 and 3 day surveys in the extensive mangrove creeks of the Indus delta, suggest that there are possibly as many as four regularly occurring different colour morphs, three of which are predominantly slaty grey or blue-grey, with one form entirely white. Occasional interbreeding between slaty and all-white individuals has long been noted (K. Eates, *JBNHS*, Vol. 31, 1926, and Parasharya and Naik, *JBNHS*, Vol. 81). The problem is aggravated by the close similarity in appearance of the Little Egret *Egretta garzetta* and the white form of *E. gularis*, and the apparent occurrence of actual hybridization between the two species in certain colonies in east Africa (Hancock and Elliot, *Herons of the World*, 1978) and at Gogha in Gujarat state (Kutch, India) (Naik *et al.*, *JBNHS*, Vol. 78, 1981). The subject is too complex to treat adequately in a book of this general nature, but authorities like James Hancock incline to the view that *E. gularis* and *E. garzetta* should be considered as one species on the basis of this narrow zone of interbreeding(*in litt.* to author, 1987). Studies have shown that all-white birds and all-slaty morphs do tend to mate with those of their kind and that such mixed slaty and white pairs are rare (K. Eates, op. cit., 1926), and discussion by F. Cooke (article on 'Polymorphism' in Campbell and Lack, 1984). However occasional cross pairing of colour morphs undoubtedly occurs and whilst the majority of white Reef Herons have olive greenish or yellow bills and more extensive olive on the tarsus than *Egretta garzetta*, birds with all-black bills have been observed in apparent Reef Heron colonies at Gogha (Naik *et al.*, *JBNHS*, Vol. 81, 1984). Much more study is required

before this problem can be resolved. My observations in Pakistan indicate that the great majority of the Little Egret population is quite distinct, not only in having an all-black bill during the breeding season, but in having more highly developed dorsal aigrette plumes and being confined in feeding to a fresh water biotope, normally shunned by *E. gularis*, and at least in one instance, breeding separately in visual contact with a nesting Reef Heron colony.

The case of the Pied Wheatear is also a reflection of the trend in this whole genus *Oenanthe* to polymorphism with a great variety of forms and species described from the Middle Eastern countries. *Oenanthe picata* occurs in three distinct colour morphs in Pakistan, the males of which are easily recognized in the field. There is evidence that the breeding range of these three forms is fairly distinct and separate and that there are also three varying colour morphs of the females, corresponding to these different areas (See Vol. 2, **Plate 33** and **Map 372** of *Oenanthe picata*). However, Paludan working in Afghanistan came to the conclusion that females of the various morphs overlapped indiscriminately (Paludan, 1959) and Scully also collected obvious examples of 'cross-breeding' between the nominate sub-species *O. picata picata* and the white-capped form *O. picata capistrata*. In the author's experience such apparently 'hybrid' forms are extremely rare and over 99.9 per cent of birds seen in the field clearly correspond to one of these three distinct morphs. For the present they are not considered as sub-species because of apparent crossbreeding and for convenience are treated merely as separate 'races' *picata, capistrata* and *opistholeuca* (Ali and Ripley, 'Handbook', Vol. 9 and Clement and Harris, 'Field Identification of West Palearctic Wheatears', *British Birds*, Vol. 80, 1987, p. 153).

Probably, the safest conclusion at this stage in our knowledge is that speciation is still very much an active and ongoing phase in the genus *Oenanthe*, and that *capistrata, picata* and *opistholeuca* show a trend suggesting that they will rapidly evolve into distinct sub-species, rather than a more homogeneous gradation of many different forms.

The term, polymorphism is also more widely interpreted as extending to light and dark plumage phases as well as normal brown-grey cryptic plumage and erythristic or reddish forms. Again Pakistan is rich in examples among the birds of prey. The Booted Eagle, *Hieraaetus pennatus* occurs in a bewildering variety of light and dark and intermediate colour forms and there is also considerable variation in the plumage of such species as *Buteo rufinus* and *Aquila rapax*. Based on my rather incomplete diary notes over many years, in Pakistan, the white-bellied Booted Eagle invariably mates with pale phase birds and likewise the darker brown-bellied birds are often seen consorting with similar dark forms. In owls of the genus *Otus*, hepatic or erythristic morphs are frequently occurring in some species but apparently not in others. In Pakistan such reddish specimens of *Otus bakkamoena* and *Otus sunia* have been noted. Likewise amongst Cuckoos, hepatic females of *Cacomantis passerinus merulinus, Cuculus canorus*, and *Cuculus saturatus* have all been noted, but it is especially interesting that in all these cuckoos it is only the females which exhibit this reddish morph, and the theory has been postulated that such individuals resemble Kestrels (*Falco tinnunculus*) in colour and appearance, a possible selective advantage in frightening off host passerine species and facilitating egg laying in the host's nest (See discussion on Polymorphism in Campbell and Lack, 1985). The grey-bellied and rufous-bellied forms of the males in the Plaintive cuckoo, were also considered as sub-species *Cacomantis merulinus passerinus* (the grey-bellied form) and *C. m. querulus* (the rufous-bellied form) (see Ali and Ripley, 'Handbook', Vol. 3, 1969), but more recent sytematists now treat them as separate species, *C. passerinus* in the north-west and *C. merulinus* in the east (S.D. Ripley, 'Revised Synopsis', 1982). See also Roberts, Grimmett and Robson's, 'Commentary on a Pictorial Guide to the Birds of the Indian Subcontinent' (*JBNHS*, Vol. 82, 1985). This separation is presumably based upon geographic allopatry, though their ranges do overlap outside of the breeding season (?) in Nepal (Inskipp and Inskipp, 1985).

GEOGRAPHICALLY ISOLATED OR ALLOPATRIC SPECIES AND THE SUPER SPECIES CONCEPT

Again this is a very complex subject difficult to discuss adequately in a general account of this nature. Essentially it rests upon the fact that due to gradual genetic mutation and change in the natural continuous process of speciation, those populations which are geographically isolated may evolve more and more widely separated characters. If the physical barrier is considerable, there is also no evidence of the possibility of interbreeding. Should they then be considered as two distinct species or merely as rather distinct separate populations of the same species? Even in the case of more widely mobile bird species with a relatively continuous distribution there can be natural barriers to any significant degree of interbreeding or genetic exchange simply because of a bird's fidelity to its own natal breeding area and the dispersal of individuals through territory maintenance (see Mayr, 'Animal Species and Evolution', 1963, and Endler, article on 'Speciation' in Campbell and Lack, 1985).

The concept of a super species is most useful when considering such problem cases. There may be two populations of birds with obvious close ancestry and similarities, which occupy widely separated (allopatric) ranges or contiguous ranges but still with very little overlap (parapatric). There may be no opportunities in

such cases to predict how they would behave reproductively with one another under natural conditions. The example of the Blue-cheeked and Blue-tailed Bee-eaters with parapatric populations in north-western Pakistan which is discussed below, being a case in point. In such instances there has to be an arbitrary decision, but by combining them within a super species definition or taxon, it does emphasize their close affinity and enables the reader for example, to extrapolate data about one species which is well known, to another much less well known species, with some confidence. There is an excellent discussion of this problem in the Introduction to volume 2 of the *Birds of Africa* (Urban, Fry and Keith, 1986).

In Pakistan there are many examples of presumed 'more ancient' birds, in terms of evolutionary ancestry, which obviously have a common ancestry but now occur in two geographically allopatric populations in Africa and the Indian subcontinent. If the species is partially a long-distance migrant there is presumably more opportunity for genetic exchange and it seems more logical to treat representatives from both regions as one species. The Pied Crested Cuckoo *Jacobinus clamator* with a population partly resident in southeast Africa and partly migrant to the Indian subcontinent, is one such example. The Hoopoe *Upupa epops*, (formerly widely treated as two species, the African population as *U. africana*), with east African migrant populations and resident Indian populations, would be another non-controversial example. But what of the case of such relatively sedentary birds as the Darter *Anhinga melanogaster* of Africa, considered as a separate species *A. rufa* on the Indian subcontinent by Ali and Ripley (see Ripley, 'Revised Synopsis', 1982, and 'Handbook', Vol. 1), but as one species *A. melanogaster* including sub-species *rufa* by K. H. Voous. The author's limited observations of Darters seen in Zimbabwe and Kenya, are that they appear darker than Indian and Pakistan birds, more rufous, less silvery grey on the neck and with less conspicuous silvery grey elongated plumes. The separation into two species does seem more logical. Another example is the King Crow, previously considered as two species (see Stuart Baker, 'Fauna Series'), *Dicrurus adsimilis* the African Drongo, and *Dicrurus macrocercus* the Black Drongo of south-east Asia. Curiously in this instance Ali and Ripley consider they should be combined as one species, whereas K.H. Voous treats them separately (Ripley, 'Revised Synopsis', 1982, and Voous, op. cit., 1977). Again the author's limited observation of the African King Crow in Zimbabwe suggests that despite the innate mimicry in the vocabulary of this species, the African population had a very distinctive repertoire of song strophes and calls, quite different from the Pakistan population and also that they showed consistently shorter and less deeply forked tails. One is tempted to speculate whether naturally occurring introductions of the African population would interbreed with the Pakistan/Indian population, especially as their territorial calls are so distinctive and the taxonomic treatment followed by Voous and argued by Charles Vaurie (1949, *Bulletin American Museum Natural History*, No. 93), would seem to add support to such a continued distinction.

The case of the African Pygmy Cormorant *Phalacrocorax pygmeus* and the south-east Asian Little Cormorant *Phalacrocorax niger* would also seem to pose similar taxonomic difficulties. In non-breeding plumage, specimens are virtually inseparable and due to the paucity of suitable wetlands in the Middle East countries, the actual areas of possible overlap of these two species are not well studied or known (M. D. Gallaher, *in litt.* to author, 1982) though certainly *P. pygmeus* extends as far west as into Baluchistan (specimen in BNHS Museum collection).

A brief digression, which may help the reader consider the practical criteria usually adopted for choice of separate species. There is a world wide complex of purely black forms of the Oystercatchers *Haematopus*, with two separate Australian and New Zealand populations, as well as the African, north American and south American populations and some isolated island endemic populations, all being considered as separate species, despite the fact that all are virtually identical in appearance (Hayman *et al.*, *Shorebirds*, 1980). The distinction here is presumably based solely upon geographic allopatry, not on questions as to whether they would in fact interbreed if they came into contact with other Black Oystercatcher populations.

We should always bear in mind the point that the final arbiter of such distinctions would be the birds themselves if their distribution ranges gradually met, and whether they then interbred freely or not.

RING SPECIES

Ring species are defined as animals, in this case birds, which have a continuous distribution often exhibiting a series of recognizable sub-species of continuing greater variation, such that the end forms, if they do overlap, do not. interbreed. In Pakistan though not altogether a resident breeding bird we have the examples of the central Asiatic Carrion Crow *Corvus corone orientalis* and the more southern Middle Eastern population of the Hooded Crow *C. c. sharpii*, with plenty of examples of hybridization between the two in the north-western border of Afghanistan (Paludan, 1959).

A much more complex ring species relates to the Herring Gull *Larus argentatus*, and the Lesser Black-backed Gull *L. fuscus*. Western Europe is considered as the overlapping area of the extreme boundaries of these two forms and here they are treated as full distinct species which do in fact breed separately. In Pakistan, where only wintering populations occur, it is possible to see a great variation in the colour of the mantle and wing coverts, of adult forms, from very dark slaty black tones as in *L. fuscus,* to intermediate lighter greys, and similar pale silvery greys as in the nominate sub-species of *L. argentatus*. Likewise, there is a variation in leg and feet colour from pale pink to sulphur yellow with intermediate muddy orange shades. Most of these birds probably belong to central Asian breeding populations of the sub-species *heuglini* and *taimyrensis*. Recent opinions amongst taxonomists, are that these are better considered as sub-species of *L. fuscus* (See Cramp *et al.*, *Birds of the Western Palearctic*, Vol. 3, 1983), rather than as Yellow-legged Herring Gulls (sub-species of *L. argentatus*) as in Ali and Ripley's 'Handbook' Vol. 3, 1969, and S. Dillon Ripley's 'Revised Synopsis', 1982).

Though *L. fuscus* is listed as occurring widely in our region (Ali and Ripley, 'Handbook', Vol. 3) the key character for separation from the dark backed form *heuglini*, is the absence of white spots on the third and fourth primary wing tips. Some individuals seen in Pakistan can definitely be assigned to *L. fuscus* on this basis (author obs. and Kent Forssgren *in litt.*), but they are in a very small minority and it is possible even to see individuals with hardly any trace of white sub-terminal primary spots.

Another perplexing problem which should be considered as belonging to a super species complex, is that of *Merops superciliosus* the Large Green or Blue-cheeked Bee-eater and *Merops philippinus*, the Blue-tailed Bee-eater. The Blue-tailed Bee-eater extends over most of the south-east Oriental region, and a small part of its population, occurring in the western extremity of its range, migrates in summer into north-west Pakistan to breed. Here it overlaps in range with the Large Green or Blue-cheeked Bee-eater which undergoes a much longer migration each spring and autumn to winter in Africa. In size and plumage characters the two species are closely similar except for the colour of the rump and tail which is blue in *M. philippinus* and more verdigris or greeny blue in *superciliosus*. There are also other minor differences in the prominence and tint of the pale borders above and below the black eye-streak. See actual species accounts for details of these differences. As the summer season progresses and the breeding birds suffer feather abrasion, plus bleaching, it is interesting and perhaps significant that in *M. philippinus* the blue tail colour fades almost to the same blue-green as in *superciliosus,* and also there is individual variation in the prominence and extent of the yellow throat patch and the pale malar streak below the black eye-band. Nesting colonies have been observed by the author around Sidhnai spill channel comprising mainly *M. superciliosus*, but with a few individuals corresponding exactly to *philippinus* and undoubtedly nesting amongst them. Likewise near Rawal lake, in a colony of *M. philippinus*, a few individuals with green tails and rumps were observed. At Lal Sohanran in Bahawalpur district both kinds occur together in considerable numbers. Unfortunately I have not been able to determine whether there are definite instances of cross-breeding in such colonies, but circumstantial evidence strongly suggests this. Daniel Marien, studying museum skins collected from north-west Pakistan and India, concluded that because of major differences in their migration and slight differences in moult pattern, they should be recognized as quite distinct and separate species (Marien, *JBNHS,* Vol. 49, 1950). If both forms evolved from a common African ancestry (Fry, *Ibis,* 1969, and Fry in Snow, 'Atlas of Speciation', 1978), then perhaps prolonged separation of *M. superciliosus* has resulted in the evolution of different migration patterns and a divergence in tail colour which has resulted in the two distinct forms which were formerly considered only as sub-species, *M. s. persicus* and *M. s. philippinus.* Ecologically they are still remarkably

similar in their addiction to a watery biotope for hunting, to a food preference for 'Odonata' (not general amongst the Meropidae), and the fact that both populations postpone the major part of their moult until after they have completed their post-breeding migration to winter quarters. Further studies of the possibility of hybridization amongst the Punjab colonies of these two Bee-eaters would prove to be of great interest, though it would not entitle us to consider them as one species. Undoubtedly the concept of a super species most closely fits the present status of these Large Green Bee-eaters, since the term implies that these species have attained a state of reproductive isolation, and they are not merely sub-species of a single variable or polytypic species (see definition in Campbell and Lack, 1985.) Field observations in Pakistan, about nesting Pied Wheatears, Reef Heron colonies and Bee-eater colonies in the northern Punjab are all worthy of much further study and likely to yield rewarding information.

The Contribution of Early Ornithologists

From the last three decades of the nineteenth century until the time that Pakistan became independent, there were a number of British Government servants who enthusiastically studied and exchanged knowledge about the bird fauna of the region. In the introduction to the first edition of the *Synopsis of the Birds of India and Pakistan* (Ripley, 1961) there is an excellent review of the accumulation of new knowledge achieved by these pioneer ornithologists and an even more thorough review of their respective contribution in the first volume of the *Handbook of the Birds of India and Pakistan* (Ali and Ripley, 1968-74).

Pakistan like many other newly developing nations has had to cope with a burgeoning human population, and the individual's economic struggle for a better life certainly diminishes the leisure time and interest which studies of natural history and of birds in particular require, and which have characterized the more prosperous western nations since World War II. Such a situation has necessitated a heavy dependence even in the present day, upon the published findings of these earlier ornithologists. This applies also to India, though to a lesser extent than in Pakistan, as the distinguished authors in their introduction to the ten volume 'Handbook Series' (Ali and Ripley, the *Handbook of the Birds of India and Pakistan*, 1968-74) pay tribute to their own heavy reliance upon the copious manuscript notes prepared by Mr. Hugh Whistler in collaboration with Dr. Claude B. Ticehurst. This was to form the basis for a comprehensive handbook of Indian birds, but with Ticehurst's untimely death in 1941 and Whistler's but a year later, their project never came to fruition. But his executors very far sightedly passed on Whistler's manuscripts to Dr. Salim Ali and these are now lodged in the very extensive library of the Bombay Natural History Society, with copies in the British Museum (Natural History).

A rather more personal tribute to some of these pioneer ornithologists including some evaluation of their reliability, therefore follows. Many of these long-dead naturalists have been my constant stalking companions in the field, and whilst poring over their writings during many a long evening, I have found in them a continued source of inspiration, and felt wonderment at their ubiquitous energy and meticulous attention to detail. Whilst some have on occasion seemed slightly dogmatic or overbearing, the majority with a refreshing humbleness, have thereby enhanced reliance upon their observations and have given special pleasure when one's own personal observations suddenly recall to mind their earlier writings.

Historically, among the earliest ornithologists of relevance to Pakistan were Major Biddulph and Dr. Scully, colleagues who stimulated each other in their bird studies and who both made careers in the 1870s and 1880s in the far northern mountain tracts where power rivalry between the Russian empire and the Chinese, even in those days, led the Imperial Indian government to try and establish a British presence in the region. Both men made extensive collections of bird skins and published their observations in the *Ibis* and these were in turn republished in the journal founded and edited by A. O. Hume, *Stray Feathers* (Biddulph, *Stray Feathers*, Vol. 9, 1881, and Vol. 10, 1881 and S. F. Scully, Vol. 10, 1881). Since their specimens are safely stored in the British Museum at Tring, their labels bearing their original field notes, and since these remote mountain regions have still largely escaped the ravaging hand of man, their findings still have great relevance today and even the few field naturalists who have accompanied mountaineering expeditions into the region in the past 50 years have been unable to make anything like such a comprehensive or significant contribution.

Coming south to the outer Himalayan ranges, there have been a good number of army officers and other officials who enthusiastically 'worked' the Murree hill ranges and adjoining territory of Kashmir. Since the author purchased a summer cottage in the Murree hills, 26 years ago, and has grown to love and know the area intimately, such early workers have had special importance. Among those first recorders were men such as Captain Cock and Captain C.H.T. Marshall (Cock and Marshall, *Stray Feathers*, Vol. 1, 1873). They concentrated on the lower slopes of the Jhelum river valley before this area was denuded of natural vegetation and developed for terraced cultivation. Sadly many of the birds which at these altitudes found suitable habitat and breeding conditions at the turn of the century have long since disappeared. Unfortunately the major publications about the current status and distribution of birds in this region have tended uncritically to document such records from the Murree hill ranges, where ecological changes even at higher elevations, though less obvious, have been no less profound. The apparent absence of many such species in the present day has been a perplexing problem, which has already been pointed out (Roberts, 'Review of Second Revised Edition of the Synopsis', *JBNHS*, Vol. 79, 1982). A Brigadier General Kenneth Buchanan was a keen egg collector who explored both Kashmir and the Murree hill range (Buchanan, *JBNHS*, Vol. 15, 1903). His manuscript notes were frequently cited by Hugh Whistler, who wrote the most comprehensive account we have of the birds of the Murree hills (*Ibis*, 1930) nearly three decades later. Colonel Rattray was another most energetic 'bird nester' who made perhaps the most extensive list of birds recorded as nesting in the Murree hills (Rattray, 1905). All these records are faithfully reiterated in Ali and Ripley's 'Handbook Series' and S. D. Ripley's *Synopsis of the Birds of India and Pakistan*. It is known that Colonel Rattray relied heavily upon local hill people for finding nests and though he and Colonel Buchanan recorded frequently that they had 'shot the female off the nest to confirm identification', there is no evidence that they preserved any specimen skins. Regrettably, egg substitutions and even nest re-location was sometimes resorted to by such hill people as a means of getting extra financial rewards from their 'foreign sahibs'! It is thus difficult to infer whether they were ever misled in their identifications as both were extremely knowledgeable and experienced ornithologists, but in the case of Col. Rattray a large number of his records are of birds never subsequently recorded by anyone else from the Murree hills. A few instances are clearly demonstrable as examples of misidentification as is recorded in the individual species accounts later in this volume. For example, his doubtful records include Lesser Spotted Eagle, Drongo Cuckoo, Golden Bush Robin and Orange Bullfinch. The Murree hill range represents in part the westernmost range extension for a number of more or less endemic Sino-Himalayan bird species. That the past 80 years have seen an inevitable shrinkage of the range and population of some hill birds is not surprising as human disturbance in such areas is now a major factor. Despite these implied criticisms, Col. Rattray was one of the first naturalists to attempt to photograph birds' nests in their natural situation and his notes are of immense value. Another very careful and reliable observer was B. B. Osmaston, a forest botanist by profession, who described for the first time the breeding habits of many Himalayan birds, especially less well known species from the inner dryer Himalayan mountain ranges (Osmaston, *JBNHS*, Vol. 14, Vol. 20, Vol. 24, Vol. 29, Vol. 31, etc.).

Descending geographically from the Himalayas to the northern Punjab, three remarkably able men have contributed to our knowledge of this area. Hugh Whistler, who served for seventeen years in the Indian police, from his earliest days was a prolific researcher and published detailed regional accounts about the birds observed in three districts where he was posted, viz., Jhelum, Jhang and Rawalpindi districts all in the Punjab (Whistler, 1914, 1922, 1930). Though he left the subcontinent in 1926 (due to his wife's ill health when a niggardly government refused to grant him compassionate leave) he continued to be most actively associated with the ornithology of the subcontinent. An energetic correspondent and acknowledged expert, he amassed a huge collection of bird skins both through his own personal efforts and through contributions from a host of enthusiastic amateurs, and this fine collection is also safely housed in the British Museum at Tring. He wrote the first *Popular Handbook of Indian Birds* of which a fourth edition has been published (ed. Kinnear, 1949, reprinted 1963) and it is still widely read. As mentioned earlier he was a contemporary and colleague of Dr. Claude B. Ticehurst, (whose name will be mentioned in reference to Sind province), with whom he hoped to collaborate on a completely revised series of volumes on the birds of India, but for his untimely death in 1942 at the age of 53.

The second of these remarkable men was Alexander Edward Jones, a military tailor and outfitter by profession who spent many winters in the northern Punjab and encouraged by his mentor Hugh Whistler, made a collection of over 3,000 bird skins beautifully prepared and mostly by his own hand. Some of these are in the British Museum but the bulk was bequeathed to the Bombay Natural History Society Museum, where they can still be recognized for their immaculate freshness and the detailed labels giving notes on the size and condition of gonads and colours of 'bare' parts. Jones published some interesting notes on the breeding of various birds of prey, and his notes on the birds he collected from around Rawalpindi, Campbellpur and Lahore districts are of special relevance (Jones, 1921), also his account of breeding birds in the Simla hills (*JBNHS*, Vol. 47, 3 parts) shows the wide breadth of his knowledge of Himalayan birds. He retired to his lifelong home at Simla where he died in 1947.

The third remarkable man was also by coincidence a police officer and a disciple of Hugh Whistler. H. Waite served through the western regions of the Punjab and particularly in the Salt Range from the World War I years to the early 1940s. He made a collection of over 2,000 bird skins of which over 1,000 came from the Punjab Salt Range, which are all in the British Museum and wrote many notes about his

Plate 10: Pioneer ornithologists of the region

(a) Herbert William Waite, C.I.E. (Punjab Police)

(b) Lieut. General Sir Phillip Christison, G.B.E., D.S.O., M.C.

(c) Kenneth R. Eates, M.B.E. (Sind Police)

(d) Alexander Edward Jones, M.B.O.U.

findings in such relatively unknown areas as Muzaffargarh district and the Suleiman hills in Dera Ghazi Khan (Waite, *JBNHS,* Vol. 37, 1934, Vol. 39, 1938, Vol. 48, 1948). He eventually returned from retirement in England to his beloved Salt Range and lived at Kalabagh on the Indus in Mianwali district where I had the pleasure of getting to know him and enjoying reminiscences of his many fascinating experiences as a police officer. He died in 1967 and before then, most generously gave me his bound copy of Ticehurst's *Birds of Sind* as well as allowing me to borrow and copy all his manuscript notes on his 2,000 specimen bird skin collection. Till the time of his death this tall gaunt old bachelor wrote letters in a copper-plate but microscopically neat handwriting.

Omitting the mountainous regions of the west it is perhaps logical to descend from the Punjab to Sind and mention Claude B. Ticehurst who served in Karachi for four years during the latter part of World War I as an army doctor. Ticehurst was a brilliant ornithologist who made a lasting contribution to our modern understanding of moult in birds and whose account of the 'Birds of Sind' published in five parts in the *Ibis* (Ticehurst, 1922-4) was a remarkable achievement after only four years of field work. He also diligently researched available information about the ornithology of Baluchistan and produced the best comprehensive account of that region, though of necessity he had to rely largely upon second-hand accounts (Ticehurst, *JBNHS,* 1926-7). He continued to be a prolific worker after return to Britain and his monograph on the difficult genus *Phylloscopus* is an essential book for anyone interested in Himalayan breeding birds, (Ticehurst, 1938). He also amassed an important collection of bird skins (about 1,500 specimens) from the Sind region all of which he made up himself and which are now lodged in the British Museum.

In Sind province one of the most interesting early ornithologists was Scrope Doig, an irrigation engineer who endured the lonely life and burning desert climate while the east Narra canal was being excavated in the 1870s. His perceptive accounts of birds, breeding habits in the swamps adjoining the east Narra were published in *Stray Feathers* (Doig, 1879 and 1880). T.R. Bell, a skilled entomologist by profession and a member of the Forest Service also made valuable observations of the bird life in the riverain forest tracts, mostly in Sukkur district and gave all his manuscript notes to Dr. Ticehurst. His manuscript notes are also lodged in the library of the BNHS. Much more recently Kenneth Eates who served throughout Sind from the early 1920s to late 1940s as a police officer, became the authority on the vertebrate fauna of Sind. His account of that fauna published in the *Sind Gazeteer* (ed. Sorley, 1968) shows the breadth of his knowledge. Eates was an avid egg collector, and after his retirement to the coastal town of Hastings in England, I had the pleasure of getting to know him. He made many new discoveries such as the first breeding records on the subcontinent of the Cream Coloured Courser (*Cursorius cursor*) and the Greater Spotted Eagle (*Aquila clanga*) and published an extensive series of notes in the *Bombay Natural History Society Journal* (Eates, *JBNHS,* 1937, 1938 and 1939). Very generously he presented me with all his old large scale maps of Sind and all his manuscript notes on his egg collection and other natural history observations(now lodged in British Museum [Natural History] at Tring). I have been able to draw on these notes with respect to the breeding habits of many Sind birds. Unfortunately a new-found sense of national pride ordained that he could not bequeath his egg collection to the Peabody Museum at Yale as he had intended. After years of fruitless searching at his behest, this collection was finally located in a congested storeroom behind Hyderabad Museum. With many specimens smashed or missing and his elaborate catalogue system in chaos, even an expert zoologist would find the task of sorting and classification well nigh impossible, so that I never had the heart to inform him of this discovery.

The western mountain regions cover a vast and little explored area ornithologically, stretching from Chitral in the north-west corner, down through the North West Frontier Province and into Baluchistan. Again the major contribution has been through army officers stationed in these frontier outposts in regions where freedom of movement was somewhat restricted even in the hey-day of British suzerainity. In Chitral, Captain Fulton and Captain Perreau collected bird skins and sought expert help in identification so that their published accounts are fairly reliable (Fulton, *JBNHS,* Vol. 16, 1904, 1907, and Perreau, *JBNHS,* Vol. 19, 1910). Fulton trekked in July right up into north-eastern Chitral and obtained several new records for the subcontinent (e.g. Azure Tit and Daurian Starling). Since his day no ornithologist has been able to visit that region, involving as it does more than 2 weeks of daily marches. Some of Perreau's recorded observations however cannot be verified from specimens and need to be regarded with caution, for example records of the occurrence of the Redwing (*Turdus iliacus*). Fulton's skins are in the British Museum and Perreau's skins (rather badly made-up) are in the Bombay Natural History Society collection. These Chitral based ornithologists are, however in a more amateur category than Captain C.H.T. Whitehead, an officer who served in Kohat, Bannu and the Kurram tribal agency and whose name will always be associated in the author's mind with the North West Frontier. Whitehead, after serving with distinction in the Boer War, served in the Frontier from about 1905 until 1914 and was killed in action a year later in France. A modest man (like Kenneth Eates) he was, as his contemporaries described him, 'as tough as nails' and certainly penetrated into remote and inaccessible areas where only an experienced mountaineer could gain access. An all round naturalist, besides discovering new records for the subcontinent of the Stoat (*Mustela erminea*) and Forest Dormouse (*Dryomys nitedula*), he was the first person to record the occurrence of the Chaffinch (*Fringilla coelebs*) and Waxwing (*Bombycilla garrulus*) on the subcontinent, besides providing the only descriptions we have of the nesting of *Phylloscopus subviridis* and *Oenanthe picata capistrata*. His published account of the birds of Kohat and the Kurram agency (*Ibis,* 1909, *JBNHS,* 1910) remains the best account available of that politically inaccessible area, and his specimens which are lodged in the British Museum, with their rather untidy handwriting, have evoked in this author, the happiest memories of exploration in the Kaghan valley. Despite the better jeepable roads of today and thirteen field trips into the area, the author cannot claim to have seen in that valley, at least five of the birds which he discovered during 2 extended camping trips.

Quetta became an important military station in Baluchistan from the 1870s and a number of army officers diligently investigated the ornithology. One of the earliest of these, of note, was Captain Barnes, stationed during one of the Afghan wars on the border at Chaman. His observations were published in *Stray Feathers* (Barnes, 1881) and his name is commemorated in a discovery of Barnes's Chat (*Oenanthe finchsii*). Much later Colonel Meinertzhagen, a contemporary of Ticehurst and Whistler, was stationed as a young captain in Quetta and in the short period of two years made an astonishingly impressive list of new breeding bird discoveries (Meinertzhagen, *JBNHS,* 1914, *Ibis,* 1920). Meinertzhagen is perhaps better known for his travels in Africa (the Giant Forest Hog is named after him) and his classic book *The Birds of Arabia*. An intellectual giant compared with many of his contemporaries, he was inclined to be contentious and arrogant, but his views of speciation and migration were ahead of his time and his contribution to ornithology is of lasting importance and value. Despite his rather quarrelsome nature he became a valued friend and mentor of Dr. Salim Ali when the latter was developing a serious interest in ornithology (Salim Ali, pers. comm. to author, 1980).

Following the surveys of Col. Meinertzhagen, the most outstanding ornithologists to work in Baluchistan, were Major C. H. Williams, an ardent egg collector who worked with his younger brother around Quetta area (Williams and Williams, *JBNHS,* Vol. 33, 1929). Later during the Second World War General Sir Phillip Christison (as he is now) was Corps Commandant at Quetta and supervised the construction of a strategic road through the Chagai desert to the border with Iran. A keen ornithologist, he was able to publish an account of the birds from this remote and hitherto unknown region and to add several new species to the subcontinent's known avifauna, e.g. Afghan Scrub Sparrow, Egyptian Nightjar and the Nightingale (Christison, *Ibis,* 1941, and *JBNHS,* Vol. 43). It was my privilege and pleasure to meet Sir Phillip in November 1985 in order to discuss some of the problems I had encountered (see species account of Carrion Crow) during my visits to study the birds of Baluchistan and he was kind enough to lend me his only surviving copy of his book *The Birds of Baluchistan* published in Quetta in 1940 and long since unobtainable.

The above account is by no means complete and is not intended to belittle or avoid reference to the many living ornithologists, both amateur and professional who have made enormous contributions to our knowledge of this region, the foremost among whom must certainly be the great Dr. Salim Ali to whom I owe an enormous debt from his writings, as well as the encouragement he has repeatedly given me in the preparation of this book.

It will be noted that nearly all these early pioneers were amateurs who had to squeeze their bird-watching into off-duty hours, and do a lot of their travelling by time-consuming horseback or on foot. Ticehurst pedalled a bicycle at high speed to get close enough to shy waders for a shot! That they did most of their bird watching with a collector's gun in hand is but a manifestation of the accepted mores of that era and for anyone not practised or trained in skinning birds, is a testimony to their diligence. Moreover the records that they left in our museums today are of permanent scientific value as well as enabling us to truly evaluate their contribution to present day knowledge. Many of them in their writings were almost lyrical in their descriptions of the habits or surroundings of the birds which they observed.

Hugh Whistler's description of the White-browed Fantail (*Rhipidura aureola*) comes especially to mind, and also Bates and Lowther's description of the characteristic mixed species foraging guilds encountered in Himalayan forest. Their attempts to verbalize bird song and calls may sometimes seem very quaint and dated, such as the often quoted human tones of the flight call of the Cotton Teal (*Nettapus coromandelianus*) described as 'fix-bayonets fix-bayonets'. They also call to mind sounds now unfamiliar to modern ornithologists. Captain Whitehead describing the sudden explosive phrases during the song of the Indian Redstart (*Phoenicurus ochruros*) likens this to 'the sound made by pouring shot into a bottle'. No doubt he loaded his own cartridges. Waite described the territorial calls of *Otus brucei* as like the sound made from pumping up a car tyre. The type of hand pump and primitive tyre valves of the 1930s models did produce a rather metronomic 'poop-poop' noise, unrecognizable from today's air-pressure hoses. That their writings have afforded so much stimulus and pleasure even up to the present day is the best memorial to their inestimable contribution.

Birds as Pests and Beneficial Agents

In many tropical and sub-tropical developing countries, where cereal food resources are a relatively scarce and precious resource, there are instances of birds inflicting serious crop losses and no account about birds in Pakistan would be complete without a realistic acknowledgement of this problem area.

Reference has already been made in the chapter on Migration to crop losses to wheat and barley, caused by migrant House Sparrows (*Passer domesticus bactrianus*) while on spring passage. Also damage to millets (*Pennisetum typhoides*) and sorghum (*Sorghum sudanese*) on autumn passage by Rosy Pastors (*Sturnus roseus*), and Black and Red-headed Buntings (*Emberiza melanocephala* and *E. bruniceps*). These losses occur most frequently in the semi-arid tracts, where crops are grown without irrigation, such as the Thar desert, the broader valleys of Baluchistan and the hilly tracts of Sind and the North West Frontier Province. They can reach serious loss levels because of the highly gregarious flocking and foraging habits of these species during migration, coupled with the characteristic that such arid-zone and rain-fed cereal crops tend to occur in scattered isolated fields and to comprise very thin plant stands of low yield potential.

During the 1970s the Agricultural Research Council of Pakistan in collaboration with the United Nations Food and Agriculture Organization carried out extensive surveys of crop losses and estimated damage levels from all kinds of vertebrate pests including birds. A considerable body of data is therefore available (*Handbook of Vertebrate Pest Control in Pakistan*, compiled and edited by T. J. Roberts, 1981). Bird losses were noted to be patchy and irregular in occurrence, in contrast to the general pattern of crop losses inflicted by insects, and to a lesser extent by mammals and rodents. Such a pattern makes over-all area estimates particularly difficult to determine, but the level of crop losses from birds were on average less than those from field rodents, and very much less than losses inflicted by insect pests. FAO surveys showed that a smaller more selective range of crops were affected, but that net losses to individual farmers were occasionally found to be very high. For example, losses to Sunflower crops *Helianthus annua* (grown for edible oil), often reached 40 per cent and in a few instances over 80 per cent, resulting in failure to harvest any crop as revealed by surveys in Rahim Yar Khan in southern Punjab and Umarkot in Tharparkar district of Sind. Losses, directly attributable to birds, to wheat crops rarely exceeded 7 per cent in the irrigated canal colonies, but in the rain-fed or arid farming zones reached as high as 40 per cent in individual fields. Similarly losses due to bird damage of 20 per cent to maize cobs were estimated in Rawalpindi district, and up to 12 per cent to citrus fruit crops in Renala Khurd (Sahiwal district), of the Punjab.

Most of these situations involve a temporary and abundant food supply in an area where birds quickly learn new feeding techniques and tend to adopt gregarious feeding habits. The Indian sub-species of the House Sparrow (*Passer domesticus indicus*) has in the past 50 years reached extraordinary population densities in the irrigated farming tracts, partly as a result of the high human population and concentration of small scattered rural villages, having houses and out-buildings of baked mud with thatched roofs. Such structures provide ideal Sparrow nesting sites.

In one detailed study of a typical farming village in Rahim Yar Khan district there were 52 households with an estimated human population of 416 persons. During March and April, 2,789 active sparrow nests were found in the same village with an additional estimated 33 per cent of the sparrow population non-breeding (due to shortage of nest sites). Sparrow damage to wheat (*Triticum vulgare*) occurs over about a six week period, commencing after the pollen anthers have disappeared and the grain develops into the milky stage, and declining as the grain hardens and ripens. Within such a period in the above village there was an estimated potential sparrow increment to the population of 11,156 juvenile birds, and a calculated total resident village population of over 16,700 sparrows, capable, between them, of consuming 50 kilograms of wheat a day (Bashir, 1980). Sparrows have learned to balance on the vertical stem of the wheat and to peck open the glumes and extract the wheat seeds *in situ*. In the early 1930s, the Agricultural Department in the Punjab already took note of the growing problem of sparrow damage and advised farmers to sow bearded or awned varieties of wheat as being relatively immune to bird attack. In the 1980s all commercial wheat varieties being cultivated were bearded and the adaptable sparrows have learned to extract the seed with equal facility. Barley (*Hordeum* sp.) by contrast is not generally eaten by sparrows because the pericarp or outer seed-coat fits so snugly over the grain. It is therefore surprising to note that in Pishin district of Baluchistan where this crop was grown in lieu of wheat, migrant flocks of *Passer domesticus bactrianus* were causing considerable damage to isolated fields (personal observation, May 1979).

The Rose-ringed Parakeet (*Psittacula krameri*) though not nearly as abundant numerically as the House Sparrow, is the second most widespread and serious pest. This is partly due to its larger size, flocking habits and the dexterity of its zygodactyl feet. Maize (*Zea mays*) grown for seed, as well as sunflower and citrus fruit crops are especially vulnerable to parakeet damage. The powerful pincer action of the 'Psittacine' mandibles enables the spadith to be ripped off unripe corn cobs, and the slippery rind of malta oranges to be cut open. Initial parakeet damage of these crops is then rapidly exploited by weaker billed sparrows and weavers (*Ploceidae*), which cannot peck through the corn cob sheath covering, and by Red-vented Bulbuls (*Pycnonotus cafer*) in the case of citrus fruit. Anyone who has kept a pet seed-eating bird will testify to the extreme palatability of sunflower seeds (*Helianthus annua*). Most commercial varieties have flower discs which bend downwards upon ripening and are therefore immune to attack from granivorous passerines. The Rose-ringed Parakeet with zygodactyl toes can cling upside-down and has no difficulty in extracting the seeds from underneath the flower head. Many farmers when interviewed reported no bird damage when they first grew this recently introduced crop but heavy parakeet damage in the second and third years, indicating the birds' adaptive and learning ability.

In Europe, freshly sown cereal is often systematically consumed by Rooks (*Corvus frugilegus*) following the corn-drill rows and it is surprising that similar bird damage has not been noted in Pakistan at this stage of cereal crop cultivation. In Multan district in one instance only, a farmer reported heavy damage to freshly sown maize and wheat by the Collared Dove (*Streptopelia decaocto*) (*Vertebrate Pest Control Handbook*, op. cit.). In Ludhiana district of Punjab in India, the House Crow (*Corvus splendens*) inflicts similar damage to newly planted wheat and is there regarded as a serious pest (Dr. Hardev Singh Toor, Ludhiana Agricultural University, pers. comm. to author, Feb-

ruary, 1975). Curiously in the Punjab part of Pakistan, where the House Crow is equally abundant and cropping patterns are similar to Ludhiana, no incidences have come to notice of similar damage by digging up newly planted cereals (See Qureshi, 'Notes on Feeding Activities of Birds in Field Areas of Lyallpur', *Pak. Jour. Zoology,* 1972). These examples of damage to newly sown cereal crops, reveal these different species learning ability.

The Rose-ringed Parakeet also feeds on ripening wheat, and is particularly fond of *Brassica* seeds (such as mustard crops) which sparrows cannot exploit. Parakeets feed in such thin-stemmed crops by hovering and biting off the whole inflorescence, which they then carry to a tree perch and consume. Generally only a few seeds are extracted before the whole seed-head is dropped to the ground where it remains abandoned.

Such food sources are, of course, very temporary and both Sparrows and Parakeets subsist for a large part of the year on grains, buds and berries, which are of no ecoomic importance. During the breeding season, House Sparrows feed their nestlings largely upon insect larvae and in Punjab villages, flocks have been watched gleaning caterpillars from fodder crops of Berseem (Egyptian Clover) and Lucerne. At such a time they are of net benefit to the farmer, but detailed studies have revealed that House Sparrows in the Rahim Yar Khan village cited above, commenced feeding grain to their semi-fledged nestlings before they had left the nest.

The Rosy Pastor (*Sturnus roseus*) is particularly abundant in Pakistan during migration and is much more omnivorous than the House Sparrow. Because its migration through southern Baluchistan and the Thar desert coincides with the spring season when the desert locust (*Schistocerca gregaria*) sometimes hatches in swarms, it was long ago realized that this Starling could be of enormous potential benefit, since flocks will actively search out and feed upon hatching locust larvae. As long ago as the 1940s the British Indian Government felt it necessary to pass special legislation to protect this species because it was often being netted for food during its spring migration. As noted in the chapter on Migration, during late summer when this same starling starts its return migration, it is often a serious pest upon Sorghum crops which have been planted in the same desert regions to take advantage of monsoon showers. Any approach to crop protection and prevention of bird damage should therefore be based upon repelling and scaring, rather than killing techniques, for most birds are beneficial at some stage in their annual feeding cycle, and there are compelling economic as well as aesthetic and auditory arguments for avoiding any control programmes aimed at attempting to destroy bird populations.

Turning to the beneficial roles played by birds, perhaps the most obvious is that of insect control. Studies of the number of moth larvae harvested daily by breeding pairs of Tits have indicated the very significant role they can play in forest hygiene and control of defoliating pests (Lack, 1954). In Pakistan to give a few random examples I have watched a small flock of Jungle Crows (*Corvus macrorhynchos*) systematically digging up and consuming the larvae of Cockchafers (Scarabaeidae) in an open forest glade in the Kaghan valley. Chafers are vegetarian and frequently serious pests of grassland and herbaceous plants. In the southern Punjab I have watched a mixed assembly of Indian Rollers (*Coracias benghalensis*), Common Mynas (*Acridotheres tristis*), and King Crows (*Dicrurus macrocercus*), vigilantly gleaning the surface of a newly sown field, of Ground Hoppers (*Chrotogonus trachypterous*), a severe potential threat to the emerging cereal shoots. In years when aphids emerge in big swarms, Swallows (*Hirundo* sp.) and Martins (*Riparia* sp.) have been noted in Multan district, to circle around the rising columns and to gorge themselves on these insects which are a potential threat to a large number of green crops.

Birds are an important agent in both cross-pollination of certain trees, and as seed dispersal agents. Generally they are rather destructive feeders on flowers which are a source of pollen or nectar food, yet because of nature's profligacy in the development of seed, such plants have obviously evolved attractive colours or accessible nectaries to encourage visits by birds, there being a net favourable balance in the resultant fertile seed production. In Pakistan it is interesting to speculate as to whether birds, with their well developed colour vision, are especially attracted to red flowers (just as most butterflies are strongly attracted to mauve and purple colours).

Several red blossomed indigenous trees rely heavily upon birds which transmit pollen from their forecrowns onto adjacent flowers and thus ensure cross-pollination. The Silk Cotton (*Salmalia malabaricum* syn. *Bombax ceiba*) has large bright crimson flowers, rich in nectar, and particularly favoured by the Sturnidae (see account of *Sturnus malabaricus*). Another beautiful flowering tree typical of dry-deciduous forest, but rare in Pakistan, is the Flame of the Forest (*Butea monosperma*) found in the Margalla hills, Abbottabad and Haripur tehsils. Its orange-pink and scarlet blooms cover every branch after the leaves are shed, and attract a large number of birds including Bulbuls and Mynas which feed on the nectar. The quicker growing Coral Tree (*Erythrina suberosa*) is comparatively rare in Pakistan occurring only along the outer Himalayan foothills but fairly widespread as a cultivated ornamental tree (Stewart, 1957). Its conspicuous clusters of velvety scarlet flowers attract migrating flocks of Rosy Pastors as well as mynas and bulbuls. In the tropical thorn forest biotope a widespread tree (which is seldom allowed to grow larger than a shrub) is the False or Leafless Caper (*Capparis decidua*) which, from February onwards, bursts forth

in a halo of delicate coral pink blossoms. These attract large numbers of Purple Sunbirds (*Nectarinia asiatica*), Rosy Pastors, and Common Babblers (*Turdoides caudatus*), all of which can be noted while feeding on the nectar to have their forecrowns dusted yellow with pollen. In the outer hill ranges and particularly on dry rocky slopes the shrub *Woodfordia fruticosa* is covered with brick red flowers throughout May and I have seen flocks of Parakeets (three different species), Purple Sunbirds and Yellow-throated Sparrows (*Petronia xanthocollis*) attracted to its nectar. The small yellowish flowers of the shrubby tree *Mallotus philippensis* are visited by many nectar and insect feeders in the foothills which act as cross-pollinators. In the Margalla hills *Pycnonotus leucogenys* and *Zosterops palpebrosa* are significant pollinators of this plant.

Many trees have evolved seeds or fruits with a soft pulpy pericarp which attract birds and mammals. Their seeds, especially the wild olive (*Olea cuspidata*), may be swallowed whole by weak billed birds eating the fruit pulp, such as the *Streptopelia* and *Columba* doves and pigeons. The kernels are voided undamaged in their faeces, thus enabling the tree to colonize or spread its seed to new areas. *Prunus cornuta*, a common tree in the monsoon influenced areas of the Himalayas, certainly benefits from seed dispersal by the Great Himalayan Barbet (*Megalaima virens*), and thrushes (such as *Turdus boulboul* and *Turdus rubrocanus*). The large edible seeds of the Chilghoza Pine (*Pinus gerardiana*), and the acorns of the Holly Oak (*Quercus ilex*) provide such an abundant harvest of seeds that they form a major item in the diet of Lanceolated Jays (*Garrulus lanceolatus*) and Nutcrackers (*Nucifraga caryocatactes*), which two species are typically associated with such forests in the inner Himalayan ranges of Pakistan. As is known, these birds consume (and thus destroy) a large proportion of the seeds they harvest, but they also cache or store surplus seeds for later winter consumption. Studies have shown that a small percentage of such caches remain un-recovered and ultimately germinate to give rise to the new dispersed tree colonies (Vander Wall et al., 1977) Plants which are aerially parasitic upon trees tend to have mucilagenous berries which depend upon birds and mammals for dispersal. The former after feeding on the sticky pulp, wipe their bills upon adjacent branches of trees where the seeds then germinate (See, S. Ali, 'Role of Sunbirds and Flower Peckers in Propagation of the Tree Parasite *Loranthus longiflorus*', *JBNHS*, Vol. 31, p. 144). Fortunately in Pakistan the Mistletoe (*Viscum album*) and *Loranthus* tree parasite are not very common or widespread (Stewart, 1957), but in the Murree hills *Loranthus* does occur and is fed upon by the Thick-billed Flowerpecker *Dicaeum agile* which consequently aids in dispersing this parasite. However in the Juniper forests of Baluchistan there is a parasitic dwarf mistletoe (*Arceuthobium oxycedri*) which is widespread and causing serious economic losses, and the seeds of this parasite are partly spread by the Missel Thrush (*Turdus viscivorus*) and the Black-throated Thrush (*Turdus ruficollis atrogularis*) which I have observed in October in Ziarat, feeding upon the berries.

INTERACTIONS WITH MAN

It is only feasible to comment briefly on a few of the areas where wild birds are exploited by man in Pakistan. The art of falconry was probably introduced into the subcontinent by the Mughal courts in the seventeenth century. It is still practised in parts of the Punjab but is fast becoming too expensive a hobby due to the scarcity of trained birds and high prices being offered by Arabs for falcons. Goshawks (*Accipiter gentilis*) are still trained to hunt hares (*Lepus nigricollis*), especially in the Punjab Salt Range, and the favoured falcons of the local gentry used to be the migratory Peregrine (*Falco peregrinus*), the Red-capped Peregrine (*Falco pelegrinoides*), and the Red-headed Merlin (*Falco chicquera*). I have also seen captive and trained Short-toed Eagles (*Circaetus gallicus*) and Shikras (*Accipiter badius*) which could not have been very effective.

In Frontier villages the Rock Partridge or Chukar (*Alectoris chukar*), is a favourite pet and cage bird, due to it aggressive nature and loud calling habits. In Quetta city a well plumaged Chukar was selling for fifty rupees in 1986.

In the Indus plains both the Grey Partridge (*Francolinus pondicerianus*) and Black Partridge (*Francolinus francolinus*) are widely kept for the same reasons. In Baluchistan the Bimaculated Lark (*Melanocorypha bimaculata*) is a very popular cage bird largely because the vehemence of its song can compete successfully with the strident noises of bazaar traffic!

There are no known instances in Pakistan of birds being captured and trained by man to aid in harvesting food resources with the exception of Hawks and Falcons, flown more for sport than to obtain food for the falconer. The professional fishing castes in Pakistan did train the Smooth Coated Otter (*Lutra perspicillata*) before it became rare due to excessive persecution, but Cormorants if captured are viewed as a potential meal. Fishermen in Dhamb village, Lasbela regularly snare Cormorants which they relish as food (author obs., in late 1970s). Hair and nylon noose snares are used by river-dwelling tribes (Mohannas and Mirbars) to capture Herons and Egrets but these birds are either used as decoys to attract other wildfowl or are eaten, and are not used to capture fish.

In the chapter on Migration, the decoying and capture of live cranes in Bannu district has been mentioned. Demoiselle Cranes (*Anthropoides virgo*) are popular pets with well-to-do Pathans, not ony for their decorative appearance but also because of their aggressive territoriality which makes them surprisingly effective 'watch dogs'. It was estimated in 1983 that approximately 600 Demoiselle Cranes and 400 Common Cranes (*Grus grus*) were decoyed and captured in that year. Most of the latter species being eaten, but a pair of Demoiselle Cranes may today fetch the equivalent of US$100 (Roberts and Landfried, 1983).

Few studies have been carried out in Pakistan on the role of birds as potential vectors of diseases transmissible to man or domestic animals and it would be fair to state that there is little positive evidence that birds are responsible for spreading any such zoonoses. Foot and mouth disease is endemic to cattle in certain regions of the Punjab and studies in Europe have indicated that birds may transmit the virus on their feet. A very unusual migrant to Karachi, the Hooded Crow (*Corvus corone sharpei*) was examined for internal parasites and found to contain nematodes (*Diplotriaena tricuspis*) and larval micro-filaria of *Diplotriaena*, but both these helminths are unlikely to be transmissible to domestic stock and even less to man (Roberts, 1977).

The concentrations of Black Kites (*Milvus migrans*), White Backed Vultures (*Gyps bengalensis*) and Griffon Vultures (*Gyps fulvus*), around municipal refuse dumps of the larger cities in Pakistan testify to the valuable role still being played by these scavengers in the 1980s.

Regrettably whenever terns and sea birds nest colonially, local fishermen have a tradition of collecting their eggs for food. I have observed this predation in nesting colonies of Slender Billed Gulls (*Larus genei*) and Gull-billed Terns (*Gelochelidon nilotica*) (Roberts, 1978) and it has been reliably reported in Indus river colonies of Terns and Pratincoles (*Sterna aurantia, Sterna acuticauda* and *Glareola lactea*) (Abdul Fateh Soomro, DFO, Kandhkot, pers. comm.). This may well account for the relative decline in numbers of these exclusively fluvial breeding species.

The recent great improvement in living standards in Pakistan, particularly in the rural areas, will, it is hoped make it easier to change attitudes about exploitation of wild birds as a food source. Already the wildlife departments of each province are regulating and restricting the keeping of caged birds of certain species and a campaign has been started to educate tribesmen about the long term benefits of limiting and restricting crane hunting. With increasing urbanization and human over-crowding, the aesthetic benefits of wild birds will hopefully be more widely appreciated, but it will ultimately and inevitably become necessary to introduce much more restrictive protection against various traditional bird trapping, decoying and egg harvesting practices.

Species' Accounts

ORDER GAVIIFORMES
FAMILY GAVIIDAE
Red-throated Diver *Gavia stellata* (Pontoppidan)

STATUS

There is only one record for the subcontinent. On 17 November 1901, a local fisherman brought a bird to W. D. Cumming at Ormara. The fisherman had knocked it with his oar as it surfaced close to his boat and claimed to have seen several (Ticehurst,

JBNHS, 1927). The skin is in the British Museum collection. The inshore coastline along the Makran is even less known and surveyed than the deeper offshore fishing waters, so the possibility of its occurrence in small numbers along the Arabian seacoast cannot be ruled out. Status VAGRANT.

ORDER PODICIPEDIFORMES
FAMILY PODICIPEDIDAE
Little Grebe or Dabchick *Tachybaptus ruficollis* (Pallas)   Map 1
Synonym *Podiceps ruficollis* (Pallas)

DESCRIPTION

Body length 25-29 cms.
Wingspan 40-45 cms.
Wing length 9.4-10.9 cms.
Bill length 18-22 mm.

The smallest grebe occurring in Pakistan, typical of the family in having a buoyant rounded body when swimming with rectrices consisting only of short downy feathers not differentiated from the rest of the flank and tail coverts. The feathers of lower back and flanks tend to look fluffy and with flank feathers and scapulars entirely concealing wings when swimming on the surface. The bill is blackish, short and sharply pointed with prominent yellow fleshy gape during the breeding season. Both sexes are alike with dark olive brown crown, back and hind neck and rufous chestnut cheeks and foreneck in breeding season. The feet are large, set well back on the body and are green-black on the tarsus, but more olive tinged on the two inner toes. All toes are broadly webbed with flap-like lobes of skin. In flight no white shows on the trailing edge of wing contra the Black-necked Grebe *(Podiceps nigricollis).* In winter plumage forepart of the neck, flanks and belly are silvery buff with greyish tinges. The iris is red-brown. Separable from *Podiceps nigricollis* in winter plumage by the latter's more contrasting black crown and silvery white not silvery buff tones of lower cheeks and foreneck, as well as slightly larger size. The downy chicks are very buoyant and pretty with silvery white lower breast and flanks, crown, neck and back with longitudinal alternating black and silvery chestnut stripes. Chicks have a greyish green bill with encircling black narrow rings and a prominent silvery white patch on forecrown.

HABITAT, DISTRIBUTION AND STATUS

Restricted to lakes, flooded borrow pits and swamps with some open water. Very rarely seen in main rivers. Uncommonly visiting estuaries and salt water lagoons on the seacoast in winter and permanently resident on the larger perennial lakes. Occasional birds recorded in the winter months on Sonmiani lagoon, Lasbela; a few occur in tidal creeks such as Ghizri and upon the Indus delta. They breed in suitable swamps in Gilgit and Baltistan in summer but migrate southwards in winter. There are huge year round concentrations on the Punjab Salt Range lakes, *circa* 200 Khabbaki, 600 on Nammal, 700 on Uchchali, nesting almost colonially. They are subject to irregular local migration when swamps dry up. Numbers on Baluchistan lakes as high as 500 on Zangi Nawar (author obs. May 1984) and lesser numbers (30+ June 1974) on Kushdil Khan lake and Hannah reservoir, augmented in summer with loosely colonial breeding noted on both lakes. An estimated 200 adults with chicks were counted on Kushdil Khan lake during a survey, 15 July 1987 (Ashiq Ahmad Khan, Pakistan Forest Institute report). They are widespread in all four provinces. Status COMMON.

HABITS

Under eutrophic lake conditions where vast quantities of wintering waterfowl congregate, have been observed to forage in definite flocks or groups of up to 15 birds and even flocks of up to 60 birds, e.g. on Ghauspur jheel (Jacobabad district, Sind) and also Uchchali lake, Punjab Salt Range. These are lakes with very little surface cover of emergent vegetation. They dive very frequently and hunt all their food swimming

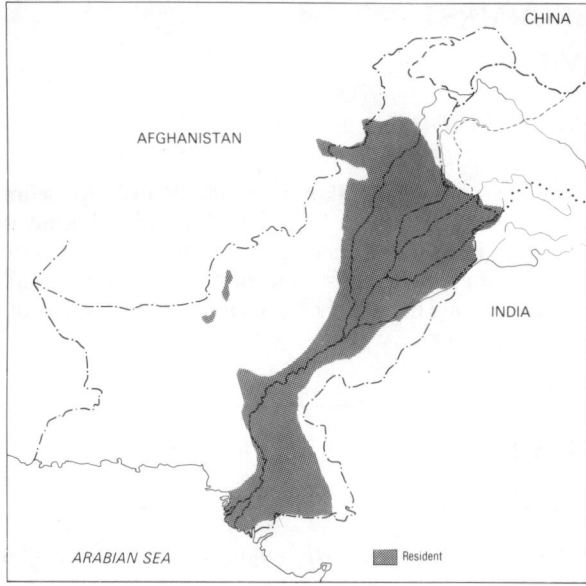

Map 1: Little Grebe *Tachybaptus ruficollis*

underwater propelled by their feet. Observed with tiny fish in bill (Thatta district, late September) but feed mainly in summer months upon aquatic insects, crustacea, tadpoles and molluscs. European studies indicate plant material not usually significant in diet.

BREEDING BIOLOGY

A very few individuals have been noted assuming breeding dress as early as 24 January (Nammal lake, Salt Range). A very vocal species compared with other grebes, with long quavering trilling calls, an important component in preparation for pair formation. In early stages of courtship aggressive chases are frequent, with birds pattering their feet across the lake surface (Hannah lake, Baluchistan, late April). Identically plumaged sexes perform a number of mutual displays. A pair parallel swimming in tandem with breasts partly raised out of the water were observed, also a pair facing each other and trilling in duet all the while twisting their hindquarters to and fro in the water with necks hunched forward towards each other (both observations on Khabbaki

lake, 31 March). Also a female noted inviting copulation on platform (possible nest site) by lying flat with neck and chin extended and flattened on surface of platform (Khabbaki lake, 1 April).

Only the immediate vicinity of the nest is defended, but the pair may start to build several platforms before final nest selected. Both sexes bring sodden vegetation which soon becomes rotten and slimy. Stems as component of structure not noted and whole nest rarely exceeds 20 cm. in diameter. The nest is capable of floating and rising with increase in lake levels (Meinertzhagen, *Ibis,* 1920, on Kushdil Khan lake, and by Bates and Lowther, 1952 on Dal lake). Courtship starts early March in lower Sind and observed in early stages late April Kushdil Khan lake, Baluchistan. Breeding season extended. Full clutch 4 to 7 eggs, earliest found 20 April (H. Waite, *JBNHS,* 1948), Kallar Kahar, Salt Range. Latest 8 July (S. Doig. *Stray Feathers,* 1879), east Narra, Sind. Eggs chalky white oval, becoming stained coffee brown to dark red-brown due to habit of both sexes of pulling loose vegetation over eggs before leaving the nest even when disturbed. May nest solitarily, but loose or scattered colonies usual on larger lakes. 30 nests all with incubated eggs 13 June (Waite, 1948), Kallar Kahar, and over 100 nests with 3 to 6 eggs 20 June (Meinertzhagen, 1920) Kushdil Khan. During a July survey of Zangi Nawar lake (Chagai) about 300 Dabchicks were present, most adults with 2 to 3 downy young in attendance, but several nests containing 5 incubated eggs were also found, and two with 4 eggs (Mohammad Idrees Chughtai, Pakistan Forest Institute, hand-written report, 1987). Both sexes incubate, and period normally 20 days and young independent after 30 to 40 days. Downy young watched (author) late July, repeatedly climbing onto parent's back each time latter surfaced from a feeding dive (Thatta district, Mirpur Sakro).

*VOCALIZATIONS

Main element is trilling described above and this is very rarely heard outside the breeding season. Trills likened to high pitched 'whinnying' vary in pitch and duration and given by both sexes. Downy chicks have a 'cheep' call like domestic chicks (Bates and Lowther, 1952).

Great Crested Grebe *Podiceps cristatus* (Linnaeus) Map 2

DESCRIPTION

Body length 46-51 cms.
Wingspan 85-90 cms.
Wing length 17.5-20.9 cms. (males), 16.8-19.9 cms. (females)
Weight 758-1,490 gms. (males), 609-1,380 gms. (females)

In Pakistan this species is normally seen only in winter plumage and can only be confused with the Red-

necked Grebe (*Podiceps grisegena*). Its long straight upright neck carriage and glistening white foreneck make it conspicuous even amongst a flock of distant ANATIDAE. Upper parts grey-brown, crown and hindneck darker brown, foreneck, lower cheeks and flanks silvery white. Bill comparatively longer than in *T. ruficollis* or *P. nigricollis* and straight, sharply pointed with no angle on lower mandible (as in *nigricollis*) or slight curve on culmen (as in *grisegena*). Bill pink to carmine at the base and horny brown at the tips and iris crimson. Feet and legs olive green with yellower tinge on toes. Downy young with longitudinal black and white stripes on head and neck, back grey-brown with less conspicuous black stripes. Individuals in breeding plumage, occasionally seen amongst winter visitors, and regularly on Baluchistan lakes from April onwards, when crown becomes black and develops occipital crest and lower cheeks develop chestnut and black tipped ear tufts and upper flanks are also tinged with chestnut. Distinguished in winter plumage from Red-necked Grebe (*P. grisegena*) by having a clear white line above its eye, whereas the dark brown crown extends down to the eye in *P. grisegena*, which also has a proportionately slightly shorter thicker neck and stouter bill. In flight it shows white on wing shoulders as well as secondaries unlike *P. nigricollis* and *P. ruficollis* but this wing pattern is shared with *P. grisegena*.

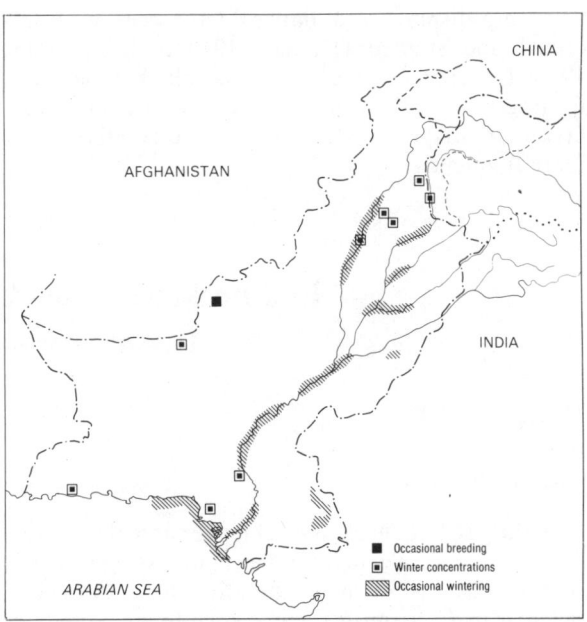

Map 2: Great Crested Grebe *Podiceps cristatus*

HABITAT, DISTRIBUTION AND STATUS

Found only on the larger deeper lakes, more frequently in the northern regions than the south. In recent years there are indications that the newly formed Rawal lake reservoir will become a major wintering ground and possibly may even attract breeding pairs. 80 were observed in 1981-2 on this lake but some disturbance resulting from commercial fish exploitation of the lake in 1982-3, resulted in a sharp drop in numbers. This was stopped on a plea by conservationists and in the winter of 1986-7 about 90 were counted on Rawal lake. In the summer of 1987 about 8 birds remained on Rawal lake until mid-May and full courtship display was watched involving several pairs. However they disappeared by the third week of May (R. F. Nana, pers. comm., 1987). Some individuals winter in salt water lagoons. 8 counted (author) 6 January on Sonmiani lagoon, Lasbela. Usually one or two each winter on lower Sind lakes such as Hadiero or Haleji. An unusually late record is 4 birds on Hadiero 23 May. (Derek Holmes, pers. comm.). Usually 2 or 3 on Salt Range lakes each winter. 12 were counted on Chashma barrage headpond (Mianwali district) 10 February. (F. Koning, pers. comm.). It is a rare visitor to Zangi Nawar lake in Chagai, Baluchistan (Christison, 1942), and an occasional pair breeds at Kushdil Khan lake, Baluchistan. 25 overwintered 1983-4 on Kushdil Khan (author obs.). Earliest sightings Punjab 10 December, latest in Sind 27 June. Status SCARCE.

HABITS

Forages solitarily and fish is a much more important item in the diet than in other grebes. Many ornitholo-

gists have described its habit of swallowing feathers, and adults have been noted feeding feathers to young. (Knights, *British Birds,* 1985). Meinertzhagen (*Ibis,* 1927) found fresh water shrimps to be the main content of stomachs in specimens secured during the summer in Ladakh (India).

BREEDING BIOLOGY

Occasional breeder on Kushdil Khan lake, Baluchistan. Three pairs with nests and eggs, one clutch of 5 eggs, found by Aiken on 12 July 1913, but only one pair which did not appear to be breeding in the subsequent year (Meinertzhagen, 1920). Three pairs were observed (author) 29 May 1979 in breeding plumage with evidence of some courtship as one pair observed facing each other with ear coverts erected, but subsequent breeding could not be confirmed. Ashiq Ahmad Khan observed one adult with two chicks during a visit to Kushdil Khan lake on 15 July 1987 (typed report, Pakistan Forest Institute). The nest is a substantial heap of aquatic vegetation, not always concealed in reeds and built by both sexes. Sometimes anchored to emergent growing vegetation, sometimes freely floating. Several nests usually built and one is used as a mating platform. Incubation period 28 days, by both parents and the eggs are elongated ovals, chalky white. Usual clutch 3 to 4 and eggs laid at two day intervals and downy young reported to be carried for up to 21 days on the parents' backs. Their elaborate mutual courtship displays and 'dances' have been well described and studied (Huxley, 1914 and Simmons, 1955). Each parent rears one or more chicks separately, feeding them bill to bill and having one individual favoured offspring which is fed preferentially and more frequently.

VOCALIZATIONS

Silent on its wintering grounds and less vocal than other grebes. It emits a loud far carrying barking call in the breeding season. Chicks are reported to be very vocal, continuously piping.

Red-necked Grebe *Podiceps grisegena* (Boddaert)

(Spelled *griseigena* in Ripley's 'Synopsis')

DESCRIPTION

Body length 40-50 cms.
Wingspan 77-85 cms.

A slightly shorter necked stockier version of the Great Crested Grebe, in winter plumage but distinguished by its brownish black crown extending down to the eye, whereas in *P. cristatus* there is a white line above the eye. The horny black bill of the Red-necked Grebe is yellow at the base contra pink in *cristatus,* and it is slightly shorter and with the culmen gently curved rather than straight. It is unlikely to be encountered in breeding plumage in Pakistan, when the reddish chestnut foreneck and absence of ear tufts, but with broad white contrasting cheek patch, make it unmistakable. In flight it shows white secondaries and wing shoulder like the Great Crested Grebe but the scapulars are wholly dark whereas in *cristatus* white extends down the scapulars.

HABITAT, DISTRIBUTION AND STATUS

One seen 14 January 1967 on Lal Sohanran lake, Bahawalpur division by author, and two companions

(Roger Holmes *et al., JBNHS,* 1967). Subsequently on 19 January in the same year, two were seen on Nammal lake in the Salt Range (Savage, 1967). These are the first records for the subcontinent and it must be considered a vagrant to Pakistan. Four were seen on Rawal lake on 15 January 1987 by M. Mallalieu (*in litt.* to author, January).

Jochen Niethammer recorded it in the 1960s in Afghanistan (*in litt.*).The Asiatic population probably winters mainly around the Caspian and few migrate as far south as the Indian Ocean. Quite a significant part of the Siberian breeding population winters around the Aral Sea.

HABITS AND BREEDING BIOLOGY

Breeds in Arctic tundra. Reported to be very vocal during breeding season and courtship rituals elaborate including 'Penguin' dances, but weed presentation is less important than in *P. cristatus.* Its wavering calls are reminiscent of 'loons' (GAVIIDAE). Its food is very similar to the smaller grebes, with molluscs, insect larvae (Odonata) and crustacea forming an important part of its diet.

Slavonian Grebe or Horned Grebe *Podiceps auritus* (Linnaeus)

DESCRIPTION

Body length 18-23 cms. (Cramp *et al.*), 33 cms. (Heinzel and Fitter)
Wingspan 59-65 cms.
Wing length 13.6-15.8 cms. (144 males),
   13.1-15.3 cms. (14 females) (Cramp *et al.*)

Only slightly larger in size than the Black-necked Grebe (*Podiceps nigricollis*) which it closely resembles in winter plumage being predominantly silvery white on the cheeks, throat and breast with blackish grey crown, hind neck and back. The chief distinguishing field points in winter are its smaller size and more dumpy appearance when compared with the Red-necked Grebe (*P. grisegena*) and similarity to the

Black-necked Grebe (*P. nigricollis*) from which it appears slightly larger headed and thicker necked, with more white about the face than *nigricollis.* Its bill is quite short and straight whereas that of *nigricollis* is more slender and slightly up-tilted on the culmen. The bill of *auritus* is pale tipped (difficult to see in the field) and recurved on the culmen, not up-tilted. The sexes are alike but the male does appear slightly larger and more contrastingly black and white in winter plumage than the female (author obs.). In *nigricollis* the black of the crown extends down to just below the eye whereas in *auritus* the black crown does not cover so much of the hind part of the head and from the rear the white cheek patches make a very distinctive pattern

(See **ill. 1**). In breeding plumage the flanks and neck become maroon-chestnut with the crown and cheeks black and the tippet of golden yellow feathers projects from behind the eyes beyond the hind neck. In *nigricollis* as its name suggests, the neck remains jet black and the golden red tippet of feathers behind the eye does not project so far backwards, whereas the forecrest is more bushy and erectile during display or excitement.

DISTRIBUTION AND STATUS

This Grebe is Holarctic in distribution, confined mainly to the boreal forest zone in the summer months and breeding on both large lakes and small ponds, provided they have sufficient cover in the form of emergent vegetation. In winter time its preferred habitat is around the seacoast in sheltered bays and estuaries, rather than inland waters. Perhaps for this reason it has not been recorded anywhere on the Indian sub-continent (See S. D. Ripley's 'Revised Synopsis', 1982).

In mid-January 1984, the author conducted waterfowl counts on the two lakes of central Baluchistan. On 17 January a single Slavonian Grebe was sighted on Zangi Nawar lake, though due to the precarious boat from which it was seen and the declining daylight, firm identification was not possible and the matter was put out of mind as one of those imponderables that surely face even the most diligent of bird watchers. A few days later (20 January) however a pair was watched at close range on Band Kushdil Khan lake, observation from a tripod-mounted telescope, enabling detailed field sketches. On both lakes identification was facilitated by the presence of Black-necked Grebes for comparison.

In January 1987, whilst conducting waterfowl counts on Khabbaki lake in the Salt Range, M. Mallalieu also saw 2 Slavonian Grebes and was able to make prolonged and careful observations (*in litt.*, January 1987 and pers. comm., February 1987).

Subsequent to the January 1987 sighting by Mallalieu and Ashiq Ahmad (Wildlife Research Officer, Pakistan Forest Institute), the author was shown in March 1987 a film taken by Naseer A. Tareen depicting Baluchistan wildlife which is to be shown on Pakistan Television. The film included close up views of a single Slavonian Grebe which was filmed in January 1986 on a small lake known as Kund just a few miles inland from the seacoast at Hab Chowki (Lasbela). Mr. Tareen was unaware of the unusual species that he was filming and identification was further confirmed by replay of the film to Rolf Passburg who is familiar with the Slavonian Grebe from the Caspian Sea region of Iran where it regularly occurs in winter (Passburg, pers. comm., 1987).

In view of its previous recorded occurrence in Iranian Seistan (Hue and Etchecopar, 1970), its occurrence in Baluchistan is not too surprising. Status RARE but possibly overlooked along the Makran coast.

HABITS

They feed by diving and swimming under water and subsist in winter mainly upon small fish and crustacea (Witherby *et al.*, 1943, and Cramp *et al.*, 1977) and in the summer by taking a higher proportion of insect larvae. Their nests are built of decaying water weeds in shallow lake water and both sexes take part in nest building and in incubation.

Black-necked Grebe or Eared Grebe *Podiceps nigricollis*

Illus. **1**
Map **3**

(Podiceps caspicus Hablizl, 1783, suppressed)*

DESCRIPTION

Body length 28-34 cms.
Wingspan 56-60 cms.
Wing length 12.7-13.9 cms. (males), 12.4-13.6 cms. (females)
Weight 265-450 gms. (fatter in non-breeding season males),
231-280 gms. (females)

Normally only observable in winter plumage when very like the Little Grebe in appearance, but is slightly larger, but with same relatively short neck and short sharply pointed bill. In profile forehead more steeply stepped and lower mandible angled to give bill a slightly up-tilted appearance. Iris a beautiful orange-red (contra reddish brown of *ruficollis*). Plumage

black and silvery white, less buff on foreneck and flanks than *ruficollis*. Individuals moulting into breeding plumage show noticeably orange-tawny flank feathers and the beginning of an occipital crest in contrast to the rounded occiput and olive buff flanks of *ruficollis* in similar plumage. There is also a noticeable whitish area behind the eye even in winter plumage. Bill blue-black, lower mandible fleshy pink (captured specimens Kharrar jheel, Renala Khurd, December). Legs greyish green, with flattened nails, all toes separately lobed with flaps of skin reminiscent of Horse Chestnut (*Aesculus*) leaf in outline. No nail on hind toe. Specimens observed Punjáb Salt Range in breeding plumage had rather iridescent tawny ear coverts and shining russet flank feathers. On Hab dam reservoir some individuals were already partially moulted into breeding plumage on 13 March. In flight, secondaries white in contrast to wholly olive brown wings of *ruficollis*.

HABITAT, DISTRIBUTION AND STATUS

Restricted to lakes in the dryer hilly regions of north and western Pakistan, it is also more adapted to a salt water and estuarine habitat in winter. The main wintering population occurs on the Salt Range lakes with average 50 to 60 birds on Khabbaki, up to 70 on Uchchali lake, usually only 20 to 30 on Nammal lake (based on 6 years' observations, December to February). 300-600 were counted by Savage in the winter of 1964-5 on Khabbaki lake and cited by Ali and Ripley ('Handbook', 1968); such a high number seems quite exceptional. 200 were seen (author) at Hannah lake (Baluchistan, December) and 30 + Kushdil Khan lake in late March. Over 100 were counted on Hab dam lake, 24 March. 1987 (S. Asad Ali, *in litt.,* March 1987). Occasional individuals sighted on Sind lakes, 5 on Ghauspur jheel, Jacobabad district in November, 1 on Khipro lake, Narra district in February (F. Koning, pers. comm.). Single birds were seen (author) January/February on Ghizri creek, Karachi (salt water estuary) in two years, 9 were seen on Sonmiani lagoon, Lasbela in January, and 9 were seen Piti creek, Indus delta on 11 November in a tight

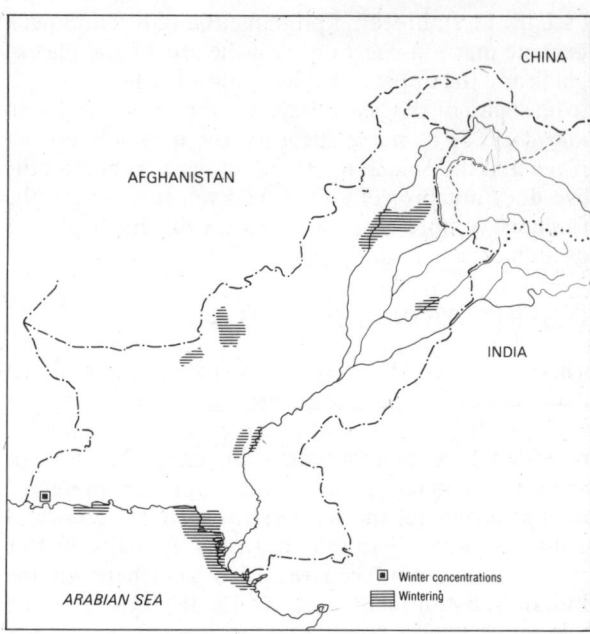

Map 3: Black-necked or Earned Grebe *Podiceps nigricollis*

swimming flock. The majority start northwards return migration at the end of February as there is a marked drop in numbers noted on Salt Range lakes.

During a visit to northern Hunza in late July a solitary bird in full breeding plumage was seen (author) on Borit lake above Gulmit (N36° 27'., E74° 52' elevation 2,500 metres/8,500 feet). It could have been an early autumn migrant or one of a breeding pair as it disappeared into reeds bordering this small lake. Status FREQUENT.

HABITS

Largely a mid-winter migrant with no records before early December, latest dates Khabbaki lake, Punjab, 30 March and Kushdil Khan lake, Baluchistan, 23 March. Feeds by diving and swimming under water like *ruficollis,* but small fish believed to be a more important component in diet for those wintering in salt water lagoons. Insects and their larvae, molluscs, and crustacea believed predominant component of diet on fresh water lakes. A specimen shot at Manchar lake had been feeding on large fresh water shrimps (Ticehurst, 1923). Tends to forage singly in winter.

BREEDING BIOLOGY

Birds seen on Nammal lake, 15 February, already partially moulted into breeding plumage, whilst birds seen on Khabbaki lake, 30 March, were still only partially showing breeding plumage. They nest colo-

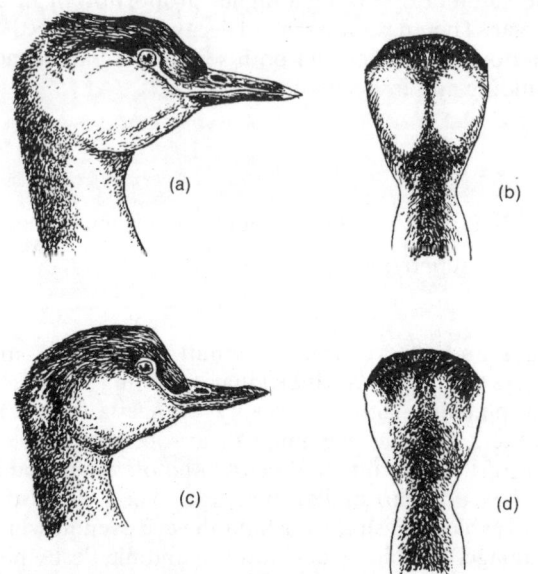

Illus. 1: Showing difference in head pattern during winter plumage, and bill shape.
    (a) Slavonian or Horned Grebe (*Podiceps auritus*)
    (b) Slavonian or Horned Grebe head from behind
    (c) Black-necked Grebe (*Podiceps nigricollis*)
    (d) Black-necked Grebe head from behind.

nially. In 1914 Meinertzhagen described a colony of 70 pairs all with eggs, discovered on Kushdil Khan lake on 20 June but the whole colony was abandoned by 11 July, with all nests deserted due to a sudden rise in the lake level and destruction of eggs. No grebes were observed on this lake by Meinertzhagen in the summer of 1913 and it is interesting that the flock of nearly 150 birds which bred during June 1914 was preceded by only 5 birds observed by him during a visit in early May (Meinertzhagen, 1920). In four summer time visits to this lake during the 1960s and 1970s and through several other reliable second hand reports, there appear to have been no observable signs of this grebe breeding in Baluchistan in the last 3 decades. Ticehurst also failed to get any information of subsequent breeding (Ticehurst, 1927) and it must be considered as a very sporadic or occasional breeder at these southern latitudes. However, during a July survey of Zangi Nawar lake in 1987, one nest containing only 1 egg was found on 8 July, the parent birds remaining in the vicinity. No other Black-necked Grebe were seen during a week long survey of the lake (Mohammed Idrees Chughtai, Pakistan Forest Institute, hand-written report, 1987).

The nests described by Meinertzhagen were typical for the species in being concealed in sedges (*Carex* sp.) and comprising more plant stems in the nest material, with a resultant much firmer anchorage to adjacent growing reed stems (in contrast to nests of *ruficollis*). Normal clutch 3 eggs, occasionally up to 5, chalky white ovals and incubated by both parents. Eggs rapidly become stained brown. Hatching asynchronous and parents remove egg shells. Eggs usually covered by weed when incubating bird disturbed off nest. Both parents tend downy chicks which have striped head and back but more white about lores and chin compared to chick of *ruficollis*. Incubation period recorded in Europe 19 to 20 days. Nesting pairs were observed in Ladakh in May on Shushal lake (4,300 metres), and the downy chicks were noted repeatedly climbing onto the backs of their parents (General S. M. Ghaur, pers. comm.).

VOCALIZATIONS

Silent in winter quarters and only vocal during nesting. Calls have been likened to that of the Slavonian Grebe (*P. auritus*) (Cramp *et al.*, 1977).

ORDER PROCELLARIIFORMES
FAMILY PROCELLARIIDAE

Slender-billed Shearwater or
Short-tailed Shearwater *Puffinus tenuirostris* (Temminck)

Synonym *Procellaria tenuirostris*

DESCRIPTION

Body length 40-50 cms.
Wingspan 95-110 cms.

Larger in size than the next species, (Audubon's or the Persian Shearwater), it is stout bodied, with long slender wings, typical of the genus and highly adapted for gliding and manoeuvrability in strong winds. Its short thick neck gives a cigar shaped appearance to the body. It has a short fan shaped tail and overall body and wing colour is a dark sooty brown on both dorsal and ventral surfaces. Its plumage is typically very dense and it emits a characteristic musky body odour. At rest wingtips extend just up to tail tip. Its bill and feet are also blackish brown with rather stout laterally compressed tarsi, an adaptation for swimming underwater. The legs are dark brown and inner two toes a lighter brown. The sexes are identical.

DISTRIBUTION AND STATUS

It is known only from one specimen in Pakistan coastal waters. This was collected in May 1899 by Mr. Walter Scott when laying a submarine cable. It was shot off the coast at Ormara (Baluchistan) following three days of gale force winds. Originally misidentified as *P. chlororhynchus* the specimen is now in the British Museum. It is believed that this petrel ranges widely during the Antarctic winter (northern hemisphere summer months) from the Sea of Japan to the southern Indian Ocean. Its status must be considered as a VAGRANT.

BREEDING BIOLOGY

It can swim strongly under water and dive straight into the sea from direct flight. A purely pelagic bird breeding in cool temperate waters off south-eastern Australia. All known breeding colonies are on islands off the coast of southern Australia with great concentrations in the Bass Strait, where an annual but strictly regulated commercial harvest of young birds is still allowed (Frith, 1976). It is a typical burrow nester breeding in colonies with egg laying highly synchronized.

Persian or Audubon's Shearwater *Puffinus lherminieri* Lesson

Synonym *Puffinus persicus* Hume *Procellariạ lherminieri* (Lesson)

(Sometimes treated as conspecific with *Puffinus assimilis* Little Shearwater, Gould)

DESCRIPTION

Body length 25-30 cms.
Wingspan 58-67 cms.

Advances in our knowledge of the PROCELLARIIDAE have been greatest in countries like New Zealand, whose scientists have first hand field experience of many remote Antarctic breeding sites. New Zealanders consider Audubon's Shearwater (·*Puffinus lherminieri*) as probably inseparable from *Puffinus assimilis*. It is a shorter winged, much smaller bird than the Slender-billed Shearwater, distinguished by white throat, cheeks and belly and under-wings except·for tips of primaries and secondaries. Rest of upper parts including whole of wings, dark grey-brown with mantle and tail blacker. Bill plumbeous to pale purplish and webbed feet opalescent white to pinkish. Outer toe and lower part of tarsus blackish, under-tail coverts are black. It is about the size of a small gull. Sexes alike.

HABITAT, DISTRIBUTION AND STATUS

A purely oceanic pelagic bird of which a population assigned to the sub-species *persicus* (Hume), is thought to breed somewhere·in the Persian Gulf. It can be regularly sighted off the Makran coast but is much less common further east around Karachi. Ticehurst records seeing a flock of about 200 in the Straits of Ormuz (extra-limital off Oman in the Persian Gulf) but did not actually see it along the Makran coast. Hume saw a few between Karachi and the mouth of the Indus on 2 March well out to sea. Captain Butler collected two specimens on 14 May 1877 off the Makran coast and the curator of Karachi Museum stated that it occurs regularly off the Indus mouth

(J. A. Murray, 1884) and the Makran coast. N. Van Zalinge watched a flock of over 60 birds diving and feeding on a surface school of anchovies in early October 1984, at a location about 32 kilometres (20 miles) from the Makran coast (*in litt.* to author), whilst working on Fisheries Department research vessel. The author has only seen it once definitely, from land at Cape Monze in very rough monsoon seas in May, as it probably feeds mainly more than 8 kilometres (5 miles) offshore. It skilfully takes advantage of every wave trough and crest in its swift gliding flight, thus it is not as easy to observe from land as Wilson's Petrel and almost certainly occurs much more plentifully than these few records indicate. Cruising offshore in small boats is not recommended in these waters during the monsoon season and modern air travel deprives us of the ọcean vessel sightings quoted by Ticehurst and Murray.

HABITS AND BIOLOGY

They are fast flying birds which can dive and swim under water well. They will plunge straight into the sea during flight. They are gregarious both in their colonial nesting colonies and under favourable feeding conditions throughout the year. The Persian breeding colonies have never been discovered but if these birds are the same as Audubon's Shearwater, they are colonial, burrow nesters and nocturnal in breeding habits with a long incubation and even longer fledging period characteristic of all Procellariiformes. Male and females are thought to relieve each other from incubation duties after intervals of 5 and 6 days, when recognition involves prolonged and excited calling to each other.

Jouanin's Petrel *Bulweria fallax* Jouanin, 1955

This species is included as it is listed as probably occurring in Indian waters in S. Dillon Ripley's 'Revised Synopsis' (1982). Jouanin's is an all-dark Petrel with a long wedge shaped tail. Its upper parts are blackish brown and under parts a slightly greyer brown. The head appears quite large and it has a fast, swooping type of flight.

It is endemic to the Indian Ocean, occurring frequently in the Gulf of Aden, Red Sea and off the south-east coast of Arabia (P. Harrison, *Seabirds*,

1983 and Cramp *et al., Birds of the Western Palearctic,* 1977, and Gallaher, Scott *et al.,* 1984).

During a Marine Fisheries survey one all-black Shearwater was clearly seen, though from a considerable distance. It could not have been *Puffinus lherminieri* as there was no white on the throat and belly. This was on 14 November 1983 about 48 kilometres (30 miles) offshore from Karachi by N. Van Zalinge and K. Forssgren, both experienced ornithologists (pers. comm., November 1983).

During another survey by the Norwegian research vessel 'M. V. Fritzov Nansen' on 1 February 1984, a flock of about 50 Shearwaters were seen fishing together by Kent Forssgren. This was about 32 kilometres (20 miles) offshore but close to Karachi. They were also all-dark underneath and showed no white on the belly.

In both these sightings the birds were far off and could not be identified but the possibility that they were Jouanin's Petrel must be considered as this is the most likely species to occur offshore in the Arabian Sea and much more likely than the slender-billed Shearwater (*Puffinus tenuirostris*).

FAMILY HYDROBATIDAE

Wilson's Storm Petrel *Oceanites oceanicus* (Kuhl)
Yellow-webbed Storm Petrel

Illus. **2**
Map **4**

DESCRIPTION

Body length 15-19 cms. (Cramp *et al.*), 19 cms. (Ali and Ripley),
   17 cms. (P. Harrison), 18 cms. (A. E. Butler)
Wingspan 38-42 cms. (Cramp *et al.*), 40 cms. (P. Harrison)
Wing length 14.7-16.3 cms. (Subsp. *exasperatus*) (Cramp *et al.*),
   13.3-15 cms. (21 specimens) (Brown and Urban)
Tarsus 35 cms. (BNHS specimen, off Bombay),
   32-36 cms. (Cramp *et al.*)
Weight 24 gms. (A. E. Butler)

This starling sized bird is a swift and powerful flyer, skimming over the sea with rapid wing beats and gliding down into wave troughs with speed and agility. Black all over except for its conspicuous white rump patch and under-tail coverts, the upper wing surface shows a slightly paler grey carpal patch becoming more noticeable in worn plumage. Under parts when examined in the hand are browner and paler than the upper surface. The white rump patch extends down to the vent and lower flanks also. Tarsi very long and black. Webbed feet have a clear lemon yellow patch between the black toes and the hind toe is reduced to a mere spicule. The iris is black and also the bill which has the nostrils fused into a short prominent tube over the base of the culmen. Its tail is short and square. The similar Black-Bellied Storm Petrel (*Fregetta tropica*) has a white patch on flanks and under-wings and has not been recorded in the Arabian Sea. The sexes are identical.

HABITAT, DISTRIBUTION AND STATUS

Some authors have speculated that it must be one of the most numerous and widespread of the earth's seabirds. Certainly it breeds further south (the major population breeds on the Antarctic continent, often on high mountain crags, not on southern latitude islands), and large numbers reach Pakistan coastal waters and can be sighted within a few hundred metres of the shore during the monsoon season. They thus perform a regular migration of over 13,000 kilometres (8,100 miles) (See chapter on Bird Migration in Pakistan), to reach their northern hemisphere summering grounds along the Makran coast.

Formerly considered only a summer season migrant to the Indian Ocean from the end of April to early November (Ali and Ripley, 'Handbook', Vol. 1, 1968). It is significant that Dr. Gibson-Hill discussing the storm petrel's occurring in the northern Indian Ocean, comments on sightings around Aden in mid-February and speculated that a portion of the population must remain as non-breeding birds throughout the southern summer (C. A. Gibson-Hill, *JBNHS*, 1948). Recent evidence obtained by observation around

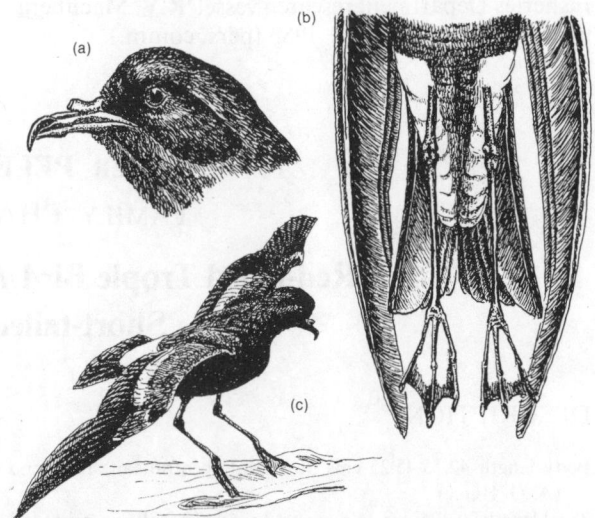

Illus. 2: Wilson's Storm Petrel *Oceanites oceanicus*
(a) Tube-nosed bill.
(b) Ventral view showing long tarsi with black toes and yellow webs to feet, also short square tail.
(c) Typical method of feeding, gliding over the water with feet touching the surface.

Map 4: Wilson's Storm Petrel *Oceanites oceanicus*

Karachi indicates that part of the non-breeding population may also spend the whole year in the northern Indian Ocean, as birds have been sighted during sea voyages around Karachi in all months except February and March (Roberts, *JBNHS,* 1984). For example 15 were sighted in a 6-hour voyage within 16 kilometres (10 miles) off the Karachi coast on 8 January. Two were sighted off Manora channel on 15 November 1983 and again 4 birds off the Makran coast on 26 November (Ras Muan) and 1 December off Astola island by Kent Forssgren during a survey by the Fisheries Department research vessel 'R.V. Machhera' in November/December 1983 (pers. comm.).

HABITS

A very intriguing little bird to watch, as it can be seen in constant movement even on the roughest days disappearing and reappearing behind each steep wave trough, often with feet pattering just on the sea's surface as though walking on water (hence the common name petrel a corruption of the disciple St. Peter). Its flight typically consists of rapid fluttering wing beats alternated with swift twisting glides always low over the sea's surface and legs extended straight out behind it. However, when water walking, it drops its long legs, holds its wings parachute fashion over its back in a 'V', and thus supported by the breeze, appears to walk on the water, sweeping its webbed feet backwards over the surface for a period of 4 to 5 seconds without fluttering its wings (November observation Hawkes Bay). It has been postulated that the bright colour of its webs may serve to startle or attract prey (*British Birds*, Vol. 73, 1980, p. 385). It has been noted, being attracted to and hovering along beside the head of a human swimmer (Passburg, pers. comm.). There are records for the Makran coast, and the Indus mouth and one was once watched feeding in calm water right inside Ghizri creek (23 October pers. obs.).

BREEDING BIOLOGY

Its single white egg is finely speckled with brown and has a prolonged incubation period, probably up to 50 days. The chick is covered with sooty black down at birth, (Ptilopaedic) takes up to 73 days to fledge and is fed by the parents by regurgitation of semi-digested oily substance. Adults are believed to feed mainly upon surface plankton or refuse.

ORDER **PELECANIFORMES**

FAMILY PHAETHONTIDAE

Red-billed Tropic Bird *Phaethon aethereus* Linnaeus
Short-tailed Tropic Bird

DESCRIPTION

Body length 42.75-45.25 cms. (excluding central tail streamers) (A. O. Hume)
Total length 90-105 cms. (Cramp *et al.*) of which half comprises tail and streamers.
Wingspan 100-110 cms. (Cramp *et al.*)
Wing length 27.3-30 cms. (A. O. Hume),
    28.57 cms. (males), 28.88 cms. (females) (E. A. Butler)
Tail length-central rectrices (elongated of adult) 60-80 cms.
Weight 481.9-567 gms. (A. O. Hume)

A rather tern-like bird with mainly white plumage and a stout down-curved bill, coloured dull red to orange-red often with dusky black commissure. The wing coverts and mantle have black crescentic or scale-like markings and outer primaries, primary wing coverts and tertials are largely black. It also has a prominent black line through its eye, curving down in a crescent in front of the eye. Its two central tail feathers are greatly elongated, white in colour except for a black quill. Like terns it has very short tarsi and graceful buoyant flight. Its legs are yellowish grey with black webs being toti-palmate like the cormorants and pelicans. Immature birds have paler yellowish bill and their legs are greyish and they lack the elongated central rectrices. Both sexes are alike. The chicks are downy white.

HABITAT, DISTRIBUTION AND STATUS

Breeds on rocky islands off the Arabian peninsula, Arabian Gulf, and North Yemen (Jennings, 1981). There have been Pakistan sightings only from January to May, mainly off the Makran coast, the first records being obtained by A. O. Hume who saw about 50 of these birds, all immatures during a 14-day cruise along the Makran coast from 14 to 28 February. He collected 6 specimens (Hume, *Stray Feathers*, Vol. 1, 1873, p. 286). Specimens were also collected between Ormara and Gwadar off the Baluchistan coast by Capt. E. A. Butler who saw about one dozen birds in all, during a sea voyage lasting for the second half of May (Butler, *Stray Feathers*, 1877). All these were immature birds without central tail plumes. It has also been seen by Sir Percy Cox off Astola island (243.2 kilometres (152 miles) due west of the Hab river) in July. More recently, on 6 November 1982, 2 were sighted about 128 kilometres (80 miles) south of Karachi from a freighter travelling from Bombay to Karachi (Walter Weitkowitz, MBOU, pers. comm., 1983). During a scientific survey off the Makran coast by the Norwegian research vessel 'M. V. Fritzov Nansen' in January/February 1984, Kent Forssgren saw a total of 5 Red-billed Tropic Birds on 20 January near Gwadar and one near Cape Monze (pers. comm., March 1984). Status SCARCE.

HABITS

A pelagic bird rarely coming within sight of land and not markedly gregarious, so that generally only single birds are seen. It appears quite tame and will fly close to ships, but rarely follows them. It feeds like terns by plunging into the sea from an aerial dive often hovering before diving. Captain Butler noted some birds catching flying-fish and they have been recorded feeding on squids. Fish are not speared but seized sideways in the bill and commonly swallowed before again taking off from the water.

BREEDING BIOLOGY

They nest colonially on all rocky islands around Oman (M. D. Gallaher, *in litt.*), also in the Persian Gulf. Pair bond is of long duration and they are believed to be monogamous with marked attachment to the same nest site. Only one egg is laid in March and unlike all other PELICANIFORMES this is richly coloured varying from white to fawn with darker purple-brown mottling. Both sexes share in incubation which lasts 40 to 46 days. The young are fed by regurgitation and take about 90 days to fledge.

VOCALIZATIONS

Described as very noisy at its breeding colony, emitting shrill grating or rattling cries. In flight even outside the breeding season it has a shrill whistle likened to the trilling of a 'Bo'sun's', pipe, also a series of screeching screams.

FAMILY SULIDAE

Masked or Blue-faced Booby *Sula dactylatra* Lesson   Illus. 3

Sula cyanops of Blanford and Oates (Vol. 4) and C. B. Ticehurst

DESCRIPTION

Body length 81-92 cms. (average 86 cms.) (P. Harrison)
Wing length 41.4-43 cms. (Stuart Baker),
    40.7-43 cms. (6 males) (Brown and Urban)
Tail length 16.8-18.2 cms. (Stuart Baker),
    16.9-18 cms. (6 males) (Brown and Urban)
Bill length 95-106 mm. (Stuart Baker),
    97-104 mm. (6 males) (Brown and Urban)
Weight 1,480-1,660 gms.

Adults are a dazzling white and resemble Cape Gannets (*Sula capensis*) and like the latter have their long wedge shaped tails entirely black, but they lack any yellow on the head and nape, distinctive of adult Cape Gannets of both sexes. The primaries and secondaries are black, the latter more extensively than in the gannet. The naked blue-black or leaden grey skin around the eye and gular area gives the face a masked appearance. Adults have a yellow iris and their legs and feet vary from leaden grey to yellow, their toes being toti-palmate. The bill is relatively straight along the culmen and under the lower mandible, dagger shaped and quite heavy and deep basally. They have no external nostril and in adults the bill is

Illus. 3: Masked or Blue-faced Booby *Sula dactylatra* showing juvenile in first winter plumage.

bright yellow in males, duller and greener in females.

Juveniles could be mistaken for Brown Boobies (*Sula leucogaster*) but differ from them in their slightly larger size, heavier straighter bills and generally lighter coloured chocolate brown back and wing coverts. They always show a whitish collar around the base of the neck or upper part of mantle except when newly fledged, and a white upper breast. In the Brown Booby the head and neck extending down to the breast is uniformly dark brown. In juvenile Masked Boobies their legs and feet are generally darker varying from slaty to orange-yellow. The under-wing pattern in immature Masked Boobies is rather greyish mottled with brown, whereas in the Brown Booby the under-wing coverts are conspicuously white and their legs and feet are pale yellowish green.

They attain their adult plumage when they are about 33 months old and as they get older the white collar on the hind neck becomes more pronounced.

HABITAT, DISTRIBUTION AND STATUS

A pelagic ocean wanderer rarely coming within sight of land and breeding extra-limitally on the Kuria Muria islands off the coast of Oman (Gallaher and Rogers, *Birds of Dhofar and Parts of Oman*, 1980).

The first record for the subcontinent was Capt. Butler's two juvenile specimens collected on 14 May 1877 when he joined a telegraph cable ship sailing to Jhask (Iran) via Astola island (E. A. Butler, *Stray Feathers*, 1877). Alan Hume (*Stray Feathers*, 1873) also saw a few immature birds off Gwadar on the Makran coast in February and Ticehurst saw several off the Makran coast near Ormara head (Ticehurst, *Ibis*, 1923, p. 460). Murray, when Curator of Karachi Museum, had a specimen brought to him on 9 July 1881. It was one out of several flocks totalling about 30 birds seen about 3 miles off Karachi coast by fishermen and it had accidentally become entangled in their nets.

More recently during a Marine Fisheries survey cruise off the Karachi coast, on 15 November 1983 two juvenile and immature Masked Boobies were watched fishing about 48 kilometres (30 miles) off Karachi seacoast by N. Van Zalinge and K. Forssgren (pers. comm., November 1983). Later more were seen off Ormara during the same survey. Another immature individual was actually watched from the shore by Kent Forssgren on 7 January 1982, fishing about half a mile off Buleji beach, Karachi (pers. comm.).

During a survey off the Makran coast by the marine research vessel 'M. V. Fritzov Nansen' in late January, February 1984, Kent Forssgren saw only one juvenile Masked Booby near Gwadar during a 13 day voyage (K. Forssgren, pers. comm., March 1984). Status RARE and probably only juvenile birds wander into Makran coastal waters.

HABITS AND BREEDING

Purely pelagic only coming to land for breeding. Feeds on fish secured by steep aerial dives plunging right under the water. Occasionally foraging in small flocks, it usually hunts solitarily. Known to feed on flying-fish and squid often in company with tropic birds. It often rests on the sea for an interval especially after a successful catch.

Birds collected by Butler had their stomachs full of flying-fish (Exocoetidae). On the Kuria Muria islands, the nearest known breeding colony, about 10,000 pairs breed from April to June on Al Oibiliya island and about 2,000 pairs on Al Hasikiya (Gallaher and Rogers, op. cit., 1980).

FAMILY PHALACROCORACIDAE

Cormorants

Great or Eurasian Cormorant *Phalacrocorax carbo sinensis* (Shaw)

Illus. **4**
Map **5**

DESCRIPTION

Body length 80-90 cms.
Wingspan 130-150 cms.
Weight 1,729 gms. (adult - Gilgit-Scully)

Recognizable by its size as the largest of our three cormorants and in breeding plumage by the noticeably yellow base of lower mandible and gular pouch. Sexes are alike except that males average larger in size. In breeding plumage the race inhabiting this region (*sinensis*) is distinguished by a more extensive prominent white thigh patch, easily visible in flight, more extensive white cheeks and throat patch and the whole of the shaggy crest, nape and sides of neck silvered with fine white scattered needle-like plumes. In some older individuals the whole crown and neck can appear silvery grey almost white. In non-breeding season these white areas disappear. The iris is pale green.

The Large or Great Cormorant has a very upright stance when on land with noticeably stout tarsi. Its toti-palmate feet are black, and the claw on the central toe is pectinate. Its neck is long and sinewy, tail comparatively long and wedge shaped and the bill which has no external nostril, is rounded on the culmen, terminating in a prominent down-curving hook. The bill is greenish to greyish horny with lower mandible and gular pouch becoming more prominently yellow in breeding season. Juveniles are more silvery brown, on upper plumage with throat and belly buffy white in winter. In adults, all the body feathers are glossed greenish, whilst the wing coverts assume a bronzy brown sheen with each feather margined in black during the breeding season. Whole plumage is duller black during winter.

HABITAT, DISTRIBUTION AND STATUS

Difficult to work out but undoubtedly some part of the population wintering in Pakistan is migratory, breeding in central Asiatic Russia. A bird shot by a local fisherman in October in Sonmiani lagoon (Lasbela) bore a Russian ring (Bachóo Ali, pers. comm., 1976) but regrettably the ring was lost before recording. Hugh Whistler (1930) observed birds in Rawalpindi district on 1 March and 18 April in flocks of 15 and 30 birds, presumably on passage, as they were not otherwise observed for the rest of the year in that district. The major population breeds in Pakistan and is locally migratory according to food availability,

especially in inland waters following the monsoon. They are equally at home on coastal waters and on inland lakes and rivers. Biddulph (1881) noted occasional birds in Gilgit in September including a flock of 5 on 12 September and Mathews (*JBNHS*, 1941) noted a few in July on Satpura lake (*circa* 2,750 metres elevation) due south of Skardu in Baltistan, wrongly cited as observation by Meinertzhagen at 3,450 metres (Ali and Ripley, 1968 and Cramp *et al.*, 1977). These northern observations are undoubtedly of migrant birds. Outside of Sind Province and the main Indus river, they are usually seen solitarily. In Sonmiani lagoon and Indus mouth there are favoured sand bars where they roost colonially during winter (*circa* 2,000 birds Pitiani creek, Indus delta, November and December in 3 different years, and over 600 birds Sonmiani in one flock, December and January in 2 different years). This wintering population disappears during the summer months, and Ticehurst thought that they were probably migrants breeding in Palearctic regions. More probably most of these birds disperse to inland fresh water colonies. Status ABUNDANT in Sind, less common elsewhere.

HABITS

They feed exclusively on fish which they catch by swimming under water using only their feet for propulsion. They frequently swim with only their head

Illus. 4: Great Cormorant *Phalacrocorax carbo sinensis*, showing the 'Chinese' sub-species in breeding plumage with white cheeks and throat and extensive area of white hair-like plumes on sides of neck.

and part of their neck visible above the surface but usually submerge by springing forward from the surface with dives seldom lasting as long as 60 seconds and usually 15 to 25 seconds duration (observed Hadiero lake). They have traditional winter roosts where they congregate in numbers, for example, Haleji lake rocky islet *circa* 300 birds. They fly from such roosts to feeding grounds with slow measured wing beats in 'V'-like formation. Individual birds spend a high proportion of the day resting on some perch (a lakeside boulder or submerged tree snag) with wings extended outwards in characteristic heraldic posture. It is thought that this may serve some special function in drying their plumage. In hot weather they cool themselves with a gular 'flutter' of their throat pouch. Birds fishing in Hadiero lake, Thatta district (where most of the fish are introduced African *Talapia* sp.), were observed being kleptoparasitized by Great Black-headed Gulls (*Larus ichthyaetus*) and were noted to emit gargling alarm or threat calls as the gulls dived upon them. They will fish in a flock when prey is shoaling but this is not a co-operative effort so much as each member out for itself. Individuals observed eagerly leap-frogging each other as the shoal swims forward as well as trying to seize captured fish from a neighbour (observed Hadiero lake 6 March a flock of 60 + all in full breeding dress). Outside creeks of Indus delta and Sonmiani, flocks of over 1,000 have been observed swimming together and feeding on a presumed fish shoal, sometimes all with bodies submerged and only necks extended above surface. Brown-headed Gulls (*Larus brunnicephalus*) swam with them as out-riders hoping to pick up escaping or injured fish but they did not behave as aggresively as *L. ichthyaetus* (observed 10 December 2 miles offshore).

BREEDING BIOLOGY

No evidence has yet been found in Pakistan of its breeding in mangroves on the seacoast of Pakistan as does its congeners *P. fuscicollis* and occasionally *P. niger*. Doig found a huge colony in a fresh water swamp in east Narra breeding in submerged Acacia *Prosopis spicigera* and Tamarisk *Tamarix aphylla* trees on Samaro Dhand (which has since dried up). In early November all nests had eggs, and some trees had 5 or 6 nests within 2 or 3 feet of each other. Nests were platforms of sticks 2 feet in diameter, rather oval (not circular) in outline, and 4 to 6 feet above the water level which was 8 feet deep. The nests had a thin lining of rushes and clutches numbered 4 to 6 eggs, being long

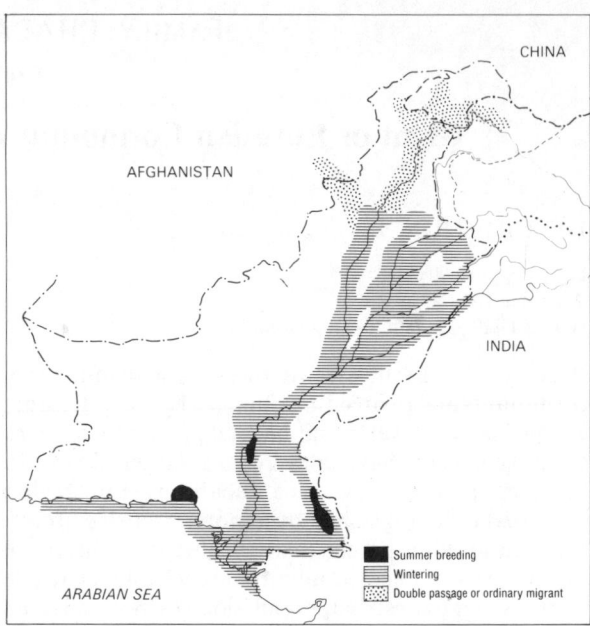

Map 5: Great Cormorant *Phalacrocorax carbo*

narrow ovals of pale greenish blue colour. The colony was estimated to be 80 metres wide and 2.5 kilometres long and to contain several thousand birds (Doig, *Stray Feathers,* 1879c). There were ibis and egrets nesting in the same colony. In the 1940s Eates found smaller colonies in the east Narra with eggs during July, August and September in mixed colonies with Shags (*P. fuscicollis*) and Pygmy Coromorants (*P. niger*) (MS unpublished). Another colony with eggs and young chicks was noted in October in nearby Saurashtra, India (Dharmakumarsinhji, 1972). Incubation period is 28 to 31 days with duties shared by both sexes. Hatching is asynchronous and chicks are black skinned and naked at first but within a week become covered with black down. They are fed by regurgitation, placing their heads inside the parent's bill. Young are fledged at 50 days.

*VOCALIZATIONS

They have quite a complicated repertoire of calls but these are mostly associated with breeding and they are much more silent in their winter quarters than *P. niger*. Their calls are mainly deep guttural sounds, with the female's voice being considered hoarser. In the nesting colony they comprise series of short rapidly repeated guttural quacking calls.

Indian [Shag] Cormorant *Phalacrocorax fuscicollis* Stephens

DESCRIPTION

Body length 60-65 cms.
Wing length 25.7-27.6 cms.

Easily overlooked as like a slightly larger version of the Pygmy Cormorant (*P. niger*). For a short period from end June to September while in full breeding plumage it is easily separable from the other two cormorants, with pure white tufts of feathers on each side of its neck behind the ear-coverts and the naked facial skin purplish black bordered by yellow on the throat. Even in winter it usually has a prominent white border to the naked skin of its gular patch which *P. niger* never shows. During the summer it also has a few white needle-like feathers on its crown and nape. Its wing coverts are noticeably bronzy green with black margins producing a handsome scale-like effect. The sexes are alike. Outside the breeding season the naked skin of the gular region becomes yellow. The legs and feet are black.

HABITAT, DISTRIBUTION AND STATUS

Equally adapted to a marine or fresh water environment this shag is much commoner in lower Sind and in estuarine waters than in the Punjab. It extends up the Indus river and its tributaries, as flocks of 60 to 70 seen Lal Sohanran lake (Bahawalpur district) near the Sutlej river. Flocks of 100 up to 200 have been regularly observed at Ghauspur jheel (Jacobabad district). Small numbers regularly seen also at Trimmu head, Jhang district but not recorded in 1920s by Whistler in Jhang district. Most numerous in Indus mangrove creeks and east Narra lakes, such as Sadori, and in lower Sind. Status COMMON.

HABITS

A gregarious species often seen fishing in flocks and roosting colonially on submerged Tamarisk bushes, where they extend their wings in typical cormorant fashion as though sun-bathing. In winter time they commonly fish in salt water creeks such as Sonmiani, Lasbela and the Indus delta and in habits are similar to others of the genus, swimming with only head and neck above the surface and diving frequently to pursue fish under water. Ticehurst records a specimen he secured in Karachi harbour having two flatfish (soles) (*Synaptura orientalis*) 12 cms. long in its oesophagus.

BREEDING BIOLOGY

No evidence has been found of this cormorant breeding in mangroves in the Indus mouth but a large colony nests during the monsoon in the mangroves at Sonmiani in company with Reef Herons (*Egretta gularis*) and Pond Herons (*Ardeola grayii*). Nesting usually starts in July but may extend up to October. They are colonial in breeding with up to 15 to 20 stick nests having been counted in one tree. Eates (MS notes unpublished) found a nesting colony in the east Narra in a lake situated in the sand hills. Their nests were built on flooded Tamarisk trees and the colony was shared with Spoonbills (*Platalea leucorodia*), White Ibis (*Threskiornis melanocephalus*) and Grey Herons

Illus. 5: Indian (Shag) Cormorant *Phalacrocorax fuscicollis*, showing white patch behind eye only developed during breeding season.

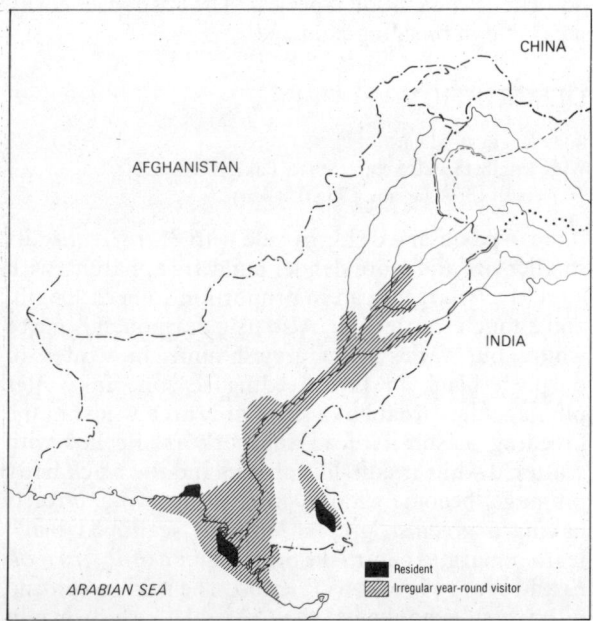

Map 6: Indian (Shag) Cormorant *Phalacrocorax fuscicollis*

(*Ardea cinerea*). This colony was discovered in September. Doig (*Stray Feathers,* 1879b) found nests with eggs in the east Narra swamps in July. Both sexes share in nest construction and normal clutch is 3 to 5, the eggs being elongated ovals of a pale bluish green

colour. Eates collected some eggs which had distinct green and purple spots on them (MS notes). The young are born naked and black, but within days become covered with brownish black woolly down. They are fed by both parents by regurgitation.

African Pygmy Cormorant *Phalacrocorax pygmeus* (Pallas)

A single specimen was collected by Capt. J.E.B. Hotson in a remote valley (Mashkai, Kharan district) of southern Baluchistan on 3 September 1909. This was wrongly listed as *P. javanicus* (or *P. niger*) by Dr. Ticehurst (*JBNHS,* 1927). The specimen, which is in the Bombay Natural History Society collection, has been identified as *P. pygmeus* (H. Abdulali, *JBNHS,* 1965). At present there is little information about possible range overlap of these two sibling species

(M. Gallaher *in litt.* to author, 1982) and in this relatively arid waterless area which remains politically inaccesible, the problem must continue unresolved.

The checklist of Afghanistan birds produced by Knud Paludan lists *pygmeus* as a possible breeder in Seistan, and *niger* as a straggler (Paludan, 1959), but Derek Scott did not come across it during a number of visits to Iranian Seistan between 1970 and 1978 (Scott, 1978).

Little or Javanese Cormorant *Phalacrocorax niger* (Vieillot)

Illus. **6**
Map **7**

Synonym (*Phalacrocorax javanicus* (Horsfield) in Blanford and Oates, Vol. 4)

TAXONOMY

In winter plumage individuals cannot be differentiated from *Phalacrocorax pygmeus* the Pygmy Cormorant of north Africa and Mediterranean countries, eastwards to Iraq. It has in the past been considered conspecific with *P. pygmeus*. See comments in chapter on The Problem of Species. The two are easily separable in breeding plumage.

DESCRIPTION

Body length 48-51 cms.
Wing length 18.1-20.5 cms. (Stuart Baker)
Tail length 13.3-14.6 cms. (Stuart Baker)

If it can be observed side by side with *P. fuscicollis*, its smaller size and more slender build are apparent, with a relatively larger head in proportion to neck length, and a much shorter bill. Also its gular pouch is never white, but varies from greyish pink in winter to purplish black in the breeding season. In winter plumage the throat has a silver grey area whilst in the breeding season its head and neck are flecked with scattered white needle-like plumes and the black body plumage becomes more glossy, the wing coverts having a greenish grey sheen with scalloped black feather margins contra the bronzy green of *P. fuscicollis* and bronzy brown of *P. carbo*. The iris is green and bill, which is hooked at the tip, is blackish brown in colour. Juveniles show in their first winter much more

extensive silvery grey on their bellies and their heads and their necks are dull rusty brown. The tail is very narrow at the base and fan-shaped.

HABITAT, DISTRIBUTION AND STATUS

This cormorant, in Pakistan appears to prefer fresh water areas, inhabiting the larger jheels and barrage head ponds in winter, but spreading out to seasonal swamps during the monsoon season. It is quite rare in northern Punjab and absent from Baluchistan except for occasional birds on Siranda jheel in Lasbela. There

Illus. 6: Little or Javanese Cormorant *Phalacrocorax niger,* in breeding plumage.

is no clear-cut proof that birds frequent the mangrove creeks in winter as this species looks closely similar to *P. fuscicollis* which favours that biotope, but the author has occasionally seen small cormorants in the mangroves which appeared to be *P. niger*. However, a colony of about 50 pairs bred in mangroves in Ghizri creek on the Karachi coast in 1983 for the first time in the memory of local fishermen. Typical winter time populations at such major wetland habitats as Lal Sohanran (Bahawalpur district) when the lake level is normal, about 3,000 contra 5 to 600 of the Indian Shag *P. fuscicollis*. In the same region, typical populations at low water levels in late February and March 300+ *P. niger,* and 30 to 40 *P. fuscicollis*. At Ghauspur jheel, Jacobabad and environs in late October there were about 7,000 *P. niger* and an estimated 1,400 only *P. fuscicollis*. Status ABUNDANT in Sind, less common elsewhere.

HABITS

The Little Cormorant frequently fishes in a flock, members eagerly swimming forward or leaping over each other as they pursue a school of fish. They normally come to the surface to swallow their prey, juggling and tossing it around until it can be swallowed head first. The smaller *P. niger* has been observed with a fish in its bill being, pursued by *P. fuscicollis* which failed to seize the former's prey. They roost communally in winter, often in *Phragmites* reed beds where they keep up a constant chatter of gargling and guttural croaking noises. In the late monsoon season in the main rice growing tracts they will often disperse into

the rice fields and are then not so gregarious and no flooded 'borrow pit' or wayside pool is too small to escape their attention. Compared with the other two *Phalacrocorax* species, frogs and tadpoles comprise a more significant item in their diet, which consists predominantly of small fresh water fish species.

BREEDING BIOLOGY

A highly gregarious breeder usually nesting in trees in association with other cormorant species and with herons and egrets. Doig found nests with fresh eggs, crowded together in partly submerged Tamarisk trees on 24 July in the east Narra, in association with *P. fuscicollis* and *Anhinga*. In Sukkur district there is a huge breeding colony on flooded Tamarisk trees on an island in the Indus river and in Shahyaki, Thatta district a smaller (*circa* 200 pairs) mixed colony in flooded *Acacia arabica* trees in association with *Bubulcus ibis* and *Ardeola grayii*. In both these colonies breeding usually commences in mid-July. Birds seen carrying sticks in the bill on arrival at nest site on 26 July but none carrying sticks on 2 August (observations same colony Shahyaki). On Haleji lake (Karachi district), a colony occupies small rocky islands from late July or early August after nesting sites are vacated by Night Herons and Purple Herons. Here, they nest in *Phragmites* reeds or *Salvadora persica* bushes. In Ghizri creek, in mangroves, a colony of about 50 pairs started nesting in late May after Reef Herons (*Egretta gularis*) had reared their young in the same patch of taller mangroves. The birds started nest building in association with *Ardeola grayii* and *Egretta intermedia*. By 8 July they had large downy young in the nests. Aggressive interactions with other species (egrets) show that both partners defend their nest vicinity area. Courtship and display depends heavily upon ritualized appeasement movements of the head and neck which are bent downwards and forwards in snake-like movements in a repeated sequence and the head and neck is occasionally bent right over the back in moments of extreme excitement. At the same time the occipital crest is raised and the gular pouch appears to be inflated making the head look, larger (observed Karachi Zoological Gardens in wild breeding colony and in the Haleji lake colony). The nest is a small platform of reed stems and sticks with some lining in the centre of thinner reed stems. The elongated oval eggs are greenish blue and a normal clutch is 3 to 5. Salim Ali suggests that nest predation by House Crows (*Corvus splendens*) has resulted in both parents immediately starting incubation of the first egg so that hatching is asynchronous (Ali and Ripley, Vol. 1, 1968), and certainly House Crows are serious predators in all heronries where they congregate in considerable numbers. No one seems to have recorded that the newly hatched young have a

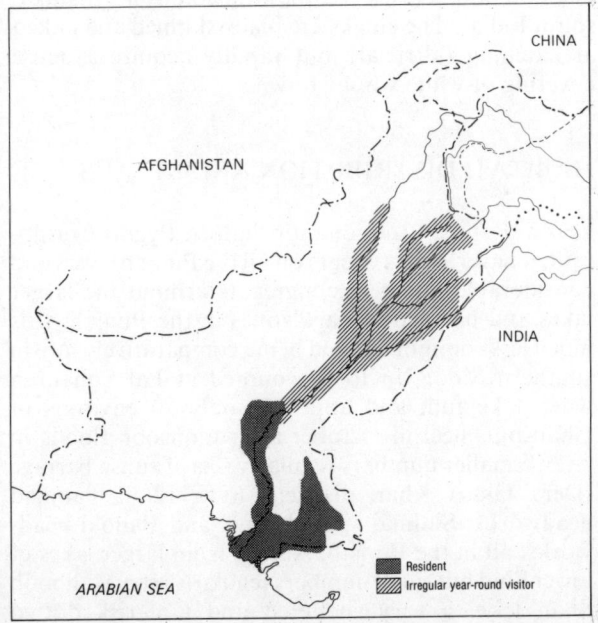

Map 7: Little or Javanese Cormorant *Phalacrocorax niger*

naked shining livid red patch on top of their skull, even after their sooty brown body down has grown. This bright red area may serve as a recognition signal. They solicit feeding by waving their wings and tickling the gular pouch or base of bill of the parents and then thrust their heads right inside the adult's bill when food is regurgitated (observed Haleji lake colony through X 50 telescope). Like all the *Phalacrocoracidae* both parents share incubation duties, and lacking brood pouches on their ventrum, put the eggs on top of their feet. Period of incubation and fledging not known.

*VOCALIZATIONS

Much more vocal than other cormorants. A winter time roost of over 4,000 Little Cormorants at Kandhkot (Jacobabad district) tape-recorded in early December at sunset sounded like the roaring murmuration of a sports stadium crowd. They are noisy in their winter time loafing and roosting areas as well as breeding colonies, keeping up an incessant variety of deep-toned grunting and groaning noises. These variable sounds are guttural and low pitched including 'gargling', 'aah-aahing', and 'kok-kok-koking'.

FAMILY ANHINGIDAE

Snake Birds

Darter or Snake Bird *Anhinga melanogaster Pennant*   Map 8

(Pennant Darter of K. H. Voous's 'Recent List')

Synonym *Anhinga rufa* (Daudin) (Ripley, 'Synopsis', Revised edition)

TAXONOMY

The Pakistan population appears truly Oriental in origin as it is commoner in the north-east than elsewhere but has spread down the Indus. Comparison with African birds shows that the Indian population noticeably differs in having the neck more silvery grey, less rufous especially in the forepart, with lanceolate wing covert and scapular feathers longer and more conspicuously margined with silvery grey. These slight differences probably only warrant sub-specific separation but no doubt there has been little genetic exchange in these two geographically separate populations of this ancient pan-tropical, family. See discussion in chapter on The Problem of Species.

DESCRIPTION

Body length 85-97 cms. (Cramp *et al.*), 90 cms. (Salim Ali)
Wing length 33.1-35.7 cms. (Stuart Baker)
Tail length 20.2-24 cms. (Stuart Baker)
Bill length 74-90 mm.

Even in distant flight they can be separated from other cormorants by the relatively longer more narrow neck, kinked even in flight and the relatively longer fan-shaped tail. Viewed closer, the straight sword shaped bill without any terminal hook, the narrow snake-like head and neck and pale streak extending down side of neck from behind the eye, are at once noticeable. The crown and hind neck are rusty brown and foreneck silvery white, more contrastingly white in the breeding season. The back, tail, and breast are black and with obscure paler slaty cross-barring on the tail feathers. The scapulars are noticeably long and lanceolate

shaped, having black centres and silvery white margins. The bill may be tipped blackish brown with the lower mandible more yellow at the base. There is no naked gular pouch and the iris is yellow. In many specimens the whole bill appears horny yellow especially in winter. The neck is kinked between the eighth and ninth vertebrae. Juveniles are more rufous buff on the head and sides of neck and silvery buff on forepart of neck and centre of belly, with no silver grey scapular plumes. Chicks appear to have some yellow bare skin on forecrown and dull white feet which later become black in adults (observed nesting colony Karnataka, south India). The chicks are black skinned and naked at hatching (altricial) but rapidly acquire a dense covering of white woolly down.

HABITAT, DISTRIBUTION AND STATUS

Less widespread and common than the Pygmy Cormorant, it nevertheless occurs on all the Punjab rivers and considerable numbers congregate around the larger lakes and barrage seepage zones in the Punjab. It is much less common in Sind being comparatively rare in southern Sind. Up to 50 counted at Lal Sohanran when lake full, and approximately 70 environs of Ghauspur jheel in October after monsoon floods in river. Smaller numbers regularly seen Taunsa barrage (Dera Ghazi Khan district), 6 to 12 at Panjnad headworks, Sidhnai spill channel and Balloki headworks, all in the Punjab. Absent from larger lakes of lower Sind but small numbers regularly seen Mahboob Shah lake, a very sheltered and tamarisk dotted swamp in Thatta district.

A few Anhinga have adapted to the salt water creeks of the Indus delta. Single birds seen in January, December, and March over a 4 year period.

HABITS

Not gregarious in hunting being usually seen solitarily, though occasionally 10 to 20 birds can be seen roosting and drying their out-stretched wings in loose association. They habitually swim when hunting with only the head and neck above water, diving frequently to pursue fish under water. Their method of hunting is unique amongst the *Phalacrocoracidae.* Instead of seizing fish prey between mandibles, they normally spear right through the body of their prey, impaling the fish, if it is small, through the upper mandible. In the case of very large fish the body is impaled on the closed bill. Then emerging with only head and upper neck above the surface, with a deft sideways flick of the neck; they remove the impaled fish and catch it between the open mandibles as it arcs through the air, (observed Panjnad headworks). They seem to prefer much smaller fish than the Great Cormorant can catch and never fail to re-catch and simultaneously swallow such prey without their bodies coming to the surface. Usually they submerge by sinking head and neck vertically and no diving is seen. They spend a large portion of the day perched on favourite trees, drying their out-spread wings and individuals appear to frequent the same favoured stretches of water for hunting each day. Studies in west Bengal indicated that 81 per cent of the diet (stomachs of 19 adult specimens examined) comprised fish (20 different species), 7.4 per cent water snakes (*Natrix* and *Enhydris*), 2 per cent tadpoles and 8.4 per cent insects mostly Odonata larvae and Hemiptera bugs (Mukherjee, 1969).

BREEDING BIOLOGY

They nest colonially with other cormorants and egrets and breeding usually commences with the onset of the monsoon. Eates found nests in submerged trees during July, August, and September usually in riverain forest, in *Acacia arabica* or *Tamarix dioica* trees in company with Little Cormorants (*P. niger*) and Little Egrets (*Egretta garzetta*). Their not very substantial nests were untidy structures of sticks with some softer water weeds in the lining. Doig (*Stray Feathers*, 1879*b*) also found them nesting in mixed Heron and Ibis colonies in the east Narra towards the end of the monsoon. The normal clutch is 3 eggs, elongated or spindle shaped and a pale greenish white colour with a chalky coating. Incubation period in the African Darter is 23 to 25 days with hatching asynchronous (Cramp *et al.*, 1977) and Salim Ali records heavy predation of downy chicks by Pallas's Eagle (*Haliaeetus leucoryphus*) (Ali

and Ripley, 1968). Young observed being fed by both parents at Karnataka (south India) put their entire heads inside the adult's mouth, the food being regurgitated, and this appeared to be a violent interaction, pushing the head of the parent bird backwards and forwards and lasting for up to eight seconds. At the onset of the breeding season they display by erecting the long silvery grey scapular plumes and this may be an aggressive posture as in *Egretta* sp. when erecting dorsal plumes. It is note-worthy that display by the Australian population (*A. melanogaster novaehollandiae*) involves wing raising only as the scapular plumes are not noticeably elongated in this species (Cramp *et al.*, 1977).

*VOCALIZATIONS

They are relatively silent once egg laying commences, but in the early summer are quite vocal especially in their favourite roosting and wing drying trees. The commonest call is a rapidly repeated rather frog-like sounding 'kek-kek-kek-ke-ke-ke-ke'. Each call lasts about 2.5 seconds and ends *accelerando.* This call is quite distinctive from that of sympatric Cormorant species.

MISCELLANEOUS

In north-west Assam, local fishermen use Darters to catch fish, the birds being perfectly tame and carried around on poles slung across the fisherman's shoulder. A ring around the bird's neck prevents the fish from being swallowed (Stoner, *JBNHS*, 1948).

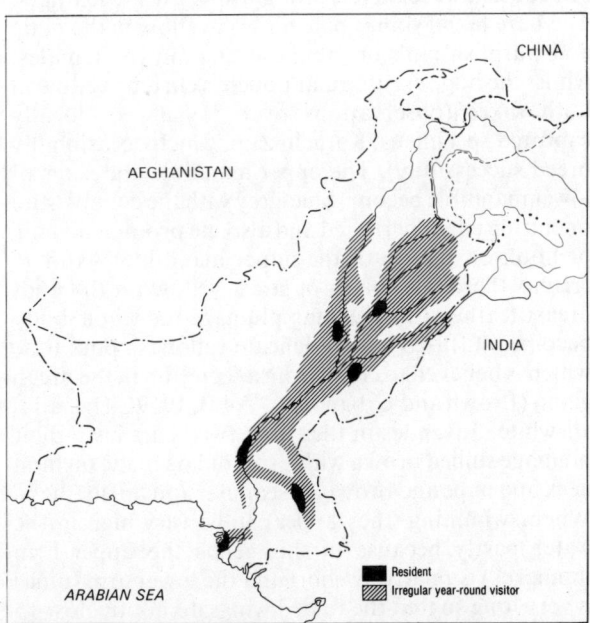

Map 8: Darter or Snake Bird *Anhinga melanogaster*

FAMILY PELECANIDAE

Pelicans

Great White or Rosy Pelican *Pelecanus onocrotalus* Linnaeus

Illus. **7**
Map **9**

DESCRIPTION

Body length 150-183 cms. includes bill 30-40 cms.
Wingspan 270-360 cms.
Wing length 70-73 cms. (Stuart Baker)
Bill length 430-450 mm.
Weight 9-12 kgs. (Brown and Urban)

A huge almost grotesque looking bird with its enormous long bill and short stubby legs, but in soaring flight it has a striking elegance, as well as when in full breeding plumage.

It can be distinguished in flight from *P. crispus* by the under-wing pattern with the primaries and outer secondaries contrastingly black with white under-wing coverts. The Dalmatian Pelican shows a more uniform greyish white under-wing pattern. When viewed closer also the legs are fleshy pink to orange and the iris (which is crimson varying to hazel brown) looks dark. *P. crispus* has leaden coloured legs and a pale lemon iris. When on the water or on land the black primaries are entirely concealed by the silky white secondary wing coverts. In winter the bill tends to be horny brown with the gular pouch pale yellow. In full breeding plumage both sexes develop a peculiar rubbery golf-ball sized knob on the forecrown at the base of the bill, and the occipital crest, which develops after the knob, becomes longer and more noticeable but consists of relatively straight spiky white feathers. The bare facial skin is pale pinkish yellow in the male and purplish pink or intense orange in the females, whilst the huge elastic gular pouch is chrome yellow in both sexes (observations over 7 years of locally captured specimens, Karachi Zoo, which occasionally breed successfully). The upper mandible and sides of lower mandible become blue-grey with the commissure or cutting edge cherry red and also the prominent 'nail' or hook on the tip of the upper mandible. As in *P. crispus* there is a splash of straw yellow on the mid-breast feathers. In breeding plumage the whole body becomes suffused with a delicate yellowish pink tone which is believed to come from a secretion in the preen gland (Brown and Urban, *Ibis*, Vol. 3, 1969). The tail is all-white. Juveniles in their first two years have their plumage sullied brown with a sort of hog mane on hind neck and nape and brownish scapular and tail feathers. When swimming, they appear to be very high in the water partly because in this genus the upper arm (humerus) is relatively short and the lower arm (ulna) is very long so that the folded wings do not fit close to the body. In juveniles and adults the feathers of the forecrown extend in a 'V' down to the culmen, whereas

in *P. crispus* the pattern is reversed with the culmen itself extending up to the forecrown in a 'V' dividing the feathered part.

HABITAT, DISRIBUTION AND STATUS

Formerly very numerous as a winter visitor to Sind (Doig described huge flocks of up to 1,000 birds in the east Narra lakes). In the 1980s only a few hundred migrate into Pakistan mostly passing through the upper Kurram valley and across north-western Baluchistan, stopping on passage fairly regularly at Kushdil Khan lake.

Occasionally a single bird can be seen in October, November, and early December on the larger lakes of the Punjab, but the main population probably winters in the shallow lakes of the Rann of Kutch or in Thatta district and the creeks of the Indus delta. They are seen less frequently and may be less numerous than *P. crispus* and generally associate in single species groups. A flock of 115 Rosy Pelicans was seen at Mahboob Shah lake, Thatta district on 8 December and 110 birds on 16 February on the same lake two years later and flocks of 68, 75 and 8 birds in Badin district on 21 February (F. Koning, pers. comm., 1974). A few birds over-winter around Haleji and Hadiero lakes (Karachi district) in association with *P. crispus* and a flock of 120 was observed in mid-October on Haleji but they did not remain more than two weeks. They are also seen at Siranda jheel and Sonmiani lagoon in tidal creeks. Most birds start northwards migration by

Illus. 7:   (a) Dalmatian or Grey Pelican *Pelecanus crispus*, male in breeding plumage.

  (b) Great White or Rosy Pelican *Pelecanus onocrotalus*, male in breeding plumage.

the third week of March, when separate flocks wheel around in thermals before gliding in a north-westerly route. On 21 March 1980 near Pipri, Karachi district, 6 flocks totalling just under 800 birds were observed on migration, and this was believed to be the major population wintering in Pakistan and north-west India. They were accompanied by a few scattered groups of Common Cranes (*Grus grus*). Their pinions made an audible buzzing sound as they circled overhead. However, during extensive IWRB surveys conducted in late January 1989 by F. Koning and Hamid Ali an important discovery was made with an estimated 20,000 pelicans thought to be all of this species, congregated on a shallow brackish lake near the southern border of Pakistan in Badin district close to the great Rann of Kutch (Koning *in litt.* to author). Earliest sightings 22 August (species unidentified, Anne Vollum, pers. comm.). Both this species and the Dalmation Pelican are sympatric on their wintering grounds. Status COMMON only in Thatta and Badin districts.

HABITS

Usually seen cruising solitarily on larger lakes, but twice seen in seepage pools fishing communally in a tight packed raft, each bird scooping up bagfuls of water in its pouch and obviously pursuing a shoal of fish as outsiders flew forward to the front and centre of the pack. They are exclusively fish eaters on Sind lakes. It has been estimated that in sub-tropical conditions adults consume about 10 per cent of their body weight or about 900-1,200 grams of fish per day (Brown and Urban, *Ibis,* 1969). They prefer to roost communally, squatting on bare open sandbars on the seacoast and lagoons, heads tucked under wing or they will roost on suitable lakes on the shores of a rocky islet. In the Rann of Kutch the stomach of one individual was full of the fish *Cyprinodon dispar* (Ali, 1960).

BREEDING BIOLOGY

Though they breed extra-limitally the discovery some 30 years ago of a colony of about 300 nesting pairs in the Great Rann of Kutch by Dr. Salim Ali is of considerable interest. A single pair of captive (pinioned) birds at Karachi Zoo has twice successfully reared young in the past 8 years (2 in 1982) and early courtship observations of these birds showed aggressive chases on water by the males, mainly of the females, but also other Pelican species. On land females were observed raising wing shoulders and quivering whole wings. The male periodically raised his bill skywards and in aggressive displays (against captive domestic geese and other Pelicans) the bill was used as a fencing weapon or from time to time extended upwards and the mandibles clapped once together making a formidable sound. In captivity only the male brought sticks to the nest which the female arranged and two plain white eggs (looking comparatively small) were laid. The chicks are naked on hatching, of a dark flesh colour with black bills and later become dark skinned turning slaty grey and then are rapidly covered with sooty black down (Brown and Urban, *Ibis,* 1969). Incubation period about 35 days (calculated at 38 days in east African colonies) (Brown and Urban, op. cit.) and shared by both sexes at Karachi Zoo. Young fed by regurgitation of semi-digested food but not observed at Karachi Zoo due to artificial feeding times of adult birds. In wild breeding colonies where large numbers nest together, the downy young when able to walk, congregate together in creches.

VOCALIZATIONS

Normally very silent but the feeding flock observed in October at Haleji lake was highly excited and occasional birds emitted a very low pitched grunt sound similar to the call of a Water Buffalo.

Map 9: Great White or Rosy Pelican *Pelecanus onocrotalus*

Grey or Spot-billed Pelican *Pelecanus philippensis* Gmelin

Body length 150 cms.
Wingspan 300 cms.
Bill length 32-35 cms.

This pelican, which can be seen alongside the migratory Dalmatian Pelican (*Pelicanus crispus*) in Sri Lanka in winter time (author obs., Yalla National park), is readily separable by its smaller size and even in winter or in sub-adult plumage, different coloured gular pouch and darker leaden grey back and wing coverts.

In Ali and Ripley's 'Handbook' (Vol. 1, revised edition, 1980, p. 29) its distribution is given as resident in (West) Pakistan. Likewise in S. Dillon Ripley's *Synopsis of the Birds of India and Pakistan* (Revised edition, 1982, p. 8) it is described as wintering in Pakistan.

There is no evidence, in fact, that it ever migrated as far north-west as Sind. Certainly Ticehurst did not meet with it (*Ibis*, 1923, p. 461) nor K. Eates in over 40 year's study of the Sind avifauna (K. Eates, pers. comm., 1971). It is also not included in the Sind Checklist (Published in the West Pakistan Gazetteer covering the Province of Sind, Appendix IV, 1969) which includes all past and present records. It is excluded from the **Checklist** of Pakistan birds.

Dalmatian Pelican *Pelecanus crispus* Bruch   Map 10

TAXONOMY

Ripley in his 'Revised Synopsis' treats *Pelecanus philippensis crispus* as a sub-species (Ripley, 1982). De Voous in his 'List of Recent Holarctic Bird Species' (*Ibis*, 1973) seems to treat *P. philippensis* as a separate species though excluding it as a Palearctic occurring bird. There appear to be ecological and other differences between these two populations as *P. philippensis* is a tree-top nester, whilst *P. crispus* nests on the ground in reed beds or inaccessible rocky islands. Whilst the Oriental population (nominate *philippensis*) nests during the winter, the Dalmatian Pelican nests from mid-April to May. In breeding plumage, the gular pouch of *crispus* males is tangerine orange and that of *philippensis* dull purple or greyish purple. Observations of the two birds feeding together (Yalla National park, Sri Lanka, late February) show a striking difference in size, plumage, and bill colour even when not in breeding plumage.

DESCRIPTION

Body length 160-180 cms.
Wingspan 310-345 cms.
Wing length 72-80 cms. (males), 68-72 cms. (females) (Stuart Baker)
Bill length 40-45 cms. (males), 36-38 cms. (females) (Stuart Baker)
Weight 10.5-12 kgs. (Brown and Urban)

Males are slightly larger than females but both are about equal in size to *P. onocrotalus*, the Great White Pelican. In winter plumage the pale iris and less extensive bare facial skin are good pointers in separating this species from the Great White Pelican and if in flight, its under-wing pattern with much less contrasting primaries and outer secondaries enables separation from a considerable distance, the whole of the under-wing appearing silvery grey. If standing, the dark grey legs are noticeable. Juveniles show more brown about the head and neck, and wing coverts than juvenile White Pelicans of the same age but are probably inseparable in the field, even in flight. See note about difference in pattern of feathers on forecrown of these two species. Old individuals from mid-January onwards begin to show a darker orange gular or throat

Map 10: Dalmatian or Grey Pelican *Pelecanus crispus*

pouch and this at once separates them from *P. onocrotalus*. In full breeding plumage they are very handsome, the whole body plumage being of the palest silky silvery white tone or grey-white. The crown and occiput is covered by a crest of curly chrysanthemum-like feathers and the gular pouch becomes mandarin orange in colour. The horny mandibles are blue-grey as in *onocrotalus* but the nail on the upper bill tip is orange, not red. Females have a less pronounced crest and their gular pouches are paler more amber coloured. The bare facial skin seems variable, as in some individuals in breeding condition this area is orange, but it may also turn a livid purple. Sub-adult birds have the orbital region and bill a leaden purplish colour and pouch greyish yellow. Spotbill Pelicans in the breeding season have dull purple gular pouches, also their feet are greyish brown. The downy chicks are white, with the forecrown naked and of an orange-pink skin colour.

HABITAT, DISTRIBUTION AND STATUS

Uses the same migration routes as the Rosy Pelican coming to Pakistan. A few are seen at Zangi Nawar lake in central Baluchistan from late March, (Christison, 1942) and also from late March to early April flocks totalling 100 to 200 birds sometimes stay at Kushdil Khan lake in Pishin district for a few days. Dalmation Pelicans have been shot on migration at Parachinar near the Kurram river on 26 March and 28 February (Colonel Dastagir, pers. comm.). First arrivals in lower Sind probably late August (see *P. onocrotalus* account) and a few birds linger on lower Sind lakes up to 15 April. Occasional individuals seen on Punjab lakes. 2 seen in October at Kharrar jheel, Renala Khurd and a single individual Lal Sohanran on 22 November (ten years' observations). A small population (20 to 40 birds) regularly winters between Hadiero and Haleji lakes in lower Sind, 100 to 150 birds in Sonmiani lagoon (Lasbela), and scattered groups throughout the Indus delta creeks (a total of 25 individuals counted during a two-day voyage). Sighting records during nine years' residence in Karachi indicate this species as about twice as numerous as the Rosy Pelican in lower Sind especially around the seacoast. Status COMMON only in Thatta district.

HABITS

Generally gregarious in winter quarters loafing by day on an isolated open sandbar in coastal regions (where human approach can be observed from afar). The Haleji and Hadiero lake populations roost on the flat top of a mid-lake rocky island which is shared, on its slopes by colonially roosting Cormorants.

When disturbed on land, birds took off laboriously by flapping their wings and slapping both feet in unison on the sand rather than by running. They became airborne after about 50 metres run. They never perch on trees as *Pelecanus philippensis* commonly does, preferring open flat inaccessible areas or rocky islands. In flight they are graceful and often soar in high circles before taking off in irregular 'V' formation. They hunt individually by swimming on the surface and sweeping their bills or whole upper neck in scythe-like fashion under the water (observed Haleji lake). They also occasionally hunt in a pack when a shoal of fish is located, carrying their wings in half raised position over their backs (observed Hadiero lake). Their diet comprises fish entirely.

BREEDING BIOLOGY

They breed extra-limitally with the main population around the Aral Sea, and like the Rosy Pelican nest on the ground in reed beds or on inaccessible bare islets, building a bulky nest of sticks and reeds. Their usual clutch is 3 or 4 eggs and incubation period 30 to 32 days and they take about 85 days to fledge. Very small young are fed by regurgitation onto the nest floor but when bigger thrust their whole heads into the parents' pouch. Curiously the captive specimens of this species have not bred in Karachi Zoo but males were observed raising their bills skywards in agonistic display and clappering the bill. This is often just one loud clap, unlike the rapid clatter of White Storks.

VOCALIZATIONS

A breeding colony in Bulgaria has been recorded emitting barking but guttural short calls like a Raven (*Corvus corax*). They are also reported to emit hissing and groaning sounds (Cramp *et. al.,* 1977).

ORDER CICONIIFORMES
FAMILY ARDEIDAE
Herons

Eurasian Bittern *Botaurus stellaris* (Linnaeus)   Map 11

DESCRIPTION

Body length 70-80 cms.
Wingspan 125-135 cms.
Wing length 33.5-35.7 cms. (males), 29.6-32.7 cms. (females)
Bill length 61.74 mm. (males), 60-68 mm. (females)
Weight 966-1,940 gms. (males), 867-1,150 gms. (females)

The sexes are alike and there is no change in nuptial plumage. They are cryptically coloured birds, with straight dagger-like bills and relatively short tarsi and comparatively thick necks tapering smoothly into the body. Ground colour a slightly reddish buff, barred and streaked with dark chestnut and black. The crown and a line behind the eye is dark chestnut, and there is a bushy nuchal crest only visible when the bird is excited. There is a whitish area around the upper throat and lower ventrum, and longitudinal lines of arrow shaped black streaks on the throat and breast. In flight the tail and wings look more tawny with black cross-barring on the wings and paler buffy wing coverts. The bill is greenish yellow, darker on the culmen and the legs are greenish yellow with very long stout toes and quite long claws, an adaptation to clambering through vertical reeds and supporting its body weight on submerged vegetation. The small area

of naked skin around the eye is green and the iris is yellow.

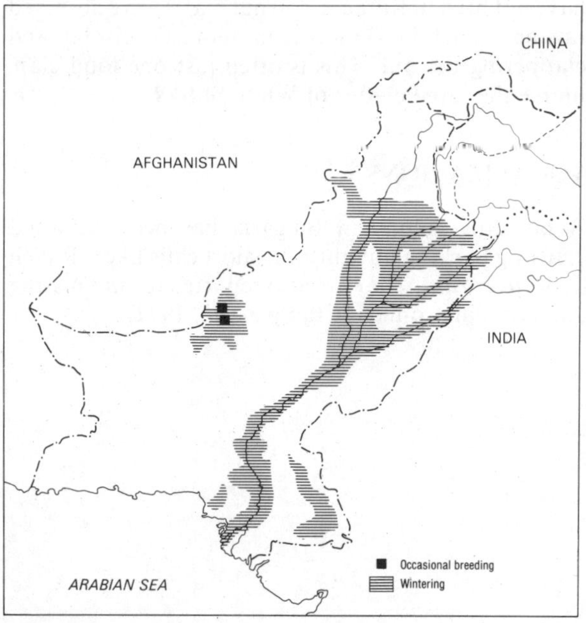

HABITAT, DISTRIBUTION AND STATUS

Wholly confined to marshy and seepage areas having extensive beds of tall reeds, either *Typha* or *Phragmites* or combinations of both. Will avoid drying out swamps and especially favours perennial water ponds with reeds growing in fairly deep water. Widespread in Pakistan mainly as a winter visitor only; and because of its secretive nature and cover haunting habits it is rarely encountered. Eates found it at Chor lake, eastern Tharparkar and Kalankot lake on the Makli hills (K. Eates, MS notes unpublished) and a specimen was killed in the seepage zone around Haleji lake (Rohil Nana, 1968, pers. comm.). It has been observed twice in the same small swamp at Kandhkot, Jacobabad district in two different successive years (author obs.) and individuals probably re-occupy the same favoured winter territory. In the Punjab, Whistler saw only one in Jhang district in three years, (Whistler, *Ibis*, Vol. 4) and Z.B. Mirza, collecting specimens for the Punjab University Museum, encountered it once only in five years in the seepage zone along Qadirabad link canal (Gujrat district). One was seen by Khabbaki lake in the Salt Range on 10 January by Mallalieu and Ashiq Ahmad (*in litt.*, January 1987). It is recorded as occurring on spring and autumn passage in Baluchistan (Meinertzhagen, *Ibis*, 1920) at Kushdil Khan lake, and Christison (*JBNHS*, Vol. 43) found a pair breeding at Malezai north of Quetta, Baluchistan in 1938. Whitehead recorded it as not uncommon in winter in the Kurram valley and around Kohat. Status SCARCE.

HABITS

Except for suitable breeding areas in Baluchistan, it is silent in its wintering quarters, but its habit of flying over reed beds at dusk often reveals its favourite haunts (observed Kandhkot, northern Sind 28 November and 13 February in two successive years). Normally very secretive, it adopts an upright concealing posture with bill held vertically and neck upstretched. It feeds principally upon fish and frogs and if opportunity avails, young Moorhen chicks, and other birds, insects and snails, snakes and eels. It hunts solitarily, often by waiting motionless clinging to reeds and with bill poised above the water.

Map 11: Great Bittern *Botaurus stellaris*

BREEDING BIOLOGY

Meinertzhagen heard them booming on Kushdil Khan lake on 19 June 1914 and thought they were breeding as next morning he flushed 'an old bird and a youngster just able to fly'. Christison on 8 June 1939 put up two young birds 'hardly able to fly', from Karak Lora near Quetta and asserted that they bred in the Malezai in 1938. This combined evidence is circumstantial, and as juvenile plumage is not very distinctive from that of the adult, actual breeding within Pakistan remains to be substantiated. Studies in Europe indicate that in a favourable swamp the male is polygamous, mating with more than one female. All the nest building, incubation and rearing of young is done by the female. The olive brown eggs take 25 days to incubate and 5 seems to be a typical clutch. The nest is always well concealed in dense reeds comprising an untidy heap of reeds in a rough circular platform at water level.

VOCALIZATIONS

Advertising calls by males are an important component in courtship and their deep resonant 'boom-boom' has given rise to its scientific name. Males will call early in the day and at dusk; a series of 3 or 4 booms followed by one to two second pauses. The origin and direction of the sound is always difficult to locate possibly because of the muffling effect of the calling bird's environment.

Little Bittern *Ixobrychus minutus* (Linnaeus)

Illus. **8**
Map **12**

DESCRIPTION

Body length 28-36 cms.
Wingspan 40-45 cms.

Unlike the Eurasian Bittern, this species is sexually dimorphic, the male being quite conspicuously patterned with biscuit coloured body and contrasting blue-black crown, mantle, tail and flight feathers. The wing shoulders are buffy cream, and the forecrown, ear coverts and upper throat are paler, almost white. There is some chestnut streaking on the flanks. The legs are olive yellow and the naked orbital skin green. In both sexes the under-tail coverts are blackish maroon. The female is more tawny buff all over but with a dark purplish brown crown and more conspicuous dark chestnut streaking on the throat. Its mantle and tail are dark chestnut contra the blue-black of the male. In plumage the female is indistinguishable in the field from the female of *I. sinensis*. These tiny bitterns have relatively short legs, long toes and relatively short thick necks fusing smoothly into the body.

HABITAT, DISTRIBUTION AND STATUS

An uncommon and local resident in Sind province, and recorded as a passage migrant in Baluchistan (Ticehurst, 1927) but it is also an occasional summer breeding visitor to that province, as Meinertzhagen collected a specimen at Kushdil Khan on 11 July, (*Ibis*, 1920), which could have been breeding and H. Waite collected a specimen on 5 June near Loralai. There are other Baluchistan records for April, May, August and September, and Christison recorded it as breeding in the Malezai, arriving in April and leaving in August (Christison, 1942). On 1 and 2 May 1984 the author encountered a roosting aggregation of about 15 birds in one particular reed bed in Zangi Nawar lake in south-western Baluchistan, some of whom may have stayed to breed. In July 1987 during a survey of Zangi Nawar conducted by Ashiq Ahmad of the Pakistan Forest Institute, and his students, a total of 5 nests of Little Bitterns were located. One on 4 July contained 4 eggs. Out of 3 found on 8 July, all within a few metres of each other, 2 were empty and the third contained 5 eggs. Another found on 9 July also contained 5 eggs. All were supported by dense *Typha* reeds and located 30 to 60 cms. above the water level (MS notes by M. Idrees Chughtai of Pakistan Forest Institute).

Illus. 8: (a) Little Bittern *Ixobrychus minutus*, adult male.
(b) Yellow or Chinese Little Bittern *Ixobrychus sinensis*, adult male.

Whitehead recorded it in Kohat (NWFP) in July and found a nest at Dandar in that month. It may be an erratic local migrant in the Punjab, as H. Waite collected a specimen on 5 June at Choa Saidan Shah (Jhelum district) where there is no suitable breeding habitat and another specimen on 28 December at Bhant in Mianwali district. Extra-limitally it is a common breeding bird in the vale of Kashmir. A specimen was shot by Scully in October in Gilgit. Its preferred habitat is tall reed beds (*Typha* and *Phragmites*) along streams (Malezai) in Baluchistan or on the margins of lakes, and seasonal swamps, especially those in which submerged tamarisks and *Populus euphratica* bushes grow up through the reeds in water of 0.5 to 1 metre depth. Observed twice, in the 1970s at Manchar in Dadu district, (author and D. Holmes) Sind and found breeding in east Narra swamps (Doig, 1878). There are Bombay Natural History Society specimens from Pithoro jheels (Sanghar district). Eates found it breeding at Karo Nai in Sukkur district, Chack Dhand near Umarkot in Tharparkar district, also at Taror Dhand near Rohri (MS notes unpublished). Status RARE.

HABITS

Will feed out in rather exposed situations in thin sedge cover (observed Manchar lake, December) and is not so secretive as *Botaurus stellaris*. Its hunting technique is to squat motionless clinging to vertical stems on the edge of a reed bed usually hunched forward with neck partially withdrawn. When approached (as at Manchar by boat) it typically adopts a cryptic stance (upright stretch and freeze). The longitudinal chestnut throat-streaks of both sexes blend effectively with the reeds. Even in this cryptic posture, it tries to rotate its body constantly to face the intruder. Its food consists of small fish, frogs and even spiders, water bugs and snails. In a study of the breeding population at Haigham marsh in Kashmir (India) regurgitations from 25 adults and 22 nestlings were analyzed and 60 per cent of their diet comprised small fish, 9 per cent amphibia, 13 per cent insects (including damsel flies, dragonfly larvae and water beetles (Dytiscidae). The predominance of fish differs from studies in central Europe where fish only formed 25 per cent of the diet (Witherby and Jourdain, 1943) (Holmes and Holmes, *Report of Oxford University Expedition to Kashmir,* 1983).

BREEDING BIOLOGY

The male advertises during the breeding season with a short rapidly repeated single croak syllabized as 'hogh' and given at the rate of 25 calls a minute (A. Voigt in Witherby *et al.,* 1943). They are apparently mono-

gamous in the breeding season, with the male doing most of the nest building, but both sexes sharing in incubation and feeding of the young. The Oxford University study at Haigham (Kashmir), reveals that nest relief was only after 5 or 6 hours during the day, both sexes taking an equal share of incubation duties. Eates found 2 nests with eggs, in July, but most freshly laid eggs from mid-August and latest date well-incubated eggs on 9 September (unpublished manuscript). Doig found a nest in the east Narra with 4 eggs on 26 May, whilst Whitehead found a nest with 7 eggs on 9 July, in Kohat district, North West Frontier Province (Whitehead, *JBNHS*, Vol 20, p. 977). In the Kashmir study, breeding was found to extend from late May to late July and the breeding density was 1 nest per 1,750 square metres, the closest nests being 16 metres apart (Holmes and Holmes, op. cit.). The nest is comparatively bulky for the size of the bird and always built in thick cover in growing reeds. A base is made at water level with flattened or bent over stems forming a platform which is then built upwards in a column which may reach 3 feet high. Nesting material is dried rushes or sometimes tamarisk twigs with quite a deep cup depression lined with green reed leaves (K. Eates, MS unpublished). Often the nest is in the fork of a flooded tamarisk. The normal clutch is 4 to 5 eggs in Sind and in the Kashmir study it was also 4 to 5 eggs (Holmes and Holmes, op. cit., 1983). The eggs are laid at 2-day intervals and are plain white, unglossy and the young at hatching are covered with bristly-looking orange or cinnamon chestnut down, with conspicuous blue eyelids.

From about 6 days of age they can leave the nest and clamber about, and by 8 days of age are agile

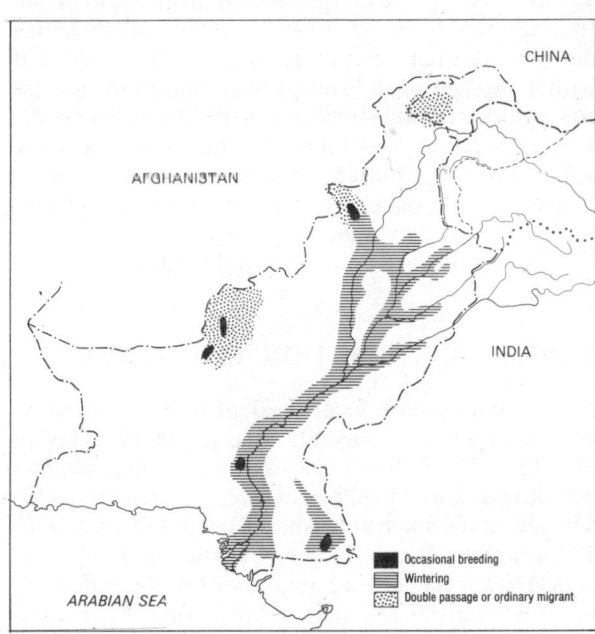

Map 12: Little Bittern *Ixobrychus minutus*

enough to be able to avoid human capture (Holmes and Holmes, op. cit.). They return to the nest to be fed when the parents call. Parents are induced to regurgitate by chicks seizing their bill sideways and then following food down to bill tip. When very small food is regurgitated onto the nest floor. (R.S. Bates, *JBNHS*, 1943). As incubation starts with first egg and hatching is asynchronous the weakest chick may starve but often the parent regurgitates food onto the nest floor, so that even if the most aggressive chick has first seized its parent's bill, another may succeed in picking up the regurgitated food (Bates and Lowther, 1952). Incubation takes 17 to 19 days (Bates and Lowther, 1952) and the chicks are fully fledged 25 to 30 days after hatching. In Sind one parent spends a large part of each day sheltering the young chicks from the sun with wings half opened and drooped forward and for the first 5 or 6 days after hatching, one parent always stays with the chicks on the nest.

VOCALIZATIONS

Bates and Lowther refer to a frog-like 'wuk' call (op. cit., 1952) often preceding flight to another patch of reeds. The advertising call of the male during the breeding season has already been described. The young when greeting their parents utter a rapid sibilant 'tu-tu-tu' (Witherby *et al.*, 1943). During the nesting relief ceremony the male has been heard uttering a low pitched 'wooff' (Holmes and Holmes, op. cit.).

Chinese or Yellow Bittern *Ixobrychus sinensis* (Gmelin) Map **13**
Synonym *Ardetta sinensis*

DESCRIPTION

Body length 30-39 cms.
Wingspan 42-47 cms.

It is slightly bigger than the Little Bittern and like the latter species it is sexually dimorphic, but females of the two species are difficult to separate even from museum skins. Adult males have a yellow-buff body with cinnamon chestnut hind neck, dark blue-grey crown, tail and flight feathers. They are brownish or olive buff on the mantle and scapulars. The legs, base of bill and orbital skin are olive yellow. The female is darker chestnut brown with indistinct streaking on the crown, hind neck and back with very conspicuous chestnut interrupted streaks (like stitching) down the side of the buffy yellow foreneck. In flight males can easily be separated from *I. minutus* by lacking the dark blue-black mantle.

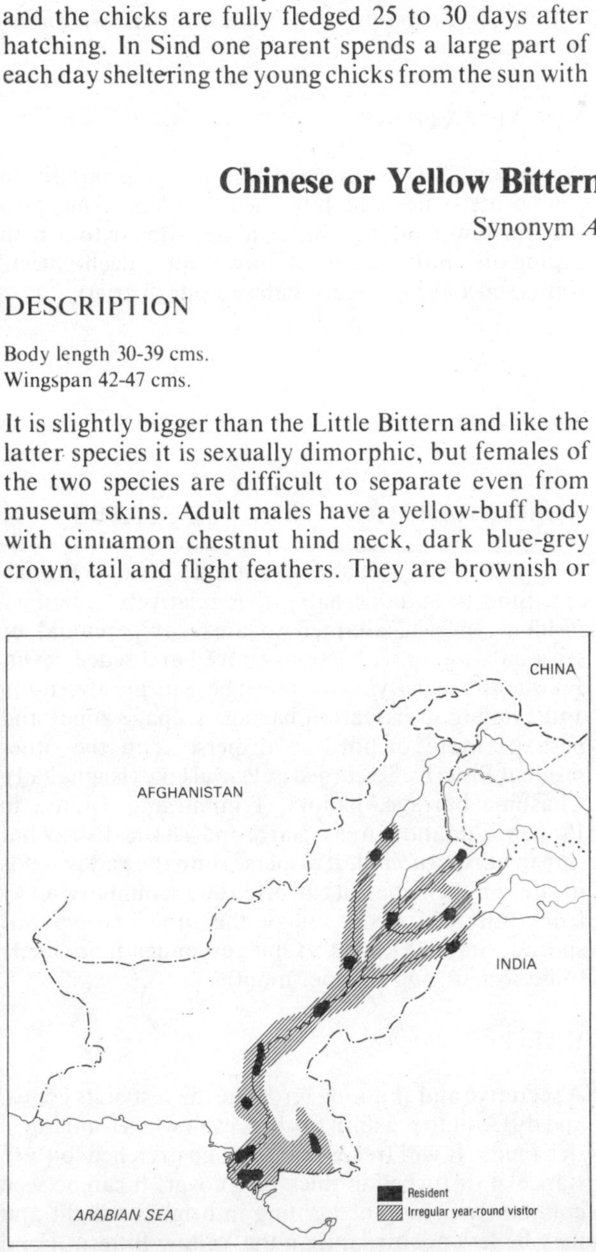

HABITAT, DISTRIBUTION AND STATUS

More widespread and commoner than *Ixobrychus minutus*, they are restricted to the same sort of habitat with dense reed cover and more or less perennial water. They have not been authentically recorded in Baluchistan but occur in the seepage zones around all the major headworks of the Punjab as well as through the swamps of Sind. They are often seen sympatrically with the Little (*I. minutus*) and Chestnut Bitterns (*I. cinnamomeus*). In Pakistan they have never been observed in mangroves or dispersing into the paddy fields as they are reported to do in Sri Lanka. Status FREQUENT.

HABITS

Similar to Little Bitterns, they are very secretive and can occasionally be seen crouching motionless on the edge of reed beds or on patches of open water, without moving for prolonged periods while waiting for suitable food prey. One bird watched while hunting repeatedly wagged its stubby tail from side to side. They are adept at clambering between vertical reeds, gripping one or

CHINA
AFGHANISTAN
INDIA
ARABIAN SEA
Resident
Irregular year-round visitor

Map 13: Yellow or Chinese Little Bittern *Ixobrychus sinensis*

more stems with their long toes. When excited, as at the approach of another Yellow Bittern, observed erecting their nuchal feathers in a shaggy crest. They may frequently be seen flying low over reed beds in the evening time and are mainly crepuscular in feeding activity. They feed on small fish, tadpoles, frogs, insects and aquatic crustacea but insects are believed to comprise a more important component of their diet than is the case with other small Bitterns.

BREEDING BIOLOGY

Doig (*Stray Feathers,* 1897*b*) found nests in the east Narra from míd-May to mid-August. Eates found nests similar to those of *I. minutus* but comparatively bulky structures, often 30 cms. across and located up to 1.5 metres above the water level (MS notes unpublished). They are made of dead reeds and sedges with quite a deep cup, and it was noted that they are always built in reeds, never in semi-submerged trees or bushes, a favourite site of the Little Bittern. The normal clutch is four eggs which are a delicate pale blue or skim-milk colour but clutches of five occasionally seen. The eggs

are more nearly spherical than those of *I. minutus*. Both sexes share in incubation and feeding of the young which is by regurgitation. Incubation period is believed to be about 22 days and the young hatch asynchronously (Hancock and Elliot, 1978) and after about ten days the chicks become very agile and frequently wander from the nest. The chicks are covered with pinkish red down. 5 were noted (author) in this state of development clambering through the reeds on 9 September at Pabooni Dhand, east Narra. Courtship is believed to involve displays with the tail being wagged sideways and the crest and scapular feathers being raised. They are more vocal than either Black Bitterns (*I. flavicollis*) or Chestnut Bitterns (*I. cinnamomeus*).

VOCALIZATIONS

A pair on 28 October were heard calling rapidly to each other, voice a high pitched 'kek-kek'. They also made a lower pitched softer noise, similar to a man urging on a horse, a sort of drawn-out 'chlech-chlech' (observed Lal Sohanran, Bahawalpur district).

Cinnamon or Chestnut Bittern *Ixobrychus cinnamomeus* (Gmelin)   Map **14**

DESCRIPTION

Body lenth 38-40 cms.
Wingspan 58-62 cms.

Smaller than the Paddy Bird, but easily distinguished from all other bitterns by its predominantly reddish chestnut or cinnamon plumage. In flight all the flight feathers and tail appear bright reddish chestnut with a darker more purplish caste to the rufous hind neck and mantle. The foreneck is pale creamy orange with two interrupted darker chestnut brown lines running down each side of the neck, like stitching, in the female and a single dark maroon streak down the centre of the neck in the male. In both sexes the nuchal feathers can be erected in a bushy crest, and are purple-maroon coloured in the male, greyish black in the female. The flank and lower breast are buffy orange in the male, with some dark brown streaking in the flanks and lower body in the female. There is a small white crescentic patch on the side of the neck and just below the ear coverts and in the pectoral region 4 or 5 pointed feathers, covering the wing shoulder when the bird is at rest, are prominently black with buff margins. Legs are bright olive yellow, bill green-yellow at base, darker brown to blackish at tip. Iris greenish gold varying to orange-yellow, the pupil contracting to a vertical slot (observed on a sunny day, Haleji lake). The naked facial skin is pinkish purple in the male.

HABITAT, DISTRIBUTION AND STATUS

Not occurring in Baluchistan but widespread in Punjab and Sind in suitable habitat. A relatively sedentary resident species, adapted to areas of perennial or seasonal swamp with extensive reed and sedge cover. Frequently seen flying over reed beds in late afternoon and evening in irrigation barrage seepage zones and perhaps more inclined to disperse than the other resident Bitterns. Recorded at Rawal lake (Islamabad), Chashma barrage, Balloki, Trimmu and Taunsa in Punjab. Common in east Narra and Thatta district but rare in northern Sind. It disperses into the paddy fields in late summer where it can often be encountered away from tall reed beds, unlike the other *Ixobrychus* species. Status FREQUENT but very much more likely to be seen during summer months.

HABITS

A secretive and skulking bird, like the rest of its genus, and difficult to see until flushed, even when foraging in rice fields. It will freeze in typical upstretched 'bittern stance' if disturbed in thick reed cover. It can be seen commonly fishing or hunting in bright sunlight and may be less nocturnal than the Yellow Bittern. Food includes flying insects Odonata, amphibians and small fish. Stomach analysis of birds from the Sunderbans,

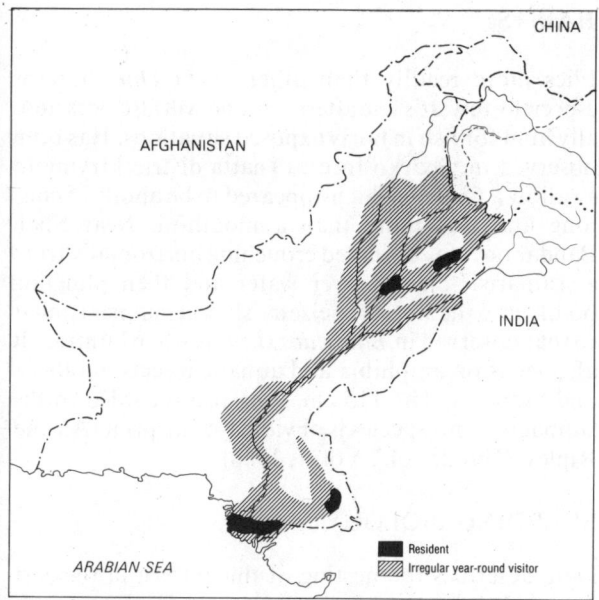

Map 14: Cinnamon or Chestnut Bittern *Ixobrychus cinnamomeus*

Bengal, indicated that amphibia comprised the bulk of its diet with insects second in importance (Mukherjee, 1971).

BREEDING BIOLOGY

During display the black and chestnut flank feathers are fluffed out conspicuously and the crest is erected. It is quite vocal in the breeding season. Will nest in thick reed beds or in submerged trees and bushes through which thickets of *Phragmites* or *Saccharum* reeds are growing. Breeding starts towards the end of the monsoon. Doig found a nest 3 August with 4 fresh eggs in the east Narra and Eates, over 20 years, only found 1 nest on 20 August with 4 downy young, in the Pithoro jheels in the Narra region. The nest, rather loosely constructed, is built up from a platform of bent-over reeds and the nest cup is generally 0.75 to 1 metre above the water surface (Eates, MS notes 1940s, and Doig, 1880). Both sexes are reported to share in nest building, incubation and feeding of the young. It is not a colonial nester. The eggs are chalky white, unmarked and the normal clutch seems to be 4. The young covered with pinkish chestnut down are very active like their congeners climbing about the reeds near their nest after about the tenth day. A fully fledged juvenile with traces of down on its head and nape was observed (author) as late as 2 November in Gharko forest, Thatta district.

VOCALIZATIONS

Three were observed in excited flight chase (a territorial dispute) on 6 October (Trimmu head, Jhang district) when one bird uttered a rapid repeated 'creck-creck-creck' reminiscent of the calls of *Ardeola grayii*. Another single note rapidly repeated 5 or 6 times sounds like 'kok-kok-kok'. Neelakantan describes the territorial call of the male as producing a hollow booming note rendered as 'gok-gok-gok'. This often results in another male advancing stealthily through the reeds to challenge the calling bird (Neelakantan, *JBNHS*, 1956)

Black Bittern or
Yellow-throated Bittern *Ixobrychus flavicollis* (Latham) (by Hancock and Elliot)

Dupetor flavicollis (Ripley's 'Synopsis')   Map **15**

Synonym *Ardetta flavicollis* Stuart Baker.

DESCRIPTION

Body length 58 cms.
Wingspan 110 cms.

Considerably bigger than the Pond Heron (*Ardeola grayii*) with a pointed heavy dagger-like bill. Males are a slaty blue-black on crown, hind neck, back, wings and tail. They are not jet black. Females are browner less slaty coloured on the back. In both sexes the front of the throat and upper breast is creamy yellow, streaked with interrupted lines of dark chestnut down each side of the neck. There are a few white edged feathers on the abdomen and the upper breast, merging into slaty colour with the feathers margined in buff.

The centre of the abdomen is whitish in both sexes. The bill is dark brown on the upper mandible, the base of the lower mandible being yellowish olive. The relatively short stout legs and feet are dark brown to greenish brown. In the breeding season the facial skin around the lores and base of the bill becomes reddish and males observed 16 October had purple orbital skin. A newly fledged juvenile observed 11 September had greyish brown plumage.

HABITAT, DISTRIBUTION AND STATUS

Similar to the rest of the genus, being confined to extensive thickets of *Phragmites, Typha* reeds, or in

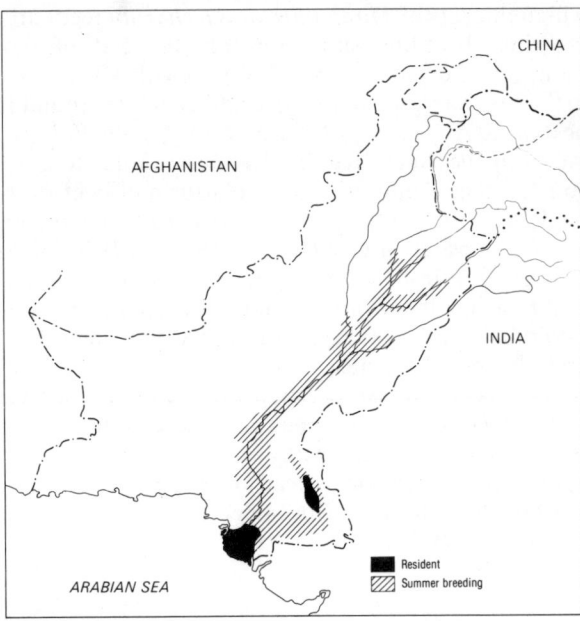

Map 15: Black Bittern *Ixobrychus flavicollis*

HABITS

Flies more readily than other *Ixobrychus* bitterns especially towards late afternoon and will also occasionally hunt for fish in fairly exposed situations. Has been observed in Gharko forest (Thatta district) trying to swallow a Catfish which appeared to be about 15 cms. long and was more than a mouthful. Near Shah Bandar one was observed crouching horizontally from a Tamarisk branch over water and then plunging bodily into the water to seize a fish, in a manner similar to that observed in *Butorides striatus* when hunting. It also feeds on amphibia and aquatic insects, molluscs and tadpoles. The 115 cm. long fish recorded in the stomach of this species is obviously a misprint (Ali and Ripley, 'Handbook', Vol. 1, 1968).

BREEDING BIOLOGY

Doig describes the nesting of this bittern in the east Narra. Nests were shallow cups made of tamarisk twigs and a few aquatic weeds about 1.75 metres from the water in dense thickets comprising tamarisk trees with reeds growing through them. Most nests were no more than 22 cms. across and 8 cms. in depth (Doig, *Stray Feathers*, 1879*b*).

All the nests he found had clutches of four eggs but Eates (MS notes) had a partially incubated clutch of 3 eggs brought to him on 20 August from the same region. This nest was well concealed in a dense reed bed. The eggs are pale greeny blue and usually laid after the start of the monsoon. Both parents are known to share incubation duties and to feed their chicks by regurgitation. The chicks wander from the nest when two weeks old, returning to be fed when called by their parents. Two young birds have been seen (author) perched in *Phragmites* reeds at Haleji lake still showing pinkish natal down on their heads and necks as late as 11 September.

seasonal inundations, submerged Tamarisk bushes with clumps of reeds and sedges growing through their branches. It is mainly a monsoon season visitor to lower Sind being almost unknown in Punjab and northern Sind. A few birds may however over-winter, as single individuals were sighted mid-December, Haleji lake in two different years (P. Conder, pers. comm. and F. Koning, pers. comm.), and on 16 January two were seen (author obs.) flighting over a reed bed at dusk. A few stray up to the Punjab as one was seen at Balloki headworks mid-August 1973 (Allan Vittery, pers. comm.). It is relatively common in the east Narra swamps (Jamrao head, Sanghar district) and in Thatta district in summer. It may well be on the increase as Ticehurst did not encounter it (op. cit., 1923). Lower Sind forms the westernmost extremity of its world distribution. Status COMMON in Sanghar and Thatta districts only.

VOCALIZATIONS

Reference to the literature is contradictory and its authentic calls do not appear to have been recorded.

Night Heron or
Black-crowned Night Heron *Nycticorax nycticorax* (Linnaeus)   Map **16**

DESCRIPTION

Body length 58-65 cms.
Wingspan 105-112 cms.

More than twice the size of the Chestnut bittern and slightly larger than the Pond Heron (*Ardeola grayii*), it is unlike the rest of the ARDEIDAE in having a comparatively large head, very short neck, and relatively short stout bill and large eye. The latter, an adaptation to its nocturnal feeding activity, is a beautiful crimson colour. Adults in breeding plumage are alike, having dark slaty purple crown, cheeks and nape and mantle, with white throat, lower cheeks and breast shading to pale creamy buff tones on the lower flanks. The wings and stubby tail are pale grey and on the nape are two long narrow white lanceolate plumes extending down the back. In winter the legs are olive green or yellow-olive. The bill is black as also the small area of facial skin around the lores. In breeding condition the legs turn salmon pink. The sexes are alike. The downy chicks have rather a greyish white bill and bare facial skin and are clothed in white hair-like down. Fully fledged nestlings have pale pinkish grey irides, and still retain a few traces of natal down on head and neck (observed Karachi Zoo). Their feet are yellow and the whole of body plumage olive brown with broad tear-drop white tips to each row of wing coverts, secondaries and narrower white streaks all over the mantle and upper breast. This spotted brown plumage of first year birds, is very striking and gradually changes to a more uniform unspotted olive brown in their second year.

HABITAT, DISTRIBUTION AND STATUS

Widespread throughout Pakistan including lowland valleys of the northern mountain areas and Baluchistan, though these populations are locally migratory in winter. They are largely sedentary in the plains. Small colonies have been observed in summer in a tall Chenar grove (*Platanus orientalis*) in Swat (Saidu Sharif) and a breeding colony of about 20 pairs in tall trees alongside the Swat river at Chakdarra. In Quetta (Baluchistan) there was a small colony on Lytton Road which were summer visitors only (Christison, 1942). This colony has since disappeared. Occasional birds have been seen in the Gilgit main valley (Biddulph, 1881). Colonies occur in the Salt Range (60 to 70 birds, Khabbaki lake, 3 year's observation), at Balloki headworks (Lahore district), Jinnah headworks (Mianwali district), Katalpur near the Ravi river (Khanewal district) and at Lal Sohanran (Bahawalpur district). In Sind there was a huge colony at Jacobabad town which was in existence at the turn of the century and was still occupied in 1982. Also at Drigh lake in Larkana district (over 700 birds), at Shikarpur (about 46 birds) and at Sanghar. A colony was observed in the mangroves of the Indus delta on Dubbi creek comprising about 80 birds and another colony breeds in the heart of Karachi city in the Zoological gardens. Status COMMON but local.

HABITS

As already indicated they are gregarious both in winter and in the nesting season. The whole colony or group will roost by day in the same favoured spot such as a canal bank tree plantation (Lal Sohanran) or lakeside tree groves (Rawal lake), or a dense reed bed (Haleji lake). At dusk they become active, flighting out to feed against the darkening sky, calling to each other with

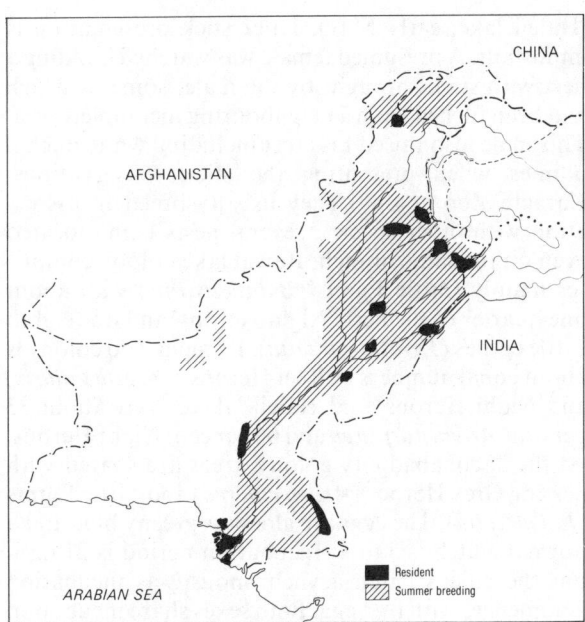

Map 16: Black-crowned Night Heron or Night Heron
Nycticorax nycticorax

their characteristic loud single 'kwok' call. They appear to hunt solitarily, or widely dispersed, despite their gregarious roosting and nesting habits, and have been seen stalking slowly in shallow pools and also standing motionless. When young are in the nest these mainly nocturnal birds also hunt by day. Food comprises mainly small fish, amphibians, especially tadpoles, larvae of dragonflies, water beetles and the occasional water snake or lizard. They will steal small nestlings whenever they are left unguarded. In a study of food habits based on analysis of stomach contents of 78 adult birds collected in the Sunderbans (Bengal, India), amphibia predominated (27.92 per cent) in the diet with mollusca second in importance (16.16 per cent), Crustacea (mostly crabs) 12.17 per cent and fishes 10.36 per cent of the diet (Mukherjee, 1971). A large traditional roost in *Phragmites* reeds at Haleji, comprising about 300 birds was eventually abandoned due to House Crows (*Corvus splendens*) numbering more than 1,500, selecting the same site for a communal roost.

BREEDING BIOLOGY

Normally nesting in tree tops, their flimsy stick nests are located quite close together, but the Haleji lake colony is located on an island on the ground in *Phragmites* reeds and in low *Salvadora* bushes. Courtship behaviour has been intensively studied in other countries. During early stages of territory establishment males were observed (author) bill rattling, a short rapid staccato noise, repeated at infrequent irregular intervals, as they stood on their territories

(Haleji lake, early May). Later stick presentation is important. A presumed female was watched building a nest with sticks brought by the male, some of which had been filched from a neighbouring incomplete nest. This male also raised his crest including white nuchal plumes when presenting the stick (observations, Karachi Zoo colony).They usually breed in association with other species, several nests being located even on the same branch. Rawal lake colony comprises mainly Cattle Egrets (*Bubulcus ibis*) with about one quarter of the nests Night Herons' and one eighth Little egret's (*Egretta garzetta*). Karachi Zoo colony is about equal numbers of Reef Herons (*Egretta gularis*) and Night Herons, and Haleji lake colony about 33 per cent *Ardea purpurea* and 66 per cent Night Herons. At the Jacobabad city colony, trees are shared with nesting Grey Herons (*Ardea cinerea*) and Little Egrets (*E. Garzetta*). The eggs are glossless greeny blue and a normal clutch is 3 to 4. Incubation period is 21 days but the chicks hatch asynchronously as incubation commences with first egg. Both sexes share incubation and feeding duties. Downy white young are fed by food regurgitated straight into their bills, but when older, food is disgorged onto the nest floor. Parents have been noted bringing whole fish in bill to young when they are older (Haleji lake, 10 May.) Even in tall trees the young wander around onto adjoining branches and in any case the nest cannot accommodate all the chicks as they grow bigger (observed Karachi Zoo). The nesting season is variable. Colonies at Haleji start breeding mid to late April, always about two weeks after Purple Herons have started nesting at the same site (observations 4 seasons). Colonies in Karachi Zoo start in March and fully fledged young have been seen out of the nest 16 April. In Jacobabad and Lal Sohanran colonies start breeding in mid to late July after the break of the monsoon (still carrying sticks 25 July.) and the Rawal lake colony starts nest building early June, (2 years' observations). After the young hatch the parents' legs change from salmon pink to orange and have re-assumed their olive green colour by the end of the breeding season.

*VOCALIZATIONS

The young at the nest keep up a constant rapid clicking noise when soliciting food. Adults utter a single loud 'quok' call in flight when they leave their non-breeding day time roost and repeat this as they fly overhead in the dark. During aggressive encounters at the nest colony they emit a loud double 'quock-quuwaark'.

Little Green Heron, Striated Heron or
Green Backed Heron *Butorides striatus* (Linnaeus)

Illus. **9**
Map **17**

DESCRIPTION

Wingspan 52-60 cms.
Body length 40-48 cms.

Smaller than the Pond Heron (*Ardeola grayii*), it has the same short neck and squat shape of the latter species, but its bill is comparatively longer and more slender than that of the Night Heron, *Nycticorax* sp. Adults are alike, with greenish black crown and short pointed nuchal crest. Lower cheeks, throat and belly are grey, sometimes with a distinctly lilac or mauve grey tinge, and the mantle is dark olive brown. The wing coverts are dark slaty green, each pointed feather margined in creamy buff and this striking pattern is the best field mark. The iris is yellow and bill black with the lower mandible turning to yellow towards the base. The toes are long and legs are stout, being olive yellow in colour but in winter some individuals have greyish green legs. The stumpy tail is slaty green-black and the flight feathers are dark grey. First year juvenile birds are more olive brown with dark streaked buffy white throat and breast and white spots on the wing coverts reminiscent of juvenile Night Herons. A bird seen on 29 July in breeding plumage had the loral naked skin area bright chrome yellow and very conspicuous and the legs were pinkish or orange flesh coloured. Its crown in sunlight was almost cobalt blue and the scapular feathers were long and lanceolate and silvery blue or lilac grey partly concealing the black and buff scaly wing coverts.

HABITAT, DISRIBUTION AND STATUS

Not an inhabitant of dense reed beds like the *Ixobrychus* genus, its particular requirements are densely vegetated channels or perennial pools, such as the mangrove creeks. It is widely but very sparsely distributed in Sind and Punjab. Single birds seen 3 times in different years at Lal Sohanran (Bahawalpur district). Occasional birds have been seen on the Chenab river creeks at Marala (E. Fernando, pers. comm.). Whistler (Ibis, 1922) only saw 1 in 3 years in Jhang district. It was only encountered twice in four winters of surveying all of Pakistan's wetlands (1963-73). At Chashma barrage on the Indus (Mianwali district) and at Ghauspur in Jacobabad district (F. Koning, pers. comm.). Derek Holmes found it relatively plentiful in the riverain forest above Sukkur (Holmes and Wright, 1968-9) and it is also not uncommon in the mangrove creeks of Sonmiani and the Indus delta. In a whole day's voyage from Rehri village to Khudi creek in the delta 12 individuals were sighted. It has not been recorded in the other parts of Baluchistan or apparently in the North West Frontier. Status FREQUENT in mangroves and riverain forest of northern Sind, elsewhere SCARCE.

HABITS

It will hunt out in the open and is not as cover-haunting as the Chestnut Bittern (*Ixobrychus cinnamomeus*), and Little Yellow Bittern (*I. sinensis*). It likes to hunt by very cautiously stalking along, head withdrawn right back into shoulders, along the water's edge and by suddenly darting its bill into the water, also by perching on a branch overhanging water and crouching downwards at 45° angle, motionless over the water. Observations of birds in the Indus delta indicate that the Mud Skipper (*Periopthalmus*) is an important food item and they have been seen plunging right into the water from such a bankside mangrove branch. They normally hunt solitarily, but they have been watched feeding within sight of each other (Ghizri creek in mangroves). In the mangroves, they also feed on crabs and shrimps, and in the Indus riverain forest of the interior, Odonata, frogs, water-beetles and small lizards, must form a more important part of their diet. Analysis of their food habits in the Sundarbans (Bengal, India) based upon stomach contents of 26 adult specimens showed that some terrestial insects are taken, such as grasshoppers and mantids. The major portion of their diet (31.8 per cent), was crustaceans followed by small fish (29 per cent) and insects forming 14.5 per cent with amphibia, mainly tadpoles, comprising 13.8 per cent of their diet and annelid worms only 3.62 per cent (Mukherjee, *JBNHS*, Vol. 68, 1971). I have seen one hunting out on exposed rocky reefs near Bombay and in the Indus on a man-made stone breastwork. They sometimes wag their tails.

Illus. 9: Little or Striated or Green-backed Heron *Butorides striatus*, adult both sexes.

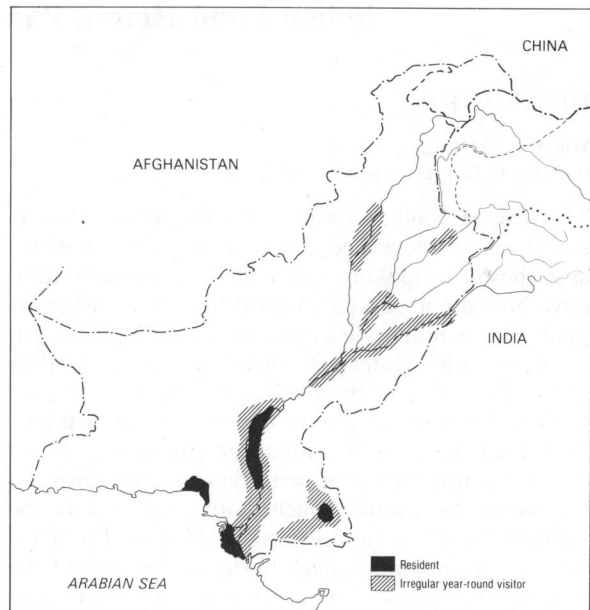

Map 17: Little Green, Striated or Green-backed Heron *Butorides striatus*

BREEDING BIOLOGY

They build rather flimsy nests of sticks which are more or less flat platforms. In the mangroves these nests are quite low down in the tree (in contrast to the topmost branches chosen by Reef Herons (*Egretta gularis*). In the riverain forest they favour a temporarily sub-merged thorn tree such as *Prosopis spicigera* and nesting normally starts after the onset of the monsoon. Eates found 3 nests in Sukkur riverain forest within a radius of 100 metres and took eggs from northern Sind in June, July and August. He also found nests in China creek, Indus delta with fresh eggs on 16, 20, 21, and 24 July in 4 different years. The normal clutch is 3 eggs which are greeny blue (Eates MS unpublished, 1940). The courtship of this rather secretive bird consists of erecting the nuchal crest, and stick presentation. The female is believed to do most of the nest building with material brought by the male. The incubation is shared by both sexes and takes 21 to 25 days (Hancock and Elliot, 1978) and typically hatching is asynchronous as incubation starts when the first egg is laid. The chicks are open eyed and sparsely covered with white down upon hatching. As is typical of the family the chicks are fed by regurgitation and soon start to clamber out of the nest.

VOCALIZATIONS

Relatively silent but has a variety of calls. When disturbed utters a double 'kyek-kyek' alarm call, not very loud. Young utter a 'tic-tic-tic' call when parent approaches the nest.

Indian Pond Heron, Paddy Bird *Ardeola grayii* (Sykes)   Map **18**

DESCRIPTION

Wingspan 75-90 cms.
Body length 42-45 cms. including bill 25 cms.

The sexes are alike in plumage even in the breeding season. A short necked squat heron, with relatively large powerful looking bill, and of a generally drab olive brown colour until flushed into flight, when its pure white tail and wings present a startling contrast. In non-breeding plumage the head and neck is streaked with olive buff and the back and wing coverts have paler yellow shaft stripes on the olive brown feathers. The throat is white becoming greyish brown on the breast and with darker brown streaking, prominent on the flanks. In normal hunched foraging posture the white abdomen, tail and wings do not show. The bill is olive yelow, becoming dark horny at the tip and the legs are olive green. The moult into nuptial plumage, which in a few individuals may commence as early as late April, produces a dramatic transformation. The whole of the head and neck becomes uniformly rich ochre yellow and unstreaked, with three not very conspicuous long narrow white nuchal plumes. The whole of the back and wing coverts are covered with purple-maroon, long filamentous plumes. The bill is jet black tipped, olive green in mid-section and dull blue basally and around the lores. The iris is bright golden yellow and immediate orbital skin also remains olive yellow. The downy chicks are whitish with buff orange down on the crown and buffy breast.

HABITAT, DISTRIBUTION AND STATUS

Very widespread and eclectic in its choice of habitat. It is a summer breeding visitor to the broader valleys of the North West Frontier Province including Swat but avoids mountainous areas and is absent from Baluchistan except for the Sibi plain and Makran coast. Throughout Punjab and Sind almost every village tank and canal or roadside borrow pit has its daily visit from a Pond Heron. They are numerous throughout the mangrove creeks and tidal creeks and can also be seen stalking over Lotus lily pads on the larger lakes, and in the main rice growing tracts they spread into all the flooded paddies. Status RESIDENT and ABUNDANT.

HABITS

Equally active, feeding by day and at night, they are solitary in hunting except where drying out pools offer a particularly rich harvest. They will often roost communally in winter, flighting from considerable distances to a favoured site such as reed beds near a

large lake, (Haleji), a grove of *Acacia arabica* at Lal Sohanran, or the *Dalbergia sisoo* trees in Ayub park, Rawalpindi. They also nest both solitarily, and occasionally colonially in mixed heronries. Their diet is very variable comprising mostly amphibia and small fish. The smallest dirtiest roadside pools will attract them. In the Indus delta they catch small Ghost Crabs (*Ocypodii*) and Fiddler Crabs (*Uca* sp.) also Mudskippers. Insects include crickets, ACRIDIDAE, Ground Hoppers (*Chrotogonus* sp.) and ants and no doubt small molluscs, worms and crustacea when available. Analysis of stomach contents of 105 adult birds from the Sunderbans, Bengal revealed a considerable quantity of unidentifiable vegetable matter 22 per cent of the total, with crustaceans 23.5 per cent being only slightly higher in importance, insects comprising 75 per cent, Arachnida only 1 per cent and Amphibia, mostly tadpoles 16 per cent. Small fish comprised only 9 per cent of the total bulk (Mukherjee, *JBNHS*, Vol. 68, 1971).

BREEDING BIOLOGY

The nesting season is very extended with occasional birds starting to breed in April and the bulk of the population breeding during the monsoon, with individual birds seen in full nuptial plumage up to the beginning of September. Courtship commences with flight chasing and aggressive displays in which the mantle and scapular plumes, and especially the crown

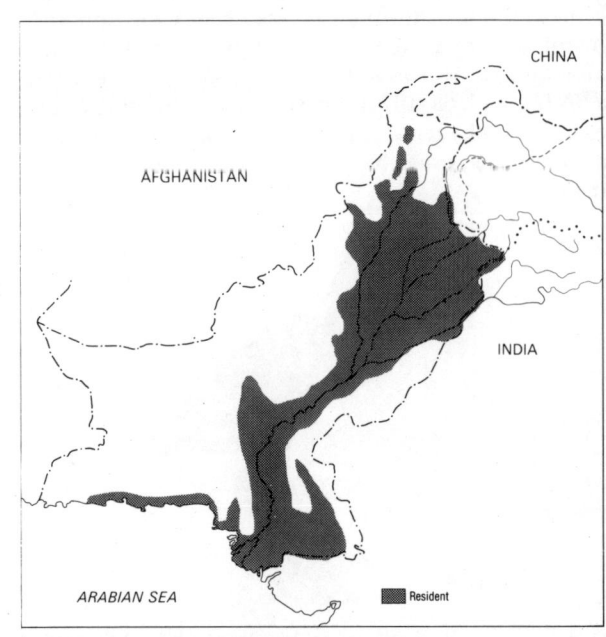

Map 18: Indian Pond Heron *Ardeola grayii*

and occiput are erected, also the breast plumes. A bird was observed on 5 May in a bamboo clump, pumping its head up and down with crest and back plumes erected. Later the same bird was noted lifting each foot alternatively in a sort of dance as it perched on a horizontal bamboo (Karachi Zoological garden). The pair bond is monogamous and the female builds the nest with material brought by the male, it being a rather substantial untidy shallow saucer of twigs and stems, often with the eggs, or chicks, visible from the ground below. The normal clutch is 4 to 6 eggs and they are pale blue. Incubation period is about 24 days, both parents sharing in this, and subsequent feeding of the young. They are tree or bush nesters and even when nesting in a mixed heronry usually select trees on the edge of the colony.

In the mangroves a nesting colony in Ghizri creek involves 6 to 7 pairs on the outskirts of a much larger Reef Heron (*Egretta gularis*) colony, and nests had eggs in mid-May. In Bahawalpur city a colony of over 30 pairs was located at the top of tall Eucalypt and Jaman trees (*Eugenia cumini*), in association with 5 to 6 pairs of *Egretta garzetta*. Nest building started in mid-April and fledged young were leaving the nest in the first week of June (D. Sterling Wyllie, pers. comm.). In Sonmiani lagoon in mangroves, a large colony was with eggs early August in association with Reef Herons (*E. gularis*), Large White Egrets (*E. alba*), Intermediate Egrets (*E. intermedia*) and Shags (*P. fuscicollis*). At Chakdarra on the banks of the Swat river, 6 pairs were noted nesting amongst a large colony of Night Herons. A solitary pair was noted in Thatta district, nesting in the same canal bank *Acacia arabica* tree, in 2 consecutive years with downy chicks in the second week of August.

*VOCALIZATIONS

Outside of the breeding season, they call frequently when disturbed or when approaching communal roosts. A rapid but spaced 'kwok-kok-kok', and also they make short croaking sounds and a more staccato and rapid 'kek-kek-kek'.

Cattle Egret (Buff-backed Heron) *Bubulcus ibis* (Linnaeus)   Map **19**

Asian sub-species: *Bubulcus ibis coromandus* (Boddaert)

DESCRIPTION

Body length 45-53 cms
Wing span 90-96 cms
Wing length 24-26 cms.
Tail length 8.3-9.6 cms.
Bill length 50-66 mm.
Weight 325-378 gms.

Similar in size and appearance to the Lesser Egret (*Egretta garzetta*) but always distinguishable from the latter by its all-yellow bill and visible gular protrusion, combined with slightly shorter neck and larger head. The sexes are alike, though males average slightly larger. It lacks the long narrow nuchal plumes of *garzetta* in the breeding season, but has instead a rather shaggy crest which becomes tawny orange in the breeding season. Its legs are slightly stouter and inclined to be olive greenish on the tibia when compared with the all-black legs of *garzetta*, also its feet are not yellow, and its bill is relatively stouter than *garzetta*. The breast plumes and mantle and scapulars develop long filamentous plumes of a golden buff, less orange than the crest and at the onset of the breeding season, the base of both mandible and bare facial skin becomes livid purple or cyclamen, with the mid-bill scarlet and only the tip yellow. By the beginning of October all traces of this nuptial plumage are lost, and the bare facial skin becomes greenish yellow. The newly hatched chick appears olive green about the face with white down on the rest of its body and the bill becomes black when the chicks are about one week old (observations Karachi Zoological gardens). Leg colour in adults can be variable with some olive yellow on the toes.

HABITAT, DISTRIBUTION AND STATUS

This is a species which has spread rapidly in Pakistan with the development of irrigation. In the 1920s it was unknown in the Rawalpindi district and rarely seen in Jhang district, (Whistler, 1922, 1930), and Ticehurst (1923) considered it only common in the better watered tracts of northern Sind. There is now a regular breeding colony on the shores of Rawal lake (Islamabad). It is a largely sedentary species, not penetrating much into the northern foothills areas but quite abundant in the major rice growing tracts such as around Sukkur, Larkana and Hyderabad. Around Jacobabad and Shikarpur it is extremely abundant and over 600 birds were counted during a 48 kilometre (30 mile) drive from Sukkur to Shikarpur. It is well adapted to grass-land areas in the riverain 'kutchas' or cultivated areas and is absent from tidal creeks or large bodies of water.

Status COMMON throughout Sind and Punjab. Absent from Baluchistan and hilly areas of the North West Frontier Province

HABITS

A much more insectivorous bird than the other ARDEIDAE, and hence the only species well adapted to foraging on dry land, grass verges of canal banks, the margins of drying out lakes etc. Their habit of associating with grazing domestic stock has often been noted and they will ride on the backs of buffaloes, so persistently that I have seen them fly and land on swimming buffaloes in a canal where no food could be expected from such an association. They are gregarious in foraging, roosting and breeding. A favoured grove of trees may be used as a traditional winter roosting site for many successive years if the birds are not disturbed. One such roost at Malir was used by some 50 birds which split up during the day into 3 to 4 feeding parties. They invariably returned to the roost about one-half hour before sunset (based on 7 year observations).

In the irrigated tracts they have adapted to 2 special hunting techniques. Flocks will congregate in newly irrigated fields. As the dry land is flooded a host of insects, worms and Crustacea flooded out of their burrows are eagerly seized by the flock. When fields are irrigated in Multan district, Ground Hoppers (*Chrotogonus trachypterous*) form the main item of their diet. They have also been observed foraging in a weed choked channel when they systematically dab-

bled with one foot extended to flush insect prey (Gharko forest, Thatta district). They are often quite vocal whilst feeding in such flocks. Stomach analysis of 318 birds collected from west Bengal in the Sunderbans, revealed a mainly insect diet (69.84 per cent of the total bulk) in which orthopterous adults formed the major part. Spiders, and ticks formed 11.11 per cent of their diet, Oligochaeta 11.11 per cent and Amphibia 7.94 per cent, and no fish were noted at all. It is undoubtedly a highly beneficial bird to agriculturalists (Mukerjee, *JBNHS,* Vol. 68, No. 3, 1971).

BREEDING BIOLOGY

Colonial in breeding, usually in trees but in Karachi Zoological gardens a colony traditionally nests in bamboo clumps. The pair bond is established at the nesting colony. Males display by constantly raising their crest and scapular plumes and by holding and seeming to 'play with' twigs. The male initiates courtship by defending a site and sometimes in the early stages sways its body slowly, lifting up each foot alternately but remaining on the same perch, (observed Karachi Zoological gardens). Aggressive encounters between nesting pairs are frequent, birds lowering their neck, ruffling out the breast and neck plumes and lunging forward with their bills. The breeding season is variable, some colonies start nesting activity in early April (Karachi) and others in late June (Rawal lake) but the majority nest after the onset of the monsoon (observed Shah Yaki, Thatta district and Jacobabad city). Normal clutch is 4 eggs of paler blue colour compared with other heron species. Both sexes share in incubation, which takes 22 to 26 days and the first egg laid hatches 5 to 6 days before the last. The young peck at, or try to grab their parent's bill to induce regurgitation and when they grow bigger they beg even more intensively, bobbing their heads up and down and waving their stumpy wings (observed Karachi Zoo colony). They often nest in association with other species. A colony near Ghulamullah in Thatta district comprised about 40 pairs alongside 20 pairs of Little Egrets (*Egretta garzetta*) located in *Acacia arabica* trees lining a canal, with young fully fledged and almost ready to fly on 8 July. The Jacobabad colony included *Ardea cinerea* and *Egretta garzetta*, and the Shah Yaki colony including *Phalacrocorax niger* and a few *Egretta garzetta*. The Rawal lake colony in one year, comprised only about 20 pairs and in addition about 15 pairs of Night Herons and 4 pairs of Little Egrets, and whilst territory defence and courtship was observed from mid-June, nest building was still noted on 15 July. The trees used for nesting at Rawal lake being only about 10 metres high contained about 9 nests per tree, and they were very flimsy structures with the eggs visible through the twigs from below. The young solicit food by keeping up a constant 'zit-

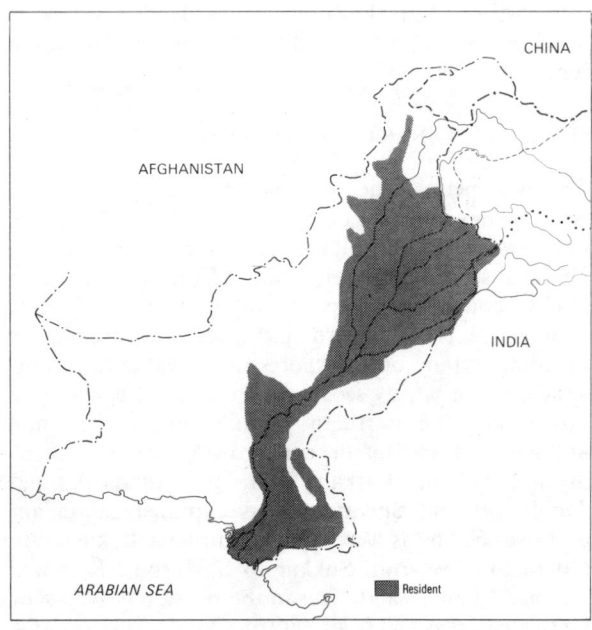

Map 19: Cattle Egret or Buff-backed Heron *Bubulcus ibis*

zit-zit' begging call, more sibilant than the begging call of the Reef Heron (*Egretta gularis*). Individual birds were still noted with bright purple-pink facial skin even when their chicks were about 14 days old, also indulging in displays with foot prancing and stick manipulation in their bills (observed 5 May in Karachi nesting colony).

*VOCALIZATIONS

Less vocal than the Paddy Bird but occasional individuals will give a low croak when at their communal roost. At the nest colony also relatively silent, occasionally uttering a rather harsh double noted call in which the first syllable is louder and of a croaking sound.

Western Reef Heron or Indian Reef Heron *Egretta gularis* (Bosc)

Synonym *Demigretta asha* (Sykes)

Illus. **10**
Map **20**

TAXONOMY

The Western Reef Heron (*Egretta gularis*) is a polymorphic species with a portion of the population (estimated at about 15 per cent around Sind coastal waters) having an all-white plumage. These white birds are almost identical in appearance to the Little Egret (*Egretta garzetta*) and show an overlap in both bill and tarsus length with that species. Some authorities consider that they should be considered as members of a super species complex (Payne and Risley, 1976), or even that they should be considered as one species (Naik and Parasharya, *JBNHS,* Vol. 81, p. 639, and *Sandgrouse* in press, and J. Hancock *in litt.* to author 1987). The basis for this contention is the observation of white and slaty forms of Reef Herons interbreeding in a colony at Gogha in Bhavnagar (Kutch, India) as well as apparent interbreeding in certain east African colonies. Also the fact that some of these white birds have largely black bills. The author's limited observations in Pakistan suggest that even in coastal regions the Little Egret (*Egretta garzetta*) is almost entirely a fresh water feeding species, distinguishable by having a relatively more

slender bill than *E. gularis* and without the yellow of the feet extending up to the bottom of the tarsus, as has been noted in all-white forms of *E. gularis*. Also the majority of white phase *E. gularis* have olive green or yellow bills. Furthermore, Reef Herons beside spending 90 per cent of their time foraging in salt water or the tidal zone, have been observed breeding separately but within sight of *E. garzetta* in a traditional nesting colony at Karachi Zoo. Until more research can be completed on this problem I prefer to treat them as separate species.

DESCRIPTION

Body length 55-56 cms.
Wingspan 86-104 cms.
Wing length 24.4-28.5 cms. (Both sexes)
Bill length 79 -89 mm. (Both sexes).
Weight 400 gms. (1 male)

A longer necked and longer legged heron than the ARDEIDAE family members hitherto described. It is roughly similar in proportions to the Lesser Egret (*Egretta garzetta*), and is particularly noteworthy in being the only species of the family in Pakistan almost wholly adapted to a salt water marine habitat, and also in exhibiting at least four quite distinct colour phases, i.e., polymorphism (see discussion, introductory chapter, on The Problem of Species). It is usually rather blue-grey all over with white throat and often a white carpal patch on the wings, highly visible in flight. The bill is slightly shorter and heavier or thicker than that of *E. garzetta,* with which it could be confused in the field, because a small proportion of the population is entirely white in plumage. The legs are dark greenish grey, often looking quite black with the feet olive yellow extending up to the bottom of the tarsus. Some wholly white birds have all the tarsi olive yellow. The bill is greenish yellow to yellow and darker horny distally. The pure white form has a yellow bill which is less slender than the black bill of *E. garzetta,* and the legs are always yellow up to the bottom of the tarsi and

Illus. 10: Western Reef (Heron) Egret *Egretta gularis* commonest ashy-grey morph. Adult in winter.

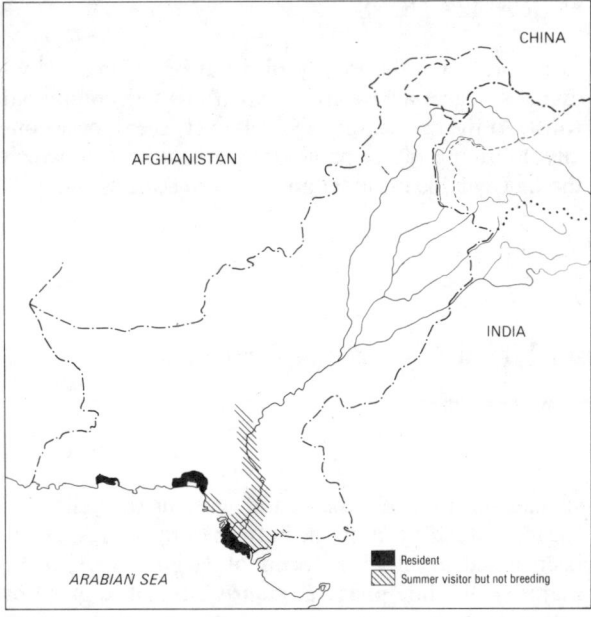

CHINA

AFGHANISTAN

INDIA

ARABIAN SEA

■ Resident
▨ Summer visitor but not breeding

Map 20: Western Reef (Heron) Egret *Egretta gularis*

look more greyish. In all colour forms the iris is greyish white and in the breeding season they develop purplish pink naked skin in the loral area and the base of the bill. Also the soles of the feet become suffused orange rather than dull olive green (observations of breeding colony Ghizri creek). In breeding plumage they develop two long slender nuchal plumes and also long lanceolate plumes on the lower breast and scapulars,some of them partly filamentous but not truly 'aigrettes', similar to those of *E. garzetta*. With respect to colour morphs there are four distinct types in Karachi waters. During a two-day voyage through the Indus delta creeks when over 1,000 individuals were examined the proportions were roughly: 70 per cent slaty grey with yellow feet and dark tarsi and sometimes a rather insignificant white carpal wing patch only visible in flight, 15 per cent white with largely yellow bills and yellow feet extending up to the base of tarsi, 5 per cent very pale ashy grey with lanceolate breast plumes grey but lower breast and vent white and also gular region white, with most of the lower tarsi showing yellow. About 10 per cent very dark purplish slaty with a prominent white carpal wing patch. An all-white adult seen in October had scattered blue-grey feathers over mantle and wing coverts.

HABITAT, DISTRIBUTION AND STATUS

A neritic species confined to seacoasts and tidal creeks along the Sind and Makran coastline with a few birds straggling inland to fresh water swamps during the monsoon season, some reaching as far as Manchar

lake in Dadu district and swamps around Hyderabad, i.e., up to about 240 kms. (150 miles) from the coast. Very abundant in mangrove creeks, up to 1,000 counted during a two-day voyage in the creeks during November and December when not breeding. Most commonly they can be seen hunting and fishing on mud flats, and lesser numbers visit tidal rock pools and rocky shorelines, but they rarely occur on purely sandy beaches. Status ABUNDANT Indus delta, COMMON Baluchistan and Karachi seacoasts, absent elsewhere.

HABITS

Colonial in breeding and usually in nightly roosts, but territorial (solitary) in feeding. Mostly diurnal in activity. Occasionally in August when rice fields are flooded Reef Herons forage in the paddy fields for frogs and fresh water fishes and can be seen in Thatta district up to 80 kilometres from the coast. In the mangroves *Uca* sp. Fiddler Crabs and *Periopthalmus*, Mud-skippers and small *Ocypoda* Ghost crabs are important food items. On rocky substrates they have been seen feeding on *Grapsidae* crabs. They hunt by slow stalking and sudden forward jabs or lunges when feeding on mud banks. Another hunting technique commonly observed is standing in shallow water at the edge of a creek, with neck withdrawn and wings half raised at the shoulder but still partially closed, as though forming a shady canopy in front of their bodies. Occasionally they stalk forward in water up to their bellies with wing shoulders hunched forward and body held in a horizontal position and neck withdrawn.

BREEDING BIOLOGY

Extensive personal observations, indicate that a few birds frequent nesting colony sites most of the year but larger numbers used them as nightly roosts 5 to 6 weeks before nesting commences. Aggressive encounters increase in intensity, with many birds erecting their scapular and breast plumes as well as their crest feathers and in high excitement erecting their two long nuchal crest feathers vertically. After pair formation much mutual preening noted particularly of each other's mantle and back feathers. Copulation frequently observable on the nest at early stages before and after first egg is laid. Male seen pulling twigs off the nest tree and presenting to brooding female, who weaved the new stick into the nest structure. The normal clutch is 4 eggs but nests with only 3 newly hatched chicks observed in about one-quarter of the colony. The eggs are rather longish ovals, of a dark greeny blue and chalky but not glossy. Newly hatched chicks have pink to carmine bills with black tips and are covered with spiky white down. In small mangrove trees usually built on topmost branches and not more than 3 nests per tree. Nests are fairly substantial stick platforms. Both sexes share incubation duties which last 23 to 26 days and this starts with the first egg laid. Heavy predation occurs in mangrove nesting colonies from local fishermen taking eggs, and House Crows (*Corvus splendens*). Young walk around on adjoining branches after about two weeks old and can fly at about 45 days after hatching. Mangrove nesting colonies are shared often with Pond Herons (*Ardeola grayii*), and Great White Egrets (*Egretta alba*) and in the more remote mangrove creeks with White Ibises (*Threskiornis melanocephalus*), Grey Herons (*Ardea cinerea*), Spoonbills (*Platalea leucorodia*) and a few Intermediate Egrets (*Egretta intermedia*) (K. Eates, MS unpublished). Inland colonies, such as in Karachi Zoological gardens, are located in tall trees (such as *Melia azedarachta* and *Mangifera indica*), in scattered groups with equal numbers of Night Herons (*Nycticorax nycticorax*) and a few Pond Herons (*Ardeola grayii*). A small colony of *Egretta garzetta* nests separately in bamboo clumps at Karachi Zoo, but in constant visual range of the nesting Reef Herons. The nesting season extends from early spring until the onset of the monsoon season. Breeding sometimes starts late March (Karachi Zoological gardens), about one to two weeks later than the Night Herons, but with fully fledged chicks observed 5 May. This colony was in occupancy by these two species in 1917 (Ticehurst) and was still used in 1986. Ghizri creek mangrove colony birds had just started laying 6 April (20 nests examined) but by 11 May most nests had downy chicks. In the Ghizri colony a pair of Brahminy Kites (*Haliastur indus*) regularly nests (and appears to be quite unmolested) in the periphery of the colony. 6 pairs of all-white Reef Herons have been observed nesting in the Ghizri colony and 16 pairs of normal Grey Egrets in one year, but in another year in Karachi Zoological gardens a white phase bird was paired with a normal grey bird and the three downy white chicks looked identical to other nearby nestlings. Regrettably it was not possible to examine this nest after the chicks fledged. A colony in Sonmiani mangroves in two years commenced nesting in mid to late July and in company with Great White Egrets (*Egretta alba*) and Shags (*Phalacrocorax Fuscicollis*). Breeding colonies studied in Bhavnagar and Gujarat, India, revealed two instances of all-white and normal grey plumaged birds paired together, in one case nest building and in the second relieving an icubating mate at the nest but the authors made no deductions about intermediate plumage forms (Naik *et al.*, 1981).

*VOCALIZATIONS

Relatively silent outside of nesting colony. Calls rather more guttural and approaching gargling tones than either Night Herons or Pond Herons. Adults returning to the nest were noted and tape-recorded emitting a gargling call. The young keep up an incessant begging call as they get bigger which can be syllabized as 'chak-chak-chak'.

Little Egret *Egretta garzetta* (Linnaeus)   Map **21**

DESCRIPTION

Body length 55-65 cms. (half of this being legs and bill)
Wingspan 88-95 cms
Wing length 24.5-30.3 cms. (males) 25.1-29.7 cms. (females)
Bill length 67-93 mm. (males) 68-89 mm. (females)
Weight 280-614 gms. (both sexes)

About the same size and build as the Reef Heron but with a slightly longer tarsus and more slender longer bill. Body all-white, legs jet black with olive yellow feet and bill wholly black in winter. In breeding plumage develop two long nuchal plumes, similar to *Egretta gularis* but longer. *E. intermedia* and *E. alba* have no nuchal plumes in the breeding season. Also the scapular and breast feathers are developed into long filamentous plumes, the famous 'Aigrettes' which during the early part of this century were commercially much in demand for decorating ladies' hats. The dorsal aigrettes extend just to the tip of the tail. In the breeding season two separate nesting colonies in Sind province had birds with bright orange to reddish pink feet at the onset of laying and the bare facial skin around the eye and lores was purplish pink, extending just to the base of the bill. By the time the chicks were hatched the majority had greenish blue bare facial skin. This colour phase may therefore be more transitory than with other breeding ARDEIDAE. The iris is yellow. Immature birds can also be seen with the usual greenish yellow feet in the vicinity of the breeding colony. Downy chicks are white in colour with blackish bills.

On 15 March 1965, in company with C.D.W. Savage and the late Roger Holmes, both experienced ornithologists, a slaty grey coloured Egret was observed feeding with a group of *Egretta garzetta,* on the Sukh Beas (old river-bed near Lahore, Punjab). It had two long nuchal plumes and a purely black bill and in size and body proportions appeared similar to the other normal white Egrets. It was presumed to have been a rare colour morph of *Egretta garzetta*. A grey

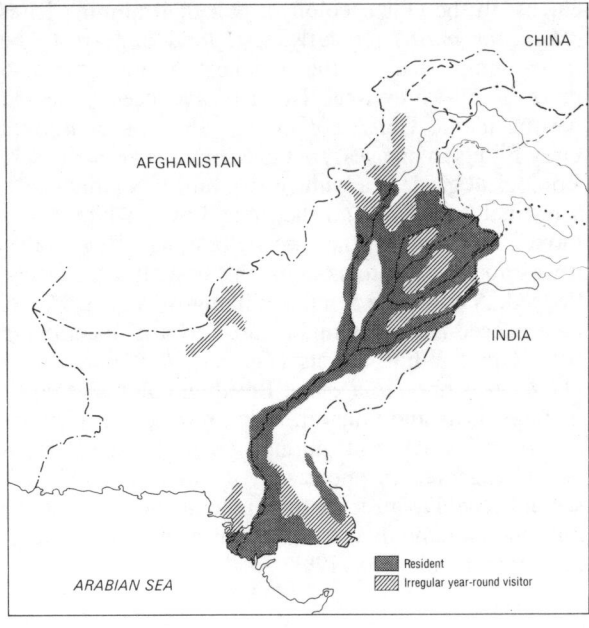

Map 21: Little Egret *Egretta garzetta*

morph occurs rather uncommonly in Africa (Brown, Urban and Newman, Vol, 1, 1982). The bird could not be secured as none of us had guns and the location was close to the main road and the international boundary with India. Being 700 miles from the seacoast, the possibility of this bird being a Reef Heron cannot be ruled out but seems improbable. None of the other white Egrets in the group had developed noticeable dorsal plumes. The Bombay Natural History Society were unwilling to publish this interesting sight record.

HABITAT, DISTRIBUTION AND STATUS

Does not normally penetrate into northern mountain areas but widespread throughout Sind and Punjab, always in association with larger bodies of water, rivers and lakes. 5 or 6 pairs were, however, found breeding at Chakdarra on the banks of the Swat river, alongside with Night Herons in May 1985 and again in July 1987. Less adapted to grassland or dry land foraging than *Bubulcus ibis*. Juveniles are locally migratory after the nesting season and a few birds may spend part of the summer on Kushdil Khan lake in Baluchistan or Zangi Nawar in Nushki district. 25 were counted on 7 May at Kushdil Khan. No breeding records for Baluchistan. It reaches dense concentrations in northern Sind around Sukkur and Jacobabad districts. Over 1,000 birds counted while driving between Sukkur and Jacobabad city in December. Normally prefers fresh water bodies but a few birds haunt tidal creeks at Karachi and these have been observed several miles down Ghizri creek in a purely marine biotope where they can be seen feeding along-

side both white and slaty plumaged Reef Herons (*Egretta gularis*). They are much more numerous in Pakistan than the other two fresh water inhabiting white egrets. Status COMMON.

HABITS

Usually gregarious in foraging and feeding and colonial in nesting and roosting. They are common around large lake shores, irrigation barrage headworks and in the major rice-growing tracts, and frequent even small roadside pools in such regions. Feed on amphibia, naiads of Odonata, fish, 'orthopterous' insects, especially *Chrotogonus* ground hoppers. Have been recorded seizing and killing small snakes and lizards. On mud flats some Annelida are taken, and surprisingly crustacea are not important in diet (Mukherjee, 1971). In rice fields insects are more important, whereas fish are the major item when feeding around rivers and lake shores. In a study based upon analysis of 138 stomachs from adult birds taken in the Sunderbans (Bengal, India) fishes comprised 60 per cent of the bulk of the diet, insects 8.3 per cent. Birds collected from cultivated tracts including rice paddies contained only 16 per cent fishes and 74 per cent insects in their stomachs. Annelida varied from 2-5 per cent and vegetable matter with sand 5-6 per cent (Mukherjee, *JBNHS*, Vol. 68, 1971).

BREEDING BIOLOGY

The Egret is colonial in breeding and often nests in association with other species. A small colony at Haleji lake on an island, breeds on rock ledges or on *Salvadora persica*, bushes in company with Purple Herons (*Ardea purpurea*) and Night Herons (*Nycticorax nycticorax*) (6 years' observations). Another wild colony in Karachi Zoological gardens nests in bamboo clumps with Cattle Egrets (*Bubulcus ibis*) and Little Cormorants (*Phalacrocorax niger*), and perhaps significantly within visual range of a nesting colony of Reef Herons (*Egretta gularis*). See discussion on taxonomy of *E. gularis*. Another colony of about 150 pairs at Shah Yaki, Thatta district, nests with *Bubulcus ibis* and *Phalacrocorax niger* in a grove of *Acacia arabica* trees. This colony is only occupied from mid-July onwards when the trees are partially submerged as a result of adjoining rice cultivation. In the *Handbook of Indian Birds* (Ali and Ripley, Vol. 1, 1968) it is stated that nesting in Sind is from July to September, but breeding commonly starts in Sukkur riverain forest in April and in Thatta district in six years' observations Little Egrets started nesting late March before the Cattle Egrets (*Bubulcus ibis*) and on Haleji lake in late May in some years, and by mid-april in others, alongside Purple Herons (*Ardea purpurea*) which had started one month earlier. A colony of 20 pairs in association with about 40 pairs *Bubulcus ibis* near Ghulamullah, Thatta district, had young fully fledged and ready to fly on 8 July. In Bahawalpur a colony started nesting in late April (in tall trees with Pond Herons (*Ardeola grayii*) and by Rawal lake Islamabad) pairs were with eggs by the third week of May. A few Little Egrets definitely nest in mangroves in association with Reef Heron and Night Heron colonies but these comprise isolated cases of one or two pairs and do not occur as extensive colonies. They are mainly adapted to a fresh water habitat and breed in proximity to such wetlands. Courtship is initiated by the males establishing a nesting territory from where they continuously utter drawn-out 'arrrgh' calls which are gargling sounds. They occasionally alternate this with slow flapping circular display flights around their territory. They make an audible noise clapping their wing shoulders together during such flights. If any other Egret flies overhead they bend forward, lower their necks and erect their dorsal (scapular) plumes in a cascading shower over their backs as well as their breast plumes (See **ill. 11**). Their nests are relatively flimsy stick platforms and these are apparently mainly built by the females from twigs brought by the male bird. Copulation was observed on 30 April. The male flew in, landing beside the nest where the female was incubating. The female stood up, moved to the side of the nest and squatted. The male immediately climbed onto her back. There was no pre-copulatory display or any vocal signals. The normal clutch is 3 to 5 eggs and both sexes share incubation duties. The eggs are glossless and greeny blue and incubation lasts 21 to 25 days. Young have been observed placing their bills inside those of their parents when food is being regurgitated (observations, 2 June, Haleji lake colony).

*VOCALIZATIONS

In winter at day time roosts or loafing grounds an occasional arrival of a newcomer elicits a chorus of low pitched guttural squawks. The most noticeable and characterstic call during the breeding season is 'gararararh', a pronounced gargling sound, uttered periodically by the male birds standing near the nest even after incubation has commenced. This gargle is unlike any call made by the genus *Ardea*. They also have a variety of shorter calls at the nesting colony, including a croaking and more staccato 'kaak-kaak' tone.

EGRET FARMING

During and immediately after World War I this was an important cottage industry in Sind province. Little Egrets were plucked for their dorsal plumes in quite a humane manner and they even bred in captivity in such Egret farms where the birds were provided with forked tree branches in which to build their nests and were sheltered from outside disturbance by rush matting. Detailed accounts were given in the *Journal of the Bombay Natural History Society*, (Birch, Vol. 23, p. 161 and Birch, *JBNHS*, Vol. 27, 1921, p. 944.). Changes in ladies' fashions abruptly diminished the demand for Egret plumes and no Egrets have been kept in captivity for this purpose since the late 1920s.

Smaller Egret or Intermediate Egret *Egretta intermedia* (Wagler)

Illus. **11**
Map **22**

DESCRIPTION

Body length 65-72 cms.
Wingspan 105-115 cms.

A pure white egret which in winter has a wholly yellow bill. It is noticeably larger than *Egretta garzetta*, and if put to flight the feet are all-black not olive yellow as in *garzetta*. Moreover *garzetta* always has a wholly black bill. The Great White Egret, *E. Alba* with a wholly yellow bill in winter, could be confused with *intermedia* though it is considerably larger than the Intermediate Egret with a longer bill. In the breeding season the bill undergoes a striking transformation, becoming jet black except for the basal area which remains bright yellow, and the bare facial skin in the loral areas becomes bright orange-yellow. In this species, while the iris turns orange briefly for the period of courtship up to egg laying no nuchal plumes or any crest are formed but both breast and back aigrettes are developed, and the filamentous tips on the back plumes extend well beyond the tail when the bird is at rest and are thus comparatively longer than similar plumes borne by *E. garzetta*. However unless displaying, the breast aigrettes are not very conspicuous. Iris is yellow, legs and feet black but greenish yellow on the tibia, in immature birds.

HABITAT, DISTRIBUTION AND STATUS

Much less numerous than *Egretta garzetta* and inclined to be sedentary. Due to their similarity in the field to Great White Egrets (*Egretta alba*) there is some difficulty in discerning their numbers and distribution,

Illus. 11: Intermediate Egret *Egretta intermedia*, male in breeding plumage, displaying at nesting site. Note absence of long nuchal plumes.

but they occur throughout Sind and Punjab in association with fresh water bodies. This species likes marshy areas and flooded grassland or weed-choked pools, but it also occurs sypatrically with *E. garzetta* on the shores of open bodies of water and irrigation barrage headworks. A small number also frequent the inland mangrove creeks and tidal channels. At Lal Sohanran (Bahawalpur district) about 500 can be seen when the lake level is high. In Ghauspur, Jacobabad district, only scattered individuals noted , compared with 30 to 40 Large Egrets (*E. alba*), in winter. In east Narra, in December, in Makhi Dhand at Sadori lake *circa* 200 and no other Egret species. No authentic records for Baluchistan except on mangrove coastline. Status FREQUENT.

HABITS

Largely sedentary and though colonial in nesting and gregarious in roosting and loafing, it hunts solitarily. Equally adapted to feed along mangrove creeks or in fresh water swamps, though it is only observed around larger bodies of water inland from the coast. In the field this Egret is often recognizable by its preferred hunting technique of cautiously and continuously stalking through weed-choked margins or along open lake shores. It rarely stands motionless to hunt like *E. alba* or *E. garzetta*. Frogs, tadpoles and naiads of dragonflies and damsel-flies are important items in its diet when foraging around fresh water swamps, but around larger lakes or estuaries fish comprise the bulk of its diet. In a study based upon analysis of stomach contents of 220 adult. specimens collected from the Sunderbans (Bengal, India), birds from the mangroves had 95 per cent by bulk of stomach contents comprising fish, 4.5 per cent crustacea and 0.5 per cent Annelida with no insects. Birds collected from cultivated tracts by contrast, had 87 per cent of stomach contents comprising insects, and only 7.5 per cent fish, 3 per cent crustacea and 2 per cent Annelida (Mukherjee, *JBNHS*, Vol. 68, 1971).

BREEDING BIOLOGY

Very little has been specifically recorded about this species, except that they are similar to other egrets in forming monogamous pair bonds for the season. In Pakistan breeding starts later than is the case of either *Egretta garzetta* or *Bubulcus ibis*, nesting activity usually not commencing until the onset of the monsoon. Birds in full nuptial plumage have occasionally been observed as early as 30 April and as late as 10 October with the bulk of courtship and nesting starting in July. Pairs indulge in mutual displays with the dorsal aigrette plumes erected vertically over their backs and bodies crouched forward with legs slightly bent and bill lowered at a 45° angle. The male frequently also presents sticks, and after the bond is established mutual preening can be regularly observed on or near the nest platform. Nesting is colonial and most often in trees, both mangroves in coastal areas, and partly submerged *Prosopis spicigera* trees, as noted at Sadori lake in the east Narra, but on Haleji lake it is often in an isolated patch of *Phragmites* reeds growing in deep water.

They ususally nest in association with other swamp dwelling birds. Eates (MS notes unpublished) found them nesting in the remote mangrove creeks (China creek, Indus mouth) with *Egretta alba*, *Threskiornis melanocephalus*, *Ardea cinerea*, and *Phalacrocorax fuscicollis*. The Sadori lake colony was in association with *Phalacrocorax niger*, *Nycticorax nycticorax* and *Egretta garzetta*. Scrope Doig also found colonies in the east Narra in flooded tamarisk trees with mixed species heronries (Doig, *Stray Feathers*, 1979b). At Haleji lake small colonies have been noted both in submerged reed beds and on islets. In the latter

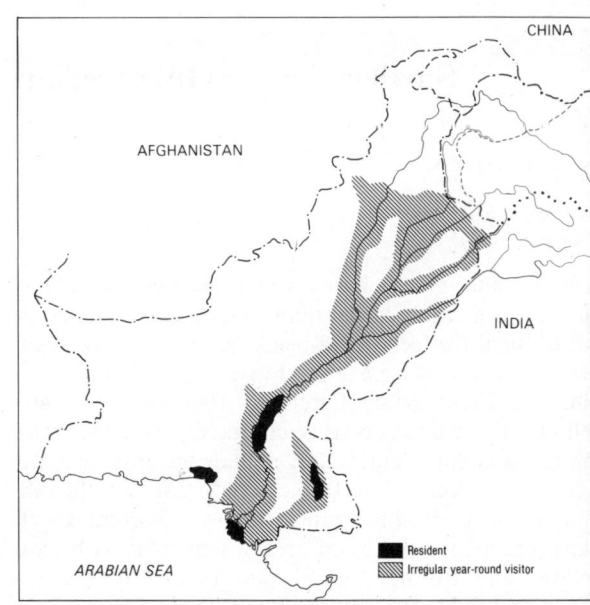

Map 22: Intermediate Egret *Egretta intermedia*

instance sites vacated by the earlier breeding *Ardea purpurea* and *Nycticorax nycticorax* are utilized, often on rock ledges or trampled down *Phragmites* reeds at water level. Their nests can be flimsy and untidy platforms of sticks or may consist largely of reed stems, some of which are picked up from the lake surface while in flight. A small colony noted in a reed bed about 100 metres from the lake shore, was in mixed association with Night Herons, with nests built on trampled-down reeds at water level and on 8 July 7 out of 10 nests located countained eggs. Both sexes share in incubation and feeding the young by regurgitation. The normal clutch seems to be quite small, 2 or 3 being recorded in Sind by Eates. The eggs are pale sea green (Stuart Baker, 1929) and incubation takes 24 to 27 days with the eggs hatching asynchronously. The young clamber about near the nest area from about 10 days and leave the nest at 3 weeks of age (Brown, Urban and Newman, 1982). Fledging period about 35 days with young ready to fly about 2 months and 3 weeks after the first egg was laid.

*VOCALIZATIONS

Relatively silent on wintering grounds but reported to emit a harsh double 'kraa-krrr' call when disturbed at the nesting colony and to emit rather hoarse buzzing sounds during greeting ceremonies at the nest (Hancock and Elliot 1978). A colony of about 20 pairs which was just commencing to nest, was tape-recorded on 25 May, and gave two types of calls, both similar to *E. garzetta* but not surprisingly, deeper in pitch and almost bovine in tone. The most distinctive call given by males on their selected nesting territory was a drawn-out ululating gargle 'ghwarr-arr-arr-arr'. The second type of call was a deep throaty 'g-a-a-a-a-rh' or 'gr-r-a-a-a-h'. This latter call appearing to be elicited more frequently during intra-specific encounters.

Large Egret, Great White Egret or Heron *Egretta alba* Linnaeus   Map **23**

Ardea alba (Ripley's 'Revised Synopsis')

Synonym *Casmerodius albus* (Linnaeus)

DESCRIPTION

Body length 85-102 cms.
Wingspan 140-170 cms.
Wing length 34.3-39.6 cms. (Brown and Urban),
    41-47 cms. (Stuart Baker)
Bill length 116-142 mm. (Stuart Baker),
    104-116 mm. (Brown and Urban)
Tarsus length 16.5-21.5 cms. (Stuart Baker),
    13.4-17 cms. (Brown and Urban)

Unless size comparison is possible, it can appear confusingly like *Egretta intermedia*. A long necked all white egret with a slender yellow (in winter) bill and long black legs. In the field it can often be identified by its larger size compared to other nearby waders, its habit of standing motionless when hunting and a noticeable kink in its neck between the eighth and ninth vertebrae. In breeding plumage, it lacks any nuchal plumes (seen in white phase *E. gularis* or in *E. garzetta*) and lacks the development of breast plumes into filamentous 'aigrettes' which at once separate it from *E. intermedia*. It possesses only dorsal aigrette plumes. The bill is wholly yellow in winter but becomes black except for the proximal region in breeding plumage. The legs which have black feet and tarsi become pinkish red or orange red throughout the tibia and down to the tibio-tarsal joint, and the naked orbital skin becomes bright turquoise, contra yellow of *E. intermedia*, during the onset of the breeding season. In winter the orbital or loral naked skin is dull olive yellow. When seen alongside *Ardea cinerea* it often appears slightly longer legged and taller. In flight it has a slower more deliberate wing beat than the other white Egrets. When viewed closely the gape which forms a black line, extends well beyond the eye (1 cm.) whereas in *E. intermedia* the gape may be less dark, more yellowish and extends only to the back of the eye. The downy chick has a yellow bill and white spiky down on its body.

HABITAT, DISTRIBUTION AND STATUS

In winter it is locally migratory according to available feeding grounds most moving southwards but a few individuals over-winter around Rawal lake and the Salt Range lakes. It regularly occurs in small numbers around Kushdil Khan lake (20 on 7 May) in Baluchistan. In the larger jheels of Punjab and northern Sind quite large numbers congregate. 80+ on Ghauspur jheel (Jacobabad district), 40 to 50 Lal Sohanran (Bahawalpur district), 100+ Taunsa barrage headpond and environs. It is well adapted to estuaries and mangrove creeks where it also breeds and is sedentary both around Sonmiani in Lasbela and the Indus mouth.

Ticehurst records it as rare outside of northern Sind (*Ibis*, 1923) and Whistler found it uncommon in Jhang (1922) and absent from Rawalpindi district. It was not unusual to be able to count 10 to 20 individuals around Haleji lake (near Karachi) in the 1980s. It has therefore

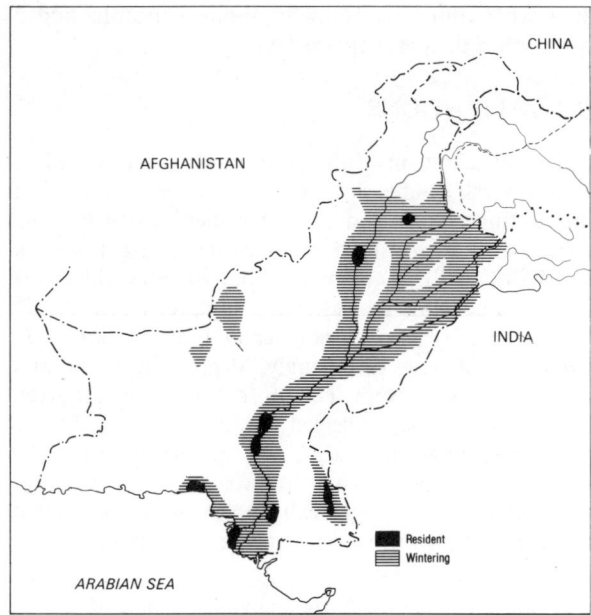

Map 23: Great Egret or Large Egret *Egretta alba*

apparently become more widespread and commoner in the past 50 years. Status COMMON.

HABITS

They often roost on the ground or loaf in congregations of 30 to 40 birds around larger lakes, but hunt solitarily. They are never found away from the main rivers or larger lakes. Their main hunting technique involves patient standing in shallows, usually in a rather erect stance with neck outstretched and held slightly forward like *Ardea cinerea* waiting for prey to come within sight. They catch mainly fish but will also seize small birds and mammals, if opportunity presents, and supplement their diet by aquatic insects and dry land orthopterous insects and even small water-loving snakes (*Natrix piscator*). Studies in the Sunderbans, Bengal, indicated that 80 per cent of the diet comprised fish belonging to 26 different species, and 6 per cent of the diet comprised mollusca. No amphibia or insects were found in the stomach contents of 70 species examined (Mukherjee, *JBNHS*, Vol. 68, 1971). They mainly hunt by daylight and are reported sometimes to travel distances of over 20 kilometres (12.5 miles) between nightly roost and feeding grounds.

BREEDING BIOLOGY

Nests colonially, always in association with other species and usually not commencing until after onset

of monsoon. On Ghizri creek in late April, occasional individuals have been seen with black bills and in full breeding plumage. Kenneth Eates only found it breeding in northern Sind around large perennial bodies of water such as the Karo Nai, in Sukkur district, in September 1930 and on 10 July in submerged *Acacia arabica* trees on Kamaro Sharif lake in Hyderabad district. Here its nests were indiscriminately intermixed with other species (Eates, MS notes unpublished).

Courtship is initiated by an increase in aggressive posturing. The dorsal aigrette plumes are frequently fanned over the back and the bird crouches forward with neck bent and erects its crest and breast feathers (which are silky and elongated but not filamentous in the breeding season). Aggressive flight chases have been noted and the vertical fanning of dorsal aigrettes is a '*garzetta* characteristic' somewhat different from the display of *Ardea* species, hence this author prefers to retain the Great Egret in the former genus. Nests can occur on flattened down reeds in reed beds or on the tops of trees. About 7 pairs regularly nest in mangrove trees in Ghizri creek in company with Reef Herons (author obs.). They were found nesting by Eates in a mixed heronry with Grey Herons (*Ardea purpurea*), Darters (*Anhinga melanogaster*), Cormorants (*Phalacrocorax spp*) and Spoonbills (*Platalea leucorodia*), at Jamrao head, east Narra. Eates also found a solitary pair nesting in an Acacia tree which was partly submerged on 10 July on Kamaro Sharif lake, Hyderabad district.

Both sexes share in incubation and feeding the young. The eggs are similar to *Ardea cinerea* in colour but fractionally smaller in size, and the normal clutch is 2 to 3 eggs. The male brings sticks or reed stems which the female weaves into the nest structure. Incubation takes 25 days and the young start clambering out of the nest after 3 weeks of age (Hancock and Elliot, 1978). They are fully fledged in 6 weeks. The young are fed in the usual manner by regurgitation. On Sonmiani lagoon in mangrove (*Avicennia*) trees, they nest together on the periphery of a much larger mixed colony of Shags (*P. fuscicollis*) and Reef Herons (*Egretta gularis*) and a few Intermediate Herons (*E. intermedia*). On Nammal lake in the Salt Range about 5 pairs nest in extensive sedge and reed beds on the eastern shore of the lake.

VOCALIZATIONS

In winter startled birds will utter a single 'kraah' call, deep and grating. Reportedly more vocal at nesting colony, uttering deep vehement cawing sounds (Cramp *et al.*, 1977).

Grey Heron or Common Heron *Ardea cinerea* Linnaeus

Illus. **12**
Map **24**

DESCRIPTION

Body length 90-98 cms.
Wingspan 175-195 cms.

Slightly bigger than the Purple Heron with a slightly thicker neck and heavier bill. Adult in breeding plumage has white face, forecrown and neck ending in long lanceolate white plumes on the breast. A broad black band extends from behind each eye terminating in a slender graceful black nuchal crest. The bill and iris is yellow and the legs and feet dark olive green. The wings and back are ashy grey and there is an interrupted line of black streaks down each side of the foreneck. The sexes are alike and in winter there is little change in plumage. Juveniles are much whiter about the foreneck with indistinct dark grey streaking, not black, on the foreneck, and the crown is more ashy extending right down to the culmen. Their backs and tails are more brownish grey, less bluish than adults. In flight the primaries and secondaries show slaty black and the carpal patches are prominently white in the Pakistan population. A pair observed at the onset of the breeding season had long silvery grey lanceolate plumes covering the wing shoulders and a magenta pink flush to their naked orbital skin and base of bill. The wing shoulder is also noticeably black in the breeding season. The downy chick has brownish grey down on its upper parts, and silky white down on throat and belly with a greyish mandible and greenish grey legs.

HABITAT, DISTRIBUTION AND STATUS

In the absence of ringing studies the proportion of wintering birds which are Palearctic migrants is unknown, but a marked passage has been noted in spring and autumn in Chitral and Gilgit valleys. Big numbers also pass on migration up the Kurram valley. The return autumn passage was noted to start at the end of September in Gilgit (Biddulph, 1881). In the Punjab they are abundant around the larger lakes and barrage headworks in winter but largely absent in summer. Ticehurst also noted a marked drop in numbers of Grey Herons along the coast and mangroves in the summer. At a rough estimate 25 per cent of the population is sedentary and 75 per cent migratory. The breeding population moves to larger permanent swamps for breeding in the monsoon season, particularly in the east Narra. In winter they are equally at home in the tidal creeks and mangroves as on inland fresh water lakes and rivers. *Circa* 500 can be seen around Ghauspur jheel and environs, 30 to 40 around Rawal lake, and smaller numbers around all the major irrigation barrage headworks in Sind and Punjab. In winter around the bare flat salt marshes of Keti Bandar up to 300 can be counted in a day's boat trip. Meinertzhagen (*Ibis*, 1920) recorded a spring and autumn passage through Baluchistan. Two were observed (author) on 7 May on Kushdil Khan lake but most pass through in March and April. Status COMMON.

HABITS

They roost colonially in winter and also often loaf on lake shores in flocks of 20 to 30. They often flight in 'V' formation between feeding grounds and winter roost, and are more active at evening time and before dawn than by day. In the Salt Range in December camping by Khabbaki lake, they were heard calling throughout the night. Their food comprises mostly small fish and amphibians but a variety of other prey when available including molluscs, crustacea, larvae of insects, tadpoles and in a marine environment crabs and mudskippers and sandfleas. An exceptional sight in October at Lal Sohanran lake, was an adult bird observed swimming strongly across a short stretch of open water, its tail cocked up comically in the air.

Birds used to be regularly snared on the larger lakes by 'Mohannas' (professional fishermen) who esteem them as a delicacy and numbers of tethered birds can be seen on their boats on Manchar lake. In the 1970s the Gilgit scouts, a locally recruited levy, wore tufts of white breast plumes on their uniform caps and these

Illus. 12: (a) Grey Heron *Ardea cinerea*, adult.
(b) Purple Heron *Ardea purpurea*, adult.

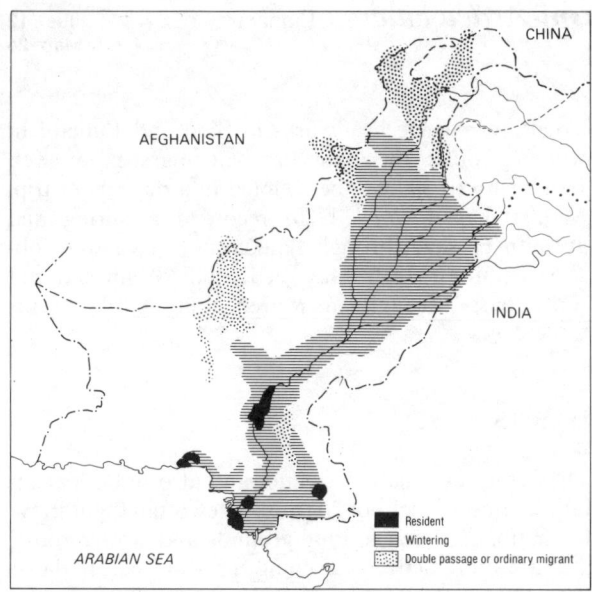

Map 24: Grey Heron *Ardea cinerea*

were regularly obtained from birds killed on spring and autumn passage. In 1987 these breast plumes were still being sold in Hunza as cap decorations to be worn on festive occasions.

They are shy and wary birds, immediately taking to wing at human approach, often with a harsh alarm croak.

BREEDING BIOLOGY

Very well studied in Europe. Very small numbers breed in the mangroves both on the Indus delta and Sonmiani, usually in association with other Egrets and Cormorants and at the onset of the monsoon. A colony nests in flooded Tamarisk trees at Kandhkot in northern Sind during the monsoon season with vast numbers of Cormorants and three Egret species. Doig (*Stray Feathers* 1879b) found a colony in the east Narra in flooded *Prosopis spicigera* thorn trees in July. An odd pair or two nests in Karachi Zoological gardens in tall trees usually commencing well before the rains but after the main Night Herons and Reef Heron colonies have got fully fledged families. Eates

found a few pairs nesting in huge Pipal trees (*Ficus religiosa*) in Jacobabad city with Night Herons. A colony used to nest (Eates, MS notes) in the 1920s in submerged tamarisks at Kalankot lake in Thatta district with Little Cormorants (*Phalacrocorax niger*) and Purple Herons (*Ardea purpurea*), and started to nest in May but this site has not been used in recent years. In some years Eates found young in the nest as late as early October.

In courtship and pair formation males initially select and defend a territory and commence giving advertising calls, a shorter sharp goose-like yelp uttered whenever another bird flies nearby. They also display by erecting their black nuchal crest and stretching their neck forwards and upwards with the bill bent slightly downwards and breast plumes partly erect. They also have a stretch display with bill and neck pointed vertically. Copulation takes place mainly on the nest, which is built by the female with material brought by the male.

The normal clutch is 3 eggs of a blue-green colour similar to those of *Egretta alba* in colour and shape. Both sexes share in incubation which lasts 25-26 days. Food is brought by both parents and generally disgorged onto the bottom of the nest. The young chicks beg with a rapidly repeated harsh 'chack-chack' call, very similar to the young of *Egretta gularis*. They also initiate regurgitation by trying to seize and stroke with their bills the parents' bills. Young take about 50 days to become fully fledged. (Cramp *et al.,* 1977) Lowther refers to a heron nest on the plains of India, in July with 3 young, which were not fed by their parents between 8.30 a.m. and 3.30 p.m. and which obviously suffered from the tropical sun. The parent upon arrival at 3.30 p.m. regurgitated small fish and then fed pieces individually to each of its 3 downy nestlings. Subsequently it remained at the nest and sheltered them with drooped wings from the sun (Lowther, 1949).

VOCALIZATIONS

Flight contact call or alarm call often throughout the year, a harsh short 'kraark'. The males advertising call only given in the early stages of breeding, has already been described, and so has the 'chack-chack' begging call of the nestlings.

Purple Heron *Ardea purpurea* Linneus

DESCRIPTION

Body length 78-90 cms.
Wingspan 120-150 cms.

A large heron with long slim neck and legs, lighter in build and of slightly smaller size than the Grey Heron but in flight very similar in size and appearance with dark slaty primaries, grey wing shoulders and mantle. The whole of the crown is blackish purple with long slender nuchal plumes and a thin dark line extending from the gape below the eye to join the nuchal plumes on the nape. The rest of the face and neck is orange fulvous with white throat and lower breast streaked with two interrupted lines of black down each side of the neck. Lanceolate scapular and breast plumes are more conspicuous in breeding plumage, when both sexes are alike and very handsome. The wing shoulder has a conspicuous bright purple-maroon patch, the mantle is also maroon with long lanceolate scapular feathers of silvery grey to white colour, falling over the grey wings. The thighs and lower belly are also purplish maroon and the breast plumes become pale slaty with fulvous mid-ribs. The legs outside of the breeding season are blackish, with olive yellow on the back of the tarsus. In the breeding season the legs are orange-brown all over and at the very onset of breeding the hind part of the legs have a salmon pink hue. The bill is yellow or olive yellow, often darker on the culmen and tip, and at the onset of breeding the entire bill and bare facial skin becomes cyclamen pink (not mentioned in Hancock and Elliot, first edition, 1978, or Cramp *et al.,* 1977) but after incubation commences, gradually assumes a more orange colour.

First year birds have more rusty chestnut on fore-part of neck and upper breast with less distinct grey streaking on sides of neck (in some individuals totally absent). In flight their wing shoulders are much browner than adults. In adults the under-wing coverts are difficult to see in flight but are a rich chestnut. They are paler fulvous in juvenile birds. The downy chicks have pinkish or creamy long soft down, pale whitish irides and dull grey-yellow feet (Plumage details recorded with X50 telescope at Haleji lake breeding colony).

HABITAT, DISTRIBUTION AND STATUS

A resident species and widespread from lower Sind to northern Punjab. They normally avoid a salt water biotope but are occasionally seen in the mangrove creeks of the Indus delta. They normally shun exposed feeding sites or open lakes, preferring reed fringed swamps and smaller lakes and seepage channels choked with *Typha* reeds. They are largely absent from Baluchistan and the North West Frontier due to absence of suitable habitat. At Lal Sohanran when the lake is full, up to 50 can be counted and over 100 on Haleji lake. They are rare on the bare margined Salt Range lakes, but can always be encountered in reed beds and seepage zones around Panjnad, Balloki and Trimmu headworks. Status COMMON.

HABITS

The only member of the true Egrets and Herons, to be very rarely encountered in a salt water habitat. They are solitary in hunting and inclined at all times to be secretive, rarely standing out in the open to hunt like *Ardea cinerea.* Their favourite hunting technique is to stand up to the thighs in weed-fringed water, neck out-stretched at a 60° angle to the body, motionless. They feed on fish, reptiles and amphibia, frogs probably being a much more important food item than is the case with the Grey Heron (*Ardea cinerea*). The two species are however sympatric in fresh water habitats and often hunt in close proximity. Snakes and dragon-flies may occasionally be taken (Cramp *et al.,* 1977). Analysis of stomachs of 70 adult specimens taken from the Sunderbans (Bengal, India) indicated 57 per cent by volume of diet was fish and only 8 per cent insects, but 20.57 per cent was reptiles, especially water snakes (Mukherjee, 1971). They can be quite vocal outside the breeding season, calling as they fly to a new feeding or resting site in a low pitched 'garrrh'. They are also quite active feeding before daylight and at dusk like *Ardea cinerea.* They have not been noted as being particularly gregarious outside of the breeding season.

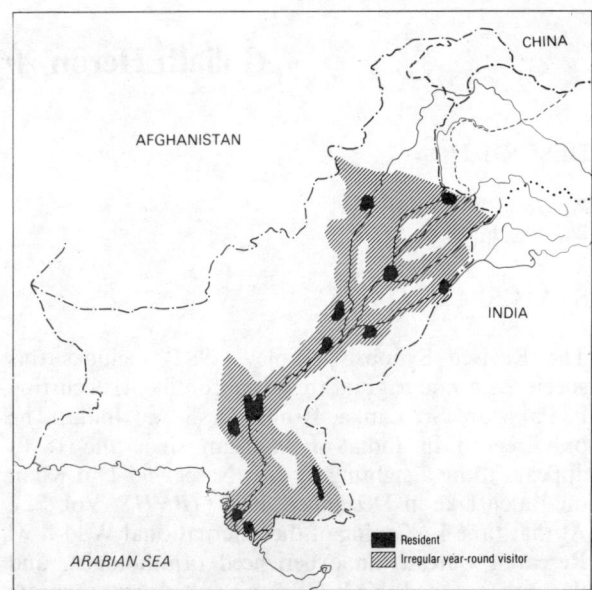

Map 25: Purple Heron *Ardea purpurea*

BREEDING BIOLOGY

A colony at Haleji lake has been observed over 6 years. Nesting commences at very variable dates but is early. In one year it started by mid-February (2 pairs feeding downy chicks on 21 February) and at the other extreme, by late April. Initially, the males stake out territories, often within a few feet of each other. Here they stand most of the day on a favoured site, periodically doing a stretch display in which the bill is pointed skywards and the pale fulvous throat appears to be slightly inflated and is prominently displayed. The bird holds this position for 4 or 5 seconds. Occasionally it pursues in flight a passing bird or aggressively displays to a closely passing or clambering bird, with crest and long nuchal plumes erected over the head, body horizontal and legs bent in a forward crouching position. When displaying, the maroon shoulder patches are prominently everted. During pair formation mutual billing or stroking of each other with closed bills, also mutual preening of pairs on the nest platform observed. The male carries sticks or reeds to the nest site which the female weaves into the nest, continuing this even after the first two eggs have been laid. The male shares in incubation duties. Copulation has been observed (11 April) immediately after the male flew in, and on the nest, the female first getting up and then crouching in front of the male. Approaching birds in flight, and when alighting, always emit a special greeting call, a repeated 'argakh-argakh-arghaal'. This colony comprises about 40 pairs located on a small island, with most nests on beaten down *Phragmites* reeds and a minority of nests on the tops of *Salvadora persica* bushes. Kenneth Eates

found colonies nesting from July to September and often in single species colonies. The Haleji colony is started by Purple Herons but always joined by Night Herons and Little Egrets. Eates came across a large colony at Kot Naro in Tharparkar district in late July with Grey Herons (*Ardea cinerea*), Large White Egrets (*Egretta alba*), Intermediate Egrets (*E. intermedia*), and Little Green Bitterns (*Butorides striatus*). The normal clutch is 3 eggs. Females have been observed (author) turning their eggs with their bills. Eggs are laid at 24 hour intervals and are pale greenish blue in colour. Incubation period is 25 to 27 days, the young hatching asynchronoulsy. They start to wander from the nest when 14 days old and are fully fledged at 6 weeks of age. Chicks have been observed inducing regurgitation by seizing and stroking parents' bills. Food is usually regurgitated onto the nest floor. Fully fledged young still unable to fly have been observed as late as 26 July in the Haleji colony. Parents spend most of the day taking turns, to shelter their downy young from the sun, with wings drooped in a characteristic stance. Males continue the 'stretch display' occasionally even after young are sprouting flight feathers.

*VOCALIZATIONS

Bill clapping or rattling not heard in this species. Both sexes have a characteristic greeting call difficult to syllabize, but a double phrase repeated 'garwak-garwak-garwak-kroh-kroh'. A flight call heard throughout the year consists of a very deep-toned guttural 'garrh'. Young chicks also keep up a monotonous begging call 'chick-chick-chick' similar to other ARDEIDAE.

Goliath Heron *Ardea goliath* Cretzschmar

DESCRIPTION

Wingspan 210-230 cms.
Body length 135-150 cms.

STATUS

The 'Revised Synopsis' (Ripley, 1982) includes this species as a rare vagrant to the subcontinent, occurring in Pakistan, Sri Lanka, Bangla Desh and India. The only record in India or Pakistan since the 1890s appears to be a sighting by Mr. Naseer-ud-Din Khan on Haleji lake in December, 1974 (*JBNHS,* Vol. 72). At that time F. Koning of the International Wild-fowl Research Bureau, an experienced ornithologist, and the author visited this lake on two subsequent separate days after being informed by Mr. Naseer-ud-Din

Khan of this sighting. We failed to locate the bird, and because a juvenile Purple Heron (*Ardea purpurea*) could be confused with this species there is a possibility of misidentification. The only other Pakistan record relates to a sighting in Baluchistan in 1870 by Blanford which was not accepted by Dr. Ticehurst due to many of Blanford's Baluchistan records being controversial (Ticehurst, 1927). It is largely a sedentary species breeding in Africa. The isolated colony in Iraq now seems to have largely disappeared (S. Cramp *et al.,* 1977). It has occurred very occasionally as a winter visitor on the Red Sea coast in Saudi Arabia (M.C. Jennings, 1981). It is not included in the Checklists of Iran (D. A. Scott, 1975) or Afghanistan (Paludan, 1959). In the light of available evidence I prefer to exclude it from the present **Checklist** of Pakistan birds.

FAMILY CICONIIDAE

Painted Stork *Mycteria leucocephala* (Pennant)   Map **26**

Ibis leucocephalus (Pennant) (Ali and Ripley's 'Handbook')

DESCRIPTION

Body Length 95-105 cms.
Wingspan 150-160 cms.
Wing length 49-51 cms. (Stuart Baker)
Bill length 252-278 mm. (Stuart Baker)

A conspicuously large white stork-like bird with long cyclamenpink legs and a heavy yellow bill which tapers to a slightly down-curved narrower tip. The sexes are alike. The whole of the crown and forepart of the face is naked and dull yellow becoming deep orange in the spring. The tail is black and the body and wings are white with black primaries and secondaries, white wing coverts and a scale-like pattern of parallel rows of lesser coverts comprising black centres and white margins. There is a band of black feathers across the breast. The most conspicuous feature is the broad white tertial feathers suffused with delicate rose pink. Juveniles have the white plumage sullied grey and the bare part of the head creamy brown with buff brown wing coverts. In Pakistan adults often become stained yellowish buff on the neck. In flight the extended neck, which is slightly down-curved, and the yellow bill and orange head can often be discerned from a considerable distance. Downy chicks have naked black faces and bill and dull greyish white body down. The base of the bill is olive yellow and also a short bar of naked skin above the eye.

HABITAT, DISTRIBUTION AND STATUS

It is more or less resident in the Indus delta, and occurs erratically along the major rivers as far as Mianwali in the Punjab, occasionally wandering to Sonmiani lagoon and the larger swamps of lower Sind after the monsoon and during the early winter. The population seems to be rapidly dwindling due to continuous robbing of the only known breeding colonies in the Indus delta by fishermen who sell the chicks to animal exporters.

Occasional records over about ten years in the Punjab, include a flock of 42 birds seen 27 March on an Indus sandbar (Jinnah barrage, Mianwali), 9 September, 20 seen roosting in *Ficus* trees by Balloki headworks on the Ravi river (Roger Holmes, pers. comm.) and about 20 at Lal Sohanran on 21 March. In Sind province, 7 seen on Manchar lake (D. A. Holmes, pers. comm.), 15 on Mahboob Shah lake, Thatta district on 8 December, 10 on 14 January in a subsequent year. During the monsoon small groups of

3 or 4 birds were occasionally sighted on the Chenab near Marala barrage (E. Fernando, 1968 pers. comm.) and 3 sighted by Dr. Mubashir Husain above the barrage in November 1984 (photographs shown to author). They are however erratic and rare visitors to the Punjab. A few remain year round on the salt marshes of the Indus delta near Keti Bandar. Status SCARCE and largely confined to the Indus delta.

HABITS

Gregarious in roosting and winter time foraging as well as during the breeding season. They forage mostly by wading slowly through shallow water, scything their opened bill from side to side in the water, often submerged to the lower cheeks, so that their food is mostly snapped up from the muddy bottom. Their food consists of frogs, fish and insects. Young at the nest colony were noted being fed mainly on fish. Reptiles and crustacea also are consumed when available. They commonly roost in trees, foraging mainly in early morning and evening. If trees are not available, they roost in flocks on open exposed sandbars or mud banks, usually squatting on the whole of the tarsus with tail just touching the ground.

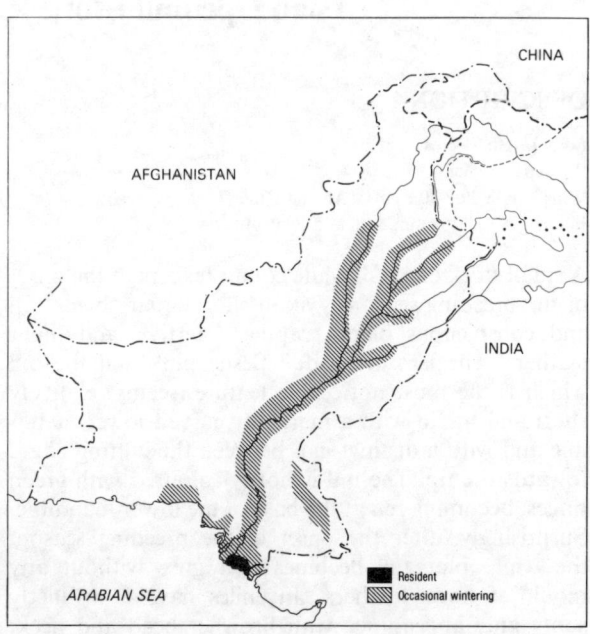

Map 26: Painted Stork *Mycteria leucocephala*

BREEDING BIOLOGY

A captive pair at Karachi Zoo performed a greeting ceremony in the breeding season in which each bird approached the other and bent forward half raising the wings which were held out but not fully opened. The under-tail coverts were ruffled outwards conspicuously and the long bill solemnly dipped downwards, repeatedly, to the accompaniment of a loud nasal sounding braying call. This vocal calling must be very occasional as it is not recorded in the 'Handbook' (Salim Ali, 1968). Kenneth Eates also describes courtship at a nesting colony in which pairs with raised half open wings did a 'high-stepping dance' and clapped their mandibles together but made no vocal sounds. Most greeting ceremonies and changeover at the nest involve an up-down display with bill opened widely and tilted upwards and occasional clapping but not rattling of the madibles (observed Delhi Zoo, wild nesting colony). In the Indus delta 4 or 5 nests were located on one tree. In the east Narra, Eates found nests in the 1940s in flooded *Acacia arabica* trees. A stick platform nest is built, material being collected by both sexes. During incubation this is continuously added to, one bird was noted bringing soggy strands of *Impomea* Water Convolvulus to its mate. The normal clutch is 4 eggs but 2 and 5 have been recorded. The dull white eggs are occasionally sparsely spotted with brown (Kenneth

Eates, MS notes), and soon become stained muddy brown. Both sexes share incubation duties. The Indus delta colony starts breeding in late September and fledged young leave the nest in late November. Eates however found birds in east Narra starting to nest in June. Actual incubation period has not been recorded. Normally the parent disgorges food onto the bottom of the nest and chicks when young are sheltered from the sun a large part of the day by one or other parent squatting over them with drooped spread wings. Young will also solicit food by trying to poke their bill into that of their parent, and also by bobbing their heads and necks up and down. Large fledged young were heard begging for food with a long drawn-out undulating tremulous wheezing call accompanied by pumping the head and neck up and down (observed at Bharatpur, India).

*VOCALIZATIONS

Mainly silent with bill clapping an important element in courtship and a weak hissing scream at nest greeting ceremonies between mates. Juveniles soliciting food at the nest have a peculiar prolonged wheezing call and a most unusual but highly audible greeting ceremony (described above) has been recorded between captive birds bowing to each other (Karachi Zoo, 8 May).

Openbill Stork or
Asian Openbill Stork *Anastomus oscitans* (Boddaert)

Illus. **13**
Map **28**

DESCRIPTION

Body length 75 cms.
Wingspan 155 cms.
Wing length 39.2-40.8 cms. (Ali and Ripley)
Bill length 153-162 mm. (Ali and Ripley)

A small stork of a dull white colour (except at the onset of the breeding season), with a short square black tail and conspicuous black scapulars, tertials and flight feathers. The legs are a dull fleshy pink and the bill which is the most noticeable feature, seems relatively short and broad with a markedly curved lower mandible and with a distinct gap between the cutting edges towards the tip. The bill is horny coloured with green tinges, becoming red at the base of the lower mandible. Surprisingly, after the onset of the breeding season, the white plumage becomes dull grey without any moult of body feathers. Juveniles have a distinctly sooty grey appearance with browner head and neck, moreover their bills are closed along the commissure

(cutting edge). The tertials in the breeding season have a green-purplish gloss. The iris is greyish white and the small area of naked skin around the eye is black.

The White Stork in flight (*Ciconia ciconia*) shows an all-white tail in contrast to the black tail of the Openbill. The downy chicks have dark brown bill, purple naked gular area and are clothed in buffy brown.

HABITAT, DISTRIBUTION AND STATUS

This stork is sedentary and fairly common in India but it no longer breeds in Pakistan. At the turn of the century both Scrope Doig (*Stray Feathers,* 1878) and Ticehurst (1923), described it as common in Sind. Ticehurst mentions a colony of 20 pairs found in a mixed heronry in the east Narra. After the monsoon juveniles range widely, a bird ringed in Thailand (Wat Phailom) was recovered 17 months later in Jessore, Bangla Desh, 1,500 kilometres due west. So it is not

Illus. 13: Asian Openbill Stork *Anastomus oscitans*

surprising that Openbill Storks still occasionally wander into Pakistan, but no evidence of regular breeding has been available since the early 1930s. Records maintained for the past twenty years include 3 sighted, July 1968, on the Chenab river above Marala barrage (E. Fernando, pers. comm.), 12 on Gungri jheel south of Sujawal, Thatta district, on 21 January 1975. (F. J. Koning, pers. comm. and IWRB Reports). A single bird on the feeder canal one mile east of Haleji lake was seen in November 1978 (Deed Ahmed Ali, pers. comm.) and another solitary bird seen at Haleji lake on 27 March 1982 (R. Passburg and author). Status RARE VISITOR.

HABITS

They are adapted to a fresh water habitat and forage mostly in shallow rivers or weed-infested tanks and jheels, where they can find herbivorous water snails, their principal food. The large olive shelled Water Snail, often known as 'Apple Snails' (*Pila globosa*) which comprises the major portion of their diet at Bharatpur, India has not ben recorded in southern Sind but other smaller fresh water whelks such as *Lymnaea acuminata*, abound. They normally forage solitarily, wading slowly through shallow water often probing in the bottom mud with opened bill and head fully submerged (observations Karnataka, south India). They roost in trees and the whole colony disperses to other wetlands after their colonial nesting. In south India they may be commonly seen foraging in flooded rice fields. Amphibia and fresh water crabs are also taken as food. In a study in the Sundarbans (West Bengal, India) stomach contents from 72 adults were analyzed and their food comprised 53 per cent Mollusca (fresh water snails, 3 per cent Oligochaeta (earth-

worms) and 4.5 per cent fish (Mukherjee, *JBNHS*, 1974).

BREEDING BIOLOGY

During the early 1930s Eates found several small colonies breeding in July and August on seepage lakes in the remote sand hill border regions to the east of the Narra canal. Here their nests, in flooded Tamarisk trees, were in association with White Ibis (*Threskiornis melanocephalus*), Grey Herons (*Ardea cinerea*), Little Cormorants (*Phalacrocorax niger*), Greater, Intermediate and Lesser Egrets (*Egretta* spp.) and Pond Herons (*Ardeola grayii*). He described colonies numbering many hundreds of these mixed species on Berwari Dhand, Laiwari and Karo Naro. Nests seen (author) in a still existing colony at Ranganathittu (Karnataka state, India) were comparatively small stick platforms, characteristically with recently added branches with green leaves still attached and even clumps of grass or water weed. Both sexes share in nest building and incubation. Clutches are 3 to 4 eggs, rarely 5 and are creamy white. The eggs hatch asynchronously, as incubation starts with first egg laid. Incubation period is about 25 days. The young are fed by both parents by regurgitation and start to wander from the nest when about 2 weeks old, returning immediately to be fed when a parent arrives at the nest. During hot weather, parent storks shelter their young from the sun with half-opened wings and one or other parent guards the chicks continuously during the first 10 days or so of their lives (Extra-limital observations, Ranganathittu). The pair bond in this colonial breeding stork is believed to last for one season only without any life-long attachment to a fixed nest site as in most *Ciconia* species.

VOCALIZATIONS

In the nesting colony, adults are reported to utter occasional deep moans, but never clatter their mandibles during greeting ceremonies as do other storks. Pairs greet each other with opened bills and an up-down bobbing of their heads and necks, accompanied by loud low pitched honking calls. Young in the nest beg for food by squealing and as they start to fledge by moaning cries. They also beg by bobbing their heads and necks up and down, tapping their bills on the sticks of the nest bowl. During copulation the pair clatter their beaks together but one individual never clatters its mandibles together as has been observed in other storks during mating (Kahl, 1970).

Black Stork *Ciconia nigra* (Linnaeus)

DESCRIPTION

Body length 95-100 cms.
Wingspan 145-155 cms.
Wing length 52 to 60 cms. (Brown, Urban and Newman)
  52-60.5 cms. (Stuart Baker).
Bill length 160-190 mm. (Brown, Urban and Newman)
  160-190 mm. Stuart Baker.
Weight *circa* 3,000 gms. (adult)

The same size and build as the White Stork (*Ciconia ciconia*), but in flight shows all-black head, neck, wings and tail with the lower breast and belly being pure white and also a conspicuous square white axilliary patch. The bill is dark red and the legs are a slightly brighter coral red. In good sunlight the black parts are glossed with green like a Mallard drake's head, with a more purple gloss on the back and wings. The sexes are alike in plumage, and both have a dark brown iris, with bare red skin around the eye. In flight the hind toe has been noted as sticking up conspicuously. Like most other storks it flies with neck fully extended unlike the ARDEIDAE. Juveniles have the lower part of their bellies dull smoky grey, contra the white of adults, and their legs are greenish yellow. The head, neck and back is dull blackish brown, not glossy. The downy chick is white all over with a yellow bill.

HABITAT, DISTRIBUTION AND STATUS

A winter migrant visitor which enters Pakistan via the northern Himalayas and on a broader front across north-western Baluchistan. Biddulph recorded flocks in Gilgit main valley each February, March and April and Dr. Scully (1887) noted a flock of over 100 in the autumn in Gilgit. A single bird was seeen in 1983 flying up the Indus river in late March in Indus Kohistan (D. Corfield, pers. comm.). It has also been noted in the

Illus. 14: Black Stork *Ciconia nigra*, note conspicuous, almost square white wing panels.

Kurram valley. On 21 June 1977, 8 adults were seen feeding on Kushdil Khan lake in Baluchistan and there is a possibility of their nesting in the vicinity (see Breeding Biology). It occurs very sporadically in winter in lower Sind and can be sighted occasionally throughout Punjab and Sind on passage. Odd birds have been sighted on Rawal lake in March in two years. On 14 February a party of 24 was seen at Sadori lake, east Narra, 12 were seen above Chashma barrage on the Indus on 28 February. (F. Koning, pers. comm.). A solitary bird was around Khadeji creek for about 8 days (just north of Karachi) during December, and another single bird sighted near Umarkot on 23 January. Lal Sohanran lake on the border of the Cholistan desert used to attract numbers on spring passage. On 13 February 1969 at daybreak 68 were counted at this lake including only 3 juveniles (Roberts, *JBNHS,* 1969). The following year at the end of February only 6 birds were noted at Lal Sohanran. Status SCARCE.

HABITS

They are extremely shy and wary and will only be seen resting in open areas where close human approach is difficult. They rest in inaccessible sand banks or spits of land along the main rivers or on large lakes. At Lal Sohanran they spend the day resting out in the desert and only come to feed around the lake margins and pools at dawn and dusk. They feed by stalking slowly through the water, probing with open bill in the mud in similar manner to Glossy Ibis (*Plegadis falcinellus*), and at Lal Sohanran they appeared to be feeding mainly on fresh water snails *Viviparia bengalensis* and *Lymnaea acuminata*. They also eat frogs, crustacea and aquatic insects and fish. In lakes such as Kushdil Khan and Rawal, fish are presumed to be the major food item as fresh water snails have not been noted in these waters.

BREEDING BIOLOGY

Though in eastern Europe this stork nests in trees in forests, often in quite dispersed separate territories, it has been recorded nesting in loose colonies on inaccessible cliffs near Zarafshan by the Aral Sea (Carruthers, 1949) and it is jut possible that one or two pairs nest on cliffs in the Tobakhakar mountains of Baluchistan (4 pairs sighted Kushdil Khan lake, 21 June 1977). J.A.W. Anderson claims to have had two young newly fledged Black Storks brought to him at Quetta in 1969 by local hunters who found a nest in that range. Paludan saw birds in Hazarajat (central Afghanistan) between 11 June and 7 August and suspected they were breeding (Paludan, 1959).

Even when nesting on rock ledges, they build a substantial pad of sticks and this has a marked depression or cup at the centre in which 3 to 5 plain white eggs are laid, at two day intervals (Cramp *et al*., 1977). Probably pair bonds are established for life. Both parents share incubation duties and caring for the young. Both sexes continue to bring material to the nest, including clumps of grass with earth attached even after the eggs have hatched. Incubation takes 35 to 36 days (Cramp *et al.*, 1977), and the chicks are constantly tended by one parent for the first 2 weeks after hatching. Young fed at first bill to bill by regurgitation, and later the semi-digested food is regurgitated onto the nest floor. Courtship is quite distinct from that of the White Stork (*Ciconia ciconia*),as bill clattering is rare and mutual displays consist of facing each other and head bobbing up and down at the same time depressing long white under-tail coverts which are also conspicuously twisted from side to side. They also emit bi-syllabic whistling noises while head bobbing.

Map 27: Asian Openbill Stork *Anastomus oscitans* and Woolly-necked or White-necked Stork *Ciconia episcopus*

VOCALIZATIONS

As in other *Ciconia* species male clatters his bill during copulation. Birds emit a hissing threat call at the nest and rather soft melodious two-toned whistles during their head tossing display. The downy nestlings also beg with continuous two-noted 'ha-chi' calls.

White-necked Stork or
Woolly-necked Stork *Ciconia episcopus* (Boddaert)

Map **28**

Synonym *Dissoura episcopus* (Boddaert)

DESCRIPTION

Head and body length 85-95 cms.
Wingspan 138-155 cms.
Wing length 44.4-49.7 cms.
Bill length 14.5-16.8 cms.

Larger than the Openbill Stork and about the same size as the Black Stork (*Ciconia nigra*) with which it looks superficially similar. However instead of the black head and neck of the former it has a downy white head and neck, except for the crown and nape which is black. Also its bill is black. There is a noticeable slaty blue area on the bare gular skin and around the eye. The body, wings and tail are black with a glossy green sheen and the belly is dull blackish brown but the breast feathers have a striking amethyst purplish gloss in good sunlight. The under-tail coverts are white. The legs are crimson and in breeding birds there is a reddish tinge to the billtip and along the commissure. The under-tail coverts are white and extend beyond

the rectrices. The iris is crimson or brownish red. Both the sexes are alike. Juveniles are dull brown not glossy purplish green but they do have white downy feathers on the neck.

HABITAT, DISTRIBUTION AND STATUS

A forest or tree haunting species and not gregarious like the other storks. It will inhabit fairly open plain areas so long as they are well dotted with groves of trees. It has never penetrated into Sind but a few have straggled westwards along the outer Himalayan foot-hills and down the rivers of the Punjab. In the 1920s Whistler found one or two pairs near the Chenab river in Jhang district, and they were probably less rare in the northern parts of Punjab at the beginning of this century than is the case today (Whistler, *JBNHS,* 1918*b*). H. Waite shot one in November 1946 in the area that is now flooded by the Rawal lake. E.

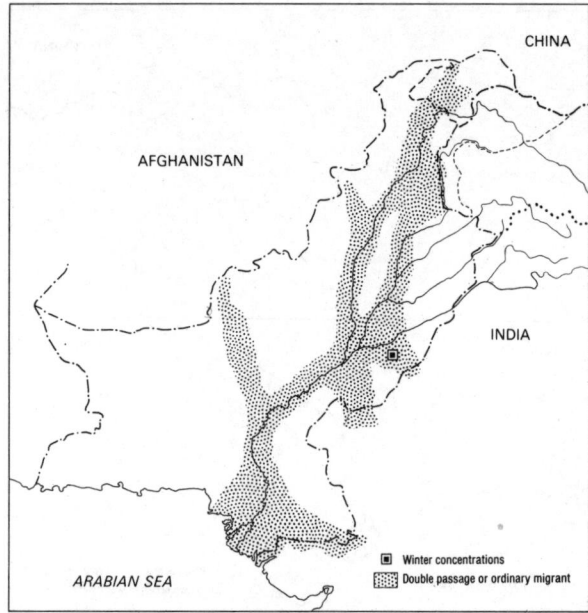

Map 28: Black Stork *Ciconia nigra*

Fernando saw one in 1968 on the Chenab river above Marala barrage which hung around from late July to early August. The author knew of a pair on the Sukh Beas just south-east of Lahore in the early 1970s and one or two pairs certainly survived on the Jhelum river around Shahpur and Bhalwal in north-western Punjab up to 1980. They are largely sedentary and must be considered very rare in Pakistan, if not extinct due to hunting by man. A locally captured bird was sold to Lahore Zoo in 1972.

HABITS

More adapted to dry land foraging than the other storks but prefer shallow marshy areas especially in irrigated forest plantations or riverain areas for foraging. They are normally not gregarious, and are believed to form permanent pair bonds. Their food consists largely of orthopterous insects, and amphibia. Fish are rarely caught except when stranded in drying out pools and they do not wade in deep water like the Openbill or Painted Storks. In Pakistan observed foraging in irrigated 'Berseen' (Egyptian Clover) crops and in drying out pools, stalking slowly along and frequently jabbing downwards with their bills to seize food prey.

BREEDING BIOLOGY

In former times it was probably fairly widespread in the northern Punjab. Whistler (1918b) recorded finding a nest on 25 December in a lofty *Acacia arabica* tree on the banks of the Chenab river near Jhang in 1917, and another nest on 23 April 1906 at Hassan Abdal in north west Punjab which had 4 incubated eggs. In northern India they nest mainly from May to July (Lowther, 1949). They are solitary nesters, though in a big enough tree, egrets may also be found nesting alongside. Both sexes share in nest building, and a fairly substantial platform of sticks and twigs is built; often with quite a deep hollow in the centre. Courtship and greeting ceremonies are similar to those of Abdim's Stork in Africa, comprising an up-down display of the head and neck accompanied by a high pitched whistling vocalization as in other *Ciconia* species. Courtship displays include bill clattering when both sexes curve their necks backwards and with bill held vertically over their backs and crown almost touching their mantle, rapidly clatter their mandibles together.

The eggs are dull white unmarked and the normal clutch 3 to 4. Both parents feed the young regurgitating food into the nest. Nothing recorded about incubating period but it is presumably about 30 days as in other *Ciconia* spp.

VOCALIZATIONS

Both sexes emit high pitched whistling calls during greeting and nest change-over ceremonies. Also both sexes use bill clattering in early courtship and possibly also as a threat gesture.

White Stork *Ciconia ciconia* (Linnaeus)

Map **29**

TAXONOMY

Vaurie (1965) describes an Asiatic (red-billed) sub-species *C. ciconia asiatica* (breeding in Turkestan), as averaging noticeably larger in size both in bill length and wing length. Most authorities (Ali and Ripley, 1968, Cramp *et al.,* 1977, C. Luthin *in litt.,* 1987) consider this population doubtfully distinct and do not accept it as a valid sub-species. Specimens collected from Baluchistan (Meinertzhagen, 1920) correspond in size to the form *asiatica.*

DESCRIPTION

Body length 100-115 cms.
Wingspan 155-165 cms.
Wing length 62-67 cms. (Stuart Baker)
    C. c. asiatica average wing length 61 cms. (14 males) Vaurie (1965)
    C. c. ciconia average wing length 58 cms. (9 males) (Cramp *et al.,* 1955)
Bill length 150-190 mm. (males), 140-170 mm. (females)
    (Brown, Urban and Newman)
Weight 2,610-4,400 gms. (males), 2,275-3,900 gms. (females)
    (Brown, Urban and Newman)

Sexes are alike. The whole head,neck, body and wing shoulders are white, also the short rounded tail. The flight feathers and tertials are black and the bill and legs are bright coral red with a small area of naked black skin around the dark brown iris. Juveniles are similarly patterned to adults, but have blackish bills well into their first year, gradually turning to red. Downy chicks have a black bill with white body down, and with pink legs which later turn grey-black.

HABITAT, DISTRIBUTION AND STATUS

A winter migrant to Pakistan. At least some of the population wintering in the north-west of the sub-continent breeds in Europe (Cramp *et al.,* 1977). A nestling ringed in Braunschweig, Germany was recovered the following winter in Bikaner, Rajasthan (Ali and Ripley, 1969). Most of the European breeding population is however, known to winter in Africa. The major population probably breeds in the USSR south of the Aral Sea in Turkestan as the race *C.c. asiatica* has been identified on passage in Afghanistan (Paludan, 1959) and a specimen assigned to *asiatica* collected in Baluchistan (Meinertzhagen, 1920). They have been observed on passage in the spring up the Kurram valley, and a fresh skin was shown to the author on 27 March 1973 at Gulistan in north-west Baluchistan.

Though it was rare in lower Sind in Ticehurst's day (*Ibis,* 1923), a small number appear regularly to winter along the Rann of Kutch Pakistan border in open marshy areas, as well as in Thatta district. 300 were seen at Ladiun lake in Thatta district, 16 November 1969 (C.D.W. Savage, pers. comm.), 57 at Mahboob Shah lake (author) 8 December, 1974 and 60 on 5 November, 1976. The potential number of White Storks migrating through Pakistan and staying for brief periods might be much greater. As Vaidya, Conservator of Forests in Bhuj, Gujarat State, India (*JBNHS,* 1986), came across a huge congregation estimated at over 3,000 birds on 2 December, 1984 on Dhand lake, located on the Great Rann of Kutch border (exact location not given but this is the region which borders Tharparkar and Badin districts of Sind). On passage 3 have been seen (author) on 31 March, 1983 at Nammal lake in the Salt Range and 5 at Sahiwal on 28 Febuary, 1970. Status SCARCE.

HABITS

Gregarious on their wintering grounds and especially on migration, though usually nesting solitarily in the breeding areas. Occasionally like Black Storks (*Ciconia nigra*) they will roost out in the desert. Kenneth Eates (MS notes) saw a party of about 50 in the middle of the day near Umarkot, some of which were sitting on their tarsi and appeared to be sunning themselves. One was seen by the author perched on the cross-bars of a

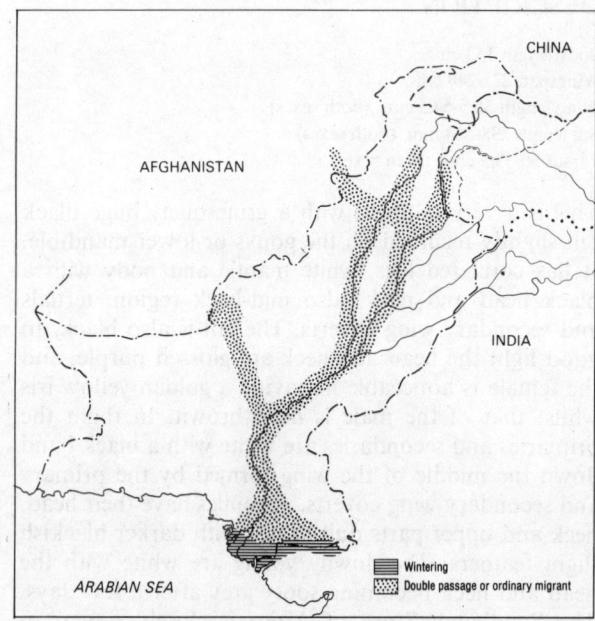

Map 29: White Stork *Ciconia ciconia*

telephone pole near Thatta and they will freely perch on trees. They spend a large part of each day like Black Storks, circling high in thermals often in association with raptors. They forage often on grassy dry land areas rather than in water, and are known to feed on orthopterous insects especially locusts. The European population, well studied, feeds on frogs, mice and insects, with more rodents in dry years, more aquatic insects in wet years (Cramp *et al.*, 1977). They also take amphibia and skinks (*Mabuya macularia*) which fairly swarm in damp grassland around Thatta lakes. In Europe they have been reported killing and eating nestlings and snakes.

BREEDING BIOLOGY

Very well studied in Europe. Their large stick nests, which are often placed on the roofs of buildings in towns and villages in western Europe, have long been associated with European folklore, and the birds are protected. They also nest on trees and cliffs and occasionally in small colonies. Nesting sites are often reoccupied year after year, and monogamous long-lasting pair bonds are formed. Both sexes share incubation and feeding duties. The plain white eggs number 3 to 5, and take 30 days to hatch, the chicks

emerging over a 5 to 10 day period, as incubation starts with the first egg. Young are fed by regurgitation and they beg by performing a head bobbing motion and with a mewing repeated call. When very small, food is regurgitated onto the nest floor, and the chicks are fed bill to bill till they learn to pick up food themselves. After they are bigger, they start grabbing their parent's bill to stimulate regurgitation.

VOCALIZATIONS

A practically voiceless bird, so that displays are silent except for the well described bill clattering in which each bird arches back its neck until its crown touches its mantle and the bill, pointed vertically, is loudly and rapidly clattered together. During copulation the male clatters his bill slightly, more slowly and softly.

On 10 December watching a flock of 10 birds circling over Chatteji lake, one individual was clearly heard clattering its bill and this was repeated twice. Such a calling during flight does not seem to have been recorded. Captive birds in Karachi Zoo frequently bill clatter with bill stretched skywards, from April to early May but other courtship behaviour has not been observed. The downy young have a cat-like begging 'mew' call.

Black-necked Stork *Ephippiorhynchus asiaticus* (Latham)   Illus. **15**

Synonym *Xenorhynchus asiaticus* (Latham)

DESCRIPTION

Body length 152 cms.
Wingspan 225-240 cms.
Wing length 56.5-64.5 cms. (both sexes)
Bill length 298-324 mm. (both sexes)
Tarsus 30-33.3 cms. (both sexes)

This is a very tall bird with a grotesquely huge black bill slightly recurved on the gonys or lower mandible. It has coral red legs, white mantle and body with a black head and neck, also mid-back region, tertials and secondary wing coverts. The tail is also black. In good light the head and neck are glossed purple, and the female is noticeable in having a golden yellow iris whilst that of the male is dark brown. In flight the primaries and secondaries are white with a black band down the middle of the wing formed by the primary and secondary wing coverts. Juveniles have their head, neck and upper parts dull brown with darker blackish flight feathers. The downy young are white with the head and neck becoming sooty grey after a few days. The Saddle-bill Stork of Africa is closely similar in appearance and like its Asiatic counterpart, the males

are slightly taller and brown eyed, the females having bright golden irides.

HABITAT, DISTRIBUTION AND STATUS

Normally sedentary but odd birds may wander after the breeding season, and occasional individuals have visited Lal Sohanran lake in the 1960s and 1970s. One pair from this area was captured in the early 1960s and exhibited at Bahawalpur Zoo. A pair was watched (author) on 30-31 January during a two day visit to Lal Sohanran in 1968 and a single bird again seen in 1970. A party of eight Black-necked storks was seen by E. Fernando on the Chenab river in Sialkot district, upstream of Marala in September 1966 and 2 were again seen on the same river in early April the following year (E. Fernando, pers. comm., 1968). A pair of Black-necked storks visited Haleji lake for about 10 days in March 1986 and one of these was photographed by Khan Mohammad Khan, Conservator, Wildlife, Government of Sind (pers. comm., May 1986).

H. Whistler saw 2 in December 1910 on the Sohan river near Rawalpindi, and Ticehurst considered them 'not uncommon' in lower Sind at the beginning of the century. A few bred in the mangroves of the Indus delta up to the late 1970s but they have not been observed or located in the past six years. Holmes and Wright saw a party of one adult and 3 immatures at Manchar lake in August 1965 and also encountered a few along the Indus delta the year before. Status must be considered no more than an OCCASIONAL STRAGGLER to Pakistan at present.

HABITS

Though they are believed to be monogamous and form stable pair bonds they seem to forage quite separately and to maintain winter feeding territories (author obs. Bharatpur, India). They are extremely shy and wary. They normally hunt by wading in water in lakes or seepage pools searching for fish and amphibia and one at Lal Sohanran was observed catching and very quickly swallowing a fish, about 15 cms. long, but have been seen catching fish up to 40-45 cms. in length, killing them by beating them on a hard surface before swallowing (C. Luthin, *in litt.,* 1987). Snakes are also consumed as well as small mammals, and one at Bharatpur sanctuary (India) was seen successfully catching and killing a Coot *Fulica atra* which was consumed (Stanley Breeden, pers. comm., 1983).

BREEDING BIOLOGY

They are thought to form a life-long pair bond, and nest solitarily, normally selecting very tall trees, and if undisturbed will re-use the same nest year after year. Kenneth Eates (MS notes unpublished) found three nests in the 1930s and 1940s, one on 11 November with 4 partly incubated eggs. near Sarkrand in Sukkur district, another with 4 downy young near Hyderabad in early December in a huge mango tree (*Mangifera*), and the third near the banks of the Indus river in a huge Pipal tree (*Ficus religiosa*) near Kotri also in November. Both sexes share in nest building which can become a huge structure from 1.5 to 2 metres across. Both sexes also share in incubation and feeding their young. The eggs are a dull creamy-white and normally 3 or 4 eggs complete the clutch. The parents

Illus. 15: Black-necked Stork *Ephippiorhynchus asiaticus,* adult female showing yellow iris.

regurgitate food onto the nest floor. E. N. Lowther, (1949) describing twenty-eight hours watching a nest from a hide, only saw the female guarding the downy chicks all day and the male never visited the nest. Also feeding was only carried out after sunset or very early in the morning. Nothing is known about incubation period or growth rate of the young.

VOCALIZATIONS

Young have been heard to utter begging calls described as 'chak' followed by 'wee-wee-wee'. Adults are silent. Courtship involves both sexes facing each other and approaching with wings out-stretched and bending forward their necks until their bills are almost touching (author obs. late March, Bharatpur, India). They then flutter their opened wings and at the same time clatter their mandibles together in a loud rattle. A similar greeting ceremony has been described on the nest with bills pointed downwards at a 45° angle and extended wingtips almost touching, accompanied by loud bill clattering (Kahl, 1970). When another Black-necked Stork flew into a feeding bird (10 February) it advanced on the newcomer, neck lowered and bill extended and rattled its bill. This threat gesture put the other bird to flight (observed Bharatpur, 1983).

Adjutant Stork or Greater Adjutant *Leptoptilos dubius* (Gmelin)

DESCRIPTION

Body length 136-160 cms.
Wingspan 210-255 cms.
Wing length 80-82 cms. (Stuart Baker)
Bill length 320-345 mm. (Stuart Baker)
Tarsus 32-33 cms.

This is a huge stork with a naked head and neck and very broad long dull yellow bill. In adults there is a peculiar pendant fleshy sack on the throat and the whole of the naked head is yellowish brown with a more pink tinge on the throat area. The back and wings are slaty grey with a white downy ruff around the lower neck and shoulders. The breast and under-tail coverts are dull white and the legs are greyish white. Unlike other storks in flight it tucks its neck in like the ARDEIDAE. The secondary wing coverts are much paler grey forming a noticeable wingbar when viewed from above in flight and this paler bar along the wing is also visible when the bird is standing (author obs., Kuala Lumpur). The Lesser Adjutant (*Leptoptilos javanicus*) lacks the throat pouch and the upper-wing surface is uniformly dark slaty. The Greater Adjutant when adult often has smeared black spots around the crown, neck and gular pouch. Juveniles have a sparse covering of downy spiky feathers on the crown and neck and also lack the pendular gular pouch. The African Marabou Stork, a close relative also looks very similar. A fossil *Leptoptilos siwalikensis* stork has been found in the Punjab Salt Range Miocene deposits.

HABITAT, DISTRIBUTION AND STATUS

In Lahore Zoo throughout the late 1960s and up to the early 1980s a captive specimen was on exhibit captured on the Ravi river near Shahdara in the late 1950s. This stork breeds mainly in Assam (India) and is believed to have become extinct as a breeding bird in Burma (C.

Luthin, *in litt.*, 1987). During the non-breeding season it wanders widely in Bangla Desh and India. Indivi-duals, particularly juveniles may wander many thou-sands of miles, mainly during the monsoon and well after the breeding season, which is from October to January. Ticehurst records a specimen exhibited in Karachi Zoo during World War I captured from Manchar lake (Dadu district). There are earlier records from around the Indus river around Rohri and east Narra (Butler, Murray and Watson cited by Ticehurst, 1923). Eates saw a party of 3 on 16 December 1930 on a sandbank on the Indus river near Sukkur. They are VAGRANTS to Pakistan occurring very irregularly.

HABITS

It is a carrion eater like its African relative and because of its large size will also take any available animal prey from snakes to lizards, as well as fish and amphibia. A specimen shot by James Murray the Curator of Karachi Museum, at Jhimpir, 50 miles north of Karachi had 6 large Spiny-tailed Lizards (*Uromastix hardwickei*) in its stomach, (Murray, *Stray Feathers*, 1878).

BREEDING BIOLOGY

Not much recorded but in the now extinct colony in Burma, they built huge stick nests both on cliffs and on the tops of tall trees (Stuart Baker, 1929). They nest colonially. The usual clutch is only 2 or 3 eggs which are white in colour. Both sexes share in nest building and rearing of the young but the period of incubation has not been recorded. Though they lack vocal muscles like other storks, besides bill clattering as a greeting ceremony, they emit loud grunts similar to the sounds made by a water buffalo (Stuart Baker, 1929).

FAMILY THRESKIORNITHIDAE

Glossy Ibis *Plegadis falcinellus* (Linnaeus)

Old name *Falcinellus igneus*

Illus. **16**
Map **30**

DESCRIPTION

Body length 55-65 cms.
Wingspan 80-95 cms.
Wing length 28-36 (7 males), 26.7-28.1 cms. (7 females)
  (Brown, Urban and Newman)
Bill length 126-141 mm. (11 males), 106-114 mm. (8 females)
  (Brown, Urban and Newman)

Slimmer and lighter in build than the White Ibis (*Threskiornis melanocephalus*) it has similar long slender neck, long legs and slender-down curved bill. The whole of the head is feathered and the general body colour is dark sooty brown, in good light with a maroon tinge on the mantle, neck and breast, with the wings glossed bronzy greenish on the wing coverts, and amethyst on the scapulars. In poor light the birds appear quite black. The bill is greyish green and a small area of naked skin around the eye and lores is leaden blue. The legs and feet are bronzy greenish, and are quite stout. The middle toe has a pectinated claw and the tarsus is scutelated in front and reticulated behind. The sexes are alike, though females average slightly smaller in size. In winter plumage there may be grey or whitish flecks on the head and neck of adult birds. In flight their head and neck remains outstretched and the feet and legs extend well beyond the square tail. Downy chicks, are covered with sooty black down and have pinkish legs and bill, the latter with a black tip and black bar in the middle of the bill.

HABITAT, DISTRIBUTION AND STATUS

Absent from mangroves or saline areas, and restricted to larger lakes, swampy areas with flooded grassland or extensive rice growing tracts. This is a largely resident species, moving sporadically in response to changed water and feeding conditions and with some influx of winter migrants from the USSR. Meinertzhagen (*Ibis*, 1920) noted flocks in Baluchistan on Kushdil Khan lake in the spring and as late as 7 July and thought the earlier flocks were on passage. Eleven were seen (author) on this lake on 7 May 1979. In lower Sind it is largely sedentary and it is comparatively rare in the Punjab. Hugh Whistler (*Ibis,* 1930) did not record it around Rawalpindi district and only saw it once in Jhang in the 1920s though H. Waite, in 20 years, collected one specimen on 7 May near Gujrat, Punjab and another on the Jhelum river at the foot of the Salt Range on 18 October. It has been seen occasionally at Lal Sohanran lake, e.g. a party of 15 was seen on 10 April. (F. Koning, pers. comm.), and very occasionally in the Punjab on spring and autumn passage, e.g. a flock of 10 at Pirawala plantation 17 April flying northwards, and on 15 September a flock of over 90 flying southwards near Chasma barrage on the Indus river (C.D.W. Savage, pers. com.).

In lower Sind it is much more frequently encountered. At Ghauspur lake in northern Sind, parties of Glossy Ibis are regular winter visitors. 80 seen 31 October, one year, 4 on 14 November in the following year. Manchar lake always has a few, 35 seen on 17 December and 25 on 21 March in another year. In Thatta district a flock of about 170 birds spends most of the year in the seepage pools around Haleji lake and a flock of 400 was seen on Ladiun lake, southern Thatta district on 14 November (D. Holmes, pers. comm.), and 130 near Badin in south-eastern Sind 17 February (F. Koning, pers. comm.). In the rice growing season they often disperse into the paddy fields and 9 were seen thus near Chauhar Jamali (Thatta district) in September. In the east Narra on 14 February 9 birds were seen by Sadori lake. Status COMMON only in Thatta district, and winter migrants appear to spread further south into India.

HABITS

A markedly gregarious species in its winter roosting and feeding habits, as well as during migration and breeding. It tends to fly in small flocks in 'V' formation even when travelling only short distances between feeding grounds. It will allow quite close approach when feeding on the ground and is not very shy. Its food consists primarily of worms, molluscs and insects such as water beetles *Hydrophilus* sp. but it will also eat frogs, tadpoles, fishes and small reptiles when

Illus. 16: (a) Oriental White Ibis *Threskiornis melanocephalus,* adult in breeding plumage. Note elongated flank feathers showing below wing, behind legs.

(b) Glossy Ibis *Plegadis falcinellus*

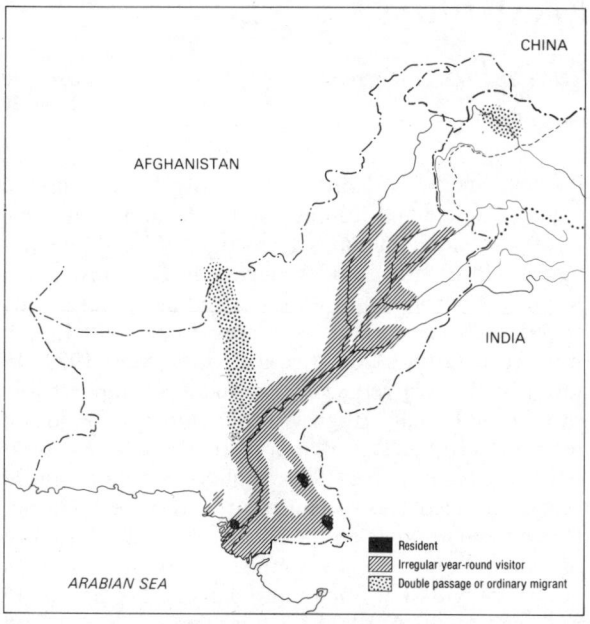

Map 30: Glossy Ibis *Plegadis falcinellus*

opportunity allows. It also feeds on ODONATA and their larvae.

BREEDING BIOLOGY

Doig found them nesting with Spoonbills (*Platalea leucorodia*) and White Ibis (*Threskiornis melanocephalus*) in thorn trees (*Prosopis spicigera*) on the edge of a remote lake in the sand dunes bordering the east Narra in June 1879. The normal clutch of eggs in this colony was three. During the Second World war K. Eates (MS notes unpublished) located a colony of about 40 nests south of the Narra canal, near Jabhrao in Tharparkar district. On 27 August nests had 2 or 3

eggs each and were located in submerged *Prosopis* trees growing around the fringes of a lake, with up to 3 nests per tree, made of sticks and not very large or substantial structures. This colony was in association with several egret and heron species. The eggs are a deep blue-green colour without markings, and 2 to 3 eggs are the normal clutch. Both sexes share incubation duties which last 21 days. The eggs despite being laid at 24 hour intervals, hatch synchronously. For the first few days the male brings food to the nest and passes this to the female who gives this to each offspring bill to bill (Cramp *et al.*, 1977). The female then leaves to forage while the male guards their offspring. The young are continuously guarded by one parent until they are 9 to 14 days old. After the first 5 to 6 days both parents bring food which they at first regurgitate bill to bill but later direct onto the nest floor (K. Eates, MS notes unpublished and Cramp *et al.*, 1977) (See differences in feeding behaviour of White Ibis chicks).

The young wander about from the nest when about 3 weeks old, clambering over branches. They may take up to 6 weeks before they can fly and even after this are still occasionally fed by their parents. The pair bond is for the season and birds have been observed mutually preening each other and rubbing each other's bills at the nest and after pair formation. Returning birds often bring a stick or grass to the nest even during incubation (Lowther, 1949). Nothing seems to be recorded about courtship.

VOCALIZATIONS

Normally silent but in winter quarters occasionally one or more individuals utter a guttural almost 'corvid' 'grarrrh' 'graarh' (heard 24 January at Haleji lake). At the nest, are said to utter a hoarse 'cough-like' call as part of their courtship ceremony (C. Luthin, *in litt.*, 1987).

Black Ibis *Pseudibis papillosa* (Temminck)   Map **31**

Synonym *Inocotis papillosus* (Temminck)

DESCRIPTION

Body length 60-68 cms.
Wingspan 90-115 cms.
Wing length 36.5-40 cms. (Stuart Baker)
Bill length 138-158 mm. (Stuart Baker)

Slightly smaller than the White Ibis (*Threskiornis melanocephalus*). Only the head and nape are devoid of feathers and the occiput is covered with bright scarlet warts or carbuncles, the rest of the face being

black. The body and wings are glossy black and the long down-curved bill is more slender than that of the White Ibis and dull blackish green in colour. The legs and feet are conspicuous brick red, a good field point in distinguishing it from the Glossy Ibis (*Plegadis falcinellus*). There is a prominent white patch on the wing-shoulder in flight but this does not always show when the bird is at rest being covered by the mantle feathers. Immature birds are a much duller sooty brown, lacking the greenish gloss of adults, and their heads are also feathered.

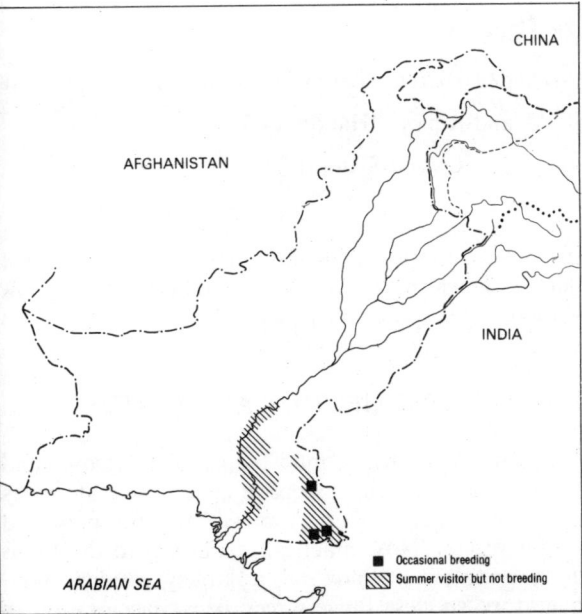

Map 31: Red-naped or Black Ibis *Pseudibis papillosa*

HABITAT, DISTRIBUTION AND STATUS

More eclectic in its choice of feeding areas, frequenting dry fallow fields or grassland as well as the drying out margins of lakes and swamps. In Pakistan it is now only an irregular monsoon season vagrant to the better watered parts of Sind along the Indus river and the eastern border regions of Tharparkar. In Ticehurst's time (1923) it was considered common and he encountered 'large flocks around most jheels'. Derek Holmes once encountered a solitary bird in September, 1964 in a tree along the Narra canal near Sukkur. A few birds more recently have been recorded around Chachro village in Mithi taluka on the eastern borders of Tharparkar. Status RARE, mainly occurring during monsoon.

HABITS

Unlike the other Ibises less gregarious in foraging and better adapted to dryland areas supplementing their diet with grains. They are quite crepuscular or nocturnal in feeding activity (author obs., late March,

Chitawan, Nepal) and will forage in muddy pools, the edges of swamps as well as in cultivated fields and stubbles. They feed on small reptiles, snakes, arachnida including scorpions, crustacea and a variety of insects, as well as small fish and amphibia.

BREEDING BIOLOGY

Not colonial in nesting and the pair chooses a tall leafy tree as a nest site often a considerable distance from water. Tree groves associated with old grave yards in Tharparkar district are particularly favoured and K. Eates found nests in such localities around Chhach village and Mithi village during the years 1945 and 1948 with the nesting season being quite variable. A clutch of 3 eggs was found on 8 February and 2 eggs on 14 August. Mr. Bell of the Forest Service found a nest with fresh eggs on 12 May. The normal clutch is 2 to 3, 4 being occasionally laid and the eggs are pale blue in colour occasionally with sparse spots and blotches of reddish brown. E. N. Lowther (1949) comments on their preference for thorny twigs for nesting material even if suitable trees are located some distance away. The nest is built mainly by the female with material brought by the male, and has quite a pronounced depression in the centre. Both sexes share incubation duties and feeding of the young by regurgitation. Incubation takes about 25 to 27 days and like most of the THRESKIORNITHIDAE the chicks hatch over a period of 5 to 7 days as incubation starts with the laying of the first egg. Very little seems to have been recorded about the courtship of this species, but they are somewhat gregarious before pair formation and males indulge in a lot of loud calling (quite unlike other Ibis species in Asia). Also stick manipulation and later during pair formation, stick presentation to the female (extra-limital observations Chitawan, Nepal).

VOCALIZATIONS

Males will advertise with loud wailing cries even during night time hours both in flight and when perched at the top of a tall tree. These calls are somewhat reminiscent of the calls of some large raptors, a wild screaming call repeated two or three times. When heard more closely the call sometimes has a rather nasal querulous tone.

Oriental White Ibis or

Black-headed Ibis *Threskiornis melanocephalus* (Latham) (Voous, 1973)

Threskiornis melanocephala (Latham) (Ali and Ripley's 'Handbook')

Threskiornis aethiopica (Latham) (Ripley's 'Revised Synopsis')

Illus. **16**
Map **32**

TAXONOMY

I am inclined to agree with the view held by the editors of *The Handbook of the Birds of Europe, the Middle East and North Africa* (Cramp *et al.*, ed., 1977) that due to long geographic separation and genetic isolation the African population *Threskiornis aethiopica* is a separate species from *T. melanocephalus*, the form breeding in south and east Asia. The African population has conspicuous black filamentous scapulars and tertials whereas in the Indo-Pakistan population these feathers are grey and rather inconspicuous. In the breeding season the Oriental White Ibis develops elongated lanceolate flank feathers, a feature not mentioned (Brown and Urban *et al.*, 1982) in describing the plumage of the African Sacred Ibis *T. aethiopica*. See comments in Chapter titled 'The Problem of Species'.

DESCRIPTION

Body length 65-75 cms.
Wingspan 112-124 cms.
Wing length 34.3-37 cms. (both sexes) (Stuart Baker)
Bill length 139-170 mms. (both sexes) (Stuart Baker)

Slightly larger than the Glossy Ibis and about the same size as the Spoon-bill, this bird has stout black legs with long tarsi and a naked black head and upper neck. Both sexes have all-white plumage though the feathers of the flank are tinged yellowish brown. The bill is long and down-curved and broader at the base than in *Plegadis falcinellus*, the culmen being more keeled and the upper mandible bearing longitudinal grooves. Juvenile birds have a covering of sooty grey down on the head and hind neck with short white feathers on the foreneck. The sexes are alike. Under the wings the humerus and ulna are naked in appearance with the naked skin bright fleshy orange in winter, turning bright crimson-red in the breeding season. The filamentous ashy or black scapular and tertial plumes develop in the breeding season and the yellowish flank feathers also become elongated and lanceolate in shape. The feathers at the base of the neck also become long and narrow in the breeding season. In juvenile birds this bare under-arm area, is black. Downy chicks are white with black heads. The throat area is however naked and pinkish purple and their bills are greyish pink.

In winter plumage adults lose their ornamental scapular plumes. The iris is red-brown and in a captive specimen examined at Korangi creek, the base of the bill was greenish grey not black, and there were a few scattered blue spots on its crown.

HABITAT, DISTRIBUTION AND STATUS

Frequents tidal creeks, mud banks in mangroves and salt marshes, as well as seepage zones alongside rivers and larger inland lakes. A marsh and wet-grassland loving species. Now practically confined to the Indus delta and lower Sind and a largely resident non-migratory species. Biggest recorded concentration in recent decades 100 birds seen 30 November 1965 at Ladiun jheel, Thatta district (Derek Holmes, pers. comm.). Usually seen in small groups. Nine seen on salt marshes at Keti Bandar on 20 February, 13 seen Dho jheel due east of Chauhar Jamali on 3 December. Occasional birds wander as far west as Sonmiani lagoon, Khinjar lake, Haleji (flock of 4 throughout November-December 1980 at Haleji lake, Karachi district). Two seen in January at Khadeji falls in hill tract north of Karachi. They probably no longer breed in the east Narra where K. Eates found breeding colonies in 1930-1 on the Karo Naro (MS notes unpublished). E. Fernando saw 2 White Ibises near

Map 32: Oriental White Ibis *Threskiornis melanocephalus*

Marala barrage on the Chenab river, north Punjab, in August 1967 (pers. comm., 1968) though there are no other Punjab records. Status, confined to lower Sind, where SCARCE.

HABITS

Locally nomadic according to water and feeding conditions and in Pakistan they appear less gregarious than Glossy Ibis outside of the breeding season. They frequently associate with grazing buffaloes, possibly benefiting from insects disturbed by the animals. Known to feed upon fresh water snails, insect larvae and particularly beetles and amphibia, tadpoles and crustacea, often wading in shallow water and submerging their whole head and neck. In coastal or salt water areas probably more fish included in diet and small crustacea. K. Eates who shot specimens which had been feeding in the tidal creeks, found their stomachs full of small crabs while those shot in the east Narra (fresh water) had been feeding on fresh water snails (K. Eates, MS notes unpublished). They gather in small flocks to roost at night in mangroves, or on land in flooded tamarisk trees. They often associate with Spoonbills and Egrets, feeding amongst the former without any signs of inter-specific aggression (author obs., Haleji lake). In summer months birds were observed to cool themselves by a rapid gular flutter (author obs. captive birds, Karachi Zoo).

BREEDING BIOLOGY

Doig (*Stray Feathers*, 1880) found this Ibis breeding in July in the east Narra district, when it was much commoner and widespread than it is today. Here they nested in a separate association on a large submerged partly fallen tree, with nests no more than 0.5 metres apart in the midst of a much bigger colony of egrets and herons. There is no recent evidence of this Ibis still breeding in the east Narra. Egg predation is heavy both by man and House Crows (*Corvus splendens*) and this sometimes results in birds attempting to breed again as late as October and November. Two small colonies known in the remoter mangrove creeks of the Indus delta (10 nests in Pitiani creek in 1972), and Khudi creek about 15 nests (1945) (K. Eates, MS notes unpublished). Breeding normally starts in July at the onset of monsoon rains. The normal clutch is 2 to 3 eggs which are white, sparsely speckled with red-brown. The nests built of twigs with a fairly pronounced cup were almost touching in these mangrove trees. The male is believed to bring most of the nest material which the female weaves into the structure. The pair bond is monogamous for that season. Courtship has not been observed but established pairs indulged in mutual neck and bill rubbing and raising and lowering their bills with neck extended as though bowing. The young are fed by regurgitation bill tip to bill tip, and when they get older, by inserting their bills right into their parents' mouths, and unlike storks, regurgitation of food onto the nest floor has not been observed (Lowther, 1949).

Incubation period is about 23 to 25 days and the young fledge in about 40 days. One parent guards the chicks on the nest continuously during the first seven to ten days but predation is still heavy from Fish Eagles (*Haliaeetus leucoryphus*) (Ali and Ripley, 1968) as well as House Crows (*Corvus splendens*) (K. Eates, MS notes).

VOCALIZATIONS

At the nest adults make wheezing noises during greeting and nest relief. Young, when small, solicit food with a repeated high pitched 'chick-chick-chick' chattering call, rather like the begging cries of young egrets. Later when partly fledged they beg with a piping squeal 'scree-scree-scree' (Lowther, 1949). At other times of the year the Ibis is very silent.

Spoonbill *Platalea leucorodia* Linnaeus   Map **33**

DESCRIPTION

Body length 85-95 cms.
Wingspan 120-135 cms.
Wing length 35-39.5 cms. (Stuart Baker)
Bill length 180-228 mm. (Stuart Baker)
Weight 1,323-1,960 gms. (3 specimens, Brown, Urban and Newman)

About the same size and bulk as the White Ibis, it can often be separated from egrets at a distance by its thicker shorter legs and much less upright body carriage. Its plumage is all-white, legs black and the long bill is also black. When viewed from the side this is slender and down-curved at the tip, but when the bird turns its head the broad rounded spatulate tip from which its name is derived at once becomes noticeable. The bill is flattened dorso-ventrally and bears fine horizontal corrugations along the upper surface and the tip is yellow in the breeding season. In nuptial plumage both sexes are alike and develop a spiky occipital crest of pale yellow feathers, extending

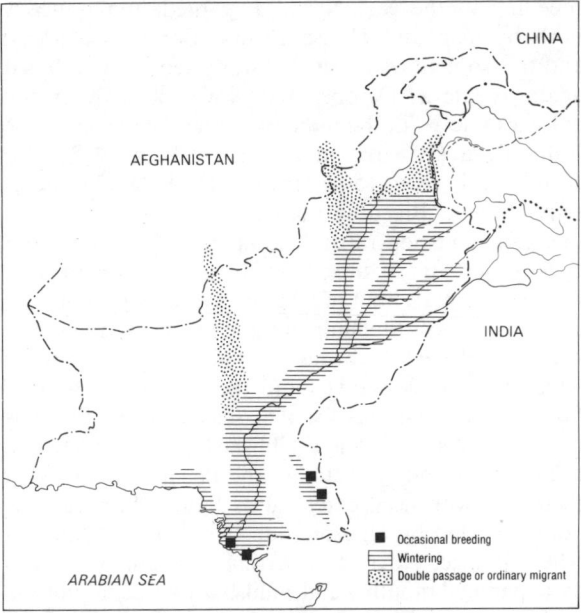

Map 33: White Spoonbill *Platalea leucorodia*

down the back of the neck, also a brighter yellow patch on the lower breast and the naked skin under the chin and throat also becomes sulphur yellow. Juvenile birds have duller grey bills with fleshy pink mottling and in flight their primaries and secondaries show small black tips. The downy chicks are white with more rounded bills, like an ibis chick, and of a fleshy pink colour. The bill becomes more flattened and develops a wider spatulate tip as the chick becomes fully fledged. In adults the iris is dark brown. In flight Spoonbills, like the Ibises, carry their head and necks fully extended at all times.

HABITAT, DISTRIBUTION AND STATUS

Associated with larger lakes and rivers, they are wholly adapted to feed in shallow water. A large portion of the population which visits Pakistan in winter is migratory and available evidence from numbers sighted on migration (Baran dam, Kurram river, late March), during winter and summer months, shows that less than 33 per cent of the population is sedentary and breeding. There is some evidence that they migrate across Baluchistan as well as up to Kurram valley and the Indus. At Kushdil Khan lake in Baluchistan, Meinertzhagen noted flocks of over 100 on 26 October and a party of 14 on 17 May. About 30 were seen (author) on 31 October in 1972 but only 3 birds on 22 March 1974. On the Indus river two flocks totalling 150 birds were seen flying high in 'V' formation above the river on 26 March (C.D.W. Savage, *in litt.)* and two flocks totalling 130 birds landed at Baran dam

reservoir on 25 March on the Kurram river during a visit by the author. Small groups of 6 or 7 birds have been seen resting besides Rawal lake, Islamabad, both in October and late March, when on passage (D. Corfield, pers. comm.). These observations have been corroborated by the ringing recoveries from two widely separated breeding colonies. One Spoonbill ringed at a nesting colony in western Turkey in Man Yas Golu, was recovered in November in Pakistan (Schuz, 1957). Another ringed near Yeysk, Sea of Azov in the USSR on 10 June 1961 was recovered two years later in June 1963 near Hyderabad.

During winter months they are gregarious and scattered flocks haunt the Indus river and its main tributaries, roosting by day on inaccessible sand banks. They also occur in the mangrove creeks in small numbers and around all the larger lakes where waterfowl concentrate. Groups of 80 and over 300 have been seen during December and January on the Indus river between Jinnah headworks and Chashma barrage in two different years. On the Salt Range lakes occasional single birds have been noted, and 7 juvenile birds on 23 March on Nammal lake. Thirteen at Trimmu headworks on 5 October, and 6 on 30 January at Lal Sohanran. In Sind bigger concentrations are more commonly sighted. 82 on Haleji lake were seen throughout December and January in one year, and 175 for a shorter visit in another year. Over 250 at Phoosani lake near Tando Bhago, southern Sind, were seen on 17 February (F. Koning, pers. comm.). In the mangroves small groups can be seen in almost any month of the year and these are probably largely sedentary, e.g. 12 seen on 12 May at Sonmiani, Lasbela, 29 seen at Ghizri creek on 30 May a year later, and 7 seen near Karachi harbour in mangroves 7 January. Status FREQUENT.

HABITS

Gregarious in foraging and roosting. They feed most actively in the early morning and at dusk and sometimes throughout the night (Ali and Ripley, 1968). Over 2 dozen birds have been watched (author), in a tight flock wading together in shallow water, bodies bent horizontally, necks extended and their bills scything to and fro in the water or mud surface. By day they roost in flocks and spend most of the time sleeping. They often rest sitting on the whole tarsus like Storks, or standing in shallow water with head and bill tucked under their wing. In flight they tend to fly in extended 'V' formation with steady shallow wing beats and frequent gliding. Their food consists of minute crustacea and shrimp larvae, and in fresh water regions, tadpoles, insect larvae, small worms and crustacea are included in their diet. They also feed on fish, small frogs, water beetles and dragonflies according to studies in the USSR.

BREEDING BIOLOGY

A few pairs undoubtedly still nest in mixed heronries in the east Narra lakes, as well as in small colonies in the mangroves of the Indus delta. They are colonial nesters and when much more numerous, Doig found a huge colony nesting on partly submerged trees in one separate colony but alongside another colony of Painted Storks(*Mycteria leucocephala*). These had well incubated eggs when discovered on 11 November. K. Eates located a colony in the Indus delta near Shah Bunder in mangrove trees on 7 July 1945 in Khalo creek and another smaller colony on Wali creek. In 1979 a breeding colony was located near the Indian border in Haidari creek which was in association with White Ibises and egrets. They had downy young on 20 July. Unfortunately local fishermen have robbed these colonies ruthlessly in recent years, to sell the half-fledged young to animal exporters. During the breeding season the parents form a monogamous pair bond and nests are built of twigs and sticks mostly by the female, with the male bringing her the material. Even after the young are hatched the male continues to bring nest material, particularly grass or earth clods, which are added to the nest lining. The normal clutch

is 3 to 4 eggs which are white with a few scattered sepia spots. The eggs during incubation soon become stained a dirty brown. One parent continuously guards the downy chicks for the first 7 days or so and both parents feed them by regurgitation. When bigger the young birds insert their heads right inside the parents' gullet in their eagerness to get food. The parent birds also shelter their young from the sun's rays for part of each day during their first 14 days after hatching. Little has been recorded about pair formation and courtship behaviour but allo-preening after pairs are formed is an important component and when excited both sexes can erect their nuchal crest feathers in an untidy fanshaped halo, behind their heads. They are relatively silent throughout courtship procedures, but during nest relief the arriving partner emits a grunting sort of call (E.N. Lowther, 1949).

VOCALIZATIONS

Young in the nest keep up an incessant begging call which is a high pitched squealing 'cheer-cheer'. At nest relief ceremonies or arrival of partner with food, the pair emit grunting calls to each other.

ORDER PHOENICOPTERIFORMES
FAMILY PHOENICOPTERIDAE

Greater Flamingo *Phoenicopterus ruber* Linnaeus
Phoenicopterus roseus Pallas of Ali and Ripley

Illus. **17**
Map **34**

TAXONOMY

The old world population, largely breeding in Africa is often treated as a separate species *P. roseus* (Ripley, 'Revised Synosis', 1982) but in his 'Recent List', Voous combines this with the north American population as one species *P. ruber*.

DESCRIPTION

Body length 125-145 cms.
Wingspan 140-165 cms.
Wing length 39.3-44.4 cms. (Stuart Baker)
Bill length 139-164 mm. (Stuart Baker)
Tarsus. 311-327 mm. (Stuart Baker)

The enormously long legs and neck of this species result in some adult male specimens standing as high as 155 cms. Females are smaller. They have white bodies with jet black primaries and secondaries and long coral pink legs. Their feet are webbed. Under

some dietary conditions the whole body and neck is suffused a delicate shell pink colour. Irrespective of body and neck colour the upper-wing coverts in flight are a beautiful deep rosy red, At rest long lanceolate scapular plumes hide these scarlet wing coverts. The iris is pale pink and the heavy looking down-curved bill is bluish pink with a jet black tip. When viewed close, it will be seen that it is the lower mandible which is large and deep and bent like a banana though with only one keel. The upper mandible is flat on the culmen and compressed dorso-ventrally, an adaptation to its specialized feeding methods. In flight the underwing coverts are a deeper vermillion-red than the upper-wing shoulders. There is much individual variation in size, with females being consistently smaller. Both sexes are alike in plumage. Juveniles are a dull sooty grey, browner on the neck and with black legs and bills. Birds in their second year assume a whiter body plumage and black-tipped pink bill but their necks remain noticeably grey-brown and the wing coverts do not hide the back primaries even when the birds are at rest. Their legs are a muddy red. The downy chicks are white with relatively short black legs and a straight normal shaped pink bill.

HABITAT, DISTRIBUTION AND STATUS

A highly gregarious species, flamingoes move around erratically according to local feeding conditions. They prefer shallow brackish lakes or coastal mud flats.

Though normally breeding extra-limitally, they can be seen in Pakistan in all months of the year. Probably the population which visits the northern Salt Range lakes, which rarely totals more than 300 to 500 birds, breeds in Afghanistan and Iran. In 1972 over 1,000 Flamingoes visited Kushdil Khan lake in western Baluchistan on passage in October (Hamid Khan, Div. Forest Officer, pers. comm.). Some of the population breeding in the Rann of Kutch, undoubtedly spend part of their non-breeding cycle in Sind. In late January 1989 a huge concentration estimated at 40,000 was discovered on the borders of the great Rann of Kutch in southern Badin district (F. Koning *in litt.*to author). They are erratically migratory, with a flock usually frequenting the creeks around Sonmiani lagoon (Lasbela) and another in Indus delta often throughout the summer and winter months depending upon monsoon conditions and rainfall in the Rann of Kutch. Flocks have also been recorded in the mangroves near Pasni, Baluchistan, and can frequently be seen on brackish lakes in the east Narra such as Sangriaro near Sanghar. The total world population is estimated at between 500,000 and 750,000 with potential breeding sites world-wide (Alan Johnson, Camargue, pers. comm., 1984).

Typical sightings and numbers: Salt Range, Uchchali lake 200, Nammal lake 30 and Khabbaki 5, Kallar

Illus. 17: (a) Head of Lesser Flamingo *Phoenicopterus minor*
(b) Head of Greater Flamingo *Phoenicopterus ruber*

Kahar 100 (mid-December), Chatteji lake near Karachi 1,300 in winter of 1981, none in 1982, Sangriaro jheel Sanghar 160 (February), Mahboob Shah lake, Thatta district 200 (November). On seacoast 250 Ghizri creek (October to March), 50 Sonmiani (March to July). Small groups are occasionally seen on other lakes such as Ghauspur (Jacobabad district) and Manchar lake. A bird was shot for food by fishermen in 1980 on Siranda jheel, Lasbela, 96 kilometres (60 miles) from Karachi. The author was able to obtain the ring from this bird which was ringed as a 'pullus' on 2 August 1977 at Ashk Island lake, Uromiyeh, Iran (N37° 27', E45° 31'). However, most ringing in Iran has been done in the huge breeding colonies on Lake Rezaiyeh in the north-eastern part of that country and some of these birds must also winter along the Arabian Sea coastal parts of Pakistan. Status in some years COMMON.

HABITS

Colonial in nesting and gregarious in feeding, sometimes remaining for periods over several months in one feeding area. They are only found around large inland lakes or along the seacoast, favouring tidal creeks with extensive exposed mud banks, and avoiding sandy strands. They will feed and sleep at intervals throughout the day but are most active at dusk and during the night, often feeding continuously through the hours of darkness (Alan Johnson, Tour. du Valat, Camargue, pers. comm., 1984). They prefer fresh water lakes with shallow margins and of a high pH value. They have frequently been seen swimming in Pakistan, a remarkable feat considering their long legs and awkwardly balanced head and neck. Swimming, flamingoes look rather swan-like with necks carried in an 'S' bend. Observed Kharrar jheel (Punjab) 5 February in about 1.60 metres depth of water, and at Sangriaro jheel (Sind) on 13 February in about 2.4 metres.

They are normally quite shy and will take to flight if a human approaches within 200 metres. A flock in flight with long trailing pink legs, and slowly beating black and vermillion wings is an unforgettable sight.

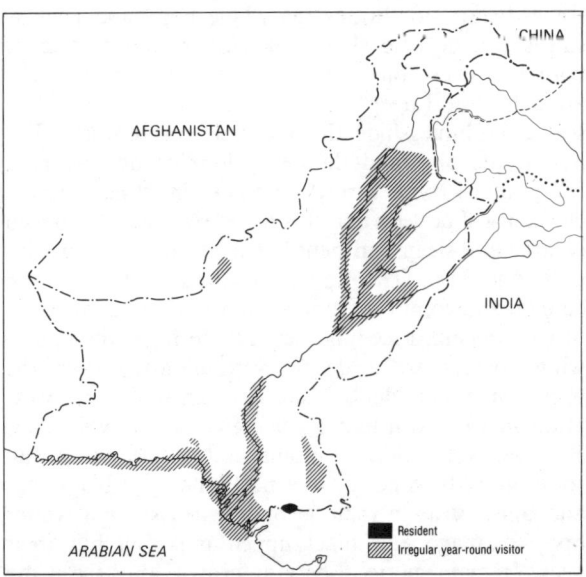

Map 34: Greater Flamingoe *Phoenicopterus ruber*

They do not fly in any particular formation but will circle on thermals and gain considerable heights when travelling greater distances. Their unique feeding methods can be observed in captive zoological specimens. The head is submerged after bending the first cervical vertebra inwards so that the crown of the head and upper mandible rest on the surface of the mud or silt under the water. The tongue acts as a pump piston, drawing water through the lamellae on the edge of both mandibles, as the bird rapidly works it up and down at the rate of about two thrusts per second with the soft gular skin visibly dilating with each stroke. This traps small micro-organisms particularly algae, micro-plankton, larvae of crustacea, particularly brine shrimps, and small insects such as CHIRONOMIDAE midges. They also eat seeds of *Juncus* and *Cyperus* sedges adapted to brackish water, and in captivity they flourish well on seeds of rice (*Oryza sativa*) and *Sorghum sudanese*. They occasionally call to each other even in non-breeding haunts with deep guttural honks and gabbling goose-like calls. Birds feeding on mud flats at Ghizri creek have been observed dabbling with one foot slightly extended, probably to flush food prey. Head remained submerged even during this procedure.

BREEDING BIOLOGY

Normally they resort to very open exposed sites where danger from predators is minimized, such as are afforded by the huge saline flats of the Great Rann of Kutch in India, where they have been known to breed since 1893 (Lester, *JBNHS.*, Vol. 8, p. 553). Here they nest in densely packed colonies usually commencing between July and October when the monsoon causes shallow flooding of the area, and the birds have some protection from ground predators. In September 1980, Syed Imtiaz Karim of the Zoological Survey of Pakistan was conducting wildfowl surveys in Badin district of Sind and reached the borders of the Great Rann of Kutch south of Gularchi where he found a breeding colony with grey downy young of varying sizes, of which he brought back 4 to Karachi Zoo (S.I. Karim, pers. comm. and *Records of Zoological Survey of Pakistan*, Vol. 10, No. 2, 1985 p. 121). This record of breeding on Pakistan's border was at approximately N24°15′, E68°45′ and 125 kilometres west of the site north of Pacham island in India where Salim Ali found them breeding in 1943 (S. Ali, *The Birds of Kutch*, 1945). It is known that initiation of nesting depends upon periodic mass displays, and that physiological synchronization is a crucial factor, with most pairs starting to nest build and lay eggs simultaneously. Occasional breeding has been recorded at Ab-e-Estada lake in Afghanistan, which is due west of Dera Ismail Khan in southern Waziristan. Human predation is a major restraint to successful breeding at Ab-e-Estada (Akhtar, 1947). At Kharrar jheel in Renala

Khurd district a number of mud nest mounds were observed in December 1967 having been recently constructed (Mountford, 1969) but there was no evidence of breeding. On Mehboob Shah lake in early November 1980 a flock of about 200 was observed carrying out initial courtship display. This seemed to be initiated by individual birds giving loud repeated honk calls and raising their bills skywards. Then small groups, more or less facing each other, stretched their necks vertically and pointed their bills skyward and opened their wings partially. They appeared to dance around a bit but without waving their wings or holding them outstretched for more than a few seconds. Subsequently, all birds in the herd appeared to move around, passing each other in different directions and calling excitedly. These two behaviour patterns viz head pivoting and then marching along in unison are essential components of courtship display in its early stages and are usually followed later by hooking (downwards bending of the head and bill whilst the neck remains upstretched). Then this is followed by wing saluting (Alan Johnson, pers. comm.). The group display observed on Mehboob Shah lake seemed to have many elements of aggression in it and was reminiscent of group displays observed in Avocets (*Recurvirostra avosetta*). Courtship can take place over many weeks even extending over several months before the flock actually selects a breeding site.

Breeding has been well described. The peculiar nests are made almost entirely of mud, being circular mounds of about 30 cms. diameter and 30-40 cms. height, surrounded by a trench resulting from excavated material. The large plain white egg is usually a singleton, but occasionally two are laid and they are pale blue with calcareous deposits on the surface. Both parents share incubation duties which last up to 28 to 31 days. Chicks are fed bill to bill in the early days by a special 'crop milk'. At 10 days they wander from the immediate nest area and join huge creches where their parents can miraculously recognize them and continue to feed them by regurgitation. At 30 days they begin to fend for themselves and are fully fledged at about 70 days after hatching.

VOCALIZATIONS

Two types of calls are commonly uttered. Loud far carrying goose-like honking not unlike the calls of *Tandorna ferruginea* and low pitched guttural gabling. The former would appear to be a contact call and the latter may be associated with various flocking and mass feeding situations. Usually they are fairly silent on the non-breeding grounds and calls may be associated mainly with mass displays or intra-specific aggressive encounters. Small young have been reported begging with high pitched 'kwick-kwick' calls, persistently repeated.

Lesser Flamingo *Phoenicopterus minor* Geoffroy

Synonym *Phoeniconaias minor* (Geoffroy)

DESCRIPTION

Body length 80-90 cms.
Wingspan 95-100 cms.
Wing length 32.9-35.4 cms. (Stuart Baker)
Bill length 100-118 mm. (Stuart Baker)
Tarsus 190-242 mm. (Stuart Baker)

It is similar in proportions to The Greater Flamingo but adult males stand only 100 cms. high compared to 140-150 for *P. ruber*. The most noticeable distinguishing feature is the slightly deeper heavy down-turned bill which looks all-black at a distance. When viewed closer there is a round dark crimson patch on the lower mandible and the base of both mandibles are suffused dark crimson-brown. The neck and body is often suffused a more distinct rosy pink than in the Greater Flamingo with one or two crimson feathers streaking the scapulars and wing coverts. The flight feathers are black. The iris is red with dark red orbital ring not pale pink as in the Greater Flamingo and the legs are also more vermillion with noticeably darker joints, and in winter time they often appear greyish pink. Juveniles are greyish brown all over with some darker brown streaking on wing coverts and the black flight feathers partially showing when the bird is at rest. The legs are dark grey and the bill dark reddish purple to brownish red. In the field both flamingoes can appear almost white due to variations in diet and they are difficult to separate unless the bill shape and darker eye can be seen. In flight under-wing pattern of adults with uniformly pink under-wing coverts and contrasting dark crimson triangular patch of axilliaries clearly separate this species from the Greater Flamingo.

HABITAT, DISTRIBUTION AND STATUS

A more specialized feeder than *P. ruber* being more adapted to saline waters and a diet of Blue-green algae and micro-organisms such as Diatoms. Ticehurst recorded seeing 4 immature birds of this species on 19

May in a flock of Greater Flamingoes at the turn of the century at a time when the latter species could be seen in Sind in thousands, (Ticehurst, 1923). A specimen was also collected from Chatteji lake in the 1870s by James Murray. There are no authentic records of sighting in Pakistan since those dates. Ticehurst writing about the years during the First World War also records that numbers were only regularly sighted on Sambhar lake in Rajputana. This is still the present day position with small numbers regularly visiting certain brackish lakes in Rajputana at different times of the year. During extensive surveys of the wetlands of Rajasthan in February 1975 a total of only 45 Lesser Flamingoes were recorded in Gujarat, former Bhavnagar state (F. Koning, pers. comm.). In January 1983 Dr. Van der Ven of the Netherland Forestry Service counted a total of just 300 again in Gujarat, and their main stronghold remains Sambhar lake, east of the Aravali hills in Rajasthan, India, where their main food, Blue-green Algae of *Spirulina* spp. is always available (M. Alam, 1982). Since going to press, an estimated 2,300 of this species were seen on the borders of the Great Rann of Kutch in southern Badin district during an IWRB survey in late January 1989 (F. Koning, *in litt.* to author).

HABITS AND BREEDING BIOLOGY

Favours salt pans and more alkaline conditions than the Greater Flamingo though both species can be seen feeding together. Apparently unadapted to exploiting plant seeds, and specialized in sieving micro-organisms from water. Often swim while foraging and scythe their heads and necks to and fro while doing so, but their method of feeding with head and bill inverted inwards is basically the same as that of *P. ruber*. Breeding in the subcontinent was not substantiated until 1974 and this was found in the former state of Bhuj (now part of Gujarat state) on the seacoast (Ali, 1974).

ORDER ANSERIFORMES

FAMILY ANATIDAE

SUB-FAMILY ANSERINAE

Swans, 'True' Geese and Whistling or Tree Ducks

TRIBE DENDROCYGNINI

Lesser Whistling Teal or
Lesser Tree Duck *Dendrocygna javanica* (Horsfield)

Illus. **18**
Map **35**

DESCRIPTION

Body length 40-43 cms.
Wingspan 82-86 cms.

Unlike most ducks, both sexes are indentical in plumage. Almost the same size as the Fulvous Whistling Teal (*Dendrocygna bicolor*) they are difficult to separate in the field but in flight the upper-tail coverts are chestnut not creamy white as in *D. bicolor*. It is a relatively small duck with an upright alert carriage especially noticeable when swimming. The whole body is rufous chestnut with black tail and flight feathers and some creamy streaks on the flank feathers but these are often indistinct and partially obsolete. The bill is dark leaden or slaty with a black nail and looks all-black in poor light. The iris is dark brown and feet are black or dark leaden grey. In flight the peculiar more rounded wing silhouette and slower wing beat give it an almost 'butterfly-like' appearance and this characteristic enables Tree Ducks, as a family to be recognized at great distance. The under surface of the wing also is dark slaty grey. When viewed close up the Greater Whistling Teal usually has more prominent buff flank feathers and paler rufous crown. The Lesser Whistling Teal has lower cheeks and throat almost white giving the crown a rather 'capped' appearance. The downy chick is described as blackish brown with large white spots on its back (Delacour, 1954).

HABITAT, DISTRIBUTION AND STATUS

This is an Oriental species found widely in south-east Asia from Pakistan in the west, to Indonesia, western Borneo, southern China, east to the Riu Kiu islands (Delacour, 1954). Adapted to fresh water swamps and lakes, preferring those with reed-fringed margins or shallow seepage pools with plenty of cover. Will feed in flooded grassland and rice fields and generally avoids open bodies of water. Not observed in mangroves in Pakistan.

It is generally a summer or monsoon visitor to the region, large numbers arriving in early May from India. Thus a flock of 400 noted on 17 May at Haleji lake, lower Sind. D. Holmes noted arrival of flocks from mid-May in the riverain forest of Sukkur in northern Sind. They straggle up the Indus river and its tributaries, but are rare in Punjab and not recorded in Baluchistan or the North West Frontier Province. A flock of 30 was seen on 6 October at Trimmu headworks on the Chenab river (Jhang district) and a pair was seen in August on Katalpur jheel, Multan district (Ainsworth Harrison, pers. comm.). Another family party with fledged young was observed in a swampy pool upstream of Taunsa barrage in late August. Its stronghold is, however, southern Sind and the rice growing tracts of Thatta district and Badin, and in this region a few scattered flocks are year-round visitors if not residents, between 50 to 200 birds being observed around Haleji lake and its adjoining seepage swamps in every month of the year (8 years' observations). Status COMMON in summer only and mainly in Sind province.

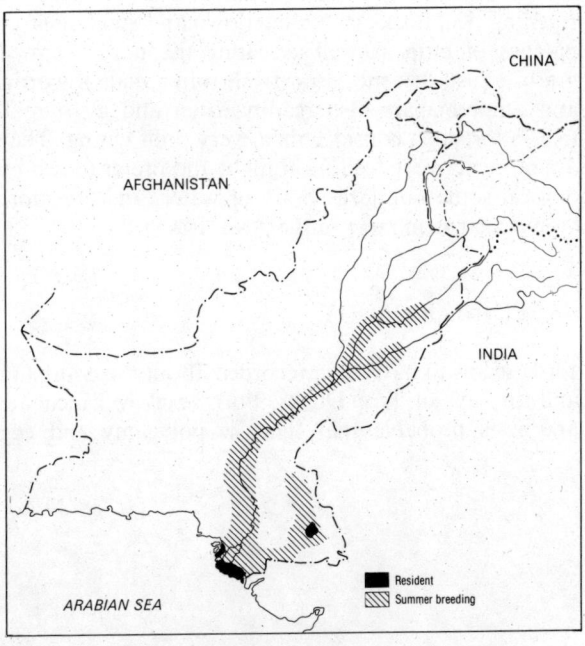

Map 35: Lesser Whistling (Teal) Duck or Tree Duck *Dendrocygna javanica*

Illus. 18: (a) Lesser Whistling Duck or Tree Duck
Dendrocygna javanica. Note paler cheeks and
less prominent creamy flank feathers.
(b) Fulvous Whistling or Greater Whistling Duck
Dendrocygna bicolor. Note darker cheeks,
and more conspicuous lanceolate creamy
flank feathers.

HABITS

Gregarious outside of the breeding season. First summer arrivals often keeping in substantial flocks of 50 to 200 for the first few weeks. They later disperse into small groups to take up breeding territories and form separate pairs. They are rather omnivorous in feeding habits and will dabble in shallow flooded vegetation as well as diving in deep water (author observation, Haleji) to feed on submerged vegetation. They eat seeds and shoots of water weeds such as *Lymnophyla heterophyla* and the rope-like strands of *Urticularia stellaris.* In Chauhar Jamali 'taluka' of Thatta district they occasionally inflict damage to ripening rice, flocks trampling down and spoiling large patches of crop as well as eating the newly formed grain. Crustacea, molluscs (fresh water snails), worms and amphibia are also readily eaten and also small fish. Sportsmen consider them very poor eating. They are more active in feeding at night and prefer to rest by day on some safe large body of water, and are more crepuscular than most of the ANATIDAE.

BREEDING BIOLOGY

Little seems to have been recorded about the courtship and displays of Tree Ducks. Both sexes will incubate and it is probable that there is polygamy and egg dumping (well known in the Red-billed Whistling Duck (*Dendrocygna autumnalis*) of the Neotropics), as nests have been found containing as many as 24 eggs (E. N. Lowther, 1949). A nest on 'Crocodile island' on Haleji lake was found containing 27 eggs in August 1986 (Khan Mohammad Khan and Syed Asad Ali, pers. comm.). Mutual preening is reported an important element in courtship.

The site can be very variable, though nests are usually located in hollow trees or cavities in tree stumps or in grass thickets growing through submerged thorny bushes. When nesting in vegetation the nest is carefully roofed over with bent grasses or sedges. The female does not line the nest with any down plucked from her breast and generally the lining is very scanty, comprising some grass and feathers. Often an old nest of a Painted Stork or bird of prey is used. The eggs, which are white to creamy white in colour, take up to 30 days (Delacour, 1954) to incubate. S. Doig (in Hume and Oates, *Nests and Eggs of Indian Birds,* 1890) found a nest with 10 unincubated eggs on 22 June placed in a creeper-festooned Tamarisk tree partly submerged in a seasonal swamp in the east Narra. He subsequently found many more nests in July and August usually 1 to 2.5 metres above the water surface and in trees covered by creepers offering concealment to the incubating bird. K. Eates found nests in Hyderabad, Sukkur and Tharparkar districts from July to September. The nests were untidy affairs lined with dried reeds, always located in thick cover, usually in dense thickets of grass and thorn bushes. The biggest clutch he encountered was 9 eggs and the sitting bird was so reluctant to fly that it could have been caught by hand (Unpublished MS). The parent bird will perform a distraction display if flushed off its eggs, often swimming around close to the intruder beating one wing on the surface and swimming in circles as though injured. Ten fledged young were observed following their single parent closely during one morning at Taunsa barrage, in a reed fringed pool on 27 August (author).

*VOCALIZATIONS

A double noted rapid twitter or whistle is constantly uttered when the bird is in flight and less commonly on the ground or swimming. Salim Ali has aptly rendered the call as 'seasick-seasick'.

Large Whistling Teal or
Fulvous Tree Duck *Dendrocygna bicolor* (Vieillot)

DESCRIPTION

Body length 50-52 cms.
Wingspan 102-110 cms.

Only slightly larger than the Lesser Whistling Teal, it can be separated in flight by the whitish cream upper-tail coverts, and when resting on the water, by its more uniform chestnut brown head and face lacking the contrasting darker crown of *D. javanica*. Also its hind neck and nape have a narrow blackish line running down the centre. The flank feathers have more conspicuous creamy falcate or curved feathers. The sexes are alike, having dull slaty black bills, leaden grey or black feet and black tail and flight feathers. The rest of the body is chestnut brown, paler and more fulvous or buff on the head. The iris is dark brown and there is no yellow orbital ring as in *D. javanica*. In flight the whole of the under-side of the wings is blackish and their relatively broad rounded wings and slow wing beat are similar to those of *D. javanica*. The downy chick is reported to be paler grey-brown than that of *D. javanica* with less contrasting white spots on the back (Delacour, 1954).

HABITAT, DISTRIBUTION AND STATUS

Like the Lesser Whistling Teal they are confined to fresh water habitats, preferring swamps and small ponds with plenty of emergent vegetation or larger lakes with submerged trees and reed thickets. Very widespread in the tropics from both north and south America, Africa and the Far East to Indonesia. This tree duck is a seasonal migrant to Pakistan though locally resident in India. In Pakistan it has not been authentically recorded north of Sind province, and even here may be considered only as an occasional visitor. The difficulty of separating this species from *D. javanica* however, makes any estimate of its population status uncertain. K. Eates definitely encountered it in 1921 on Khinjar lake, Thatta district, and several years later in Hyderabad district, all winter

sightings when specimens were secured during a duck shoot (unpublished MS). Ticehurst records sightings by Sir Evan Jones around Hyderabad and by S. Doig in the east Narra and by Colonel Butler in the winter on Manchar lake (Ticehurst, 1923). On 26 June the author observed two pairs of this duck on Haleji lake which could be compared to groups of Lesser Whistling Teal perched in *Phragmites* reeds nearby. It must be an irregular visitor at all seasons, seldom coming to breed. Status RARE.

HABITS

Similar to those of *Dendrocygna javanica* and the two species are quite sympatric in Bangla Desh and in north-eastern India. They are gregarious, but usually less numerous and occurring in much smaller flocks than the Lesser Whistling Teal. However the author has seen flocks of 300 and 400 on the larger jheels of Sylhet district in Bangla Desh. They are well able to dive and are believed to feed largely on aquatic vegetation and seeds, as well as rice grains, fresh water molluscs, larvae of water beetles, tadpoles and worms.

BREEDING BIOLOGY

Nests in disused raptors' nests, or in natural cavities or on the ground in thick clumps of sedges and submerged bushes. The eggs are ivory white, numbering 6 to 8, and the incubation period is recorded as 30 days. The pair bond is more lasting than in dabbling ducks, with the male definitely sharing in incubation duties.

VOCALIZATIONS

Has a similar double noted whistle most frequently uttered as a flight call but also when perched in trees when other ducks flight overhead. It is a stronger call than that of *D. javanica*, the call carrying further and being higher pitched.

SUB-FAMILY ANSERINI
TRIBE ANSERINI

Mute Swan *Cygnus olor* (Gmelin)

DESCRIPTION

Body length 145-160 cms.
Wingspan 208-238 cms.

A fine looking bird with all-white body plumage, short rounded tail and long wings tapering to a point. The bill is orange-red with a prominent black knob at the base of the upper mandible and covering the forecrown. The eye is dark brown with a narrow black loral triangle extending from the eye to the base of the upper mandible. Feet are black. Males and females are alike though the males develop a more prominent black knob during the breeding season.

HABITAT, DISTRIBUTION AND STATUS

Very few authentic records since the beginning of this century. On 26 January 1911, Mr. P. Lord shot one on the Sohan river near Rawalpindi. Meinertzhagen reported 'large numbers' visiting Kushdil Khan lake in 1901 in winter and single individuals in February 1911 and November 1913 (Meinertzhagen, 1920). Ticehurst records that a flock was seen on Manchar lake in January 1878 and that in 1900 there were several birds shot and many more seen, 8 at Kotri on the Indus on 13 January, one in mid-March on Manchar lake and 10 at Lakhi at the end of March. I have been unable to trace any authentic records in Pakistan since 1913. Status RARE VAGRANT.

Whooper Swan *Cygnus cygnus* Linnaeus
Alpheraky's Swan *Cygnus cygnus jankowskii* (Ripley's 'Revised Synopsis')
Bewicks' Swan *Cygnus cygnus bewickii* Yarrell (Ripley's 'Revised Synopsis')   Illus. **19**
Cygnus columbianus bewickii (K. H. Voous, 1973)

TAXONOMY

Recent studies showing the great individual variation in bill pattern of Bewicks' Swan (Bateson, Lotwick and Scott, 1980) and increased knowledge about clinal size variations in eastern Siberian breeding populations, have resulted in considerable disagreement between taxonomists as to the classification of Swans. K. H. Voous in his 'List of Recent Holarctic Birds Species' (*Ibis,* 1973) separates *Cygnus cygnus* the Whooper Swan as a single species having no distinct sub-species. Some authorities combine *Cygnus cygnus* with the Trumpeter Swan *C. buccinator* of north America. Voous treats *Cygnus columbianus* as a super species, combining Bewicks' Swan as a sub-species *C. columbianus bewickii*, mainly Palearctic in breeding, and the nominate Whistling swan *C. columbianus columbianus* breeding in Nearctic regions. Alpheraky's Swan *C. columbianus jankowskii*, which is recorded as occurring in Pakistan in Sind and Baluchistan, is considered a sub-species of Bewicks' Swan on the basis of the greater amount of yellow at the base of its upper mandible, it being otherwise equal in size (S. Delacour, Vol. 1, 1954, and Dement'ev *et al.*, Vol. 4, 1952). Examination of skins in the Bombay Natural History

Society's collection by Abdulali (*JBNHS,* Vol. 65, p. 420), indicated that birds attributable to this sub-species were shot in Campbellpur, Attock district, Punjab, as well as in the Rann of Kutch (India). A single bird visited Haleji and Hadiero lakes in lower Sind during December 1984 and was eventually trapped by local fishermen. Photographs taken by the author confirmed that it was like a Bewick's Swan in smaller size and with a shorter straighter neck whilst the amount of yellow on its upper mandible was more extensive than is typical of *C. bewickii* (see **Fig. 19**). The authors of the *Handbook of the Birds of the Western Palearctic* (Cramp *et. al.*, 1977) consider Alpheraky's Swan to be inseparable from Bewick's Swan (Vol. 1, p. 385). S. Dillon Ripley in consultation with Jean Delacour, in his 'Revised Synopsis' has treated both Alpheraky's and Bewick's Swan as races of the Whooper Swan *C. cygnus jankowskii* and *C. cygnus bewickii* (Ripley, 'Revised Synopsis', 1982) but this view is not taken by any of the standard works (K. H. Voous, 1977, C. Vaurie, 1965, Cramp *et al.*, 1977, and Ali and Ripley, 1969). It would seem prudent to treat all old records of both Bewick's and Whooper Swans on the subcontinent, with caution, unless supported by skins.

DESCRIPTION

Whooper Swan
    Body length 145-160 cms.
    Wingspan 218-243 cms.
Alpheraky's Swan
    Body length 117-124 cms.
    Wingspan 118.7-204 cms.(both sexes) (Dement'ev, *et al.*)
    Wing length 49-55 cms.
    Bill length 90-99 cms.

A huge snow white bird with straight neck and smoothly sloping upper mandible, bright yellow basally and black at the tip, and yellow area extending in a wedge along the base of the upper mandible beyond the nostril. The legs and feet are black, and iris dark brown. The yellow area of the bill extends in a narrow loral triangle of naked skin to the eye. The amount of yellow in the upper mandible is more restricted in *bewickii* whilst in the Whooper Swan, *C. cygnus* the yellow area extends narrowly down each side of the commissure to at least half the total bill length. The sub-species *jankowskii* has more extensive yellow than is typical of the Bewick's population wintering in Britain (author obs.). In size *bewickii* is smaller than *C. olor* and also *C. cygnus*, with a relatively shorter neck, usually carried rather straight. Smaller bill and the steeper forehead of Bewick's, give it a different, more 'docile' head shape.

HABITAT, DISTRIBUTION AND STATUS

Up to the mid 1960s Whooper Swans were reported to be occasional winter visitors to Seistan in the Un-i-hirmand swamps of eastern Iran (Savage, *Wildfowl Trust 16th Annual Report*, 1965), so their occasional occurrence in Baluchistan is to be expected. Five adults were seen in Hamoun-i-Puzak (Afghanistan) in January 1976 (D. A. Scott, *in litt.*) and 6 by Lindon Cornwallis in the southern Caspian (Iran) in January 1977 (D. A. Scott, *in litt.*). However the only authentic records for the whole subcontinent were, until recently, of 3 birds shot in the early 1900s. Meinertzhagen records a swan shot by Mr. A. B. Aitken on Kushdil Khan lake on 17 December 1913 which was a female. A painting of the bill and head was sent to the British

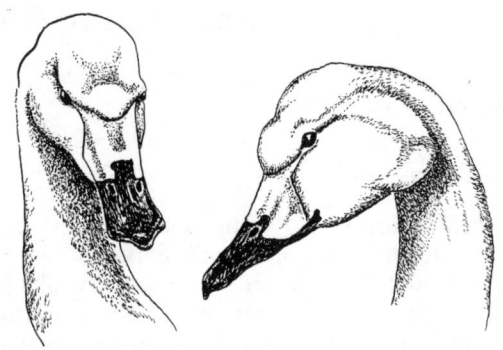

Illus. 19: Alpheraky's Swan *Cygnus columbianus jankowskii* (based on photograph of captive specimen).

Museum where it was identified as *Cygnus minor jankowskii*. Its weight was 6.57 kg. Wing length, closed, 50 cms. Unfortunately this specimen lodged in the MacMahon Museum was destroyed in the 1935 Quetta earthquake. Ticehurst records a Whooper Swan shot on 31 January 1904 by a Mr. Crerar at Kambar lake, Larkana district. This skin is still in the Bombay Natural History Society Museum. He also recorded a swan shot on 2 December 1907 by a Mr. McCullock near Jacobabad which was identified as *Cygnus cygnus bewickii* Yarrell. In the winter of 1981 Mr. Inayatulla Arbab, Sub-divisional Officer (Wildlife) of the Baluchistan Forestry Department reported that three swans visited Zangi Nawar lake in south-western Baluchistan N29°25'., E65°47'. On 7 and 8 February 1983, Mr. Ashiq Ahmed, Wildlife Management Specialist of the Pakistan Forest Institute visited Zangi Nawar after again receiving reports that wild swans had reappeared (pers. comm., 1983). There were 8 swans which reportedly stayed for about 12 months and he identified them as Whooper Swans with the aid of two field guides (Heinzel, Parslow and Fitter, and Ali and Ripley, Vol 1, 1969), being able to observe them through a telescope at about 150 metres distance (Ahmed, *JBNHS*, Vol. 82). In the winter of 1983-4 about 15 swans visited Zangi Nawar lake in early December but stayed only a few days.

In all there have only been about 15 authentic records of Whooper, Bewicks' or Alpheraky's Swans on the subcontinent since the 1870s. Status RARE VAGRANT.

White-fronted Goose *Anser albifrons* (Scopoli)

DESCRIPTION

Body length 65-78 cms.
Wingspan 130-165 cms.

A typical 'Grey Goose' with pale margined broad scalloped feathers on wing mantle, darker brown head and neck and generally grey-brown body plumage. The bill is pink and legs are orange and a restricted area of the forecrown and front of face is white. The Eastern Greylag (*A. anser rubrirostris*) though having a similar coloured bill, has pink not orange legs, and it also lacks any white patch at the base of the bill and on the forecrown. The White-fronted Goose also has darker and more extensive brown cross-barring around its breast though older specimens of the Greylag also develop noticeable dark cross-bars. In flight the rump of this species is dark grey-brown with a whitish border whereas in flight the Greylag shows a pale grey rump.

HABITAT, DISTRIBUTION AND STATUS

Breeding in the Arctic tundra of the USSR and north America a small population regularly winters around the Caspian Sea and Black Sea coasts of northern Iran and northern Iraq. An occasional bird straggles south-eastwards into Pakistan where they have mainly been encountered on the Indus river or its tributaries. Dr. Ticehurst between 1917 and 1919 handled two specimens shot on Manchar lake which he identified as corresponding exactly with the nominate sub-species. The last definite record was a sighting in January 1968 by C. D. W. Savage on the Indus near Sukkur (pers. comm.). Status in Pakistan RARE VAGRANT.

Lesser White-fronted Goose *Anser erythropus* Linnaeus

DESCRIPTION

Body length 53-66 cms.
Wing length 120-135 cms.

A small neat bird with a shorter neck and smaller bill than any of the other 'Grey Geese', it is dusky brown all over with conspicuous white forepart of face extending on the forecrown right up to the top of the crown. In *Anser albifrons* the white area is less conspicuous, usually being restricted to a narrow band over the forecrown. Its bill is pink and legs orange like *Anser albifrons* and when fully mature there are conspicuous dark blackish chestnut bars around the lower breast. If viewed close enough, a prominent orange orbital ring of bare skin at once distinguishes this species from the larger White-fronted Goose and this is present in immature birds also, which lack the white forehead and dark belly bars, and so this is perhaps the best field character. Also in flight it can readily be separated from the Greylag (*Anser anser*) even if seen as a solitary bird, by the wing shoulder being a much darker grey and the secondary wing coverts showing a narrow paler bar when the upper surface of the wing is revealed. In the Greylag the wing shoulders are strikingly pale grey.

HABITAT, DISTRIBUTION AND STATUS

Breeding in slightly more southern latitudes across the Arctic than *A. albifrons*, from Norwegian Lapland to Kolyma in Siberia, there are wintering populations (*circa* 5,000) which regularly occur around the south coast of the Caspian Sea (D. A. Scott, *in litt.*), and in the Sea of Azov. Formerly abundant in the Seistan swamps of south-east Iran in the mid-1960s (Savage, *Wildfowl Trust 16th Annual Report*, 1965), none were sighted in the Seistan basin in the 1970s despite regular surveys (D. A. Scott, *in litt.*). Only the occasional straggler is likely to reach the Indian subcontinent. Dillon Ripley in the first edition of the 'Synopsis' lists recordings from Sind and the NWFP, also from widely separated localities across northern India. Neither Hugh Whistler nor Claude Ticehurst recorded any sighting of this species within Pakistan and undoubtedly in the past there has been confusion between this species and *Anser albifrons*. A. O. Hume, in 1871, shot 3 specimens of *erythropus* on the Jhelum river on 27 November (*Stray Feathers*, Vol. 1, p. 269, 1873). He saw 5 birds altogether during a 4-month survey of the Indus, 2 more birds being sighted near Sukkur and even at that early date when other wild geese were abundant on the Indus he considered the Lesser White-front to by very rare (*Stray Feathers*, 1873, op. cit.). Paludan (1959) includes *Anser albifrons* in his checklist of Afghanistan but not *A. erythropus*. An experienced wild fowler, Mr. Ainsworth Harrison believes he indentified this goose among a small flock of Greylag Geese on 21 January 1962 on the Chenab river upstream of Marala headworks (pers. comm., 1964). Such a sight record has to be treated with caution. C. D. W. Savage records that there were eight sightings within Pakistan between 1945 and 1965 (Savage, 1968). Status RARE VAGRANT.

Greylag Goose *Anser anser* (Linnaeus)
Eastern Greylag Goose *Anser anser rubirostris* Swinhoe

Map **36**

DESCRIPTION

Body length 80-90 cms.
Wing length 152-180 cms.

The largest of the 'Grey Geese', the eastern population averages slightly larger in size than the European population with a distinctive rose pink not orange bill, and generally paler grey-brown breast plumage. The feet are pink, the head and neck dusky brown and also the scapulars which are handsomely margined whitish buff. In flight the rump patch is pale silvery grey as also the wing shoulders with a darker brown bar along the medium secondary wing coverts. Mature or older birds develop blackish chestnut crescents and short bars on the lower breast and belly. The sexes are alike.

HABITAT, DISTRIBUTION AND STATUS

A winter migrant to the subcontinent, from widely dispersed breeding grounds stretching from the southern borders of the Arctic tundra zone and across the boreal forest of the USSR down to the central Asian steppes. Numbers still breed around the northern shores of the Caspian and Aral seas.

Earlier writers comment in stirring phrases about the large flocks of this magnificent bird visiting Pakistan in winter. Meinertzhagen records flocks of Greylag spending the whole winter on Kushdil Khan

lake. Ticehurst records the earliest arrivals around southern Sind in the first week of October when 'large skeins' could regularly be seen flying towards the Indus mouth. It was never recorded around Rawalpindi in the 1920s (Whistler) and Whitehead noted it migrating up the Kurram valley in small numbers and considered it very rare in Kohat district (Whitehead, 1910). Even in 1922 Whistler noted that it was much less common than formerly in Jhang district, a few being encountered along the Chenab river. In 1926 Dr. Salim Ali visited Manchar lake in Sind and wrote (*JBNHS*, 1927) graphically about the huge flocks of both Greylag (*circa* 10,000 seen on 1 January.) and Barheaded Geese, which regularly wintered there. A Russian ring recovered from a Greylag Goose shot on the Kabul river near Nowshera in March 1979 was sent (by author) to Moscow via the Bombay Natural History Society but no response has been obtainable as to the origin of this bird, however one ringed in July 1959 by the Russians at Chatyr-kul lake in Kirghiz N40° 40′., E75° 18′ was shot on the river Indus 48 kilometres (30 miles) east of Dera Ghazi Khan by Ainsworth Harrison on 23 December 1960 (*JBNHS*, Vol. 59, No. 3, p. 964).

Like all wild geese the Greylag is intolerant of human disturbance and this has largely accounted for its decline throughout the region, as few corners of this crowded country are now free of regular and frequent human intrusion. Records maintained for the past 20 years indicate that Greylags regularly migrate down all the northern Himalayan valleys from the Shyok and Indus in Baltistan to Gilgit, Hunza and Yasin valleys to the west. The late Mir of Hunza in the 1940s used to obtain bags of over 100 in a day's shoot during migration (pers. comm.). A very small number spends the winter on the Salt Range lakes, usually in Uchchali (maximum number 200, 25 January 1976), and a few may spend part of the winter in the barrage headpond and sheltered islands upstream of Rasul barrage (on the Jhelum river) and occasionally Taunsa barrage (on the Indus near Dera Ghazi Khan). In Sind in recent years, odd birds have been recorded on the lakes of Sanghar district, e.g. Sangriaro, Khipro and Ghauspur in Jacobabad, but never more than 2 to 3 individuals at one time. Careful enquiries from local hunters on several visits indicate that none have been reported visiting Manchar lake in recent years nor Kushdil Khan lake in Baluchistan. During more than 20 winter time boat trips in the Indus delta, Greylag Geese were seen only once in November 1968 near Keti Bandar. During comprehensive wetland bird counts of Pakistan organized by the IWRB a typical total population count was 305 in two flocks both on

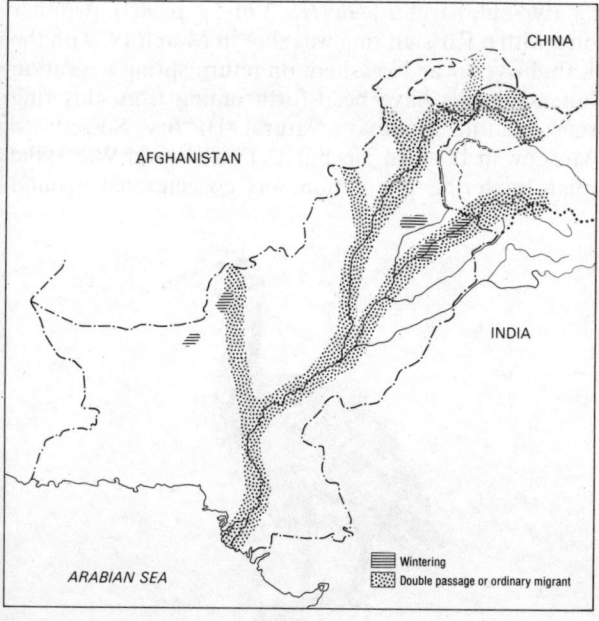

Map 36: Greylag Goose *Anser anser*

Punjab lakes and none seen in Sind (January/February 1973). In 9 years' observations of the main wetlands around Karachi, 10 Greylag were recorded, once only, on Hadiero lake (near Thatta town) in January 1981 (Khan Mohammed Khan, Sind Wildlife Board, pers. comm.). The annual concentration of these birds in the flooded marshes of Bharatpur sanctuary in India and their tameness in proximity to humans is surely one of the wildlife spectacles of the present day. Several thousand Greylags still winter regularly in the Seistan basin (Iran/Afghanistan) just west of Baluchistan (D. A. Scott, *in litt.*). Status locally FREQUENT, allowing for its gregarious habits.

HABITS

They are markedly gregarious except when nesting and they migrate in diagonal ribbons and 'V' skeins, keeping up a constant chatter of gabbling calls, also when feeding, especially if one group of birds flights to a different area. They are more catholic in choice of resting areas than the Barheaded Geese, preferring to swim on large lakes while resting whereas the Barheaded Goose seems more confined to the main rivers and prefers to roost on dry land such as inaccessible sandbars. In shallow lakes they frequently forage on submerged water weed, tipping up with their prominent white vents and under-tail coverts reminiscent of a domestic goose. They largely feed at night, flighting to crops and grassy areas to crop the green vegetation. They are mainly grazers and herbivorous in diet. During a mid-winter visit to Khapalu (elevation 2,700 metres) in south-eastern Baltistan, a pair of Greylags was seen for two days (8-9 February) haunting quiet stretches of the Shyok river. Both could fly strongly and showed no external sign of injury. Daytime temperatures at that time ranged around 0° centigrade.

Bar-headed Goose *Anser indicus* (Latham)

Illus. **20**
Map **37**

DESCRIPTION

Body length 70-78 cms.
Wingspan 150-165 cms.

Smaller than the Greylag Goose (*Anser anser*) they show a uniform blue-grey back and wing shoulder in flight with all dark slaty black flight feathers. If seen close, the bright yellow-orange bill and white face are noticeable. The sexes are alike. The foreneck is dark grey-brown becoming paler on the upper breast. The face is white with two prominent bands of white running down each side of the neck, the hind neck being dark chocolate brown or sooty brown coloured. There are two jet black crescentic bars running across the rear part of the crown and a little lower across the occiput. The tail is dark blue-grey with white tips and the flanks are handsomely scalloped with burned umber brown. Under-tail coverts and lower flanks are pure white and the mantle is pale blue-grey, the wing coverts being scalloped with white margins. The upper-tail coverts are white, the feet are orange-yellow like the bill which has a prominent black nail. The iris is dark brown.

HABITAT, DISTRIBUTION AND STATUS

A winter migrant visitor to the subcontinent following much the same routes as the Eastern Greylag (see species account) through the northern Himalayas. They breed mainly in the high mountain plateaux of Tibet and central Asia, a few breeding within Indian territory in Ladakh.

There have been several recoveries of birds ringed by the Russians in Kirgiz SSR, at Lake Chatyr-kul (N40° 40′., E75° 15′). An 8-year old bird was killed migrating up the Gilgit river by the Raja of Punial on 10 April and a second bird (a juvenile) was shot on the Indus river near Dera Ghazi Khan also ringed from Chatyr-kul, Kirghiz (*JBNHS*, Vol. 59, p. 964). Another bird with a Russian ring was shot in March 1979 on the Kabul river near Nowshera on return spring migration but no details have been forthcoming from this ring sent via the Bombay Natural History Society to Moscow in 1980. In Ticehurst's time (World War I) the main wintering population was concentrated around

Illus. 20: Bar-headed Goose *Anser indicus*

Map 37: Bar-headed Goose *Anser indicus*

Manchar lake, and even in 1926 Dr. Salim Ali saw thousands of this species there. In the present time no Bar-heads visit Manchar lake but a flock of varying size winters on the headpond and seepage lakes around Taunsa barrage near Dera Ghazi Khan, 157 in January 1973, 400 in 1972 and 300 in December 1968 (F. Koning and C. D. W. Savage and own observations). This is the main concentration of this species in Pakistan and numbers occasionally build up to over 1,000 in spring migrations (e.g. February 1968). Small numbers regularly come down the Chenab river and sometimes stay on the sandbars upstream of Marala barrage (40 seen 9 March). They also occasionally visit the Salt Range lakes (6 on Khabbaki lake on 3 January) and Kharrar jheel near Renala Khurd,

Punjab (68 on 5 February) and on the Indus below Panjnad headworks on sandbars in the river where numbers of adult birds were netted and ringed successfully during the 1960s by Lieut. General Marden.

They have not been recorded from Kushdil Khan lake in Baluchistan, and they are rare in Sind and apart from small groups seen on the headpond above Guddu barrage (author obs.) in the northern border area, there are no records. Holmes and Wright saw none during 1963-5 when travelling widely in the province. Accumulated evidence particularly the IWRB wildfowl counts during the early 1970s indicate that it is less numerous than the Greylag in Pakistan and largely confined to one wintering area. Status locally FREQUENT to RARE allowing for its gregarious habits.

*HABITS, BREEDING AND VOCALIZATIONS

Like the Greylag it is a very shy and wary bird on its wintering grounds. It feeds mostly by night in cropland or grazes on river banks and roosts during the day on inaccessible and exposed sandbars in the riverain areas. In former times it has been reported as inflicting serious damage to gram or chickpea crops (*Cicer arietum*).

A few small breeding colonies survive (outside of Pakistan), on the high plateau lakes of north-eastern Ladakh, where their behaviour is very tame and fearless of humans in contrast to their behaviour on their wintering grounds. In 1983, a colony of about 100 pairs were nesting on hummocks on the marshy fringes of Tso Moriri lake (General S. M. Ghaur, pers. comm., 1983).

They are inclined to be less vocal than the Eastern Greylag in their wintering grounds but occasionally if a new flight of birds arrives the whole herd will join in a chorus of deep-pitched cackling. It is almost like a human voice in tone and considerably lower pitched than the calls of *Anser anser*.

TRIBE TADORNINI

Shelducks and Sheldgeese

Ruddy Shelduck or Brahminy Duck *Tadorna ferruginea* (Pallas)   Map 38

Synonym *Casarca ferruginea* (Pallas)

DESCRIPTION

Body length 61-67 cms.
Wingspan 121-145 cms.

This large and very handsome duck exhibits slight sexual dimorphism, the males averaging larger in size and with differences in head and neck pattern during

the breeding season. Their heads are pale creamy, merging to chestnut, with the neck and rest of the body a rich orange-chestnut. The bill, legs and feet are black. In flight the wing shoulders show pure white, with glossy green-black flight feathers and tail. The secondaries or speculum being particularly bright iridescent green. In breeding plumage the male has a narrow black collar half-way down its neck, and its head looks slightly larger and darker than that of the female, which is often pure white on the crown and forepart of the face. In flight the under-wing coverts are wholly white. At rest the white wing shoulder is usually wholly concealed. The male loses the black neck-ring in winter.

HABITAT, DISTRIBUTION AND STATUS

Widely distributed from southern Europe across north Africa the Middle East, Tibet, China, and central Asia (Kazakhstan to Transbaikal), with some birds wintering in Iran, Iraq and the Indian sub-continent. In its winter quarters usually keeps in flocks and is quite gregarious, and will be found from the mangrove estuaries of the seacoast up to the northern lakes and river areas. The biggest concentrations noted were 156 on 16 November, near Keti Bunder on the Indus delta, on salt-marsh grass flats. 597 were counted resting on water on Taunsa barrage headpond, on the Indus river near Dera Ghazi Khan, on 5 February 1972 (C. D. W. Savage and F. Koning, pers. comm.). In the late 1960s and early 1970s flocks of shelducks frequented the salt pans fringing the Ravi river between Sidhnai and Shorkot in Multan district. Over 15 years' observations, flocks totalling 60 to 70 were noted, with a maximum of 141 seen together on Sidhnai spill channel on 22 November 1964. A flock of 110 was noted on salt flats near Pindi Battian in Sheikhupura district in Punjab throughout February in 1967.

They are winter migrants to the Indus plains, most birds being first sighted from early November (earlier sighting Sind, in Karachi, 29 October). Most have departed by late March. The latest record is of a party of 7 on the old bed of the river known as the Sukh Beas, 67 kilometres (42 miles) north-east of Lahore on 21 April.

Most wintering birds are migrants from breeding grounds in Kirghiz SSR and Trans-Caspian area. They also breed on the high mountain plateaux of Tibet and Ladakh in India. A bird shot near Lahore in October 1959 had been ringed as a juvenile on 9 July 1959 in Russia on Lake Son-Kul in Kirghiz province (N41°50'., E75°00') (*JBNHS*, Vol. 59, No. 3, p. 964). In mid-February, under conditions of ice and snow, two small parties of Ruddy Shelduck were seen (author) on the Indus and Shyok rivers in Baltistan. Some birds haunt their breeding grounds throughout the year (Delacour, 1954). In Ladakh, by certain lakes where they breed they also remain the year round (Brigadier Moti Dhar, pers. comm., 1983). A few pairs probably breed along Pakistan's northern border region where small streams and lakes provide suitable breeding conditions. Dr. Schaller while searching for *Ovis ammon polii* in the Hindu Kush range in extreme

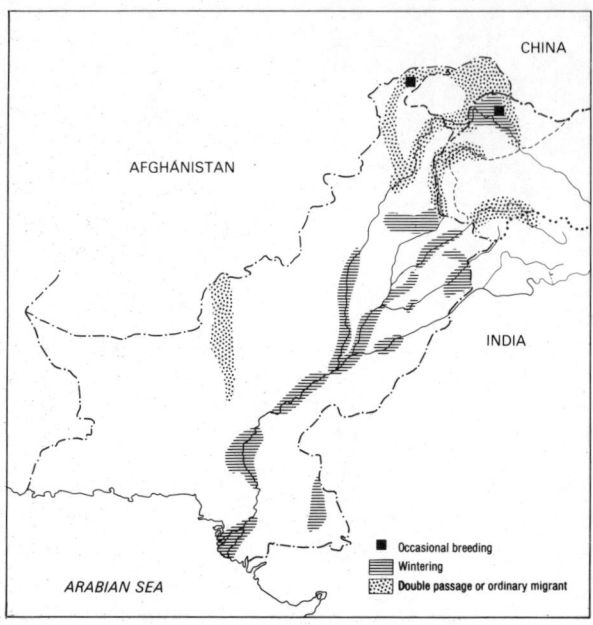

Map 38: Ruddy Shelduck or Brahminy Duck *Tadorna ferruginea*

north-east Chitral in June, encountered several pairs in the vicinity of a lake at the head of one of the tributaries of the Karumbar river and these were almost certainly breeding in the vicinity (exact location N36°37'., E73°40') (Major Amanullah Khan, pers. comm., 1985). Bowes-Lyon reported seeing breeding pairs north of Mastuj and towards the Wakhan in northern Chitral in May and June, during a botanical expedition in the late 1950s. Mathews encountered a flock of 60 on the Indus near Skardu on 24 July (*JBNHS*, 1941). Outside of Pakistan it is a common breeding bird and winter visitor to western Iran, but mainly a winter visitor in small numbers to south-east Caspian and Seistan (D. A. Scott, *in litt.*).

HABITS

In Pakistan it is markedly gregarious as the above records show, and small groups always seem to keep in contact with loud calls when flighting and feeding. They prefer to roost by day in bare open regions where approaching predators or humans can be easily observed, as they are very wary on their wintering grounds. They are therefore more often encountered on brackish lakes free of any marginal reed cover, or open salt marsh flats, when near the coast, rather than mangrove creeks. They sleep during the day by choice, lying on bare open ground, head tucked under wing. 156 were noted thus, on the banks of the Indus near Keti Bandar. They feed mostly in late evening and early morning, being mostly herbivorous, grazing like geese. They are also reported to eat molluscs, crustacea and insect larvae, and because of their predilection for alkaline waters, their flesh is not considered good eating. Near Sidhnai spill channel they were observed pecking at the succulent leaves of the salt tolerant *Suaeda fruticosa*. They have been recorded as gleaning cereal grains from stubbles, as well as eating amphibia spawn, tadpoles and small fish. They frequently call to each other when feeding or foraging, their calls being very far carrying.

BREEDING BIOLOGY

Pair bonds are monogamous and are certainly formed before arrival on their breeding grounds (observed late March in Chitawan, Nepal). Delacour believes that they form life-long pair bonds. Only the female incubates, and she does not sit until the full clutch is laid so that hatching is synchronous. 8 to 9 glossy white eggs are laid, and the female plucks her own breast down for the nest lining. Breeding is often loosely colonial in Ladakh (Brigadier Moti Dhar, pers. comm., 1983). They usually build their nest in a rock cleft or ledge in the mountain range closest to the lake or marshy ground where they feed. Such nests can

therefore be quite a considerable distance from the area to which the newly hatched chicks have to be escorted.

Incubation lasts 28 to 29 days, during which the male bird remains near the nest site. He also helps to tend the chicks, accompanying the young until they are fully fledged, even though they are able to feed themselves from the time of hatching. Little has been recorded about actual courtship in wild birds, but vocal calling with responses between the two sexes are an important element, the female having very different calls.

*VOCALIZATIONS

On wintering grounds males utter a loud nasal trumpeting reminiscent of a goose honking. This call is often given in flight and can be heard from a distance of at least 2 kilometres. The flight call of the female is deeper and sounds like 'angh-angh'. In Chitawan, Nepal, females were observed giving a rapidly repeated gargling call to which the male responded with deeper bi-syllabic calls 'chuk-wah-chukwah'.

Shelduck or Common Shelduck *Tadorna tadorna* (Linnaeus)   Map **39**

DESCRIPTION

Body length 58-65 cms.
Wingspan 110-130 cms.

A large duck, bigger than a Mallard (*Anas platyrhynchos*). They are conspicuously plumaged in black (with a green gloss), chestnut and white. The head and neck are glossed green in good sunlight and the scapulars also. The flight feathers are black with the secondaries white-tipped and glossed green. The upper tail coverts and tail are white with black tips, the wing shoulders, lower neck and belly are snow white. There is a broad dark chestnut band around the breast and upper mantle, with the under-tail coverts chestnut, and the mid-belly has a longitudinal black bar extending from the chestnut breast band down to the vent. The legs are pale flesh pink and bill darker carmine pink. In the breeding season the male develops a prominent pink knob on the forecrown and the female's bill is distinctly paler. The iris is brown. Females are about equal in size to the males.

HABITAT, DISTRIBUTION AND STATUS

This is a Palearctic breeding species wintering south to north Africa, the Middle-East, southern China and

Japan as well as the Indian subcontinent. They are winter visitors to Pakistan, much less frequently encountered than *Tadorna ferruginea*, being one of Pakistan's less common ducks.

They have been recorded on spring passage in the Kurram valley, and north-western Baluchistan (Kushdil Khan lake) and even around Turbat (Hotson in Nihing valley), but there is no definite evidence of their migrating through the northern Himalayas. Normally they are only encountered in ones and twos on larger inland lakes, e.g. 2 on Khinjar lake on 21 March (J. Vieilliard, pers. comm.). A few usually overwinter on the Salt lakes, the highest number being recorded over about 15 years being 26 on Uchchali lake. Ticehurst considered it very rare in Sind, his few records were from Manchar lake and some seen on the Larkana lakes (Ticehurst, 1923). Recently quite large numbers have been sighted on Karachi coastal waters. About 80 were counted on Siranda lake, Lasbela (less than 2 miles from the sea) on 6 February (Roberts, *JBNHS*, 1980). A flock of 35 was seen in Karachi harbour near Mauripur salt works on 19 February 1975 and F. Koning counted 283 in the harbour in January of that year. 126 were counted on Hadiero lake near Thatta town, lower Sind on 7 February 1973 and 74 were counted on the Sukh Beas 40 miles out of

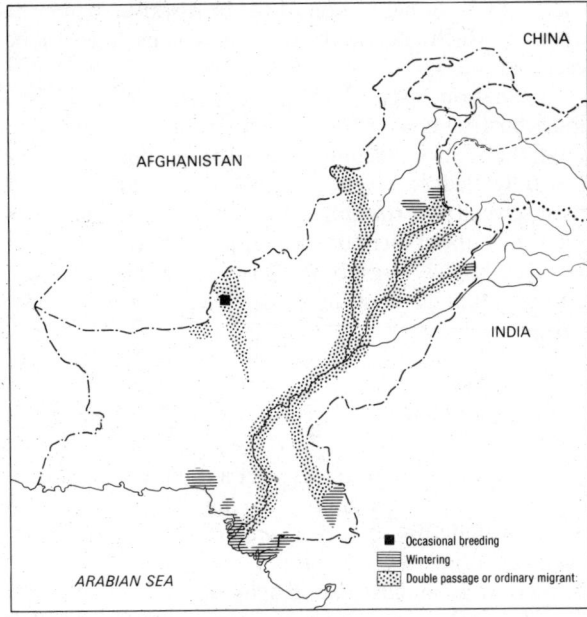

Map 39: Shelduck *Tadorna tadorna*

on land as does *Tadorna ferruginea* and are much more dependent upon animal food than the Ruddy Shelduck. They often feed by wading in shallows, scything their bills to and fro in the shallow water or on the mud surface, for micro-organisms like Diatoms, larvae of crustacea, and brine shrimps which are important items in their diet as well as molluscs and insects when available.

BREEDING BIOLOGY

Formerly bred in considerable numbers in Seistan in Afghanistan and Iran, just west of Baluchistan, but these marshes have an erratic area due to irregular rainfall and snow melt. Numbers however, still breed on Lake Ab-e-Istada in Afghanistan. Occasional birds occurred on Kushdil Khan lake in Baluchistan during the summer months (Meinertzhagen saw 6 on 19 May, 1914) but it was not until 1937 that a pair was recorded breeding successfully in a hole in the bank of Kushdil Khan lake, Baluchistan (Christison, 1942). Eleven young were fledged. In 1939 another pair nested on the north-eastern end of the lake. No Shelducks have been seen on Kushdil Khan lake during several summer visits during the 1970s and they must be sporadic breeders. Pair bonds are monogamous and probably life-long. They are burrow nesters, the female shaping the nest hollow which is lined with her own breast down. The creamy white eggs are incubated by the female alone and the normal clutch is 8 to 10. Incubation period is 29 to 31 days. Both parents tend the young for the first 15 to 20 days after hatching, though the ducklings feed themselves from the time of hatching and are very active.

Lahore in late December 1968 (Rhys Davies, pers. comm.). Large numbers, estimated around 900 were also seen on Sandho lake on the border of the Rann of Kutch about 39 kilometres (19.5 miles) south of Badin (F. Koning, pers. comm.). It is therefore possible that quite a substantial population winters in remote areas along the Sind coast and Rann of Kutch and may have escaped detection. Mr. Koning estimated the total wintering population in this region at between 2,000 and 3,000 birds. Status SCARCE.

HABITS

Throughout its range a bird of brackish and saline wetlands, most likely to be encountered in Pakistan on such lakes, as Uchchali in the Salt Range, and Siranda in Lasbela and the greatest concentrations have recently been recorded along the seacoast. In feeding habits they are more prone to feed in water than by grazing

VOCALIZATIONS

As in the TADORNINAE calls of males and females markedly dissimilar. Males have a loud whistling tone often uttered in flight. Their whistles are quite melodious and inflected, suggesting smaller passerine birds. The female emits a much deeper pitched and nasal call rendered as 'gaga-gagaga'.

Cotton Teal or Quacky-Duck *Nettapus coromandelianus* (Gmelin)

White-Quilled Pygmy Goose in Australia

Illus. **21**
Map **40**

DESCRIPTION

Body length 31-34 cms.
Wingspan 50-60 cms.

A very small duck, smaller than the Common Teal (*Anas crecca*), with a rather short black bill, deep at the base like a goose. In breeding plumage the male has a largely white face, neck and belly with a shiny iridescent green-black crown not reaching down to the eye, and black glossed green back, wing shoulders and tail. A narrow black collar extends around the base of the neck becoming wider in the forepart. There is a whitish patch on either side of the upper-tail coverts and the under-tail coverts are black. The feet are black, the hind toe being well developed and unlobed and all toes having strong claws. These adaptations enable it to perch well on branches. In winter the male moults into an eclipse plumage similar to that of the female which has a more extensive dark crown and back, but these areas are dark brownish grey and not glossy except on the wing shoulder which has some green gloss. The face and breast and flanks are greyish, with some darker speckling on the lower neck, and the breast having a more buffy tinge. There is a broad dark grey-brown line through the eye. In flight both sexes show a white band through the wings formed by the secondaries and base of the black-tipped primaries, but this white area is more limited in the female. Juveniles look like females with more brown on the flanks and a less distinct paler eye stripe.

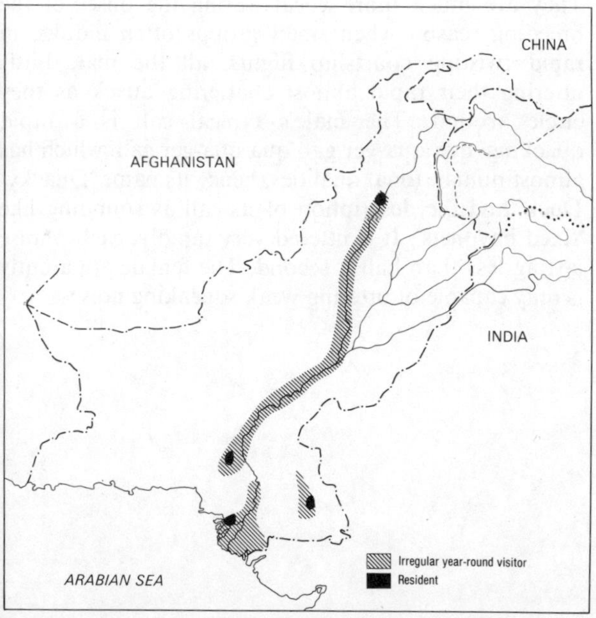

HABITAT, DISTRIBUTION AND STATUS

It occurs throughout south China, south-east Asia, Indonesia and north-east Australia. This sub-tropical duck is a resident in the subcontinent, but was considered rare in the north-west at the turn of the century. Ticehurst considered it rare in lower Sind, recording between 1908 and 1915 about 7 sightings, or authentic cases of birds shot during duck shoots. It occurred at that time, mainly around Thatta district and the tamarisk-dotted swamps of Sujawal. It was not encountered in the Punjab in the 1920s by Whistler (*Ibis*, 1922 and *Ibis*, 1930) or H. Waite (*JBNHS*, 1948). K. Eates writing about its occurrence in Sind from the 1920s to 1940s, only saw it twice in 30 years and did not believe that it bred (MS notes unpublished).

It appears to have increased considerably since his day as small flocks seem to be resident on several lakes in lower Sind and occasional groups have travelled up the Indus and been recorded in the Punjab.

They like best, lakes with plenty of cover especially submerged weeds and lotus lilies (*Nelumbium* sp.) on their surface. A population of between 300 and 400 live year-round on Hadiero and Haleji lakes approximately 80 kilometres (50 miles) north-west of Karachi, e.g. 300 counted 17 November 1973 on seepage pools bordering Haleji and about 150 on 6 March 1980 on Hadiero lake. In the Punjab they are largely summer visitors, but numbers over-winter, 10 seen on the seepage pools beside Jinnah barrage on the Indus at Mianwali on 23 January 1981. In summer they are regularly seen on Salt Range lakes, 8 on Nammal lake on 6 September 1967, 2 seen on Kallar Kahar lake in August 1968 (C. D. W. Savage, pers. comm.). 23 were seen on Lal Sohanran on 22 November 1972. In Sind they also occur on Manchar lake and some of the east Narra lakes (24 counted on 20 March on Manchar lake, J. Vieillard, pers. comm.) and 20 on Drigh lake in Larkana district on 25 January, 8 on Sonari lake, Sanghar on 19 January. (F. Koning, pers. comm.). Numbers are augmented in the monsoon season in lower Sind and it is not certain how many of these

Map 40: Cotton Pygmy Goose or Cotton Teal *Nettapus coromandelianus*

Illus. 21: Cotton Pygmy Goose or Cotton Teal *Nettapus coromandelianus*, male in breeding plumage.

breed, as small flocks can be seen throughout the summer months. Status COMMON only in Thatta district of Sind.

HABITS

They are perching ducks which nest in cavities, usually in trees, and perch freely on branches. In winter they consort in small groups and will forage throughout the day, preferring areas of water, choked with rope-like strands of the water weed *Urticularia stellaris*, as well as *Limnophyla heterophyla* common on Manchar lake, and in seepage pools near Jinnah barrage they were seen feeding upon *Vallisneria spiralis* water weed. They will frequent very small seepage pools and seem to forage mainly while swimming. Their short deep bills seem to be an adaptation for feeding on seeds, especially those of *Nelumbium* Lotus lilies and seeds of water weeds such as *Hydrilla verticillata* and of *Scirpus roylei*. They will also eat young shoots, rice grains and fresh water crustacea, insect larvae and worms. In a study carried out in the Sunderbans, West Bengal, stomach contents of 43 birds revealed only 68 per cent of their diet was vegetable matter comprising aquatic weeds, whilst 18.5 per cent was mollusca, 7.5 per cent insects (aquatic), 18.5 per cent small shrimps and fish fry (Mukherjee, *JBNHS*, Vol. 71, 1974).

Though their legs are set further forward under the body than in the typical dabbling ducks, they have very short tarsi and are awkward on land and appear to forage entirely while swimming and by dipping the head and neck under water, never by diving. They fly fast and strongly however and typically utter rapid low pitched chattering contact calls in flight. They can, however, dive well if frightened (to avoid predation), when unable to fly.

BREEDING BIOLOGY

About 10 per cent of males start to assume full breeding plumage from mid-March and by early May, all are in breeding dress which they do not begin to lose until mid-November in Pakistan. Courtship involves a lot of flight chases usually of more than one male chasing females and breeding seems to be initiated by group displays. Little has been recorded about actual pair formation or displays but the bond is believed to be of short duration and only the female incubates. The nest is lined with a bit of grass or feathers but not with any down. Nests in holes in buildings have been reported, but usually a tree is selected close to water. The eggs are plain white and the normal clutch is 6 to 14 but one suspects that like others of this tribe of perching ducks, bigamy is frequent with a second female dumping her eggs in an already occupied nest hole as 22 have been recorded in one nest (Stuart Baker). Only the female cares for the ducklings which hatch synchronously and have been reported being pushed out of the nest cavity and fluttering to break their fall as they neared the ground (Ali and Ripley, 1968).

*VOCALIZATIONS

They are much more vocal during the onset of the breeding season when small groups often indulge in rapid twisting courtship flights, all the male birds uttering their rapid almost chattering quack as they circle around. The male's typical call is a rapid quacking 'qua qua-ger gab qua qua-ger gab' which has almost human tonal qualities, hence its name 'Quacky-Duck' and the description of its call as sounding like 'fixed bayonets'. It is uttered very rapidly, each phrase lasting less than half a second. The female apparently is only capable of uttering weak squeaking noises.

Nakta or Comb Duck *Sarkidiornis melanotos* (Pennant)
Knob-billed Goose of Africa
Sarkidiornis melanonotus (Pennant) of Ticehurst

DESCRIPTION

Body length 70-78 cms. (males), 50-60 cms. (females)
Wingspan 140-160 cms. (males)
Wing length 33.9-40.6 cms. (males),
    28-30.9 cms. (females) (Stuart Baker)
Weight up to 2,610 gms. (males),
    1,925-2,300 gms. (females) (Hume)

There is marked sexual size dimorphism in this species of perching duck which is predominantly plumaged in black and white. The males are about as big as the Bar-headed Goose (*Anser indicus*). The bill is black and the legs and feet are dark plumbeous. The wing coverts are glossed green and the secondaries and tertials are glossed bronzy purple. The back is duller and greyer and the head and neck are freckled with black spots. A half collar of black feathers extends around the pectoral region leaving the centre of the breast white and the under-tail coverts are buffy yellow margined around the vent by a black band. The male in the breeding season develops a rounded fleshy fin-like projection on its upper mandible which is laterally compressed. In the winter this shrinks and becomes more of a rounded knob. The female is similarly coloured but much smaller in size and lacking the knob at the base of the upper mandible. Also she lacks the black pectoral bands and her breast and flanks are dull buffy colour, less white. In flight the wings are entirely black, both above and below, and the wings look rather broad and rounded.

HABITAT, DISTRIBUTION AND STATUS

A resident species in India, rare in south India and inclined to be locally migratory according to water and feeding conditions. Although rare in the north-west of the subcontinent, it has been recorded earlier in this century as rare, but locally resident in southern Sind. In 1874 on 14 July a female with a fully formed egg in her oviduct was shot near Lahore at a place where the Bari Doab canal was being constructed. There were also Nakta ducks exhibited in Lahore Zoo in the 1920s obtained from a local nest and reared artificially (Wright and Dewar, 1925). It prefers well wooded countryside and fresh water ponds or tanks that are well covered with emergent aquatic vegetation with trees nearby. It was probably resident in very small numbers until the 1930s in Thatta district and south-eastern Sind. Ticehurst (*Ibis,* 1923) records birds being killed by duck hunters in Badin 18 January 1879, 8 being seen at Sujawal on 27 December 1911 and another shot in the same area on 12 February, 1912.

One was recorded in February 1918 by Gibson near Tando Bago, Hyderabad district, Sind (Gibson, *JBNHS,* 1918). The Bombay Natural History Society Museum has a specimen shot near Umarkot on 26 December 1917. Eates did not come across it during over 25 years working in Sind. Status in Pakistan today, probably EXTINCT.

HABITS AND BREEDING BIOLOGY

Very little seems to have been recorded about this interesting duck in India. They are largely herbivorous in diet and will take rice and cereal seeds, seeds of sedges and young green buds and shoots of water weeds such as *Potamogeton indicus*, also insect larvae, and molluscs available in marshy areas.

The much larger sized males in this species indicate the possibility that pairing may commonly be polygamous or bigamous, as up to 47 eggs have been found in one nest (Livesey, *JBNHS,* 1921). The pair bond is usually very temporary and copulation is always by forced rape, as in the African *Sarkidiornis*. Courtship consists of a male approaching the female in the water with breast and wings slightly raised and the head wagged slowly from side to side. Generally the female responds by trying to escape.

Nesting takes place during the monsoon months and usually a mango tree is selected by the female who does all the incubation and care of the young. She selects a hole, or natural hollow in a tree or the cavity formed between dividing main branches. Some twigs and grass are placed in the nest bottom. (Lowther, 1949). The eggs are ivory white in colour and the normal clutch is 7 to 15. They are reported to take 30 days to incubate. The male has been noted as visiting the nest tree daily but never approaching close to the nest cavity and the female is very wary and circumspect in approaching the nest (Lowther, op. cit.).

J. A. W. Anderson reported that in the late 1950s a pair of Nakta ducks nested in a hole in the ruined building on Lasbela island in the Indus river near Sukkur.

VOCALIZATIONS

According to Lowther the male emits a loud honking call, probably a summoning call when he circles around the nest tree, but Delacour only recorded a low hissing from captive pairs which bred successfully. In winter time females occasionally utter a weak grunting call, but both sexes are relatively silent most of the year.

TRIBE ANATINI

Dabbling Ducks

Eurasian Wigeon or Wigeon *Anas penelope* Linnaeus   Map **41**

DESCRIPTION

Body length 45-51 cms.
Wingspan 75-86 cms.

The sexes are about equal in size and in flight the pale silvery grey body and leading edge of wings with broad white bar formed by wing coverts is at once noticeable in the drakes. Females lack this white band on the upper-wing, but their wing shoulder looks greyish. Both sexes have a glossy green speculum and dark blackish grey primaries. On the water the male's high crowned chestnut head with creamy yellow forecrown and pale grey bill is diagnostic. The bill has a black nail visible from a distance. The breast is pinkish vinaceous, the flanks vermiculated grey and the lower belly is pure white, under-tail coverts are black and the scapular feathers are lanceolate grey and black divided by a white mid-rib. Legs and feet are grey. The female has a duller grey bill and her flank feathers look noticeable rufescent and less streaked than the flanks of female Mallard (*A. platyrhynchos*) or Common Teal (*A. crecca*). In silhouette swimming Widgeon have a rather peaked forehead and pointed tail.

HABITAT, DISTRIBUTION AND STATUS

This is a winter visitor to Pakistan from the Palearctic region, spending these months on large fresh water lakes, with a distinct preference for more open bodies of water or along the main rivers. It also frequents some of the tidal creeks and salt marshes of the Indus delta. Numerically it is about the fifth most numerous duck species wintering in Pakistan based upon IWRB surveys. Average wintering population 10,000 to 20,000 birds (e.g. 17,670 actually counted during January and February, 1973 and 12,000 the previous winter). The Salt Range lakes often hold considerable numbers, e.g. 600 on Nammal, 1,100 on Uchchali and 1,500 on Kallar Kahar in one winter. In Sind over 4,000 annually visit Hadiero and Haleji lakes where they are usually the dominant duck species. Earliest arrival dates noted were 26 October and the latest 15 April but most have left by late March.

They have been shot in Gilgit on passage and a bird ringed on Manchar lake was recovered in Siberia (N56°., E70°.). Numbers winter each year on Kushdil Khan lake in Baluchistan and several flocks of 100 to 300 winter on different creeks in the Indus delta. Status ABUNDANT.

HABITS

Quite vocal, even in winter quarters the male's whistle being unmistakeable. They are usually very active feeding at night, occasionally dipping down to graze on submerged vegetation in the typical manner of dabbling ducks depending upon water depth. But more commonly they feed from the surface swimming with head only partially submerged. On the Salt Range lakes such as Khabbaki their principal food is *Myriophyllum* sp. and *Vallisneria spiralis*. In the salt marshes around Keti Bandar they come on land at night to graze. Their diet is largely vegetarian in winter and they are quite tolerant of brackish lakes or halophytic plants such as *Sueda fruticosa* and *Salsola foetida*.

BREEDING BIOLOGY

Extra-limital breeding right across the steppe and taiga zones of northern Europe and the USSR but not extending right into the Arctic tundra.

*VOCALIZATIONS

In winter time males can always be heard giving their ringing 'whee-oo' call. They emit this in flight as well as on the water. Females have a distinct purring call less easy to detect in a large gathering of waterfowl. The call which may be continuously repeated sounds like 'krrr-krr'.

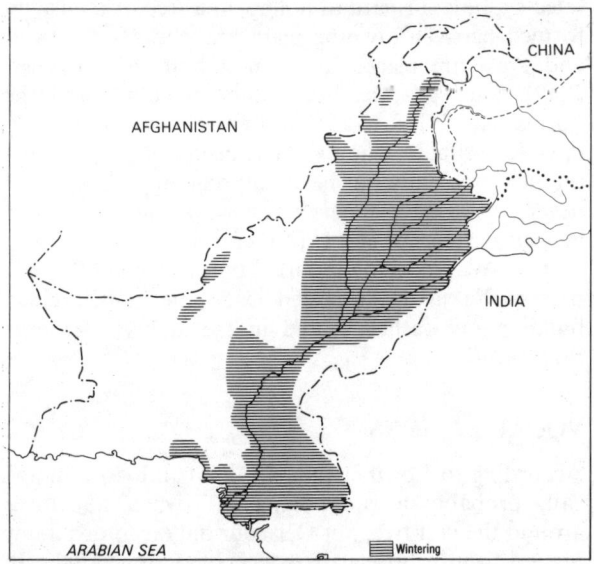

Map 41: Northern Pintail *Anas acuta*, Common Teal *Anas crecca*, Mallard *Anas platyrhynchos*, Widgeon *Anas penelope*, Northern Shoveler *Anas clypeata* and Gadwall *Anas strepera*

Falcated Duck or Falcated Teal *Anas falcata* Georgi

DESCRIPTION

Body length 48-54 cms.
Wingspan 76-82 cms.

Similar in size to the Gadwall, to which it is thought to be closely related, but the drakes are very decorative and striking in appearance with high peaked crown of a bronzy maroon and bottle green rear part of head, extending from behind the eye into a bushy nuchal crest. The throat is white, framed by a marrow black line and the breast is grey with delicate scale-like markings and flanks vermiculated grey. Wing shoulders pale greyish white with long down-curving falcated scapular plumes jet black narrowly bordered with white. The female is difficult to separate from a female Wigeon, both having a dark chestnut crown. But the female Falcated Teal lacks the darker tail of the Wigeon and has the whole of the breast mottled with brown (white in the mid-belly region of the Wigeon).

STATUS

Breeds in eastern Siberia from the upper Yenesei south to lake Baikal and eastwards to Amur and Ussuriland. Normally they winter in Japan, Korea and eastern China. In the Indian subcontinent it is a straggler, occurring mainly in Assam and Bangla Desh where it is apparently quite rare today (D.A. Scott, *in litt.*, 1987). The author kept as a pet a drake purchased in December 1969 in Chittagong bazaar.

In Pakistan there are about five authentic records. One shot (an adult male) in the east Narra (Sanghar district) in Sind on 10 January 1901 (*JBNHS*, Vol. 14, p. 149) and a Sind specimen in the BNHS collection. The Game Book of the Amir of Bahawalpur had a record of one being shot in the early 1930s at Jajjian Abbas in December and a Major General McLeod shot one at Feroza, in December 1879, in Bahawalpur district. It was a female. There is also a record of one shot in the 1920s in Jhelum district. Status RARE VAGRANT.

Gadwall *Anas strepera* Linnaeus   Map 41

DESCRIPTION

Body length 46-56 cms.
Wingspan 84-95 cms.

In the field it looks the same size as the Mallard (*Anas platyrhynchos*) but is in fact slightly slimmer bodied and relatively longer winged. Compared with other dabbling ducks the drake could be mistaken for a female, being inconspicuously coloured. He is however, much more silvery grey than a female Mallard (*A. platyrhynchos*). A good field point is the quite extensive black of upper and under-tail coverts extending to rump and vent, coupled with grey wing shoulders and rather blackish breast, and an all-black or dark leaden bill. The head is mealy brown, with narrow black scaling on the upper breast. Mid-belly is white, and legs are orange-yellow. There are long pale grey scapular feathers which usually conceal the speculum or secondaries when the bird is at rest. In flight the drake shows a conspicuous speculum pattern, of pure white outer secondaries bordered by a black band (the secondary wing coverts) and a chestnut band (the median secondary wing coverts). This bright chestnut red band often shows as a small patch when swimming. The female has a paler face and upper throat with a darker brown crown and narrow dark eye-streak. Her breast is more darkly streaked than a typical female Mallard and shows a lot of white on the flanks when swimming, which the female Mallard does not show. Also in this species the bill shows a yellow panel on the lower half of the upper mandible, both in females and males during eclipse plumage. In flight the female also shows grey wing shoulders and a white panel on the outer secondaries, in contrast to the dark speculum of female Mallard or Wigeon (*Anas penelope*) with which it could be confused.

HABITAT, DISTRIBUTION AND STATUS

They are winter migrants from the Palearctic to the subcontinent, breeding in the middle latitudes or taiga zone of central Asia. Largely confined to fresh water lakes or swamps, occurring whenever there are concentrations of waterfowl during the winter, especially preferring less alkaline lakes in which there is a good deal of aquatic or emergent vegetation. However, it has been twice noted in October on sandbars in Ghizri creek (on the Karachi seacoast). It is very widespread throughout Punjab and Sind and a few usually winter on Band Kushdil Khan lake in Baluchistan. Ticehurst

(1923) noted that it was the commonest duck in Sind in winter, and Whistler recorded it (*Ibis*, 1922) as abundant in Jhang district, Punjab, in winter. It must have suffered a greater decline than other species during the past 60 years as it is now less common than any of the other dabbling ducks which visit Pakistan except the Spotbill (*A. poecilorhyncha*), and less common than the *Aythia ferina* or *A. fuligula*. Total numbers counted during IWRB surveys of all the major wetlands were 2,700 in 1972 and 4,000 in 1973 (cf. over 12,000 *A. penelope* and *A. platyrhynchos* in the same year). It is commoner in Sind than Punjab according to these surveys (F. Koning, pers. comm.) and there is evidence of a very sharp decline in numbers occurring during the 1930s (C. D. W. Savage, 1968).

Earliest arrivals come in late October and a few birds may linger up to May in Rawalpindi district (Major Stockley, *JBNHS*, Vol. 28) and in Kohat (Whitehead, 1910). According to shooting records they migrate through Gilgit, the Hunza and Kurram valleys. Birds ringed on Manchar lake (Dadu district, Sind) have been recovered in Omsk region of Novosibirsk, USSR chiefly between latitudes N50° and 60°. Status COMMON.

HABITS

Even in large mixed concourses of waterfowl they seem to recognize each other, as Gadwall generally feed in small groups of con-specifics. They will also feed in association with Coots (*F. atra*), presumably picking up weeds dislodged by the diving coots (D. A. Scott, *in litt.*). Like other dabbling ducks they frequently tip up their tails and feed on submerged vegetation or seeds, when their black under-tail coverts make them quite conspicuous. They feed mainly on seeds, shoots and other plant parts, particularly from submerged water weeds such as *Limnophila heterophyla* in Sind and *Vallisneria spiralis* and *Hydrilla verticillata* in Punjab.

*VOCALIZATIONS

On thir wintering grounds females can often be heard giving a 'quack-quack-quack' decrescendo-type of call quite similar to that of the female *A. platyrhynchos*. The drake has a deeper pitched croaking call (heard in March, Haleji lake) and during the breeding season drakes have a repertoire of fluting whistling calls.

Baikal Teal *Anas formosa* Georgi

DESCRIPTION

Body length 36-42 cms.
Wingspan 60-72 cms.

A small duck, only slightly larger than the Common Teal (*A. crecca*). Drakes are beautifully patterned on the head with buffy yellow at the base of the bill and mid-cheeks, divided by a narrow black line curving down from behind the eye to the throat. There is also a narrow white line above the eye. The crown and chin are black and the nape and sides of the neck are glossed with green and purple. The breast is pinkish buff, spotted with black and with a conspicuous vertical white bar in the pectoral region. Flanks and wing shoulders grey with white belly and graceful curving falcated scapulars of black and cinnamon and creamy white in longitudinal bands. The speculum is bronzy green, edged with black and white and the bill and legs are bluish slaty. The female looks very much like the female of *A. crecca* but with a more distinctive face pattern, including a round white patch at the base of the bill, a dark streaked crown and an interrupted darker line through the eye to the nape and the rest of the face light creamy, speckled with brown.

HABITAT, DISTRIBUTION AND STATUS

Breeds within the boreal forest zone in north-east Siberia, wintering largely in Japan and south-east China, where it is now becoming quite rare (D. A. Scott, *in litt.*, 1987). It is only a straggler even to the north-eastern corner of the subcontinent and there are only two authentic records for Pakistan. One from Sind, shot in 1878 by Colonel Le Messurier (*Stray Feathers*, Vol. 8, p. 324) and one from the Punjab about 50 years ago. Status VERY RARE VAGRANT.

Common Teal *Anas crecca* Linnaeus Map **41**

DESCRIPTION

Body length 34-38 cms.
Wingspan 58-64 cms.

A small duck. Males have dark chestnut maroon heads with a central panel of bottle green extending from just in front of the eye and curving down the hind neck. This green area is framed by a narrow creamy buff line. The breast is creamy buff spotted with brown, and the wing coverts and flanks are finely cross-barred grey, under-tail coverts black with round yellow buff patch on the rear flanks. Speculum is iridescent green with secondary coverts forming a narrow creamy border. The scapular plumes are grey and black but not very long or conspicuous except for a broad white border which shows well when at rest.

Bill and feet are dull leaden grey to black. The female looks like a miniature female Mallard with a distinctly dark crown and line through the eye, but showing some white on the belly and the throat which the female Mallard lacks. The speculum in flight is green not purple-blue but is narrowly bordered by white lines as in *A. platyrhynchos*.

HABITAT, DISTRIBUTION AND STATUS

Breeds widely in Asia from the borders of the Arctic tundra to steppic latitudes and is a winter migrant to Pakistan. Three recent ringing recoveries include an adult male ringed 24 October 1965 at Kharrar jheel in the Punjab, shot in Altai, USSR on 5 May 1967 near Topchicka N52°50'., E83°08', and of a female ringed at Chany lake, Novosibirsk, USSR on 30 July 1967 and shot in November of the same year near Shekhupura, Punjab. A first year male ringed on 19 November 1965 in Thatta district of Sind was shot at Pavlogradsk near Duvanovka on 5 September 1966, N54°13'., E73°34'. The Baraba Steppe is one of the principal breeding grounds in the Soviet Union for Teal wintering in Pakistan (C. D. W. Savage, *in litt.*).

It is a dabbling duck which prefers to frequent shallow seepage pools on the fringes of large bodies of water, especially if they have plenty of tamarisk bush cover. They will also congregate by day on large open bodies of water where there is hunting pressure.

They are one of the earliest autumn migrants amongst the ANATIDAE. On 31 August a flock *circa* 30 seen Muzaffargarh, Punjab, 27 August, flock of 50+ on Kushdil Khan lake, Baluchistan. By the third week

of September flocks can be frequently encountered in the flooded rice fields in lower Sind (*circa* 20 on 21 September near Chauhar Jamali, Thatta district). They are widespread in Punjab, Sind and wherever there is suitable swampland or water, in Baluchistan and the NWFP. They are by far the commonest duck in winter in Pakistan. Because of their preference for thick cover and even of small shallow ponds and seepage zones in the periphery of larger rivers or lakes, they are very difficult to census. In the IWRB surveys of all major wetlands in Pakistan, numbers were always regarded as under-estimates but 88,400 were counted in 1973, and 37,500 in 1972. Status VERY ABUNDANT.

HABITS

They often feed throughout the night and by day roost in big flocks on more open bodies of water. They can adapt to a wide variety of habitats and water of varying degrees of salinity or acidity, being able to subsist on a wide range of foods including rice grains, seeds of water weeds and sedges (*Scirpus* and *Carex* spp.) and of halophytic plants such as *Salsola foetida* and *Suaeda fruticosa*. They will also eat small molluscs, insect larvae and water beetles. They commonly feed by wading in shallow water or by swimming, with bill immersed, dabbling under the water surface or in the mud. In deeper water will regularly 'up-end' to get at submerged vegetation alongside larger Pintail (*A. acuta*) and other species (observed Haleji lake seepage pools). They are very swift flyers, twisting and turning with amazing agility, often low over the water surface in tight flocks, before all suddenly plunging in to land. On Ghauspur jheel, Jacobabad district, which in a good monsoon year covers an area of about 260 hectares, 8 or 9 separate feeding flocks can be seen totalling up to 6,000 birds (observations, 14 February 1975).

*VOCALIZATIONS

Females in their daytime resting places often utter high pitched soft and rapid quacking sounds (tape recorded in December, Chatteji lake). Such calls are rapid enough at times to be likened to a soft clucking almost approximating human voice tones. Males utter an inflected whistle call, 'kerick-kerick'.

Mallard *Anas platyrhynchos* Linnaeus

synonym *Anas platyrhyncha*

Map **41**

DESCRIPTION

Body length 50-65 cms.
Wingspan 81-98 cms.

A large duck, with the male handsomely patterned with dark bottle green head and neck and maroon breast divided from the dark green neck by a narrow white ring. The body is mainly grey and its tail white with black rump and vent. The bill is greenish yellow and feet orange. The female has a dull olive brown bill, but similar orange legs and both sexes have an iridescent cobalt blue speculum narrowly bordered by black and then white. The female has a dark brown crown and line through the eye, with the buff cheeks and flanks heavily streaked with darker brown. The belly is dull grey-brown not white. For distinction from female Gadwall (*Anas strepera*) see description of that species.

HABITAT, DISTRIBUTION AND STATUS

The Mallard is an abundant Holarctic breeding species from the Great Lakes region of Canada westwards to Alaska and across the USSR to western Europe. It is a winter migrant to Pakistan with no evidence of breeding, though an occasional pair still probably breeds on the Hokra jheel in the main vale of Kashmir (Bates and Lowther, 1952). Birds ringed on Manchar lake in Sind and on Jhajjia Abbasian in Bahawalpur have been recovered in the Novosibirisk region of the USSR between N52°., and 56°. First arrivals are recorded in early September in Baluchistan but do not seem to reach Sind till mid-October, and they leave quite early, usually by about the the middle of March. They occur all over Pakistan but sometimes in a rather patchy distribution. Considerable concentrations have been counted, e.g. 300 on Warsak dam headpond, North West Frontier Province, up to 700 (27 January.) on Rawal lake, Islamabad and on Rasul barrage headpond (Jhelum district) 500+ on 7 December. In Gilgit a few birds spend the whole winter in willow swamps along the river but there is a marked influx of birds on spring and autumn passage. In Sind they are not very common, but normally 400 to 500 winter on the three large lakes north-west of Karachi, appearing quite often after the beginning of the new year. They have

also been seen in the mangrove creeks but usually in quite small numbers. The total wintering population is certainly much greater than that of Gadwall (*Anas strepera*), but they are about fifth in abundance amongst the dabbling ducks. IWRB surveys in February 1972 gave a total count from all major wetlands of 11,700, and in February 1973 about 7,000. The latest date that birds have been seen was 9 April on Rawal lake. The Game Book records of the Amir of Bahawalpur indicate that during the 1930s and early 1940s this was the commonest duck on Bahawalpur jheels. Total bags of up to 3,000 ducks used to be secured during two days shoots organized on these lakes. Status ABUNDANT but mainly in the Punjab.

HABITS

They will rest by day on large open bodies of water such as Chashma barrage headpond (1,200 on 2 February 1973) but usually prefer bodies of water with cover from submerged tamarisk bushes, or well bordered by reeds for feeding. They are usually seen keeping in small groups of conspecifics and their presence is often first revealed by the loud ringing calls of the female. In their wintering grounds they feed mainly upon seeds and submerged vegetation such as shoots and leaves of water weeds (*Hydrilla verticillata* and *Vallisneria spiralis*). As the weather warms up in February an increasing amount of insect food is included in their diet, such as CHIRONOMIDAE and Odonata larvae, small molluscs and water beetles. They feed by swimming and submerging their bills or heads in the water, as well as by up-ending where submerged weeds offer better feeding opportunities. They will also flight into stubble and flooded grassland to feed at night, when they peck at the vegetation or dabble in the mud as they walk. In Pakistan they are very shy and wild and are strong flyers.

*VOCALIZATIONS

In winter females have a loud ringing decrescendo cry 'quack-quack-quack-quack' getting quicker and fainter after the first two calls. The males have a more nasal or wheezy quack which is less frequently repeated than the female's call.

Spotbill Duck *Anas poecilorhyncha* J.R. Forster   Map **42**

DESCRIPTION

Body length 50-65 cms.
Wingspan 81-96 cms.

About the same size as the Mallard, it often looks bigger on the water, having a rather upright neck carriage. They are readily distinguished from other dabbling ducks by the rather pale silvery buff forepart of their necks, coupled with the pure white flash of their tertial feathers, which show even when at rest. Both sexes have orange feet. Their crowns look blackish with a broad dark eye-streak and speckled whitish cheek and broad pale supercilium. The breast is spotted and the flanks scalloped with white margins to the brown feathers. The bill when viewed closely is the striking feature of this duck being black with the distal one-third a bright chrome yellow. There is a prominent black nail and a less conspicuous round fleshy orange scarlet patch at the base of the upper mandible. Females are like the males in plumage though slightly smaller and duller in colour. In flight the bottle green speculum bordered by black and white is conspicuous, and also the white tail. The tips of the greater wing coverts are also white giving a broad white bar to the middle of the wing.

HABITAT, DISTRIBUTION AND STATUS

A partly Oriental and Palearctic faunal species of which the nominate sub-species inhabits the whole of the subcontinent, whilst another race lacking the red spot at the base of the bill, inhabits Burma and the Shan states. It also occurs as far north as Korea and Manchuria (Delacour, 1956) and the whole of Japan and China. In Pakistan they are largely summer monsoon visitors with a small residual population over-wintering in south Sind. They migrate up the Indus in summer and can be found in a few scattered pairs breeding in the swampy pools and seepage channels around Balloki headworks (on the Ravi river), Trimmu headworks (on the Chenab river), Katalpur jheel and Sidhnai spill channel in Multan district. In Sind small flocks have been seen throughout December, January and early February on the Sanghar lakes (Sadori and Soonahri) and larger numbers on Haleji lake (*circa* 50 in November 1982). Numbers are supplemented by summer visitors and pairs disperse to smaller seepage pools and canal-side borrow pits in Thatta district to breed. They favour fresh water areas with plenty of emergent sedges such as *Scirpus roylei* and *Juncus bufonius*, and also reedy margins. Maximum number counted were 68 on Haleji lake, Thatta district in November 1976. Absent in the NWFP, but in surveys conducted by Ashiq Ahmad of

the Pakistan Forest Instute in Baluchistan, several Spotbilled Duck were seen on Kushdil Khan lake in the early spring and in July 1987, 15 adult birds were seen with an unrecorded number of young, establishing the fact that they had bred (Ashiq Ahmad, Wildlife Research Officer, typed report, 1987). Status COMMON in Thatta district only.

HABITS

Not markedly gregarious but keeps in small groups of upto a dozen birds outside of the breeding season. Mainly feeds by swimming with head submerged or only bill submerged. In deeper water feeds by up-ending like Mallard *Anas platyrhynchos*, with which it seems to share a preference for the same type of habitat. It is mainly herbivorous, feeding on seeds and young shoots of water weeds such as *Scirpus roylei*, *Hydrilla verticillata* and *Potamogeton indicus*. It feeds on rice in the late monsoon season, also the rhizomes of water lilies, and to a lesser extent upon fresh water snails, water beetles and other insect larva. It flies strongly, and before the breeding season small parties often move around looking for suitable feeding grounds. It will also feed by walking on marshy ground pecking at vegetation such as *Cyperus eleusinoides*.

BREEDING BIOLOGY

They breed during the monsoon season, usually when temporary ponds are full of water and vegetative

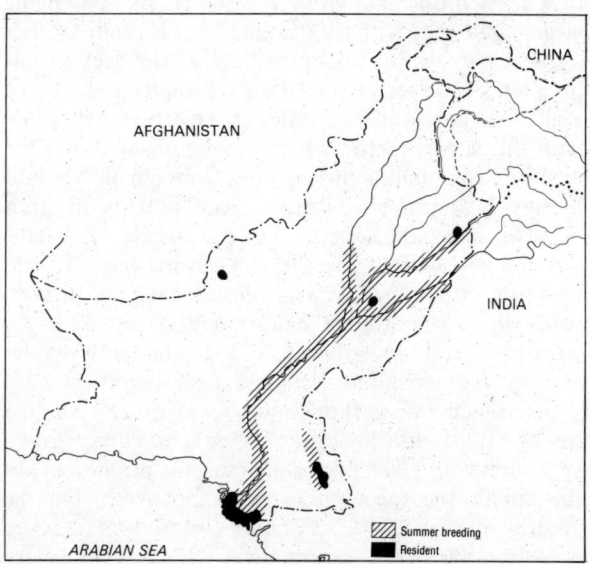

Map 42: Spot-billed Duck *Anas poecilorhyncha*

growth is vigorous. Nest sites are on the ground in sedges or grass thickets, clumps of tall *Saccharum* being favoured. They often nest on islets or sheltered banks of deep pools where bushes provide good cover. Eggs have been found on 7 June in the east Narra (Doig) and K. Eates found a clutch of 7 unincubated eggs on 20 July on Chach Dhand north of Umarkot and 11 eggs at Khinjar lake in early August. The eggs are a pale greenish grey, very similar to those of a Mallard and the normal clutch is 7 to 9. The pair bond appears to be fairly stable, pairs have been noted as early as mid-March and the male remains in the vicinity of the nest while the female incubates and after hatching, assists in protecting the young. The nest itself is quite well made, being lined with dried grasses and down, and the young hatch synchronously after about 26 days from the last egg being laid. Courtship has elements similar to other dabbling ducks.

In the spring there is much flight chasing with one or more males pursuing a female, or an unattached male aggressively chasing a mated pair trying to drive off the male. On the water, threatening postures include neck lowered and outstretched and swimming towards the victim (obervations, Mirpur Sakro in April). Displaying males have also been watched swimming up to females and jerking their heads very upright with neck held vertically and breast high in the water (observations Haleji, 31 May). Downy chicks were found at Katalpur jheel in mid-August (Ainsworth Harrison, pers. comm.), and full fledged young still following the mother in line, at Balloki headworks in late August (Roger Holmes, pers. comm.).

VOCALIZATIONS

When flushed in non-breeding season they fly off with a rapid repeated and short 'quack-quack-quack' very similar to a female Mallard (*Anas platyrhynchos*). The drake has a more nasal wheezy quack, again reminiscent of a drake Mallard.

Northern Pintail *Anas acuta* Linnaeus

Map **41**

DESCRIPTION

Body length 51-66 cms.
Wingspan 80-95 cms.

Slightly smaller and slimmer in build than the Mallard (*Anas platyrhynchos*), the Pintail has a noticeably longer and more slender neck and in the males a long pointed tail. The drake with a snow white neck and breast, is conspicuous from a distance, its body being largely pale grey with a chocolate brown head, extending in a narrow line down the back of the neck. It has long lanceolate scapulars with black shafts and narrow white margins and the outer-tail feathers are white with the central rectrices black and elongated into the well-known pintail. Females differ from female Mallard in having a rather uniformly pale buffish chestnut head without a darker crown or eye streak. They have rather a pale buffy breast and the legs in both sexes are blackish to plumbeous. The bill is also grey in both sexes with a longtitudinal black patch at the base of the culmen and on the nail in the drake. The more slender neck and pointed tail are good field marks in both sexes in separating the Pintail from other dabbling ducks. In flight the drake reveals a glossy green speculum with black and white margins posteriorly. In the female the speculum is more brownish. On the anterior margin of the secondaries in both sexes there is a narrow buffy yellow line which is not very conspicuous in flight.

HABITAT, DISTRIBUTION AND STATUS

A Holarctic species which occurs over nearly the whole of the USSR and which winters in western Europe, the Nile delta and southern Asia. It is a winter visitor to Pakistan found in all regions. It enters over the Himalayan ranges from late September and also observed on passage in Baluchistan. Earliest arrivals noted on 2 September on Ghizri creek, Karachi coast. Most birds have left Sind by early March and migrating flocks have been seen (author obs.) crossing the Murree hills high-up as early as 17 March though a few stragglers can be seen up to 22 May (Rawal lake) and 30 March (lower Sind lakes). Normally quite gregarious, and preferring large open bodies of fresh water, where they can be seen in concentrations of 2,000 to 3,000 on the larger Sind lakes such as Drigh and Langh in Larkana district, with up to 1,500 in such places as Chashma barrage headpond, Taunsa barrage headpond and lakes such as Hadiero in Thatta district or Ghauspur jheel in Jacobabad district. IWRB surveys indicate that this is the second-most numerous of the dabbling ducks which occur in Pakistan. The February 1972 census total based on counts of all the major wetlands, was less than 10,000, but the February 1973 census showed 28,400. A large flock winters each year on tidal creeks in the Indus delta, and small flocks have been seen at Keti Bandar on the salt marsh flats. They are more abundant in Sind than in the Punjab. Status ABUNDANT.

HABITS

Highly gregarious in winter with flocks tending to separate into all-male or all-female bands. They are shy ducks, easily disturbed and when put to flight commonly circle till they reach a considerable height and then head off in search of another safer haven. They are often active feeding by day, as well as by night and most commonly forage by swimming and up-ending, their black tails being depressed to maintain balance as they search the lake bottom for food. Their longer necks are believed to give them an advantage over other dabbling ducks for this foraging technique. They are adaptable in diet subsisting mainly upon seeds, tubers and rhizomes, and during the mid-winter months on such water weeds as *Potamogeton indicus*, *Cyperus eleusinoides* and *Vallisneria spiralis*. They also consume molluscs, insect larvae and water beetles when such food is easily available especially in the spring.

VOCALIZATIONS

In their winter quarters they are largely silent, though wintering birds in Britain commonly make a quiet 'purring' (burp whistle) in courtship. In their breeding areas, they have a repertoire of calls not dissimilar from other dabbling ducks. The males have a thin high pitched squeaky call and the females utter a variety of quacking calls; shorter and more staccato than those of *A. poecilorhyncha* or *A. platyrhynchos*.

Garganey *Anas querquedula*

Garganey Teal — Australian name. Old name *Querquedula querduedula*   Map **43**

Mistakenly sometimes called Blue-winged Teal in the Indian subcontinent, but not to be confused with the American Blue-winged Teal *Anas discors* of that name.

DESCRIPTION

Body length 37-41 cms.
Wingspan 60-63 cms.

Slightly larger than the Common Teal (*Anas crecca*),

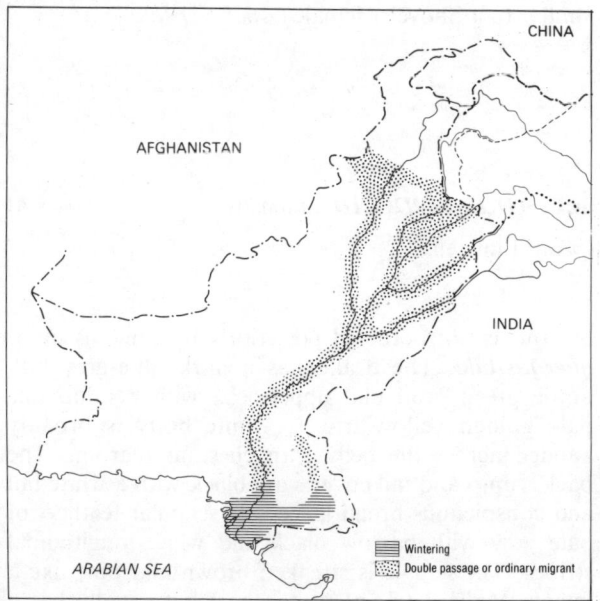

Map 43: Garganey *Anas querquedula*

and much smaller than the Mallard, these dabbling ducks often come to notice in mixed flocks of *Anas crecca* when their pale grey-blue wing shoulders flash in the sun. In breeding plumage the male is unmistakable with dark chocolate brown head and broad white sickle shaped line from eye and around the ear coverts. Its breast and back are brown with pale grey contrasting flanks and conspicuous lanceolate scapular plumes which are blue-grey and black with a narrow white rib through the black area. The under-tail coverts are mottled brown and buff. In flight the speculum is bottle green bordered by white; in the female this area is a duller browner green. The female is very like a Common Teal but has much more white on the belly and in flight shows the same pale grey wing shoulders. However, immature females lack the pale shoulders and are then easily confused with female Common Teal. Also the female shows a more conspicuous dark eye-stripe and broad supercilium than other female dabbling ducks. In both sexes the bill and legs are blackish grey. Unlike most other dabbling ducks visiting the subcontinent, males often retain their eclipse plumage until the early spring, much later than other ducks.

HABITAT, DISTRIBUTION AND STATUS

This winter migrant to the subcontinent breeds in mid-latitudes across Asia and eastern Europe avoiding the northern tundra areas and cold mountainous regions. The population which winters in the region comes from a wide area in the USSR with ringing recoveries in Rajasthan (India) from as far north as Kiev region (N51° 19'., E30° 14') and as far east as Leningrad (N60° 30'., E32° 50' which were shot in Maharashtra state, India. Birds ringed on Manchar lake have been recovered north of lake Balkash (N77°., E53°).

A few Garganey are recorded in Baluchistan on passage (Kushdil Khan lake in March to May and October to November). Very few Garganey over-winter in Pakistan, the majority wintering further south in India and especially in Sri Lanka. Huge numbers however pass through on both autumn and spring passage, being especially numerous from mid-March to early April in the northern Punjab where birds may tarry for several weeks on suitable bodies of water. In the lakes of lower Sind, spring migrants begin to arrive in numbers from mid-February and stay till late March. It is, like *Anas crecca*, one of the earliest migrants in autumn, birds being noted on 27 August on Kharrar jheel, Renala Khurd, Punjab and on 21 September in flocks over rice fields at Mirpur Sakro, Gharo taluka, southern Sind. In March it is often the only species remaining on suitable lakes, 1,200 on Hadiero lake on 10 March 1979, over 2,000 seen at Haleji lake on 27 February 1977 and over 1,000 on Haleji lake on 7 April 1974. On Rawal lake *circa* 40 seen 30 April and 21 May, 4 Garganey in two separate years.

Small numbers over-winter in southern Sind but they have not been noted in the Punjab, e.g. 200 on 14 December on Haleji lake and during the IWRB wetlands surveys, a total of 700 during February 1973, all in southern Sind and scattered over 9 different lakes. They will settle on the smallest pools and in rice fields and are not restricted to large bodies of water but they have not been noted anywhere along the seacoast. Status ABUNDANT only on passage.

HABITS

Seems to have a preference for foraging in shallow water and is tolerant of quite alkaline pools, rarely being seen on land or feeding in deeper water by up-ending. Usually they swim about with their bills and heads partly submerged. They apparently include a high proportion of animals in their diet when available and this includes a liking for fresh water snails such as *Planorbis* and *Lymnaea acuminata*, larvae of midges CHIRONOMIDAE, also crustaceans and worms. They also feed on vegetable matter including seeds, buds and roots especially of aquatic weeds such as *Lemna minor*, *Potamogeton indicus* and *Limnophila heterophila*.

*VOCALIZATIONS

Mass displays of the Garganey seem to be quite usual as flocks congregate in March on some of their last staging grounds in Pakistan and it seems probable that some pair formation develops before final migration. In these mass displays which take place on the water, drakes swim around indicating increasing excitement and give head jerking or pumping displays and also finally throwing back of head onto shoulders followed by ceremonial drinking. Some individuals also raise one wing showing their bright speculum in a wing-flap display (observed Hadiero lake, 10 March). During such a season males become quite vocal emitting short rapid strophes of their peculiar clicking call somewhat like the sound produced by rapid stroking of a bar across wooden slats or rails. Each strophe only lasts 1 or 1.5 seconds and may be repeated while in flight or on the water. Females are relatively silent but are reported to give a crescendo-like series of 'quacks' similar to a Shoveler female (*Anas clypeata*).

Shoveler or Northern Shoveler *Anas clypeata* Linnaeus Map **41**

Old name *Spatula clypeata* Linnaeus

DESCRIPTION

Body length 44-52 cms.
Wingspan 70-84 cms.

The rather long and broad spatulate shaped bill is the best diagnostic feature. When swimming the white necks and breast of the drakes are very conspicuous. In flight the males show pale blue wing shoulders and females a grey-blue wing shoulder, in both cases similar to Garganey (*A. querquedula*), and the Shoveler, like the latter, has a bottle green speculum but this is not bordered posteriorly by white as in *A. querquedula*. The drake has a dark blue-grey bill, bottle green head and upper-neck with a noticeable pale golden yellow iris. Its white body is broadly banded across the belly with chestnut maroon. The back, rump and tail coverts are black with a white tail and conspicuous broad lanceolate scapular feathers of pale grey with narrow black and white longitudinal stripes. The female is streaked brown and buff like a female Mallard (*A. platyrhynchos*) but usually has a less distinct dark crown and of course the green not blue speculum and spatulate bill at once distinguish it. The males retain their eclipse plumage often up to early January but can be distinguished by having a yellow iris in contrast to the females' dark brown iris, also their flanks usually have a more rufescent tawny colour. The legs and feet are orange in both sexes.

HABITAT, DISTRIBUTION AND STATUS

A widespread winter migrant visitor to the subcontinent being one of the earliest migrants to arrive, birds being shot on 30 August in Gilgit and as early as 15 September in lower Sind. Like the Garganey (*Anas querquedula*) it is one of the last to depart, 21 May on Rawal lake (1980), and in another year over 200 on 10 April (D. Corfield, 1981) on the same lake. Ringing recoveries include a female ringed on 14 November 1965 at Ghauspur jheel in Sind (netted from flooded grassland by C.D.W. Savage and author) with Common Teal, and shot in Yakut near Suntar in the USSR on 30 May 1967 (N62° 10'., E77° 37'), also a male ringed at Manchar lake, Dadu district, and recovered in Barabinsk district of Siberia (N55°., E76°). They are common on passage in Baluchistan, the Kurram valley and northern Himalayan areas. Adapted to a large variety of conditions in winter they occur on the main rivers, large lakes including alkaline lakes, and on small shallow muddy pools and flooded grassland areas. Their preferred feeding areas are shallow muddy margins of drying out pools or at night time in flooded grassland. In Pakistan they are the third most abundant duck species after Teal (*A. crecca*) and Pintail (*A. acuta*) according to IWRB wetland surveys in the early 1970s. In February 1972 over 18,000 were counted and in February 1973 just under 22,000 when all the major wetland were surveyed. Large numbers can be seen in Punjab, e.g. 2,500 on Uchchali lake, Salt Range (January, 1981), in late March 1,100 on Taunsa barrage headpond on the Indus. In Sind on Ghauspur jheel (Jacobabad district) 2,300 in mid-February as well as on Sanghar lakes such as Soonahri (over 4,000, 14 February) and the

Larkana lakes (3,000 on Drigh and over 2,000 on Langh lakes late February). Status ABUNDANT.

HABITS

They are inclined to feed in stagnant or muddy water, and as their bodies have a musky smell they are not popular amongst duck hunters. They are usually quite gregarious in their winter quarters, feeding exclusively by swimming or walking in very shallow waters and sieving plankton and the larvae of micro-crustacea with their peculiarly adapted bills. They can be seen swimming with neck outstretched and submerged or quite commonly with the whole head and neck submerged, with the bill being swept sideways to and fro in scything motions. Small floating seeds, plant debris, molluscs and insect larvae are also included in their diet and brackish lakes often provide highly eutrophic conditions for micro-plankton. They are rarely seen feeding in tidal creeks but occasionally rest by day on broad sandbars and mud flats (200 seen on Ghizri creek with Gulls and Waders in December). They feed quite frequently throughout the hours of daylight, but only at night in areas where they come onto land, such as the wet grassland margins of pools. They often feed in the company of Teal in such marshy areas. Very occasionally they can be seen up-ending to feed and on Uchchali lake in about 2 metres of water they were observed feeding by swimming in dense tight packed rafts, all with heads submerged, a habit not noticed on any other occasion.

VOCALIZATIONS

Relatively silent but male calls are often drowned out in the cacophony of other waterfowl noises, as they have a rather soft nasal 'kraay-kraay' call and shorter wheezy 'chik-chik' calls. Females have quite different voices uttering a series of decrescendo quacks of 4 to 5 calls similar in tone to a female Garganey (*Anas querquedula*).

Marbled Teal *Marmaronetta angustirostris* (Menetries)

Synonym *Anas angustirostris* (Menetries)

Illus. **22**
Map **44**

DESCRIPTION

Body length 39-42 cms.
Wingspan 63-67 cms.

It is larger than a Common Teal (*A. crecca*) and about the same size as the Garganey (*A. querquedula*), but in flight and on the water usually easy to recognize

because of its pale sandy buff colouration with evenly coloured pale buff wings, lacking any speculum and a broad dark brown band through the eye. The sexes are alike in plumage, but the male has a prominent nuchal crest in the breeding season. Pale blotches on the mantle and wing coverts give a marbled effect to the plumage. The crown and lower cheeks are pale creamy with brown and white fine cross-barring on the nuchal crest. The bill and feet are blackish, and in the female there is a paler grey-green triangle at the base of the upper mandible, lacking in the male (obs. breeding birds at Zangi Nawar). The iris is dark brown. From a distance this duck looks noticeably 'black eyed', due to the pale silvery buff head and broad dark brown eyestripes and often appears to have a large rather 'shaggy' head.

HABITAT, DISTRIBUTION AND STATUS

It is one of Pakistan's rarest ducks, with only very small numbers occurring as winter visitors on the lakes of the Indus plains. Substantial numbers can only be seen in years when water conditions are favourable, in one location, viz., Zangi Nawar lake (surface area after good rains *circa* 2,700 hectares) in Chagai district of Baluchistan. Over 200 were counted on 18-19 January 1984 and an estimate 150 breeding pairs in early May of the same year. The small Indus plains wintering population is probably locally nomadic. Some breed in Transcaspia (USSR) and recently it has been found nesting in Sinkiang (D.A. Scott, *in litt.*, 1987). A few stay to breed in scattered localities in south-western Pakistan. On Siranda jheel in Lasbela near the seacoast, breeding of 1 to 2 pairs has twice been recorded (K. Eates, 1945 and Frank Ludlow, 1915). During the past 10 years it has not been possible to find evidence of breeding on Siranda. A few pairs still breed in Khairpur district of Sind in quite small swamps and oxbow lakes near the river (Holmes and Wright, *JBNHS*, 1968-9 and author's obs., 1980). Breeding has

Illus. 22: Marbled Teal *Marmaronetta angustirostris*. Male with partially raised nuchal crest and in display posture.

also been recorded on Kushdil Khan lake in some years (Meinertzhagen, 1920). Only on Zangi Nawar lake is there a substantial resident population.

During 5 winter surveys in the 1970s of nearly all the major wetland areas by IWRB teams, there were only 2 sightings of this duck. 7 on Soonahri lake (Sanghar district) in 1974 and 50 near Kandhkot on 10 February 1973 in a reed fringed pond. In thirty years the author has only 3 wintering records from the Indus plains, 2 near Mirpur Sakro (Thatta district) 11 October 1981 and 6 Ghauspur jheel 30 November 1979 and a pair evidently breeding, near Khairpur in April 1980.

It appears to have been more abundant formerly. During the First World War Dr. Ticehurst considered it 'pretty common' in parts of Sind, particularly on Manchar lake, the Larkana lakes and Pithoro jheels in the east Narra. It is also reported as abundant on Manchar lake in the 'Handbook' (Ali and Ripley, 1969). Despite 4 IWRB surveys of Manchar and numerous visits by other ornithologists, e.g. C.D.W. Savage, John Wright, Derek Holmes, and Jacques Vieillard, during the 1960s and 1970s there have been no recent sightings on Manchar. There are no records at all of this duck from the Punjab. Status RARE.

HABITS

It prefers bodies of water with plenty of cover in the form of reeds or inundated Tamarisk bushes and avoids open water. It is well adapted to quite alkaline lakes and will also frequent small brackish drying out pools and shallow tamarisk-studded seasonal inundations. They are erratically migratory as none were seen on Kushdil Khan lake during visits on 27 October, 7 June and 21 May in three different years, but 4 were seen on 18 July by C.D.W. Savage though he found no evidence of breeding. The author saw one on Sonmiani lagoon (sea water) on 31 March 1978. They are probably omnivorous in feeding, taking seeds, vegetable matter and aquatic animals when available. A pair was watched near Khairpur in a small shallow pool swimming with head submerged and bill making audible dabbling noises in the muddy bottom. On Zangi Nawar lake in the Chagai they feed by upending

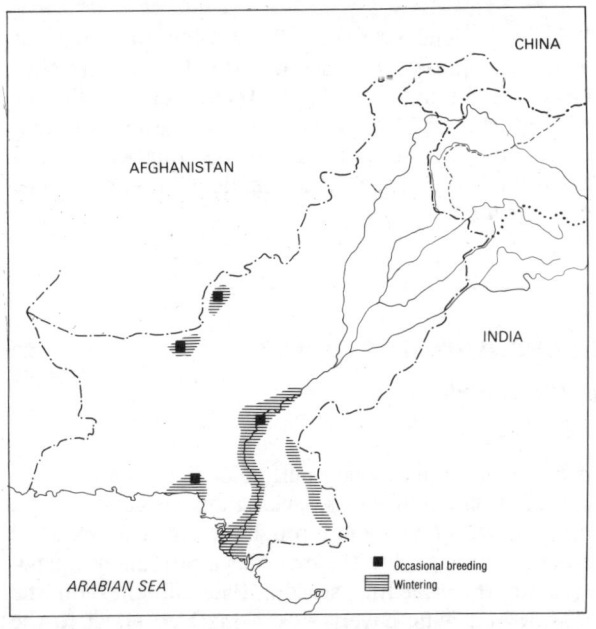

Map 44: Marbled Teal *Marmaronetta angustirostris*

while swimming in water of about 1.5 metres depth and were probably consuming the water weed *Ruppia maritima*, the only plant growing there. It has been recorded as feeding on annelid worms, molluscs and aquatic insects in the USSR (Dement'ev and Gladkov, 1952), as well as being mainly vegetarian in diet in India, taking seeds and shoots of aquatic plants, (Ali and Ripley, op. cit.). A captive group kept by the author flourished on a mixture of cereal, millet and sorghum seeds thrown into water with pieces of dried fish meal.

BREEDING BIOLOGY

A pair watched in Khairpur, 11 April 1980 was obviously about to breed. The male displayed frequently by erecting his nuchal crest and stretching straight upwards his head and neck and then suddenly jerking his head downwards until it was pressed down between wing shoulders. They seemed quite tame during courting and allowed approach to within about 70 metres. It was usually not possible to hear any vocalizations during this display. On Zangi Nawar lake on 5 May pair formation was only complete in about 75 per cent of the estimated population of 300 birds. Much courtship flight chasing was noted with the pursuing male actually jerking back his head at times while in flight, in the typical display posture.

Actual breeding records include a duck with 14 ducklings in August 1913 found by Mr. Aitken on Kushdil Khan lake in Baluchistan. This duck gave a wing-injury distraction display. Frank Ludlow's local native egg collector brought him 2 clutches from Siranda jheel on 14 June 1915; one a clutch of 12, the other 9 eggs. The nests were in a *Juncus* sedge clump on small islands in the lake with the entrance roofed over by bent rushes. Local game watchers living at Zangi Nawar (now a sanctuary) reported that nests were always built well inside dense clumps of *Phragmites* reeds and were usually unapproachable by boat. John Wright found evidence of breeding with 3 pairs seen on a small lake near Tando Musti Khan in March 1965. Siranda jheel in Lasbela only occasionally has enough water to provide suitable breeding conditions and none were observed there during several visits during the late 1970s. There is heavy human predation of this bird when it does breed. Brigadier Christison reported local hunters collecting their eggs in Zangi Nawar lake (Christison, 1942) and Ticehurst recorded that local hunters collect the eggs on Siranda jheel (Ticehurst, 1927).

Only the female appears to care for the young after hatching and in captivity incubation takes 25 to 27 days, the ducklings hatching synchronously and being able to feed themselves from that time.

VOCALIZATIONS

Captive males uttered a weak squeaking call when head-jerking, during display (noted Khanewal, April 1970 but not tape recorded). Females were not heard calling even during courtship.

TRIBE AYTHYINI

Pochards

Red-Crested Pochard *Netta rufina* (Pallas)   Map **45**

DESCRIPTION

Body length 53-57 cms.
Wingspan 84-88 cms.
Wing length 25-27.8 cms. (males),
    24.3-26 cms. (females) (Dement'ev)
Bill length 44-54 mm.
Weight 1,000-1,200 gms. (males),1,050-1,210 (females) (Dement'ev).

One of the handsomest of ducks, and larger than other diving ducks of the genus *Aythya* with slightly longer tarsi and more upright stance on land. The drake which is in full breeding plumage by early November, has an orange-chestnut head darker brown on the cheeks and lower part, with a slightly erectile bushy crest of brighter orange. His iris is crimson and bill bright cerise red. The breast is black as well as the back, rump and tail coverts. The flanks are white and the wings and tail greyish brown unstreaked. Females have blue-grey bills, brown irides and a large round head with conspicuous white throat and cheek patches, framed by dark brown crown extending over the eye into the loral region and down the hind neck. Their breast and flanks are pale buffy brown with some indistinct vertical barrings. Wings and rump dull light brown. The male has orange-red feet and legs and the females' are orange-fleshy or pink, with the webs blackish in both sexes. In flight both show dull white secondaries and primaries narrowly tipped with brown. The drake also has the leading edge of the wing prominently white.

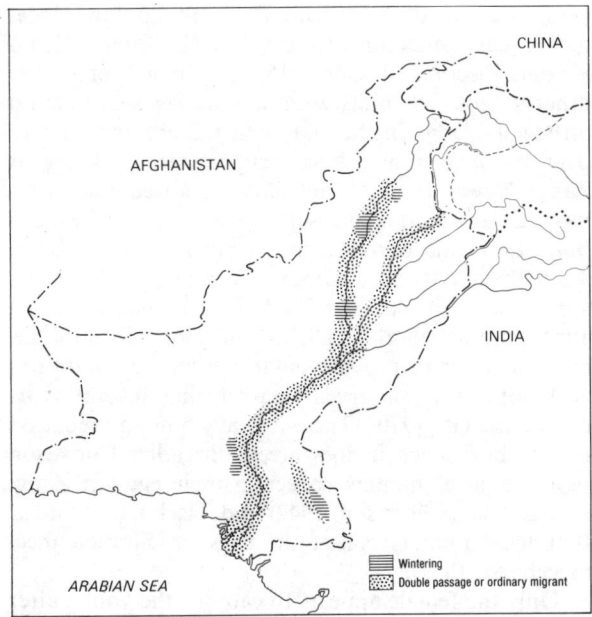

Map 45: Red-crested Pochard *Netta rufina*

HABITAT, DISTRIBUTION AND STATUS

The Red-crested Pochard has a limited breeding range in the warmer steppic latitudes in central Asia and Turkestan. It is a winter migrant visitor to Pakistan which has now become rather rare. Birds ringed on Drigh lake, Larkana were recovered in Russian Turkestan (N37°21′., E66°20′), and another ringed in Bahawalpur was recovered near lake Baikal (N55°., E105°). They are gregarious in their wintering grounds and prefer larger lakes, having fairly deep water and open areas. They generally avoid very small weed fringed pools and are not seen on seacoast. In recent years, none have been seen on Manchar lake where they used to occur in thousands (Salim Ali, 1927). One or two flocks usually spend the winter on the headponds or deeper seepage pools upstream of Chasma barrage (Mianwali district) or Taunsa barrage (Dera Ghazi Khan district) both on the Indus. 175 were counted on Chashma and only 12 on Taunsa in 1974

(F. Koning, pers. comm.). 200 on Chasma barrage headpond on 23 January 1981 in about 5 separate flocks. In early January 1987 during severe winter weather several thousands were estimated at Chashma barrage amongst a total concentration of perhaps 70,000 ANATIDAE (Ashiq Ahmad, Widlife Investigation Officer, *in litt.*, January 1987). A few winter on the remoter deeper lakes of Khipro in the eastern borders of Sind. 65 on Wassawari jheel seen 15 February and on Bakar lake in Sanghar district 100 to 200 each winter. Odd individuals are occasionally sighted on other lakes both in Punjab and Sind. They are rare in Baluchistan but a few are occasionally sighted on Kushdil Khan lake and Zangi Nawar. They appear to arrive quite late, few being seen before mid-October and all have gone again by mid-March. Status SCARCE.

HABITS

A largely herbivorous duck which obtains its food mainly by diving but also by dipping its head and neck beneath the surface.

They usually feed in small groups when surface dipping but in open deeper water quite commonly in flocks, birds diving in quick succession when one individual starts to feed, (observed Chashma barrage in deep water). In the Khipro lakes *Myriophyllum intermedium* water milfoil is abundant, and in the seepage ponds around Chashma and Taunsa, *Potamogeton indicus* and *Saggitaria guayanensis*. They are known to feed upon seeds, buds and roots of these aquatic weeds, also occasionally aquatic insects such as water beetles, amphibia spawn, crustaceans and small molluscs.

VOCALIZATIONS

They are largely silent in their winter quarters but during the breeding season have a variety of vocalizations, including a hoarse quite loud but short call used both as a contact or warning/threat call, rendered as 'baht'. Females during the breeding season utter a soft repeated 'guk-guk-guk-guk' call.

Common Pochard *Aythya ferina* (Linnaeus)   Map **46**

DESCRIPTION

Body length 42-49 cms.
Wingspan 72-82 cms.

A short necked stocky duck with a comparatively large head and bill, the forehead sloping smoothly back to

form a rather high crown. The male has a pale grey body with black breast, rump and tail coverts. Head, chestnut red with a pale blue-grey bill having a prominent black nail. The iris is crimson. In eclipse plumage similar to female but head more golden brown and back greyer. The female is yellowish

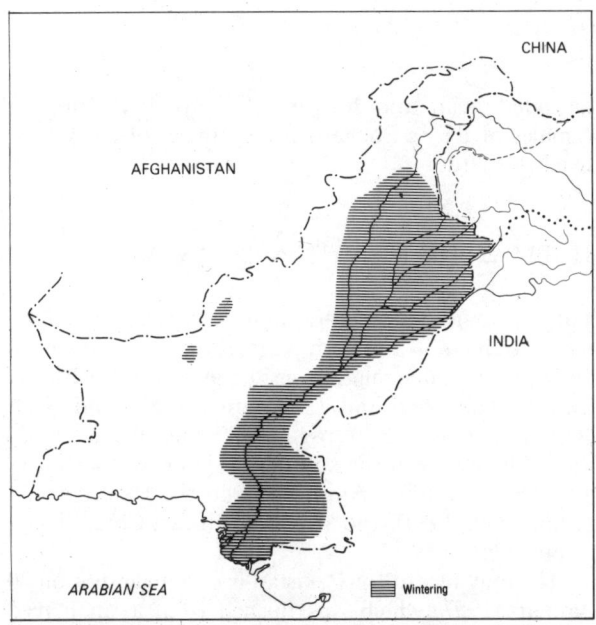

Map 46: Common Pochard or Pochard *Aythya ferina*

brown, not streaked or scaled with darker brown, and often having a hoary or whitish area around the base of the bill and cheeks. On the wing both sexes show a greyish speculum and no white which distinguishes them from the other Pochards found in Pakistan. They typically fly in tight compact flocks, usually higher in the sky than other ducks and with a very fast wing beat. The legs are slaty blue with a prominent lobe on the hind toe. Drakes are usually fully moulted into breeding dress by mid-November.

HABITAT, DISTRIBUTION AND STATUS

A purely Palearctic breeding bird not extending to North America, and preferring southern warmer more steppic latitudes for nesting. They are abundant winter visitors to Pakistan wherever there are lakes with large enough areas of open water and sufficient depth for bottom feeding by diving. They avoid shallow reedy pools or water covered with surface or emergent vegetation. In Pakistan it is the commonest of the diving ducks with flocks of over 500 being a common occurrence on the larger lakes of the Punjab such as Uchchali, and river irrigation barrage headponds such as Jinnah barrage. It is equally common in Sind. Typical counts, Lal Sohanran, Bahawalpur 350 on 12 January and 1,500 on 22 November in another year. Khabbaki lake, Salt Range 300 on 21 November and

1,200 on 28 December in another year. Flocks regularly winter on Kushdil Khan lake in Baluchistan and in Sind, up to 1,000 may be seen on Haleji lake near Karachi and scattered small flocks totalling about 800 on Manchar lake, 3,000 on lake Phoosani in Badin district. IWRB wetland surveys showed a total count of 11,900 in February 1972 and 27,600 in February 1973. They are the third or fourth most abundant duck species in winter according to water and feeding conditions, being about equal in numbers to the Shoveler (*Anas clypeata*). Earliest arrivals seen in Sind, first week of November and latest seen on 31 March on Haleji lake. In Baluchistan, the northwards migration in spring does not start until late April (Meinertzhagen, 1920). Stray birds have been seen in May, June and July around Rawalpindi and Kushdil Khan lakes, but no evidence of breeding. One duck ringed in Bahawalpur on Jhajjia Abbasian was shot in central Siberia, USSR at N76°., E84°. Status ABUNDANT.

HABITS

Very gregarious in winter and typically they feed entirely by keeping to open water and swimming together in packs or rafts, diving to swim to the bottom to forage. They use only their feet to propel themselves under water and if the lake is deep they spring slightly out of the water at the commencement of a dive. They feed largely upon submerged water weeds and their seeds, especially Pond Weed *Potamogeton indicus* and in more alkaline lakes such as Hadiero *Myriophyllum* sp. (Water milfoil). They have also been recorded as eating *Lemna* duckweed and animal matter such as larvae of insects, molluscs, annelid worms and amphibia tadpoles (Dement'ev and Gladkov, 1952). They are quite awkward on land, and because of their relatively narrow pointed wings, they require some effort to take off from water, pattering rapidly along the surface, but once airborne flying very strongly with rapid wing beats. On Haleji lake, Sind they feed mostly during the day but they are not disturbed on this lake which is a sanctuary. On Khabbaki lake in 1965 a flock of 300 spent the day sleeping or preening but were heard actively feeding most of the night.

VOCALIZATIONS

Silent on their wintering grounds and calls confined to courtship when the male utters a rather soft nasal 'wiwierr'.

Baer's Pochard *Aythya baeri* (Radde)

DESCRIPTION

Body length 46 cms. (Ali and Ripley) 40 cms. (King)
Wing length 21-23 cms. (males),
    18.6-20.3 cms. (females) (Delacour)
    20.8-24. cms. (males),
    19.3-21.5 cms. (females) (Stuart Baker)
Tail length 6.7-7.2 cms. (males) (Delacour)
Bill length 48-50 mm. (males),
    47-48 mm. (females) (Delacour)
    39-42 mm. (Stuart Baker)

Similar in appearance to the White-eyed Pochard (*Aythya nyroca*) but distinguished in the male by having a black head and neck. This black area is glossed with green, and this gradually merges into rich maroon-chestnut on the breast. In eclipse plumage the male looks like the female and both have the head and neck a dull but very dark brownish black and a paler fulvous chestnut patch on the face near the base of the bill. They also average noticeably larger in size than the White-eyed Pochard.

Like *Aythya nyroca* the bill is slaty blue with a black tip and in the male the iris appears straw yellow or white. In the female the iris is white. The pale chestnut patch on the face, the darker, browner and more rounded head and longer bill help to distinguish females of Baer's Pochard from females of the White-eyed Pochard.

HABITAT, DISTRIBUTION AND STATUS

Largely confined to central and eastern China, Japan and extreme north-eastern parts of the subcontinent of India, as a winter migrant visitor as well as to Burma and central Thailand (Delacour, 1959, and S.D. Ripley, 'Revised Synopsis', 1982 and King, *et al.*, 1975). It breeds in the southern regions of the Soviet Far East, including Amurland and the Ussuri region (Flint, *et al.*, 1984), and also in north-east China (D. A. Scott *in litt.*, 1987).

The only record for Pakistan is of a male shot on 30 January 1957, which was flushed from a small reed covered lake near Gujrat, Punjab by Brigadier Haider Jang. The head, wing and foot of this specimen was preserved in the British Museum (Natural History) at Cromwell Road. They also confirmed its identity (Brigadier Haider Jang, *in litt.* to author, 20 July 1986). Status accidental VAGRANT.

Ferruginous Duck or
White-eyed Pochard *Anthya nyroca* (Guldenstadt)

Map **47**

Old name *Nyroca rufa* (Stuart Baker, *Fauna of British India*)

DESCRIPTION

Body length 38-42 cms.
Wingspan 63-67 cms.

A slightly slimmer, smaller diving duck than the Common Pochard (*A. ferina*) often with tail carriage higher (not touching the water as in *A. ferina*) and thus reminiscent of a dabbling duck. The best field point is the generally uniform dark brown body with snow white under-tail coverts and vent. The male is dark maroon-chestnut on head and flanks with a conspicuous pale bluish white iris and dark greyish black bill, paler blue at the tip and with a black nail. The wing coverts and back are darker brown, less rufescent. The female looks closely similar but has a brown eye and her flanks are rather paler and more buffy brown than chestnut, with some indistinct dark barring. The head is comparatively large as in other diving ducks, with a rather high crowned profile and short neck. In flight, could be confused with *Aythya fuliga* as both have white flight feathers, brown tipped forming a continuous white bar, but in the Ferruginous Duck, this wingbar is much more conspicuous. The female Ferruginous Duck when swimming shows conspicuous white under-tail coverts and white extending in an oval patch up the belly whilst the female Tufted Duck usually has black or dark brown under-tail covert, but some individuals (of *fuligula*) occasionally show a lot of white on the under-tail coverts. The legs and feet are dark slaty grey with the hind toe well lobed. In winter, drakes often show discernible vertical buffy bars on the flanks.

HABITAT, DISTRIBUTION AND STATUS

Largely a winter migrant to Pakistan and much less common than the other Pochards, and only seen while on passage on such open water as the Punjab Salt Range lakes or Kushdil Khan lake in Baluchistan. It favours secluded reed and sedge-fringed pools and lakes with plenty of cover. It is not normally encountered on rivers and has not been noted on the seacoast or in mangrove creeks which habitat it frequents in India.

It is most plentiful in lower Sind where an estimated one or two thousand probably winter. IWRB surveys of all major wetlands revealed only 65 in February 1972 and 90 in February 1973, but because of its predilection for reed beds and cover, many are overlooked. Earlier records indicate that it was much more numerous in former times. Ticehurst found it one of the most numerically abundant species in Sind during the First World War. Numbers are probably augmented on spring migration, by birds coming from further south, and as they are one of the later species to depart for their wintering grounds, they then become more conspicuous. For example, 24 seen Lal Sohanran on 17 March, 10 on Rawal lake, Islamabad in October and 6 on 10 May the year following. 40 seen at Trimmu barrage headpond and seepage pools on 10 September and 'commonest species' on Nammal and Khabbaki lakes in the Salt Range on 8 April. (C.D.W. Savage, pers. comm.). Odd birds may still be encountered in early May, e.g. Ticehurst in the Bolan pass (*Ibis*, 1927).

There is a good deal of evidence (see below) that about 15 pairs of this duck remain to breed each year on Zangi Nawar lake in the extreme south-western part of Baluchistan. This lake is well covered by scattered reed beds and during a visit from 3-5 May about 15 pairs were encountered (author) mainly hiding in reed beds, but indulging in courtship display. In the USSR this duck breeds in southern steppic latitudes, avoiding the boreal forest zone further north. It breeds in the Kirghiz steppes, the Seistan basin (Iran/Afghanistan), and also in Tibet. Autumn migrants return very early, a party of 4 birds being seen on Nammal lake in the Salt Range on 8 August 1968 (C.D.W. Savage, pers. comm.). 20 were seen on Chatteji lake (Thatta district) in mid-October but were on passage as they only stayed for a few days. A bird ringed on Haleji lake, Karachi district in January was recovered three years later in Kazakhstan in the Syr Daria region at N44° 34′., E66° 07′. at the end of April. Status FREQUENT but localized, even in suitable habitat.

HABITS

Except when on passage, not a very gregarious species, often being seen in pairs or small groups. They feed by diving in deeper lakes, and are largely herbivorous in diet, but they tend to prefer shallow water, and will forage swimming on the surface, provided it is thickly covered with water plants and emergent vegetation and in such conditions they forage by swimming with neck extended and head submerged (observed Trimmu headworks on Chenab). Their diet consists largely of seeds of water lilies *Nymphaea* sp., also *Potamogeton indicus* and Duckweed (*Lemna minor*). They also take stems of plants, rice grains and when available amphibia especially tadpoles, molluscs, water beetles and

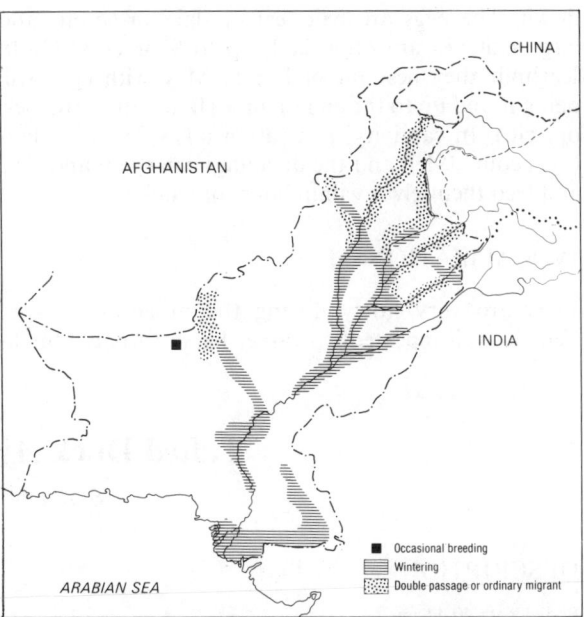

Map 47: Ferruginous Duck or White-eyed Pochard *Aythya nyroca*

insect larvae. They are sometimes difficult to flush, preferring to hide in dense reed beds and they are quite shy ducks, feeding principally at night.

BREEDING BIOLOGY

Due to their habits of hiding in dense reed bed cover, observations in early May on Zangi Nawar were difficult to make but pair bonds seemed to be fairly well established with males swimming closely around females as though to cut off their escape and females calling repeatedly and rapidly. Males on such occasions stretched their heads and necks upright while swimming. Local game watchers asserted that these Pochard breed on Zangi Nawar (Mohammed Rafique and Noor Mohammed, pers. comm., May 1984). Their nests, they claim, are always on the water surface in dense *Phragmites* thickets and the eggs are often covered from the hot sun by down, feathers or wet weed when the female is not incubating, an observation which corroborates the behaviour noted by Bates and Lowther of the duck breeding in the vale of Kashmir (Bates and Lowther, 1952). During a survey from 4-12 July 1987 by Ashiq Ahmad, a female with 15 ducklings (wing quills sprouting) was found on Zangi Nawar lake on 10 July. (Mohammad Idrees Chughtai, handwritten report Pakistan Forest Institute). Bates and Lowther found them fairly common on the Hokra jheel and some were found in 1967 nesting on the Dal lake, Kashmir valley (John Tyak, pers. comm.). Their nests are not very substantial pads of rushes, lined with finer grasses or leaf strips and as incubation progresses, quantities of the female's own dark brown breast down. The eggs are pale 'cafe-au-lait' in colour and large clutches are often laid, up to 9 or even 11. In Kashmir they nest rather late in May with eggs still being found up to the end of June (Bates and Lowther, op. cit.). In captivity, incubation takes 25 to 27 days (Delacour, 1959) and the ducklings can swim and dive and feed themselves within hours of hatching.

*VOCALIZATIONS

They are very noisy during the breeding season. Females call repeatedly a rather harsh throaty 'durrh-durrrh-durrrh' or 'gurrrh-gurrrh'. This call is similar to that of *Aythya ferina* but slightly higher in pitch and less growling in tone. At Zangi Nawar, they also called (author obs.) during flight when the male was closely following behind. On the water, females emitted between 15 to 20 such calls in rapid succession.

In their winter quarters they are largely silent, but a drake was heard calling on 10 September at Trimmu irrigation barrage on the Chenab river. This was a high whistling call, less melodious than a Widgeon's whistle (*Anas penelope*) and considerably weaker but having a similar inflection 'whee-whoo'.

Tufted Duck *Aythya fuligula* (Linnaeus)   Map 48
Synonym *Nyroca fuligula*

DESCRIPTION

Body length 40-47 cms.
Wingspan 67-73 cms.

These short necked, rather compact diving ducks are often overlooked in Pakistan amongst the greater numbers of Common Pochard (*A. ferina*), because the drakes often retain their eclipse plumage until well on into January. In breeding plumage the drake, with black back and rump, and tail offset by shining white flanks, is easily distinguished. The head, neck and breast are also black, glossed with purple in good light and the slender curving crest down the back of its neck, often blows in the wind. The pale golden iris is also conspicuous from a distance. In both sexes the head has a high crowned silhouette, less backward sloping than in *A. ferina* and the bill is slaty blue which distinguishes it from females of *A. nyroca* and *A. ferina*. The female is dark brown almost black on the head and back and sometimes shows some white on the under-tail coverts (but never as clear or extensive as in *A. nyroca* females) and some white on the forepart of the face at the base of the bill (but rarely as clear and extensive as in *Aythya marila*, the scaup). The flanks usually have indistinct whitish buff vertical scalloping and in breeding plumage this becomes markedly tawny. The male in eclipse plumage, though brown always looks darker than the female and his flanks tend to look grey rather than brown. Both sexes have blue-grey feet and in flight look very similar to *A. nyroca*, with brown tipped white secondaries and buffy white primaries tipped dark brown. They can be separated from both sexes of *A. nyroca* in flight, usually having dark brown or black under-tail coverts, and the female Tufted Duck even in winter has a small down drooping nuchal crest like the male. Also, their head shape is different being more rounded and flatter on the crown, less peaked than in *A. nyroca*.

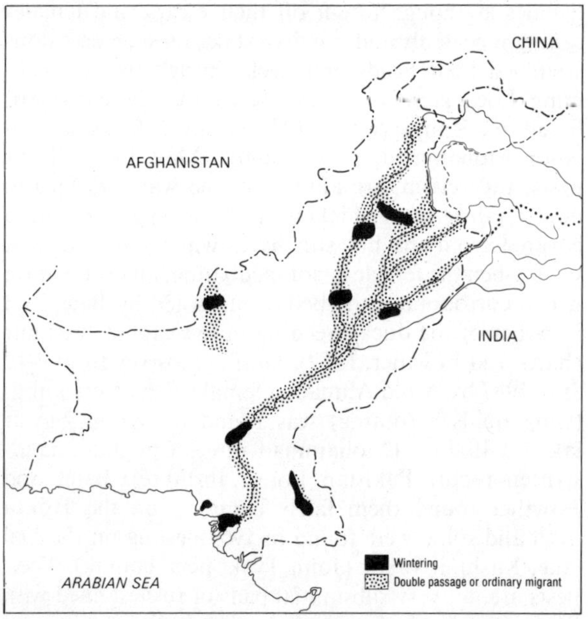

HABITAT, DISTRIBUTION AND STATUS

A Palearctic breeding duck, absent from North America but breeding from western Europe across Asia mainly in the boreal zone from sub-Arctic latitudes down to the fringes of the steppe. They winter in Pakistan in relatively large numbers, concentrating on favourable waters, and they tend to prefer deeper lakes with large stretches of open water and avoid smaller pools with emergent vegetation. They arrive in Baluchistan in late September, numbers often being seen on Kushdil Khan lake up to early April.

In the Punjab they are seen on passage on Salt Range lakes, e.g. 41 on Nammal lake on 23 March, 150

Map 48: Tufted Duck or Tufted Pochard *Aythya fuligula*

on Khabbaki lake on 21 February. Large numbers over-winter on such deep water bodies as Chashma barrage headpond (300 on 11 February 1972, F. Koning, pers. comm.) and Kharrar jheel, Renala Khurd, Punjab 1,500 seen on 16 March 1970 (J. Vieillard, pers. comm.), with only 400 on the same lake earlier that year on 11 February.

Perhaps the biggest concentrations are to be found on Khinjar lake in Thatta district lower Sind where 1,200 were counted in 1973 and 2,700 in 1972 and 1,350 in January 1987 (Zoological Survey Data). According to IWRB surveys of Pakistan's wetlands they are about sixth or seventh in abundance amongst wintering ducks, being much less common than *Aythya ferina,* but though unevenly distributed, they are occasionally more numerous than Mallard (*Anas platyrhynchos*) in some years.

Most have migrated north by early April. Latest date seen 30 April on Rawal lake, Islamabad. A group of 5 birds was seen by the author on Phandar lake, Gupis district of Gilgit on 15 November at about 3,200 metres (10,500 feet) elevation. There are no known ringing recoveries of this species from the subcontinent. Status COMMON but local in distribution.

HABITS

As noted above they tend to congregate in Pakistan in considerable numbers on large bodies of water which offer suitable feeding conditions. Here they will be seen in widely scattered flocks numbering 30 to 50 birds. Often they are plentiful on lakes where the Common Pochard is entirely absent and this separation is accounted for by their different feeding habits, the Tufted Duck eating mainly animal foods whilst the Pochard prefers vegetable foods. They are also much more diurnal in feeding activity than the Common Pochard, which enables them to exploit more animal matter in their diet. In winter their diet is often predominantly molluscs, crustaceans, and insects (Cramp *et al.,* 1977). They also consume some seeds and vegetable matter particularly in the spring and summer months. They obtain their food by diving and are less subject to human disturbance when they congregate on very large bodies of water such as at Chashma barrage headpond (covering an area of nearly 20 kms. long by 2 kms. broad) or Khinjar lake (approximately 300 square kms.). When they are disturbed by shooting they fly very high in large flocks and often do not return for some time.

VOCALIZATIONS

Birds in flight frequently utter a harsh low 'krr-kurrr' call almost like a rattling growl, which can be heard above the characteristic whistling of their wings. Birds sometimes give this call on the water and it can regularly be heard from disturbed flocks flighting overhead. On their breeding grounds the drakes have a whistling call similar to that of *A. nyroca* and a rapid call rendered as 'buck-buck-buck'.

Scaup or Greater Scaup *Aythya marila* (Linnaeus)

Old name *Nyroca marila* (Linnaeus)

DESCRIPTION

Body length 42-51 cms.
Wingspan 72-84 cms.

Very like the Tufted Duck, they are compact, large headed diving ducks, but the drakes in breeding plumage combine a black head and breast with a grey back in constrast to the black back of *A. fuligula.* Also in a good bright light, the head is glossed bottle green rather than purple. Females are much harder to separate in the field, but generally in winter plumage the female scaup shows a much clearer and more extensive white patch around the base of the bill, and its back and wing coverts are margined a paler more silvery buff. Also there is no trace of a nuchal crest, always present in females of *A. fuligula*. In flight both sexes show white secondaries tipped dark brown and greyish white primaries tipped darker, and the females are doubtfully distinguishable in flight, except for slightly paler grey wing shoulders and generally heavier build with broader bill. Both sexes have blue-grey bills with a black nail, and their legs and feet are dull slaty grey coloured with the webs distinctly blacker.

HABITAT, DISTRIBUTION AND STATUS

The most northerly of the *Aythya* diving ducks· in breeding, being Holarctic in distribution and largely confined to Arctic tundra zones. In its wintering grounds it is almost wholly adapted to a marine environment, but numbers spend the winter around the Caspian and Black seas, and the lakes of Kazakhstan. There are no recent records for the Persian Gulf or Red Sea, though Dr. D.A. Scott did see a pair on

Haun-al-Hammar, near Basrah, Iraq in January 1979 (*in litt.*, 1987). It may straggle to the Arabian Sea however as there have been scattered records in Pakistan. Three were collected on the Indus river. A female shot 3 November 1881 near Attock, another 23 kilometres (14 miles) upstream of Attock and a third near Hasan Abdal on the Jubbee river, and Christopher Savage obtained two certain records during the 1950s and 1960s (Savage, 1968). On 24 December 1948 a Mr. J. H. Butler shot a drake in breeding plumage on Nammal lake, Mianwali district (*JBNHS*, Vol. 48, No. 1, p. 117). On 27 March 1982 a female was seen on Haleji lake by the author and Rolf Passburg, and it is certainly possible that it is overlooked because of its similarity to *Aythya fuligula*, coupled with its winter

biotope being typically offshore in coastal waters. Status VAGRANT.

HABITAT, DISTRIBUTION AND STATUS

On sea in winter they are often gregarious, associating in large flocks. They feed by diving, being propelled under water solely by their feet. Their food consists largely of molluscs with a much lesser intake of crustaceans and occasionally small fish. They are silent in their winter quarters but on their breeding territory there is a marked dichotomy in vocalizations, with males having low crooning calls and females louder more guttural 'karrr' and 'querrr' calls.

TRIBE MERGINI

Scoters, Sawbills and Sea Ducks

Long-tailed Duck or Old Squaw *Clangula hyemalis* (Linnaeus)

DESCRIPTION

Body length 40-47 cms.
Wing length 73-79 cms.
Tail up to 13 cms. in males.

The drake is unmistakable with long black central tail feathers in all plumages. In winter the male has a largely white head, neck and flanks with a circular dark brown blotch on the sides of the neck, broad brown breast band and blackish brown rump. In eclipse plumage the forecrown, lower cheeks and neck are mottled black and brown with chestnut margined feathers on the wing coverts and scapulars, having dark brown centres. Females lack the long tail, but have white on the head and neck in winter with a dark crown and neck patch similar to the males, and in summer plumage a more restricted greyish eye-patch with the crown dark brown extending down the nape and forming a collar around the lower neck. On the back its brown plumage is scalloped with buffy chestnut. Both sexes have rather short deep bills of dull grey, that of the male being tipped flesh pink with a black nail. In flight, the wings are dark brown all over except for the scapulars of the male in winter

plumage which are white, long and lanceolate in shape.

HABITS AND VOCALIZATIONS

The Long-tailed Duck breeds throughout the Holarctic, close to the pack-ice in Arctic tundra latitudes, and normally winters in northern Pacific waters off Alaska and the Kamchatka peninsula. There is also a huge wintering population off north European coasts, with millions recorded on migration in the Baltic (D. A. Scott, *in litt.*, 1987). A few have been recorded in recent times as wintering in the Caspian and Black seas. In the past, occasional birds have straggled down to Pakistan region. A female was shot on Kushdil Khan lake in January 1938 (M.B. Reeve, *JBNHS*, Vol. 40, p. 334), also one in October 1938 at Chaman (A.E. Dredge, *JBNHS*, Vol. 37, p. 549). One was shot in Sind in 1936 (*JBNHS*, Vol. 43, p. 487). Ticehurst records that it had been shot on the Indus 'several times at Attock', in north-west Punjab (Ticehurst, 1923). There have been no records within the past 50 years. Status RARE VAGRANT.

Common Goldeneye *Bucephala clangula* (Linnaeus)

Map **49**

Old name *Glaucion clangula* (Linnaeus)

DESCRIPTION

Body length 42-50 cms.
Wingspan 65-80 cms

In eclipse plumage the male has a dark brown head with a round greyish white patch below and in front of the eye. Its lower neck is white and breast and flanks are grey-brown, barred darker. The pale golden iris is noticeable in both sexes, and the feet are conspicuous in having yellow legs and toes with black webs. In breeding plumage the male has pure white flanks and the head looks black, but bottle green in good light, with a round pure white patch at the base of the bill and in front of the eye. The mantle, rump and tail are black and the scapulars are white with diagonal black stripes. Females lack the white face patch and have dark brown head and backs, with a narrow white band along the scapulars and silvery grey flanks barred with brown. In flight the female shows dark brown wingtips (primaries), with white secondaries having a narrow brown line through the middle, and grey wing shoulders bordered by dark brown on both sides. Females have more orange-yellow feet with grey webs. The best field character in flight is the conspicuous white wing patch formed by the secondaries, not extending into the wingtip as in *Aythya* diving ducks. This white patch is often partly visible when the bird is perched.

HABITAT, DISTRIBUTION AND STATUS

This duck breeds throughout the Holarctic in both North America, western Europe and the USSR, being confined to coniferous forest as it nests in hollow trees. Birds breeding in north central USSR, winter mainly around the Black and Caspian seas, the Aral Sea and Tadzhikistan. It appears to be largely absent in winter from south-east Iran or the Seistan basin (D. A. Scott, *in litt.*) and Pakistan birds are perhaps more likely to be overshoots from the central Asian (Tibetan plateau) wintering population. A few have been recorded in recent years in Pakistan on the major rivers, particularly on the Indus near Tarbela dam (1973, Edward Sicely, *in litt.*), and on the Chenab river near Marala barrage. In 1966, 7 were seen on 10 February and 4 in November in the same place by Edward Fernando. One, an immature male was shot by him and examined by the author. In Baluchistan one was shot on Kushdil Khan lake on 12 December 1940 and 2 specimens dating from 1916 obtained in January from Zangi Nawar were in the Quetta Museum (Christison, 1942). R. N. Slater in 1879 and 1880 saw small groups of Goldeneye on the Indus near Ghazi Ghat (D. I. Khan)

and said that 3 or 4 occurred there every winter but gives no dates (*Stray Feathers*). R. M. Stocker collected 6 or 7 birds on the Indus near Attock in December (*Stray Feathers*, Vol. 10, p. 515). Ticehurst (1923) also mentions 3 occurrences in Sind in February and March without giving dates. It seems probable that a few Goldeneye overshoot their Caspian winter grounds and spend the winter in tidal creeks or lagoons along the Makran coast, as most sightings are of small parties of birds in spring and autumn when they are probably on passage. Status in Pakistan RARE.

HABITS

Edward Fernando noted that they swam strongly, keeping close to the river bank and diving frequently. They feed on crustacea, molluscs and small fish and probably insect larvae. On their breeding grounds they dig up tubers of water plants such as *Sagittaria* and *Potamogeton* which they obtain by diving under water, and feed on fresh water snails and insect larvae. They are silent outside of the breeding season but produce rythmic whistling noises with their wings in flight during courtship as well as emitting loud 'zeee-zeee' calls (Cramp *et al.*, 1977).

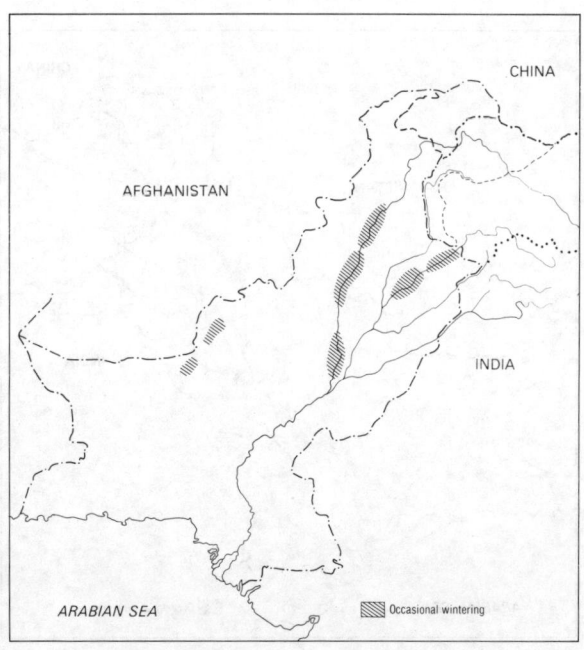

Map 49: Common Golden-eye *Bucephala clangula*

Smew *Mergus albellus* (Linnaeus)

Map **50**

Old name *Mergellus albellus* (Linnaeus)

DESCRIPTION

Body length 38-44 cms.
Wingspan 55-69 cms.

A small diving duck with narrow bill slightly hooked at the tip, often swimming low in the water and diving frequently like a typical Merganser. Slightly bigger than the Common Teal (*Anas crecca*) they show a high crown and steep forehead in silhouette and a slightly bushy occipital crest more pronounced in the males. In eclipse plumage the male has a dark chestnut red head with white throat and lower cheeks similar to that of the female. The narrow bill is leaden grey with a black hooked tip. The breast and flanks are grey with some white indistinct barring and the mantle is blackish. The loral area and a narrow band extending through the eye looks almost black. Females in flight show brown speculum narrowly bordered with white and a broad white band mid-wing (the median wing coverts).

The male is very striking in breeding plumage, being largely pure white, except for a small round black patch between the eye and the bill, and a black line extending through the base of the side of the crown. Flanks are vermiculated pale grey, the mantle is jet black and the scapulars are white with two narrow black lines circling the upper and lower breast. The rump and tail is dark grey. The drake has a crimson iris and dark plumbeous grey feet, whilst the female has brown eyes and grey feet, with the webs distinctly black. Males begin to show their breeding plumage

from late November but others can be seen in late February still in partial moult.

HABITAT, DISTRIBUTION AND STATUS

This duck breeds only in the Palearctic boreal forest zone in relatively cool northern latitudes, but it requires forest near water, and it nests in tree holes. Wintering populations occur in the wetland areas of the central Asian steppes and southern Caspian, and the Seistan basin (327 in Iranian Seistan, January 1977, 43 in Afghani Seistan in January 1976) (D. A. Scott, *in litt.*, 1987). Small numbers migrate into Pakistan, possibly only in some years of more severe winter temperatures. They are usually only seen for a few weeks from late January to early February. At this time they occur on the Indus river or Salt Range lakes, usually in flocks, but in some years as only a few scattered groups of 4 or 5 birds. In the 1930s H. Waite counted flocks of up to 200 on the headpond of Jinnah barrage at Kalabagh (pers. comm.). On Nammal lake *circa* 300 were seen in late January 1968 and 150 in 1969 in early February (C.D.W. Savage, pers. comm.). Frederick Koning counted 72 on Nammal lake on 2 February 1973 and 57 on 10 February 1972. Very few birds straggle down to Sind but on Ghauspur jheel in northern Sind after a good monsoon, 32 were counted by the author on 4 and 5 November 1971. In Lal Sohanran, Bahawalpur over a period of 10 years' regular visits, only 2 were sighted on 13 January 1967.

A drake was shot on 19 May on Kushdil Khan lake, Baluchistan (Christison, 1942) but in some winters none are recorded, though in occasional years there are parties of up to 20-30 birds. Karim Dad Junejo, a keen wildfowler saw it only once on the Sanghar jheels in eastern Sind, during 20 years of keeping shooting records though he several times saw small flocks on Manchar lake in the 1960s (pers. comm.). Status in Pakistan, an irregular winter visitor and SCARCE.

HABITS

They often keep in tight flocks when feeding and all seem to dive simultaneously or in quick succession (observed Ghauspur jheel in water of about 2 metres depth). Such simultaneous diving might suggest an advantage in securing startled prey when fish are the principal food source. In winter their food consists mainly of fish and they are largely diurnal in feeding. On their breeding grounds insects become an increasingly important component of their diet as the water gets warmer. On their wintering grounds they are silent. Males during courtship are reported to make a rapid clicking call.

Map 50: Smew *Mergus albellus*

Red-breasted Merganser *Mergus serrator* Linnaeus

Map **51**

DESCRIPTION

Body length 52-58 cms.
Wingspan 70-86 cms.

Slightly smaller and slimmer built than the Goosander (*Mergus merganser*) with a slimmer bill but in eclipse plumage they need to be seen well to separate them. The drake in breeding plumage has a bottle green head with two stiff nuchal crests and differs from the Goosander in having a chestnut and black breast band. Lower neck is pure white and also the wing coverts and scapulars with the mantle black and rump and tail grey and also the flanks vermiculated pale grey. The bill is crimson and legs orange. In eclipse plumage the male has a dark chestnut and brown streaked mantle in contrast to the much greyer back of *M. merganser* and the hind neck is chestnut all the way to the mantle whereas in the Goosander the chestnut extends only half way down the neck, the lower part being pale grey. In flight the primaries are black and the secondary and median coverts white with two parallel narrow black bars across the wing. In flight the Goosander shows a pale grey back and more white on the secondaries, but the females are difficult to separate at a distance or in bad light.

HABITAT, DISTRIBUTION AND STATUS

Breeding across the Holarctic region, in the northern boreal and sub-arctic areas, this is a duck well adapted to a marine environment, and large numbers migrate in winter to the southern Caspian shores, but only small numbers reach the Persian Gulf coast and coast of Persian Baluchistan (D. A. Scott, *in litt.*, 1987). In Pakistan a small number winter off the Makran coast and can occasionally be encountered close inshore around Karachi seacoast as well. Ticehurst (1923) records one shot on 24 November 1875 off Chahbar on the Makran coast and another obtained by Yerbury in Karachi harbour. 2 were shot on Kushdil Khan lake in Baluchistan, one on 17 February 1934 and the second on 19 November 1939 (Christison, 1942), also one was shot on Kushdil Khan lake in 1902 (Nurse, *JBNHS*, Vol. 14, p. 400). On 20 December 1974 close to Buleji fishing village outside Karachi 3 were watched fishing close inshore, and a drake in breeding plumage off Cape Monze on 2 April 1982 was sighted by Rolf Passburg and the author. These were the only sightings by the author during nine year's residence in the area. One was seen close to Buleji in 1955 (Leslie, *JBNHS*, Vol. 53, p. 708). Its Status in Pakistan must be considered RARE.

HABITS

They swim with the body low in the water and tail usually touching the surface. Rough seas do not seem to bother them unduly but they feed frequently close inland. They dive frequently to hunt their food under water, often staying under for as long as a minute. They are not usually seen consuming fish prey on the surface, so must be able to swallow most of their smaller prey under the water. The three birds watched at Buleji were frightened away by a motor boat and took to flight but an hour later they flew back to the same spot evidently having failed to find such a good feeding area. There are no authentic records of this duck on inland waters or on passage across Pakistan.

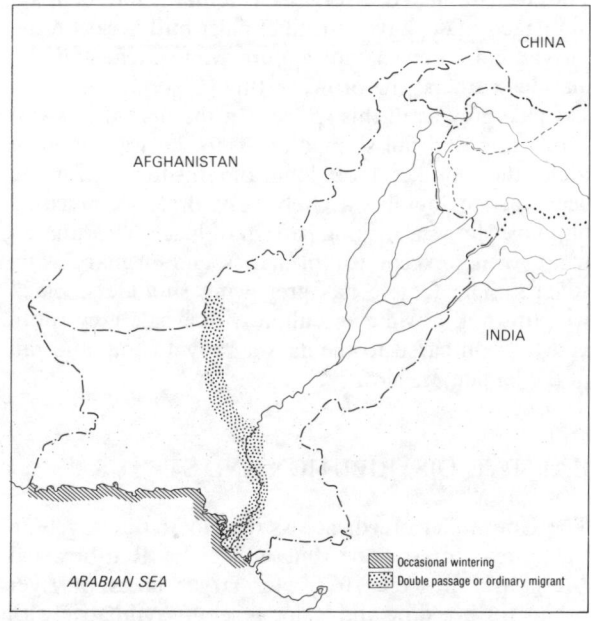

Map 51: Red-breasted Merganser *Mergus serrator*

Goosander *Mergus merganser* Linnaeus
Common Merganser (North America)

Old name *Mergus castor*

Map **52**

DESCRIPTION

Body length 58-66 cms.
Wingspan 82-97 cms.
Sub species *orientalis* — Wing length 26.2-28 cms. (males),
    23.5-26 cms. (females) (Dement'ev)
Bill length 47.5 mm. (males), 40-47 mm. (females)
Weight 1,700-2,060 gms. (males), 1,400-1,700 gms. (females) heavier
    in spring) (Dement'ev)

The male in breeding plumage has a dark bottle green head and upper neck but no stiff nuchal crest as in *Mergus serrator*. The back is black with prominent white wing shoulder and scapulars, grey rump and tail. The narrow saw bill is crimson with a down-curving black nail at the tip and the breast and flanks are creamy white. Females are smaller than males and have chestnut heads with a stiff nuchal crest and are thus confusingly like *M. serrator* females. However, their chin and forethroat are clear white, and the chestnut of the head only extends down to the nape, the lower part of the neck being pale grey posteriorly. The Red-breasted Merganser female and eclipse-plumaged male, have indistinct paler buff areas on the throat and foreneck not a pure white area, and the mantle feathers are browner (in *M. serrator*), much less pale grey as in this species. In the field they swim with rump and tail sloping down to the water and in flight their bodies look long and fusiform with the neck carried low like a Grebe. The drake in breeding plumage shows a striking pattern of black back and all white wings except for primaries and primary wing coverts. The female has grey wing shoulders, black wingtips and a white speculum with a pale grey rump and tail compared to the darker brown rump and tail of the female *M. serrator*.

HABITAT, DISTRIBUTION AND STATUS

The Goosander breeds across the Holarctic, largely in the boreal forest zone, but unlike the Red-breasted Merganser it seems to prefer larger lakes or river basins for breeding and is not generally sympatric with that species. It nests in tree cavities and in winter, part of the population migrates to the Pacific seacoast and the Caspian and Aral seas. There is a huge winter distribution across the coast of western Europe and both sides of the Atlantic. A small population breeds in the high plateau regions of Tibet, and in the Altai and Turkestan regions of the USSR.

In Pakistan it is a very rare winter visitor, and it may be presumed that the population which breeds in central Asia, occasionally visits the Indus river and its tributaries. It has been sighted occasionally off the Makran or Karachi coast and more frequently on the Indus river and its main tributaries. E. Fernando shot a male in eclipse plumage on the Chenab river near Marala barrage in November 1967 and on 30 December 1966 C.D.W. Savage saw a party of 13 in the same area. In 1970, 3 probably all females were seen on the Indus near Tarbela on 19 February (C.D.W. Savage, pers. comm.). Other sightings include a fine drake in breeding plumage on Nammal lake, Salt Range on 28 March 1968 (author) and 3 on Rawal lake which remained for 3 weeks from mid-February to early March (Michael Porter, *in litt.*). The author saw 3, including 2 drakes, one in breeding plumage and the other sub-adult, on 17 January 1981 on Rawal lake, Islamabad with David Corfield. It has been recorded in Sind, including 2 in December 1965 on the Indus at Sukkur by J.O. Wright (pers. comm. to author and Holmes and Wright *JBNHS*. Vol. 65, 1968). It is an irregular visitor to the lakes of Baluchistan, one being shot in November 1938 on Kushdil Khan and one on Zangi Nawar on 2 December 1939 (A.F. Christison, *JBNHS,* Vol. 43, 1942). Further evidence that this population may breed north of Turkestan, comes from records by Major Biddulph (1880) that he came across it on passage in Gilgit, and by Perreau that it was fairly

Map 52: Goosander or Common Merganser *Mergus merganser*

well known in Chitral in March while on northward passage. Status in Pakistan SCARCE.

HABITS AND VOCALIZATIONS

They are more adapted to rivers, even fairly turbulent mountain torrents, than is *Mergus serrator,* in their wintering grounds and are much less addicted to sea coast estuaries and lagoons, preferring large inland fresh water lakes and reservoirs. Beyond these slight differences in habitat preference they are largely fish eaters, obtaining their food by diving underwater, often hunting in small groups, in exactly the same manner as the Red-breasted Merganser. They are reported to probe with their bills under stones in turbid water and in their breeding grounds vary their diet with aquatic insects, amphibia and molluscs.

They are silent in their winter quarters but during courtship and display the male emits not very loud twanging calls, rendered 'uig-a'. also a high pitched bell-like note (Cramp *et al.,* 1977). Females emit a much harsher 'karr-karr' call.

TRIBE OXYURINI

Stiff-tailed Ducks

White-headed Duck or
Stiff-tailed Duck *Oxyura leucocephala* Scopoli

Illus. **23**
Map **53**

DESCRIPTION

Body length 43-48 cms. including tail 8-10 cms.
Wingspan 62-70 cms.
Wing length 14.7-16 cms. (both sexes) (Dement'ev)
Tail length 11-12 cms. (males) 9.5 cms. (female) (Dement'ev)
Bill length 46-50 mm. (both sexes) (Dement'ev)
Weight 737 gms. (1 male, Salt Range, 22 February) (W.A. Whitehead)

A small diving duck with a long narrow very stiff tail, short thick neck and bill which is swollen at the base. Their heads are large and bodies stocky. The male in eclipse has a conspicuous white face with black on the ridge of the crown and forming a narrow hind collar. The rest of its body is barred with yellow-buff and dark grey-brown, with the tail coverts rather chestnut and noticeably short, and the narrow dark brown tail having very stiff quills. Normally they swim with the tail partly submerged but occasionally this is erected at a 45° angle and carried in this posture for some minutes. They have also been observed to sleep with their tail cocked up (D.A. Scott, *in litt.,* 1987). The swollen bill is dull leaden grey but in breeding plumage it changes dramatically to bright pale blue (colour change commences from mid-February), and the brown area on the crown and hind collar diminishes so that the dark eye is entirely framed by white. The breast and rump take on a rufescent tinge and the back becomes more yellow-buff. The female, similar in size with an equally long stiff tail has less white about the face, the brown crown area extending from the base of the bill and through the eye to the nape, with another speckled brown parallel band extending from the nape to the back of the ear coverts. The female's bill is dull leaden grey and her body striated in buff and dark brown, but her upper-tail coverts are more chestnut.

In both sexes the feet are leaden coloured with the hind toe well lobed and the tarsi set very far back on the body. In flight the wings are more greyish brown with no marked speculum pattern, and their wing beats are extremely rapid and whirring. Females captured in late December had still not moulted their tails which were very abraded and worn. (Plumage details recorded from mist-netted birds Khabbaki lake, Salt Range).

HABITAT, DISTRIBUTION AND STATUS

This Palearctic counterpart of the *Oxyura,* Stiff-tailed ducks, is uncommon throughout its very restricted Mediterranean and steppic breeding range. Almost the whole migrant population which winters in the subcontinent, remains on one or two lakes in the Punjab Salt Range so that Pakistan has a special responsibility for the conservation of this species.

They are bottom feeding ducks which require lakes of a fairly high pH and having a food growth of submerged water weeds. A few very scattered records are known of individuals wintering on Sind lakes. Karim Dad Junejo, an experienced wildfowler, captured an injured adult male specimen on Soonahri

Illus. 23: White-headed Duck *Oxyura leucocephala,* male in breeding plumage.

Map 53: White-headed Duck *Oxyura leucocephala*

lake in Sanghar on 10 January 1972, the first he had recorded in twenty years of duck hunting on the Sind lakes. Whitehead recorded an occasional individual near Kohat in the NWFP, and occasionally flocks have been seen on Kushdil Khan lake in Baluchistan on spring and autumn passage, though in most years none were sighted (Meinertzhagen, 1902). One was shot on 17 February 1934 and another on 19 November 1939, both immature birds, on Kushdil Khan lake (*JBNHS,* Vol. 43, p. 487). The total wintering population in Pakistan is around 700 to 800, most of which winters on Uchchali or Khabbaki lakes in the Soon valley of the Punjab Salt Range. Before it dried up, in the mid 1970s as many as 700 were counted on Kharrar jheel in Renala Khurd district, 32 were counted on Kharrar lake in January 1987 (Zoological Survey records). Occasional flocks were seen on other Salt Range lakes, e.g. 90 on Kallar Kahar on 22 January 1972 and on the same date 210 on Khabbaki. In January 1987, 467 on Uchchali and 70 on Jhasar (M. Mallalieu and Ashiq Ahmad, *in litt.,* January 1987). Latest dates seen, 25 April on Nammal lake, but most birds migrate

north by mid-February. The population which winters in Pakistan is believed to breed largely in Kazakhstan in the USSR (C.D.W. Savage, pers. comm.).

HABITS AND BREEDING BIOLOGY

Jacques Vieillard who was familiar with the breeding of this species in Morocco, believed that conditions were closely similar on Khabbaki lake for this species to nest in Pakistan, but several visits in late April have failed to reveal any sign of breeding activity among the few that linger. They nest in emergent reeds and often in fairly thin *Juncus* cover, utilizing old nests of Moorhens (*Gallinula chloropus*) and Coots (*Fulica atra*), or the platforms of Dabchicks (*Tachybaptus ruficollis*). No down is used in the nest lining and only the female incubates and tends the ducklings which can dive and feed themselves from hatching.

In winter they have been observed as feeding individually, preferring the shallower lake margins where they dive continuously, usually remaining submerged only for 5 to 7 seconds and diving again within moments of surfacing. They feed on submerged plant material and seeds, especially *Vallisneria spiralis* water weed, and *Cyperus eleusinoides* both found abundantly on Khabbaki and Nammal lakes. Savage found seeds of *Melilotus indicus* and *Rupia rostellata* (wrongly identified as *Ruppia maritima*) in the crop of one bird. They almost certainly feed on micro-crustacea and copepods in such alkaline water, but these would not be traceable in crop or stomach contents, being rapidly digestible. On their wintering grounds they are almost impossible to put to flight, and for bird ringing studies could easily be induced to swim into mist nets. It is probable that they are actually flightless in December and January, as a captured bird showed newly growing flight feathers. On Kharrar jheel, on 4 and 5 January on the other hand, several were seen flying around very strongly.

VOCALIZATIONS

During the hours of darkness in late February, White-headed Ducks were heard (author obs.) emitting low grunts and these were thought to be males. Females responded with more goose-like short calls.

ORDER ACCIPITRIFORMES
Birds of Prey
FAMILY ACCIPITRIDAE

Crested Honey Buzzard *Pernis ptilorhynchus* (Temminck)
Honey Buzzard (*Pernis apivorus*)

Plate **12**
Map **54**

TAXONOMY

The author is only familiar at first hand with the population inhabiting north-west India and Pakistan from which there seems to be no consistent feature for separating *Pernis apivorus* the European Honey Buzzard from *P. ptilorhynchus* the Oriental Crested Honey Buzzard.

Reference to the taxonomic literature only tends to confirm the belief that this is best treated as a super species complex, if not conspecific. Most keys differentiate the two species upon tail pattern whilst admitting that this is not a consistent feature. Ali and Ripley ('Handbook', 1969) separate the two sub-species of *P. ptilorhynchus ruficollis* and *orientalis* on the basis of difference in broadness of tail bars, the identical key given by Stuart Baker ('Fauna Series', 1928) to separate *P. apivorus* from *P. ptilorhynchus*. The Pakistan population has a very slight nuchal crest, only visible when the bird is excited and from examination of the series of skins in the British Museum (Tring) the size and shape of the nuchal feathers of *P. apivorus* are the same as *P. ptilorhynchus* collected from northern and central India. Vaurie (*Birds of the Palearctic Fauna*, 1965) splits *Pernis* into two species but admits that their ranges overlap in Siberia where they actually inter-breed (p. 148, op. cit). Dr. Dillon Ripley in his 'Revised Synopsis' (1982) treats them as two species but adds a footnote that some authors consider them conspecific and cites Brown and Amadon Vol. I, 1968, who treat the Oriental or Asiatic populations *ptilorhynchus* as sub-species of *P. apivorus*. K.H. Voous in his *List of Recent Holarctic Bird Species*, 1973, treats them as separate species but also cites Brown and Amadon and the preference of some authors to treat them as conspecific.

The author's opinion is based upon comparison of voice recordings made of the Pakistan population and European recordings of *P. apivorus* which are indistinguishable, also the fact that their ecological requirements and breeding behaviour do not seem to differ significantly, whilst the extreme polymorphism exhibited by this species coupled with variation in tail pattern between immature and adult specimens adds to the confusion. Adult birds seen in all regions of Pakistan do, however, show a consistent tail pattern, with one broader dark sub-terminal tail bar and two narrow bars higher up the tail partly obscured by the tail coverts.

DESCRIPTION

Body Length 52-68 cms.
Wingspan 135-150 cms.
Wing length 43.2-48.3 cms.
Tail length 24.2-26.7 cms. (Stuart Baker).

The Honey Buzzard has a relatively longer, narrower tail than buzzards of the Genus *Buteo* which gives it an overall greater average body length but its much smaller head and narrower neck at once distinguish it in the field from either the Common or Long-legged Buzzards, particularly noticeable when the bird can be observed perching. In flight the tail pattern is diagnostic, blue-grey above with dark blackish cross-bars, a very broad one sub-terminally and two narrower ones at the base of the tail, the latter being partly obscured by the tail coverts. Though more silvery white, this tail pattern is clearly visible from underneath and the under-wings are pale silvery grey with parallel rows of spots and darker carpal patches, only confusable with the under-wing pattern of *Circaetus gallicus* (see **Plate 12**). When viewed perching the head and neck are distinctly blue-grey in adults with the iris golden yellow and the bill appearing relatively weak and

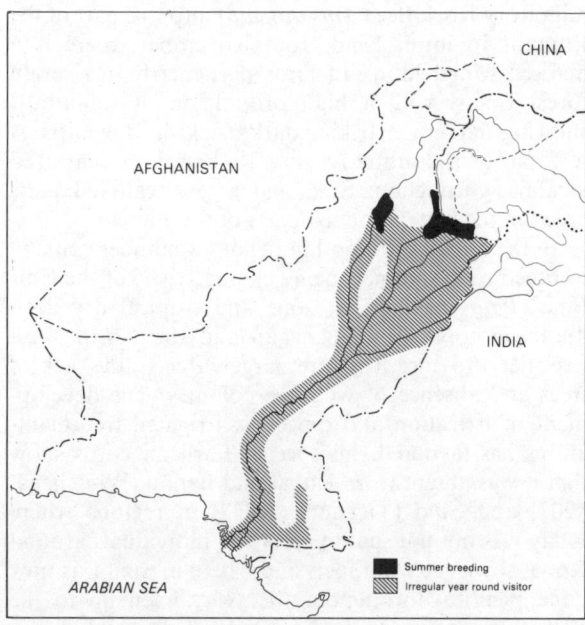

Map 54: Crested Honey Buzzard *pernis ptilorhyncus*

narrow between the culmen and the gonys. The rest of the body is darker brown-grey to umber brown with the throat and breast varying from white, to rufous, to darker brownish grey. The lower flanks in adults usually show rufous cross-barring and the upper breast some vertical dark brown streaking. Immature or sub-adult birds have a dark brown iris and heavier vertical streaking all over the breast and lower belly. Pale morphs may have the whole of the throat and upper breast pure white unmarked and darker morphs have a grey-brown breast concolorous with the wings and back, and without distinct darker streaking. The tarsi are relatively long, naked and yellow, as also the cere which extends well down the upper mandible leaving only the hooked tip black. When at rest the wings sometimes show a darker bar formed by the tips of the secondary wing coverts though in rufous forms this may show as a paler wingbar. In Pakistan the darker morphs are less commonly observed than a paler grey or white breasted form though all colour phases have been noted. Many authors have commented on the small scale-like feathers of the forecrown lores and ear coverts; an adaptation against the attacks of angry Hymenopterous insects.

HABITAT, DISTRIBUTION AND STATUS

Subject to local migration movements, its exact year round distribution in Pakistan is difficult to work out. In the northern regions it is only a summer visitor with a marked arrival in considerable numbers from late February to early March throughout the Punjab, particularly noticeable in the vicinity of irrigated forest plantations. Its arrival coincides with that of the migratory Rock Bee (*Apis dorsata*) into this part of the Punjab. In lower Sind from November, there is a marked winter influx of birds particularly in riverain forest tracts, with a high proportion of sub-adults showing the rather striking dark blackish brown iris. A few pairs undoubtedly stay to breed in scattered localities throughout Sind and a few scattered pairs breed in the better wooded parts of the Punjab.

In the Himalayan foothill regions a number come in to breed each summer, being characteristic of the Chir Pine (*Pinus roxburghii*) zone and tropical dry deciduous biotope. In Baluchistan and the North West Frontier Province it is rare, largely due to the lack of trees and absence of wild bee colonies. The development of irrigation and especially irrigated tree plantations has favoured this species. Earlier records show that it was absent from Kohat and Bannu (Whitehead, 1907) and Sind (Ticehurst, 1922) in regions where today it is not unusual to see a few individuals around Kohat town each summer, and where in Sind it is now quite plentiful throughout the year. Even up to the 1930s Eates considered it rare in Sind. In the 1970s as many as one dozen individuals could be flushed while walking through Pirawala forest plantation, and on 20 June at Pir Patho riverain forest in Thatta district, lower Sind, 11 individuals were located in one afternoon within a small area of about half a square mile. Status in Pakistan, locally migratory with some breeding, FREQUENT.

HABITS

Rather secretive birds, often spending much of the day concealed in the leafy foliage of a tree. They will however often allow close approach while perched. During the middle of the day they often circle overhead on thermals and this is especially noticeable in their summer breeding haunts. In winter there is some evidence of gregariousness with considerable numbers being located within one small patch of forest (see above under Status and Distribution). One or two individuals often roost in the author's garden at Malir outside Karachi during the month of December, but due to harrassment by House Crows (*Corvus splendens*) they tend to use such sites only in late evening, dispersing by early morning. Their hunting technique often consists of perching and watching for long periods and then flying low through the trees to another vantage point. Individuals have twice been noted hovering briefly and then plunging vertically into a grass thicket or clump of bushes. They are adapted to feed on bees, their larvae and combs and an individual was watched carrying in its bill a lump of bee's comb, not in its feet as would seem more normal for raptors (observation Manga valley, Murree foothills). They will catch lizards and small birds, also large insects such as Bush Crickets and Mantids. Wasps of *Polistes* Genus are favoured in lower Sind in early autumn and the Rock Bee (*Apis dorsata*) in the Punjab during March and April. Other authors have recorded them consuming berries and robbing birds' nests and presumably a greater variety of non-Hymenopterous insects are taken in the winter months when bees are scarcer. On 15 November in Malir the author watched one tear down a large comb of the Yellow Wasp (*Polistes hebraeus*). This was carried to a convenient tree perch nearby and systematically pecked open. The developing larvae appeared to be consumed with quite a quantity of adherent comb.

BREEDING BIOLOGY

They construct quite substantial nests which are usually in trees and both sexes carry material to the nest. An individual was seen carrying a freshly broken branch with leaves in its talons on 9 April near Bhegari embankment north of Sukkur (Sind). The nest cup is itself comparatively small and copiously lined with smaller branches bearing green leaves. The normal clutch in Sind was found to be only 2 eggs and they closely resemble small *Milvus migrans* eggs in shape and colour being richly blotched with darker red-brown against a rufous background (K. Eates, MS notes based upon 6 nests examined). Eggs can be found from late March up to early May and are usually located quite high up in Eucalyptus or Shisham trees (*Dalbergia sissoo*). A nest seen at Kalabagh, Mianwali district on the Indus was in a Eucalyptus, whilst in Marala barrage canal colony (Sialkot district) a nest was located in the topmost fork of a Pipal tree (*Ficus religiosa*). In this latter instance there were downy young in the nest in July. In the Murree foothills near Tret at 914 metres (3,000 feet) elevation a pair was frequenting a newly completed nest judging from the amount of fresh green leaves incorporated into the structure in a Long-needle Pine (or Chir) (*Pinus roxburghii*) on 20 May. Stuart Baker describes the eggs as being quite variable in pattern, some having a paler buff ground colour and others with mauve under-markings. Incubation period varies from 32-35 days with the female believed to take the greater share

of incubation duties. Chicks take about 44 days to fledge and are fed by both parents. Courtship is said to involve both sexes circling around high in the air with the male usually above and following an undulating or swooping flight path (Brown and Amadon, 1968). During courtship the male will also characteristically climb steeply upwards at the end of a swoop and at the apex will rapidly quiver its wings over its back before again swooping downwards and forwards. This display seems to be rather unique amongst raptors.

*VOCALIZATIONS

What is probably a mate attracting or territorial call is often given in the spring by a bird perched in a tree. It consists of a repeated rather drawn-out scream which sounds inflected or di-syllabic 'whe-yeeuw, whe-yeeuw'. They also call in flight and on such occasions each call is usually of shorter duration and a more melodious 'piha'. They are relatively silent in the non-breeding season. In the Punjab males start this territorial calling from early March.

Black-winged or Black-shouldered Kite *Elanus caeruleus* (Desfontaines)

Plate **12**
Illus. **24**
Map **55**

DESCRIPTION

Total length 31-35 cms.
Wingspan 75-87 cms.
Wing length 26-27.8 cms. (males), 26.3-28.7 cms. (females)
Tail length 11-13.5 cms. (both sexes) (Brown, Urban and Newman)
Weight 200-270 gms. (males),
219-293 gms. (females) (Biggs *et al.*, in Brown and Urban)

A small graceful looking raptor of a pale dove grey colour with white throat and belly and forecrown and black tertiary wing coverts and shoulders highly visible in flight on the upper surface as well as when perching. The primary wingtips are also darker grey, the bill black with yellow cere and legs, and a large ruby red iris. Juveniles have a browner grey mantle

and upper breast, with a scaly pattern formed by the white margins to mantle and nape feathers and whitish tips to the primary and secondary wing coverts. Its long slender wings beat slowly in a graceful almost tern-like pattern in flight and it characteristically hovers with relatively slow wing beats held high over its back whilst seeking prey. The absence of black wingtips and short tail distinguishes this bird from the larger adult male Pallid Harrier which can be encountered in exactly similar terrain. Also its primaries are never splayed while gliding as is the case with *Circus macrourus*.

HABITAT, DISTRIBUTION AND STATUS

Probably the commonest and most widespread raptor in the Indus plains occurring from the Potohar and Salt Range down to Karachi. It is equally at home in desert border regions, being frequently encountered in Lasbela and the Tharparkar desert areas. It is a summer visitor to the North West Frontier Province occurring up to 1,400 metres (4,700 feet) elevation. In parts of Sind such as Kandhkot tehsil in the north, and Thatta district in the south it is usual to encounter 10 or 12 individuals during a drive of 112 to 160 kms. (70-100 miles). It avoids mountainous regions and is only a straggler to central and northern Baluchistan. In Pakistan, it prefers open lightly wooded areas, with

Illus. 24: Black-shouldered Kite or Black-winged Kite *Elanus caeruleus.*

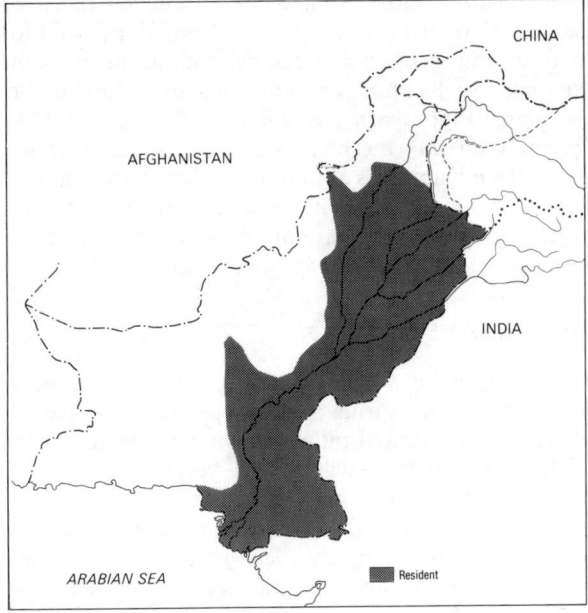

Map 55: Black-winged Kite *Elanus caeruleus*

BREEDING BIOLOGY

The pair bond may last for much of the year, but no very spectactular nuptial displays have been noted during the breeding season. A new nest is built each year which is often quite a shallow small structure. Two nests in mango trees (*Mangifera indica*) have been found in Malir, and a third in an *Acacia arabica* tree on a canal bank near Thatta. Eates found them breeding almost colonially in scattered *Prosopis spicigera* trees in the Hab valley with six nests located along a three mile stretch of dry wadi. Laying usually begins from late April extending upto early July, the normal clutch being 3 to 4 eggs, rarely 5. Most eggs are laid in May and Eates found that they would lay again if the first clutch was taken. The eggs are creamy buff in ground colour with blotches and streaks of dark brown and purplish brown (K. Eates, MS notes). Both sexes vigorously defend the nesting areas and have been seen chasing off House Crows (*Corvus splendens*), the female taking the major share of incubation duties. Incubation lasts about 26 days and the young are fledged and ready to leave the nest 30 to 35 days after hatching. Three fully fledged but still flightless young were found in the nest at Khar in the Hab valley on 26 August 1985 (Rolf Passburg, pers. comm.). Normally the male brings most of the food to the nest and the female feeds the young. Hatching is asynchronous resulting in considerable variation in the size of the young.

*VOCALIZATIONS

Outside of the breeding season they are very silent. During the nesting season both sexes have been described uttering a variety of calls, including purring or wheezing cries given by the female while incubating, and high pitched whistles by both sexes as warning or alarm calls, (Cramp *et al.*, 1980). But the most typical vocalization is that given by the male, who calls for prolonged periods throughout the day, and particularly in the evening, while perched in a tree, both to advertise the nesting territory, and also possibly as a part of courtship. This is a short upward inflected whistle, repeated at measured intervals, 'toowhit-toowhit' or 'pleewit-pleewit'. Often this continues uninterrupted for three or four minutes. Being rather soft and melodious in tone, it is unlike the calls of typical raptors, and more reminiscent in timbre of some of the smaller *Charadrius* plovers.

Two juvenile birds perched side by side near the roadside 6 January when approached gave rather hoarse hissing and drawn-out calls, which were not very loud. These were presumably alarm cries and do not correspond clearly to any of the vocalizations described by other observers (Cramp *et al.*, Vol. 2, 1980).

fallow land or scrub desert, avoiding forested regions or extensive swamps but in the eastern Himalayas it often frequents forest glades. Status COMMON.

HABITS

They spend a large part of the day perched on telegraph or power lines or exposed trees from where they can watch for larger insects and reptile food prey. K. Eates an ornithologist who worked in Sind Province during 1930s and 1940s noted its preference for reptiles and that both *Calotes versicolor* and *Acanthodactylus cantoris* lizards were frequent food items. (MS notes unpublished).

The author has seen one eating a *Meriones hurrianae* Desert Jird, and capturing a White Wagtail (*Motacilla alba*), after hovering over some *Juncus* sedge and plunging to the ground. They also catch locusts and crickets on the ground as well as Desert Larks (*Calandrella* spp.) and diurnal rodents. In hunting they hover closer to the ground than the Kestrel (*Falco tinnunculus*) characteristically does, with rather slow wing beats raised high over their backs and can thus be distinguished at a distance from a hunting Kestrel which commonly employs a shallow rapid fluttering wing beat and takes more advantage of air currents. Generally they are solitary in hunting but can occasionally be seen in pairs outside of the breeding season. Ali and Ripley (1968) report their habit of roosting colonially with up to 20 birds converging on one site but this has not been recorded by any observers in Pakistan.

Black Kite *Milvus migrans migrans* (Boddaert)
Eared or Large Indian Kite *Milvus migrans lineatus* (Gray)
Pariah or Indian Kite *Milvus migrans govinda* (Sykes)

Map **56**

DESCRIPTION

Body length 55-60 cms. (up to 66 cms. in sub-species *lineatus)* including tail 20 cms.
Wingspan 160-180 cms.
Wing length 42.6-46.3 cms. (males), 44.8-48.2 cms. (Females) sub-species *migrans* (Brown, Urban and Newman)
Weight 630-928 gms. (males), 750-941 gms. (females) (Brown, Urban and Newman)

A rather variably streaked dark brown bird with long narrow wings and tail giving it a particularly buoyant flight. The head and neck is generally dirty buff to rufous with dark shaft streaking and the long tail is forked with distinct darker brown cross-barring. Three races are recognized, the nominate one being uniformly darker brown on body and wings and lacking paler carpal patches on the underside of the wings but with a whitish buff head and neck.

The Eared or Lineated Kite is the largest race, usually showing distinctly darker ear coverts and a more rufous crown and neck with darker shaft streaks and prominent pale shaft streaking on mantle and breast, as well as conspicuous paler carpal patches when viewed in flight from underneath. The sub-species *govinda* is generally smaller in size with less contrasting dark ear coverts and some streaking on the breast, with less uniformly dark plumage than is typical of the nominate race and less conspicuous carpal patches. There is, however, considerable inter-gradation in plumage and young birds of all three sub-species show more conspicuous pale tips to wing coverts and streaking on body plumage. The long tarsus is yellow and naked and the bill blackish horny with a greenish yellow cere. The iris is dark brown.

HABITAT, DISTRIBUTION AND STATUS

The sub-species *govinda* is largely sedentary and confined to the Indus plains, being more numerous in the vicinity of towns and villages where food supplies are more easily available. A small population of *govinda* is resident year round in Baluchistan in the vicinity of towns' The nominate sub-species *migrans* is largely a summer breeding visitor to Baluchistan, the North West Frontier Province, Chitral and Gilgit, whilst the sub-species *lineatus* is a winter visitor to the plains being most often encountered in the vicinity of drying out jheels and canals. It regularly occurs around Murree town in the summer, though evidence of breeding is not available as the sub-species *govinda*

also occurs there in summer and its breeding range extends mainly from northern Kashmir towards Tibet. Black kites are still relatively common throughout Sind and the Punjab, and flocks of up to 400 birds wheeling in a wide straggling circle can be observed over parts of Karachi city at eventide, and flocks of 150 to 160 over Rawalpindi city. Breeding populations of the nominate sub-species are thinly but widely distributed in Baluchistan and the NWFP, often in association with suitable rocky gorges or cliffs which are preferred nesting sites for this much less commensal population. They are also summer breeding visitors to the main valleys of Baltistan, Gilgit and Chitral. Status ABUNDANT.

HABITS

They are gregarious in their roosting and foraging areas during the winter, and all 3 sub-species tend to be locally migratory according to food availability, with large congregations assembling in Thatta district of lower Sind at the end of the monsoon to exploit the vast numbers of stranded fish when the rice canals dry out. They are adept at picking up refuse with their feet both in harbours and from the ground and will often transfer food thus seized to their bills for consumption

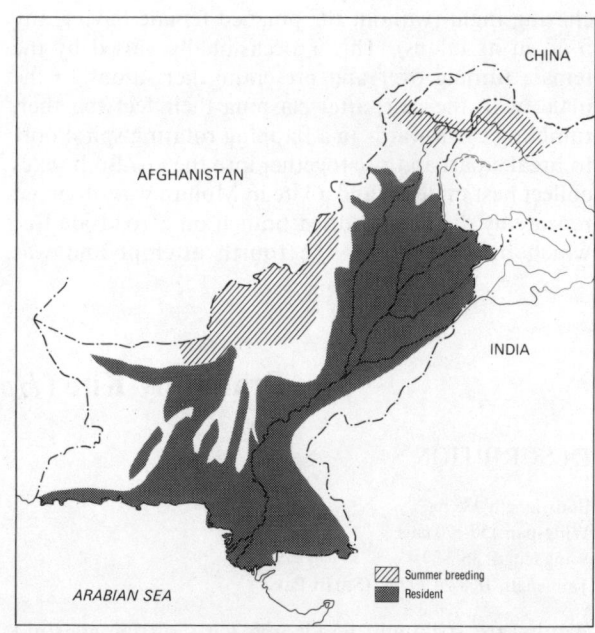

Map 56: Black Kite *Milvus migrans*

while sailing around in the air. They are dependent mainly upon weak or sickly animal prey or carrion and human garbage but there are authentic instances of birds swooping down and seizing healthy domestic chickens as well as large healthy fish which were basking near the surface of a lake. They will also take reptiles, mole crickets and *Chrotogonus* grasshoppers. They are amazingly dexterous in aerial manoeuvrability, swooping down in crowded thorough-fares to seize food and avoiding speeding traffic as well as overhead telephone wires. In certain localities such as Delhi where food is served in the open, individual kites become adept at seizing morsels from the human customers and the author had this experience in the late 1940s and Guy Mountfort in Dacca, Bangla Desh (*in litt.*, 1987). Such behaviour is learned, not innate as it has not been recorded in Pakistan.

BREEDING BIOLOGY

Breeding occurs over a very extended period in the sub-species *govinda* with birds beginning to call territorially in early winter and many nests with eggs being found in January up to mid-March. In the mountain areas breeding occurs from March to April with first arrivals being noted in early March in Gilgit, when birds were already apparently paired up. The pair bond duration is not known but certainly there is evidence of favoured nest sites being reoccupied year after year, e.g. a cliff nest site at Rohri in Sukkur district used for four years in succession (K. Eates, MS notes) which would lend support to the establishment of long lasting pair bonds. Courtship often commences in November with two birds indulging in repeated chasing flight (without the pursued female having any food in its talons). This is occasionally varied by the female turning over and presenting her talons to the male, with the pair, after clasping their feet together, tumbling earth-wards in a flapping rotating spiral only to break apart and rise together into the air. Both sexes collect nest material and a kite in Multan was observed repeatedly diving at a dried branch on a roadside tree which broke off after the fourth attempt and was

subsequently borne away to the nest in its bill after transfer from its feet. Most nests are untidy platforms with odd pieces of cloth, dried dung cakes or even paper adorning the nest rim. Nests with fresh eggs have been found throughout January in the vicinity of Karachi, but a nest in Multan district was still being built on 12 March. Most sites are in the fork of a tall tree, but ledges on buildings, power pylons, date palms and natural cliff ledges have all been noted in Pakistan. *Milvus migrans migrans* in Baluchistan normally nests on cliffs, well away from towns or villages and an incubating bird on her nest was found at Hannah gorge, Uruk valley, outside Quetta on 15 April. The eggs are laid at 2 day intervals and are variable in marking, usually with a white ground colour and sparse purplish red blotches and reddish brown speckles. The female is believed to do most of the incubating with the male bringing food to his mate and later assisting in feeding young. The normal clutch is 2 to 3 eggs, rarely 4 and the parents can be quite aggressive when young are in the nest. Eates records a pair nesting in his compound which hit the head of an unwary visitor, knocking off his spectacles. In a study in Delhi, 46 per cent of clutches comprised 3 eggs and 43 per cent only 2 eggs. Nest building by both partners was prolonged, lasting from 24 to 68 days with incubation being done largely by the female which took from 29 to 35 days, the eggs hatching asynchronously. Chicks left the nest on average, 64.3 days after hatching and could fly at 42 days of age. Out of 18 nests studied, only one chick survived to fledging in the majority of cases (Desai and Malhotra, *Ibis*, 1979).

*VOCALIZATIONS

The plaintive rather high pitched wailing of Kites is a familiar noise in the towns of Pakistan, particularly at the onset of the breeding season. The typical call is a repeated long drawn-out 'kwee-errrr'. In courtship flight chases and when excited they have a more rapid 'kee-yik-yik-yik' call. Outside of the breeding season they sometimes emit a drawn-out whistle 'wheeeu' which is probably a contact call.

Brahminy Kite (*Haliastur indus*) (Boddaert)

Plate **12**
Map **57**

DESCRIPTION

Body length 48 cms.
Wingspan 150-170 cms.
Wing length 36.2-39.4 cms.
Tail length 18.8-20.7 cms. (Stuart Baker)

Adults are strikingly handsome with rufous chestnut tail, wings and lower belly and white head, neck and

breast, narrowly streaked with black shaft stripes. In flight the dark blackish tipped primaries and paler chestnut secondaries make a bold and characteristic pattern with the tail being relatively shorter and rounded at the tip, in contrast to the longer forked tail of the Black Kite. They have relatively long wings, angled at the carpal joint like *Milvus migrans* and because of their similar wing shape, the all-brown sub-adult birds can be confused with that species. However even in juvenile birds the head and neck is a paler rufous tone than the rest of the body and the rounded tail is not cross-barred (as in *Milvus migrans*), whilst the secondaries from underneath are much paler chestnut and sharply defined from the darker primaries. They are less buoyant in flight than the Marsh Harrier (*Circus aeruginosus*) and the chestnut under-wing coverts are distinctive. Adults have yellow feet, tarsi and cere. The iris is dark brown and the bill blackish horny. When at rest the white plumes of the neck extend down to the upper mantle and on the breast right down to the tibia.

HABITAT, DISTRIBUTION AND STATUS

Only really common in lower Sind and in the mangroves; it also occurs sparsely along the Indus river system as far as Panjnad headworks in Bahawalpur division and is fairly plentiful in the rice growing tracts of Jacobabad district, Sind. It occurs around Taunsa barrage in Dera Ghazi Khan, and has been noted occasionally at Lal Sohanran in Bahawalpur. However, through most of the Punjab it is uncommon or absent. Its stronghold is the Indus delta where 30 to 40 individuals can be encountered in a day's voyage along the mangrove creeks. In the summer it spreads through the rice growing areas of Thatta and Badin districts. It has not been recorded in the interior of Baluchistan or the North West Frontier Province and is never found far from major river systems or large lakes. Status

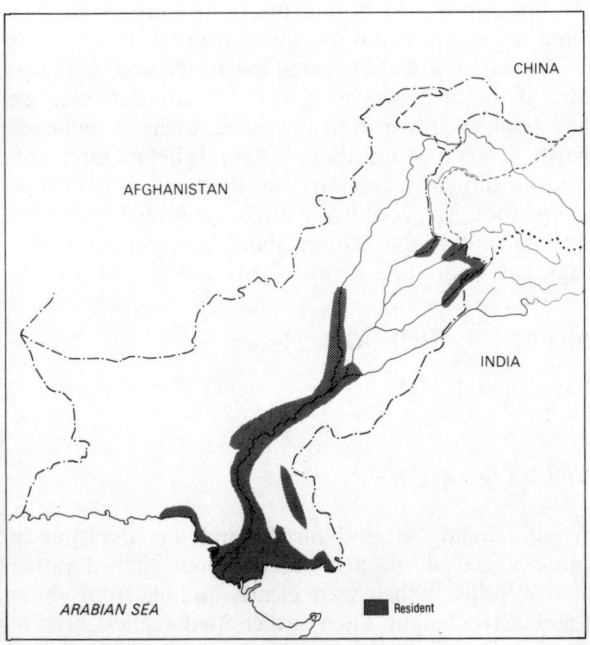

Map 57: Brahminy Kite *Haliastur indus*

COMMON only in Thatta district, elsewhere RARE and confined only to the river system.

HABITS

Like the Black Kite it often searches for food with slow flapping flight, traversing 30 to 40 metres above the ground and they can be very dexterous in swooping down to seize prey in their talons without landing. I have watched them catching *Periopthalmus* mud skippers in this fashion as well as quite large fish basking near the surface in tidal creeks. In the rice growing season, frogs form a substantial portion of their diet, and one successfully carried off in its talons a shot-gun wounded Teal (*Anas crecca*) at Haleji though it is not known how the bird was killed. Salim Ali talks of their scavenging habits in the sea harbours of India but this has not been noted amongst Sind birds. They appear to be sedentary in Pakistan and in the Indus delta are quite gregarious, a nightly roost of over 80 birds being counted in Khai creek in mid-November, 1983.

BREEDING BIOLOGY

The nesting season is variable as the following records will show. A female found, incubating eggs 20 February, nest in mangrove tree *Avicennia officinalis* in Sonmiani lagoon. Nest located on periphery of Reef Heron colony (*Egretta gularis*) in mangrove on Ghizri creek, eggs found mid-March one year and with two recently fledged but flightless young on 30 April the subsequent year (author obs.). K. Eates found nests with eggs on 3 January in a mangrove tree and another nest on 3 March with two eggs well incubated near Jati, and a third with half-fledged young in a nest near Sukkur on 15 April. A pair was seen nest building in November and this can continue for upwards of two months before egg laying commences. The nest is a comparatively small structure built of sticks and is built anew each year with both parents defending their territory quite vigorously when young are in the nest. Lowther (*A Bird Photographer in India,* 1949) recorded parents feeding their young about once an hour and frogs were the main food item.

Courtship seems mainly to involve both sexes wheeling around in circles and emitting their very characteristic and unique mammal-like 'meeow' call, which can be given from a tree perch or while circling around the nest site. Incubation takes 26 to 27 days and the normal clutch is 2 eggs, rarely 3 with the female doing most, if not all, of the incubating. The male certainly brings food to the nest for his brooding mate. The eggs are greyish white feebly spotted and blotched with reddish brown. Their nest is not decorated with rags or other rubbish, like *Milvus migrans* being a comparatively smaller and neater structure. In Gharko riverain forest (Thatta district) on 20 June three nests were located within 500 metres of each other in tall *Terminalia* trees, each with two fully fledged young, capable of flight and perched in the vicinity of the nest tree. The young begged vociferously for food, hopping from branch to branch to harass their parents. They are capable of flight about 8 weeks after hatching.

*VOCALIZATIONS

When disturbed at the nest they will utter a repeated high pitched short scream reminiscent of the Arctic Skua (*Stercorarius parasiticus*) but during courtship and nesting their call is a quavering curiously mammalian sounding 'meeah' unlike the call of any other raptor.

White-tailed Sea Eagle *Haliaeetus albicilla* (Linnaeus)   Plate **13**

DESCRIPTION

Body length 70-90 cms.
Wingspan 200-240 cms.
Wing length 55.2-64 cms. (males), 62.1-71.5 cms. (females)
Tail length 25.4-33.1 cms. (males), 27.6-33 cms. (females)
Weight 3,075 gms. (males), 4,080-6,920 gms. (females) Brown, Urban and Newman)

This huge fish eagle averages about 10 per cent bigger than Pallas's Fish Eagle (*H. leucoryphus*). Both sexes are similar in plumage but females are up to 25 per cent heavier. Though almost equal in wingspan with Pallas's Eagle, they look much bulkier in body proportions. Adults have dirty buff to greyish white heads and necks, paler almost white on the crown, and a pure white tail which is wedge shaped when closed, but short and rounded when gliding. The rectrices are not square-tipped as in *H. leucoryphus* but have pointed tips. Their body plumage is dull umber brown with paler buff margins to the mantle and wing coverts. The bill and the cere are yellow as also the feet, with the tibia well covered by dark chestnut bushy 'trousers'. Juveniles are harder to distinguish from juvenile Pallas's Fish Eagles, though the tail feathers show varying amounts of white mottling with darker margins, and the overall body plumage is a darker brown whilst the heavier deeper bill is noticeable. Sub-adults have dark bluish bills even up to their third year and they look very like a vulture in outline when sailing overhead. At rest both juveniles and adults show rather prominent lanceolate neck hackles.

HABITAT, DISTRIBUTION AND STATUS

Some of the Russian breeding population, predominantly juvenile birds, migrate down the Indus in severe winters. A pair of adults was recorded in January 1968 and again in February 1969 near Chashma barrage (Mianwali district, Punjab) (Roberts and Savage, *JBNHS*, 1970). On Hadiero lake in Thatta district, Sind, from late December to early January 1980, a sub-adult bird was observed and one or two birds generally turn up at Ghauspur jheel in Jacobabad

district each winter. In 1971, one adult and 3 immature birds were recorded. Ticehurst records seeing an adult on the Hab river, Lasbela district on 14 November (Ticehurst, *Ibis*, 1922). An adult was noted on the Quetta rubbish dump by Christison in February (*JBNHS*, 1942) and a sub-adult bird on Zangi Nawar lake in Baluchistan during January 1984 visit (author). Status RARE winter visitor.

HABITS

Mainly fish eaters, they have been observed frequenting larger lakes or the main Indus river in Pakistan and have not been noted off the seacoast. They seize fish basking near the surface without actually diving in as does an Osprey and they are quite adept at catching waterfowl. An adult eagle was watched catching a swimming Teal (*Anas crecca*) and a sub-adult catching a Coot (*Fulica atra*). In both instances the prey was swimming on an open lake, and was repeatedly swooped upon, the eagle returning with ponderous wing beats and again swooping down over the same individual. The teal appeared exhausted and unable to dive at the fifth swoop. It was secured after the eagle had plunged down onto the water where it remained partly submerged for about 8 seconds before taking off without difficulty and carrying the hapless teal firmly in one foot. The coot was entirely consumed except for its legs and wings, within about ten minutes of the eagle carrying it to a rocky hill overlooking the lake.

BREEDING BIOLOGY

Extra-limital.

VOCALIZATIONS

A pair circling around on thermals at Ghauspur in January, called to each other in low pitched rather hoarse 'yelps', which were clearly audible from about 1,500 metres height. Each bird emitted a quick series of 10-15 'yelps', the calls of the larger female being lower pitched and more reminiscent of a dog.

Pallas's Fish Eagle,
Ring-tailed Fish Eagle *Haliaeetus leucoryphus* (Pallas)

Synonym *Haliaeetus leucogaster* (Gmelin)

Plate **13**
Map **58**

DESCRIPTION

Body length 76-84 cms.
Wingspan 200-250 cms.
Wing length 55.6-57.8 cms. (males),
    55.8-59.8 cms. (females) (Ali and Ripley)
Tail length 27.1-27.5 cms. (males),
    27.4-29.1 cms. (females) Ali and Ripley.

A huge eagle, approximating in size *Aquila chrysaetos* with a rather distinctive dark purplish brown body and dirty buff head and neck paling almost to white on the crown and lores. The tail is pure white with a very distinctive dark brown broad sub-terminal band. The bill is deeply compressed and plumbeous as is the cere. The nape and neck feathers are lanceolate. The legs and feet are dull yellowish white. Sub-adult birds have allbrown tail and head and neck but when viewed closely the base of the tail feathers are mottled white and the head and neck is paler fulvous with distinctive neck hackles. In flight silhouette the neck of both adults and sub-adults looks rather long and thin, and the bill is noticeably heavy and deeply compressed laterally when compared with *Aquila* spp. They are faster in wing beat and more graceful in flight than *H. albicilla*, and the tail is relatively longer and rounded at the tip not wedge shaped. Females are slightly larger than the males.

HABITAT, DISTRIBUTION AND STATUS

This resident fish eagle has become increasingly rare in Pakistan especially in the Punjab and northern regions and even on the lower Indus it is now very sparsely distributed. Around Ghauspur jheel and Kandhkot in Jacobabad district the greatest concentration probably occurs, with about 5 pairs within a ten mile radius. In Thatta district there are four recent deserted eyries known to the author. The reasons for this decline are not known but in at least two instances too much human disturbance in the vicinity of the nest site was undoubtedly a contributory factor. During a survey of all Pakistan's major wetlands in 1974, only 26 breeding pairs were located and the total Pakistan population was estimated at less than 40 pairs (F. Koning, *JBNHS*, 1972). They are fresh water eagles, being absent from tidal creeks or mangroves and can occasionally be seen along the Indus up to Mianwali in the Punjab but are more likely to be encountered in the vicinity of large lakes. They are absent from Baluchistan and the North West Frontier Province except in the latter province where it borders the Indus. They

may be locally migratory, as W.H. Mathews encountered 2 on the Indus in July in Baltistan as well as 1 on the Deosai alpine plateau in late June (*JBNHS*, 1941). They are occasionally sighted over the Salt Range lakes. Status SCARCE.

HABITS

Mainly a fish eater, this eagle is also adept at capturing waterfowl. They catch fish mainly by swooping upon prey basking near the surface and do not plunge into the water like Ospreys, but they seem to be very successful as I have seen one carry off a Carp which must have weighed about 2.5 kg. In hunting for waterfowl the usual technique is to glide very low over the water relying upon surprise to snatch an unwary bird on the surface, and at Haleji lake, Purple Gallinules (*Porphyrio porphyrio*) have twice been observed (author) falling victim in this manner. H. Waite (*JBNHS*, 1948) described an eagle hovering over Dabchicks (*Tachybaptus ruficollis*) at Kallar Kahar lake and successfully catching them by suddenly stooping. They will repeatedly swoop over a swimming bird in this manner until it is too tired to dive again. They also take a heavy toll of nestlings from colonies of Little Cormorants (*Phalacrocorax niger*), egrets

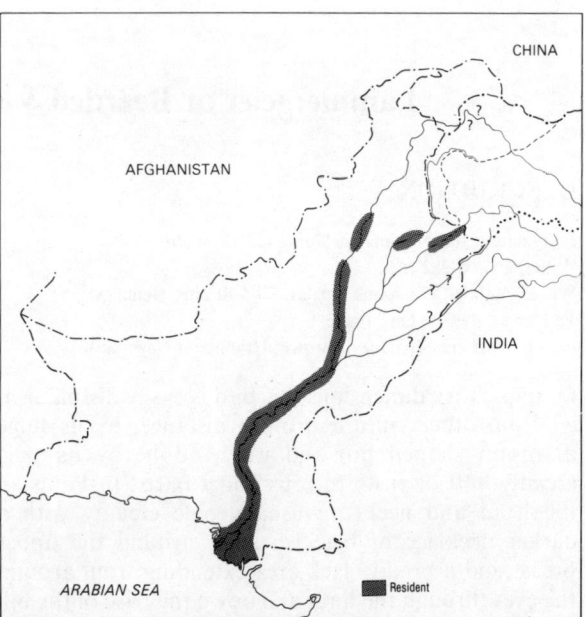

Map 58: Pallas's Fish Eagle *Haliaeetus leucoryphus*

and ibises and Lowther recalls a pair robbing almost all the offspring from a Painted Stork (*Mycteria leucocephala*) colony (Lowther, 1949). Fish prey may be eaten on the ground or on a tree perch. E.N. Lowther also records seeing the remains of 5 turtles on the nest of one pair that he was photographing.

BREEDING BIOLOGY

The pair bond is often life-long with the same single nest site being used year after year, usually an enormous accumulation of sticks in an isolated tall tree. Near Sujjawal bridge an eyrie has been built in a tall electricity pylon and this (at the time of writing) has been occupied for 3 years. .

Both sexes hang around the nest site, perching together in the nest tree for some weeks before nest repair or rebuilding commences. This is usually from late October to early November when both sexes strike at drying tree branches to break them off and bear them away in their feet to the eyrie, as well as smaller stems with green leaves which are used to line the nest cup. An eyrie at Lal Sohanran was decorated or lined with fresh green branches of *Calotropis procera* and *Tamarix dioica* and another bird was observed breaking off a stem with green mango leaves which was borne away to the eyrie. Copulation has been seen throughout November and the first week of December in lower Sind and probably extends over 4 or 5 weeks. Courtship involves mutual soaring around in close formation and a lot of loud calling. A pair has been noted (early February) duetting on the nest with bills raised to the sky and necks thrown back. The normal clutch is 2 to 3 eggs which are pure white, unspotted. K. Eates found 3 eggs more usual than 2. The female does most of the incubation which lasts about forty days with the male bringing food regularly to his brooding mate. The female will sit tight when approached, often lowering her head and neck to conceal her presence from the ground. The white downy chicks hatch from mid-December to early January and remain in the nest vicinity until late March when they are able to fly but still receive food from their parents (observations of eyrie at Haleji lake over 9 year period). The eggs hatch asynchronously 2 to 3 days apart. It is rare for more than 2 young to survive to fledging. As the young grow in size the female begins to leave them and also hunt for food to bring to the nest. In the first 2 weeks after hatching one or both parents continuously guard the chicks, as marauding House Crows (*Corvus splendens*) which are attracted to the eyrie by the prey remains, are a real threat.

*VOCALIZATIONS ·

The wild cry of this Eagle carries for distances of over a mile and is regularly uttered during the early part of the breeding season both while flying and from tree perches. It is a high pitched ringing continuous 'qwark-qwark' rising and falling in pitch so that it is evocative of the creaking of un-oiled wooden cart wheels. This cry is distinctive and cannot be confused with any other eagle. Recordings made of a pair duetting on the nest platform reveal that the cries of the female are very similar to her mate and only slightly lower pitched.

Lammergeier or Bearded Vulture *Gypaetus barbatus* (Linnaeus)

Plate **11**
Illus. **25**
Map **59**

DESCRIPTION

Body length 100-115 cms. including tail 42-44 cms.
Wingspan 266-282 cms.
Wing length 74.3-78.7 cms. (males), 71.5-81 cms. (females)
Tail length 42.7-46 cms. (males),
    43.7-46.9 cms. (females) (Brown, Urban and Newman)

Of impressive dimensions this bird is easily distinguished from other vultures from a distance, by its huge diamond shaped tail and when adult, by its pale creamy buff or rusty fulvous under parts. In Pakistan the head and neck are usually pale creamy with a darker necklace of blackish spots around the upper breast and a broad black area extending from around the eyes through the lores and down the base of the bill in the peculiar bristles which gave rise to its English name. The crown and mantle are silvery brown with lanceolate feathers becoming greatly elongated over the scapulars and bearing almost white shaft streaks and darker brown margins. The wings are comparatively long and slender and uniformly grey-brown. The axilliaries from underneath are conspicuously creamy white. The tibia is heavily feathered creamy to orange-buff and the naked tarsus is greyish blue. The bill is horny brown to black and the iris golden yellow. Sub-adult birds have paler brown to whitish forecrown and ear coverts with the rest of the head and neck darker brown and the same colour as the wings with paler margins to the mantle and wing covert feathers, giving a spotted effect. In adults, the tuft of black bristles protruding below the gonys of the lower mandible can often be detected when the bird flies past in aerial silhouette. Males and females are very alike in both size and plumage.

HABITAT, DISTRIBUTION AND STATUS

A largely sedentary and resident species though occupying very extensive territories. It is an inhabitant of mountain regions, as well adapted to the arid and bare crags of south and central Baluchistan, as to the green alpine pastures of the Himalayan highlands. It occurs even in lightly forested areas and appears largely sedentary throughout its range, remaining at sub-zero temperatures even in the inner ranges of the Karakorams and Hindu Kush. A pair appeared resident around Dunga Gali and used to haunt the adjoining mountain ridges throughout the summer in the early 1960s but they have since disappeared from the Murree hill range and only occur as occasional visitors. They are sparsely distributed through northern Baluchistan including the Bolan pass, the Takht-i-Sulaiman range in Southern Waziristan and the Safed Koh range. They can regularly be encountered around the Lowari pass in Dir, the northern valleys of Chitral and throughout Gilgit and Baltistan. Judging from earlier accounts, this magnificent bird is much rarer in Baluchistan than in former times, though it is still maintaining its status in the far northern regions of Pakistan. In Sind it is only an occasional visitor to the higher hill ranges such as Kirthar in January and February and it is not common or resident as implied in Vol. 1 of the 'Handbook' (Ali and Ripley, 1968). Status SCARCE.

HABITS

The comparatively long narrow wings and broad tail give it great aerial manoeuvrability and they characteristically hunt by quartering up and down across the contour of steep mountain faces, hugging every col and scree slope. They hunt in a seemingly effortless glide requiring minimum wing beats and availing of every up-draught, their long tails bending from side to side to assist in balance and steering. They are largely scavengers and of timid disposition, rarely challenging vultures at a carcass but preferring to sit and wait until

Illus. 25: Lammergeier or Bearded Vulture *Gypaetus barbatus,* showing appearance of 'beard' in adult of both sexes.

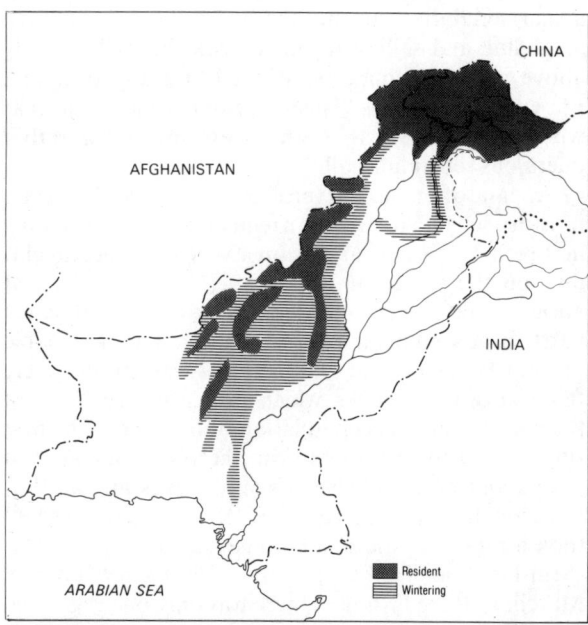

Map 59: Lammergeier or Bearded Vulture *Gypaetus barbatus*

the clean-picked skeleton remains. They are specialist bone feeders and will carry aloft heavy bones dropping them time and again in a deliberate attempt to break them open on the rocks below. (R.L. Fleming, *JBNHS* Vol. 52, p. 933-4). An adaptation of this habit is the bearing aloft of the Four-toed or Afghan tortoise (*Testudo horsfieldi*) observed in Baluchistan, which is devoured after its carapace is fractured by dropping onto rocks. An adult has been watched taking evasive action when chased by a much smaller Jungle Crow (*Corvus macrorhynchos*) in the Kaghan valley, and also another on the slopes of Musa-Ke-Masala carrying a large piece of dried animal hide which it twice dropped on the rocks and retrieved in a futile attempt to break this up. Another adult which swept overhead on the Burzil pass within about 80 metres, made a loud buzzing noise with its pinions as it glided past. They hunt solitarily and are extremely hardy, roosting on ice covered cliffs at sub-zero temperatures in regions such as Gilgit and Baltistan. They are capable of swallowing small bones and quite substantial bone fragments (up to 25 cms. long) whole (Brown, Urban and Newman, Vol. 1. 1982,). When taking off and landing on level ground, they do not need to run forward (author obs. Kargah valley), as do the other vultures (*Gyps* and *Aegypius*).

BREEDING BIOLOGY

They are believed not to reach sexual maturity until about 7 years of age and to form life-long pair bonds (Cramp *et al.*, 1980). Courtship includes some spectacular aerial dives around the site of the eyries as well as gliding and sailing together, with the male usually above and the female occasionally turning over and presenting her talons. This is the only time of the year when they are vocal, both sexes emitting a rather querulous screaming call.

A pair often has several eyrie sites within their territory which are used in irregular rotation. The only nest seen by the author was on a very bare vertical cliff face on the slopes of Rej Khan (Hazara district) at about 3,500 metres located on a horizontal ledge partly protected by an overhang and not approachable on foot beyond about 200 metres. Nests are relatively flat platforms of sticks which may often be built up from a substantial accumulation of branches. The nest rim is adorned with dried skin, sheep's wool and even plastic and other rubbish. Usually 2 eggs are laid but occasionally only one. Eates (MS notes unpublished) took 6 eggs from mountain ranges surrounding Quetta (Murdar, and Takhatu) on 14, 16, 17, 19, and 29 March. In three instances there was only one egg in the nest. The eggs are a dirty white ground colour with brick red blotches. Hatching is asynchronous, eggs being laid at 3 to 5 days intervals, with an incubation period of about 55 days. The weaker sibling invariably

is killed or dies of starvation and only one chick is reared, which is very weak and feeble for the first 29 days. Both sexes bring food which is regurgitated for the chicks in the initial stages and one parent broods the very young chick to protect it from the cold throughout most of the day. Observations in Africa indicate that only the female incubates but that the male shares in brooding the young chick and that both sexes bring food to the young. The period from hatching to full fledging varies from 110 to 122 days (Brown, Urban and Newman, 1982.) and flight feathers start to sprout from the thirtieth day. The parents do not leave the chick alone at the nest, until it is about 60 days old. Despite the long dependence of the chick on its parents, up to 170 days total, pairs will breed or attempt to breed annually.

VOCALIZATIONS

During display gives a shrill screaming call, likened by Hugh Whistler to a 'loud squealing'. The young at the nest give a rapid piping begging call often when the approaching parent is still at some distance in the air (Bates and Lowther, 1952.)

Egyptian or Scavenger Vulture *Neophron percnopterus* (Linnaeus)   Plate **11**
Map **60**

DESCRIPTION

Body length 60-70 cms.
Wingspan 155-180 cms.
Wing length 44.3-50.6 cms. (Stuart Baker)
Tail length 22.8-26.3 cms. (Stuart Baker)
Weight 1,584-2,180 gms. (Brown, Urban and Newman)

A small vulture with relatively long narrow wings and a diamond shaped tail. Adults have the forepart of the face bare of feathers and yellow, as is the cere which extends two-thirds down the length of the relatively narrow hooked bill. Its bare head is surrounded by a shaggy ruff of spiky feathers of a dirty buff colour. Adults after their fifth year have the whole body plumage white except the primaries and secondaries which are jet black. The legs and feet are yellow in the plain's population and more reddish in the Himalyan population. Sexes are equal in size. Juveniles in their first years are dark umber brown all over, with forepart of the head whitish and naked. From their second year they show a few white paler feathers in the mantle and wing coverts and their bellies and under-wing coverts are much paler brown with the tail varying to isabelline. The rump is much paler almost white when viewed from above, in all sub-adult plumage stages. The head and neck hackles are also

brownish and generally darker than the mantle and breast feathers. The iris varies from dark brown to reddish.

HABITAT, DISTRIBUTION AND STATUS

Resident in the plains and locally migratory in the northern mountain regions. In some towns in India they are quite commensal, roosting on tall buildings, and in Pakistan small numbers will frequent municipal refuse tips and slaughter houses, but they generally loiter only around the city outskirts. They are extremely widespread and adaptable, being encountered in remote desert regions such as along the Makran coast, and mountain regions which are sparsely populated by humans. A population breeds in the northern dryer Himalayan regions of Chitral, Gilgit and Baltistan, as well as the mountain ranges of Baluchistan, drifting in winter down to the main valleys and possibly the adjacent plains. In the Indus plains they are seldom encountered away from large towns but varying numbers congregate around slaughter houses and rubbish dumps, favouring bare open places on the outskirts of towns, such as disused brick kiln sites or uncultivated saline plains. They are more typically found in desert areas than the other vulture species. An

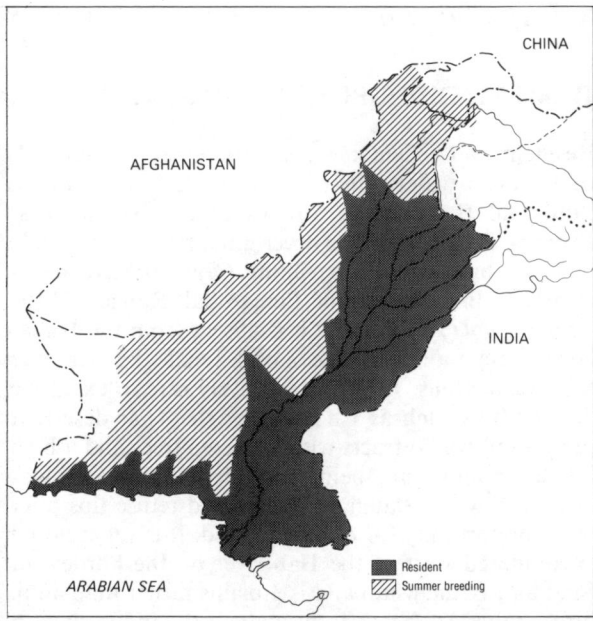

Map 60: Egyptian Vulture or Scavenger Vulture *Neophron percnopterus*

adult was observed in early May quartering over a snow covered ridge in Dir at 2,896 metres (9,500 feet) elevation. Status LOCALLY COMMON.

HABITS

In a few places, they develop the habit of roosting colonially, flighting in at dusk from great distances. Power pylons near the Pipri railway shunting sidings some 20 miles west of Karachi are a traditional site with upwards of 80 to 100 birds congregating to roost on the cross-bars of 2 or 3 adjacent pylons. They are carrion eaters and scavengers and will take precedence at carcasses over House Crows (*Corvus splendens*) but give way to the much larger White-backed Vultures (*Gyps bengalensis*). They are too weak billed to be able to exploit carcasses of domestic animals as efficiently as the larger vultures and hence subsist partly on human refuse such as rotten fruit and vegetables, excrement and the occasional beetle. They spend a large part of the day sailing in thermals and travelling great distances from their roosting grounds in search of food prey.

BREEDING BIOLOGY

The pair bond is probably life-long with birds not normally starting to breed until their fifth year. Courtship consists of the pair circling together high in the air and less frequently one bird engaging in spectacular swooping dives followed by steep climbs. Where migratory, the pair has been noted arriving together at its traditional nesting site. In Pakistan 2 eggs are the normal clutch and these are normally laid from late March to April in the Punjab Salt Range (H. Waite, 1948) much later than in the other vultures. Both sexes assist in nest building and incubation. Nesting is usually solitary and preferred sites are cliff ledges, often inaccessible but trees are used in the plains. Two occupied nests have been found by the author on the top of much-lopped *Prosopis spicigera* trees, one in the Hab valley and the second in the desert border east of Rahim Yar Khan. Another nest with the bird incubating was found in a date palm tree (*Phoenix dactylifera*) at Keti Bunder on the Indus delta (J. Vieillard, pers. comm.) in the third week of March. Eates (MS notes) collected over 100 eggs from Sind with 90 per cent of clutches comprising 2 eggs. He considered the peak laying period to be the last week of March and the first week of April. In Chitral (Perreau, 1910), nests have been found with eggs in May. Eates found a colony of 6 nests located over a one-quarter mile stretch of sandstone cliffs at Khadeji falls, 30 miles north-east of Karachi. Nests were located in wind-fretted holes. He never observed colonial nesting where tree sites were used. The nest itself is a flattish structure, more substantial when in a tree, and often decorated with rags, dried pieces of dung and other rubbish. It will be reoccupied year after year if the birds are not disturbed. The eggs are handsomely marked having a dirty white ground colour blotched with grey and dark blackish brown varying to reddish brown. The egg-shell texture is chalky and pitted. Incubation takes about 42 days with hatching asynchronous. Only one young survives the fledging period which is up to 77 days. Initially the young chick is fed by regurgitation of food onto the nest floor and one or both parents continuously shelter and protect the young for the first 40 days. After that age the young is left alone by day but both parents roost each night in the nest vicinity.

VOCALIZATIONS

Salim Ali in the 'Handbook' states that it is a remarkably silent bird with no calls recorded, but Brown and Amadon note that mewing and hissing sounds are utterred at the nest.

MISCELLANEOUS

K. Eates refers to persistent Sindhi folkore as to the efficacy of the egg yolk contents against snake bite and scorpion stings. The contents are soaked on a piece of cotton wool and placed over the bite area and this is supposed to extract and absorb the venom.

Oriental White-backed Vulture *Gyps bengalensis* (Gmelin)

Plate **11**
Map **61**

DESCRIPTION

Body length 75-85 cms. (including tail 20-25 cms.),
Wingspan 180-210 cms.
Wing length 53.5-57.8 (both sexes) (Stuart Baker)
Tail length 21.7-23.2 cms (Stuart Baker)

Sexes similar in size and colouration, but about 25 per cent smaller than *Gyps fulvus*. It is generally a darker bird with blackish grey dorsal plumage and distinctive white under-wing coverts contrasting with dark grey to black remiges. The mid-back region is pure white in adults and there is a conspicuous white downy ruff where the neck joins the body, but the white rump is not always visible when the bird is in flight, and this is not developed until birds are about 5 years of age. The head and neck are naked and blackish grey, sometimes with a sparse cover of white down on the lower neck. Sub-adult birds are much harder to separate from Griffon Vultures though they attain pure white under-wing coverts after their first year. The body is however much paler with buffy feathers on the breast similar to *Gyps fulvus* and the neck ruff consists of pale buff lanceolate feathers, not white down. In both adults and sub-adults the legs and feet are greyish black and the bill is black or greenish grey with the cere black. The iris is yellowish brown. In flight the short rounded tail and relatively broad and wide wings aid in identification, coupled with the relatively small head drawn into the shoulders, which distinguish vultures from Eagles (*Aquila* spp.).

HABITAT. DISTRIBUTION AND STATUS

Resident and sedentary. The commonest of our vultures, now widely distributed throughout Punjab, Sind and the broader valleys of the North West Frontier Province. In the dryer mountainous areas it is replaced by the Griffon vulture (*Gyps fulvus*), particularly in Baluchistan, the Punjab Salt Range and the Tribal agencies. Similarly it is absent from the Himalayas though a few birds can occasionally be seen over the Murree hills. It is comparatively rare in extensive desert tracts such as Cholistan or the Thar desert. It prefers cultivated tracts with scattered trees and a high human population, being attracted to larger towns and cities where slaughter houses and refuse tips offer more opportunity for obtaining food. It is very seldom encountered west of the Hab river on the borders of Sind and Baluchistan and is a plains rather than a hill bird. Status ABUNDANT throughout the Indus plains.

HABITS

Gregarious both in traditional roosting and loafing sites and in breeding colonies, though foraging singly. They have a high aspect ratio in wing design (i.e. surface area in comparison to wing chord). This enables them to glide with minimum effort for very long distances and to gain height very effectively by circling around thermal updraughts. They are believed to travel enormous distances daily (probably well in excess of 200 miles) by such means, searching for suitable food in the shape of animal carcasses. Their acute vision enables them to detect carrion, usually by such aids as the activity of dogs and crows in relation to the suspected animal carcass, and also by remaining in visual contact with other vultures. They are immediately attracted to any bird which proceeds to descend earthwards. In this way a host of birds will appear from all directions of the compass, coming into human vision with uncanny speed from a sky which previously appeared empty of sailing vultures. They are purely carion eaters and once a carcass is discovered will crowd around both onto and inside it, jostling for position and fighting over portions of the carcass in their eagerness to gorge as much as possible. Where the carcass is small and the congregation of vultures large, the White-backed vulture will frequently try to drive off rivals with wings raised over its back and by jabbing with its bill, but I have not noticed the foot raising gesture adopted by the Griffon Vulture (see next species account). They also will deliberately fly and land on the backs of other feeding vultures, hanging on like a circus rider until the victim is forced to retreat from the carcass.

Map 61: Oriental White-backed Vulture *Gyps bengalensis*

They will share a carcass with Griffon Vultures, but invariably give precedence to Cinerous Vultures (*Aegypius monachus*), but often keep hungry 'pie' dogs (feral dogs) at bay, as well as Egyptian Vultures (author obs.). There are records of this species attacking the carcass of a freshly dead bullock with such speed and efficiency that no meat was left on the skeleton within about 40 minutes of commencing their feast (Ali and Ripley, 1968). Gorged birds sit around on the ground near a carcass often for 5-6 hours before they can take off again and they are ungainly on the ground, requiring space to run forward with flapping wings before they can become airborne. They frequently choose a tree grove as a traditional nightly roost and this is often in the vicinity of their traditional nesting colonies also. Their faeces will eventually kill such trees which are favoured for communal roosting. An interesting behavioural trait is that of sunning or standing with wings outstretched in ·heraldic pose usually after feeding on a carcass. This is believed necessary to restore feather shape or camber after prolonged gliding and is never resorted to after a night's rest which allows feathers to regain their normal shape (C. Houston, *Ibis,* 1980).

BREEDING BIOLOGY

It is presumed that the pair bond is life-long as they show traditional attachment to a nest colony and nesting trees, though a new nest is usually built each year. Courtship consists mainly of mutual soaring, usually with the male bird slightly above and behind and following closely every twist and turn of the bird below. This is alternated with spectacular steep descents towards the nest site the buzzing sound of the wind in their primaries audible from distances of up to 0.8 km. Such swooping dives sometimes terminate in the pair landing on the nest tree or sometimes spiralling upwards to continue soaring together. Nest construction or repair starts in October or November with both sexes bringing branches to the stick nest. The nest itself is comparatively small in relation to the size of the bird. Branches with green leaves are particularly favoured and these are often broken off by the bird perching and tugging with its bill rather than by swooping down and striking with its feet as observed in most other raptors. The nest itself is comparatively small in relation to the size of the bird and both sexes

spend much time perched together on the nest platform before the single egg is laid. This is unmarked, plain chalky white, but a few eggs (1 in 10 out of K. Eates' collection) have faint lavender under-markings and some red-brown spots. The lining of the egg shell is yellow-brown not greenish as in almost all other birds' eggs. Both sexes share incubation duties but they are easily disturbed and will fly off the nest if approached. In some colonies 5 to 6 nests will be built in close proximity in the same tree. I have found small colonies comprising only 5 to 6 nests, but some of the traditional breeding sites such as on the Bhalwal/Shahpur road, contain over 80 pairs, and over 50 pairs in the Rasul barrage canal colony. Breeding can be spread over a considerable period, as nests with eggs and well developed downy young have been noted in mid-January and in late February, some with downy young, with other nests containing nearly full grown and fully fledged chicks. Incubation takes about 50 days and the young are not fledged until nearly 3 months after hatching. On 25 February 4 or 5 nests examined in the Changa Manga forest colony, all contained almost fully grown and fully fledged young. Young exercising their wings and nearly as big as their parents have been seen in colonies on 25 March.

The newly hatched young are fed bill to bill with food first regurgitated onto the nest floor. Birds in sub-adult plumage have been observed at nest colonies near Khinjar lake, bringing food in their crops, and it is presumed these were breeding birds, not helpers as A.E. Jones made similar observations (cited by Ali and Ripley, 'Handbook', Vol. 1). All food is brought to the nest in the crop and regurgitated.

*VOCALIZATIONS

At a carcass, excited birds emit prolonged hoarse hisses and cackling 'kek-kek-kek' calls, and at the nesting colony a rapid chattering 'kek-kek-kek' is also probably a threat call most frequently uttered upon the too close arrival of a neighbour. Pairs also indulge in short bouts of hoarse strangulated screaming when together on the nest. These calls rise to a squeal when the bird is very excited. The sound is a bit reminiscent of the preliminary braying of a donkey and such calling is associated with the early part of the breeding cycle.

Long-billed Vulture *Gyps indicus* (Scopoli)

Plate **11**
Map **62**

DESCRIPTION

Body length 80-95 cms.
Wingspan 205 cms.
Wing length 56-65 cms. (Stuart Baker)
Tail length 23.8-27.4 cms. (Stuart Baker)

It is about the same size as the White-backed Vulture (*Gyps bengalensis*) but much paler in colouration, the breast and belly being a rather isabelline pale brown with broad pale shaft stripes on the feathers. The neck is rather black, naked and scraggy in appearance and the bill is relatively narrower in depth than the length of the cere, hence its name. The tail and flight pinions are blackish brown and the rump pale buff. There is a ruff of downy white feathers at the base of the neck in adults. In general body colouration they resemble a small edition of *Gyps fulvus* but the blacker less down covered neck and relatively slender bill distinguish it from the Griffon Vulture. Sub-adult birds are impossible to separate in the field from sub-adult White-backed Vultures, both having more white down on the nape and neck and a ruff of pale buff lanceolate feathers at the base of the neck. The bill is greenish horny with the cere dirty greenish and the iris brown. The legs and feet are ashy or greenish grey. The population of Griffon Vultures breeding in the hills of the Punjab Salt Range was originally wrongly described as a sub-species of the Long-billed Vulture (*Gyps indicus jonesi*) (Whistler, 1927). Observations of Salt Range breeding vultures by this author in the 1960s, showed that they were indistinguishable from Griffon

Vultures (*Gyps fulvus*) in all respects, a view conveyed to Dr. Ripley, who subsequently placed *G. indicus jonesi* as a sub-species of *G. fulvus* (see footnote S.D. Ripley, p. 54, 1982 'Revised Synopsis',). At present there are two separable sub-species, *G. indicus indicus* breeding on cliffs or ruined buildings in the plains and the other *G. indicus tenuirostris*, a tree nester, averaging larger in size and breeding only in the north-eastern Himalayan foothill zone.

HABITAT, DISTRIBUTION AND STATUS

Virtually absent from Pakistan except for a few pairs nesting on cliffs on the rocky outcrops in Nagar Parkar area, in the extreme south-east corner of Sind. Paul Goriup found this vulture quite numerous around the cliffs of the Kharunjhar hills (December, 1980) and the author saw 4 specimens on a buffalo carcass near Koshki in Badin district in December 1978. Status RARE RESIDENT.

HABITS

Feeds on carrion and shares a similar biotope with *Gyps bengalensis* except for a predilection for roosting and nesting on cliffs or ruined forts rather than trees. They congregate at carcasses in considerable numbers like other *Gyps* species.

BREEDING BIOLOGY

They are colonial in nesting and K. Eates found small colonies in Tharparkar district at Kasbo, Kalinjhar and Vorao on rocky outcrops up to 325.28 metres (1,100 feet) elevation. Egg laying started from the second week of December up to the first week of January. One colony comprised only 3 nests, another 6 nests and the third 16 nests. The nests were comparatively flimsy structures comprising only a few sticks and refuse such as sheep's wool and bits of animal hide and rags placed on a rocky ledge. Only 1 egg is the normal clutch with a dull white ground colour and rather sparingly spotted with red-brown (Eates, MS notes unpublished). Both sexes share incubation duties and feeding the young on regurgitated food.

VOCALIZATIONS

Not accurately described but Brown and Amadon refer to hissing and cackling calls characteristic of the Genus.

Map 62: Long-billed Vulture *Gyps indicus*

Eurasian Griffon Vulture *Gyps fulvus* (Hablizl)

Plate 11
Map 63

TAXONOMY

The population breeding in the hills of the Punjab Salt Range was originally described by Whistler as a new sub-species of the Long-billed Vulture *Gyps indicus jonesi* (Whistler, 1927). The author pointed out (*in litt.*) to Professor Dillon Ripley that this population was in fact indistinguishable from the Griffon Vulture *Gyps fulvus*, a view accepted in the revised edition of the 'Handbook' by Ali and Ripley, Vol. 1, 1969 and S.D. Ripley, 'Revised Synopsis', 1982, on p. 54.

DESCRIPTION

Body length 95-105 cms.
Wingspan 24-29 cms. (both sexes)
Wing length 70-75 cms. (Salt Range specimen) (A.E. Jones)
Tail length 30-31 cms. (Salt Range specimens) (A.E. Jones)
Weight 7-8 kgs. (adults)

Much larger than the Oriental White-backed Vulture *Gyps bengalensis* and of a generally paler sandy colour on the wing coverts, mantle and breast. Individuals seen roosting in company with the Cinerous Vulture *Aegypius monachus* were only slightly smaller. The semi-naked head and neck is finely covered with white down. In breeding birds this neck down appears white and thicker but some winter birds observed around

Karachi had grey almost naked necks with sparse buff coloured down. The breast is streaked with lanceolate feathers having darker centres. Tail and remiges are blackish brown and upper-wing coverts creamy chestnut. The bill is horny brown and cere darker plumbeous. The legs are blackish grey and the neck ruff in adults is conspicuously white, and comprises pale buff lanceolate feathers in birds up to their fourth year. In flight silhouette from underneath the wing coverts are uniformly isabelline or buffy brown contrasting less sharply with the darker grey black primaries and secondaries. Adults after their fourth year show whitish bars in the region of the humerus and ulna. The streaking on the breast and the dark crop patch are often readily seen as the bird soars overhead. The tail is rounder tipped and slightly less wedge shaped than the tail of *Aegypius monachus*. Sub-adult birds are darker in body plumage and lack white bars on the under-wing coverts. They cannot be separated in the field from sub-adult Himalayan Griffon Vultures.

HABITAT, DISTRIBUTION AND STATUS

A resident species, performing local migrations to the plains in winter. This vulture is widespread in the dryer and lower hill tracts of Pakistan where it is resident and breeds. In winter time numbers migrate to the Indus plains. It is more likely to be encountered than *Gyps bengalensis*, in extensive desert tracts in winter, such as the Thal and Cholistan deserts and is relatively common in lower Sind in winter. In the northern Himalayas it is replaced by the Himalayan Griffon, but in summer in the Murree hills it can be encountered with the White-backed *G. bengalensis* and Himalayan vultures *G. himalayensis*. Breeding throughout Baluchistan, and parts of Sind Kohistan and the Punjab Salt Range, it is comparatively rare in the northern parts of Baluchistan during mid-winter. Status in Baluchistan, NWFP and the Salt Range, COMMON and in Sind FREQUENT.

HABITS

A gregarious species throughout the year, in its winter roosts and nesting colonies, though individuals forage singly. They travel vast distances daily in their search for carrion upon which they subsist. Like the White-backed Vulture, they will congregate with other vulture species, very quickly once a carcass has been located. At a large carcass, they will feed in a jostling mass alongside White-backed Vultures (*Gyps bengalensis*), but if the prey is a small animal they will generally

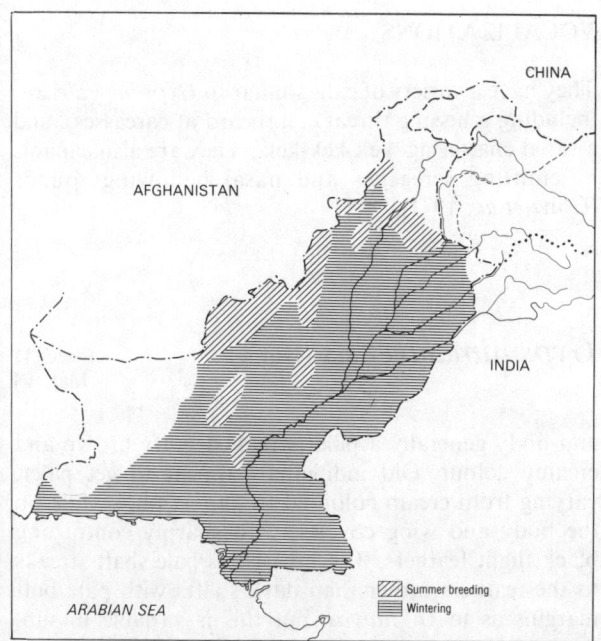

Map 63: Griffon Vulture *Gyps fulvus*

drive off the smaller *G. bengalensis*. The normal method of attack is with wings raised over the back, and head lowered with bill out-stretched. They then advance on their opponent with loud hissing. Another gesture frequently observed when approaching a smaller carcass is the raising of one foot with toes stiffly out-stretched, which seems to be an effective threat gesture, as other feeding vultures back away to allow such a bird to reach the carcass and commence feeding.

Like the other *Gyps* vultures, they spend a lot of the time roosting or resting, and observations show that in 24 hours, 16 hours may be spent roosting and 8 hours cruising around at an average air speed of 65-80 kms. (40-50 miles) per hour, looking for food. They thus commonly cover up to 480 kms. (300 miles) in a day. Obviously many individuals can go without food for many days, but they are capable of consuming up to 2 kgs. of animal flesh (a quarter of their body weight), when a suitable carcass is encountered (U.N. Glutz von Blotzheim, K.M. Bauer, and E. Bezzel, 1971).

They differ mainly from *Gyps bengalensis* in their presumed Palearctic origins and adaptation to a montane habitat. Though they migrate down to the warmer foothill and plains regions in winter, where they are sympatric with the White-backed Vulture, they breed exclusively on mountain cliffs, spending the spring and summer months in the arid steppic mountain regions of Baluchistan, the NWFP and the Attock hills and Salt Range of the Punjab.

BREEDING BIOLOGY

Nesting commences in early January in the warmer or lower altitude breeding colonies, and as late as March in the higher mountain ranges of Baluchistan. Stuart Baker ('Fauna Series') notes that colonies are smaller than for *G. bengalensis*, with from 12 to 20 nests being usual. A colony in the Khair-i-Murat range of Campbellpur district, Punjab comprised about 40 pairs (author), and one on the northern face of Takhatu

range in Baluchistan was estimated to have about 25 pairs (author). The pair bond is of long duration and may be life-long. Courtship starts about 5 or 6 weeks before egg laying, and involves the pair rising together on thermals and then indulging in spectacular swooping glides at a steep angle earthwards, either wingtip to wingtip, or one just above and behind its partner. A nest pad is built on their chosen cliff ledge, with branches, leafy twigs and bits of rubbish. H. Waite took eggs from the Punjab Salt Range from early January to early February (Waite, 1948), and Eates took the clutch of a single egg from five nests on 28 January from the Kalla Chitta hills further north. Nesting in the higher hills of Baluchistan starts later. Eates took 3 single eggs between 11 March and 23 March from Koh-i-Murdar range and the Urak valley, both bordering the main Quetta plain. Marshall found newly hatched young on 25 April (cited by Meinertzhagen, 1920). Nests are usually located in wind fretted hollows or inaccessible precipice faces and comprise a scanty lining of *Artemisia* and *Juniperus* branches in Baluchistan with more substantial stick structures in the Punjab Salt Range. Both sexes take part in nest building, incubation of the single egg and care of the young. Incubation takes about 52 days and the brooding bird sits very tight during sub-zero weather which can occur in Baluchistan right up to early March. The eggs are plain white, unmarked of a rough chalky texture. The downy white young are fledged in 90 days from hatching but remain for up to 120 days on the nest ledge. They solicit food from their parents by vigorous wing flapping as they get older (observed Khair-i-Murat).

VOCALIZATIONS

They have a variety of calls similar to *Gyps bengalensis*, including a hissing threat call (heard at carcasses), and a rapid chattering 'kek-kek-kek'. They are also capable of emitting screeches and nasal bellowing sounds (Glutz *et al.*, 1971).

Himalayan Griffon Vulture *Gyps himalayensis* (Hume)

Plate **11**
Map **64**

DESCRIPTION

Body length 115-125 cms.
Wingspan 261.6-306.3 cms. (Dement'ev *et al.*)
Wing length 75.5-80.5 cms. (Stuart Baker)
Tail length 36.5-40.2 cms. (Stuart Baker)
Weight 10-12 kgs. (Dement'ev *et al.*) 8-10 kgs. (Tibet) (Schafer)

Giant version of the Griffon (*Gyps fulvus*) with a heavier covering of white down on the head and neck,

and body generally a paler streaked buffy brown and creamy colour. Old individuals appear to get paler, varying from cream coloured to almost pure white on the body and wing coverts, with sharply contrasting black flight feathers. The breast has pale shaft streaks to the feathers rather than dark shafts with pale buff margins as in *G. Fulvus*, but this is variable in subadults and the two species are probably inseparable in the field unless size comparison can be made. In overhead flight the under-wing coverts are more uniformly pale creamy and lack the cinnamon tones with white centres, characteristic of the under-wing coverts of *G. fulvus*. The iris is dull yellow, bill horny yellowish to greenish, being paler than in *G. fulvus*, with the cere pale brown. The legs and feet are greyish white.

The downy chicks are clothed with a thick buffy yellow down (Bates and Lowther 1952). Almost pure white adult specimens (except for black tail and wings) have twice been observed on the Lowari pass in Dir and the Babusar pass in Chilas. The sexes are alike in size and colouration.

HABITAT, DISTRIBUTION AND STATUS

Not uncommon in the Himalayas where it is resident and relatively sedentary; in this respect it is quite distinct from *Gyps fulvus*. In winter it hunts along the main valleys, ascending in summer to the highest alpine slopes where nomadic sheep and goat flocks are more likely to provide food. They reguarly visit the Murree hill range and will visit carcasses in company with Griffon and White Backed Vultures. They can be seen all over Chitral, Gilgit, Hunza, the Kaghan and Neelum valleys and Baltistan.

They do not descend in winter to the plains, and are not reliably reported from Baluchistan. Status FREQUENT to COMMON throughout the northern mountain regions.

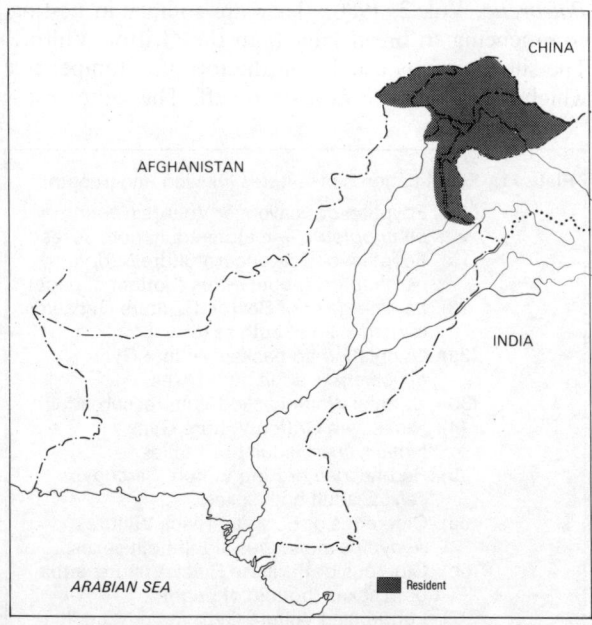

Map 64: Himalayan Griffon *Gyps himalayensis*

HABITS

It is no uncommon sight to see one of these vultures soaring over snow covered ridges and slopes during the winter months, and in summer sailing in arcs high above 6,096 metres.(20,000 feet) high mountain peaks in Baltistan. They are gregarious in their nightly roost and nesting places and probably more inclined to hunt for food in closer visual proximity with other Griffons than would be normal amongst plain's inhabiting vultures. They are exclusively flesh eaters, relying upon carrion and do not haunt mountain villages in the same way as the Bearded Vulture often does. They will assemble in quite large numbers at a carcass behaving very much as other *Gyps* species but they are shyer than the plain's species often hanging around at some considerable distance from a dead animal for upwards of 2 or 3 days before any individual has the courage to start feeding (observations on buffalo carcass, Dunga Gali and pack horse at Burawai). They are fond of sunning themselves on rock slabs, often lying flat with the wings held slightly apart from the body and because of the instinctive reaction of one vulture to join another bird when it is seen landing, 3 or 4 birds can be encountered together sunning themselves or roosting on a warm day.

BREEDING BIOLOGY

Because of their long fledging and incubation periods, nesting commences early, even in the far north, with nests being occupied from the second week of February when the surrounding slopes are still snow-covered. The eggs are usually white unspotted and the normal clutch is a singleton. Occasionally as in other *Gyps* species, an egg is laid with streaks and blotches of reddish brown. They nest colonially, selecting hollows or sheltered ledges and their nests are quite bulky stick structures placed on inaccessible cliffs. Bates and Lowther only knew of relatively small colonies, the largest comprising no more than six nests (op. cit., 1952). Traditional nesting sites are reoccupied year after year and the surrounding cliffs become white with their excrement. It is probable that adults are not usually mature until their sixth year and the whole breeding cycle extends up to 7 months of the year as captive birds were observed copulating in January.

VOCALIZATIONS

Not accurately recorded, but known to consist of a variety of hissing, grunts and cackles similar to *Gyps fulvus* (Brown and Amadon, 1968).

Eurasian Black Vulture or
Cinerous Vulture *Aegypius monachus* (Linnaeus)

Plate **11**
Map **65**

DESCRIPTION

Body length 100-110 cms.
Wingspan 250-295 cms. Sexes alike
Wing length 73.5-80.4 cms. (males), 75-84.5 cms. (females)
Tail length 33-41 cms. (both sexes) (Brown, Urban and Newman)
Weight 7,000-11,500 gms. (males), 7,500-12,500 gms. (females)
   (Europe) (Brown, Urban and Newman)

A uniform dark chocolate brown all over with relatively short wedge shaped tail and huge broad wings of high aspect ratio (surface area in relation to length). Their bills are deeply compressed and blackish horny with pale violet cere and pale pink or bluish pink naked face and nape. The crown is bare with a rather high pointed occiput giving a distinctive profile when compared with the rounded crown of *Gyps* vultures. There are spiky lanceolate feathers ascending the rear part of the neck to the occiput and the chin and throat, also an area around the eyes and lores is covered with dark chocolate down giving the face a characteristic masked pattern. The feet and legs are dull white to yellowish coloured. In flight the uniformly dark brown tail and flight feathers and wing coverts, when viewed from above or below, coupled with whitish feet make this vulture unmistakeable. It is slightly larger than *Gyps fulvus* with which it is sympatric in Pakistan. Sub-adult birds are a slightly paler brown on the wing shoulders and neck ruff. Nestlings are covered with brownish grey down.

HABITAT, DISTRIBUTION AND STATUS

It has become fashionable to exhibit this, the largest and handsomest of Eurasian vultures, in European and American zoos and this has led to a constant demand from animal exporters in Pakistan which appears to have definitely led to its decline. It breeds very sparsely throughout Baluchistan and the tribal areas of the North West Frontier Province. There is some migration of birds to lower Sind in winter, as well as to the Thar and Cholistan deserts and the Potohar plateau around Rawalpindi. In the late 1950s upwards of 30 Cinerous Vultures could be counted on Karachi municipal refuse dumps in Ghizri creek area and it was from here that animal trappers started netting birds. In the 1980s, at most 10 or 12 Cinerous Vultures come in winter to the outskirts of Karachi, the escarpments of Khadeji falls being their favourite winter roost, also the North West end of Hab dam reservoir. 15 to 20 birds could still be seen in summer around Quetta rubbish dumps in the early 1970s and single birds have been recently recorded every second or third year in the vicinity of Rawalpindi refuse tips. They perch on trees or on the bare ground and can take off from flat ground without needing to run more than 2 or 3 steps. They leave their wintering grounds by mid-February and return to their nesting haunts on the higher Juniper studded crags on Pakistan's western borders. A resident species performing local migrations down to the plains in winter. Status SCARCE.

HABITS

A carrion feeder, like *Gyps fulvus,* it will take precedence over the latter species at a small carcass such as a sheep (observed Cholistan desert), but will feed alongside Griffon Vultures at a larger carcass such as a dead camel (observed near Kachlak, Baluchistan). These birds are gregarious in their winter roosts, favouring traditional sites where thermals assist them in ascending to optimum hunting levels. Typically they roost on the ground choosing a flat open escarpment if possible, close to a steep hill or cliff where air currents help their take-off. Though they nest in trees they appear to avoid perching to rest or roost in trees, in contrast to *Gyps* species which will readily perch in trees.

BREEDING BIOLOGY

Pair bonds are long lasting and probably life-long (Cramp *et al., Handbook of Birds of the Western Palearctic,* Vol. 2, 1980). They are solitary in nesting, commencing to breed later than the Griffon Vulture. The site chosen is usually on the top of a Juniper tree which is growing out of a steep cliff. The same nest is

---

Plate 11: Flight patterns of vultures (viewed underneath)

  (1b) Egyptian or Scavenger Vulture *Neophron percnopterus* — mature adult, both sexes.
  (1b) Egyptian or Scavenger Vulture *Neophron percnopterus,* both sexes (bottom of page)
  (2) Lammergeier or Bearded Vulture *Gypaetus barbatus,* adult both sexes.
  (3a) Oriental White-backed Vulture *Gyps bengalensis,* adult both sexes.
  (3b) Oriental White-backed Vulture, sub-adult.
  (4) Himalayan Griffon Vulture *Gyps himalayensis,* adult both sexes.
  (5) Red-headed or King Vulture *Sarcogyps calvus,* adult both sexes.
  (6a) Cinereous or Eurasian Black Vulture *Aegypius monachus,* adult both sexes.
  (6b) Cinerous or Eurasian Black Vulture, adult both sexes (bottom of page)
  (7) Long-billed Vulture *Gyps indicus,* adult both sexes.
  (8a) Eurasian Griffon Vulture *Gyps fulvus,* adult both sexes.
  (8b) Eurasian Griffon, adult both sexes.

Plate 11

(1a)

(2)

(3a)

(3b)

(4)

(5)

(6a)

(7)

(8a)

(1b)

(8b)

(6b)

Plate 12

(1)

(2)

(3)

(4)

(5)

(6)

(7)

(8a)

(8b)

(9)

(11a)

(11b)

(10)

(12)

repaired and reoccupied each year if unmolested and becomes an enormous structure. A single egg is laid which has a white ground colour, covered over most of its area with buffy patches and overlaid with dark brown and maroon stains. Some eggs show more white ground colour. C.E. Williams describes one nest more than 1.8 metres (6 feet) deep and 1.5 metres (5 feet) in diameter. The platform was lined with some grass and the nest which was decorated with animal skins and bits of rag and dung (Williams and Williams, *JBNHS,* 1929), was described as very foul smelling. Delme-Radcliffe took a nest near Quetta at 3,200 metres (10,000 feet) on 5 April. T. E. Marshall who took several nests around Quetta, describes all as being large flat structures often 2 metres wide, and built on the crown of Juniper trees growing out of cliffs. One had a week-old nestling covered with sooty grey down on 16 April and on the same date another nest with the egg about to hatch. He could actually hear the chick chirping inside the egg. A third nest found early in May had a 10-day old chick in it (T. E. Marshall, *JBNHS,* 1911).

K. Eates (regrettaby!) took 52 eggs from various eyries in Baluchistan. All were at elevations of 3,048 metres (10,000 feet) or higher and collected between 10 March and 27 March. 38 were on cliffs and 14 on the crown of *Juniperus macropoda* trees. 34 of these eggs came from Koh-i-Murdar range which is relatively bare of Junipers which might account for the high proportion of nests on cliff ledges. Most of these were collected during the 1930s and early 1940s and about one-third were white all over and unspotted (K. Eates, MS notes unpublished). Incubation period is estimated at 52 to 55 days with the young fledging at 60 days but

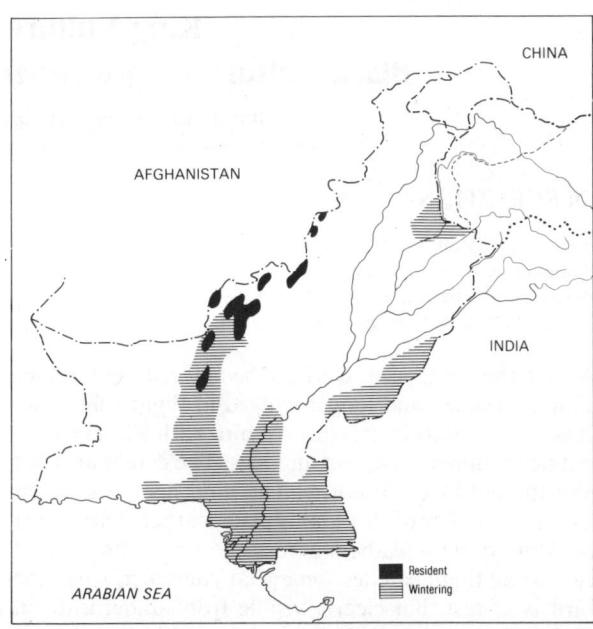

Map 65: Cinerous or Eurasian Black Vulture *Aegypius monachus*

remaining on or near the nest until able to fly at about 100 days of age (Brown and Amadon, 1968). These intervals may be underestimated, (see below).

Barnes describes a fully fledged young bird taken from a nest on the Khojak range near Chaman, at the end of May (*Stray Feathers*, Vol. 9, 1880). Another chick taken in late May from an eyrie in a tall pine tree at the foot of the Wana plain in South Waziristan, was described, when a few days old as covered with brownish grey down, having a light pink cere, yellow irides and creamy white feet and legs. Though adapting well to captivity, it was unable to stand until one month old, and did not take its first short flights until 5 months old, but flapped its wings and jumped in the air, as though exercising, for about 2 weeks before making its first successful flight (Major Hughes, *JBNHS* 1917). Both parents share incubation duties and tending the young. Incubation takes up to 55 days (Dement'ev *et al.*, 1951) and the young take up to 120 days to be able to fly and leave the nest (Cramp *et al.*, 1980). Both parents feed their young mainly by regurgitation onto the nest floor. As soon as they can shuffle, the young go to the edge of the nest to void their excreta (Major Hughes, op. cit., 1917).

VOCALIZATIONS

Stuart Baker described birds at the nest emitting quite loud 'squalling and roaring noises'. Brown and Amadon (1968), state that they occasionally utter croaks and hisses at a carcass, but there are no recent descriptions of their calls and they appear to be a relatively silent species.

Plate 12: Flight patterns of smaller eagles, buzzards and hawks (viewed underneath)

(1) Short-toed or Snake Eagle *Circeatus gallicus*, typical adult.
(2) Crested Honey Buzzard *Pernis ptilorhyncus*, typical adult.
(3) Bonelli's Eagle *Hieraaetus fasciatus*, pale morph adult.
(4) Long-legged Buzzard *Buteo rufinus*, adult.
(5) Mountain Hawk-eagle *Spizaetus nipalensis*, adult.
(6) Desert Buzzard *Buteo buteo vulpinus*
(7) Osprey or Fish Hawk *Pandion haliaetus*
(8a) Booted Eagle *Hieraaetus pennatus*, adult pale morph (viewed underneath)
(8b) Booted Eagle, adult darker morph (viewed from above)
(9) Brahminy Kite *Haliastur indus*, adult (from below)
(10) White-eyed Buzzard or Buzzard Eagle *Butastur teesa*, adult (from below)
(11a) Black-shouldered Kite or Black-winged Kite *Elanus caeruleus*, adult (from above).
(11b) Black-shouldered Kite, adult (from below).
(12) Shikra or Indian Sparrow Hawk *Accipiter badius*, adult male.

King Vulture, Red-headed or
Black Vulture *Torgos calvus* (Scopoli) (Ali and Ripley's 'Handbook')
Synonym *Sarcogyps calvus* (S.D. Ripley's 'Synopsis')

Plate **11**

DESCRIPTION

Body length 84 cms.
Wingspan 200 cms. (both sexes)
Wing length 60-62.5 cms. (Stuart Baker)
Tail length 22.6-25.7 cms. (Stuart Baker)

About the same size as *Gyps bengalensis* but rather slimmer bodied and longer winged in flight silhouette. It is a much darker grey-brown bird with a livid naked red head, upper neck, feet and legs. The cere is also red and the bill blackish horny. The hind neck is covered by a spiky collar of short lanceolate feathers with a ruff of white downy feathers across the upper breast and two white thigh patches somewhat concealed when the bird is at rest, but clearly visible from underneath in flight silhouette, and very conspicuous when the bird is taking off. A thin white line along the underside of the wings at the base of the remiges is also a good field mark. There are lappets or loose flaps of red skin on either side of the neck in adults. The iris is red-brown and sub-adults have the head and neck a duller fleshy brown with some white down on the crown and the body plumage is a lighter brown colour with the flight and tail feathers darker blackish brown. The front of the neck is fringed by white down as in adults. Downy chicks develop neck lappets but the naked part of the face and neck is a 'pinkish leaden colour' (Whistler, MS notes).

HABITAT, DISTRIBUTION AND STATUS

This is a largely sedentary and resident species, more adapted to well wooded or forested foothill areas as well as the plains in regions of higher rainfall, so that it has never been numerous in the dryer north-west. Recorded as sparsely distributed in the lower Punjab in the 1920s (Whistler *Ibis*, 1922), it was commoner about the Rawalpindi plateau (Whistler, *Ibis*, 1930) and by no means rare around Lahore (Currie, *JBNHS*, 1916a). Ticehurst recorded it as fairly common in the better watered parts of Sind (*Ibis*, 1923) and Waite recorded it as a scarce resident in the Salt Range (*JBNHS*, 1948). Nowadays it appears to be a rare straggler with occasional individuals being seen (Roger Holmes and author) around Lahore, Kasur and in Tharparkar district, but there is little evidence that any resident population remains along the main Indus river. The author has about 3 records of sightings over the past 30 years including one or two birds at Karachi municipal refuse dumps, and one near Mianwali in the Salt Range and one near Khanewal. It has not recently been encountered around Rawalpindi or the Salt

Range. Paul Goriup (pers. comm.) observed a pair at Nagar Parkar in 1980. Status RARE and evidently much declined since the beginning of this century.

HABITS

A solitary hunter adapted to exploit carcasses with other large vulture species as well as smaller carrion such as rats. It is never gregarious at roosts or feeding assemblies around a large carcass and has the reputation of being rather timid.

BREEDING BIOLOGY

A long lasting or permanent pair bond is formed and the birds perform very spectacular courtship displays, flying together in graceful circles, their entire bodies almost touching and also with the female underneath turning on her back and grabbing the feet of her suitor, the inter-locked pair performing cartwheels through the air before breaking apart again.

They breed solitarily, nesting in tall trees and the same nest may be used year after year. They also commence egg laying much later than the *Gyps* genus vultures; March being the usual season. A. J. Currie found three nests in March and April in the vicinity of Lahore (op. cit. 1916) and K. Eates found a nest with well fledged young on May 12 from the Hab valley, as well as a nest being built (with fresh green leaves added) at the end of February near Sujawal (Thatta district) and another one with fresh eggs in mid-March near Sukkur (MS notes) A single egg is laid, plain white unmarked and both sexes take part in nest repairing, incubation and tending the young. Incubation is said to last 45 days. Waite found nests quite low down in stunted *Acacia modesta* trees, but on steep hill-sides in the Salt Range with eggs in mid-March (op cit. 1948).

VOCALIZATIONS

Said to emit hisses and grunting calls at a carcass, like other vultures, and a raucous 'roaring' during copulation (Salim Ali, 'Handbook', 1969).

MISCELLANEOUS

A fossilized *Torgos* vulture dating from the Miocene sediments named *Torgos tracheliotus* (like the African Lappet Faced Vulture) has been found in the Punjab Salt Range.

Short-toed Eagle *Circaetus gallicus* Gemlin

Plate **12**
Illus. **26**
Map **66**

DESCRIPTION

Body length 62-67 cms.

Wingspan 185-195 cms. Females are slightly larger than males.

Wing length 52-53.6 cms. (males),
    53-57.1 cms. (females) (Stuart Baker)

Tail length 28.7-33 cms. (both sexes) (Stuart Baker)

Weight 1,180-2,000 gms. (males),
    1,304-2,324 gms. (females) (Cramp *et al.*)

A medium sized eagle, with a comparatively long tail and thick neck with large rounded almost owl-like head. It is considerably larger than a Long-legged Buzzard (*Buteo rufinus*). Of a predominantly grey-brown colour dorsally and white below, the only other species with which it can be confused in Pakistan is the Honey Buzzard (*Pernis ptilorhynchus*) or Bonelli's Eagle (*Hiraaetus fasciatus*). It is considerably larger than the Honey Buzzard and the large rounded head of the Short-toed Eagle is diagnostic from a distance. The eyes are large with pale yellow iris and the tarsi, long unfeathered and olive green to greyish green in colour. The toes are comparatively short, an adaptation to grasping serpents. The throat and upper breast are grey-brown, lower breast and belly pure white with some broad cross-barring on the flanks. The strongly hooked bill looks relatively weak compared with the head size and the cere is bluish white. The mantle, wings and tail vary from brown to greyish with darker grey cross-bars on the tail and darker black-brown secondaries. The primaries are tipped dark grey and the tail has three darker cross-bars being relatively long and rounded at the tip. Individuals vary with the head and throat being quite dark grey-brown and the under-wing coverts being more heavily spotted in dark phase birds. Bonnelli's Eagle has a much smaller head and relatively slender neck and is usually slightly bigger and darker on the upper surface with under-wing coverts wholly dark brown and unspotted and only one very broad terminal tail bar.

HABITAT, DISTRIBUTION AND STATUS

A sparsely distributed sedentary resident. Very widely but locally distributed from the lower hill ranges of Baluchistan and Sind Kohistan, through the eastern desert border regions of Cholistan and Tharparkar. It is thinly distributed in the Punjab Salt Range and the Margalla hills but absent from the irrigated Indus plains or the northern Himalayan regions. The only area where it is comparatively common is in southern Sind in the rice growing tracts. A day's drive in Thatta district south of Sujawal will usually result in the sighting of 4 or 5 individuals. There is an odd pair resident along all the higher hills of the Punjab Salt Range in *Olea cuspidata* forest and birds have been seen around Mashelakh reserve north-west of Quetta as well as around Ziarat valley at 2,484.4 metres (8,000 feet) elevation in Juniper forest. Specimens have been observed around Kohat and Tarbela dam also. As many as 6 individuals have been seen simultaneously from the road between the Makli hills and Hadiero lake just south of Thatta. It's status is FREQUENT.

HABITS

A competent glider and soarer, doing most of its foraging on the wing. In the summer it subsists largely upon snakes and K. Eates records seeing one carrying a 4 foot long (1.2 metres) Rat Snake (*Ptyas mucosus*) in its feet. In Ziarat an eagle perched on a large boulder was seen to seize a basking Rock Lizard (*Agama nupta*) by merely shuffling sideways and striking with its foot (author obs.) It frequently interrupts its gliding to hover with wings held high over its back and uses this technique in searching for prey, plunging suddenly to the ground. Another individual was seen flying low

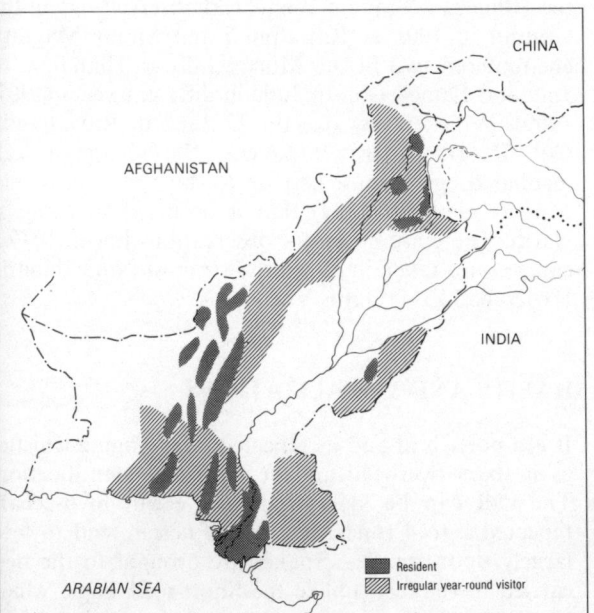

CHINA

AFGHANISTAN

INDIA

ARABIAN SEA

▨ Resident
▧ Irregular year-round visitor

Map 66: Short-toed or Snake Eagle *Circaetus gallicus*

Illus. 26: Short-toed or Snake Eagle *Circaetus gallicus*, typical plumage of adult, both sexes.

overhead with the distinctive tail of a partially swallow-ed *Calotes versicolor* lizard still protruding from its mouth, probably carrying this to feed its incubating mate. It will consume insects, fledgling birds and amphibia when occasion provides but the preferred food is snakes. It is gregarious but pairs often remain together outside of the breeding season.

BREEDING BIOLOGY

Their nests are hard to find being comparatively small structures, generally concealed inside a leafy tree and not too near the top of the tree crown. Both sexes share in nest building and an old nest may be re-used and repaired or a new one is built. Branches with green leaves as well as grass are favoured for lining the nest cup which is quite deep so that the incubating birds are generally invisible from the ground. They generally lay eggs from mid-January to February and K. Eates found a fresh egg on 14 February in an *Acacia arabica* tree in Duber forest reserve near Rohri in northern Sind. Only the female incubates, and she is a tight sitter, being brought food by the male whilst brooding. The normal clutch is a singleton and it is plain white unmarked. The incubation period is about 47 days with the chick fledging after 80 or 90 days from

hatching. A parent bird on arriving at the nest was see to pull out a snake from its crop using its foot. The normally stand on the protruding reptile's tail and pu it out of their gullet by jerking the head and body u and back (G. Mountfort, obs. in Spain and *in litt.* t author). The downy chicks seem to know instinctivel to swallow the snake head first. Partially consume young *Varanus griseus* lizards have also been seen o the nest.

It is not known if the pair bond is life-long, as ofte one individual initially starts defending a territory an a fresh nest is built by most pairs. Courtship include mutual soaring and one individual of either sex ma occasionally perform a sort of undulating or swoopin 'sky-dance' display comparable to that of the Hone Buzzard but lacking the 'butterfly-like' hover at th apex of the flight path.

*VOCALIZATIONS

Usually silent outside of the breeding season but young male bird called constantly on 16 November. I the nesting season both sexes call repeatedly from perch or in flight, a clear sharp whistling call 'piee-on Another variant 'mioouk-mioouk' was given by female perched on a power pylon.

Crested Serpent Eagle *Spilornis cheela* (Latham)

DESCRIPTION

Rather variable in size
Body length 50-74 cms.
Wingspan 110-180 cms. Females average slightly larger.
Wing length 46.8-50.7 cms. (Stuart Baker)
Tail length 29.5 31.5 cms. (Stuart Baker)

A large generally dark chocolate brown eagle with broad rounded wings and a very characteristic pattern when viewed from underneath with a broad pale creamy bar through the middle of the tail and a conspicuous white bar along the base of the secondaries and through the middle of the pinions. Generally some white spotting is visible on the under-wing coverts which are a paler chestnut brown. Immature birds are paler all over and the white spots on the belly are only fully developed in adults. Tarsi are long yellow and bare of feathers. The head is comparatively rounded with a conspicuous nuchal crest or hood of short pointed feathers with white spots at their base which can be erected in a fan shaped hood when the bird is excited. The naked lores and cere are bright yellow as also the iris.

HABITAT, DISTRIBUTION AND STATUS

Essentially a forest bird and adapted to higher rainfall

areas, it is but a straggler within Pakistan territory. A few authentic records have been made of individual wandering along the foothill or Siwalik range belov the Himalaya. Captain Whitehead observed one in th autumn of 1905 at Rawalpindi and Major Magrat encountered two in the Murree hills at Thandiani i July and Dunga Gali in June in different years (1907 1909). Whistler saw one on 25 June in Rawalpindi Col. R. H. Rattray (*JBNHS,* 1904) took a wel incubated egg from a nest on 6 May. In Sind Jame Murray claimed (1884) that it occurred in the eas Narra. There are no recent observations but in 1975 a captive bird taken in Sanghar district was on exhibit i Hyderabad Zoo. Status VAGRANT.

HABITS AND VOCALIZATIONS

It is a noisy bird and its repeated clear ringing whistle as it soars overhead are the best field identification The call can be syllabized 'hwu-eeeah hwu-eeeah repeated 2 to 4 times. It is a tree nester, said to fee largely upon reptiles. Snakes are brought to the nes carried in the feet, unlike the Short-toed Eagle whic characteristically swallows its prey before going to th nest.

Marsh Harrier *Circus aeruginosus* (Linnaeus) Map **67**

DESCRIPTION

Body length 48-56 cms.
Wingspan 115-130 cms.
Wing length 38.5-40.5 (males),
    39-43 cms. (females) (Stuart Baker)
Tail length 23.4-24.5 cms. (males),
    23.8-25.8 cms. (females) (Stuart Baker)
Weight 405-667 gms. (males),
    620-740 gms. (females) (Dement'ev *et al.*)

The largest of the harriers inhabiting Pakistan with comparatively broader tail and wings, though *Circus cyaneus* the Hen Harrier is only fractionally smaller. Both sexes are predominantly dark chocolate brown but adult males have a contrasting pale grey tail and flight feathers except for dark tipped primaries. Females have noticeable pale creamy to white crown, throat patch and leading edge to the wings. Sub-adults however lack these creamy markings and could be confused with *Milvus migrans* except for the more buoyant flight with pronounced dyhedral of wing and absence of forked tip to the tail. Adult males have paler chestnut under-wing coverts and the breast is streaked creamy and dark brown. The bill of all harriers is strongly hooked but small and weak. The iris is hazel to dark brown and the tarsi comparatively long, naked and bright yellow in colour. The cere is yellow.

HABITAT, DISTRIBUTION AND STATUS

Confined to reed beds, lake and swamp environs, they are the most common and widespread of the harriers, visiting Pakistan as winter migrants. They avoid estuaries or coastal mangroves though an occasional individual can be sighted in the Indus delta. They do not regularly frequent the main rivers, but in early autumn, birds spread out over the rice growing tracts and may then be encountered far from larger swamps or lakes. They are migratory with a passage northwards to Russian breeding grounds being noted in Gilgit from mid-March to the end of April. Return passage has been noted in late August (Biddulph, 1881). First arrivals in the plains can be noted as early as mid-September and a few stragglers stay around the larger swamps of Sind up till mid-April. In Baluchistan odd birds stay throughout the summer on Zangi Nawar and Kushdil Khan lakes with some circumstantial evidence of occasional breeding. In a January visit to Zangi Nawar the author counted over 50 Marsh Harriers, but only 2 remained by early May and local game wardens stated that they did not breed there (Roberts, *JBNHS*, Vol. 82, 1985). In Sanghar district in November up to 20 Marsh Harriers were counted on Sadori lake, an area of about 800 hectares. Status COMMON.

HABITS

It is believed that in their southern breeding range many adult males remain near their breeding grounds, as this phenomenon would help to explain the curious fact that only about 5 per cent of winter sightings in Pakistan are of adult males, the majority being adult females or sub-adult birds which again appear to be mostly females.

Marsh Harriers in winter spend a large part of the day perched on reed beds or low submerged bushes around their feeding grounds but they will occasionally indulge in high soaring on thermals particularly at the onset of warmer spring weather. While hunting they typically fly within 5 metres of the ground or reed tops, sometimes gliding very swiftly and even lower over more open stretches, ready to pounce on an unwary bird. Their food includes waterfowl, small passerines, amphibia and the occasional small reptile. They are quick to exploit wounded duck as any human hunter will testify and I have observed many birds trying to pirate waterfowl from eagles and consuming fish which they have robbed from Black Kites. When hunting over lakes it is noticeable how Coots (*Fulica atra*) and all species of waterfowl react to their approach and take evasive action. One was watched swooping over Purple Gallinules (*Porphyrio porphyrio*), but this appeared to be more in play as no physical attacks ensued. However on one occasion a female was seen to plunge onto an unsuspecting

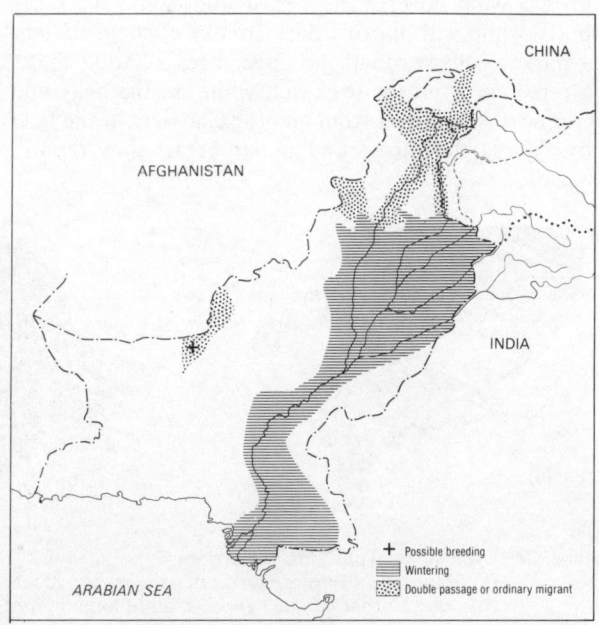

Map 67: Marsh Harrier *Circus aeruginosus*

swimming coot (*Fulica atra*) which it held under water with its own body partly submerged for over one minute. The attack was not successful as the bird eventually relinquished its prey and the Coot swam and flew away quite actively. When hunting over rice fields frogs provide a major part of their diet.

BREEDING BIOLOGY

Nests are built on the ground in dense reed beds and the pair is quite territorial. Brigadier Christison thought that several pairs bred on Zangi Nawar and Malezai in Baluchistan (*Ibis,* 1940) and he was shown old nests which might have belonged to this species but the great majority breed extra-limitally. There is some evidence of polygyny with several females nesting within the territory of one male. The normal clutch is 4 to 6 unmarked pale bluish white eggs and the female does all the incubation with the male bringing her food at the nest. The nidicolous young are fed by both parents.

*VOCALIZATIONS

Occasionally in winter a female will emit repeated long drawn-out weak whistling calls, most usually when sitting in a reed bed. These may be food soliciting calls. On 19 February a male was watched displaying at Hadiero lake, in the vicinity of a foraging female. He rose high into the air and called in a short high pitched 'kik-kih-kih'. He then swooped upon the female who turned over and presented her claws to him.

Hen Harrier *Circus cyaneus* (Linnaeus)

Illus. 2
Map 6

DESCRIPTION

Body length 44-52 cms.
Wingspan 100-120 cms.
Wing length 34.1-35.7 cms. (males),
    37.5-39.2 cms. (females) (Stuart Baker)
Tail length 21-22.1 cms. (males),
    24.6-25.5 cms. (females) (Stuart Baker)
Weight 300-400 gms. (4 males),
    480-600 gms. (3 females) (Dement'ev *et al.*)

Adult females are difficult to separate in the field from other *Circus* species having a conspicuous white rump and dark brown body with three cross-bars of dark brown on the pale brown tail, and paler grey-brown wing coverts.The breast and facial discs are buffy white with narrow dark streaks condensing into a darker collar around the upper breast. Adult males are pale grey on the back and white on the belly and can be distinguished from all other harriers in the field by having the throat and upper breast grey (contra pure white in *Circus macrourus*), coupled with the three primaries being black up to their bases, giving a more extensive looking black wingtip. Also the trailing edge of the wing forms a narrow darker grey line particularly noticeable from the underside, a conspicuous field point not always well shown in field guides. Their comparatively larger size and habitat preference are also useful pointers. Sub-adult birds have a much more orange-tawny breast which is unstreaked.

HABITAT, DISTRIBUTION AND STATUS

A winter migrant visitor to Pakistan, breeding extra-limitally. It is most likely to be encountered in the Himalayan outer foothill areas, such as Margalla hills, the broad *Saccharum* studded stretches along the Jhelum and Chenab rivers below Mangla dam and Marala barrage. Occasional birds occur throughout

(a)   (b)   (c)

Illus. 27: Typical facial patterns of Harriers.
    (a) Montagu's Harrier *Circus pygargus*, above left-adult female, below right-1st. year male.
    (b) Hen Harrier *Circus cyaneus*, adult female - note more streaks and spotted crown and cheeks without dark·mask lines.
    (c) Pallid Harrier *Circus macrourus*, left-adult female, right-1st. winter male. Note conspicuous darker double·mask circlet and absence of streaking on breast of male.

CHINA

AFGHANISTAN

INDIA

ARABIAN SEA

▨ Wintering

Map 68: Hen Harrier *Circus cyaneus*

Punjab and Sind, frequenting uncultivated or semi-desert tracts. A few over-winter in the main valleys of Gilgit and Chitral and the Potohar plateau around Rawalpindi. It is the most common harrier in Baluchistan particularly on spring and autumn passage. On 31 March six were counted driving between Quetta and the entrance to the Bolan pass. Whitehead noted

autumn passage in September and October and return passage from late March to early May in Kohat district. Ticehurst did not record this harrier in Sind though the author has one record of an adult male hawking over open salt marsh flats at Keti Bunder. Status SCARCE.

HABITS

They are adapted to subsist on a variety of diurnal rodents, lizards, ground feeding passerines and ortho-pterous insects. Characteristically they hunt low over the ground surface, alternating glides with slow measured wing beats. A specimen which haunted the Margalla hills from December to March quartered over more open bush studded ridges as well as abandoned cultivation at the bottom of side valleys. They roost on the ground at night, often gregariously. At Quetta airfield one was observed carrying *Meriones libycus* Sand Jird in its feet and this colonial and diurnal rodent is probably an important prey item in Baluchistan.

BREEDING BIOLOGY

Nesting is extra-limital, breeding occurring south of the tundra zone in the USSR, preferably on heath land or rank grass. The male arrives first on the nesting territory and immediately commences spectacular display flight dancing. These birds nest on the ground like other harriers.

Pallid Harrier *Circus macrourus* (Gmelin)

Illus. **27**
Map **69**

DESCRIPTION

Body length 40-48 cms. including tail 16-18 cms.
Wingspan 95-120 cms.
Wing length 33.2-36 cms. (males),
    36.3-38.6 cms. (females) (Stuart Baker)
Tail length 20.1-22.1 cms. (males),
    22.9-24.7 cms. (females) (Stuart Baker)
Weight 311-374 gms. (4 males),
    402-550 gms. (17 females) (Dement'ev *et al.*)

Smaller than the Marsh Harrier *C. aeruginosus* or Hen Harrier, *C. cyanus*, adult males can be distinguished from other harriers by their wholly white undersides, lacking the grey throat of *C. cyaneus* and grey upper breast of Montagu's Harrier, *C. pygargus*. Also absent are the white upper tail coverts (conspicuous in *Circus cyaneus*), coupled with absence of dark wingbar characteristic of *Circus pygargus*. The dorsal surface is a very pale grey and the black wingtips are restricted to the distal half of the first 3

primaries being more limited in area than the wingtips of either male Montagu's or Hen Harriers. Juveniles have a rich tawny throat and belly unstreaked and their pale creamy faces are framed by two dark brown crescents in a mask-like pattern, the second or outer crescent lying just in front of the wing shoulder and being totally absent from sub-adult *C. pygargus* (see **ill. 27**). Generally sub-adult Hen Harriers (*C. cyaneus*) have the least contrasting face mask of all three species. Adult females are probably impossible to separate reliably in the field from *C. pygargus* and *C. cyaneus*.

HABITAT, DISTRIBUTION AND STATUS

A winter migrant visitor to Pakistan breeding from the Caspian to Tian Shan range in central Russia. First arrivals in lower Sind have been seen as early as the last

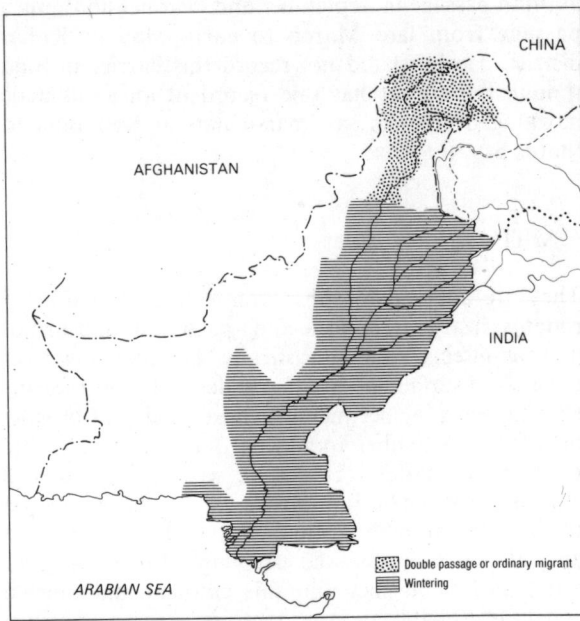

Map 69: Pallid Harrier *Circus macrourus*

border regions of Thar and Cholistan deserts. It will exploit more arid areas than those favoured by the other harriers. Status COMMON to FREQUENT.

HABITS

A typical harrier in its buoyant flight and habit of quartering up-wind within 5 metres of the ground or crop surface, as it searches for food prey. It avoids well wooded areas and usually hunts solitarily. Its food is opportunistic and *Lacerta* lizards are important in sand dune areas, whilst larks *Galerida* and *Calandrella* spp. are important in mountain steppe areas. Where rodents are plentiful these provide the principal food source. In the USSR and Africa it has been reported as feeding upon grasshoppers and locusts when these are abundantly available. It roosts at night on the ground and sometimes several individuals will congregate together at such a roost.

BREEDING BIOLOGY

Breeds extra-limitally. Nesting habits similar to other harriers with the male arriving first on the breeding territory where he performs spectacular flight displays. Only the female incubates and the 4 or 5 eggs are laid on a platform nest made in a patch of weeds or rank grass, over a 10 day interval, the chicks hatching asynchronously.

VOCALIZATIONS

Very silent on its wintering grounds. Adults utter a high pitched chattering call 'kik-kik-kik' in the nest area.

week of August and there is a marked return passage early in April. Biddulph noted a passage through Gilgit from the end of August and returning again at the beginning of April. It is absent from Baluchistan in the winter but has been noted there on passage between 15 March and 15 April. It is the most widespread and common harrier in Sind and Punjab after the Marsh Harrier and is most likely to be encountered in the so-called 'barani' or non-irrigated wheat growing tracts such as the plains of Bannu and Dera Ismail Khan, the Potohar plateau of the Salt Range, and uncultivated tracts in between, extending down to sandy islands in the Indus delta and the

---

Montagu's Harrier *Circus pygargus* (Linnaeus)

Illus. **27**
Map **70**

DESCRIPTION

Body length 43-47 cms. with tail 16-18 cms.
Wingspan 105-120 cms.
Wing length 32-38 cms. (males),
  32.5-39 cms. (females) (Dement'ev *et al.*)
Tail length 21.3-24.1 cms. (both sexes) (Stuart Baker)
Weight 258-288 gms. (5 males),
  340-380 gms. (4 females) (Dement'ev *et al.*)

The smallest of the harriers, they look particularly graceful on the wing, as they glide twisting and turning for balance, low over the ground. Males are pale grey all over with black primaries and a conspicuous

narrow black wingbar formed by the tips of the secondary wing coverts There is usually some streaking in the region of the tibia and lower flanks and the whole of the upper breast is grey with no abrupt change to pure white on the lower breast as in *C. cyaneus*. Also the rump is grey not white and cross-barring on the outer tail feathers usually rather conspicuous. Females are similar in plumage to *C. macrourus* with a conspicuous white rump and long narrow cross-barred tail. Sub-adults are a darker tawny colour than in the other harriers with no paler white ring around the ear coverts as in *C. macrourus*. In the field sub-adults can be separated from *C.*

macrourus by the absence of a second dark ring circling the facial discs but adult females are doubtfully separable.

HABITAT, DISTRIBUTION AND STATUS

A winter visitor breeding extra-limitally in Turkestan, the Altai and west central USSR. It is probably the least commonly encountered of the harriers wintering in Pakistan but occurs in all provinces in a variety of treeless open biotopes. An adult male frequented the rocky escarpments and euphorbia dotted (*Euphorbia caducifolia*) gravel plains around Dhabeji in Thatta district one winter (1981-2), and another frequented the grassy meadows fringing Uchchali lake in the Salt Range. Birds on passage have been seen quartering uncultivated tracts dotted with *Saccharum* cane-grass clumps alongside the Ravi and Chenab rivers in Punjab. Another was seen on the Gadap plain in February within 3 kms. (2 miles) of the Karachi seacoast. In April on spring passage these birds are most likely to be encountered, as confirmed by Whistler in Jhang (1922), Whitehead in Kohat (1907) and the author's observations at Khairpur 10 April and at Sidhnai spill channel near the Ravi river on 13 April. Status SCARCE.

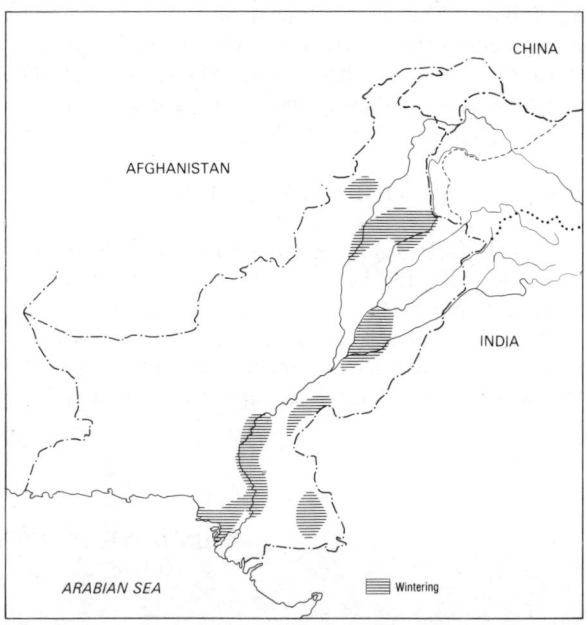

Map 70: Montagu's Harrier *Circus pygargus*

Pipit (*Anthus spinoletta*) in the Punjab Salt Range. Orthopterous insects and reptiles are also recorded as being taken (Brown and Amadon, 1968).

HABITS

They avoid cultivated crops and well-wooded areas, hunting in the same manner as the Pallid (*C. macrourus*) or Hen Harrier. (*C. cyaneus*), flying low over the ground and often quartering systematically up-wind over an open stretch of grassland, returning down-wind to cover an adjacent strip. They catch small passerines and rodents stooping swiftly to the ground when prey is spotted. Birds are probably more important than rodents in their winter quarters and one male was watched stooping upon and catching a Water

BREEDING BIOLOGY

Breeds extra-limitally but more often near marshes or swamps than *C. cyaneus* or *C. macrourus*. Nests are built on the ground, often several females nesting in close proximity. The males may be bigamous and they arrive about 10 days ahead of the females on their nesting territories, where they immediately start aerial displays. The male brings food to the female while egg laying and incubation is in progress, and characteristically this is transferred to the female in an aerial pass.

Pied Harrier *Circus melanoleucus* (Pennant)

DESCRIPTION

Body length 46-49 cms. (Ali and Ripley)
Wing length 34.4-36.7 cms. (males),
    35.5-38.7 cms. (females) (Stuart Baker) 34.4-36.7 cms. (males),
    (Brown and Amadon)
Tail length 19.7-21.7 cms. (males) 21.1-24 cms. (females),
    (Stuart Baker) 19.7-21.7 cms. (males), (Brown and Amadon)
Tarsus 7.6-8 cms. (males) 8.1-8.8 cms. (females) (Stuart Baker)
Weight 265-325 gms. (males), 455 gms. (1 female) (Brown and
    Amadon)

The male is unmistakable, differing from other harriers in his all-black head, neck and upper breast,

with silvery white belly, under-wing coverts and leading edge of wings. There is a broad black band through the wings formed by the median wing converts. The rest of the wings and tail are grey with the primary wingtips black as well as the trailing edge to the secondaries. The male also shows a conspicuous white rump in flight. Females are difficult to separate from females of the Hen Harrier (*Circus cyaneus*) and Pallid Harrier (*Circus macrourus*) but generally look much paler on the underside, being white with a few brown streaks on the lower belly instead of buffy brown with dark streaks as in females of *cyaneus* and *macrourus*, but this is a variable character and not reliable. The coverts along the forearm are whitish buff, a pattern similar to that of the female Marsh Harrier (*C. aeruginosus*) which always has a dark rump.

HABITAT, DISTRIBUTION AND STATUS

The Pied Harrier is a breeding bird in the Soviet Far East, including the Lake Baikal region, Amur region and south into north China (Flint *et al.*, 1984).

It is a winter migrant mainly to the north-eastern part of the Indian subcontinent and continuing down the eastern coast as far south as Sri Lanka. It is quite common in Assam, Bangla Desh (author obs.) and is only considered a rare straggler to Uttar Pradesh and the central part of Madhya Pradesh (Ali and Ripley, 'Handbook', Vol. I, 1968).

There is only one sight record for Pakistan. On 29 December 1985 Mark Mallalieu saw an adult male at Uchchali lake in the Salt Range. It was hunting around the lake all day and identification was unmistakable (Mallalieu *in litt.*, January 1986). This sighting was independently corroborated by Ashiq Ahmad, Wildlife Investigation Officer, Pakistan Forest Institute who camped for 3 days at Uchchali lake at the beginning of January 1986 (Ashiq Ahmad, pers. comm., June 1986). Status ACCIDENTAL VAGRANT.

Goshawk *Accipiter gentilis* (Linnaeus) Map **71**
Synonym *Astur palumbarius*

DESCRIPTION

Body length 40-50 cms. (males),
    up to 62 cms. (females) with tail 18-22 cms.
Wingspan 125-140 cms. (males), up to 165 cms. (females)
Wing length 29-32.3 cms. (males),
    35.3-36.2 cms. (females) sub-species *schvedowi* (Stuart Baker)
Weight 556-600 gms. (2 males),
    1,000 gms. (1 female) sub-species *schvedowi* (Dement'ev *et al.*)

A very large hawk, similar to a Sparrow Hawk in general build and colouration. Adults are blue grey to dark brown above with a long tail having three darker cross-bars and the flight feathers blackish. The ear coverts are dark grey with a pale whitish line or supercilium above the eye. The throat is white and breast also, but it is finely cross-barred with dark grey, the bill is powerful with a hooked tip and bright yellow cere. The tarsi are yellow and bare of feathers and the iris is bright yellow. Immature birds are heavily streaked on the nape and breast with vertical tear-drop dark brown marks against a buffy yellow background and are browner on the crown, back and tail than adults. In flight pattern the broad wings are finely cross-barred on the remiges with dark grey striations or spotting on the under-wing coverts, in overall pattern closely similar to a large *Accipiter nisus*. The sexes are alike in plumage.

HABITAT, DISTRIBUTION AND STATUS

A rare winter migrant to Pakistan favouring the better wooded areas and older records indicate that it was formerly more plentiful. In Biddulph's day, juveniles migrated in autumn through the main valleys of Gilgit and were trapped by falconers (Biddulph 1881). David

Mallon saw one in early November 1987 in the Hunza valley near Passu (*in litt.*, 1987). Similarly juvenile birds regularly migrated down the Chitral valley in autumn and were trapped by falconers for sale on the plains (Prince Burhan-ud-Din, pers comm.). Enquiries made during the 1970s indicated that Goshawks are nowadays very seldom encountered on passage in these northern areas. In the 1920s and 1930s observers such as Hugh Whistler and H. Waite considered it very rare in the Punjab. The author saw an adult female at Rawal lake, Islamabad on 8 March 1974 and two more

Map 71: Goshawk or Northern Goshawk *Accipiter gentilis*

have been recently sighted by D. Corfield in the same vicinity on 14 March 1980 and 2 November 1982. Z. B. Mirza collected one on 11 January 1964 at Jallo plantation north of Lahore for the University Museum. Another Goshawk, obviously an escaped bird was seen (author) in Malir outside Karachi in 1979 as it had jesses on one leg. In September 1983 Kamal Islam encounterd a fine adult male in coniferous forest in the Neelum valley, Azad Kashmir (pers. comm., 1983). There is no evidence that it breeds within Pakistan territory. Status RARE winter visitor.

HABITS

Traditionally much favoured for the sport of falconry, particularly by Punjabi landowners, who train them to take hares and Grey Partridges. They are such bold and persistent hunters that they are highly prized for local falconry and can be trained to hunt gazelles as well as bustard. Fortunately they are not used by Arab falconers whose financial resources have stimulated so much falcon trapping. They are basically woodland haunting species preferring to shelter in leafy trees for a part of the day, launching forth suddenly to pursue their quarry with great speed and agility, twisting between the trees, and in the case of ground quarry running on the ground into thorny thickets to pursue their escaping prey (observed by trained Goshawks catching hare (*Lepus nigricollis dayanus*) in Jabba valley, Mianwali district). Wild birds in their winter quarters probably feed mainly upon *Streptopelia* and *Columba* pigeon species.

BREEDING BIOLOGY

Said to form a life-long pair bond and to reoccupy the same territory each year using 3 or 4 traditional nest sites. The male does most of the nest building, a large stick platform. The pair engage in loud vocal duets in the early morning. The normal clutch is 3 to 4 eggs, bluish white ovals unspotted and the female does most of the incubation which takes 35 to 38 days. Young start to hunt for themselves after they are about 50 days old.

VOCALIZATIONS

Silent outside the breeding season. Both sexes are noisy at the nest site, particularly in the early morning when their repeated calls 'ki-weer, ki-weeer' are reminiscent of *A. nisus*.

Besra Sparrow Hawk *Accipiter virgatus* (Temminck)

DESCRIPTION

Body length 31-36 cms. with tail 14-16 cms.
Wingspan 58-65 cms.
Wing length 16.5-17.5 cms. (males),
    19.7-21 cms. (females) sub-species affinis (Stuart Baker)

Could be confused with a small male *Accipiter nisus* or *A. badius* but if seen clearly there is a narrow black median line down the centre of the white throat which is entirely absent in *A. nisus,* and this line is usually a pale grey in *A. badius*. It is also a much darker coloured hawk dorsally with three blackish cross-bars on the tail, more conspicuous than is usual in specimens of *A. badius*. It also averages smaller in size than the Shikra, and males tend to have the upper breast region uniformly chestnut without distinct cross-barring, whilst the lower breast is more conspicuously barred with rufous. Females are larger in size and browner in dorsal colouration with some dark streaking on the upper breast which is never present in *A. badius*. Specimens are never as pale as *A. badius* and the prominent tail bars and heavier cross-barring on the breast are more similar to *A. nisus* than *A. badius*.

HABITAT, DISTRIBUTION AND STATUS

It is typically an inhabitant of tropical rain forest in adjoining south-east Asian countries and the only region where it is likely to occur is in the higher rainfall forest of the Murree hills. There are 5 specimens of Sparrow Hawk collected by H. Waite from the Galis down to Murree foothills, which were all identified by Hugh Whistler as *Accipiter nisus melanoschistos*. Three pairs frequently seen by the author in two different years at Nathia Gali and Dunga Gali, which were probably breeding were also *A. nisus*, lacking any median throat stripe. Bates and Lowther did not encounter it in Kashmir (1952). However Whistler and Kinnear reported that a specimen taken from Murree was the Besra Sparrow Hawk (*JBNHS*, Vol. 38, 1936, p. 435). It is unlikely to be more than a straggler to the north-western region and until more evidence is forthcoming must be considered a VAGRANT to Pakistan.

Eurasian Sparrow Hawk *Accipiter nisus melaschistos* (Hume) Map **72**

Accipiter nisus nisosimilis nisosimilis (Tickell)

DESCRIPTION

Body length 28-38 cms.
Wingspan 55-70 cms.
Wing length 21.2-21.9 cms. (males),
    24.5-26 cms. (females) sub-species *melaschistos* (Stuart Baker)
Tail length 13-17 cms.
Weight 142-147 gms. (7 males),
    198-270 gms. (7 females) sub-species *melaschistos* (Scully)

A small hawk with long narrow barred tail and rather broad rounded wings. Adult males are dark slaty grey on crown, back and tail, with a narrow white supercilium and three darker cross-bars on the tail. The larger female is browner grey on the upper parts with the flight feathers darker blackish grey. The throat and breast is white finely cross-barred with rufous in adult males and blackish grey in females. Sub-adult birds are browner with a less conspicuous white supercilium and dark brown vertical streaks on the upper breast. The tarsus is yellow and naked, the cere is yellow and the strongly down-hooked bill is blackish. The irides are golden yellow in juveniles and adults. The race *melaschistos* which breeds in Baluchistan is found in winter on the plains. It is reputed to be a much darker bird than *nisosimilis*. Females are about 25 per cent larger in size than the males.

HABITAT, DISTRIBUTION AND STATUS

A. n. melaschistos is a summer breeding visitor to the higher mountains of Baluchistan associated with juni-per scrub forest. It also breeds in coniferous forest in the Himalayas. *A. n. nisosimilis* is a regular winter visitor to the Indus plains arriving about the end of October, and most likely to be encountered in the better wooded tracts of Punjab and Sind. H. Waite collected this sub-species in winter in the Salt Range and Rawalpindi district. *A. n. melaschistos* is also a regular summer visitor to Gilgit, breeding in the main valleys. C. H. Whitehead recorded it as rare but breeding in the Safed Koh range. Status FREQUENT in Baluchistan, less commonly met with throughout the northern mountain regions. FREQUENT winter visitor to Indus plains.

HABITS

A swift and stealthy hunter, capable of flying within a few inches of the ground surface or along a cliff face, as it relies upon surprise to encounter unsuspecting prey. It spends a large part of the day perched in concealment in a tree waiting for unwary prey to come into view and its victims comprise principally birds, from *Streptopelia* doves to small passerines. At times they circle around on thermals and can be seen soaring at considerable heights. They invariably hunt solitarily. Ticehurst writes that the female, called 'Bashah' is popular amongst Sindhis for falconry. In winter time one or two can always be encountered in the Indus mangrove creeks where they live off the waders. Prey is usually seized off the ground and consumed on some suitable vantage point close to where it is killed.

BREEDING BIOLOGY

In Baluchistan confined to the higher hills, arriving in May when pairs immediately start nest building. The male brings most of the materials. Often an old magpie's nest (*Pica pica*) is used as the foundation, the nest comprising an untidy platform without a deep depression in the centre. In the northern Himalayan regions an old nest of a Jungle Crow (*Corvus macrorhynchos*) is often used. They do however occasionally build a completely new nest. These are always situated in trees. Williams describes one in Hannah gorge, north of Quetta with four incubated eggs located in an apricot tree on 15 June whilst a pair was watched building a nest in the top of a Juniper tree at Ziarat on 10 May. The normal clutch is 4 eggs, white in ground colour, boldly marked with brick red smears and blotches. Biddulph found a nest in Gilgit at 3,048 metres (10,000 feet) on 23 June, 9.1 metres (30 feet) up

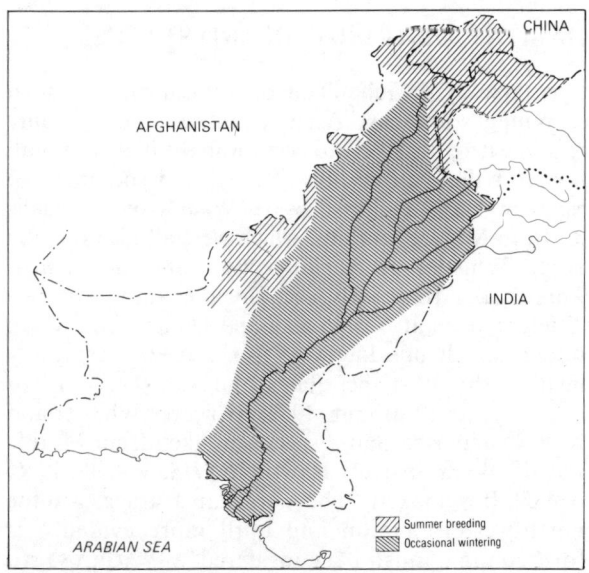

Map 72: Eurasian Sparrow Hawk *Accipiter nisus*

in a Blue Pine (*Pinus wallichiana*) with four well incubated eggs. The female does nearly all the incubation with the male bringing food for her to the nest and subsequently bringing most of the food whilst the nestlings are still small and in need of the parents' shelter and protection. Incubation takes 30 to 31 days and the young are fledged in 24 to 30 days but cannot hunt for themselves for another month after this, during which they are fed by both parents.

VOCALIZATIONS

Both sexes are very noisy in the vicinity of the nest, often giving away its presence. On the intrusion of a human or potential predator the typical warning call is a chattering sharp 'kek-kek-kek'. An advertising call or territorial call given at the beginning of the nesting season is a more drawn-out 'kew-kew-kew'. They also give a whistling call 'whee-oo' when arriving with food.

Indian Sparrow Hawk or
Shikra *Accipiter badius cenchroides* (Severtzov)
Accipiter badius dussumieri (Temminck)

Plate **12**
Map **73**

DESCRIPTION

Body length 29-32 cms. (males), 30-38 cms. (females)
Wingspan 60-70 cms.
Wing length 17.7-19.6 cms. (males), 20.9-22.1 cms. (females)
    (sub-species *cenchroides*) (Stuart Baker)
Weight 136-167 gms. (males) (BNHS collection)

A smaller paler version of the Sparrow Hawk *A. nisus* with adult males having almost white ear coverts, pale blue-grey crown mantle and tail with darker blackish primaries and three darker grey tail bars with only the terminal one being conspicuous. Females are larger with much browner plumage on crown and mantle. The throat in both sexes is white unmarked, except for a narrow dark median line and the breast in males is finely cross-barred with pale chestnut. The female has heavier darker chestnut barring on her breast and immature birds are darker brown with vertical streaking on the breast and spotting on the feathered tibia. The tibia feathering of adults is plain white, or with only faint obsolescent barring. The smaller size, paler grey upper parts and a median dark throat stripe, are the best field marks to distinguish this species from *A. nisus*. Typically the iris is orange in adults, yellower in juveniles, the legs are yellow and the cere greenish yellow, and the bill is plumbeous horny blue. In under-wing flight pattern adults show hardly any barring or spotting on the flight feathers and wing coverts. The race *cenchroides* is larger and paler and *dussumieri* is smaller and darker ashy above.

HABITAT, DISTRIBUTION AND STATUS

A summer breeding bird in the broader valleys of Baluchistan, and the outer Himalayan hill ranges where it is typically associated with sub-tropical pine forest. It is resident and largely sedentary throughout the Indus plains being more partial to irrigated forest plantations and better wooded tracts, avoiding pure desert facies. Status COMMON in Sind and Punjab, uncommon summer breeder Baluchistan and plains of NWFP.

HABITS

This little Hawk often invades garden areas and the outskirts of towns and is more adaptable to urban areas than the equally common White-eyed Buzzard (*Butastur teesa*). Typically it hunts by waiting concealed inside a leafy tree, and then pouncing in swift low flight upon some prey spotted on the ground nearby. Like all the *Accipiter* hawks it relies upon stealth and surprise to catch its food, seeking prey by swift low flight along the edge of tree avenues, around the contours of bush studded hills or low over the tops of

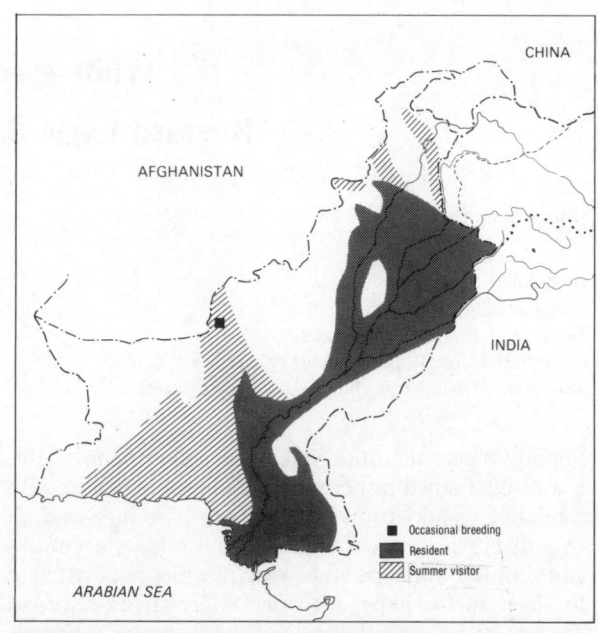

Map 73: Shikra or Indian Sparrow Hawk *Accipiter badius*

tall crops. Breeding pairs in the vicinity of the author's home exhibited great skill in spotting and catching *Calotes versicolor*, Garden lizards as well as Palm Squirrels (*Funambulus pennanti*), both major food items. They have also been seen successfully pouncing upon House Sparrows (*Passer domesticus*), a Praying Mantis and in the olive forest of the Kala Chitta, catching White Cheeked Bulbuls (*Pycnonotus leucogenys*) at a drinking hole. They also take field rats, doves and amphibia as opportunity occurs.

They are solitary outside of the breeding season, becoming very vocal in mid-winter before and during establishment of pair bonds.

BREEDING BIOLOGY

Pair bond is monogamous during breeding season, with aerial displays often a conspicuous feature in the early stages, both sexes swooping after each other, diving and tumbling in mutual acrobatics. The nest is built afresh each year, generally in a tall leafy tree with both sexes collecting sticks which are carried to the nest in the feet, but the female places twigs, brought by the male and does most of the actual nest construction. (observations Khanewal bungalow garden). This is contrary to the statement that only the female collects nesting material (*Birds of the Western Palearctic*, Stanley Cramp, ed. 1980). One nest blown down in a storm, was replaced by an entirely new construction, in the same *Terminalia* tree within 4 days in the first week of March. The normal clutch is 4 eggs, rarely 5 and these are pale bluish grey often unmarked in Sind

eggs, but with darker lavendar blotches and scattered black speckles in a few instances. Incubation is by the female alone and takes 33 to 35 days. A pair was seen carrying food to partly fledged young 15 April. Eates found many nests of *dussumieri* in Sind from April through to June with most eggs laid in April and the average clutch size 4 eggs.

The race *cenchroides* was observed on passage through Quetta in October by Meinertzhagen and Williams found several nests in an avenue of Chinar trees at Pishin, all with young towards the end of June (Meinertzhagen, *Ibis*, 1920 and Williams and Williams, *JBNHS*, 1929). On 20 May on the Margalla hills at nearly 1,219 metres (4,000 feet) a female Shikra was watched building a nest in a *Pinus roxburghii* tree. She collected more than one branch before going to the nest, carrying these in the left foot and landing with difficulty on her right foot, with much wing flapping to maintain balance. *Calotes* and *Acanthodactylus* lizards were common food items brought to the young in the Khanewal garden nest.

*VOCALIZATIONS

The typical advertising call is a two-noted high pitched 'ti-whik, ti-whik, ti-whick'. Individuals call loudly for several minutes on end in the early morning and again in the evening with short bouts irregularly throughout the day. This call is similar to one of the calls of the Drongo (*Dicrurus adsimilis*). An excited female was tape recorded rapidly uttering a single phrased 'kee-kee-kee' call.

White-eyed Buzzard or

Buzzard Eagle *Butastur teesa* (Franklin)

Plate **12**
Map **74**

DESCRIPTION

Body length 31-36 cms.
Wingspan 80-82 cms.
Wing length 27.8-29.6 cms. (males),
    29.4-31.4 cms. (females) (Stuart Baker)
Tail length 15.1-16.9 cms. (both sexes) (Stuart Baker)

Slightly bigger than the Shikra (*Accipiter badius*), this is a heavy bodied rather ungraceful looking hawk with a relatively short rounded tail, long bare tarsi and an overall greyish brown colour. Adults have a conspicuous milky white iris and a square white neck patch at the base of the nape, with two white streaked broad vertical stripes on either side of the throat. The crown, ear coverts and rest of the neck is grey-brown with

some darker streaking. It is otherwise a very variable bird with sub-adults being generally paler on the body and more rufous in tone whilst some adults have a large amount of silvery-brown to buff white areas around the head, breast and wing shoulders. Generally the tail is more rufous and indistinctly cross-barred. In flight the paler grey wing shoulders and dark brown body are conspicuous from above and from underneath. The wingtips are darker with indistinct barring on the primaries and the under-wing coverts are a darker brown than the silvery brown flight feathers. Sub-adult birds have a brown iris, with pale creamy tips to the wing coverts giving a spotted effect. The white pattern on the throat and nape, only well developed in adults, is the best field mark. The cere is yellow as also the legs and feet.

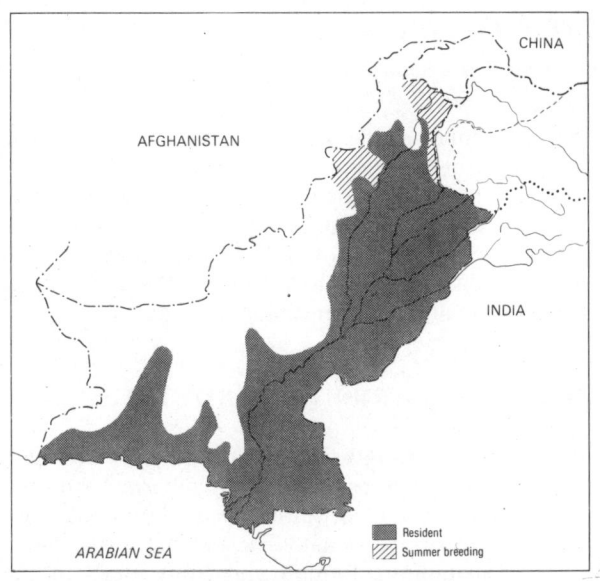

Map 74: White-eyed Buzzard or Buzzard Eagle *Butastur teesa*

HABITAT, DISTRIBUTION AND STATUS

One of the commonest and most widespread raptors in the Indus plains, extending up to the sub-tropical Chir pine zone (*Pinus roxburghii*) of the foothills in summer, including the hills around Balakot in Hazara district, Lehtrar at 1,066.8 metres (3,500 feet) in Rawalpindi district, and the Miranzai and Samilzai valleys in Kohat district (Whitehead, 1911). It even extends into the deserts of Cholistan and 15 were counted in a 32 km. (20 mile) drive up to and beyond Fort Abbas. It is as well adapted to irrigated cultivation and well wooded areas, as it is to fairly treeless scrub desert facies. It may have increased with the spread of cultivation as H. Waite (*JBNHS*, 1948) only encountered it at Sakesar on passage in the Salt Range in the 1930s and early 1940s, whereas today it is widespread in the Salt Range. It avoids extensive mountainous areas and appears to be absent from Baluchistan except along the Makran coastal regions, and also absent from the higher hills of the NWFP. Status ABUNDANT resident with some local summer migration.

HABITS

They shun the immediate vicinity of towns or villages but are generally quite tame allowing close human approach. They do most of their hunting by perching on a tree or telephone pole and watching, and are comparatively sluggish in habits for a raptor. Their principal food is orthopterous insects and small reptiles. When a field is being irrigated they will often settle on the ground nearby to spring upon mole-crickets GRYLLOTAPIDAE and Ground Hoppers (*Chrotogonus* sp.), as they are flushed out by the water. One has been seen eating a Desert Gerbil (*Meriones hurrianae*) in Pirawala plantation and another carrying a *Calotes* garden lizard to its nestlings in Malir. They have never been observed trying to chase birds and rely mainly upon a swift pounce to the ground from some nearby perch, to secure a variety of mammalian, and herpetero fauna as well as amphibia and insects. They occasionally soar in thermals, but their flight is usually rather direct and low, like a Sparrow Hawk's (*Accipiter nisus*).

BREEDING BIOLOGY

Most pairs breed from mid-March to the end of May with peak egg laying in April. They become very noisy during the nesting season, both sexes calling loudly at all times of day. The nest, built by both sexes, is always high up in a tree and consists of sticks similar in size and outward appearance to a House Crow's nest. The normal clutch is 3 eggs but Eates occasionally found only two incubated eggs in nests located on 21 April and 17 May. There is no lining in the nest except finer twigs and the eggs are greyish to greenish white unmarked, very rarely with pale red-brown flecks (Eates, MS unpublished). Only the female incubates and she sits very close, even when the nest is approached by a human intruder. In the Punjab they prefer small groves of trees by cattle sheds or tree-lined canals. The male brings food to his mate at the nest and incubation takes about 19 days. The pair bond is long lasting as a pair was haunting a tree grove in Badin and calling excitedly in late January and another pair was observed copulating on 12 February near Rahim Yar Khan; this took place on top of an isolated *Acacia* tree in which there was no nest. No particular displays have been noted, except frequent soaring over the nest site.

*VOCALIZATIONS

Mainly in the early morning and evening both birds call loudly, a repeated di-syllabic scream 'ki-weeahr, ki-weeahr' repeated 5 or 6 times in succession.

Desert Buzzard and other sub-species *Buteo buteo vulpinus* (Gloger)   Plate 12
Buteo buteo burmanicus (Hume) *Buteo buteo japonicus* (Temminck and Schlegel)

TAXONOMY

This is such a variable species that many different forms have been described as separate sub-species. I follow Ripley's 'Revised Synopsis' (April, 1982) in combining the Desert Buzzard (*Buteo vulpinus*) as a sub-species of *Buteo buteo* (see also Vaurie, *Birds of the Palearctic Fauna,* 1965), and in combining *B. buteo burmanicus* and *japonicus* as one sub-species. The author inclines to the view that when more is understood about this group *Buteo jamaicensis* the Red-tailed Hawk of north America and *Buteo hemilasius* the Upland Buzzard may be conspecific, and simply geographically isolated populations, with *Buteo buteo refectus,* a 'large Himalayan race' described by L. A. Portenko (1935) as being another link.

DESCRIPTION

Body length 51-55 cms.
Wingspan 102-120 cms. (males), 117-127 cms. (females)
Wing length 35-37.7 cms. (males),
       37.8-39.2 (females) (sub-species *vulpinus*) (Stuart Baker)
Tail length 18-19.1 cms. (males),
       18.2-20.1 cms. (females) (sub-sp. *vulpinus*) (Stuart Baker)
Weight 600-675 gms. (4 males),
       710-1,175 gms. (5 females) (sub-sp. *vulpinus*) (Dement'ev *et al.*)

With relatively broad wings and broad rounded tail, they are usually seen soaring around on thermal currents, so that a description of the under-wing pattern is appropriate. In *vulpinus* the tail is pale chestnut, varying to pale isabelline, often with the outer tail feathers partly white and usually with cross-barring on the central rectrices, visible from above and below. The under-wing characteristically shows dark chestnut carpal patches and dark tips to the secondaries and primaries with interrupted spotted bars through all the flight feathers, and the under-wing coverts generally rather rufescent. The head, neck and breast is usually pale buff or chestnut with some dark streaking on the breast. The principal character of the sub-species *vulpinus* is its rufescence, relatively smaller size, and cross-barred tail. *B. japonicus* is larger and either predominantly dark umber brown all over, or in paler phase birds variably marked with paler buff streaks especially about the neck and breast. The head may be largely whitish buff in some instances. The tail is invariably cross-barred darker brown and generally the lower flanks are heavily streaked dark brown with finer streaks on the breast. In flight silhouette the under-wing coverts may be entirely dark brown in dark phase birds or silvery white with the leading edge of the wing pale brown, in contrast to the more rufescent tones of the wing converts of *vulpinus*. The legs and feet are yellow as also the cere, in all sub-species, and the strongly hooked bill is blackish. Compared with the Honey Buzzard (*Pernis* sp.) with which they might be confused they have shorter tails and less distinct spots and bars on the under-wing. Females are about 20 per cent larger than males.

HABITAT, DISTRIBUTION AND STATUS

A winter visitor to Pakistan occurring almost anywhere in open country. Specimens of *B. japonicus* have been seen in Baluchistan in October and March and both sub-species are occasionally encountered in the Indus plains throughout Punjab and Sind, both in the irrigated tracts and in scrub-desert biotope. Recent sight records include the Sibi desert plain (13 April), Khaur at the foot of the Salt Range (26 January), Kushdil Khan lake, Baluchistan (30 October), Bahawalpur (14 October), and the Mansehra plain, Hazara district, (22 February). Biddulph considered it an uncommon winter visitor to Gilgit, and Whitehead also recorded it as a rare winter visitor in Kohat district. In their winter quarters specimens corresponding to *japonicus* have most frequently been encountered in Punjab and Baluchistan with *vulpinus* occurring both in Sind and eastern border regions of the Punjab. Status FREQUENT.

HABITS

They will sit for hours on a telephone pole or prominent hillock or tree and take off with heavy laboured wing beats, but soar gracefully when hunting and can hover without moving their wings, balancing by twists of the tail when there is sufficient wind. They will perch on trees and on the ground and take a variety of prey including field rats, reptiles and small birds. Most of their prey is taken on the ground to which they descend not very swiftly, with wing raised over the back and feet extended.

BREEDING BIOLOGY

Breeding believed to be extra-limital. Silent on their breeding grounds. Specimens of *Buteo buteo vulpinus* have been seen in Ziarat on 17 May 1980 and on 9 May 1979 near Zindra, both in Baluchistan. These birds were soaring around all day and not apparently on migration so the possibility of the occasional breeding pair cannot be ruled out.

Long-legged Buzzard *Buteo rufinus* (Cretzschmar)

Synonym *Buteo ferox* (Gmelin)

Plate **12**
Map **75**

DESCRIPTION

Body length 50-65 cms. with tail 19-24 cms.
Wingspan 126-143 cms. in males, 140-148 cms. in females.
Wing length 42.5-45.9 cms. (males),
    44.8-49.6 cms. (females) (Cramp *et al.*)
Tail length 20.7-24.4 cms. (males),
    22.3-26.2 cms. (females) (Cramp *et al.*)
Weight 590-1,281 gms. (males),
    945-1,760 gms. (females) (Cramp *et al.*)

Slightly larger than the Common Buzzard (*Buteo buteo*) it is mainly distinguished by having a pale chestnut unbarred tail, and a good deal of white or creamy buff around the head, neck and shoulders. There is however wide variation with both light and dark forms. In flight pattern from underneath, the wingtips are darker and there is a prominent dark carpal patch with the under-wing coverts being rufous and the flight feathers much whiter generally with indistinct bars in the distal portion of the secondaries. The tibia region is always dark chestnut, a useful field point, with the lower breast streaked and the upper breast and throat much paler, often nearly white. In some birds when viewed from above the outer tail feathers are largely white. Sub-adult birds show faint barring on the tail. The tarsus is yellow as also is the cere, the iris hazel brown and bill horny black. Females are about 20 per cent larger than males.

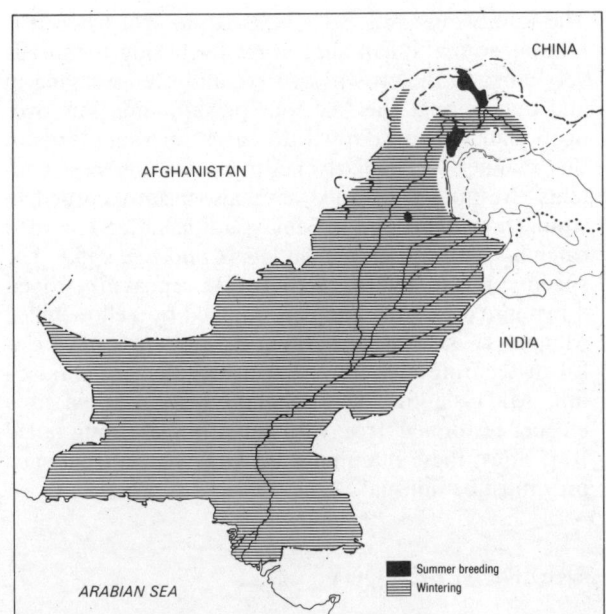

Map 75: Long-legged Buzzard *Buteo rufinus*

HABITAT, DISTRIBUTION AND STATUS

In winter this buzzard is very widespread and comparatively common throughout Pakistan. It particularly frequents desert regions and broader stony plains on the west of the Indus basin, but is less common in winter in mountainous regions. It is less frequently to be encountered in well cultivated and wooded areas but in regions such as the Punjab Salt Range, Indus delta, Cholistan desert border and Sind Kohistan it is possible to see 8 or 9 individuals in half a day's car journey. They are winter migrant visitors to these regions, first arrivals being no d from mid-August in Baluchistan and the third week of August on the Indus plains. A very small fraction of the population stays to breed in the northern mountain areas of Pakistan, whilst the majority of birds travel further north to central Asiatic breeding grounds. In the sand dunes along the Karachi seacoast a few stragglers can still be seen up to the end of March and one was seen in Ghizri creek on 9 April, but most have gone north by mid-March. Status COMMON winter visitor, rare breeder in northern mountains.

HABITS

They will settle on trees or on the ground and spend quite a part of the day thus, perched on some vantage point. In breezy weather they will hang motionless in the air as well as soaring around while hunting for food. In still calm air they cannot hover. When taking off, their flight appears slow and laboured but they will circle effortlessly without a wing beat for hours on end, once they have attained sufficient height. In sand dune desert areas an important food source is the colonial burrowing Sand Jird (*M. hurrianae*), in the Punjab and Sind, *M. libycus* in Baluchistan. They also catch lizards particularly *Agama* rock lizards in the dryer hilly tracts. They are also probably capable of catching hares and ground feeding birds such as larks.

BREEDING BIOLOGY

In the higher Himalayan valleys, the occasional pair can be encountered in June and July and is presumed to be breeding. A pair was seen on the Lowari Pass in Dir in mid-May in *Abies* fir forest at 2,895.6 metres (9,500 feet), and at Burawai in the Kaghan valley, Hazara district a pair was watched from 12 to 17 June haunting a Juniper studded cliff at about 3,353 metres (11,000 feet). A pair was also haunting the forested upper slopes of Manshi forest on 11 to 13 June 1986 and appeared to have a nest in the vicinity.

They generally nest on cliffs in preference to trees and pairs come back to traditional nesting cliffs though they may build a nest on a different site each year. Nests are lined with a lot of fresh green leaves and are often bulky structures, built by both sexes. Males are said to perform an undulating display flight similar to other *Buteo* species. Eggs taken from nests in Kashmir were well covered with rich deep brown blotches and indistinct under-markings of grey on a white background. The normal clutch is 3 eggs, occasionally 2 or 4. Nigel Hacking observed a pair in display flight on the southern escarpment of the Punjab Salt Range in February and actually saw them copulating, so that it is possible that they bred in that area. There are one or two old records of breeding (Cock's near Nowshera, April 1872, and Theobald, Punjab Salt Range) which could never be properly authenticated but which were either *Buteo rufinus* or *Buteo buteo vulpinus* (Whistler, *Ibis*, 1930).

VOCALIZATIONS

Silent outside the breeding season, but both sexes have a weak, wailing or mewing call similar to *Buteo buteo*.

Black Eagle *Ictinaetus malayensis* (Temminck)   Plate **13**

DESCRIPTION

Body length 69-81 cms.
Wingspan 175-265 cms.
Wing length 52-55.3 cms. (males),
    53.8-56.8 cms. (females) (Stuart Baker)
Tail length 28.5-31.2 cms. (both sexes) (Stuart Baker)

A rather long winged large eagle with a long square tipped tail and relatively small bill. The body plumage is dark brown (blacker in the population inhabiting Malaysia and Sri Lanka), with the tail showing distinct grey cross-bars. The feet and cere are bright yellow the claws being relatively straight compared to the strongly recurved claws of other raptors. The claw on the inner toe is remarkably short. It is typically a bird of mountain forest regions only and can usually be identified from its habitat, large size and dark colouration. Usually a small white carpal patch shows in under-wing silhouette and occasionally indistinct darker barring on the primaries. Juvenile birds have more buffy breasts with dark shaft streaks and paler rufescent heads and necks with paler shaft streaks on the nape. At rest the long pointed wings and tail give it a rather kite-like appearance, but the tarsus is covered with close feathering down to the toes like *Aquila pomarina*.

HABITAT, DISTRIBUTION AND STATUS

An inhabitant of tropical forests including lower altitude Himalayan mixed deciduous forest, they only come into Pakistan as occasional wanderers and just into the Murree hills. In 23 years of visits to Dunga Gali, the author has sighted lone individuals in 3 years, two in 1965, one in 1979 and one in 1982. In the latter year an individual was watched hunting in the Kao valley for several days in succession during early May, whilst the 1965 sighting was at the height of the monsoon. At the turn of the century it was an occasional breeder in the Galis, Col. Buchanan took a nest on 29 April 1899 at Changla Gali and a nest with one downy chick was reported the previous year. There have been no recent records of breeding. Status RARE and irregular STRAGGLER.

HABITS

The long wings and tail give great control when it is hunting so that it can glide at relatively slow air speeds very close to the crowns of trees and even weaving in and out amongt the tree tops or skimming low over open glades. It is adapted to search for small birds in this manner, particularly nesting birds and nest contents. Nestling field mice have also been recorded in stomach contents and in Malaysia it has been recorded catching bats and cave Swiftlets (*Collocalia* spp.). The specimen watched in Kao forest repeatedly dived down into a clearing and was mobbed by Yellow-billed Magpies (*Cissa flavirostris*) but did not appear successful in securing any prey. Extra-limitally on Maxwell hill, Malaysia one was seen (author) perched in a creeper-festooned tree with an arboreal Long-tailed Rat. Since these mammals are largely nocturnal such prey must be unusual.

BREEDING BIOLOGY

Extra limital. Said to perform spectacular stooping displays during courtship but pair bond believed to be life-long. The nest recorded in Changla Gali had only 1 egg, apparently the normal clutch for the species. The nest is usually located at the top of a tall deciduous tree and re-used, both sexes adding fresh green leaves and branches to the structure.

VOCALIZATIONS

The single bird haunting Kao valley occasionally uttered a single loud mournful whistle, not very dissimilar from *Spilornis cheela*. It was syllabized at the time as 'klee'.

Lesser Spotted Eagle *Aquila pomarina* (C.L. Brehm)

DESCRIPTION

Body length 60-65 cms.
Wingspan 134-159 cms.
Wing length 47-50.5 cms. (males),
    49.3-50.8 cms. (females) (Stuart Baker)
Tail length 23-24.8 cms. (both sexes) (Stuart Baker)

Only slightly smaller than the Greater Spotted Eagle, they are difficult to separate in the field when in adult plumage. Juveniles show fewer rows of spots on the wings and these are smaller than in *A. clanga*. They are also more rufescent tawny on the nape and generally not so dark. In flight pattern from underneath adults show paler brown under-wing coverts whilst *A. clanga* has a uniformly dark brown under-wing. On the upper surface *A. pomarina* usually shows paler whitish bases to the primaries. The long narrow closely feathered tarsi when at rest are one of the most distinctive field points. Also in flight silhouette it looks slightly longer tailed and less broad winged than *A. clanga* (see Porter, Willis *et al.*, 1974 for details of differences in flight pattern).

HABITAT, DISTRIBUTION AND STATUS

This eagle has not been definitely recorded in Pakistan. It is a regular winter visitor to Bharatpur sanctuary in neighbouring Rajasthan so might occur in the Punjab. Colonel Rattray (*JBNHS*, 1905) claimed to have found this eagle's nest in the Murree hills near Dunga Gali and he shot the female, but this record is not accepted by'Ali and Ripley ('Handbook', Vol. 1, 1969). It should be looked out for. One adult bird was sighted by Rawal lake (Islamabad) on 3 March 1988 by 3 very experienced Swedish ornithologists (*Bird Watching in Pakistan*, Hirschfeld, Kjellen and Ullman, privately published, March 1988, Malmo, Sweden).

Greater Spotted Eagle *Aquila clanga* (Pallas)

Plate **13**
Map **76**

DESCRIPTION

Body length 65-72 cms.
Wingspan 155-182 cms.
Wing length 48.6-50.1 cms. (males),
    54.2-56.5 cms. (females) (Stuart Baker)
Tail 24-26 cms. (males), 25-26.7 cms. (females) (Stuart Baker)
Weight 1,665-1,925 gms. (males),
    1,770-2,520 gms. (females) (Cramp *et al.*)

A large very dark brown almost blackish eagle with the tarsi feathered right down to the toes and the primary wingtips extending to the tail tip when at rest. The bill is deep and strongly hooked on the upper mandible with a bright yellow gape and cere. In flight silhouette, the short rounded tail barely extends beyond the trailing edge of the broad wings and this is the best field character to distinguish it from *Aquila rapax*. The overall dark colour and prominent pale whitish rump are further guides, also the down droop of the wings when gliding and viewed from in front. Juveniles in their first winter have prominent buff tips to the secondaries, lesser median and tertial wing coverts, forming three or more rows of spots which are large and conspicuous. Sub-adults in their second winter have creamy white tips to the wing coverts only, which makes them confusingly like juvenile Tawny Eagles, *A. rapax*. The Greater Spotted Eagle is considerably smaller than the Tawny Eagle and the adults overall dark colour with prominent white tips to the upper tail covert are diagnostic. Juveniles or sub-adults could only be confused with *Aquila pomarina* which only has prominent pale spots on the major secondary wing coverts and is rather smaller in size with noticeably close feathered narrow tarsal shanks, extending below the bushy thighs. *Aquila clanga* tends to have the tibia feathers merging more gradually and extending further down the tarsus.

HABITAT, DISTRIBUTION AND STATUS

In winter this eagle is the most likely raptor to be encountered around large lakes, barrage headponds and swamps. It is not uncommon and up to six individuals can be seen in the sky at one time around the larger lakes of Sind and Punjab. It also haunts the larger canals, and a few over-winter in the mangroves of the Indus delta. It has not been recorded in Baluchistan. Though Salim Ali in the 'Handbook' describes it as resident in West Pakistan, the majority of birds are migratory, disappearing during the summer, probably to their Russian breeding grounds. However a few do remain to breed, at least in the better forested areas of the east Narra canal and the Indus river. Status COMMON.

HABITS

They can often be seen perched on low tamarisk bushes on the edge of jheels and swamps where waders and duck are feeding. They will take waterfowl and fish when opportunity occurs, but amphibia are probably a staple item in the diet. On several occasions birds have been seen eating a fish but it is not known how such prey was secured and it is probable that it was pirated in most instances.

BREEDING BIOLOGY

Kenneth Eates recorded the first nesting of this species on the subcontinent when he discovered a female incubating a single egg on 1 March 1936 four miles from Bhagar lake in Sukkur district, Sind. He subsequently discovered 2 more nests in 1937, 1 in 1938 and 3 in 1939, mostly in Sujawal and Sanghar districts of Sind (Eates, 1937e). His nests were usually located in well wooded areas not far from lakes and eggs were found from April till June, the clutch varying from 1 to 2 and the nest described as a large platform of sticks often in an *Acacia arabica* tree (but in two instances in Mango trees (*Mangifera indicus*). He described the eggs as white, blotched with purple-brown and fine speckles of red-brown. He found the remains of a Purple Gallinule (*Porphyrio porphyrio*), a Collared Dove (*Streptopelia decaocto*) and a Tiger Frog (*Rana tigrina*) on various nests, as well as unidentified fish. He also observed that the male brought food to his incubating mate, swooping low over the nest and allowing her to pluck the food from his talons without rising from the nest (MS notes unpublished). The 'Handbook' (Ali and Ripley, Vol. 1, 1968) states that this species breeds in Baluchistan, Sind and Punjab and that nesting occurs from May to June. There is no known record of nesting in the Punjab as yet, though this is by no means improbable. Baluchistan lacking

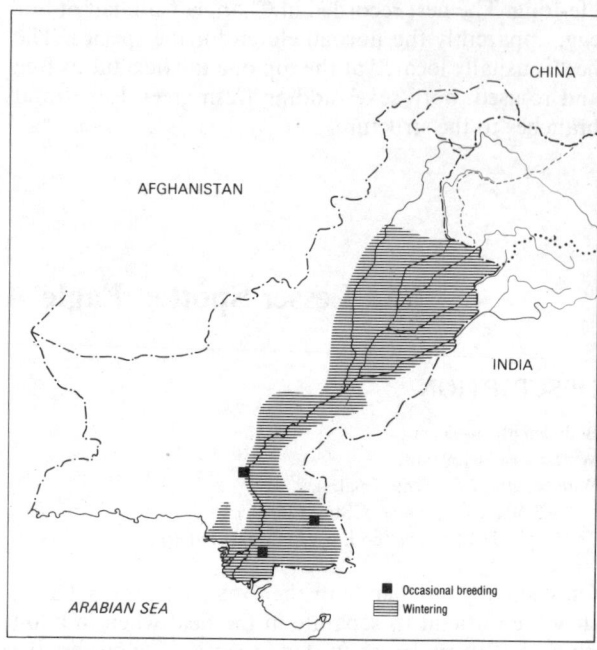

Map 76: Greater Spotted Eagle *Aquila clanga*

either suitable wetland habitat as well as being practically devoid of trees is unlikely to provide suitable breeding habitat for this species. The author supposing this species only a summer breeder, stumbled upon a nest in Thatta district quite by accident, about 10 miles south of Mirpur Sakro. Located at the top of an *Acacia arabica* tree growing on the bank of an irrigation channel in an area subject to seasonal inundation and rich in both waterfowl, fishes and amphibia. When discovered on 6 January 1984, it contained a single downy young which on 24 January was judged from its sprouting quills and size to be about 5 weeks of age so that hatching occurred sometime in the third week of December. Both parents spent a large part of the day patrolling the air space over the nest and keeping watch on their offspring, and the male was observed displaying over the nest site on two occasions. In another area of Thatta district in early November at Mahboob Shah lake a male had previously been watched displaying before a female, so there is some circumstantial evidence that nesting may commence any time during the winter months as well as the summer. The display consists of a series of swooping or undulating plunges, with wings half closed, followed by a steep upward swoop to an almost vertical stall. These displays even include some complete aerial somersaults and are often performed after nesting commences when another raptor crosses the territory or during courtship, while the female (larger in size) is circling around nearby. The male calls also during these displays. Incubation in this species is

42 to 44 days and period of fledging up to 65 days (Brown and Amadon, 1968).

*VOCALIZATIONS

Their call is a high pitched quite musical yelp often given when a conspecific is observed flying by or when two birds are soaring together on thermals. Birds may call when perched in a tree or while soaring overhead, just before or after displaying. The call can be likened to a high pitched yapping and a fairly rapidly repeated 'kyew-kyew-kyew'. The voice of the female is similar but slightly lower pitched and hoarser. They are quite vocal both during winter and summer.

Tawny Eagle *Aquila rapax vindhiana* (Franklin)
Steppe Eagle *Aquila rapax nipalensis* (Hodgson)

Plate **13**
Illus. **28**
Map **77**

DESCRIPTION

Body length 65-77 cms.
Wingspan of *vindhiana* 169-180 cms.
Wingspan of *nipalensis* 174-260 cms.
Wing length 49.7-53 cms. (males),
    51-55 cms. (females) sub-species *vindhiana* (Cramp *et al.*)
    56.5-61 cms. (males) 60-64.5 cms. (females) sub-species *nipalensis*
Weight 2,520-3,500 gms. (males),
    2,300-4,850 (females) sub-species *nipalensis*

Formerly the Steppe Eagle was considered a separate species. (Dillon Ripley 'Synopsis' 1961 and Ali and Ripley 'Handbook' Vol. 1, 1968, first ed.) They are nowadays considered as conspecific. Generally the Steppe Eagle is a larger version of the Tawny with darker brown body plumage in sub-adult stages, but more prominent creamy white wingbars formed by the tips of the greater and median secondary wing coverts. The Tawny Eagle is often a very pale creamy buff in juvenile and sub-adult plumage with the flight feathers contrasting blacker, but there is great variation with intermediate colour forms, and it is impossible always in the field to identify immature Tawny eagles and separate them from sub-adult Imperial Eagles (*A. heliaca*). The tail is longer and narrower than *A. clanga* and the wings are relatively longer and narrower. Uniformly pale yellow-buff forms of *vindhiana* are fairly easy to recognize as also sub-adult forms of *nipalensis* with their pale wingbars. The feet and cere are yellow in all sub-species, the cere being darker leaden coloured in juvenile specimens. The legs are covered with bushy feathers to the bottom of the tarsus. The tail usually shows indistinct cross-barring. The deep set yellow iris and strongly hooked bill give this eagle quite a fierce expression.

HABITAT, DISTRIBUTION AND STATUS

A. r. vindhiana is resident and a breeding bird throughout the remoter less settled parts of Punjab, Sind and the lowland areas of Baluchistan and the NWFP. It definitely avoids mountainous areas but is typically a bird of the desert, being frequently observed in the Sibi, Thal and Thar deserts over 48 kms. (30 miles) from the nearest cultivation.

The Steppe Eagle is a winter migrant visitor from central Asia and it is inclined to be gregarious around the larger lakes where waders and waterfowl abound. On the Potohar plain outside of Rawalpindi there is a traditional roosting site of this eagle where up to 38 birds have been counted at one time. They roost on the ground on low sandstone ridges, and from late November this roosting site has attracted large numbers of Steppe Eagles for the past six years since it was discovered. Tawny Eagles occur sparingly in Baluchistan in the plains areas and the Chagai district, individuals being sighted throughout the year, but there is no evidence of their breeding in that province. They have adapted quite well to irrigated cultivation also. Status COMMON.

HABITS

The Tawny Eagle has the reputation of being a scavenger and certainly it will frequent refuse dumps alongside vultures. In winter time it often resorts to colonial roosts especially in the vicinity of refuse

Illus. 28: Steppe Eagle *Aquila rapax nipalensis*. Typical appearance of 2-3 year old bird, both sexes.

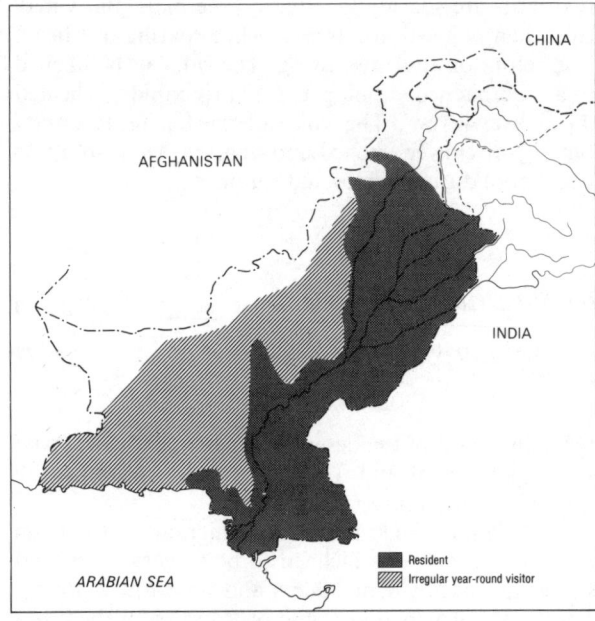

Map 77: Tawny Eagle *Aquila rapax*

dumps or slaughter houses. One such site, near an isolated poultry farm in the desert to the north of Karachi, usually has between 20 and 35 Tawny Eagles, most of which roost on the cross-bars of adjacent power pylons, but many roost also on the ground in the vicinity. They are also adept at robbing other birds of prey of their quarry, often frequenting wetland areas in winter time solely for this purpose. In desert regions they like lizards especially the Spiny-tailed *Uromastix*, and they will even eat locusts (K. Eates, MS notes). They can hunt mammals, often perching motionless for hours on a sand hill or bush near burrow colonies of *Meriones hurrianae*, Desert Jirds, which they catch by a sudden pounce (observed Rahim Yar Khan on borders of Cholistan desert). Analysis of 30 regurgitated pellets from a Tawny Eagle's roost by the Hab Dam in January 1989 revealed that Coots (*Fulica atra*) were the principal food item but also a few Pochard (*Aythya ferina*) and Tufted Ducks (*Aythya fuligula*) were also occasionally taken (F. Koning, *in litt.* to author).

BREEDING BIOLOGY

A male displaying in its territory on 3 March first of all called intermittently from a prominent perch at the top of a tree—each series of calls lasting about 2 seconds. It then flew in a wide circle swooping low and then rising steeply for 40 to 50 metres. A rather ponderous flight dance and nothing like the spectacular and similar displays by *Aquila clanga*. Courtship and display can occur sporadically from December onwards, when males have been observed pursuing females in swooping flight, diving down upon the female from behind, also both sexes circling each other a few metres apart, calling loudly. Talon presentation has also been observed, without any actual exchange of food prey by an apparently displaying or courting pair. The nest is usually built right on the top of a tree being a platform of thorny branches devoid of any shelter. An unusual nest site, built on the cross-arm of a power pylon was obviously in use early March 1987, near Gharo, Thatta district. Egg laying commences early, sometimes in January but in the northern Punjab more usually from the end of February. Often the platform is 1.2 or 1.5 metres (4 or 5 feet) across and the centre is lined with some green leaves or grass clumps. The normal clutch is 2 to 3 eggs which are quite glossy in texture, a dull skim-milk blue, sparingly blotched with brown and purple. H. Waite (*JBNHS,* 1948) found a nest with 2 well incubated eggs in a Shisham tree (*Dalbergia sissoo*) at Kallar Kahar in the Salt Range. K. Eates found nests in every district of Sind with eggs as early as 24 December and as late as 6 March and about one-fourth of these were clutches of 3 (out of over 60 eggs collected, MS notes). A nest seen by the author on 11 April had only one large fledged chick which the female was sheltering from the hot sun even at 4.30 p.m. This was on Bhegari Bund 22.5 kms. (14 miles) north of Sukkur at the top of an *Acacia arabica* tree and there was a House Sparrow's nest of grass tucked into the base of the eagle's nest, both parents busily carrying caterpillars to their brood as the hen eagle sheltered her chick within 60 cms. (2 feet) of them.

*VOCALIZATIONS AND BEHAVIOUR

During aerial pursuits and courtship they are quite vocal. The female utters a very low pitched hoarse barking. A male when performing a swooping flight display uttered a more rapid but guttural yelping 'keh-keh-keh', both sexes giving calls of much deeper pitch than *A. clanga*. During an aggressive encounter birds were heard uttering a more throaty guttural bark. This occurred when one was feeding on a fish on a flat open sandy plain and a second bird swooped low over it and landed within 2 metres. The attacker then jumped at the feeding bird which turned over and presented its talons. Both remained with widespread wings and feet interlocked looking a bit ludicrous sitting flat on their tails with heads and necks well drawn back from each other and bills half open. Thereafter both birds sprang into the air and the fish was at least, temporarily abandoned (observations made with X50 telescope, Haleji lake, December).

Imperial Eagle *Aquila heliaca* (Savigny)

Plate **13**
Map **78**

DESCRIPTION

Body length 72-83 cms.
Wingspan 190-210 cms.
Wing length 54-60.9 cms. (males),
    58.9-63.4 cms. (females) (Cramp *et al.*)
Tail length 26-29.2 cms. (males),
    28-31.1 cms. (females) (Cramp *et al.*)
Weight 2,450-2,718 gms. (5 males),
    3,160-4,530 gms. (4 females) (Cramp *et al.*)

A fine looking eagle considerably larger than *Aquila clanga* and slightly larger than *Aquila rapax nipalensis*. Adults are easy to distinguish as they have conspicuous pale buff, sometimes almost white hackles on the hind neck and extending over the crown, whilst one or two tertial and scapular feathers are pure white. The overall body colour is a very dark brown and the tail usually a more greyer brown, with a broad darker grey terminal bar.

Sub-adults are much harder to differentiate from other *Aquila* species as they have paler buffy brown wing shoulder and under-wing coverts, often an isabelline brown similar to the juvenile *Aquila rapax*, also the crown and nape are not pale gold as in adults. Adult specimens have often been seen with pale buff almost white crown and *Aquila chrysaetos* is never this pale on the head and neck. Also the tail of *A. chrysaetos* in adults is cross-barred on its upper portion and much paler, almost white in sub-adult specimens.

Immature Imperial Eagles have noticeably streaky neck and breast plumage, often with two pale wing bars on the secondaries and secondary wing coverts like *A. rapax nipalensis*. An adult female shot by C.H.T. Whitehead weighed 3.6 kg. (8 lbs.) It is believed that adults around 5 years of age attain the golden buff head and neck but that conspicuous white scapular feathers are not acquired until the seventh or eighth year. The tarsus is well feathered down to the toes which are yellow like the cere. Occasional sub-adults can be seen with creamy feathers on the leading edge of the wings as well as on the nape and hind neck, such a pattern closely resembling that of a female Marsh Harrier (*Circus aeruginosus*). One such was seen on Manchar lake by the author and another similarly marked was shot by Ticehurst. Females are about 10 per cent larger in size than males.

HABITAT, DISTRIBUTION AND STATUS

This is an inhabitant of open plains or desert areas in its winter quarters, avoiding high mountainous regions and often being attracted to the vast assemblage of waterfowl which congregates around major lake and wetland areas. It has never been recorded in the higher regions of northern Baluchistan or the northern Himalayan regions. They are believed to be largely winter migrant visitors from a central Russian breeding population. Formerly a few pairs did breed in the north-west part of the subcontinent and there is some evidence also of a local breeding population in the arid broken foothill country of southern Baluchistan. Surprisingly H. Waite never recorded it in the Salt Range though the author once saw an adult at Khabbaki lake in December.

In its winter quarters solitary individuals are usually encountered. They regularly occur in the major desert tracts of Cholistan and Tharparkar. Judging from Ticehurst's accounts of Sind it is much rarer than formerly (op. cit., *Ibis*, 1923).

Occasional pairs evidently did breed formerly. Stuart Baker ('Fauna Series', 1928) refers to an authentic nest from Dera Ismail Khan and another from the Baluchistan-Sind hills. Ali and Ripley ('Handbook' 1968) refer to two authentic records from Jhelum in the Punjab and Hansi in (east) Punjab, India, both at the turn of the century. Eates' manuscript notes refers to four nests taken on the Lasbela Sind boundary. Two of these from the Hab valley each contained two eggs, collected on 24 February 1944 and 30 November 1944 and a third with one egg on 10 March 1946 near Moidan. Status SCARCE and largely a winter migrant.

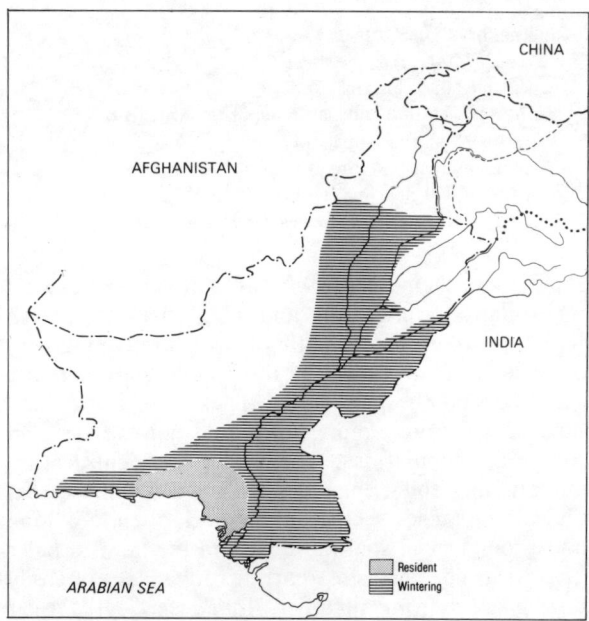

Map 78: Imperial Eagle *Aquila heliaca*

HABITS

Even in their winter quarters a pair has been observed in high soaring display flight calling to each other. They are quite vocal in the presence of conspecifics. They have the reputation for sluggishness, spending a large part of the day perched on a tree or other suitable vantage point. For two successive winters one adult roosted on the same power pylon over a 6 or 8 week's period on the periphery of Karachi airport in open uncultivated thorn scrub and the author, driving daily to work became very familiar with its individual markings. It would take off and leave the area sometimes after about 9 a.m. when warm air provided thermal currents. They are adept at robbing other birds of prey and will consume fish as well as birds and mammals and are not averse to feeding upon carrion when available. Salim Ali records *Uromastix* lizards and Russell's Viper in the stomach of two different individuals. Observations around Punjab lakes indicate that this eagle lives largely by piracy.

BREEDING BIOLOGY

They are tree nesters, and in their northern breeding grounds occupy a permanent nesting range, repairing and re-using several nest sites in irregular rotation. K. Eates refers to four nests collected in Kalat and

Lasbela districts of Baluchistan along the sides of broad stony valleys in Kalat and the Hab regions. They were all located on the tops of stunted thorn trees growing out of crags or cliffs. This is a region where there are no tall trees. The nests were not particularly large and were noted as being smaller than those of *A. rapax*, he found in Sind but they were well lined with dried grass and sometimes green leaves and were more compact than Tawny Eagles' nests. One nest contained 2 eggs, the other three a single egg, and they were located on 24 February, 16 November, 30 November and 10 March, the most recent one in 1946. He described the parent birds as being rather timid, merely circling around when the nest was robbed. The eggs are not glossy, but dull white with sparse blotches of purplish brown and some grey under-markings. Both sexes incubate the eggs and as in most eagles the female principally feeds the young while the male does the hunting.

VOCALIZATIONS

During soaring displays in November one bird uttered a rather hoarse dog-like call rendered 'gowk-gowk-gowk'. Another individual with food prey, when attacked by a second eagle uttered a more rapid but similar hoarse low pitched 'krok-krok-krok'.

Golden Eagle *Aquila chrysaetos* (Linnaeus)

Plate **13**
Map **79**

DESCRIPTION

Body length 75-88 cms.
Wingspan 204-220 cms.
Wing length 63-65.5 cms. (males),
    66-70 cms. (females) Himalayan population (Stuart Baker)
Tail length 31.5-33.5 cms. (males),
    35-36.5 cms. (females) (Stuart Baker)
Weight 4,000-4,100 gms. (males),
    6,350 gms. (1 female) sub-species *daphanea*

Bigger than the other *Aquila* eagles found in Pakistan with relatively longer tail and wings than *A. heliaca* which it also exceeds slightly in size. Adults look very dark brown almost black in the population inhabiting Pakistan and the pale golden hackles of crown and hind neck are often more rufous and not so pale as in European populations. The best field identification guide is its habitat, plus generally paler white carpal and rump patches. The head and neck often look long and prominent in soaring flight and the tail is a paler grey with indistinct cross-barring. Juveniles are easily distinguished from all other *Aquila* species by their prominent extensive white base to primaries and tail. Females are about 10 per cent larger in size than males.

HABITAT, DISTRIBUTION AND STATUS

An inhabitant of high mountain regions occurring throughout the Karakorams, Hunza, Gilgit and Chitral

Plate 13: Flight patterns of larger Eagles (viewed from above).
(1a) Tawny Eagle *Aquila rapax*, typical pale morph adult.
(1b) Steppe Eagle *Aquila rapax nipalensis*, typical mature adult.
(1c) Steppe Eagle *Aquila rapax nipalensis*, typical sub-adult.
(2a) Golden Eagle *Aquila chrysaetos*, mature adult.
(2b) Golden Eagle *Aquila chrysaetos*, sub-adult.
(3) Asiatic Imperial Eagle *Aquila heliaca*, mature adult.
(4) Black Eagle *Ictinaetus malayensis*, adult.
(5a) Greater Spotted Eagle or Spotted Eagle *Aquila clanga*, mature adult.
(5b) Greater Spotted Eagle or Spotted Eagle *Aquila clanga*, 2nd. year bird.
(6a) White-tailed or Grey Fish Eagle *Haliaeetus albicilla*, mature adult.
(6b) White-tailed or Grey Fish Eagle *Haliaeetus albicilla*, 3rd. or 4th. year old sub-adult.
(7) Pallas's or Ring-tailed Fish Eagle *Haliaeetus leucoryphus*, mature adult.

Plate 13

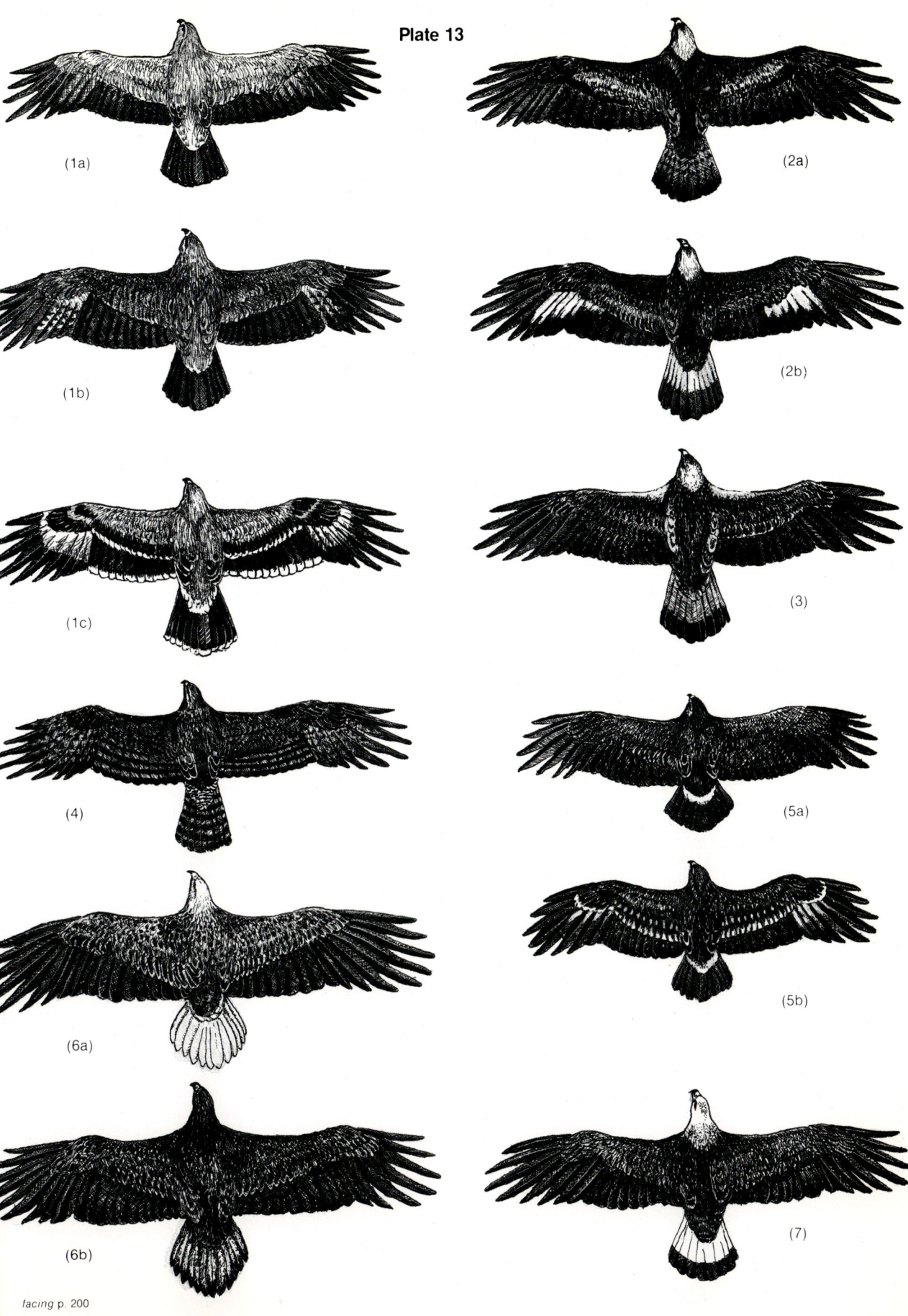

(1a)

(2a)

(1b)

(2b)

(1c)

(3)

(4)

(5a)

(6a)

(5b)

(6b)

(7)

Plate 14

(1)

(2)

(3a)

(3b)

(4a)

(4b)

(5)

(6a)

(6b)

(7a)

(7b)

(8a)

(8b)

(9)

along the main mountain chains. They also inhabit moister alpine slopes of Azad Kashmir and Hazara district and occur on all the higher ranges of the North West Frontier Province and northern Baluchistan. They are largely sedentary and I have watched them soaring over the Shyok river in Baltistan in mid-winter when the maximum day temperatures at valley floor level were about 0°C. Occasionally sub-adult birds may wander to the plains in winter. Recent Sind records include an immature bird which the author watched on 25 December on the Chaukandi ridge of hills (under 100 metres elevation), just north of Malir. It appeared exhausted and only flew a few metres each time it was flushed. F. J. Koning also saw an immature specimen soaring over the Indus delta on 13 January. On 17 March (1984) a sub-adult Golden Eagle was seen searching the cliffs at Karchat on the Kirthar hills, Sind, at less than 180 metres elevation, and another Golden Eagle was seen on 14 February 1985 in the same region (Khan Mohammad Khan, pers. comm.) no doubt attracted by the presence of newly born Ibex. It is noteworthy that 3 of these records, spanning just under ten years, are all of sub-adult birds which probably wander widely. Status SCARCE resident.

Map 79: Golden Eagle *Aquila chrysaetos*

HABITS

Usually seen solitarily and on the wing either soaring high overhead or searching the ground below as they glide over the steep slopes seemingly quite slowly and close to the ground, with hardly a wing beat, maintain-

ing balance and air speed by delicate movements of their wingtips. Meinertzhagen has recorded one while shooting Chukar Partridges (*Alectoris chukar*) in Baluchistan, diving into a covey as they flew across a wide valley and successfully striking a bird in mid-air. In Hazara district they have been seen catching unwary Long-tailed Marmots (*Marmota caudata*) as they sun themselves atop a large rock (Imtiaz, Chowkidar, Burawai Rest House, pers. comm., 1982). Most prey is in fact struck on the ground. They have been recorded as hunting Snow Pigeons and Snow Cocks (*Columba leuconota* and *Tetraogallus* species) in the northern regions. According to Bates and Lowther they rarely succeed in capturing the Himalayan Snowcock as it is such a wary bird (*Breeding Birds of Kashmir*, 1952). There are also records of them taking Hares in Baltistan. In Europe mammals are reported to be a preferred food over birds (Brown and Amadon, Vol. 2, 1968). The Cape Hare (*Lepus capensis*) is fairly plentiful in Baluchistan in the higher valleys frequented by the Golden Eagle. On 4 February 1985 one was actually watched seizing a newly born Ibex (*Capra hircus*) at Karchat, which it succeeded in carrying off in its talons with considerable difficulty, 'beating its wings like a helicopter' as it rose from the cliff face (Syed Asad Ali and Khan Mohammad Khan, pers. comm., February 1985). I have watched one in Kargah nullah, Gilgit, being mobbed by a Kestrel (*Falco tinnunculus*) as it sailed around at high altitude, and Koning saw a Greater Spotted Eagle (*Aquila clanga*) dive upon and mob a juvenile soaring around over the Indus delta.

Plate 14:
- (1) Black-breasted or Rain Quail *Coturnix coromandelica*, male.
- (2) Snow Partridge *Lerwa lerwa*, both sexes.
- (3a) See-see Partridge *Ammoperdix griseogularis*, female.
- (3b) See-see Partridge *Ammoperdix griseogularis*, male.
- (4a) Western Tragopan *Tragopan melanocephalus*, male.
- (4b) Western Tragopan *Tragopan melanocephalus*, female.
- (5) Himalayan Snow Cock *Tretraogallus himalayensis*, both sexes.
- (6a) Koklass Pheasant *Pucrasia macrolopha*, male.
- (6b) Koklass Pheasant *Pucrasia macrolopha*, female.
- (7a) Himalayan Monal *Lophophorus impejanus*, female.
- (7b) Himalayan Monal *Lophophorus impejanus*, male.
- (8a) White-crested Kalij Pheasant *Lophura leucomelana hamiltonii*, male.
- (8b) White-crested Kalij Pheasant *Lophura leucomelana hamiltonii*, female.
- (9) Cheer Pheasant *Catreus wallichii*, female.

BREEDING BIOLOGY

They are vocal during the nesting season uttering dog-like 'yelps' and pairs indulge in soaring displays over their territory. In May over the Neelum valley, Azad Kashmir, Kamal Islam saw a female turn over on her back and the pair clutching each other's feet, came tumbling through the sky (pers. comm., 1983). Another type of display of undulating dives and upward swoops, carried out by both sexes individually, which is believed to serve as a territorial display and threat to conspecific intruders (Harmata, *Raptor Research*, 1982). Williams describes 3 nests in the mountains around Quetta, Baluchistan from which he took eggs (Williams and Williams, *JBNHS*, 1929). Each nest was a huge platform of sticks built on the top of a juniper tree growing out of a cliff face. The nest also incorporat-ed bits of animal skin and rags and the depression in the centre being lined with grass. Two eggs is the normal clutch and on one occasion there was a chick and an infertile egg. Breeding starts in early March depending upon the severity of the winter. Eates through his collector got a clutch of two fresh eggs on 17 March and one partly incubated on 14 April from Murdar and Takhatu mountain ranges. The pair bond is believed to be life-long, often with two or three traditional nest sites being used. Incubation is mainly by the female and takes 43 to 45 days, with the period up to fledging a further 65 to 70 days. There are no descriptions of nests found in the Himalayas. Major C. H. Williams mentions that local shepherds in Baluchistan always try to destroy the eggs of this eagle, if the nest is accessible, since they consider it a threat to their young lambs and kids (op. cit., 1929).

Bonnelli's Eagle *Hieraaetus fasciatus* (Vieillot)
Synonym *Nisaetus fasciatus* (Ripley 'Synopsis,' first ed.)

Plate **12**
Map **80**

DESCRIPTION

Body length 68-72 cms. including tail 23-29 cms.
Wingspan 150-180 cms.
Wing length 46.5-50.7 cms. (males),
    47.8-52.3 cms. females (Cramp *et al.*)
Tail length 24.6-29.1 cms. (both sexes) (Cramp *et al.*)
Weight 1,712-2,386 gms. (Brown, and Amadon)

Larger than the Booted Eagle (*H. pennatus*) but with the same general proportions with a relatively long narrow tail, square tipped and long wings. Adults are variable in colour but there are several rather distinctive field characters. Generally they are pale buff or whitish all over the under parts from throat to the heavily feathered tarsi and this at once separates them from all the *Aquila* eagles and also from the Short-toed Eagle (*Circaetus gallicus*) because the latter always has a grey-brown throat and upper breast contrasting with the rest of the white under parts. There is fine dark shaft streaking of the breast feathers in adult Bonnelli's whilst sub-adult birds have quite rufous chestnut patches on the breast and belly.

In flight silhouette the head and neck look rather slender and long compared with the larger almost owl-like head and shorter neck of *Circaetus gallicus* and the broad dark sub-terminal bar at the tip of the tail is conspicuous from underneath whereas the other 3 or 4 narrower tail bars are only visible on the upper surface of the tail. Its upper parts are rather a grey-brown in adults, quite dark in tones, but with a conspicuous whitish patch in mid-back region. Some-times this shows as a paler streaky white collar around the top of the mantle. In contrast to the under-wing pattern of *Circaetus gallicus* the under-wing coverts are often quite dark brown, darker than the second-aries, whereas they are silvery white with parallel rows of dark spots in the Short-toed Eagle, and in the Booted Eagle (*H. pennatus*) the under-wing coverts are always paler than the flight feathers, the reverse of the pattern in Bonnelli's. Usually the leading edge of the wings of Bonnelli's adults when viewed from underneath, is paler creamy contrasting with the darker median under-wing converts. Sub-adult birds are rusty to fulvous on the breast with bolder shaft streaking and their upper plumage is more brown, less grey in tone than in adults. In their second year, their under parts become paler buff with bolder shaft streaks than is typical in adult birds, also the under-wing coverts may be buff or pale brown less dark than in adults. The sexes are alike both having orange to buffy yellow irides and bright yellow toes. The tarsus is fully feathered to the base of the toes.

HABITAT, DISTRIBUTION AND STATUS

Most typically a bird of the warmer dryer hill ranges of southern Baluchistan particularly the Chagai and Ras-Koh hills (Christison, 1942), the NWFP (resident around Kohat, Whitehead) and the Punjab Salt Range. Absent from the Himalayas and the well settled cultivated tracts, but occurring sparsely throughout the Indus basin, especially in the vicinity of larger lakes in winter. It also frequents desert border regions. It is a cliff nester by preference, but utilizes trees in parts of Sind. Status widespread but SCARCE.

HABITS

The pair bond is probably life-long and the male at least shows attachment (author obs.) to the nest site even outside the breeding season. They are strong flyers and active hunters, believed to subsist mainly on birds and mammals and to shun carrion and rarely attempt to take reptiles in contrast to *H. pennatus* or *A. rapax*. The author has twice been lucky to see one of a pair swoop down and take a hare. Once in Chak Jabbi, *Lepus nigricollis dayanus* (in the Salt Range), and once above Karchat in the Kirthar hills, Sind Kohistan *Lepus capensis*. They also take Grey Partridges (*Francolinus pondicerianus*), doves, and around lakes, waders and coots. Salim Ali ('Handbook', Vol. 1) records them as taking prey as large as the Painted Stork (*Mycteria leucocephala*) and a wounded Greylag Goose (*Anser anser*). Christison records them as frequently stooping to seize wounded Chukar (*Alectoris chukar*) when out shooting in the Chagai. Pairs often keep together outside the breeding season and occupy a favourite roosting or daytime perching site. This was observed in a lone tall tree on the banks of the east Narra canal and on the summit of a rocky ridge in the Kirthar hills during January and February visits.

BREEDING BIOLOGY

In hilly tracts cliff eyries are preferred, from sea level (Cape Monze near Karachi and Ormara, Makran) up to 2,134 metres (7,000 feet) in some of the Baluchistan hill ranges. Generally the nests are quite bulky stick structures and 3 or 4 nests are built within close proximity of each other on the same cliff site, each nest

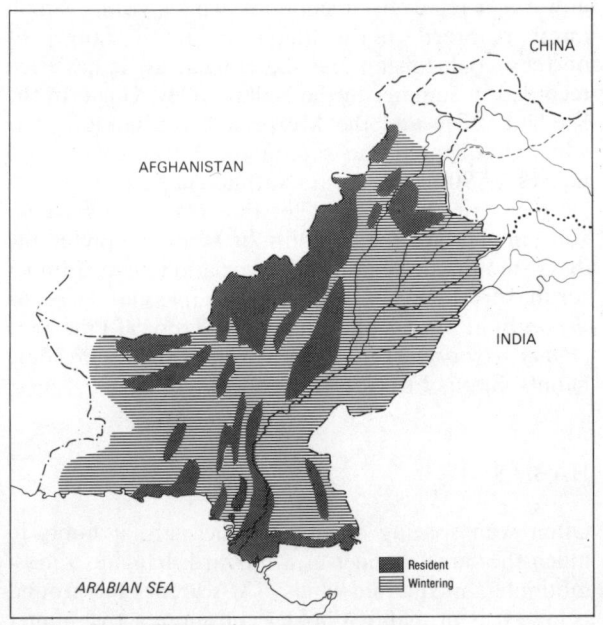

Map 80: Bonelli's Eagle *Hieraaetus fasciatus*

being repaired and used on an irregular basis. The breeding season is extended, being from December to February in Sind but later in April and May in northern and central Baluchistan where Marshall (*JBNHS*, 1903) found a nest with one egg on 14 May and another with 2 eggs on 26 May. Eates found thirteen nests in Thatta district and in the east Narra, most of these were in large trees, often very bulky structures. Out of 13 clutches taken, 7 were of 2 eggs and 6 of only 1 egg. He described the nests as being well lined with small twigs and grass and if in a tree, often a leafy one with the nest partly concealed and not right on top as in *Aquila rapax* (K. Eates, MS notes). A nest site in low hills near Cape Monze, discovered by the author in 1979, appeared to have been long used. It contained three very bulky stick nests in wind fretted hollows on a limestone cliff and different nests had been repaired and utilized in different years. According to available evidence, in one year only the male frequented the site, perhaps having lost his mate, but the following year, a new mate was evidently acquired. Egg laying usually starts in late December at this site and two eggs are the normal clutch. These are greyish white in colour, rather poorly marked with some light red blotching (K. Eates, MS notes). The female does 90 per cent of the incubation with the male perching nearby to warn of danger or defend the eyrie for they are bold in diving upon and attacking even a human intruder. Incubation takes about 37 to 40 days (Cramp *et al.*, Vol. 2, 1980) and fledging another 8.5 weeks, so that the young are usually not strong enough to leave the nest until the end of March. In 1983, two fully fledged and well grown young were seen perching on the sides of the nest at Cape Monze on 20 March. The female was perched close by but flew away silently and did not return. When attempts were made to photograph the nest from an adjacent cliff, the young retreated to the back of the rock ledge on which the nest was located and crouched out of sight. The male does most of the hunting and bringing food to the eyrie, whilst the chicks are continuously guarded or brooded by one or other parent during the first week or so after hatching. Both parent birds tend to be somewhat secretive while breeding and they do not soar over the nest site, or perch on prominent rock pinnacles, both of which habits are characteristic of the species when they do not have young in the nest.

VOCALIZATIONS

Not as vociferous as the Booted Eagle but pairs will call during soaring displays. One was heard on 17 January at Khadeji falls giving a repeated fluting call, a long inflected whistle 'eee-ou, eee-ou'.

On another occasion in late September a pair was pursuing each other over the Makli hills (Thatta district) when one bird gave rather short harsh low pitched calls, syllabized at the time as 'kik-kik-kik'.

Booted Eagle *Hieraaetus pennatus* (Gmelin)

Plate **12**
Map **81**

DESCRIPTION

Body length 45-53 cms. with tail 16-18 cms.
Wingspan 100-121 cms.
Wing length 34.2-37.8 cms. (males),
    37.4-42.5 cms. (females) (Brown, Urban and Newman)
Tail length 18.6-20.4 cms. (males),
    19.8-21.5 cms. (females) (Cramp *et al.,* Vol. 2)
Weight 510-770 gms. (males),
    840-1,250 gms. (females) (Cramp *et al.*)

Superficially like a Black Kite with relatively long narrow wings and tail but the tail is rounded at the tip and shorter on average than that of a typical *Milvus migrans.* It is generally slightly smaller than the Long-legged Buzzard (*Buteo rufinus*). It is a very variable bird in colour with three principal morphs, a very pale one in which the head is slightly buffy with the upper breast somewhat rufous but the rest of the body and under-wing coverts being white or greyish white with scattered dark shaft streaks. The contrasting dark blackish primaries make this pale morph very distinctive in under-wing pattern. Another morph is rather uniformly dark umber brown all over but with the tail generally paler chestnut and unbarred, and always two conspicuous white or creamy patches at the base of the forewing where it joins the body, clearly visible in flight when the bird wheels and can be seen from above. The most common morph is an intermediate phase in which the body is tawny with darker streaks and paler tips to the upper tail coverts and secondary wing coverts. In under-wing pattern this form shows darker flight feathers and paler brown under-wing

coverts, the most reliable diagnostic character in the field, as also small pale whitish patches at the base of each wing, visible on the dorsal surface. The general flight silhouette is of a small buzzard but with a longer tail. The tarsus is feathered down to the toes unlike that of *Buteo* species, hence its name. All colour morphs occur in Pakistan. An almost white specimen was seen in the Uruk valley, Baluchistan and another near Kalam in Swat, Kohistan. A uniformly dark umber specimen was seen at Malir near Karachi. A more typical form, such as one, noted on the summit of Miranjani in the Galis had the back umber brown with paler buff wing shoulders, whilst the under-wing coverts were creamy buff and the flight feathers blackish grey. Some specimens show faint barring on the tail. Very pale specimens look like miniature Scavenger Vultures in colour and wing pattern. The cere and feet are yellow and the strongly hooked bill horny black. The iris varies from orange to buff. Capt. Biddulph shot a specimen which weighed 680 grams (1.5 lbs.). Females are about 10 per cent larger than males.

HABITAT, DISTRIBUTION AND STATUS

A winter migrant to the Indus plains where it is more likely to be encountered in regions closer to the hilly country to the west of the Indus, than in the irrigated canal colonies. Specimens have been noted in the Punjab Potohar, Salt Range and around Attock, also the Thal and in Sind, Dadu and Thatta districts. It is however a relatively uncommon winter visitor. Small numbers breed in the higher mountain ranges of northern Baluchistan and the Himalayas. It has been recorded in summer in the Naltar valley, Gilgit, in the Kaghan valley and the Murree hills including a pair which frequented the summit of Miranjani 3,017.5 metres (9,900 feet) near Nathia Gali in 1969 and another pair seen in the Chir Pine (*Pinus roxburghii*) zone in the Murree foothills. It seems to prefer the dryer mountain regions in association with Juniper scrub forest (in Baluchistan), or Deodar (*Cedrus deodara*) in Swat Kohistan and the Tropical Pine zone (*Pinus roxburghii*) in Hazara district, in its breeding haunts. Status FREQUENT, locally migratory.

HABITS

Often seen soaring around on thermals, it hunts in much the same manner as a buzzard, hanging almost motionless in suitable winds as it searches the ground below. It is probably a bolder and more active hunter than *Buteo* species and will dive readily with half

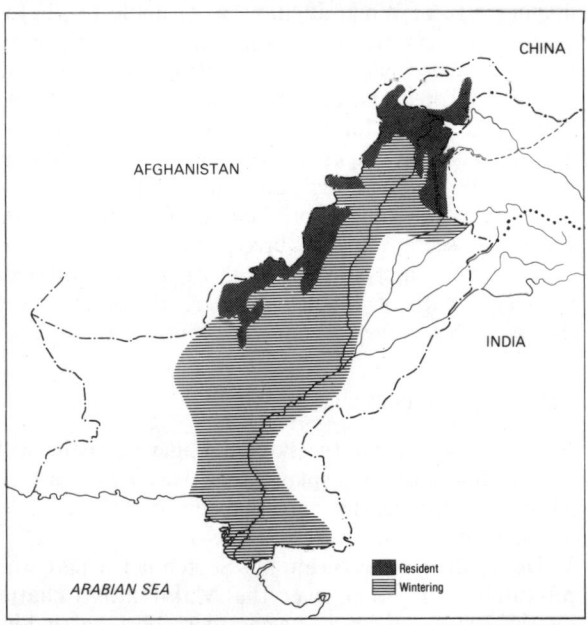

Map 81: Booted Eagle *Hieraaetus pennatus*

closed wings to swoop upon flying birds. A specimen in the Uruk valley plunged down into an apricot (*Prunus armeniaca*) orchard, twisting and diving between the trees more in the manner of a Sparrow Hawk (*Accipiter* spp.) than a *Buteo*. Its food prey is variable like most raptors, including birds, mammals and reptiles. In the Juniper forests of Ziarat, Baluchistan, in summer the staple food is probably *Agama caucasia* the Northern Rock Agama which it often hunts by perching on a strategic rock waiting to pounce on lizards which emerge to sunbathe. Evidence from feather pluckings in a small 'Tangi' (sheltered ravine) near Chautar, east of Ziarat which was in the territory of a pair of Booted Eagles indicated that it was probably this Eagle which had killed Mistle Thrushes (*Turdus viscivorus*) and they have been recorded (Ali and Ripley, 'Handbook', 1968) as taking Rock Partridges or Chukar (*Alectoris chukar*) which occur in the same habitat in Baluchistan. Pikas (*Ochotona* spp.) are probably hunted also in much the same manner as Rock Agamas, by waiting on the ground ready to pounce from a nearby vantage point. Pair bonds are long lasting and several authors refer to the habit of pairs hunting cooperatively (Ali and Ripley, 1968, and Cramp *et al.*, 1980).

BREEDING BIOLOGY

Nests were described by C. H. T. Whitehead from the Kaghan valley and Col. Rattray from Murree. Generally they are located at the top of a tall tree and are fairly large structures. Dr. Scully found in Gilgit, nests in tall Chinar trees (*Platanus orientalis*) in the main valleys (*circa* 1,524 metres (5,000 feet) elevation). Rattray's nest contained two well incubated eggs on 20 March and was located at about 2,134 metres (7,000 feet) elevation. Scully obtained a fledged nestling near Gilgit on 12 July. Whitehead's nest from the Kaghan was found in June at 3,048 metres (10,000 feet) with the female bird sitting, but on an infertile egg. A.E. Jones describes a nest found at 1,800 metres (6,000 feet) near Simla with two highly incubated eggs on May 10. It was located 23 metres up in a huge Deodar tree (A. E. Jones, *JBNHS*, 1938, p. 568).

The author has twice been privileged to watch the aerial display of this bird. In early May above Baba Kharwari in Baluchistan when a male repeatedly stooped with wings closed like a falcon upon the female, bringing himself up short with a spectacular upward stall each time, and calling loudly a two-noted whistle syllabized at the time as 'pi-wee-pi-wee-pi-wee'. The second occasion on 28 May on the alpine slopes of Musa-ke-Masala in Hazara district, only the male, alone was visible. He flew directly into view, calling loudly till high over a steep forested amphitheatre, here he suddenly commenced absolutely vertical plunges with wingtips pressed close against the flanks, falling with the velocity of a stone, an estimated 200 metres or more only to rise as suddenly and steeply into the air without any discernible change in half-closed wing position. At the end of the momentum of each almost perpendicular rise, and just stalling, it levelled out in flight and then repeated the plunge. After some minutes and about six looping plunges, it disappeared as suddenly as it had arrived. The normal clutch is 2 eggs, white or greyish white in ground colour, unmarked or sparsely spotted with faint reddish brown marks. There is obviously wide variation in choice of nest sites from isolated fir trees at the upper limit of tree growth down to the valley floors of the inner Himalayan ranges. Only the female incubates, with the male assuming the usual role of food provider. The young are believed to leave the nest about 90 days after incubation starts.

VOCALIZATIONS

They are noisy on the nesting territory but silent in winter quarters. One displaying male at Ziarat repeatedly called in a screaming whistle reminiscent of *Spilornis* species, syllabized at the time as 'pi-weepi-wee, pi-wipee wipee'. A male watched displaying high over the Machiara valley (Azad Kashmir) on 2 April called for about 3 minutes, each phrase lasting about 3 seconds, and being very evocative in cadence and structure to the song of the small Cuckoo *Cuculus poliocephalus*. Another bird watched at Lehtrar soaring over *Pinus roxburghii* forest occasionally emitted a rather rapidly repeated and short ringing call like the 'kik-kik-kik' of a *Dendrocopos* woodpecker. These two types of calls are closely similar to calls described by other observers (Brown and Amadon, 1968, Cramp *et al.*, 1980).

Hodgson's Hawk Eagle or
Mountain Hawk Eagle *Spizaetus nipalensis* (Hodgson)

Plate **12**
Map **82**

DESCRIPTION

Body length 72 cms.
Wing length 47.5-49.1 cms.
Tail 28.3-29.8 cms. (Stuart Baker)

A confusingly variable bird but only likely to be encountered in Himalayan forest zones. About the size of Bonnelli's Eagle (*H. pennatus*) but with comparatively longer tail and shorter rounded wings. In flight silhouette from beneath the throat looks white and lower belly chestnut with the under-wings silvery white and a prominent black trailing edge, also 3 dark crossbars along the secondaries. The tail is also crossbarred and the general pattern similar to *Pernis ptilorhynchus*, but it is a distinctly bigger and heavier bird than the Honey Buzzard, particularly about the head and bill. When seen perched, there is a prominent nuchal crest of long black streaked feathers and the upper throat is heavily streaked, with rufous crossbarring on the thighs and lower belly. Sub-adult birds are paler buffy colour on the head and body, lacking the chestnut cross-barring. The tarsi are feathered down to the toes as in *Hieraaetus* species. Some juvenile birds have an almost white head and body.

The upper parts are usually dark grey-brown not blue-grey.

HABITAT, DISTRIBUTION AND STATUS

Nowadays only likely to be encountered in well forested slopes around the lower Neelum valley, Jhelum valley or Murree hills. H. Waite collected a specimen on 14 June at 1,800 metres (6,000 feet) elevation near Murree in 1947 (specimen in British Museum) and Lt. Col. Buchanan found a nest at Changla Gali on 26 August 1900 and collected the female (Stuart Baker, 'Fauna Series', 1928).

In 24 years of visits to the Murree hills the author only twice encountered this eagle. A sub-adult bird on 23 May 1979 hunting along the ridges below Mukshpuri peak and a sub-adult bird on 2, 3 June 1986 on the summit of Mukshpuri peak. This bird when sailing around above the valley below was repeatedly mobbed by Jungle Crows (*Corvus macrorhynchos*), which it seemed to ignore. Status RARE and probably only irregular visitor.

HABITS AND BREEDING etc.

They will spend long periods perched on a tall tree watching and waiting and then pounce swiftly on suitable prey, their short rounded wings giving them manoeuvrability in trees. They are said to be capable of taking Koklass Pheasants (*Pucrasia macrolopha*) as well as smaller Himalayan forest birds. The bird encountered two days running in 1986, was perched on an isolated Silver Fir and on the second day, after being apparently driven away by mobbing Jungle Crows, returned and swooped low, or 'dive-bombed' low over my head, a territorial type of behaviour. Their nests are placed high up in a tree such as a Deodar (*Cedrus deodara*) growing out of a cliff, and the same nest may be re-used year after year being re-lined with fresh green leaves. The female alone incubates and the normal clutch is one egg. the laying season usually being April and May. Like all forest adapted raptors they are quite noisy relying partly on calls to maintain contact. A. E. Jones described its call as 'a shrill metallic whistle which might easily be imitated by a toy flute'. Brown and Amadon (op. cit., 1968) describe its call as 'klu-weet-weet'.

Map 82: Hodgson's or Mountain Hawk-eagle *Spizaetus nipalensis*

FAMILY PANDIONIDAE

Osprey *Pandion haliaetus* (Linnaeus)

Plate **12**

DESCRIPTION

Body length 55-58 cms. including tail 14-21 cms.

Wingspan 145-170 cms.

Wing length 45-51 cms. (males), 47-51 cms. (females) (Brown, Urban and Newman)

Tail length 18.7-21 cms. (males), 19.4-23.2 (females) (Brown, Urban and Newman)

Weight 1,120-1,740 gms. (males), 1,208-2,050 gms. (females) (Brown, Urban and Newman)

This specialized fish-eater is intermediate in size between the Long-Legged Buzzard (*Buteo rufinus*) and the Greater Spotted Eagle (*Aquila clanga*) in company with which it can occasionally be seen in Pakistan. It has relatively long wings, angled at the carpal joint, giving it a rather characteristic flight silhouette. The lower face and breast are shining white with a slight bushy crest on the nape which is streaked buff and dark brown, as also the crown, and a dark brown band through the eye. Its tail is fairly long and prominently cross-barred, and in under-wing flight pattern the secondaries are barred, the primaries tipped blackish brown and the primary wing coverts are blackish with the secondary wing coverts white. A circlet of black streaks frames the upper breast. The head and bill appear relatively small, though when perched the pale golden iris looks large and owlish. The cere is dull greenish blue and the legs are pale greenish grey with long white feathered tibia and naked tarsus. Immatures are similar to adults but with paler margins to the upper wing coverts giving a rather spotted or scaly effect.

HABITAT, DISTRIBUTION AND STATUS

A widespread winter visitor throughout the Indus basin closely associated with major rivers or lakes and jheels. It is much commoner on the seacoast and Indus delta and comparatively rare further north. It is virtually absent from Baluchistan and the North West Frontier Province due to absence of suitable large bodies of water. It regularly occurs at Rawal lake, Islamabad and visits the Punjab Salt Range lakes. There are always one or two individuals haunting the main irrigation barrage headponds. In southern Sind they are common, 15 were counted during a six hour voyage through Ghizri creek and Ibrahim Hydari on 25 October and 12 were once counted together, roosting on the ground around the drying out margins of a seepage pool at the north end of Haleji lake in early March. The odd individual remains throughout the summer. Two were seen on 8 July near Mirpur Sakro and another individual frequented Haleji lake throughout one summer. Most, however have left for their northern boreal forest breeding grounds by early April. Status COMMON.

HABITS

They will, when hunting, systematically beat up-wind, traversing a lake or tidal creek, wheeling back to the end of the lake again and beating slowly up-wind on another traverse. They catch fish by hovering, often before plunging which they do feet first with legs extended and frequently submerging totally under water. They must have remarkably keen eyesight as 3 different Ospreys were watched in Ghizri creek within the space of one hour successfully catching large fish in water that was not only rather turbid but also choppy due to a light breeze. Characteristically they emerge from their dive and if a fish is secured, grasping it with the fish's head forward and one foot firmly clasping the fish along its back and behind the other foot. One Osprey with a particularly large struggling fish, flew only a few hundred metres to a nearby sand bar in the Indus delta where it landed and appeared to be grappling with the fish. A local fisherman immediately plunged into the sea, it being very shallow and ran to the island to disturb the Osprey off its kill, hoping to frighten the bird and thus secure the fish, but it took off with the fish securely gripped in its talons despite the man's proximity. Characteristically Osprey will shake their whole bodies after a dive while still in flight to get rid of surplus water from their plumage. Not all dives are successful, however as one Osprey was watched, diving five times in succession without catching anything on Ghauspur lake. Occasionally Ospreys will soar around on thermals even doing so carrying partly consumed fish prey. They are fond of perching on telegraph or telephone poles to rest, in areas close to their feeding grounds, or on the open ground and seem to prefer such perches to trees. There is no evidence that they ever attempt to capture waterfowl, being specialized fish eaters.

BREEDING BIOLOGY

Extra-limital though, Ospreys do nest in comparable latitudes on the southern shore of the Caspian in Iran and on some islands in the Persian Gulf, but there is no evidence of breeding by birds which over-summer in Pakistan. In the USSR they normally build a huge bulky nest in the topmost fork of a tall tree, often returning to repair and add to the same nest. The normal clutch is 2 to 3 eggs and the incubation period is 35 to 38 days, largely carried out by the female with the male bringing food to his brooding mate.

*VOCALIZATIONS

They only occasionally call on their wintering grounds but in October at Haleji lake a pair of birds has twice been seen chasing each other and performing a mutual soaring display with one bird repeatedly calling, a high almost chirruping whistle. Another Osprey perched on a partly submerged snag called loudly when another Osprey flew overhead in early March. This was a loud but similar toned whistle-like call 'chewk-chewk-chewk'.

ORDER FALCONIFORMES
FAMILY FALCONIDAE
Eurasian Kestrel *Falco tinnunculus* (Linnaeus)

Map **83**

DESCRIPTION

Body length 32-35 cms. including tail 12-15 gms.
Wingspan 71-80 cms.
Wing length 23-26.3 cms. (males),
   24.5-26.4 cms. (females) (sub-species *interstictus*) (Stuart Baker)

This is a delicate looking falcon with relatively long slender tail and long pointed wings usually discernible from all angles. Its most notable characteristic in the field is its ability to hover in the air with shallow rapid wing beats while searching the ground below. The body is a reddish chestnut with black spots, smaller on the wing shoulder and with brownish black flight feathers when viewed from above. The throat is whitish and breast fulvous buff with scattered dark shaft streaks on the feathers. The crown, hind neck and also the tail of the male is blue-grey with a single broad blackish sub-terminal band on the bottom of the tail whilst the female has a chestnut tail closely cross-barred and her crown is chestnut with very small narrow streaks. Both sexes show a black moustachial streak extending in a curve from below the eye down either side of the throat. The iris is dark brown and the cere and feet, which are unfeathered on the tarsi, are yellow. There is also an area of yellow naked orbital skin around the eye. In flight silhouette when viewed from below the under-surface of the tail and wings is silvery grey with cross-barring obsolescent except for the broad black terminal tail bar visible in both sexes. The under-wing coverts are a buff to fulvous with black streaking. Similar in size to *Accipiter nisus* with which it could be confused, the more pointed wingtips of the Kestrel and less prominent barring on wings and tail are usually sufficiently distinctive. Females are slightly larger than males.

HABITAT, DISTRIBUTION AND STATUS

Largely resident and breeding throughout the mountain tracts of Pakistan with some winter migration of part of the population to the plains. This wintering population generally avoids well wooded and cultivated tracts, preferring scrub desert and the 'barani' or dry farming foothill zones such as the Punjab Salt Range and the plains areas around Dera Ghazi Khan, Bannu and Indus Kohistan. It is quite common and widespread throughout Baluchistan down to the Makran coast, and the Himalayas in mountainous regions, with a substantial part of the population remaining in the hills. Birds have been observed in Gilgit in mid-December and Baltistan in the Indus and Shyok valleys in early February and around Quetta in January. Winter migrants to the plains may be noted from the end of September and remain up to the end of March. In the absence of any ringing data, migration patterns remain speculative but some central Asian breeding birds, characteristically paler than the nominate sub-species (which is the typical breeding bird in the Himalayas), have been collected in winter in Punjab and Sind. Whistler noted a marked northward passage in March in Rawalpindi district (Whistler, 1930 and Ticehurst, 1923). Brigadier Christison noted marked migration of Kestrels in spring in the Chagai on the south-western borders of Baluchistan. They were moving slowly eastwards, birds flying short distances, low to the ground and averaging only about 13.8 kms. (8 miles) per hour due to frequent stops. He was able to count between 50 and 60 within his range of vision on some occasions and the passage continued for about 10 days from 14 April. A specimen which he collected from the migrants was typical of the race *Falco tinnunculus tinnunculus* (Christsion, *Ibis*, 1941). It is tempting to speculate that these were birds which had wintered in Africa returning to breeding grounds north of the Himalayas. Status in Pakistan COMMON.

HABITS

They do not stoop on prey like the larger falcons but rely largely upon a hovering technique with relatively gentle descents to the ground to catch prey on the ground. Their food is very variable according to local conditions and abundance. Ticehurst (*Ibis*, 1923) shot a bird in Sind which had been feeding exclusively upon locusts. Captain Biddulph (*Stray Feathers*, 1881) records as many as 20 Kestrels in Gilgit valley, hovering over the ripe wheat fields when they are being harvested to catch field mice. The author has, on several occasions seen Kestrels carrying Rock Agama Lizards (*Agama himalayana* and *A. caucasia*) in their talons both in the Kaghan valley and in Ziarat, Baluchistan and a Garden Lizard (*Calotes versicolor*) in the feet of one by Rawal lake, Islamabad. In the desert areas besides locusts, they catch ground roosting birds such as Pipits and Larks as well as orthopterous insects and Dung Beetles SCARABIDAE. Probably in the warmer regions of Pakistan with less grass cover and absence of Microtine voles, a staple food in Europe, their diet in Pakistan consists predominantly of small birds, reptiles and insects. In the Hannah valley due east of Quetta, above a rocky gorge, a flock of 8 Kestrels was watched on 26 March circling around and catching with their feet unidentifiable insects which they then transferred to their bills and ate in flight.

BREEDING BIOLOGY

Courtship can be prolonged, a pair occupying its nesting territory and calling to each other for 2 or 3 months before nesting commences. Males have been seen presenting *Agama nupta* Rock Lizards to the mate perched on a crag below Bhagnotar, Murree hills as part of courtship. Both sexes frequently call from a prominent tree perch or cliff as well as when sailing around in the air, but most commonly one of the pair gives its characteristic tremulous advertising or invitation call, when perched. Nesting is usually on a cliff ledge with little or no nesting material being placed on the ledge and most eggs are laid between April and early June. A nest with 4 eggs was found on 17 April in an earthern cliff on the slopes of Sakesar in the Salt Range (H. Waite, 1948). Whistler (*Ibis*, 1930) found nests with eggs in the Murree hills in May and Meinertzhagen (*Ibis*, 1920) recorded complete clutches laid on 14 May and one with 2 eggs on 21 May, around Quetta valley Baluchistan.

A nest found by the author contained downy young on 19 May in the Mana valley at about 2,134 metres (7,000 feet) elevation in north-eastern Baluchistan. It was located in a wind-fretted hollow in limestone cliff, there being no visible nest material beyond remains of food kills beneath the chicks. An apparently occupied nest was however located in a Deodar Tree (*Cedrus deodar*) found by the author in early June near Drosh in southern Chitral at about 2,134 metres (7,000 feet) elevation which was in an old crow's nest and Bates and Lowther record Kestrels utilizing a Jungle Crow's nest (*Corvus macrorhynchos*) in a pine tree in Kashmir (op. cit., 1952). Cliff sites are however much more

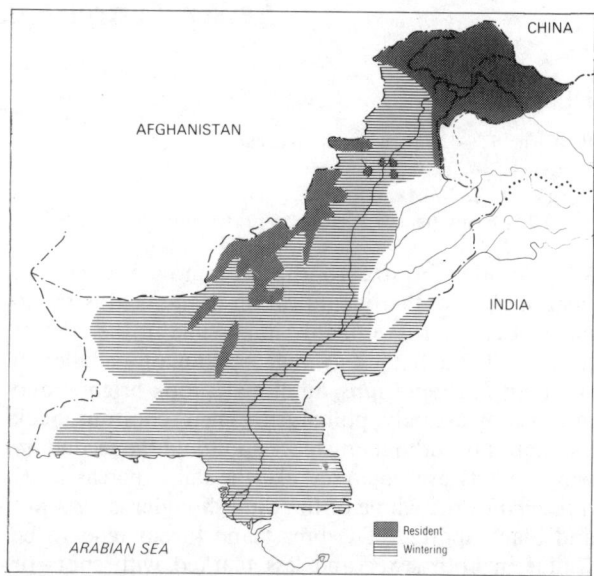

Map 83: Common or Eurasian Kestrel *Falco tinnunculus*

usual in Pakistan and indeed almost every suitable cliff site is occupied by a breeding pair in the narrower side ravines and valleys of Baluchistan, Chitral and Gilgit. The normal clutch is 4 eggs but 2 or 3 incubated eggs have been recorded. Usually only the female incubates, (Cramp *et al.*, 1980), the male regularly bringing food to her. The eggs are usually dull red in ground colour covered all over with darker red-brown smears, streaks and blotches. Incubation lasts 28-9 days and the chicks hatch asynchronously as the eggs are laid over a period of 5 to 8 days. Copulation may continue for some weeks before egg laying commences.

*VOCALIZATIONS

The most characteristic call is a querulous drawn-out 'wriih' or 'whreeah' given by either sex as an invitation or advertising call usually when perched near a potential nest site. A pair was heard calling together for some ten minutes on 17 May near Sojh in the Kaghan valley, and this duetting is very characteristic of the early nesting stages. They were perched on separate fir trees (*Abies pindrow*), one bird constantly uttered the 'vree-vree' tremulous call, the other responding with a short sharp and rapid 'kek-kek-kek'. This latter type of call is often uttered by one bird when the pair is soaring or when approaching the nest. On one occasion two females were observed chasing each other near Lake Saif-ul-Maluk (Kaghan valley) when the pursuer uttered a slightly quicker more excited 'kik-kik-kik' call. Copulation has been observed following the 'vreeh-vreeh' call which in this instance was given by the female. They can be quite crepuscular in activity and a male (?) was heard uttering the 'vreeh-vreeah' call repeatedly, well after dusk had descended, at Hazar Ganji in the Chiltan hills outside Quetta.

Lesser Kestrel *Falco naumanni* (Fleischer)

DESCRIPTION

Body length 29-32 cms. with tail 11-12 cms.
Wingspan 58-72 cms.
Wing length 22.2-24.4 cms. (males),
    23.9-24.6 cms. (females) (sub-species *pekinensis*) (Stuart Baker)

Very similar in size to the Common Kestrel (*F. tinnunculus*) so that identification in the field is by no means easy and is probably impossible in the case of females. Though they average fractionally smaller in size than *F. tinnunculus,* adult males look brighter and more contrastingly plumaged. Their chestnut back and nape are without any black spots and the secondary wing coverts are blue-grey like the tail, whereas in *F. tinnunculus* the whole of the wing shoulder is chestnut and black spotted. The breast and throat tend to be whiter in both sexes and less marked with spots or streaks on the flanks. Females are barred on the tail and black spotted over the body like female *F. tinnunculus.* Two reliable characters which separate the Lesser Kestrel from the Common Kestrel are the elongated central pair of rectrices and the pale horn coloured claws. The claws of *F. tinnunculus* always being black. In soaring flight the different tail pattern often shows up well despite the central tail feathers only extending fractionally (observed in migrating flocks of *F. naumanni* in Zimbabwe).

HABITAT, DISTRIBUTION AND STATUS

A rare straggler to the Indian subcontinent, mainly in peninsular India as there is a migration passage between east African wintering grounds and breeding grounds in eastern Siberia. It is generally highly gregarious during migration but no such passage has even been noted or recorded through Pakistan. However stray birds may be over-looked as a female shot in Gilgit by W.M. Conway is in the British Museum. The label gives no details as to original location, or collection date, and indeed was originally wrongly identified as *F. tinnunculus.* A population breeds in Turkestan and regularly occurs on migration in south-west Iran, so it should be looked out for in Baluchistan. Status UNKNOWN, presumed VAGRANT.

HABITS AND BREEDING

Very gregarious in breeding, roosting and on migration. It is largely insectivorous and catches a lot of flying insects in aerial sallies as well as hunting low over the ground and hovering before picking up insects on the ground.

Red-headed Merlin or Turumtee *Falco chicquera* (Daudin)   Illus. **29**
Map **84**

DESCRIPTION

Body length 31-36 cms. with tail 12-13 cms.
Wingspan 60-70 cms.
Wing length 19-20.1 cms. (males),
    22-23.2 cms. (females) (Stuart Baker)

A very beautiful little falcon, with sexes alike in colouring though the female is usually slightly larger with heavier darker cross-barring on the breast. Generally they are very pale blue-grey on mantle and wings with flight feathers blackish and the grey tail narrowly white tipped with a broad dark sub-terminal bar. The crown down to the nape is reddish chestnut, the cheeks and throat are pure white, the rest of the breast and feathered tibia being finely cross-barred white and dark grey. A darker chestnut moustachial streak extends down from below the eye, each side of the throat. The legs and feet, cere and naked orbital ring are all-yelow. The iris is brown. Juvenile birds have the chestnut head and neck streaked black and heavier black streaks on the upper breast and bars on the lower flanks. The back and wing shoulders are also barred with black. In flight they look very like a large Merlin with short narrow tail and strong direct flight with narrow pointed wings.

HABITAT, DISTRIBUTION AND STATUS

Resident and believed to be largely sedentary, they are very widely but thinly distributed in the Indus plains and outer foothill areas of NWFP and Baluchistan. Occasional birds are seen in the desert border regions of Tharparkar and Cholistan, but it is entirely absent from the higher plateaux and mountain regions of Baluchistan. Whistler (*Ibis*, 1930) only saw it twice around Rawalpindi and Whitehead (*JBNHS,* 1911) noted it around Peshawar and Mianwali but considered it rare around Kohat. It seems to like groves of trees on the edge of desert as a pair haunted the rest house garden at Tajjal rest house on the east Narra and another frequented the rest house garden at Fort

Illus. 29: Red-necked Falcon or Red-headed Merlin *Falco chicquera*, adult, both sexes.

Abbas. The author has records of pairs seen around Rawal lake, near Gujranwala, outside Karachi at Malir, Lal Sohanran near Bahawalpur, Ghauspur jheel in Jacobabad district and on the Sibi desert plain. It is probably less rare in Sind than Punjab, but like all the facons occurring in Pakistan, with the single exception of the Kestrel, it appears to have declined drastically in numbers since the 1940s, due in some measure at least, to the falcon trade.

Characteristically hunting cooperatively, the pair uses combined tactics to cut off escape routes of flighting birds which they are pursuing. They are active and quick hunters swooping down, in low and direct flight to flush flocks of ground feeding birds and then relentlessly pursuing their intended quarry with twists and turns. Their main food prey is probably birds and most of these are taken on the wing. I have watched one hunting a flock of Ashy-crowned Finch Larks (*Eremopterix grisea*) which were first flushed by the swooping bird and then one individual was hotly pursued as the little flock dispersed in their panic. Another successfully swept down upon feeding Little Stints (*Calidris minuta*) and killed one in the air. They must occasionally swoop upon birds in mock attack as one was watched pursuing Plain Sand Martins (*Riparia paludicola*) flying over Haleji lake though these Martins when pressed could easily out-manoeuvre and outfly their attacker (observed in December in late afternoon). In a rocky gorge of the Baran nullah to the west of Thanabulla Khan in Dadu district, where a pair has its territory, one was watched launching from its perch and swooping upon a Mallard Duck (*Anas platyrhynchos*) as it flew up the gorge. This attack was presumably a form of 'play', the quarry being too big for this little falcon to attack normally as food. *Streptopelia* doves are an important food item and I have seen one carrying a Pied Bush Chat (*Saxicola caprata*) to present to its mate. Over the year the pair is remarkably faithful to its nesting territory, usually continuing to remain within a few miles of the nest site, outside of the breeding season.

They probably will take small mammals and reptiles occasionally and have been reported hunting bats as they emerged at dusk (Mian Mohammad Sharif, in Rahim Yar Khan, pers. comm., 1971 and Fry, 1964). K Eates (MS notes unpublished) noted them feeding on locusts when these were plentiful.

BREEDING BIOLOGY

Invariably nesting in trees, they often repair an old crow's nest, or a kite's nest and as far as is known never construct a completely new nest. Acacia trees are favoured, but *Casuarina* and even date palms have been used in Sind (K. Eastes, MS notes), and they prefer fairly well wooded areas. The author has come across a nest in a mango orchard (*Mangifera indicus*) near Chauhar Jamali, Thatta district and another in a Neem tree (*Melia azederach*) in the east Narra. The normal clutch is 4 eggs, rarely 5 and these are laid from late February till early April. Records of 23 clutches taken in Sind by K. Eates showed a mean of exactly 4 with only 4 clutches of 5 eggs. Earliest date clutch completed 12 February, latest date 20 March (Eates, MS notes unpublished). In Ticehurst's day (*Ibis*, 1923) a pair even bred in Burns' Public Gardens in Karachi city but increase of traffic and the spread of the city has long since made this locality unsuitable. He notes fully fledged young on 16 April in this Karachi nest. The eggs vary from pinkish red to yellowish stone in ground colour, thickly freckled with dull brownish red. Usually only the female incubates (Brown, Urban and Newman 1982) despite the statement in Ali and

Map 84: Red-headed Merlin or Red-necked Falcon *Falco chicquera*

Ripley's 'Handbook' (1969). The male is the main food provider during incubation which lasts about 33 days (Brown and Urban, 1982) and courtship feeding also occurs even before nesting has started. The pair is very noisy at its nest site, the male calling loudly when he brings food, and they also indulge in duetting from perches near their nest site at the beginning of egg laying (obs., Malir). Incubation starts with the first egg laid and hatching is asynchronous.

VOCALIZATIONS

A rapid twittering call has been heard at the nest and also shrill chattering screeches, the latter appeared to be a threat or warning call when House Crows (*Corvus splendens*) were being driven from the nest. It also has an advertising or territorial type of call similar to a Merlin's (*Falco columbarius*) consisting of a high pitched squeal.

Merlin *Falco columbarius* (Linnaeus) Map 85

DESCRIPTION

Body length 25-30 cms. with tail 8-10 cms.
Wingspan 50-62 cms.
Wing length 19.6-20.4 cms. (males),
    22-22.4 cms. (females) (Stuart Baker)

About the same size as the Red-headed Merlin (*Falco chicquera*) and of similar body proportions and character, being a bold and dashing falcon despite its small size. The upper surface is much darker than *F. chicquera* in the male, the crown being slaty dark blue-grey on the upper parts and with finely black streaked, buffy or rusty fulvous under parts. The grey tail is cross-barred blackish with a broader sub-terminal band and white tips, and the primaries are also blackish. The eye is large and dark and there is no distinct moustachial streak. Females are about 10 per cent larger with darker brown-black wings and tail lacking the slaty tones of the male and with more contrasting creamy white bars on the tail and broader black tear drop streaks on the lower flanks and tibia. The legs, feet and cere are yellow. There is generally a faint whitish streak above the eye. When at rest the wingtips do not quite reach to the tip of the tail.

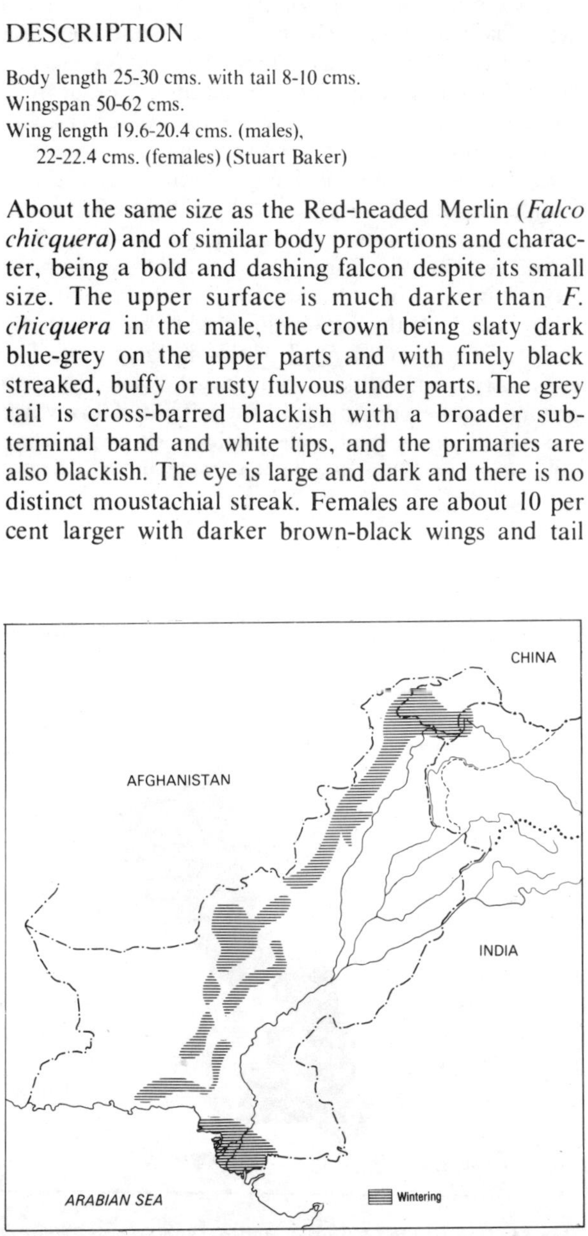

HABITAT, DISTRIBUTION AND STATUS

The Merlin is a Palearctic winter visitor to the north-western part of the subcontinent, occurring very sparingly in all four provinces. It seems to inhabit any type of fairly open country including cropland, scrub desert and mountain valleys. Brigadier Christison recorded it as a rare winter visitor around Ziarat and Loralai in north-eastern Baluchistan (*JBNHS*, 1942) and Captain Biddulph (*Stray Feathers*, 1881) considered it an uncommon winter visitor to Gilgit, sticking mainly to the side valleys and narrower gorges or ravines. Whitehead (*JBNHS*, 1911) recorded it as a fairly common winter visitor around Kohat and Whistler (*Ibis*, 1922) considerd it a regular but uncommon winter visitor around Jhang district in the Punjab. Recent records (author) include fairly regular sightings around Dhabeji and Mirpur Sakro in Thatta district of Sind, also at Taunsa barrage in Muzaffargarh district of the Punjab and Rahim Yar Khan district in southern Punjab. Being a migrant visitor, single birds are always seen, in contrast to the resident *F. chicquera*, which is usually in pairs. The earliest dates noted (author's diary) are the end of October and latest dates end of February. Status RARE.

HABITS

They hunt mostly small birds which they strike down in flight, hunting by swift low and stealthy flight along the edges of tree avenues bordering cropland, alongside canals and other areas on the edge of cover and around bush-dotted sand hills. Ticehurst (*Ibis*, 1923) saw one kill a Pied Wheatear (*Oenanthe picata*) outside Karachi

Map 85: Merlin *Falco columbarius*

and the author saw one flush a flock of Short-toed Larks (*Calandrella cinerea*) from a fallow field, chasing them low over a cotton field just at dusk and striking a bird within a few metres of the ground which it immediately took to an embankment and started to consume. This method of hunting is typical for the species which rarely stoops from a great height, preferring the quick low stealthy attack, striking a fleeing bird in direct flight. They spend considerable parts of the day perched on some vantage point such as a telegraph pole or bush on a rise in the ground and watch out for prey from such vantage points. They have been recorded catching dragonflies in other areas, and no doubt will take small mammals, but ground-feeding birds seem to be their principal prey.

BREEDING etc.

Extra-limital. They are usually silent outside of the breeding season. They breed in open heath land or sand dunes, nesting on the ground.

Northern Hobby *Falco subbuteo* Linnaeus

Map **86**

DESCRIPTION

Body length 30-36 cms. with tail 8-10 cms.
Wingspan 82-92 cms.
Wing length 25-27.5 cms. (males), 27.7-28.6 cms. (females)
 (Himalayan population) (Stuart Baker)
Weight 131-232 gms. (males),
 141-340 gms. (females) (Cramp *et al.*)

Larger than the Red-headed Merlin or Holarctic Merlin (*F. chicquera* and *F. columbarius*) with a slim and graceful build and relatively longer more pointed wings. At rest the wingtips extend just beyond the tail tip (author obs., Dunga Gali). They are consummate flyers and gliders, often hunting high in the air. About the same size as *Falco tinnunculus* they are predominantly dark brown to slaty brown on the crown and mantle with pale whitish throat and breast heavily streaked with black spots. They have a prominent rufous chestnut feathered tibia and under-tail coverts. The long slim tail is slaty brown with outer feathers showing narrow rufous cross-bars. The throat is bordered on each side by blackish moustachial stripes. The dark brown eye is rimmed with a yellow orbital ring as also the cere.

Females are about 10 per cent larger and much browner on the crown and back, sometimes almost black. Juveniles are like the females but with pale buff margins to the feathers giving a scale-like pattern to the back and wing shoulders. There is usually an indistinct paler creamy supercilium or line above the eye.

HABITAT, DISTRIBUTION AND STATUS

It is a summer breeding visitor to the northern mountain regions and uncommon irregular winter visitor to the Indus plains. Much less of a forest dwelling bird than its Oriental counterpart, *Falco severus*, it will typically be associated in its breeding haunts with the outer Himalayan forest at lower elevations, including the sub-tropical 'Chir' (*Pinus roxburghii*) zone, but will inhabit quite open country denuded of forest such as around the lower Kunhar valley at Balakot, and even inner Himalayan valleys such as Shangrila, south of Skardu in Baltistan (P. Fabian pers. comm., July 1984) and Besal in the inner reaches of the Kaghan valley. In winter it can turn up anywhere in more open country, having been recorded in Lasbela and occasionally in various parts of Sind. H. Waite (*JBNHS*, 1948) recorded it on autumn passage around the summit of Sakesar in the Salt Range and Christison noted it around the higher peaks of Ziarat in north-eastern Baluchistan (*JBNHS*, 1942). Meinertzhagen had records of two seen at Ziarat on 24 July and one collected near Quetta on 1 October, (Meinertzhagen, *Ibis*, 1920). Ticehurst saw 1 or 2 in the

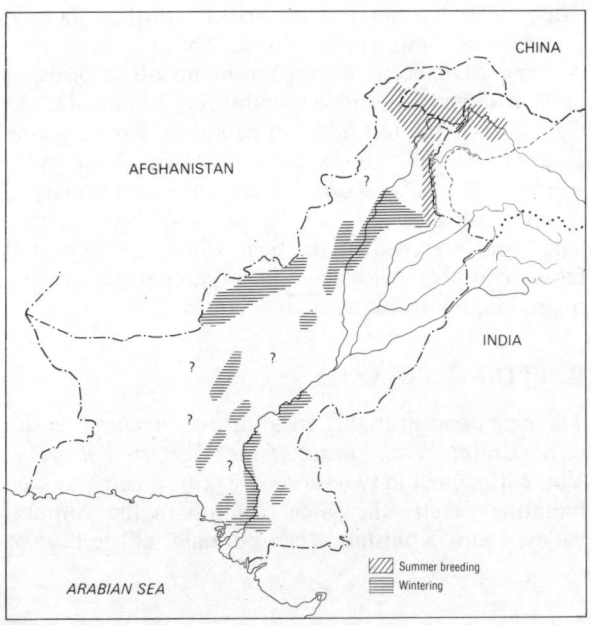

Map 86: Hobby *Falco subbuteo*

hilly areas to the north of Karachi each winter (*Ibis*, 1923). The author in ten years' residence in lower Sind noted only one on 13 December 1974 near Morobulla Khan, Thatta district. However it is not uncommon in its northern breeding haunts, particularly the outer spurs of the Murree hills, where 5 or 6 birds can sometimes be seen in the air together, exploiting insects in a favourable air current. Status FREQUENT and migratory but a rare winter visitor.

HABITS

Typically the Hobby hunts by very swift flight, its long narrow wingtips scissoring the air like a swallow, and interspersed with short swift glides. It is particularly active in hunting at dusk. Another characteristic technique involves fairly leisurely sailing around on thermals and circling through thermal up-draughts, catching larger insects such as Odonata and Coleoptera carried up in such air currents. In many areas, individuals seem to specialize on bats (Microchiroptera), and this was observed (author) in a specimen haunting the old garrison church at Khanspur in the Murree hills, as well as being noted by Ticehurst (*Ibis*, 1923) in the birds wintering around Karachi. Sometimes they are quite gregarious in hunting insects, as the author watched 5 circling around near Bhurban in October in the Murree hills, feeding largely upon Odonata. The insects were seized by the feet and transferred to the bill for consumption in flight. The fast flying Sheath-tail Bats (*Taphozous nudiventris kachhensis*) which roost colonially, often attract Hobbies in winter and in Rahim Yar Khan the author saw one making regular evening meals, overtaking them easily with one quick swerve. They can, in fact, catch swallows and martins (HIRUNDINIDAE family) on the wing (Cramp *et al.*, Vol. 2, 1980)

They have been recorded robbing other birds of prey of their food and one individual around Dunga Gali learned the technique of capturing Rock Agama Lizards (*Agama nupta*), as it was twice seen flying from a cliff-face with one of these large lizards clasped in both feet close to its belly like a torpedo carrier. They are largely silent in their winter quarters and from available evidence, rather gregarious during migration, but in winter hunt solitarily.

BREEDING BIOLOGY

The pair bond probably lasts for one breeding season only (Cramp *et al.*, *Birds of the Western Palearctic*, Vol. 2, 1980) but in two successive years a pair was seen haunting exactly the same cliff top in the Alipurai valley, Indus Kohistan. They normally utilize the old nest of a tree-nesting bird. In well wooded areas they utilize the abandoned nest of a Jungle Crow (*Corvus macrorhynchos*) or any large stick nest located high up in a tall tree. Captain Biddulph (*Stray Feathers*, 1881) records them as arriving in the main valleys of Gilgit to breed at the end of April. Colonel Buchanan found a nest below Changla Gali with 3 unincubated eggs on 20 June at about 1,524 metres (5,000 feet) elevation and noted that this nest was well lined with goat's hair and wool, but as far as is known neither bird in the pair add any material to the old nest which they adapt for their eggs. Buchanan's nest was located at the top of a *Pinus wallichiana* Blue Pine (Whistler, *Ibis*, 1930). They seem to be late breeders, as a pair was watched indulging in aerial displays and haunting the summit of Mukhshpuri in early June with apparent intention to breed, but after about one week they deserted the area. Bates and Lowther record nests with eggs in late June and July and confirm that they are late nesters in Kashmir (Bates and Lowther, 1952). They did not come across any cliff nests in Kashmir but cliff sites are occasionally used in the USSR (Dement'ev *et al.*, 1951). It is probable that the pair of Hobbies frequenting Shangrila in Baltistan (Paul Fabian, pers. comm.), a relatively treeless region, chose a cliff site and this appeared to be the site of the pair seen (author) in the Alipurai valley in 1985 and 1986, as there were no trees in the vicinity of this cliff.

The pair perform marvellous aerobatical displays during the nesting season and the male indulges in courtship feeding, calling loudly when he approaches with food. Three to four eggs are the normal clutch of a dull yellowish buff to pale brick red ground colour, densely speckled with darker red-brown. The female does most of the incubation (Whistler, *JBNHS*, 1928a). The eggs hatch asynchronously after about 28 days incubation and during this period the male often calls the female off the nest to feed her, performing a spectacular aerial food pass.

VOCALIZATIONS

They call in relatively rapid bursts of short syllable type calls, usually some tonal variation of 'tew-tew-tew' or 'kiu-kiu-kiu' and both sexes have a similar call except that the male is usually higher pitched. Whistler described their call as a harsh plaintive and rising 'tee-tee-tee' (Whistler, 1928a). A pair haunting the summit of Mukhshpuri mountain in the Murree hills during early June often indulged in spectacular courtship flights during which both uttered rapid 'kiu-kiu-kiu' calls in bursts of 4 to 6 calls. The male was also heard calling while hunting bats even after sunset.

Oriental Hobby *Falco severus* (Horsfield)

DESCRIPTION

Body length 27-30 cms. with tail 9-11 cms.
Wingspan 55-82 cms.

A slightly smaller version of the Hobby (*Falco subbuteo*), but distinguished from the latter by its overall darker upper parts being slaty black on crown and mantle and with the whole of the breast a deep rufous chestnut. Sub-adults have the chestnut breast heavily streaked black, but in adults the breast is plain chestnut, unmarked. The dark chestnut breast at once distinguishes this bird from *Falco subbuteo*.

HABITAT, DISTRIBUTION AND STATUS

It is typically a forest bird, extending over most of south-east Asia and the eastern Himalayas, including Nepal where it breeds.

In Ripley's 'Synopsis' both first and second editions (1961 and 1982), its distribution is given as extending from 'Pakistan (Murree?) and Kashmir (?) eastwards'. I have not been able to trace any evidence for this statement and all the records of Hobbies (Rattray, 1905, Whistler, 1930 and author's obs.) noted in the Murree hills or anywhere else within Pakistan territory have been of *Falco subbuteo*. For the present time therefore, I prefer to exclude it from the **Checklist.**

Sooty Falcon *Falco concolor* (Temminck)

Illus. **30**
Map **87**

DESCRIPTION

Body length 33-36 cms. with tail 12-15 cms.
Wingspan 85-110 cms.
Wing length 26.4-28.3 cms. (males),
   27.3-29.7 cms. (females) (Cramp *et al.*)

Very similar to a Hobby (*Falco subbuteo*) in size and build, with long slender wings and swift graceful flight. Adults are variable with light and dark morphs, the latter being confusingly similar to Eleonora's Falcon

(*Falco eleonorae*). The most typical adult phase is, however much paler than *F. eleonorae*, being a uniformly ashy grey-brown bird, concolorous in both breast and under-wing pattern, but dorsally showing slightly darker wing and tail tips. Juveniles are streaked with buff and dark grey on the body and under-wings and the tail shows lighter buff cross-barring. The iris is dark brown and the cere, orbital ring, legs and feet tend to be rather orange in colour, becoming almost reddish in some older birds, particularly in darker morph birds. Some sub-adult birds are creamy white about the throat and lower cheeks with heavy streaking on the breast, at which stage they are probably indistinguishable from *F. subbuteo*. Their toes are particularly long and well developed. When at rest the long slender wingtips extend just beyond the tail tip.

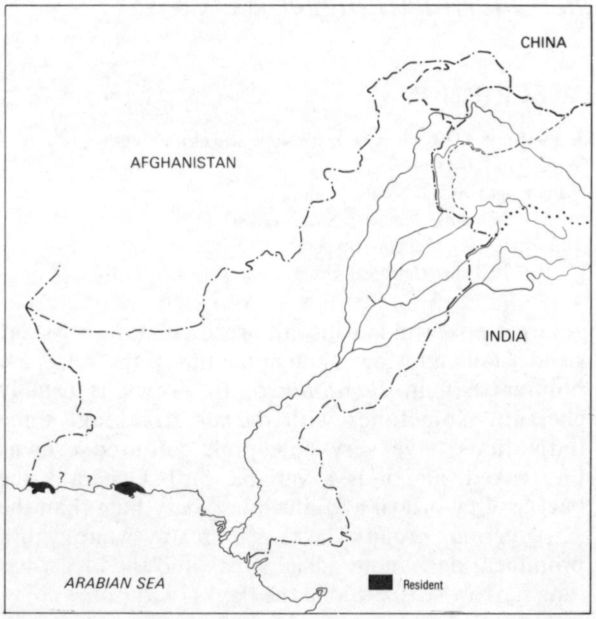

HABITAT, DISTRIBUTION AND STATUS

A very specialized falcon in its ecological adaptations as it is similar to Eleonora's Falcon in its dependence upon migratory birds for a large part of its food prey and breeding only towards the onset of autumn when such migrants are most abundant and when their own nestlings require to be fed. They also are adapted to highly arid coastal areas in the main Palearctic, African migration routes (Gallaher, *Journal Oman Studies*, 1979).

In Pakistan it only occurs as a summer breeder along the Makran coast probably from about Ormara and westwards. W. D. Cummings when working in this region at the beginning of this century, during the laying of a sub-marine cable, collected several speci-

Map 87: Sooty Falcon *Falco concolor*

Illus. 30: Sooty Falcon *Falco concolor*, mature adult.

mens. An adult male shot on 31 August 1912 from near Ormara was deposited in the Quetta Museum (destroyed in the 1936 earthquake). Two more collected in early May 1901 are deposited in the British Museum, also collected from near Ormara and Cummings reported seeing another in August of that year in the hills near Chabbar further west (Ticehurst, *JBNHS*, 1927b). Status RARE and confined to Makran coast as a summer breeder.

HABITS, BREEDING etc.

They are very like Hobbies in their aerial hunting methods, being extremely swift and manoeuvrable so that most prey is taken on the wing. They are also largely crepuscular in hunting activity and will feed upon bats and insects when these are available, as well as a variety of migrant birds. In Oman they have been recorded as feeding upon the Cuckoo (*Cuculus canorus*), Hoopoe (*Upupa epops*), Whitethroat (*Sylvia communis*), Wilson's Storm Petrel (*Oceanites oceanicus*), Red-tailed Wheatear (*Oenanthe xanthoprymna*) and Spotted Flycatcher (*Muscicapa striata*) (Gallaher, 1979). All these species are summer migrant visitors to Pakistan likely to occur on passage along the Makran coast.

They nest on the bare ground making no nest lining, but choosing a hollow, sheltered from the fierce sun either under a rock overhang or a clump of *Euphorbia* cactus. The normal clutch is 2 eggs which on the islands off the Persian Gulf have been found from mid-August up to early September (Gallaher and Woodcock, *The Birds of Oman*, 1980). During a survey in July and August 1978 in the Gulf of Oman it was found that on favourable islands, pairs nested within 200 metres of each other. Only the female incubated and tended the eyrie whilst almost all the food hunting appeared to be done by the male. The clutch size varied from 2 to 3, more commonly the former and the first hatched chick was observed on 27 August (Gallaher, 1979)

Laggar Falcon *Falco jugger* or *Falco biarmicus iugger* (J.E. Grey)   Map **88**

TAXONOMY

The larger heavy bodied falcons form a very difficult group taxonomically and the Lagger has been variously classified by different experts. Dillon Ripley in his 'Synopsis' (1961) and 'Revision' (1982) treats it as a sub-species of the Lanner *Falco biarmicus*, i.e. as *F. biarmicus jugger* and this is followed in Ali and Ripley's 'Handbook' (1969) whereas Brown and Amadon (1968) treat it as a distinct species *Falco jugger* and this seems to be preferred and followed by De Voous in his 'Recent List' (*Ibis*, 1977).

It is ecologically distinct from most of the *Falco peregrinus* group, and in its browner colouration and tendency to have a paler and variably coloured crown and nape it seems to conform more with the *Falco biarmicus/cherrug* group. The author prefers this latter treatment as followed by Ali and Ripley and to consider it as a geographically distinct sub-species of *Falco biarmicus*.

DESCRIPTION

Body length 43-46 cms. about the same size as a Peregrine
Wingspan 100-110 cms.
Wing length 30.5-32.8 cms. (males),
    32.3-36.4 cms. (females) (Stuart Baker)
Tail length 16.7-17.5 cms. (males),
    16.9-19.8 cms. (females) (Stuart Baker)

A larger powerful falcon with broad chest, rather short slender tail and wings, though the tips of the wings are blunter than in *F. subbuteo*. Its crown is usually chestnut, sometimes with darker streaking, some individuals have very pale pink coloured crowns unstreaked, and it is a variable bird. Generally the back is grey-brown and much less slaty blue than the *F. peregrinus* group. The face is creamy white with a prominent dark moustachial streak and the breast has long narrow streaks along the flanks, with some cross-barring on the tibia or thighs. Sub-adults have darker, heavier streaking on the breast and the crown is usually dark streaked and brown rather than chestnut. In all forms there is a yellow orbital ring and cere and the legs and feet are yellow. The wing coverts and mantle feathers are usually margined paler, giving a handsome scale-like pattern. When at rest the wingtips do not reach the tail tip. Females are about 15 per cent larger in size. The tail is cross-barred with whitish bars on the inner webs of the outer tail feathers but these white areas are much less extensive than in the tail of the Saker (*Falco biarmicus cherrug*).

HABITAT, DISTRIBUTION AND STATUS

The Lagger is resident throughout the dryer less intensively cultivated tracts of Indus plains, but does not normally extend into the higher mountain regions of Baluchistan or the Himalayas. It is a bird of much more arid and open tracts than those inhabited by the Peregrine. It will however be found along canals with their tree-lined embankments (e.g. in Badin district of Sind), as well as in extensive areas of scrub desert or low hills and erosion gullies as in Dadu district of Sind. It does not shun extensive sand dune tracts. It avoids true mountainous areas or densely human-populated areas. It is more or less sedentary wherever it occurs, being widely but very thinly distributed in southern Baluchistan, Sind, Punjab and the lower plains areas of the NWFP. Whitehed considered it an uncommon resident around Kohat (*JBNHS*, 1911). Recent sightings (author) include pairs on the rocky escarpments of the Salt Range, the Thal desert, Trimmu headworks in Jhang district, Karchat, Dadu district, Lasbela in Baluchistan and on the Bannu plains. A pair haunted the mangroves of Sonmiani Hoar during the early 1970s. In Christison's day a pair nested at Baleli near Quetta, but he considered it very rare in northern Baluchistan (*JBNHS*, 1942). Status now SCARCE and becoming RARE.

HABITS

They often remain in pairs throughout the winter and roost together on favoured lookout posts. Thus, from November to February 1971-2 a pair haunted a brick kiln chimney near Mailsi and another pair roosted nightly for about 14 days on a power pylon near Khanewal, both localities in the southern Punjab. They will fly with strong rapid wing beats, alternating with glides and swoops low over the ground, or less commonly circle high overhead and stoop like a Peregrine, but generally they are much less active than the latter species, preferring reptile and rodent prey which can be caught on the ground. They will take Rock Pigeons (*Columba livia*), Doves (*Streptopelia* spp.) and the Indian Roller (*Coracias benghalensis*) in the vicinity of their roosts, such as bridges and brick

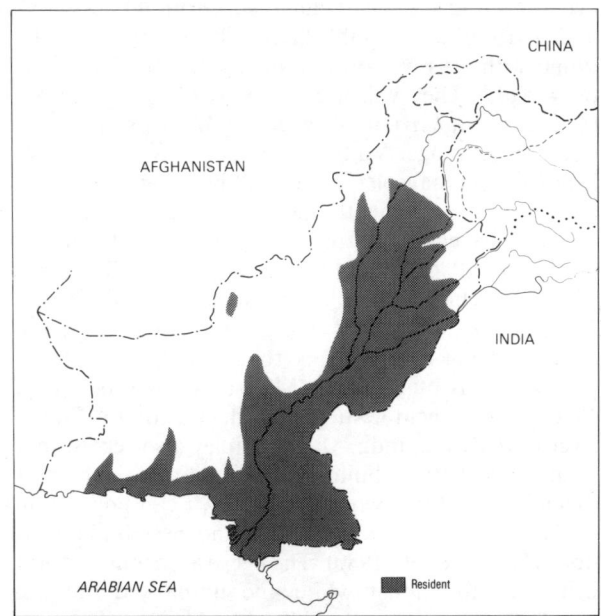

Map 88: Laggar Falcon *Falco jugger*

kiln chimneys, but in desert areas Spiny-tailed Lizards (*Uromastix hardwicki*) were noted to be a favourite food item, as well as feeding on locusts (*Schistocerca gregaria*) when these are plentiful, and Desert Jirds (*Meriones hurrianae*) (K. Eates, MS notes). Earlier this century Ticehurst noted that this falcon often adapts to an urban environment, nesting on tall public buildings or monuments in Karachi and preying upon feral pigeons (*Ibis*, 1923). They still nested on certain public buildings in Karachi up to the late 1940s, some sites such as Frere Hall Museum having been traditionally occupied by a pair for known periods of over 30 years (K. Eates, MS notes). The increased traffic and noise of modern-day Karachi has changed this pattern of behaviour and none have been noted within our larger towns in recent decades. Though trained for falconry they are not considered as bold or active hunters, and are used by Arab falconers mainly as a decoy bird to attract Peregrines or Sakers.They normally attack their prey flying low and swiftly over the ground according to Arab falconers and are less inclined to 'wait on' and stoop like the Saker (*Falco cherrug*).

BREEDING BIOLOGY

In Pakistan they nest from January to April, but most commonly during February and March. E. N. Lowther in central India records a nest in a tall *Salmalia* tree with 3 young ten days old on 3 March. (Lowther, 1949) and Kenneth Eates who took no less than 56 clutches from different parts of Sind, found 5 eggs on four occasions and the mean clutch size from 40 nests was 3.85 (MS notes unpublished). The earliest nest he found with eggs was on 27 January and the latest was on 4 April. They will use old nests of House Crows (*Corvus splendens*) or Black Kites (*Milvus migrans*) in trees, or Scavenger Vulture's nests on cliff sites or bare ledges on old mausoleums and tall buildings. Out of 40 nests taken by Kenneth Eates, 3 were on ledges in disused Scavenger Vulture's (*Neophron*) nests and 14 were in old Black Kite's (*M. migrans*) nests, all in trees, 10 were on cliffs, 11 on old tombs and 2 on buildings in Karachi city. Where cliff sites are used both sexes will actually bring a few sticks to the ledge though no proper nest is built (Eates, MS notes). Whistler (*Ibis,* 1930) records them nesting in earth cliffs on the Sohan river near Rawalpindi. Ali and Ripley also record that both sexes share in building and repairing of the nest ('Handbook', 1969) whereas in the African population of *F. biarmicus*, it is stated that no nest building is done (Cramp *et al.*, 1980). The eggs vary from uniform brick red to almost white and unmarked but are usually very handsomely blotched and spotted (Eates, MS notes). Both sexes incubate but the female does

the larger share and both sexes are very territorial in defending the nest site, attacking even vultures that soar within the vicinity. Incubation period is unrecorded but estimated at usually between 30 and 31 days for the African *F. biarmicus* (Brown Urban and Newman, 1982). The period from laying to first flight of the young is considered to be about 60 days (Brown and Amadon, 1968). Young out of the nest and able to fly were still being fed by their parents on 8 May, who called noisily during this process (K. Eates, MS notes). Courtship includes spectacular swoops by the male onto the circling female and much calling together and mutual soaring.

VOCALIZATIONS

Relatively silent except during the breeding season. At the nest site both sexes occasionally emit a shrill prolonged scream 'whi-eee-eee'. This wailing call may act as a food summons to the female, and is also used during courtship and when food is being brought for the young.

Saker Falcon *Falco cherrug* J.E. Gray
Shangar Falcon *Falco cherrug milvipes* Jerdon

Illus. **31**
Map **89**

Synonym *Falco biarmicus cherrug* J.E. Gray, Ripley's 'Revised Synopsis', 1982
Falco biarmicus milvipes Jerdon Ripley's 'Revised Synopsis', 1982

TAXONOMY

S. Dillon Ripley in his *Synopsis of the Birds of India and Pakistan* treats the Saker as part of a *Falco biarmicus* super species group including the Laggar Falcon, *Falco biarmicus jugger*. This is followed by Ali and Ripley in volume 1 of the 'Handbook'.

K.H. Voous in his 'Recent List' (*Ibis*, 1977) splits *Falco cherrug* including the Saker and the Shangar falcons into a separate species from *Falco biarmicus*, the Lanner. Brown and Amadon (Vol. 2, 1968) also treat the Saker as a separate species *Falco cherrug* from the Lanner *F. biarmicus*. There is considerable size variation in all the geographically separated populations of these larger falcons of both the *F. cherrug*, *F. biarmicus* and the *F. rusticolus* (Gyr falcon) groups. Also if juvenile specimens of museum skins of *F. cherrug* the Saker, *F. jugger* the Laggar, and *F. biarmicus* the Lanner, are laid side by side, there appear to be no distinctive features by which they can be readily separated, with much variation in plumage colour and pattern, including both crown and tail patterns which have been used to separate these three species in the past.

With the limited amount of museum material available from our region the author inclines to the view that they are better regarded as part of a *F. biarmicus* super species complex.

DESCRIPTION

Saker *Falco cherrug*

Body length 50-56 cms. (Ali and Ripley)
Wing length 34.8-37 cms. (males),
　39-41.2 cms. (females) (Stuart Baker)
Tail length 19-20 cms. (males),
　20.7-21 cms. (females) (Stuart Baker)
Weight 950 gms. (female) (Dement'ev)

Shanghar *Falco cherrug milvipes*

Body length 50-58 cms. (Ali and Ripley)
Wing length 34-35.1 cms. (males),
　37.4-43.5 cms. (females) (Stuart Baker)
　34.8-38 cms. (13 males), 38.7-41.1 cms. (11 females) (Dement'ev)
Tail length 18.8-23.6 cms. (Stuart Baker)
Weight 750-990 gms. (males),
　975-1,150 gms. (females) (Cramp *et al.*)

The Saker is a much browner bird than the Peregrine, lacking the blue-grey or slaty tones of the latter. Its tail instead of being cross-barred tends to show large pale spots, not extending right across both webs. The crown is pale often white or pale buff with dark streaks. Adults have dark sepia brown centres to the feathers of back and wing shoulder, each feather narrowly margined paler rufous. The long pointed wings have blackish flight feathers, spotted with white on their inner webs and edged with buff. There is usually a rather indistinct dark moustachial streak (much less definite than in the Peregrine) and a dark streak from behind the eye to the nape. The under parts are white, spotted with dark brown, more heavily streaked with dark brown in immature birds.

The Shanghar differs from the Saker in having the whole of the back barred rufous and dark sepia brown rather than spotted. Also the tibia and lower flanks tend to be barred rather than streaked darker. Also its crown is usually darker more rufous buff, rarely showing white. It is usually a bigger bird than the nominate sub-species or Saker.

As mentioned above there is much variation in the plumage of these falcons. The crown varies from white to pinkish buff and the tail varies from having distinct cross-bars to completely unbarred dark brown in some individuals. The white breast is less prominently streaked in old birds but with more prominent flank streaking. Sub-adults have the crown and ear coverts streaked white and dark brown and a more definite dark moustachial streak extending from the base of the eye. In younger birds the flank feathers are uniformly dark ashy without prominent pale margins to the feathers, but sometimes these feathers are margined with paler rufous buff. In both sub-adults and adults the conspicuous oval white spots on both webs of the tail feathers, form the best field mark. The iris is dark brown, the legs and feet, cere and orbital ring are dull yellow.

Illus. 31: Saker Falcon *Falco cherrug*, typical mature bird, both sexes.

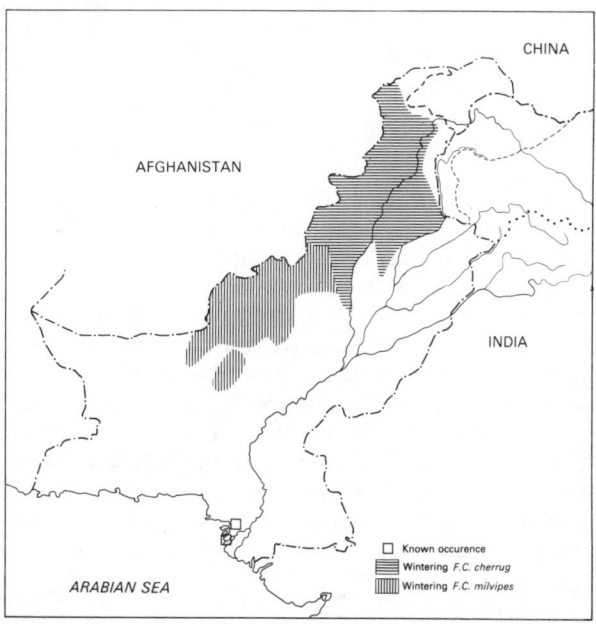

Map 89: Saker Falcon *Falco cherrug* and Shanghar Falcon *Falco cherrug milvipes*

HABITAT, DISTRIBUTION AND STATUS

A winter visitor both in the mountain regions and the foothill country around Mianwali, Kohat and Attock districts. It is now a very uncommon visitor more likely to be encountered in Baluchistan or the NWFP, but may occur in the desert border regions of Punjab and Sind as well as in the Indus delta. The population which winters in Pakistan breeds in central USSR from Trans-Caspia to Kazakhstan. Recent records for the Saker include an adult sighted in mangroves in January in the Indus delta (author with F. Koning), an adult seen (author) at Beleli customs post on 31 October north of Quetta and another at Mastuj in northern Kalat on 3 March. (J. Vieillard, *in litt.*) and another near Chashma barrage on the Indus on 26 February. Ticehurst only noted one record from the Kirthar hills (Ticehurst, 1923) and Whistler did not record it around the Potohar (Whistler, 1930) loess plateau of Rawalpindi district. However one was recorded spending most of the winter around Rawal lake in 1986-7 (M. Mallalieu *in litt.* to author, January 1987). In Baluchistan the Shangar (*milvipes*) has been collected from around Quetta in October and January (Ticehurst, *JBNHS*, 1927). Meinertzhagen also collected an immature male at Kushdil Khan on 17 May (Brit. Mus. coll. and Meinertzhagen, *Ibis*, 1920). This falcon can be trained to hunt Houbara (*Chlamydotis undulata*) and Gazelle (*Gazella gazella*) and hence is much sought after by Arab falconers. This has undoubtedly led to their rapid disappearance from Pakistan where a good condition female would fetch as high as US$ 50,000 in 1983. Status SCARCE, becoming very RARE.

HABITS

Always seen solitarily, and usually in bare open countryside such as saline flats or bare gravelly plains. They will hunt, like Peregrines circling high over their prey and killing their quarry after a steep aerial dive, binding onto a bird and carrying it down to the ground. This is the technique used by trained birds handled by falconers, but in the wild a lot of its hunting is done by swift low direct flight, repeatedly striking at a fleeing bird from behind. They are said to feed upon Spiny-tailed Lizards (*Uromastix hardwicki*) and Sand Rats and Jerboas (*Allactaga* spp.) (Dement'ev and Gladkov, 1951). Rodents (*Meriones* sp.) are more easily available than reptiles in winter in the cold steppic plateau regions of Baluchistan. In Baluchistan, one was observed (author) mobbing by diving upon a pair of Eagle Owls (*Bubo bubo*), perched in a rocky gorge, whilst another individual watched at Chashma was repeatedly mobbed in its turn by a House Crow (*Corvus splendens*) much smaller than itself.

BREEDING BIOLOGY

Extra-limital. They will make no nest, laying their eggs on a bare earth ledge, or use the nest of another bird such as an abandoned Scavenger Vulture's nest (*Neophron*). They nest usually on cliffs but occasionally will use a tree. They are said to avoid human artefacts such as ruined buildings, a favoured nest site for other falcon species. 3 to 5 eggs are usually laid from late March. The female does most of the incubating, with the male gathering most of the food prey.

Peregrine *Falco peregrinus* (Tunstall)

Shaheen Falcon or
Black Shaheen *Falco peregrinus peregrinator* (Sundevall)   Map **90**

TAXONOMY

A population of smaller falcons with chestnut red crown and blue-grey mantle occurs as a sparsely resident breeding bird in the dryer hilly tracts of Pakistan and this is classified as a sub-species *Falco peregrinus babylonicus* by Dillon Ripley in his 'Revised Synopsis' (1982) as well as by Ali and Ripley in the 'Handbook Series' (1969). The author prefers to include *babylonicus* as a sub-species of the *Falco pelegrinoides* super species group. See next species. The migratory Peregrine occurring in Pakistan belongs to the sub species *japonensis* Gmelin, (Ripley, 1982) described as having a white breast with dark cross-bars and pale grey upper parts. Earlier ornithologists have considered the population visiting Pakistan to be no different from *F. peregrinus callidus* Latham (see Ticehurst, 1923). Both these sub-species are typically fairly large pale birds.

DESCRIPTION

F. peregrinus
Body length 40-48 cms. with tail 11-13 cms.
Wingspan 100-110 cms.
Wing length 29.1-32 cms. (males),
    34.8-36.7 cms. (females) (Cramp *et al.*)
Weight 582-750 cms. (males), 925-1,300 gms. (females) (Cramp *et al.*)

F.p. peregrinator
Body length 38-46 cms.
Wingspan 90-100 cms.
Wing length 26.5-29.5 cms. (males),
    31.2-34.2 cms. (females) (Stuart Baker)

In the field it can at once be distinguished from *Falco jugger* by its darker blue-grey upper parts and breast with horizontal dark barring, instead of vertical streaks. Immature birds however, have vertical streaks not bars, on their breast. In adults there is quite often a fulvous or buffy tone to the lower breast and flanks. It has a prominent black moustachial streak and a plain slaty black crown without any white streaks. The female is about 15 per cent larger. There is no white on the tail as in *Falco cherrug*. It is a typically broad shouldered, short tailed compact falcon, and its direct rapid wing-heating flight and narrow pointed wings alternating with short glides, is suggestive of all the larger falcons. The iris is dark brown and orbital ring, cere, legs and feet are dull chrome yellow. In under-wing flight silhouette the tail shows narrow cross-barring and the breast and under-wing coverts have darker spotting or cross-barring than the flight feathers.

The Shaheen is very similar in size and colouration except for the breast and lower belly and under-tail coverts which are rufous chestnut with black cross-barring.

HABITAT, DISTRIBUTION AND STATUS

The Shaheen (*peregrinator*) is a largely resident species breeding in peninsular India and possibly the eastern Himalayas in the outer ranges, favouring always areas with steep cliffs and escarpments. There is some local migration outside of the breeding season. Whistler

(*Ibis,* 1930) records a sighting by Magrath in 1908 of a single bird in Thandiani on 2 July and his own observations of 2 birds on 16 February over the outer Murree foothills and at Murree on 5 August, both in 1926. The author has one sight record only, after 5 continuous stormy days at Dunga Gali on 7 July 1966, of a juvenile bird. Meinertzhagen shot a female at Kushdil Khan lake in Baluchistan on 26 October but later told Ticehurst that he was doubtful about his identification (Ticehurst, *Birds of Baluchistan,* 1927*b*). Dr. Ticehurst did not record it in Sind. Ali and Ripley in the 'Handbook' (1968) describe it as resident and breeding in Pakistan, listing localities such as the Thal desert, Kohat district and Chitral. The author has never come across any recent evidence of this sub-species of the Peregrine breeding anywhere in Pakistan and believes that earlier records stem from confusion with *babylonicus.* The present day status of this falcon in Pakistan must be considered an occasional straggler, largely in the outer Himalayan regions and not breeding within the region covered by this book.

The Peregrine Falcon (*F. peregrinus*) is a winter visitor occurring very widely in the Indus plains, most likely to be encountered around large jheels or irrigation barrage seepage zones, where waterfowl provide ample food resources. Occasional individuals can be encountered along the mangrove creeks and Karachi seacoast in winter, often roosting each evening on a favoured power pylon. C.H.T. Whitehead (*JBNHS,* 1911) considered it as a passage migrant around Kohat and uncommon. Whistler (*Ibis,* 1922) considered it a rare winter visitor around Jhang. Recent sightings (author) include birds seen on the coast at Cape Monze, and it seems to like frequenting seacoast cliffs, the Indus delta (where it roosts on navigation beacons and power pylons), Karchat in the Kirthar hills, Ghauspur jheel in Jacobabad district, Lal Sohanran jheel in Bahawalpur and Rawal lake near Islamabad. It is very much sought after by Arab falconers who will pay a higher price for large females in good plumage, even higher than for the larger *Falco cherrug* (i.e. over US$60,000 in the mid 1980s). Consequently it is becoming increasingly rare.

The Shaheen is largely a sedentary and endemic population breeding in peninsular India and its status in Pakistan is not well understood and thought to be that of an OCCASIONAL STRAGGLER only. The Peregrine is a regular winter visitor to the Indus plains, but nowadays heavily persecuted by falcon trappers and becoming increasingly scarce. Status SCARCE.

HABITS

Ticehurst (*Ibis,* 1923) describes the Peregrine as a typical duck hawk. It certainly frequents regions where waders and waterfowl abound and normally occurs solitarily in its winter quarters. It is a very bold and aggressive hunter, striking most of its prey in the air, making steep dives and plunges, or pursuing its prey with fast low flight, twisting and turning with every evasive flight of quarry. The author has watched the Shaheen stooping over Rock Pigeons (*Columba livia*) which shared the stone breastwork of a dam at Krishnarajasagar in Karnataka, south India and this pigeon is also a favourite food prey in Pakistan. It seldom strikes prey on the ground and therefore birds form the principal food source with mammals being relatively uncommon.

BREEDING BIOLOGY

Very well studied and known. They breed extra-limitally preferring inaccessible ledges on cliffs and often utilizing the same site year after year. Virtually no nest lining is made and the normal clutch is 3 to 4 eggs laid over a period of eight to nine days. The male does a small share of the incubation but mostly assumes the role of food provider, bringing food to his mate as a part of the courtship ritual also.

VOCALIZATIONS

Silent in their winter quarters. During the breeding season and when alarmed or excited utter a variety of short quick staccato calls 'hek-hek-hek' or 'chik-chik-chik'. A female Shaheen at the nest site uttered a prolonged 'chirr-r-r' call (Ali and Ripley, op. cit.).

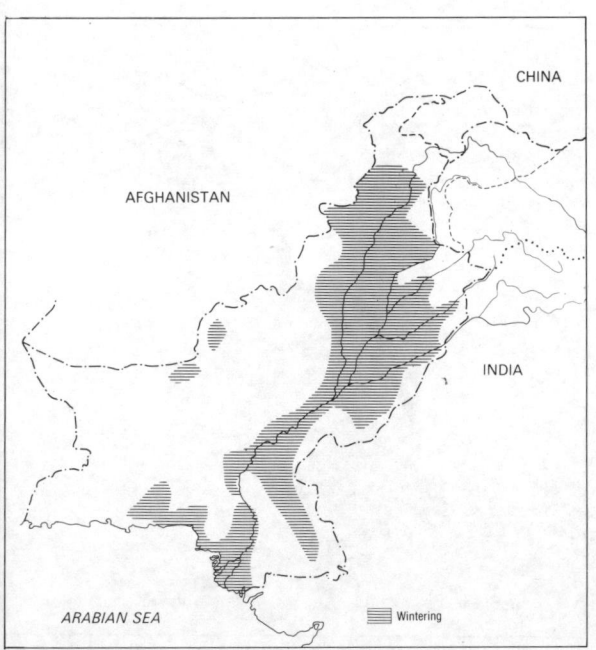

Map 90: Peregrine Falcon *Falco peregrinus*

CHINA

AFGHANISTAN

INDIA

ARABIAN SEA

▤ Wintering

Red-capped Falcon *Falco pelegrinoides* Temminck
Red Shaheen or
Red-capped Falcon *Falco pelegrinoides babylonicus* (P.L. Sclater)
Red-capped Falcon or
Barbary Falcon *Falco peregrinus babylonicus* of Ali and Ripley   Map **91**

TAXONOMY

In the *Birds of the Western Palearctic* (Cramp *et al,* 1980) (of which de Voous is a senior editor), the Barbary Falcon (*Falco pelegrinoides*) is taken as a super species largely inhabiting Africa with a western population occurring from Iran eastwards, described as *Falco pelegrinoides babylonicus* Sclater 1861. The author prefers this new classification for this largely sedentary, smaller falcon with its distinctive head pattern. Vaurie considered it also a sub-species of *Falco pelegrinoides* and Dr. Ticehurst with great perception writing in 1923 said that he was 'exceedingly doubtful' whether it should be considerd a race of the Peregrine (Ticehurst, *Ibis* 1923).

DESCRIPTION

Body length 38-46 cms.
Wingspan 90-100 cms.
Wing length 27.3-28.4 cms. (males),
    32-33.8 cms. (females) (Stuart Baker)
Weight 330-398 gms. (males),
    513-765 gms. (females) (sub-species *babylonicus*) (Cramp *et al.*)

A small neat 'peregrine' with relatively pale blue-grey upper parts and having a rufous crown and nape with narrow dark moustachial streak and darker blackish forecrown and line bordering nape and ear coverts. The wingtips are noticeably darker and the tail has obsolescent cross-bars. Juveniles are much browner with dark streaking on the breast and flanks and chestnut border to the mantle and wing covert feathers. Adults have almost white throats and breasts with more buffy rufous belly and flanks, having indistinct cross-barring on the flanks and tibia. The iris is dark brown, and the cere, orbital ring and legs are sulphur yellow. They look very like Red-headed Merlins (*Falco chicquera*) except for larger size and rufous lower belly and the broad buffy nuchal collar.

HABITAT, DISTRIBUTION AND STATUS

It appears to be an uncommon, locally migratory species on the dryer hilly tracts of western Pakistan where it occasionally breeds. Meinertzhagen (*Ibis,* 1920) considered it resident in the higher mountains

around Ziarat, Baluchistan and he shot a female on 23 July and saw others there in June. Christison considered the Red-capped Shaheen as resident in small numbers in the hills of Nushki, Koh-i-pusht and Raskoh in the Chagai district of Baluchistan (*Ibis,* 1941). Whistler (*Ibis,* 1930, and specimen British Museum) shot an adult female in Rawalpindi on 4 February 1926 and believed that it bred in the outer foothills around Taxila and Margalla. C.H.T. White-head considered it less common than the Laggar Falcon (*Falco jugger*) but still fairly common in the foothills of Kohat and Bannu (Whitehead *JBNHS* 1911). Perreau (*JBNHS,* 1910) encountered it in lower Chitral including one seen in late April. Ticehurst (*Ibis,* 1923) considered it a not uncommon winter visitor to Sind particularly in the northern part of the province. Reliable sight records of this falcon in recent times are few but a specimen was trapped by local falconers in Jacobabad district of Sind in October 1983 and another was seen by the author, trapped near Ahmadpur Sharkia in Bahawalpur district in the late 1960s. It has been reliably sighted on the Rawalpindi plateau (Potohar) in 1974 (author) and again in 1981

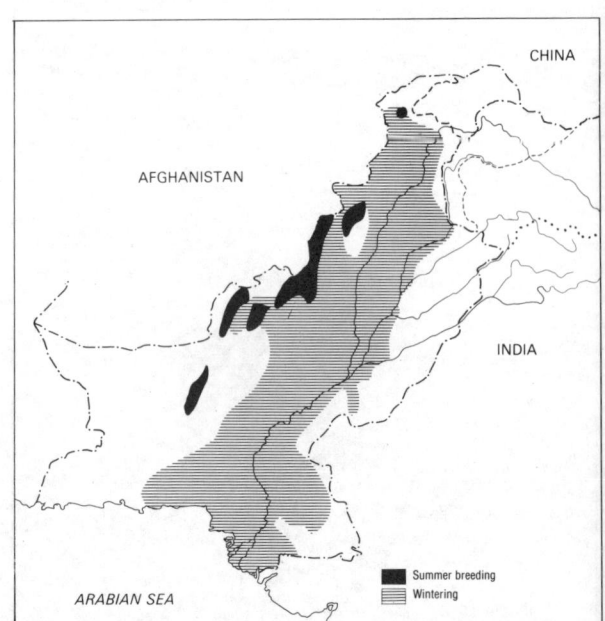

Map 91: Red-capped Falcon *Falco pelegrinoides*, including Barbary Falcon *Falco pelegrinoides babylonicus*

(D. Corfield, pers. comm.). In Sind it is a winter visitor from November to February and it apparently breeds sparingly in Baluchistan and the NWFP. Status SCARCE resident locally migratory.

HABITS

Similar to other large falcons, it is a very swift flyer of great manoeuvrability and skill when hunting. It is believed to take mainly birds, preferring to strike them on the wing rather than on the ground. M. Bell of the Forest Service recalls one which repeatedly swooped upon a Great Stone Plover (*Esacus magnirostris*) which had young on a sandbank of the Indus river near Sukkur. The Stone Plover resisted each swoop with spread wings and up-raised bill, refusing to take to the wing and the falcon eventually landed briefly on the ground beside it before taking off in 'apparent disgust' (quoted in Ticehurst *Birds of Sind*, 1923). They are also reported to be quite crepuscular in hunting and feed on bats where these are easily available (Ali and Ripley, 'Handbook', 1969).

BREEDING BIOLOGY

Two young were taken (Stuart Baker, 1928) from an eyrie on the Gomal pass west of Dera Ismail Khan at 650 metres altitude. J. Anderson reported having taken eyasses from several nests in Kalat in southern Baluchistan (pers. comm., 1969). No details are available about the dates of these encounters but they were all cliff sites in relatively hot arid mountain regions. No other authentic information is available about the breeding of this sub-species.

ORDER GALLIFORMES

SUB-FAMILY PHASIANAIDAE

Snow Partridge *Lerwa lerwa* (Hodgson)

Plate **14**
Map **92**

DESCRIPTION

Body length 37-40 cms.
Wingspan 50-55 cms.
Tail length 11-13 cms.
Weight 453-623 gms.

About the size of a Rock Partridge (*Alectoris chukar*) but having a slightly longer tail and slimmer build. The powerful digging bill is crimson-red, and the legs are a slightly more orange-red. The body is greyish chestnut, being finely cross-barred all over the head, neck and upper parts with dark blackish brown against a white background, with the wing shoulders and scapulars being washed with rufous. The breast is dark chestnut, broadly streaked along the flanks with white, and with a narrow vertical white line down the centre of the breast. The under-tail coverts are chestnut, cross-barred blackish. The sexes are alike but the male has a blunt spur on the hind part of the tarsus. The iris is reddish brown with a small naked area of red skin around the eye. In flight a narrow white trailing edge to the wing is conspicuous. The downy chick has a black bill and is feathered halfway down the tarsus. Its head and neck is greyish, darker on the ear coverts and crown. The back is mottled black and chestnut and the breast is rufous (Meinertzhagen, *Ibis*, 1927).

HABITAT, DISTRIBUTION AND STATUS

It occurs from Hazara district in the NWFP in the west, extending into neither Afghanistan (Paludan, 1959) nor the USSR (Flint *et al.*, 1984). However, its range extends across the Himalayas eastwards to Arunachal Pradesh (Inskipp and Inskipp, 1985). Adapted to high alpine slopes, this Himalayan endemic is rarely encountered below 3,300 metres (11,000 feet) elevation in the western part of its range, and in Pakistan has now become very rare and local. It is associated with steep rock-strewn slopes interspersed with patches of creeping *Juniperus squamata* and *Bergenia* species, right up to the limit of permanent snow. It is therefore sympatric with the Snow Cock *Teraogallus* in such regions, though possibly preferring better vegetated and less dry mountain aspects. In Gilgit it occurs in the headwaters of the Naltar valley above 3,900 metres (13,000 feet) where a survey was conducted by Ghulam Rasul (DFO Wildlife, Northern Areas) in 1984. In the upper reaches of the Khunjerab valley, it is found sharing rocky outcrops with the Snow Cock. Here, there is an estimated population of 30 to 40 birds (Jalal Uddin, Game Watcher, Khunjerab National Park, pers. comm., 1987). More recently in the winter of 1987 a covey of 10 birds was found haunting the steep slopes above Rama lake on the flanks of Nanga Parbat (Astor district) and regrettably two specimens were shot and they have been preserved by Ghulam Rasool, DFO Wildlife, Gilgit. Captain Whitehead (*JBNHS*, 1914) found nests with eggs between 3,800 and 4,250 metres (12,500 and 14,000 feet) elevation in the Kaghan valley and Col. Buchanan found a nest on the Safed Koh range above Parachinar at 3,352.8 metres (11,000 feet). In the northern reaches

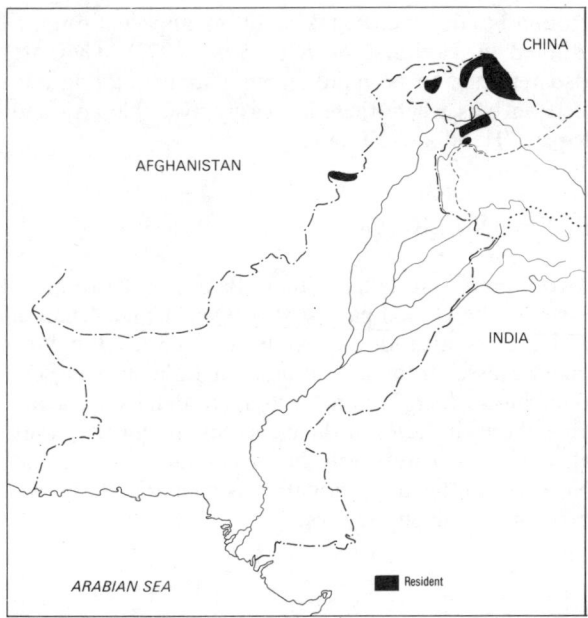

Map 92: Snow Partridge *Lerwa lerwa*

of the Hunza valley it occurs up to 5,182 metres (17,000 feet) in summer and there are reliable sightings from the Shimshal valley (north-east Hunza), in the 1960s by Wing commander Shah Khan (pers. comm., 1969) and from the Khunjerab by Z.B. Mirza (pers. comm., 1973). Curiously, W. H. Mathews (*JBNHS*, 1941) hunting for Ibex in the Deosai plateau and Tseri valley in Baltistan did not encounter it, nor did the author while camping in August (1946) on the Deosai plateau. It still survives on some of the highest alpine slopes in Azad Kashmir also, particularly above Machiara on the right bank of the Neelum valley (Sheikh Abdul Qayyum, Chief Game Warden, pers. comm., 1984). Status RARE and LOCAL.

HABITS

A typical rock partridge in its ability to run well over the ground and jump from stone to stone, and its habit of foraging and roosting in coveys or small groups. Fleming describes them as quite noisy as they assemble to roost at night, huddling together in small caves or hollows under rocks (*Birds of Nepal*, 1976). Meinertzhagen examined 10 stomachs from specimens collected in Ladakh and found that they had fed on lichens, moss, vegetable shoots and grit (*Ibis*, 1927). Due to their inaccessible habitat, very few naturalists have encountered them in the western Himalayas but Whitehead (1914) commented upon their ridiculous tameness which gave away the presence of their nest and wrote that, consequently, they are regularly robbed of their eggs by Gujar shepherds. Meinertzhagen (who badly needed meat for his porters) was able to collect every bird in a covey of ten as they repeatedly allowed him to approach closely after every shot (op. cit., *Ibis*, 1927).

BREEDING BIOLOGY

The pairs do form a lasting bond, in that the male will guard the incubating female and help to tend the downy nestlings for some days, but the female makes the nest scrape and does all the incubation. The nest is located in a hollow, usually sheltered under an overhanging ledge and well lined with moss and leaves (Whymper, *JBNHS*, 1910). The normal clutch is 3 to 5 eggs of a pale olive to greyish buff ground colour, finely speckled with reddish brown all over. Whitehead (*JBNHS*, 1914) found well grown downy chicks on 2 July, as well as a nest with fresh eggs on 3 July in the Kaghan valley, implying that females will relay after an early breeding failure. One nest was found at 4,200 metres (14,000 feet). There is no description of the males' courtship display.

VOCALIZATIONS

They have a low whistling call, repeated on a shriller, higher key when alarmed (Meinertzhagen, *Ibis*, 1927). Also a series of cackling calls when settling to roost (Fleming *et al.*, 1976).

Himalayan Snowcock or
Ram Chukar *Tetraogallus himalayensis* G. R. Gray

Plate **14**
Map **93**

DESCRIPTION

Body length 72 cms. (Ali and Ripley)
  76 cms. (large male, Kashmir) (A.C. Bruce)
Wingspan 111 cms. (Kashmir) (A.C. Bruce)
Wing length 28-31.2 cms. (both sexes) (Stuart Baker),
  32-34 cms. (6 males), 27.5-31.5 cms. (11 females) (Dement'ev)
Weight 3,628.7 gms. (large male, Kashmir) (A. C. Bruce),
  2,670-2,700 gms. (Whistler)
  2,200-3,100. gms. (males), 2,000-2,570 gms. (females)
  (Dement'ev).

A giant partridge in shape and body proportions with a powerful down-curved olive grey bill and stout orange to vermillion legs and feet. A small area of bare yellow or orange skin surrounds the eye. The iris is brown or greyish brown. The body is predominantly ashy-grey with conspicuous long loose white under-tail coverts and a white throat and upper breast bordered by two vertical dark chestnut lines running down the sides of the neck. The upper part of the mantle has a broad buffy and black speckled collar and the lower back is conspicuously patterned longitudinally with pale creamy streaks, forming interrupted lines. The flanks also have similar creamy streaks forming irregular parallel lines and these are narrowly bordered with dark chestnut. The primaries are white, tipped blackish grey, with a narrower white wingbar extending across the secondaries, this pattern being conspicuous in flight. There is a conspicuous pattern of very dark maroon (almost black) square spots encircling the white upper breast and this white area is sharply divided from the dark grey lower breast. There is a blunt spur on the legs of the male, which otherwise appears the same as the slightly smaller females.

HABITAT, DISTRIBUTION AND STATUS

Where the Himalayan Ibex *(Capra ibex sibirica)* occurs, the Himalayan Snowcock is at home. It haunts the highest, semi-arid uplands, whether precipitous valleys or the higher spurs of major mountain peaks, rarely being found below 3,600 metres (12,000 feet) and more typically between 3,900 and 4,570 metres (13,000 and 15,000 feet). In the vast inner mountain areas of northern Pakistan, it is relatively plentiful at these altitudes. Dr. Schaller found it fairly common at about 4,877 metres (16,000 feet) elevation in Arkari and Besti nullahs in Chitral (pers. comm., 1972). In the Safed Koh range it is less common but has been shot as low as 3,000 metres (10,000 feet) in autumn, when berries tempt them to lower altitudes (Col. Dastagir, pers. comm., 1969). They occur sparsely in Swat Kohistan but their stronghold is Gilgit, Hunza and Baltistan. W.H. Mathews records seeing as many as 40 in one morning above Tseri village at 3,900 metres (13,000 feet) in western Baltistan

(JBNHS, 1941), and the author encountered small family groups of 5 to 8 birds on two successive days, in late November under snowfall conditions while stalking Ibex in the Shingai-gah nullah in western Gilgit at just over 4,250 metres (14,000 feet) elevation. In a study on the slopes of Minapin Glacier in Hunza in August 1984, the largest flock encountered was 43 birds (J. Mayers, Cambridge Karakoram Expedition, 1984). It occurs sparingly in the higher mountain ranges of central Afghanistan and Nuristan (Paludan, 1959) but has not apparently been recorded in Iran. It also occurs in the mountains of central Asia and Kazakhstan (Flint *et al.,* 1984). In the Himalayas it occurs eastwards to central Nepal where it is rare in the high inner ranges and its range overlaps with the Tibetan Snowcock *(Tetraogallus tibetanus)* (Inskipp and Inskipp, 1985). Status COMMON in suitable habitat.

HABITS

Gregarious outside of the breeding season occurring in small coveys and in winter sometimes in larger parties of up to 20 or 30 birds. In the 'breathtaking' altitudes where the human intruder is forced to pause (for breath) every few paces, the Himalayan Snowcock characteristically avoids encounter by running uphill. They are reluctant to fly, but when pressed they then take off down the slope, usually hurtling at terrific speed and covering a very considerable distance before alighting again. In a study

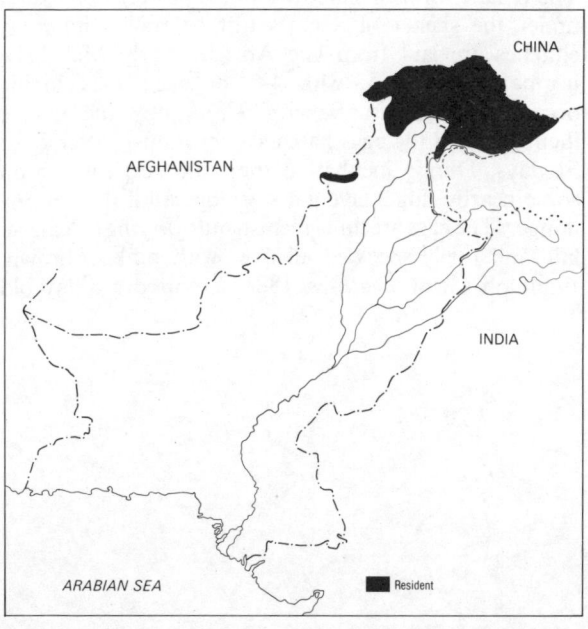

Map 93: Himalayan Snowcock *Tetraogallus himalayensis*

of the population inhabiting the slopes above Minapin glacier in Hunza, James Mayers found that they retreated at night to roost in rocky outcrops at higher elevations, travelling down by walking over screes in the early morning to feed on alpine meadows, with peak activity periods in the early morning and late afternoon. Their diet from faecal analysis, indicated that they are largely herbivorous with grasses *(Poa* and *Alopecurus* spp.) and sedges *(Cyperaceae* sp.) being predominant food items in their diet during late July/August (J. Mayers, Cambridge Karakoram Expedition, 1984, Report).

They remain at high altitudes even in winter, where snow lies on the ground, but are able to subsist by taking advantage of windswept slopes or ridges where the vegetation is exposed. They subsist upon all kinds of vegetable matter including berries of *Ephedra,* shoots of *Artemisia maritima* and, when the ground is not frozen, they dig for bulbs, particularly those of *Gagea* species (observed by author above Saif-ul-Maluk). However grasses, including their flowering seed heads form their staple diet. In the USSR grasses, juniper berries, the bulbs of wild onions, and (in winter) fallen spruce seeds, are all included in their diet (Dement'ev, Vol. 4, 1952). During his Hunza study Mayers witnessed five attacks on resting and roosting Snowcocks by Golden Eagles *(Aquila chrysaetos).* None of these were successful: the Snowcocks rapidly glided downhill once flushed and were much faster than their pursuers (Mayers, 1984).

BREEDING BIOLOGY

Major Biddulph *(Stray Feathers,* 1881) noted that in summer the coveys broke up and they consorted in pairs. The female forms a shallow unlined nesting depression under the shelter of a grass tuft or rock, and most clutches are laid from late April to early May. The normal clutch is 5, with 4-6, or even 7 occurring occasionally (Stuart Baker, 1928). Only the female incubates and the eggs hatch synchronously after 28 to 31 days. During incubation the male keeps watch on some nearby ridge, giving a warning call if there is any danger. The eggs are dull greenish buff varying to 'cafe au lait' and finely speckled all over with pinkish brown. Biddulph *(Stray Feathers,* 1881) captured a 3-day old

chick in Gilgit on 28 May. It had silky down, greyish white underneath, but rufescent on the crown and back with blackish-brown stripes on the crown and auricular region, and spots on the body. On a climb up Sitaram, the highest peak in the Safed Koh (4,750 metres/15,500 feet) Lieutenant Fairbrother came across a brood of 9 downy chicks on 21 June. Despite both parents being with the chicks, he was easily able to catch the young by hand *(Stray Feathers,* 1880, p. 207).

The display of the male has similar elements to many Tetraonidae. The body is lowered in a crouching posture and the neck extended with all the feathers erected, making the white and chestnut bordered upper breast very conspicuous. The tail is erected over the back and the white under-tail coverts fluffed out so that they become very conspicuous. The wings are also partly opened and lowered to the sides. He circles the female thus, periodically running onto a rock or mound to throw back his head and give his loud whistling call (Meinertzhagen, *Ibis,* 1927). The male seizes the hind neck of the female in his bill during copulation and this area becomes bare of feathers and bruised during the breeding season (Meinertzhagen, op. cit.)

*VOCALIZATIONS

During the breeding season and particularly in the early summer, the males call frequently, especially from about 6 a.m. onwards and again in the late afternoon. They have been heard calling in the Kaghan valley in the first week of May and as late as the end of July on the Khunjerab pass. They ascend a prominent rock or spur before calling and their cry is an almost human sounding inflected whistle ended on two shorter-rising whistled notes which can be syllabized as 'cour-lee-whi-whi'. This rather unvarying stereotyped call is repeated at intervals for 1 to 5 minutes, and sometimes for as long as 15 or 20 minutes. The first part is reminiscent of a Curlew *(Numenius arquata).* They also utter a 'chok-chok-chok' call which often accelerates into a rapid chatter, reminiscent of a Chukar, which is probably a contact or warning call. Their whistling advertising or courtship call can be heard over distances of over 3 kms. on a calm day. They will also utter shorter alarm whistles when disturbed and while in flight.

Chukar or Rock Partridge *Alectoris chukar* (J. E. Gray)   Map **94**

DESCRIPTION

Body length 34-39 cms.
Wingspan 50-54 cms.

A densely plumaged, neat and rounded looking partridge, blue-grey on the crown but a pinkish brown colour on the hind neck and back. When at rest, the smooth contoured plumage practically conceals the tail and flight feathers which are chestnut in colour. The legs are vermillion and the bill orange-red. The forecrown, lower cheeks and throat are buffy white prominently encircled by a thick black border forming a collar at the base of the throat and passing through the eye. The iris is orange-brownish and the eye surrounded by an orbital ring of naked red skin. The rump and lower breast are more blue-grey and there is a small chestnut patch behind the eye encircled by the black line which extends from the forecrown and through the eye. The lower flanks are barred vertically by the handsome scalloped pattern of black, edged with orange-chestnut. The lower belly and under-tail coverts are pale chestnut. There is a small blunt spur on the tarsi of the males but otherwise both sexes are identical in size and plumage. Juveniles are more sandy buff, less rufous on the back with narrow creamy shaft streaks and black vermiculations on the wing coverts. They lack the black circlet through the eyes and around the throat and have only one or two black and chestnut crescentic marks on the flank. Males weigh up to 765 grams (27 oz.) and females 540 grams (19 oz.). The population inhabiting the northern mountain regions averages larger in size than specimens from the Punjab Salt Range (author obs. and Stuart Baker, 1928).

HABITAT, DISTRIBUTION AND STATUS

Very adaptable to all kinds of arid, rocky and hilly country ascending to the higher mountain valleys of the inner Himalayan ranges. Now very rare on the lower more accessible foothill ranges, a few still survive in the Kirthar range in Sind, in the Salt Range around Sakesar and on the Margalla hills in Punjab. It occurs in the higher valleys of Swat and Indus Kohistan, and throughout Baluchistan, although mostly between 2,100 metres (7,000 feet) and 3,900 metres (13,000 feet) elevation. It can be associated with degraded foothill scrub comprising *Dodonea viscosa* and *Reptonia buxifolia* as low as 600 metres (2,000 feet) elevation (e.g. the Samana range in NWFP, and up to 3,300 metres (11,000 feet) amongst *Juniperus macropoda* forest in Baluchistan. It is tolerably common in Chinji forest reserve in the Salt Range. In the Kurram valley and Waziristan it is still widespread from about 900 metres (3,000 feet) up to 2,400 metres (8,000 feet) on the slopes of the Safed Koh range. In summer it has been seen at 3,300 metres (11,000 feet) in the Lutko nullah (Chitral)

(T. Braham, pers. comm.) and up to 3,900 metres (13,000 feet) in the Shingaigarh nullah in western Gilgit (author's observation 16 November after light snow-fall). Males were calling from the same cliffs as Himalayan Snowcock on the Khunjerab pass at 4,500 metres (15,000 feet) in late July (author obs.). In the Shyok valley in eastern Baltistan at 2,700 metres (9,000 feet) with snow on the ground, flocks were encountered in mid-February. This is largely a sedentary species which avoids forest cover. There is a good deal of evidence from local hunters (Ticehurst, *JBNHS,* 1927, and Williams and Williams, *JBNHS,* 1929) and observations on the ground, that populations of this bird fluctuate markedly, numbers building up to remarkably high densities after favourable rains and feeding conditions, and then sometimes declining drastically either due to disease or some factor associated with the stressing conditions induced by drought.

HABITS

This bird is gregarious outside the breeding season, and where really plentiful, coveys in the autumn can often number 20 to 30 birds. These may well be family groups as a pair can rear large broods. In Baltistan W. H. Mathews typically noted in August, coveys of 4 adults and about 30 younger, smaller birds (*JBNHS,* 1941). In the Hazarganji reserve (Chiltan range, Baluchistan) in 1984 after a year of good rainfall the author encountered

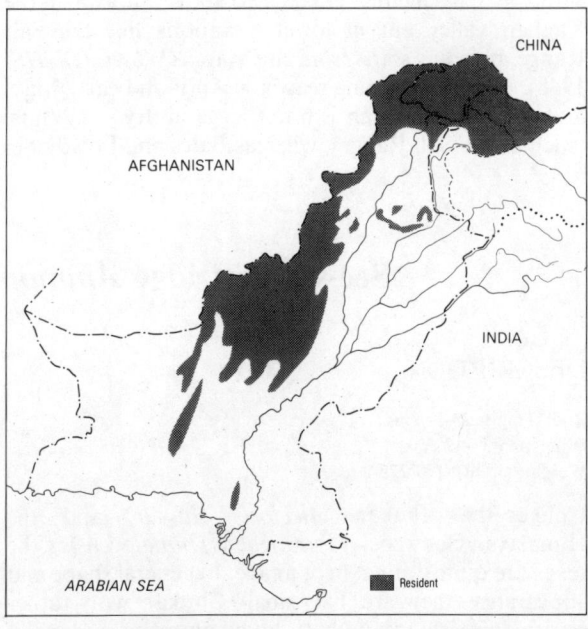

Map 94: Chukar *Alectoris chukar*

two coveys each of over 40 birds. They are particularly active in early morning and evening when the male birds run rapidly over the ground, jumping up steep rocks until they reach some rocky prominence of ridge-top from where they give their typical rapid calls. They are strong runners, reluctant to fly unless disturbed, and well able to hop over big boulders and travel fast on the ground over hilly terrain. If alarmed they will burst into flight, with rapid whirring wing beats alternating with glides, usually hugging the contours of the hillside. When coveys are flushed, the birds often split up and fly off in different directions, relying on calls from the ground to re-establish contact when danger has passed. Their food consists largely of vegetable matter, including seeds, leaves, berries and bulbous roots. In seasons of good rains, grass seeds are the staple diet, but in hard winters they dig extensively, leaving conspicuous excavations on bare slopes where they have been foraging for bulbs and rhizomes. They require water daily and often descend at about 8 a.m. to a suitable stream for drinking. In Baluchistan both Golden Eagles *(Aquila chrysaetos)* and Bonnelli's Eagles *(Hieraaetus fasciatus)* have been observed (Meinertzhagen, *Ibis,* 1920 and Christison, *Ibis,* 1941) preying upon Chukar. In both Baluchistan and the NWFP they are very popular as cage birds, often to be seen in a crowded noisy bazaar where they seem to settle down and call lustily at eventide.

BREEDING BIOLOGY

Nesting occurs over a long period according to latitude and altitude. Some birds in the Himalayas ascend to alpine pastures and do not start breeding until late June. A hen bird with 8 downy chicks was seen (author) on 8 July at 3,300 metres (11,000 feet) above Burawai in the Kaghan valley but in lower elevations like the Salt Range, breeding starts from late March (Waite, *JBNHS,* 1948). The main nesting season is April and early May, and the normal clutch is 6 to 9 eggs in dryer habitats, (such as the Salt Range), whereas Bates and Lowther in

Kashmir, (op. cit., 1952) a heavier rainfall area, found four clutches varying from 15 to 19 eggs. Williams (*JBNHS,* 1929) found most nests in Baluchistan from the end of April up to July and noted that when the spring rains failed, hardly any birds bred, but that when there were good rains, some pairs had two broods. The nest is a mere scrape in the ground partly sheltered by the root of a bush or an overhang with a clump of grass. It is scantily lined with bits of dried grass and the eggs are glossy and of a dirty cream colour marked with minute pale brown speckles all over. Only the female incubates, but the pair bond is normally monogamous and the male remains on guard in the vicinity of his brooding mate. E. M. Nicholson encountered a female with downy young on 5 July near Harboi in Kalat region of southern Baluchistan, and she performed a broken wing distraction display (Nicholson, MS notes unpublished). The downy chicks are very attractive with chestnut crown and a dark line extending from behind the eye and the back mottled in greyish brown and black. The breast is pale creamy fawn (observations, Burawai, Kaghan valley). Observations on captive breeding Chukar indicate that the male takes little part in tending the young after they hatch, but that if a second clutch is laid, he will look after the first brood. The male does however defend the immediate nest site. Incubation takes 22 to 24 days, hatching is synchronous, and the young are capable of feeding themselves immediately.

*VOCALIZATIONS

At evening time and again in the early morning males are very vocal. These calls can also be heard as dusk is descending. The call is rapidly repeated staccato 'chu-chuk-chuk' with variations. Sometimes commencing 'chuk-a-ka, chuk-a-ka, chuk-a-ka' and ending 'ka-ka-ka'. It is far carrying and somewhat reminiscent of a domestic hen calling but more rapid. When very excited the male also gives higher pitched squeaking croaks like those of *Francolinus*.

See-see Partridge *Ammoperdix griseogularis* (J. F. Brandt)

Plate ·14
Map **95**

DESCRIPTION

Body length 23-27 cms.
Wingspan 39-42 cms.
Weight (adult) 190-225 gms.

Unlike the Chukar *(Alectoris chukar)* and the Himalayan Snowcock *(Tetraogallus) himalayensis),* the sexes are quite distinct in plumage. In general shape and appearance they are like small Chukar with rather cryptic pinkish sandy buff body plumage and paler sandy buff under-parts. The male has blue-grey head and

neck, with a broad white line through the eye framed above by a black line, broader on the forecrown, narrowing behind the eye. There is a less distinct blackish grey area on the lower ear coverts. The sides of the neck are spotted in a pattern of small grey and white scales or spots. The upper breast is warm pinkish buff and the lower belly creamy with longitudinal parallel lines on the flanks, of black margined with chestnut. Instead of lying vertically as in the Chukar, these dark lines lie in a more horizontal plane. Females are more rufous on the crown with fine grey cross-barring on the back and a white throat and sides of neck, speckled with dark grey. The breast is pinky buff with darker cross barring, lacking the flank bars of the male. The rump of the female bears noticeable black shaft streaks forming spots. Both sexes have an orange bill and dull yellow legs varying in some areas to olive green (e.g. in Sind Kohistan). The iris is bright yellow and there is usually no trace of a spur. At rest the short rounded wings are largely concealed by the wing coverts. The short rounded tail is also hidden by the sandy buff under-tail coverts, but when put to flight, shows conspicuously chestnut outer tail feathers.

HABITAT, DISTRIBUTION AND STATUS

Though sympatric with *Alectoris chukar* in many parts of its range, the See-see is essentially a bird of much lower altitudes, being absent from the northern inner Himalayan ranges, and better adapted to warmer southern latitudes. They are found throughout Baluchistan in suitable terrain, in Sind Kohistan, the Punjab Salt Range and the NWFP. Due to persistent persecution in the latter province they are now quite rare and local. If not persecuted they can become very numerous and attractively tame, entering the environs of farm yards and threshing floors (observed Chashma Spring in North Waziristan). They are resident and sedentary, from the hills bordering north-west Karachi (Khadeji falls), through the Kirthar and adjoining ranges. They even occur in the sand dune habitats in areas around Nushki in southern Baluchistan, and in the low gravelly hills of the Tobakakar range in western Baluchistan. They occur throughout the Salt Range, including the Kalla Chitta range in the Punjab, in some of the hills north of Kohat in the NWFP and in the Cherat hills south of Peshawar and in the Khyber agency. They have survived in greater numbers than the Chukar in the more hilly tracts of the Salt Range. There are no records of their occurrence in the Himalayas. They will occasionally descend to the flat plains at the foot of the Salt Range (author obs.) as well as to the narrow cultivated valleys within the hills

HABITS

Usually seen in pairs or at the most as 2 to 4 birds, and not occurring in large coveys like the Chukar, even where the population is quite dense. They descend to valley streams, sometimes more than once a day, to drink and appear to be thirsty little birds. They can become very tame when not molested and a bird was seen enjoying a dust bath on a threshing floor near Barshore in Pishin district where women folk were working a few yards away. Their food is largely vegetable matter including green shoots and leaves, grass seeds, berries and probably some insects. A specimen shot by Whistler in the Salt Range had its crop full of the green leaves of a creeping Trefoil *(Trigonella occulta)*. Dement'ev and

Gladkov record ants and beetles in their diet (op. cit., 1952). They are essentially sedentary birds not moving to higher altitudes in the summer, as do the Himalayan populations of the Chukar.

BREEDING BIOLOGY

They apparently breed as monogamous pairs. The only courtship display noted was when a male circled a female trailing only one wing facing her side and with rather upright stance (obs. captive birds, Lahore Zoo, 1968). The female makes a shallow hollow or scrape for her nest, usually well concealed between stones, or under the shelter of a larger rock or bush. It appears to have some lining of bits of dried grass and twigs and the eggs number 8 to 14, with a maximum of 16 in the Salt Range (Williams and Wiliams, 1929 and Waite, 1948). They breed from March to April in the Salt Range, although Meinertzhagen *(Ibis,* 1920) found nests up to 12 May in Baluchistan at higher elevations. The male remains in the vicinity of the brooding female who does all the incubating. The eggs are of a uniform creamy white colour to pale buff, unmarked with any spots. The incubation period is over 21 days. On 2 July 1952 near Khuzdar in Baluchistan E. M. Nicholson (MS notes unpublished) found a female with downy young. The adult feigned injury wth a broken wing so convincingly that his motor driver tried to catch it. The author encountered a family party on 13 July at Khadeji falls in which both the male and female were tending the chicks and called them to safety; these chicks were newly fledged but refused to fly, running with agility up a steep

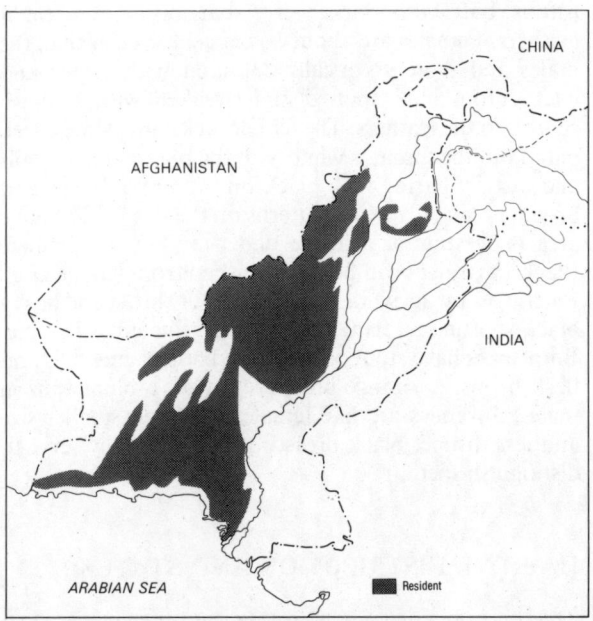

Map 95: See-see Partridge *Ammoperdix griseogularis*

boulder strewn slope. During the early part of the breeding season males have been observed frequently fighting together 'long and fiercely' (Major C. H. Williams, op. cit., 1929). They also call throughout the day from rocky promontories (author obs.) but most incessantly in the early morning and evening.

*VOCALIZATIONS

When put to flight their short-rounded wings produce a peculiar high whirring noise which probably aids as a flight signal to others in the vicinity. The onomatopoeic name 'See-see' does not well represent this wing sound. Their advertising call is an inflected short phrased 'khooit-khooit-khooit', monotonously repeated, a bit reminiscent of a high pitched Chukar but rising at the end. They also give softer whistle-like contact calls. In the breeding season males behave very like Chukar (*Alectoris chukar*), chasing and attacking other approaching males aggressively and running over the ground to various prominent boulders and ridge tops, where they ascend to the skyline and call for 3 or 4 minutes continuously.

Black Partridge or Black Francolin *Francolinus francolinus* (Linnaeus)   Map 96

DESCRIPTION

Body length 33-36 cms.
Wingspan 50-55 cms.

They average slightly larger than the Grey Partridge (*F. pondicerianus*) but resemble them in their stout plump bodies and rather short tail. The male is handsomely patterned being jet black on the lower part of the head except for a white patch on the ear coverts and having the whole of the breast black with the flank feathers boldly patterned with white margins in a chevron pattern. The crown is rufous buff, black streaked and there is a broad maroon-chestnut collar around the lower neck framed below on the mantle by a black and white spotted area. The lower back and central tail feathers are cross-barred white and black and the under-tail coverts and vent are chestnut. The wings are rufous buff with darker cross-barring on the flight feathers. Females are about 10 per cent smaller than the males and more cryptically coloured with crown and back rufous buff, spotted and streaked with blackish centres to the feathers. The female lacks any white cheek patch but the throat is white with the breast creamy buff and heavily barred with black on the upper breast and broader black scale-like patterns on the flanks. The nape area is chestnut as also the under-tail coverts. Female Black Partridges can be distinguished from female Grey Partridges by the white not ochraceous throat and heavy black spotting on flanks, rather than fine cross-barring. Both sexes have stout down-curved horny grey bills and flesh brown to orange-brown legs with a blunt spur in males. Juveniles are like females but their smaller size and less distinct black breast markings usually serve to distinguish them.

HABITAT, DISTRIBUTION AND STATUS

The most favoured game bird on the Indus plains, their numbers have declined very rapidly in the past 50 years, probably due to loss of suitable habitat as well as over-shooting. They avoid very open or bare hilly country and are practically absent from the northern mountain regions and most of the NWFP and Baluchistan. They are very dependent upon good ground cover in the form of scrub jungle for nesting and though they have adapted to shelter and forage in crops, particularly cotton and sugar cane, such areas are only used temporarily. In the 1940s they were extremely plentiful in Sind but improvement in agricultural practices and particularly the levelling and clearing of land and cutting down of scrub jungle has led to their virtual disappearance from many areas.

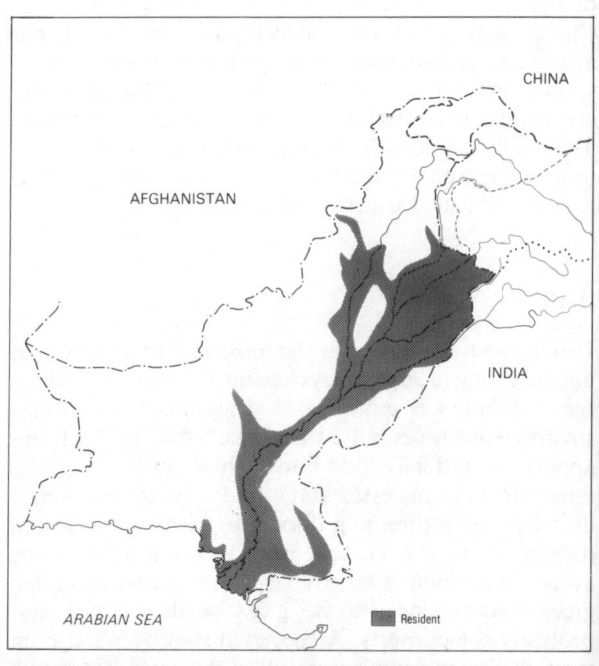

Map 96: Black Francolin or Black Partridge *Francolinus francolinus*

Relict populations survive around irrigated forest plantations in the Punjab (particularly Lal Sohanran, Bahawalpur, Pirawala, Khanewal) and the riverain forest regions in Sind. A few survive in the bush-studded erosion gullies and ravines of Lasbela especially around the Jal-e-Maidan near Uthal in Baluchistan (Deed Ahmad Ali, pers. comm., 1977) and there is a good population (author obs.) in the Margalla hills north of Islamabad extending up to Tret at 1,200 metres (4,000 feet) elevation. They are absent from the Salt Range hills (Waite, 1948 and author obs.), this habitat being too open for them. Whitehead (*JBNHS,* 1911) noted that small numbers occurred on the lower hills of the Kurram valley where the Dwarf Palm *(Nannorrhops ritchieana)* provided good cover. A few birds now live on the periphery of Karachi airfield (author obs.) where young tree plantations and tall grass has provided an ideal undisturbed habitat. They are entirely absent from the main desert tracts such as the Thal or Cholistan but occur in *Saccharum* thickets in the east Narra. Status FREQUENT but only locally COMMON.

HABITS

They can occur sympatrically with *Francolinus pondicerianus* which is also a bird quick to exploit good vegetative cover, but in the Margalla hills the Grey is confined to the lower slopes. Similarly in the Indus plains the Black Partridge only shares the better watered or tree sheltered regions where *F. pondicerianus* also occurs. Normally they are not highly gregarious, usually only foraging in twos and threes. At night they roost on the ground (unlike *pondicerianus),* but males like to climb part way up a tree or stump when giving their call. They are not as aggressive as *pondicerianus* and hence cannot be so easily decoyed by a captive bird and snared, as is done with the Grey Partridge. They are largely herbivorous but will eat a variety of insects and even human faeces. A study in Sind indicated that seeds and vegetable matter formed the bulk of their diet but a large number of ants (Formicidae), beetles, and fewer numbers of flies (Diptera), and wasps (Hymenoptera) were included. In winter mustard *(Brassica campestris),* a cultivated crop, was a favoured food (Faruqi *et al.,* 1960). In a study at Faisalabad (formerly Lyallpur) stomach contents comprised grains, weed seeds and various insects such as locust hoppers (Husain and Bhalla, *JBNHS,* 1937). They are largely sedentary.

BREEDING BIOLOGY

When the weather begins to warm up in February, males start calling, especially in the early morning and evenings. Where the population is dense such as around Rawal lake, as many as 4 individuals can be heard within an area of 150 hectares, and in Pirawala irrigated forest near Khanewal, as many as 6 calling males could be heard within a 40 hectare area in the 1970s. The pair bond is monogamous but only the female nest-builds and incubates. Display mainly consists of the male circling the female with tail fanned, wingtips scraping the ground, and neck upstretched. Breeding can be over an extended period but often does not start till the onset of the monsoon, because of their need for good ground cover. Nests are usually well concealed in thickets of grass or thorn bushes. Whistler (*Ibis,* 1930) found most nests with eggs in June in the Punjab, but in Sind some birds start to breed in March: Ticehurst found chicks hatched on 4 April near Dadu (*Ibis,* 1924). There is a second nesting period after good monsoon rains, and Doig (*Stray Feathers,* 1879) found nests with eggs in September and October. Eates also noted birds laying again in September and considered them as double brooded in a good monsoon year (MS notes unpublished).

Out of over 50 nests found by Eates in Sind, 5 to 7 eggs were the normal, with 9 once, and 8 on a few occasions. The eggs are hard-shelled and glossy, varying from olive buff to pale stone coloured. The incubation period is 18 to 19 days with hatching synchronous (A. Lee, pers. comm., based on captive birds, 1980), and the chicks able to feed themselves immediately. In the wild, males have been seen in attendance with a female and chick.

*VOCALIZATIONS

Males will call any time of the year, and have been heard just after sunrise in November and January but they call most incessantly from February onwards, often at intervals throughout the day. The call is stereotyped, short and very distinct but difficult to describe. Each call lasts just under 1.5 seconds varying to 1.75 seconds and is uttered at irregular intervals every 15 to 30 seconds. It comprises a far carrying rather grating call of two joined couplets, 'chick-ghweeh, (pause) geek-ka-geek'. The first phrase as staccato and grating in tone with the last two notes drawn-out, followed by a brief pause and then a more rapid 3-noted 'long, short, long', equally grating, almost insect-like in tone.

Indian Grey Partridge or
Grey Francolin *Francolinus pondicerianus* (Gmelin)

Map **97**

DESCRIPTION

Body length 33-35 cms.
Wingspan 48-52 cms.

Slightly smaller than the Black Partridge *(F. francolinus)*, but of the same plump game-bird build with rather short stubby tail and long thin neck and upright carriage when running on the ground. Both sexes are alike in plumage, being grey-brown and chestnut on the upper part of the body, the individual feathers having creamy white shaft streaks and dark brown borders and creamy cross-bars also. There is an indistinct broad creamy supercilium, the cheeks are rufous and the throat is a warm yellow-buff encircled by a narrow black gorget. The breast and belly are buff with fine irregular cross-barring. The outer tail feathers are chestnut and the flight feathers are cross-barred brown and pale buff. The bill is leaden coloured, the legs dull red or brownish red and the male has a blunt spur on the tarsus. The distinctive buffy and black bordered throat pattern without large black spots on the breast or flanks distinguish this Francolin from the otherwise very similarly coloured female Black Partridge *(F. francolinus)*.

HABITAT, DISTRIBUTION AND STATUS

Much better adapted to arid conditions than the Black Francolin, they are more widely and evenly distributed throughout the Indus plains, penetrating into the major desert tracts as well as the arid broken foothill country to the west of the Indus. Avoiding intensively cultivated and heavily populated areas they are most plentiful in undisturbed tropical thorn forest habitat and occur throughout the lower hills of Makran and Lasbela, Sind including the Thar desert, Punjab including the Salt Range and Thal desert and the NWFP in the lower protected hills such as those around Cherat and in parts of Kohat. They avoid the higher hills and are absent from around Quetta, but share the same habitat with the See-see *(Ammoperdix griseogularis)* in Kohat, Cherat and the Salt Range. They share the same habitat with *Francolinus francolinus* in irrigated forest plantations and the outer slopes of the Margalla hills. They can be encountered on the upper slopes of the Kirthar hills at 900 metres (3,000 feet) elevation as well as deep into sand dune desert such as around Yazman in Cholistan. Surprisingly a few maintain a foothold around the outskirts of Malir city, sheltering in undeveloped building plots which have become infested with *Prosopis juliflora* (Mesquite), and during the author's residence in Malir it was a pleasure to hear their cheery calls at sunrise against the background noise of heavy vehicular traffic. Status COMMON.

HABITS

Despite constant persecution from sportsmen and bird trappers this hardy bird seems able to survive in proximity to man. They are noted for being pugnacious and can easily be attracted by a captive 'call-bird' and netted. They are extremely popular as cage birds partly because of their loud and ringing calls and also their aggressiveness. Records of fees collected by the Sind Wildlife Management Board during the 1970s (when the author was a member), for permits to keep captive birds, indicate that several thousand are still trapped each year in that province alone.

Where undisturbed they keep in small family parties, feeding and roosting gregariously. They have the unusual habit, for a francolin, of roosting in trees at night. Often two birds will perch in close bodily contact. The dense green foliage of *Salvadora oleoides* being particularly favoured in the Punjab. They can go without water much better than Sandgrouse (Pteroclididae) and even See-see *(Ammoperdix)* and seem well adapted to exploit insects as well as berries of *Ziziphus* and *Capparis aphylla*, grass seeds, green leaves and shoots. They also dig up small lizards such as *Hemidactylus* and desert *Lacerta*'s, as well as the succulent rhizomes of grasses. This may help them to exploit a dryer biotope than *F. francolinus*. In a study conducted at Faisalabad

Map 97: Grey Francolin or Indian Grey Partridge *Francolinus pondicerianus*

(formerly Lyallpur), stomachs from 10 specimens taken between March and July contained grains, seeds, weed seeds, vegetable matter, as well as locust hoppers, ants, termites, beetle grubs: most of the latter are injurious to agriculture (Husain and Bhalla, *JBNHS,* 1937).

BREEDING BIOLOGY

They form a monogamous pair bond, but the female does all the incubation. Nesting is mostly in the spring, eggs being found in March and April, but a few pairs nest in September and October after the monsoon rains. Eates (MS notes unpublished) considered that in Sind they started breeding earlier than the Black Partridge: occasionally he found nests in February, but mostly from mid-March. The nest is always well concealed inside a clump of grass growing up through a thorn bush. The hollow, scraped out by the female, is generally lined with some pieces of grass and leaves. The biggest clutch Eates found was of 9 eggs (unpublished MS notes). The eggs are glossy, pointed at one end and vary from pale brownish to pale buff unmarked.

Incubation takes 18 to 19 days and the chicks hatch synchronously. Both parents tend the young chicks after hatching. On 1 June at sunrise in Dhaman-i-koh ravine at the foot of the Margalla hills, 3 families with fledged chicks smaller than Button Quails *(Turnix* spp.) were seen picking up grit within a 400 metre stretch of road. In each two adults were in attendance. Unlike *Alectoris,* these family parties do not join together but forage separately.

*VOCALIZATIONS

One of the more familiar sounds in rural areas, the males call all the year round, mostly at sunrise and sunset. They start off with some subdued short 'check-check' or 'chak-chak-chak' calls, working up to a rapidly repeated 3-noted call, aptly rendered by Indian writers as 'khateeja-khateeja-khateeja'. They also have a softer more whistling 'kila-kila-kila' call, probably a contact call. In periods of high excitement as just before flight, both sexes give a high whirring 'khirrr-khirrr' call which doesn't carry so far.

Common Quail or Grey Quail *Coturnix coturnix* (Linnaeus)   Map **98**

DESCRIPTION

Body length 16-20 cms.
Wingspan 32-35 cms.

A compact short necked little game bird with rounded body and short tail. Both sexes are streaked warm ochre and black without any cross-barring on the body. The male has a distinctive head pattern with three black crown stripes separated by mesial and superciliary creamy stripes. The sides of the face are white with darker chestnut ear coverts and a black gorget framing the white throat with a black chin and line down the centre of the throat usually with a second inner circlet of spots. The upper breast is rufous buff with pale shaft stripes and the flanks are browner with blackish spots and white shaft streaks. The flight feathers are barred dark brown and rufous buff, and this field mark is helpful in separating the Common from the Rain Quail *(Coturnix coromandelica),* which has plain chestnut wing quills. Females are like males in general cryptic colouration but they lack the black and white throat pattern and have heavy black streaking on the upper breast. In both sexes the bill is horny brown coloured and the legs yellow to fleshy brown with no spurs on the tarsi. The short wings are heavily cambered and their flight is by rapid whirring wing beats.

HABITAT, DISTRIBUTION AND STATUS

Unique amongst the Galliformes in being migratory, surprisingly little is known about their movements. However in the northern Himalayan regions including Gilgit and Chitral a regular spring and autumn passage is noted, with a few birds staying to breed in the broader valleys in the summer. There is a similar marked double passage in spring and autumn throughout the Indus plains, the north-west providing the main passage route for birds which spread throughout the Indian subcontinent to spend the winter. Though there are no ringing recoveries, their passage through certain localities is quite predictable. The autumn passage starts from late August and they are numerous in irrigated cropland throughout the Punjab during September and are still numerous in lower Sind till early October. The return spring passage starts in late April and continues through May. Along the coast of Karachi and on sandy islands in the delta, migratory quail can be flushed, often in emaciated condition (R.F. Nana, pers. comm.), from late September and through October. It is thought that there is considerable east-west movement of these birds along the coast. While on passage they are usually encountered as scattered birds or in twos and threes throughout suitable arable areas, which they seem to prefer to broken hilly country or open desert. Some quail on return spring passage stay to breed and nests have been found in Sind, the Punjab and Baluchistan. In northern Hunza around Sost males were heard calling in tall wheat in late July and local farmers confirmed that they regularly nested in the standing wheat crops in July and August (author obs., 1987). The odd bird also overwinters in Punjab and Sind, indicating a rather gradual and dispersed movement during migration, perhaps with a tendency to linger if feeding conditions are attractive. It is thought that most of these spring and autumn migrants are birds that breed in eastern Europe and central Asiatic Russia. Christison (*Ibis*, 1941) noted a passage of quail in early March and again in August September, through the Chagai. Many migrating quail are trapped in late March as they pass through Bannu district, sheltering in the ripening wheat (author's observations) and Whitehead (*JBNHS*, 1911) found a few birds staying to breed in April and May in Kohat district. Status still COMMON though not as abundant as indicated by earlier writers.

HABITS

In the spring quail can be heard calling and birds can often be flushed from crops. But they are remarkably secretive and furtive birds in their habits and provide little or no opportunity for continuous observation. When flushed they fly rather straight but for short distances, dropping suddenly back into the cover of crops often after only 100-200 metres. They do not congregate in coveys and seem to be encountered singly

or at the most 2 or 3 scattered individuals. They are believed to travel mostly at night while on migration, a habit well known and exploited by Punjabis who put call birds (captive male quails in cages) out in the crops at evening time knowing that numbers of migrating birds will be attracted during the night to settle in the vicinity. In this manner large numbers are subsequently netted and sold in the villages for eating. This technique was also used to obtain good bags during quail shoots; though this has not been a popular sport in modern times, the quarry being considered too small for the present day cost of cartridges. Whistler records himself going out after tea and shooting 42 quail in one evening near Khanna in Rawalpindi district (*Ibis*, 1930) and Ticehurst recalls 2 guns bagging 128 in three hours shooting at Dadu (western Sind) on 14 March 1913. Evidence of their gradual decline is available from the fact that netted birds are rarely offered for sale nowadays, the few secured by local trappers being often sold at higher prices for the aviary trade. Their food consists largely of seeds, including fallen cereal grains in stubbles, as well as grass seeds, Chicory (*Cichorium* sp.), and insects including ants and ground living insect larvae. Captive quails were noted (author) foraging like domestic chickens by scratching and jumping backwards and digging assiduously with their bills. A study in western Rajasthan, India showed 90 per cent of their food by weight was weed seeds including grasses and legumes, 18 per cent cultivated grains, mostly millet and only 8 per cent insects and Arachnida (Mukherjee, 1963).

BREEDING BIOLOGY

Nesting is inclined to be sporadic with rather scattered authentic records. The calling of males occurs throughout spring passage and is not evidence of actual breeding. Whistler (*Ibis*, 1922 and 1930) did not actually record finding any nests in the Punjab but Waite found many between the Jhelum river and the foot of the Salt Range escarpment, mostly in mid-March (Waite, *JBNHS*, 1924). Whitehead (1911) found nests in April in Kohat, as did Williams (*JBNHS*, 1929) in Baluchistan around Quetta, Mastung and Mach. K. Eates found 5 or 6 nests in Sind. Lindsay-Smith (*JBNHS*, 1913) had clutches brought to him in late May, well incubated from wheat fields in Faisalabad (Lyallpur) district. Nests in Baluchistan were found in lucerne crops and in thick grass on the fringes of a small irrigated tree plantation (Williams and Williams, *JBNHS*, 1929). In Sind (Eates, MS notes) nests were invariably in the riverain tracts in open clearings where the 'Blue Pea' or vetch had been planted as a crop. The female is a tight sitter and in most written accounts the nest has been accidentally discovered during grass cutting operations. The normal clutch seems to be 5 to 7, the largest clutch found by Eates comprising 10 eggs, but in another instance only 5 were found 'well incubated' (K. Eates, MS notes). The

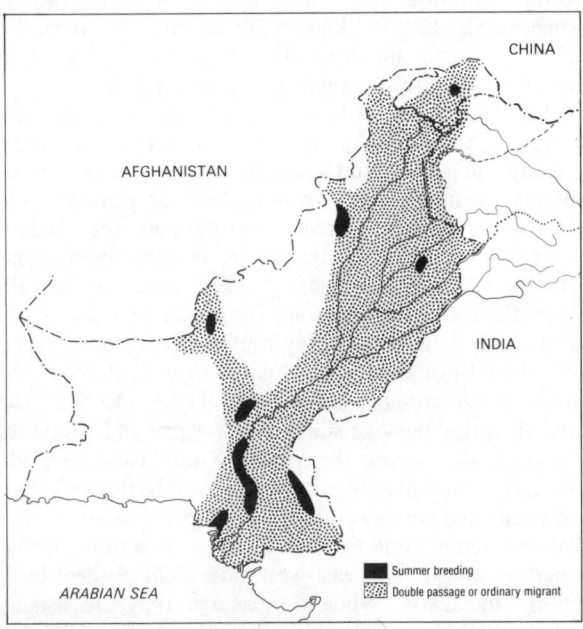

Map 98: Common quail *Coturnix coturnix*

Summer breeding

Double passage or ordinary migrant

CHINA

AFGHANISTAN

INDIA

ARABIAN SEA

nest is always cleverly concealed, in a hollow or under a clump of vegetation with a few feathers in the lining. The eggs are variable in pattern, with finely speckled and heavily blotched specimens sometimes occurring in the same clutch. They are reddish brown to yellowish buff in ground colour, pointed at one end and glossy in texture. Most eggs in Sind are laid in April with one nest found on 1 May (K. Eates, MS notes). Williams found nests around Quetta from 9 April to 29 May, Capt. Biddulph (*Stray Feathers,* 1881) in Gilgit found a nest with 11 eggs on the point of hatching on 26 June. Only the female incubates for the 17 to 20 days necessary. Eggs are laid at 24 hours intervals but they hatch synchronously. The chicks feed themselves and can flutter short distances off the ground from 11 days of age.

*VOCALIZATIONS

Disturbed birds in flight emit a short squeaky alarm call or whistle. Calling males utter the rather melodious 3-phrase liquid whistling call 'whit-te-tooh' or 'kwik-whi-khwik'. It is a relatively low pitched whistling tone with the first syllable very short and the last syllable emphasized and slightly more sustained. It is not a loud call and can be uttered throughout the year, usually from thick cover, but is most commonly heard during the breeding season. Two males recorded in Hunza in July called in bouts of 3 to 6 calls, repeated at approximately 1 second intervals. When heard at very close range the last part of the call has a distinctly squeaky 'un-oiled' intonation.

Rain Quail or Black-breasted Quail *Coturnix coromandelica* (Gmelin)   Plate **14**
Map **99**

DESCRIPTION

Body length 16-18 cms.
Wingspan 31-33 cms.

Slightly smaller than *Coturnix coturnix,* but closely similar in body proportions and cryptic plumage pattern, with rounded plump body and short stubby tail. It is rather more white about the head and throat with a distinctive pattern of black chin and vertical line joining two encircling black stripes extending from behind the eye and around the throat in a gorget, aptly described by Whistler ('Handbook', 1949) as an anchor mark when viewed from in front. The crown and back is more rufescent, less buff than in the Common Quail, but with similar pale creamy shaft streaks. The most noticeable feature is the extensive patch of black in the mid-breast region and heavy black streaking along the flanks. The flight feathers are plain chestnut, unbarred in contrast to those of *Coturnix coturnix*. The female lacks the black face mask of the male, and is probably inseparable in the field from the female Common Quail except for its slightly smaller size and unbarred flight feathers. Both sexes have legs and feet pale fleshy to fleshy grey, with the bill horny brown in the female and more leaden coloured in the male. The iris is hazel brown.

HABITAT, DISTRIBUTION AND STATUS

This is an Oriental endemic species largely confined to the subcontinent (it occurs all over India, Sri Lanka and in the dryer parts of Burma). It breeds during and after the monsoon, following an irregular migration dispersal. In Pakistan it is a monsoon season visitor to the plains, coming into the irrigated crops and grassy areas during late June and breeding from August through to

September, dispersing eastwards and southwards by early October as the dryer weather approaches. It is most plentiful in the north-eastern part of the Punjab particularly in Lahore, Gujranwala and Sialkot districts, being rare in southern Punjab. H. Waite (*JBNHS,* 1948) did not encounter it in the Salt Range, nor Whistler (*Ibis,* 1930) in the Rawalpindi plateau regions. Curiously C. H. Whitehead (*JBNHS,* 1911) records that a few Black-breasted Quail were shot during May each year around Kohat, but he did not encounter them later in the summer. Many are netted for food during the summer

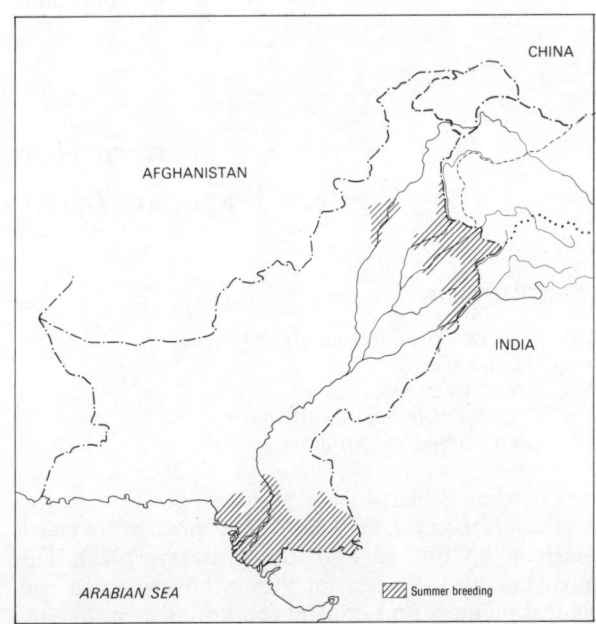

Map 99: Rain Quail or Black-breasted Quail *Coturnix coromandelica*

and can be seen, being offered for sale in the Tollington poultry market, Lahore, from early August till the end of September. They frequent the same habitat as Common Quail, liking thick green crops or grassland and avoiding bare, open ground or forest plantations. In Sind they are irregular visitors in years of good rainfall and abundant grass growth, but in a dry year they are practically absent. They breed in suitable grass areas in such years (Ticehurst, *Ibis,* 1924) and penetrate to the border areas of Lasbela in the Hab valley. They are absent from the dryer hilly tracts or extensive desert regions such as cover most of Baluchistan and the NWFP. Status FREQUENT.

HABITS

Perhaps slightly more gregarious than *Coturnix coturnix,* with small parties of 3 to 6 being flushed together. They are similar in habits, keeping to thick ground cover, particularly standing crops and tall grassland. They are very skulking, preferring to run, and when flushed fly low over the crops for relatively short distances plunging suddenly into cover again and running so that they can rarely be flushed twice. They feed mainly on grass seeds and other weed seeds, but in captivity flourish on seeds of Millet *(Pennisetum typhoides).* According to Fleming (1976) they also feed on the ground-dwelling insects and their larvae as well.

BREEDING BIOLOGY

Ticehurst quotes Le Messurier finding nests in 1875 at Sapoora near Karachi on 8 August after good monsoon rains (op. cit., 1924). One clutch of 5 eggs was found and females shot with oviduct eggs on the same date, whilst downy chicks were encountered on 1 September in the same year. K. Eates (MS notes unpublished) found a clutch of 3 eggs on 29 July on the Moach plain just north-west of Karachi, and a nest containing 6 eggs on 2 August in different years. The nests were well concealed in thick grass and scantily lined with dried grass and leaves and were very similar to *Coturnix coturnix* nests often being built in the depressions formed by the hoof-prints of cattle. The eggs are yellowish white to brownish buff in ground colour, glossy in texture and speckled with dark brown, although some are blotched more boldly. They are, in fact, small editions of the Common Quail eggs. It is thought that only the female incubates, but that the pair bond is lasting, with the male assisting in the tending of newly hatched chicks, which can feed themselves from the time of hatching.

The incubation period is 18 to 19 days and one observer noted that the chicks remained in the vicinity of the nest for some days after hatching, the female constantly calling to her young. Courtship is believed to be similar to *Coturnix coturnix* with the male circling the female, wing dropped and partly spread and neck upstretched with feathers ruffled.

VOCALIZATIONS

Its call is quite different from *Coturnix coturnix* (Smythies, 1953) being a double whistle best rendered as 'which-which' (Fleming) or 'whit-whit' (Smythies, 1953) and constantly repeated in the early morning and evenings. This double whistle is repeated 3 to 5 times in rapid succession, then there is usually a pause of some minutes.

Western Horned Tragopan or
Western Tragopan *Tragopan melanocephalus* (J. E. Gray)

Plate **14**
Map **100**

DESCRIPTION

Body length 68-72 cms. including tail 19-25 cms.
Wingspan 80-100 cms.
Wing length 25.7-29 cms. (males),
 (22.5-25 cms. (females) (Stuart Baker)
Bill length 17-20 cms. (Stuart Baker)

It is a relatively large stout pheasant, bigger than the Koklass *(Pucrasia)* or Kaleej *(Lophura),* with males weighing up to 2.15 kgs. (Stuart Baker, 1928). The tragopans are all noted for their richly patterned and spotted plumage and brightly contrasting areas of bare skin around the throat and eyes. The Western Tragopan is perhaps the most soberly clad of these 5 species of Sino-Himalayan pheasants. In shape they have noticeably long broadly graduated tails, short rounded wings and relatively short but stout bills. The male has a grey-brown back and wing coverts, vermiculated black and buffy brown, dotted all over with round white ocelli, whilst the lower breast and belly are black with larger round white spots becoming smaller on the upper breast. The tail is black with the outer feathers barred on the inner webs with fragmented pale chestnut bars. The upper-tail coverts extend well down each side of the tail and are pale hazel chestnut colour with prominent white ocelli framed by black chevrons. These ocelli are very conspicuous when the bird is retreating on the ground. The sides of the head around the eye, have an extensive area of bright red skin and there is a bushy crest of slender black feathers tipped blood red posteriorly. The most striking aspect of the male are the fiery orange-vermillion and iridescent hackles on the foreneck, offset by blood red hind neck and small maroon-red patch on the wing shoulder. The naked throat area is blue. Many of the feathers on the breast and lower flanks are blood red below the white spot.

In the breeding season, the gular skin area can be expanded in display into a peculiar bib-like flap of skin. This becomes flesh pink with a median purple line and two cobalt blue triangular patches on either side of the median line. This lappet shrivels up outside of the breeding season becoming invisible. There are also two small fleshy horns, coloured blue, on either side of the crown which are erected when the cock bird displays. The legs are pink to fleshy grey with quite a sharp spur. The female lacks any red plumage on the neck or white ocelli on the body. She is cryptically plumaged grey-brown all over with black shaft streaks on the mantle and white centres bordered black to some of the breast and flank feathers. Her throat region is paler buff and wing shoulders more chestnut, whilst the upper-tail coverts have the same prominent tail pattern of chestnut black and white ocelli as the male. The legs are more brownish pink and there is a restricted area of bare red skin around the eye. The legs are stout with well developed toes.

HABITAT, DISTRIBUTION AND STATUS

A very hardy Himalayan pheasant adapted to survive on the coldest north-facing slopes in summer and never, in the western part of its range, descending in winter below the forest line at 2,134 metres (7,000 feet) elevation. In

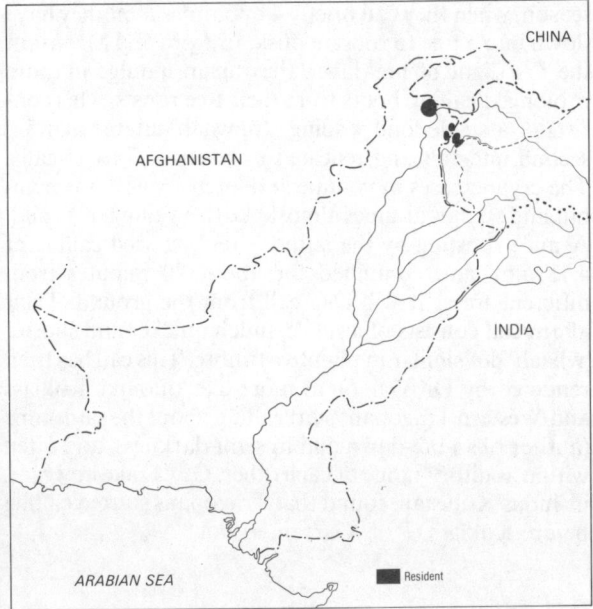

Map 100: Western Tragopan *Tragopan melanocephalus*

Azad Kashmir, it is characteristically associated with mixed coniferous forest, Blue Pine *(Pinus wallichiana)* and Brown Oak *(Quercus semecarpifolia)* and it occurs in mixed Brown Oak *(Quercus semecarpifolia)* and Deodar *(Cedrus deodara)* forests in Indus Kohistan. In summer it will forage up to the edge of the tree line at 3,350 metres (11,000 feet) but prefers to haunt sheltered cols and slopes where there is a thick ground cover of *Viburnum* spp., *Spirea laureola* and *Berberis* sp. A small, viable population survives around the tributaries of the Duber valley in Indus Kohistan. In a survey carried out from 8 May to 27 June 1987 a total of 26 different calling males were heard in Indus Kohistan, with an estimated population of 30 to 40 pairs in the watershed of the Duber valley, and smaller populations in the upper reaches of the Patan and Kayal valleys. Birds were encountered up to the upper limit of the tree line. (Guy Duke and Paul Walton, hand-written Report to WPA, July 1987). A further survey on the east bank of the Indus in May/June 1989 in the mid-Palas valley resulted in the surprising discovery of a population estimated as high as 200 pairs (G-Duke and J. Eames, 'Report to ICBP', Cambridge, 15 August 1989). Another smaller more threatened remnant survives in the mountain valleys forming the watershed from Makrhrah peak down to the Kaghan valley. In this area of Hazara district, birds were seen at 3,352 metres (11,000 feet) by Mr. Abdul Rahman Khan, Conservator of Forest (Wildlife) in the Kanshian nullah and the Bichla watershed (pers. comm., 1973). Only 1 or 2 were seen in Nuri forest in May 1984 during surveys by Richard Grimett and Craig Robson. In Azad Kashmir, Ainsworth Harrison found them at higher elevations above Pir Hasimar and Pir Chinasi in Kotli district, at the head of the tributary valleys off the left bank of the river Jhelum (per. comm., 1971). The main population, however survives in the Machiara valley, a tributary of the Neelum river (Kishenjanga) on its right bank, where an estimated population of 70 to 80 birds are protected (author's obs., 1984). A smaller population occurs above Salkala higher up the Neelum valley in an area comprising the furthest buttress slopes of the Kaz-i-nag range. A total population of 20-30 birds are estimated to survive in this latter area (Kamal Islam, pers. comm., October 1983). There are presumed to be more birds surviving on the Indian side of the cease-fire line above Salkala. In the late 1960s, occasional birds visited the northern slopes of the Galis in winter where Ainsworth Harrison shot them (pers. comm., 1969). Status RESIDENT and RARE. The IUCN lists this pheasant as endangered in its *Red Data Book*.

HABITS

Besides their extreme rarity, they are very shy and nervous birds so that little has been recorded of their habits. In winter they descend to about 2,150 metres (7,000 feet) but keep within the forest even when the ground is carpeted with snow (Pir Khan, Forest Guard, Machiara, pers. comm., 1984). In summer they occur mainly between 2,743 and 3,352 metres (9,000 and 11,000 feet) elevation on northern aspects, particularly where good shrub and bush cover is available as in moister cols and ravines. They are not gregarious, usually only single birds being flushed at a time. At the approach of dusk, and even after dark they fly up into trees to roost, but feed mostly on the ground, being active in foraging from late afternoon and in the early morning. Caecal droppings of the Tragopan, seen in March and late May in Machiara (Azad Kashmir) are very distinctive being always blackish and of a liquid consistency, in sharp contrast to the more normal fibrous tubular extrusions which they also produce in common with all game birds and which are not distinguishable from the droppings of the Koklass pheasant. Craig Robson (pers. comm., 1984) noted the same semi-liquid black caecal dropping from Tragopans located in Duber valley, Indus Kohistan, about 70 kms. north-west of Machiara. Their diet is known to comprise mainly young buds and leaves, but they are partial to the red berries of *Skimmia laureola* and in the autumn to the acorns of *Quercus semecarpifolia,* and in July and August, the berries of *Viburnum nervosum* (obs. author Machiara and pers. comm., Pir Khan, retired Forest Guard, Machiara).

Like many pheasants, the cock bird will respond to a sudden noise with an alarm crow. Kamal Islam, whilst studying the Tragopan, used this technique to locate individuals, striking a hollow log to elicit their characteristic 'whaa' alarm cry (pers. comm., October 1983). The author's impression from encounters in Machiara sanctuary, is that they prefer to run uphill when disturbed, flying only when danger is imminent. Their short high cambered wings make more noise in flight than the Koklass or Kaleej, and produce such a distinctive noise as to be instantly recognizable for that species.

BREEDING BIOLOGY

Breeding occurs during the month of May extending up to early June. Though recent publications tend to repeat the statement that they often nest in trees, the only nest found by Kamal Islam, was a scrape on the ground (pers. comm., 1983) and another nest found on the ground is described in Stuart Baker ('Fauna Series', 1928). The pair bond is monogamous but the female does all the incubation whilst the male helps to tend the chicks. The incubation period is 28 days (Delacour 1951). A nest with 4 unincubated eggs was found on 12 May 1984 at Machiara. It was located on a narrow earth ledge on a rock face at about 3,350 metres (11,000 feet) (Sheikh Abdul Qayyum, Chief Game Warden, Azad Kashmir,

pers. comm., 1984). The eggs are pale buff in ground colour, feebly freckled with dark brown; clutches are believed to be quite small, varying from 3 to 6. A female with only two chicks was sighted by Kamal Islam in June.

The display has only been described at first hand by Delacour and is typically lateral with the male twisting his body with partly fanned tail sideways towards the female and dropping the inner wing. This display is without erection of the throat bib or 'horns'. It usually culminates in a full frontal display when the cock bird will suddenly stop and with all plumage fluffed out, slowly raise and lower its half spread wings, shaking the head and neck spasmodically. During this period, the horns and throat lappet suddenly expand and contract several times. It then shakes its plumage and the wattles contract with astonishing rapidity as it assumes normal feeding (Delacour, *Pheasants of the World,* 1951). There have been no aviary or zoological collections of Western Tragopan since the Second World War and the World Pheasant Association has made repeated efforts to obtain a pair for captive breeding. About 4 males and 2 females have been captured from Azad Kashmir, and 1 male from the Kaghan valley, all of them as single individuals, over a period of about eleven years, but none of these has long survived captivity or in some cases, tran-shipment to Europe.

*VOCALIZATIONS

Unlike the Koklass *(Pucrasia)* or Monal *(Lophophorus)* pheasants, in which the females have quite different calls from the male, both sexes of the Western Tragopan utter very similar distinctive nasal bleating calls. Males are inclined to be more silent except during the breeding season, when they call briefly 4 or 5 times after they have flown into a tree to roost at dusk. In April and May from the first light of pre-dawn, they again indulge in more prolonged calling bouts from their tree roosts. This consists of a single loud wailing 'khuwaah' uttered at 1.5-2 second intervals and repeated in bouts of 7 to 15 calls. The cry increases in volume and intensity as it is drawn-out and sounds at times almost like the wailing of a child. A male roosting by the author's tent, started calling at 4.15 a.m. and continued for about 20 minutes from different trees. It will also call from the ground. Their alarm call consists of a single, much quicker and shorter 'whaah' but similar in plaintive timbre. This call has been rendered by Dr. Garson as more like 'quoink'. Koklass and Western Tragopans start calling about the same time (author obs.) pre-dawn and in semi-darkness, and often within auditory range of each other. Guy Duke however, in Indus Kohistan, found that Tragopans started calling before Koklass (P. J. Garson, *in litt.*).

Koklass Pheasant *Pucrasia macrolopha biddulphi* (Marshall)

P. m. castanea (Gould)

Plate **14**
Map **101**

DESCRIPTION

Body length 58-64 cms. including tail 20-27 cms.
  in males, 16-17 cms. in females
Wingspan 55-70 cms.
Wing length 23.3-24.9 cms. (males. Kashmir) (Stuart Baker)
Weight 1.3-1.4 kgs (males), 1.0-1.1 kgs. (females) (Whistler)

A medium sized rather slim pheasant with a long conspicuous crest on the nape (up to 10 cms. long in male) and a medium length, wedge shaped tail. Four different sub-species have been described, mainly based upon the amount of chestnut on the mantle and breast. The race *castanea* occurring in Chitral and Swat having the most extensive dark maroon chestnut areas, with the race *biddulphi* confined to Gilgit and Hazara districts, having no chestnut in the mantle and the nominate sub-species *macrolopha* with silvery grey mantle and flanks, occurring in the Murree hills. Like most pheasants there is considerable individual variation in plumage pattern and after examining the series of skins in the Bombay Natural History Society collections and the British Museum at Tring, the author feels doubtful about the validity of such sub-species. Generally the population inhabiting the inner dryer mountain ranges north and west of the Murree hills, has more chestnut in the body plumage but the tail pattern is so variable as to lack consistency with the identification keys usually given (see Stuart Baker, 'Fauna Series', 1928 and Ali and Ripley, 'Handbook', 1969).

The cock bird has a bottle green head and neck with a conspicuous white oval patch below the ear coverts, a pale hazel brown crown with long pencil-shaped blackish green plumes (9 cms. only), extending in a drooping crest on either side of the crown. These plumes are erected during display. The centre of the breast is dark liver chestnut and the rest of the body is silvery grey with long lanceolate feathers on the mantle and flanks having black centres. The wing coverts and back have more buffy yellow margins to the feathers and there are broad arrow-shaped upper-tail coverts margined with pale grey and having darker brown centres extending on either side of the tail almost to its tip. The plumbeous grey legs are quite long and stout with a prominent curved spur. The rump is generally paler grey and under-tail coverts are chestnut. The flight feathers are buffy brown with paler outer webs. Females have a less distinct nuchal crest, and are chestnut with darker brown streaks on the breast and darker brown mantle and ear coverts. They have a white patch on the side of the neck similar to the males and creamy buff throats with black scale-like markings on the sides of the throat. The tail is also long straight and graduated.

HABITAT, DISTRIBUTION AND STATUS

The least endangered of our Himalayan pheasants, the Koklass is widely distributed in coniferous forest areas between 2,100 and 2,700 metres (7,000 and 9,000 feet) elevation from the higher Murree hill peaks, through Azad Kashmir, Hazara district, Swat and Chitral. A few birds survive precariously in the Deodar and Ilex Oak scrub forest *(Quercus ilex)*, of the higher peaks in the Malakand agency and also in north-eastern parts of Dir in similar biotope. It occupies somewhat lower elevations than the Monal *(Lophophorus impejanus)*, preferring forest glades for foraging rather than the alpine slopes on the upper edge of the forest. In Machiara (Azad Kashmir) it shares the same habitat and feeding areas as the Western Tragopan *(Tragopan melanocephalus)*. A census carried out by Z. B. Mirza (1978) in the Neelum valley in 1977 revealed 88 calling birds within a 6.5 square mile area or the equivalent of 5.25 birds per square kilometre. In the Galis between Nathia Gali and Ayubia, there were 59 calling males in a square mile area in April 1978, equivalent to 22.75 birds per square kilometre (Severinghaus, *WPA Journal,* 1978-9). This compares with a density as high as 20 calling males per square kilometre in the best habitat around Simla (Himachal Pradesh, India) (Dr. P. J. Garson, pers. comm., 1987, Gaston *et al.,* 1981, Gaston 1980, and Khan and Shah, 1982). Status COMMON.

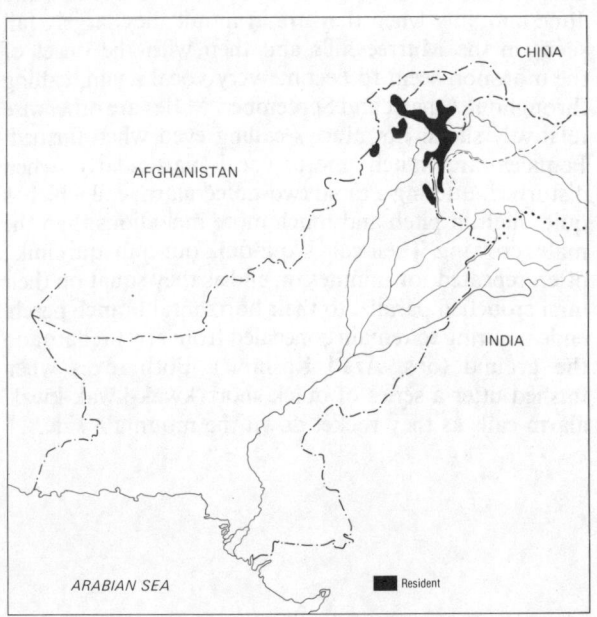

Map 101: Koklass Pheasant *Pucrasia macrolopha*

HABITS

Koklass frequent the steepest and most precipitous forested areas and are now only found in relatively high numbers in the western part of the Murree hills as well as the lower Kaghan valley. They are extremely shy and skulking and can rarely be watched behaving naturally by the human observer, usually slinking away rapidly through the undergrowth or bursting up through the trees uttering their loud alarm cries before rocketing down the slope, diving and twisting between the trees with incredible speed and agility. In winter and early spring they sometimes congregate in loose groups and the author once flushed 7 birds together in early May below Nathia Gali, but in the summer they are usually encountered in pairs. Nesting starts in April, continuing through till late June. They forage mostly in early morning and late afternoon, favouring open grassy glades where they feed on leaves of grass and other forbes. They will eat berries in season, particularly those of *Lonicera quinquelocularis* and *Viburnum nervosum*, larvae of insects and various invertebrates, scratching in pine needle litter like a typical pheasant (author obs., Kao Forest, Dunga Gali). At dark they fly into trees to roost, often perching about half way up and close to the main trunk and it is from such perches that the males start to crow with first light of day. Many writers have noted their habit of calling at all times of the day when startled, particularly by a sudden rumble of thunder, so frequent in the outer Himalayas during April and May. The display of birds in captivity has been well described (Wayre and Harrison, 1972) and is a typical lateral presentation. The cock bird circles a female with his tail fanned open and completely twisted on its side so that the upper surface faces the female. His whole body stance is somewhat upright and tilted inwards, not crouching, and the neck and back feathers are ruffled and raised with the inner wing slightly dropping. The pale chestnut crown remains slicked down but the long blackish crest feathers on either side become raised vertically 'like the ears of a rabbit', and the white cheek patches are also fluffed up so that they are visible from the front view also (J. O. Harrison and Philip Wayre, *Outdoorman*, April 1972). Two nests found by the author below Dunga Gali were snug, well rounded hollows but with only a scanty lining of dried leaves and on very steep banks; in one case sheltered under tree roots and the other under a bush. The normal clutch is about 6 eggs but 5 well incubated eggs were found by Col. Buchanan on 27 May on the slopes of Miran Jani (Murree hills) (in Whistler, *Ibis*, 1930). Another nest of

10 fresh eggs was found by him on 6 May at Changla Gali (Whistler, *Ibis*, 1930) but such a large clutch seems to be unusual. Only the female incubates, though the cock is reported to be monogamous. Incubation period by captive birds at the Pheasant Trust (in Norfolk, U. K.) varied from 25 to 26 days (Wayre, 1969). The eggs are pale buff thinly speckled all over with small reddish brown specks. The chicks hatch synchronously and the egg shells are left just on the periphery of the nest (author obs.). I have twice encountered females with downy chicks at Dunga Gali. Once on 19 June a bird with only 3 chicks about a week old which were a golden tawny colour with darker brown eye streak and stippled on the back and wings with reddish brown. These chicks could fly for short distances like all Phasianidae chicks, which rapidly develop small flight feathers. There was no evidence of the male in attendance on either occasion, but Bates and Lowther (1952) report that the male does help in tending the young brood.

*VOCALIZATIONS

Males call at dawn almost throughout the year and this calling is more prolonged and reaches its peak from mid-May to early June when females are egg laying. At that time birds start crowing at 4.15 a.m. with the first light of pre-dawn. It is a rather harsh and hoarse 'ka-ka-kaaah-kah' repeated at intervals of 15 seconds up to one full minute and other males in the vicinity take up the call. This is an advertising or territorial call given by individual males from the same approximate vicinity each day and while the bird is still perched in its night time roosting tree. They will also fly to the ground and continue calling. They call less frequently just at dusk also and occasionally in the middle of the day. By late June and July when they are in moult they largely fall silent in the Murree hills and then with the onset of the monsoon seem to become very vocal again, calling throughout August and September. Males are otherwise relatively silent, not always calling even when flushed. Females are much more vocal particularly when disturbed, uttering a rapid two-noted alarm call which is quite high in pitch and much more melodious than the males crowing. Their call is 'qui-quik qui-quik qui-quik', often repeated for minutes on end as they squat on their tarsi crouched parallel to their horizontal branch perch, endeavouring to remain concealed from any predator on the ground (obs. Azad Kashmir). Both sexes when flushed utter a series of quick short 'kwak-kwak-kwak' alarm calls as they rocket down the mountain side.

Himalayan Monal or
Impeyan Pheasant *Lophophorus impejanus* (Latham)

Plate **14**
Map **102**

DESCRIPTION

Body length 68-73 cms.
Wingspan 100-110 cms. Females slightly smaller
Tail length 21-23 cms.
Wing length 28.9-32 cms.
Weight 2,260 gms. (males) (Stuart Baker), 2,120 gms. (females)

This is a rather large heavily set pheasant. The male is truly resplendent with black throat, breast and belly, offset by metallic iridescent wings of bluish purple and bluish green, a pale golden chestnut square tail, with iridescent green upper-tail coverts and white rump to mid-back region. The head bears a striking crest of glossy metallic green spatulate feathers with wire-like black naked vanes, devoid of barbs. The head and ear coverts are golden metallic green shading to coppery red on the sides of the neck and a more golden copper on the lower nape. The bill is strongly down-curved and powerful. The female is slightly smaller and totally different in colouration. She has a short bushy crest on the nape and an area of bright blue naked skin around the eye. Her chestnut tail is tipped white with darker brown cross-bars and the whole of the rest of the body is rufous brown, finely barred with darker brown except for a white throat area. In both sexes, the iris is brown and the legs are olive brown, the scutes edged with black. The male has a blunt spur on the tarsus, he also has a dark cobalt blue patch of naked skin around his eye.

HABITAT, DISTRIBUTION AND STATUS

The Monal spends a large part of the year on the upper limit of the coniferous tree-line, and in summer forages on alpine meadows up to 4,877 metres (16,000 feet) elevation. It is widely distributed wherever there is some coniferous forest in the Himalayas, occurring in Chitral, Dir, Swat, Indus Kohistan, Hazara district and Azad Kashmir and Gilgit. But in all these regions it is very thinly distributed due to constant persecution by local hunters. The gorgeous male is highly prized for its plumage and the crest is used as a cap badge in the northern regions, giving the wearer some social status. Its large size is also a welcome addition to the cooking pot, worth the effort of a whole day's climb. It was the regular cap badge of the Gilgit Scouts, until this local levy was disbanded in the 1970s. In the Galis (Murree hills) a small population survived until the early 1950s (Hamid Khan, pers. comm., 1960 to author), but it is now extinct. It is now very rare in just 2 or 3 of the forested valleys of Gilgit around Chilas, Naltar and Astor. The largest surviving population is probably now found on both slopes of the mountainous ridge which separates the watershed of the Neelum and lower Kaghan valleys, particularly on the flanks of Jamgarh peak, and in the Bunja and Kala nullahs on the Hazara side. It still survives precariously in Deodar forest in the upper reaches of Birir nullah in Chitral, as well as in Swat in the Jabbah valley and the forested slopes north of Mankial. Captain Whitehead recorded it as occurring above 2,743 metres (9,000 feet) in the Safed Koh and as being fairly numerous (1911). Philip Wayre (Norfolk Pheasant Trust) saw 10 in one morning in 1971 (pers. comm., 1971) above Manshi Forest Reserve on the slopes of Musa-ke-Masala, and the author flushed a total of 7 birds during a morning's climb up Rejkhan on 13 June 1985 in the same area. At Kadir Gali on the eastern side of the Kaghan valley in early spring 1985, the author saw a total of 7 birds between 6 and 8 a.m. Status RARE.

HABITS

Monal roost at night in coniferous trees, usually at around 2,400 metres (8,000 feet) elevation (author obs. Kaghan valley). At dawn and again late in the afternoon, they ascend to the adjacent green grass ridges on the edge of the forest and above the tree-line to feed, usually between 3,000 and 3,500 metres (10,000 and 11,000 feet). Where relatively plentiful they are gregarious and can be seen at sunrise feeding in threes and fours. They will occasionally scratch for their food as the Koklass or

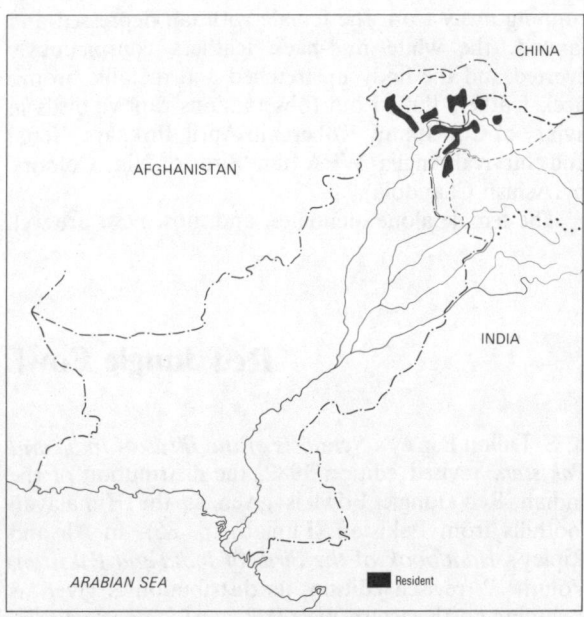

Map 102: Himalayan Monal *Lophophorus impejanus*

Kaleej typically does, but usually spend a lot of their time digging in the ground with their powerful bill, often in one place for a prolonged period. In the winter and early spring they feed on the bulbs of *Gagea,* and the rhizomes of *Iris hookeriana,* also probably *Eremurus* tubers. In summer they eat green shoots, seeds and insects. One was observed on Kadir Gali ridge eating the purple flowers of *Iris hookeriana.* They are able to dig in quite deep snow and do not descend in winter below the snow-line, being very hardy birds. They are always very alert and cautious, running swiftly over the precipitous ground if they sense danger.

BREEDING BIOLOGY

Several naturalists have written rather incomplete accounts of the cock's display (Bates and Lowther, 1952 and Major Roden in Stuart Baker 'Fauna Series', 1928). The full display appears to comprise three different elements. The initial phase starts with a direct frontal approach often from as far as 10 metres distance from the feeding hen bird. It consists of spreading and fanning the tail, and raising both wings over the back, the secondaries spread in a broad fan and the head and neck lowered to expose the shimmering golden green mantle feathers to maximum effect. The tail may be briefly raised over the back but is more commonly depressed on the ground. When highly excited this approach is accomplished with the bird jumping up and down on its flattened tarsi whilst shivering its wings. Following this, if the female does not move away, there is sometimes a lateral display with the cock bird circling around and twisting his partially fanned tail towards the female whilst his head and neck are upstretched but with the bill pointing downwards. The third phase consists of running away from the female with tail depressed and fanned, the white mid-back feathers conspicuously everted and the body upstretched and metallic bronze neck feathers fluffed out (observations captive birds in aviary of Col. Jimmy Roberts in April, Pokhara, Nepal and Survival Anglia/WPA film 'Birds of Nine Colours' by Ashish Chandola).

The female alone incubates, and most nests are well concealed hollows scratched out under a fallen log or the shelter of a rock. The nest may be lined with some dried leaves and is often shielded by over-hanging grass and branches. The normal clutch is 4 to 6 eggs, but Bates and Lowther found a clutch of only 3 eggs on the point of hatching. Eggs are laid from early May till early June. The period of incubation is about 28 days. A hen bird with five chicks just able to fly was seen on 2 June at 3,749 metres (12,300 feet) elevation on the slopes of Musa-ke-Masala (Kaghan valley). When disturbed she escorted them by short flights down the slope (Philip Jones, pers. comm., 1975). A female with 3 or possibly 4 chicks, still in down, was flushed in Machiara Sanctuary on 1 June 1984 at about 3,200 metres (10,500 feet) on a precipitous forested slope. The chicks remained hiding while the hen bird kept up loud whistling alarm calls for a period of 4 or 5 minutes. The Jungle Crow *(Corvus macrorhynchos)* is a serious predator of the eggs of this pheasant, even when they are cleverly concealed.

*VOCALIZATIONS

When alarmed, calls loudly as it flies away, usually down hill with a swift whirring flight. This call is a series of rapidly emitted high pitched screams. It also has soft contact calls when foraging, sounding like 'kluk-kluk'. The female with chicks (described above) flew a short distance and then, on the ground, uttered a series of high pitched whistling alarm calls (tape-recorded) for a period of about 5 minutes, some longer 'kleeh-wick-kleeh-wick', alternating with shorter more urgent 'kwick-kwick' cries.

During the breeding season males call at dusk from their roosting tree, for upwards of 4 to 6 minutes and again in mid-morning after they have finished feeding and descended to the safety of the forest. During these calling periods they utter upward inflected melodious whistles reminiscent of the Curlew *(Numenius arquata)* 'kur-lieu' or 'kleeh-wick'. The female is also capable of uttering similar calls. These two-noted whistles are alternated with single-noted higher pitched 'kleeh' calls. A typical song pattern is 'kur-lieu-kleeh-kur-lieu-kleeh-kleeh-kurlieu-kurlieu', each call spaced 3 or 4 seconds apart.

Red Jungle Fowl *Gallus gallus* (Linnaeus)

In S. Dillon Ripley's *Synopsis of the Birds of India and Pakistan,* revised edition 1982, the distribution of the Indian Red Jungle Fowl is given as the Himalayan foothills from Pakistan (Punjab) (p. 85). In Ali and Ripley's *Handbook of the Birds of India and Pakistan,* Volume 2, revised edition, its distribution is given as including north-eastern West Pakistan (compact edition, 1983, Vol. 2, p. 103).

I can find no evidence that the Red Jungle Fowl has ever existed in Pakistan, even in the Murree foothills or the extreme southern portion of Azad Kashmir in Mirpur district. Writing about the birds of the Murree hills and Rawalpindi district in 1930, Hugh Whistler comments that the Red Jungle Fowl does not occur and that Magrath's second-hand report of this bird at Thandiani *(JBNHS,* Vol. 18, p. 298) cannot be accepted

(Whistler, *Ibis,* 1930, p. 272). Indeed this misapprehension probably stems from the fact that local hunters and hill people know and call the White Crested Kalij Pheasant 'Jangli Murghi' meaning 'Jungle Fowl' and the Kalij occupies the same low altitude biotope where the Jungle Fowl would be likely to occur in Pakistan.

In 1969/70 during a wildlife survey conducted by the National Council for Wildlife Preservation some forest department officials in the Azad Kashmir Government claimed that the Jungle Fowl occurred in their area. Subsequent investigations revealed that these claims were also based upon confusion with the Kalij *(Lophura leucomelana)* (Z. B. Mirza, pers. comm., 1970, and T. J. Roberts, *Pakistan Journal of Forestry,* Vol. 20, No. 4, October 1970). For the present time it is excluded from the **Checklist** of Pakistan birds.

Kalij or Kaleej Pheasant or
White-crested Kaleej *Lophura leucomelana hamiltonii* (J. E. Gray)
Nepal Kaleej *Lophura leucomelana leucomelana* (Latham)

Plate **14**
Map **103**

DESCRIPTION

Body length 65-73 cms. (Ali and Ripley)
Wing length 21.6-24.9 cms. (males), 20.3-21.5 cms. (females)
    sub-sp. *hamiltonii* (Stuart Baker)
Tail length 22.8-32.7 cms. (males) (Stuart Baker)
Weight up to 1.25 kgs. (males), 0.567-1.02 kgs. (females)
    (Stuart Baker)

This is a rather long legged, slim bodied pheasant, slightly larger than the Koklass *(Pucrasia macrolopha).* Both sexes have a medium long curved tail, laterally compressed and reminiscent of a bill-hook blade in outline. The tail feathers are downward pointing but very broad basally. Both sexes also have a shaggy hairy crest which is erectile when excited. The male is gun metal blue-black on the head, neck and mantle with a brighter blue gloss visible in sunlight, and the rump feathers are prominently margined white. The breast is pale ashy grey streaked white with the feathers on the flanks lanceolate and having darker brown centres. The bill is greenish white, the upper mandibles being powerful and down-curved. Both sexes have a livid scarlet patch of naked skin around the eye and the legs are brownish with fleshy or purplish tinges. Males have a prominent spur on the tarsus. The hair-like crest varies from dirty white to dark ashy even in the population of *hamiltonii* which inhabits Pakistan. The female is grey-brown all over with the mantle and wing coverts margined creamy white giving an attractive scalloped pattern. The tail and flight feathers are more chestnut brown, flecked blackish and are relatively less elongated and down-curved than in the male. The hairy crest is dirty buff to grey and the ear coverts and throat are more creamy white with the breast feathers boldly patterned with creamy borders but in a rounded scale-like pattern, unlike the lanceolate feathers of the male. The outer tail feathers, visible only in flight are glossed blue-black. A row of over-lapping broad spear shaped tail covert feathers, extend down each side of the laterally compressed tail, each one conspicuously margined whitish. The iris is orange-brown. The nominate sub-species *leucomelana* has a dark brown crest and less prominent white scale pattern on the rump. Birds seen in the Murree hills with dark brown head crests, often intergrade between these extremes.

HABITAT, DISTRIBUTION AND STATUS

A resident endemic species most typically associated with foothill country, extending from Wild Olive *(Olea cuspidata)* and 'Phalai'*(Acacia modesta)* scrub forest, up to the *Pinus roxburghii* sub-tropical pine zone. Because of the relatively high level of human settlement and disturbance in this zone, they have been much persecuted by local hunters, and the present population is very

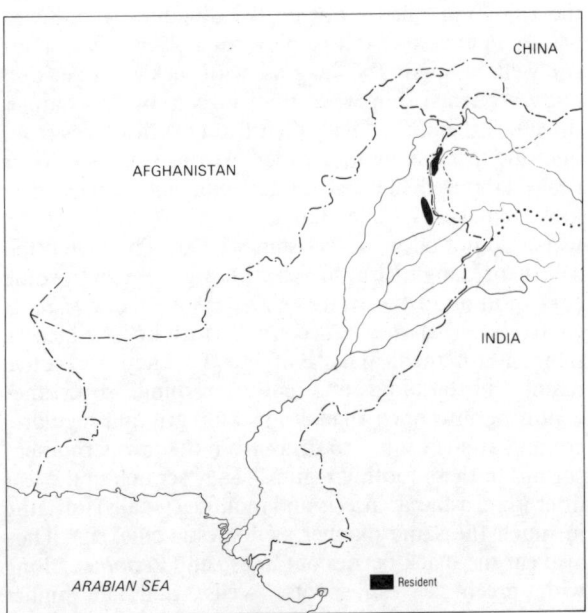

Map 103: Kalij Pheasant *Lophura leucomelana*

sparse and fragmented. A few may still survive on the lower slopes at the head of the Siran, and the mouth of the Kunhar valley in Hazara district, and they definitely occur along the lower reaches of the Neelum valley, especially the Saidpur valley and Machiara valley and on the Jhelum river at Pir Chinasi. Its main surviving population now extends from the Margalla hills, through to Lehtrar and at Panjar and Kahuta. They occur in thick 'jungle' at the foot of the Margalla hills as low as 457 metres (1,500 feet) elevation just west of Islamabad, and have been flushed at Ghora Gali at 2,081 metres (6,500 feet) elevation. There is no recent evidence of their having extended into other parts of the north-west such as Chitral or Swat.

The resettlement of hill farmers away from the Margalla hills bordering Islamabad and the creation of the Margalla National Park has led to a small increase in the population in this region. Recent evidence of their continued survival around Kuwai in the lower Kaghan valley is lacking. They still occurred in the 1960s in this part of the Kaghan valley (Bashir Hussain, Forest Guard, Kuwai, pers. comm.). They must be considered as endangered in Pakistan. Status SCARCE.

HABITS

Adapted to live in dense undergrowth at lower elevations, usually never far from hill streams or springs. They are seldom seen but in early morning are very noisy and vocal when undisturbed, perhaps relying upon such means to maintain social contact. In the creeper-festooned thorny bushes of the Margalla ravines under the sweet scented *Carissa opaca* bushes, 3 individuals have been heard calling within auditory range of each other (May, 1985), and within an area of 20 hectares. In the adjoining 'quarry ravine', 4 individuals were heard within an area of about 60 hectares in June, 1986. They run well, hiding in the tall grass with neck extended and body crouched, but when undisturbed and no danger threatens them, they will often fly short distances from one side of a ravine to another. At this time with their heavy laboured flight and their long curved tails, they look remarkably like Jungle Cocks *(Gallus gallus),* which do not occur in Pakistan. The local hillmen often call them 'Jangli Murghi' which has given rise to some confusion as to the status of the Jungle Cock *(Gallus gallus)* (see that species account). More knowledgeable hillmen call the Kaleej 'Ban kukar'. They are active mainly in evenings and early morning, sometimes venturing into open roads to pick up grit and regularly coming at such times to drink from the few permanent springs in these foothill regions. They scratch in the leaf litter for crustacea, insects and mollusca (snails) foraging in much the same manner as domestic chickens. They also eat the black berries of *Carissa* and *Ziziphus,* along with green leaves, shoots, seeds, and, at higher elevations, the acorns of *Quercus incana* in season. At night they fly into trees to roost.

BREEDING BIOLOGY

The pair bond is not as transient as in many Galliformes, and pairs have been flushed (author) together in May and early April. Males can be heard display calling throughout the summer. The male displays when close to a female by circling around her, with drooped wings and body tilted sideways and tail fanned outwards. A drumming noise is produced by males as an advertising display, by a very rapid beating of the half open wings against the sides of their body, but this is usually extremely difficult to observe in the thick cover they inhabit. A. E. Jones was wrong in considering the drumming as partly vocal (quoted by Ali and Ripley, 'Handbook', 1969), as close observation of captive birds has confirmed (K. Howman, *in litt.* to author, 1985). The pair bond may be monogamous or bigamous, only the female making the nest scrape on the ground and incubating the eggs. However, instances have been recorded of the male bird solicitously tending and looking after the downy chicks (H. W. Watson, *JBNHS,* 1914). The nest is well concealed under a bush or overhanging roots, but is scantily lined with grass or dried vegetation. Col. Buchanan found a clutch of 8 eggs below Changla Gali on 14 July. (Whistler, *Ibis,* 1930). Another nest near Tret with 10 eggs was found by Major Amanullah Khan in 1983 in early May: these eggs were laid at 24 hour intervals, the clutch was completed on 13 May, and the eggs hatched synchronously after 27 days incubation. The normal period is considered to be 24-25 days (Wayre, 1969). The first egg was pipped at 10 p.m. and 8 chicks had dried down and were quite active by next morning, with all 10 hatched by 12 noon (Major Amanullah Khan, pers. comm., 1983). The eggs look like small domestic pullet's eggs, varying from whitish to pale buff, unmarked and not very glossy. Local people invariably rob and eat eggs if a nest is discovered. Due to overgrazing in the hills by goats, another nest near Tret was destroyed by water run-off after a heavy thunderstorm (Amanullah Khan, pers. comm.). Eleven day old chicks (seen by author) had downy rufous golden bodies with a dark line through the eye and spots on the crown; their flight feathers had sprouted and were barred chestnut and buff. The female calls them constantly with soft 'chicks'. Predation of chicks is high, as a female with only 4 downy chicks was located (author) below Malach in the Haro nallah on 17 August and they were estimated as being about 8 days old.

*VOCALIZATIONS

Their varied repertoire of calls is difficult to describe. Both sexes will utter long sequences of rather low pitched clucks alternating with high pitched squealing chirrups with crest erected and indicating an excited state (observations based on wild caught female kept by author). Such calling sessions are usually at dusk or early morning. The drumming (described above) only occurs in the breeding season and is produced by the male, usually hidden in thick cover. The squeals and chuckling calling sessions occur mostly during the breeding season extending up to September. A female with chicks was tape-recorded in late June at dusk giving a prolonged series of short high pitched clucks and chirrups and as Whistler had so aptly described them, they are reminiscent of the squeaks of a Guinea Pig *(Cavia porcellus).*

The male, when in the presence of the female, indulges in long calling sessions and these are heard usually from early morning till about 9 a.m. and less frequently in the late afternoon. The short low pitched clucks 'terook-terook-terook', are repeated for one to two minutes on end becoming more rapid as excitement increases and alternating with higher pitched more rapid and squealing 'ter-tr-r-r-eeeh, tuk-tuk-tuk-treeh' calls.

Cheer Pheasant *Catreus wallichii* (Hardwicke)   Plate **14**

DESCRIPTION

Head and Body length 61-118 cms. including tail
Wingspan 80-95 cms.
Wing length 23.5-26.9 cms. (males),
    22.3-24.5 cms. (females) (Stuart Baker)
Tail length 38.8-58.4 cms. (males), 31.7-46.8 cms. (females)
    (Stuart Baker)
Weight 1.2-1.6 kgs. (males), 0.9-1.25 kgs. (females) (Rattray)

Rather a large drab coloured pheasant, bigger than either the Koklass or Kaleej. Both sexes have long barred tails and conspicuous hair-like crests on their napes. The male is mottled with buff, black and white, giving an overall rather grey effect to the plumage and there is a conspicuous bare crimson facial patch. His crown and crest are blackish brown and the long narrow tail is buff with broad grey cross-bars, narrowly bordered by black on either side of the grey bands. The breast is creamy with some blackish barring on the flanks and the wing coverts have creamy tips forming curved pale bars. The rump and upper-tail coverts are more chestnut in tone. The female has a bare brick red facial patch, less crimson than that of the male. She looks superficially like the male but her upper throat and neck is scaled dark brown, with creamy borders to the feathers and her lower breast and belly is a rich chestnut with creamy margins to the feathers. The long tail is cross-barred with broad light brown bars bordered by dark blackish brown. In both sexes the iris is orange-brown, the bill is horny to whitish, and the legs plumbeous greyish. The male has a prominent spur. The female is slightly smaller than the male.

HABITAT, DISTRIBUTION AND STATUS

This Himalayan pheasant is quite sedentary showing little or no altitudinal migration. It is generally associated with the outer hill ranges of the Himalayas, typically in the barer, unforested slopes but favouring very precipitous terrain with some shrub and tall grass cover. In Pakistan it is on the western limit of its overall distribution and would appear from earlier records (Whistler, *Ibis,* 1930 and Rattray, *JBNHS,* 1905), never to have been very plentiful. Its sedentary habits and

addiction to more open treeless slopes, away from closed canopy forest had made it particularly vulnerable to local hunters, and in Pakistan it has been gradually exterminated from all its more accessible haunts, largely during the 1950s and 1960s. In the Murree hills and particularly the lower ranges around Tret and the Margalla hills it was hunted even up to the early 1960s (J. Ainsworth Harrison, pers. comm., 1971). It was reliably recorded around Bansra Gali (Hazara district) in the late 1960s (Mohammad Mumtaz Malik, Conservator Wildlife NWFP. pers. comm. to author, 1984). A small party was found at 2,400 metres (8,000 feet) near Dunga Gali by Major Venour who collected a specimen on 25 July 1906 (Venour, *JBNHS,* 1907), showing that given suitable steep precipitous terrain and good ground cover they can be found from 600 metres up to 2,600 metres (2,000 up to 8,500 feet).

It used to occur in the lower Kaghan valley around Kuwai where it was known as 'Reyal' or 'Reyar' but appears to have become extinct since the 1970s (Nizamuddin, Shikari, Shogran, pers. comm). In 1979 one was shot in Kamil Gali (west of Shahran, above the Nadi valley) by a local hunter at about 1,500 metres (5,000 feet) (Shafiq-ur-Rehman, Game watcher, pers. comm., 1986). Since the late 1970s there have been no reported sightings anywhere in these regions though it was seen in the late 1960s behind Garhi Habibullah in the Pir Chela hills and Trakama pass (Mohammad Jamdad Khattak, DFO, Abbottabad, pers. comm., 1985). In January 1983 a single bird was located in the upper reaches of Machiara valley on the west bank of the Neelum river (Kamal Islam, pers. comm.) and higher up the Neelum valley a potentially larger population still survives on the outermost flanks of the Kaz-i-nag range above Salkala. In this locality, in December 1977 between 1,829 and 2,285 metres (6,000 and 7,500 feet) in light snow, 20 Cheer were flushed with dogs (Z. B. Mirza, WWF Report). In September 1986 James Burt (World Pheasant Association Research Officer) surveyed Pir Chinasi forest area in the southern part of the Jhelum valley region of Azad Kashmir and established that a very small population, possibly no more than 6 or 7 birds still survived in this pocket of forest (pers. comm., March 1987). The area is however subject to heavy human pressures both from wood cutting and grazing and has no sanctuary status. The World Pheasant Association has, over the past several years generously donated Cheer Pheasant eggs to Pakistan for a captive breeding scheme. So far about 4 introductions have been attempted since 1980, mainly in the Margalla hills, but also in Dunga Gali and Malkandi forest in Kaghan valley. Aviary bred eggs from England have been flown to Pakistan and after rearing, fully grown young pheasants have been released. Unfortunately there has been a very high rate of predation of these birds after release, by Hill Foxes *(Vulpes vulpes)*, Leopard Cats *(Felis bengalensis)*, Himalayan Palm Civets *(Paguma larvata)* and Yellow-throated Martens *(Martes flavigula)*, all of which have been trapped in the vicinity of the Cheer breeding enclosures. Efforts are continuing for this re-introduction.

Status in Pakistan RARE and in view of current hunting by local hillmen, in imminent danger of extinction.

HABITS

They are gregarious, living throughout the year in small family parties. This, coupled with their adherence to a fixed territory and their noisy calling habits early in the morning before dispersing from their night time roost (Young *et al.*, 1987) have made them particularly vulnerable to hunters. They roost at night in trees. With their strong down-curved upper mandible they are well equipped to dig and feed upon tubers, subterranean rhizomes and roots to supplement their diet in winter time. In the summer, seeds and berries of various sorts are consumed and earlier in the spring young shoots and leaves. They are said to be very difficult to flush without the use of dogs (Ainsworth Harrison, pers. comm., 1969), preferring to run through the undergrowth to escape detection or crouching in the thick undergrowth, relying upon their cryptic plumage for concealment until

almost stepped upon. When flushed they fly noisily and strongly, plunging down the steep hillsides, twisting between the trees.

BREEDING BIOLOGY

The pair bond is reported to be lasting throughout the breeding season with the male helping to tend and protect the newly hatched chicks. A nest scrape is made on the ground by the female alone and this is usually cleverly concealed under overhanging vegetation and in the shelter of a rock or bush. They normally nest on very steep precipitous ground. Only the female incubates, taking about 26 days. The normal clutch is 9 to 14 eggs and they are a dull creamy white to pale greyish buff, sparsely freckled with reddish brown. Occasionally entirely unmarked eggs are found. Most clutches are laid from late April to early June. Captive birds were observed displaying in a manner somewhat similar to the Common or Ring-necked Pheasant *(Phasianus colchinus)*, which is a lateral display with the inner wing half opened and dropped and the tail fanned and twisted sideways. But unlike Monal, which has its neck upstretched during lateral display, the body is more crouched forward with the neck extended horizontally (observation captive birds, Pokhara, Nepal).

*VOCALIZATIONS

The full territorial or advertising call is a rather loud and pleasant 'chir-a-pir, chir-a-pir-chir-chir' (Ali and Ripley, 'Handbook', 1969). They also have a number of social contact calls particularly at dusk when going to roost, and exhibit chorusing behaviour 10 to 20 minutes before sunrise (Young *et al.*, 1987). These include high piercing whistles 'chewewoo', interspersed with short calls 'chut', usually repeated at the beginning and end of bouts of whistling, as well as short staccato harsh notes in combination with the whistles (Young *et al.*, 1987).

Indian Peafowl *Pavo cristatus* Linnaeus   Map **104**

DESCRIPTION

Body length 92-122 cms. (without train) (males),
    80-90 cms. (females)
Wingspan 240 cms.
Train 140-160 cms.
Weight up to 5 kgs. (males), 2.75-4 kgs. (females)

A long legged and long necked bird, comparable in size to a rather rangey domestic turkey. The iridescent dark blue head, neck and breast of the resplendent cock is familiar to all, offset by a featherless white cheek patch and the fan shaped crest of short wiry black quills topped

by small spatulate iridescent green tips. The mantle and back is brilliant golden bronzy green, each feather margined with black, the lower belly is dull black, wing coverts barred black and grey and the flight feathers are largely pale chestnut, unbarred. In flight the under-wing coverts are black. The tail is dully grey-brown and wedge shaped, and except during moult, this is normally completely hidden under the greatly elongated and decorative upper-tail coverts which form its well known train. The train consists of bronze green feathers, nearly all ending in the ocelli which are the famous Peacock's 'eyes' — each one roughly pear shaped and broadest at its base comprising an inner purplish black indented pupil surrounded by azure blue, and the whole framed by a coppery disc narrowly edged, again with an outer rim of green and dark bronze. The long powerful thighs are buff and the legs and feet are horny brown. Females have a crest similar to the male but the crown and nape is rufous brown and the rest of the neck and back grey-brown, with a metallic green gloss to the lower neck. The belly is dull white, flight feathers are chestnut and the tail dark brown with whitish tips. There is some scale pattern to the flank feathers which have glossy green tips. In both sexes the iris is dark brown and the actual white cheek patch consists of bare skin. Males have a strong spur on the tarsus.

HABITAT, DISTRIBUTION AND STATUS

The Peacock exists very precariously in the extreme north-eastern border regions of the Punjab as well as the south-eastern corner of Sind. In the 1940s and up to independence it was fairly widespread in the better watered parts of Sind (author obs., 1947) such as around Mirpurkhas, Khipro, the Makhi Dhand in Sanghar district, and the Thar desert. It was also quite common in Hyderabad district. K. Eates considered that this whole population was feral and that it had been introduced by the Mirs of Talpur after the Kalhora dynasty was overthrown (unpublished mansucript notes). After independence it was rapidly exterminated by local hunters who had no religious compunctions about shooting a bird, regarded as sacred by the Hindu villagers living in Tharparkar. It survives today still in good numbers in the extreme southern border regions of

Tharparkar around Islamkot and Nagar Parkar where predominantly Hindu communities still afford it protection. Since this population extends widely throughout the Thar desert and into Rajasthan it seems probable that it was indigenous and not in fact introduced by the Talpur Mirs. Up in the Punjab there used to be a small feral population around Renala Khurd, and there is still quite a large feral population at Kallar Kahar in the Salt Range, given some protection because of their habit of roosting by the local Mosque. In the foothill regions of Sialkot district and further north in the Jhelum watershed, around Panjar and Kahuta a small genuinely wild population survives very precariously. However they are regrettably still hunted ruthlessly and even as late as 1983, the Punjab Wildlife Protection Ordinance still allowed wild Peacocks to be trapped. The author has twice encountered wild peacocks in the Lehtrar valley, the first on 14 December 1969, and again on 10 April 1971 (4 birds on the latter occasion) at about 900 metres (3,000 feet) elevation. A wild caught bird was secured near Tret in 1975, and local hunters still maintain that they survive in the region (Abdul Razzaq, resident of Tret, pers. comm., 1983). Status RARE in the true wild state.

HABITS

At night they are noisy, uttering their loud 'mee-haw' screams before flying up into a tree to roost: this makes them particularly vulnerable to local hunters. They feed mainly in the early morning and late evening and are quite omnivorous. Many writers have noted their propensity for killing snakes. At Renala Khurd while out riding one early morning, Mrs. Anne Cranfield observed a circle comprising some 6 or 7 peacocks, pecking at a coiled snake in the middle (pers. comm., 1957).

They will also feed upon flower buds and green shoots, insects, seeds and small lizards. In Lehtrar valley they were observed flying up into tall Chir pines *(Pinus roxburghii)* to roost at dusk. Usually they consort in small parties of 3 to 4 females with one cock bird. They usually live near water, and in the Murree foothills regularly come to streams to drink in the early morning. They are quite shy and furtive and despite their large size, are surprisingly adept at avoiding detection and slipping through the undergrowth.

BREEDING BIOLOGY

The dance of the male is too familiar to require detailed description. A feral free range population kept by the author at Khanewal in a citrus orchard, was observed to breed either in May or after the onset of the monsoon rains in July and August. The males each had a particular territory or dancing ground where they would fan their trains in the early morning and late afternoons, quivering their erected upper-tail over their backs with

Map 104: Indian Pea Fowl *Pavo cristatus*

an audible rustle of the quills and lifting their feet up and down in a sort of dance. One individual repeatedly displayed towards a pet rabbit, ignoring female peacocks feeding nearby! Females usually laid under a pile of brush-wood (cotton sticks or citrus prunings) with no lining to the hollow scraped out of the bare earth. The eggs are creamy buff, unmarked and the usual clutch is 4 to 6 eggs. Occasional eggs are finely speckled with red-brown like a domestic turkey's eggs (Eates, MS notes). This feral population in Khanewal had been reared from birds captured in Kaloi, Tharparkar district of Sind. Wild nests were also found near Diplo and Mithi in the 1950s in Tharparkar, Sind, but it is not known if peacocks still survive in these areas now. These Sind nests were found in July and August. K. Eates found several nests and only once recorded a clutch of 5 eggs, 3 or 4 eggs is usual. The cock is polygamous and takes no part in tending young, usually having a harem of 2 to 3 breeding females, though recent studies of a feral population suggest that both sexes are promiscuous and that the male does not defend a harem of peahens but rather a small breeding territory (Rands *et al.*, 1984). The incubation period is about 28 days and the female is very attentive in looking after her chicks.

*VOCALIZATIONS

The loud trumpet like scream of the male, mostly to be heard at dusk and occasionally throughout the night, is well known. It is an inflected cat-like 'mee-awh'. They also have an alarm or excited call comprising a fairly rapid 'kok-kok' or 'ko-ko-ko'. The female gives this call when warning her chicks of possible danger. They also emit loud short but almost explosive 'baap' calls, which are closely similar to the honking sound of old-fashioned air-bulb car horns and this appears to be a warning or danger call.

ORDER GRUIFORMES
FAMILY TURNICIDAE
Little Bustard Quail or
Indian Little Button Quail *Turnix sylvatica* (Desfontaines)   Map **105**
Synonym *Turnix dussumieri* (Temminck)

DESCRIPTION

Body length 15-16 cms.
Wingspan 25-30 cms.

A very diminutive, quail-like bird with short neck, plump rounded body and short but sharply pointed tail. The feet are distinguished by having only three front toes and no hallux. It has a blackish crown with a broad median whitish line down the centre and also a creamy white supercilium and darker streak through the eye. The hind neck is ferruginous, scalloped with buff, and the upper parts are barred rufous and black. The lower breast and belly is creamy whitish and the upper breast is reddish biscuit in tone with black scale-like spots on the sides of the breast and neck, margined whitish. Specimens collected (by E. Fernando in 1968) from Marala barrage on the Chenab river in Sialkot had a plumbeous bluish bill. The legs are also pale blue-grey and the iris is yellow. The female is similar in plumage but about 15 per cent larger in size than the male. Sometimes the bill is whitish and the legs fleshy white (Salim Ali, 'Handbook', 1969).

Its smaller size and unbarred flight feathers and less prominently streaked crown, help to distinguish it in the field from *Coturnix coturnix* the Common Quail, and its more blue-grey bill and feet with pointed tail separates it from *Turnix tanki*.

HABITAT, DISTRIBUTION AND STATUS

This is a resident species in the grassy plains areas of the subcontinent, which enters Pakistan as an erratic summer breeding visitor, during and immediately after the monsoon period. Like other Oriental rainy season visitors it is most likely to be encountered in the north-east corner of the Punjab and the south-eastern corner of Sind. In the latter province Ticehurst recorded it as a rainy season visitor occurring locally and only in small numbers, mostly around the east Narra and around Hyderabad and Jacobabad (Ticehurst, *Ibis,* 1924). K. Eates considered it an occasional monsoon migrant visitor to lower Sind, and he shot specimens in November 1944 on the Bidook plain, Lasbela, close to Karachi and also on the Gadap plain east of the Hab river (MS notes unpublished). In Sialkot, Gujranwala and Rawalpindi districts of north-east Punjab it also occurs irregularly in the summer. Husain collected a specimen from Faisalabad (formerly Lyallpur) on 14 April but considered it rare (Husain and Bhalla, *JBNHS,* 1937). On 29 November Whistler records finding the remains of two unidentifiable Turnix species, taken from the crops of two harriers which he shot in Rawalpindi district *(Circus macrourus* and *Circus cynaeus)* (Whistler, *Ibis,* 1930). H. Waite never collected any specimens but heard them calling around Nurpur in the Salt Range during August and September. The author encountered one in Multan district at Khanewal while combining wheat on 21 April and E. Fernando collected specimens (now in Stuttgart Museum) at Marala in *Saccharum* Cane grass thickets on the banks of the Chenab river in October. Status FREQUENT to SCARCE.

HABITS

They are weak flyers, preferring to run through cover when disturbed, but if flushed their short rounded wings produce a characteristic whirring noise and they usually fly a very short distance close to the top of the crop or ground-cover, before plunging back into the vegetation. They scratch on the ground for insects and seeds and feed on a variety of grass and weed seeds, cultivated cereal grains, young green shoots and ants. Studies of the Spanish population indicated that insects comprised at least 50 per cent of the diet (Cramp *et al.,* 1980), but in India, the bulk of their diet is probably grass seeds which they glean from the ground. A specimen collected near Faisalabad had 'grains, weed seeds and insects' in its stomach (Husain and Bhalla, 1937). They are not very gregarious and are usually flushed solitarily. When they alight, they have a characteristic very upright posture quite unlike that of *Coturnix,* but this is difficult to observe in the normal grassy cover which this species frequents.

BREEDING BIOLOGY

This Button Quail has bred successfully in captivity and observations of behaviour of a number of captive females, confirm that the usual sexual roles are reversed and that the female practices successive polyandry (Wintle, 1975). She chooses a second mate immediately after the first clutch is completed. The normal clutch is only 3 or 4 eggs and these are incubated solely by the male. Courtship seems to consist exclusively of the female giving advertising calls. These consist of surprisingly loud and resonant purring and throbbing calls. This accompanied by the female pumping her chest or upper breast up and down, in a vibrating manner, the head and neck drawn into the shoulders and the bill pointed downwards and closed all the time. She is reported to twist from side to side while 'drumming' and certainly the calls have a remarkably ventriloquial effect, their origin and direction being difficult to pinpoint. They call most actively in the early morning and late evening. Each call lasts about 10 seconds and is uttered at intervals of about 15 seconds. When a male comes into view the female then displays by performing a strange swaying walk accompanied by exaggerated slow steps

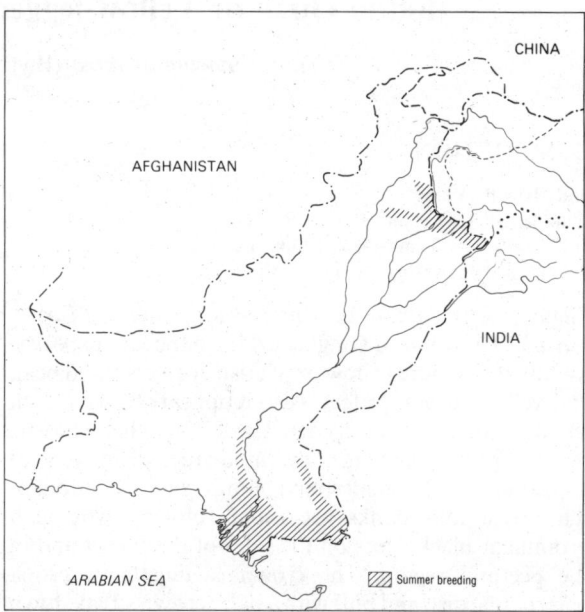

Map 105: Little Bustard Quail *Turnix sylvatica*

(M.W. Ridley, *WPA Journal,* No. 8, 1983). Both sexes help to construct the nest which is generally located in thick grass, which consists of a roofed over canopy made from bent and woven grass stems beneath which only a shallow scrape is made in the ground with one lateral entrance. The eggs are greyish white in ground colour, finely speckled with light brown and some black spots also. The cock bird is reported to sit very tight. Incubation is exceptionally short, taking only 12 to 14 days (compared with 17 to 20 days in *Corturnix coturnix).* The nidifugous chicks are like other Gruiformes in that they depend upon bill to bill feeding by the parent bird (in this case the male) for the first few days after hatching. At this stage the male presents an entirely insect diet to the chicks. In this respect they differ from the Galliformes which can feed themselves immediately after hatching. The young develop and mature remarkably rapidly and can feed themselves entirely after they are 20 days of age and become sexually mature at 5 to 6 months of age (Bell and Bruning, *International Zoo Year Book* 1974).

VOCALIZATIONS

The soft drumming calls of the female have been described under breeding biology. H. Waite (*JBNHS,* 1948) recorded them as calling all day long in the 'Bajra' fields *(Pennisetum* or Millet) around Nurpur in the Punjab Salt Range during August and rendered these calls as a resonant booming 'hoon-hoon-hoon'.

Button Quail or Yellow-legged Button Quail *Turnix tanki* Blyth

Turnix maculatus tanki (Blyth) of Stuart Baker's 'Fauna Series'

DESCRIPTION

Body length 15 cms.
Wing length 7.9-8.9 cms. (Stuart Baker)
Tail length 3.5-4.0 cms. (Stuart Baker)
Weight 34-36 gms. (males), up to 45 gms. (females)

Slightly larger than the Little Button Quail *(Turnix sylvatica)*, it can be distinguished from the latter species, usually by its more yellow or whitish horn coloured beak and yellow to orange feet. Also it appears to have even less of a tail than its congener. The feet are characteristic of the genus, lacking any hind toe. The general appearance is of a small round plump quail-like bird, but the breast and flanks are more rufous tawny with prominent black spots on the sides of the throat and in the pectoral region. The crown is indistinctly cross-barred, blackish and buff with a paler crown streak down the centre, and a darker line through the eye with a pale buff supercilium and cheeks. The back is greyish brown cross-barred with black and the lower belly and vent are more whitish than in the Common Quail, also the chin and throat are whitish and unmarked. The female is about 15 per cent larger than the male and has a broad rufous chestnut collar on the back and sides of neck with greyish tones on the mantle and scapulars with black spots and the wing shoulders showing creamy buff unspotted. In the hand its straighter bill, with less down-curved upper mandible, indicates that it has no close relationship with the Galliformes.

HABITAT, DISTRIBUTION AND STATUS

In the Punjab it appears a more frequent visitor than in lower Sind. It is an irregular local migrant visitor to Pakistan, mainly when grass and general vegetative cover is increased after spring rains or during the monsoon rainy season. E. Fernando purchased 5 specimens (shown to author) from a local hunter on 5 November 1967, who had live-trapped them in rank grass near the Chenab river at Marala barrage. Captain Whitehead (*JBNHS*, 1911) recorded them as regular summer visitors in small numbers to the 'grass' farm at Kohat where they were occasionally found breeding. In lower Sind they occur irregularly and only after good monsoon rains, and Ticehurst (*Ibis*, 1924) considered them less common than *Turnix sylvatica*, but mentions the fact that both species commonly share the same habitat, as Le Messurier shot them together on 8 August at Sapoora near Karachi. K. Eates never came across *T. tanki* in Sind but records that S. A. Strip shot this species in the Moach and Gadap grasslands in August and September after a good monsoon year, in the mid 1930s (MS notes unpublished). He also recorded that E. Fardy

shot them farther to the south-east in Badin district. Status FREQUENT.

HABITS

Reputed to be very secretive and hard to flush. When put to flight it bursts suddenly forth with rapid whirring of wings producing a high thrumming noise and it settles down after a very short flight into thick cover, whence it immediately starts running and cannot be flushed a second time (Fleming and Fleming, 1976). It is omnivorous in diet, consuming seeds, green shoots and a substantial quantity of ground dwelling insects, their larvae and soil-born crustacea and in general habits are similar to others of the genus.

BREEDING BIOLOGY

Said to have an extended breeding period from March to November, with the female initiating courtship, presenting food to the male (Seth-Smith, 1903) and also by emitting the characteristic rapid throbbing or drumming calls of the genus. Probably both sexes select the nest site which is in thick cover with the nest itself, a scantily grass lined scrape on the ground, and only occasionally partly roofed over with stems of grass and other vegetation. Only the male incubates the eggs and the female is believed to be successively polyandrous

Map 106: Yellow-legged Buttonquail *Turnix tanki*

Illus. 32: Yellow-legged Buttonquail *Turnix tanki*, female.

(Stuart Baker, 1930). Most eggs are laid after the monsoon when grass cover is at its maximum and the normal clutch is 4 eggs. These are more handsomely marked than *T. sylvatica* eggs, being glossy, strong shelled and slightly pointed at one end. They are greyish white to pale limestone in colour, finely stippled all over with yellowish or grey-brown, and with pale mauve markings and a scattering of bolder blacker spots and blotches (Stuart Baker, 1930). Only the male tends the newly hatched chicks, which emerge synchronously after about 13 days (Stuart Baker, 1930) up to 16 days (Ali and Ripley, 1969) incubation. They have to be fed bill to bill by the male for the first 5 or 6 days after hatching and he alone tends them solicitously. A male has been recorded giving a broken wing distraction display to protect his downy chicks (Ali and Ripley, 1969).

VOCALIZATIONS

The advertising call of the female which is the basis of courtship is similar to that of *Turnix sylvatica* and difficult to describe. It consists of short 10 to 15 second bursts of sound of rapid throbbing or purring nature not unlike the 2 stroke engine throb of a very distant motorcycle. This is made by the female visibly inflating her upper chest and throat. Its direction and origin is hard to determine. The female is said to turn and twist as she calls and these postures may help to attract an approaching male. Fleming (1976) renders the calls as 'off-off-off'. But this appears to be a second type of call, more subdued but equally carrying as the drumming calls and in the opinion of E. Lowther (1949) the 'pook-pook' call is only given by the male. During courtship females are said to be very pugnacious, attacking and driving off any rival hens that approach while she is 'booming'.

FAMILY RALLIDAE

Slaty-legged Crake or
Philippine Banded Crake *Rallina eurizonoides* (Lafresnaye)

DESCRIPTION
Body length 25 cms.
Wing length 12.2-13.2 cms. (Stuart Baker)
Tail length 5.5-6.4 cms. (Stuart Baker)
A dark rufous chestnut rail with whitish chin and throat, warm cinnamon chestnut breast, and the whole of the flanks, belly and under-tail coverts, barred with black and white. It has slaty grey legs and feet, the toes being long and well developed and without any webs. The bill is olive green, darker brown on the tip and the iris is crimson.

STATUS

In Stuart Baker's 'Fauna Series' (Vol. 6, 1929), its distribution is given as extending across the sub-Himalayas from the North West Frontier to Assam (p. 16). This is repeated in Ali and Ripley's 'Handbook', (Vol. 2, 1969), but in S. Dillon Ripley's 'Revised Synopsis' (1982) it is described as occurring throughout the subcontinent, with only India and Sri Lanka (specifically listed).

The author has seen and tape-recorded this crake in its wintering grounds in Sri Lanka but has never heard it calling in Pakistan. Neither H. Waite nor H. Whistler came across it in the foothill regions of the Punjab and no evidence has come to light of an actual specimen having been collected from our region. It is therefore excluded from the **Checklist,** but these very secretive birds often escape notice, and it should be looked out for.

Water Rail *Rallus aquaticus* Linnaeus

Map **107**

DESCRIPTION

Body length 23-28 cms.
Wingspan 38-45 cms.
Bill length 3-4 cms.

Because of its extremely secretive habits one is rarely afforded a prolonged view but its long slender red bill slightly down-curved at the tip, serves to distinguish it from any other rails or crakes inhabiting the region. Both sexes are alike in plumage having slaty blue-grey head, cheeks, neck and breast. The crown, hind neck, and upper parts are chestnut streaked black, the feathers of the mantle and wing coverts actually having blackish brown centres and rufous margins. A noticeable feature is the white under-tail coverts with boldly patterned flanks and ventrum of vertical white and black bars. The legs are stout and quite long with long well developed toes, and are fleshy brown in colour. Juveniles lack the slaty grey cheeks and breast being dull fulvous buff with only indistinct flank barring. The downy chicks are wholly black with white bills. The iris in adults is crimson.

HABITAT, DISTRIBUTION AND STATUS

A winter migrant visitor to Pakistan, most likely to be encountered in the Punjab and relatively uncommon in Sind. Dr. Sculley *(Stray Feathers,* 1881) considered it a spring and autumn passage migrant through Gilgit and secured a specimen (Brit. Mus. coll.) on 25 April.

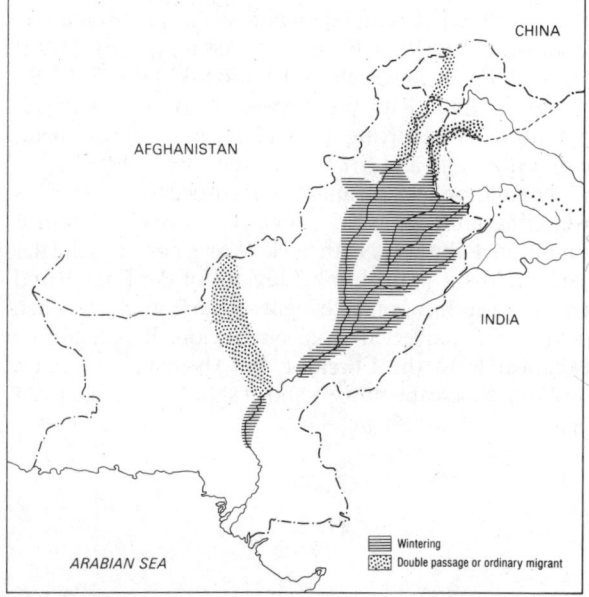

Map 107: Water Rail *Rallus aquaticus*

It is perhaps most plentiful in the few marshy areas of the Punjab Salt Range and the seepage zone upstream of Rasul barrage on the Jhelum river and in the water-logged *Typha* swamps of the northern Thal desert adjacent to the Jhelum river. H. Waite (hand-written record of skin collection) collected specimens from 13 October to 1 March from the Salt Range (Kallar Kahar lake, Dulla on the Soan, and from Gujrat and Muzaffargarh districts of the Punjab). It is much less common in southern Punjab, but the author has several sight records from Sidhnai spill channel (Multan district), Trimmu headworks (Jhang district) and Lal Sohanran (Bahawalpur district).

In Sind, Holmes and Wright *(JBNHS,* 1968) noted it around Sukkur in seepage zones alongside the main canals in early November but considered it rare. The author has only one record from Sind of a bird flushed from flooded *Tamarix* bushes at Wassowari lake in Sanghar district on 15 February 1973. Ticehurst obtained no definite Sind records but noted British Museum specimens which had been collected from Shikarpur south-west of Jacobabad district *(Ibis,* 1924). Whitehead *(JBNHS,* 1911) collected one specimen from Dhand tank in Kohat district and Meinertzhagen collected two specimens from Baluchistan on 1 March and 23 November. But due to lack of suitable marshy habitat it occurs rather sparingly in Baluchistan and the NWFP. During a visit to Zangi Nawar lake in the Chagai desert of south-east Baluchistan on 17/18 January 1984, the author heard many birds calling in the evening and on the second evening estimated a total population of as many as 50 Water Rails all around the lake which comprised a chain of lagoons extending over a distance of about 12.8 kilometres (8 miles) (Roberts, *JBNHS,* 1985). Near Khaur in the Salt Range, Nigel Hacking found five frequenting a small swamp of less than 1 acre area, which is an indication of how plentiful they can be in that region (pers. comm., 1974). Status SCARCE to FREQUENT in Salt Range.

HABITS

They are mostly crepuscular, or nocturnal, in feeding activity and very secretive by daylight. They are quite agile and run swiftly if they have to cross open exposed areas. When foraging undisturbed they flick their tails up and down like a Moorhen *(Gallinula chloropus).* They like to probe in mud or shallow muddy water and their bills are well adapted for this, their diet comprising mostly aquatic insects and their larvae (Odonata and Coleopetra). Mollusca (fresh water snails), and worms (Annelida). They will also eat seeds of sedges, berries, small amphibians, spiders and green shoots and succulent roots according to availability (Dement'ev, *et al.,* 1951, and Cramp *et al.,* 1980). They maintain contact by loud calls and are especially vocal at dusk or if suddenly disturbed, even in their winter quarters. They can swim and fly strongly but seldom engage in either activity being agile on their feet and capable of penetrating the densest tangles of reeds and submerged branches. The bird flushed on Wassowari lake (see above) was in a typically open habitat of submerged grass and *Tamarix dioica* and was almost stepped upon by the camel I was riding.

BREEDING BIOLOGY

Both sexes take part in nest construction and incubation of the eggs (Bates and Lowther, 1952). Most breeding is extra-limital, the sub-species *korejewi,* Zarudny, breeding in the USSR in Turkestan and in north-eastern Iran. It does however, regularly breed in the main vale of Kashmir so may be found to breed occasionally in some of the more suitable swamps, such as Gakuch or Singhal on the Gilgit river, or Bara on the Shyok river in Baltistan, areas which the author has been unable to explore in summertime.

Pair formation is probably initiated by the male calling and later by pursuit of female with tail raised over back and white under-tail coverts everted. Later in display the male raises his wings to show the prominently barred flanks and again fluffs out the undertail coverts (Cramp *et al.,* 1980) The nest is always well concealed in dense reed cover and is an untidy circular pad of reeds about 15.2 cms. (6 ins.) in diameter but usually not very substantial (Osmaston, *JBNHS,* 1927 and Bates and Lowther, 1952). The normal clutch is 5 eggs and incubation takes 19 to 20 days (Cramp *et al.,* 1980). The eggs are quite glossy and pale cream in ground colour, not very thickly spotted and blotched with reddish or purplish brown and also some under-markings of lavendar grey. The young chicks stay in the nest for the first 4 or 5 days where they are fed bill to bill and constantly brooded by one parent.

VOCALIZATIONS

In the dense reeds which they inhabit it is not surprising that loud calls play an important role in social contacts and biology. There is a well known loud squealing call, usually repeated 4 or 5 times which has been likened (Osmaston, 1927b) to the squeal of a piglet 'wheah-wheeh-wheeh', the first phrase usually being louder and higher pitched and gradually dying away or the reverse and increasing in crescendo and higher pitch. These calls can be heard most frequently at dusk and are regularly given by birds even in their winter quarters. They probably denote alarm or territorial advertising and are believed to be given by both sexes. The male is also believed to emit a courtship call in the breeding season comprising a softer more trilling 'tyick-tyick-tyick' (Cramp *et al.,* 1980). Bates and Lowther (1952) liken this to a Moorhen's call and render it as 'ker-ick, ker-ick'.

Spotted Crake *Porzana porzana* (Linnaeus)

Illus. **33**
Map **108**

DESCRIPTION

Head and body length 22-24 cms.
Wingspan 37-42 cms.

Noticeably larger than the Little *(P. parva)* or Baillon's *(P. pusilla)* crakes but still quite a diminutive bird when compared with the Moorhen *(Gallinula chloropus).* In shape and body proportions similar to both the previous species mentioned with blue-grey face and supercilium, throat and upper breast, and the lower flanks having vertical bars of black and brown and white. It is at once distinguishable by having small white scattered spots over the sides of the throat, neck and breast and even on the grey supercilium. The bill is yellowish green, darker on the tip and reddish to orange at the base of both mandibles. The crown and upper parts are rufous to

Illus. 33: (a) Spotted Crake *Porzana porzana.* (b) Baillon's Crake *Porzana pusilla*
(c) Little Crake *Porzana parva.* All drawn to same scale. Note two-toned bars on flanks of Spotted Crake and White spots on crown and mantle. Note dark streaked crown and white 'smears' on mantle of Baillon's Crake with conspicuous flank bars and paler less conspicuous barring on flanks of Little Crake with unstreaked crown.

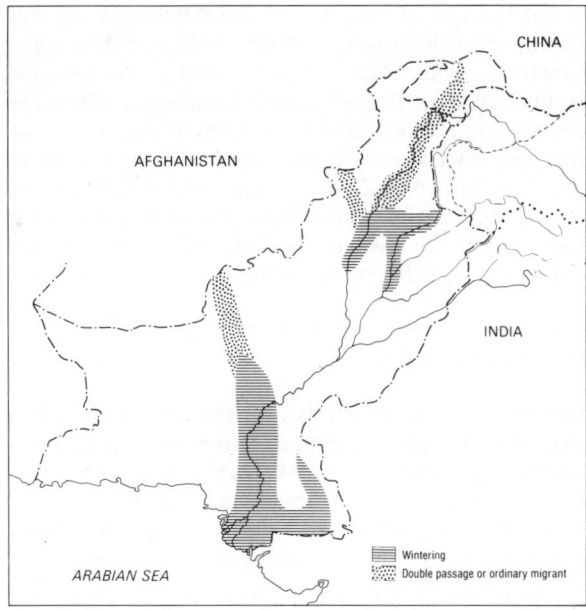

Map 108: Spotted Crake *Porzana porzana*

olive brown, boldly streaked with black and with some white spotting on the hind neck and mantle as well. Scapulars, lower back and rump are narrowly streaked with creamy white. The vent and under-tail coverts are buff and not prominently barred, a useful field point when only the retreating bird is glimpsed. The vertical flank bars are actually chestnut with narrow black borders and separated by broader irregular white bars. The sexes are similar in plumage. The legs are dull yellowish green. In flight the white spots on the wing coverts show well, which feature is lacking in *P. parva* and *P. pusilla*. Juveniles lack grey on the throat and breast, being whiter and have cinnamon tawny on the sides of the neck and flanks but have fairly prominent chestnut and white flank bars. The iris in both sexes is red-brown. The all-black downy chicks have black legs and feet and a dark tipped bill with a reddish yellow base.

HABITAT, DISTRIBUTION AND STATUS

It is very similar to Baillon's *(P. pusilla)* and the Little Crake *(P. parva)*, all being winter migrant visitors to Pakistan, and in occurring often in the same swamps alongside its congeners *P. pusilla* and *P. parva*. It is widely scattered in suitable reedy sheltered swamps in Sind and Punjab and has been noted on passage in Baluchistan, NWFP and Punjab. Dr. Scully *(Stray Feathers,* 1881) collected them in Gilgit during April only, but Major Biddulph *(Stray Feathers,* 1881) had 2 specimens brought to him collected near Gilgit in mid-April and the second in early July. Meinertzhagen *(Ibis,* 1920) collected one near Quetta on 31 October and noted specimens in the Quetta museum collected in October and November in previous years. Whitehead *(JBNHS,* 1911) recorded it as a spring migrant in fair numbers

through Kohat district. In the Punjab Whistler collected one in Jhang district on 8 February and another at Kallar Kahar in the Salt Range on 24 February. (Whistler, 1920 and 1930). Z. B. Mirza collected it from Balloki headworks in 1970 (Mirza *Pak. Journ. Zoology* 1973) Ticehurst noted autumn arrivals as early as 12 September in southern Sind (Ticehurst, 1924). Holmes and Wright also found this species around Sukkur *(JBNHS,* 1968). Khanum and Qadri from Karachi University collected specimens from Thatta district in the winter of 1971-2 (Khanum and Qadri, *Pak. Jour. Zoology* 1972). The British Museum has a specimen collected from Dost Allee jheel in Larkana district, Sind. The author has observed it at Mahboob Shah lake in Thatta district where several birds were calling to each other in 1979 but on many subsequent winter visits failed to locate it again. Also in a seepage area between the main railway line and canal south-west of Khairpur in early April 1980, where Baillon's Crake was also observed at the same time. It is so secretive that sightings are very few and until more evidence is available it must be considered uncommon. Status SCARCE.

HABITS

Presumed to be typical of the genus as nothing special has been recorded. Probably largely nocturnal or crepuscular in feeding activity, shy and skulking, being able to walk on lotus lily pads and submerged water weeds, and swim when required in order to hunt for insects, larvae of Odonata and fresh water snails *(Lymnae).* It also feeds upon Annelida, minute crustacea, Arachnida and seeds of sedges (*Carex* sp.) and rice *(Oryza sativa)* and other plant material.

BREEDING BIOLOGY

Presumed to be extra-limital. J.L. Peters (1960) lists it as breeding in the British Isles and north-west Kashmir. Since the beginning of this century it has apparently not bred in Britain (Cramp *et al.,* 1980) and there is no evidence as yet, of its breeding in north-west Kashmir, though Major Biddulph (1881) speculated that it might breed in Gilgit on the basis of a specimen shot by a local hunter in early July. Its breeding biology is similar to that of *Porzana pusilla,* with perhaps a more substantial nest cup being built which is partly roofed over by a canopy of plant stems. (Cramp *et al.,* 1980) Clutches are often large (8-10 eggs) with both sexes sharing incubation duties and the first hatched young remaining in the nest until all the eggs are hatched.

VOCALIZATIONS

Silent in its winter quarters. Males have an advertising call in the breeding season consisting of loud whistle-like squeaks, rather ascending in pitch and spaced at one second or lesser intervals, 'kwit' or 'whitt'. These calls may be repeated 20 to 50 times (Cramp *et al.,* 1980).

Little Crake *Porzana parva* (Scopoli)

DESCRIPTION

Head and body length 18-20 cms.
Wingspan 34-39 cms.

This is a tiny very secretive bird with stubby up-turned tail and large feet and long legs, like a miniature Water Hen *(Gallinula chloropus)* in general body outline. It is confusingly similar to Baillon's Crake *(Porzana pusilla)* and males in non-breeding plumage are probably indistinguishable in the field unless very good views are obtained of the crown and under-tail coverts. They are slightly bigger than Baillon's Crake with noticeably longer wings and tail (15 per cent longer). The males are slaty blue-grey on head throat and breast, with the crown and back chestnut and streaked with black on the mantle and wing coverts only. Baillon's Crake *(P. pusilla)* has much more distinct chestnut and black streaking on the crown, in both sexes. *P. parva* has paler creamy streaks on the back and the scapulars are prominently edged with pale buff. The lower flanks and under-tail coverts are barred with white and darker grey, but much less distinctly than in *P. pusilla*. The legs and toes are dirty green and the bill is generally bright green with the base of the upper mandible red, but this red spot is not easily seen in the field and is a feature shared with *Porzana porzana* the Spotted Crake which also occurs in similar biotopes in Pakistan. The iris in both sexes is crimson. Females are unlike males in having a whitish throat and the lower breast merging into a pale cinnamon fulvous colour, with indistinct brown vertical bars on the flanks and more blackish grey barring on the vent and under-tail coverts. Juveniles are like the females. In flight the wing feathers are dark brown, unbarred in both sexes. In size they are hardly bigger than *Motacilla maderaspatensis* (author obs.) which is often seen alongside. Chicks are downy black all over with stubby pale straw coloured bills and black legs and feet.

HABITAT, DISTRIBUTION AND STATUS

They are confined to swamps having dense reed cover but will often frequent quite small patches of suitable habitat. They are winter visitors to Pakistan occurring very sparsely in Sind and Punjab in suitable fresh water swamps, but are never found in mangrove swamps. They often appear relatively common in certain localities and are strangely absent from other equally suitable areas. In the borrow pits alongside Sidhnai spill channel (Ravi river, Multan district) a number over-winter and the author with D. Holmes and J. Wright counted 5 females and 3 males on 21 March in a 4 hour visit to this area in 1966. Birds have been noted in Multan area in all months from 6 September up to 21 March. The author heard several birds calling including a female in the seepage zone at Balloki headworks on 26 February 1987. The

only other records from northern areas lie outside of the Punjab. Neither Whistler *(Ibis,* 1930) nor Waite *(JBNHS* 1948) encountered it in the northern Punjab where it must undoubtedly occur as an irregular winter visitor. In Sind it occurs regularly on some of the larger reed fringed lakes of Larkana district, such as Drigh, and there are a series of skins in the British Museum from Dost Allee lake, in Larkana district. Ticehurst himself never encountered it. (op. cit. *Ibis,* 1924) during several winter visits to the lakes of lower Sind. Peter Conder only made one positive identification, a sighting on 9 February 1980 at Haleji lake, Thatta district (pers. comm.).

In Baluchistan numbers were noted in small tamarisk studded swamps at the mouth of the Bolan pass and there was a specimen in the Quetta museum collected from near Quetta (Ticehurst, The Birds of British Baluchistan, *JBNHS,* 1927b). Dr. Scully (1881) collected 3 specimens (Brit. Mus. coll.) in Gilgit between 5 October and 2 November and considered it a passage migrant through Gilgit. During Guy Mountfort's WWF expeditions to Pakistan (Mountfort, 1969) both the author and Col. Mountfort saw this species in a small tamarisk swamp at 1,829 metres (6,000 feet) elevation on 3 November at the mouth of the Karumbar (Ishkoman) valley in Gilgit and again on 4 November in a *Phragmites* filled swamp on the Gilgit river below the mouth of the Darkot (Yasin) river at 1,981 metres (6,500 feet). The second bird was calling loudly and afforded satisfactory views for positive identification. Status SCARCE to RARE.

Map 109: Little Crake *Porzana parva*

HABITS

Though sometimes active by day they are mainly crepuscular and can often be seen (dimly) feeding out in the open close to reed beds just after darkness falls. They are not usually vocal in their winter quarters though one on autumn passage in Gilgit was heard calling. A male was observed repeatedly jerking its tail up and down as it probed in the mud for food. One, when flushed, settled in an isolated patch of *Typha angustifolia* reeds in a borrow pit and refused to be put to flight again despite two of us wading in and quartering the patch covering no more than 400 square metres. Undoubtedly they can stay submerged with only their bills above water and thus escape detection. They feed largely on animal matter (Mason and Lefroy, 1912), including aquatic insects, water beetles and small flies (Diptera), also worms (Annelida) and water snails (Mollusca). A lesser amount of seeds of sedges are also consumed including *Carex* and *Sparganium*, water lilies *(Nelumbium* spp.) etc.

BREEDING BIOLOGY

Extra-limital but breeding habits similar to those of Baillon's Crake *(P. pusilla)* except that roofing over of the nest has not been reported for *parva*. Both sexes take part in brooding the young and feeding them for the first few days until they are able to feed themselves.

*VOCALIZATIONS

A female bird was tape-recorded on 4 April 1985 in a *Phragmites* reed bed on the side of Rawal lake (Islamabad). The bird was only a few feet away and by subsequent play-back, showed itself, permitting identification. It started by giving short high pitched 'py-ook'-'py-ook' calls, each spaced about 2 seconds apart. A series of 5 to 10 such calls then terminated in a rapid but decelerating trill of slightly lower pitched calls 'piou-pi-pi-pi-pi-pi-pi-pe-pe'. The last part of this rapidly stuttering call falling slightly in pitch like a mechanical bell or clockwork toy running down. This calling continued more or less without interruption for about 4 minutes.

The male is said to have a rather lower pitched more nasal 'quek' or 'quak' repeated at one call per second or slightly faster but without ending in the trill of the female (Cramp *et al.,* 1080). A male bird was heard calling in Gilgit in 1967 and field notes made at the time describe this call as a rather spaced and measured 'kweck-kweck-kweck'.

Baillon's Crake *Porzana pusilla* (Pallas)

Illus. **33**
Map **110**

DESCRIPTION

Body length 17-19 cms.
Wingspan 33-37 cms.

This is the tiniest of the *Porzana* crakes, smaller in size than the Little Crake *(P. parva)* but very similar in general appearance, being rather long-legged with a laterally compressed body and a short up-cocked tail, short thick neck and long toes. The male is superficially similar to *P. parva,* having dark slaty blue-grey face, throat and upper breast, whilst the crown and mantle is chestnut but heavily streaked with black, and on the mantle and wing coverts there are conspicuous and irregular creamy white streaks. The flanks have more prominent vertical barring of white and blackish grey than in *P. parva* and the bill is bright green, paler horn coloured at the tip and lacking any red patch at the base of the upper mandible and gape. The legs vary from olive brown in winter to a distinctly pinkish grey or dirty flesh colour in the breeding season. The iris in both sexes is orange to scarlet. The black streaked crown is a feature shared by both male and female which helps to distinguish it from *P. parva,* also the much more distinct barring on flanks and under-tail coverts. Females have white throat changing to pale fulvous on the sides of the neck and grey on the lower breast. There are broad vertical bars on the flanks which are dark grey and white becoming blacker on the vent. There is an indistinct grey supercilium and grey ear coverts in contrast to the chestnut ear coverts of the female *P. parva.* The males have more conspicuous white streaking on the back than *P. parva.* Also in both sexes the wings are relatively shorter in relation to body length than in *P. parva.* Juveniles look very like females of *P. parva* being cinnamon chestnut on the breast with much less conspicuous vertical barring on the flanks.

HABITAT, DISTRIBUTION AND STATUS

Confined to swamp areas having a lot of reeds or margins of lakes and pools with emergent vegetation. They are definitely more numerous and widespread in winter in Pakistan than *P. parva* and a few may possibly stay to breed in the northern mountain valleys though most of the population are winter visitors. H. E. Barnes (*Stray Feathers,* 1880) while stationed at Chaman on the borders of north central Baluchistan noted an autumn passage of many birds in September, most of them so reluctant to fly that they could be caught by hand as they crouched inside small bushes. Meinertzhagen (*Ibis,* 1920) obtained specimens at Quetta on 24 and 31 August and when they appeared numerous and were probably on autumn passage. Large numbers appeared to be over-wintering on Zangi Nawar lake in the Chagai district in south-western Baluchistan, estimated over 100, 17-18 January 1984 (Roberts, *JBNHS*, 1985). C.H.T. Whitehead (*JBNHS*, 1911) records them as passing through Kohat in large numbers from late April to 20 May, but he did not note them on autumn passage whereas he encountered this species in September around Rawalpindi. In Gilgit, Major Biddulph ('Stray Feathers', 1881) collected a male in breeding condition on 20 May near Buni. In their winter quarters they have been noted in widely scattered parts of Punjab and Sind. H. Waite (*JBNHS*, 1948) collected 11 specimens betwen 3 September and 22 May most of them in April and October from Gujrat district, Jhelum and Muzaffargarh districts. He noted many in swamps formed by borrow pits at Sarai Alamgir in Gujranwala district (MS notes on his personal collection). Dr. Salim Ali (*JBNHS*, 1941) collected a specimen from Bahawalpur on 9 February 1939 and the British Museum has a specimen from Manchar lake in Sind where Ticehurst also saw it on 26 December 1919 (*Ibis*, 1924). He also collected two from a small swamp close to Karachi on 4 September. He also collected one in a very small swamp which is created by a small spring seeping out of limestone cliffs near Rehri fishing village on the mouth of Gharo creek on 6 February. This small swamp still exists but recent exploration by the author has produced only *Tringinae* waders and wagtails. Status SCARCE.

HABITS

Very similar to the Little Crake, they are secretive in habits and do not emerge to feed in open areas until after

Map 110: Baillon's Crake *Porzana pusilla*

dark. The author has noted them at Bharatpur India, swimming well, with head jerking to and fro like a Moorhen *(Gallinula chloropus)*, they also twitch their tails up and down as they forage on floating water weeds or upon muddy embankments and individuals have been regularly sighted (author at Bharatpur), at dusk near the same patch of reeds indicating that they have favourite foraging areas even if they are not territorial in winter quarters. They are believed (Bates and Lowther, 1952) to be more adapted to less tall vegetation, such as *Juncus* sedges and rice fields, than *P. parva*, and to avoid very tall reed beds. Their food is similar to *P. parva*, comprising mainly small invertebrates including aquatic insects, snails, Annelida, Arachnida, but they also consume seeds of rushes, sedges, wild rice and water weeds and appear less exclusively insectivorous than *Porzana parva* (Mason and Lefroy, 1912).

BREEDING BIOLOGY

Dr. Scully (*Stray Feathers*, 1881) felt certain that it bred in Gilgit around rice paddy cultivation and found it a regular summer visitor to the main valleys and a male in breeding condition was collected on 20 May. Conclusive evidence from Gilgit is still lacking, but only slightly to the east, in more southerly latitudes this species regularly breeds in the vale of Kashmir. Bates and Lowther (1952) considered it a common and plentiful breeding bird in Kashmir on suitable lakes such as the Hokra jheel and marshes around Sumbal. Laying starts from early May and continues till July and large clutches are produced; 6 to 8 eggs being usual (Philips, *JBNHS*, 1946, part 3, and Bates and Lowther, 1952). The nest, made on the ground, is well hidden in weedy vegetation usually not in tall reeds but on the embankment of a lake on short sedges and reeds and even in rice fields, and the nest is invariably partly roofed over by bent plant stems carefully pulled into position by the breeding pair. Both sexes share in nest building and incubation. The eggs are slightly glossy, of a creamy yellow to pale brown ground colour, streaked and freckled quite profusely and boldy with reddish brown. There is circumstantial evidence that they may be double brooded as Bates and Lowther found slightly incubated eggs as late as 11 August. Both parents brood the young chicks which hatch asynchronously and depend upon bill to bill feeding by their parents for the first few days. Incubation period is 14 to 16 days (Cramp *et al.*, 1980).

*VOCALIZATIONS

Accounts of its voice given in *Birds of the Western Palearctic* (1980) refer to previous confusion between this species and *Porzana parva* and describe the main advertising call of the male as quite different from *P. parva* being stuttering or hard-creaking sounds 'trrrr-trrrr' each lasting 2 to 3 seconds and distinct from the 'crek-crek' call described for *P. parva*. By contrast, Bates and Lowther (1952), who spent many hours watching these crakes on their nest describe the call as a high pitched 'crake' which is repeated slowly at first and gradually at increasing speed. The author has often heard these measured single 'crake-crake' calls in swamps when the light was failing, but by good fortune on 24 April 1983 at Balloki headworks on the Ravi river in bright early morning sunshine a crake was heard calling territorially in some dried and trampled down *Typha* reeds. Tape recordings were made and played back as close as possible to the actual spot whence the calling emanated. Within seconds and at a distance of barely 6 metres, an adult male Baillon's Crake emerged, its pinkish legs and prominent blackish flank bars clearly seen, all-green base to bill, leaving no doubt as to its identity. At the time the calls were written down as 'crake-crake-crake' slightly shriller than one of the typical calls of *Fulica atra*. One hour later, play-back of the tape, again produced a curious and possibly aggressive, adult Baillon's Crake from an area about 0.4 kilometres distant and this seems to be therefore a fairly typical advertising or territorial call for the species. It is also worth speculating that at this late spring date, in this Punjab region that Baillon's Crake might occasionally breed in the seepage zones around Balloki headworks. Ali and Ripley in their 'Handbook', (1969) also describe the call of this species as a single loud high pitched 'crek', each call rather well spaced in time but gradually increasing in tempo. Clearly more studies need to be carried out on the vocal repertoire of these two *Porzana* species.

Ruddy Crake or Ruddy-breasted Crake *Porzana fusca* (Linnaeus)   Map 111

Synonym *Amaurornis fuscus* (S. Dillon Ripley's 'Synopsis' and Ali and Ripley's 'Handbook')

DESCRIPTION

Body length 21-23 cms.
Wing length 36-40 cms.

About the same size as the Spotted Crake but averaging slightly smaller, this is a typical crake in its skulking secretive habits, short neck and laterally compressed body, large feet and stubby up-tilted pointed tail. Its crown is deep maroon-cinnamon coloured, and its cheeks, neck and upper breast and wing coverts are more olive dark brown. The lower flanks and under-tail coverts are barred with olive grey and buffy white. The bill is brownish green to horny brown and the legs are orange-brown to red-brown. The sexes are alike. In general body proportions, it has a slightly longer bill than the other crakes and reminds one of the White-breasted Waterhen *(Amaurornis phoenicurus)*. Juveniles are darker brown than the adults on the crown and hind neck, with more white about the cheeks and sides of the neck but have similar barring on the lower belly and flanks. The iris is crimson in adults.

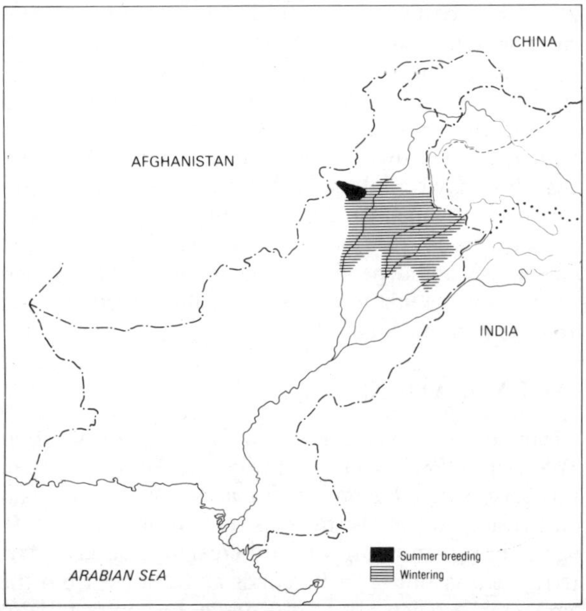

HABITAT, DISTRIBUTION AND STATUS

It inhabits small reed choked borrow pits and swamps, particularly in areas of rice cultivation as well as ox-bow swamps in the riverain areas. An Oriental endemic species which apparently occurs irregularly only in the northern regions of the Salt Range and the plains regions of western NWFP. It has not been recorded in Sind or Baluchistan and may be a partial local migrant in Pakistan which comprises the north-western periphery of its overall range. Whitehead considered it quite numerous in Kohat district and found it breeding in a marsh near Dandar in the lower Kurram valley. H. Waite collected specimens (Brit. Mus. coll.) from Bhant in Mianwali district on 25 December and Jagga and Pahrwal both in Gujrat in January. M. Mallalieu found it in April and May in the reed beds below Rawal lake dam, near Islamabad (pers. comm., June 1986). It was seen several times in small swamps fringing the Indus river at Kundian (Mianwali district) by C. Savage and the author during 1968 and 1969 on different dates between January and 29 March. Sighted (author) at

Map 111: Ruddy Breasted Crake *Porzana fusca*

dusk feeding in the open at Balloki headworks on the Ravi river 64 kilometres (40 miles) south-west of Lahore on 28 March 1981. Surprisingly Whistler (*Ibis,* 1930) did not record it in Rawalpindi district. It is a fairly plentiful breeding bird in the vale of Kashmir to the north. Status RARE to SCARCE in the north-western foothill regions of Pakistan only.

HABITS

Active mainly at dusk and in early morning when it can occasionally be seen feeding on the muddy fringes of canals or streams, if close to reed bed cover. It flies reluctantly, preferring to run into cover if disturbed. When foraging it flicks its tail up and down and jerks the head in a typically Moorhen *(Gallinula chloropus)* fashion. It will swim well, but usually avoids deep water and seems to prefer rice fields and the sedge choked depressions on the fringes of permanently inundated swamps. It is, like its congeners, very shy and secretive and seldom seen during daylight. It feeds on a variety of small invertebrates especially aquatic insects, mosquitoes and their larvae, and insect pupae, as well as snails, and some vegetable matter such as succulent roots, seeds and sedges and marsh plants (Mason and Lefroy, 1912). A specimen collected near Delhi had 18 vermiform insect larvae, 1 spider, 3 Pond Skaters (Coleoptera) and 2 small snails in its stomach (Donahue, (*JBNHS,* December 1967).

BREEDING BIOLOGY

In Kashmir the breeding was well described by Bates and Lowther (op. cit., 1952). Most nests are fairly shallow cups of reeds and the dry leaves of aquatic plants, located on lodged rice plants, or in thick weeds on embankments between paddy fields or in shallow submerged sedges. Most nests are partly concealed by a bent over lattice work roof of plant stems. In Kashmir they do not breed in dense tall stands of *Typha* or *Phragmites* reeds as *Rallus aquaticus* does, but near Kohat in a smaller marshy area they were found nesting in clumps of reeds or sedge grass (Whitehead, *JBNHS,* 1911). 6 to 9 eggs are laid, which are pale creamy in ground colour, boldly spotted all over with red-brown and purplish grey. In Kashmir eggs were mostly found from late June to early August when the rice fields provide more cover, but near Kohat Whitehead found most eggs laid from mid-June to mid-July with 7 the maximum clutch size (Whitehead, 1911, op. cit.) Both sexes share incubation duties and tending the newly hatched young. The chicks hatch asynchronously and are dependent on their parents for feeding for the first few days.

VOCALIZATIONS

On their breeding grounds, they call in late evening and early morning, and this is a metallic 'tewk-tewk' repeated every 2 or 3 seconds, usually followed by a trill or bubbling call likened to that of a Dabchick (Bates and Lowther, op. cit.).

Corn Crake *Crex crex* (Linnaeus)

DESCRIPTION

Body length 27-30 cms.
Wingspan 46-53 cms.

Larger than the Water Rail *(Rallus aquaticus)* and about the size of the Grey Partridge *(Francolinus pondicerianus),* it is larger than a Moorhen *(Gallinula chloropus)* and chiefly distinguished from other rails and crakes, by its comparatively short deep bill, and preference for a dryer grassland habitat. Both sexes are rather alike, having pale chestnut buff crown and back with black spots on the crown and heavy black streaking on the mantle. Cheeks and sides of neck are buffy yellow with bright chestnut wing coverts and dull white flanks barred vertically with chestnut. In the breeding season males have some blue-grey about the face but this is much less distinct in the female. The bill is more fleshy brown than green and the legs and feet are also fleshy brown. The iris is orange-brown. It has the typical laterally compressed body of a rail and stout legs with short up pointed tail and secretive skulking habits of the family.

HABITAT, DISTRIBUTION AND STATUS

This crake has been collected only once within our area, a female specimen shot by Dr. J. S. Scully (*Stray Feathers,* 1881 and Brit. Mus. coll.) on 8 October 1881 while on autumn migration in Gilgit main valley. It was in a maize field. As they breed in central Siberia and the Russian steppic regions of Latvia and Estonia, the occasional straggler might pass through the northern border regions of Pakistan on passage to their wintering grounds in east Africa. They have been recorded on passage further west in the Red Sea and Persian Gulf. Status VAGRANT.

Brown Crake *Amaurornis akool* (Sykes)

Crimson-legged Crake of La Touche and Chinese authors

DESCRIPTION

Body length 28 cms. (Fleming), 25 cms. (King *et al.*),
  28 cms. (Ali and Ripley)
Wing length 11.4-13.1 cms. (Stuart Baker)
Tail length 5.4-6.3 cms. (Stuart Baker)
Weight 114-170 gms. (males),
  110-140 gms. (females), (A.O. Hume)

This is one of the larger crakes being roughly the same size as the Water Rail *(Rallus aquaticus)* and noticeably larger that the Ruddy Crake *(Porzana fusca)*, with both of which, it is often sympatric. It is about the shape and build of a small edition of the White-breasted Waterhen *(Amaurornis phoenicurus)* with the same rather short heavy bill and with the crown and upper parts dark olive brown whilst the cheeks, breast and belly are ashy grey. The flanks and under-tail coverts are also olive brown and the immediate chin and throat area are paler greyish white. In adult birds the bill is green, darker at the tip, and the irides are crimson, whilst the legs which are stout with long tarsi and toes, become livid purple in the breeding season. Outside of the breeding season the legs are brownish pink and juvenile birds have brown irides. Females are slightly smaller than males but otherwise similar.

HABITAT, DISTRIBUTION AND STATUS

The Brown Crake occurs widely across India, Bangla Desh, south-eastern China including Hong Kong and northern Vietnam. It is an inhabitant of sub-tropical or warmer wetland regions both in the plains and the Himalayan foothill zone.

It is apparently rare in the main vale of Kashmir as Bates and Lowther (1952), as well as many other keen ornithologists (Phillips, 1946, and Osmaston, 1927) who worked the lakes and reed-beds of that area, never came across it, but there is a specimen in the British Museum collection from Kashmir (labelled Noshera—a locality untraceable on the maps). It has not been recorded from within Pakistan until the summer of 1987 when M. Mallalieu observed a group of 2 or 3 frequenting a reed-bed on the western shores of Rawal lake between 18 July and 15 August (*in litt.* to author and published in *Oriental Bird Club Bulletin*, No. 6, 1987).

H. Waite while posted to Ambala in the Indian part of Punjab (about 480 kms. 300 miles south-east of Islamabad) collected a series of 6 from Chandigarh, Jughadhri and Tajuwala, as well as 2 from the foothills around Kangra further north (approx. 320 kms. 200 miles south-east of Islamabad) (specimens in British Museum and hand-written notes of skin collection). It also occurs just to the south-east of Pakistan in Saurashtra (Gujarat State, India) where Dharmakumarsinhji (1972) found it breeding on a secluded village tank. Its occurrence near Islamabad at the foot of the Himalayas is therefore not surprising though it does represent a new western extension of its range. Status RARE and localized resident.

HABITS

They are mostly active in feeding at dawn and dusk when they will emerge into the open. Like all the rails they tend to be very shy and secretive and will run for cover if they think they have been observed. Individuals watched by the author in Chitawan, central Nepal as well as at Bharatpur were regularly seen in the early morning and evening time on open patches of short grass or drying out fringes of pools with tail cocked up and wagging as they walked, just like the White-breasted Waterhen, pecking at the base of water weeds or grass clumps and apparently gleaning snails and insects. They are also said to feed largely upon Annelid worms when these are plentiful (Mason and Lefroy, 1912). They will fly up into trees to roost and are agile in clambering through and up between tall reeds, like all the family. Ali and Ripley (1969) also report that they consume the seeds of marsh plants. Their flight appears feeble with long legs and toes dangling but they can run with speed.

BREEDING AND VOCALIZATIONS

They nest in dense reed beds, mainly during the monsoon season making a not very substantial pad of rush blades and sedges with a slight hollow in the centre. Their eggs described as richly coloured being pale yellowish or salmon buff, boldly marked by well defined spots and blotches of dark reddish brown (Stuart Baker, 1929). 4 to 6 is the usual clutch and both sexes share in incubation duties and are reported to be very close sitters.

Dharmakumarsinhji (1972) describes the call as shrill and reminiscent of the Little Grebe *(Tachybaptus ruficollis)*. This may be the same call described by Osmaston as a long drawn-out vibrating whistle, gradually descending in scale and by Fleming (1976) as a high rippling trill lasting 3 or 4 seconds.

White-breasted Water-hen *Amaurornis phoenicurus* (Pennant)   Map 112

DESCRIPTION

Body length 32 cms.
Wingspan 45-54 cms.

About the size and build of the Moorhen *(Gallinula chloropus)* but with a slightly longer bill and usually longer necked and leaner appearance. The forepart of the face, including cheeks and forecrown and the entire throat and breast are white. The crown, hind neck and back are grey-green or slaty grey but with distinct greenish olive cast in certain lights. The under-tail coverts and vent are rufous chestnut. The iris is crimson, the bill sage green and the large feet and long legs are dull chrome yellow or bright olive yellow. In the breeding season the base of the upper mandible is reddish. The flank feathers tend to be sullied grey. Both sexes are alike in plumage.

HABITAT, DISTRIBUTION AND STATUS

Less aquatic than the Moorhen *(Gallinula chloropus)* it still requires permanent swamp areas with fairly dense bush cover or reed beds, though it will forage on quite dry land and at some distance from the nearest water. It is resident but somewhat erratically distributed throughout Punjab and Sind from small colonies in the seepage streams below Rawal lake outfall at Islamabad in the north, down to the roadside borrow pits of Thatta district in the south. In recent years small colonies have been noted (author) in Changa Manga irrigated canal forest plantation (Kasur district), and Daphar irrigated forest plantation (Sargodha district), alongside the Lower Bari Doab at Renala Khurd, and Balloki headworks, also in Faisalabad alongside canal seepage zones, and in Shahdara outside Lahore. In fact the seepage zones alongside major canals are a favourite micro-habitat for this species. It is quite plentiful in the major rice growing tracts of Dadu, Larkana, Sukkur in northern Sind and Thatta district, spreading out into the rice fields during the late summer. It occurs along the Chenab river in Muzaffargarh district (Punjab) but was never encountered (author) along the Ravi river in Multan district during over 20 years of observations. Status FREQUENT but irregular in Punjab. COMMON in Sind. Largely absent from Baluchistan and the NWFP.

HABITS

Inclined to be crepuscular, coming out into canal side roads and village paths to forage at dusk. They constantly jerk up their tails as they walk and are very agile in clambering up reeds and into thorny bushes. They can also be quite active foraging during the daytime. They are normally solitary in feeding and subsist upon Mollusca (snails), worms (Annelida), aquatic crustacea and insects and their larvae, as well as a lesser proportion of vegetable matter and seeds. Stomachs of 5 birds collected from Bihar, (north-east India) between July and September showed that the diet was partly vegetable matter including bulbous roots or water weeds, grass and weed seeds and grit and a great variety of insects, ants *(Phidole)* and Coleoptera (Tenebrionidae, Carabidae), *Myllocerus* weevils, crickets, *Apis florea* and Molluscs (Mason and Lefroy, 1912). They are noisy birds, calling for prolonged periods in the summer months and usually clamber up reeds to some vantage point before starting to call.

BREEDING BIOLOGY

They form stable pair bonds during the breeding season, which is after the onset of the rains. Both sexes are believed to share in nest construction and incubation duties. The nests, made of dried reeds, rushes and plant stems, are extraordinarily difficult to find, looking like a collection of rubbish caught up in thick tangles of reeds or grasses growing up between *Tamarix* or *Prosopis* bushes. They may be built on the lower branches of such bushes but they are usually close to or on the ground. K. Eates (MS notes unpublished) found a nest on 18 July with five unincubated eggs and clutches of up to 7 eggs have been recorded. The eggs are creamy or pinkish

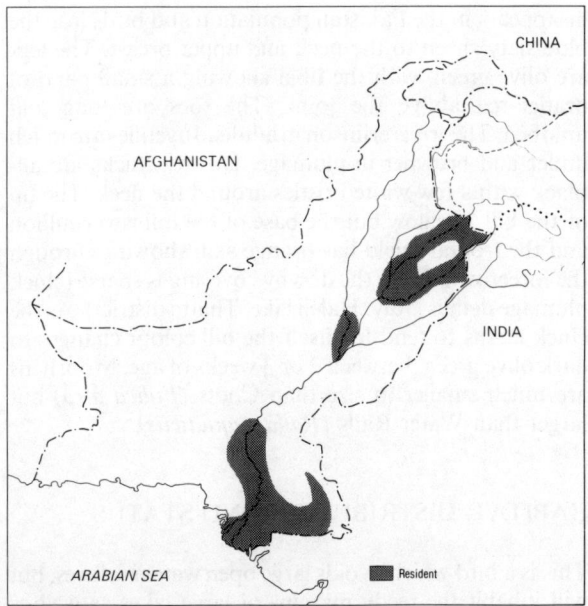

Map 112: White-breasted Waterhen *Amaurornis phoenicurus*

white in ground colour, boldly streaked and blotched with darker red-brown. The downy chicks are all-black, including their bills, and are adept at clambering through the vertical reed stems and diving into the water to hide when the nest is approached. They remain in the nest vicinity, however, for the first 5 or 6 days after hatching and the male also assists in escorting and protecting the young family.

*VOCALIZATIONS

The very loud and weird calls of this rail are much more often heard than the bird itself is seen. They start calling from early February (author obs.) and continue through to October, mostly from late afternoon and even during the hours of darkness, and again in the early morning and always from thick cover with the bird itself well hidden from view. Often a pair will duet or call together. A calling session may last from 1 minute up to 15 continuous minutes on end. Usually starting with ululating throaty 'kaargh-kaargh' calls and then breaking into 'kurrwarh-kurrwagh-kurrwagh' calls 'krrr-kwok-kwok-krrr-oowark-oowark' the last two calls rapid and inflected upwards. Sometimes a single noted 'kuk' is repeated monotonously for minutes on end even after dusk. These calls are believed to have a territorial function helping space out males and even pairs, and help to avoid encounters with unpaired birds, as they appear intolerant of conspecifics and individuals have been observed (author) both at Gharko and Haleji driving off an approaching bird.

Moorhen, Waterhen *Gallinula chloropus* (Linnaeus)   Map **113**

DESCRIPTION

Body length 30-33 cms.
Wingspan 49-53 cms. Smaller than European bird.

Except when adults are seen in good sunlight they look drab black all over with conspicuous white under-tail coverts. Adults have slaty black heads and necks with wings dark olive brown and flanks a paler slaty grey, the flank feathers tipped white forming an interrupted thin white line. The bill extends to a frontal shield over the forehead and in adults the bill is olive green in winter but becoming yellow tipped and bright scarlet on the base and frontal shield. Both sexes are similar in plumage. After the breeding season the white flank stripe disappears in the Pakistan population and birds lose the blue slaty sheen to the neck and upper breast. The legs are olive green, with the tibia showing a small band of orange-red above the joint. The toes are long and unlobed. The iris is crimson in adults. Juveniles are much duller and browner in plumage. Downy chicks are all-black with a few white bristles around the neck. The tip of the bill is yellow but the base of the bill is vermillion and the frontal shield has orange skin showing through the forecrown where the downy covering is sparse (chick plumage details, July, Haleji lake, Thatta district). As the chick learns to fend for itself the bill colour changes to dark olive green, between 2 or 3 weeks of age. Moorhens are much smaller in size than Coots *(Fulica atra)* but larger than Water Rails *(Rallus aquaticus)*.

HABITAT, DISTRIBUTION AND STATUS

This is a bird which avoids large open water surfaces, but will inhabit the reedy margins of large lakes as well as smaller reed choked borrow pits and canal side seepage zones. It is resident and very widely distributed in suitable habitat throughout Pakistan. The author has watched them in mid-February at —12°C in the Shyok valley of Baltistan at 2,682 metres (8,800 feet) as well as at 45°C in the furnace heat of May in the southern Punjab at Lal Sohanran. It is sparsely resident in Gilgit and Baltistan and Baluchistan and the NWFP but it is comparatively rare in both these latter provinces (it occurs in large numbers however, on Zangi Nawar in Baluchistan and breeds in the Kurram valley). In Sind and Punjab it is much commoner, occurring abundantly around the seepage zone of all irrigation barrages, as well as along the ox-bow swamps and lakes of the river

Map 113: Common Moorhen or Common Waterhen
Gallinula chloropus

system. It is not found in areas of only seasonal inundation preferring permanent swamps and regions of sufficiently stable water levels to afford permanent reed cover. Ali and Ripley ('Handbook', Vol. 2, 1969) considered that the winter population is augmented by migrants from the USSR. Status ABUNDANT.

HABITS

It swims much more commonly while seeking food than does *Porphyrio porphyrio* and is more diurnal in feeding activity than *Amaurornis phoenicurus,* thus avoiding to some extent, competition with these two related species with which it shares the same habitat. It dives to feed much less commonly than *Fulica atra* and prefers to pick insects, arachnids and other invertebrates off the leaves and stems of emergent aquatic vegetation as well as foraging on the muddy margins of lakes and ditches. Compared with its European counterpart it appears in Pakistan as being shyer and more secretive. It climbs well and often perches in trees. Its diet is omnivorous, but usually with a greater proportion of vegetable matter and plant seeds than animal matter (Mason and Lefroy, 1912). A bird watched at Haleji lake was pecking at and apparently swallowing seeds of *Polygonum.* It also eats Annelida, larvae of aquatic insects, flies *(Diptera)* and small tadpoles (Cramp *et al.,* 1980). The habit of constantly flicking its tail as it forages has often been recorded and during display (author obs.) it everts its white under-tail covert by fanning them outwards.

BREEDING BIOLOGY

The pair bond may be long lasting and they may frequently be double brooded as there seem to be two peak breeding periods from spring and again during the monsoon season (K. Eates, MS notes). Eates found a nest with 7 eggs on Siranda jheel, Lasbela on 7 July while another nest was located by the author in the seepage zone adjacent to Haleji lake in mid-May. The nest is usually built over water and concealed in reeds, being a bed of bent down rushes and reed stems. Sometimes it is quite a bulky structure built up from the water table, and it always has a well defined cup. Both sexes assist in nest building but a presumed male was observed bringing reed stems in its bill while its mate was already sitting on the nest (observed 12 May, Haleji lake). The normal clutch is 8 to 9 eggs but Bates and Lowther (1952) found 5 incubated eggs in Kashmir and as many as 12 have been recorded. The eggs are a pale pinkish stone colour and are thinly speckled all over with dark reddish brown blotches mainly around the broader end. Both parents share incubation duties equally and bring food to the young chicks which normally hatch synchronously (Cramp *et al.,* 1980). They are fed, at least partially, for up to 2 or 3 weeks after hatching and will swim freely like their parents. They are often seen with the male (slightly larger than his mate) bringing up the rear to the little flotilla, both parents assiduously herding and clucking to their chicks (10 July observation at Haleji lake, with brood of 7 downy chicks estimated 7 or 8 days old).

Bates and Lowther noted a distressed parent, when the nest sight was disturbed, swimming around in the cover of reeds and slapping the water. This is done with one of the feet (Cramp *et al.,* 1980) and not with the wings as assumed previously (Bates and Lowther, 1952).

*VOCALIZATIONS

The usual contact call or advertising call emitted from thick cover and throughout the year, is the well known loud, 'kurrik'. An alarm call is a more rapidly repeated 'keh-keh-keh' or a softer more staccato 'kik-kik-kik'. They are more vocal in the evenings and early morning and also during the breeding season, but the variety of their calls and the constancy with which they can be heard throughout the year, indicate the probability of long lasting pair bonds.

Purple Gallinule, Purple Coot, Purple Swamp Hen *Porphyrio porphyrio* (Linnaeus)   Map **114**

DESCRIPTION

Body length 45-50 cms.
Wingspan 90-100 cms.
Wing length 24.4-27.1 cms.
Leg length 8.8-9 cms.

Large bulky looking rail, bigger than a Coot *(Fulica atra),* with stout legs and huge feet and the heavy bill with arched culmen extending to a broad square frontal shield. The sexes are alike. The bill is scarlet and legs are slightly duller pinker red with a dusky black collar on the tibio-tarsal joint. In bright sunlight the scapulars and wing coverts show turquoise green, with the head and neck a very pale blue-grey. The rest of the body plumage is purplish slaty. The under-tail coverts are white and a conspicuous signal when the bird flicks up its tail. Juveniles are more greyish to sooty black on the head, neck and breast with dull leaden bills. In the breeding season the mid-breast region of adults also shows a lot of blue-green or aquamarine gloss and the top of the frontal shield shows a rounded slightly raised heel extending up to a ridge on the top of the crown. The iris is red. Downy chicks are black with legs and feet appearing disproportionately large which are brown, not pink. Their bill is red with a white tip and there is a bare red patch on their crown, with scattered white bristles on the back (K. Eates, MS notes). They also have a distinct claw on the meta-carpal joint of the wing.

HABITAT, DISTRIBUTION AND STATUS

More or less sedentary but subject to irregular local migratory movements especially during and after the monsoon when seasonal rains provide new feeding areas and shelter. They are commoner in Sind but small colonies exist in the seepage zones of all the major irrigation barrages in the Punjab including Islam headworks on the Indus at Kalabagh, Mianwali district. A few usually occur at Kallar Kahar in the Salt Range and there are numbers at Trimmu head on the Chenab river, and at Lal Sohanran and at Panjnad in Bahawalpur. Occasionally seen on the Sidhnai spill channel (Multan district). There are large numbers in the reedy portions of Manchar lake and other Sind lakes, with perhaps the largest concentration in Pakistan occurring on Haleji lake where an estimated 1,200 spend the entire year. Fluctuating numbers haunt the fringes of Zangi Nawar lake in southern Baluchistan (Christison, *Ibis*, 1941 and author obs. 1984), but they have not been noted on Kushdil Khan, nor in the NWFP. Though they have penetrated to the main vale of Kashmir they have not been noted in the northern mountain valleys of Pakistan. Status COMMON.

HABITS

Rather gregarious in their habits, they can often be observed feeding or roosting in small groups of a dozen or so birds. They will swim across open water, but do so rarely, preferring to forage on lotus lily pads and floating weeds or to clamber through reed beds. They appear to feed mainly on aquatic vegetation probably taking a lesser quantity of animal feed than other rails. Their principal food in Pakistan is the pithy stems of *Scirpus* and *Phragmites* reeds. They characteristically pull up submerged rhizomatous stems and grasp them firmly between the hind toe and three bunched front toes of one foot which is then raised towards the bill (author obs.). The hard outer cortex is skilfully stripped off by the bill and the inner pith consumed. This results in the production of fibrous green faeces easily recognized from this species, suggesting rather rapid and incomplete digestion. They have been recorded as eating birds' eggs, molluscs and various insects (Cramp *et al.*, 1980). They appear to be quite diurnal in habits and to prefer to forage on the ground, or on floating vegetation, never feeding like *Fulica atra* or *Gallinula chloropus*, while swimming. They are quite vocal throughout the year, indulging in a variety of contact and threat calls. K. Eates reared one as a pet (pers. comm. to author in 1967) and noted that it became very territorial, recognizing his own servants, but running up to and pecking strangers who entered his compound. Aggressive encounters between feeding birds can often be observed. They are preyed upon by many eagles and the author has several times seen *Haliaeetus leucoryphus*, secure and carry away this species, as well as *Aquila heliaca*. A Marsh Harrier *(Circus aeruginosus)* which swooped upon Purple Gallinules was however warded off quite aggressively with wing flapping and bill striking movements.

BREEDING BIOLOGY

They nest over an extended period where suitable dense reed cover is available, and downy chicks following their parents can be observed at Haleji lake in lower Sind from mid-May till mid-October. K. Eates found nests with eggs on 20 April at Haleji and Motiwar near Sujawal 17 April also 5 eggs at Sujawal on 10 July (MS notes unpublished). The nest is usually quite a bulky structure, built up from water level on a platform of bent over reeds and it is always concealed in dense reeds. The final nest cup may be as high as 76 cms. (30 ins.) above the water table but often with sloping sides made from material dragged to the nest site. The female does most of the building of the nest but both sexes carry nest material and share incubation duties. The young are tended by both parents and are fed bill to bill for at least 3 or 4

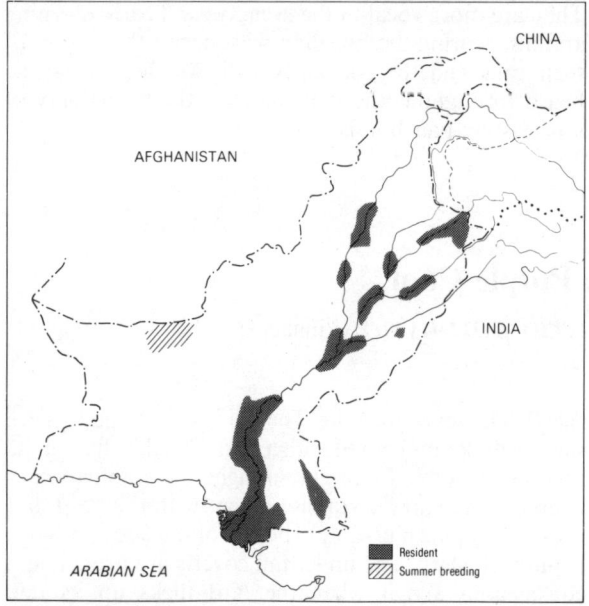

Map 114: Purple Gallinule or Purple Swamphen *Porphyrio prophyrio*

weeks after hatching and even when they are able to forage for themselves. When either parent gives a warning call the downy young will crouch and if necessary submerge their whole bodies (observed Haleji lake). The eggs are like large Moorhen eggs *(Gallinula chloropus),* having a pinkish buff ground colour and scattered with quite bold purplish brown spots and speckles with some secondary grey under-markings. The normal clutch is about 5 eggs but from 3 to 7 have been recorded (K. Eates, MS notes, Bates and Lowther, 1952). Hatching is synchronous after an incubation period of 23 to 25 days.

*VOCALIZATIONS

They have a variety of loud and complicated calls difficult to describe. Commonly one calling bird will elicit a whole chorus of rejoinders, and the combined wailing, squawking and cackling of the group adds to the peculiar charm and excitement of a day spent alongside a Pakistani jheel. One of its characteristic calls is the repeated wailing of almost human tone which breaks into a sort of whistling piping. Another call is a guttural gurgling alternated with short raucous 'kiuks'. It also emits loud trumpet like short calls 'puk-puk', but the social significance of all these various calls is difficult to interpret as the calling bird is so commonly obscured from sight when most vocal. The louder trumpeting type calls are thought (author obs.) to denote alarm or warning and the repeated wailing is more of an advertising or territorial song (Cramp *et al.,* 1980). They can be heard giving such territorial song calls throughout the year, and often females will chorus back in shorter higher pitched wails.

MISCELLANEOUS

Fossil *Porphyrio* bones have been found in the Siwalik beds of the Punjab Salt Range in Miocene sediments.

Watercock or Kora *Gallicrex cinerea* (Gmelin)

Illus. **34**
Map **115**

DESCRIPTION

Body length 34-43 cms.
Wingspan 68-86 cms.

They are longer legged than the Common Coot *(Fulica atra).* The female is noticeably smaller than the male and both are about the same size as the Common Coot *(Fulica atra).* In summer the male bird looks all-black but in a good light only the head and neck is plain black and there is faint vertical grey or dull white barring on the lower flanks and across the under-tail coverts whilst the back is slaty grey with the wing coverts and tertails margined with pale chestnut. The most striking feature is the bright scarlet bill, with a frontal shield over the forecrown extending into a round backward sloping horn or projection on the top of the crown. The bill tip is bright yellow and the legs are red as also is the iris. In non-breeding plumage the legs become olive green and this colour often persists until early July. The horn on the head shrivels and the base of the bill becomes yellowish with only a small triangular horny shield. It then becomes similar to the female, in which the back is dark brown, the feathers margined fulvous and the throat and belly are pale buffy brown with wavy darker bars. In general shape it is a typical rail with stout legs, long toes and rather laterally compressed body with up tilted pointed but short tail.

HABITAT, DISTRIBUTION AND STATUS

This widely distributed Oriental species occurs from China, the Philippines to the Celebes in the east. In Pakistan it is only a rainy season breeding visitor largely confined to lower Sind but a few birds also invade the north-east corner of the Punjab, descending down the Ravi and Chenab rivers. It is confined to reedy margins of roadside and canal borrow pits and seepage zones, until the rice growing season, when it disperses into the paddy fields and low lying flooded areas. In Sind the

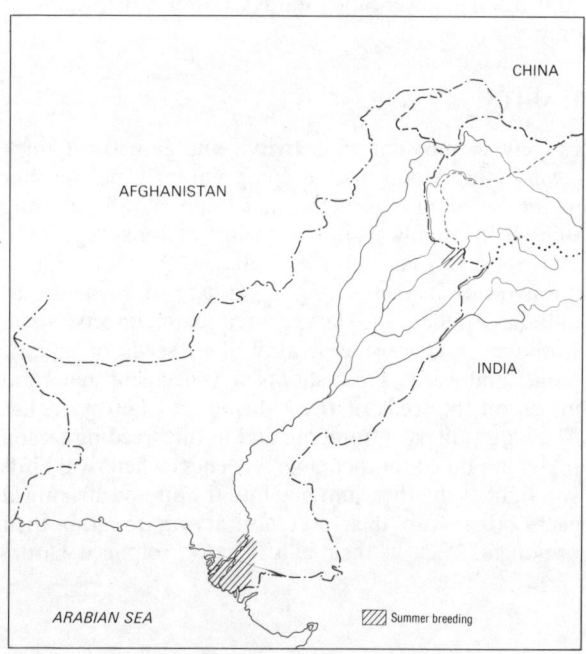

Map 115: Watercock or Kora *Gallicrex cinerea*

Illus. 34: Watercock or Kora *Gallicrex cinerea*, male in breeding season.

earliest arrivals have been noted on 16 May, but they are not widespread until early July when the rice is being transplanted. Then they occur throughout Thatta district and less commonly in the rice growing tracts of Hyderabad and Badin districts. After breeding and by late September they disperse again, migrating south-eastwards back into India. E. Fernando encountered them in the reed beds upstream of Marala barrage in Sialkot district in summer only (pers. comm., 1969) and the late Roger Holmes watched a female at close range on 22 September 1967 (pers. comm. to author) in reed beds upstream of Balloki headworks on the Ravi river (Lahore district). Z.B. Mirza also collected a specimen from Balloki (*Pak. Jour. Zoology* 1973). In Ticehurst's day they were rare in Sind (*Ibis*, 1924) and have undoubtedly increased due to the great increase in rice cultivation in lower Sind. Status COMMON in lower Sind only.

HABITS

Largely crepuscular in activity, emerging from thick cover to forage only in the evening and early morning but in the breeding season the male will call all day long though invariably from the shelter of rank vegetation and sedges. Like all the Rallidae they are rather omnivorous in diet taking advantage of such fare as tadpoles, pupae and larvae of aquatic insects, small Mollusca, and crustacea, as well as seeds of aquatic sedges and weeds, green shoots and succulent stems and in season the seeds of rice (Mason and Lefroy, 1912). They are solitary in foraging and in the breeding season males are noted for their aggressiveness when rival birds will fight each other, jumping into the air and striking at each other with their feet and trying to grab their opponents neck in their bills. These prolonged battles

have no doubt given rise to its trivial name of 'Water-cock'.

BREEDING BIOLOGY

A secretive bird, despite its constant and loud booming advertising calls, so that little has been recorded about its breeding. It is believed that the pair bond is stable and that the male assists in nest building and incubating. Nests are found from late July onwards in lower Sind and a hen bird with downy young was noted at Gujjo in early September 1967 (Jack Coles, pers. comm.) K. Eates found nests with incubated eggs on 20 July and partly incubated eggs on 16 July. Nests are well hidden in reed beds like that of the Purple Gallinule *(P. porphyrio)* and consist of a large rather untidy pad of rushes and sedges with a shallow depression a few inches above the water level. Most nests are built over water where the reeds emerge from a water depth of 70-95 cms. (Eates, MS notes). The eggs are variable in ground colour, from pale pink to yellowish stone with rather longitudinal brick coloured blotches and spots of reddish brown. The normal clutch seems to be 5 or 6 eggs but up to 9 have been recorded. Eates describes the young chicks as being covered in black down on the back and brown down on the belly and having a black bill with an ivory white tip to both mandibles. Legs are black and wing claw on the carpal joint is present. There is no trace of a frontal shield to the bill in the downy chicks.

*VOCALIZATIONS

The advertising call of the male is heard incessantly during the monsoon season in the rice fields of Thatta district and this call obviously plays a vital role in courtship as well as in spacing out rival males. It is a very resonant and regularly repeated hollow sounding pumping sort of call, preceded by shorter and rather Paddybird-like *(Ardeola)* 'quoks'. These are given with the head raised and neck up stretched and sound like 'kakh-kakh-kakh', uttered at a measured pace of about one per second, and this is then followed by a much louder and resonant 'qhumb-qhumb-qhumb' uttered at a faster tempo and like a pumping noise. These calls are produced by the bird crouched forward with head lowered and neck out-stretched, and gular region visibly inflating with each call (author obs. Haleji). The bill points groundwards and the red horn which is fleshy at its tip actually bends over forwards and trembles as the bird's whole body distends with each call. Earliest calls heard were 6 June and latest heard, 9 September. A bird watched calling, appeared to have its bill closed. I have recorded these two types of call continuously alternating without a break for over 15 minutes.

Black or Eurasian Coot *Fulica atra* Linnaeus

Map **116**

DESCRIPTION

Body length 36-38 cms.
Wingspan 70-80 cms.
Wing length 18.5-22 cms.
Tarsus length 5.6-6.4 cms.

A large stocky rail with no white on the under-tail coverts and an all-black body. The bill is pinkish white with a pure white round frontal shield on the forecrown. The body plumage is sooty black on head and neck, greyer on the flanks and back. In flight there is a narrow whitish band to the trailing edge of the secondaries. The iris is crimson and the feet greenish grey with long curiously lobed webs on the toes, a unique feature amongst the Rallidae. The sexes are alike. In flight the long toes trail behind the tail tip and the rounded wings beat in rapid whirring strokes. The tibia shows a smear of orange-red just above the joint. There is no seasonal variation in plumage.

HABITAT, DISTRIBUTION AND STATUS

An abundant winter migrant, concentrating mainly on the larger lakes and irrigation barrage headponds of the Indus basin. Part of the population migrates across Baluchistan and migrating Coots have been noted at Chaman in March (Ticehurst, *JBNHS*, 1927*b*) and through Chitral in late April and early May (Perreau, *JBNHS*, 1910) and in Gilgit they are common on passage in spring from the first week of March till mid-April and again in autumn (Scully, *Stray Feathers*, 1881). It is an occasional sporadic breeder on some of the larger Sind lakes and likewise on Band Kushdil Khan lake (Meinertzhagen, *Ibis*, 1920) in Baluchistan but this is exceptional, the entire population normally departing in late spring to northern breeding grounds in the USSR. Though still an abundant visitor, loss of habitat to agricultural drainage schemes in its breeding areas, has undoubtedly greatly reduced the winter migrant population to Pakistan. In his account of Manchar lake in Dadu district, visited in December 1926, Dr. Salim Ali writes of huge rafts of Coots numbering so many thousands of birds that they could only be counted in acres (*JBNHS*, 1927). Today barely 2 or 3,000 normally winter on Manchar (Roberts, *JBNHS*, 1967). Extensive wetland surveys in the early 1970s and waterfowl censuses revealed the biggest concentration in the whole of Pakistan on Khinjar lake (a huge lake of approximately 240 square kms. (90 square miles) surface area). Here an estimated 71,000 were counted in 1972 (Fred Koning, IWRB Report), and an estimated 79,670 on 13 January 1987 (IWRB Wetlands Report for Pakistan prepared by the National Council for Wildlife, Islamabad). Other favoured wintering areas such as Sangriaro lake in Sanghar district of Sind which had an

estimated 10,000 birds or over, and 8,000 on Dungrio lake (census estimates, mid-January 1972). Coots winter on Zangi Nawar in southern Baluchistan in huge numbers (the author estimated 66,000 on 17/18 January 1984), but Christison (*Ibis,* 1941) found little evidence of breeding on that lake. The author counted over 300 on Kushdil Khan lake on 22 June but no evidence of breeding in that year, but Christison did find a solitary nest in 1938 on that lake which was the only breeding record for Baluchistan that he could find (*JBNHS,* 1942). During a May visit to Zangi Nawar there were still about 2,000 birds and local game watchers asserted that 600 or 700 pairs nested each year on the lake (Roberts, *JBNHS,* 1985). During frequent summer visits to the main lakes of lower Sind (by the author) in only 3 out of 10 years a total of 3 to one dozen Coots have been noted over-summering (usually in company with Purple Gallinules), but no evidence of breeding was obtained. The last date noted for birds on spring passage was 7 April at Rawal lake in Rawalpindi district. Each winter during boat trips around the tidal lagoons of Ghizri and Keamari, a few individual Coots are encountered. These are always healthy looking and are observed diving far from land and appear to be subsisting satisfactorily upon marine organisms, probably mainly molluscs but also including small fishes. Status ABUNDANT.

HABITS

Despite aberrant behaviour noted above, in which an occasional individual adapts to a marine environment

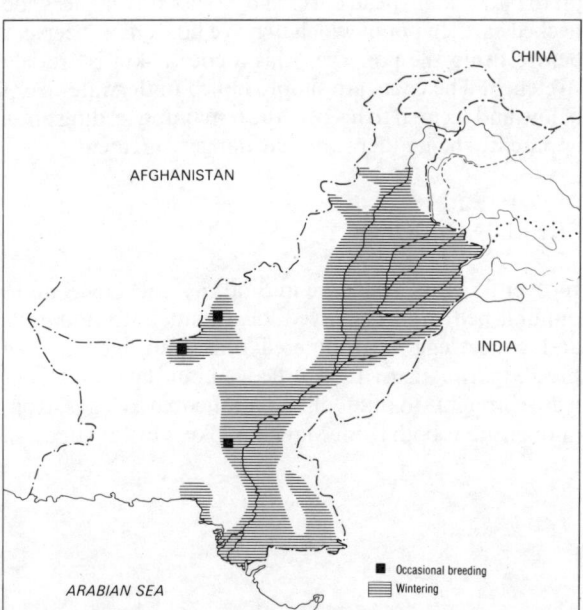

Map 116: Black or Eurasian Coot *Fulica atra*

and animal diet, most coots are highly gregarious in their wintering grounds, subsisting largely upon submerged water weeds which they obtain by diving. They will also forage by sieving material from the surface and less commonly by walking on floating vegetation and along the margins of swamps and lakes. They are perhaps more highly adapted than any other Rallidae to feed upon aquatic algae and parts of dicotyledonous water plants, including reeds and stems. In a study conducted in the Sunderbans, west Bengal, stomach contents of 36 birds were analysed revealing that 63.5 per cent of their diet was vegetable matter, mostly submerged weeds, with 6.2 per cent Annelid worms, 13 per cent fresh water Gastropods, 7 per cent aquatic insects and 6.7 per cent small fishes (Mukherjee, *JBNHS*, 1974). Even when feeding on land they tend to do so gregariously in small groups. They will feed both by day and night, but are mainly active in early morning and evening, and of all the Rallidae are best adapted to swimming and feeding on open water and are consequently the least secretive and most tolerant of human presence and disturbance. They are preyed upon by many raptors which congregate in winter on the larger lakes of Pakistan and form a major item in the diet of *Aquila clanga* and the author has observed both *Haliaeetus albicilla* and *H. leucoryphus* successfully capturing coot in Sind. At Bharatpur, Rajasthan India, the Black-necked Stork (*Xenorhynchus asiaticus*) has been noted as successfully catching and preying upon Coots (S. Breeden, pers. comm., 1983). They are still a traditional food item of Sindhi 'mirbars' (professional fishing tribes), who drive them with boats at night into flight nets, suspended on bambo poles out in the middle of large lakes (observed Manchar, 1967). They are also stalked in the early morning, when busy feeding, by a 'mirbar' often wading up to his neck, his head covered by a specially made, wide necked earthen pot in which two eye holes have been cut before firing the pot. Over this a coot's skin is crudely stretched! The coots are simply pulled underwater from below and secured to his belt, the remaining feeding birds apparently being unaware that danger threatens.

BREEDING BIOLOGY

In over 35 years' residence in Sind, K. Eates (MS notes unpublished) who employed local hunters to find nests and collect eggs, encountered only four instances of breeding. In August 1936, 4 fledged but flightless chicks were brought to him plus 3 unincubated eggs from another nest, both from Manchar lake. On 15 August in the same year a nest containing 5 eggs was obtained by him on the same lake and finally on 28 July 1941 local hunters brought him a clutch of 3 unincubated eggs from the same lake. In 1939, Brigadier Christison recorded one pair of Coots hatching out 5 young in the last week of June on Kushdil Khan lake in central Baluchistan and the same pair hatching a second brood on 18 August using the same nest (*JBNHS*, 1942). Until recently these were the only authentic breeding records traceable. Varying numbers breed each summer on Zangi Nawar lake, Baluchistan (Roberts, *JBNHS*, 1985 and Ashiq Ahmad, 1987). A parent bird with 4 downy chicks was seen as late as 6 July. (Mohammad Idrees Chughtai, 1987). In the main vale of Kashmir they breed on the more sheltered jheels but only in small numbers (Bates and Lowther, 1952).

The normal clutch size in Kashmir was found to be 8 with variation between 5 and 12, and eggs were found from May to July and they are commonly double brooded with, in at least one instance both parent birds watched, brooding 3 fresh eggs whilst alternately swimming out to feed 4 fledged youngsters (Bates and Lowther, 1952). The eggs are glossless, pale buff or pinkish stone coloured, widely and regularly marked with a scattering of small black blotches and spots. Breeding has been well studied in Europe, both parents defending the nest territory aggressively and taking an equal share in incubation and feeding the young which are fully fledged at about 55 days and can feed themselves entirely from about 30 days after hatching (Cramp *et al.*, 1980) Kenneth Eates gives a detailed description of chicks in down (secured 15 August, Manchar lake). The bill is white tipped but scarlet around the base extending up to the crown. There is some naked purple skin above the eye and there are yellow bristles around the neck and face with the down around the face enclosed in scarlet quill sheaths. The feet are black.

*VOCALIZATIONS

In winter time they can be heard calling throughout the night (observations 5 December, Khabbaki lake, Salt Range). The commonest call is a short single staccato 'kewk' rather querulous in tone. Aggressive calls comprise a series of shorter higher pitched calls 'ieu-ieu-ieu'. A variety of other social interaction calls have been described, comprising softer chuckling and clicking noises.

FAMILY GRUIDAE

Cranes

Sarus Crane *Grus antigone* (Linnaeus)

DESCRIPTION

Height about 150-156 cms.
Wingspan 265-272 cms.

A huge pale grey bird with dingy red, naked head and upper neck and long fleshy red legs and feet. The relatively stout straight bill is greenish horny coloured with the immediate top of the crown also ashy green. The toes are unwebbed and the inner secondary feathers are elongated and gracefully down-curving concealing the wing-tips and tail.

HABITAT, DISTRIBUTION AND STATUS

A sedentary species which pairs for life, they are now extinct as a breeding bird in Pakistan. K. Eates however saw two pairs in Larkana district on Drigh lake in 1929 and 'mirbars' (professional fishermen) took a pair of fledged chicks from there in the same year (MS notes). J. Anderson's father found a pair breeding in Larkana on one of the larger lakes as recently as 1939 (pers. comm., 1971). In Ticehurst's day (World War I) he saw one in October near Mirpurkhas and mentions a specimen in the Karachi Museum from Jhimpir but considered it rare even at that time (*Ibis,* 1924). In August 1968, E. Fernando saw a pair on the Chenab river above Marala barrage (pers. comm.). Comparatively numerous across the Indian border in Rajasthan their absence from Pakistan is testimony to their vulnerability to, and intolerance of, human interference. Status VAGRANT.

Common Crane *Grus grus* (Linnaeus)

Illus. **35**
Map **117**

DESCRIPTION

Wingspan 220-245 cms. Males are slightly larger than females
  (observations on numerous captive pairs).
Height 110-120 cms.

They are tall elegant grey birds with long legs and graceful, dark tipped and curved filamentous plumes covering the wingtips and tail, formed by the elongated tertials and inner secondaries. The head and neck are darker grey, almost black, with a broad white band extending from behind the eye down the side of the neck where it gradually merges into grey. The forecrown and lores have naked bluish black skin with a broad warty red band extending over the mid-crown. The relatively short bill is greenish horny, paler white or yellow basally. The iris is usually pale golden but some adults have reddish brown and even crimson irides, irrespective of the season (observations *circa* 40 captive birds).

HABITAT, DISTRIBUTION AND STATUS

They are regular winter migrants, passing mainly down the Indus, and entering via the Koh-i-Safed range and Kurram valley in the NWFP but migrating flocks have also been observed in early March flying up the Chenab river above Marala barrage in Sialkot district. Due to their intolerance of human persecution only occasional small parties are observed resting on passage within Pakistan. In the late 1970s, 5 were noted at Sidhnai on the Ravi river 6 October, and 21 cranes on Mahboob Shah lake, Thatta district 5 December and 2 at Haleji lake, Karachi district on 4 October (author obs). The majority of birds however, when reaching near the coast, turn eastwards and winter in the Great Rann of Kutch and Saurashtra in Indian territory. The only area where they do over-winter in Pakistan is at Sandho and Jabho lakes in Badin district in the Rann of Kutch right on the Indian border, where 105 and 15 respectively were counted in late January 1973 (F. Koning, pers. comm.). In the Kurram valley spring migration is usually noted from early March to early April, with great regularity (local crane hunters' statements to author, 1982-3). In the autumn they can be seen over Thatta district as early as mid-September and some birds even follow the Hab river. 120 were seen just west of the Hab on 1

Illus. 35: Common Crane *Grus grus*

October (Deed A. Ali, pers. comm., 1980). H. Waite noted them on passage over Mianwali district as early as 15 August (*JBNHS*, 1948). In 1983 on 14 October a strong west to east passage of six flocks totalling over 650 birds were observed (author) over lower Thatta district from mid-day to evening. The largest flock was over 200 birds and smallest comprised 26 birds. About 10 per cent only of these flocks could definitely be identified as juveniles. This is the largest number reliably recorded on any one day in this region in recent decades. In Christison's time a small flock of Common Cranes over-wintered each year (15 in 1939 and 40 in 1940) on Zangi Nawar lake in south-western Baluchistan (*Ibis,* 1941), and a migratory flock of 55 birds was seen over Ziarat on 22 March by the author. Local hunters however only knew of the Demoiselle *(Anthropoides virgo)* as a regular passage migrant visitor to Kushdil Khan lake (1972 and 1976 enquiries). Surveys of crane hunting in the Kurram valley conducted in 1982 and 1983 indicate that up to 750 Common Cranes (and as many *Anthropoides virgo)* are annually captured alive or killed by local crane hunting camps, located along migration routes, and that in early 1983 there were over 5,170 captive cranes in the NWFP, mostly held by villagers in Bannu district. All hunters interviewed admitted that numbers of migratory cranes had diminished markedly in the past several decades and the annual catches were also greatly decreased. Efforts are being made (1983-4) to alert the local Provincial Government authorities to this hunting pressure and steps are under way to try and limit and regulate such hunting which is an age long cultural tradition and passion amongst these strongly independent frontier tribesmen. (see Roberts and Landfried, 'International Crane Workshop', Bharatpur,

India, 1983, and S. E. Landfried publications in various journals, 1983). Present day status SCARCE even on passage and no permanent wintering flocks occur within Pakistan except in the extreme south-eastern border regions.

HABITS

Gregarious in its wintering range, usually resting in very open exposed situations near large bodies of water, sandbars in rivers etc. and flighting out to feed on cropland only in the early morning and evening, roosting during the night and again during the hotter hours of day on a site where human intruders cannot approach unseen. On migration it soars to great heights circling on thermals and then by alternate flapping and gliding flight in broad extended 'Vee' formation, lead birds taking it in turn to occupy the most arduous leading flight position.

They are omnivorous in diet, subsisting mainly upon green shoots and leaves, seeds of grasses and cultivated crops. especially cereals. They also consume insects, mostly ground dwelling Orthopterous spp. when available. Captive birds allowed to forage on free range were observed pecking at unidentifiable objects on the ground, in stubble crops, as well as eating young shoots of Mustard *(Brassica campestris)* in a growing crop. A small flock over wintering at Sultanpur jheels, outside Delhi (India) was noted to return in small groups of 3 to 8 just at sunset, the entire flock assembling close together on the drying out lake margin, whence there was no hidden ground or tree cover within about 1 kilometre.

BREEDING BIOLOGY

Extra-limital. Pair bonds are life-long and hunters can often capture pairs because of the succourant behaviour of the other partner when one bird is captured. Captive pairs are used as decoys and in the spring can be induced to give their bugling duet calls in unison at a command from their owners. However these boreal forest nesting birds rarely breed successfully under captive conditions in the NWFP, whereas *Anthropoides virgo* does so, not infrequently (see the species account).

*VOCALIZATIONS

The voices of the two sexes are different in pitch and one of the only reliable means of differentiating them. While on migration they have guttural but high pitched contact calls, a 'kraah-krraah', also a more grating 'krrrr'. In the breeding season males will give a very far carrying trumpeting series of calls which comprise a series of four bugle-like cries 'dee-da-deh-dee' increasing in speed and crescendo. The male raises his bill skywards while thus calling and erects his rump plumes. The female responds with shorter harsher quacking calls in unison with the male's bugling. Due to the long convolutions of their tracheal pipe, these calls have great resonance and will carry for distances of 4.8 to 6.2 kms. (3 to 4 miles).

Map 117: Common Crane *Grus grus*

Siberian Crane or Great White Crane *Grus leucogeranus* Pallas   Map **118**

Synonym *Bugeranus leucogeranus*

TAXONOMY

Dr. George Archibald of the International Crane Foundation considers that this species should be placed in the same genus as the Wattled Crane *(Bugeranus carunculatus)* of Africa on the basis of similar duetting calls (G.W. Archibald, 1975. Proceedings of the International Crane Workshop, and S.D. Wood, *The Wilson Bulletin*, No. 3, Vol. 91, 1979, pp. 384-99).

DESCRIPTION

Body 60-65 cms.
Wingspan 230-260 cms.
Height 120-140 cms.

Alan Octavious Hume rightly referred to this magnificent snow-white bird as the 'Lily of Birds'. It is larger than *Grus grus,* with a relatively longer and more slender bill (an adaptation to its specialized feeding habitat). The legs and bill are crimson, extending over the forepart of the face to the back of the crown in a bare red naked area. The primaries and base of the primary wing coverts, are jet black; only revealed when the wings are spread as the rump is cloaked by long curving white tertial feathers. Juveniles in their first winter have rusty orange heads and necks, without the naked red face patch and the wing coverts and scapulars are brushed with rusty buff.

HABITAT, DISTRIBUTION AND STATUS

In 1983, 36 birds wintered at the famous Bharatpur Sauctuary in Rajasthan, India. It was noteworthy that about 80 per cent of these were family groups with juveniles, and very few adults without young were seen (pers. observation, February 1983). They leave their Russian breeding grounds on the Ob river in the western Siberian tundra region of the USSR in early autumn but often do not reach Bharatpur until mid-December or later. The areas where they spend the intervening 10-12 weeks while on autumn migration and the fate of non-breeding or sub-adult birds remains a mystery. A few scattered observations indicate that Ab-i-Estada lake in central Afghanistan is an important resting or staging area during such migration. This means that the small population of 30 or so birds must overfly Zhob district of Baluchistan and Multan area in the Punjab. There have been no reliable sightings of this crane in Pakistan since Hume's record of two small parties seen in 1875 (quoted in Ali and Ripley's 'Handbook'). Dr. Landfried interviewing Crane hunters in 1981, ascertained from one old hunter that white cranes had been seen in the Kurram valley in the early 1960s and that several had

actually been captured and eaten in about 1964. In the author's lifetime experience however, such second-hand reports have to be treated with caution as Great White Egrets *(Egretta alba)* are also occasionally killed and eaten during these crane camps. Hunting of migrating cranes also occurs in Zhob district of Baluchistan. Out of the Ob river population, 2 or 3 pairs also still winter on the southern Caspian marshes (ironically gaining some protection from the area being a duck-hunting preserve.). The main population, comprising about 200 birds breeds in Yakutia in north-eastern Siberia and winters on the great marshes of Lake Poyang in China. It is listed in the IUCN *Red Data Book* as severely threatened with extinction and obviously risks further decimation from hunters during migration across Pakistan, though the species is legally protected from hunting in Punjab, Sind and the NWFP. Status RARE, even on passage.

HABITS

They feed almost exclusively upon vegetable matter obtained by probing in marshy or shallow flooded areas and do not forage on dry land as do *Grus grus* and *Anthropoides virgo.* They have been observed at Bharatpur feeding mainly upon the hard round tubers of

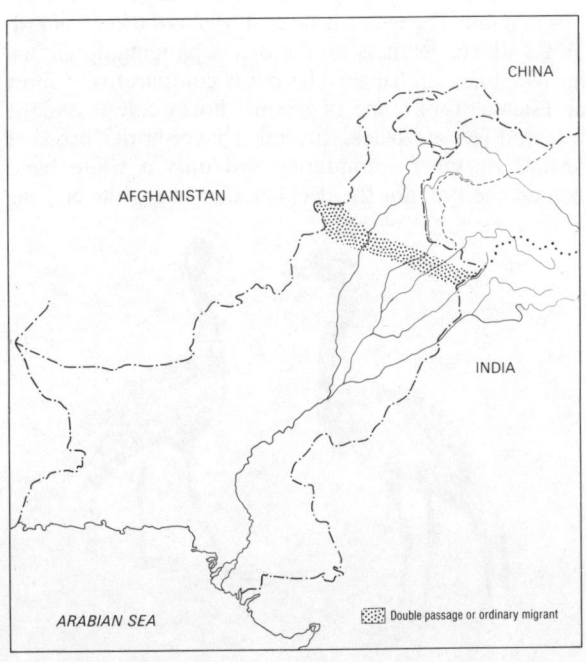

Map 118: Siberian Crane *Grus leucogeranus*

the Sedge *Cyperus rotundus.* In their breeding grounds more animal food is consumed including Lemmings *(Lemmus obensis),* amphibia, small fish and lizards (Dement'ev and Gladkov, 1951). At Bharatpur they seem capable of probing in water up to 40 cms. depth and swallowing some food without raising their heads; it is therefore very difficult by observing live birds to ascertain their diet. Juveniles are occasionally fed bill to bill and remain by their parents throughout their first winter (observed Bharatpur, 8-9 February), this no doubt reinforces the parent offspring bond. When such bill to bill feeding was noted the item was invariably a sedge tuber. Each family group maintains its own winter feeding territory, but they roost at night all together and this is the only time in winter that they can be heard calling, usually upon the arrival of more birds at the roost.

BREEDING BIOLOGY

Extra-limital.

*VOCALIZATIONS

Much less vocal than *Grus antigone* or *Grus grus.* Birds were heard giving intention calls before flighting to their evening roost. Their call is a rather rapid alternating of two types of calls giving a yodelling effect. The first a high pitched tremulo 'kerrh' then a more falsetto 'khoon'. Greeting calls are a rapid 'khee-koon-khee-khoon' of higher pitch than other cranes, rather musical and pleasant to the human ear much less bugling than the Sarus or Common Crane's calls. It could also be syllabized 'heehaw' but the tones are purer and more high pitched than that of a donkey braying.

Demoiselle Crane *Anthropoides virgo* (Linnaeus)

Illus. **36**
Map **119**

DESCRIPTION

Body length 50-53 cms.
Wingspan 165-185 cms.
Wing length 45-53 cms.
Bill length 65-70 mm.
Height 90-100 cms.

A comparatively small crane, but very elegant with its bright crimson eye and white down-curving ear tufts. Placed in the same genus as the Stanley Crane of South Africa, they have pale blue-grey bodies with slender lanceolate down-curving tertials which conceal the wing tips and tail. The whole head and neck is darker blackish grey with the feathers on the breast hanging down in a narrow blackish fringe. The bill is comparatively short and slender for a crane, of greenish horny colour, and the legs and feet are black. Juveniles have shorter browner tertials or inner secondaries and only a white band behind the eye, not the peculiar hair-like white curving

Illus. 36: Demoiselle Crane *Anthropoides virgo*

plumes of adults (observed 6 month old captive bred birds).

HABITAT, DISTRIBUTION AND STATUS

This crane breeds in steppic latitudes in central Russia from the plains north of the Black Sea eastwards to outer Mongolia. It is a widespread winter visitor to the Indian subcontinent which enters via the Indus watershed in Pakistan. It has declined dramatically in numbers in this century, even if it is still more plentiful than *Grus grus.* The main wintering population now is probably to be found in Gujarat state in India, which includes the Kathiawar peninsula and Rann of Kutch. Here in the winter of 1982-3, Dr. J. A. Van Der Ven estimated the total population at about 25,000 birds (pers. comm., February 1983) scattered over more than 20 different night time roosts. A. A. Phillips writing at the turn of the century described migrating flocks on their staging grounds in the Punjab which covered a broad band of ground 50 metres wide and 2.4 miles long, and totalling many hundreds of thousands of birds This has been quoted by Stuart Baker ('Fauna Series', 1929), and Ali and Ripley ('Handbook', 1969) as evidence of their continued abundance in present times. In Pakistan a few small flocks tarry in the eastern desert border regions of Cholistan and Thar (author obs., Rahim Yar Khan district and Karim Dad Junejo, Sanghar, pers. comm.) while on migration, but there are no long term wintering populations nowadays. Migration is over a more extended front than *Grus grus* though flocks of both species often follow the same routes and travel together. Crane hunters who visit Zhob district report (1983) large numbers of Demoiselle, but no Common Cranes, whilst in late March and April the *Grus grus* is the predominant species migrating through the Kurram valley. They also migrate in smaller numbers through Gilgit and the Hunza valleys, passing southwards in late August (Major Biddulph, 1881). In the Punjab they are much more commonly seen on migration than *Grus grus,* whereas in southern Sind only *Grus grus* is encountered (author obs. over 7 year period). Flocks can be seen crossing southern Punjab in a north-westerly direction from mid to late March (28 year's author observations). In Baluchistan flocks cross Pishin district in mid to late March and flocks have been seen (author) on passage at Baran dam on the Kurram river (Bannu district) on 29 March. The return passage is from late September to early October. A flock of 26 was seen on 4 October and 36 on 6 October in two consecutive years at Khanewal, Multan district. Highly prized as pets, they are still ruthlessly hunted on migration by the Wazir tribesmen of Bannu district. Status in Pakistan FREQUENT on migration only.

HABITS

Like the Common Crane they are highly gregarious on their wintering grounds, as well as while on migration and sometimes associate with *Grus grus.* They rest by day on exposed sandbars on the shores of large lakes or in river beds where humans cannot approach them unseen, and fly out to forage in green crops in the early morning and late afternoon. They are very shy and wary and difficult to approach. They often rest in remote desert areas also by day, coming to fingers of cultivation extending into the edge of desert to feed in the early morning (observed Cholistan, Rahim Yar Khan district, late February 1957). In India they have been reported as feeding largely on cereal seeds gleaned from stubbles, but they also eat green shoots and buds of plants and insect larvae and pupae. Captive birds on free range in Bannu district brought *Chrotogonus* grasshoppers to their newly hatched chicks.

BREEDING BIOLOGY

Pair bond is life-long and they nest on the ground in semi-arid open grassland areas defending a large territory from conspecifics. Unmated captive cranes of opposite sex when kept on free range in Bannu district, established pair bonds and started egg laying after 7 and 8 years in captivity (pers. com. Crane owners, Bannu). Often eggs for the first year or two are infertile or the chicks do not survive, but after several years of failures some pairs have bred consistently in captivitiy, often successfully rearing both chicks, an event seldom achieved in the wild. The normal clutch is 2 eggs and they are glossy, olive buff in ground colour with yellow-brown blotches and smaller spots. Both sexes share in incubation but mainly the female at night time, with the male defending the immediate nest vicinity. In Bannu district breeding pairs normally lay in May and in the least disturbed or farthest corner of the walled

Map 119: Demoiselle Crane *Anthropoides virgo*

compound of the homestead. Domestic chickens, dogs and even the house owner are attacked if approaching within 15 feet of the nest, the male bird flying upwards and striking down with feet and bill on the head and shoulders of an approaching human (observed in Bannu). Incubation takes about 30 days and the eggs hatch asynchronously. For the first 3 days the chicks are too weak to accompany their parents out of the compound into the surrounding fields, and are brought grasshoppers which are fed to them bill to bill. A detailed survey in February 1983 revealed 97 instances of captive bred young surviving more than 12 months based on the previous 5 or 6 years and there were 4 or 5 cases of 2 chicks being successfully reared. They remain as a family group throughout the first year and the male was observed erecting his black breast plumes and lowering his head and neck in aggressive threat posture whenever a local villager approached too close. During courtship dancing, observed in March adults were observed erecting their white head crest over the sides of their heads like miniature curved wings, but their white crests are sleeked down during threat postures.

*VOCALIZATIONS

A pair will call in unison or duet, the female's call being lower pitched. Their calls are much lower pitched and guttural than those of *Grus grus,* lacking the shrill bugling quality of the latter. Birds on migration or while in flight call to each other with a 'kraak-kraak' fairly high pitched call, not unlike that of *Grus grus,* which is equally far carrying and resonant. This can often be heard before the birds are visible in the sky.

FAMILY OTIDIDAE

Bustards

Little Bustard *Tetrax tetrax* (Linnaeus)   Map 120

Tetrax tetrax orientalis (Hartert)

DESCRIPTION

Body length 40-45 cms.
Wingspan 105-115 cms.
Weight 907-975 gms. (India and France) (Males, Brit. Mus).
  700-750 gms. (females) (Dement'ev and Gladkov)

The smallest of the bustards, but still quite a stout pheasant-sized bird, with comparatively long legs and straight long neck and rather short bill. Outside the breeding season, both sexes are predominantly white on the belly with the whole head, neck and back a pale sandy buff finely vermiculated with grey and black spotting. The long scapulars and tertials conceal the wingtips when at rest, but in flight these are boldly patterned in white with black borders to the secondaries and all-black wingtips. In the breeding season the male develops a blue-grey throat and cheeks, with a jet black neck prominently patterned by a white throat chevron and broader white collar below. The legs are olive yellow with very short stout toes and no hind toe, typical of all the Otididae. The iris is pale yellow, becoming more orange in adult males and the bill is olive to blue-grey, darker on the tip. In the field the absence of any black on the neck of both sexes in winter time, plus the more extensive white on the secondaries and 'inner' primaries serve to separate this from the larger Houbara *(Chlamydotis undulata)*. Males are slightly larger than females. The sub-species *orientalis* which occurs in this region (Stuart Baker, *Game Birds of India*, Vol. 2, 1921) is characterized by having a more greyish tinge to the upper parts and wings averaging longer *(Birds of Africa*, Vol. 2, 1986). Males have a whistling wing beat owing to their short emarginated primary and a rather duck-like flight, not the measured wing beats of other species.

HABITAT, DISTRIBUTION AND STATUS

This is a steppic bird from middle latitudes, but appearing to prefer undulating plains on the borders of hills or stony slopes and attracted to open grassland or low green crops. They are irregular winter migrants which have never been more than rare and erratic in occurrence within our limits, though Stuart Baker refers to bags of up to 20 being obtainable in some winters in Baluchistan and the NWFP ('Fauna Series', Vol. 6, 1929). There are a few old records from Gilgit, for example Major Biddulph records one shot on a stony plain on 27 March about 10 kilometres (6 miles) from Gilgit town whence a pair had been noted the year before *(Stray Feathers*, 1881). Whitehead noted it as occurring very rarely around Kohat and Bannu *(JBNHS*, 1911). H. Waite did not record it in the Salt Range, but more recently Rohil F. Nana (pers. comm.) encountered two birds in the Kala Chitta hills in a broad grassy valley in January 1976. It may still occur irregularly in Lasbela and the Makran as Cumming obtained one at Ormara in 1904 (wings preserved in the British Museum) and J. Anderson claims to have shot it in Lasbela in the mid 1960s. More recently, Afsar Mian during studies of the Houbara Bustard in the Chagai desert, reported sighting a single bird in December 1985 near Yakmuch. James Murray, Curator of Karachi Museum, had one in his collection which is still preserved in Karachi University Museum but its origins are unknown, and Ticehurst cites one being shot near Karachi and identified by a Mr. Strip 'one exceptionally cold winter in January', presumably at the turn of the century (op. cit., 1924). More recently keen sportsmen from Swabi district of the NWFP claim that this species regularly occurs between Buner in Swat and Gobati canal in a sandy and grassy stretch of northern Swabi district, and that 12 to 20 birds are shot there in a good year (F. Rohil Nana, pers. comm., 1980). It was also recorded fairly often around the 1920s and 30s on the Jamrud plain near Peshawar. Present status RARE and an erratic winter visitor.

Map 120: Little Bustard *Tetrax tetrax*

HABITS

The diet consists predominantly of vegetable matter, with the occasional insect being eaten (Coleoptera and Orthoptera). They will peck at young green shoots and leaves including *Artemisia maritima*. They are gregarious in both breeding and foraging so that where one occurs several will usually be encountered.

BREEDING BIOLOGY

Extra-limital. The jumping display of the males has been well studied in Portugal and includes preliminary calling and erection of black neck-ruff followed by foot thumping and then repeated butterfly-like leaps into the air during which the all-white under surface of the wings are prominently displayed and the bird flaps its wings, emitting a whistling sound, and lands back on the same spot. Males display from a ridge or prominent piece of high ground usually in visual contact with other displaying males. Males are polygamous and take no part in incubation or care of young. Females, which are slightly larger than males, mate with vigorously displaying males and tend to nest in territories of dominant males to avoid the attention of other males. Clutch size is 4 to 6 eggs.

Houbara *Chlamydotis undulata* (Jacquin)

Illus. **37**
Map **121**

McQueen's Bustard *Chlamydotis undulata macqueenii* (J. E. Gray)

DESCRIPTION

Body length 55-65 cms.
Wingspan 135-170 cms.
Wing length 41 cms. (Stuart Baker)
Tail length 18 cms.
Weight up to 1,700 gms. (males),
    up to 1,900 gms. (females) (Stuart Baker)

In size between *Tetrax* and *Ardeotis*, with a rather long thin neck and noticeably large yellow iris. They are long-legged cursorial birds with the whole of the upper parts rufous sandy, finely vermiculated with grey and black. Females are somewhat smaller but similar in plumage. The broad square tipped tail has grey cross-bars edged with black and the whole of the lower breast and body is white. The crown is black with narrow white down-curved straggly crest feathers. The most conspicuous features are the bands of filamentous black feathers down each side of the neck. These feathers are black basally, white in the middle and black tipped, they are wholly white on the lower part and even longer. The crest, the long slender feathers along the sides of the neck and the upper breast feathers are erectile during display. The lower breast is blue-grey forming a bib of longer feathers. The legs have three very short cushioned toes and no hind toe, and are greenish yellow and the bill is rather compressed dorso-ventrally, greenish yellow at the base and black or horny tipped. In flight the primaries are white, black tipped and with secondaries largely black, with white on the carpal and secondary wing coverts and indistinct black wingbars. The sub-species *macqueenii* is larger and paler than the nominate north African population with more black on the crown and a darker grey chest bib.

HABITAT, DISTRIBUTION AND STATUS

The Houbara, like all the Otididae, has declined dramatically in numbers during the past 30 years and has attracted world-wide concern because of hunting pressure from Arab falconers. Breeding mostly in central Asiatic Russia mainly in the Kuzil Kum desert region south-east of the Aral Sea, they are perhaps the best adapted of all the Otididae to an arid semi-desert environment. Some breeding also occurs in the more remote south-western border regions of Baluchistan. The sub-species *macqueenii* is migratory, entering north-west Pakistan and crossing in a broad front through Baluchistan and part of the NWFP. Some birds winter in Rajasthan, India, but a large part of the population remains in the desert border areas of Cholistan and Thar as well as in Lasbela and Chagai, desert plateaux of Baluchistan. The first arrivals are usually noted early in October and birds leave their eastern winter range from late February, beginning to return north-westwards. They are usually all gone from Baluchistan's border areas by early April.

Illus. 37: Houbara Bustard *Chlamydotis undulata*. Note wing tips completely covered by tertials and short thick cushioned toes with absence of hallux.

Recent studies on their breeding grounds in the Soviet Union indicate an overall reduction of the total population from 1956 to 1979 of about 70 per cent (Alekseev, 1980). Another Russian study noted a dramatic decline in numbers starting from 1968, by no coincidence the year when large scale hunting by Arab visitors to Pakistan was started (T. Ponomareva, 1979).

Winter rainfall is localized and erratic in the desert regions of Pakistan and Houbara movements depend heavily upon availability of food and freedom from disturbance. The latter has been the most serious factor affecting the population in recent years as population densities are so low that food supplies are not a limiting factor. In February and March 1980, P. Goriup of ICBP conducted a survey on their Pakistan wintering grounds and estimated a potential population upwards of 20,000 birds, but even at that time noted that as many as 3,000 birds were killed annually by increasingly organized and numerous Arab hunting parties. More recently, the estimates of all the favoured Houbara wintering areas in Sind indicate population densities as low as 3.86 birds per 10 square kilometres in 1979-80 and 3.3 birds per 10 square kilometres in 1981-2. This means taking the total area of the province, a density of just over 1.75 birds per 10 square kilometres (Mohammad Ibrahim Surahio, Project Officer, Sind Wildlife Management Board Official Report, 1983). S. B. Mirza similarly estimated a 20 per cent decline over 9 years in the Houbara population over-wintering in Rahim Yar Khan district of the Punjab. Status SCARCE becoming RARE.

HABITS

They are largely nocturnal in feeding where disturbance is high and there is evidence that they migrate largely by night and in suitable tracts, travelling partly on the ground covering 1 to 2 kms. per night (observations of local hunters, Baluchistan, corroborated by M. I. Surahio in Sind and Professor Afsar Mian in Baluchistan).

Where numbers are higher they are gregarious in feeding, and they are strongly attracted to crops of mustard *(Brassica campestris)*. In 1947 the author saw a flock of 22 birds in a mustard crop at about 6 a.m. in February on the borders of Chak 95, 24 kilometres (15 miles) west of Rahim Yar Khan. Even an extensive search over many hundreds of square kilometres revealed at most 1 or 2 birds within this same region in 1983. In winter, they feed mainly on vegetable matter and particularly succulent shoots, but will catch and eat Coleoptera, Orthopterous insects and consume large quantities of grit. In Sind Kohistan tracts, they have been noted favouring the berries of *Zizyphus nummularia* and flowers of *Capparis aphylla* as well as succulent leaves of *Haloxylon* and *Grewia populifolia* (Neale Taylor, pers. comm., 1982).

29 stomach contents from Houbara were collected by Afsar Mian during early and late winter visits to Arab

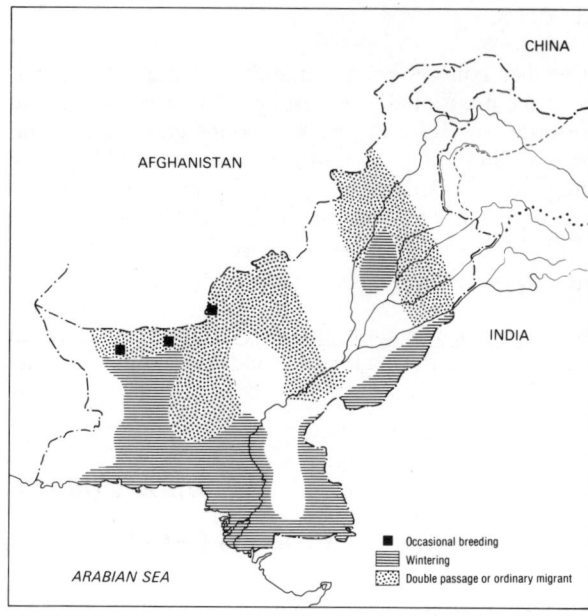

Map 121: Houbara Bustard *Chlamydotis undulata*

Occasional breeding

Wintering

Double passage or ordinary migrant

hunting camps in 1984 from Chagai and Kharan districts. When examined visually by the author, it was striking to note the high proportion of insect remains in the samples from late winter feeding birds. In 2 samples, almost 90 per cent comprised animal matter. Crude visual examination suggesting that they had fed almost exclusively upon dung beetles Scarabidae, and Tenebrionid beetles of medium size with a smaller proportion of ants *Camponotus* spp. Detailed analysis by Afsar Mian later showed 74 per cent of plant material and 25.3 per cent insect remains from the early winter sample and 51.3 per cent of Coleoptera in the late winter samples, with 7.8 per cent Formicidae and 42.4 per cent plant material. These late winter samples were taken from birds hunted in Kharan. Plant material comprised mostly young shoots in early winter, with more seeds and flowers in late winter, the predominant plant species being *Anabasis* sp., *Tribulus terrestris, Haloxylon ammodendron,* and fourthly *Alhagi camelorum.* Strongly aromatic plants such as *Artemisia, Salvia* and *Atriplex* were taken in much smaller quantity (Afsar Mian and Mohammad Rafiq, *Annual Technical Report to WWF Pakistan,* 1985-6).

Houbara are exceedingly shy and wary, with keen vision, and will normally run away when a human is sighted, only taking to the wing if suddenly disturbed. Hunters have long noted that they can be approached by camel back or on local bullock carts and have used this subterfuge to get near enough to their quarry. When approached very close they sometimes retreat into the shelter of a bush and freeze by squatting on the ground, neck outstretched and this habit makes them particularly vulnerable to falconers when equipped with fleets of desert adapted vehicles. If pursued by a falcon, they will forcibly eject mucilagenous faecal matter into the attacking falcon and this tactic sometimes enables them to escape. They are strong and swift flyers though their deliberate wing beats and large size give a deceptive impression of slowness.

BREEDING BIOLOGY

The males are believed to be polygamous and display on well spaced out territories. They are aggressive towards each other and will erect their tails and fan them open over their backs with their wings, spread and dropped crouching forward like a Stone Curlew *(Burhinus oedicnemus)* during such agonistic encounters (this has been observed in Pakistan in February by Neale Taylor 1982, and General J. Marden, 1960s, pers. comm.). The courtship display is quite different, involving the erection of the neck ruff and gradual sinking of the head with head crest also erected, back into the shoulders. The lower long white plumes in the pectoral region are everted vertically and completely hide the head and upper neck, giving a prominent white splash of colour visible from distances of 400 to 500 metres. The males perform runs and walks whilst in this posture. Only the female incubates and tends the young, which are nidifugous but have to be fed bill to bill for the first few days after hatching. No nest is made, the eggs being laid in the merest scrape under the shelter of a bush. The normal clutch is 3 to 4.

There has recently been much interest focussed upon evidence of bustards breeding in Baluchistan and it is becoming increasingly apparent that a few birds do remain each summer to breed in the Makran and Chagai desert regions. Dr. Mohammad Yaqub Bhatti, Director of Animal Husbandry found a Houbara egg in 1968 in the Mashlaq Reserve north of Quetta (reported in the *Outdoorman*, February 1972 issue). Dr. Ticehurst quotes Cummings working near Ormara in the 1880s and Hotson surveying around Panjgur, both encountering Houbara in June and having fledged young brought to them (*JBNHS*, Vol. 32). On 9 April 1983, one Houbara Bustard egg was collected from Yakmach east of Noakundi, in south-western Chagai, Baluchistan (shown to author by Deputy Inspector General of Forests). Brigadier Christison also wrote that breeding in the Makran is recorded from time to time (op. cit., *Ibis*, 1941). The egg from Yakmach was glossy in texture and dark olive in ground colour, blotched with darker brown and some smears of grey. Also 2 eggs taken from somewhere in Baluchistan were flown to the Dubai Wildlife Research Centre in 1983 (Paul Goriup, pers. comm.). In late March and April 1985, by the efforts of local shepherds and the incentive of large monetary rewards, about 6 newly hatched Houbara chicks and 6 or 7 eggs were collected from Kharan, Khuzdar and the Chagai regions of Baluchistan for captive breeding projects in Saudi Arabia (Naseer A. Tareen, pers. comm., 1987); Mr. Tareen took cine films of 2 eggs and 3 chicks at the collection camp. Again in early May 1987, Mr. Tareen saw nearly 50 Houbara (presumed potential breeding birds) and filmed females with chicks in Kharan and Turbat districts *(in litt.* to author, June 1987). These may be potential migrants to the USSR, or more probably part of a regularly breeding population contiguous with Iran.

VOCALIZATIONS

Houbaras are generally silent, but may emit a short call during display. Females and chicks remain in contact with high pitched peeping calls. A gruff bark is sometimes given when alarmed.

Great Indian Bustard *Ardeotis nigriceps* (Vigors)

Illus. **38**
Map **122**

Choriotis nigriceps (Vigors) Dillon Ripley 'Revised Synopsis'

Eupodotis edwardsii (Blanford and Oates)

TAXONOMY

Due to close morphological similarities with the Great Arabian Bustard *(Ardeotis arabs)* and the Kori Bustard *(Ardeotis kori)* of Africa, and the even closer similarity in display and appearance with the Australian Bustard *(Ardeotis australis)*, most authorities consider that the Great Indian Bustard *(C. nigriceps)* should belong to the genus *Ardeotis* and that there is no basis for retaining a separate genus *Choriotis* (as in S. Dillon Ripley's 'Synopsis' and Ali and Ripley's 'Handbook Series'). The authors of the *Birds of Africa* (Vol. 2, 1984) in fact consider that *Ardeotis* should be regarded as a super species group.

DESCRIPTION

Body length 122 cms. (males), 92 cms. (females)
Wingspan 250-275 cms.
Height of standing male 100 cms.
Weight upto 12 kgs. averaging 9.5 kgs. (males), 6.5-7 kgs. (females)

Males are much larger than females. They have comparatively longer bills than the genus *Otis* and the bill is flattened dorso-ventrally. The crown is black with a short shaggy nuchal crest. The rest of the face and neck is white, finely striated with black, but appearing white from a distance and the neck feathers are rather loose giving it a thick appearance. The back is dark brown-buff finely vermiculated with black. There is an interrupted collar of black around the lower breast and the belly is white. The tertials are long, extending to the tip of the broad square ended tail. The legs are stout, greenish yellow and there is no hind toe. The wing shoulders are black and spotted with white and in flight most of the flight feathers are dark brown merging to dark grey on the secondaries. These and the primaries are tipped with white. The outer tail feathers also are tipped white.

HABITAT, DISTRIBUTION AND STATUS

It is listed as a species endangered with extinction in the IUCN *Red Data Book* and Dharmakumarsinhji conducted surveys on behalf of WWF and estimated a total population throughout its range in India of no more than 1,260 in 1969 and 745 in 1978 (World Wildlife Fund Project 453 reports, 1970-8). It is a resident endemic species on the subcontinent, adapted to semi-desert and grass-steppe conditions, and undergoes local migratory movements when rainfall creates favourable feeding conditions. In the 1970s small numbers were occasionally sighted in the border desert regions both in northern Cholistan and in the Thar desert. In 1969 a party of 18 birds wintered east of Khipro in the Thar desert (Karim Dad Junejo, pers. comm., 1972). In 1969, east of Bhawalnagar in Cholistan, a group numbering about 20 stayed from March to the end of September in the desert beyond Fort Abbas (W. A. Kermani, pers. comm., 1969). In 1965, 3 were shot a few miles east of Derawar in Cholistan (General J. Marden, pers. comm., 1967). One was also shot in 1960 in August at Gadra in the extreme north-eastern corner of Tharparkar district. Status RARE VAGRANT to Pakistan.

HABITS

They are gregarious in feeding and roosting when not disturbed. They prefer areas of open rolling country or thorn-bush dotted plains, especially where tall grasses

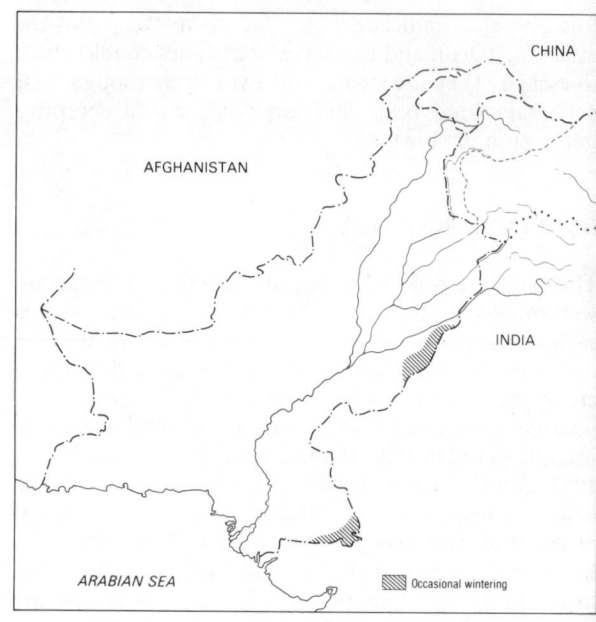

Map 122: Great Indian Bustard *Ardeotis nigriceps*

prevail after monsoon rains. They are wholly cursorial, roosting on the ground and travelling large distances on foot to feed if not disturbed. They consume a variety of vegetable matter, seeds of millets and sorghums, green shoots and tender leaves and small reptiles including snakes and lizards, also insects especially Blister Beetles *(Cantharis* sp.) and locusts *(Schistocerca* sp.). Studies in Karera, Madhya Pradesh revealed that the bulk of their diet in summer was seeds of 'jowar' *(Sorghum sudanense),* large ants *(Camponotus* sp.), and *Chrotogonus* grasshoppers (Asad R. Rahmani, pers. comm.). They feed mainly in the early morning and late afternoon and like all the Otididae are very wary and difficult to approach.

BREEDING BIOLOGY

The males are polygamous and take no part in incubation or care of the young. The female lays but one egg in a mere hollow scraped in the bare ground. The display of the male which takes place incessantly during and immediately after the monsoon season involves a rather static posture with the tail bent right back over the rump and the neck and head upstretched. The air sacs at the base of the neck are inflated so that the feathers below the uptilted bill, spread outwards in a huge white puff ball and the breast feathers hang down between the legs, almost touching the ground in a grotesque greyish white wobbling bag. The wings remain folded during this period and the male does not walk around like the displaying Houbara. Kenneth Eates took several eggs of this bustard in the late 1940s from nests found around Nagar Parkar in extreme south-eastern Sind. His local

Illus. 38: Great Indian Bustard *Ardeotis nigriceps*

collector found a nest on 22 August 1948 near the village of Maṅasiri, located in tall grass and another on 27 September 1948, 128 kilometres (80 miles) east of Diplo near the Indian border. Eates described the eggs as being olive drab in ground colour, sparingly smeared with reddish brown and a few small purple spots scattered here and there (MS notes).

Studies by Asad R. Rahmani indicate a 30 day incubation period. The chicks are precocial and nidifugous and the female immediately takes them from the nest site to areas of better cover. The chick will stay with its mother for up to 6 months though it is capable of feeding independently from about 1 week of age. However the mother has been seen to feed a four-week old chick bill to bill ('Study of the Ecology of the Great Indian Bustard', Annual Report 1981-2, *JBNHS*), and young birds may remain with their mothers until the next breeding season.

*VOCALIZATIONS

Males give a deep booming moan while displaying (Stuart Baker, *Game Birds of India*, Vol. 2). They also have an alarm call when taking to flight, described as between a bark and a bellow and syllabized as 'hook' (Dharmakumarsinhji, *Birds of Saurashtra*, 1954, and Stuart Baker, 'Fauna Series', Vol. 6).

Great Bustard *Otis tarda* Linnaeus   Map **123**

DESCRIPTION

Body length 75-105 cms.
Wingspan 190-260 cms.
Weight 8-16 kgs. (males), 3.5-6 kgs. (female)
    6.35 kgs. (female, Peshawar - Roos-Keppel)

A huge swan sized bird, males being considerably bigger than females. Males have a blue-grey head, and cinnamon chestnut lower neck, back and tail, the wing coverts and mantle being closely barred black and the tail having wider spaced more prominent cross-bars. The outer tail feathers are white as are the secondaries on the wing, and the belly. The under-tail coverts comprise loose and long white feathers. In flight the flight feathers are black tipped with a greyish carpal area and the secondaries and secondary wing coverts make a broad conspicuous white patch. In non-breeding plumage the cinnamon area on the lower neck is reduced in area whilst older males in the breeding season develop conspicuous long white filamentous feathers, slanting like whiskers from the base of the bill and across the lower face. Females are not only much smaller but have noticeably thinner necks and more grey on the wing joints, less white on the secondaries, and more distinct but narrower barring on the tail. In both sexes the legs are dark olive brown, very stout and with only 3 toes. The bill is horny black at the tip and the iris is brown.

HABITAT, DISTRIBUTION AND STATUS

It has never been more than a rare vagrant to the subcontinent with most records from Peshawar area. In Mardan district, NWFP, 7 were seen in 1911, and 2 young females shot in January of that year. Another was shot near Peshawar on 1 December 1917. (Sir George Roos-Keppel *JBNHS*, 1918). Brigadier Christison records one shot near Zhob in Baluchistan on 2 April 1940 and 2 captured near Nushki and kept as pets by the Political Agent in 1937 (Christison, 1941). Ticehurst records one out of a party of four, shot on 1 January 1911 near Jacobabad. There is also a record of one being shot in Chitral by Capt. Lyall at the turn of the century. Undoubtedly the Russian breeding population from which these stragglers entered Pakistan is nowadays greatly reduced due to the extensive disappearance of undisturbed breeding habitat, and the chances of recent visitations are much less. Recent records include a bird shot in the late 1970s on the Jamrud plain which has been preserved as a mounted specimen by a local landowner

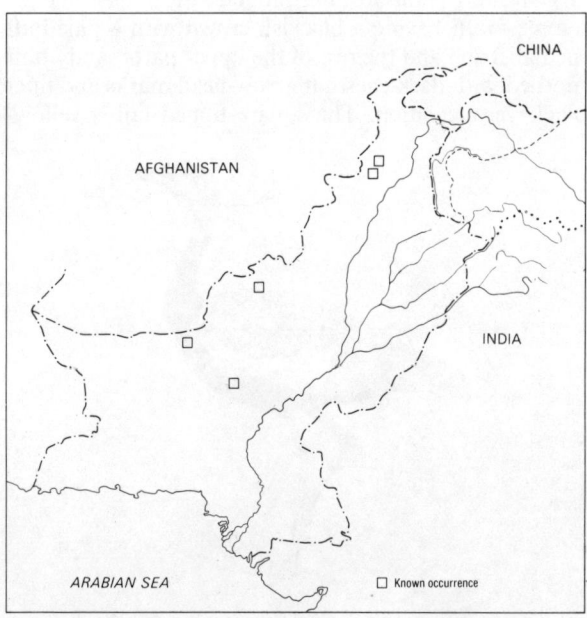

Map 123: Great Bustard *Otis tarda*

(Arshad Khan, Professor of Zoology, Peshawar University, pers. comm.). During surveys to study the Houbara Bustard, Afsar Mian saw 2 in October 1984 near Padak and the following year a single bird in Bisemah, southern Kharan (Afsar Mian, *in litt.* to *JBNHS*). Status VAGRANT.

HABITS AND BREEDING BIOLOGY

They are quite gregarious and despite their bulk take to the wing and fly more readily than *Ardeotis* bustards, often flighting from nightly roosting places to feeding areas. They subsist largely upon green vegetable matter but will consume cereal seeds, Orthopterous insects and field mice. In Poland they have been recorded eating *Microtus* voles, and one in England was observed catching and eating a Long-tailed Field Mouse *(Apodemus)*, but seeds of wild plants and green shoots form the bulk of their diet.

Males seem to display within fairly close range of each other in a 'dispersed lek' system and they are polygamous. The display is a spectacular eversion of the under-wing coverts and tail pressed over the back with wings dropped and much white plumage predominating, at the same time air sacs in the sides of the neck are inflated to a grotesque balloon. The remarkable thing is that this whole bodily transformation takes place in a matter of seconds and the cinnamon brown bird becomes an almost spherical mass, predominantly white when viewed from the side. Females are attracted to this wholly visual display and come to these traditional display sites for copulation. Two to three eggs comprise the usual clutch and the nest is a bare scrape on the ground. Males take no part in incubation or rearing of the young.

Lesser Florican or Likh *Sypheotides indica* (J. F. Miller)

Illus. **39**
Map **124**

DESCRIPTION

Body length 40-48 cms. (males) 44-52 cms. (females)
Wing length 18-20.4 cms. (males),
    20.9-24.7 cms (females) (Stuart Baker)
Tail length 8.2-11.4 cms. (males), 11.4 cms. (females) (Stuart Baker)
Weight 395-560 gms. (males), 510-740 gms. (females)

Only slightly larger than the Little Bustard *(Tetrax tetrax)* it is a much longer legged bird than the latter but averages lighter in weight. Females are larger than males, and both sexes have a noticeably slim neck and rather large head. In non-breeding plumage the male is like the female, both having a blackish crown with a pale buff medial stripe and the rest of the upper parts sandy buff mottled with dark chestnut arrow-head marks and finer black vermiculation. The square tipped tail is yellow-buff with black bars and some mottling. The chin and throat are white turning to creamy buff on the foreneck with black streaks, and the lower belly is whitish. The legs are yellowish horn, the iris pale yellow and the bill is yellow, darker along the culmen. In the summer the male assumes a striking breeding plumage which comprises an all-black head, neck and breast except for a small white throat patch and collar around the base of the neck. He also develops a long crest of black wire-like feathers up to 10 cms. long with spatulate tips which extend horizontally from the nape curving slightly upwards at the tip. In flight the primaries and secondaries are dark brown banded with ochraceous yellow and the entire wing shoulder in both sexes is conspicuously white.

HABITAT, DISTRIBUTION AND STATUS

The Lesser Florican is a specialized inhabitant of open tracts of grassland and is an endemic Oriental species wintering in the Deccan grassland plains and migrating north-westerly in the monsoon to breed in north-west Gujarat state. Due to shrinkage of suitable grassland habitat in Gujarat state and the vulnerability of breeding birds to disturbance, they have become much reduced in numbers. Recent surveys have revealed that there is now little more than 6,000 hectares of suitable habitat left in the Kathiawar peninsula (still regarded as the breeding stronghold of the species). In this survey, conducted September 1981, covering all likely breeding areas, a total of only 21 displaying males were located and they must be considered in critical danger of extinction in Kathiawar (Goriup and Karpowicz, *ICBP Annual Report,* 1981-2, and International Bustard Symposium, Peshawar, October, 1983).

Illus. 39: Lesser Florican or Likh *Sypheotides indica*, male in breeding plumage.

Map 124: Lesser Florican or Likh *Sypheotides indica*

In the early decades of this century part of this monsoon breeding dispersal into Gujarat state (India) also spilled over into southern Sind and the grass plains of southern Lasbela in Baluchistan. By the 1940s they were very scarce and irregular rainy season visitors and K. Eates records the last record known to him, of 3 males shot in September 1937 by Colonel C. B. Rubin around Band Murad Khan near the Hab river. Dr. S. H. Rizvi flushed a male florican in September 1958 in the same area near Band Murad Khan beyond Mangho Pir, and in 1964 in the same place he shot a female, mistaking it for a Stone Curlew *(Burhinus oedicnemus)* (Dr. Rizvi, pers. comm., 1980). In July 1986 several birds were seen in a patch of tall open grassland 32 kms. (20 miles) north-east of Lahore, and near Alloo village in Kasur district. A female from these 1986 visitors was shot, and identified by the author from the mounted specimen, as a Lesser Florican. In July 1987 several more were seen west of this area in a patch of open grassland on the old bed of the Sukh Beas river (Rohil F. Nana, pers. comm., 1987). This rather surprising northward extension of their range may well be the result of fast disappearing suitable nesting habitat, and persistent drought in Gujarat state (Sankaran and Rahmani, 1986). Status VAGRANT.

HABITS AND BREEDING BIOLOGY

Observations of foraging birds indicate that they are more insectivorous in diet than other bustards. They subsist partly upon grass seeds and young green leaves, but mainly upon insects especially *Acrididae* grass-hoppers and *Cantharidae* beetles, and caterpillars (Dharmakumarsinhji, 1972, Sankaran and Rahmani, 1986). When more plentiful they were quite gregarious in habits. Males arrive before females on their breeding grounds in June and July and stay until about early October. They are very conspicuous at this time as they display on a favoured piece of high ground in open tall grassland. The display is aerial, and takes place mainly during early morning and evenings. The male repeatedly jumps into the air to a height of 1 to 1.5 metre and emits a rattling sound at the top of the spring and then flutters back to the ground, exposing the white wing shoulders to the maximum and delaying the return to earth. During this aerial leap the head and neck are bent back over the body and the spatulate head plumes are twisted forward over their heads. The whole jump lasts no more than 2 or 3 seconds and makes the bird highly visible, not only to females in the vicinity but also to unsports-man-like human hunters! (Details from R. S. Dharmakumarsinhji, *Sixty Indian Birds*, 1972, and photographs taken by Paul Goriup, 1982.)

K. Eates found nests in Junagadh (India) and notes that 4 eggs is a normal clutch and found unincubated eggs 28 August and 16 September. The eggs are dark olive in colour, rather oval in shape and not pointed, with reddish brown smears at one end. Males are polygamous and only the female incubates and looks after her young.

VOCALIZATIONS

The rattling sound of the male is audible up to 500 metres, and may possibly be produced by specialized feathers as in a snipe 'drumming', by its highly emarginated primaries beating the air (P. Goriup, *in litt.*, 1987). When foraging they are reported to utter a high pitched whistling 'like wind passing through a crack in a window during a strong gale' (Dharmakumarsinhji, 1972).

ORDER CHARADRIIFORMES

FAMILY JACANIDAE

Pheasant-tailed Jacana *Hydrophasianus chirurgus* (Scopoli)

Illus. **40**
Map **125**

DESCRIPTION

Body length 44-47 cms. (males in breeding plumage) including·
    tail 15 cms. 50-54 cms. (females) including tail.
Body length 31 cms. (both sexes) in winter plumage.
Wing length 18.2-24.2 cms.
Tail length 20-32.5 cms. (in breeding plumage)
Bill length 25-29 mm. (Stuart Baker)

Actually a slightly smaller bird when seen alongside the Bronze-winged Jacana *(Metopidius indicus)*. In the monsoon season both sexes assume similar breeding plumage which is very striking. The forepart of head and neck and the entire wing shoulders being white with the breast and belly dark chocolate brown and long down-curving central tail feathers jet black. The mantle and rump are dull brown and the brownish white tertials are elongated, curving upward when the wing is closed, like the prow of a boat. In adult birds the first primary feather tip becomes spatulate distally with an almost barbless quill or rachis. The rear part of the crown and a dividing line down the side of the neck is black and nape and hind neck is bright straw yellow, somewhat iridescent, the feathers bristling outwards like fur." The iris is dark brown (becoming yellow in winter), and the slender slightly down-curved bill is bluish. This breeding plumage is assumed in early May and lasts up to early October. In winter the long tail feathers are lost and the mantle becomes sandy buff and the whole of the breast becomes white instead of black with a narrow black breast band and a white supercilium with the whole of the crown and hind neck dark brown, whilst the black line bordering the now brownish yellow neck patch becomes broader and extends through the eye. The bill also becomes grey-green. The upswept long tertials,

however persist. Like all the Jacanidae the toes are enormously elongated with very long straight claws, enabling them to walk over floating vegetation and particularly lotus lily leaves *(Nelumbium nelumbo)* without sinking. The hind claw is in fact not as long comparatively as in the heavier Bronze-winged Jacana. The legs and feet are blue-grey to greenish grey. The female is slightly larger than the male and to add to their peculiarities there is a spur on the carpal joint. In flight the wings are wholly white except for the tips of the primaries, the primary wing coverts and outermost secondaries.

HABITAT, DISTRIBUTION AND STATUS

For preference they inhabit permanent swamps and lakes where there is a good growth of submerged vegetation but in winter will resort to more open bodies of water as long as they contain a good submerged growth of *Hydrilla verticillata* and *Vallisneria spiralis* water weeds. In summer they disperse widely into the rice fields. They are resident but subject to local erratic wanderings according to the condition of seasonal inundation wetlands. They are widespread and common in the major rice growing tracts of Sind especially Hyderabad, Larkana and Thatta districts. Also in Shekhupura, Gujranwala and Sialkot in the Punjab. Individuals have occasionally turned up in Gilgit (Scully, *Stray Feathers,* 1881) and Bund Kushdil Khan lake (Meinertzhagen, *Ibis,* 1920) in Baluchistan, but they are mostly summer breeding visitors in the Punjab and resident in Sind. The Rev. Storrs Fox (*JBNHS,* 1937) recorded finding one on 31 May, 1933 on one of the storage tanks in Murree, 2,100 metres (7,000 feet), which was presumably a bird migrating up the Jhelum valley. On 29 May 1986, the author encountered one near Besham (Indus Kohistan) on the roadside, which was probably migrating up the Indus valley to Gilgit. On a typical large lake such as Hadiero in southern Sind or Ghauspur in Jacobabad district, the wintering population may be as high as 150 individuals. Status COMMON.

HABITS

Somewhat gregarious in winter quarters, usually foraging in loose association with conspecifics within some portion of larger wetland areas. They can swim well, and do so readily, but their preferred method of foraging is by slow measured tread across floating water

Illus. 40: Pheasant-tailed Jacana *Hydrophasianus chirurgus,* breeding plumage, both sexes. Note peculiar spatulate tipped 1st. primary and upward curving tertials.

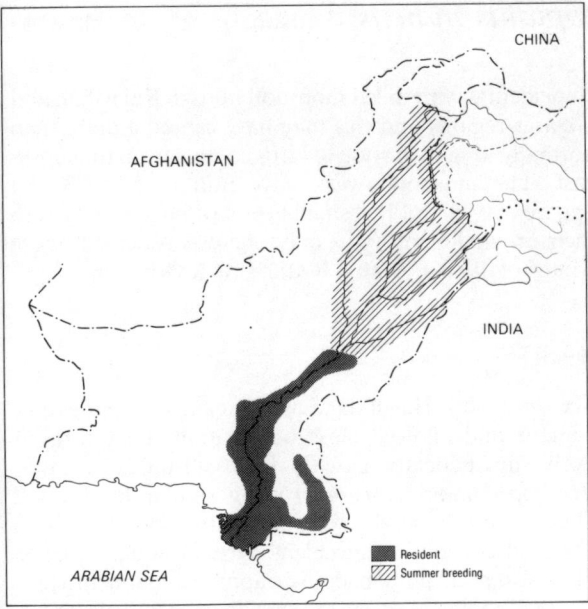

Map 125: Pheasant-tailed Jacana *Hydrophasianus chirurgus*

weeds and lily pads, whence they peck at insect larvae, the spawn of molluscs and amphibia and feed largely upon fresh water snails and small bivalve molluscs. They also eat seeds and succulent parts of water plants. Mason and Lefroy examined the stomach contents of two birds and found these to contain remains of small bivalve molluscs *(Corbicula orientalis)* and parts of water snails *(Vivipura* and *Ampullaria) (The Food of Birds in India,* 1912). They are not particularly shy or fearful of humans and forage actively throughout the day.

BREEDING BIOLOGY

Females are polyandrous, mating with a succession of males. Each breeding male guards and stays with his chicks for 50 or 60 days until his moult into winter plumage is complete and the chicks also are fully fledged, so that they look alike in size and plumage, thus only one brood can be reared by each male per year (pers. comm. J. S. Serrao, BNHS Research Project, 1963). Breeding takes place after the onset of the monsoon when courtship is initiated by a number of flight chases and much calling. The female helps to construct the nest and defend it until she has completed laying and will vigorously drive off potential predators such as House Crows *(Corvus splendens)* (observations author, Haleji lake). The normal clutch is 4 eggs and the nest is always built on floating vegetation, usually not very well hidden, in scattered or emerging reeds. Often it is a very flimsy construction consisting of a few pieces of *Hydrilla* water

weed pulled together in a pad or bent and entwined rushes (such as *Cyperus rotundus* etc.). The eggs are very glossy and olive brown with a purplish bronze sheen, and are markedly pyriform in shape. If the nest is observed too closely, the male, who does all the incubation will drag them under his chin and transport them across the water surface to a new location (Peter Jackson, observed near Delhi, pers. comm.). Copulation takes place on vegetation near the nest site, the female solicits the male by tipping forward at an angle of 45°. He often stands on her back for over one minute even though only attempting copulation once (observed Gharo, Thatta district in late July). The male also jumps over the female's back from side to side as an invitation to mating (J. S. Serrao, pers. comm.). Eggs are laid at 24 hour intervals and the male starts incubating immediately, possibly as a protective measure. The newly hatched chicks are creamy white all over at hatching, but later become more chestnut dorsally and develop blackish spots on the crown and back whilst the throat and belly remain white unspotted. They can feed themselves from the time of hatching but the male attends them solicitously and has been observed giving a broken wing display to distract a human intruder. They have also been observed (Vir, Thatta district in late September) actually carrying their downy young pressed between their flanks and closed wings, the long toes of the chicks dangling visibly below the wings. This may be a characteristic behavioural pattern of all Jacana species [observed and filmed for the African Jacana *(Actophilornis africana)* (Urban *et al.,* Vol. 2, 1986) and described for the Australian Lotus bird *(Irediparra gallinacea)* Frith, ed. 1979], and is a strategy by the male to protect his offspring from predators while he forages for his own needs.

Incubating males, partly fledged chicks and small downy chicks have been observed on the same date (8 July) on Haleji lake. Many Jacanas nest inside paddy fields in Sind and males with downy chicks in attendance have been seen (author) as late as 30 September. K. Eates' detailed description of a chick in down is as follows 'legs grey-brown with edges of scutes on tarsus, yellow. Bill whitish horny. Head yellow-brown more chestnut on back. A black line extends from behind the eyes to nape and occiput is black extending down hind neck. Upper part of wings black bordered by chestnut and back spotted with black'.

*VOCALIZATIONS

In winter quite vocal, giving contact calls when flighting to new areas, the call is a distinctive plaintive 'queeear' sound. In the breeding season the territorial song is a cat-like 'mee - ooph' often accelerating to 'ooph-ooph-ooph' and alternated with quicker shriller 'meyu-meyu-meyu-meyu'.

Bronze-winged Jacana *Metopidius indicus* (Latham)   Illus. **41**

DESCRIPTION

Body length 28-31 cms.
Males slightly larger than females.
Wing length 14.5-19.8 cms. (males), 15.2-18. (females)
Bill length 31-39 mm. (Stuart Baker)
Tail length 3.4-5.2 cms

This Jacana has a much heavier bill than the Pheasant-tailed Jacana, and is a slightly larger bird. It has the enormous long legs and long toes typical of the family and is at once recognizable by its dark blue-black head and neck with broad white line extending from above the eye to around the nape. In bright sunlight the hind neck and mantle reflect a beautiful deep purple light and the neck and flanks are iridescent green. The back and wing coverts glisten bronzy buff with a golden sheen in good light. The under-tail coverts and vent are dark vinaceous chestnut. The legs are dull greenish grey and the long claw on the hind toe is absolutely straight. In flight the wings show no white. The bill is greenish yellow with a red spot at the base and a blue-green frontal shield which also becomes red in the breeding season. This is rather less in area than that of a Moorhen *(Gallinula chloropus)* and can be lifted away from the forehead. Juveniles in their first winter have a dull chestnut crown darkening to a blackish band extending down the back of the neck. There is a greyish white superciliary line in front of the eye and a dark blackish line extending from behind the eye, with the ear coverts white. The throat and breast are a distinctive cinnamon orange and the lower belly and under-tail coverts pure white with some vertical grey barring in the lower flank region. The wings are glossy bronze like those of adults. The legs are olive grey and the bill yellower than that of adults (observation of 5 juveniles, Haleji lake 28 December).

HABITAT, DISTRIBUTION AND STATUS

It is a resident species adapted to live in lakes, swamps and even small village tanks, provided these are well covered with emergent vegetation particularly lily pads such as Lotus *(Nelumbium nelumbo)*. It is often sympatric with the Pheasant-tailed Jacana and is largely sedentary. In the monsoon season and particularly after poor rains it is subject to local migratory movements. It had not been recorded anywhere in Pakistan until the winter of 1980 when there must have been an influx of birds into lower Sind because one was seen on a weed-fringed canal at Ghulamullah in Thatta district in January 1980 (Anne Vollum, pers. comm.) and later on 9 February 1980 two were watched on Haleji lake.

After this, birds were seen in one lotus carpeted corner of the lake continuously up to the summer of 1983 and 5 juveniles were watched in the early winter of 1981, evidence of probable breeding. Their arrival was coincidental with failed monsoon rains in Rajasthan and Gujarat regions and this may have caused a more than normally long migration of birds coming from the southeast. The monsoon was above normal in 1983 and apparently part of this small group of birds returned to their original territory as only one was observed in the winter of 1983-4. Status RARE VAGRANT.

HABITS

The birds on Haleji lake were relatively fearless of humans and allowed close observation. They foraged freely in association with Pheasant-tailed Jacanas *(Hydrophasianus chirurgus)* using similar techniques. They subsist largely on animal matter but have been observed consuming succulent parts of aquatic plants. They also peck at lily-pads in an apparent manner to feed on small molluscs or egg spawn adhering to their under-surface and no doubt eat insects and their larvae. They are adapted to walk on weed-choked water surfaces and like other Jacanas can walk on floating water-lily leaves. They can swim well when necessary and if disturbed will hide by crouching submerged between the lily pads rather than taking to flight.

BREEDING BIOLOGY

They nest during the monsoon from June to September and their breeding biology is thought to be similar to that of *Hydrophasianus chirurgus,* but they generally build a more substantial nest pad of rushes twisted round and round in a circular pad and composed of reeds and water weed. The eggs are also very striking, being the same size and pyriform shape as that of *Hydrophasianus* but even more glossy in texture with a rufous brown to pale olive

Illus. 41: Bronze-winged Jacana *Metopidius indicus*. Note heavier bill and longer hind claw, than in Pheasant-tailed Jacana.

ground colour and the whole surface criss-crossed by squiggly black lines. The normal clutch is four eggs and these are incubated solely by the male, the female being polyandrous and assisting only with initial nest-building and laying of the clutch. Kingsley Kefford (*JBNHS*, 1945) records watching a male carrying downy chicks to safety under one wing. In this instance the nestling was carried some 10 metres, its dangling legs clearly visible below the parent bird's closed wing. This appears to be a behavioural trait of the whole family (See account of Pheasant-tailed Jacana), enabling the male to forage for himself while giving total protection to his offspring against predators.

*VOCALIZATIONS

The call is described as a wheezy, repeated 'seek-seek-' (Fleming), or a 'piping squeal' (E. N. Lowther). Two birds tape-recorded during an aggressive encounter uttered an accelerating rather rapidly repeated and grating 'geek-geek-geek-geek' which rose in pitch as one bird flew at the other.

FAMILY ROSTRATULIDAE

Painted Snipe *Rostratula benghalensis* (Linnaeus)

Illus. **42**
Map **126**

DESCRIPTION

Body length 23-26 cms.
Wingspan 50-55 cms.
Wing length 11.5-13.6 cms.
Bill length 41-47 mm. (Stuart Baker)
Weight 99-140 gms. (males), 125-180 gms. (females)

Very like a Common snipe *(Gallinago gallinago)* in size and shape, but with slightly longer legs and shorter bill which is noticeably down-curved and slightly thickened along the distal third. Males are slightly smaller than females and drabber in colour with an olive brown face and upper breast, with the eye prominently framed by a broad whitish spectacle mark extending behind the eye and a buffy white median stripe down the centre of the crown. The lower breast is framed by a dark blackish circlet below which, is a broader white circlet extending upwards in a pectoral band. The rest of the under parts are white. The wings and back are metallic bronzy grey-green with prominent buff spots in rows along the coverts. The flight feathers are blackish brown with rows of buff spots and the wing is much more rounded than that of true snipes *(Gallinago)*. The short tail is cross-barred buff and greenish grey. Females have the head and breast a rich maroon-chestnut with similar white eye-ring and creamy mesial line on the top of the crown and black band framing the lower breast. Their wings and back lack the pale buff spots conspicuous in the male. The edge of the mantle in both sexes is bordered with two pale creamy lines lying along the top of the wing coverts and in flight these show as two longitudinal back stripes. In flight legs dangle like a rail and are coloured olive green, sometimes more yellowish olive in females. The feet are not webbed and the iris is dark brown.

HABITAT, DISTRIBUTION AND STATUS

Adapted to swampy borrow pits and reed-fringed ponds or stagnant drainage channels choked with reeds, they are resident and widespread throughout the Indus basin, but curiously local in occurrence and subject to erratic wanderings when marshes dry out. They tend to occur in small colonies and can be very sedentary where swampy areas are permanent. They have been recorded recently around the seepage outlet of Rawal lake, in certain small

Illus. 42: Greater Painted-snipe *Rostratula benghalensis.* Female on left and male on right.

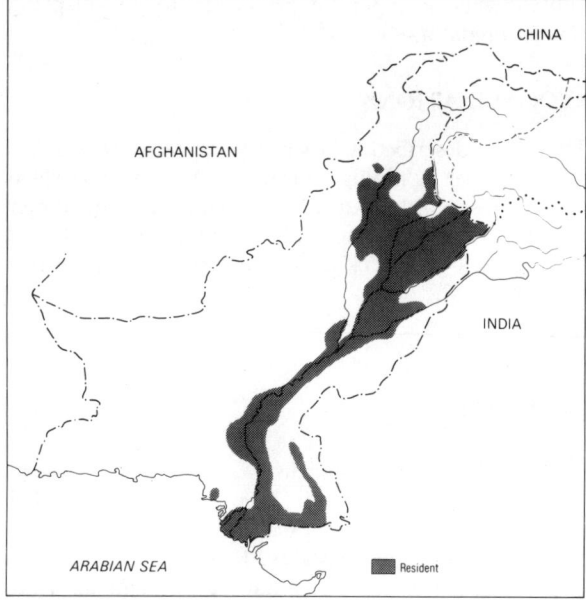

Map 126: Greater Painted-snipe *Rostratula benghalensis*

scything their bills to and fro (Ali and Ripley 'Handbook').

They have a remarkable display used in threat and courtship accurately described by Hugh Whistler. The tail is raised and fanned and the wings spread wide and bent forward to beyond the top of the beak. The bright patterned wings and back are thus shown to maximum effect and this startling visual display is accompanied by a growling and hissing call ('Popular Handbook', 1949).

BREEDING BIOLOGY

Females are polyandrous, the male doing all the incubating and caring for the young. K. Eates found nests in Sind from May to September and the clutch was commonly only 2 eggs (incubating male flushed off 2 eggs in 3 separate nests). However Whistler (1949) gives the normal clutch as four eggs. The nest is well concealed in a clump of *Saccharum* grass or *Phragmites* reeds and comprises a pad of moist reeds and sedges slightly hollowed in the middle. The eggs somewhat pointed at one end, are not very glossy and are yellowish in ground colour scrawled and blotched with blackish red or very dark brown. The chicks are brooded on the nest for the first few days and fed by the male also, bill to bill but they are active and can swim and run around within hours of hatching. The downy chicks are creamy white with chestnut patches on crown and back framed by narrow longitudinal black lines as well as a parallel black line running through the eye and down the side of the neck and over the sides of the back.

The advertising call of the female which can be heard throughout the night during the monsoon, consists of a soft hollow sounding 'hoop-houp-houp' which is far carrying and rather ventriloquial in direction and this probably attracts males.

A male with 4 partly fledged chicks was seen in flooded grassland near Hyderabad by Derek Holmes on 5 June (pers. comm.) and Eates found another male with downy young near Sukkur on a small island in a borrow pit in late August (MS notes). N. Van Zalinge found a male with 4 downy chicks on 22 August in the seepage below Hab dam *(in litt.,* August, 1985). Studies in Tamil Nadu, south India indicate that in favoured habitat nesting can be almost colonial with 10 nests located within an area of 3,300 square metres and 13 metres being the closest distance between nests. Courtship was noted to involve repeated and prolonged chasing, culminating in the female entering relatively open dry ground where she squatted and invited copulation (W. D. Wesley, paper given at BNHS Centenary Seminar, Bombay, December 1983).

*VOCALIZATIONS

The advertising female makes two types of calls. First a series of inflected resonant 'kerr whooup-kerr whooup-

swamps in the vale of Peshawar, at Kallar Kahar lake in the Salt Range, and in the seepage zones around most major irrigation barrages. They are also quite widespread in the major rice growing tracts. In recent years they have been observed at Manchar lake, Khoski in Badin district, around Mirpur Sakro in the rice fields, in the seepage zone below the Hab river dam and even on the tidal creeks at Ghizri in Karachi. They have also been seen feeding in very saline brackish pools near Kandhkot in Jacobabad district. F. Koning saw 10 in January within an area of less than 3 hectares in a seepage zone on the right bank of the Indus below Chashma barrage (Mianwali district), and the author saw 5 together in a small reed bed at Balloki headworks on the Ravi in early April. They are largely absent from the hilly tracts and northern valleys though breeding occasionally in the main vale of Kashmir. Status FREQUENT

HABITS

Largely crepuscular or nocturnal in feeding, and except in winter time when ponds dry out, very seldom seen as they remain hidden in vegetation. A bird watched (near Ghauspur, Jacobabad district), feeding out in the open around drying out tamarisks, probed with its bill in the mud like a snipe and continuously wagged its tail up and down like a Common Sandpiper *Actitis hypoleucos*. Another at Manchar on the banks of the feeder channel swam across and disappeared into the reeds on the other bank. They feed on a variety of aquatic insects and their larvae, Mollusca and Annelida, as well as weeds of aquatic plants. They sometimes feed in shallow water by kerr whoup' with emphasis on the last syllable, immediately followed by a higher pitched shorter duration 'khoonk-khoonk-khoonk', but human consonants are an impossible medium, and this may well be the same call as 'cook-cook' by Ali and Ripley (1969). Both types of calls can be monotonously repeated in alternating series but the higher pitched 'khoonk' is usually repeated in a longer series up to 30 or 40 times. Calling is mostly just at dusk and for the first hour or two after darkness falls and again before dawn. These calls can be heard any time of the year, but are most frequently uttered in the summer and early autumn. A male with chicks has been heard giving the hissing threat call.

FAMILY HAEMATOPODIDAE

Oystercatchers

Old-world or Eurasian Oystercatcher *Haematopus ostralegus* Linnaeus Illus. **43**
Map **127**

DESCRIPTION

Body length 40-45 cms.
Wingspan 80-86 cms.
Wing length 24-26.1 cms. (Stuart Baker)
Tail length 9.9-11.4 cms. (Stuart Baker)
Bill 7-9 mm. long.

A large stoutly built shore bird (heavier bodied than the Bar-tailed Godwit *(Limosa lapponica)* with a thick neck, high forecrown and long straight orange-red bill which is laterally compressed and paler yellower at the tip. It is black above with white under parts and a black head, neck and upper breast. In winter plumage there is a wide curving area of white extending around the lower cheeks from the throat. This is not confined to juvenile birds (see Ali and Ripley 'Handbook', Vol. 2, p. 203). In flight a white bar through the wings shows prominently and a broad black terminal bar to the white tail and rump. The legs and feet are coral pink with unwebbed toes. The wing coverts are often bleached quite brownish in the autumn. The sexes are alike.

HABITAT, DISTRIBUTION AND STATUS

They are winter migrant visitors to the seacoast throughout Makran and Sind, avoiding narrow inland tidal creeks or fresh water swamps and preferring wide flat mud banks and sand banks where a plentiful supply of bivalve molluscs and marine worms provide food. The only records of migration are of birds sighted by Col. Meinertzhagen at Kushdil Khan lake in Baluchistan on 29 April and 17 May *(Ibis,* 1920). These were possibly Seistan breeding birds (see Paludin, 1959). There is also one solitary record for Wular lake in Kashmir (Ward, *JBNHS,* p. 946). A small part of the wintering population also over-summers along the Karachi coast but these are non-breeding birds. Seen on 20 June in Sonmiani lagoon *circa* 20 scattered birds, also about 15 on a sand bank in scattered twos, Indus Delta 29 July. Status ABUNDANT only on the seacoast.

HABITS

Quite gregarious in wintering quarters, flocks can be seen roosting or resting on sand bars in mixed company with gulls and other waders. They call frequently during flight to new feeding areas, having pointed wings and a strong rapid wing beat. They are shy and wary around Karachi seacoast and do not allow close approach. Their food appears to consist mainly of inter-tidal marine molluscs and their wedge tipped bills are adapted for opening hard-shelled bivalves as well as gastropod molluscs. They also feed on marine worms being quick to see them extruding their casts and running towards the hole to stab. On rocky headlands, as at Cape Monze, observed pecking off Limpets (Patellidae), and also digging out small crabs *Graspus strigosus*. In Ghizri creek near Karachi on the more extensive mud flats, there is usually a wintering population of between 200 and 500 Oystercatchers in scattered groups and all these leave by mid-May. In April they all have bright black nuptial plumage and wholly black throats (observation 22 April over 200 birds). They return from end September, numbers building up during October and at this time all have rather brownish bleached wing coverts and mantle, and white throats. Juveniles with duller bills and grey

Illus. 43: Eurasian Oystercatcher *Haematopus ostralegus,* showing summer plumage without white collar on lower neck.

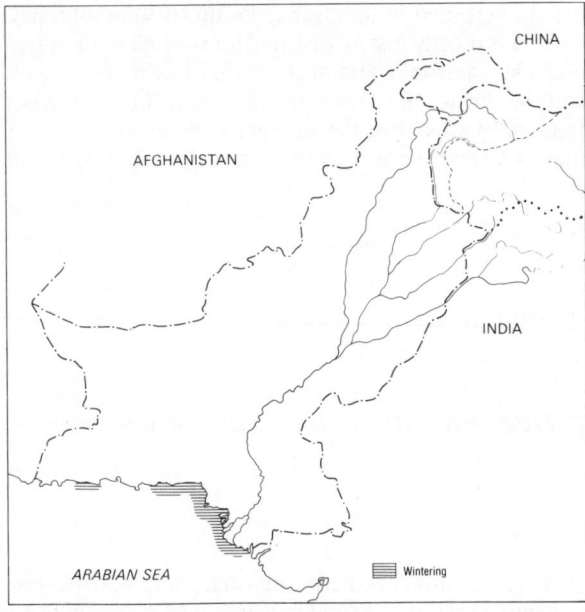

Map 127: Eurasian Oystercatcher *Haematopus ostralegus*

legs can occasionally be discerned in October and November. Surprisingly no recent records have been obtained of any birds on passage through the interior of Pakistan.

BREEDING BIOLOGY

Extra-limital. Very well studied in Europe.

*VOCALIZATIONS

There is a discernible dialectical difference in the calls of Oystercatchers around the Karachi and Makran coast, compared with similar tape recordings made of Oystercatchers in North Wales. The well known shrill piping call is most commonly mono-syllabic in Pakistan and shriller than the flight and contact calls recorded from North Wales which are commonly di-syllabic and less frequently mono-syllabic. They can be rendered as 'teeep'. They also utter a rapid 'peep-peep-peep' when alarmed and taking to flight. The full piping display calls so commonly heard around Anglesey have never been heard in Pakistan even amongst birds just prior to spring migration.

FAMILY IBIDORHYNCHIDAE

Ibisbill *Ibidorhyncha struthersii* Vigors

Illus. **44**

DESCRIPTION

Body length 41 cms.
Wing length 65-80 cms.
Bill length 6-7 cms.

A very distinctive plover-like bird in flight, but more suggestive of a curlew on the ground with long slender down-curved bill, black facial mask and earthy grey back, wing coverts and neck. The crown and forepart of the face is black, margined posteriorly with white, turning to blue-grey on the nape and hind neck. The belly is white with a broad circular breast-band of black framing the grey neck and upper breast. The grey tail has faint darker ashy cross-barring. The bill and legs and feet are dull crimson and the iris is crimson.

HABITAT, DISTRIBUTION AND STATUS

This strange bird is a specialized feeder upon aquatic insects and their larvae, and is confined to mountain streams or rivers having rather wide shingle banks, typically flowing in fairly level stretches where the water is not too turbulent and the river forms several channels with intervening island shingle banks.

It is an endemic species confined to the high plateau mountain regions of central Asia, extending from Kazakhstan in the west, south to Kashmir and eastwards along the inner Himalayan ranges to Nepal and Sikkim (Flint *et al.*, 1984, and Fleming and Fleming, 1976). It prefers high elevations for its breeding, often being found between 3,300 and 4,300 metres (10,000-14,000 feet).

It does not seem to have penetrated much, into the colder dryer inner mountain ranges of the north-west, not having been recorded from Chitral or Hunza and there are only three records from our area. An immature female bird was collected by Biddulph on 6 August 1880 near Gilgit (Biddulph, *Stray Feathers,* Vol. 10, p. 275). There is some altitudinal migration in winter, especially to rivers which do not freeze over and H. Waite collected

Illus. 44: Ibisbill *Ibidorhyncha struthersii*

an adult female in the Sohan valley near Kahuta (just to the east of Islamabad), at about 600 metres (2,000 feet) on 15 January 1931 (H. Waite, pers. comm., 1966 and specimen in British Museum). During a January 1986 visit to Baltistan to photograph Snow Leopards, R. Roberts (no relation of author) came across a party of 5 birds feeding in the vicinity of hot springs in the mountain stream near Chu-tran village up the Shigar valley. They were no doubt attracted by the unfrozen water and small crustacea observed swimming in that locality, as the ground was snow carpeted and the other stretches of the stream frozen over at that season (Roberts, *in litt.*, 1986). Status VAGRANT.

HABITS AND BREEDING

Their long curved bills are well adapted for probing around and under submerged stones at the edge of streams and they are specialized feeders on aquatic insects. In the early morning or when water temperatures are low, such insects remain hidden under stones or in the silt mud, and the Ibisbill feeds by probing. When the water warms up, and there is typically a very considerable diurnal temperature variation in these high extremely dry altitudes, the Ibisbill wades into the river and feeds by picking up floating insects, typically standing in water up to its belly and remaining for long periods at favoured feeding stations. In a study of their feeding habits in Langtang Kola, Nepal, by Pierce, their diet was studied by analysis of faeces and regurgitated pellets and during 7 days in mid April, their diet comprised approximately 40 per cent of Stone-fly larvae, 34 per cent Mayfly larvae, 7.89 per cent Caddis Fly larvae, 9.2 per cent Caddis Fly adults and 2.63 per cent fish. It was found that when the water temperature rose above 5°C., sufficient invertebrates drifted in the water and the Ibisbills picked them up off the water after wading into the stream (R. J. Pierce, *Ibis*, Vol. 128, January 1986).

Breeding starts from late April and the nest site is a mere bowl, hollowed in the sand and lined with several hundred small pebbles tossed in by the birds themselves (Pierce, 1986, op. cit.). Four pyriform eggs are laid normally and both sexes incubate. Two types of calls have been recorded, short (up to 40 times) piping calls like an Oyster Catcher *(Haematopus ostralegus)* and more drawn out 'kleep' or 'klew-klew' more reminiscent of a Green Shank *(Tringa nebularia)* (Pierce, op. cit., and Osmaston, *Ibis*, Vol. 1, 1925).

FAMILY RECURVIROSTRIDAE

Black-winged Stilt *Himantopus himantopus* (Linnaeus)

Illus. **45**
Map **128**

DESCRIPTION

Body length 35-40 cms.
Wingspan 67-83 cms.
Wing length 24-25 cms. (males), 22.7-23.6 cms. (females)
Bill length 54-68 mm. (females), 60-69 mm. (males), (Stuart Baker
Leg length 25 cms. with exceptionally long tarsi 11.5-14.5 cms.

The extremely long legs especially the tibial portion, make this bird very distinctive. They have rather large rounded heads, long necks and all-white bodies with wholly black wings. The bill is long, thin, straight and black. The legs are coral red and the iris is bright red. In the breeding season males generally have dusky black markings extending from behind the eye over the nape and part way down the hind neck, but males with all-white heads have also been recorded (Goriup, *British Birds*, 1982, p. 12). Females may be wholly white on head and neck or dusky grey with as much variation as is exhibited in the males. In non-breeding plumage both sexes show more extensive grey on nape and hind neck. Juvenile birds have brownish black backs, with more extensive ashy grey areas on crown, ear coverts and hind neck. Juveniles also have plumbeous or pinkish grey legs. In flight the all-white tail and rump shows up against the sharp pointed black wings and the crimson legs extend from the tibio-tarsal joint onwards, beyond the tail tip.

HABITAT, DISTRIBUTION AND STATUS

A resident bird with some local migration in winter to more southerly latitudes and also dispersal according to availability of suitable feeding areas. A bird ringed at Bharatpur Rajasthan, in 1967 was shot at Balkh in Afghanistan in 1970 (*JBNHS*, Vol. 68, April 1971). They are quite highly adapted to alkaline or highly brackish pools, favouring very shallow water on the margins of drying out lakes and ponds, often quite small areas such as man-made borrow pits and tanks. They do not utilize marine or sea-shore areas nor fresh water rivers but prefer rather open exposed areas of marsh and lakeshore. Resident and breeding throughout the Indus watershed, and undoubtedly one of Pakistan's most widespread Charadriformes. Up to 1,600 congregate around Ghauspur jheel (Jacobabad) in winter, 1,100 on 16 February 1972 and over 1,600 on 15 February 1971 (F. Koning Surveys), 200 on Kallar Kahar in the Salt Range and over 200 on the small pond by Mauripur Air

Illus. 45: Black-winged Stilt *Himantopus himantopus*, both sexes in typical non-breeding plumage showing extensive grey on hind crown and nape.

Force Base, Karachi. Occurs on Band Kushdil Khan in Baluchistan (40+ on 31 October, 1971), as well as Siranda jheel in Lasbela (Baluchistan). Water logging in many parts of Sind and Punjab, and increased salinity due to canal irrigation has undoubtedly favoured the spread and increase of this species. Status ABUNDANT

HABITS

They wade in shallow water and probe for food with bills fully submerged often up to their eyes, and occasionally wade in deeper water up to their breasts. They can swim but do so very rarely. Their food comprises mainly animal matter including probably minute crustacea, larvae and nymphs or aquatic insects, copepods, Annelida and to a lesser extent seeds of rushes and sedges.

This is a gregarious species throughout the year and loosely colonial in nesting also. They are noisy during evening time particularly as the warmer spring weather approaches and they can be heard calling all night. They are not particularly shy or nervous, becoming in certain spots accustomed to feeding alongside heavily traffic-laden roads and on the outskirts of towns when they learn that the pedestrian traffic is not a threat to them. If nervous in the proximity of humans they will bob their heads and whole body up and down. They roost head tucked under wing and standing on one leg in shallow water (typically), and also occasionally on dry land sitting on the whole tarsus.

BREEDING BIOLOGY

Loosely colonial in nesting and thus can be highly dependent upon local water conditions. They normally nest from May to early July, with peak breeding in June. Important elements of courtship seem to involve group displays to synchronize breeding, and much aerial chasing and calling plus unison tandem running of pairs, not adequately described in Vol. 3, *Birds of the Western*

Palearctic. Pairs have been seen in close parallel, running with rapid steps through shallow water, head and neck lowered and bill extended. This is often followed by birds facing each other with upright threat or display posture and with both sexes repeatedly calling and often followed by female soliciting copulation (observed 29 January near Khinjar lake). Courtship also observed 13 November with attempted copulation, the female was noted to crouch forward with head and neck lowered (soliciting). The male ran quickly on either side of her, and probed nervously in the mud. He then flew up onto her back sitting on his whole bent tarsi. During group displays all members, comprising from 6 or 7 up to 15 birds, call excitedly while bowing inwards or running past each other with rather upright posture. During this excited calling, individuals frequently fly short distances over others in the group (observed in March, Ghizri creek, and Puran near Khoski, Badin district). Copulation is apparently only successful if carried out with the female standing in shallow water (Goriup, *British Birds*, 1982). There is an excellent account of breeding behaviour and courtship of isolated pairs breeding in Portugal by Paul Goriup *(British Birds,* 1982, pp. 12-24). A colony comprising about 50 pairs was found near Shorkot road (Jhang district) in May 1964 in drying out mud and another on Siranda jheel in Lasbela comprising 15 nests on 24 May. (Roberts, *JBNHS,* 1970). K. Eates also found nesting colonies on Siranda jheel on 19 June, and H. Waite found a colony of 35 nests on Kallar Kahar in the Salt Range in late June. Meinertzhagen found them breeding on Kushdil Khan lake with 3 eggs per nest on 10 June. The full clutch is usually only 3 eggs but 4 are also sometimes laid. Nests are usually about 50 metres apart and may be in shallow

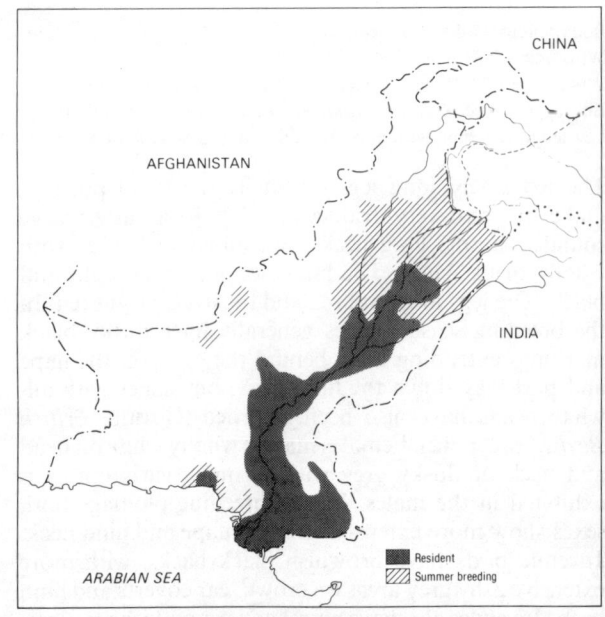

Map 128: Black-winged Stilt *Himantopus himantopus*

water or on a grass clump in shallow water. Often they comprise a bit of mud and weed scum scraped together to form a shallow saucer. Where the ground is quite dry such as on an island in Siranda jheel, nests were made of a pad of dried weeds *(Haloxylon* and *Salsola)*. The eggs are pyriform in shape, of pale buff ground colour rather widely blotched with coalescing black blotches, not unlike the eggs of a Red Wattled Plover *(Hoplopterus indicus)*. Both sexes share incubation duties and tending the young. Dharmakumarsinhji reported regular egg wetting by both sexes during incubation and injury feigning when a human approached the nest with eggs (op. cit., 1972). The nidifugous chicks are able to feed themselves from the time of hatching. Incubation takes 22 to 25 days (Cramp *et al.,* 1983).

*VOCALIZATIONS

Being gregarious throughout the year they have a complicated variety of calls all rather similar in plaintive coot-like *(Fulica atra)* timbre, but varying in length and time. Flocks in spring indulge in choruses of short quickly repeated 'kwit-kwit-kwit' calls. The alarm call or warning cry is a more staccato 'kek-kek-kek-kek'. A more drawn -ut 'kurwait-kurrwait' perhaps denotes more of a contact and less of an alarm call as it is often uttered by individual birds flighting from one part of their feeding grounds to another. In the spring and early summer they are often very vocal throughout the hours of darkness.

Pied Avocet *Recurvirostra avosetta* Linnaeus

Illus. **46**
Map **129**

DESCRIPTION

Body length 42-45 cms.
Wingspan 77-80 cms.
Wing length 22-23.5 cms.
Bill length 84-91 mm. (Stuart Baker)

About the same body size as the Black-winged Stilt *(Himantopus himantopus),* with an equally long slender black bill, but this is prominently curved upwards along its distal half and the legs are pale blue, or bluish grey, being much shorter than those of the Stilt, though equivalent in length to similar large waders such as

Limosa sp. The body plumage is predominantly snow white with the crown and hind neck black and narrow black bands over the back formed by the scapulars and the outer wing coverts. The wingtip is also black, whilst the tail, rump and secondaries as well as wing shoulder are all pure white. Its flight pattern is particularly striking both from underneath, with all-white neck, body and wings except for black tips and dorsally with an almost butterfly like pattern of two broad black bands down either side of the white back, framed by white wings with curving black wingbars and solid black tips. The sexes are alike. The iris is brown.

HABITAT, DISTRIBUTION AND STATUS

Mainly seasonal visitor to Pakistan occurring in small numbers throughout the year, often with a noticeable increase from spring to early summer. An occasional breeder on Kushdil Khan lake in western central Baluchistan and on Siranda jheel in Lasbela. In the Punjab it is a rather uncommon and sporadic visitor. Occasional birds seen as far north as Rawal lake (Islamabad), e.g. 3 on 22 May 1981 and 8 on Kallar Kahar in the Salt Range on 1 May, 25 seen on 1 September Ghazi Ghat (Muzaffargarh district) and over 200 seen Kharrar jheel, Renala Khurd in October (Chris Finney, pers. comm., 1968). In Sind and Baluchistan its stronghold is in areas along the border of the Rann of Kutch and brackish pools close to the seacoast and all along the Makran coast. 170 were counted on 22 March Khinjar lake (Thatta district) (J. Vieillard, pers. comm.), 14 in brackish pools Kandhkot (Jacobabad district) 14 November and 800 on Sandho lake 10 January 1971 on the borders of the Rann of Kutch (F. Koning, pers. comm.). They appear rather specialized feeders preferring alkaline and shallow or drying out pools, hence their erratic but widespread occurrences. They are

Map 129: Pied Avocet *Recurvirostra avosetta*

quite well adapted also to exploit tidal creeks and estuarine mud flats. Ghizri creek usually has an over-wintering population which totalled over 200 birds on 22 April 1983 and 22 January 1982. On 18 May a flock of 75 frequented a salty pool in the sand dunes between Ghizri and the open sea, but by the end of May or early June all these birds had disappeared and it is thought that they mainly breed in the Rann of Kutch across the border in India. However, smaller groups are occasionally sighted in any month. 50 were seen on a pool near Clifton beach on 21 July 1974 and 3 in seepage pools adjacent to Haleji lake on 10 July 1976. On 25 May, 17 Avocets were seen on Kushdil Khan lake, Baluchistan (J. Anderson, pers. comm., 1972) and on 22 May 1974 the author observed a small group of 6 Avocets by that lake, displaying. Status ERRATIC but FREQUENT year round visitor, only sporadically breeding.

Illus. 46: Pied Avocet *Recurvirostra avosetta*, both sexes.

HABITS

They are colonial in both breeding and in foraging like *Himantopus*, and characteristically feed by wading in shallow water and scything their bills to and fro. They will also feed while swimming. 40 birds were seen on Mauripur pool outside Karachi on 2 April feeding as they swam. They will also pick food items from the surface of water or in oozing wet mud. They are often seen in association with Greater Flamingoes *(Phoenicopterus ruber)* and Black Winged Stilts *(Himantopus himantopus)*, like those two species being rather dependent on warm shallow and saline waters for their feeding. They subsist mainly upon minute crustacea, copepods, water-fleas and larvae and pupae of Diptera and also marine Annelida and to a lesser extent tiny Mollusca. They are, much less vocal outside of the breeding season than *Himantopus*.

BREEDING BIOLOGY

Well studied in Europe. Communal courtship is probably an important component in synchronizing breeding activity particularly for a species which breeds in rather transient biotope of shallow water with a very high evaporation rate. Dr. Salim Ali discovered the first breeding in India when he located a colony of over 500 pairs in April 1945 in the Great Rann of Kutch *(JBNHS,* 1945, p. 420). Kenneth Eates found 3 pairs nesting on 10 June 1940 in Siranda jheel, Lasbela. One nest had 4 incubated eggs and the others with 2 and 3 eggs respectively. They were located on a small island in the lake alongside nests of *Charadrius alexandrinus* and *H. himantopus* (MS notes unpublished). When terns and gulls nested on Siranda lake in 1979 no Avocets were noted by the author. Meinertzhagen found one pair

nesting on Kushdil Khan lake in 1913 but none in 1914 and gives no dates or any other details. Two full clutches of 4 eggs found on 10 and 17 June on this lake (Ali and Ripley, 'Handbook', Vol. 2, p. 334) actually refer to Dr. Ticehurst's records of Black Winged Stilts' nests *(JBNHS,* 1927*b).*

Communal displaying observed by the author in late May near Ghizri creek, involved a lot of excited calling and pairs running in unison closely side by side with neck lowered and bill held horizontally. This is followed by the pair facing each other and lowering their bills to the ground surface and calling rapidly 'kloo-kloo-kloo'. No copulation was observed (18 May early morning 1979). The interesting feature of this display is that the flock totalling just over 70 birds had been in the vicinity since April but they left the area about ten days later, presumably to start breeding in more suitable habitat.

Usually the full clutch of 4 eggs is laid on the bare mud or clay, with very little nesting material collected. Both sexes incubate and wet the eggs with their body plumage regularly during incubation, which is reported to last 23 to 25 days. The eggs are very similar in appearance to those of *H. himantopus*, pointed at one end, of varying shades of pale olive brown in ground colour with rather widely spaced bold black splotches and smears. They are larger in size however, than Black Winged Stilt eggs. Salim Ali recorded numbers of parent birds in the Rann of Kutch colony giving distraction-cum-threat displays, with wings held out sometimes close over their backs and sometimes spread on the ground. Nests at this stage contained incubated eggs.

*VOCALIZATIONS

In flight and on the ground a clear liquid whistle 'klewit-klewit', alarm call a more strident rapidly repeated 'kleet-kleet-kleet'. Usually only vocal in the breeding season.

FAMILY DROMADIDAE

Crab Plover *Dromas ardeola* Paykull

Illus. **47**
Map **130**

DESCRIPTION

Body length 33-35 cms.
Wingspan 75-78 cms.
Wing length 20.9-22.5 cms. (males),
     20.1-21 cms, (females), (Stuart Baker)

It is about the same build and size as an Avocet with equally long legs, with which it could be confused at a distance. The most noticeable field character is the rather large head and massive looking black bill with a pronounced gonys or arch on the lower mandible, reminiscent in shape of the bill of *Esacus magnirostris*. They are wholly white on the head and neck unlike the Avocet *(Recurvirostra)*. The body is white with black tipped wings and a black saddle patch on the back. The black wingtip extends to the outermost secondaries and over the primary wing coverts also. The legs are blue-grey or bluish white and the toes are partially webbed and the iris is dark brown. There is a minute black patch in front of and behind the eye, which is not shown in most illustrations (Cramp *et. al.,* Vol. 3, 1983). The sexes are alike. Juveniles are speckled grey-brown on the crown and hind neck and mantle. When seen alongside Oystercatchers *(Haemantopus)* they are much longer legged and smaller bodied. Downy chicks are grey on the crown and back with a whitish nuchal band and their under parts are dirty white (Meinertzhagen, 1954).

HABITAT, DISTRIBUTION AND STATUS

A shore bird confined to feeding in the inter-tidal zone in sub-tropical latitudes and occurring very sporadically along the west coast of the subcontinent including Cape Comorin (a specimen collected 1980 by the Bombay Natural History Society bird ringing team). In Pakistan it occurs very sparsely along the Makran coast and occasionally in the Indus Delta, but it seems likely that numbers pass through on migration to wintering grounds in the Rann of Kutch. Studies by the 1984 Oxford University expedition to the Gulf of Kutch Marine National Park revealed a wintering population of between 2,500 and 5,000 Crab Plovers (Palmes and Briggs, *Forktail*, Vol. 1, 1986). During a voyage through the Indus delta mangrove creeks, a flock of 12 was seen on 15 November, 1967 on a broad mud flat and two days later this had increased to 30 birds. Several extended winter time boat trips in this region in subsequent years have failed to locate any more. On Sonmiani lagoon in Lasbela, single individuals have been sighted 9 May 1976 and 12 May 1974. Also for a few days in December 1971, a flock of 17 haunted the pool at Mauripur. On 11 September 1984 a party of just over 70 were seen on the

mouth of the Hab river by N. Van Zalinge (pers. comm., 1984), about 25 per cent of these birds were in juvenile plumage. On revisiting the place with companions on 14 September, only 1 Crab Plover remained. H. H. Dharmakumarsinhji discovered regular wintering flocks of 50 to 70 birds on the rocky coastline of Jamnagar in the Gulf of Kutch, Gujarat state (pers. comm., 1980), though at the time of publishing his book *(Birds of Saurashtra,* 1954) he considered them very rare in Kutch with only 1 or 2 sightings of single birds. Status RARE.

HABITS

Individuals at Sonmiani foraged like Grey Plovers, *(Pluvialis squatarola),* with whom they associated, running suddenly forward then stopping abruptly and tilting the whole body forward to pick up food. They feed largely on small crabs, but will tackle quite large crabs, shaking the victim's whole body until the claws and legs are dismembered (Palmes and Briggs, 1986, op. cit.). In Bhavnagar they have been observed catching Mudskippers *(Periopthalmus)* (Dharmakumarsinhji, pers. comm., 1981). In Sonmiani and the Indus delta they fed on very sticky mud banks and were in association with Bar-tailed Godwits *(Limosa lapponica)* and Grey Plovers *(Pluvialis squatarola).* They can be quite nocturnal in feeding activity according to the state of the tide as their food prey of crabs is more accessible

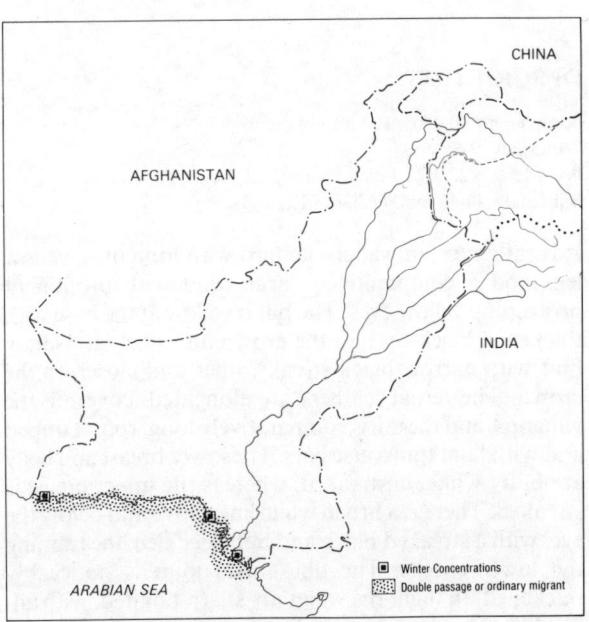

Map 130: Crab Plover *Dromas ardeola*

Illus. 47: Crab Plover *Dromas ardeola*, adult winter plumage showing grey mottling on nape.

during the brief period of maximum low tide. Where plentiful they are highly gregarious at traditional roost sites (up to 1,200 birds on Bhaidar island in the Gulf of Kutch) but during feeding gradually disperse to about 50 metres between individuals (Palmes and Briggs, 1986, op. cit.).

BREEDING BIOLOGY

In mid-September 1981 Paul Goriup encountered nearly 80 Crab Plovers on the shore-line and along the seacoast of the Gulf of Kutch in the former Jamnagar state in Gujarat state, India. There were a number of grey-backed juveniles and it was observed that these were still being fed by their parents. These young birds were able to fly well and there has never been any record of breeding on the subcontinent but the possibility should definitely be investigated. They nest in burrows which the birds excavate themselves in sand banks or under rocks or slabs of coral. The nest tunnel is up to 180 cms. long and rises at the end to widen into a nest chamber. Usually only a single egg is laid which is dull white, unspotted and relatively enormous for the size of the bird. There is no information on the period of incubation but both sexes have been observed bringing food to the nest burrow and the nidicolous chicks appear to be semi-altricial being unable to stand after hatching and remain in the nest burrow requiring to be fed for sometime. Their eggs are quite unlike any other Charadriforme eggs as indeed are most of their nesting habits.

VOCALIZATIONS

Babbling calls, a rapid 'techuk-techuk-chuk-chuk-chuk' uttered mainly at night (M. D. Gallaher, *Birds of Oman*). Also on mud flats in breeding season a sharp 3-noted whistling call 'kew-ki-ki' or 'ki-tewk' (Cramp *et al.*, Vol. 3, 1983).

FAMILY BURHINIDAE

Stone Curlew *Burhinus oedicnemus* (Linnaeus)   Map **131**

DESCRIPTION

Body length 40-44 cms. with bill 4 cms.
Wingspan 77-85 cms.
Wing length 22.3-22.7 cms. (males), 21.7 cms. (1 female)
Bill length 44-46 mm. (Ali and Ripley)

It is rather a slim wader-like bird with long olive yellow legs and a comparatively large head with prominent protruding yellow eyes. The bill is yellow at the base with thickened black tip and the crown and back are sandy buff with narrow black streaks, finer and closer on the crown. The tertial feathers are elongated, covering the wingtips, and the tail is comparatively long, round tipped and with faint transverse bars. The lower breast and belly are buffy white, unstreaked, whereas the upper breast is streaked. There is a broad white line above and below the eye, with a streaked black and buff speckled line framing the lower cheeks. The tibio-tarsal joint is noticeably thickened. In flight the wings are sharp pointed, with all-black remiges having two white widely separated patches on the base of the primaries, and white bars along the secondary wing coverts framed by parallel buff and black lines. At rest this pattern on the wing coverts shows as horizontal black and white bars framed below with vertical black streaks. There is no hind toe. The sexes are alike.

HABITAT, DISTRIBUTION AND STATUS

This is an inhabitant of scrub desert and extensive sand dune desert, with shy secretive habits so it may be more widespread than available records indicate. It is apparently absent from the Salt Range (Waite, 1948) but occurs regularly on some of the broad stony undulating hills around the mouth of the Khyber pass, the Jamrud plain and parts of north-eastern Kohat in the NWFP. It is virtually absent from the cultivated alluvial plains of Punjab and Sind but occurs sparingly in the Cholistan and Thar deserts, and in Sind it is most likely to be encountered on the broad arid valleys between the hill ranges west of the Indus river or on barren sandy islands in the Indus delta and along the Makran coast. A pair usually haunts the low hills on the north shore of Hab dam lake. It can be very local in occurrence with 2 or 3 pairs found in one vicinity but absent from other equally suitable tracts. Because of erratic rainfall patterns it is subject to local migrations. Meinertzhagen collected a pair on the Popalzai plain near Chaman and Christison recorded it as a sparse resident around Dalbandin and Zangi Nawar in the Chagai (Meinertzhagen, *Ibis*, 1920 and Christison, *Ibis*, 1941). A flock of 5 birds was seen on 21 January 1983 near Thano Bulla Khan (Hyderabad district) by Neale Taylor (pers. comm.) and Eates once flushed as many as 12 birds from a small patch of desert near Sarhad in Sukkur district in December. Status SCARCE RESIDENT

HABITS

They are almost wholly nocturnal in feeding activity and their large eyes probably indicate a high degree of visual acuity in darkness. In the winter they occasionally associate in small groups or flocks of 5 to 10 (see account of Distribution and Status). K. Eates who watched one foraging noted that it suddenly ran forward very fast, body crouched and seized a grasshopper. They feed largely upon Coleoptera and Orthopterous insects, including especially *Tenebrionid* beetles and *Chrotogonus* grasshoppers and the large black ants *Camponotus compressus*. They have also been recorded as eating worms, small reptiles and occasionally seeds. A specimen collected near Delhi had a large millipede (Diplopoda) and an adult beetle (unidentified) in its stomach (Donahue, *JBNHS*, Vol. 64). When camping out on sandy islands in the Indus delta the author has often seen their distinctive tracks (no hind toe) fresh in the dew criss-crossing the sand dunes within 20 feet of where he and his companions had been lying asleep so they are apparently much bolder in the dark. Such tracks indicated that they had been systematically searching around the base of scattered grass clumps, probably for feeding grass hoppers.

BREEDING BIOLOGY

A nest was found with two fresh eggs on a bare clay flat but under the shelter of a *Suaeda fruticosa* bush near Lal Sohanran (Bahawalpur district) on 4 April (Alan Vittery, pers. comm.). Kenneth Eates found many nests in Sind chiefly in March and April. They usually comprised the merest scrape unlined, in the shelter of a grass tuft or small bush and the parents run crouching away from the nest, never giving its presence away, but after the eggs are well incubated or the nidicolous chicks are hatched they will give a broken wing distraction

Map 131: Stone Curlew *Burhinus oedicnemus*

display or even an aggressive display running towards the human intruder with wings spread and carpal joints twisted forward to the ground, tail raised and fanned, and crown and neck feathers ruffled (author obs. Delhi zoo). Both sexes share in incubation which takes 24-26 days (Cramp *et al.*, 1983), and tending the chicks, and probably the pair bond lasts for the whole breeding season. The normal clutch is 2 eggs, not pointed at either end being broad ovals and they are greenish buff or stone coloured with secondary grey markings, blotched and smudged with blackish brown (Eates, MS notes unpublished). Newly hatched chicks rely upon crypsis for concealment and will lie flat, head and neck extended and motionless when a predator is in the vicinity. For the first few days the downy chicks are brought food by their parents and remain with them at least until they are two months old and after becoming fully fledged.

VOCALIZATIONS

They call at night throughout the year (heard Indus delta, 12 November and 13 December) but only occasionally outside of the breeding season. They have a variety of rather mournful yodelling whistles reminiscent of a Curlew *(Numenius arquata)*, but with the second syllable often more drawn-out 'currleeuw currleeeh' etc. They also have shorter repeated 'ker-vic-ker-vic-ker-vic' calls, more often heard in the summer breeding season.

Great Stone Plover *Esacus recurvirostris* (Cuvier)

Illus. **48**
Map **132**

Great Stone Curlew *Esacus magnirostris* (Vieillot) Dillon Ripley 'Revised Synopsis'

DESCRIPTION

Body length 48-52 cms.
Wingspan 90-100 cms.
Wing length 25.2-27.3 cms.(both sexes)
Bill length 74-84 mm. (Salim Ali)

A long legged, curlew-sized bird, with a dispropor-tionately heavy looking bill and huge goggle eyes. The sexes are identical. They have earthy brown or buffy grey crown, neck and mantle with only very faint dark streaks on the mantle, yellow iris and black rimmed eye, surrounded by a narrow white ring extending behind the ear coverts. This white area through the eye, framed above and below by black, with nearly the whole of the ear coverts being black and with a narrow black malar streak. The secondaries are paler more greyish buff with the wing shoulders darker almost black and separated by a narrow white line. The legs are olive yellow, the feet lacking any hind toe, and the long bill is yellow on its basal half, black tipped with the lower mandible shaped like the keel and prow of a boat (a pronounced gonys). In flight the outer tail feathers are black, tipped with a narrow white bar above, and the remiges are also black with white on the inner webs of the first three primaries and white tips to the inner primaries, the darker wing shoulders and pale panel of the secondary wing coverts showing up prominently.

HABITAT, DISTRIBUTION AND STATUS

Whilst the Stone Curlew *(Burhinus oedicnemus)* is widespread in desert areas, the Great Stone Plover is closely tied to the vicinity of water, its preferred habitat being grassy flats along river banks, salt marsh grass in delta areas, sandy or shingly islands in rivers and the shores of lakes but preferably ones with stony drying out shores and the shores of lagoons on the coast. It occurs very sparsely throughout the Indus and its tributaries and in Sonmiani lagoon, Lasbela and particularly the salt marsh flats around Keti Bandar in the Indus delta. 5 were seen (author) on 9 March on the banks of the Chenab river upstream of Marala barrage in Sialkot district, and twelve on 11 February 1971 on the banks of the Indus upstream of Chashma barrage on the Indus river (F. Koning, IWRB Surveys). In March 1980, 3 were seen frequenting a stony stream bed which drains into Rawal lake (D. Corfield, pers. comm.). H. Waite collected specimens (Brit. Mus. coll.) on the Jhelum river near Tahlianwala. In Sind a few breed around the estuary of the Hab river and 5 were seen (author) on mud flats on the shores of Miani Hoar (lagoon) in Lasbela. Status SCARCE RESIDENT.

HABITS

Very similar to the Stone Curlew *(Burhinus oedicnemus)* in being largely nocturnal in activity. They often occur in pairs during early spring and autumn so the pair bond may be of lasting duration. In winter they are more gregarious but groups of more than 3 to 5 are seldom seen as they are not sufficiently numerous in Pakistan. Like the Stone Curlew they are swift runners, foraging on the ground and seeking their prey visually. They subsist on a variety of animal food including molluscs, especially gastropods, amphibia, insects and beetles and an occasional small lizard or bird's egg. Dharmakumarsinhji (1972) records one finding and swallowing whole, the eggs of a Kentish Plover *(Charadrius alexandrinus)*. Crabs are believed to form the major part of their diet in estuarine areas.

BREEDING BIOLOGY

These birds are very tolerant of heat and unsheltered situations and their nests are often in quite exposed situations on a sandbank in the river or on bare rock. The nest itself is a shallow scrape usually not even shaded by an adjacent bush or grass clump. The usual clutch is two eggs and in Sind, Eates found the earliest nests in the

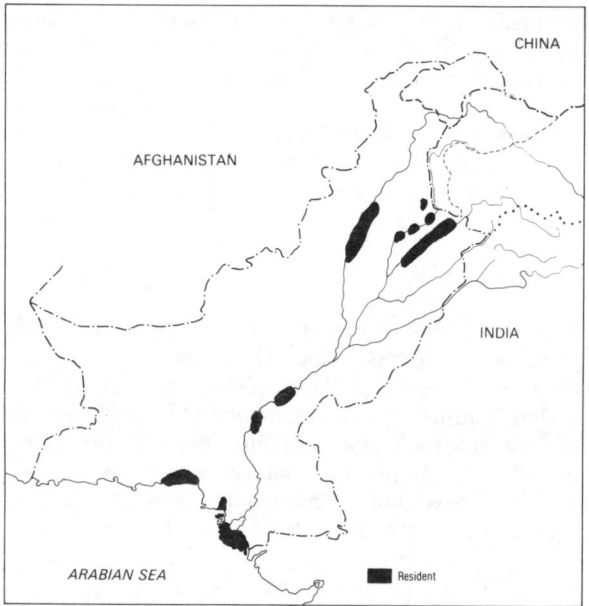

Map 132: Great Thick-knee or Stone Plover *Esacus recurvirostris*

Illus. 48: Great Thick-knee or Stone Plover *Esacus magnirostris*. Note very large eye and lack of dark streaking on upper plumage.

Sukkur and on shingle banks near perennial pools on the Hab river. The eggs are pale creamy in ground-colour and the full clutch is 2 eggs (Stuart Baker, 1929), they are heavily streaked with black and with some scrawling lines also. Eates describes nests as often being located amongst debris left by high tides on creeks. Both parents take equal shares in incubation and protecting the young after hatching. Incubation lasts 28 days. The downy chicks even on the first day of hatching can swim well (S. G. Neginhal, *JBNHS,* 1982). Predators are a constant problem for such ground nesting birds (see account of *Falco pelegrinoides).* House Crows *(Corvus splendens)* were watching Corfield when he found and photographed the eggs of this bird on the shores of Rawal lake. Consequently they were destroyed within hours, probably by this species (pers. comm., 1981).

third week of March with most laying in mid-April. Swinhoe found a nest on the Hab river on 24 May (in Ticehurst, 1923). However, David Corfield (pers. comm.) found a nest on the northern shores of Rawal lake with a completed clutch on 6 June 1981. Eates (MS notes unpublished) latest date of finding a nest was on Kotrato island in Korangi creek on 20 June. Nests were found by him on sandy islands in the Indus river near

VOCALIZATIONS

When disturbed or alarmed, occasionally they emit a harsh 'craak'. On their nesting territory at night, they give wild wailing cries 'kree-kree-kree' (W. A. Phillips, 1939.) The warning calls to downy chicks are a deep whistling. Another alarm call described as 'kill-ick-kill-ick-kill-ick' (Neginhal, op. cit.).

FAMILY GLAREOLIDAE

Cream-coloured Courser *Cursorius cursor* (Latham)   Map **133**

DESCRIPTION

Body length 19-22 cms.
Wingspan 51-57 cms.
Wing length 16-17.1 cms. (males), 16.2-16.6 cms. (females)
Bill length 30-32 mm. (males), 29-33 mm. (females) (Salim Ali)

A sleek plumaged cursorial bird, with very upright stance, rounded head and constricted neck, the upper body plumage being earthy brown or sandy buff, the wing coverts entirely concealing the blackish brown wing tip (primaries and primary wing coverts). The outer tail feathers and trailing edge of the secondaries are tipped white, with a narrow blackish band anteriorly. The head pattern is striking. A dark brown eye is followed by a black line curving down the back of the neck to join in a 'V' point on the nape. Above this is a broad white supercilium and another narrow black line with the mid-crown and occiput clear blue-grey. The forecrown is chestnut and the sides of the neck also have warmer cinnamon tones. The peculiar beak is dark brown sharply down-curved and pointed. The legs have long tarsi and short toes without any hind toe. They are coloured dull white. In flight the entire under-wing is conspicuously black except for the narrow white trailing edge to the secondaries and a narrow chestnut band

along the forewing. The lower ventrum is whitish buff. Juveniles lack the black eye streak being cinnamon chestnut on the crown and upper breast with grey spotting on the back and wings. Both sexes are alike.

HABITAT, DISTRIBUTION AND STATUS

Rather thinly but widely distributed winter visitor to the more arid treeless regions of the Punjab. In southern Sind and Baluchistan it is resident and breeding, typically associated with the outskirts of small hamlets in the broader low altitude valleys or clay flats in the Thar desert and the Hab valley and Lasbela. Occasional birds have been seen in Rawalpindi district on the Potohar plains and it was collected by H. Waite (Brit. Mus. coll.) at Jand in Attock district. There appear to be no records from the NWFP. In the Chagai desert of western Baluchistan Christison (*Ibis,* 1941) considered it a common passage migrant in spring and autumn, but over 3 years did not note any over-summering. The author saw two birds near Zangi Nawar lake in early May 1984, in sand dune desert. Three pairs were seen on 7 March 1968 in Dasht valley 32 kms. (20 miles) east of Quetta (J. A. W. Anderson, pers. comm.). Whistler

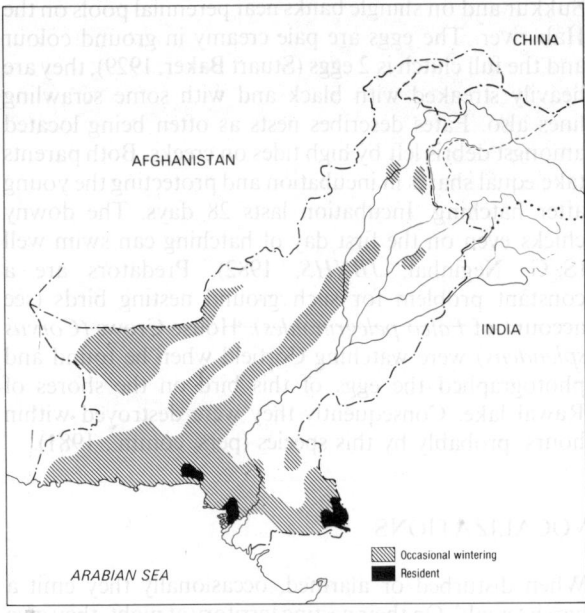

Map 133: Cream-coloured Courser *Cursorius cursor*

found it a regular winter visitor to the salty clay flats in Jhang district (Whistler, 1922). A family group was seen (author) in sand hills by a village at Hathungo near Khipro, Sanghar district on 16 February, and they can be regularly encountered around Karchat below the Kirthar hills and in the Hab valley and the Gadap plains north-west of Karachi. After the monsoon and breeding season, they sometimes gather in small flocks and even associate with Indian Coursers (*Cursorius coromandelicus*). The highest number seen were 17 on 6 February 1977 on Chaukandi limestone ridge just north of Karachi. Only 5 birds were seen in the same area later on 22 November in the same year. They occur in sand dune regions as well as stony desert and rocky outcrops, not being dependent on special edaphic differences but only requiring open treeless barren desert country. They are most likely to be encountered near small villages where goats or cattle are penned at night, the dung attracting beetles and other insect food prey. Status SCARCE.

HABITS

Diurnal in feeding activity and gregarious at all times of the year even during the breeding season. They run forward rapidly when foraging, their feet appearing almost as a blur and bend the whole body forward to dig beneath plant roots or probe around stones, and appear to hunt visually. They feed mainly upon insects, Dung Beetles (Scarabeidae) and especially ground living *Tenebrionid* beetles, and orthopterous insects and small molluscs (Mason and Lefroy, 1912) also ants *(Camponotus compressus)* and when available, larger prey such as Fringe-toed lizards *(Acanthodactylus*

cantoris) and scorpions. They will also eat seeds and small molluscs, but insects and their larvae are believed to comprise the main part of their diet, most of which are obtained by digging in the hard dry ground. When alert or watching an intruder they have a very upright stance with neck and breast almost in one vertical plane. Their flight is strong with measured wing beats like a *Vanellus* plover, but their wingtips are more pointed.

If not molested they will become very tame in the precincts of herders' settlements and remote hamlets bordering the desert and in fact are usually very tame, preferring to run rather than take to the wing.

BREEDING BIOLOGY

Eates found nests in the Hab valley and in Tharparkar district, and noted that more than two adults were usually seen in the nest vicinity indicating that they are gregarious even during nesting. He found 2 eggs (the usual clutch number) on the point of hatching on 3 May 1937. The nest is a mere scrape in the sand with no lining and in an open situation with no bushes nearby. The egg is comparatively larger than that of the Indian Courser *(C. coromandelicus)* (MS notes unpublished).

Studies in North Africa indicate that both sexes share incubation (Urban, Fry *et al.*, 1986) but that the male mostly remains on watch in the nest vicinity and later helps to tend the chicks which have to be fed for the first 2 or 3 weeks, though they become self feeding before fledging (Cramp *et al.*, 1983). Both parents bring them whole and sometimes quite large insects. No reports of deliberate egg wetting have been noted with this species but in midday heat the female crouches over the eggs rather than sitting on them. High ground temperatures in the nesting season may account for the female being very reluctant to leave her eggs and returning to them even when disturbed, if the intruder moves away a short distance (K. Eates, MS notes unpublished). The eggs are pale yellow buff, finely and evenly spotted with brown. A territorial display flight was noted by Paul Goriup on the Canary Islands in which the male flies diagonally up in the air with shallow rapid wing beats, like a butterfly, and then volplanes down uttering an ululating or warbling plover-like cry (Goriup, pers. comm., 1981). A fully fledged juvenile observed in late August, ran with head retracted and crouching its body low but its parents remained in upright alert stance though they ran away with it (observed Hab valley).

VOCALIZATIONS

The display call heard by P. Goriup was rendered by him (pers. comm., 1981) as 'wheck-wheck-cowrr'. They have two types of calls given in flight. When disturbed or alarmed they call in a liquid ventriloquial 'whit-whit' (D.B. Shirt in ICBP *Bustard Studies*, No. 1, 1983) and also a contact or excited flight call described as a double-noted whistle 'quiddle' rather reminiscent of the Indian Sandgrouse *(Petrocles exustus).*

Indian Courser *Cursorius coromandelicus* (Gmelin)

Map **134**

DESCRIPTION

Body length 23-26 cms.
Wingspan 58-60 cms.
Wing length 14.3-16.3 cms. (both sexes)
Bill length 23-30 mm (Salim Ali)

It is very similar to the Cream Coloured Courser in build and general appearance, though averaging slightly larger and distinguished by having a rich liver chestnut crown and white throat grading into rusty orange on the lower throat and a darker cinnamon chestnut breast, in contrast to the blue-grey crown and sandy buff breast of *Cursorius cursor*. Its bill is the same courser shape, being black, sharp pointed and down curving and it has a similar head pattern with black lines running from the lores through the eye and meeting in a 'V' on the lower nape, framed above and below by white lines. The back and wing shoulders are sandy brown, slightly darker in tone than that of *C. cursor* and the black flight feathers are entirely concealed by the wing coverts when the bird is at rest. The chestnut breast darkens to maroon with the lower belly white but having a black patch between the legs. The rump and vent are also whitish. The outer tail feathers are tipped white, with narrow black bands, the central tail feathers being concolorous with the back. The iris is dark brown and the legs and feet are dull white with short thick toes and no hallux. In flight the whole underside of the wings is black but the dark chestnut breast and black patch between its thighs at once distinguishes it from *C. cursor*. The wing beats are very marked, like a plover, but the wings are long and pointed at the tip and it is a fast and powerful flier when alarmed. The sexes are alike. Juveniles lack the black and white head-stripes and have rather spotted plumage on crown and mantle with white throat and rufous chestnut breast.

HABITAT, DISTRIBUTION AND STATUS

They haunt a similar habitat to the Cream Coloured Courser *C. cursor* but are perhaps better adapted to more mesic conditions, such as the drying out margins of jheels', old river beds and fallow fields. They are locally nomadic, according to rainfall conditions and food availability, but are resident and breeding. They are rather scarce and erratic in distribution throughout Punjab and Sind, over-lapping with the Cream Coloured Courser in lower Sind. In Thatta district both species can often be encountered in the same locality in winter. It is rare west of the Indus. A favourite haunt is the Sukh Beas old river bed east of Lahore and some of the grass flood plains next to the Jhelum river and Qadirabad drainage channel in Shekhupura district. K. Eates, (MS notes unpublished) considered it widespread in lower Sind, particularly Badin district and Thatta district around Jangshahi. H. Waite (*JBNHS*,

1948 and Brit. Mus. coll.) collected specimens near Mithan Kot in Dera Ghazi Khan district, Dina in Jhelum district and Kunda in Muzaffargarh district, mainly in the winter months. In the late 1960s about 10 used to haunt the grasslands of Walton airfield in Lahore but probably this region is nowadays too disturbed by traffic. After the monsoon during September and October, they congregate in small flocks, and can often be seen in the dry rocky plateau country of south-western Thatta district. 8 were seen on 2 October Chaukandi ridge 1976, 5 seen 13 October near Jangshahi (Kamal Islam, pers. comm.) and 8 seen 21 October near Pipri north of Karachi in 1983. Status SCARCE.

HABITS

Pair bonds seem to be quite long lasting as they are usually encountered in pairs from about March through to August and then in the post monsoon and early winter they congregate in small flocks of 5 to 10 birds. They are diurnal in feeding activity and typically forage by running forward in fast spurts and bending down the whole body (necessary because of their long legs) to dig and peck vigorously in the ground. Their main food is ground living *Tenebrionid* beetles and the soil born larvae and pupae of insects as well as ants and orthopterous insects such as mole crickets *(Gryllotalpa)* and larvae of grass-hoppers (Mason and Lefroy, 1912). Whistler ('Handbook', 1949) also records them as eating

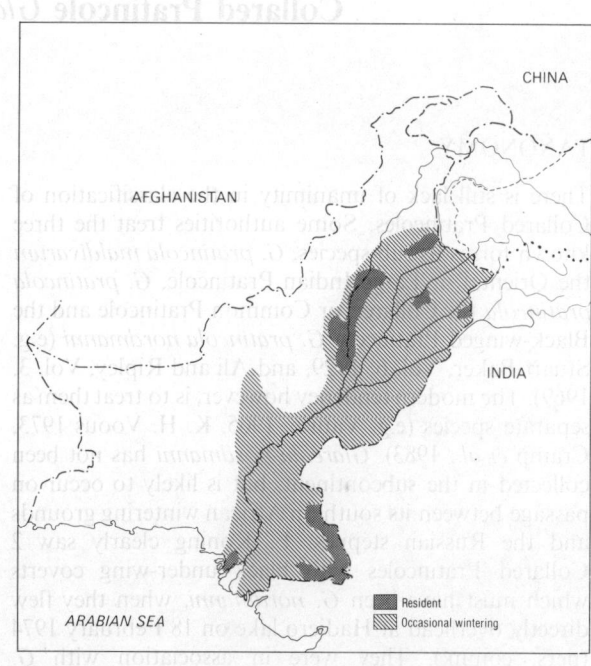

Map 134: Indian Courser *Cursorius coromandelicus*

ants (Formicidae), weavils (Curculionidae) and small molluscs. They have the typical rather bold-upright stance of *C. cursor* when alert and danger threatens and are very silent in their winter quarters. Both Cream Coloured and Indian Coursers have been regularly seen in the early winter months, frequenting the same bit of ground in the low hills around Dhabeji and Jangshahi in Thatta district in the autumn. (author obs., and Neale Taylor, pers. comm.).

BREEDING BIOLOGY

In lower Sind the peak egg laying season is the third week of April. K. Eates over 30 years found altogether 3 nests in March, 32 in April and 19 in May with 12 nests in June and 9 in July (MS notes). Eggs are laid on bare ground with no obvious nest scrape or lining material and the normal clutch is 2. The eggs are rather spherical in shape, buff or cream coloured but heavily marked all over with inky grey and overlaid with spots, lines and scratches of black or dark brown. They are almost perfectly camouflaged against the bare drought cracked or flaked earth where they are laid, as is the downy chick. Both sexes tend and protect their young and Eates noted the parents giving low subdued alarm calls to which the chick immediately responded by crouching with head and neck flat on the ground and freezing. The chicks are

pepper and salt coloured with rather short close down. Holmes and Wright found a downy chick in late May near Badin (op. cit., *JBNHS*, Vol. 65, 1968). Scrope Doig (*Stray Feathers*, 1879b) found it nesting alongside Collared Pratincoles *(Glareola pratincola)* on 4 May in the east Narra. When disturbed at the nest, Eates noted that the parent birds emitted no alarm cries, nor attempted any distraction displays. Dharmakumarsinhji thought that in Gujarat state there was another post monsoon breeding season (Dharmakumarsinhji and Lavkumar, op. cit., 1972). There is no clear evidence as to whether the male shares any incubation duties, but Dharmakumarsinhji implies that both sexes share these duties. The young chicks are partly fed by their parents for the first week or two after hatching, and remain dependent until after fledging.

VOCALIZATIONS

A remarkably silent species even during the breeding season. Dharmakumarsinhji while photographing an incubating bird heard it emit a 'low grunt' (*60 Indian Birds*, 1972) and Whistler reports that when disturbed it rises with a 'distinctive note' (Popular 'Handbook', 1949), which Salim Ali describes as a low 'chucking'. Parties of birds flushed by the author on several occasions were never heard emitting any calls.

Collared Pratincole *Glareola pratincola* (Linnaeus) Map **135**

TAXONOMY

There is still lack of unanimity in the classification of Collared Pratincoles. Some authorities treat the three known forms as sub-species, *G. pratincola maldivarum* the Oriental or Large Indian Pratincole, *G. pratincola pratincola* the Collared or Common Pratincole and the Black-winged Pratincole *G. pratincola nordmanni* (e.g. Stuart Baker, Vol. 6, 1929, and Ali and Ripley, Vol. 3, 1969). The modern tendency however, is to treat them as separate species (e.g. Vaurie, 1965, K. H. Voous 1973, Cramp *et al.,* 1983). *Glareola nordmanni* has not been collected in the subcontinent, but is likely to occur on passage between its southern African wintering grounds and the Russian steppes. F. Koning clearly saw 2 Collared Pratincoles with black under-wing coverts which must have been *G. nordmanni,* when they flew directly overhead at Hadiero lake on 18 February 1974 (pers. comm.). They were in association with *G. pratincola.*

DESCRIPTION

Head and body 23-25 cms. with tail 10-12 cms. (outer feathers)
Wingspan 60-65 cms.
Wing length 17.6-20 cms.
Tail length (outer most feather 10.2-11.9 cms. Stuart Baker)
Bill length 15-16 mm.

With large head, rather flattened crown and short thick neck it is similar in outline to other Pratincoles. The wingtips are long and slender extending just short of the tail tips. Above it is greyish brown to sandy buff, with large dark eyes prominently ringed with white. The lower cheeks and throat are fulvous yellow, framed by a narrow black circular gorget and the upper breast is slightly ochraceous buff, merging gradually into white on the lower belly. The very similar Oriental Pratincole *(G. maldivarum)* is much darker grey-brown above and in flight lacks any white trailing edge to the secondaries. Also its tail is not so deeply forked as in the Collared Pratincole. The legs and feet are dusky black with comparatively longer tarsi than in *Glareola lactea*. The black down-curved bill is red at the gape. The wing quills and tail are black, the latter much more deeply forked than in *Glareola maldivarum* or *G. lactea*. The outer rectrices are white on the outer web and the rump is conspicuously white. Juveniles look very different, their plumage being spotted all over with dark tips to the feathers edged prominently with creamy white. The throat is whitish with the upper breast spotted blackish and there is no black gorget line. In flight the under-wing pattern shows a narrow trailing white edge and the whole wing is dark chestnut with narrow blackish brown borders and the wingtip dark brown.

HABITAT, DISTRIBUTION AND STATUS

Summer breeding visitors to Pakistan, from a wintering population in east Africa. They are rare stragglers up into the Punjab, being seen on the Ravi at Balloki in several years (R. S. Holmes and Z. B. Mirza, pers. comm.). They are mainly to be encountered in lower Sind and especially around the larger lakes and the salt marsh flats of the Indus delta and along the Rann of Kutch borders in Badin district. Some are always to be seen around Manchar lake in Dadu district. They are less restricted to rivers than *G. lactea*, preferring wide open bare flats created by drying out seasonal lakes or swamps and tidal creeks subject to periodic flooding. Earliest dates seen 18 February, latest dates 17 September. Status COMMON in lower Sind only, in summer.

HABITS

By late March pair formation and breeding activity has usually commenced and most have departed by the end of August. They are probably less exclusively aerial in foraging, collecting some of their food by running over the bare ground. They are gregarious at all times of the year and particularly active hawking the air in flocks at dusk. Their diet comprises a wide variety of larger adult insects, moths (Heterocera), also flying White Ants (Isoptera) and Coleoptera. During the heat of the day they generally rest on the ground, squatting in the hollows made by camel or buffalo hoof-prints, almost invisible against their sandy background, until the eye catches the opened black bill of the panting bird.

BREEDING BIOLOGY

Before Haleji lake was made into a reservoir its early summer drying out provided ideal breeding conditions and Ticehurst (*Ibis*, 1923) describes visiting a breeding colony there. Nesting commences from late March to early April. Derek Holmes found them quite common in June around Sir creek south of Jati in Thatta district, also between Sehwan town and Manchar lake (*JBNHS*,

Map 135: Collared Pratincole *Glareola pratincola*

Vol. 65, 1968). K. Eates found nests with eggs on 10 April near Jangshahi and on 12 May near Keti Bander on the Indus delta, young had all hatched and their wing quills were sprouting (MS notes). The normal clutch is 2 to 3 eggs, and usually the nest, a mere scrape in quite bare ground, is sheltered in a hoof-print depression. The eggs are yellow-buff in ground colour with under-markings of grey and densely spotted and blotched with black. Eggs are variable, and Scrope Doig (1879*b*) found many with pale greenish olive ground colour varying to light stone. Both sexes share incubation duties and feeding the downy young by regurgitation. Ticehurst describes the mass distraction display of parent birds in a nesting colony he visited, once the chicks had hatched. Doig (1879*b*) describes scattered birds lying spread-eagled in a static type of distraction display. They are very active soon after hatching and will run from an intruder and hide cleverly in some animal's hoof-print (Ticehurst, *Ibis*, 1923). Studies in the Middle East indicate young are only fed by regurgitation for the first week and that they are fully fledged at 30 days (Cramp *et al.*, 'Handbook', Vol. 3, 1983).

VOCALIZATIONS

Very vocal in defence of their nest colony, uttering tern-like calls 'kirr-kirri' which change to louder screaming type calls when a colony is disturbed. They also have more complicated sequences of calls in the early breeding season involving short staccato calls repeated 'kit-kit-kit' immediately followed by a more rapid trilling 'kerrichtik-kerrichtik'.

Eastern Collared Pratincole or

Oriental Pratincole *Glareola maldivarum* (after C. Vaurie 1965, K.H. Voous, 1973)

Large Indian Pratincole *Glareola pratincola maldivarum* J.R. Forster

(after Ali and Ripley, Vol. 3, 1969)

DESCRIPTION

Body length 24-25 cms.
Wing length 17-19 cms.
Outer tail length 7.1-8.5 cms. (c.f. 10-12 cms.
     of *Glareola pratincola*)
Bill length 13-15 mm. (Stuart Baker).

When at rest the wingtips in this species often extend slightly beyond the tail tip. The upper parts are a darker sandy brown than in the Collared Pratincole and there is no white trailing edge to the wing formed by the tips of the secondaries which are a conspicuous field point in *G. pratincola pratincola*. Moreover the mantle of *G. pratincola* is a paler isabelline brown colour. The tail is not so deeply forked as in the Collared Pratincole and shows comparatively less white. It is otherwise similar with deep rusty maroon under-wing coverts, a conspicuous white rump and lower belly. The yellow ochre chin and throat patch is encircled by a narrow black ring. The short black bill is quite strongly down-curved on the culmen and the base is crimson-red, with a very wide gape when the bill is opened. The irides are dark brown and the legs and feet brownish black, the toes being unwebbed with the hind toe very weakly developed.

HABITAT, DISTRIBUTION AND STATUS

Not clearly known. It breeds widely in India in the north-east, from Madhya Pradesh, Bengal and Assam. Also in Bangla Desh and Burma (Stuart Baker, op. cit., 1929, Ali and Ripley, op. cit., 1969). It apparently breeds sporadically westwards, occasionally extending its range into Pakistan where the only definite record is of a few pairs found breeding in a colony of *Glareola pratincola pratincola* in the east Narra, Sind in the 1870s where Scrope Doig collected specimens which are now in the British Museum at Tring (Doig, *Stray Feathers*, Vol. 8, 1879*b*). In the USSR it is a very rare breeding bird in south-east Transbaikal and is considered mainly a migrant from south-east Asia (Dement'ev *et al.*, 1951).

There are no recent records of authentic sightings in Pakistan. The specimens collected by Scrope Doig in Sind were nesting in the first week of May on a flat, recently dried out, salt plain dotted with small sedge clumps. The usual clutch size is 3 eggs and these are laid in a bare scrape, hollowed out on the ground, usually in the slight shelter of an animal's hoof-prints, the digging of wild boar or a dried cake of cattle dung (Doig, op. cit.).

Small Indian Pratincole or Swallow-plover *Glareola lactea* Temminck   Illus. **49**
Map **136**

DESCRIPTION

Head and body length 16-18 cms.
Wingspan 42-48 cms.
Wing length 14.2-16 cms.
Bill length 9-10 mm. (Stuart Baker)
Tail length 5-5.7 cms. (Stuart Baker)

They have comparatively large heads with flattened crowns and short necks, and long pointed wings extending beyond the tail tip when at rest. In flight they look remarkably like large swallows. Their legs are black with comparatively short tarsi though they run fast and well along the ground. The short sharply down-curved bill is black with a wide fleshy red gape and the large eye is dark brown framed by a whitish eye-ring. The crown is dark olive brown, the rest of the neck, back and wing coverts are paler earthy or sandy brown. The short tail is forked with white upper tail coverts and outer webs to the feathers and white tips, but the central tail feathers are black. The ear coverts are yellowish white and the breast is white with a tinge of lemon about the throat and upper breast. A black loral line extends from the eye to the base of the bill, and the wingtips are black. In flight the under-wing pattern is distinctive with black under-wing coverts and axilliaries and white remiges-except for the outer quills which are blackish, but the tail looks all-white from underneath. From above, the tail shows white with a broad black tip. When they squat on the bare ground they become almost invisible.

HABITAT, DISTRIBUTION AND STATUS

It is more or less confined to the Indus river and its tributaries in the Punjab and upper Indus in Sind with

Map 136: Small or Little Pratincole *Glareola lactea*

only exception being the klepto-parasitic Jaegers (Stercorariidae). This led to reduced tarsal length, exaggeratedly wide mouth gape with short hook tipped culmen as in Hirundines, adapted for crushing the hard chitinous exoskeleton of insects before swallowing while in flight.

They are markedly gregarious throughout the year being tied closely to fresh water river systems and requiring open exposed areas for colonial nesting, well protected from predators, such as islands in rivers or wide expanses created by drying out temporary inundations. Their diet is purely adult insects and this includes Diptera of many kinds. Coleoptera and Hemiptera. They are particularly active feeding at dusk, often in loose flocks, sometimes skimming low over the water surface, at other times rising high with twisting zig-zagging flight like a swallow. They will occasionally feed on the ground also while resting, running forward swiftly to peck at an object. There have been no ringing studies on this species in the subcontinent and their erratic movements are not understood but after breeding they sometimes desert an area and in Pakistan seem to disappear entirely from the Indus in the mid-winter months.

occasional breeding in the east Narra and the Indus delta. It will occasionally haunt large lakes particularly if there are wide expanses of drying out mud or grassland around their margins. In Pakistan it is mainly a summer breeding visitor and being highly gregarious is often seen in considerable flocks, but lesser numbers do over-winter, particularly in the southern reaches of the Indus. Ticehurst recorded them in Karachi district as late as 4 November. (Ticehurst, *Ibis,* 1923). Whitehead recorded them as spring and summer visitors, common along gravel banks of the Kurram river in the NWFP during April and May (op. cit., 1911). The author saw flocks of 30 to 40 birds hawking round the Baran dam on the lower Kurram river at Bannu in late March. Winter records include 150 seen 15 February, 1971 at Ghauspur jheel (Jacobabad district) and again 350 on Ghauspur jheel on 16, January 1987 and 70 near Jatti in Thatta district on 28 January, 1987 (F. Koning, *in litt.)* and 13 on 22 February, at Lal Sohanran in 1971 (Koning, IWRB Surveys). Odd parties can be seen hawking over larger lakes, 5 or 6 seen Rawal lake on 27 September (D. Corfield, pers. comm.) and over 60 around Sidhnai spill channel 25 February. Sometimes they congregate in large numbers. Well over 500 were counted (author and J. Vieillard) wheeling in the sky and resting on a sandbar below Jinnah barrage, on the Indus near Mianwali on 29 March. Status ABUNDANT.

HABITS

Perhaps unique amongst the whole order of Charadriformes in being adapted to hunt all their food in air spaces rather than on the ground or in the water, the

BREEDING BIOLOGY

Around Marala barrage on the Indus, colonies started breeding from mid-March (E. Fernando, pers. comm., 1969), and in lower Sind, from late March to April. Derek Holmes found a colony of about 50 pairs on an island in the Indus below Hyderabad, on 30 March, with some downy chicks already hatched by the second week of April (pers. comm., 1967). Huge numbers breed on sandbars between Sukkur and Guddu barrages during April and May, according to flood conditions and such colonies are usually in association with Indian River Terns *(Sterna aurantia)* and Black-bellied Terns *(Sterna acuticauda).* The normal clutch is 3 eggs, but often only 2, and both sexes share incubation duties (Lowther 1949 and K. Eates, MS notes). Nests are a scrape hollowed in the sand, occasionally decorated with a few shells and rootlets and the eggs are chalky textured not glossy,

Illus. 49: Small or Little Pratincole *Glareola lactea.* Note wingtip extends well beyond tail, and throat is unmarked in contrast to other Pratincoles.

creamy to pale greenish buff, finely and lightly marked with reddish brown, and purple spots and streaks. They tend to be very variable in marking. The colony is usually in one area and at some distance from other tern species and individual nests will be about 5 to 10 metres apart. Courtship and pair-bond formation probably takes place for several weeks before actual nesting. Pairs have been observed at Sidhnai from 28 March up to 13 April but no nests could be located. Display includes a lot of flight chasing and mutual aerial flights with repeated calling. On the ground they run towards each other and when approaching close, face to face bow their heads down with bill extended and the tail cocked upwards. Also the wings are raised. These postures have elements of aggression or distraction also, as birds have been observed chasing away rivals with wings raised over back and in distraction displays one or both wings are often raised. On 2 April at Sidhnai, distraction display of one bird was impressively executed, it tumbled over the ground dragging one wing then the other as though mortally injured. Other observers have described mass distraction displays when a nesting colony is visited with over 20 birds dragging themselves over the ground hopping on one leg and lying half on one side with wing out-stretched (Whistler, 'Handbook', 1949, E. N. Lowther, *A Bird Photographer*, 1949). Aggressive birds,

when approaching each other were observed fluffing out their neck feathers and with head rather upstretched, unlike the head bowing of courting pairs (noted at Sidhnai, 1981).

Incubation takes about 17 to 18 days and the chicks are precocial and nidifugous, covered with close grey down above, mottled buffy and dusky grey and white below with a narrow black line extending from behind the eye. Both parents feed their chicks by regurgitation (probably boluses of only partly digested insects). They also regularly wet their eggs with their belly plumage during incubation (observed Dharmakumarsinhji (1972) in Gujarat state).

*VOCALIZATIONS

E. N. Lowther (1949) records an incubating bird uttering a rapid 'tuk-tuk-tuk' like a House Gheko *(Hemidactylus)*. In flocks and while on the ground they frequently call during the breeding season. The advertising type of call is very like that of the Little Tern *(Sterna albifrons)* 'temik-temik-temik', or 'ke-terrick-ke-terrick' with rising intonation. When foraging in huge flocks at dusk they have shorter contact calls sounding like 'trrit-trrrit-trrrit'.

FAMILY CHARADRIIDAE

Plovers

Little Ringed Plover *Charadrius dubius* Scopoli

Illus **50**
Map **137**

DESCRIPTION

Body length 14-15 cms.
Wingspan 42-48 cms.
Bill length 15-17 mm. (Salim Ali)
Tarsus length 2.3-2.6 cms. (Salim Ali)

Slightly smaller than the Ringed Plover *(Charadrius hiaticula)* to which it is closely similar, being chiefly distinguished by the lack of narrow white bar on the upper wing surface during flight, and a prominent chrome yellow orbital ring during the breeding season, which *hiaticula* lacks. When viewed closely the wholly black bill (orange at the base in *C. hiaticula*) is another helpful diagnostic point but in the breeding season, *C. dubius* also shows some yellow at the base of the lower mandible. In winter plumage it is closely similar to the Kentish Plover *(C. alexandrinus)*, but can be distinguished from the latter by having a slightly darker and more extensive breast band and in being slightly smaller in size.

These little plovers have swift direct flight and on the ground tend to crouch low while running, in rather a

furtive manner. Their faces and under parts are white with a prominent wide black breast band in the breeding season and a broad black line over the crown extending through the eye and across the ear coverts. There is a white area on the immediate forecrown and a narrow white line behind the black crown band. The rest of the nape, hind neck and back is pale sandy buff. In flight the outer tail feathers show white. The legs are brownish yellow sometimes with an olive caste and lack any hind toe. In the Common Ringed Plover *C. hiaticula* the legs always show some orange tones even in winter. The fleshy yellow orbital ring is supposed to be more prominent in the sub-species *jerdoni* which is also smaller in size than *curonicus*. The white throat collar extends right around the back of the neck at all seasons.

In winter plumage the crown is entirely grey-brown as also the ear coverts and the breast band becomes browner and more limited in area but not as indistinct as in winter plumage Kentish plovers. Juveniles only have a dark band in the pectoral area like winter plumage Kentish plovers *(C. alexandrinus)*, but lack the prominent white forehead of the latter species, or any white wing band in flight. The sexes are alike and the iris is dark brown.

HABITAT, DISTRIBUTION AND STATUS

Less of a marine coast dwelling species than *C. hiaticula*, preferring the drying out margins of lakes and seasonal pools as well as the banks of rivers, whether shingle or sandy banks. However they are regularly to be seen in small numbers along the tidal creeks and inner channels around Karachi and the Indus delta. The sub-species *jerdoni* is resident and breeding whilst the sub-species *curonicus* is a winter visitor, arriving from the beginning of August and departing again by early April (Ticehurst, 1923). It breeds sparsely but widely throughout Sind, Punjab and on the few suitable river-beds in Baluchistan and the NWFP. Eates found nests along the Hab river on shingle banks, and the author found it nesting on the Khadeji stream just north of Karachi. It nests on islands in the Indus in Sukkur district as well as along the Chenab and Ravi rivers. In Baluchistan breeding pairs have recently been encountered (author) on the Loralai stream at Mekhtar and the Surkhab valley in Pishin. It breeds along the Sohan river in Rawalpindi district and the dry torrent beds draining into Nammal lake in the Salt Range. Whitehead (*JBNHS*, 1911) found them breeding on gravel banks in the Kurram river from May onwards in Kohat district and Major Biddulph (*Stray Feathers*, 1881) only noted them in Gilgit on passage in April and May. They have also been recorded on passage through Chitral main valley in April (Perreau, *JBNHS*, 1910). Small groups can be encountered throughout winter in the Punjab around drying out pools, larger lakes and along rivers. Status COMMON.

HABITS

Entirely ground feeding, living on invertebrates picked

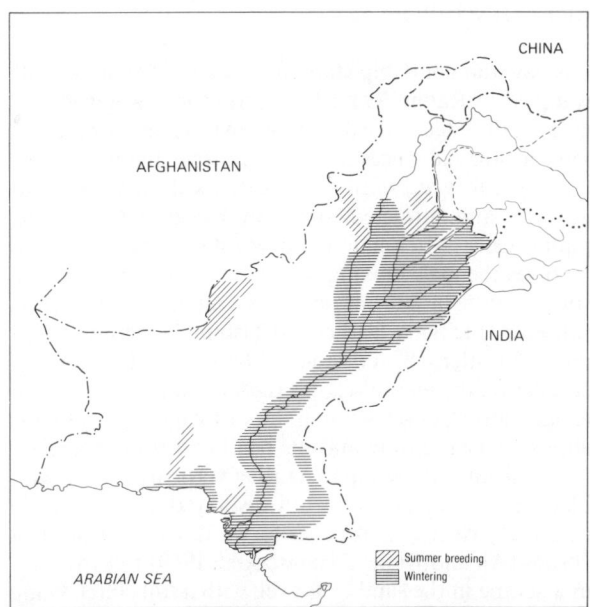

Map 137: Little Ringed Plover *Charadrius dubius*

up by visual hunting, often with characteristic swift runs forward and then pauses to search fresh ground. They also feed in shallow water and an individual was watched (author, Ravi river, Sidhnai) extending one foot and dabbling with it before bending down to pick up food. They are usually seen in loose association with conspecifics, but are not markedly gregarious and frequently share the same feeding habitat as *C. alexandrinus*. When one individual is flushed it is often joined by others of the same species and they fly fast and low over the ground, wheeling in unison. Their food is mainly insects of which flies Diptera form a major item especially Muscidae for which it is often attracted to refuse, and Chironomidae. Beetles are also important, including Weevils (Curculionidae), with occasional worms and small crustacea.

Illus. 50: (a) Little Ringed Plover *Charadrius dubius*, adult in breeding plumage, both sexes.
(b) Kentish or Snowy Plover *Charadrius alexandrinus*, adult in breeding plumage, both species drawn to same scale.

BREEDING BIOLOGY

Display and courtship starts from early March. A male watched in Rahim Yar Khan district in a water-logged area on the edge of the desert, on 9 March made repeated semi-hostile advances to unresponsive females, body crouched forward with flank feathers fluffed and head lowered and attracted attention by its rapid trilling piping calls. After pair bond is established, courtship involves the male moving very close to his mate with tail splayed in a broad fan and body often tipped slightly sideways, the flank feathers are also prominently fluffed out (Whistler, 'Handbook', 1949). At the time of copulation the male also runs right up to the female with exaggerated 'goose-stepping' lifts of each foot till he is almost marking time and his head and neck draws in. The male then draws up to a rather vertical position with black breast-band prominently displayed and the female stationary close by, crouches forward. Copulation then ensues (Ali and Ripley, 'Handbook', 1969) Eggs are laid in a scrape in the sand or gravel, with a full clutch being four eggs. K. Eates found partially completed clutches on the Hab river on 16 May, 21 May and 3 June (MS notes). A pair with downy chicks was seen by the author on the Surkhab stream (Baluchistan) on 21 June. Breeding usually commences later than for *C. alexandrinus* in Baluchistan. At Kushdil Khan lake on 2 April many pairs were still displaying and courting while feeding at the water's edge whilst two pairs of Kentish Plovers had young or full clutches of eggs. The eggs are broad ovals sharply pointed at one end, of a pale greenish grey to buffish stone colour with fine spots and speckling of brownish black with wiggly hair-like lines. Incubation takes 24 to 25 days (Cramp *et al.*, 1983). Both parents incubate and brood the young chicks after hatching, but they are capable of feeding themselves from the start.

*VOCALIZATIONS

In winter time when put to flight often utter short peep-like whistles as alarm cry. Disturbed at the nest a similar but more drawn-out and louder 'cheeoo' warning call is given often by birds running on the ground. A rapidly repeated piping whistle is uttered during the breeding season and a harsh chattering call when extremely agitated or threatened.

Common Ringed Plover *Charadrius hiaticula* Linnaeus   Map **138**

DESCRIPTION

Body length 18-20 cms.
Wingspan 48-57 cms.
Wing length 12-13.8 cms.
Tail length 5.2-6.4 cms. (Stuart Baker)
Bill length 13-15 mm.

A large version of *Charadrius dubius* with a broad black pectoral band and white throat and belly. The crown, back and mantle are sandy brown and the head pattern is similar to that of *C. dubius*, with a black line from base of bill through lores and then extending in a broad band over the crown as well as behind the eye in a broad line across the cheeks. There is also a broad black band round the breast extending in a narrower black collar over the hind neck, but separated by a white line from the grey-brown nape and hind part of crown. In flight it is distinguished from *C. dubius* by a prominent white wing-bar and by its more orange-red legs and orange base to the bill extending over the upper mandible as well. In flight the tail looks darker brown towards the tip and the outer tail feathers are white. There is no yellow orbital ring (conspicuous in *C. dubius*) in the breeding season. Both sexes are alike in plumage. In winter the young look very like juveniles of *C. dubius* with some white on the forecrown and the breast band more brownish and interrupted in the centre. The legs are yellowish in winter.

HABITAT, DISTRIBUTION AND STATUS

A regular winter visitor to seashore and estuarine areas around Karachi, which has probably been overlooked.

Map 138: Common Ringed Plover *Charadrius hiaticula*

Ali and Ripley ('Handbook', 1969) describe it as a rare straggler to the subcontinent. A flock of 15 was seen at Mauripur near the salt-works on 19 February 1973 (F. Koning, pers. comm.). The following year the author with F. Koning, saw 7 on a very muddy bank adjacent to mangroves in Karachi harbour on 7 January 1974. Ecco Smit counted nearly 100 on Ghizri creek pools 29 August 1977 in company with over 300 Little Ringed Plovers. A party of 3 was seen by a tidal creek behind Clifton beach (Karachi) on 4 October 1980 by Paul Goriup. In all cases birds were observed feeding closely as well as flying and identity was certain. Derek Holmes was also quite certain he saw one on Khinjar lake in early May on a seepage pool in 1965 and another on 16 May near Jhol in the same year (pers. comm.). Dr. Scully collected one specimen in Gilgit, an immature female, on 11 October, 1879 (skin in British Museum).

It has probably been overlooked in the past and can be considered a regular winter visitor in small numbers around Karachi seacoast. Status SCARCE.

HABITS

It is more likely to be seen feeding in oozy mud and mangrove embankment situations than *C. dubius*, which seems to prefer a more sandy substrata. It feeds on marine worms, crustaceans and molluscs, hunting mainly by sight and also employing a foot dabbling technique in wet sand.

VOCALIZATIONS

Flight call note a liquid two-noted 'too-i' 'too-i' also 'queep'.

Kentish or Snowy Plover *Charadrius alexandrinus* Linnaeus

Illus. **50**
Map **139**

DESCRIPTION

Body length 15-17 cms.
Wingspan 42-45 cms.
Bill length 19-22 mm. (males), 18-21 mm. (females), (Salim Ali)
Tarsus length 26-29 mm (Salim Ali)

Slightly bigger than the Little Ringed Plover *(C. dubius)* but often feeding in similar habitat and easily confused in winter plumage. They have black bills and legs with grey-brown or sandy brown crown, nape and back, the forepart of the face, throat and breast being white. In breeding plumage the nape and crown of the male is a pale pinkish chestnut, and there is a black smear behind the eye and over the forecrown with the forehead and a line above the eye remaining white. Females in breeding plumage are duller their crowns being concolorous with their grey-brown mantles. There is a short half collar of black in the pectoral region. The eyes are large with dark brown iris and as in other *Charadrius* plovers there is no hind toe. In winter plumage the crown becomes concolorous with the back and black marks behind the eye and in the pectoral region become greyish brown.

HABITAT, DISTRIBUTION AND STATUS

A largely resident species in lower Sind, locally migratory as a summer breeder in Baluchistan, arriving in early April and leaving early September, (Meinertzhagen, *Ibis*, 1920). It breeds sporadically also further north as far as Rahim Yar Khan and even occasionally north-east of Lahore, and it is a fairly wide-spread winter visitor to Sind and Punjab in suitable habitat. It particularly favours bare clay flats, salt pans and sandy margins of larger lakes and seasonal ponds and tanks, particularly if they are drying out with wide damp margins. It also occurs along streams and rivers in winter. In winter seen on margins of Salt Range lakes, especially Uchchali lake, also on salt flats and seepage pools alongside Sidhnai spill channel, Multan district. Occurs throughout the year, around seasonal ponds in the old bed of the Sukh Beas north-east of Lahore. Its real stronghold however is on the seacoast along sandy beaches and tidal creeks with sandy banks. Here it can be seen in loose flocks in winter and breeds in summer. It has not been noted on migration through Gilgit but in the NWFP a spring passage has been noted around Bannu and Kohat (Whitehead, 1911 and Magrath 1908b). Curiously one adult in breeding plumage was seen on 14 July 1975 by the author at 2,438 metres (8,000 feet) elevation on a gravel bank on the Kunhar river at Naran (Hazara district). Status ABUNDANT in lower Sind and Makran coast, elsewhere erratically distributed.

HABITS

Gregarious in winter quarters though feeding as dispersed individuals. They commonly congregate to roost when the state of the tide is unfavourable, or at night, on some bare arid stretch of sand, sometimes in association with *Charadrius mongolus*. They feed largely by sight and by picking invertebrates off the ground surface, most usually over dry sand or clay and sometimes in shallow water. They are less inclined to feed in the inter-tidal zone than the Sand Plovers *(C. mongolus and C. leschenaultii)*. They have been observed foot dabbling in

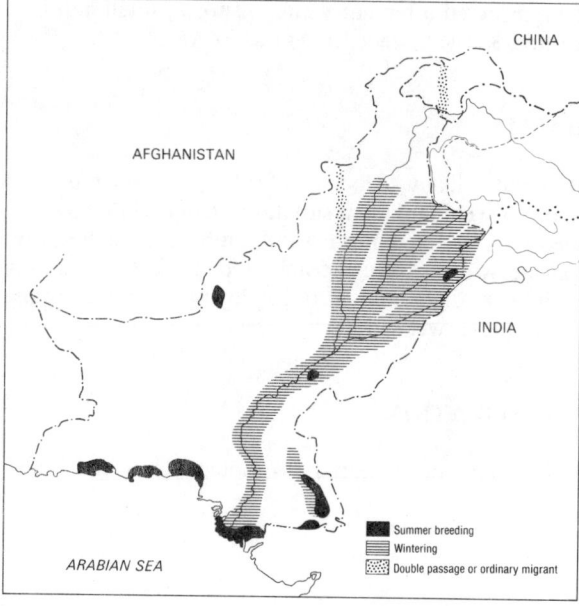

Map 139: Kentish or Snowy Plover *Charadrius alexandrinus*

wet sand. They often forage in sand dune areas beyond the high tide line and in inland waters are more adapted to saline conditions than *Charadrius dubius*. Their food consists largely of crustaceans mostly minute crabs, and worms such as Nereidae. Inland they also consume ants, flies (Diptera) and adult and larval beetles predominate. In Karachi, Ticehurst (*Ibis*, 1923) noted that crabs predominated in the stomach of specimens shot. When disturbed they fly low to the ground often being joined by conspecifics and forming a compact flock, but they are not very shy and often prefer to run away from a human intruder rather than take to the wing.

BREEDING BIOLOGY

A favourite breeding haunt in Karachi is among the sand dunes between Ghizri creek and the open seacoast and on the islands in the Indus delta. Here pairs start nesting in mid-April. Nests have also been found by the author at Siranda jheel in Lasbela and behind Clifton beach, usually with 2 to 3 eggs, only once with four. The nest rim is often decorated with small pieces of shell and protected in the lea of a ridge of wind-blown sand, caught up by a dried *Salsola* bush or piece of refuse (an old rubber sandal in one case). Parents with downy chicks, well able to run, have been encountered near Mauripur salt-works 6 June. A nest was found in a clump of *Juncus* sedge with

2 young obviously recently hatched on the drying out margins of a small jheel west of Rahim Yar Khan on 10 April 1974.

On the north shore of Kushdil Khan lake in western Baluchistan 7 nests were found on 21 June 1977. These nests were all built in buffalo or cattle hoof-prints on flat open drying out mud. Nests were decorated or lined with a very few bits of dried grass or weed and contained mostly 2 or 3 eggs, the latter being the usual full clutch number. Each nest was more than 50 metres from the adjacent nest. On 2 April 1985 in the same location a pair with three newly hatched young (down still wet) was located, the nest scrape as is typical, being in a completely open exposed position. R. Holmes discovered 2 downy chicks with agitated parents, north-east of Lahore by a drying out pool on a bare salty expanse in early May 1967 (pers. comm.). The downy chicks are white on the lower face and breast with buffy grey crown and back and there is a narrow white collar encircling the neck between the speckled crown and mantle. Eggs are variable but usually rather greenish olive or stone grey with a mixture of hair-line squiggles and bold blotches of blackish brown. Around Karachi, it has been noted that eggs laid in May were wetted by both sexes with their breast plumage at the time of incubation change-over. At Clifton beach and Rahim Yar Khan parents were seen giving a broken wing distraction display, this was done while the bird ran rapidly away. Individuals soon took fright and flew away landing about 150 metres distance and then calling to the chick with soft 'toowhit-'toowhit' whistles. Even at Kushdil Khan lake where the breeding pairs all seemed to have only partially incubated eggs, several birds gave a distraction display on the ground with one opened wing dragging and body tipped. Both sexes share incubation duties which last 24 to 27 days. The chicks are self-feeding but both parents tend them and brood them for the first few days after hatching. Along the coast of Karachi they nest in association with Saunders Little Tern *(Sterna saundersi)* often within 100 metres. At Siranda jheel they nested close to Black Winged Stilts *(H. himantopus)*. The latest nesting record is one found 7 July 1945 by Vernon Jones near Clifton oil pipe-line with 2 eggs (Eates, MS notes). Out of over 50 nests taken by Eates around Karachi only 3 contained 4 eggs and over 30 had 3 eggs.

*VOCALIZATIONS

When flushed they often call and this is not a single note like *C. dubius* but a double-noted very plaintive whistly 'wit-wit' or 'wit-wit-wit'. When calling to downy chicks and extremely agitated the call is 'tooweet-tooweet'.

Lesser Sand Plover or
Mongolian Sand Plover *Charadrius mongolus* Pallas

Plate **15**
Map **140**

DESCRIPTION

Body length 19-20 cms.
Wingspan 45-48 cms.
Wing length 12.4-12.9 cms.
Tail length 4.4-5.3 cms.
Bill length 16-18 mm. (maximum 20 mm.) (Stuart Baker)

Intermediate in size between *C. dubius* and *C. hiaticula,* it is rather a drab little plover when seen in its winter plumage. In Pakistan at the end of spring passage and in late summer many can be seen in breeding plumage, when it is beautifully marked. Plumage descriptions in recent bird publications are sometimes conflicting or inaccurate as there is variation between eastern and western breeding populations. In winter the crown, hind neck and mantle is grey-brown to sandy buff. The throat cheeks and under parts are white with an indistinct pectoral band only slightly 'warmer' in tone than its mantle, and interrupted by white in the mid-breast region. There is a whitish superciliary streak and the ear coverts are streaked buffy brown. In flight an indistinct but variable narrow white bar shows across the wings. The bill is black and noticeably shorter and less heavy than in the Greater Sand Plover *(C. leschenaultii).* The legs and feet can be dirty yellowish or greyish green. It looks closely similar to the Kentish Plover *(C. alexandrinus)* except for slightly larger size and lack of narrow white collar on the hind neck. In breeding plumage the legs become jet black, the crown rusty chestnut and a broad bright orange-chestnut pectoral band surrounds the upper breast sharply divided from the white throat. There is no trace of any black line dividing the white throat area and the chestnut breast band in the population wintering or migrating through Pakistan (see incorrect description Ali and Ripley, 'Handbook', 1969 and incorrect illustration in the *Pictorial Guide,* BNHS, 1983, Plate 40 and correct illustration in Flint *et al.,* 1984). The nominate sub-species *C. mongolus mongolus* which shows this black line, evidently does not winter in Pakistan. The forehead from base of bill is black in a broad band passing through the eye and rapidly narrowing to a point through the ear coverts. In full breeding plumage there is no white anywhere on the forecrown. The race which winters in the north-west of the subcontinent has been assigned to *C. mongolus atrifrons* (Wagler), largely on the basis of a slightly longer bill and shorter wing than the nominate sub-species (see Cramp *et al.,* 1983). Juvenile birds in their first winter have an even more indistinct darker area in the pectoral region and some darker streaking on the wing coverts. A few scattered individuals showing breeding dress can be seen as late as 25 October.

HABITAT, DISTRIBUTION AND STATUS

It winters along the Makran and Sind coast and is undoubtedly the most numerous of all the waders in this region. Except rarely, while on passage, it is never seen on inland waters, and is most abundant on open mud flats but will forage on rocky shores and sandy beaches and on mangrove mud banks as well. It is generally abundant from late September with some post breeding arrivals from early August, and most have departed again by late May. A good number however over-summer in non-breeding plumage. Small flocks of 15 to 20 birds can be seen in Ghizri creek, along Clifton beach and in Sonmiani lagoon throughout June and July. Twenty were noted 6 June 1980 in a mangrove creek behind Hawkes Bay of which all were in full breeding dress; these were probably late migrants. In mid-winter flocks of 300 to 700 can commonly be seen roosting in dry areas when the tide is high.

Holmes and Wright noted a sizeable spring passage in the spring of 1965 along the more saline jheel margins of lower Sind. On 16 May, over 150 were noted at Jhol, and smaller numbers up to 28 May. 100 were counted at Hadiero lake and 50 at Jhol on 30 May and the last was seen near Hyderabad on 12 June (op. cit. *JBNHS,* 1968). It has not been noted on passage elsewhere but W. H. Mathews saw a small party on a shingle bank on the

Map 140: Lesser Sand Plover or Mongolian Plover
Charadrius mongolus

Indus river near Skardu, Baltistan in late July and again in early August 1941 (*JBNHS,* 1941). They breed on the high plateaux of Tibet and in eastern Ladakh so might occasionally breed in Baltistan. Col. Meinertzhagen found several pairs on the southern borders of the Deosai plateau at 4,000 metres (13,000 feet) in August and from their behaviour felt certain that they had young in the vicinity. This area is just inside Pakistan territory (Meinertzhagen, *Ibis,* 1927). Status VERY ABUNDANT.

HABITS

Very gregarious in their wintering grounds. They roost in compact groups when tidal conditions interfere with feeding, often on dry sand dune areas or gravelly ridges well away from the tide's edge. They feed in typical plover fashion, running forward rapidly in little spurts, body crouched forward and then suddenly pecking at the surface. Birds when disturbed fly in tight flocks, turning and twisting in unison as they wheel low over the mud flats. Their food is probably variable covering a broad range of small molluscs particularly bivalves according to C. B. Ticehurst (1923), crustacea, especially small crabs and neritic worms. Individuals were watched successfully capturing an unidentified fly (Diptera) which was running over the mud in considerable numbers at Hawkes Bay. It also feeds on sand-fleas (Amphipods).

BREEDING BIOLOGY

Before departure northwards, individuals develop full breeding plumage and have greatly enlarged gonads (Ticehurst, *Ibis,* 1923) indicating readiness to breed immediately upon arrival at summering grounds. At this time they start courtship display and a good deal of intra-specific aggression. On 15 May at Ghizri creek, there was much territorial or advertising calling and birds would crouch with head lowered and neck tucked in, running in unison (presumably when of opposite sexes), or one would chase another with wings raised vertically over back and neck extended, bill low to the ground (presumably a rival male). The very fascinating calls are described below. In Ladakh just east of Pakistan territory they breed on dry open steppes near mountains but on the drying out margins of small marshes or lakes formed by drainage from adjacent hills. These are semi-desert regions, bare of trees and often at altitudes from 4,000 up to 6,000 metres (13,000-19,480 feet). They are colonial in nesting apparently in suitably extensive habitat (Dement'ev and Gladkov, 1951). Stuart Baker describes the eggs as pointed at one end and like those of *C. hiaticula* with the normal clutch 3 eggs and the nest a mere hollow scraped in the ground ('Fauna Series', Vol. 6). Downy chicks can be found in Ladakh in mid-June and they will crouch motionless when their parents give warning calls. Parent birds also give a broken wing distraction display (General S. M. Ghaur, Bombay, pers. comm., 1983).

*VOCALIZATIONS

On their wintering grounds, birds when flushed and forming flight flocks will occasionally utter rather dry sounding short calls 'chitik-chitik'. At the onset of the breeding season they develop a very unique call or display song. This is remarkably like the grating and accelerating 'tuk-tuk' of *Caprimulgus asiaticus* and can be syllabized as 'trit-it-it-it-turkhweeoo' with the first part a rapid high pitched clucking call, breaking vehemently into a harsh grating and inflected end phrase. The whole song lasts hardly more than 1 second and may be repeated 4 or 5 times within the space of half a minute. This type of song call is not mentioned in the *Handbook of Birds of the Western Palearctic* (Cramp et al., 1983).

Large Sand Plover *Charadrius leschenaultii* Lesson

Greater Sand Plover (K. H. Voous)

Geoffroy's Sand Plover (Stuart Baker)

Plate **15**
Map **141**

DESCRIPTION

Head and body length 22-25 cms.
Wing length 12.8-14 cms.
Tail length 5-5.7 cms.
Bill length 23-25 mm. (Stuart Baker)

Only fractionally larger than the Lesser Sand Plover *(C. mongolus).* Both species in winter plumage can easily be confused. It is however slightly longer-legged and with a distinctly longer and heavier bill (see especially winter identification of these 2 species, *British Birds,* Sinclair and Nicholls, 1980). In winter plumage the bill is black and legs vary from greyish yellow to greyish green. The forepart of the face, throat and under parts are white and the crown, ear coverts, a short pectoral band and the whole of the back are sandy buff or greyish brown. The forecrown and a broad line over the eye is white and the mid-breast area is white, the pectoral band being incomplete. In flight they are like *C. mongolus* with white outer tail feathers and an indistinct narrow white wingbar and cannot be separated. After examining perhaps thousands of both species, often seen together, the author considers that difference in bill size and shape is the most reliable character.

In breeding plumage they develop a bright orange-chestnut breast band extending up the sides of the neck which is sharply divided from the white throat but this is narrower and extends less far down the breast than in *C. mongolus*. The crown is generally more concolorous with the mantle and not so chestnut as that of *C. mongolus* in breeding plumage, and the black face mask is narrower, being divided in front of the eye into a loral line and a narrow forecrown stripe with some white on either side of the forecrown. These white spots show up very clearly and at once distinguish this plover from *C. mongolus* when seen in Pakistan in breeding plumage. The legs become black and like others of the genus lack any hind toe. The iris is dark brown. Both sexes are alike in plumage.

HABITAT, DISTRIBUTION AND STATUS

A winter visitor to the seacoast along the Makran and Sind, favouring sandbars and open mud flats, but also occurring in muddy areas near mangroves and on rocky headlands. They often feed in association with *C. mongolus* but are always much less numerous. Autumn arrivals still in breeding dress have been noted as early as 29 July in the Indus delta near Ghabi Dero, but most arrive from late August. The spring migration starts after birds moult into breeding dress from late

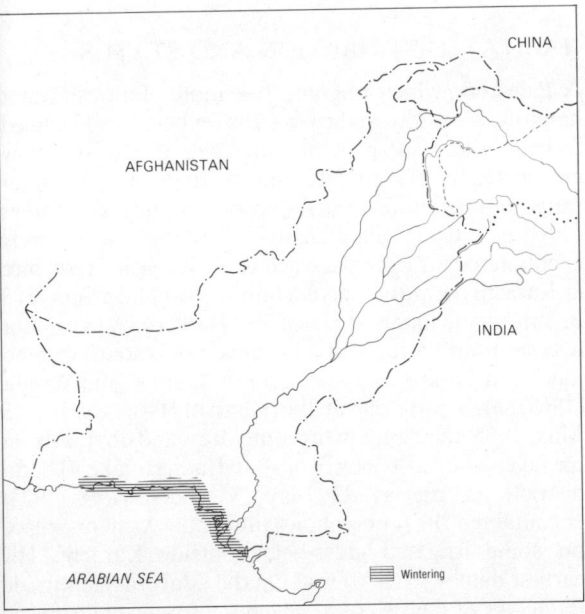

Map 141: Greater Sand Plover or Geoffrey's Sand Plover
Charadrius leschenaultii

March to early April and by 22 April, about 15 to 20 per cent are in full breeding dress. Most have left by mid-May. A very small number over-summer and remain in non-breeding dress. Birds are occasionally encountered in Baluchistan on migration. Marshall collected one in March near Quetta (Meinertzhagen, *Ibis*, 1920) and Christison collected one from a flock of 5 in north Chagai region on 21 April 1939, and another from a flock of six in March 1938, west of Quetta (*JBNHS*, 1942, Vol. 43). Otherwise they are never recorded on passage inland nor even on saline lake shores away from the coast. Status ABUNDANT winter visitor.

HABITS

Gregarious but usually seen in small groups of 10 to 20 birds amongst larger congregations of waders. Will forage in exactly the same manner and types of habitat as the Lesser Sand Plover *C. mongolus*, but better adapted to muddy areas than the Kentish Plover *C. alexandrinus*. They are similar in habits, with swift flight low over the ground, forming tight flocks, wheeling and twisting in unison. In feeding they adopt similar techniques to *Charadrius mongolus* or *Pluvialis squatarola*, running forward rapidly and bending down suddenly to pick up food. Individuals watched on the Indus delta fed on baby *Ocypoda rotunda* crabs and in another area were observed catching quite large Fiddler crabs *Uca annulipes*, seeming to be cautious about the large claw, repeatedly picking it up and dropping it until the claw came off (observations 11 December and 18 April). They also feed upon beetles (Buprestidae), larvae of flies and marine worms (Ali and Ripley, Vol. 2, 1969). Birds collected in Turkmenia USSR had beetles and ants in their stomachs (Dement'ev and Gladkov, 1951).

BREEDING BIOLOGY

It breeds further north than *C. mongolus* but over a wider longitudinal range, from the south-eastern Caspian region through Turkmenia to north-west Mongolia (Flint *et. al.*, 1984). Isolated breeding has been recently recorded from north-western Afghanistan (G. Niethammer, *Ibis*, 1967). Like others of the genus, both parents share incubation duties, and the nest is a scrape in the ground with 3 eggs being commoner than 4. They are however, solitary, not colonial in breeding. Dement'ev and Gladkov describe the eggs as similar to large eggs of *C. alexandrinus* (op. cit., Vol. 3, 1951). Despite observing many birds in full breeding plumage just before spring migration, the author only once saw pre-courtship display. This consisted mostly of aggressive chasing of conspecifics often approaching with both wings held vertically over the back and neck out-stretched and bill held parallel to the ground. This is preceded by rapid calling, the phrases being more melodious and twittering than those of *C. mongolus*. Birds were also observed chasing each other in flight and calling at the same time (21 April, Ghizri creek).

*VOCALIZATIONS

Displaying birds on 21 April were recorded giving repeated and rapid short bursts of a high twittering trill. This varied in tone and inflection becoming more grating when the bird was more excited, 'chirrrrr' or 'dhrrr-dhrrr-dhrrr'. These calls are more melodious than those of *C. mongolus* lacking entirely the final harsh phrase. In winter time birds when flushed utter very short trills 'trrrt' as an alarm or warning call.

Lesser or American Golden Plover
Eastern Golden Plover *Pluvialis dominica* (P.L.S. Muller)   Map **142**

Synonym **Pacific Golden Plover** *Pluvialis dominica fulva*

DESCRIPTION

Head and body length 23-26 cms.
Wingspan 60-72 cms.
Wing length 16-16.5 cms.
Tail length 6-6.4 cms.
Bill length 22-27 mm. (Stuart Baker)

The sub-species which occurs in Pakistan *fulva* is markedly smaller than the Grey Plover *(P. squatarola)* which occurs sympatrically. It has a rather upright stance with large rounded head and narrow neck and long slender dark grey legs. In winter plumage there is a distinctly buffy tinge to the crown and back which is grey-brown as is the crown with buffy speckling and whiter margins to the wing coverts with a few buff spots. The ear coverts and upper breast are more darkly patterned with streaks and spots and the forecrown and a line above the eye looks paler. The iris is dark brown and the bill, which is more slender than that of *P. squatarola*, is also black. Compared with the Eurasian Golden Plover *(Pluvialis apricaria)*, it is greyer and less golden spotted on the back in winter plumage but its yellow tones about neck and mantle help to separate it from the colder grey-white tones of *P. squatarola*. The most reliable field character is the dull grey axilliaries and under-wing coverts revealed on the under surface when in flight (axilliary jet black in *P. squatarola* and pure white in *P. apricaria*). In breeding plumage which can occasionally be seen from April onwards the belly, throat and face become jet black whilst the crown and back are conspicuously spangled with golden buff like *P. apricaria*. Juveniles in their first year are much more yellow tinged on the crown and back and could not be mistaken for *P. squatarola*. There is no hind toe which *P. squatarola* does have. Generally its appearance in winter plumage is more leggy and slimmer necked than the Eurasian Golden Plover *(P. apricaria)* and also it shows a more distinct whitish supercilium. For a good account of specific identification see Pym *(British Birds, 1982)*.

HABITAT, DISTRIBUTION AND STATUS

A Palearctic winter migrant. It is more of a fresh water haunting species than the Grey Plover being also inclined to be gregarious in its winter quarters. In Pakistan they are probably on the western-most fringe of their winter range and therefore occurring very sparingly, sometimes only as solitary individuals. They are also rarely encountered far from the seacoast. In ten years' residence at Karachi the author has encountered inland groups of 3 or single individuals at Haleji and Hadiero lakes on four occasions and a flock of 8 in saline pools south of Dho jheel near Ladiun in Thatta district. Holmes and Wright (1968) saw a party of 8 at Jhol south of Hyderabad on 16 May, 1965; they were in breeding dress and obviously on passage. They also saw 3 or 4 at Hadiero lake (Thatta district) as late as 23 May. Ticehurst *(Ibis, 1923)* encountered them only 3 or 4 times, always in one place on some irrigated grass-fields outside Karachi. His earliest date was 25 August. On the tidal grass alongside the banks of Ghizri creek, which is a fresh water stream, just outside Karachi a few always over-winter and in mid-May 1977 over 70 were counted in a flock, all in

Map 142: Eastern Golden Plover or Lesser Golden Plover *Pluvialis dominica*

breeding plumage and on 17 May in 1980 there were only 3 in this locality. J. Vieillard saw a flock of 77 in the Indus mouth in 1970 (pers. comm.) and R. Passburg watched 3 feeding in mud near a mangrove creek at Sandspit on 30 December 1982. They are occasionally seen with Grey Plovers on the tidal mud flats, (2 seen 25 October, 1 on 3 November). Scully (*Stray Feathers,* 1881) observed them only on autumn passage through Gilgit and collected two males on 27 September and 3 October. The only other inland record is a bird collected by H. Waite (Brit. Mus. coll.) on 28 September near Sarai Alamgir, Gujrat district in the Punjab. Status SCARCE winter visitor.

HABITAT, DISTRIBUTION AND STATUS

Extra-limital and confined to the Kamchatka peninsula

in eastern Siberia so that the majority of the wintering population comes down the eastern sea-board of the subcontinent. When in flocks, shy and difficult to approach but a solitary individual at Hadiero lake was tame and allowed two of us to walk within 10 metres. They feed mainly upon insects and their larvae and minute crustaceans, hunting by sight. They breed in the Arctic tundra, both birds sharing incubation which lasts 27 days and 4 eggs are the normal clutch (Dement'ev and Gladkov, 1951).

*VOCALIZATIONS

Birds flushed utter a rather mellow inflected call softer than that of *Tringa erythropus* but similar in sound, 'too-ee'.

Greater Golden Plover *Pluvialis apricaria* (Linnaeus)

DESCRIPTION

Head and body length 26-29 cms.
Wingspan 67-76 cms.
Wing length 18.1-19.4 cms.
Tail length 6-7.5 cms.
Bill length 21-26 mm. (Stuart Baker)

Slightly larger than the Lesser Golden Plover, but probably inseparable in the field unless seen in flight when the shining white axilliaries and under-wing coverts are clearly distinctive from the greyish brown under-wing coverts of *P. dominica.* Another useful field character when in partial breeding plumage, is the white vent and under-tail coverts, which are blackish in *P. dominica.* In winter plumage it is distinguishable from *P. squatarola* by a buffish yellow caste to the upper plumage, the speckling on the wing coverts being noticeably buff, less grey than in *squatarola.* Also when compared with the Grey Plover *squatarola,* its bill is more slender and crown darker contrasting more sharply with the paler grey stripe above the eye. Legs are dark greenish grey and iris dark brown. Lower belly greyish white, upper breast mottled with darker grey. In breeding plumage it is prominently spangled with golden spots and black, with a black belly and cheeks, bordered

narrowly by white. When on the ground its large rounded head, short constricted neck and upright stance are similar to *P. squatarola.*

HABITS AND BREEDING BIOLOGY

A Palearctic winter migrant. Brooks shot one near Sehwan, Dadu district of Sind on 24 January 1878 and Blanford shot one at Gwadar on the Makran coast in January 1872. Ticehurst records one shot on 7 January 1919 in Karachi, by a Captain Hanna, which is in the Bombay Museum (Ticehurst, 1923). There have been other records in the eastern part of the subcontinent, but no recent authentic records from Pakistan. On 29 July 1977 while on a boat trip in the Indus delta, the author saw on a mud bank a solitary large Golden Plover in breeding plumage and its plumper build with more conspicuous golden speckling at once suggested that it might be *apricaria.* When eventually put to flight the under-wings looked bright white, but in the glaring light off the water one cannot be certain. Status VAGRANT but no doubt the few visitors to the region are easily overlooked, especially when in winter plumage.

Grey Plover or Black-bellied Plover *Pluvialis squatarola* (Linnaeus)   Map **143**

DESCRIPTION

Head and body length 27-30 cms.
Wingspan 71-83 cms.

It is slightly bigger than the Eastern Golden Plover *(P. dominica)* but juveniles often have quite a yellowish wash to the upper body plumage and could be mistaken in winter plumage for the Golden Plover. It is always easily identifiable in flight because the axilliaries show as a jet black patch against the white under-wing coverts and pale grey spotted flanks. In breeding plumage the upper wing coverts and back are spotted white and black without any trace of yellow or buff) gold and the crown is also white closely spotted with black. Also in breeding plumage the lores, ear coverts, throat and belly are black broadly margined with white over the eye, down the sides of the neck and in the pectoral region. The legs are dark plumbeous grey and the feet have a small hind toe not present in *P. dominica.* In winter plumage the rather short heavy bill and large dark eye, pick it out from other waders. It is then mostly white on the throat and mid-breast region, speckled grey on crown, back and wing coverts with paler speckling on the sides of neck and breast.

HABITAT, DISTRIBUTION AND STATUS

A Palearctic winter migrant. Whilst never highly gregarious nor seen in flocks like *Pluvialis dominica,* it is nevertheless very common along the tidal creeks and mangroves of the Sind and Makran coast. It is occasionally seen on inland waters as single individuals. On 5 February 1975 two were seen on Kharrar jheel at Renala Khurd (Punjab). Four were seen on Ghauspur jheel (Jacobabad district) on 16 February in another year. They prefer the margins of open lakes with no reed cover when encountered inland, but the inter-tidal zone on the seacoast is their preferred habitat. In a day's voyage around the Ghizri creeks, up to 300 individuals can be counted, always feeding in association with other species of waders but occupying a variety of habitats, especially wet muddy banks near mangroves, as well as rocky headlands and sandy beaches. A few birds remain all summer on their wintering grounds. 4 were noted on Ghizri creek on 30 May in non-breeding plumage. A few individuals can be noted from the end of March beginning to moult into breeding plumage and by 22 April out of over 200 birds counted on Ghizri creek, 80 per cent had partially or wholly moulted into breeding plumage. The majority migrate to their breeding grounds from late April and by early May they have all departed. The latest date birds were seen in breeding plumage was on 12 May when over 30 were counted around the mangrove channels of Sonmiani lagoon in Lasbela.

Most autumn arrivals are seen on the coast again from end September onwards but a few birds in breeding dress can be encountered from early August, and on 29 July 1977 during a trip through part of the Indus delta four were seen, all in breeding plumage. Status COMMON to ABUNDANT around mangrove creeks.

HABITS

Mostly noted foraging on areas recently exposed by the tide, but wade in shallow water to forage on the margins of inland lakes or ponds. They tend to stand rather erect and upright for periods, then run forward with body held more horizontal and peck at the ground dipping down suddenly to the surface. They appear to do most of their food seeking by sight and are diurnal in feeding. In flight they are like the smaller *Charadrius* plovers, tending to form small flocks, and to fly very swiftly low over the ground, turning and twisting as they go, exposing the prominent black patch under the wing. Their food in winter consists largely of small crabs, but small molluscs, insects and Annelida are also taken. They seem to associate while feeding indiscriminately with large waders such as *Numenius* sp. as well as *Calidris* sp. and perhaps this indicates only their preference for areas where food supplies are plentiful.

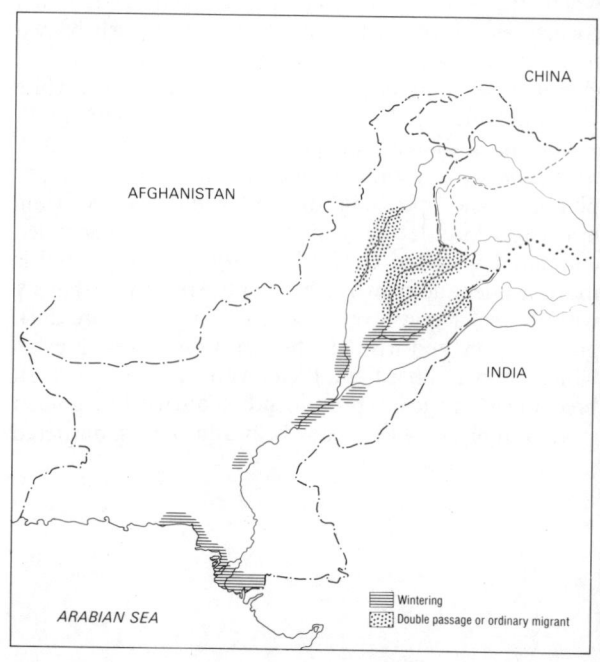

Map 143: Grey or Black-bellied Plover *Pluvialis squatarola*

BREEDING BIOLOGY

They breed in the high Arctic tundra of the USSR. Their nesting habits are similar to those of *P. apricaria*, the nest being made on the ground and both sexes taking equal share in incubation. The normal clutch is 4 eggs and incubation takes 26 to 27 days. The chicks are self-feeding from the time of hatching.

VOCALIZATIONS

They can be heard calling in the spring quite often while on the ground as well as in flight. Their call is a pleasant sounding tri-syllabic 'keee - owheee' and variations on this, in a repetitive wailing tone, are said to form part of their courtship and territorial song when breeding.

Yellow-wattled Lapwing *Hoplopterus malabaricus* (Boddaert)

Illus. **51**
Map **144**

Vanellus malabaricus (Boddaert) of Dillon Ripley's 'Revised Synopsis', 1982

DESCRIPTION

Body length 24-28 cms.
Wingspan 65-69 cms.
Wing length 18.4-20.2 cms. (Stuart Baker)
Tail length 8-8.9 cms.
Bill length 26-28 mm. (Stuart Baker)

A slightly smaller, slimmer version of the Red-wattled Lapwing *(H. indicus)* but very similar in body proportions and general appearance. Long bright yellow legs and an olive to sandy brown back are offset by a white belly and tail, with the latter having a sub-terminal black band. The upper breast is sandy buff with a slightly greyer tone than the mantle, bordered at the bottom of the buff area with a narrow black line. The crown and occiput are black bordered behind the eyes by a white line extending in a 'V' down the nape, the chin is also black. The most noticeable feature is the broad flaps of yellow fleshy skin at the base of the black tipped bill. This wattle extends below the eye in a roughly triangular flap and in front of the forehead in a smaller more square shaped flap. The slender straight bill is black tipped, yellow at the base and there is a narrow yellow orbital ring. In flight the wingtips including outer first

secondaries are black, with the secondaries and outer secondary wing coverts white. Both sexes are alike. This species has no hind toe whereas *Hoplopterus indicus* has a minute hind toe. The iris is silvery white to pale yellow. The brown breast and neck easily distinguishes this Lapwing from the Red Wattled Lapwing.

HABITAT, DISTRIBUTION AND STATUS

A bird of dry fallow land, abandoned cultivation and bare flat gravelly plains or low hills avoiding both very saline and marshy areas such as are favoured by the Kentish Plover or the Red Wattled Plover *(C. alexandrinus* and *H. indicus).* This is an endemic Oriental species which only comes into the extreme southern portion of Pakistan around Lasbela and lower Sind as a summer breeding visitor. It has never been recorded north of Hyderabad or beyond the extreme south-eastern corner of Lasbela. A few birds linger in Thatta district until mid-January but these are unusual and there is a marked influx of birds from late February through March at which time they immediately start territorial calling or advertising. Birds first heard calling at Malir 20 February, and 25 February in three consecutive years. After the monsoon from about August onwards they gather in small flocks often on bare open sandy patches. The largest flock seen was 20 birds on the periphery of Karachi cantonment on 28 September and 15 seen 19 August in the Hab valley. Records of late staying individuals were 20 December Chaukandi tombs and a flock of 4 seen 3 January on the periphery of Karachi airport and a pair seen at Dhabeji 15 January. Status FREQUENT only in lower Sind.

Illus. 51: Yellow-wattled Lapwing *Hoplopterus malabaricus,* both sexes.

HABITS

They often share the same habitat with the Cream Coloured Courser *(Cursorius cursor)* in Sind. They are gregarious outside of the breeding season, but are usually already in pairs upon arrival in Pakistan in March and

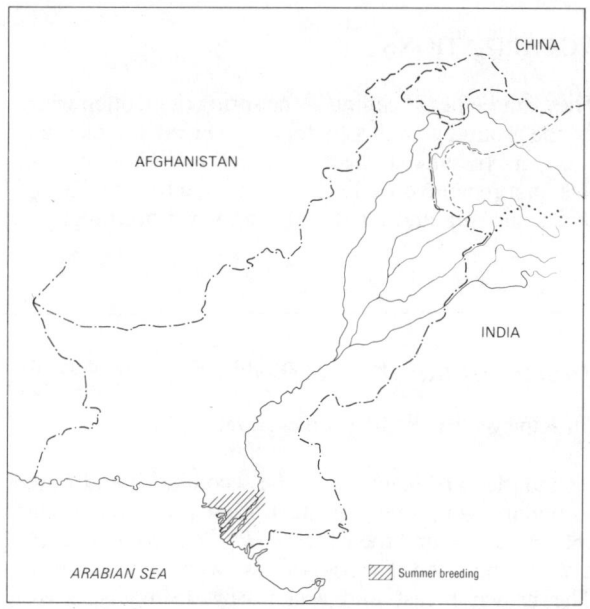

Map 144: Yellow-wattled Lapwing *Hoplopterus malabaricus*

BREEDING BIOLOGY

Very secretive about their nest site, running furtively away from it as soon as danger threatens. They call noisily over their territory, and breeding pairs advertise their presence by their agitated manner. The 'nest' is hollowed out in bare open ground often quite exposed and with barely any lining, beyond a few grass stems, being a mere scrape hollowed in the sand. Four eggs is the normal clutch. K. Eates (MS notes unpublished) found the earliest completed clutch on 27 April with most eggs being laid in mid-May to the third week of May and the latest nest with eggs on 10 June. The eggs are pale sandy or clay coloured often with hardly any markings. A few eggs are handsomely marked with large blotches of dark brown and spots and streaks of inky purple. Both parents share incubation duties, and egg wetting during incubation by the parents has been observed (Jayakar and Spurway, 1965). The chicks will freeze when their parents give them warning cries. They are nidifugous and precocial, being able to feed themselves from the time of hatching.

*VOCALIZATIONS

Their territorial calls at nesting time are vey distinctive, a plaintive 'tee-whit, tee-whit' followed by a rapid repeated 'te-wit-wit-wit-te-wit-te-wit-wit-wit-wit'. Slightly shriller than that of *H. indicus* but equally far carrying and given both from the ground and in flight. Outside of the breeding season they are more silent than the Red Wattled Lapwing but occasionally utter a two-noted call 'dee-wit dee wit' which probably is a warning cry.

eggs may be found from late April to early June. They are largely nocturnal in feeding activity running rapidly forward and pausing to search the ground surface. Their food includes beetles, orthopterous insects, and larvae of insects. Stomachs of eight specimens examined in Bihar, India all contained nothing but remains of small molluscs, chiefly *Melania tuberculata* (Mason and Lefroy, 1912).

Red-wattled Lapwing or
Plover *Hoplopterus indicus* (Boddaert), K.H.Voous 'Recent List'

Illus. **52**
Map **145**

Vanellus indicus (Boddaert) Dillon Ripley's 'Revised Synopsis'

DESCRIPTION

Head and body length 32-35 cms.
Wingspan 80-81 cms.
Wing length 21.2-23.3 cms. (Stuart Baker)
Tail length 10.7-11.6 cms
Bill length 32-34 mm.

Longer legged than the Black-bellied or Eastern Golden Plovers, they have no distinct winter plumage, or marked difference between the sexes. Their backs are grey-brown or olive brown unstreaked or spotted, with a narrow white trailing edge to the secondaries and all-black flight feathers. The upper tail coverts are white as is the tail with a broad black sub-terminal band. The head is black on crown and throat with a large white patch on

the cheeks extending down the neck to the flanks and a black patch covering most of the upper breast. The long legs are yellow with a conspicuous fleshy crimson orbital ring which extends over the sides of the forehead in a small red wattle, the whole pattern of the eye-ring and wattle suggesting the outline of the small metal detachable cap from a tin of aerated beverage. Downy chicks are distinctively patterned with white face and under parts, crown and back greyish buff speckled all over with black with the black spots tending to form lines on the sides of the crown. There is a broad black band extending around the back of the nape from behind the eye and a black patch in the pectoral region sometimes extending right around the throat. Another distinctive black line extends around the flanks under the wings and the sides of the tail. The legs are olive grey. There is a small horny spur developed at the bend of the wing during the breeding season. In the winter of 1984, Dr. Mubashir Hussain saw and photographed a partial albino specimen (shown to author) near Trimmu headworks, it retained the black head and wingtips, but the rest of its body was wholly white.

HABITAT, DISTRIBUTION AND STATUS

Not markedly gregarious but remarkably ubiquitous throughout the irrigated tracts of Sind and Punjab and occurring wherever there are ponds, roadside borrow pits or seepage areas alongside canals, the margins of rivers and streams. They avoid seacoast or marine habitats but may occur alongside streams draining into the sea where they are affected daily by tides as long as there is some salt marsh grassland habitat. Not usually recorded in Baluchistan except along the Makran coast and absent from mountainous regions in the north though it has been recorded (Meinertzhagen, 1920) in Quetta and Loralai in stony streambeds and in Gilgit as a straggler in summer. Scully collected one on 24 April and saw one in June the previous year *(Stray Feathers*, 1881). Whitehead *(JBNHS*, 1911) also considered them summer visitors to the Kurram valley. Wherever they occur in Sind and Punjab they are resident. Status ABUNDANT.

HABITS

Attracted to water, they are most plentiful in marshy areas, the margins of lakes and alongside canals, but occur around surprisingly small wetland habitats, especially village ponds (tanks) and roadside borrow pits where seepage has occurred. They are often seen in pairs and behave quite territorially even in winter months. A pair called loudly and circled over my head, until leaving

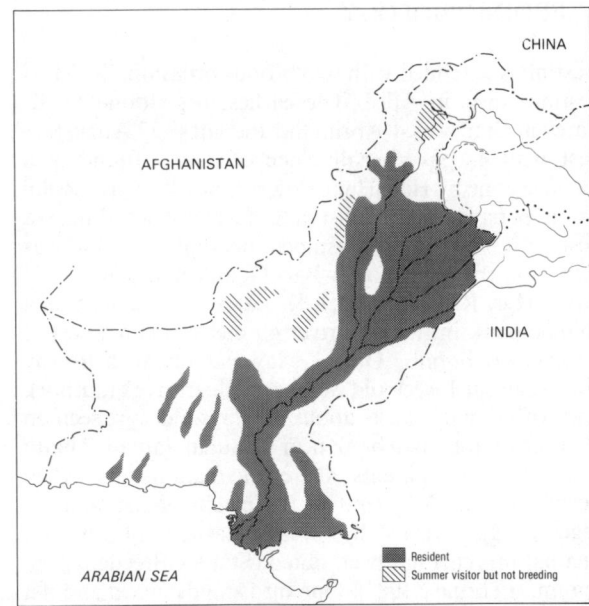

Map 145: Red-wattled Lapwing *Hoplopterus indicus*

Illus. 52: Red-wattled Lapwing *Hoplopterus indicus*, both sexes.

the vicinity, on 24 January, Sanghar, Sind. Occasionally they will congregate in loose small associations, e.g. 15 loafing and roosting alongside a roadside pond near Gharo, Thatta district observed 12 October. Such associations are however, uncommon. They are quick to exploit the food supply created when cultivated land is freshly irrigated and insects are flooded out of their burrows. They will also forage, wading in shallow water and probing with their bills but this is less common than surface feeding, and hunting visually, running forward rapidly to bend down and peck at the ground like other Charadriidae. Mason and Lefroy found mainly insects in 9 stomachs examined, including *Chrotogonus* grasshoppers, larvae of Click Beetles, Elateridae, ants and a few weed seeds and one crustacean (prawn); they also found Diptera larvae (Tipulidae) in some individuals (op. cit., 1912). In a study of the stomach contents of 174 birds collected from the Sunderbans (west Bengal), the diet was wholly animal matter, including 63 per cent insects, 28 per cent fresh water Mollusca, 5 to 6 per cent spiders, 1.9 per cent snakes and 1.2 per cent Annelid worms (Mukherjee, *JBNHS*, Vol. 72, 1975).

Birds examined in Afghanistan had fed on beetles including water beetles, and weevils (Curculionidae) (P.J.K. Burton, in Cramp *et al.*, 1983). They are more nocturnal in feeding than diurnal and can be heard calling throughout the night whereas they tend to be quieter during daylight hours. They are wary of humans but if unmolested will feed close to busy footpaths and thoroughfares. They have a curious habit of standing on roads during the day when not feeting but possibly they are attracted thence by the need for obtaining grit.

BREEDING BIOLOGY

Nesting is extended with two broods occasionally. Most nesting starts in April. The earliest nest found by K. Eates in Sind was 4 April, and the latest 27 August. A nest with 4 eggs partially incubated was found in a seepage zone by Haleji lake (lower Sind) 19 April (Rohil Nana, pers. comm.), whilst in Sahiwal district, Punjab a nest with 3 eggs (and presumed incomplete clutch) was found on the main Lower Bari Doab canal bank on 24 April. (Dr. R. Orr, *in litt.*). K. Eates found a nest on a shingle bank in the Hab river 40 metres from a nest of *Charadrius dubius*. On 14 May parents with downy chicks, about 1 week old were seen Ghizri creek (author), and with downy chicks about ten days old were seen on 19 June on the canal bank near Chauhar Jamali (Thatta district). Both parents share incubation and when nesting in mid-May and at the height of the summer regular egg wetting by breast feathers of bird on incubation relief has been noted (Stanley Breeden, pers. comm., February 1983). One bird stands guard and if a predator (man included) approaches, gives special short warning calls, a high pitched brief 'tit'. The incubating bird has then been observed to crouch lower for concealment or if danger is imminent to slip unobtrusively off the eggs and after running crouched some distance to join its mate in the air in mobbing or warning flight (Dharmakumarsinhji, op. cit., 1972). House crows are serious predators of the eggs. The nest is usually a scrape in the ground with the rim built up a bit by pebbles, goat faecal pellets, and sometimes with a few grass stems in the nest bowl. The pyriform shaped eggs point inwards, and are brownish grey, sometimes with an olive tint and covered with blackish blotches all over. The chicks are self-feeding but tended and protected by both parents. Behavioural repertoire of this plover is typical of the plovers. Aggressive encounters between roosting or loafing birds are more likely in the spring. Three birds 3 April at Sidhnai were noted running with very short rapid steps up to their opponent and stretching upright with breast puffed out and neck extended. Courtship or pair formation includes display flights with tail fanned and shallow rapid wing beats accompanied by loud calling over the nesting territory. On the ground the male also approaches the female with tail fanned and then stands very erect with neck up-stretched and breast puffed out.

Hatching is synchronous. D. Corfield found a nest in shingle on the shores of Rawal lake with 3 eggs. In early May when he found one chick with dry down and the second just emerged, he returned 30 minutes later with a camera, but the third chick had hatched as evidence of egg-shells showed and the parent bird had safely taken all three some hundreds of metres away (pers. comm., 1983). Young chicks will squat in the cover of some plant or bit of debris when the parents utter a warning note (observed Ghizri creek, May).

*VOCALIZATIONS

The loud querulous cries of this Lapwing are one of the most familiar country-side sounds. At dusk pairs often call to each other 'cheet-cheet-chit-tit-toowhit-tit-tit-toowhit'. It has been rendered as 'did he do it' or 'pity to do it' and these phrases once read do seem to evoke its wild calls. They are very vocal during the breeding season, protesting at any intrusions into their nesting territory. Alarm calls are a more rapid call 'treent-treent-tit-treent-tit-treen'. A territorial song often heard at dusk and after is a repeated 'tint-tint-tint', sometimes ending in 'ti-too-whit'.

Sociable Plover *Chettusia gregaria* (Pallas)
Sociable Lapwing *Vanellus gregarius* (Pallas) of Ali and Ripley   Map 146

DESCRIPTION

Head and body length 27-30 cms.
Wingspan 70-76 cms.
Wing length 19.6-20.4 cms.
Tail length 8.4-9.1 cms. (Stuart Baker)
Bill length 27.5 mm.

Intermediate in size between the White-tailed Lapwing *(Chettusia leucura)* which is smaller, and the Red-wattled Lapwing *(H. indicus)* which is bigger, with slightly shorter bill and shorter legs than the White-tailed Lapwing. At all seasons they have a noticeably dark crown and nape which is black in breeding plumage, speckled dark brown in winter, and this bordered below by a broad white line from the forecrown and running across the nape to meet in a white 'V'. In breeding plumage a black line extends below from the lores and through the eye. This is very similar to the head pattern of *Cursorius coromandelicus* with which it agrees in general body colour also, but the latter has a noticeably down-curved bill and white legs. The Sociable Plover has black legs and bill which is fairly short and straight. The back and wing coverts are grey-brown or sandy buff and the sides of the neck have a warm ochraceous tinge, darkening on the lower belly to deep blackish chestnut in the breeding season. In non-breeding dress the throat is white and the upper breast is pale greyish buff, mottled with darker brown. The lower belly is white. In flight the primaries and primary wing coverts are all-black, with white secondaries and upper tail coverts. The tail is black with the outer tail feathers white at the base and white throughout the outer webs. Juveniles have the crown and mantle streaked buff and blackish but their black legs distinguish them from juvenile White-tailed Lapwings.

HABITAT, DISTRIBUTION AND STATUS

It is somewhat similar to the Yellow-wattled Lapwing (*H. malabaricus*) in its habitat preference for dryer land. It usually forages on fallow fields, waste land or scrub desert though it will be seen on grassy swards on the edges of lakes and rivers occasionally, close to other plovers. It is a winter migrant to the north-west of the subcontinent which appears to have become very rare now in Pakistan. At the beginning of the century, Ticehurst considered it fairly common in northern Sind (*Ibis*, 1923) but it was never encountered in the mid 1960s by Holmes and Wright (*JBNHS*, 1968) during three years of touring in that region. F. Koning over four winter visits covering all the major wetlands only encountered one flock of 6 on 14 February 1972 on ploughed land in Jacobabad district (IWRB Reports and pers. comm.). Paul Goriup saw and photographed one in breeding plumage 11 December near Nagar Parkar, in the Thar desert in 1980. In the Punjab, Whistler (*Ibis*, 1922) during World War I found it not uncommon in winter in young wheat in the riverain areas. H. Waite (*JBNHS*, 1948) encountered a flock of 12 on 17 October 1947 on the shores of Khabbaki lake and again a flock on 11 November on the banks of the Chenab river near Jhang. There have been no recent sightings in the Punjab. Whitehead recorded it as a passage migrant in April through Kohat and Bannu districts (1911). A few were recorded by Captain Fulton (*JBNHS*, 1904) on the grassy banks of the Chitral river, with a spring passage in May when flocks of 4 or 5 could be encountered including one in the Bumboret nullah at the end of May at 1,800 metres (6,000 feet) elevation. The only Baluchistan records are from Chaman on the Afghanistan border by Barnes who considered it 'very uncommon' (*Stray Feathers*, 1880). Major Biddulph only saw 1 or 2 each year in Gilgit main valley when they passed through at the beginning of April in full breeding dress (*Stray Feathers*, 1881).

Though it is described as 'regular and fairly common in Pakistan' in the 'Handbook' (Ali and Ripley, 1969), lack of recent authentic sightings indicates that its present status is RARE.

Map 146: Sociable Plover *Chettusia gregaria*

HABITS AND BREEDING BIOLOGY

It is restricted in its Russian breeding range to the western steppic regions, mainly to dry areas of wormwood *(Artemisia)* betwen Kuibyshov and Barnaul on the Kalunda steppe of the USSR (Knystautas, 1987). It is probable that recent agricultural developments have effected the breeding population of this species adversely, which is now listed in the *Red Data Book of the USSR* (Knystautas, op. cit.). Their status in Pakistan as winter visitors has likewise obviously changed. They nest in small colonies in dry areas on the ground. In winter quarters they are also gregarious feeding in ploughed fields and on dry grassland. Their food comprises mostly beetles, grass-hoppers (Acridiodae) and coleopteran larvae (Curculionidae) (Dement'ev and Gladkov, 1951). They are undoubtedly beneficial to agriculture in their feeding habits. They feed by picking up items from the ground surface, sometimes probing between clods and around the base of plants.

VOCALIZATIONS

Not recorded on the subcontinent, but described as a short rather rasping 'reck-reck' or 'ket-ket' (Cramp *et al.*, 1983).

White-tailed Lapwing *Chettusia leucura* (Lichtenstein) K. H. Voous, 1977   Map **147**

Vanellus leucurus (Lichtenstein) Dillon Ripley's 'Revised Synopsis', 1982.

DESCRIPTION

Head and body length 26-29 cms.
Wingspan 67-70 cms.
Wing length 16.9-17.8 cms.
Tail length 7.3-7.8 cms. (Stuart Baker)
Bill length 35-38 mm.

Only slightly smaller than the Sociable Plover *(Chettusia gregaria)* or equal in body size, but noticeably longer legged. It is about the same size as the Yellow-wattled Lapwing *(Hoplopterus malabaricus)* but easily separable because of the very different habitat which they frequent. It is a sandy brown or buff olive brown on the head and back becoming greyer on the lower neck and breast. The lower breast and belly and the tail are all-white. Its legs are bright chrome yellow and the bill blackish with no yellow at the base nor any facial wattles. The iris is dark brown and there is no orbital eye-ring as in the wattled plovers. In winter plumage, the forehead and lower cheeks are whitish and the upper breast pale grey-brown. In breeding plumage the only change is a slight darkening of the grey upper breast and the head. In flight it is very distinctive with all-white tail and upper-tail coverts, black wingtips but prominent white primary wing coverts and secondaries, also there is a narrow black line separating the buff wing shoulders from the outer secondary wing coverts. Lack of any head crest, rather slim build and upright carriage characterizes this elegant plover in the field. The sexes are alike. Juveniles which can be seen in the autumn have brown tips to the tail feathers and black tips to the feathers of the mantle and scapulars so they look rather like a winter plumage Reeve *(Philomachus pugnax)*, but with much longer legs.

HABITAT, DISTRIBUTION AND STATUS

An inhabitant of fresh water marshy areas and lake shores, often sharing the same habitat with Redshanks *(Tringa totanus)* and Marsh Sandpipers *(Tringa stagnatilis)* in Pakistan and never occurring in such dry areas as the Yellow-wattled Plover *(Hoplopterus malabaricus)*, nor such limited restricted damp areas as the Red-wattled Plover *(H. indicus)*. With one localized exception, it is a winter migrant to Pakistan, being an early visitor, numbers being seen from early September in lower Sind and departing again about mid-March. Latest date seen 22 March. It is rare in the Salt Range and the NWFP possibly due to lack of suitable habitat. Whitehead noted it as rare in winter around Kohat but more frequently encountered around Bannu *(JBNHS,* 1911). H. Waite *(JBNHS,* 1948) occasionally saw it at Kallar Kahar in the Salt Range and Khinjar jheel in Muzaffargarh district. It is quite common on the seepage zones of all the major irrigation headworks in the Punjab such as Balloki headworks, Sidhnai spill channel, Trimmu headworks and Jinnah barrage on the Indus. In Sind it is most plentiful in the east Narra and around Sanghar where loose flocks of 6 to 7 birds can be encountered in suitable wetlands. A flock of 15 was seen at Dho Dand near Ladiun, Thatta district, 9 January. At Ghauspur, 38 were counted around the lake shore in 15 February 1971. A few are regularly seen, but less commonly, in southern Sind, Mahboob Shah lake, and Haleji etc. In the first week of May 1984 about 10 pairs were discovered on Zangi Nawar lake in extreme south-western Baluchistan. Their courtship displays and territorial behaviour were observed over three days, which indicated that they intended breeding there and this was corroborated by the local wardens who patrol the lake. (Roberts, *JBNHS,* 1985). During a summer survey of Zangi Nawar in July 1987 by Ashiq Ahmed, a pair was located with a nest containing two eggs on 9 July. It was situated on a bare open stretch of mud on the lake shore (field notes by Mohammad Idrees Chugtai, Pakistan Forest Institute). On previous winter visits to this lake (author), this plover was not recorded. They also occur on passage through Baluchistan and 30 were counted on the shores of Kushdil Khan lake on 30 October. They were also noted by Delme Radcliffe on passage through Quetta valley in spring and autumn

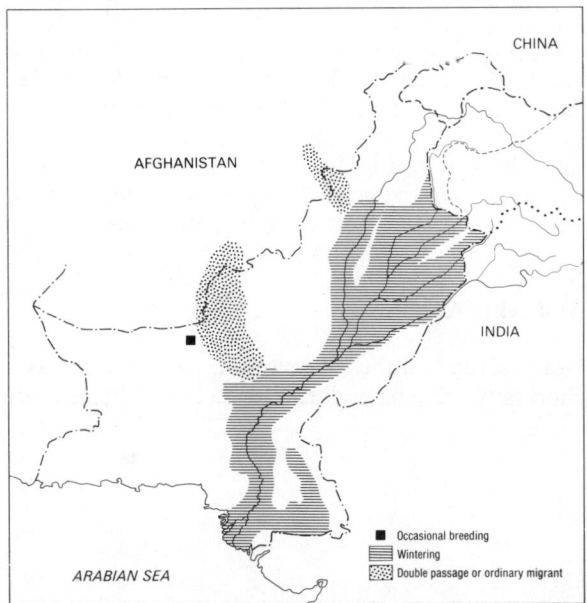

Map 147: White-tailed Lapwing *Chettusia leucura*

(Ticehurst, *JBNHS,* 1927*b*). It has not been recorded in Gilgit. Status COMMON.

HABITS

Observed flighting to dryer ploughed fields at dusk to roost in a loose flock (9 birds Ghauspur jheel). They are mostly diurnal in feeding activity and where numbers are sufficient, gregarious in their wintering grounds, occurring in groups of 3 or 4 up to 7 to 8 birds. They more frequently seek food in shallow water and foot dabbling with one leg extended is a common technique. Their food includes insects, chiefly water beetles (Hydrophilidae) and grasshoppers, but also worms (Annelida) and crustacea and small molluscs. They have been noted probing in mud with their bills. Birds on migration on Kushdil Khan lake, Baluchistan, appeared nervous and in the presence of humans some bobbed their bodies forward, remaining standing in the same positions (30 October, observation).

BREEDING BIOLOGY

Colonial in nesting in Iraq (Baker, 'Fauna Series', 1929) but also nests solitarily. Breeding mostly extra-limital except for a newly discovered colony in the Chagai district of Baluchistan (Roberts, 1985). Its nesting habits are like other plovers', a scrape in the ground with little or no lining, serves for a nest and 3 to 4 eggs, pyriform in shape, pointed ends inwards, comprising the clutch.

Intra-specific aggressive behaviour has been observed in the spring in the lower Sind. One individual will bend its body horizontally with head and neck lowered and run forward calling at the same time till the opponent takes to wing (observed 14 October, Mirpur Sakro). On Zangi Nawar lake in Baluchistan where pairs were defending small islets on the lake on which they were presumed to be starting to nest, other waders such as *Tringa glareola* were driven away by crouching forward, neck outstretched and running up to the intruder. Male birds were noted on these islets assuming a very upright posture and calling repeatedly perhaps 20 to 30 times when in visual contact with a conspecific.

The territorial display observed at Zangi Nawar consisted of the male bird, in shallow fluttering flight, slowly rising then volplaning to the ground again calling all the while. As it descended, the calls were accelerando.

*VOCALIZATIONS

Rather silent in winter quarters but when disturbed or going to roost in evenings utters a short querulous two-noted call 'dzee-whik' usually not repeated. Wintering birds in an aggressive encounter uttered a short shrill 'wik-wik-wik' call and this has also been heard when birds are flying around. During the breeding season, they become much more vocal, males calling repeatedly in their territory, 'teeuwhit-teeuwhit'. A higher pitched and more melodious call than that of the Red-wattled Plover, and somewhat reminiscent of the calls of the Gull-billed Tern *(Gelochelidon nilotica).* In display flight the calls are more rapid and excited, 'ti-toowhit-ti-toowhit' rising in pitch and tempo to a crescendo as they fly upwards on a shallow trajectory and then volplane to earth again.

Green Plover, Lapwing or Peewit *Vanellus vanellus* (Linnaeus) Map **148**

DESCRIPTION

Body length 31 cms.
Wing length 22-23.6 cms.
Tail length 10.8-11.9 cms.
Bill length 23-27 mm. (Stuart Baker)

About the same size as the Sociable Plover *(Chettusia gregaria)* but rather shorter legged than the *Chettusia* or *Hoplopterus* Plovers. Slightly larger than the Grey Plover *(Pluvialis squatarola).* They have a conspicuous plume on the back of the crown of upswept narrow black feathers. The back is dark olive brown in winter with a broad black breast band and the top of the head extending to the crest is also black. The rest of the body is white except for the under-tail coverts which are pale chestnut and a broad black band through the tail. The face also has black around the eye, usually extending in a few spots behind the eye and in a diagonal line below the cheeks. The short straight bill is black and the legs and feet are orange-brown. In breeding plumage the whole of the forepart of the face, chin and throat also becomes black. There is generally a buffy tinge to the hind neck and sides of the crown in winter plumage. In good light the back shows bronzy green with a purple sheen on the wing shoulder. In flight the flight feathers are all-black except for the wingtips which are white and the wings are noticeably broad tipped giving the bird a very characteristic flight outline. The sexes are alike in plumage.

HABITAT, DISTRIBUTION AND STATUS

A Palearctic winter migrant occurring not very commonly in the northern Punjab and NWFP, mainly on grasslands near seepage areas, and edges of lakes and also in gravelly river-beds and occasionally on dry fields. A few over-winter along the Gilgit main river and in Chitral main valley. They come down to the plains by late October often leaving again by early February. Their favoured habitat is around the grasslands created by seepage zones upstream of the major irrigation barrages in the Punjab, for example Balloki and Sidhnai on the Ravi, Panjnad and Jinnah barrages on the Indus. In Sind it is rare especially in the southern part. In 10 years of observations an occasional bird has been seen around Haleji lake and roadside ponds near Gharo. A favourite wintering ground is the grassland around Ghauspur jheel, 202 counted 10 November 1979 and over 450 on 9 December 1968. Flocks have been seen on bare open grassy plains near the Jhelum river (40 or more) at Mandi Bahauddin 15 January, 30 at Sidhnai on 13 November on drying out mud flats. A few always haunt the shores of the Salt Range lakes in winter. In Baluchistan, a flock of 40 was seen on 30 October around Kushdil Khan lake, probably still on passage. Meinertzhagen considered them regular winter visitors in flocks to Baluchistan (*Ibis*, 1920), but Christison only noted a few around Zangi Nawar lake in the Chagai in winter (*Ibis*, 1941). Whitehead (*JBNHS*, 1911) recorded it as a common winter visitor around Kohat and Kurram valley. A few individuals were regularly seen (author) around the refuse tips in the Potohar plain 10 miles north of Rawalpindi in November.

Major Biddulph recorded that flocks of 15 to 20 could be encountered in March on spring migration but that all had gone north by 25 March (*Stray Feathers*, 1881). It is never seen in the big flocks which are a feature of western Europe in winter. Status locally COMMON.

HABITS

Largely diurnal in feeding. Runs forward in short spurts and bends down to peck at the ground, walking more slowly on rough ploughed land with head lowered as if searching. Food consists of all kinds of invertebrates, especially larvae and pupae of insects, worms and small

Map 148: Northern Lapwing or Green Plover *Vanellus vanellus*

molluscs. Around lake margins in grass they appeared to be catching flies (November, Ghauspur lake).

BREEDING BIOLOGY

They breed extra-limitally nesting solitarily in the northern steppe regions of the USSR. The nest is a bare scrape often on open ground and 3 to 4 pyriform shaped eggs are laid, both sexes sharing incubation duties.

VOCALIZATIONS

Though very vocal on their nesting territory they are rather silent in winter quarters, an occasional individual uttering the loud squeaking 'peea-whit' (heard Gilgit airfield, 12 November).

FAMILY SCOLOPACIDAE

Sandpipers and their allies

SUB-FAMILY CALIDRIDINAE

Great Knot or Eastern Knot *Calidris tenuirostris* (Horsfield)

Plate **15**
Map **149**

DESCRIPTION

Body length 27-29 cms.
Wingspan 62-66 cms.
Wing length 16.5-18.5 cms.
Tail length 6.3-6.9 cms.
Bill length 39-47 mm. (Stuart Baker)

About 15 per cent larger than the Red Knot *(Calidris canutus)* with a slightly longer and heavier bill, and a predominantly grey colour without any buff tones often visible in the plumage of *C. canutus* when in winter dress. It is a noticeably stocky bird with short neck and tarsi, but compared to *C. canutus* appears more upright, less rounded in body outline with a smaller head. The crown is streaked with black and there is indistinct grey mottling on the sides of neck and pectoral region. The mantle is grey streaked with black and the lower rump and upper-tail coverts are a clearer white, more indistinctly scalloped with grey than in the Red Knot *(C. canutus).* For a good account of differences and identification see Marchant *(British Birds,* 1986). The throat and breast are white. In body size it could be confused with the Bar-tailed Godwit *(Limosa lapponica),* or Grey Plover *(Pluvialis squatarola),* both occurring in Pakistan in the same areas and being about the same size and greyish colour, but the Great Knot's legs are shorter and stouter than *L. lapponica's* and its wholly black bill is stouter at the base lacking the pink colour of the godwit's bill which is much longer. In general body proportions and bill length it looks like a Reeve *(Philomachus pugnax)* but its tail lacks the cross-barring of *P. pugnax* or the dark mesial line down the middle of the white upper-tail coverts. In flight the white rump, grey-brown unbarred tail and a white wingbar are helpful field characters. The legs are greenish dusky. In summer plumage it is more boldly patterned in black and white, with dark spotting on the flanks and pectoral region and mantle, and lacks any of the chestnut red on the face and breast which at once distinguishes the Red Knot *(Calidris canutus).*

HABITAT, DISTRIBUTION AND STATUS

A purely maritime species in its winter quarters, only being seen in the inter-tidal zone along the Karachi and Makran coast in winter time. It breeds in eastern Siberia in the sub-Arctic zone and is an uncommon visitor to Pakistan, inhabiting usually the more extensive and remoter mud flats and estuaries. A few visit Ghizri creek mud flats each winter, where they associate with Bar-tailed Godwits and Grey Plovers. Ben King saw a flock of about 70 on a sandbar near the mouth of Ghizri creek on 15 March 1981, but the author over ten years and numerous visits to this creek has only twice located small groups of 6 birds, 23 October and 23 March 4 birds. Around Sonmiani lagoon in Lasbela it is commoner with small flocks of scattered individuals being seen from October through till late May, and the maximum number 14 in March (author obs.) and the latest date seen, 4 on 5 June (Ticehurst, *Ibis,* 1924). Individuals showing some of their breeding plumage, particularly the heavy black spotting on the breast have been noted alongside birds in winter plumage, both on 23 October and 20 May. They have not been recorded even on migration on any inland fresh waters, but on 10 January 1974, a flock of about 22 birds was seen on Sandho lake, a shallow salty expanse of water, on the Rann of Kutch on the southern border of Badin district adjacent to India (F. Koning, pers. comm.)

Curiously, none have been recorded along either coast of peninsular India, but they occur regularly in winter along the Orissa coast, and in Bangla Desh as well as occurring throughout the south-west Pacific from the

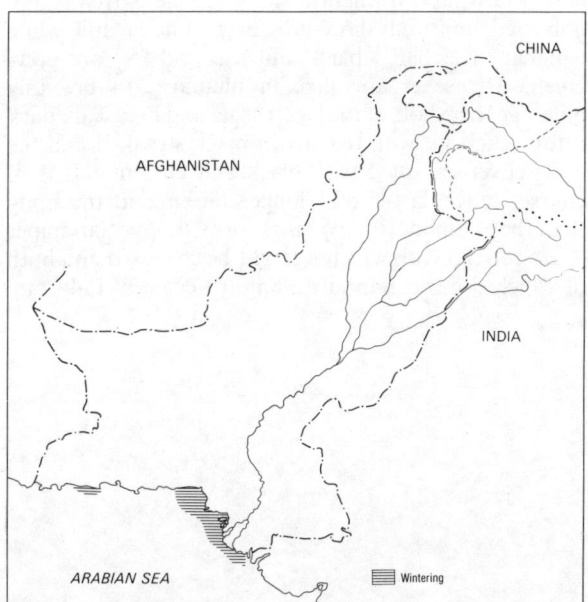

Map 149: Eastern Great Knot *Calidris tenuirostris*

Philippines to Indonesia and Australia. Even in Dr. Ticehurst's days (World War I) he considered it the least common of all the waders visiting Karachi seacoast. Status RARE.

HABITS AND BREEDING BIOLOGY

They nest on the more exposed rocky ridges of the sub-Arctic zone in north-eastern Siberia. (Hayman, Marchant and Prater 1986). They are gregarious in their winter quarters, but forage individually often in association with Bar-tailed Godwits *(Limosa lapponica)* and Grey Plovers *(Pluvialis squatarola)*. They probe in the mud while feeding (observed Dambh village) and presumably subsist principally upon marine worms and small molluscs. Normally they are encountered in small flocks of 6 to 12 birds.

VOCALIZATIONS

Little has been recorded, but on its breeding grounds its voice has been described as 'typical of the Plovers and unlike other Sandpipers' (Dement'ev and Gladkov, Vol. 3, 1951). Wintering birds have been described as giving a flight call 'queet-queet' somewhat similar to the call of the Black-tailed Godwit (Yamashina, 1961).

Red Knot *Calidris canutus* (Linnaeus)

DESCRIPTION

Body length 23-25 cms.
Wingspan 57-61 cms.
Wing length 16.2-17.4 cms.
Tail length 5.0-6.6 cms. (Salim Ali)
Bill length 3.2-3.5. cms.

Smaller than the Great Knot *(C. tenuirostris)*, it is a rather stout bodied bird with short legs and a relatively short bill which is comparatively shorter than that of the Great Knot *(C. tenuirostris)* and is broad at the base and slightly thickened distally. In winter plumage it is pale grey with some vertical barring along the flanks, and scalloped spots on the lower neck. The face and under parts are whiter. In flight it shows a pale barred rump, unbarred uniformly grey-brown tail and a dull white wingbar. The bill is black and legs and feet are grey-green. The sexes are alike in plumage. In breeding plumage the whole of the face, throat and breast are dark rufous chestnut with the crown black streaked and the wing coverts spotted with black and chestnut. It is of course, much larger and longer looking in the body than the Dunlin *(C. alpinus)* or Curlew Sandpiper *(C. testaceus)* with which it might be confused and both of which occur commonly along the coast of Pakistan.

HABITAT, DISTRIBUTION AND STATUS

It must be only a rare straggler to the subcontinent. Colonel Meinertzhagen *(Ibis,* 1920) collected a specimen on 26 March on Kushdil Khan lake, Baluchistan. The only other record for the region is of a bird collected on 15 December 1928 in Sri Lanka. The Baluchistan specimen was a female and was alone, no other Knots were seen by Meinertzhagen *(Ibis,* 1920). Status VAGRANT.

HABITS AND BREEDING BIOLOGY

Breeds in the high Arctic being more abundant in the Nearctic or north American Arctic islands than in Asia. In winter they are normally gregarious, feeding in the inter-tidal zone and being confined to the seacoast. They feed by probing in the mud, upon a variety of molluscs, both bivalves and gastropods, and small crabs in their European feeding grounds (Cramp *et al.,* 1983).

VOCALIZATIONS

A liquid whistling 'quick-ick' when put to flight also a muttered 'knut' as a contact call.

Sanderling *Calidris alba* (Pallas)

Calidris albus Dillon Ripley, 1982

Plate **15**
Map **150**

DESCRIPTION

Body length 20-21 cms.
Wingspan 40-45 cms.
Wing length 11.8-12.9 cms. (Stuart Baker)
Tail length 5.2-6.2 cms.
Bill length 23-29 mm. (Salim Ali)

It is fractionally larger bodied than the Dunlin *(C. alpinus)* or Curlew Sandpiper *(C. ferruginea)* and in winter plumage most distinctive amongst the maritime waders because of its pale grey unstreaked crown and back and predominantly white face with a touch of black at the wing shoulder. Its short black bill at once distinguishes it from Dunlin and Curlew Sandpipers alongside which it is frequently seen on the tide line. The legs and feet are also jet black. In flight a broad white wingbar, and the blackish primary wing coverts show up well, also the central tail feathers are darker greyish brown and also the middle of the rump. In breeding plumage, the mantle becomes much more conspicuously patterned with black and rufous margined feathers and there is a pale chestnut buff collar around the nape and in the pectoral region, mottled with black. In Pakistan all birds show full breeding dress by mid-May. The feet lack any hind toe or halux, probably an adaptation to their special foraging techniques.

HABITAT, DISTRIBUTION AND STATUS

It is a winter migrant to Pakistan spending the winter all along the coast from Karachi and throughout Baluchistan. It typically avoids mangrove creeks or very muddy estuaries, preferring sandy beaches and a firmer substrate or occasionally rocky coastlines. It arrives in good numbers along the Karachi seacoast from its breeding ground in September and most have gone north again by mid-May but a few small flocks have been noted throughout the month of June (author), along Clifton beach, Karachi. It has never been noted on inland waters while on migration. Status ABUNDANT on seacoast only.

HABITS

The Sanderling is instantly recognizable amongst the waders which throng the shoreline by its active habits. It characteristically forages by running swiftly along the tide's edge stopping suddenly to pick up minute crustacea and larvae of diptera brought in by each wave or to probe in the sand. Shallow sandy beaches are it favoured habitat and it generally avoids the sticky mud typical of mangrove shores. Sanderlings forage individually but can be seen in scattered groups and between 30 and 40 can generally be counted along one mile of Clifton beach at Karachi. Birds seen on the rocky headlands at Buleji and Cape Monze were probably resting and were not noted foraging actively. They will roost gregariously at high tide, but are among the most active of waders and Ticehurst noted their habit of continuing to hunt right up to the turn of the full tide when other 'probing' waders had long since gone to roost.

BREEDING BIOLOGY

Breeds in the high Arctic tundra, preferring gravelly or stony sites with scanty vegetation. Studies in the Canadian Arctic indicate that females commonly lay two clutches, the first of which is incubated by the male alone who also looks after the chicks, while the female incubates her second clutch (of four eggs).

VOCALIZATIONS

Its flight call is a liquid 'twick-twick'. This is a rather shrill and sharp call, syllabized by some authors as 'plick-plick' (Brunn and Singer, 1970).

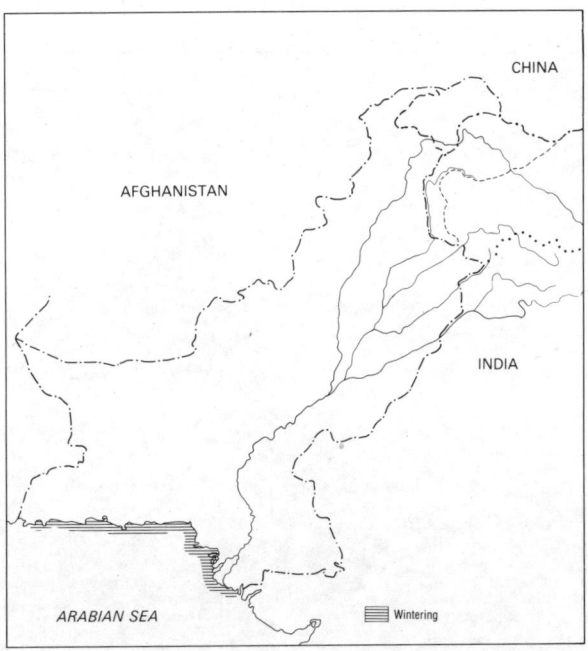

Map 150: Sanderling *Calidris alba*

Little Stint *Calidris minuta* (Leisler)

Plate **15**
Map **151**

DESCRIPTION

Body length 12-14 cms.
Wingspan 34-37 cms.
Wing length 9-10.2 cms.
Tail length 3.4-4.2 cms.
Bill length 17-20 mm. (Salim Ali)

This tiny sandpiper is about the same size as a Wagtail *(Motacilla sp.)* when seen alongside. With short straight bill, compact body and stout legs it looks from a distance, bigger than it is. In winter plumage it is grey above and white below with darker mottling (feather shafts) on the mantle and crown and some grey mottling in a restricted area around the sides of the neck. It is the same size as Temminck's Stint *(C. temminckii)* but distinguished from the latter by having jet black not olive green legs and when spread the outer tail feathers are pale grey not white as in *temminckii*. It looks like a diminutive Dunlin *(C. alpina)* in general porportions and body colouration, especially in flight. In breeding plumage the feathers of the back and wing coverts become black with rufous margins and the foreneck and upper breast are tinged dull rufous with dark brown spotting. The sexes are alike. In flight a narrow white wingbar and dark centre to the rump are conspicuous.

HABITAT, DISTRIBUTION AND STATUS

This is a highly gregarious species on migration and in its winter quarters. The bulk of the population spends the winter on the seacoast but large numbers can be observed on inland waters and seepage pools during spring and autumn passage. First arrivals have been noted (author) in lower Sind (Haleji lake) as early as 21 July when still in breeding plumage. Ticehurst (*Ibis*, 1924) noted them as arriving in numbers by the end of July in Karachi mud flats. During a 13 hour voyage through the Indus creeks on 29 July only one flock of Little Stints, all in breeding dress, was encountered. It has been recorded on spring passage around Kohat and in the Kurram valley (Whitehead, 1911) and in Gilgit on autumn passage in September (Scully, 1881). It has also been noted as plentiful on spring passage (mid-April) and autumn (October) in Baluchistan on Kushdil Khan lake (Meinertzhagen, 1920). Flocks totalling about 400 were noted 17 March at Lal Sohanran lake (Bahawalpur division). A considerable number of birds over-winter in the Punjab but they are much less common than Temminck's on inland fresh-waters. Flocks have been recorded in November and March in Muzaffargarh district, and in early January at Taunsa barrage on the Indus (Dera Ghazi Khan district). H. Whistler considered them as common as Temminck's Stints in winter in Jhang district (*Ibis*, 1920). Most birds have assumed their summer breeding dress by the third week of April and most have left by mid-May, but stragglers have been seen as late as 9 June (Siranda jheel, Lasbela). A few scattered individuals were noted on Sidhnai spill channel (southern Punjab) as late as 9 May. In mid-winter, around the mud flats of Ghizri creek five to ten thousand could be counted in the 1980s during a half day's boat trip. In the late 1960s as many as 2000 could be counted during a walk along the oil pipeline at Clifton. Status very ABUNDANT along the coast from Karachi to Baluchistan. It typically avoids mangrove creeks or very muddy estuaries preferring sandy beaches and a firmer substrate or occasionally rocky coastlines. In good numbers along the Karachi seacoast from its

HABITS

A remarkably active little bird usually seen on the drying out margins of small pools, roadside borrow pits, tidal creeks and estuarine mud flats, usually in small flocks or scattered groups. They forage most commonly by picking at the surface and darting about actively on mud banks. Less commonly they feed by probing with their bills, and will wade in shallow water (observed Ghizri creek, Karachi). They are generally encountered in company with Dunlin *(C. alpina)* and Curlew Sand pipers *(C. ferruginea)* and are often quite tame and quicker to return to settle and feed near a human intruder than their larger congeners. They become more vocal and quarrelsome as spring advances, keeping up a constant twittering and frequently chasing each other when approached too closely by another feeding bird. They typically fly in tight compact flocks when disturbed, turning and wheeling in unison and travelling very swiftly close to the water or ground surface. Their

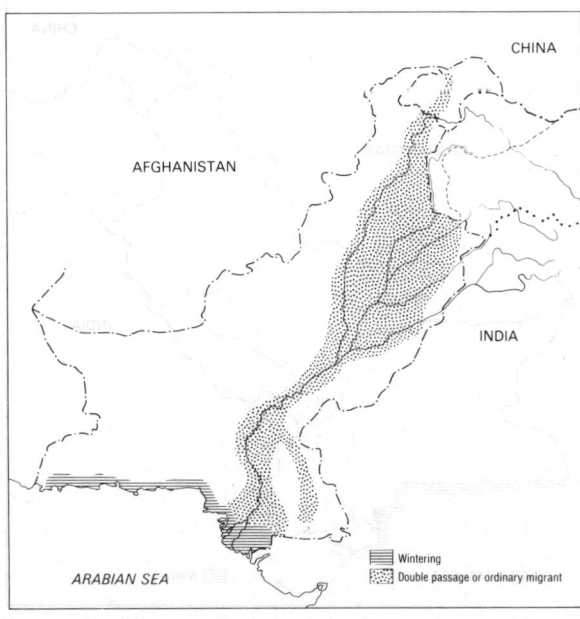

Map 151: Little Stint *Calidris minuta*

food consists of small crustacea, larvae of diptera and adult flies, and small molluscs. Ticehurst (*Ibis*, 1924) noted tiny marine shells in their stomachs, and in a study in Bihar district, India they had consumed *Planorbis* snails, water beetles *(Copelatus pugnax)* and other unidentified insects (Mason and Lefroy, 1912).

BREEDING BIOLOGY

They breed exclusively in the Palearctic high Arctic coastal regions both of Scandinavia and the USSR. Ticehurst noted that adults in worn breeding dress arrived back on the Karachi coast from the end of July and that juvenile birds appeared, three to four weeks later. There is no evidence of any significant population of non-breeding birds which over-summer on Pakistan's coast. Like other high Arctic breeding Calidridinae they are probably double clutched, the female alone incubating and tending the second clutch, whilst the male

rears the first, thus taking advantage of an abundant Arctic summer food supply and compensating for the numerous ground predators. On 5 May (1977) on Ghizri creek, Little Stints were in single species flock concentrations of 150 to 200 birds and were indulging in much excited displaying and aggressive posturing. Pairs approached each other with wings raised over backs, head and neck lowered and extended and then jumped at each other attempting to strike with their feet. All the while the whole flock kept up a constant rapid twittering.

*VOCALIZATIONS

By late April birds are very vocal, keeping up constant rapid twittering cries which vary in pitch, rising and falling. A flock calling simultaneously as it feeds on the mud flats in early May emits a massed high pitched trilling chorus of quite melodious twitters. In winter, they emit low 'tirrt-tirrrt' calls in flight.

Temminck's Stint *Calidris temminckii* (Leisler)

Plate **15**
Map **152**

DESCRIPTION

Body length 13-15 cms.
Wingspan 34-37 cms.
Wing length 9-10 cms.
Tail length 4.5-4.8 cms. (Salim Ali)
Bill length 16-20 mm. (Stuart Baker)

Whilst *C. minuta* calls to mind a miniature Dunlin in winter plumage, this diminutive sandpiper with its grey upper-breast is more like a small Common Sandpiper *(Actitis hypoleucos)*. They have slightly longer wings than the Little Stint, extending to the tail tip when at rest, and are mainly distinguished from *C. minuta* by the darker grey upper parts and grey upper-breast with olive yellow not black legs and the outer tail feathers showing pure white in flight in contrast to the wholly grey outer tail feathers of *C. minuta*. In winter plumage the crown and mantle is more uniformly mottled dark grey and less speckled than that of *C. minuta* in similar plumage. Both sexes are alike. The bill is short, straight and stout at the base. In breeding plumage the back becomes more boldly patterned with black centres to the feathers and rufous margins but the upper parts are generally less bright rufous and more brown than in *C. minuta* and the upper breast becomes buffish brown with darker spots. (based on birds seen Sidhnai spill channel, early May).

HABITAT, DISTRIBUTION AND STATUS

A winter migrant to the subcontinent, being very numerous throughout the inland lakes and inundation

areas, in the Punjab plains. No small roadside borrow pit or canal bank seepage pool is too small to escape the attention of this little wader. In lower Sind it is comparatively rare being usually encountered as single individuals or in parties of only 3 or 4 birds, but in the Punjab Salt Range and Sargodha and Muzaffargarh

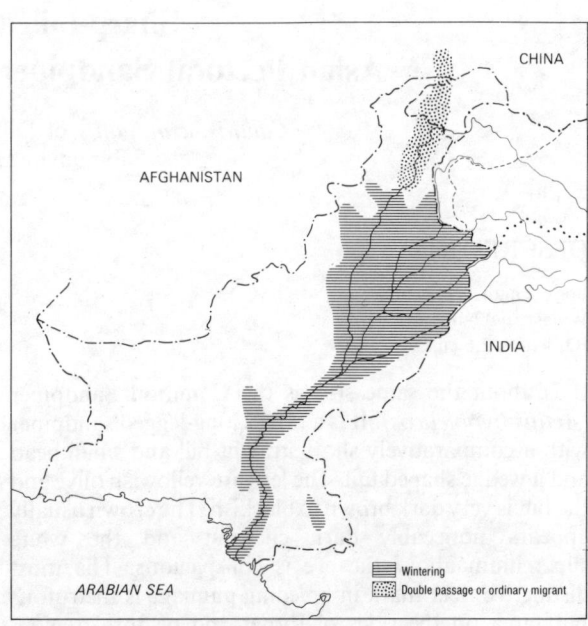

Map 152: Temminck's Stint *Calidris temminckii*

districts it is extremely plentiful with groups of 10 to 20 birds frequenting almost every wayside pool. Occasionally an individual bird will be seen in company with other waders in the brackish tidal creeks of the Indus delta, Ghizri creek and the coast, but generally they prefer sweet water habitats. Frequently they can be encountered feeding in newly irrigated cropland some distance from permanent swamps. There is a marked spring passage through Gilgit from the middle of May onwards (Biddulph, 1881). Earliest arrivals have been noted (author) in lower Sind on 28 August in rice fields and roadside seepage pools near Chauhar Jamali (Thatta district). Ticehurst (*Ibis,* 1924) noted arrival in mid-August of first birds, being adults in worn breeding dress with juveniles coming in a month later. None have been noted over-summering and generally all have migrated north by the first week of May. Status ABUNDANT mainly in Punjab and in winter.

HABITS

Less active than *C. minuta* being slower and more deliberate in foraging, and more inclined to frequent vegetated wetlands, also to avoid the more brackish exposed mud flats, favoured by the Little Stint. They feed mainly by picking items from the surface and have not been noted probing in mud, but will dip their bill into shallow water. Their diet includes mainly Diptera larvae and Coleoptera but they also take small molluscs and Annelida. They will fly in close formation in swift wheeling packs when disturbed, often climbing high into the sky and zig-zagging downwards as they approach a new and undisturbed feeding place.

BREEDING BIOLOGY

They breed solitarily in the Arctic regions of Scandinavia and the USSR but avoid such exposed situations as are used by the Little Stint, breeding often on the edge of the boreal forest zone. They are believed to be successively bigamous. The first mated male, after a pair bond of only 1 week's duration, incubating the eggs and tending the nidifugous chicks, whilst the female mates with a second different male and then incubates and herself rears, her second clutch. Meanwhile her first mate later in the summer seeks a second mate after his chicks have become independent, which is within 15 to 18 days of hatching. Some females lay up to 3 clutches within a period of about 21 to 23 days, averaging 4 eggs per clutch (Cramp *et al.,* 1983).

VOCALIZATIONS

Its flight call is slightly different from that of *Calidris minuta* being a more prolonged twitter 'trrrrit', or a trilled 'tirrirtrir'. The flight call of *C. minuta* is a shorter and sharper note.

Sharp-tailed Sandpiper or
Asian Pectoral Sandpiper *Calidris acuminata* (Horsfield)

Calidris acuminatus of Dillon Ripley's 'Revised Synopsis'
Synonym *Eriola acuminata*

DESCRIPTION

Body length 17-22 cms.
Wingspan 42-48 cms.
Bill length 2.6 cms.

It is about the same size as the Common Sandpiper *(Actitis hypoleucos)*. It is a rather long-legged sandpiper with a comparatively short straight bill and small head and a wedge shaped tail. The legs are yellowish olive and the bill is very dark brown to blackish. The crown usually appears noticeably dark chestnut and the white supercilium above the eye is conspicuous. The most distinctive field mark in breeding plumage is the rufous buff area on the sides of throat and pectoral region, speckled with black and the narrow dark margins to the flank and lower breast feathers forming small chevron patterns of a scaled effect. The sexes are alike and in winter the rufous colouring around the upper breast becomes more of a grey-brown and the black chevron pattern of the breast feathers disappears. It then looks like a long-legged Dunlin, but has a distinct whitish supercilium lacking in the Dunlin, and a much shorter bill. In flight it lacks any distinct wingbar, and the tail looks wedge shaped due to the longer central rectrices. The mid-rump and central tail feathers are blackish grey. The best field points are the all-white under parts with buff or grey-brown darker pectoral area, short bill and wedge shaped tail. The Dunlin with similar rump pattern, shows a conspicuous wingbar and is much greyer on the upper parts, in winter plumage.

HABITAT, DISTRIBUTION AND STATUS

It is only a straggler to Pakistan recorded on passage, as it breeds in the tundra zone of Russian Siberia and normally winters in the southern Pacific region being a regular visitor to New Zealand, Australia and the Solomon Islands. A specimen was secured on 1 August 1880 by J. Biddulph *(Stray Feathers,* Vol. 10, 1887) and W.A. Phillips collected a specimen on 18 September 1958 in Sri Lanka. On 5 September on Kharrar jheel (Renala Khurd) in 1970, Allan Vittery watched one at close range for several hours, feeding amongst a huge flock of Marsh Sandpipers *(T. stagnatilis)* (pers. comm., 1973). The Gilgit bird was in company with a flock of Ruffs. Because of a lack of skilled observers in Pakistan, it seems probable that occasional birds may pass down the western seaboard of the subcontinent more regularly than these two records suggest. Status VAGRANT.

HABITS AND BREEDING

They breed on the Chukchi peninsula of north-eastern Siberia. In winter they frequent tidal lagoons, estuarine mud flats and fresh water lakes. In the USSR they have been reported as feeding upon amphipoda, small crustaceans and molluscs, and Diptera (Dement'ev and Gladkov, Vol. 3, 1951).

VOCALIZATIONS

In their winter quarters they emit two types of call. A rather sharp shrill twittering reminiscent of a Swallow *(Hirundo rustica)* 'tree-trit' and a shorter metallic softly repeated whistle 'pleep-pleep' (Cramp *et al.,* Vol. 3, 1983).

Curlew Sandpiper or Curlew Stint *Calidris ferruginea* (Pontoppidan) Plate **15**
Map **153**

Synonym *Calidris testaceus* (Pallas) Dillon Ripley's 'Revised Synopsis', 1982

DESCRIPTION

Body length 18-19 cms.
Wingspan 42-46 cms.
Bill length 3.4 3.9 cms. (males),
 3.8-4.2 cms. (females) (Sind, Ticehurst)

The population which winters in Pakistan is closely similar in body size and bill length to the wintering population of Dunlin *(Calidris alpina)* with which it is confusingly similar. It does however have a comparatively longer neck, more upright carriage and slightly longer bill and legs, when both species can be observed feeding together, and in general body outline more resembles a miniature Curlew *(Numenius).* In winter plumage it is grey with darker streaking on crown and back, white below except for some grey and brown mottling on the breast, with a prominent all-white rump and white wingbar. The Dunlin *(C. alpina)* can at once be separated in flight, by the dark band running down the centre of the rump and upper-tail coverts. The legs and feet are black to dark grey and the bill is black. The sexes are alike in plumage. In summer they moult into a rich vinaceous chestnut colour over the whole of the lower face, throat and belly with indistinct darker barring along the flanks. The crown becomes blacker and the mantle is also stippled with black and chestnut, each feather being narrowly margined white. The throat area also remains whiter and there is a white ring above and below the eye which is often not noticeable in museum skins, but very conspicuous in the field. This is not well shown in the illustration in the *Shorebirds Identification Guide* (Hayman, Marchant and Prater, 1986).

HABITAT, DISTRIBUTION AND STATUS

A purely maritime species in its winter quarters, with first autumn migrants arriving along the Karachi coast from the last days of July and first week of August. They have been noted on migration in Gilgit, specimens being collected from 2 August up to 7 September (Biddulph, *Stray Feathers,* 1881). D. Corfield noted a flock of 18 on 5 March 1980 at Rawal lake, Islamabad and on 8 April 2

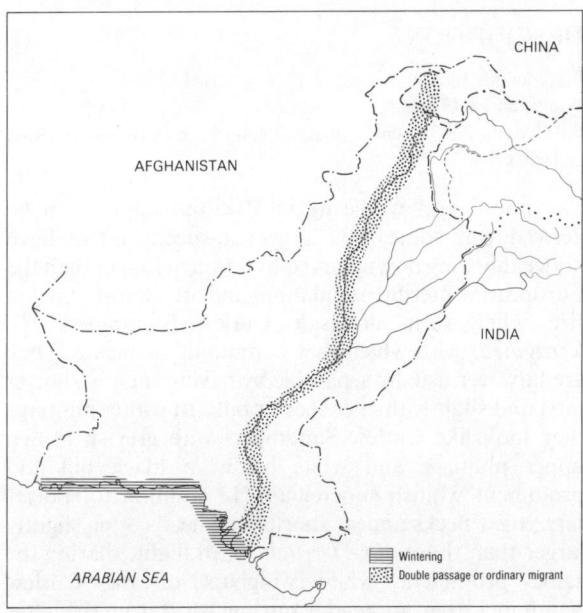

Map 153: Curlew Sandpiper *Calidris ferruginea*

birds (pers. comm., 1981). H. Waite also collected a specimen (Brit. Mus. coll.) on autumn passage at Khanna in Rawalpindi district on 21 September. Meinertzhagen noted them as numerous on spring passage from 16 May onwards at Kushdil Khan lake, in western Baluchistan, all birds being in full breeding dress (*Ibis,* 1920). They are one of the most numerous waders in winter time along Pakistan's coastline, and around Karachi are certainly more abundant than the Dunlin (*C. alpina*). They are not really common until August and the juveniles do not arrive until the first week of September (Ticehurst, 1924). Adults begin to lose their breeding plumage towards the end of August. A few birds probably over-summer in non-breeding dress as they have been seen (author) 9 June on Siranda jheel in Lasbela and on Ghizri creek, Karachi on 12 June.

Surveys (author) in the Indus delta and creeks indicate that about half the population is developing its chestnut red breeding dress by mid-April, whereas by 23 April only about 10 per cent of the mid-April population is still on the winter grounds and of these, about 80 per cent are now in full breeding dress. Status ABUNDANT along seacoast only.

HABITS

A highly gregarious species usually seen feeding in scattered association with both its congeners and Dunlin (*C. alpina*) and Little Stints (*C. minuta*). They prefer open mud banks or exposed flats with some sandy substrate or salt marsh grassland areas and are not usually seen in the narrower mangrove covered creeks. Occasionally they are seen feeding on purely sandy beaches in company with Mongolian Sand Plovers (*C. mongolus*) and Sanderlings (*C. alba*). They feed by probing in soft mud or jabbing more quickly into the surface and often feed in shallow water, submerging their whole heads. Their diet during winter comprises small mollusca, marine worms and the larvae, pupae and adult forms of insects including beetles and Diptera (Cramp *et al.,* 1983).

BREEDING BIOLOGY

Its breeding range is more restricted than that of *C. alpinus* with which it so commonly associates in winter. It breeds along the coastal lowlands of the Arctic from the Siberian mainland to offshore Arctic islands.

VOCALIZATIONS

Its flight call is a melodious twitter 'chirririp-chirririp', less grating than the calls of *C. alpina*. Flocks when disturbed often take off silently and their flight call is only occasionally heard.

Dunlin *Calidris alpina* (Linnaeus)

Plate **15**
Map **154**

DESCRIPTION

Body length 18-20 cms.
Wingspan 38-43 cms.
Bill length 2.9-3.6 cms. (males), 3.4-3.9 cms. (females) (Sind Ticehurst).

The population wintering in Pakistan appears to be derived from some of the larger sub-species as they have noticeably long down-curved bills (much longer than the European wintering population) and are almost equal in size when seen alongside Curlew Sandpipers (*C. ferruginea*) with which they commonly associate. They are however usually separable by having slightly shorter tarsi and slightly thicker shorter bills. In winter plumage they look like Curlew Sandpipers with greyish brown upper plumage and white below, a black bill and prominent whitish supercilium. In addition to shorter tarsi, their necks appear shorter and heads often slightly larger than those of *C. ferruginea*. In flight, sharing the same prominent white wingbars of the Curlew Sandpiper, they are readily distinguished from the latter by the rump pattern which has white borders, clearly separated by a blackish grey centre. *C. ferruginea* shows an all-white rump in flight. The legs and feet are black. The breeding plumage is very striking, as it is the only wader visiting the region with a large black patch on the middle of the lower belly, coupled with quite a chestnut and black streaked pattern on the head and crown. The upper-breast and cheeks are white with conspicuous fine black streaks in parallel lines extending around to the hind neck. The sexes are alike. In winter plumage the black belly patch disappears entirely but this is conspicuously visible from mid-April and again in adult birds when they first return in the autumn.

HABITAT, DISTRIBUTION AND STATUS

A winter migrant visitor to Pakistan, being highly gregarious and often encountered in company with Little Stints (*C. minuta*) and Curlew Sandpipers (*C. ferruginea*). Along the seacoast it is extremely common but it also occurs in small numbers throughout the winter on the sandbars and ox-bow lakes along the Indus and the main Punjab rivers. First arrivals on the sea-coast may be noted from early August and the return spring passage commences again from mid-March. They have been noted on autumn passage through Baluchistan (Kushdil Khan lake) on 10 November and in the spring in smaller numbers, between 16 and 20 May. (Meinertzhagen, 1920). D. Corfield noted 8 birds in partial breeding plumage on 9 May 1980 at Rawal lake, Islamabad (pers. comm.). Ticehurst noted older birds as first autumn arrivals around Karachi, with juveniles coming in 3 or 4 weeks later. 200 were counted (author) in Kharrar jheel (Renala Khurd, Punjab) on 16 March, most in breeding plumage. Spring migrants also recorded 17 March. Lal Sohanran lake (Bahawalpur) and Nammal lake (Salt Range) on 21 March. Flocks with Little Stints were seen on sandbars in the Indus around Kandhkot, Jacobabad district 9 January. Latest date seen 30 May on Ghizri creek, Karachi. H. Waite collected specimens 30 December in Mianwali district (Brit. Mus. coll.) and H. Whistler noted a few on seepage pools in Jhang district throughout the winter (*Ibis*, 1922). Observations over 7 years around Karachi indicate that most Dunlin start northward migrations before Curlew Sandpiper. Birds begin to show traces of breeding dress from late March but on 22 April a few Dunlin still in winter plumage could be seen amongst mixed flocks of waders on Ghizri creek. On 4 May at Sonmiani lagoon (Lasbela) all Dunlin were in full breeding plumage. Ticehurst believed that a few birds over-summered along the seacoast and that these were in breeding dress unlike the majority of Palearctic waders which remain on the seacoast in summer. Status ABUNDANT.

HABITS

Gregarious in their winter quarters, feeding typically by wading in shallow water or walking along the mud and probing with the bill, less commonly pecking at the surface. Their typical feeding attitude is a rather hunched-up or crouching stance. They often feed in close packed flocks and jab very rapidly into the mud (observed Sonmiani lagoon). They will feed along the tide's edge on sandy beaches as well as muddy shores and a few always frequent Clifton beach, Karachi. When disturbed or seeking fresh feeding grounds, they commonly congregate in single species flocks, flying in close formation, twisting and wheeling in unison and at

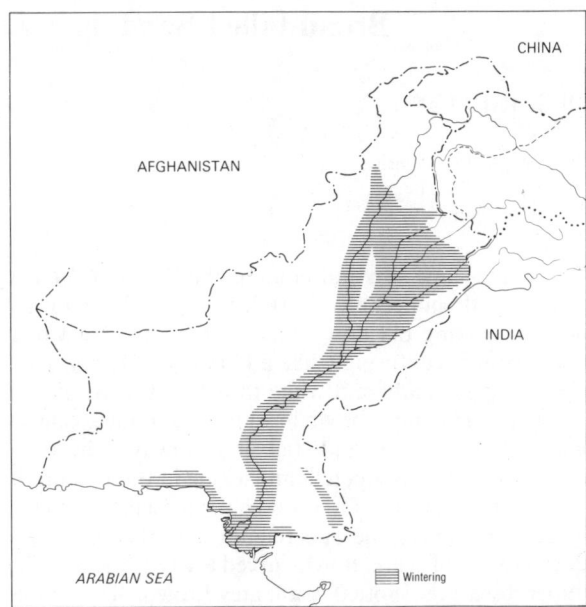

Map 154: Dunlin *Calidris alpina*

great speed, usually fairly close to the ground, their white bellies flashing in the sun as the whole flock simultaneously banks. Their food comprises marine worms, small bivalve molluscs and crustaceans (Amphipoda), gastropod snails (Hydrobia) and insects and occasionally seeds (Cramp *et al.*, Vol. 3, 1983).

BREEDING BIOLOGY

Breeds circum-polarly from Alaska and eastern Greenland across northern Europe and Siberia. It chooses moist boggy ground and tussock-tundra for nesting but occupies a wide latitudinal range from boreal forest and even extending into temperate latitudes up to the high Arctic zone. The pair bond is monogamous.

VOCALIZATIONS

The flight call is a shrill rather prolonged 'treeep' or 'treeer'. When feeding they sometimes utter a more soft purring babble 'pur-r-r-r' (Yamashina 1961, Ali and Ripley, 1969).

Broad-billed Sandpiper *Limicola falcinellus* (Pontoppidan)

Plate **15**
Map **155**

DESCRIPTION

Body length 16-18 cms.
Wingspan 37-39 cms.
Bill length 3 cms.

About the same size and build as the Dunlin *(Calidris alpina)* with noticeably short legs and a long rather heavier looking bill down-curved at the tip. In winter plumage it is confusingly like a Dunlin *(C. alpina)* with greyish upper body and white throat and belly, and in flight similar pattern of white wingbar and dark centre-band down rump and tail, but always shows a conspicuous white supercilium lacking in the Dunlin. It usually shows traces of a second paler mesial strip on the forecrown, above the supercilium and the bill looks distinctly swollen and down-curved towards the tip. The upper breast is spotted with grey-brown. The bill is black, sometimes with a greenish caste and the legs and feet are olive varying to yellowish grey. The sexes are alike in plumage. In summer they moult into a very distinctive plumage with a snipe-like head pattern of dark crown divided by two mesial and two superciliary whitish stripes and a dark line running through the lores and extending behind the eye. There are also creamy lines running along the edge of the mantle and the top of the wing coverts adding to the general snipe-like appearance. The sides of the neck and upper breast are heavily spotted with black and the back is patterned with black feathers margined rufous buff, like scales (based on observation of birds' Ghizri creek, 10 May). In winter plumage it tends to look more contrasting black and white like a Sanderling *(C. alba)*, than the more brownish grey Dunlin *(C. alpina)*. Juveniles in their first winter have a pectoral band of buffy rufous with finer black streaking than that of adults.

HABITAT, DISTRIBUTION AND STATUS

A winter migrant visitor to Pakistan occurring only along the seacoast, with a definite preference for wet muddy embankments and creeks, never being encountered on hard sandy beaches. Their similarity to Dunlin *(C. alpina)* in winter plumage has probably resulted in many birds being overlooked, but they are numerically much less common than the other inter-tidal feeding waders and more likely to be encountered in twos and threes rather than in flocks like the Sand Plover *(Charadrius mongolus)* and Sandpipers *(C. ferruginea* and *alpina)* with which they commonly associate. First autumn arrivals can be seen from the last few days of July around Karachi creeks when adults in worn breeding dress return whilst immature birds arrive in the first week of September. There seems to be an influx of birds on passage during the spring months with numbers building up on the creeks of the Indus delta from mid-March till mid-April when they commence their northward migration.

On a half-mile stretch of Ghizri creek, Karachi, over 400 were counted on 17 April, 1977. By 15 May a survey in the same area revealed no more than 30 to 40 birds, all in full breeding dress and evidently ready to migrate to their breeding grounds. Birds have been noted as late as 4 October still showing the rather black upper plumage and spotted breast of their summer breeding plumage. Latest date seen 30 May (author) but Ticehurst (1924) encountered a few birds as late as 19 June which were not in breeding dress and he believed that a small number regularly over-summered around the seacoast. There are few records for the whole subcontinent of this wader being seen inland or on migration (Ali and Ripley, 1969), but M. Mallalieu saw one on 21 August 1987 on the shores of Rawal lake *(in litt.,* 1987). Status FREQUENT to SCARCE.

HABITS

Not as gregarious in winter as other waders, being usually encountered foraging singly though often in association with other wader species, especially Little Stints *(Calidris minuta)*. In spring however they tend to congregate in small flocks separate from other species. Their feeding method is rather slow and deliberate with both picking motions at the surface and deep probing

Map 155: Broad billed Sandpiper *Limicola falcinellus*

observed, also occasional rapid runs forward (observed Ghizri creek on mud recently exposed by tide). Their short legs and rather black and white appearance are helpful in distinguishing them from the slightly larger sized Dunlin *(C. alpina)*. Ticehurst *(Ibis, 1924)* found small gastropod shells in the stomachs of birds shot near Karachi, and elsewhere on their wintering grounds they feed mainly upon marine worms, insects including beetles, and small crustaceans (Amphipods). They also take some seeds of grasses and sedges (Cramp *et al.*, 1983).

BREEDING BIOLOGY

Its breeding distribution is more or less limited to the boreal forest zone in Scandinavia, Finland and Estonia, and its preferred nesting habitat is open wet bogland. As far as is known the pair bond is monogamous and they are loosely colonial in breeding.

VOCALIZATIONS

Its flight call is a low pitched double-noted 'chiprit' or 'chr-reek' or 'crtirr' (Cramp *et al.*, 1983 and Dement'ev and Gladkov, 1951).

Ruff (and Reeve) *Philomachus pugnax* (Linnaeus)

Plate **15**
Map **156**

DESCRIPTION

Body length 26-30 cms. (males), 20-24 cms. (females)
Wingspan 54-58 cms. (males), 48-52 cms. (females)
Bill length 3.5 cms.

A confusingly variable wader, both in size, variations in breeding plumage and leg colour. Males are about 10 to 15 per cent larger than females. In winter plumage, the form in which nearly every bird is likely to be sighted in Pakistan, the legs are generally yellowish red to pinkish red and the bill yellowish to olive at the base with a darker tip. Females are about the same size as a Redshank *(Tringa totanus)* but with comparatively longer legs. Males are slightly bigger than a Greenshank

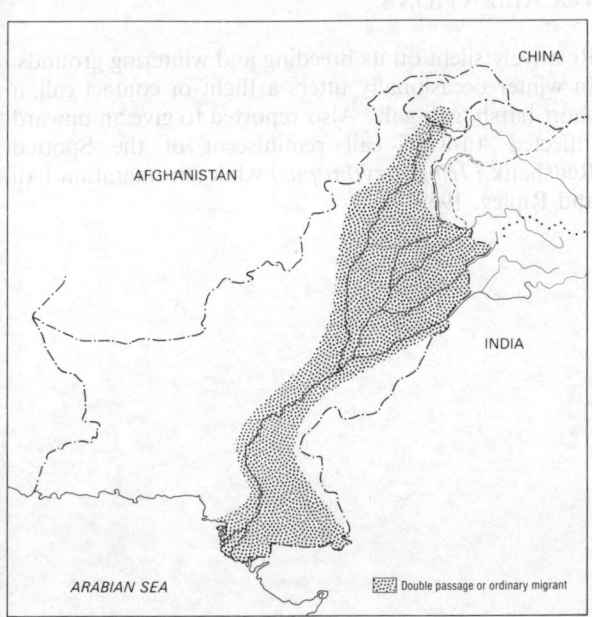

Map 156: Ruff (& Reeve) *Philomachus pugnax*

(Tringa nebularia). Both sexes have a noticeably upright carriage, long neck and short bill, slightly down-curved at the tip. In winter plumage the back and wing coverts are black margined with buff producing a handsome scale-like pattern untypical of any other winter plumage wader likely to be encountered in Pakistan with the exception of the much smaller and rare Sharp-tailed Sandpiper *(Calidris acuminata)*. The breast is suffused greyish brown with the throat and lower belly whiter. In flight a narrow white wingbar is visible and the rump is white with a narow blackish grey centre stripe and darker terminal tail band. In breeding plumage the male is characterized by conspicuous ear tufts on the nape and neck ruff of feathers highly variable in colour. Sometimes the head and neck ruffs are contrasting in colour, sometimes uniform in pattern. They vary from pure white through cross-barred white and cross-barred chestnut to all-black and no two birds seem to be alike. The forepart of the face becomes bare of feathers and warty with highly coloured skin varying from yellow to brick red or olive grey. This colour is independent of the colour of the head feather tufts. Individuals with olive green legs, pink legs and vermillion-orange legs, and with dark brown bills and black bills, pink at the base have all been noted in Pakistan during full winter plumage. About one in 20 male birds seen on spring passage show the initial development of the neck and head feather tufts. Occasionally some will start to moult into breeding plumage from late January. White or pale buff and white are the commonest colour morphs for these neck ruffs noted on spring passage. In a flock comprising all male birds observed on Khinjar lake 1 February, 19 showed development of white head and neck ruffs. A male seen on 5 April near Mianwali (Punjab) had purplish brown neck ruffs developing. Another on 13 February at Sidhnai (Punjab) had conspicuous black and white feathers developing on the head and neck. On 10 April near Rahim Yar Khan out of a flock of 30, three males were noted with partly developed head and neck ruffs,

one pale orange-chestnut, a second barred black and white and a third pure white.

HABITAT, DISTRIBUTION AND STATUS

An abundant passage migrant through Pakistan in spring and autumn with the bulk of the population passing through the Punjab in September and early October and again in March and early April. A few birds linger in lower Sind all winter, but the majority travel further south into the subcontinent, many wintering in Sri Lanka. It is highly gregarious on migration and in winter quarters, and is most frequently seen on roadside pools and borrow pits, being rarely encountered on the coast. It has been noted (Scully, 1881) on passage through Gilgit in September and through Kohat and the Kurram valley from late February to mid-May (Whitehead, 1911). On 18 March near Quaidabad in the Thal, over 1,500 were counted (author) in 2 flocks along small seepage pools and roadside pools and practically all of these appeared to be females in non-breeding plumage. On the Sukh Beas, in Lahore district on 21 April over 300 were counted in company with small groups of Spotted Redshanks *(Tringa erythropus)*. Earliest arrivals noted by Ticehurst in lower Sind were on 6 August *(Ibis.,* 1924). In Baluchistan they have only been noted on spring passage when large flocks were seen by Kushdil Khan lake (Meinertzhagen, 1920). A bird ringed at Bharatpur (Rajasthan, India) was recovered in Norilsk in the Krasnoyarsk region of west Siberia. Status ABUNDANT, mainly on spring passage.

HABITS

In Pakistan they have not been noted on the seacoast or on tidal creeks, except where there is a flow of river water coming down the creek and they appear to have a definite preference for fresh water or inland lakes and pools. They will forage on grassland near to lake shores and also by wading. Often they feed gregariously in dense flocks, all birds probing with their bills together

(observation Kandhkot, Jacobabad district, early October). They feed on a variety of insects and their larvae, including Chironomidae (midges), Coleoptera and Diptera. They also take small molluscs (water snails), Annelid worms, Amphipoda and occasionally seeds of grasses and sedges. They forage both by day and by night and are often very active throughout the night when they first return on autumn migration (observed Kandhkot, Jacobabad district). In Pakistan they are only seen in large dense flocks while on passage as they break up into smaller groups as the winter season advances.

BREEDING BIOLOGY

European studies of the lek displays and promiscuous breeding system of these interesting waders have been well reported. Males fight and display on a particular piece of high dry ground and the females visit these display areas for copulation only, carrying out all the nesting, incubation and protection of chicks unaided by the males. The nidifugous chicks become self-feeding 5 or 6 days after hatching.

The high preponderance of male Ruffs on spring migration showing predominantly white crown and neck ruffs in Pakistan, is interesting, as it is now known that these tend to be peripheral or non-displaying males which cannot occupy the centre of a lek and which must adopt opportunistic methods to be able to mate with any females attracted to the darker coloured displaying males (see A. J. Hogan-Warburg, *Ardea,* 1966, and J. G. Van Rhijn, *Behaviour,* 1973).

VOCALIZATIONS

Relatively silent on its breeding and wintering grounds. In winter occasionally utters a flight or contact call, a short harsh 'gah-gah'. Also reported to give an upward inflected 'tu-whit' call reminiscent of the Spotted Redshank *(Tringa erythropus)* while on migration (Ali and Ripley, 1969).

SUB-FAMILY GALLINAGININAE
Snipes and Dowitchers
Jack Snipe *Lymnocryptes minimus* (Brunnich) K. H. Voous's 'Revised List' Illus. **53**
Map **157**

Capella minima (Brunnich) of Dillon Ripley's 'Revised Synopsis', 1982.

DESCRIPTION

Body length 17-19 cms.
Wingspan 38-42 cms.
Bill length 4 cms.
Weight 56.7 gms. (*JBNHS*, Vol. 40, p. 331)
 34-70 gms. (September to May) (Cramp *et al.*)

The size of a Starling *(Sturnus vulgaris)* this is the smallest snipe occurring in the subcontinent, often recognizable in the field by its habit of crouching and not taking flight until almost trodden upon. In flight the generally small size, narrow wedge shaped all-dark tail lacking bright rufous and the relatively shorter bill are helpful in separating this species from the Common Snipe *(Gallinago gallinago)*. In the hand or if observable on the ground, it appears much darker than the Common Snipe, with no mid-coronal paler creamy streak. The lower parts are white, tinged with buff on the neck and breast with heavy short streaks and no vertical fine dark barring on the flanks, conspicuous in *G. gallinago* and *G. stenura*. It has a purplish or metallic green sheen to the mantle and wing coverts, lacking in other snipe which visit the region. The outer tail feathers lack any white as in the Common Snipe and central pair of rectrices are pointed and elongated (unlike the tails of *G. Gallinago* and *G. stenura*) (see **ill. 53**). There are two pairs of longitudinal pale creamy back stripes obvious both in flight and when at rest. The head looks comparatively large with a large dark brown iris, legs and feet are stout and long, being yellowish grey or pale olive

Illus. 53: (1) Pintail Snipe *Gallinago stenura.* (2) Common or Fantail Snipe *Gallinago gallinago.* (3) Jack Snipe *Lymnocryptes minimus* (4) Solitary Snipe *Gallinago solitaria.* Showing upper surface of spread tails, (5) Fantail or Common Snipe. (6) Pintail Snipe. (7) Jack Snipe. (8) Solitary Snipe. Showing crown pattern and bill shape. (9) Common and Pintail Snipe. (10) Jack Snipe.

green. The bill is straight slightly broader at the tip when viewed dorso-ventrally, being black tipped and greyish yellow basally. The sexes are alike.

HABITAT, DISTRIBUTION AND STATUS

It breeds from northern Scandinavia in the west throughout the forest tundra zone of the USSR, overlapping in range with the Common Snipe *(Gallinago gallinago).* The Jack Snipe is a winter migrant to Pakistan, probably more frequently encountered on passage than throughout the winter, with evidence that most of the wintering population enters the subcontinent from the north-west. Meinertzhagen noted autumn passage through northern and central Baluchistan starting around mid-October and the return passage in mid-April when he encountered as many as 100 in a day *(Ibis, 1920).* It has not been recorded in Gilgit, but Whitehead noted a few as wintering in the Kurram valley and larger numbers passing through up to mid-April, but always as scattered individuals and in much smaller numbers than *Gallinago gallinago* (Whitehead, 1911). In winter it is widespread but uncommon through the Punjab from the Potohar plateau, down to Rahim Yar Khan district. Whistler noted it from 4 February till 2 April around Rawalpindi, most birds being seen while on passage. In Sind, Ticehurst noted it as much less plentiful than the Common Snipe and arriving in autumn much later than *G. gallinago.* Each October at Sidhnai spill channel, it was noted (author obs.) as being fairly common in rice field stubbles, particularly areas which were still inundated and covered with *Juncus* sedge. It has also been seen at Rahim Yar Khan in November (author) and December and around Marala barrage throughout the winter in suitable marshes (E. Fernando, pers. comm.). Status SCARCE

HABITS

When flushed has comparatively slow and weak flight, not zig-zagging like the Common Snipe. Prefers flooded vegetation, especially grass and sedge infested rice stubbles and the grassy margins of permanent swamps. Such areas must be wet and muddy but may often be quite limited in area and it will be attracted to quite small isolated marshy patches of ground provided there is sufficient vegetative cover. It usually forages solitarily and is thought to be active mainly in the early part of the

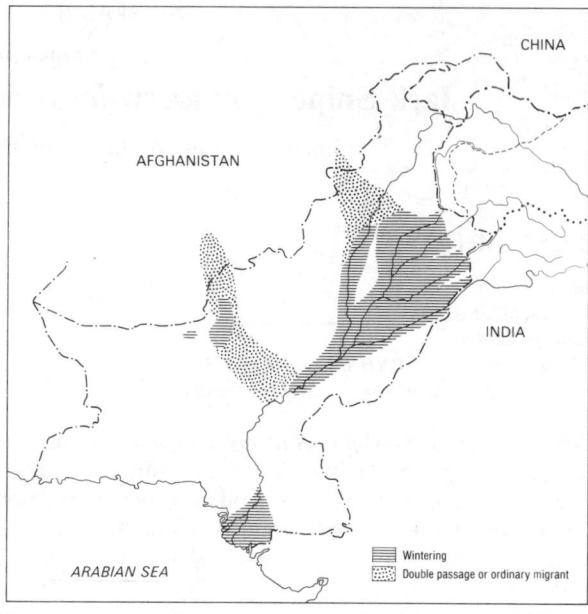

Map 157: Jack Snipe *Lymnocryptes minimus*

night and before dawn, probing in the mud like all its genus and feeding principally upon annelid worms and the larvae and adults of aquatic insects including water beetles (Hydrophilidae) and flies (Tipulidae) and small molluscs. It also eats seeds of marsh plants.

BREEDING BIOLOGY

They nest in the sub-Arctic and northern boreal zone in open tracts of coarse bog, from Fenno-Scandinavia across the USSR. The pair bond is thought to be of brief duration with the female carrying out all the incubation, but this has not been well studied and probably mating is fairly promiscuous at the beginning of display and courtship, with males later on assisting in rearing young after they are hatched.

VOCALIZATIONS

Usually rather silent even when flushed, unlike the Common Snipe. A specimen at Rahim Yar Khan emitted an alarm call which was weaker and less grating in tone but otherwise similar to the flight call of *Gallinago gallinago.*

Common Snipe or Fantail Snipe *Gallinago gallinago* (Linnaeus)

Illus. **53**
Map **158**

DESCRIPTION

Body length 25-27 cms.
Wingspan 44-47 cms.
Bill length 6-7 cms.

About 25 per cent larger in size than the Jack Snipe *(Lymnocryptes minimus)* and separable from the latter by its much longer bill, mid-coronal pale creamy stripe and larger size, with prominent white trailing edge to wings, conspicuous in flight, and largely absent in the Pintail Snipe *G. stenura.* It has a rufous chestnut and black barred tail whilst the Jack Snipe has a dark tail. When viewed on the ground it is also distinguishable from the Jack Snipe by the vertical black barring on the greyish white flanks, those of *L. minimus* being streaked not barred. It has two prominent dark stripes through the eye and down each side of the crown. The eye is large, high set with a dark brown iris, the legs are olive green and the bill yellowish horny at the base, blackish at the tip. The mantle and scapular feathers are barred chestnut and black and margined pale buff forming two irregular creamy lines down either side of the back. The upper breast is buffy with dark blackish wavy or scale-like bars merging into the purer white unmarked lower belly and flanks with vertical straight black bars. The chief distinguishing feature from the similarly sized Pintail Snipe *(G. stenura)* is the white trailing edge to the wing in flight and the more distinct creamy lines down the centre of the back which are more broken up in the Pintail Snipe. *G. stenura* has a very distinctive tail when seen in

the hand and less clearly defined pale tips to secondaries and creamy margins to the wing scapulars, also its flank bars tend to be more broken up and less extensive. The sexes are alike in plumage.

HABITAT, DISTRIBUTION AND STAUTS

It breeds throughout the northern boreal zone in north America, Europe and Asia. It is a widespread winter migrant visitor throughout Pakistan. The upper Kurram valley is an important migration route, as military officers serving in Parachinar used to regularly conduct snipe shoots and records of their bags in the Game Book preserved at the Mess (author obs.) revealed that a total bag of 200 birds between 5 and 6 guns was not unusual during the latter part of the month of January up to early February, obtainable by only three or four hours of walking the rice paddy stubbles. Whitehead noted that before January there were much smaller numbers observed in this valley and considered this the start of a spring migration (Whitehead, 1911). In Gilgit area a few birds over-winter in boggy areas alongside the rivers where contour irrigation channels produce small seepage areas. Four were flushed (author) in early December 1969 over a distance of less than 1 km. There is a not very marked spring migration in April and again from the end of August. In Baluchistan first autumn arrivals are noted from mid-August with a marked return passage in March (Meinertzhagen, 1920). The largest bag recorded in the Game Book of the Staff College at Quetta was 166 in a one day's shoot in March 1914 (Meinertzhagen, *Ibis,* 1920). Ticehurst records first arrivals in autumn along the Makran coast as well as in lower Sind, during two successive years, on 24 August and 28 August (Ticehurst, 1927*b*). In Sind the earliest arrivals come in by mid-August, but they are not really plentiful until late September. 5 were flushed on 23 August around Gharo in Thatta district while carrying out rice crop surveys over a small area of about 6 hectares (author). Latest dates recorded in Sind 29 April. (Ticehurst, 1924), in Multan district at Sidhnai, 14 April (author). The favoured habitat of snipe is shallow flooded rice stubbles or margins of lakes which are drying out, or where the ground is well covered with short vegetation. They are normally rather secretive by day and will only forage out in the open at night time but can occasionally be seen during daytime in exposed situations such as upon islets in shallow lakes or channels of major rivers (observed Jhelum river at Rasul in Jhelum district) and in seepage zones around Haleji and Chatteji lakes, lower Sind. As evidence of their former abundance, Ticehurst records the maximum hunting bag for lower Sind, of 307 snipe in 6 hours' shooting by 3 guns in 1923. Though not particularly gregarious, snipe tend to congregate in big

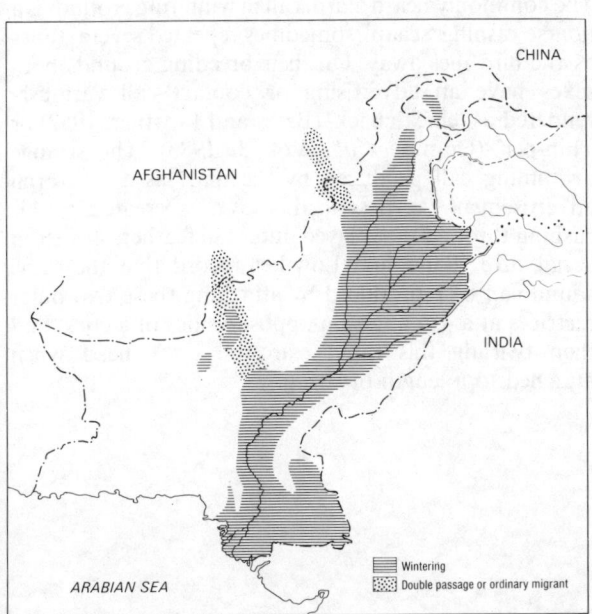

Map 158: Common or Fantail Snipe *Gallinago gallinago*

numbers during the daytime, where shelter and cover is good and adjoining areas are rich feeding grounds. Such aggregations were described as wisps of snipe in older sporting books. At Nalwah, in a water-logged and brackish area of stunted *Tamarix aphylla* on the edge of the Cholistan desert in November 1976, the author flushed 46 birds during about 2 hours' random exploration of an area of barely 80 hectares, and no doubt much higher densities occur in areas closer to extensive swamps. Snipe shooting is no longer a popular form of hunting since independence (1947), and therefore there is no recent evidence of population densities to reveal whether numbers visiting Pakistan in winter are being maintained or are declining. But in their wintering grounds within the past several decades, conditions have obviously improved with the increase in irrigation, rice cultivation and water-logging problems, all of which create favourable habitat for snipe, as well as the lack of any significant hunting pressure. Status COMMON.

HABITS

Largely crepuscular or nocturnal in feeding activity, obtaining most of its food by probing in mud with its long bill as well as by picking at the surface and by sweeping its mandible tips along leaf and sedge stems (observed Kandhkot Jacobabad district). Mason and Lefroy examined stomach contents of 13 individuals collected throughout the migration season from October to April, in Pusa district of Bihar (India) and found that the diet consisted chiefly of Annelida worms and aquatic insects, including *Elaterid* larvae, Tipulidae larvae, *Mesomorpha villiger* and the caterpillars of Geometrid moths. In March a high portion of their diet included molluscs, *Planorbis* snails and occasionally crustacea (water shrimps). Also a few seeds including sedges and leguminous weeds (Mason and Lefroy, 1912)

Undisturbed birds will forage actively during daylight and when they note danger or think they are being observed they will at once squat on their tarsi and freeze. In good cover, they rely on this natural concealment, but if the intruder approaches too close they will suddenly spring into the air uttering their harsh, short alarm cries. At dusk in favoured habitat, undisturbed snipe can often be seen flying overhead at considerable heights probably seeking new feeding grounds, and they characteristically swerve or bank rapidly from side to side even on such high trajectory flights, travelling with rapid continuous wing beats at considerable speed.

BREEDING BIOLOGY

There are no known records of the Common Snipe breeding within Pakistan territory, but it does breed sparsely around some of the larger swamps of the main vale of Kashmir. Similar large marshy areas do not occur in adjacent mountainous regions of Pakistan. Most of the winter visiting population in Pakistan must breed mainly on wet moorlands in the temperate and boreal zones of the USSR. The territorial advertising and courtship flights of this species have been well studied and recorded even by European observers of the country-side, before much was understood about breeding biology. The males only, perform high circling aerial flights, usually preceded as they mount into the air by a sharp high pitched double call 'chick-chack chick-chack'. After circling round a sudden dive is made with the outer tail feathers splayed which produce bleating or drumming noise as the air is forced through their stiffly vibrating outermost rectrices. Both sexes are somewhat promiscuous at the beginning of the nesting season, where extensive suitable nesting habitat attracts many individuals and the one female may copulate with several different males. When one female starts nesting, usually one male will stay on that territory and probably both parents assist with feeding the newly hatched chicks, bill to bill, though all the nest building and incubation is carried out by the female alone. The nest is cleverly concealed in a tuft of sedges or vegetation and the glossy pyriform eggs are remarkably obliterative with their surroundings.

*VOCALIZATIONS

The commonly heard alarm call in wintering grounds is a hoarse rasping 'scaap', sometimes repeated several times as the bird flies away. On their breeding grounds both sexes have an advertising or contact call variously rendered as 'chick-chack' (Bates and Lowther, 1952) or 'chip-per' (Cramp *et al.,* Vol. 3, 1983). The strange drumming call produced by the male as a territorial advertisement, as described above, is created by air rushing through the splayed outer tail feathers in a steep aerial dive. Bates and Lowther record that the same sound can be reproduced by attaching these two outer rectrices at a 45° angle on opposite sides of a cork, and then twirling this rapidly around one's head when attached to a length of string.

Pintail Snipe *Gallinago stenura* (Bonaparte)

Illus. **53**
Map **159**

Asiatic Snipe (Russian authors)

DESCRIPTION

Body length 25-27 cms.
Wingspan 44-47 cms.
Bill length 5.5-7 cms.
Weight 134.6 gms. (*JBNHS*, Vol. 40, p. 331)

Identical in size to the Common Snipe *(G. gallinago)* but with the bill averaging slightly shorter. In the field it cannot reliably be separated from the Common Snipe, though the trailing edge of the wing normally does not show such a marked white margin. It is slightly slower and heavier in escape flight (author obs., Chitawan) and its wingtip is slightly blunter (Cramp *et al.,* 1983). The only sure identification, is in the hand, when the 7 outer tail feathers on each side are peculiarly long and thin. Both sexes are alike. The body is overall rather a dull dark chestnut brown vermiculated with buff, with less distinct longitudinal creamy lines on the back and edges of scapulars, when compared with the Common Snipe *(G. gallinago)* and Jack Snipe *(L. minimus).* The crown is dark blackish brown with two mesial paler buff stripes and a broad central pale creamy crown stripe and a dark line through the eye. The upper breast is chestnut with black scale-like marking and the lower belly is white with black vertical barring on the flanks, but this barring tends to break up into spots and to be less extensive than in *G. gallinago.* The tail is white tipped and chestnut with obsolete black barring. The upper-wing coverts often look duller buff spotted rather than white tipped and overall are paler than in the Common Snipe *(G. gallinago),* also its creamy supercilium is broader in front of the eye than is the case with the Common Snipe. In flight the under-wing coverts are much darker with more conspicuous black cross-barring when compared to the Common Snipe. The eye is large with iris dark brown. The stout legs are greyish green and the bill is thickened at the tip, and blackish brown with its base pale green. The tail comprises 26 to 28 feathers, the outer 7 or 8 on each side being reduced to stiff 'pins'.

HABITAT, DISTRIBUTION AND STATUS

It breeds in the northern half of west Siberia, Transbaikal region and across eastern Siberia (Flint *et al.,* 1984). The difficulty of distinguishing this species in the field from the much commoner *Gallinago gallinago* has resulted in a paucity of reliable records. It is a winter migrant visitor, much less plentiful than the Common snipe *(G. gallinago)* and is thought to be largely a double passage migrant through Pakistan. Certainly, abundant records indicate that it is the dominant and most plentiful species wintering in southern peninsular India

and also in the north-eastern parts of the subcontinent. Ticehurst noted that it occurred in lower Sind from September to November when it was not uncommon and then again in the spring. In mid-September it was about as plentiful as the Common Snipe *(G. gallinago)* but by December when other snipe were abundant it was largely absent. A solitary record of one shot near Thatta in December by a Captain Hanna is recorded (Ticehurst, 1924). It has not been reliably recorded in winter anywhere else in Pakistan including on migration in the northern areas. Status UNKNOWN.

HABITS

Individuals flushed in the Terai of Nepal, looked noticeably darker in flight than the Common Snipe, and were also more sluggish being reluctant to fly as far when flushed (pers. obs., 1982). On their wintering grounds they typically frequent a slightly dryer habitat than that favoured by *Gallinago gallinago*. Specimens shot on migration in Sind, had the larvae of Coleoptera in their stomachs (Ticehurst, 1924). They are reported to have a preference for insects and molluscs and feed upon a variety of other insects and their larvae and are less dependent upon worms than the Common Snipe (Dement'ev and Gladkov, Vol. 3, 1951).

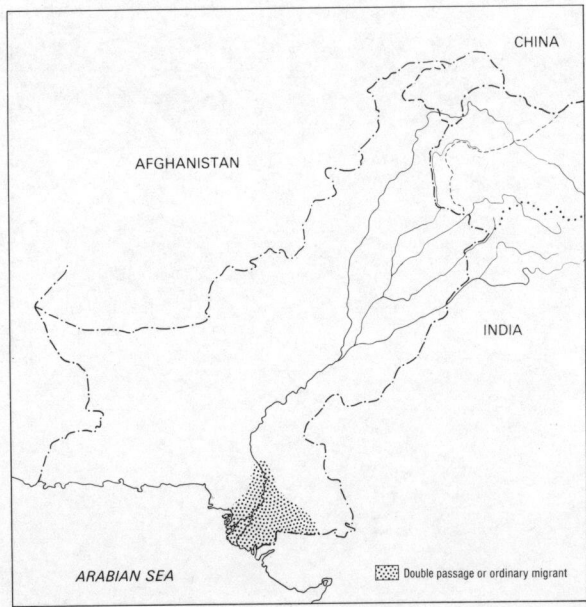

Map 159: Pintail Snipe *Gallinago stenura*

BREEDING AND VOCALIZATIONS

It breeds in more northern latitudes of the USSR than the Common Snipe *(G. gallinago)*, nesting in both tundra and boreal forest zones, in grassy bogs formed around small woodland streams. In the beginning of the breeding season, groups of males will perform courtship flights together. These often commence at dusk and continue in darkness and their 'song' (?) consists of an intermittent spaced 'chuin-chuin-chuin' rising in pitch and becoming more rapid until it terminates in a metallic trill (Dement'ev and Gladkov, 1951). Breeding is solitary and in other aspects believed to be similar to that of the nesting of the Common Snipe.

On their wintering grounds, disturbed birds are often silent when flushed, but will occasionally utter a similar short rasping alarm call, as in *G. gallinago* rendered as 'scaap' (Cramp *et al.*, Vol. 3, 1983).

Solitary Snipe *Gallinago solitaria* Hodgson

Hermit Snipe (Russian authors)

Illus. **53**
Map **160**

DESCRIPTION

Body length 30-32 cms.
Wingspan 51-56 cms.
Bill length 6-6.7 cms.

Smaller than a Woodcock *(Scolopax rusticola)* this is a large snipe, considerably bigger than the Great Snipe *(Gallinago media)*, (which has only been recorded as a straggler on the eastern seaboard of the sub continent). It is doubtfully distinguishable in the field from the Wood Snipe *(G. nemoricola)*, though the latter averages shorter in both wing and bill length and the two species inhabit a very different type of biotope as the name Wood Snipe suggests. When seen alongside the Common Snipe *(G. gallinago)* it is a much larger paler and greyer bird and when flushed its slower heavier flight and tameness, at once distinguish it (obs. author, Gilgit). The head is typically patterned like that of the other snipe with a large dark eye situated high in the skull with a blackish line through the eye and along the sides of the crown, separated by an indistinct or faint paler mid-coronal streak, but having two more prominent creamy white mesial lines. The upper breast is buffy with light rufous speckles, with the rest of the lower breast and belly prominently cross-barred black against a pale grey background, like a Woodcock *(S. rusticola)*. This cross-barring is more extensive than on either *G. stenura* or *G. gallinago*. The back and mantle show two pale creamy white lines running longitudinally. The tail is bright chestnut, tipped white, with a narrow black sub-terminal bar. It is more wedge-shapped when open than on *G. stenura* or *G. gallinago*. The scapulars are noticeably white bordered and the back stripes are almost pure white also, in contrast to the more creamy yellow back stripes of the Common and Jack Snipe. The legs are dull yellowish green or dirty yellow and the bill is horny black at the tip, olive green basally (description based on birds observed, Gilgit in November and December in two successive years).

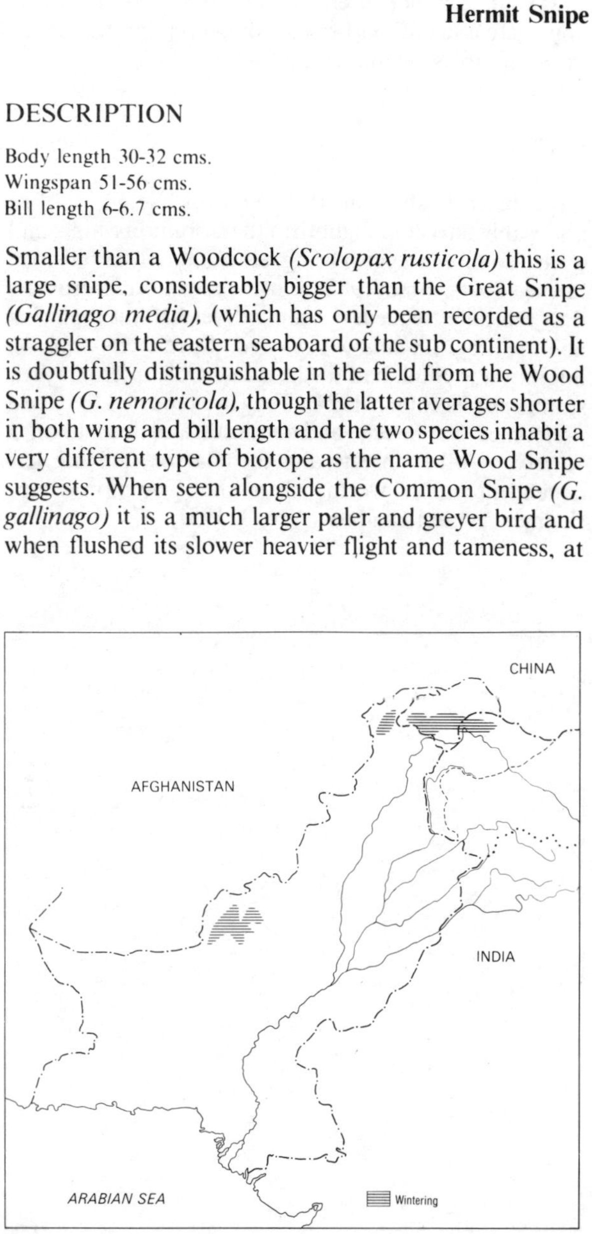

HABITAT, DISTRIBUTION AND STATUS

A breeding bird in the high mountain ranges of southern Asia, probably adapted to dryer steppic montane areas. Known to breed in the Tian Shan and Altai ranges in the USSR and in the Soviet far east, and in the inner higher mountain ranges of the Himalayas and Karakoram. It is typically an inhabitant of high alpine valleys in mountain regions. It is largely sedentary or performs short altitudinal migrations in winter down to surrounding valleys. All Pakistan records are of wintering birds. The author watched one on 20 February 1983 at Bara on the Shyok river in north eastern Baltistan, feeding on the banks of an unfrozen stream. Others have been seen from 10 November till 15 December in 1969 and 1970 in marshy grassland bordering the Gilgit river, east of the

CHINA

AFGHANISTAN

INDIA

ARABIAN SEA

Wintering

Map 160: Solitary Snipe *Gallinago solitaria*

own. Major Biddulph noted them as uncommon in
suitable valleys of Gilgit from 1,524 metres to 2,750
metres (5,000-9,000 feet), in winter and spring *(Stray
Feathers,* 1881). They undoubtedly breed on the higher
plateaux of Baluchistan in the Kaliphat and Zarghun
ranges, as a young bird was shot in Kowas Tangi near
Ziarat (*JBNHS*, 1910, pp. 541-2). Meinertzhagen noted
5 being shot during the winter of 1913 mostly during
October, between 1,524-2,750 metres (5,000-9,000 feet)
in valleys adjacent to Quetta (Meinertzhagen, *Ibis*, 1920)
It has also been shot at Zangi Nawar lake near Nushki on
3 October 1938 and near Pishin in October 1940
(Christison, *JBNHS,* 1942). Ticehurst records one shot
near Loralai on 13 March 1925 which at that late date
must have been close to its breeding haunts in the nearby
mountain ranges in north eastern Baluchistan. He also
mentions the earliest record in Baluchistan of a bird shot
on 29 September. There appear to be no records from
Chitral, though it occurs in adjacent regions in
Afghanistan. Status RARE RESIDENT.

HABITS

They have a curious habit of bobbing their whole bodies
up and down while feeding, rather like a Common
Sandpiper *(Tringa hypoleucos).* They feed by probing in
wet grass and mud. Birds observed in Gilgit were feeding
during mid-morning and were allowing approach to
within 10 metres. When flushed they flew with slow
heavy flight compared to other snipe species, generally
settling within 100 metres. When approached, their first
reaction is to squat and freeze with feathers slicked
down. Their diet appears similar to other snipe, with
larvae of insects particularly Coleoptera and Tipulidae
being obtained by probing in mud, as well as small snails
(Mollusca) and particles of sand, also the remains of
Diptera (flies) (Dement'ev and Gladkov, 1951). They are
reputed to be solitary in habits outside of the nesting
season.

BREEDING BIOLOGY

No nests have been discovered on the subcontinent (Ali
and Ripley, 1969), but in the Munu-Sardyk Range of
Mongolia, USSR, breeding has been well described. As
no account is given in the 'Handbook Series' (Ali and
Ripley, 1968-74) I quote extensively. The nest itself is a
well concealed hollow in a damp willow thicket, with no
significant lining and the clutch comprises 4 pyriform
eggs, their pointed ends inwards. In colour they vary
from light greenish brown to cinnamon buff with
blotches and large round markings of black, brown and
dingy violet (Dement'ev and Gladkov, 1951). Fledged
juveniles have been found in August and incubated eggs
in late June in outer Mongolia. They nest up to 3,000
metres (9,840 feet) elevation in the USSR and prefer
wide pebbly beds of streams and glacial rivulets. At the
commencement of breeding, males display territorially
by flying in wide circles and periodically diving
downwards with tail feathers spread and wings half
folded. This dive is accompanied by a 'sharp jarring
sound' which is probably caused by the vibration of the
narrow outer tail feathers as it is intermittent and
coincides with each stoop of their aerial dive. They
periodically also emit loud cries, likened to the call of a
Willow Ptarmigan, and have been rendered 'zhzhzh-
chok-chok-chaaa' (op. cit., 1951). The birds flushed
(author) in Gilgit occasionally uttered a single alarm call
which was a harsh low 'krek', distinctively lower and
more croaking than the alarm call of *Gallinago
gallinago.*

Woodsnipe *Gallinago nemoricola* Hodgson

Synonym *Capella nemoricola* (Hodgson) of Ali and Ripley

DESCRIPTION

Body length 31 cms.
Wing length 13.3-14.1 cms. (Stuart Baker)
Tail length 6.3-7.4 cms. (Stuart Baker)
Bill length 61-67 mm. (Stuart Baker)

This is a larger snipe than either the Common Fantail *(G.
gallinago)* or Pintail Snipe *(G. stenura)*, but it is
confusingly similar in appearance to the Solitary Snipe
(G. solitaria) being more or less barred all over the belly
and ventral region as is *solitaria*. Though similar in size
to the Solitary Snipe, it averages shorter in wing length
and in bill length, the wing never exceeding 15 cms. in
length, whereas in *G. solitaria* the wing varies from 15.3
to 16.9 cms.

In the field it looks darker about the face than *solitaria*
and with less conspicuous white borders to the scapulars,
the 'back stripes' being more buffy and broader than in
solitaria (Inskipp and Inskipp, 1985). In flight the under
parts look more dusky without the whiter central belly
area conspicuous in the Solitary Snipe.

HABITS

Normally found in well wooded foothill country and
even in winter quarters preferring a forest habitat in

contrast to the Solitary Snipe which frequents stream beds and marshy areas in comparatively open and mountainous regions.

STATUS

A Major Barton claimed to have shot a specimen of this snipe near Mardan in 1887, basing his identification upon the description given in Hume and Marshall's 3

volume work *(Game Birds of India, Burma and Ceylon* 1879-80), but he did not preserve the specimen (Barton *JBNHS*, Vol. 14, p. 607).

There are no other records of its occurrence so far west, and it has only regularly been reported from Dalhousie in Himachal Pradesh (76° E) and eastwards in the Himalayas (Ali and Ripley, 'Handbook', Vol. 2 1969). For the present time, I prefer to exclude it from the **Checklist** of Pakistan birds.

SUB-FAMILY SCOLOPACINAE

Woodcocks

Eurasian Woodcock *Scolopax rusticola* Linneaus   Map 16

DESCRIPTION

Body length 33-35 cms.
Wingspan 56-60 cms.
Bill length 6.5-8 cms.
Weight 329.5 gms. *(JBNHS,* Vol 40, p. 331)

With beautifully patterned cryptic plumage, the Woodcock looks superficially like a large Snipe, with long straight bill and large dark eye set high up near the crown. It does however, appear larger in the head, shorter and thicker in the neck and more robust in body than any snipe and of course it is comparatively larger in size even than *Gallinago solitaria*. The most distinctive field mark is the dark blackish crown and nape, with

three paler buffish cross-bars in contrast to the longitudinal pale brown stripes in Snipes; also its more rufous tawny upper parts. The short, fan shaped tail, is dark grey tipped, and blackish basally, with the rump and upper-tail coverts chestnut, bearing darker striations. The forecrown, lower face and breast are rufous buff with narrow dark barring particularly conspicuous as vertical bars in the flank region. The flight feathers are chestnut, narrowly cross-barred dark brown, and when at rest the back is vermiculated with black and rufous chestnut with creamy spots bordering the feather tips and some black spots on the wing coverts. Shorter in the leg than a snipe, with orange-brown legs and feet (seen on Gilgit specimen), varying to leaden grey or pinkish (seen on Shahran specimen). It is feathered down to the tibia-tarsal joint unlike snipe in which the base of the tibia is naked. The bill is swollen at the tip and flexible, being blackish brown with the base of the lower mandible reddish to purplish pink. In flight and at rest there are no conspicuous longitudinal creamy lines on the back as in snipe. The sexes are alike in plumage.

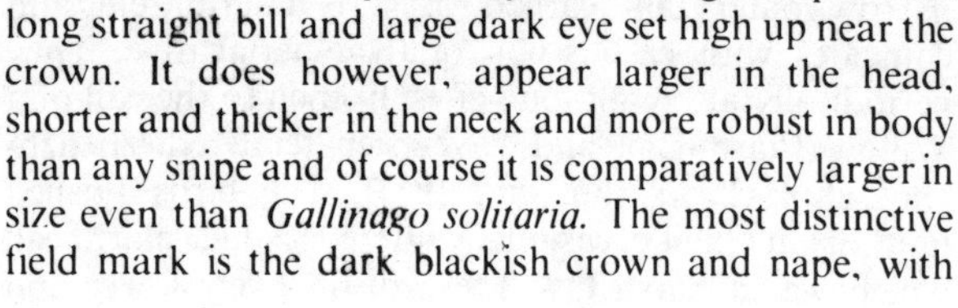

HABITAT, DISTRIBUTION AND STATUS

A sparsely distributed breeding bird in the moister temperate-Himalayan forest tracts, characterized by mixed coniferous and deciduous forest. It appears to have become extinct as a breeding bird in the Galis in the 1950 s but a small population survives in Azad Kashmir on the forested slopes of the Neelum valley (Pir Khan, pers. comm.). Kamal Islam watched one roding in May 1983 at Machiara (pers. comm., 1983) also in Hazara district in Manshi and Malkandi reserve forests on either side of the lower Kaghan valley, where the author has seen them in several years. In winter they descend to adjacent valleys and a few may straggle across the Indus plains, to winter in the Sholas of the Nilgiri hills in southern peninsular India. At the turn of the century it

Map 161: Eurasian Woodcock *Scolopax rusticola*

was described as breeding freely around Changla Gali in the Murree hills (Rattray, *JBNHS*, 1905). Whitehead noted that 5 or 6 were shot each winter in the more sheltered gardens of Kohat (Whitehead *JBNHS*, 1911) and a few may still survive as breeding birds in the Safed Koh range, as 6 Woodcock were shot in the gardens of the Parachinar rest house during the winter of 1968-9 (Col. Dastagir, pers. comm., 1969). In Chitral Capt. Fulton described it as resident throughout the year but uncommon (*JBNHS*, 1904). In Baluchistan it was a scarce but regular winter visitor around Quetta from October to February (Meinertzhagen, *Ibis* 1920), but Christison did not encounter it in the 1940s. There are reports of a few Woodcock still descending to the Uruk valley in winter around the seepage zones created by the water catchment tanks (Abdul Rahman, Shikari, pers. comm., 1973). Major Biddulph described it as a winter visitor to the valleys of Gilgit, being more numerous during the severe winter of 1877-8 *(Stray Feathers*, 1881). On 6 November 1968 during Guy Mountford's World Wildlife Fund expedition to Gilgit one was seen (author) just outside Gilgit town, so a small population probably still survives to breed in the few forested valleys of Gilgit. A stray individual was found in a citrus orchard at Khanewal, Punjab in December 1962. Repeatedly flushed, it flew with rather slow hesitant flight between the trees, settling each time within 3 or 400 metres. Ticehurst *(Ibis,* 1924) also records one stray bird being shot in Lyari gardens (Karachi) on 4 November 1877, but the majority of the Pakistan breeding population undergoes local and short migrations to adjacent valley or foothill regions with those that winter in the Nilgiri hills, probably overflying north and central plains' regions of the subcontinent. Status RARE but resident.

HABITS

They are mainly crepuscular and solitary in feeding activity, frequenting the soggy banks of mountain streams or muddy areas around hillside springs. They feed by probing in a semi-circle with their bills as they squat on their tarsi, before shuffling forward to a fresh location. Their diet comprises largely of *Lumbricoides* earthworms and small mollusca and larvae of insects especially coleoptera including Elateridae, Scarabidae, Circulionidae etc. Their gizzards always contain a quantity of gravel, and in one Indian specimen, fragments of black Coleoptera (Mason and Lefroy, 1912).

While feeding they often dabble in the mud with one foot extended (observed Shahran, early June). The bill is usually inserted for about half its length sometimes at a slight angle. While paddling with foot extended they sometimes flap their wings over their head (observed Shahran, Hazara district). The significance of such wing flapping is not understood but the foot dabbling presumably flushes subterranean food prey, their movements then being more easily detected through the

sensitive bill tip. A bird captured near Naran in the Kaghan valley cocked up and fanned its tail, which was probably a threat display (Major Amanullah Khan, photographs).

BREEDING BIOLOGY

Males perform their distinctive territorial or courtship flights at dusk, starting at sunset flying in wide circles with slow fluttering flight high over the tree tops. As darkness falls these roding flights get closer to the ground or tree tops. Being a shy secretive bird, its presence is often only detected at dusk when roding starts. In the amphitheatre of Manshi forest at Shahran, in 3 different years roding was observed by one or two individuals over forest located at about 2,591 metres (8,500 feet) elevation and started only after the sun had set and dusk was rapidly gathering. It was noted from the end of May until mid-June but during an early July visit to the same area no roding was observed. These male birds alternately fluttered rapidly and glided, circling quite swiftly through the sky and their bills could be noted opening wide with each call, their legs and feet also dropping as they called. These calls are an alternating and rapid double-noted squeak sounding like 'seek-seek' and a frog-like croak.

In the Murree hills on 15 May on the slopes of Mukhshpuri mountain, two males (?) were observed flight chasing each other rapidly at dusk, presumably an aggressive encounter by one bird with a second male which had already started roding. They uttered rather high pitched whistling squeaks as they flew past. Studies in Europe indicate that males are successively polygamous, the males seeking females on the ground by roding and successively fertilizing up to 4 different females. Only the female incubates and the male deserts her, as soon as the clutch is completed, to seek a second mate (Hirons, 1980). Colonel Rattray (1905) describes a nest found at Changla Gali in the Murree hills, as a shallow cup in leaf litter on a steep bank, with four eggs which were broad ovals in shape. They are pale creamy in ground colour with reddish brown blotches and smudges and under-lying markings of lavender grey. Nesting may start from early May onwards and Col. Rattray noted them roding as late as 26 July. There are conflicting accounts of female woodcock transporting their young between their thighs while flying (See Ali and Ripley, 'Handbook', 1970, p. 294, and account given by Col. R. S. P. Bates, *JBNHS,* 1942, p. 69). In *New Dictionary of Birds,* edited by Landborough Thomson (1964) it is also stated that the female has been authentically recorded transporting its young in flight to a safe place, but in the *Handbook of the Birds of the Western Palearctic* (Vol. 3, 1983, p. 453) the authors seem to imply that such a behaviour does not occur and has been wrongly interpreted from their distraction display. However carrying of young in flight, pressed between the thighs (tibia) and belly has several times been witnessed by naturalists in the Himalayas, e.g. Davidson in Stuart Baker, *Game Birds of India*, (Vol. 2, 1921, p. 14) and R.S.P. Bates, *(JBNHS*, Vol. 43, 1942) and it has been reliably observed in Scotland. (C. L. Mckelvie *Book of the Woodcock*, 1986). Incubation takes 22 days and hatching is synchronous, the nidifugous chicks requiring to be fed bill to bill by the parent for the first few days (Cramp *et al.,* op. cit.).

*VOCALIZATIONS

During roding flight the male has a distinct song (described above) comprising two to five slow croaking calls followed by a shrill 'tseek' or sometimes a bi-syllabic 'twissick'. These calls are uttered at 2 to 3 second intervals while flying around at dusk. A bird roding at Dunga Gali only uttered a single croak followed each time by a sibilant 'tseek'.

A bird disturbed while feeding uttered a number of rapid croaking calls while on the ground. It then launched into a loud sequence of squeaking and groaning sounds difficult to describe adequately (observed Shahran forest rest house, early June). Another bird while feeding uttered rapid and loud quacking calls (Shahran, Kaghan valley). During distraction display females have variously been described as uttering harsh croaks like a Nutcracker *(Nucifraga caryocatactes)* also hissing and yapping calls (Bates and Lowther, 1952).

SUB-FAMILY TRINGINAE

Godwits, Curlews and Sandpipers

Black-tailed Godwit *Limosa limosa* (Linnaeus)   Map **162**

DESCRIPTION

Body length 40-44 cms.
Wingspan 70-82 cms.
Bill length 7.5-12 cms.

The Black-tailed Godwit is distinctive amongst waders likely to be encountered in a fresh water habitat, because of its large size, and very long straight bill. It is about equal in size to a Whimbrel *(Numenius phaeopus)*. In winter plumage grey-brown on head, neck and back, with the lower breast and belly white and a prominent white rump. As females are slightly larger than males, there is often considerable size variation in a flock, but amongst other waders they are distinctive by their very long necks, upright carriage and comparatively long leaden grey or dull grey-black legs. Some individuals show white mottling on the forepart of the face and a white supercilium but this is variable and generally the mantle and wing coverts are not conspicuously dark streaked but of a more uniform grey, in contrast to the Bar-tailed Godwit *(Limosa lapponica)* in similar winter plumage. In flight they are unique amongst waders and easy to recognize from a distance with a pure white rump and tail having a broad black bar at the tail tip extending up the middle rectrices, and with a broad white bar through the wings with a black trailing edge to the secondaries and solid black wingtips and black primary wing covers. The bill is livid pink basally and tapers to a darker horny black point. The iris is dark brown. The very similar Bar-tailed Godwit is separable by its preference for a salt water habitat, with a more streaked grey and white back and relatively shorter legs. Also its bill has a more pronounced upward curve and in flight the very broad white wingbar is unique to *Limosa limosa*, amongst waders in our region.

In breeding dress, which is often partly visible before birds migrate northwards from late March, the whole of the head, neck and upper breast becomes a reddish cinnamon colour with some darker barring on the flanks, and the mantle is more rufous chestnut with black barring on the feathers. However, a few grey feathers remain in the back. Both sexes are similar in breeding plumage though the female is duller in colour.

HABITAT, DISTRIBUTION AND STATUS

In Pakistan it frequents fresh and inland water, in contrast to the seacoast-haunting Bar-tailed Godwit *(Limosa lapponica)*. A winter migrant visitor which is highly gregarious both on passage and on its winter feeding grounds. It mainly spends the cold weather months on the larger lakes and wetlands of the interior of Sind province, being seldom encountered, even on passage, in the Punjab and also very seldom encountered on the seacoast (but see below). In 3 years' observations, D. Corfield only once encountered a flock of 9 birds on Rawal lake, Islamabad on 1 March 1980 (pers. comm., 1983). The first few autumn arrivals in Sind may be encountered from the last week of August, but not in any numbers, until the second half of September. On Manchar lake in 1927, Salim Ali recorded flocks totalling 10,000 or more (*JBNHS*, 1927), but nowadays such large concentrations are never encountered. The biggest number recorded by the author, was about 3,000 at Ghauspur jheel in north-western Sind on 14 February 1971. The northward migration starts from mid-March.

On 28 March 1980 at Khinjar lake, Thatta district, a flock of about 40 was seen, which included 7 birds with pinkish chestnut feathers appearing on the face, neck and upper breast, and these birds were about to start their spring migration. A few scattered flocks often over-summer in lower Sind, in non-breeding plumage, haunting the seepage pools around Haleji lake until the end of June and thereafter spreading into the rice fields. A flock of 50 was noted 1 June 1974 at Haleji and 25 on 16 June 1975 and 22 on 8 July in rice fields near Mirpur Sakro, Thatta district. Records of birds in Punjab are probably mostly of flocks on passage but over 200 were seen on 5 February 1973 at Taunsa barrage on the Indus (F. Koning, pers. comm.) and 2 birds on Mailsi spill channel on the Sutlej river on 2 and 4 October, and at Sidhnai on the Ravi river on 6 September. A single bird was seen (author with P. Goriup) on Clifton beach, Karachi on 16 January 1981. There are no records of birds on migration from Baluchistan but Dr. Scully collected a male in breeding dress and a female from Gilgit, and noted them on passage during the first half of April and in the autumn in the third week of September *(Stray Feathers,* 1881). Neither Whitehead nor Whistler recorded it around Kurram and Kohat in the NWFP nor around Rawalpindi district of the Punjab. Status ABUNDANT only in lower Sind.

HABITS

They feed often in deeper water than other waders, taking advantage of their long legs and may be seen up to their bellies in water, plunging their whole nead and neck beneath the surface. They commonly feed in flocks, probing in the mud or other scummy shallows of a drying out pond, their bills making shallow rapid probes alternated with occasional deep stabs up to the base of the bill. Their food includes small crustacea and water beetles (Ticehurst, 1924), also larvae of beetles and Diptera flies, mollusca, annelid worms and earthworms (Lumbricidae). Some seeds and leaves are also occasionally taken including rice (Mason and Lefroy, 1912). On their breeding grounds they feed predominantly upon grassland insects especially beetles, Carabidae and Curculionidae and around lake shores larvae of Dytiscidae and Hydrophilidae (Water beetles) (Dement'ev and Gladkov, 1951). In flight they often travel in 'V' formation and travel at considerable height, twisting and turning through the air so that their silvery white under-wing pattern and black barred upper-wing pattern alternatively flashes in the sunlight. They are

Map 162: Black-tailed Godwit *Limosa limosa*

normally rather silent in their winter quarters but in late March as migration time approaches, they become noticeably aggressive, with birds sparring with each other, wings raised over their backs and sometimes jumping at each other striking with their feet and the whole flock making a soft twittering call (observed Haleji lake 7 April).

BREEDING BIOLOGY

Extra-limital, they nest semi-colonially and have a stable monogamous pair bond, both sexes sharing incubation duties and caring for the young, though the nidifugous and precocial chicks can feed themselves from the time of hatching. They breed from temperate to boreal zones in lowland grassy or moorland areas.

*VOCALIZATIONS

Often silent in winter or occasionally uttering tri-syllabic flight calls 'wit-wit-wit' (Ali and Ripley, 1969). During feeding, prior to migration, they utter short brief 'kik-kik-kik' calls (Haleji lake, 7 April). On their breeding grounds, the displaying male has a peewit-like *(Vanellus vanellus)* loud plaintive call (Cramp *et al.,* 1983).

Bar-tailed Godwit *Limosa lapponica* (Linnaeus)   Map **163**

DESCRIPTION

Body length 37-39 cms.
Wingspan 70-80 cms.
Bill length 8.5-10 cms.

It is slightly smaller than a Whimbrel *(Numenius phaeopus)* and shorter legged than the Black-tailed Godwit *(Limosa limosa)* but overlaps in body size with the latter. It is ecologically separated from the fresh water inhabiting Black-tailed Godwit, so that identification is usually not difficult, but it can also be separated from *Limosa limosa* in winter plumage, by its more streaked appearance on the back and wing coverts, and its tail which is cross-barred grey and white in contrast to the solidly black tail of *L. limosa*. In flight the rump shows prominently white but the wings are all grey-brown without the conspicuous white panels of the Black-tailed Godwit. Also its bill is slightly up-tilted towards the tip, whereas that of the Black-tailed Godwit appears straighter. When feeding on the ground its plumage looks more uniformly grey-brown than that of a Curlew or Whimbrel. The bill is fleshy pink basally and black towards the tip, with the legs dark olive grey to leaden grey and the iris dark brown. The sexes are not alike in contrast to *Limosa limosa,* when in breeding plumage. The male assumes a cinnamon chestnut body plumage extending from the head to the vent with some narrow black shaft streaks on the feathers of the crown, nape and flanks. The Black-tailed Godwit by contrast, in breeding plumage still has the lower breast and belly white with dark vertical barring on the flanks. In breeding plumage the female becomes darker on the mantle and scapulars,

like the male, the feathers bearing black centres and chestnut barring, whilst its head and neck become buff, finely steaked with black and the lower breast and belly remains white. A whitish supercilium is usually prominent in both sexes in winter plumage and in breeding females. The cross-barred white tail is visible in all plumages from above and below when in flight and is the most reliable field point.

HABITAT, DISTRIBUTION AND STATUS

Like the Black-tailed Godwit a winter migrant visitor, but not as highly gregarious on its wintering grounds, and almost wholly confined to the seacoast and inter-tidal zone in estuaries and lagoons where large expanses of sand banks or mud banks are periodically inundated by the tide. It occurs very rarely on inland waters. H. Waite collected specimens (Brit. Mus. coll.) on autumn passage on Nammal lake in the Salt Range on 2 October, and on 14 October at Ghauspur jheel in Jacobabad district the author saw a solitary Bar-tailed Godwit amongst a flock of Black-tailed Godwits.

In one day's voyage around Ghizri creek in mid-winter and on the more open sandbars of the Indus delta up to 100 Bar-tailed Godwits may be counted, but they are less numerous than most other waders. They are generally encountered in loose groups of 5 or 6, feeding in association with Curlew *(Numenius arquata)* and Grey-Plovers *(Pluvialis squatarola)*. Adults in worn breeding dress are amongst the earliest autumn arrivals and have been noted in Karachi creeks from the fourth week of July (Ticehurst, *Ibis* 1924). They depart again by the end of April and during boat trips around the Indus delta lagoons and Ghizri, it was estimated that more than 50 per cent were well moulted into breeding dress on 18 April 1980, and on 22 April 1983, 66 per cent were in full breeding plumage. A small number of birds always over-summer, particularly around Sonmiani lagoon, and these retain their winter or non-breeding dress. A flock of 8 birds was noted on 23 June 1973 by Clifton oil pipe-line Karachi and scattered groups of 2 or 3 birds at Sonmiani on 24 May in winter plumage. On a visit to Sonmiani on 4 May a few Bar-tailed Godwits in breeding plumage were still around and these birds were presumably late migrants. On a boat trip on Ghizri creek on 22 April over 200 Bar-tailed Godwits were counted, in an area where not more than 50 or 60 birds were noted a month earlier, indicating an influx of migrants from further south. Status COMMON, only on seacoast.

HABITS

They probably probe in the mud at shallower depths than the Black-tailed Godwit, as noted in Ghizri creek,

Map 163: Bar-tailed Godwit *Limosa lapponica*

where the head is often cocked sidways while probing. Ticehurst noted that their stomachs contained bivalve molluscs and small crabs (Brachypoda) (op. cit., 1924). Birds shot in India contained small Mollusca sandworms (Annelida) and minute Acephalae (Jellyfish) (Mason and Lefroy, 1912). On their breeding grounds larvae of fresh-water insects, worms and molluscs form a more important part of their diet (Dement'ev and Gladkov, 1951). In flight they are very swift and like some of the smaller waders, turn and twist in synchronized manoeuvres with great agility.

BREEDING BIOLOGY

Their nesting range is in more northern latitudes than *Limosa limosa,* being confined to the high Arctic extending from Alaska to eastern Siberia across to Norwegian Lappland. They nest in tundra bog and peat land and the pair bond is monogamous with the male doing more than half of the incubation and both sexes tending the chicks, which are self-feeding from the time of hatching (Cramp *et al.,* 1983).

VOCALIZATIONS

In its winter quarters its flight call is a low piping whistle (Ali and Ripley, 1969). In a large flock its voice has been likened to the rapid and melodious 'chattering' of Jackdaws (Dement'ev and Gladkov, 1951). The call is also described as a low barking 'kurruc' or 'yak' (Cramp *et al.,* 1983). Its flight call has also been described as 'kirruc-kirruc' (Heinzel and Parslow 1972). The courtship flight song of the male has been described as sweet rapid notes 'a-wik-awik-awik' (Cramp *et al.,* 1983).

Whimbrel *Numenius phaeopus* (Linnaeus) Map **164**

DESCRIPTION

Body length 40-42 cms. including slender down-curved bill, which is 7-10 cms. long.
Wingspan 76-89 cms.

It is a larger wader about the same size as a Godwit *(Limosa)* but with slightly shorter legs and neck and more bulky body. Similar in outward appearance and colouration to the Curlew *(N. arquata)* it is chiefly distinguished from the latter by averaging 30 per cent smaller in size with a shorter bill and a dark brown crown with conspicuous creamy white longitudinal stripes. There is a broad mid-coronal whitish streak and two narrower creamy white superciliary streaks. The legs and feet are bluish grey and bill is fleshy pink at the base, horny black distally. The body plumage is boldly streaked in sandy brown and buffy white, with arrow marks on the flanks, narrower dark streaks on the throat and cross-barring on the flight feathers and tail with wavy barring on the wing coverts and mantle. In flight the rump shows prominently white.

HABITAT, DISTRIBUTION AND STATUS

A winter migrant visitor to Pakistan, over-flying most of the country but arriving in considerable numbers and conspicuously noisy in the rice fields of lower Sind from mid-July onwards. It is in fact one of the earliest autumn migrants. Later in the autumn it resorts to the mud banks of the mangrove creeks and coastal mud flats, and it is never encountered away from the seacoast. Ticehurst *(Ibis,* 1924) considered that the bulk of the population passed southwards, but a substantial population over-

winters and over 100 have been counted during a 2 day voyage through the Indus delta creeks on 10 December and over 120 on a day's voyage in Sonmiani lagoon (Lasbela) on 12 May. During late August and early September they are particularly numerous along the sandy and rocky shores of Karachi and the Makran coast west from Hawkes Bay through Cape Monze to Sonmiani, being seen in flocks of 15 to 20. A few birds over-summer in the mangrove creeks. Two groups of 5

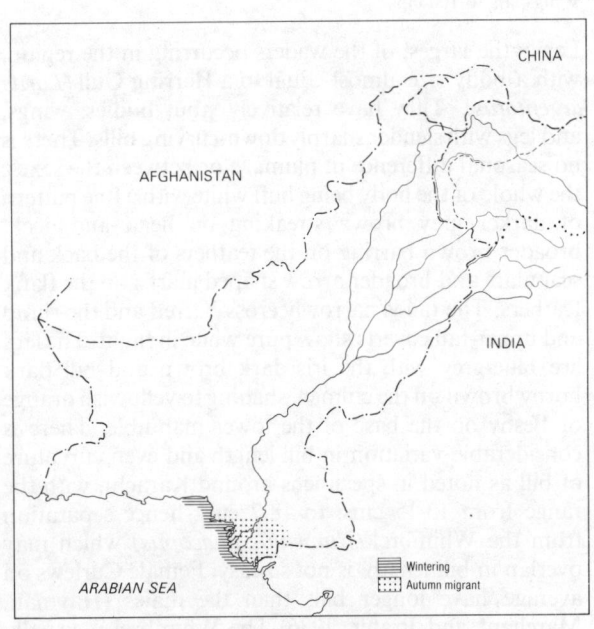

Map 164: Whimbrel *Numenius phaeopus*

and 3 Whimbrel were seen in the rice fields around Mirpur Sakro on 8 July and it is presumed that these were returning breeding birds. There is no record of birds on passage in the northern areas (Ticehurst, *JBNHS,* 1927) in contrast to that of the Curlew *(N. arquata),* and available evidence indicates more of a west to east passage through the Middle East and Iran from the breeding population of central Siberia and Fenno-Scandinavia. Status ABUNDANT only along Makran and Sind seacoast.

HABITS

More gregarious than the Curlew *(Numenius phaeopus)* specially on southern migration. They have a preference for feeding in sticky mud rather than sandbars around Karachi and unlike the Curlew perch freely in the mangrove trees at high tide. They usually forage in company with Curlew and Oystercatchers *(Haematopus ostralegus).* They have been observed picking items from the surface and making shallow rapid probes with their bill. They feed upon *Uca* Fiddler crabs, (observed Indus delta). Also other small crabs such as

Thalamita crenata and *Sesarma longipes* (Ali and Ripley, 1969). They also consume Mollusca and insects and worms (Annelida).

BREEDING BIOLOGY

Circum-polar in the boreal and sub-Arctic zones, choosing boggy moorland or tundra for nesting. They nest solitarily with the male sharing in incubation duties and caring for the newly hatched chicks, which are nidifugous and self-feeding from the time of hatching.

*VOCALIZATIONS

Quite a noisy bird on arrival at its winter quarters, with both sexes uttering their loud rapidly repeated whistle when in flight. It is a single note given so rapidly as to be almost a whinny 'pee-pee-pee-peep'. Sometimes this whistle is repeated only 4 or 5 times, occasionally in a trill of 7 or 8 notes. In the breeding season they utter a more drawn out inflected whistle like a Curlew *(N. arquata)* (Cramp *et al.,* 1983).

Eurasian Curlew *Numenius arquata* (Linnaeus)   Map **165**

DESCRIPTION

Body length 50-60 cms. including very long down-curved
   bill 13-17 cms.
Wingspan 80-100 cms.

Easily the largest of the waders occurring in the region, with a body size almost equal to a Herring Gull *(Larus argentatus).* They have relatively stout bodies, wings, and legs with slender sharply down-curving bills. There is no seasonal difference in plumage or between the sexes, the whole of the body being buff white with a fine pattern of darker grey brown streaking on head and neck, broader brown barring on the feathers of the back and scapulars and broader arrow shaped marks on the flank feathers. The tail is narrowly cross-barred and the rump and upper-tail coverts show pure white in flight. The legs are blue-grey with the iris dark brown and bill dark horny brown on the culmen, shading to yellowish orange or fleshy on the base of the lower mandible. There is considerable variation in bill length and even curvature of bill as noted in specimens around Karachi, with the range from 13-15 cms. to 18.7 cms. hence separation from the Whimbrel *(Numenius phaeopus)* which may overlap in bill length, is not so easy. Female Curlews on average have longer bills than the males (Hayman, Marchant and Prater, 1986) The Whimbrel is usually separable when seen alongside (as frequently occurs in

the Indus mangroves) by its shorter less curved bill, and smaller over-all body size. The Curlew also lacks any distinctive dark crown stripes as in the Whimbrel.

HABITAT, DISTRIBUTION AND STATUS

A winter migrant visitor to the region with very few birds remaining on inland fresh water lakes and river sandbars, the majority wintering along the seacoast. It has been recorded on passage in Chitral (Perreau, *JBNHS,* 1910) at the end of April, and occasionally through Baluchistan, in April on Band Kushdil Khan lake shores, one or two being seen on 9 April and 12 August and 30 November (Meinertzhagen, *Ibis,* 1920). David Corfied saw a ;single bird at Rawal lake, Islamabad on 15 March 1981 and one on 22 March 1982. H. Waite *(JBNHS,* 1948) also noted a regular autumn passage over the Punjab Salt Range and collected specimens (Brit. Mus. coll.) 13 October at 15 Kallar Kahar and 24 October at Uchchali lakes. It is abundant around the more extensive tidal mud flats on Sonmiani and Dambh villages in Lasbela and throughout the Indus mangrove creeks and lagoons. In a 3 day boat voyage through the mangroves of the Indus in mid-November an estimated 5,000 were seen (author and R. Passburg), and in one day's voyage in December about 1,150 were counted. In lower Sind first arrivals are found in the rice fields from early August as well as around the larger fresh water lakes, but most travel to the seacoast to spend the winter.

In the Punjab a few scattered individuals remain throughout the winter in the larger wetlands. 3 were at Panjnad headworks 14 November, 2 at Sidhnai spill channel 22 November, 2 at Nalwah on the edge of the Cholistan desert near Rahim Yar Khan on 29 January and 3 birds on the sandbars around Mailsi syphon on the Sutlej river 2 January, and 4 a year later on 5 January in the same locality. In northern Sind around Ghauspur jheel a few Curlew always over-winter. A flock of 12 seen 30 November 1974 and 32 on 5 November 1969. A few birds undoubtedly over-summer but these may be supplemented with autumn migrants from as early as late July as over 30 were noted in a boat trip to Ghabi Dhero in the Indus delta on 29 July 1977 and on 22 July a flock of 7 was seen near Jati in Thatta district in rice fields. Status ABUNDANT on the seacoast in mangroves, otherwise scarce.

HABITS

Not as gregarious as the Whimbrel *(Numenius phaeopus)* and usually encountered in loose groups of 5 or 6 birds feeding amongst Bar-tailed Godwits *(Limosa lapponica)* and Oystercatchers *(Haematopus ostralegus)*. On first arrival and during migration they keep in small flocks and often fly in 'V' formation, e.g. 32 birds between the Indus and a nearby lake on 4 November. Compared with wintering birds in western Europe they are definitely not so gregarious in Pakistan. Observations over many years indicate that in Punjab and the northern Sind where they are rare, small numbers can be predictably encountered at certain localities which might indicate fidelity of individuals to particular winter feeding territories. Like their European counterparts they are shy and wary and particularly active feeding at night, often calling to each other throughout the hours of darkness (observations while camping Indus delta, and mist-netting results at Ghauspur lake). They can be seen walking along over the mud, stabbing several times rapidly with their bills and then occasionally rapidly inserting their whole bill up to head (observation Indus delta). They feed upon small crabs *(Uca* sp., and *Ocypoda* sp.) (Ali and Ripley, 1969), bivalve molluscs (Mason and Lefroy, 1912) and annelid worms in the inter-tidal zone. Inland, insects form a higher portion of the diet especially larval and adult beetles (Carabidae, Histeridae, Hydrophilidae, Tenebrionidae etc.) (Cramp *et al.,* 1983). Depending upon circumstances they feed both during daytime and

Map 165: Curlew *Numenius arquata*

at night, and can be seen roosting on banks or sandbars when the tide is high in mixed company with other waders.

BREEDING BIOLOGY

It nests on moorland and boggy heathland in temperate and boreal latitudes and extending into the steppe zone in the USSR. The pair bond is monogamous, the male usually sharing equally in incubation and both parents brooding the small chicks which are self-feeding.

*VOCALIZATIONS

Familiarity with the breeding bird in North Wales, as well as wintering Curlew in Pakistan indicates that they are quite vocal throughout the year though the bubbling song of a breeding bird is uttered less frequently during the winter, it can be heard in Pakistan throughout these months during many different situations including uttered by birds on the ground. The main contact or alarm call is the well known repeated 'cur-lee cur-lee'. When very alarmed it may utter a more rapid 'cur-cur-cur' and these cries often have a wheezy tone. The beautiful sounding liquid bubbling call rises and falls with a vibrato, effect and may last for up to 1½ seconds.

Spotted or Dusky Redshank *Tringa erythropus* (Pallas)

Plate **15**
Map **166**

DESCRIPTION

Body length 29-31 cms.
Wingspan 61-71 cms.
Bill length 5-6 cms.
Tarsus 5.3-6.1 cms.

An elegant wader, likely to be mistaken in winter for the commoner Redshank *(T. totanus)* to which it looks closely similar but having a distinctly longer more slender bill and longer legs than the latter. If put to flight it usually at once identifies itself by the very distinctive two-noted call note and is separable from the Redshank in lacking the clear white secondaries and tips to inner primaries, prominent in both summer and winter plumage on *T. totanus*. Its upper body in winter plumage is a rather pale grey with white rump and cross-barred tail and upper-tail coverts. The mantle is less olive in tinge than *T. totanus* and it is also usually more upright in carriage. The bill is black at the tip, red basally and the legs and feet are orange-red. There is a whitish supercilium and the throat, foreneck and breast are white with obsolete grey speckling along the sides. In breeding plumage they are strikingly different, the whole head and neck and upper breast moulting into a uniform dark sooty brown with fine white speckling on the back and flanks which are otherwise the same dark sooty colour. Both sexes are alike in plumage. In full breeding plumage a white eye-ring is noticeable and the legs and feet are a darker red than those of *Tringa totanus,* also the bill is slightly down-curved at the tip, whereas that of *T. totanus* is perfectly straight (obs. Sukh Beas, Lahore district, 21 April).

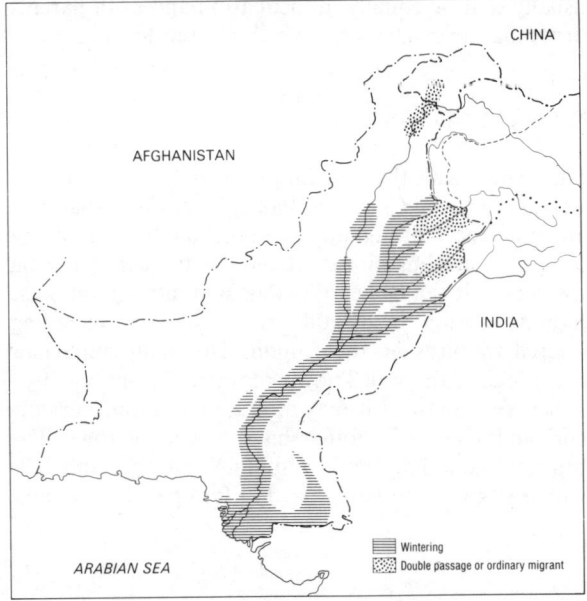

Map 166: Spotted or Dusky Redshank *Tringa erythropus*

HABITAT, DISTRIBUTION AND STATUS

A winter migrant visitor to the region, which is much less common in Pakistan than the other 'shanks' or 'sandpipers'. First arrivals in the Punjab are noted in early September and they are on return spring migration again in late April. One was shot out of a flock of 8 near Kushdil Khan lake, Baluchistan by Meinertzhagen on 26 October *(Ibis,* 1920) and one was collected on migration in Gilgit on 23 April (a male partly moulted into breeding plumage) by Biddulph *(Stray Feathers,* 1881). The author saw several flocks of 10 to 20 birds in partial and full breeding dress, on the Sukh Beas 67.2 kms. (42 miles) north-east of Lahore on 21 April and H. Waite collected a specimen at Khanki headworks in Gujrat district on 10 May (Brit. Mus. coll.). Birds in full breeding plumage have also been noted on 2 April and 10 April on roadside pools in Rahim Yar Khan district and a group of 12 was noted as late as 8 May at Haleji lake, Thatta district. They are generally later than Redshanks in commencing their return spring migration. During the winter they are mostly seen in ones or twos in company with other waders, frequenting seepage pools near canal headworks, the drying out margins of lakes and even highly brackish roadside pools throughout Sind and Punjab. They have been noted on the Indus creeks on salt marsh grass flats (e.g. at Keti Bandar) in mid-November but generally they are only found on inland waters, though they do not shun very brackish and salty drying out pools. Twenty birds were seen (author) in one small pool 12 February 1973 near Sadoori lake in Sanghar district and 30 were seen together on 10 January 1974 on Sandho lake on the Indian border of Badin district (F. Koning, pers. comm.). It is a gregarious species in its winter quarters and large flocks have been encountered some years at the beginning of the winter at Kandhkot and Ghauspur in Jacobabad district. Three flocks of 120 to 160 birds each seen (author) 14 November and 9 December in successive years, at which time their feeding behaviour is very distinctive, but such aggregations have not been noted elsewhere. There seems to be a definite increase of birds during spring passage, possibly of birds wintering further south. Status FREQUENT.

HABITS

Even single birds have been noted feeding by swimming in deeper water and scything with their bills (observations Khinjar lake in seepage zones). At Ghauspur a close-packed flock fed in shallow water in a tight group, massed together, their heads plunging in with such rapid bobbing fashion as to suggest a sewing machine, as they prodded in the ooze with their bills. Compared with other waders they are much more ready to swim while foraging and also to feed in close packs.

Their food comprises a variety of animal invertebrates including larvae and adults of water beetles, flies molluscs and worms. They have been recorded with 'their gullets full of small fish', in India (Inglis, *JBNHS*, 1903), and Water Snails *(Limnae* and *Planorbis)* formed over 50 per cent of their diet in Russian birds (Dement'ev and Gladkov, 1951). Two or three Spotted Redshanks (otherwise rare in the region) regularly occurred over 4 successive winters on one small roadside brackish pool near Jangshahi, Thatta district until the pool dried out in 1983, indicating the possibility of this species having territorial fidelity to a suitable restricted wintering ground. On numerous equally suitable pools and lakes in the vicinity, Spotted Redshanks were normally never encountered. Flocks at Kandhkot exhibited remarkably coordinated flock behaviour, both when wheeling and twisting in flight unison and when wading and starting to feed, simultaneously probing with rapidity, and also when a swimming flock turned in unison and headed for deeper water at the approach of a Marsh Harrier *(Circus aeruginosus)*.

BREEDING BIOLOGY

Extra-limital in sub-Arctic regions on the fringes of the lightly wooded tundra. They nest solitarily and breeding behaviour seems to have been little studied, but the male does a large part of the incubating as well as brooding and tending of self-feeding young.

*VOCALIZATIONS

The flight call is very distinct and comprises an inflected 'two-wit', only repeated after several seconds' interval. This is used as a warning or alarm call and as a contact call. On its breeding grounds, it emits a creaking or grinding series of whistle-like notes (Cramp *et al.,* 1983).

Redshank *Tringa totanus* (Linnaeus)

Plate **15**
Map **167**

DESCRIPTION

Body length 27-29 cms.
Wingspan 59-66 cms.
Bill length 3.7-5 cms.
Tarsus 4-5.5 cms.

It is a long legged, neat wader with medium length straight pointed bill. Slightly shorter legged and noticeably shorter in the bill than the Spotted Redshank *(Tringa erythropus)* and considerably smaller than the Green-shank *(T. nebularia).* Its most noticeable features are the bright vermillion-red legs and feet and in flight, broad white trailing edge to the wings. It has dusky olive grey upper parts, spotted on the wing shoulder with some white and pale buff and greyish white throat and upper breast, with some indistinct darker spotting. The scapulars are cross-barred with dark grey. In flight the secondaries and tips of inner primaries form a broad conspicuous white bar to the trailing edge of the wing, a feature lacking in *Tringa nebularia* or *Tringa erythropus.* The white rump extends up between the scapulars in a white cigar shaped pattern. The white rump pattern is similar to the Greenshank, Spotted Redshank and Marsh Sandpiper, none of which show all-white seondaries. In summer plumage the wing coverts and scapulars are more streaked and heavily barred and the breast becomes more heavily spotted. In winter plumage they are more uniformly grey-brown on the upper parts. The bill is dark horny brown at the tip, red on the basal half of the lower mandible. The iris is dark brown. The sexes are alike in all plumages. Immature birds have more yellow-orange coloured legs, but are longer legged than the Terek Sandpiper *(Xenus cinereus)* which also lacks the white rump.

HABITAT, DISTRIBUTION AND STATUS

A common and widespread winter migrant visitor to the region with the main population concentrating along the seacoast, but also a few occurring in small numbers around all the larger inland waters and tolerant of both sweet and saline pools. Autumn migrants usually arrive a bit later than the Greenshank *(T. nebularia)* but have been noted as early as 21 July while still in breeding dress, with heavy spotting on the breast, (Haleji lake, 1979). It

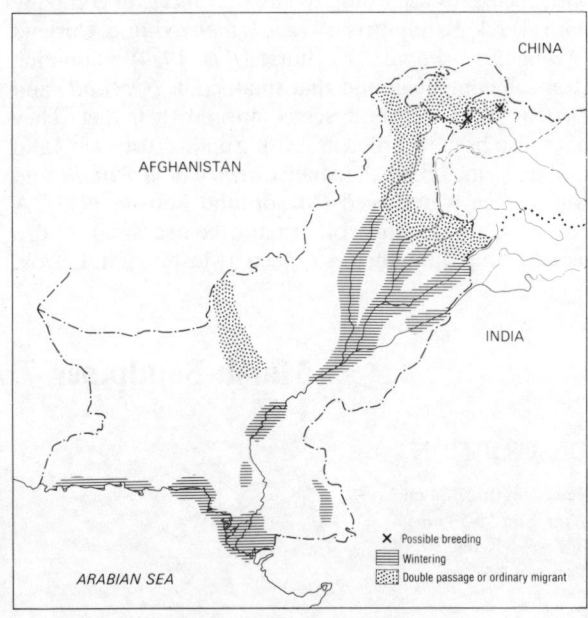

Map 167: Common Redshank *Tringa totanus*

is not sighted very often in the northern areas on passage, but Dr. Scully collected a specimen in Gilgit on 10 April and saw others in the first week of September *(Stray Feathers,* Vol. 10, 1881). In Baluchistan they have been noted in mid-May and again in late October on passage (Meinertzhagen, 1920). Two were seen Kushdil Khan lake 31 October 1972 (author). In the Punjab Salt Range lakes they are usually only sighted during autumn and spring passage (H. Waite, 1948) but small groups of 2 or 4 birds have been noted on Sidhnai spill channel (Ravi river), Lal Sohanran (Bahawalpur) and Panjnad headworks in November and early December over several years. On the seacoast they avoid sandy beaches but in the mangroves on mud banks they are very numerous. On a one day boat trip over 2,000 can be counted along the Indus delta creeks (author's diary, 10 December 1983). They are equally common in the mangrove creeks of Sonmiani lagoon in Lasbela and presumably further west along the Makran coast. Most birds have flown north by the end of April and latest date seen 22 April on Ghizri creek. Status COMMON.

HABITS

Definitely not as gregarious as the Dusky or Spotted Redshank, especially on migration and normally seen in twos or threes, they are more adapted to brackish waters and the seacoast than either *T. erythropus* or *T. nebularia.* They are nervous birds, bobbing their whole body and head up and down when observed closely. They will occasionally forage in deep water with their breast partly submerged but have not actually been observed swimming. They favour sticky mud banks amongst the mangroves, probing and picking rapidly with their bills according to circumstances, in company with Terek Sandpipers *(Xenus cinereus)* and Curlews *(Numenius arquata).* Ticehurst *(Ibis,* 1924) examining stomach contents found that small crabs *(Ocypoda* and *Uca* sp.) and soft molluscs comprised their diet. They have also been recorded as feeding upon crustaceans and molluscs, including the snail *Corbicular orientalis* and some green water weed (Mason and Lefroy, 1912). A considerable quantity of grit and coarse sand is also usually found in their stomach (Mason and Lefroy,

1912). Inland, larvae of Coleoptera, dragon fly larvae (Odonata), flies and worms (Annelida) comprise more of their diet. (Dement'ev and Gladkov, 1951, Cramp *et al.,* 1983).

BREEDING BIOLOGY

It breeds in Ladakh (India) just to the east of Pakistan territory (Meinertzhagen, *Ibis,* 1927) and it is quite probable that a few pairs breed on the Deosai plateau of Baltistan where Mathews heard them calling at night in mid-July *(JBNHS,* 1941). Several pairs were also seen by Meinertzhagen near Skardu haunting a marshy spot at 2,000 metres (6,500 feet) on 18 August and he thought that they might have bred there *(Ibis,* 1927). In these regions it usually breeds from 3,629 metres (12,000 feet) upwards in tussocky grassland near mountain streams. A clutch of four pyriform eggs is laid in a nest on the ground which is a small hollow and scantily lined with grass blades or other rotten vegetation on a mound of moss or grass tussock. Incubation shared by both sexes, takes 24 days and the precocial nidifugous chicks hatch synchronously and are self feeding from hatching. Often only the female tends them as they develop. Most of the population however breeds in central Asia, in Kazakhstan and the Pamirs (Dement'ev and Gladkov, 1951). A bird ringed in Bharatpur (Rajasthan, India) was recovered on its breeding territory in the Altai, near Blagoveschchenka, USSR (Ali and Ripley, 1969). At the beginning of pair formation, males perform display flights over their nesting territory, a sort of alternating fluttering upwards and gliding downwards with depressed wings in an undulating flight path, uttering a rapidly repeated inflected whistle call 'tui-tui' (Cramp *et al.,* 1983).

*VOCALIZATIONS

On their wintering grounds when disturbed they are quite noisy, usually uttering a series of rapidly repeated calls 'teu-hew-hew' or 'teu-hu', both on the ground before taking flight and also when on the wing.

Marsh Sandpiper *Tringa stagnatilis* (Bechstein)

Plate **15**
Map **168**

DESCRIPTION

Body length 22-24 cms.
Wingspan 55-59 cms.
Bill 3.6-4.5 cms.
Tarsus 4.7-5.7 cms.

In winter plumage the generally rather white head and neck and grey upper plumage make it look very similar

to a miniature Green-shank *(Tringa nebularia).* It is longer legged than a Wood sandpiper *(T. glareola)* but about equal in size and body shape with a very slender needle-like black bill. In flight, like the Greenshank *(T. nebularia)* shows the same barred tail and cigar shaped white rump patch between the scapulars, also without any white bar on the wings. In breeding plumage the mantle and wing coverts become more buffy, less grey and are heavily spotted with black as also the hind neck and upper breast. In winter plumage the upper parts are a cold grey without the olive tones of Wood Sandpipers or their larger cousins, Redshanks. Their legs are grey-green and iris dark brown, but sometimes in the spring the legs turn bright yellow-olive. If size comparison is not possible, the needle-like and straight bill usually serves to distinguish it from the up-turned and heavier looking bill of the Greenshank *(T. nebularia).*

HABITAT, DISTRIBUTION AND STATUS

A winter migrant to the region, with the bulk of the population passing through to winter in the more southern regions of the subcontinent. They are one of the earliest migrants to arrive in the autumn. 20 seen in a loose flock still in breeding dress on Haleji lake, 21 July. E. M. Nicholson records an autumn passage through Kalat and southern Baluchistan from 15 August (typescript notes, 1952). Allan Vittery saw a flock of 130 on Kharrar jheel, Renala Khurd in late August, 1973 (Vittery, pers. comm.). A small number over-winter in the more extensive marshy and wetland areas of the Punjab, especially the Sukh Beas, Kasur district, Sidhnai spill channel (Multan district) and around Taunsa barrage in Dera Ghazi Khan district. In the spring there is a marked northward passage in company with Wood Sandpipers in the fresh water creeks such as Hab river and Ghizri on the Sind coast. Over 200 seen Ghizri creek 1 April, but by 25 April all had left, only the Wood Sandpipers *(T. glareola)* remaining. They normally do not frequent such small and isolated roadside pools and tanks as *T. ochropus* or *T. glareola* and are seldom seen on the tidal mud flats, but Ticehurst did see some on the coast *(Ibis,* 1924). Latest date seen (author) 10 April Rawal lake, Islamabad. Status COMMON.

HABITS

Except on migration not very gregarious, the flock noted by Allan Vittery being exceptional (see above). They are graceful and delicate feeders darting rapidly along the water's edge picking at floating objects and off vegetation. They often forage in company with dabbling ducks and other waders (observed Chatteji lake in flooded Tamarisk). Food items include insects and their larvae, molluscs and crustacea. Ticehurst found small shells of *Gastropods* in their stomach (op. cit., 1924). Russian studies show few terrestial beetles in 12 stomachs examined, their diet mostly comprising water beetles *(Corixa),* larvae of Dytiscidae and *Planorbis* water snails (Dement'ev and Gladkov, 1951).

Map 168: Marsh Sandpiper *Tringa stagnatilis*

BREEDING BIOLOGY

Birds ringed at Point Calimere, Madras in south India during December were recovered in the USSR in Novosibirsk region in early May on their breeding grounds (Ali and Ripley, Vol. 2, 1969). It breeds in fresh water swamps from middle latitudes in steppe and temperate forest zones, generally avoiding alkaline or bare saline flats and preferring well vegetated ground cover. It often breeds colonially and in company with terns *Chlidonias niger* and *C. leucopterus,* and Redshanks *(Tringa totanus)*; several dozen nests being found within an area of 2 hectares (Dement'ev and Gladkov, 1951). Both sexes share incubation and tending of the young.

VOCALIZATIONS

Not as vocal on wintering grounds as the Greenshank *(T. nebularia)* or Redshank *(T. totanus),* but birds on first arrival in autumn occasionally utter flight calls, which are a soft fluting reminiscent of a Wood Sandpiper *(T. glareola)* 'tee-yew-tee-yew-tee-yew'. During early courtship they sing while fluttering over their territories, and this is more of an ululating call 'tyurlyu' (Cramp *et al.,* 1983).

Greenshank *Tringa nebularia* (Gunnerus)

Plate **15**
Map **169**

DESCRIPTION

Body length 30-33 cms.
Wingspan 68-70 cms.
Bill length 5-6 cms.
Tarsus 5.5-6 cms.

This is a long legged rather large wader, bigger than the Redshank *(T. totanus)* and distinguished by its long strongly made up-turned bill and olive green legs. The iris is dark brown and bill is also green at the base, blackish on the distal half. In winter plumage the head and neck are white, faintly spotted with dark grey and darker on the crown. The rest of the under parts are white and the mantle and back is rather a cold greyish tone. In flight the back shows prominently white between the scapulars as also the upper-tail coverts and tail which are cross-barred dark grey, and it is confusingly like the smaller Marsh Sandpiper *(T. stagnatilis)*. In breeding plumage there is only a very slight change (whereas *T. stagnatilis* looks quite buffy rufous on the back) with the back and wing coverts becoming more darkly spotted and blotched with black and brown and more distinct black streaking on the neck and upper breast. The sexes are alike. Only *Tringa stagnatilis* and this species among the Sandpipers, have the whole forepart of the crown and head white in non-breeding plumage.

HABITAT, DISTRIBUTION AND STATUS

A winter migrant visitor to Pakistan, which can occur in every month of the year, and which is one of the earliest returning migrants at the end of the summer and the latest to depart in the spring. Birds have been seen in Muzaffargarh and Multan districts of Punjab on return migration as early as 21 August and in the rice fields around Sujawal in lower Sind as early as 28 July. The main autumn passage continues through September, e.g. 9 birds seen Rawal lake, 29 September, 1980 (D. Corfield, pers. comm.). They are usually not very gregarious preferring to forage singly, but 87 were counted in 4 or 5 groups at Lal Sohanran on 17 March in very hot weather, and were obviously on passage. They are quite common along the main river systems, around the larger fresh water swamps and seepage pools near large canals or irrigation headworks, and they will be equally at home in brackish or saline pools. Though not nearly as common as *T. totanus* on the seacoast, a few over-winter on the coast both on sandy beaches and along the mangrove creeks. In the Punjab they are numerically more common than the Redshank *(T. totanus)*. A few also over-summer along the coast, 2 being seen at Mauripur on the salt pans on 6 June and Corfield saw occasional individuals around Rawal lake in every month of the year (pers. comm., 1983). They

occur in suitable areas in the NWFP and Baluchistan, also during the winter months 30 were seen (author) on Kushdil Khan lake on 30 October 1972. Birds on passage have been collected from 10 to 17 August (Biddulph, Vol. 9, 1881) and in early September (Biddulph, 1882) from Gilgit. The return passage takes place mainly during the latter part of April through the Punjab. Status ABUNDANT.

HABITS

As noted above they are not normally gregarious in winter. They are usually shy and wary, and if observed closely will bob their heads and necks and whole body nervously up and down. Very eclectic in their choice of habitat, from small roadside borrow pits, to river banks and estuarine shores. They are active foragers often catching prey by running through water with part of the head immersed in the water, as well as probing in mud and picking at the surface. Their diet includes a lot of small fish fry (Dement'ev and Gladkov, 1951 and Mason and Lefroy, 1912). They have been recorded swallowing a frog also (Ali and Ripley, 1969). Their main diet however is probably insects especially aquatic larvae of Diptera, Coleoptera, Odonata nymphs, crustacea and annelid worms, and water snails *(Hydrobia)* etc. On the mangroves, small crabs appeared to be successfully caught with lunging movements of head and neck.

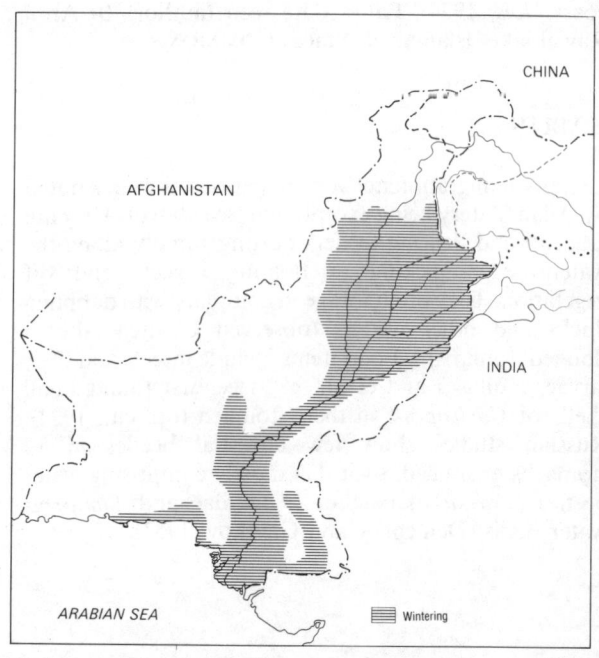

Map 169: Greensank *Tringa nebularia*

BREEDING BIOLOGY

In northern Scotland the breeding of this species has been very well studied by D. Nethersole Thompson (pub. Poyser, 1979). They breed over a wide range of latitude from the tundra and taiga forest zone down to the steppes, usually favouring moorland with boggy areas and streams and avoiding forest. They nest solitarily, both parents sharing incubation duties. The young are precocial and nidifugous.

*VOCALIZATIONS

In winter quarters they usually give voice to a ringing triple call 'teeu-teeu-teeu' when flushed or disturbed. It drops in pitch on the last note and is rather mournful and when familiar, is easily distinguished from the call of the Redshank *T. totanus* which is more markedly bi-syllabic. This call is also used as a contact call between two birds and can be given on the ground as well as in flight.

Green Sandpiper *Tringa ochropus* Linnaeus

Plate **15**
Map **170**

DESCRIPTION

Body length 21-24 cms.
Wingspan 57-61 cms.
Bill length 3-3.8 cms.
Tarsus 3-3.7 cms.

Approximately the same size as the Wood Sandpiper *(T. glareola)* but with slightly shorter legs. It is noticeably longer legged and larger than the Common Sandpiper *(Actitis hypoleucos)*. It is a very dark coloured wader compared to others of the genus, its rather black and white appearance helping to distinguish it at once from the Wood Sandpiper. It has olive brown upper plumage and is mainly white below, finely streaked with dark grey on the throat and upper breast. In flight it shows no white wingbar, but the rump and tail look noticeably white showing very distinct broad sub-terminal dark tail bars on the central rectrices, in contrast to the even,

slender barring throughout the tail of *T. glareola*. In flight also, the whole of the underside of the wings look blackish grey which is the best distinguishing feature from the similar appearing Wood Sandpiper. Its bill is dull greenish olive, black at the tip and straight, tapering to a fine point. The iris is brown and legs and feet are dull olive green, less yellowish than the legs of *T. glareola*. The sexes are alike in plumage. In the breeding season there is very little change but the upper parts show more noticeable buff speckling but less conspicuous than in *T. glareola*. The face and neck in breeding plumage are also more boldly streaked with blackish brown. When flushed it often zig-zags like a snipe and then climbs quite high. The flight call is melodious and very distinctive. Status COMMON.

HABITAT, DISTRIBUTION AND STATUS

A winter migrant to the region, which is well distributed throughout Pakistan in winter, and commoner than other waders in the NWFP and Baluchistan because of its predilection for small streams, isolated water tanks and limited marshy areas which provide too small a habitat to attract other sandpipers. First arrivals are noted in Sind from the end of July (author obs. and Ticehurst, 1924). In Gilgit it is very common on passage throughout April and again from the middle of August to the end of September (Scully, 1881). A pair was seen (author) for 2 days on 22-23 July on the shores of a lake on the Shandur pass (Gilgit/Chitral border) at 4,200 metres (14,000 feet) indicating that autumn passage starts very early. In Baluchistan a few over-winter, ascending mountain streams in the spring to 2,400 metres (8,000 feet) but there is no evidence of breeding (Meinertzhagen, 1920). Around Rawal lake, Islamabad a marked passage is noted from the end of September and again throughout April, but one bird was noted 23 June. (D. Corfield, pers, comm., 1983). Ticehurst also saw occasional individuals over-summering in Sind, e.g. 1 on 21 June (*Ibis*, 1924 p. 123). 2 birds were seen on Khadeji creek (Karachi district) on 13 July (author) and

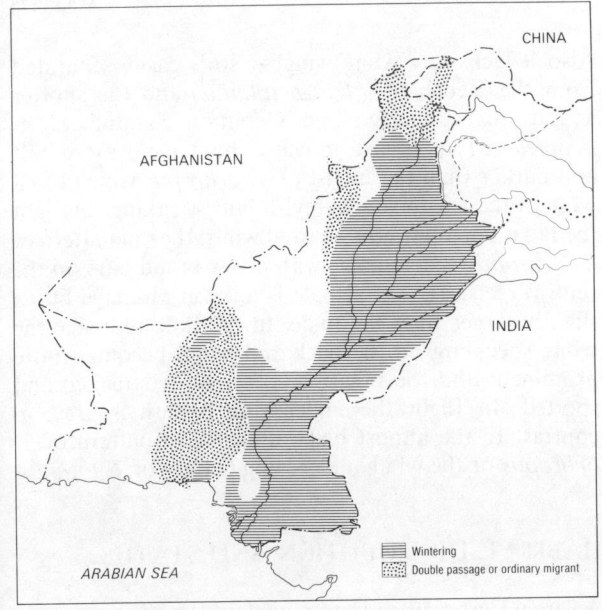

Map 170: Green Sandpiper *Tringa ochropus*

1 on 18 July in another year in a roadside pool near Chauhar Jamali in Thatta district. It frequents the rice fields when it first returns in late July and early August in Larkana and Thatta districts of Sind, also seen in rice fields in Gujranwala district 15 August. It does not seem to occur on the seacoast. E.M. Nicholson noted a passage through Kalat in southern Baluchistan of very exhausted birds during early September (unpublished notes). It is never seen in such large groups as the Wood Sandpiper even on migration and is probably not as numerically abundant as *T. glareola*. Status COMMON.

HABITS

A solitary feeder and preferring rather sheltered locations for feeding. The streams draining into Rawal lake with wooded sides are a favourite habitat in winter, also small tamarisk fringed road and canal-side seepage pools which do not harbour any other waders. It is active both by day and night in feeding (observation 9 p.m. January, Khoski, Badin district). It feeds on the edge of water picking up bits and sometimes by wading. It has been observed dabbling in the mud with one foot (Sidhnai spill channel). It feeds on a variety of insects and their larvae including Coleoptera *(Opatrum, Myllocerus)* also Mollusca (*Planorbis* and *Melania* snails) (Mason and Lefroy, 1912). Mosquito larvae *(Culicidae)* have been recorded in Faisalabad (Husain and Bhalla, *JBNHS*, 1937). It also eats small shrimps, annelid worms and small fish (Cramp *et. al.*, 1983). When observed or nervous it sometimes reacts by bobbing the tail and rear part of the body up and down, like a Common Sandpiper *(A. hypoleucos)*.

BREEDING BIOLOGY

Unique amongst the genus because of its habit of using an old nest in a coniferous tree for egg laying, usually that of a Thrush *(Turdus* sp.). It breeds in the Palearctic zone from the boreal forest down to the northern temperate zone in swamps, bogs and marshes usually near rivers or streams.

Captain Fulton found numbers in July between 2,700-4,200 metres (9,000-14,000 feet) at the head of the Turikho valley in Chitral, but could not prove breeding *(JBNHS,* 1904). They also occur in Tadzhikistan in summer but as yet there is no evidence of breeding this far south in the USSR (Dement'ev and Gladkov, 1951). The pair bond is monogamous and both sexes share incubation duties. They perch freely in trees during this season, but in their winter quarters show no attachment to wooded or forest areas.

*VOCALIZATIONS

In its winter quarters the commonly heard call is a melodious triple whistle given as a flight or alarm call and a contact call. It can be rendered as 'tlee-wit-wit' or sometimes more upward inflected 'klui-weet-weet'. On its nesting territory it utters short songs 'k-too-wit-tit-tit-toowit-too-wit'.

Wood Sandpiper *Tringa glareola* Linnaeus

Plate **15**
Map **171**

DESCRIPTION

Body length 19-21 cms.
Wingspan 56-57 cms.
Bill length 2.5-3.2 cms.
Tarsus 3.2-4 cms.

Similar in size to the Green Sandpiper *(Tringa ochropus)* but slightly longer legged than *ochropus* and bigger than the Common Sandpiper *(Actitis hypoleucos),* and smaller than a Redshank *(T. totanus).* It is olive brown on the upper parts with noticeable buffy white speckling on the mantle and wing coverts, and the upper breast and throat are pale buffy white. There is a noticeable pale buff superciliary streak and darker line through the eye. The sharp pointed bill is straight and black with a dull olive green base. The iris is dark brown and the legs are usually bright olive green, in the spring quite olive yellow. In flight it shows a square white rump patch and tail like the Green Sandpiper *(T. ochropus)* and so is easily separated from the Redshank or Marsh Sandpiper which have white extending up the back in a cigar shape.

Also it lacks any white wingbar so is easily separated from the Redshank *(Tringa totanus)* and the shorter legged but similarly sized Common Sandpiper *(A. hypoleucos).* Its breast, in winter plumage is more buff and darker than the breast of *T. ochropus* which has a whiter breast with some greyish buff streaking. In flight the tail is narrowly cross-barred whilst the tail pattern of *T. ochropus* is distinctive with heavy broad bars on the central rectrices. The female is alike in plumage but is slightly larger than the male. In breeding plumage the white speckling on the back and wings becomes more prominent and the breast becomes more streaked and spotted. In flight the under-wing pattern is grey, in contrast to the almost black under-wing pattern of *T. ochropus* or the white under-wing pattern of *T. totanus.*

HABITAT, DISTRIBUTION AND STATUS

A very widespread and common double passage migrant through Pakistan, with a sizeable population over-wintering. Unlike the Green Sandpiper *(T. ochropus)* this species has never been recorded over-summering in the plains. They are attracted to all types of wetlands, from relatively small village tanks to large lakes, the seepage zones around irrigation headworks and boggy grassland around the margins of lakes and particularly drying out rice stubbles. They seem attracted to shallow flooded areas with good vegetative cover, similar to *Gallinago gallinago,* and generally avoid very brackish or salty waters. However on migration they can be encountered on the seacoast and large numbers congregate to feed on the salt marsh grasslands of Ghizri creek and the Indus delta before their spring migration to their breeding grounds. 2 were seen on a rocky promontory on Buleji beach on the seacoast near Karachi on 30 April. Ticehust *(Ibis, 1924)* noted them in flocks of 20 to 30 in lower Sind in the first half of September. Odd birds can still be seen on passage in Punjab up to the beginning of May. Dera Ghazi Khan district, 3 seen 5 May. Over 100 counted along a 40 hectare stretch of Ghizri creek on 22 April and only 12 seen 10 May in the same year. On passage they pass through Gilgit in big numbers from the end of April up to mid-May, later than *T. ochropus* and there is a marked return passage from the beginning of August (Biddulph *Stray Feathers,* 1881 and Scully, *Stray Feathers,* 1881). In Baluchistan it is also mostly a double passage migrant in April and May and again during August (Meinertzhagen, *Ibis* 1920), however a few over-winter (author obs.) around Zangi Nawar lake in Chagai district. Birds ringed in India in Manjhaul, Bihar and Bharatpur, Rajasthan in March and October have been recovered in the USSR near Ushakovo, Tyumen region 57° 49′N., 68° 04′E and Mukhtuye district, Yakutian 60° 30′N., 116° 10′E (Ali and Ripley, 1969). Status COMMON.

HABITS

Only gregarious on migration and adapted to feed in shallow water or recently water-logged areas. They can swim well (seen Rahwari head on Narra canal, Sanghar district). Their food includes a majority of small molluscs. *Planorbis* species, *Melania tuberculata* also water beetles Hydrophylidae, Odonata nymphs, bivalve molluscs (*Corbicula orientalis* and the occasional ant *(Camponotus compressus),* (Mason and Lefroy, 1912). In the USSR annelid worms were taken rarely, and aquatic insects *Corixa* and *Entomostraca* were found in stomachs (Dement'ev and Gladkov, 1951).

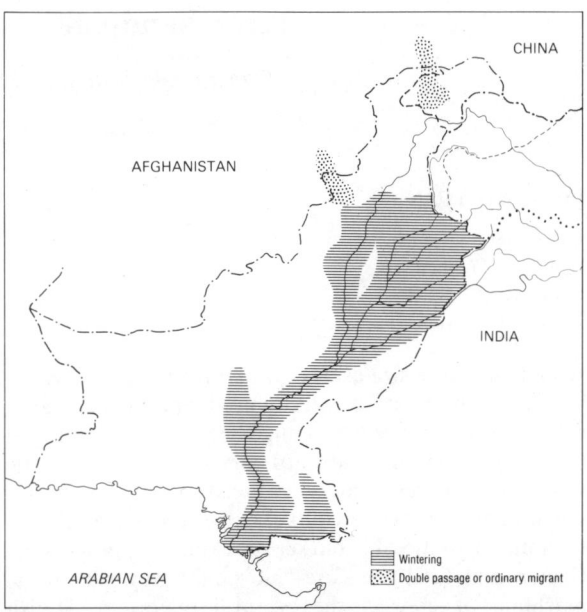

Map 171: Wood Sandpiper *Tringa glareola*

BREEDING BIOLOGY

It breeds over a wide latitudinal range in northern Europe and the USSR from the sub-Arctic down to temperate zones, preferring coniferous forest or hillocky tundra. Curiously an isolated population nests in the Pamirs where it is common at 3,750 metres altitude (Dement'ev and Gladkov, 1951). The female often leaves the male to care for and tend the young which are nidifugous and self-feeding. Both sexes share incubation duties and they usually build a nest in the form of a shallow cup in a sedge or heather tussock, but occasionally they will utilize an old thrush's nest in a tree.

*VOCALIZATIONS

The flight call, once familiar, is distinctive. It is a rapid, repeated whistle, usually repeated as three phrases, 'chip-ip-ip'. It is more metallic and not so melodious as the flight call of *T. ochropus.* It has also been rendered as 'fi-fi-fi-fi' (Dement'ev and Gladkov, 1951). During territorial and breeding display flight it has a rapid warbling call.

Terek Sandpiper *Xenus cinereus* (Guldenstadt)

Tringa terek (Latham) Dillon Ripley's 'Revised Synopsis'

Synonym *Tringa cinereus*

Plate **15**
Map **172**

DESCRIPTION

Body length 22-24 cms.
Wingspan 57-59 cms.
Bill length 4.4-5.2 cms.
Tarsus 2.6-3.2 cms.

About the same size as *Tringa ochropus* or *T. glareola* but one of the most easily recognized sandpipers because of its noticeably short orange legs (see below) and relatively long slightly upward curved bill. There is little difference between winter and breeding plumage in this species and the sexes are alike. Upper parts grey-brown without any noticeably darker streaking or any spotting. Crown and hind neck are pale sandy grey with some marbling of grey in the pectoral region, a whitish supercilium, throat and breast. In breeding plumage the mantle and scapulars become finely streaked with black feather shafts. This shows conspicuously as a black line along the edge of the scapulars when the bird is at rest. In flight the rump and tail are uniformly grey-brown, but the trailing edge of the wing shows a white bar, not quite so broad as in the Redshank *(T. totanus)*. The base of the secondaries and wingtips are darker blackish brown. The iris is dark brown and the base of the bill shows orange, the tip being blackish and slightly upward curved. In breeding dress the whole of the long slender bill becomes blackish. Pakistan wintering birds have much more orange-red legs compared with the yellower legs of

Scandinavian birds (see illustration in Hayman, Marchant and Prater, 1986). The neck is short and the whole bird with its short tarsi looks rather squat. It is very active in foraging however, running in spurts over the mud.

HABITAT, DISTRIBUTION AND STATUS

This migratory wader is confined to the seacoast during the winter, preferring mangrove creeks or exposed sand and mud flats. First arrivals are noted from early August and they mostly leave again by mid-May. Latest date seen 30 May in excited twittering flocks at Ghizri creek. They have rarely been noted inland on fresh water even on passage, except one record in India at Bharatpur, Rajasthan in September (Salim Ali, 1969). However, on 23 May 1987, M. Mallalieu encountered a party of about 8 birds resting on the shore of Rawal lake which he was able to photograph. All were in full breeding dress with all-black bills. It is numerically much less abundant in the mangroves than the Redshank *(T. totanus)* or Lesser Sand Plover *(Charadrius mongolus),* but is very widespread and common so that during a day's voyage in the Indus delta creeks during November and December over 100 could be counted. They also occur along the Makran coast and are common on the mud flats at Sonmiani. They can be seen in loose or scattered groups of 10 to 12 birds but usually forage singly though in association with other waders. Status COMMON to ABUNDANT in Indus delta and Makran coast mangrove lagoons.

HABITS

They like sticky mud and frequent all the inner mangrove creeks as well as the more open tidal flats. They feed in a very active manner darting hither and thither and running with blurred legs and probing deeply in the mud with their bills. They feed principally on small crabs (Brachypoda) (Ticehurst, 1924) but have been seen swallowing quite large crabs whole (author), also small bivalve molluscs and annelid worms. Ticehurst noted their habit of washing their food prey before swallowing it (op. cit., 1924). They have the habit of bobbing their head when foraging also rocking the whole body like a Sandpiper, probably a sign of nervousness when approached. It has been noted (author) that while feeding they will aggressively chase away conspecifics.

BREEDING BIOLOGY

They nest solitarily and on the ground, mainly in the

Map 172: Terek Sandpiper *Xenus cinereus*

taiga and boreal forest zone extending into the sub-Arctic. They avoid mountainous or rocky areas, preferring river margins, coastal flats and willow scrub. There are conflicting reports of the role of the sexes in incubation but the chicks are self feeding from hatching and for the first few days are tended by both parents.

*VOCALIZATIONS

It becomes more vocal in the spring, flocks or groups when flying around, often uttering short twittering warble-like calls 'twit-a-wit-a-wit' or 'tee wee wit-wee-wit'. On its breeding grounds it has a melodious tri-syllabic whistling call (Cramp *et al.,* 1983).

Common Sandpiper *Actitis hypoleucos* (Linnaeus)

Plate **15**
Map **173**

Synonym *Tringa hypoleucos* Linnaeus, Dillon Ripley's 'Revised Synopsis'

DESCRIPTION

Body length 19-21 cms.
Wingspan 38-41 cms.
Bill length 2.3-2.7 cms.
Tarsus 2.3-2.6 cms.

A small sandpiper with relatively shorter legs and smaller body than the *Tringa* genus. In the field it very characteristically has a low flickering type of flight along the water's edge, and on the ground constantly bobs its head up and down, and rocks its body or wags its tail. Also when flying low over water it typically alternates rapid wing beats with short glides on stiff down-pressed wings, less characteristic of other *Tringa* waders. It is a uniform olive grey on the crown, neck and back without any paler speckling or darker streaking, and the upper breast is also uniformly grey-brown, sharply divided from the white lower breast and belly. There is a narrow whitish supercilium. The legs are dark olive to greenish grey and the straight narrow pointed bill is olive green basally, horny brown at the tip. In flight the rump and central tail feathers are concolorous with the olive brown wings and there is a narrow white wingbar conspicuous on both upper and under surfaces of the wing. The outer tail feathers are white with indistinct barring. The dark grey-brown upper breast sharply contrasting with the white lower breast, is one of the best field marks in winter. The sexes are alike, and breeding plumage is similar to winter plumage except for the upper breast becoming slightly darker.

HABITAT, DISTRIBUTION AND STATUS

A winter visitor throughout Pakistan, with considerable numbers breeding along the mountain streams and rivers of the northern areas. A few birds over-winter in the mountains, along the shingle banks and pools of the Gilgit and Chitral rivers (author obs.), and there is a marked passage through Gilgit from 15 May and again from late August (Biddulph, *Stray Feathers* 1882). It has been noted on passage through Chitral from end of April to May (Fulton, 1904) and in Baluchistan. H. Waite

(*JBNHS,* 1934) collected a specimen at Fort Monro on 25 July and one was seen (author) at Ziarat, Baluchistan in a dry stony nullah on 29 April. Meinertzhagen thought a pair was breeding at 2,700 metres (9,000 feet) in the Uruk valley near Quetta, but could not locate the nest (op.cit. 1920). On the plains it first arrives from the beginning of August in the northern Punjab (H. Waite, 1948) and in lower Sind (Ticehurst, *Ibis,* 1924). It departs again in early May. It frequents inland fresh water and the seacoast, and inland is often encountered around small isolated village tanks, alongside canals and roadside borrow pits, away from large bodies of water. It also occurs on the margins of brackish lakes and drying out pools where conditions are quite saline. On the seacoast it is very typically an inhabitant of the more secluded muddy creeks and mangroves, rarely being encountered on the wider more open tidal flats. It is

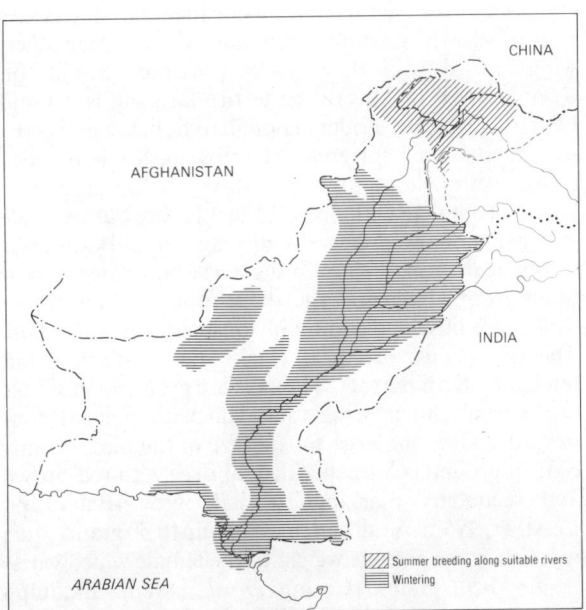

Map 173: Common Sandpiper *Actitis hypoleucos*

generally solitary in occurrence but so plentiful that 60 to 70 can be counted during a day's cruising in the Indus delta creeks. It breeds in the Kaghan valley, Swat river, including the Ushu and Utrot tributaries, Deosai plateau and northern reaches of Chitral, from alpine streams at 3,629 metres (12,000 feet) down to rushing torrents at 2,285 metres (7,500 feet). Status COMMON.

HABITS

Not gregarious, invariably encountered singly outside of the breeding season. Their habit of bobbing their tails and rocking their bodies while foraging has been described above. They seek food prey mainly by picking from the surface of mud and in shallow water and especially items washed in on the water's edge, whether a lake shore or tidal creek. Their food comprises a wide range of adult and larval insects, including beetles (Carabidae, Elateridae, Curculionidae, Hydrophilidae) and Diptera flies (Culicidae, Muscidae, Tipulidae). Also small molluscs, annelid worms, spiders and crustaceans (Dement'ev and Gladkov, 1951, Ali and Ripley, 1969 and Cramp *et al.,* 1983). It is very active when feeding, stabbing with its bill, probing between pebbles and running forward suddenly. Its flight with alternating rapid fluttering wing beats and glides low over the water, is very distinctive.

BREEDING BIOLOGY

Males have a territorial display consisting of repeated singing from a prominent boulder or islet in the river where they have established their nesting territory, alternating with occasional low rapid fluttering flights and glides over their territory while they call. Both sexes also call excitedly from the ground and chase each other, wings raised over their backs (observed Shojh, on Kunhar river 15 May). The territorial song is a rapid twittering which is strident enough to be heard above the hiss of mountain torrents. At Burawai, Kaghan valley (3,000 metres elevation) in late May, pair formation was already complete with one bird incubating but the male still occasionally giving his display song. Both sexes share incubation, the nest being a well concealed hollow in the ground often underneath the roots of a tamarisk or *Indigofera* bush and commonly on an islet in midstream. The nidifugous, precocial chicks are self-feeding but tended by both parents. The eggs are pyriform in shape, the normal clutch being four, laid with pointed ends inwards. They are large for the size of the bird, creamy buff in ground colour speckled all over with red-brown and secondary markings of pale grey (Bates and Lowther, 1952). A full clutch weighs up to 50 grams often more than the average weight of the female which varies from 45-51 grams (Cramp *et al.,* 1983). Biddulph collected juvenile birds in Gilgit from 7 August which had no brown mottling on the throat or breast, as in

adults, and had fine barring on the wing coverts of black and chestnut *(Stray Feathers,* July 1882). On the lower Deosai plateau they nest at 3,600 metres (12,000 feet) elevation, by small mountain streams usually in the lea of peat hummocks. In the Jora valley above Burawai in the Kunhar valley, 9 pairs of Sandpiper were encountered (author) along about 4.8 kms. (3 miles) of stream, and one pair nested within 20 metres of Citrine Wagtails *Motacilla citreola,* on the same boggy islet which was devoid of any bushes and in a bare alpine habitat at about 3,500 metres (11,750 feet) elevation.

*VOCALIZATIONS

In winter time its flight or contact call is a thin high pitched 'swee-wee-wee' or 'seep-seep-seep'. On their breeding grounds both sexes have rapid twittering or piping calls of varying tempo and pitch, 'tweedidi-deedidi-deedidi-deedidi'. The first phrase is often slightly downward inflected and emphasized. The female will join in a chorus of rapid single piping notes at the same time.

Plate 15: Waders & Shore-birds in winter plumage

(1) Greater Sand Plover *Charadrius leschenaultii*
(2) Lesser Sand Plover or Mongolian Sand-plover *Charadrius mongolus*
(3) Great Eastern Knot *Calidris tenuirostris*
(4) Sanderling *Calidris alba*
(5) Little Stint *Calidris minuta*
(6) Temminck's Stint *Calidris temminckii*
(7) Curlew Sandpiper *Calidris ferruginea*
(8) Dunlin *Calidris alpina* (the race wintering in Pakistan usually has a bill length equal to that of the Curlew Sandpiper)
(9) Broadbilled Sandpiper *Limicola falcinellus*
(10) Ruff *Philomachus pugnax* (many individuals are more contrastingly black and white in winter plumage, the bird illustrated is the more common buff morph).
(11) Spotted or Dusky Redshank *Tringa erythropus*
(12) Common Redshank *Tringa totanus*
(13) Marsh Sandpiper *Tringa stagnatilis*
(14) Greenshank *Tringa nebularia*
(15) Green Sandpiper *Tringa ochropus*
(16) Wood Sandpiper *Tringa glareola*
(17) Terek Sandpiper *Xenus cinereus*
(18) Common Sandpiper *Actitis hypoleucos* (note wingtip shorter than tail-unusual in waders)
(19) Ruddy Turnstone or Turnstone *Arenaria interpres*
(20) Rednecked or Northern Phalarope *Phalaropus lobatus*

Plate 15

(1)

(2)

(3)

(4)

(5)

(6)

(7)

(8)

(9)

(10)

(11)

(12)

(13)

(14)

(15)

(16)

(17)

(18)

(19)

(20)

facing p. 360

Plate 16

(1)

(2)

(3)

(4)

(5)

(6)

(7)

(8)

(9)

(10)

(11)

(12)

(13)

(14a)

(14b)

facing p. 361

Turnstone or Ruddy Turnstone *Arenaria interpres* (Linnaeus)

Plate **15**
Map **174**

DESCRIPTION

Body length 22-24 cms.
Wingspan 50-57 cms.
Bill length 2-2.5 cms.
Tarsus 3.5 cms.

A short-billed and relatively short legged shore-bird or wader, about the same size as a Terek Sandpiper *(Xenus cinereus)* and larger than a Dunlin *(Calidris alpina)*.

They are noticeably pied about the head and breast in summer plumage, being more uniformly mottled brown and black in these areas in winter. Their relatively short legs are vermillion or organge and their straight bills, stout at the base, are black. In winter plumage the head is streaked brown and white with blackish ear coverts and pectoral band mottled with white in middle of the breast. The lower breast and belly is white. In breeding plumage the back and wing coverts are bright orange chestnut and black and the crown becomes white, speckled with black, with a black half collar around the white neck and a line through the eye and extending over the ear coverts. In flight there is a white wingbar contrasting sharply with the black flight feathers, and two broad white bars extending over the shoulders and down each side of the wing base. The back and rump are white with a broad sub-terminal black tail bar and also black edges to the upper tail coverts. The sexes are alike. The flight pattern with white rump and broad black bar on tail and across white rump, with white panels along the base of each wing, is unique amongst small waders making this bird easy to distinguish amongst a host of other waders.

HABITAT, DISTRIBUTION AND STATUS

A not very numerous winter visitor to the Sind and Baluchistan seacoast. Rocky shores are not very widespread along the coast of Pakistan though in its European wintering grounds this bird is characteristically associated with such a habitat. It occurs on tidal mud flats, occasionally on open sandy shores and on the broader mud banks of the Indus creeks, and Makran lagoons. They are especially numerous on the rocky promontaries at Buleji beach and Cape Monze but will regularly be seen around the wider creeks of the Indus delta and the shores of Sonmiani lagoon, usually in company with Dunlin *(Calidris alpina)*, Bar-tailed Godwits *(Limosa lapponica)* or Sand Plovers *(Charadrius mongolus)*. They arrive from their breeding grounds from about mid-August and most have departed again by mid-May. By the third week of April birds seen around Ghizri creek were mostly in full breeding dress, though many have been seen in the same area as late as 15 May. Five or six were noted on 13 June in Sonmiani lagoon and these were probably birds which over-summer and do not breed. In mid-winter during a half day's voyage around Sonmiani lagoon and creeks (Lasbela) 30-40 individuals can be counted. Occasionally they are seen in small flocks e.g. 15 roosting on rocks at Buleji. They have not been noted inland on fresh water bodies during migration in Pakistan though there are one or two records of single individuals from India (Lucknow, Patna, Manipur) (Ali and Ripley, 1969). On 15 November 1983 one perched on a fishing vessel 48 kms. (30 miles) off the Sind coast (N. Van Zalinge, pers. comm.). They are great travellers in winter and can be encountered on remote islands throughout the south-west Pacific. Status FREQUENT.

HABITS

They can swim well and there is evidence that they can obtain some form of food in this manner as observed in the Lakshadweep Islands (A. Hume, *Stray Feathers,* 1876). This may account for their ability to exploit even remote oceanic islands in winter (observed author, Solomon Islands, October 1965). Where plentiful they often feed in flocks, and cooperatively and have a variety of tactics according to circumstances. They will flip over shells and seaweed with their short bills, also peck at the surface, jab and probe with their bills, particularly in rock fissures or amongst detritus on the high tide mark. They feed upon a variety of insects, sandfleas and diptera

---

Plate 16: Heads of Terns in Breeding Plumage (sexes are alike).

(1) Sandwich Tern *Sterna sandvicensis*.
(2) Gull-billed Tern *Gelochelidon nilotica*.
(3) Swift or Great Crested Tern *Sterna bergii*.
(4) Bridled Tern *Sterna anaethetus*.
(5) Lesser Crested Tern *Sterna bengalensis*.
(6) River Tern *Sterna aurantia*.
(7) Common Tern *Sterna hirundo*.
(8) Black-bellied Tern *Sterna acuticauda*.
(9) White-cheeked Tern *Sterna repressa*.
(10) Little Tern *Sterna albifrons*.
(11) White-winged Black Tern *Chlidonias leucopterus*.
(12) Caspian Tern *Sterna caspia*.
(13) Whiskered Tern *Chlidonias hybridus*.
(14a) Indian Skimmer or Scissorbill *Rhynchops albicollis*.
(14b) Indian Skimmer or Scissorbill *Rhynchops albicollis*, cross-section through mid-point of bill.

flies, molluscs (Limpets *Patella* sp. at Cape Monze), small crabs *(Graspus strigosus),* at Cape Monze and on Buleji beach rocks, and annelid worms (Ali and Ripley, 1969, Dement'ev and Gladkov, 1951 and Cramp *et al.,* 1983). Occasionally they have been seen feeding off dead animals washed in by the tide and also plant material and seeds during the breeding season.

BREEDING BIOLOGY

They breed circumpolarly in high Arctic latitudes throughout North America, Scandinavia and the USSR. The pair bond is monogamous and they nest on the ground, normally solitarily. The female carries the greater share of incubation but the male often remains alone to tend and protect the young while the female departs a few days after her eggs hatch., (Cramp *et al.,* 1983). A few Turnstones breed near Bukhara, Zarafshan on lake shores according to D. Carruthers *(Beyond the Caspian,* 1948).

VOCALIZATIONS

Its flight call in winter is a rattling sort of chuckle or twitter which sounds similar to the flight call of *Charadrius leschenaultii.* On 22 April pairs were engaged in aggressive encounters chasing each other with head lowered and neck extended and they emitted twittering

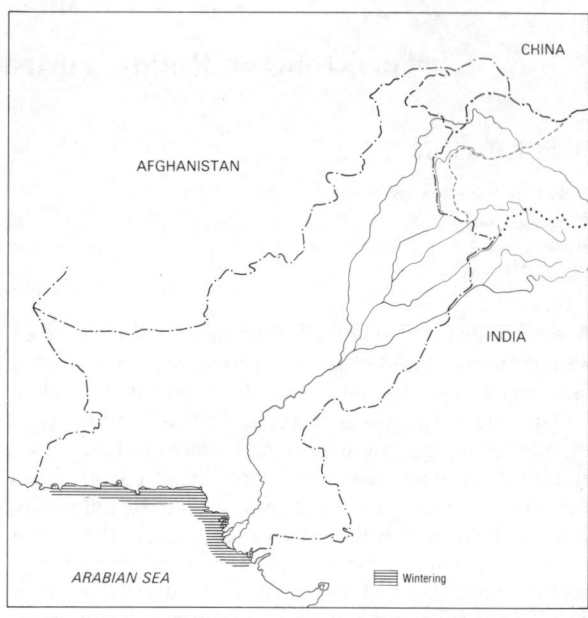

Map 174: Ruddy Turnstone *Arenaria interpres*

warbles or purring, rattling type extended calls. These birds ruffled their feathers occasionally when running up to an opponent but their tails were not fanned or raised in any way (observations Ghizri creek, 1981).

SUB-FAMILY PHALAROPODINAE

Phalaropes

Red-necked Phalarope or
Northern Phalarope *Phalaropus lobatus* (Linnaeus)

Plate **15**
Map **175**

Synonym *Lobipes lobatus*

DESCRIPTION

Body length 18-19 cms.
Wingspan 32-41 cms.
Bill length 1.9-2.3 cms.

These small waders are generally seen swimming and with rather small heads and slender necks and short needle-like bills slightly down-curved distally, they are quite distinctive. In winter plumage they are whitish around the head, neck and breast with a dark grey almost black eye stripe and crown and some pale grey on the hind neck. They are slightly smaller in size than a Dunlin *(Calidris alpina)* and in winter plumage might be mistaken for a Sanderling *(Calidris alba)* but the black streak behind the eye and very slender bill are helpful field points. In the spring they moult into quite bright breeding plumage, the females being about 5 per cent

larger and more brightly coloured but not as distinctive from the male as is the case with the breeding plumages of other phalarope species. Both sexes are grey on the crown, but hind neck of the female is blue-grey and the male duller brown-grey, and the upper breast is also grey with a conspicuous round white patch from the lower cheeks and the throat, and a broad reddish chestnut patch behind the eye and extending in a collar down the side of the neck and framing the white throat. The mantle feathers become lanceolate with black centres, edged with rufous and buff and the lower belly is white. In flight there is a narrow white wingbar, the upper tail coverts and central rectrices are dark brown-grey and in winter plumage the scapulars are black streaked. The iris is dark brown and the legs and feet are blackish brown and the bill is dark horny brown. The toes have slight fleshy lobes and small webs (hence the specific name).

Map 175: Red-necked Phalarope *Phalaropus lobatus*

HABITAT, DISTRIBUTION AND STATUS

This species is highly gregarious and pelagic in its winter quarters which are off the seacoast of Baluchistan and Sind where it usually, can only be encountered some miles from the shore, often in flocks of 30 to 100 or more birds. It starts migration from its tundra breeding grounds from late August to mid-September, and usually makes the flight to the seacoast non-stop, but occasional birds can be encountered on inland lakes or fresh water ponds on passage. Recent records in Punjab include 4 October. Kharrar jheel, Renala Khurd 2 birds (Rhys Davies, photograph, 1965), Muzaffargarh 1 September 1970 near Ghazi Ghat (author), and in lower Sind 6 birds in roadside pools 12 September 1980 near Mirpur Sakro, Thatta district (R. Passburg, pers. comm.). Captain Fulton (*JBNHS.* 1904) collected a single specimen on 14 September near Drosh in Chitral. The earliest autumn arrival recorded 19 August 1977 on Ghizri creek (Ecco Smith, pers. comm.). There are more records of birds sighted during spring passage, including 3 seen on 22 May 1975 on Rawal lake, Islamabad, including one in breeding plumage (B. Amstutz and author). H. Waite (*JBNHS,* 1934) collected 2 on 25 September 1930 at Fort Munro in the Sulaiman hills. Derek Holmes saw 10 in a seepage pool by Khinjar lake on 28 May, 1965. On Siranda jheel, Lasbela on 16 May 1976, 10 were seen feeding in the lake including several in partial breeding plumage (author) and 9 on 8 May 1977 around Ghizri creek (Ecco Smith, pers. comm.). Meinertzhagen *(Ibis,* 1920) notes birds collected from Baluchistan in May and September and he shot one on 28 March. Movement to inland pools and lakes within a

few miles of the open sea, prior to migration in the late spring seems to be a regular pattern for this species. From September and May they normally occur up to 32 or 48 kms. (20 or 30 miles) offshore, though occasionally coming close inshore. On 22 January 1983 at the mouth of Ghizri creek about 9.5 kms. (6 miles) offshore, 5 or 6 flocks of 15 to 20 Phalaropes were encounterd. They constantly flew around to join other feeding flocks and did not seem particularly afraid of our boat. On 13 November 1983 Nicholas Van Zalinge encountered a flock of about 30 some 11.2 kms. (7 miles) off Manora head outside Karachi and when 32 kms. (20 miles) offshore towards the Indus mouth several more small groups. During fishing trips along the Makran coast from 26 November to 1 December small flocks were frequently encountered between 32 and 48 kms. (20 and 30 miles) offshore as far as Astola island (Van Zalinge, pers. comm.). Status COMMON but only offshore.

HABITS

They fly low over the sea surface in swift fluttering flight and in their feeding grounds are highly gregarious. They swim buoyantly, with jerks of head and neck like a Waterhen and their tails held high. When feeding in inland pools they spin around in the water and sometimes scythe with their bills from side to side in the water surface layers. They will also feed by wading and walking in the mud, jabbing or darting forward to catch food prey off the surface (author obs. Pribiloff islands). They seem very tame if encountered on inland pools during migrations, more so than most other waders on migration. Their food when feeding pelagically has not been identified but off the Arabian coast comprised minute gastropod molluscs (Cramp *et al.*). Inland their food includes insects and their larvae especially aquatic beetles and *Corixa* bugs, and Amphipoda and larvae of Empididae in coastal areas (Dement'ev and Gladkov, 1951).

BREEDING BIOLOGY

Only the male incubates the eggs but pair bonds may be briefly monogamous as females have been noted helping to tend chicks. Females are aggressive in courting males and successive polyandry is common, the female laying a second clutch for another male. They nest in loose colonies on boggy ground throughout Arctic tundra from north America through Asia and Europe, often in heath and scrub birch habitat some miles inland from the seacoast.

VOCALIZATIONS

On their wintering grounds occasionally utter low short scratchy notes 'kirk-kirk', and warning calls 'tirric-tirric' or 'twick-twick'. Usually very silent in their winter quarters. During the breeding season the female has a territorial song 'wit-wit-wit' during display and advertising flight.

FAMILY STERCORARIIDAE

Skuas and Jaegers

Pomarine Skua or Jaeger *Stercorarius pomarinus* (Temminck)

Illus. **54**
Map **176**

Pomatorhine Jaeger (Ali and Ripley's 'Handbook', Vol. 3)

DESCRIPTION

Body length 48-70 cms.
Wingspan 124-138 cms. (Urban *et al.*)
Central tail feathers in adults up to 20 cms. in length
Central tail feathers of breeding adult extend up to 10.5 cms. beyond outer tail feathers.

A confusingly variable bird with pale or dark colour morphs at all ages and varying amounts of darker barring around breast and lower flanks. They are slightly larger than Brown-headed Gulls *(Larus brunnicephalus)* and also slightly larger than the Parasitic Skua *(Stercorarius parasiticus)*, with both of which it may associate in Sind coastal waters. They are long, slender winged, gull-like birds with stout barrel shaped bodies and of a blackish brown colour on their upper parts showing whitish carpal wing patches in flight and barring on the upper-tail coverts. In winter plumage the central tail feathers of adult birds as well as immatures, only extend a short distance beyond the tail tip and are difficult to see. The crown is usually darker than the rest of the head and in pale morphs the lower part of the head, neck and breast are white or buffy white with yellowish tinges on the hind part of the neck. Juvenile birds are heavily barred on the whole of the breast with broader bars towards the rump. The bill is horny blue, strongly hooked at the tip of the upper mandible and with an angled gonys like a Gull *(Larus)*. It is slightly

heavier and thicker in appearance than the bill of *Stercorarius parasiticus*. The webbed feet and legs are black. In breeding plumage the central rectrices become longer and are spatulate tipped, with the rachis (quill) somewhat twisted in birds after they attain the age of 4 years. About two-thirds of the birds sighted around Karachi coastal waters were sub-adult or immatures, and about one-half were dark morphs (sample size 26 birds over 3 year period). It requires experience to distinguish this species from sub-adult Arctic Skuas *(Stercorarius parasiticus)* and readers are referred to Svensson, 1984 and Cramp *et al.*, 1983, and Olsen and Christensen *(British Birds,* Vol. 77, 1984, pp. 443-50).

HABITAT, DISTRIBUTION AND STATUS

The 'Handbook', Volume 3 (Ali and Ripley, 1969), records only one definite occurrence of this species for the whole region; a single bird off the coast of Sri Lanka in 1912, and describes it as an accidental vagrant. It is confusingly like the Parasitic Skua *(Stercorarius parasiticus)*, and so may have been mistaken for that species, as recent observations suggest (Passburg, Forssgren and author) that it is not rare along the Karachi seacoast, where it is more likely to be encountered than *S. parasiticus*. A maximum number of 7 have been sighted in one day's boat trip off the Indus

Illus. 54: (a) Arctic Skua or Parasitic Jaeger *Stercorarius parasiticus.* (b) Pomarine Skua or Jaeger *Stercorarius pomarinus.* Showing typical appearance of adult or sub-adult winter plumage birds. Note bolder barring on ventral region and especially on rump of Pomarine Skua, with less extensive pale carpal patches on under-wing of Arctic Skua and more streaked, less conspicuous dark breast band. Note heavier bill of Pomarine Skua.

mouth, and individuals have been positively identified in all winter months, as well as during September and April and June by several independent observers, and good photographs of swimming and flying birds have helped to confirm identification (Walter Weitkowitz, November 1982, N. Van Zalinge, December 1983, K. Forssgren, May 1980, pers. comm., also author with R. Passburg numerous observations, December to April, and in June). Their migrant status is not clear, as they may be mostly non-breeding sub-adult birds following feeding tern flocks according to conditions, rather than regular winter migrants. Status SCARCE.

HABITS

More pelagic in habits than the gulls and terns on which they depend for food, usually resting on the sea and well offshore. They very occasionally roost on sandbanks on the periphery of roosting flocks of terns and gulls, but when disturbed tend to fly out to sea, in contrast to the terns which wheel around and return to land to settle. Though they typically roost or rest, out on the sea they are generally stationed near the open mouths of creeks or by promontaries where cross currents provide good fishing conditions for the terns upon which they klepto-parasitize.

In summer they associate with feeding flocks of White-cheeked Terns *(Sterna repressa)* and in winter with feeding flocks of Sandwich Terns *(Sterna sandvicensis).* On one occasion two birds were resting together on the sea about 3.2 kms. (2 miles) off Cape Monze and when one chased a tern the other joined in the pursuit, so that they hunted cooperatively, though it was evident that only one bird could gain the disgorged fish prey. Their food consists entirely of fish and possibly squid *(Sepia* sp.) klepto-parasitized from terns and gulls, when they are in Pakistan waters. However they have

Map 176: Arctic Skua or Parasitic Jaeger *Stercorarius parasiticus* and Pomarine Skua or Jaeger *Stercorarius pomarinus*

been observed catching fish directly from the sea by dipping down to the surface (Urban *et al.,* 1986) and on their summer breeding grounds to feed largely upon *Microtine* rodents (Cramp *et al.,* 1983).

BREEDING BIOLOGY

They nest in the Arctic zone around the pole in coastal tundra regions. Both parents feed their young by regurgitation. They are silent in their winter quarters.

Parasitic Jaeger or Arctic Skua *Stercorarius parasiticus* (Linnaeus)

Illus. **54**
Map **176**

Richardson's Skua of Stuart Baker's 'Fauna Series'.

DESCRIPTION

Body length 43-56 cms.
Wingspan 110-125 cms.
Wing length 29-34.5 (males), 29-34 cms. (females) (Urban *et al.*)
Central tail feathers of breeding adult 19.8 cms.

About the same size as a Slender-billed Gull *(Larus genei)* and averaging slightly smaller than the Pomarine Skua *(S. pomarinus).* It has long graceful wings and a gull-like appearance. There are light and dark colour morphs, with some birds uniformly dark brown above,

darker brown almost black on crown and nape and with a slightly paler brown breast. Light morph birds may be pale grey-brown on the breast varying to almost pure white with the hind neck and lower cheeks showing yellow tones, and with indistinct grey-brown mottling on the hind neck and in the pectoral region. They retain in all colour morphs, a conspicuous dark brown crown extending down to the eye. Immature birds are more yellow-brown on the upper parts with dark brown barring on back and breast, but these bars are narrower and less distinct than in the Pomarine Skua. The bill is bluish black and legs black. A sub-adult bird found dead (Clifton beach, 24 January) had leaden blue tarsi and toes, with black webs. In breeding plumage the central rectrices are pointed and extend beyond the tail tip 6.5 to 10.5 cms. In flight paler carpal patches are not usually as noticeable on upper-wing surface as in *S. pomarinus*. Pale morphs show varying tinges of yellow on hind neck, some being wholly white without any yellow tinge (observed roosting bird, Sonmiani). Many of the birds observed off Sind and Makran coast are sub-adult with varying amounts of barring on the flanks. Pale morph specimens (based on 7 years' diary records around Karachi) appear to be in a minority.

HABITAT, DISTRIBUTION AND STATUS

Small numbers visit the coastal waters of Sind and Makran, mainly from the post monsoon months until the following spring, with some evidence of an annual spring passage along the coast. On 13 March 1981, 12 birds all believed to be this species, flew past Cape Monze in a westerly direction over a 3 hour period, all appeared to be adult pale morph birds (R. Passburg with X40 telescope, pers. comm.). They associate with roosting and hunting flocks of terns *(S. bergii and S. sandvicensis)* and Slender-billed Gulls *(Larus genei)* at the mouths of rivers, the major creeks and lagoons such as Pitiani and Khudi creeks on the Indus, the Hab river and Sonmiani lagoon, and birds usually as single individuals, have been sighted in December, January and February each winter (9 observations) in 3 annual surveys by boat. On 12 December 1983 one pale phase adult bird was seen roosting close to 2 Pomarine Skuas and a huge flock of Sandwich Terns and mixed gulls at the mouth of Khudi creek, Indus delta. Summer sightings include one harrying White-cheeked Terns *(Sterna repressa)* on 6

June 1980 off Cape Monze and a dark morph bird on 13 June 1981. During a sea voyage along the Makran coast to Iran, Captain E. A. Butler observed about 12 between Pasni and Gwadur between 14 and 17 May *(Stray Feathers,* 1877a). Due to difficulty of separating this species from *S. pomarinus* when observed out at sea it is difficult to determine the exact status of these two Skuas in the Arabian Sea, but recent records (N. Van Zalinge, R. Passburg, K. Forssgren and author) suggest that it may be less common than *S. pomarinus* in the winter months. Status FREQUENT.

HABITS

They can often be watched chasing terns from the Karachi coastline, where the barren Soorjar hills jut out into the sea at Cape Monze, and cross currents off the headland seem to attract flocks of fishing terns. They are remarkably persistent and adroit in pursuing their victims, twisting and turning with speed and grace. They have also been observed chasing Slender Billed Gulls *(Larus genei)* in Sonmiani lagoon. Their food consists almost entirely of parasitized fish and squid *(Sepia* sp.) taken from gulls and terns, their initial approach to their victim often being low over the water approaching the bird from behind.

BREEDING BIOLOGY

It breeds in the high Arctic on barren tundra and moorland, right around the northern hemisphere, and often quite far inland or down to the edge of the boreal forest zone. It is silent at sea but noisy on its breeding grounds.

FAMILY LARIDAE

Gulls and Kittiwakes

Sooty Gull, Hemprich's Gull or Aden Gull *Larus hemprichii* (Bruch)

Illus. **55**
Map **177**

DESCRIPTION

Body length 43-46 cms.
Wingspan 105-118 cms.
Wing length 32-34.8 cms. (Stuart Baker)

This is a medium sized gull, about equal to the Brown-headed Gull *(Larus brunnicephalus)* but with a proportionately longer and heavier bill. It looks heavier bodied than the Slender-billed Gull *(L. genei)* when seen alongside, though similar in body length, and it is

noticeably lighter in build than the Herring Gull *(L. argentatus).* It is a very distinctive gull at all seasons due to its uniformly dark sooty grey or dark brown wings and mantle when in flight. The head is darker blackish brown than the back, with some greyish mottling on the forecrown in winter plumage. The upper-tail coverts and tail are white as also the lower belly. The wingtips are darker, almost black and in flight the trailing edge of the secondaries forms a conspicuous white bar enabling this bird to be distinguished at a distance from Skuas

(Stercorarius sp.), which in pursuing other gulls, behave in a similar manner. In breeding plumage the head becomes darker sooty brown with a narrow but conspicuous white ring framing the upper part of the eye and a broad white collar on the sides of the neck, narrower at the back, reminiscent of the Stonechat's collar *(Saxicola torquata)* which is sharply distinct from the sooty brown head above, but merges more gradually on its lower side into the grey-brown of nape and breast. This collar disappears immediately after breeding and birds observed around Karachi up to the end of April are still in non-breeding plumage without the white half collar. The sexes are alike in plumage. Immature plumages have not been well studied, but juvenile birds have a dark sooty brown tail and some paler margins to the mantle and scapular feathers and the upper breast is dark brown. Ticehurst considered that the dark tail band was not entirely lost until the fourth year *(Ibis,* 1924). The legs and feet are slaty olive or dusky olive in specimens observed along the Sind and Makran coast, but in breeding birds have been described as pale dirty yellow (Butler, *Stray Feathers,* 1877a). The black of the head extends in a 'V' down the nape to join the mantle (note errors in leg colour and head pattern, Plate 34, Ali and Ripley 'Handbook', 1969). The bill of adults is pale green to yellowish green with a black band from the gonys and partly across the upper mandible and the whole bill tip is reddish, sometimes horny white on the extreme tip.

Their flight is strong and graceful with angled wing beats and they frequently chase and successfully pirate fish prey from other gulls and terns. The whole of the under-wing is dark grey-black in flight with small white spots on the inner primary wingtips, sometimes obsolescent.

HABITAT, DISTRIBUTION AND STATUS

This gull is usually only encountered along the Sind and Makran seacoast in the summer months. Occasionally birds may be seen on the channel buoys or offshore

Map 177: Sooty Gull or Hemprich's Gull *Larus hemprichii*

around Manora head and Karachi harbour from March to October (earliest date 26 March latest date 21 October based on 10 years' observations) but there is a marked influx of birds all along the coast from early May, with numbers building up throughout June. None were seen on Sonmiani lagoon 25 April during a six hour boat trip, but were numerous in the same region on 9 May. Flocks can be seen resting on the sandy beaches up to the end of June but thereafter they disperse to breed. The main nesting colony is on Astola island off the Baluchistan coast, more than 320 kms. (200 miles) west of Karachi, but scattered pairs breed on the sand dune islands of the Indus creek mouths during the monsoon. During 4 exploratory voyages throughout the length of Pakistan coastline at distances between 24 and 48 kms. (15 and 30 miles) offshore from early November to early December 1983 many gulls of 6 species visited the vessel (which was on a fishing research survey) but no Sooty Gulls were observed (N. Van Zalinge, pers. comm.). The statement in the *Birds of the Western Palearctic* (Cramp *et al.,* 1983) presumably repeated from Ticehurst *(Ibis,* 1924) that it is a 'common winter visitor to coastal Pakistan' is therefore misleading. Juvenile birds and adults in non-breeding plumage have been noted at Sonmiani lagoon (Lasbela) as early as 28 August. Status COMMON in summer.

HABITS

Contrary to what has been written about its habits 'as essentially an inshore gull' in Arabia (Meinertzhagen, 1954) it is the most pelagic of the gulls frequenting

Illus. 55: Sooty or Hemprich's Gull *Larus hemprichii.* Adult in breeding plumage, late summer. Note conspicuous white tips to secondaries.

Pakistan's coastline, and is never seen inland (as are all the other species) or on fresh water or in the tidal creeks Before breeding it becomes quite gregarious, with small flocks resting on the beaches during June, largest noted 21 birds 24 June at Clifton beach, Karachi. They habitually klepto-parasitize Slender-billed Gulls *(L. genei)* and Sandwich and Lesser Crested Terns *(Sterna sandvicensis* and *S. bengalensis),* as frequently observed off Clifton beach, and at Sonmiani. They have also been noted following returning fishing boats at Dambh village, Baluchistan and one was seen eating a Ghost Crab *(Ocypoda rotunda)* on Clifton beach. They are also reported to feed on turtle hatchlings in coastal regions of Arabia (Urban *et al.,* 1986). They are no doubt opportunistic feeders like most gulls, but more skilful at klepto-parasitism than other Laridae.

BREEDING BIOLOGY

When Captain Butler visited Astola island on 29 May thousands of Sooty Gulls already in pairs, were roosting on the island with what appeared to be many nest scrapes but no actual egg laying had commenced *(Stray Feathers,* 1877a). They still nest colonially on Astola island 'N25°08., E63°50') many birds starting to lay in late June. They also nest on Zeila island off the Somali coast. The normal clutch is 2 or 3 eggs (Meinertzhagen, 1954). Incubation takes about 25 days (Fogden, *Ibis,* 1964). The eggs are pale stone or olive drab in ground colour, marked with large but sparse blotches of dark and light brown and some secondary marks of lavender. Most nests are built in the shelter of clumps of bushes *(Suaeda fruticosa)* or in the shelter of boulders and rocks as protection against the fierce sun in June and July. The nest is a hollow in the sand with a scanty rim of grass roots and twigs (K. Eates, MS notes). A few pairs still nest solitarily on islands in the Indus delta (Natha Khan, pers. comm., 1980) and in their African breeding haunts they more typically nest solitarily (Urban *et al.,* 1986). K. Eates was shown a nest in the Indus delta on 30 August 1946 with 2 eggs, and 4 more nests in September 1953 on islands at the mouth of Malir and Ghizri creeks. These were on barren sandy windswept islands. Three had downy young but the fourth nest contained 3 fresh eggs. These nests had hardly any lining. On 8 August 1941, K. Eates sent a local fisherman to Astola island where most nests had downy chicks but he brought back 31 eggs partly incubated (K. Eates, MS unpublished). Fishermen from Pasni still visit uninhabited Astola to plunder gull eggs, and the population visiting Sind coastline nowadays is clearly much smaller than in Ticehurst's time (World War I), when he described flocks of 100 (Ticehurst, 1924) roosting on the mud banks.

VOCALIZATIONS

Silent outside of the breeding season but Captain Butler *(Stray Feathers,* 1877a) described the breeding colony on Astola as keeping up 'their incessant peculiar mournful cries'. Their call note is a loud rather drawn-out 'kioow' similar to *L. argentatus* (Cramp *et al.,* 1983), repeated usually 12-16 times, becoming progressively shorter and lower pitched (Urban *et al.,* 1986). On their non-breeding grounds they are relatively silent gulls.

Great Black-headed Gull *Larus ichthyaetus* Pallas

Illus. **56**
Map **178**

DESCRIPTION

Body length 59-70 cms.
Wingspan 149-175 cms.
Wing length 4.9-5.2 cms. (females),
    4.6-4.9 cms. (females) (Cramp *et al.,* 1983)
Bill length 5.9-7 cms. (Cramp *et al.,* 1983)
Weight up to 1,700 gms. (specimen Pasni, Baluchistan,
    Bishop, *Stray Feathers,* 1875)

Could be confused with *L. argentatus* with which it overlaps in size but the bill appears longer and heavier looking and the wingspan also averages slightly more. Males are bigger than females with a wingspread up to 175 cms. compared with 156 cms. for *Larus argentatus heuglini* (Dement'ev and Gladkov, 1951). In winter plumage adults are white on the head with some dark speckling on the hind crown and nape and usually a dark crescent around the ear coverts extending in front of and behind the eye, forming a distinct mask. The mantle is pale silvery grey, with tail all-white, and wingtips black with the outermost tips white. Usually the wing-beats look slower and heavier than *Larus argentatus* in flight and also the base of the primaries of adults show pure white rather than silvery grey, which extends down to the black wingtips of the Herring Gull. Immature birds are streaked and mottled with dark grey-brown except on the mantle, but show more white on the belly than sub-adult *L. argentatus*, invariably with a denser band of brown mottling in the pectoral region and sometimes extending right around the upper breast. The tail has a broad blackish brown band which becomes narrower in the second summer by which time the mantle and wings are largely pale blue-grey. Older birds have more restricted black areas on the wingtips and immature birds in their first three years tend to show more white about the throat and forecrown with distinct dark

Map 178: Great Black-headed Gull *Larus ichthyaetus*

spectacle marks around the eye and around the nape and ear coverts. In breeding plumage adults assume a black (very dark brown) head, in shape and pattern as extensive as that of *Larus ridibundus* (the Black-headed Gull) and with two conspicuous crescentic white rims above and below the eye. The bill becomes orange basally with a broad black band across the gonys and both mandibles, with the extreme tip scarlet. Adults showing almost complete moult into nuptial plumage have been noted as early as the end of January and are commonly seen with black heads from mid-February. Sub-adult birds have the bill yellow not orange with the tip more dark brown with a black sub-terminal band. Juveniles have the legs and feet greyish brown but they turn paler grey, eventually becoming dirty yellow (observations Haleji lake, March, and Dement'ev and Gladkov, 1951). With its bright plumage and orange black-tipped bill this huge gull in breeding plumage is a striking and impressive sight.

HABITAT, DISTRIBUTION AND STATUS

A winter migrant visitor to Pakistan, a few lingering around the larger fresh water lakes or river barrage headponds, with small numbers also wintering along the seacoast. It is by no means common but because of its wide ranging habits and large size it is often seen and recognized during its wanderings. On Khinjar lake in Thatta district it is the predominant gull, though usually seen as solitary individuals, with a wintering population of 20 to 30. Fifteen were counted on Chashma barrage headpond 9 February 1973 and 31 on 17 January 1981

and 20 on Ghauspur jheel and surrounding flooded areas on 16 February 1974. Odd individuals have been seen in October while on passage, around the Salt Range lakes (Uchchali and Khabbaki). Odd birds have been seen (author) in the Indus mouth and along the Indus river within 16 kms. (10 miles) inland of the coast (16 November and 7 January) and at sea, as far as 32 kms. (20 miles) off the coast from Karachi in late November (N. Van Zalinge, pers. comm., 1983). The latest dates seen (author) in lower Sind 17 March and on passage in Baluchistan 29 March (Meinertzhagen, *Ibis,* 1920). Major Biddulph also collected a juvenile in its first year plumage on 26 August in Gilgit (*Stray Feathers,* July 1882). A juvenile bird ringed in Kazakhstan (USSR) on Lake Taldy-Kurgan, Azakol, Srednivist, N46°05'., E81°44' on 17 June 1980, was recovered on 27 February 1981 at Mahim off Bombay (J. S. Serrao, pers. comm., Bombay, 1981). Status FREQUENT.

HABITS

As might be expected from such a large gull it is quite predatory in habits. Several were seen harrying, by aerial hovering, a fishing flock of Great Cormorants *(Phalacrocorax carbo sinensis)* on Hadiero lake (Thatta district) in February. They dived upon the Cormorants as they emerged with fish in their bills and in at least one case succeeded in snatching a *Tilapia* fish. In another incident a Marsh Harrier was feeding on an unidentifiable bird in a reed bed on the edge of Haleji lake (Thatta district) when a Great Black-headed Gull mobbed it with repeated swoops until the harrier took to the air with its prey and dropped it. The gull however had hardly commenced feeding when a Black Kite *(Milvus migrans)* robbed it by the same tactics, the gull attempting to carry the remains of the bird in its bill until forced to drop it. In the USSR it has been reported to feed mainly

Illus. 56: Great Black-headed Gull *Larus ichthyaetus.* Adult in breeding plumage.

on fish but also insects (beetles and locusts), Jerboas and Yellow Wagtail *(Motacilla flava)* nestlings (Dement'ev and Gladkov, 1951). On Khinjar lake birds cruising alongside fishermen's gill nets and baited lines, appeared to be able to successfully steal captured fish.

BREEDING BIOLOGY

Nests in colonies around inland seas or brackish lakes in central Asia eastwards to north-west Mongolia, e.g. Syr Darya, Aral Sea and Caspian Sea islands (Dement'ev and Gladkov, 1951). Egg laying starts in early April.

*VOCALIZATIONS

Rather silent in winter quarters but an alarmed bird at Keti Bandar uttered a low pitched barking squawk 'whe-ow' similar to that of the Great Black-backed Gull *(Larus marinus)*. Sub-adult birds when mobbing Cormorants several times gave excited calls 'qoh-qoh', in tone very like a human voice saying 'go'. On their breeding territory they have a quavering laughing sound rendered 'kyauu-kyauu-kyauu' (Dement'ev and Gladkov, 1951).

Black-headed Gull *Larus ridibundus* Linnaeus   Map **179**

DESCRIPTION

Body length 34-37 cms.
Wingspan 91-106 cms.
Wing length 29.5-31.5 cms. (males),
        28.5-30.2 cms. (females) (Stuart Baker)
Bill length 2.9-3.6 cms.

In winter plumage confusingly like the Slender-billed Gull *(Larus genei)* with which it frequently consorts, but separable with experience by relatively shorter and darker tipped bill and darker legs and usually a more pronounced smudge of grey on the ear coverts. However many winter birds have brighter red legs than during the breeding period, and many Slender-billed Gulls acquire a dark grey ear spot in winter. This gull is about 15 per cent smaller than the Slender-billed Gull *(L. genei)* and more than 20 per cent smaller than the Brown-headed Gull *(Larus brunnicephalus)*. In flight it is a slender graceful gull with long pointed wings, which have an all-white narrow triangular tip, with only a narrow black line along the leading edge of the primaries, and a broader black edge posteriorly, formed by the black primary feather tips. In flight the under-wing tip looks dark grey, whereas that of *L. brunnicephalus* shows black with prominent white inner spots and *L. genei* shows a more constrasting white and dark panel to the under-wing tip. The mantle is pale silver grey with all-white rump, tail and head except for some grey smudges around the eye and on the ear coverts. The legs are bright scarlet in winter and the dark red bill looks slender without the marked gonys of the larger gulls *(L. argentatus* and *L. fuscus)*. The tip usually looks horny black. In breeding plumage the head becomes dark brown with white crescentic rings above and below the eye, and the legs become dark crimson. Most of the population commences its northward migration while still in winter dress; e.g. out of a flock of about 400 on Haleji lake on 27 March only 4 birds with black heads were noted. First year birds can be seen throughout their first winter with black tail bands and brown mottling

diagonally across their upper-wing surface and blackish brown secondaries. Their legs are also yellowish at this age. Immature Slender-billed Gulls at this age have clearer white primary wing coverts, which area is still mottled with dark brown in first year Black-headed Gulls.

HABITAT, DISTRIBUTION AND STATUS

In Pakistan, they are only encountered in the winter months and on larger bodies of water inland or around the seacoast. First arrivals in autumn, have usually not been noted until early September, and in the spring they depart well before the Brown-headed Gulls *(Larus*

Map 179: Black-headed Gull *Larus ridibundus*

brunnicephalus). A flock of over 3,000 obviously on passage, seen (author) on Kushdil Khan lake 23 March 1974, and 30 to 40 usually spend the winter on this isolated lake in north-western Baluchistan. Flocks gather on the seepage pools around Haleji and Khinjar lakes by the end of March and are usually gone north by the first week of April (9 years' observations). Dr. Scully (*Stray Feathers*, 1881) collected a bird on migration in the Gilgit valley on 2 May. Meinertzhagen (*Ibis*, 1920) noted 'thousands on the Quetta plains' in the first week of March, and found that few remained after the end of April. Captain Perreau (*JBNHS*, 1910) noted that a few pass up the Chitral valley from March to May. The latest date seen on the Arabian seacoast was one in breeding dress on 24 June in a tidal pool near Karachi, with Slender-billed Gulls. About 12 were also seen (author) in breeding dress on Kushdil Khan lake, Baluchistan on 22 June 1971. A few individuals over-winter on the larger inland lakes of Punjab and on the barrage headponds, e.g. Taunsa barrage, Chashma barrage, Uchchali lake and Lal Sohanran reservoir, but are more likely to be encountered in large numbers only on passage. At Ghauspur jheel in Jacobabad district of northern Sind, however, over 4,000 came in from the Indus river and surrounding marshes each evening to roost, 4-6 February 1974 (F. Koning, pers. comm.). The main population, however, spends the winter along the seacoast and large numbers can be encountered round the Sind and Baluchistan creeks and lagoons where they can be seen roosting in flocks on the sandbars. Status COMMON to ABUNDANT.

HABITS

Gregarious on migration and on their wintering grounds where food supplies are abundant, though scattered individuals will be seen on many of the larger inland lakes. They are versatile and opportunistic feeders. Around the coastal fishing villages they hang around for offal, and flocks follow the fishing trawlers as they haul their trawl nets up and down the creeks. They also mingle freely with *Larus brunnicephalus* and *Larus genei*, roosting on sandbars with both species and wheeling around when nets are being hauled in. They can also be seen hunting for small molluscs and crabs on the sandbars and mud flats at low tide. At Ghauspur flocks of 60 to 70 birds were circling around high up, swooping on unidentified insects. They also can occasionally be seen walking over newly ploughed land close to lake and river banks, searching for insects. A wide variety of foods are taken including garbage, insects both adults and larvae, and molluscs.

BREEDING BIOLOGY

Well studied in Europe. They nest colonially on fresh water bodies usually in close association on small islets. The pair bond is monogamous and the downy chicks are fed by regurgitation.

*VOCALIZATIONS

Silent on their wintering grounds, but when in feeding flocks and on passage they are more vocal but still comparatively weak voiced compared with birds in their breeding haunts. When disturbed at their roosting places in winter a few individuals will occasionally utter rather high pitched throaty or quavering 'kwarr' calls. Eleven different types of calls have been differentiated in detailed studies in Europe (M. Moynihan, 1953 and 1956, *Behaviour*, Vol. 6 and Supplement No. 4.).

Brown-headed or Tibetan Gull *Larus brunnicephalus* Jerdon

Illus. **57**
Map **180**

DESCRIPTION

Body length 42-46 cms.
Wingspan 110-120 cms.
Wing length 33-34.8 cms. (Stuart Baker)

A medium sized gull about equal in size to the Sooty Gull *(Larus hemprichii)* and noticeably heavier bodied and larger than the Black-headed Gull *(L. ridibundus)* as well as being slightly larger than *L. genei*. The chief distinguishing field mark of both adults and juveniles is the wingtip pattern which is black, both above and below, broken by a prominent white band on the primary feather tips forming almost a solid circle when the wing is half closed. This compares with the less extensive black wingtips of *L. canus*, which has a smaller white spot on the black primary tips; and which is not likely to occur in our region and the largely dark grey under-wing tip of *L. ridibundus* and triangular white tips when viewed from above of *L. genei* and *L. ridibundus* with narrow black posterior edge.

In breeding plumage both sexes assume a dark brown head, the rest of the body and tail being white with silvery grey mantle. The head when viewed closely (author obs. Sonmiani, May) is more grey-brown and paler than *L. ridibundus* in similar plumage, but with a noticeably blacker collar, broader around the throat and narrower over the nape where it divides sharply from the white neck. The legs and feet are dark red and the bill which is more slender than that of the larger gulls, e.g. *L. argentatus,* is dark red often with a darker brownish tip. The winter plumage of adults involves loss of the dark brown head and the legs also usually become paler more orange-red. As in other allied gulls in winter, the head is not wholly white, there being a smudge of grey-black on the cheek behind the eye and varying amounts of dark streaking on the crown and nape. The irides are confusingly variable being sometimes almost white in both adults and juveniles (observed Ghizri creek 18 April), varying to red-brown and straw yellow, but usually the irides of adults are paler in the breeding season and brownish in immature birds. Immature birds have a broad dark brown sub-terminal tail bar and brown wingbars formed by mottling on the secondary and primary wing coverts and the whole of the head and neck show varying amounts of grey brown mottling. They can usually be separated from immature Black-headed Gulls *(L. ridibundus)* by the more extensive brownish black wingtips with white tips to the inner primaries. Immature birds have yellow legs and sometimes more orange-red bills than adults, with dark brown tips.

HABITAT, DISTRIBUTION AND STATUS

This species breeds around high altitude lakes in central Asia including Ladakh, adjacent to Pakistan on Tsukr and Tso-morari lakes. It is a winter visitor to coastal regions of Pakistan, with small numbers lingering on larger fresh water bodies inland and around the Indus irrigation barrages and headponds. On the seacoast it is common both around fishing villages and in the tidal creeks, extending westwards to Lasbela and Sonmiani lagoon. In the spring they linger later than Black-headed Gulls *(L. ridibundus),* being plentiful throughout April, and have been seen in considerable numbers on

Sonmiani lagoon up to 12 May. First arrivals are not usually noticed before October, but by 10 November over 100 were seen during a 6 hour boat trip around the Indus creeks. Winter sightings of scattered individuals occur on the Punjab Salt Range lakes, Chashma barrage headpond (Mianwali), Taunsa barrage and Ghauspur jheel (author) and Drigh lake (F. Koning, 1973) in upper Sind. They are less numerically abundant than Black-headed Gulls on the open sea but in the mangrove creeks and lagoons are often more numerous than other gull species and large numbers hang around the fishing villages at Ibrahim Hydari, Rehri and other hamlets on the Indus delta. There are no records from the interior of Baluchistan. Ticehurst (Ibis, 1924), during World War I, considered it to be the least common of the gulls around Karachi, whereas in recent times (1980s) it is more plentiful along the coastal creeks than the Black-headed Gull *(L. ridibundus)* in winter, and numerically much more abundant as a winter visitor, than *L. hemprichii* is as a summer visitor. While camping at the mouth of Khudi creek on 25 October an estimated 200 *L. brunnicephalus* were roosting with other terns and gulls on a sandbar. The dominant species was *L. genei (circa* 500). Status ABUNDANT.

HABITS

Due to its larger size it is quite successful at piracy and food robbing and is an aggressive species in mixed feeding flocks. On 18 April by a fish-net trap (weir) on Ghizri creek, a flock of *Sterna bengalensis* was catching fish adroitly as they bumped into the net, while Brown-headed Gulls perched on the net supporting poles or swam around and successfully attacked and robbed many of the Lesser Crested Terns of their food prey. All this accompanied by much loud calling and attempts by the gulls to catch fish themselves, swimming and dipping

Illus. 57: (a) Head of Slender-billed Gull *Larus genei.* Adult in breeding plumage.
(b) Slender-billed Gull in winter plumage showing wingtip pattern and blackish mask behind eye.
(c) Brown-headed Gull *Larus brunnicephalus.* Adult in winter plumage showing wingtip pattern.

following fishing boats as they return from the open sea. In their breeding grounds they have been noted feeding on small fish which abound in the snow-melt streams which drain into the otherwise highly brackish, and fishless lakes in Ladakh where they breed (Brigadier Moti Dhar, pers. comm., 1983). In their Mongolian breeding haunts they also feed on crustaceans particularly *Artemia* species (Dement'ev and Gladkov, 1951). Inland on fresh water bodies insects, gastropods and Mollusca form part of their diet and they have also been observed hawking winged termites (Salim Ali, 'Handbook', 1969). Winter time observations suggest that it is a less gregarious species than *L. ridibundus*.

BREEDING BIOLOGY

In Ladakh, egg laying starts in early July and a substantial nest pad is built where the ground is in boggy situations, but is a mere scrape scantily lined if on dryer ground (Meinertzhagen, *Ibis,* 1927). The pair bond is monogamous, normal clutches being 3 eggs, usually ochre in ground colour with dark reddish and blackish brown spots and bold blotches and sometimes violet grey under-markings (Stuart Baker, 1929). They are colonial nesters and both sexes share incubation duties.

*VOCALIZATIONS

Aggressive or excited calls of birds around a fish weir were lower pitched and more guttural than calls of *Larus ridibundus* in similar circumstances but otherwise similar in general timbre. It was written at the time 'krreeah-krreeah'. However they have a very varied repertoire of calls like all the Laridae including more reedy high pitched begging calls and drawn-out cries rising in pitch and becoming thinner.

Map 180: Brown-headed Gull *Larus brunnicephalus*

their heads and necks under water. On several occasions in different years Brown-headed Gulls have been watched hovering over, and trying to land amongst feeding Plumbeous Dolphins *(Sotalia plumbea)* as though in a deliberate attempt to benefit from pieces of squid (*Sepia* sp.) or fish which appeared near the surface as a result of the Dolphins feeding. Their skill in flying to and anticipating where the Dolphins next surfaced in the turbid water, indicating probable previous practice. They will scavenge on refuse and garbage like other Laridae, and are always to be seen in winter around Karachi inner harbour. Flocks have been noted

Slender-billed Gull *Larus genei* Breme

DESCRIPTION

Body length 42-46 cms.
Wingspan 92-110 cms.
Bill length 3.8-4.2 cms. (cf. 2.7-3cms. for *Larus ridibundus*)

It is a longer bodied bird than the Black-headed Gull *(L. ridibundus)* and averages about 15 per cent larger than the latter species, but it is nevertheless a rather lightly built and graceful gull with a long neck and long slender bill as its name suggests. In winter they are very similar to Black-headed Gulls with all-white bodies and tails, silver grey backs and wings and some darker grey marks on the ear coverts and occasionally a few spots on the nape. Some birds in winter have pure white heads. In

summer they are wholly white on the head and the breast assumes a pale pinkish flush, more easily discerned when the bird is resting on the ground than when flying overhead. The bill is red with no pronounced gonys on the lower mandible and the legs and feet are vermillion to bright red, but can be variable. Generally they are more orange, even yellow-orange in non-breeding plumage and darker red during the breeding season. The irides are also variable but in breeding birds tend to be white or very pale yellow, with the iris always much darker in juvenile birds. The wingtip pattern is similar to *Larus ridibundus* with the leading edge of the wingtip white extending over the primary wing coverts to form a white triangle (the 5 outer primary feathers contra four outer primaries in *ridibundus*). The whole of the rest of the wing is pale silvery grey, slightly paler than *L. ridibundus*. The trailing edge of the wingtip is narrowly black and the under-wingtip tends to show a contrasting pattern of a white panel on the leading edge and dark grey-brown panel on the inner primaries, less uniformly dark grey over the whole wingtip than is typical with *L. ridibundus* when seen from below. The longer neck, more sloping forehead and relatively longer bill are helpful field points when the bird becomes more familiar.

HABITAT, DISTRIBUTION AND STATUS

In Pakistan a purely coastal and neritic species occurring in large numbers in the lagoons and estuaries along the Makran coast and in the wider creeks of the Indus delta. They can also be frequently sighted out to sea and up to 32 kms. (20 miles) off the coast when they usually associate in loose flocks or fly in lines, until a shoal of small Clupeidae fish is sighted near the surface. In the post monsoon season small numbers come onto inland lakes within a few miles of the coast, while large numbers can be seen offshore along the open coast at all times of the year. Because of difficulty in separating it from the Black-headed Gull in winter, it may often be overlooked inland, but numbers have been seen in December, January and February on Haleji lake, approximately 72 kms. (45 miles) from the coast and single birds as far inland as Khinjar lake, 96 kms. (60 miles) inland. They breed colonially on brackish lakes or inland seas, and judging from their abundance, there must be several suitable locations along the largely uninhabited Makran coast. The one known location in Pakistan is Siranda

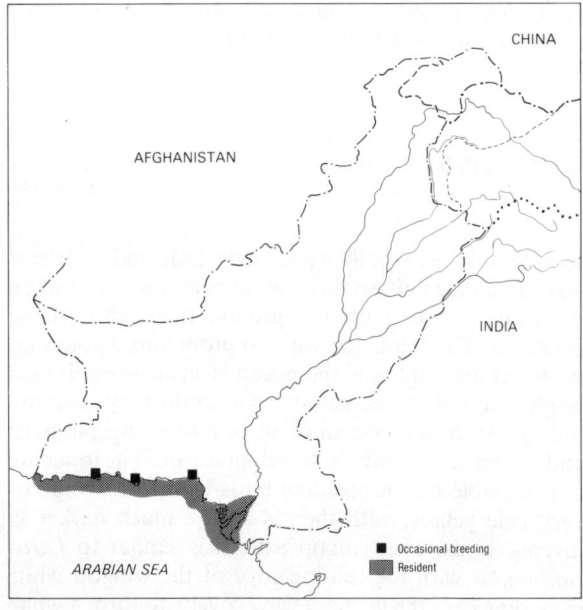

Map 181: Slender Billed Gull *Larus genei*

jheel in Lasbela (see below under Breeding) where breeding is irregular and dependent upon suitable water conditions.

A Russian ring, recovered in December 1985 by a fisherman, from a bird caught near Ghizri creek (and purchased by the author) suggests that some proportion of our seacoast wintering population breeds on salty inland lakes in Kazakhstan. At the time of writing, the source of this ring had not been indicated by the Russian authorities, but the nearest known large breeding colonies are in Kazakhstan (Flint *et al.,* 1984). Status ABUNDANT.

HABITS

An opportunistic feeder occasionally seen in the outer harbour, picking up refuse tipped from vessels and also seen following fish trawlers, but it seems to be more successful at catching fish than *Larus ridibundus* and more inclined to forage out in the open sea than the latter species. They are gregarious in their breeding, winter roosts and feeding patterns. On 21 October off Clifton beach an estimated nine hundred birds of this species were following shoals of small fish Clupeidae, probably *Sardinella* species, some swimming in the shoal and others flying overhead, plunging into the water and dipping with their heads and necks under the sea. Sooty Gulls *(Larus hemprichii)* were busy klepto-parasitizing on the successful.

They can be seen, throughout the winter months, flying along the Karachi coast at evening time from west to east to join conspecifics in a large night time roost on the exposed sandbars at the mouth of Ghizri and Khudi creeks. Over 500 were counted coming to roost, from about 5 p.m. on a sandspit at Khudi creek in early December. During the breeding season a variety of insects, crustaceans (*Artemia* sp. and *Gammarid* sp.) and fish are reported to be consumed (Dement'ev and Gladkov, 1951).

BREEDING BIOLOGY

Despite the thousands of this gull which must haunt the Sind and Baluchistan coasts, relatively small numbers breed on Siranda lake, the only known colony of the species on the subcontinent and in ten years of the authors nearby residence, breeding was only confirmed twice, in both cases in mid-summer following a year of above normal monsoon rains when the inundated area of this shallow brackish lake was well above normal. In 1945, K. Eates found a colony of about 200 pairs on 9 June on a large bare island in the lake. On this date about 80 per cent of the nests contained full clutches of 3 well incubated eggs, and the remainder were newly hatched chicks. He noted that the island was shared by a colony of Caspian Terns *Sterna caspia,* which were separate and at some distance and had started breeding before the gulls (MS notes unpublished). On 24 May 1979 the author received reports of gulls and terns breeding on Siranda and wrote an account of a colony of some 60 pairs on a bare island (Roberts, *JBNHS*, Vol. 77, 1980). Each nest is a quite substantial pad of dried stems, *Salsola* and other material with quite a few feathers, gull and flamingo noticeable in the cup. The nests were 1 to 2 metres apart, fairly evenly and densely packed on the higher part of the island and most contained downy chicks, so that laying must have commenced in mid-April. The eggs were comparatively smaller than Caspian Tern eggs, seen on the same day and were rather pale in ground colour for a gull's egg, creamy white to buffy stone colour, speckled and blotched with sepia brown with secondary smudges of grey and under-lying darker lavendar marks. About 80 per cent of the clutches and broods comprised 2 not 3. The eggs are protected from the fierce sun by both parents who wet them with their breast feathers and also perform a fairly rapid incubation change-over. Eates reported young being fed by regurgitation and that their food was small sprats (*Sardinella* spp. ?).

Chicks in down have black bills and feet, pale creamy white down on upper-parts, with white throat and under parts. They are mottled with black on the crown and back, with some smaller black spotting in the pectoral region and flanks. The whole crown looks blackish from a distance with a spectacle mark of black around the eye. The newly hatched chicks are semi-altricial and semi-nidifugous and both parents shelter them continuously from the sun's rays for the first two or three days.

It was apparent from the Siranda lake visit, that nesting in this species is very synchronous for the whole colony. In 1979 when visited 2 weeks later the whole colony had been deserted due to the lake level having dropped by evaporation and the island being exposed to predation. No chicks or adult gulls remained, only addled eggs. The whole colony must have deserted *en masse*. Natural predation by foxes (*Vulpes vulpes pusilla*), numerous in the vicinity, as well as egg robbing noted (author) by local fisherman were also hazards which the colony faced. Caspian Terns on the other hand had just started nesting on another island further out in the lake.

*VOCALIZATIONS

Calls are similar to those of *Larus ridibundus* in type but are deeper in tone and have a nasal inflection. They do not sound as high pitched or melodious as those of *Larus ridibundus*, often sounding hoarse and dropping in pitch. Two types of calls were uttered, a more excited calling 'ka-ka-ka' and a more drawn-out 'kraaah-kraaah'.

Common Gull or Mew Gull *Larus canus* Linnaeus   Map **182**

DESCRIPTION

Body length 40-44 cms.
Wingspan 114-126 cms.
Bill length 3.2-3.6 cms.

The wingtip pattern of this gull when adult is very like that of the Brown-headed Gull (*Larus brunnicephalus*). Though it is slightly shorter bodied than the latter and longer in the wing, in the field they could appear similar in size. The wingtip is all-black with prominent white 'windows', but these are smaller in area and not so conspicuous as in *L. brunnicephalus*. Its white head and neck are slightly streaked grey in winter. The most distinguishing field points are its relatively shorter yellow green bill and dark iris and mewing cries. The Brown-headed Gull has a more slender, slightly longer dark red bill. The legs and feet of the Common Gull are olive yellow.

HABITAT, DISTRIBUTION AND STATUS

It is not included in the 'Revised Synopsis of the Birds of the Subcontinent' (Dillon Ripley, 1982), nor in the

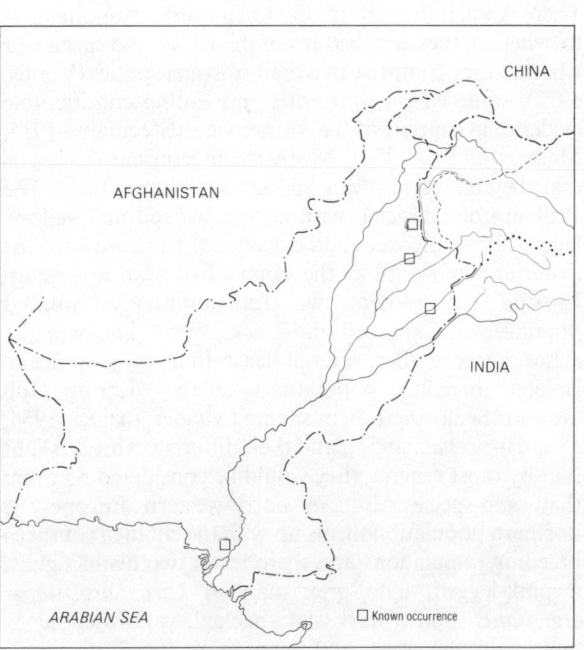

Map 182: Common or Mew Gull *Larus canus*

'Handbook', Vol. 3 (Ali and Ripley, 1969). However occasional birds undoubtedly reach Pakistan in winter and have hitherto been mistaken for *Larus brunnicephalus* or otherwise overlooked among immature gulls which present special difficulties in identification. Allan Vittery saw one on the Ravi river at Balloki headworks in 1972 and this is recorded in *Birds of the Western Palearctic*, Vol. 3, p. 793. Fred Koning while working for the IWRB in Pakistan saw one on Rasul barrage headpond on the Jhelum river early in February 1974 and subsequently 2 more, one adult and one immature on Khinjar lake in January 1976. In all cases he was attracted initially by hearing their unmistakable call. These sightings are also published in IWRB reports of his Pakistan survey. More recently it was sighted again on Khinjar lake on 4 April 1984, by Richard Grimmett and Craig Robson (see Roberts,

Grimmett and Robson, 1985, *JBNHS,* Vol. 82, p. 569). In the winter of 1986/87 a single bird spent several months on Rawal lake (Islamabad) (M. Mallalieu, *in litt.,* 1987).

Most of this gull's breeding population in Europe and western USSR, winters in the Mediterranean and north African region to the Persian Gulf, whilst the east Siberian population *L. canus kamtschatschensis* winters in south-east Asia as far south as Formosa, but the western Siberian breeding population *L. canus heinei,* also has an isolated breeding population at Tabriz in north-east Iraq and this species winters in the Black and Caspian Sea regions down to the Persian Gulf (Dement'ev and Gladkov, 1951). It is highly probable that stray wanderers from this population may regularly enter northern parts of Punjab as the above records show. Status RARE VAGRANT.

Herring Gull *Larus argentatus* Pontoppidan

Silver Gull of Russians

Lesser Black-backed Gull *Larus fuscus* Linnaeus

Map **183**

TAXONOMY

Along the coast of Pakistan in winter, a bewildering variety of large gulls occur, belonging to the *Larus argentatus/fuscus* group, which show all gradations in both leg and mantle (back and wing coverts) colour. There is still difference of opinion among taxonomists as to whether they are better combined as one species or whether they comprise two really distinct species (Voous, 1977), some earlier authorities presenting considerable evidence in support of the former view (Stegmann, 1934, Meinertzhagen, 1954). More recent consensus view is that they represent a ring species (see chapter titled 'The Problem of Species'), with darker backed and yellow-legged forms increasing in occurrence from west to east (Vaurie, 1965) until at the point of overlap in western Europe they form two reproductively isolated populations (Campbell and Lack, 1985). The problem arises because there are at least four geographically isolated breeding populations of the Herring Gull around the northern hemisphere (Meinertzhagen, 1954, Dement'ev *et al.,* 1951), and their differences are so slight that by most criteria, they would be considered no more than sub-species. But in north-western Europe the northern population meets up with the chain of southern breeding populations and there result two distinct gulls, a pink-legged, light grey mantled *Larus argentatus argentatus,* and a dark slaty-backed, yellow-legged *L. fuscus.* In this western European region *L. fuscus* is normally a spring migrant visitor which starts to breed some two or three weeks after *L. argentatus,* so that they

are reproductively isolated. It is, however, noteworthy that there have been increasing reports of hybridization between these two species, occurring in Europe (K. Paludan, 1951, F. Goethe, 1957, M. P. Harris, 1970 and Cramp *et al.,* 1983).

Map 183: Herring Gull *Larus argentatus*

The important aspects for the region covered by this book are that all populations are closely similar in size, habits and behavioural traits, and the treatment given hereunder is intended only to present a coherent view of a complex picture taking into account most recently accepted systematic views. It must be recognized that field sightings of sub-adult birds along Pakistan's coast cannot reliably be assigned to any particular sub-species or population, and only when more ringing studies can be conducted will the true picture emerge. Studies of wingtip patterns of museum specimens collected from different regions, do however, show some consistency.

Two populations from the southern species breeding 'arc' (see Meinertzhagen, 1954) undoubtedly occur.

Larus argentatus cachinnans Pallas

DESCRIPTION

Body length (average) 55-67 cms.
Wingspan 138-155 cms.
Wing length 43-47 cms. (males) (Meinertzhagen)
Bill length 52-63.5 mm. (males) (Cramp *et al.*)

This is a large gull. It has a slightly darker grey mantle and wing coverts than the nominate race or western European population and has yellow legs and feet. It breeds around the Black, Azov, Caspian, and Aral seas and salt lakes of Kazakhstan and Turkestan. Specimens collected from Sind and Makran coast exactly correspond with this population (Vaurie, 1965, Dement'ev and Gladkov, 1951 and British Museum specimens). Based on sightings of adult birds on the ground (author diary notes over 10 winters), between 10 and 20 per cent of the wintering population in Pakistan belongs to this sub-species.

Larus argentatus mongolicus Sushkin

Wing length 42.5-45.2 cms. (males),
42.7-45.7 cms. (females) (Dement'ev)

This is another large sub-species with a somewhat darker grey mantle than *cachinnans* (but still considerably lighter than *heuglini*), and legs which are usually pink or bluish grey (Dement'ev and Gladkov, 1951 and author obs.). This population breeds around Lake Baikal, the lakes of the Altai mountain range and in northern Mongolia. Specimens corresponding to this sub-species have been collected from Karachi in February and along the Arabian seacoast up to Baghdad in the north-west (Dement'ev and Gladkov, 1951). Sightings of adult birds on land (when the legs are clearly visible) indicate that 20-30 per cent of the population belongs to this species. On inland waters records are inadequate but indications are that it forms a substantial part of the population.

Turning now to the northern arc of breeding gulls, these comprise, putatively only two sub-species.

Larus fuscus taimyrensis Buturlin

Wing length 44.3-47.7 cms. (males) (Cramp *et al.*, 1983)

This is a medium sized gull being dark grey on the wings and back, but paler in colour than *heuglini* or the nominate sub-species *fuscus*. The legs and feet are usually yellow, but can vary to fleshy yellow or pink. The wingtips are distinctly black with small white spots on the first two primaries, only weakly developed. This population breeds east of *heuglini* in regions of north and central Siberia extending to the Arctic coast, the Taimyr peninsula and with a second disjunct population breeding in northern Kazakhstan and south-west Siberia (Dement'ev and Gladkov, 1951). Specimens belonging to this population have been collected from Karachi (Brit. Mus. coll.), and along the Makran coast and off Bombay. It is estimated that amongst adult birds seen on the ground and categorized sub-specifically 30-50 per cent of the coastal population belongs to this sub-species which has never been seen on inland waters except while on passage.

Larus fuscus heuglini (Bree)

Wing length 44.7-46.9 cms. (males) (Cramp *et al.*, 1983)
43-46.5 cms. average 45 cms. (Ticehurst, 1924)

A medium sized gull with a dark slaty grey mantle (similar exactly in tone to *Larus fuscus graellsii* of western Europe). It has yellow legs and feet with a few individuals showing fleshy yellow and even flesh pink leg tones (Ticehurst, 1924 and author obs.). The black wingtips are extensive, extending to the seventh and sometimes eighth primary, with usually only one white spot on the tip of the first primary. This population breeds in northern Siberia from the Kanin peninsula, lower Pechora and lower Ob valleys, eastwards to where it inter-grades with *taimyrensis* (Dement'ev and Gladkov, 1951). It winters in Iran, the Makran coast (Brit. Mus. coll.) and Sind. Sightings of adults are confined to coastal areas, and are estimated to comprise between 15-25 per cent of the adult population.

Sub-adult Herring and Slaty-backed Gulls

These are very numerous in winter in Pakistan especially on inland waters where they comprise the bulk of large gulls seen. On the coast also they appear to comprise about two-thirds of the population. They usually have black bills, bodies heavily mottled with dark brown, if in their first winter, but with grey mantles in their second winter, and with a dark brown sub-terminal tail band with some obsolete cross-barring at the base of the tail feathers and darker blackish brown primaries and secondaries. Their legs and feet are usually bluish grey varying to pale pink. In older individuals the base of the bill shows yellow and the tail bar is narrower.

HABITAT, DISTRIBUTION AND STATUS

Summarizing the above, large gulls are common along the seacoast in winter, with the majority of birds being sub-adult and roughly half the population dark slaty backed, and half the population paler grey but noticeably darker than that of European Herring Gulls. A few birds can be seen as late as the end of May in Karachi harbour and Sonmiani lagoon and occasional individuals have been sighted in June. It is quite usual to see pale grey and dark slaty birds resting side by side on sandbars in the Indus creeks and the majority of such groups have yellow legs and feet. When more than 5 or 6 dark backed birds are seen roosting together, usually at least one bird shows fleshy orange tints to its legs. These large gulls can be encountered in November and December quite commonly between 24 and 32 kms. (15 and 20 miles) off the coast and they often follow fishing boats (N. Van Zalinge, 1983, pers. comm.). They are not as gregarious as the smaller gulls *(ridibundus, genei* and *brunnicephalus)* nor even seen in such large flocks as the latter three species. A flock of *L. argentatus cachinnans* was noted on migration on 7 April at Kushdil Khan lake in Baluchistan (Meinertzhagen, 1920). A few *Larus argentatus* type gulls spend the winter on Kushdil Khan lake each year. Observations on Rawal lake Islamabad between 1980 and 1983 indicate autumn passage is quite late, birds being seen from 31 October and throughout November and again around 17 March. A single bird was seen on 8 June (D. Corfield, pers. comm.). Status ABUNDANT in winter.

HABITS

Opportunistic and omnivorous feeders, taking fish scraps and refuse, following ocean going vessels into the harbour and also wading in the surf to hunt *Ocypoda* crabs. They are known to eat molluscs, crustacea, insects, fish and annelid worms. They are shyer and more wary than gulls around European towns, and have not been observed on Karachi municipal refuse tips, where other scavenging bird species abound, nor settling on buildings or the ground close to humans in the fishing villages, preferring to swim offshore and forage on the mud flats.

BREEDING BIOLOGY

Extra-limital. Colonial, ground nesters. Breeding has been studied and described in detail in Europe. Chicks are fed by regurgitation and are looked after by both parents, the pair bond being monogamous and stable.

VOCALIZATIONS

On wintering grounds relatively silent. Occasionally utters call notes or excited calls varying in intensity and attenuation 'keeow-keeow' or 'kew-kew' and 'kaw-kaw'. Calls lower pitched and more throaty generally than for nominate *L. argentatus.* Young birds have high pitched drawn-out whistle given in flight and on the ground and this is probably a begging call.

Lesser Black-backed Gull *Larus fuscus* Linnaeus

DESCRIPTION AND STATUS

The nominate sub-species of the Lesser Black-backed Gull *(L. fuscus fuscus)* is darker in mantle colour than Slaty-backed Gulls breeding in Siberia and north-eastern USSR. It has proportionately more slender or narrower wings than *Larus argentatus* (Cramp *et al.,* 1983). In the *Handbook of the Birds of India and Pakistan* (Ali and Ripley, Vol. 3, 1969) and in the 'Revised Synopsis' (Dillon Ripley, 1982) both *Larus argentatus heuglini* and *Larus fuscus fuscus* are listed as occurring in Pakistan and peninsular India in coastal waters. There is no evidence that the nominate sub-species *Larus fuscus fuscus* occurs in the coastal waters of the Arabian Sea or in Pakistan *sensu strictu,* as this population breeds in north-western Europe in Scandinavian countries, hardly extending eastwards into the USSR and in winter migrates to the eastern Mediterranean and down the west and east coasts of Africa. The consensus view of recent studies and experts, is that all the dark-backed gulls wintering off the coast of

Pakistan and India belong to the sub-species *heuglini* and *taimyrensis (Birds of the Western Palearctic,* Vol. 3, ed. Cramp., 1983). The author agrees with this view. During 3 weeks' winter survey of Pakistan's coastal sea-bird population (with the research vessel 'Dr. Fritzov Nansen)', Kent Forssgren who is an expert on the breeding birds of the Baltic and especially *Larus fuscus fuscus* , was of the view that all the Slaty-backed Gulls in this region, whether *heuglini* or *taimyrensis* appear to be more logically assigned to sub-species of *Larus argentatus* on the basis of voice, relative wing length and width, lighter mantle colouration and habits (Forssgren, pers. comm., 1984). His observations reinforce the view that *Larus fuscus fuscus* should be excluded from the **Checklist** of birds for this region and would tend to support the view already held by some authorities that the bulk of the Holarctic breeding population is best treated as a *Larus argentatus* super species complex with only *Larus fuscus* in north-western Europe having achieved full specific status, having crossed the barrier of geographical separation and still remaining distinctive.

Great Black-backed Gull *Larus marinus* Linnaeus

STATUS

Larger than *Larus ichthyaetus* with an all-white head, flesh pink legs and much darker mantle than *Larus fuscus heuglini*. This very large gull could occur occasionally along the seacoast. On 16 November 1899 Col. R. Meinertzhagen shot one on Deoli jheel in Rajasthan, a brackish lake 88 kms. (56 miles) south of Nasirabad. Though the specimen has not been preserved it was recorded with adequate diagnostic evidence (Ali and Ripley, 'Handbook', 1969 and Meinertzhagen,

JBNHS, 1900). The author during an early February (1975) boat trip around Sonmiani, Lasbela saw a conspicuously large adult gull with pink legs and very dark grey, almost black mantle which was probably *Larus marinus*. On being disturbed however it flew out of sight so did not afford prolonged or close observation, but it originally attracted attention by its more contrasting black and white plumage and size, and would not have raised doubts as to its identity, if its occurrence had not been so unlikely. Status a possible VAGRANT.

FAMILY STERNIDAE

Terns and Noddies

Gull-billed Tern *Gelochelidon nilotica* (Gmelin)

Plate **16**
Map **184**

DESCRIPTION

Body length 35-42 cms.
Wingspan 86-102 cms.
Wing length (very variable) 30-33 cms.
　(Sind specimens) (Ticehurst)
Bill length 3.5-4 cms.

This is a large tern with a noticeably short and comparatively deep bill, arched on the culmen and a tail which is not deeply forked. In breeding plumage the cap is glossy black extending well down the hind neck in a

'V'. The mantle is pale grey with no dark on the wingtips and the wings are white underneath. The bill and legs are black. Sexes are alike. In winter plumage the whole of the crown is white, unlike most other terns, (with a streaked hind crown), with a dark line below and behind the eye, and occasionally some faint spotting on the lores and around the nape. Juvenile birds have a speckled black nape with blackish brown spotting on the back and in rows on the wing coverts. This is rapidly lost in their first winter and their plumage becomes like adults'. They look like large Whiskered Terns *(Chlidonias hybridus)* in general colouration but actually have the tail more deeply forked. They are bigger in body than the Indian River Tern *(Sterna aurantia)* but with shorter outer tail feathers and not so deeply forked a tail as the latter. In sub-adult plumage the legs and feet have a dark red tinge. The toes are fully webbed. They are pure white on the rump contra *Sterna aurantia*. Their slower wing beats and more graceful appearance in the air help to separate them from the similarly sized Black-headed Gull *(L. ridibundus)*. As in all terns the sexes are alike.

HABITAT, DISTRIBUTION AND STATUS

A widespread tern in the region, occurring sparsely along all the major rivers and lakes during the winter months, with some migration to the Caspian and Aral seas and steppe regions of south-eastern USSR in the summer through the northern mountain valleys. It is common in coastal areas inshore, being a neritic species favouring mangrove creeks or tidal mud flats rather than the open seas. Irregular breeding also occurs on islands in the main Punjab rivers and on inland brackish lakes near the coast. In Baluchistan it occurs in small numbers on spring passage in April and May with birds being seen in

Map 184: Gull-billed Tern *Gelochelidon nilotica*

early and mid-May on Kushdil Khan lake in 1975 and 1976 and also on June 22, 1979, and there is evidence of occasional breeding on the larger lakes (Roberts, *JBNHS,* 1985). It has been seen on passage at the end of April and during early May in Chitral main valley (Perreau, *JBNHS,* 1910). Whitehead also recorded it on the Kurram river in NWFP in late April and May (*JBNHS,* 1911). Specimens on return migration have been collected also in Gilgit, an adult on 1 August already moulting into winter plumage and a juvenile bird on 3 August (Biddulph, July 1882). In winter it can occasionally be seen on the Punjab Salt Range lakes and numbers on the Indus right up to Kalabagh in February and March. In the creeks of the Indus delta it is the commonest species of tern during the autumn and winter months. During a twelve hour boat trip in the Indus creeks on 29 July it was the most numerous tern species. In years of suitable water levels, it nests colonially on Siranda lake in Lasbela (Roberts, *JBNHS,* 1980) and the author found a small breeding colony in May 1984 on Zangi Nawar lake in the Chagai. H. Waite (*JBNHS,* 1917) found it nesting on the Chenab river in June not far from the Rivaz bridge. Status COMMON.

HABITS

Not gregarious when foraging and will usually be seen singly, but they usually roost in mixed flocks with other terns and gulls. They are great wanderers, particularly after the monsoon, when they can be seen hunting over rice fields, roadside and canal seepage pools and borrow pits. On the coast they often hunt over mud banks and sand flats not over the water. This tern never plunges bodily into the water, its normal hunting technique being to dip down to the surface and to pick food items off the water or land surface. On Ghizri creek they have been seen successfully catching *Uca* sp. (Fiddler) Crabs by this method. Their food includes insects, especially orthopterous insects when feeding inland. In the USSR besides orthopterous insects, they feed on *Lacerta* lizards, spiders (Arachnida) and beetles (Scarabidae) (Dement'ev and Gladkov, 1951). They also catch locusts in sand dune areas near the coast (observed Ghizri), and can be quite beneficial in this manner. Fishes are not a major item in their diet.

BREEDING BIOLOGY

Courtship pursuit flights have been observed in Ghizri creek from late May with excited calling and attempts at fish presentation, so this starts before actual arrival on the breeding grounds. The advertising or courtship calls are very distinctive, perhaps indicating no clear relationship with other terns. Kenneth Eates found an 'enormous breeding colony' on Siranda jheel when he visited it for the first time on 9-10 June 1945. They were breeding on 3 small islands in company with *Charadrius alexandrinus* and *Sterna albifrons*. At that date most pairs had downy young but a few were still with eggs and he noted that 3 was the normal clutch. Slender-billed Gulls were later in nesting in that year (MS notes unpublished). On 24 May 1979 they were again observed breeding on Siranda (Roberts, *JBNHS,* 1980). Their nests, hollows in the sand, were decorated with bits of dried weed and shells (unlike the nests of *Sterna caspia*). Seventeen pairs in two small colonies were noted, all with 2 eggs and they had apparently commenced breeding after Slender-billed Gulls *(Larus genei)* nesting nearby. The downy chick is white on throat and breast, ochraceous buff on the head, wings and back with small scattered spots and a bluish area around the eye. The legs are brownish red and bill black with a fleshy pink base (specimen collected 12 June, Siranda jheel). The eggs are variable in colour from pale limestone buff to greenish grey in ground colour, boldly blotched with black and reddish brown and with secondary markings of grey and lilac. Both sexes share incubtion duties, and egg wetting by the bird taking over incubation duties was noted at Siranda. Incubation period is of 23 days (Dement'ev and Gladkov, 1951).

*VOCALIZATIONS

Very silent in winter time, but during courtship has an advertising call, comprising a 'plover-like', musical 'ker-wick-wick', 'kerr-wick-kuwick-kuwick', uttered quite rapidly while in flight. In winter time they can very occasionally be heard uttering shorter 'wick-wick' calls when excited. When approaching a small nesting colony, birds uttered a much shorter more grating threat or alarm call (author obs. Zangi Nawar).

Caspian Tern *Sterna caspia* Pallas

Plate **16**
Map **185**

Hydroprogne caspia (Pallas) (of Dillon Ripley's 'Revised Synopsis', 1982)

DESCRIPTION

Body length 50-59 cms.
Wingspan 127-140 cms.
Wing length 38-42.1 cms. (Stuart Baker)
Bill length 6.3-7.3 cms.

This is the largest tern throughout the world in the family Sturnidae, distinguished by its bright coral red and heavy bill, and comparatively short forked or almost square tail. The mantle is pale grey with the primaries darker brown-grey along the trailing edge of the wingtip. The tail and upper-tail coverts are white. The legs and feet are black. In breeding plumage the top of the head is black with a slight crest on the occiput. In winter plumage the whole crown is streaked black and white sometimes with a blacker area on the lores and behind the eye. Sometimes the bill tip is dusky. Juveniles have paler more orange bills. The under surface of the wingtips is dark grey. Often in flight, the tail looks pointed, with the neck rather short and thick and the body fusiform in shape. Like all the terns when hunting, it typically flies with the bill angled downwards, and as in all terns the sexes are alike.

HABITAT, DISTRIBUTION AND STATUS

A resident species, but subject to local movements which have not been studied. It inhabits both coastal areas and in winter time spreads up the Indus river and frequents larger lakes and irrigation barrage headponds. Some of the population is also probably migratory, travelling to central Russian breeding grounds around the Black and Caspian seas and Azov Sea. An adult in breeding plumage was seen at Rawal lake Islamabad, 22 April (D. Corfield, pers. comm., 1982) and an immature bird on the Sohan river near Rawalpindi in late autumn (Whistler, *Ibis*, 1930).

A large flock was seen on Kushdil Khan lake on 30 April remaining there till 17 May (Meinertzhagen, *Ibis*, 1920). It has also been noted (author) on 7 May on Kushdil Khan lake in 1974. On the Karachi seacoast they can be seen throughout the winter and up to the end of June. The maximum number recorded by me on any inland lake; *circa* 40 on Ghauspur in Jacobabad district, 4 November. In June and July they presumably disperse to breed and have been recorded nesting on Siranda jheel, Lasbela in early June 1979 (Roberts, *JBNHS*, 1980) and on Astola island near Pasni on the Makran coast. Status COMMON.

HABITS

This tern is mainly a fish feeder preferring to hunt over deeper water. It can hover when searching, and hunts by diving steeply into the sea, often disappearing entirely beneath the surface. Meinertzhagen noted them picking refuse off the surface of the sea, and robbing other terns and gulls as well (Meinertzhagen, 1954). In the Baltic Sea small colonies nest on islands, but quite often these birds fly inland and hunt on inland fresh water lakes where the bulk of their food is Perch (K. Forssgren, pers. comm., 1981). They have also been recorded as catching swimming crabs (Portunidae) and prawns in the Indian Ocean (Ali and Ripley, 1969). They commonly roost in mixed company with other gulls and terns on lake shores and sandbars on the coast, when not seeking food, but are usually seen hunting solitarily.

BREEDING BIOLOGY

On 19 June and again on 25 June a pair was observed in courtship flight chases and calling excitedly, though fish presentation was not actually observed. This was on Clifton beach some 96 kms. (60 miles) east of their nearest known breeding grounds. They nest colonially, sometimes in small aggregations close to other Terns *(Gelochelidon nilotica)* and Gulls *(Larus genei)*, and sometimes in big mono-specific colonies. About 150 nesting pairs were found on one island in Siranda jheel on 12 June (Roberts, *JBNHS*, 1980), about 98 per cent of these nests had only 2 eggs and only 2 out of 150 nests were noted with 3 eggs. K. Eates found a colony of about 50 pairs of Caspian Terns nesting on Kajar island in the

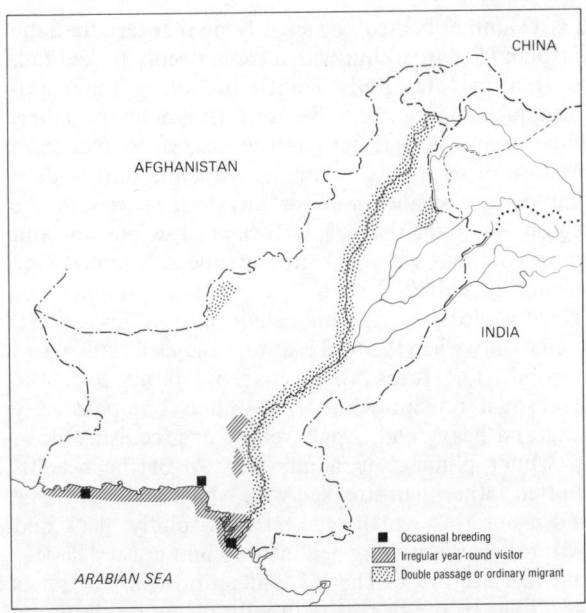

Map 185: Caspian Tern *Sterna caspia*

Indus delta, near the Indian border, in the early 1950s. They had both eggs and newly hatched downy chicks in late July (Eates, unpublished notes). They were also found (K. Eates) nesting in huge numbers with Sooty Gulls *(L. hemprichii)* on Astola island, Makran coast in June. In all instances their colonies are some little distance from other breeding sea birds, and the nest scrapes located on bare open sand. The nests on Siranda lake, being within 1.5 metres of each other, consisted of scrapes in the sand barely lined with a few stems and dried pieces of Saltwort. The eggs are more pointed at one end than Slender-billed Gulls' eggs (seen on same day) and larger and vary from olive buff to greenish stone ground colour with dark brown spots and blotches, and grey under-markings. The adult birds dive boldly on human intruders to the colony and make a lot of noise with their alarm cries. Both sexes share incubation and the downy chicks are fed by regurgitation. Studies in the Baltic Sea with colour marked birds indicate that the pair bond is life-long for older successful breeding pairs, but that young birds after unsuccessful breeding usually form new pair bonds

the following year (K. Forssgren, MS in preparation). In winter, individuals disperse widely and 'paired' birds will be quite separate, but return to a traditional breeding site where the old pair bond is re-established. In Mr. Forssgren's study a 30 year old male was still breeding successfully. Usually at least one parent will accompany the young to their wintering grounds, where the young are occasionally still fed and frequently give prolonged begging calls (observed annually at Hadiero lake, October till early December).

*VOCALIZATIONS

Begging call of juvenile birds on their wintering grounds is a drawn-out thin whistle not unlike that of young *Larus argentatus*. Their alarm call is a loud harsh 'kuwow-kuwow'. A contact call by flying birds, especially in the spring and autumn is a low pitched grating or rasping sound 'e-erragh-e-erragh' not very loud considering the bird's size. A variation of this can also be given in louder form as an aggressive or threat call.

Swift or Great Crested Tern *Sterna bergii* Lichtenstein
Large Crested Tern

Plate **16**
Map **186**

DESCRIPTION

Body length 46-55 cms.
Wingspan 108-122 cms.
Wing length 34-37.6 cms. (Stuart Baker)
Bill length 6-6.8 cms.

It is a slimmer bodied and slightly smaller tern than the Caspian (*S. caspia*) but with a more deeply forked tail, so that in total body length including outer tail feathers it may exceed the body length of the latter. This largely pelagic tern is the easiest to recognize because of its dark grey mantle which is darker than that of any smaller gulls or any tern species in the region. In flight the deeply forked all-white tail and rump contrasts with dark grey mantle and upper-wing surface including the tips. In breeding plumage the crown is glossy black with a slight nuchal bushy crest, visible only when the bird is at rest and especially when excited. The lores and a narrow band over the forecrown remain white. The bill is comparatively long and heavy and is pale yellow or greenish yellow. In winter plumage in adults the top of the head is spotted rather than streaked with white, while the nape and a line through the eye remains solidly black and with the feathers elongated into a slight crest. The legs and feet are black. The iris is deep brown. The sexes are alike. In the field it is usually recognizable by its larger size, pale yellow bill and dark ashy grey mantle.

Its wings appear long and narrow and it is a graceful flier. The sexes are alike.

HABITAT, DISTRIBUTION AND STATUS

Resident along the Sind and Baluchistan coast, this tern is generally more common than *Sterna bengalensis*, the Small Crested Tern. It is only very occasionally seen in the wider creeks and channels, preferring to hunt in the open sea. A few always rest on the outer channel buoys outside Karachi harbour, and can always be seen roosting in company with other terns and gulls on the outer standspits and beaches around the Indus mouth and Sonmiani lagoon. Individuals were occasionally seen from a fishing boat when it was trawling about 40 kms. (25 miles) off the Karachi coast in early December and more frequently seen at 24 kms. (15 miles) offshore (N. Van Zalinge, pers. comm., 1983). They can be seen throughout the autumn and winter, fishing near the tide-races and outer sandbanks of the mouth of the Hab river. They have never been noted on inland waters or on migration inland. They have been seen roosting on the beach at Clifton in mid-June and also roosting with other terns in Sonmiani lagoon on 9 June. They were also noted as being common all along the Makran coastal waters from Gwadur to Pasni in early December (N. Van

Zalinge, pers. comm., 1983). They breed in May and June both on Astola island and islands in the Indus delta, close to the border with India. Status COMMON.

HABITS

This tern often fishes in groups with other Terns, e.g. *Sterna bengalensis* and *Sterna sandvicensis*. They normally fly at a considerable distance above the sea surface and when looking for fish hover and sometimes descend lower, check and hover again. Their normal method of securing prey is to plunge slantingly into the sea with wings closed, usually submerging the entire body momentarily. Their food comprises mainly fish, but they probably also take some mollusca (e.g. squids). They are quite sociable and will roost and fish in company with other tern species and often come and perch on larger fishing trawlers or merchant ships (observed December, N. Van Zalinge).

BREEDING BIOLOGY

They breed colonially on the ground, their nests being a slight hollow in the bare sand with no lining. Captain Butler found a colony of about 100 pairs on 8 August on Astola island and as they each had only one egg, breeding had just started. He noted that they nested very densely, each nest being hardly half a metre from its neighbour, and considered that predation by Hemprich's Gull (*L. hemprichii*), which was also about to start breeding on the same island, probably accounted for their nests being so close packed. During a later visit on 9 June colonies had 3 eggs per nest. The eggs are very variable in ground colour, no two being quite alike, varying from white to pale cream, pink to rich salmon buff in ground colour with dark red-brown speckles and blotches and a few grey under-markings (Butler, *Stray Feathers*, Vol. 5, 1877a). Meinertzhagen has postulated that variation in egg pattern for this species and other densely colonial nesting sea birds, is probably an adaptation aiding in individual egg recognition by incubating birds (Meinertzhagen, 1954). In June 1941, K. Eates was shown a small colony of 30 pairs on an island at the mouth of Kajar creek in the Indus delta, near the

Map 186: Swift or Large Crested Tern *Sterna bergii*

(present day) border with India. In 1943 on 27 August a colony of about 200 was located on this same island, and also a much smaller colony on an island near Nandi-Kohar also near to the Rann of Kutch and Indian border. All the nests in June had only 1 or 2 eggs but in 1944 he found a colony comprising 40 nests with fresh eggs in the first week of July, (Eates, MS notes). Downy chicks are greyish yellow above, white below with dark brown spots on the back and head and streaks about the wings. Both sexes share incubation duties which take 20-22 days (Witherby and Jourdain, 1943), whilst period of fledging is 30-35 days (Cramp *et al.*, 1983).

VOCALIZATIONS

A bi-syllabic call is reminiscent of that of the Common Tern (*Sterna hirundo*) being slightly hoarser 'kurr-rik-kurr-rik'. At the nest colony Butler described birds as giving loud screaming calls, probably alarm cries.

Lesser Crested Tern *Sterna bengalensis* Lesson

Plate **16**
Map **187**

DESCRIPTION

Body length 38-44 cms.
Wingspan 89-109 cms.
Wing length 27.1-29.5 cms. (Stuart Baker)
Bill length 5-5.7 cms.

This is a large tern, about equal to the Sandwich Tern (*Sterna sandvicensis*) in size and slightly smaller than the Swift Tern (*S. bergii*). It has a pale grey back and wings, no different in colour from the Sandwich Tern but is distinguished by its bill which is relatively long and heavy and bright orange or cornelian yellow in colour. In breeding plumage the crown is black extending to the nape where the feathers are slightly elongated and form a bushy crest. The black cap also extends right down to the base of the bill in contrast to the Large Crested Tern which has a narrow white band over the forecrown at the base of the bill. Its bill remains the same colour throughout the year, being sometimes paler, more yellow-orange in young birds. The legs and feet are black and the sexes are alike in plumage. In winter in adults, the forecrown is almost wholly white and the black cap becomes reduced to the occiput region, consisting of black and white streaks, extending from behind the eye and around the nape. The tail is deeply forked and white as also the rump. When seen roosting alongside *Sterna sandvicensis* it appears equal in size, wing length and in the grey colour of the mantle. The sexes are alike.

HABITAT, DISTRIBUTION AND STATUS

A purely marine tern which occurs all along the Baluchistan and Sind coasts, generally being seen more commonly in the spring, summer and autumn months. It often fishes far out to sea and has been observed 42 kms. (20 miles) from the Karachi coast, fishing in a loose flock (N. Van Zalinge, 1983, pers. comm.). Ali and Ripley (1969) repeat Ticehurst's statement (*Ibis*, 1924) that it is the commonest tern in Karachi harbour and coastline. In recent years, it has not been noted very often inshore and is much less numerous than *Sterna sandvicensis* or *Gelochelidon nilotica* in the harbour or tidal creeks, and my own records of observations over a ten year period, 1973-83, indicate that it is less commonly encountered than *Sterna bergii*. It is resident the year round and at least some birds breed on islands in the Indus delta. It very occasionally can be seen in the larger wider creeks of the Indus delta and frequently fishes inside Sonmiani lagoon. Status COMMON.

HABITS

Quite gregarious in roosting and in hunting. A small flock was watched on 18 April flying around and over a fish net trap in Ghizri creek, successfully catching fish that got crowded near the net. In the same area on 25 October when fishermen were hauling in their nets, this species hovered over the net, dipping down to seize fish near the surface. Normally they hunt by hovering over the sea and plunging in from a considerable height. They are often attracted to places where fish are shoaling and hunt in mixed company with *Sterna sandvicensis*. They are mainly fish eaters, but presumably also take squid and prawns according to availability.

BREEDING BIOLOGY

Like *Sterna bergii*, this tern seems to breed towards the end of the monsoon. K. Eates, through his native egg-collector, discovered 3 small nesting colonies on bare sandy islands near the mouth of Kajar and Mall creeks in the southern part of the Indus delta, bordering the Rann of Kutch. On 27 August a colony of about 200 pairs was found on Kajar island most with clutches of 2 or 3 eggs, well incubated. This is in contrast to clutch sizes in the Middle East, where 1 or 2 eggs was normal (Cramp *et al.*, 1985). Eates's collector stated that they bred in that region from mid-August up to mid-September in previous years. The nests were mere hollows or scrapes in the sand, with at most a few grass

Map 187: Lesser Crested Tern *Sterna bengalensis*

stems on the rim and the eggs varied from buff, to limestone to terracotta in ground colour, with small dark brown speckles and larger blotches, occasionally with scrawled lines (Eates, MS notes unpublished). Very little seems to have been written about their breeding habits, but it is known that they nest colonially, laying their egg in nest scrapes close together, and that both parents share incubation. Probably eggs are protected from over heating, by wetting with breast feathers as the August sun temperatures would be lethal within minutes to an unprotected developing embryo. Incubation takes 21 to 26 days

and the fledging period is 32 to 35 days (Cramp *et al.*, 1985).

*VOCALIZATIONS

An excited flock, while fishing uttered rather high pitched cries of the same type as *Sterna sandvicensis* but their cries were not quite so grating as the latter. A bi-syllabic 'kirrik-kirrik' and a thin high pitched creaking 'kreek-kreek-kreeek' were tape recorded at this time.

Sandwich Tern *Sterna sandvicensis* Latham

Plate **16**
Map **188**

DESCRIPTION

Body length 38-45 cms.
Wingspan 91-112 cms.
Bill length 5.2-5.8 cms.

It has a much more slender bill than *Sterna bengalensis* or *Gelochelidon nilotica*. The bill is black with the extreme tip horny yellow or ivory. This is rather a pale looking tern of large size, with the upper-wings and back silvery grey and the tail not very deeply forked. Its crown is jet black in the breeding season, extending down the nape in a bushy crest. The legs and feet are black. Both sexes are alike. In winter plumage the forecrown is white becoming speckled black and white in mid-crown, and streaked more heavily black around the nape. The wingtips are quite pale without noticeably darker tips. It can be distinguished in winter from the Gull-billed Tern by its slimmer wings, its relatively longer and more slender down-curved black bill and more deeply forked tail. When seen roosting alongside the Lesser Crested Tern they are equal in body length and size. The sexes are alike.

HABITAT, DISTRIBUTION AND STATUS

Since Ticehurst's day the status of this tern appears to have changed, as he considered it a winter visitor 'in fair numbers' in some years only (Ticehurst, 1924). In recent years, records confirm that it is one of the commonest terns both in winter and early summer along the seacoast, especially around the Indus creeks and westwards to Sonmiani in Lasbela. The nearest known breeding areas are around the shores of the Black and Azov seas and the eastern shores of the Caspian Sea in the USSR (Flint *et al.*, 1984), and its status on Pakistan's seacoast is not clear. A bird ringed in June 1975 at the Krasnovodsky Reserve on the Caspian Sea (USSR) was recovered off the coast of

Kerala, south India (Ambedkar, *JBNHS*, 1985). Another with a Russian ring was recovered on 17 September 1983 on Rameswaram island, Tamil Nadu, though the location where it was ringed was not reported (Lal Mohan, *JBNHS*, 1986).

They have been noted (author) as common around Karachi seacoast from October right through to late June. A flock of 15 was seen on 20 June off Clifton beach. It is especially plentiful along the coast of Cape Monze, the mouth of the Hab river, Buleji beach etc., during April, May and June and is usually the most abundant tern in the wider creeks of the Indus delta during October to December. On 1 February 1983 a

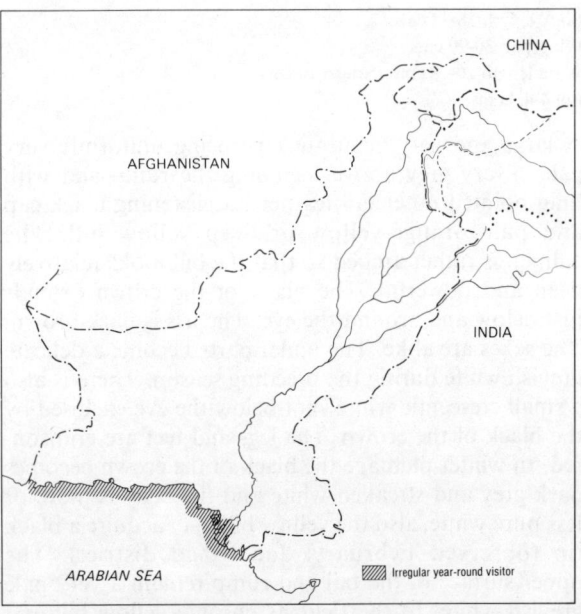

Map 188: Sandwich Tern *Sterna sandvicensis*

flock of at least 700 Sandwich Terns was seen about 5 miles off Karachi seacoast fishing in a shoal of anchovies (K. Forssgren on research vessel 'D. Fritzov Nansen,' pers. comm.). It is a purely maritime species and can be encountered out at sea beyond the sight of land. It is more common than any of the other tern species in the Indus delta in winter and along the open seacoast in spring. Birds seen (author) in late May were in full breeding plumage calling excitedly, and behaved as though courtship chasing had commenced, but over 30 birds seen on 13 June were still in winter plumage. Status ABUNDANT non-breeding visitor except during monsoon months.

HABITS

This tern dives for fish, hunting over the open sea, often in association with flocks of other Sandwich Terns and Lesser Crested Terns. A flock of about 200 was seen roosting on a sandbar on 8 December at the mouth of Pitiani creek, with 3 Skuas roosting alongside (*Stercorarius pomarinus*). They often follow fishing boats and hover around when the trawl nets are being hauled up. They don't usually make as deep dives as

Large Crested Terns (*Sterna bergii*), watched fishing in the same vicinity and often hunt closer to the water's surface. In their Caspian Sea nesting grounds they are reported as feeding on orthopterous insects and also crustaceans (*Idotea baltica*) (Dement'ev and Gladkov, 1951).

BREEDING

Colonial in compact groups with nests close together. Breeding is extra-limital. It has been well studied and described in Europe. Juvenile plumaged birds such as are common in October in parts of the British seacoast have not been noted in September or October off the Sind coast.

VOCALIZATIONS

They occasionally call in their winter quarters and have been seen indulging in noisy chases in April and May which appeared like courtship behaviour. The flight call is a bi-syllabic rather high pitched 'kirrik-kirrk', grating in tone.

Indian River Tern *Sterna aurantia* J.E. Gray

Synonym *Sterna seena* Sykes

Plate **16**
Illus. **58**
Map **189**

DESCRIPTION

Body length 38-44 cms.
Wingspan 80-90 cms.
Wing length 26-28 cms. (Stuart Baker)
Bill 4-4.3 cms.

A large and very beautiful tern being uniformly very pale silvery grey above including the rump and with long pointed outer tail feathers, a glistening black cap and pale orange-yellow or deep yellow bill. The culmen is rather arched so that the bill looks relatively deep and powerful. The black of the crown extends just below and around the eye. The iris is dark brown. The sexes are alike. The under parts become a delicate greyish white during the breeding season. There is also a small crescentic white spot below the eye enclosed by the black of the crown. The legs and feet are crimson-red. In winter plumage the black of the crown becomes dark grey and streaked white and the belly is more or less pure white, also the yellow bill may acquire a black tip (observed February, Jacobabad district). The upper surface of the tail and rump remain a very pale greyish white. In the field its chrome yellow bill and long tail streamers are helpful pointers. It is also

generally the most vocal and noisy of inland tern species. Juveniles have the forecrown and a broad supercilium white and the back and wing covert feathers are margined narrowly with blackish grey and white outer borders, giving a scale-like pattern. The downy chicks are grey with black spotting on crown and back.

HABITAT, DISTRIBUTION AND STATUS

An Oriental species confined to the Indian subcontinent, Burma and the Malaysian peninsula. It is a fresh water adapted tern inhabiting the larger rivers during the breeding season, but wandering widely in winter to large inland lakes and in lower Sind adapted to hunt along the larger irrigation canals. It is but rarely seen in the tidal waters of the Indus creeks (1 observed at Keti Bander, 20 February) and is virtually absent from the hilly tracts of Baluchistan and the NWFP. Birds have been seen frequenting the desert lakes of Khipro and Sanghar in the winter (eastern borders of Sind) and on Baran dam near Bannu (NWFP) in late March, but typically they are seldom encountered far from the major rivers of the Punjab and Sind. Status COMMON.

Illus. 58: River Tern *Sterna aurantia*

HABITS

Colonial in breeding and also often in their daily foraging and roosting outside of the nesting season. The pair bond appears to be of long duration as they are frequently encountered hunting and travelling in pairs throughout the cold weather months. They typically hunt by beating upwind with deep regular strokes of their long pointed wings, producing a rather typical jerky flight, 8 to 10 metres above the water surface. When they see food prey they plunge sharply into the water if after fish, or less commonly dip to the surface to pick smaller prey. They will congregate in noisy feeding flocks wheeling around and they fly upwind over the water surface and when fish are caught they swallow these dexterously in flight, flipping them around in their bills until the prey goes in head first. They feed principally on small fish but will also take crustacea and aquatic insects. They roost on sand banks at the water's edge, often in mixed association with other tern species and hunt only during the day, often roosting at intervals during daytime as well.

BREEDING BIOLOGY

They become more gregarious at the approach of the nesting season and also more vocal, foraging birds frequently engaging in bouts of excited calling. Pair bonds are established by courtship feeding of fish by the male and also by flight chasing. Fish are usually presented to the female on the ground, and beforehand the male with the fish in his bill, gives a particular type of advertising or presentation call (observations Mirpur Sakro, 28 March). Dharmakumarsinhji describes courtship in this species with the male presenting a fish while raising and lowering his tail and bowing his head forward (Dharmakumarsinhji, 1972).

Breeding is invariably on sandbars or islands on the major rivers and may be somewhat extended due to varying water levels in the river and loss of earlier clutches due to flooding. Whistler (*Ibis*, 1922) found it breeding on the Chenab river in Jhang district in May

and June, whereas K. Eates (MS notes unpublished) found colonies on the lower Indus, near Karachi in April. H. Waite found a colony with downy chicks on the Jhelum river on 12 May near Tahlianwalla (MS notes). Scrope Doig found them breeding on the east Narra mainly in June and July (*Stray Feathers*, 1879b) but in one year as late as August (Ticehurst, 1924).

Nests are usually located quite close together on bare open sand and consist merely of hollows formed by the birds rotating their bodies. In south India nest rims decorated with a few small stones have been recorded. Colonies are normally found in association with Black-bellied Terns (*Sterna acuticauda*), Small Indian Pratincoles (*Glareola lactea*) and Little Terns (*Sterna albifrons*), on sandy islets in the Indus river and usually these are in a separate part of the island from the other species. The normal clutch is 2 to 3 eggs and these are most likely to be found from late April to early May. The site chosen for such colonies varies from year to year as might be expected, since the river channels so frequently gouge new courses and erode or create new sandbars in their beds. The eggs are less variable than those of *Rynchops albicollis*, being pale buff or stone coloured, sometimes of a greenish tinge and with spots and blotches of reddish brown and dark brown and some grey under-markings. They are broad ovals in shape. Both sexes share incubation duties which is believed to take 18 to 19 days (Ali and Ripley, 'Handbook', 1969). The downy chicks are buffy brown with darker spots on the crown and back, more contrastingly marked than the chicks of the Skimmer. Both parents have been noted dipping low over the water surface to deliberately wet their bellies and then

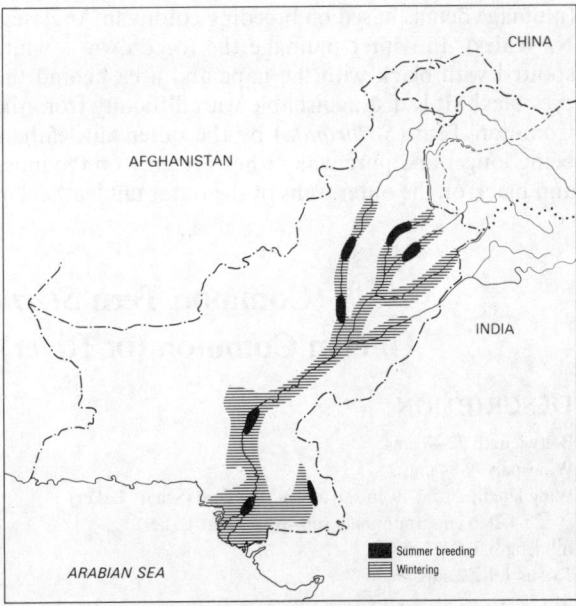

Map 189: River Tern *Sterna aurantia*

transferring this moisture to their eggs and also to their downy chicks after hatching (Lowther, 1949). When danger threatens, such as human intruders, parent birds call to their young which crouch with head and neck flattened and rely on crypsis to escape detection. The adult birds will swoop and dive upon the human invader calling loudly all the while (Dharmakumarsinhji, 1972). Dharmakumarsinhji noted that chicks would not accept 'dried-out' fish prey until the parent bird had returned to the water and dipped the fish, after which it was readily swallowed (op. cit., 1972).

*VOCALIZATIONS

Contact calls in flight consist of fairly short staccato shrill calls 'kiuk-kiuk' quite high pitched and melodious, not grating. During display, a more extended rapidly repeated bi-syllabic call is given 'kierr-wick kyerrwick-kierrwick' often accelerating in a crescendo. It is not greatly dissimilar from the calls of the Common Tern (*Sterna hirundo*) during the breeding season, but may be described as less grating and more melodious.

Roseate Tern *Sterna dougallii* Montagu

DESCRIPTION

Body length 36-43 cms.
Wingspan 76-79 cms.
Wing length 21-22.6 cms. (Stuart Baker)

A long tailed and slender winged tern of graceful proportions, about equal in size or slightly smaller than the Common Tern (*Sterna hirundo*), noticeably smaller than the Gull-billed Tern (*Gelochelidon nilotica*) and slightly bigger (10-15 per cent) than the White-cheeked Tern (*Sterna repressa*). It has noticeably long tail 'streamers' (outer rectrices) and a very pale white appearance alongside *S. repressa*. In winter time the bill is all-black. In the breeding season after incubation starts, the base of bill gradually appears red. The crown and nape are glossy black during the breeding season and the legs and feet crimson red, also there is a faint rosy flush to the breast and belly which gradually fades as the breeding season progresses (plumage details based on breeding colony in Anglesey, N. Wales). In winter plumage the forecrown is white spotted with black with the nape and area behind the eyes black. It is distinguishable with difficulty from the Common Tern (*S. hirundo*) by the outer tail feathers being longer and pure white (they are grey on the inner and black on the outer webs of the outer tail feathers in

S. hirundo), and by its paler white wingtips, the primary feathers of the Common Tern showing dark grey on the underside. It is also easily distinguished by its distinctly harsher more grating calls. At rest the tail tip extends clearly beyond the wingtips.

HABITAT, DISTRIBUTION AND STATUS

It has not been authentically recorded from Pakistan. It is a maritime species haunting coastal waters and will not be found inland or along rivers. It has been recorded breeding on islands off the west coast of India in the 1940s, e.g. Vengurla rocks and Rameswaram island, but it has not been possible to confirm breeding in recent years (Humayun Abdulali, pers. comm., 1981). It has also been recorded breeding on islets off the coast of Sri Lanka. The nearest known breeding grounds to the Karachi coast are at Masirah off the Omani coast in the Arabian Sea, where a small colony was discovered in June 1974 (and one specimen lodged in the British Museum) and again a few pairs were noted in 1979 on the same island (M. Gallaher *in litt.*, 1982 and Gallaher *et al.*, 1984). It may well occur in small numbers and irregularly off the coast of Pakistan in winter time when it would be difficult to separate if from *Sterna hirundo*.

Common Tern *Sterna hirundo hirundo* Linnaeus
Tibetan Common (or River) Tern *S. hirundo tibetana* Saunders

Plate **16**
Map **190**

DESCRIPTION

Body length 32-40 cms.
Wingspan 79-84 cms.
Wing length 24.5-27.9 cms. (sub-sp. *tibetana*) (Stuart Baker)
　　25.4-28.6 cms. (nominate sub-sp.) (Stuart Baker)
Bill length 3.4-3.9 cms.
Tarsus 1.9-2.2 cms.

A medium sized tern with deeply forked tail, slender wings and graceful appearance. In breeding plumage

this tern normally has a crimson bill with a black tip and bright red legs and feet. However the population breeding in eastern Siberia *S. hirundo longipennis* has an all-black bill throughout the year. The crown is jet black extending from the base of the bill to the nape and down to the top of the eye during the breeding season. The back and wings are pale blue-grey with the rump, upper-tail coverts, and central tail feathers white. The outer tail feathers are grey, their outer webs being narrowly blackish. It is distinguished from *S. repressa* in winter plumage by its white rump and slightly larger size. In winter plumage the bill appears entirely black even in the nominate sub-species but the red feet are helpful. The forecrown in winter is white, gradually becoming black on the mid-crown and nape. The throat is white but breast in summer sometimes has a greyish tinge. The wingtips show the first primary largely black. At rest the wingtips do not· extend beyond the tail tip, being about equal, in contrast to the very similar Roseate Tern (*S. dougallii*). The Tibetan River Tern in breeding plumage is darker grey on the mantle with more conspicuous dark markings on the wingtips, and the throat white, contrasting with the rest of the breast and belly which are pale grey. The bill is red, but black on the distal half.

HABITAT, DISTRIBUTION AND STATUS

This tern is only likely to be encountered on the seacoast, but a small population *S. hirundo tibetana* nests on the high plateau lakes and streams of Tibet, the Pamirs and Ladakh. Some specimens seen on the seacoast in early May had all-black bills, also a party of 3 roosting birds on 7 October and these individuals might have belonged to the sub-species *longipennis*. *S.h. tibetana* has also been noted in Gilgit on 23 August on southward passage (Biddulph 1882) and David Corfield saw one on Rawal lake on 2 July 1983 which was presumably also on passage. Dr. Mrs. Jean Orr saw several on the Shyok river at Khapalu (Baltistan) during May 1980, presumably on passage to their breeding grounds (pers. comm., 1982). Two were also seen on the Indus river at Skardu in early August (W.H. Mathews, *JBNHS,* 1941). As regards the nominate sub-species, its status is by no means clear as they are rarely sighted outside of the early summer months (Ticehurst, *Ibis,* 1924 and author obs.) at which season they are not uncommon along the Karachi and Makran seacoast. Meinertzhagen noted them on Kushdil Khan lake in May and early June, 1914 when they were plentiful (*Ibis,* 1920), but there are no subsequent records from northern Baluchistan. Ticehurst considered it only a summer visitor to Karachi coastline (*Ibis,* 1924). Recent confirmed identifications (author) include a few winter as well as summer sightings, e.g. on 19 February (a single bird in non-breeding plumage), 7 October and 8 November (2 birds in each case). A tern which was either this species or *S. dougallii* was observed about 19 kms. (12 miles) off the Sind coast on 7 December 1983 (N. Van Zalinge, pers. comm.). Birds have also been reliably identified on 19 February and 30 April and many sightings (Van Zalinge, Passburg and author) during May and throughout June, but not in any other months. Latest summer sighting 29 July. Birds seen 30

April were in full breeding plumage with pale grey breasts and basal two-thirds of bill bright red. Similarly birds noted on 30 May included individuals in worn plumage with nearly all-black bills and some in full breeding plumage with long tail streamers and red on the bill showing clearly. Birds seen on 7 October were in very worn plumage. They are mostly seen in the summer months in company with *Sterna repressa*. They can be· seen fishing in the mangrove creeks, roosting on the rocks on the open seacoast and also 3-5 kms. offshore fishing amongst flocks of other terns. It is presumed that all visitors to the coast are non-breeding birds. Birds seen on 8 November were also in worn plumage and were about 9-11 kms. offshore, fishing with *Sterna repressa* and *S. sandvicensis* (Walter Weitkowitz, pers. comm., 1983). Status SCARCE except in early summer.

HABITS

On passage the Russian and Tibetan breeding populations are adapted to follow river systems. They hunt mostly fish prey which they obtain by plunging into the water in a steep dive.

BREEDING BIOLOGY

The Tibetan Tern (*S. hirundo tibetana*) nests colonially in marshy ground where snow-melt streams debouche onto the high plateau valleys. Eggs are found chiefly in

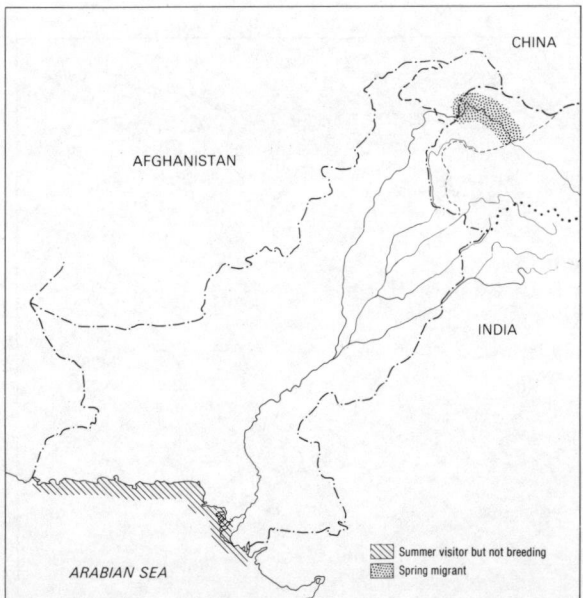

Map 190: Common Tern *Sterna hirundo* including *Sterna hirundo tibetana*

June. They are reported to be very bold in defending their eggs or young, diving on the intruder repeatedly (Ludlow, *Ibis,* 1928). The normal clutch is 2 to 3 eggs, with both parents sharing incubation duties and the chicks being fed by regurgitation. A few pairs still nest on the lakes of eastern Ladakh (Brigadier Moti Dhar, pers. comm., 1983), but they have never been recorded breeding in Baltistan.

VOCALIZATIONS

Their rather drawn-out grating calls have been rendered 'kriyayaya'. When excited or threatening they utter shorter 'kek-kek-kek' calls or a more rapid 'kikikikik'. A roosting flock on 30 May broke into loud 'kikikik' calls when disturbed and executed spectacular aerial circuits and dives.

White-cheeked Tern *Sterna repressa* Hartert

Plate **16**
Map **191**

DESCRIPTION

Body length 30-36 cms.
Wingspan 69-79 cms.
Wing length 22.7-25.4 cms. (Stuart Baker)

Slightly smaller than the Common Tern (*Sterna hirundo*) and noticeably smaller than the Sandwich Tern (*S. sandvicensis*), this is a slender bodied tern with a shorter less deeply forked tail than the Common Tern (*S. hirundo*). In winter its bill is largely black and legs are orange-red. In the breeding season the bill is bright red basally with the culmen and tip black and the legs and feet are coral red. The most noticeable field point both in summer and winter plumage are the tail and rump which are grey and concolorous with the wings. In breeding plumage the crown from base of bill to nape is jet black, the lores and ear coverts are white,

but the throat and sides of neck extending to the whole of the breast are dusky grey almost a vinaceous grey. In winter the breast is white. Thus in the summer the cheeks show up contrastingly white against the dark throat and sides of neck and the black cap. Immature birds have an all-black bill and their darker grey mantle and rump (like adults') are helpful identification points.

HABITAT, DISTRIBUTION AND STATUS

Curiously, Ticehurst (*Ibis*, 1924) never met with this tern which is considered a summer visitor to the Sind and Baluchistan coasts (Ali and Ripley, 'Handbook', 1969). This is rather a pelagic tern which is not often seen from the land until the early summer. It is however very numerous along the open seacoast from March to June and flocks can be seen fishing just beyond the surf, in full breeding plumage. It has not been noted inside the main creeks or lagoons of the coastline, except once in Ghizri creek on 29 June after stormy weather (Ecco Smith, pers. comm., 1977). Earliest dates seen (author, 7 years' records) 20 March and flocks have been noted off Buleji beach and Cape Monze throughout May and June. They can often be seen resting on the beach in small flocks. Captain Butler noted them as common all along the Makran coast during a cruise to Astola island between 13 and 17 May (*Stray Feathers*, 1877a). Recently they have been recorded off the Sind coast from late October, through to early December both outside the harbour off Manora head (Walter Weitkowitz, 8 November 1983) and in considerable flocks off the Rann of Kutch about 24 kms. (15 miles) out to sea in early December and around Astola island (Makran coast) in mid-November (N. Van Zalinge, pers. comm., 1983). They breed abundantly in the Persian Gulf and on Masirah island off the Arabian peninsula. They also breed on Vengurla rocks off the coast of Maharashtra. No breeding has actually been recorded within Pakistan territory. Status COMMON in summer only.

Map 191: White-cheeked Tern *Sterna repressa*

HABITS

They are gregarious in breeding and in their hunting. They appear to subsist entirely on small fish and are adept at manoeuvering in strong head winds, plunging bodily (author obs.) into the sea after their quarry in steep dives. Apparently the east African population seldom obtains food by diving into the water, getting its food prey mostly by contact dipping in the surface water (Urban *et al.*, 1985).

BREEDING BIOLOGY

Breeding as far as is known is outside of Pakistan territory. Colonial in nesting with eggs being laid in a bare scrape or hollow in the sand. Over 3,000 pairs nest on Sheetvar island in the Persian Gulf. The usual clutch is only 2 eggs of variable colouration, whitish cream to olive buff with small spots and speckles of pale reddish to dark reddish brown with under-markings of grey or lilac. In the Persian Gulf they commence egg laying in mid-June (Meinertzhagen, 1954).

VOCALIZATIONS

Birds fishing in a flock off Cape Monze were heard (author) calling to each other and this was similar to the calls of *Sterna dougallii* being more grating and lower pitched than similar calls of the Common Tern (*S. hirundo*).

Black-bellied Tern *Sterna acuticauda* J.E. Gray

Synonym *Sterna melanogaster* of Stuart Baker

Plate **16**
Map **192**

DESCRIPTION

Body length 30-35 cms.
Wingspan 70-80 cms.
Wing length 22.1-24 cms. (Stuart Baker)
Bill 3.2-4 cms.

This tern with its long tail streamers, bright orange bill and dark belly is one of the more striking species amongst this elegant family. They are smaller in size than the River Tern (*Sterna aurantia*) being closer in size to the Whiskered Tern (*Chlidonias hybridus*), but readily distinguished from the latter by the longer outer tail feathers, resulting in a deeply forked tail, and its vermillion or orange-red bill. The iris is dark brown and the legs and feet are orange-red (darker than the bill), in contrast to the darker coral red of the legs and feet of *Sterna aurantia*. In full breeding plumage the entire lower breast and belly becomes deep chocolate brown varying to black, with the upper breast and sides of neck dark grey. The crown extending from the base of the bill to the nape is glossy black, and the back and wings are pale silvery grey, the tail being paler grey. Both sexes are alike. In non-breeding plumage the forecrown becomes white and the nape streaked black and white and the black area on the belly disappears for a short period only after breeding. Thus its breast and belly is white with some grey in September but by November the lower belly shows some blackish area increasing by December to the whole of the vent and belly. It is confusingly like the Whiskered Tern (*C. hybridus*) in breeding plumage, showing narrow white cheek and neck patch, but the under-tail coverts are concolorous with the black belly whereas this area remains white in *C. hybridus*. The tail is paler grey in *S. acuticauda* with noticeably longer outer tail feathers which are also white, whereas the tail of *C. hybridus* is the same grey tone as its mantle and wing coverts.

HABITAT, DISTRIBUTION AND STATUS

Like the River Tern *S. aurantia* this is an exclusively fluvial or fresh water species, being particularly adapted

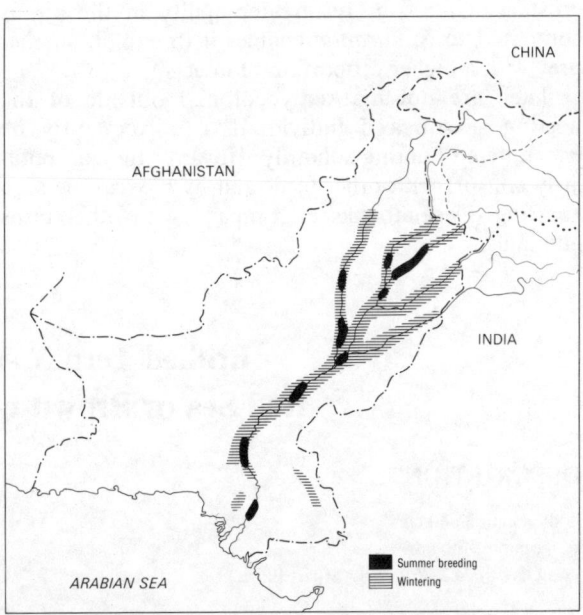

Map 192: Black-bellied Tern *Sterna acuticauda*

to feed and nest along rivers. It is more frequently encountered on other fresh water bodies such as lakes and ponds, than its larger cousin (*S. aurantia*) and only very rarely descends to tidal or brackish creeks near the coast. In Pakistan it is commonest in northern Sind and throughout the Punjab, being rare in lower Sind and practically absent from Baluchistan and the North West Frontier. An occasional bird can be seen on the Indus near Sujjawal bridge in lower Sind, whereas around the Punjab irrigation headworks at places like Trimmu, Taunsa, Chashma and Balloki, it is the commonest tern species. It wanders erratically in winter time and then numbers haunt the larger lakes which have open water and are not heavily weed fringed, e.g. Nammal lake in the Salt Range, and Manchar lake in Dadu district of Sind. H. Waite collected 2 (Brit. Mus. coll.), on the Leh river near Rawalpindi in January and February. It is less numerous throughout the plains than *S. aurantia*. Status FREQUENT.

HABITS

It is more insectivorous in diet than the River Tern and will hunt food prey by dipping down to the surface and even hawks dragonflies (Odonata) and beetles at dusk, flying over land adjacent to lakes or rivers. It will also catch fish by diving bodily into water and catches most of its food in a similar fashion to the River Tern, beating upwind 8 to 15 metres above the water surface with graceful measured strokes of its wings, sailing around to start again at the same spot downwind. Its food comprises a variety of Coleoptera, Odonata, Isoptera (flying termites), as well as fresh water crustacea and fish. Its greater agility in the air as compared to *S. aurantia* enables it to exploit smaller prey with a higher proportion of insects.

They are not markedly colonial outside of the nesting season and individuals can frequently be encountered hunting solitarily. However they do regularly consort with other terns and by day can be seen roosting on sandbanks in company with other terns and gulls.

BREEDING BIOLOGY

Very similar to that of the Indian River Tern with whom it usually associates during the breeding season. It nests colonially on the ground, on bare sandy islands in the main rivers. The nest itelf is a scrape hollowed out in the sand usually unlined except for a few strands of dried grass. The clutch comprises two or three eggs variably coloured creamy buff to greenish and dark buffy, spotted and blotched with reddish brown and with pale grey under-markings. K. Eates (MS notes unpublished) found a colony of about 20 pairs nesting on an island in the Indus near Sukkur and located several hundred metres apart from a larger colony of *Sterna aurantia* located on the same island. On 27 April this colony contained downy chicks in about one-third of the nests and well incubated eggs in the remainder. E. N. Lowther (1949) describes a colony on the Jumna river near Delhi in early May, in which the parent birds were observed wetting their eggs and downy young with their breast feathers. This colony was also located a few hundred metres apart from a mixed colony of *S. aurantia* and *R. albicollis*. H. Whistler (*Ibis*, 1922) found colonies on the Chenab river from late March to May always on river sandbars in company with Little Terns (*Sterna albifrons*) and River Terns (*Sterna aurantia*). The incubation period is believed to be 15 to 16 days (Ali and Ripley, 'Handbook', 1969) and the male will bring food to his mate in the early stages of incubation and egg laying whilst later sharing incubation duties. This species also occasionally nests solitarily (Whistler, 'Handbook', 1949).

VOCALIZATIONS

Their contact call is a shrill but quite melodious short sharp 'krek-krek'. It is lower pitched than the cry of the Indian River Tern (*Sterna aurantia*), and not so strident. When excited or angry they utter a more rapid and repeated 'kek'kek-kek' call.

Bridled Tern *Sterna anaethetus* Scopoli
Red Sea or Brown-winged Tern of Ali and Ripley

Plate **16**
Illus. **59**

DESCRIPTION

Body length 35-40 cms.
Wingspan 65-78 cms.
Wing length 24.2-26.1 cms. (Stuart Baker)

A dark sooty brown tern with deeply forked tail. Its upper parts are rather grey-brown on the mantle,

darker sooty brown on the wings with crown and nape black. Its under parts are white. The forehead is white extending a little above and beyond the eye and there is a greyish white collar between the black of the nape and the grey brown mantle. A black loral stripe runs through the eye and joins the black of the nape. Outer tail streamers are conspicuously white and this shows well when they are flying. In flight the leading edge of the upper surface of the wings also shows white up to the carpal bend and this forms a conspicuous mark. The face pattern with white forehead, short superciliary stripe and white throat, divided by a black line through the eye gives rise to its common name 'bridled'. The rather slender bill is black as also the legs and feet and the iris is dark brown. The sexes are alike.

Immature birds have the top of the head mottled white and greyish margins to the feathers of the mantle and wing coverts.

HABITAT, DISTRIBUTION AND STATUS

This is a pelagic tern which wanders widely over the Arabian Sea and Indian Ocean from its breeding grounds on the Persian Gulf and off the Arabian coast and in the Red Sea. It occurs only very occasionally in coastal waters far from its breeding haunts, but is occasionally sighted along the Makran and Sind coasts and can be seen fairly regularly along the extreme western seaboard of Baluchistan from Gwadur to Pasni. Captain Butler in his sea voyage along the Baluchistan coast on 13-17 May notes that he did not encounter it until opposite Gwadur whereafter it became common but not as abundant as it was when they entered the Persian Gulf (*Stray Feathers*, August, 1877a). N. Van Zalinge during three coastal surveys, from 13 November to 15 December 1983 between 16 and 32 kms. (10 and 20 miles) offshore all along the coast saw 2 resting on some driftwood about 6.5 kms. (4 miles) off Karachi harbour on 14 May and 4 on 30 November near Astola island (pers. comm., 1983). W. Weitkowitz saw a flock of 15 to 20 some 48 kms. (30 miles) south of Karachi on 29 August 1983. There is evidence that some numbers breed on Vengurla rocks off the Maharashtra seacoast, India and these southern birds may have belonged to that breeding population. In 10 years of observations (author, Passburg and Van Zalinge) from land, there are but 2 recent records. One seen fishing on 21 June 1974 over some tidal pools at Clifton beach (author) and 3 seen off Cape Monze on 20 May 1980 (R. Passburg pers. comm.). Status in Pakistan's coastal waters, RARE.

HABITS, BREEDING etc.

It feeds exclusively on crustaceans and fish which occur in deep water zones, outside of the breeding

Illus. 59: Bridled Tern *Sterna anaethetus*. Note prominent white leading edge to wings and white outer tail feathers. Also white area extending backwards over eye which is lacking in the very similar (extra limital) Sooty Tern (*Sterna fuscata*).

season. It has the habit of resting on the open ocean more frequently than the more neritic or coastal terns, but is reluctant to float on the sea being especially attracted to roost on the rigging of ships or to perch on floating driftwood. In Captain Butler's account (*Stray Feathers*, 1877a) specimens were caught by sailors at night while they were roosting on the rigging of his vessel. Three of the sightings off the Makran coast by Van Zalinge (see above) were of birds resting on floating driftwood.

They hunt food by hovering over the surface and dipping to the surface and also by plunging and diving into the water. They apparently are able to hunt successfully at night time also, having been recorded with bathy-pelagic (deep sea) fishes (Gonostomtidae family) in their stomach 3 hours after sunset in the Pacific. These fishes only come to the surface to feed after darkness falls (Morzer Bruyns and Voous, 1965). They nest colonially in dense packed masses with nests being less than 0.5 metres apart, laying in May and June in the Persian Gulf. The normal clutch is only 1 egg and nesting may be prolonged up to August with a replacement clutch being laid by unsuccessful parents (Meinertzhagen, 1954).

VOCALIZATIONS

The call of this tern has been rendered as 'wide-awake'. Its call has also been described as a staccato 'wrep-wrep' (Cramp *et al.*, 1985).

Little Tern, Least Tern *Sterna albifrons* Pallas
Sterna albifrons albifrons
Sterna albifrons pusilla

Plate **16**
Illus. **60**
Map **193**

TAXONOMY

In the 'Handbook' (Ali and Ripley, 1969), two sub-species are described from the Indo-Pakistan region, *S. albifrons albifrons* with shafts of the first three primaries dark brown, breeding only in Makran on salty lakes, and north-west India on rivers, mainly in Gujarat state. The second, *S. albifrons sinensis* with the first three primaries white, mainly a winter visitor to the coast but with at least one known breeding colony on Uttan Washi, a small island off Bombay. In the 'Revised Synopsis' (Dillon Ripley, 1982) the sub-species *sinensis* has been eliminated and instead *pusilla* has been included. *Sterna albifrons pusilla* being ascribed to the population which breeds inland on the rivers of north-west India and Pakistan, whilst the nominate sub-species is described as breeding along estuaries and brackish marshes in Pakistan and Gujarat state (India), presumably not very far from the seacoast. In the absence of ringing studies no further light can be thrown on this problem as winter plumage birds cannot be separated, and abound along the seacoast, in tidal lagoons and mangrove creeks, and may belong to any one of several described sub-species. The entire population wintering in Pakistan has an all-black bill and is confined to seacoastal regions and no Little Terns are ever seen on inland fresh waters or rivers in these cold weather months. In the spring an influx of birds on inland fresh waters can be noted, which are then in breeding plumage and which travel up the Indus and its tributaries and breed on sandbars in these rivers. These birds cannot be separated from *Sterna albifrons albifrons* (Ticehurst, 1924 and Eates, notes unpublished). The formerly designated sub-species, *S. albifrons saundersi* (Stuart Baker, 'Fauna Series', 1929 and Ali and Ripley, 'Hand-

book', 1969), with black shafts on the first three primaries and which only breeds on the seacoast, has now been separated as a distinct species (D. Ripley, 'Revised Synopsis', 1982).

DESCRIPTION

Body length 20-28 cms.
Wingspan 50-54 cms.
Wing length 16-17.5 cms. (sub-sp. *pusilla*-Pakistan specimens)
    (Stuart Baker)

This is a very diminutive tern with long slender wings and a characteristic rather deep beat of its wings in flight. The tail is slightly more deeply forked than in *Chlidonias* terns and it is chiefly recognized by its pure white breast in breeding plumage and its yellow bill. It is slightly smaller in size than the Whiskered Tern (*Chlidonias hybridus*). Its wing beats are noticeably quicker than larger fresh water terns. In breeding plumage the bill is yellow with only the extreme tip black and in Pakistan many individuals, perhaps the majority, have a wholly yellow bill. The crown is glossy black extending to the nape, but a narrow band over the forecrown and extending over the eye is white. This white forecrown patch usually extends backwards just over the eye, whereas in *S. saundersi* the white forecrown is more restricted in area, barely reaching the eye (See **ill. 60**).

The legs and feet are orange-yellow. The upper-wing surface and back is light grey with the tail and rump white. In winter plumage the black cap becomes streaked with white and the forecrown is more exten-sively pure white. All *Sterna albifrons* sub-species, after breeding, moult into rather similar plumage in which the dark on the outer shafts of the primaries becomes browner and less distinct, and the bill turns

(a)

(b)

Illus. 60: (a) Typical head of Little Tern *Sterna albifrons albifrons* breeding on inland river banks in Pakistan. Note more extensive white area on forecrown and very reduced black tip to bill. Many individuals have a wholly yellow bill during the breeding season.
　　　　　(b) Typical head of Saunders' Tern *Sterna saundersi* in breeding season. Note more extensive black tip to bill and white forecrown not extending backwards over eye (both heads copied from photographs of Pakistan specimens).

dark brown and eventually black, which makes it very difficult to determine which breeding populations remain along the seacoast.

HABITAT, DISTRIBUTION AND STATUS

In winter they are gregarious and can be seen in all months in small flocks fishing off the open seacoast and particularly in the mangrove creeks of the Indus delta and in the harbour and along the Makran coast. A flock of 12 Little Terns was seen fishing together on 29 July in the Indus creeks, sub-species unknown. Usually not more than 5 or 6 are seen fishing or roosting together during the winter months, often in company with other tern species.

From early April these terns suddenly arrive alongside the larger fresh water lakes of lower Sind and they then gradually travel up the Indus and its tributaries to their breeding grounds. 30 were, however, still seen on 1 June around Haleji lake (Thatta district). In the southern Punjab on the Ravi river near Abdul Hakim pairs of Little Terns were first noted between 12 April and 15 April in 3 consecutive years. Nesting starts in May on the Chenab in Jhang district (Whistler, 1922), a pair was noted (author) on 19 May near Jhelum town, on the river. They are not as numerous on the Punjab rivers as *Sterna aurantia* or *Sterna acuticauda* during the breeding season. Status FREQUENT.

HABITS

This tern usually hunts 10 to 20 metres above the lake or sea surface and dives steeply into the water after its food prey. They take mainly small fish along the seacoast but on fresh water bodies may be seen dipping to the surface as well as plunging under the water and they may then be seeking aquatic insects. They are more vocal than many other terns especially at the onset of spring and can frequently be heard calling while they are hunting for food.

BREEDING BIOLOGY

Courtship initially involves attempted fish presentation by males accompanied by much excited calling and flight chases between pairs. When nesting on rivers they are usually in small colonies, often mixed with nests of River Terns (*Sterna aurantia*) and Scissor Bills

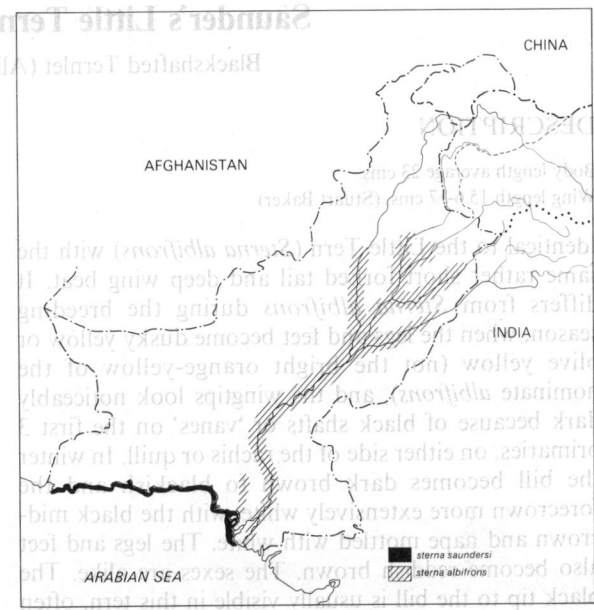

Map 193: Little Tern *Sterna albifrons* and Saunder's or Black-shafted Tern *Sterna saundersi*

(*Rhynchops albicollis*). The nest scrape is made in the bare sand and may have a few shells or stems decorating the rim. The male often brings fish to his brooding mate until the chicks are hatched. The normal clutch is 2 to 3 eggs, occasionally only one and both sexes share incubation duties. K. Eates found small colonies of this tern on the Hab river between Band Murad Khan and Mochko and on Siranda jheel he found a colony of over 100 pairs on 9 June 1945. The eggs he considered to be darker in ground colour than those of Saunders Tern (*Sterna saundersi*) (K. Eates, MS notes). Vernon Jones found two clutches each of 3 eggs on the Hab river on 18 May. The parent birds are reported to wet their chicks from their breast feathers when they come to relieve the other partner from brooding duties during daylight hours (Dharmakumarsinhji and Lavkumar, 1972).

*VOCALIZATIONS

Their rapid high pitched calls are reminiscent of the Pied Kingfisher (*Ceryle rudis*). They call 'kerrik-kerrik-kerrik' also a rapid more bubbling 'ke-ker-kewik, kerkewic-kerkewik' when in a flock together.

Saunder's Little Tern *Sterna saundersi* Hume

Illus. **60**
Map **193**

Blackshafted Ternlet (Ali and Ripley's 'Handbook')

DESCRIPTION

Body length average 23 cms.
Wing length 15.6-17 cms. (Stuart Baker)

Identical to the Little Tern (*Sterna albifrons*) with the same rather short forked tail and deep wing beat. It differs from *Sterna albifrons* during the breeding season, when the legs and feet become dusky yellow or olive yellow (not the bright orange-yellow of the nominate *albifrons*), and the wingtips look noticeably dark because of black shafts or 'vanes' on the first 3 primaries, on either side of the rachis or quill. In winter the bill becomes dark brown to blackish and the forecrown more extensively white, with the black mid-crown and nape mottled with white. The legs and feet also become reddish brown. The sexes are alike. The black tip to the bill is usually visible in this tern, often extending over the distal one-quarter (author's photographs), whereas in *S. albifrons* breeding on Pakistan's rivers, the entire bill is quite often yellow.

HABITAT, DISTRIBUTION AND STATUS

A marked influx of this tern can be observed along the seacoast from early April. They breed in small dispersed colonies along the coast, including Sonmiani area. Their winter movements are not known but probably they join part of the population which winters on the larger creeks and lagoons along the seacoast. They are a purely maritime species. They have now also been found breeding in the Kathiawar peninsula and on islets off the north-western coast of Sri Lanka. Status around Karachi and Lasbela coast ABUNDANT. By September these terns have disappeared from their breeding haunts.

HABITS

They fish in creeks, over the open sea and usually in pairs or in small groups outside of the breeding season. They feed upon small fish and crustacea, which they obtain by hovering and plunging vertically into the water.

BREEDING BIOLOGY

They come back to traditional nesting sites and nest in loose association, in that 4 or 5 nests are usually located over an area of 100 to 150 hectares. The pair bond is long lasting and involves much courtship chasing and fish presentation by the male to the female. The nest is a hollow scrape in the sand with a scant lining of broken shells. The normal clutch is 2 eggs not 3, and both sexes share incubation duties. A colony at Ghizri near Karachi under observation during two summers, was 0.8 kms. from the open sea, located between low sand dunes, the two closest nests 200 metres apart. Egg laying started late May in 1976 and early June in 1977. The male was observed presenting fish on 7 June while the female was brooding 2 eggs, suggesting that the female does take a larger share of incubation. The eggs are warm buff in ground colour with small spots and squiggles of dark brown and with secondary grey markings. In the 1940s over 40 pairs nested in the Ghizri colony (K. Eates, MS notes) but at the present time, 6 pairs are the maximum number discovered in any one year around Ghizri. There is much more human disturbance now in this area, plus nest and chick predation by feral dogs and House Crows (*Corvus splendens*). On an outer sandbar of Sonmiani lagoon about 20 pairs were nesting (author obs.) in June 1975. Downy chicks have flesh coloured legs and feet and a greyish pink bill with a darker tip. The down is yellow buff on the upper parts and dull white underneath. There are darker brownish longitudinal lines on the crown and brown spots and patches on the back and flanks.

*VOCALIZATIONS

Similar to the calls of *Sterna albifrons*. When coming in to relieve an incubating mate both sexes call in a rapid high pitched 'kiwick-kiwick-kiwick' and 'keer-i-wick, keer-i-wick, keer-i-wick'.

Whiskered Tern *Chlidonias hybridus* (Pallas)

Plate **16**
Illus. **61**
Map **194**

DESCRIPTION

Body length 24-26 cms.
Wingspan 66-71 cms.
Wing length 22-24.2 cms. (Stuart Baker)
Bill length 31-33 mm. (males),
    26-29 mm. (females) (Indian specimens) (Cramp *et al.*)

This small tern is intermediate in size between the Least Tern (*Sterna albifrons*) and the Black-bellied Tern (*S. acuticauda*), and it can be seen in the same habitat with both these species in summer. It has rather a shallow fork to its tail and in summer breeding plumage, is very attractive. It then has a dark red bill, jet black cap and the mantle and upper-tail coverts are uniformly silver grey. The tail however remains white. The whole of the belly, except the under-tail coverts up to the throat becomes dark grey-black, shading to solid black on the lower belly with the cheeks and sides of neck contrasting sharply white. The legs and feet are dark red or brownish red. The webs on the feet are deeply emarginated unlike the webs of *Sterna* genus. When at rest the wingtips extend well beyond the tail. The sexes are alike. In winter plumage the whole of the throat and belly becomes white and the bill dark brownish red almost black. The black crown becomes flecked with grey and white, with the forepart of the head wholly white. This tern is chiefly recognizable by its rather squarish tail lacking long tail-streamers, and in winter, with its all-white crown it looks rather like a miniature Gull-billed Tern *Gelochelidon nilotica* except for more extensive black streaking on its nape and its relatively more slender bill. In summer plumage the whole of the under-wing and under-tail coverts remain white. It is readily separable from *Chlidonias leucopterus* even when in the early stages of moulting into breeding plumage because of its white under-wing coverts contra jet black of the latter.

HABITAT, DISTRIBUTION AND STATUS

This is a year round and common visitor to both inland lakes and rivers and the coastal regions of Pakistan, especially exposed tidal mud flats and creeks during the winter. Breeding within Pakistan has never actually been proved. In the monsoon season they spread throughout the major rice-growing tracts and larger lakes, and birds in breeding plumage can be seen (author obs. 7 years throughout Thatta district) in July, August and September in Sind. However no evidence of breeding has been found. It is also common on all the inland larger lakes, irrigation barrage headponds and rivers during the winter and early summer months. Nesting colonies regularly occur extra-limitally in the main vale of Kashmir and

occasionally (Lowther, 1949) on the plains of north-eastern India. Status ABUNDANT but mainly as an erratic double passage migrant.

HABITS

This tern is very graceful in hunting, its normal method of foraging being to dip down to the surface to pick up food either from the water, exposed mud banks or water lily leaves. They have never been seen diving into the water like other fresh water inhabiting terns (*Sterna aurantia* and *S. albifrons*). Their food comprises aquatic insects including Odonata, water beetles and also small molluscs and crustacea. Ticehurst (*Ibis,* 1924) noted them as feeding upon 'Libelluline' dragonflies. In a study conducted in the Sunderbans (west Bengal), stomach contents of 13 birds revealed 6.6 per cent of their diet consisting of

Illus. 61: (a) White-winged Black Tern *Chlidonias leucopterus* in breeding plumage viewed from below.
(b) White-winged Black Tern viewed from above. Note all-black underwing coverts with white undertail coverts and prominent white upper wing coverts and rump.
(c) Whiskered Tern *Chlidonias hybridus* viewed from below.
(d) Whiskered Tern viewed from above. Note uniformly grey remiges (flight feathers). Only wingtips are grey in *leucopterus*. Note ventral area almost black and cheeks showing prominently white. From above note grey rump and uniformly grey upper wing coverts.

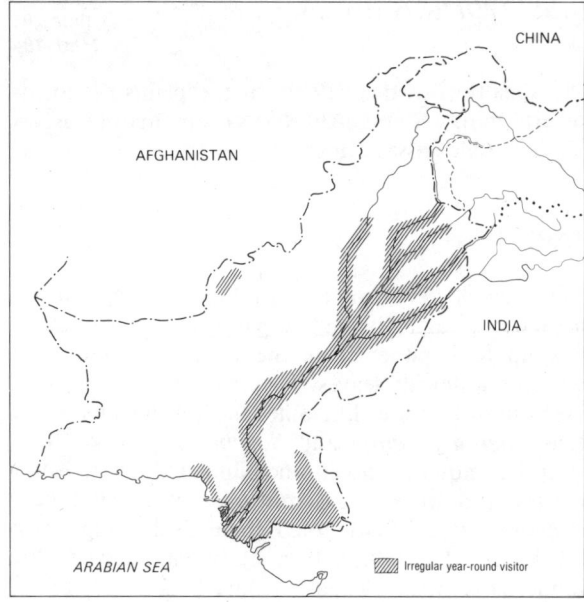

Map 194: Whiskered Tern *Chlidonias hybridus*

frequently whilst thus foraging. They have also been seen (author) successfully hawking in the air, adult dragonflies over newly transplanted rice, twisting and swooping with speed and dexterity (Jati, Thatta district, July).

BREEDING BIOLOGY

In Kashmir they nest colonially on fresh water lakes having good surface cover of emergent plants. Nests are untidy structures made from long reed stems woven and attached to floating vegetation. Adults have been observed frequently robbing other nests for nesting material (Bates and Lowther, 1952). The normal clutch is 3 eggs and due to high egg predation the season is a prolonged one with eggs being laid from mid-May till late July. The eggs are variable in ground colour from pale grey to greenish or yellow-brown with blotches and spots of dark purple-brown. Both parents nest build and share incubation duties. The newly hatched chicks can swim well and are mottled on the back with yellow buff and black.

*VOCALIZATIONS

Their contact or flight calls are peculiarly grating and dry creeky sounds and are repeated as they fly around, varying in pitch and duration. They sometimes call in December and January on their wintering grounds. Nesting colonies are reported to be very noisy, uttering harsh 'kreek-kreek' calls (Bates and Lowther, 1952).

frogs and tadpoles, 20 per cent small fishes, 20 per cent crustaceans and 53.3 per cent insects including damsel flies and dragonflies (Mukherjee, *JBNHS,* 1975). They often hunt in groups, working their way systematically upwind beating low over the marsh or lake surface, and then circling around to the windward end and starting again. During the summer months they call

---

White-winged Black Tern *Chlidonias leucopterus* (Temminck)
C. leucoptera of Dillon Ripley's 'Revised Synopsis'

Plate **16**
Illus. **61**
Map **195**

DESCRIPTION

Body length 22-24 cms.
Wingspan 64-68 cms.
Wing length 19.1-22 cms. (Stuart Baker)
Bill length 25-28 mm. (males), 23-26 mm. (females) (Cramp *et al.)*

In the field it looks about the same size as the Whiskered Tern (*Chlidonias hybridus*) and in winter plumage is difficult to distinguish, though the latter (*hybridus*) has a slightly longer bill and less contrasting darker remiges. White-winged Black Terns have a more distinct black spot behind the eye, black crown and nape. They average slightly smaller in size than *C. hybridus*. First winter birds are quite different to first winter Whiskered Terns with obvious dark primaries, contrasting with pale medium and greater wing coverts and a darker grey mantle. In full breeding plumage it is a strikingly different tern with all-white tail and rump,

jet black body including the whole of the head, neck and belly and the wings with dark silvery grey flight feathers, merging gradually to pure white on the leading edge of the wing coverts. *C. hybridus* by contrast, has an all-grey mantle and rump and the upper surface of the wings is uniformly grey without any pure white on the wing coverts. When viewed from underneath *C. leucopterus* has conspicuously black under-wing coverts and white remiges shading to grey on the wingtips. *C. hybridus* by contrast has all-white under-wing coverts with dark grey secondaries and primaries when viewed from beneath, as well as the contrastingly white cheek patch. In breeding plumage the bill is dark blackish crimson and the legs and feet are bright orange-red. In winter plumage the black on the body, head and neck entirely disappears. The bill becomes brownish black and the nape, and only a small area behind the eye is streaked black. The under-wing coverts are also pure white. Its tail is not deeply forked and is more square ended when compared with *C. hybridus* (author obs. Haleji lake). Also in winter plumage the whiter rump of *leucopterus*, slightly shorter bill and more extensive white area around back of neck, help to distinguish this species from *C. hybridus*.

HABITAT, DISTRIBUTION AND STATUS

It is a fresh water tern which associates in winter with the Whiskered Tern (*Chlidonias hybridus*) and so has probably been overlooked. It mainly occurs as a rare winter visitor to the north-east coast of India and the 'Handbook' (Ali and Ripley, 1969) cites only 3 records for the whole western seaboard of the subcontinent, including two in Jasdan, Saurashtra in Gujarat state. In Pakistan it has recently been discovered as a regular spring passage migrant with specimens in partial or full breeding plumage being sighted (J. Vieillard, pers. comm.) on Manchar lake on 2 March 1970, and 5 or 6 being seen on Haleji lake in Thatta district from late April till early May in each year between 1977 and 1985 (Roberts, 1978 and 1981). The latest dates seen on Haleji 29 May and earliest date 18 February (John Walmsley, pers. comm., 1972). On 14 May 2 were seen (author) on a brackish pool alongside Ghizri creek about 1 mile from the open sea, indicating that birds might have wintered further south along the coast of India. Status a RARE spring passage migrant.

HABITS

Similar to the Whiskered Tern, they feed on fresh water pools and lakes with slow wing beats and graceful buoyant flight, dipping down suddenly to the surface to pick insects, rarely if ever entering the water. They hunt systematically beating upwind over a small pool or inlet circling around to the windward again and again, traversing up against the wind. They are commonly seen doing this in company with *C. hybridus*. Their food presumably comprises largely aquatic insects and their larvae, especially Odonata and water

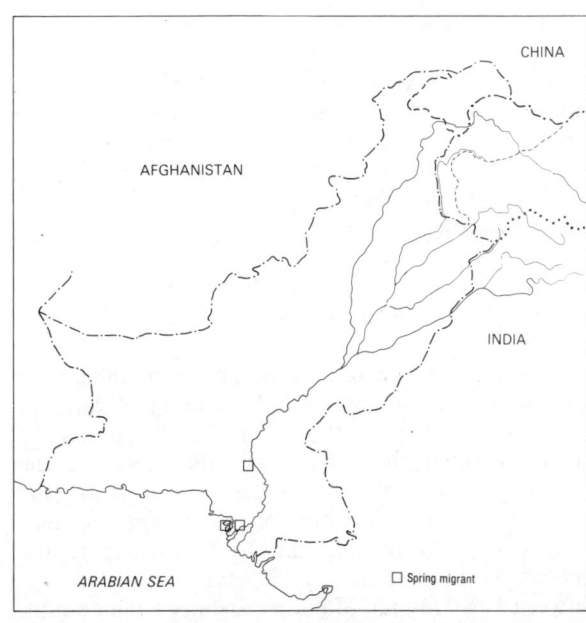

Map 195: White-winged Black Tern *Chlidonias leucopterus*

beetles. In the USSR they feed primarily upon insects including Coleoptera, Gryllidae and Diptera also less commonly small fish and amphibia and Araneae (Dement'ev and Gladkov, 1951).

BREEDING BIOLOGY

It nests extra-limitally in the Black Sea and Caspian Sea and the Sea of Azov extending into the steppe country as far north as Moscow. It nests colonially, building flimsy reed platforms on dead floating aquatic vegetation.

VOCALIZATIONS

Calls heard at Haleji lake were less grating than those given by *Chlidonias hybridus* and were a quickly repeated 'kick-kick-krrick'. Its call has also been rendered as 'karr' (Dement'ev and Gladkov, 1951).

Common or Brown Noddy *Anous stolidus* (Linnaeus)

DESCRIPTION

Body length 41-45 cms.
Wingspan 79-86 cms.
Wing length 27.1-30 cms. (Stuart Baker)

This tern is a smoky chocolate brown colour all over its body, head and wings with the forecrown white,

shading to grey on the mid-crown and greyish brown on the nape. The tail is both wedge shaped and indented, the central and inner two pairs of rectrices being shorter than the outer ones, and of a darker blackish brown colour. The remiges are also a darker blacker brown than the wing coverts. It has black legs and feet, with very short tarsus and the toes fully webbed. The iris is brown. There is no change in this sub-tropical species between summer and winter plumages. The sexes are alike. It is readily distinguished in flight from *Sterna anaethetus* which may occur in the same waters, by its all-brown under parts and its wedge shaped tail, the latter having a deeply forked tail with the outer rectrices conspicuously white, as well as the leading edge of the wings.

HABITAT, DISTRIBUTION AND STATUS

This is a purely pelagic or oceanic tern inhabiting sub-tropical latitudes and only coming to land to breed. It extends from the eastern Pacific, across the Indian Ocean to the mid-Atlantic and breeds on the Lakshadweep islands in the Indian Ocean. It appears to occur as a straggler, only off the extreme western coastal areas of Baluchistan. In his voyage to the Persian Gulf, Captain A. E. Butler noted a few between Pasni (on the Makran coast in extreme south-western Baluchistan) and Jhask in Iran. This was at the end of May (*Stray Feathers,* 1877*a*). W. D. Cummings also caught 5 alive at Ormara on 4 May 1901 after they had taken refuge in tamarisk bushes following a severe cyclone (Ticehurst, *JBNHS,* 1927).

One of Butler's specimens is now in the British Museum collection at Tring. During a 13 day survey all along the Baluchistan and Karachi coast from 20 January to 2 February 1984 no Noddys were noted though *Sterna anaethetus* was encountered (K. Forssgren, pers. comm.). Status VAGRANT to north-west coast of Baluchistan.

HABITS

They range widely over the oceans, being able to subsist on small fish shoaling near the surface and apparently they rarely need to plunge into the sea to catch their food prey, relying upon under-water predators which frequently induce such small fish to leap bodily out of the water.

BREEDING

There are large colonies off the north coast of Somalia and one on Mait island in the Gulf of Aden, so birds from this population may wander up to the Persian Gulf and along the Makran coast. They nest off the Somali coast in June.

FAMILY RYNCHOPIDAE

Skimmers

Indian Skimmer or Scissorbill *Rynchops albicollis* Swainson

Plate **16**
Map **196**

DESCRIPTION

Body length 38-43 cms.
Wingspan 102-114 cms.
Wing length 34.4-39.8 cms. (Stuart Baker)
Bill length 5.8-7.5 cms. (upper mandible)
    7.8-10 cms. (lower mandible) (Stuart Baker)

It is, in size, larger than the Indian River Tern (*Sterna aurantia*), being about the size of a House Crow (*Corvus splendens*). A very striking large tern-like bird, patterned in dark brown and white. The wings are extremely long and extend when at rest, well beyond the tail tip. Its short, almost square tail is white with the upper-tail coverts and inner webs of the central rectrices dark brown, matching the colour of the upper-wing surface. The head looks grotesquely large, partly an adaptation to its peculiarly deep laterally compressed bill. The head, neck and under parts are pure white except for the crown and nape which are dark blackish brown. The forecrown is white and in flight a narrow white band shows along the trailing edge of the wings. The legs and feet are

vermillion-red with short tarsi considering the size of the bird. It waddles awkwardly when walking. The iris is dark brown, the black of its crown extending just down to and around the eye. The bill is long, strongly arched downwards on the culmen, and coloured vermillion or dark orange with a yellow tip. When viewed laterally the lower mandible is blunt tipped and extends well beyond the upper, whilst both mandibles are compressed laterally into thin scissor-blades, the palate of the upper being grooved to fit the lower (See **plate 16**). There is a series of curved grooves along the side of the lower mandible (See **plate 16**) which are thought to counteract the tendency while skimming, for the bill to be pushed downwards in the water (Urban *et al.,* 1986). Juveniles have horny brown, straighter bills, more tern-like in appearance, and their upper parts are less dark brown, the feathers bearing fulvous creamy margins, giving a scale-like pattern. In flight in both adults and immatures, the under surface of the wings is wholly white and their rather slow, shallow wing beats make them very distinctive from other terns. The sexes are alike.

HABITAT, DISTRIBUTION AND STATUS

This is an exclusively Oriental species, confined to the Indian subcontinent, Burma and the Indo-Chinese archipelago. It is highly adapted in its feeding methods to a fluvial habitat, being rarely encountered away from large rivers and only frequenting inland lakes or tanks and ponds while on passage. In Pakistan it is a summer breeding visitor which has declined markedly in the past 50 or 60 years, probably because of the shrinking of its undisturbed habitat due to irrigation barrages on the main river systems and the gradual encroachment of temporary cultivation in the river bed and river island tracts, as seasonal flooding has become less of a hazard. At the turn of the century Hume noted it as tolerably common between November and February all along the Indus (Hume, *Stray Feathers*, 1873). Holmes and Wright who surveyed the same parts of the Indus during the mid 1960s never encountered any Skimmers except during the monsoon season and there are no recent winter sightings from any other part of Pakistan.

In lower Sind erratic spring migrants may be noted from mid-February through early April and they travel up the main Indus river, spreading up to its main tributaries in the Punjab from early April. Thus a party of 18 was seen on the mouth of the Indus on 23 March (Jacques Vieillard, pers. comm., 1970). Individuals on passage have been seen (author) on Haleji lake (72 kms. (45 miles) north-west of Karachi) on 20 February, 15 March and 27 March between 1980-3. A party of 3 was watched fishing on 13 April on the Ravi river near Sidhnai, Multan district (author obs.), and

12 were seen upstream of Chashma barrage on the Indus (Mianwali district) on 1 April 1970 (C.D.W. Savage, pers. comm.). Dr. Mubashir Hussain saw and photographed a flock of 14 Skimmers on 27 December 1983 at Trimmu headworks. One was seen on Rawal lake (Islamabad) on 15 April 1975 (Bruce Amstutz, pers. comm.) and Whitehead saw a flock of about 30 on the Indus near Kalabagh in April (*Ibis*, 1909). Status RARE.

HABITS

A very specialized feeder, normally obtaining all its food from the water surface, usually near the edge of river channels and lagoons, often where the water depth is only 3 or 4 cms. Characteristically they forage by flying steadily and slowly along the edge of a river or poolside embankments with the mouth open and lower mandible immersed at the tip and constantly skimming the water surface. They have a very favourable wing-loading ratio, with extremely buoyant flight so that they can maintain a fairly constant ground speed and elevation with very shallow wing beats, requiring minimum lift. In fact when winds are strong they cannot hunt successfully (author obs. Sidhnai) as there is too much lift and they then rest on some sandbar, head and body pointing into the wind. They prefer to feed at dusk and dawn and even during the hours of darkness when the hot winds and duststorms of the summer months subside and the air is relatively calm. They hunt like other surface feeding terns by systematically quartering the water surface, beating upwind and then turning round to the same starting point downwind. Their food comprises a high proportion of small surface dwelling crustacea and insect larvae and when available small fish. Stomachs examined by Stuart Baker (1929) contained only a small quantity of thick oily material. Hugh Whistler found one with its crop crammed full of small fish (in Ali and Ripley, 'Handbook', 1969).

BREEDING BIOLOGY

The pair bond is monogamous and they are colonial in breeding. Nesting usually takes place from April to June with the timing being highly dependent upon river levels and the breeding activity of other terns. H. Waite found them nesting on islands in the Chenab river in the first week of May near Chuha Mal (Gujrat district) in the 1940s (pers. comm., 1967). E. Fernando found them nesting on an island upstream of Marala barrage on the Chenab river (Sialkot district) also in early May (pers. comm., 1969). K. Eates found colonies varying from 20 to 40 pairs in late April on islands near old Ali Wahan and Jherruck on the Indus river in 1933 and 1936 (MS notes). Normally both

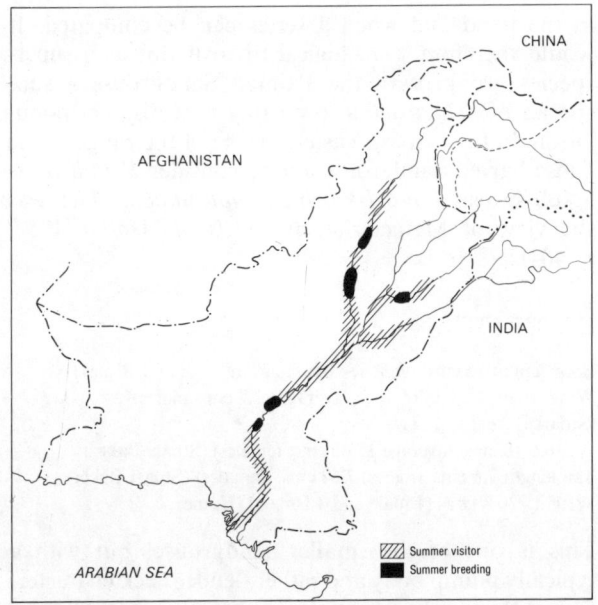

Map 196: Indian Skimmer or Scissorbill *Rynchops albicollis*

pairs start making nest scrapes a week or so before selecting a final site and the nest itself is a bare hollow in the sand, unlined and in an open exposed situation. Often several nests of the same species are located within a few metres of each other (10 metres, Dharmakumarsinhji, 1972) but slightly apart from the other tern colonies. They nest on such islands in association with *Sterna aurantia*, *Sterna acuticauda* and *Glareola lactea*. The eggs are very variable in ground colour from salmon buff to greenish stone and greyish white, blotched and streaked with dark brown and some blue-grey under-markings (Eates, MS notes). The female does most of the incubating but the male does relieve her frequently (Dharmakumarsinhji, 1972). Both parents wet the eggs and chicks with water from their breast feathers (Lowther, 1949). The incubation period has not been recorded but is about 21 days in the African Skimmer (*Rynchops flavirostris*) (Urban *et al.*, 1986). The normal clutch is 3 eggs but occasionally 4 or even 5 are found (Lowther, 1949) and clutches average larger in size in this species than the other sympatrically breeding river terns. The downy chicks

are yellowish grey or sandy coloured with scattered darker brown blotches on the crown, back and wings and thus blend well with the sandy ground. They will crouch flat, selecting hollows or footprints in the sand when their parents give warning cries. E. N. Lowther noted a Skimmer presenting a fish to another individual on the ground, when incubation in a breeding colony was well advanced (op. cit., 1949), so this indicates that fish presentation is important in the pair bond relationship.

VOCALIZATIONS

The contact calls of this bird are relatively grating and harsh and have been rendered as a nasal 'kap-kap' (Whitehead, 1909). When disturbed at their nesting colony they utter continuous twittering cries (Whistler, *Popular Handbook*, 1949). Stuart Baker describes the call as a shrill chattering scream ('Fauna Series', 1929). Compared with *Sterna aurantia* they are much less vocal.

ORDER **PTEROCLIDIFORMES**
FAMILY PTEROCLIDIDAE
Sandgrouse

Close-barred Sandgrouse *Pterocles lichtensteinii* Temminck

Plate **17**
Map **197**

Lichtenstein's Sandgrouse *Pterocles lichtensteinii arabicus* of Stuart Baker

Arabian Close-barred Sandgrouse *Pterocles indicus arabicus* Neumann of Ali and Ripley's 'Handbook'

TAXONOMY

De Voous in his 'Recent List' (*Ibis*, 1977) implies that *Pterocles indicus* (Gmelin) the Painted Sandgrouse is usually treated as a separate species, and this is following Stuart Baker (*Game Birds of India*, Vol. 2, 1921 and in his 'Fauna Series', Vol. 5, 1928), who treats *Pterocles lichtensteinii* as a separate species from *Pterocles indicus* the Painted Sandgrouse, in both volumes. S. Dillon Ripley in the 'Revised Synopsis'. (1982), and Ali and Ripley in their 'Handbook' (Vol. 3, 1969), by contrast consider the Close-barred Sandgrouse as a sub-species of the Painted Sandgrouse (*Pterocles indicus arabicus*).

As the following accounts will show, in Pakistan there is as yet no evidence that the ranges of the Close-barred and the Painted Sandgrouse overlap, but the author's field experience suggests that they are remarkably similar in their ecological requirements, voice and peculiarly crepuscular drinking habits. The males of each species can be clearly separated in the field and the hand, whilst females are probably only separable

in the hand and when a series can be compared. It would therefore seem logical to treat this as a superspecies group, with the Painted Sandgrouse a subspecies of *lichtensteinii*, occurring as a disjunct population on the extreme eastern fringe of the range of the Close-barred Sandgrouse and to consider *P. indicus* as a sub-species *Pterocles lichtensteinii indicus*. This was the view of Meinertzhagen (*Birds of Arabia*, 1954, p. 464).

DESCRIPTION

Body length 26 cms. (Ogilvie-Grant) 27 cms. (Ali and Ripley)
Wing length 17.1-18.66 cms. (average 17.9 cms. males)
(Stuart Baker)
16.6-18 cms. (average 17.65 cms. females) (Stuart Baker)
Tail length 7.6 cms. (males) 7.62 cms. (females) (Stuart Baker)
Weight 226.8 gms. (1 male and 1 female) (Hume)

This is one of the smaller sandgrouse, but with a typically plump body and rather slender neck characteristic of the family. The male is ochraceous buff all over, finely barred with black, with the wing coverts and the nape a paler creamy white, closely barred with black. Across the mid-crown there is a thick black band running from above each eye and another black band running down either side of the forecrown but not extending to the immediate forecrown region which remains white. The tail is pointed but relatively short without the central rectrices extended into a pintail. The mid-breast region is deeper ochraceous yellow with two encircling blackish bands, the upper one more brown, the lower blacker. The outer secondaries and flight feathers of the wing coverts are tipped ochraceous. The under-wing coverts and axillaries in flight show pale whitish grey. The bill is red-brown or dirty red and noticeably more slender than in other Sandgrouse species. The legs are fully feathered along the front on the tarsus whilst the feet and hind part of the tarsi are brownish yellow or orange-yellow. As in all the Sandgrouse of the genus *Pterocles,* the hind toe is vestigial. The iris is red-brown with dull yellow naked skin surrounding the eye.

Females are like males, but lack the chestnut breast band with blackish bars across the breast, noticeable in the male and their heads and necks are more finely streaked and spotted without the black and white bars across the crown, so conspicuous in the male. Like the male the chin and throat are creamy white or whitish, finely streaked and barred with black whereas in the Painted Sandgrouse male, the chin and throat are a rich ochraceous buff unstreaked. Females of the Close-barred Sandgrouse are paler than females of *P. indicus* particularly on the head, neck and wing shoulder and they have only 14 tail feathers whereas

P. indicus has 16. Furthermore the naked portion of their legs and feet are orange-brown or yellow-brown whilst the feet of *P. indicus* are more olive yellow.

Juveniles are like females without the prominent black and white bars on the crown or thick bars on the breast. Ticehurst and Meinertzhagen who examined and compared specimens from Sind and Saudi Arabia could find no distinct differences between the two populations (Ticehurst, *Ibis,* 1923, p. 647). Status locally FREQUENT.

HABITAT, DISTRIBUTION AND STATUS

This sandgrouse is very closely associated with a stony or rocky type of scrub desert in areas of low hills, a habitat which it shares to some extent with other sandgrouse, but it is never found in rolling sand hill country nor in flat clay salt pans, where other sandgrouse regularly consort, and this characteristic is so well recognized, that local hunters and villagers know it by the name of 'hill' or 'mountain' sandgrouse ('Pahari but-teetar'). It also avoids higher altitude mountain plateau regions, being absent from northern Baluchistan.

It extends in range across the hilly or rocky portions of the Sahara eastwards through Sudan, Ethiopia and extending southwards to Kenya (C. Vaurie, 1965) and eastwards through southern Baluchistan in the Makran and the hilly tracts of Sind, but not on the east side of the Indus river. In the 1960s it was not uncommon in the low rocky hills just north and north-west of Karachi around the outskirts of Malir and around Khadeji falls (author obs.). It is still widespread in the Kirthar range and westwards through the Hab valley and Pab hill range and throughout the central regions of the Makran but Christison considered it rare further to the north-west in the Chagai (Christison, *Ibis,* 1941). There are specimens collected from Panjgur, and Kolwas district in southern Baluchistan and from Ormara on the Makran coast (Ticehurst, *JBNHS,* 1926c, and Brit. Mus. coll.). It has been recorded also as far north as Kharan. Wherever it occurs it is resident. Status locally FREQUENT.

HABITS

Not as highly gregarious as the other sandrouse species, but sharing similar habits with the Painted Sandgrouse. Typically they are encountered in small parties of 3 to 4 birds, or in pairs during the breeding season, which extends over most of the summer months. They are rather distinctive in a number of other respects in that when flushed they rarely fly very far and can easily be marked down and with circumspection approached quite closely. On foot in open ground, I know of no other sandgrouse which can be

Map 197: Lichtenstein's or Close-barred Sandgrouse *Pterocles lichtensteinii* and Painted Sandrouse *Pterocles indicus*

approached so easily. Perhaps their cryptic plumage and the broken nature of the terrain which they favour renders them less conspicuous and vulnerable to predators. Their other unique habit, shared with the Painted Sandgrouse (*P. indicus*), is in only flighting to traditional drinking places after dusk has fallen and continuing until a full hour after darkness. Most other Sandgrouse habitually drink quite late in the morning, or if twice a day, again in the late afternoon. While camping at Karchat in the Kirthar National Park pairs and parties could clearly be heard calling as they flew over our tent for one hour after darkness. These habits have been noted by other observers both in Pakistan (Ticehurst, *Ibis,* 1923.) and elsewhere in Africa and the Middle East (Meinertzhagen, 1954, and Cramp *et al.,* 1985).

Their diet comprises almost entirely of seeds. Crops of birds shot by Ticehurst in the Kirthar were crammed full of the seeds of the xerophytic Acacia (*Prosopis spicigera*) (*Ibis*, 1923, p. 647). In Arabia, Acacia seeds were also recorded as their favourite food and when these were exhausted *Cassia* and Asphodel were consumed (Meinertzhagen, 1954.). Gallaher also recorded the seeds of *Asphodelus tenuifolius* in the crop of a female collected in Oman (Gallaher, *Jour. of Oman Studies,* 1977).

BREEDING BIOLOGY

The breeding season is rather extended as is typical of the Pteroclididae, depending upon local rainfall and food availability, but this species breeds mostly in the hot summer months in Pakistan. Dr. M.H. Rizvi found a nest near Karchat (Dadu district) in the first week of June, the very hottest season in the whole year for that area. He was able to approach the incubating female close enough to take cine film which was kindly shown to the author. At that time there were 3 eggs and the nest was a mere hollow scraped by the bird's breast in the soil, but it was slightly sheltered or shaded in the lea of a bush. The female bird whilst sitting, panted with bill open and enhanced evaporative cooling by a rapid gular flutter. It is certain that both parents incubate very closely and do not leave the eggs exposed for long periods during daylight hours as the heat of the sun would be sufficient to kill the developing embryo. The male has well developed incubation patches and as in other sandgrouse species, incubates during the night, the female by day (G. L. McLean in Campbell and Lack, 1984 and Meinertzhagen in Landsborough Thomson, 1964).

K. Eates found several nests in the Hab valley around Khar, and on the Surkhan escarpment just west of Landhi. The complete clutch seems to be 3 eggs, less commonly only 2 and he found this number in nests on 4 May, 20 May and 26 May in different years. The eggs are rather elongated ovals of a salmon buff or pink ground colour, heavily blotched and spotted with reddish brown and pale purple under-markings forming a ring or zone at one end (K. Eates, MS notes unpublished). It seems highly probable that both parent birds bring water on their belly feathers to cool the eggs and also to give drinks by feather sucking and regurgitation to their chicks after hatching, as this has been observed for other sandgrouse species (Meinertzhagen, in Landsborough Thomson, 1964). The chicks are well insulated at birth from the hot sun by thick filamentous down and can feed themselves.

VOCALIZATIONS

When flushed, birds utter a rather low pitched rapid and guttural call almost like a clucking. Tape recordings were stolen, but field notes at the time described the call, probably an alarm call, as 'geg-geg-geg'.

Flighting birds going to drink also have typically lower pitched and more grating calls than other sandgrouse species, but they are bi-syllabic and similar to other sandgrouse in their rather liquid tones 'gwittoo gwittoo'. I find Ticehurst's description of their call as like a Sparrow's 'chirrup', quoted by other writers (Ali and Ripley, 1969) rather puzzling and have not heard this call (Ticehurst, 1923). Generally when flighting their calls are not repeated so rapidly and are briefer than for example, those of *P. exustus*.

Painted Sandgrouse *Pterocles indicus* (Gmelin)

Plate **17**
Map **198**

Formerly *Pterocles fasciatus* Gray

See taxonomic position of *P. indicus* under account of *Pterocles lichtensteinii*

DESCRIPTION

Body length 28 cms. (Ali and Ripley),
    26.7-28.6 cms. (Stuart Baker)
Wing length 15.8-18.4 cms. (males),
    15.2-17.5 cms. (females) Stuart Baker
Tail length 8-10.1 cms. (males) Stuart Baker,
    8.2-9.5 cms. (females) (Hume)
Weight 198-226 gms. (males) (Jerdon),
    180-191.3 gms. (females) (Hume)

This is another small Sandgrouse, averaging only slightly larger in size than *Pterocles lichtensteinii*. It is very similar in appearance to *P. lichtensteinii*, but the male differs in several easily seen respects. Whilst the upper parts are closely barred with black and dark brown, the whole of the breast, throat, cheeks and a broad collar around the hind neck or upper mantle, is plain yellow ochre without any barring. In *lichtensteinii* this throat and breast area is spotted and streaked on the hind neck and cross-barred on the breast. The lower breast bears two broad dark bands, the top one maroon or chocolate, the lower more blackish and separated by a broader band of pale yellow-buff, lighter in tone than the ochraceous upper breast. This pattern is similar to that of the breast of *lichtensteinii*, but in the latter species the dark bands are much narrower and both are more blackish, the upper one less noticeably maroon. The forecrown is white with a very broad black band in front of the eye and across the top of the forecrown. In *lichtensteinii* this black band slopes backwards in a 'Vee', but in the Painted Sandgrouse it is straight and does not slope back towards the eye. The rest of the belly, lower back and tail are cross-barred black. Specimens from Pakistan are definitely a paler ochraceous colour, drabber and closer in body tones to specimens of *lichtensteinii* when compared with specimens from peninsular India.

Females lack the black coronal band on the forecrown and the ochraceous throat and breast of the male. The whole of their throat, neck and breast is finely cross-barred against a paler buff background, except for the immediate area of the chin and upper throat which is unstreaked. When compared with the closely similar female of the Close-barred Sandgrouse, the back is a warmer darker buff or chestnut buff tone, and usually the upper throat of *lichtensteinii* is speckled or finely streaked darker. Juveniles are like females but more closely and finely streaked and cross-barred.

In both sexes the bill is rather more slender than in other Sandgrouse species, and is coloured dull red whilst the legs and feet are feathered down the front of the tarsus, but a yellowish olive or greeny grey colour on the bare parts in contrast to the orange-brown colour of the feet of *lichtensteinii*. However this feature is not constant as some museum specimens have the feet colour described as dingy brown. Both sexes have 16 tail feathers in contrast to *lichtensteinii* which has only 14 tail feathers. The iris is brown.

HABITAT, DISTRIBUTION AND STATUS

Typically it is found in areas with scattered thorn bushes, *Acacia modesta* and *Ziziphus nummularia* and *Capparis aphylla*, avoiding cultivation and keeping to a stony substrate, but in low hilly tracts, not in the higher mountains.

The Painted Sandgrouse is a truly Oriental faunal species occurring as an isolated or disjunct population well to the north and east of where the Close-barred Sandgrouse occurs, with no overlap. It extends through the dry hilly and rocky areas of north central India, Rajasthan and down through peninsular India as far south as Karnataka (Mysore) and eastwards to Bihar (Ali and Riply, 'Handbook', 1969). In Pakistan it does not occur anywhere south of the Punjab Salt Range and immediate foothills nor west of the Indus in Sind where the Close-barred Sandgrouse occupies a very similar habitat in Lasbela and Sind hilly tracts.

This is rather a rare and local Sandgrouse in Pakistan, being confined to low hilly areas and stony rocky escarpments in the Punjab Salt Range, northwards to Attock and again on the west side of the Indus around the low hills around Kohat, the lower Kurram valley and even as far as the foothills flanking the Koh-i-Safed range (Col. Dastagir, pers. comm., 1968). Whitehead found it rare around Kohat but records one of a pair shot at Shinwari (1,160 metres/ 3,700 feet) (Whitehead, *JBNHS,* 1911, p. 968). Major Barton also recorded small numbers in the Buner foothills east of Mardan in the early part of the century and sent a skin to the Society (F.J.H. Barton, *JBNHS,* 1902, p. 606). H. Waite considered them resident in fair numbers in stony broken ground at the foot of the Salt Range and also in the hills behind (Waite, *JBNHS,* 1948, p. 112) and the author also encountered them at the foot of the hills north of Khushab, when 5 or 6 birds flighted over on two consecutive evenings at dusk (March, 1982). Status locally FREQUENT.

HABITS

Very similar to the Close-barred Sandgrouse (*Pterocles lichtensteinii*), indicating their close phylogenetic relationship. They are not as gregarious during the day and while feeding, as other sandgrouse species, and are usually encountered only in twos and threes, except when coming to traditional drinking places and even here, in Pakistan do not congregate in large numbers.

Their diet is mostly hard seeds, predominantly those of grasses but they will also eat young shoots (Dharmakumarsinhji, *Birds of Saurashtra*, 1954). Hume examining the crops of shot birds, found only hard seeds (in Stuart Baker, *Game Birds of India*, Vol. 2, 1921). But Stuart Baker also mentions termites as being eaten in certain seasons, (probably when abundant after summer rains) (Stuart Baker, op. cit., Vol. 2, 1921).

Like the Close-barred Sandgrouse they rely on crypsis for protection from predators and often crouch until almost stepped upon. When flushed they do not fly far and their trajectory, though quite strong and fast is fairly close the ground, birds usually pitching down suddenly within 200 or 300 metres where they can be marked down. This is in sharp contrast for example, to the behaviour of *Pterocles orientalis* which takes to flight before it can be approached within about 100 metres (author obs.).

They also share with *P. lichtensteinii* the habit, untypical of other Sandgrouse, in only drinking once, after dusk has fallen, when they flight to traditional drinking places. Here they typically land some distance from water and crouch motionless for some time before approaching to drink.

Stuart Baker also refers to another rather unusual trait, in that this Sandgrouse has been noted consorting regularly to traditional places for dust bathing (G.O. Allen in Stuart Baker, 1921, op. cit.).

BREEDING BIOLOGY

In peninsular India, Stuart Baker stated that eggs have been found in every month of the year, but that in heavy rainfall periods, particularly during the monsoon, it largely ceases to breed (Stuart Baker, 1928).

Dharmakumarsinhji (1954) in Saurashtra (Kutch, India) considered that most breeding occurred in the spring and early summer from February up to May with a few breeding after the monsoon in late August up to October. In the Punjab Salt Range, H. Waite found most nests from March to May and occasionally found nests in June and early October (Waite, *JBNHS*, 1948).

The nest, typical of the whole family, is a mere scrape hollowed out by the bird's breast, with a few bits of grass placed in the bowl, but one nest found by Waite (*JBNHS*, 1925) had a 'neatly made pad of grass' in the hollow. The full clutch is normally 3, occasionally 2 eggs and they are quite richly coloured, being pale cream to salmon buff in ground colour, sparingly spotted with purple-red and secondary blotches of lavendar or reddish grey (Stuart Baker, 1928). The female incubates by day and the male by night. H. Waite (*JBNHS*, 1925) shot a female off the nest on 16 March during daytime. Other details of breeding have not been recorded, but the young are able to feed themselves from hatching, being nidifugous and no doubt protected by both parent birds until they are fledged and able to fly independently to drinking places.

*VOCALIZATIONS

Like all Sandgrouse, birds flighting to drink call to each other but not as incessantly as in other species. Their calls are low pitched rather guttural and similar in tone to the calls of *P. lichtensteinii*. Birds tape recorded near Katha Saghral at the foot of the Salt Range, gave rather grating not very far carrying calls as they flew overhead 'qutchikuh-qutchikuh' or 'ko-khwi-chot' rapidly enunciated, the last two notes rising then falling and more vehement. This same flight call is rendered by Salim Ali (1969) as 'chirik-chirik'. When flushed they give very similar alarm calls to *P. lichtensteinii*, a croaking low pitched 'geg-geg-geg'.

Coronetted Sandgrouse *Pterocles coronatus* Lichtenstein
Crowned Sandgrouse

Plate **17**
Map **198**

DESCRIPTION
Body length 28 cms.
Wingspan 65-73 cms.
Tail 12-13 cms.
Weight 240.97 gms. (Stuart Baker, and 'Game Birds', 1921).

Like all the Sandgrouse this species has a very short stout bill, a short thick neck and full rounded breast. This is another small sandgrouse without elongated

central tail feathers. The male is the only sandgrouse with the whole of the lower breast and belly pinkish isabelline brown without any markings. The rather similar male Spotted Sandgrouse has an oval black patch between the thighs. Its face pattern is also distinctive with a rufous brown crown encircled around the hind neck by a blue-grey band. The cheeks are ochraceous becoming more rufous isabelline in a collar around the lower neck. There is a narrow black bib in the region of the chin and two curving black lines on either side of the forecrown extending from the lores to above the eye but not touching the eye. The forecrown and forepart of the face is whitish giving more contrast to these black patches. The wing coverts and mantle are beautifully marbled with darker sandy brown and paler cream. The flight feathers are dark brown tipped creamy, and the tarsus is covered anteriorly in pinkish buff feathers, the hind part of the leg and the toes being bare and a dirty grey colour. The short bill is blue-grey and the iris dark brown. The female has the same conspicuous ochre yellow cheeks and sides of the neck, but the whole of the crown, nape and back is finely barred and flecked with dark brown. The upper breast is also closely cross-barred with grey and dark brown, the lower belly being more pinkish brown with some darker spots along the flanks. In both sexes the central tail feathers are slightly elongated so that the tail tapers to a point but lacks the fine pintail of species such as *P. exustus* or *P. senegallus*. The tail feathers are cross-barred in the female and pinkish brown in the male. In flight, from the underside the belly is ochraceous, and the wing coverts look pinkish white, much paler than the buffy yellow underwing coverts of *P. senegallus* or blackish of *P. indicus*.

HABITAT, DISTRIBUTION AND STATUS

This sandgrouse seems to be an inhabitant of really barren and windswept stretches of desert, more arid than those regions favoured by *Pterocles exustus*. It is comparatively local in distribution being confined to regions west of the Indus, which probably indicates its Ethiopian faunal origins. There are some local migratory movements in winter time down to the foothills and broad flat valleys between the hill ranges of Sind Kohistan and a marked passage has been noted through the Bolan pass down to the Sibi plain in early winter (Williams and Williams, 1929). It is however, largely sedentary, subject to local migrations according to food availability. Its stronghold is in the wide clay flats and sand dune tracts of southern Baluchistan from Khuzdar in Kalat and down, through the Makran (Hoston, cited by Ticehurst, *JBNHS*, Vol. 32, 1927). It also occurs in the extreme south-western corner of the Chagai in the sand dune desert north of Koh-i-Sultan range (Christison, *Ibis*, 1941) and along the regions bordering Afghanistan further north, around Chaman, where it has been found breeding (Williams and Williams, *JBNHS*, 1929). In Sind it can generally be encountered (author obs.) in the wide flat valleys between the hill ranges of Surjana, Sumbak and Kirthar in Dadu district. Flocks come to drink in the pools of the Baran river, particularly near the Surjana pass. After good monsoon rains small parties come to drink from seepage springs and pools on the Moach plateau south of the Super-highway between Karachi and Hyderabad. A permanent spring near Manjand village 77 kilometres (48 miles) north of Hyderabad in Dadu district, was visited by flocks totalling between two and three hundred in March 1985. Surprisingly, they have not been recorded anywhere in the Punjab including Cholistan, nor in the Tharparkar desert regions of Sind. Where it does occur it can be encountered in small parties varying from 5 to 15 birds and appears to be quite plentiful. In Sind Kohistan they are often found feeding in the same localities as *Pterocles exustus* and *Pterocles senegallus*. Status FREQUENT to COMMON.

HABITS

They are gregarious like others of the family and have regular drinking places to which they flight, starting to arrive around 8.30 a.m. This sandgrouse does not come to drink again in the evening at least during the winter months (author's observations at two known traditional drinking places near Karchat, and Manjand in Sind Kohistan). This agrees with observations made by Whitaker (*Birds of Tunisia*, 1905) regarding the species in north Africa and Meinertzhagen in Oman and Saudi Arabia (op. cit, 1954). Ali and Ripley (Vol. 3, 1969) claim that it also comes to drink in the evening and this warrants further confirmation. Birds coming to the Karchah Chashma in Dadu district on 20 March were noted still arriving 3½ hours after sunrise (10 a.m.) and many individuals walked right into the water partly submerging their breasts which suggests that

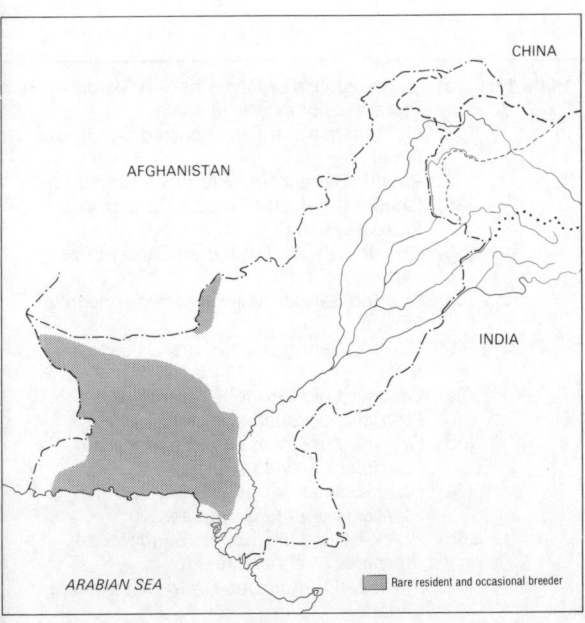

Map 198: Crowned or Coronetted Sandgrouse *Pterocles coronatus*

Rare resident and occasional breeder

they may already have started breeding as this action is regularly used to wet and cool eggs as well as provide water to their chicks. They are believed to travel long distances, up to 64 or 80 km. (40 or 50 miles) between these drinking places and their daytime foraging areas. They have quite an upright carriage on the ground, like a Grey Partridge (*Francolinus pondicerianus*) and drink by submerging their heads and then raising them with opened bill pointed skywards. When coming to drink at Manjand they flew in groups of 5 or 6 birds, with occasional flocks of up to 20 or 30 birds, and were very cautious about approaching the drinking place, settling some 3 or 400 metres away upon first arrival. Their diet consists almost entirely of seeds including ephemereal grasses (*Bromus tectorum*) and the nuts of halophytic Chenopodiacae such as *Suaeda fruticosa*. In Saudi Arabia stomachs were found containing 'a variety of hard seeds' (Meinertzhagen, 1954). Also birds shot just after sunrise near Khinjar lake on 22 June had their crops full of seeds (Ticehurst, *Ibis*, 1923).

BREEDING BIOLOGY

In Pakistan it breeds in Baluchistan in the windswept sand dune tracts near Chaman and between Gulistan and Saranan (Williams and Williams, 1929). It must undoubtedly breed in southern Baluchistan around Khuzdar, Turbat and Noakandi as birds have been seen in these places throughout the summer months also around Koh-i-Sultan and the Kacha range in the Chagai (Christison, *Ibis*, 1941). The breeding season is probably quite extended as Ticehurst shot a female near Khinjar lake (Thatta district) on 22 June with a fully developed egg in its oviduct (*Ibis*, 1923). Nests with eggs were found by Barnes on 30 April near Chaman, and another one on 27 May 1908, in the same area (Barnes, *Stray Feathers*, 1880 and Stuart Baker, *Game Birds of India*, 1921). Williams (1929) found nests in May and June between Saranan and Chaman. No attempt at lining is made, the eggs being laid in a bare hollow scraped by the parent bird's rotating breast in the sand, often in a very open exposed situation. The normal clutch is 3 eggs, but sometimes 2 are laid. They are broad ellipses in shape, of a glossy close texture, and a creamy white to pale limestone ground colour with dark brown spots and streaks of yellow-brown and sometimes under-markings of pale purple (Williams and Williams, 1929). K. Eates found one nest in Sind close to the Baran river near Jangri. It contained 3 eggs on 12 May 1948, this was on a bare stony slope (MS notes). Both sexes share incubation duties and it is believed that males bring water to their incubating mates and that both sexes give water to their chicks by feather wetting as well as by regurgitation. The male incubates by night and the female by day and the period of incubation has not been

recorded but is probably just over a full four weeks (G.L. McLean in Campbell and Lack, 1984). The nidifugous chicks can feed themselves from the time of hatching but depend for water on their parents, who bring this daily on their soaked belly feathers.

*VOCALIZATIONS

Their flight calls are quite distinctive being rather higher pitched than *P. exustus* and their bi-syllabic calls 'quittoo-quittoo' are more rapidly elided and higher in pitch than the two-noted calls of *P. senegallus* and are interspersed with shorter calls 'quit-quit-quit-quidu-ke-quidu' which are less grating than those of *P. exustus*. Like most sandgrouse the call is very far carrying, often being audible before the birds are in sight. In my experience they are just as vocal and noisy when coming in to drink as other sandgrouse species, despite what other authors have written (Meinertzhagen, 1954 and Ali and Ripley, 1960).

Plate 17: (1a) Lichtenstein's or Close-barred Sandgrouse *Pterocles lichtensteinii*, male.
(1b) Lichtenstein's or Close-barred Sandgrouse, female.
(2) Painted Sandgrouse *Pterocles indicus*, male.
(3a) Chestnut-bellied or Indian Sandgrouse *Pterocles exustus*, female.
(3b) Chestnut-bellied or Indian Sandgrouse, male.
(4a) Spotted Sandgrouse *Pterocles senegallus*, female.
(4b) Spotted Sandgrouse *Pterocles senegallus*, male.
(5a) Crowned or Coronetted Sandgrouse *Pterocles coronatus*, female.
(5b) Crowned or Coronetted Sandgrouse *Pterocles coronatus*, male.
(6a) Black-bellied or Imperial Sandgrouse *Pterocles orientalis*, female.
(6b) Black-bellied or Imperial Sandgrouse *Pterocles orientalis*, male.
(7a) Pin-tailed Sandgrouse *Pterocles alchata*, female.
(7b) Pin-tailed Sandgrouse *Pterocles alchata*, male in summer plumage.

Plate 17

(1a)

(2)

(1b)

(3a)

(3b)

(4a)

(4b)

(5a)

(5b)

(6a)

(6b)

(7a)

(7b)

facing p. 408

Plate 18

(1)
(2)
(3)
(4)
(5)
(6)
(7)
(8)
(9)
(10)
(11)
(12)
(13)
(14)

facing p. 409

Spotted Sandgrouse *Pterocles senegallus* (Linnaeus)

Plate **17**
Map **199**

DESCRIPTION

Body length 33-36 cms.
Wingspan 71-80 cms.
Central tail feathers 12.7-16.7 cms.

Larger than either the Chestnut-bellied Sandgrouse *P. exustus*, or the Coronetted Sandgrouse *P. coronatus*, its name is derived from the spotted plumage of the female. Both sexes have the central pair of rectrices elongated into a pintail and when viewed from underneath in flight the pale creamy unmarked wing lining and sandy buff body at once separate it from *P. exustus*. An elliptical longitudinal black bar in the mid-belly region and between the thighs often shows up well in flight. Males lack any encircling breast bands and have pale isabelline crown and mantle with a rich ochraceous orange throat, ear coverts and hind collar, framed above and below by bluish grey supercilium and neck collar. The tail is isabelline brown with the point black and the rectrices are black with buff tips.

The female has a similar ochre yellow cheek and throat patch, but her crown, mantle and rump are boldly spotted with black against a sandy buff background. The upper breast is also spotted and the wing coverts are chestnut with parallel rows of black spots. The tail is cross-barred and black at the tip. Like the male, there is a narrow black band extending from between the thighs to the vent. The bill is bluish grey or pale leaden and the toes and naked rear of the tarsus, greyish blue. The forepart of the tarsus is feathered. The iris is dark brown. Juvenile birds have the central black belly streak but the rest of the belly and lower breast is covered by small crescentic black marks. The upper parts are sandy yellow with blackish crescentic bars and streaks on the feathers. Their central tail feathers are also not so elongated and are barred to the tips (Meinertzhagen, 1954).

HABITAT, DISTRIBUTION AND STATUS

An abundant, but erratically occurring, winter visitor to the main deserts of Cholistan in Punjab and Thar in Sind. It avoids very hilly or high mountainous country and is absent from northern Baluchistan or the Punjab Salt Range, but it is quite adapted to a stony terrain and low foothills. It occurs sporadically throughout the foothill areas of the Makran as well as along the foothill areas of Sind where barren plains extend westwards up to the Indus river. Thus, it occurs around Manchar lake and Manjand in Dadu district, the Sibi desert, especially around Jacobabad and in Larkana ditrict around Nagar Parkar it is numerous in December and January, as well as in the deserts east of Dharki and Khairpur in northern Sind and appears to be sparsely resident in the region. In Baluchistan it is mainly a passage migrant but odd birds have been shot in December and January around Quetta and Said Hamid to the north-west (Meinertzhagen, 1920). Christison noted a huge spring and autumn passage through Chagai of birds which breed in southern Afghanistan (Christison, *Ibis,* 1941). Hotson encountered it in May at Nihing in the Makran and at Surab in August in southern Kalat (cited in Ticehurst, 1927*b*), and a few pairs undoubtedly breed in southern Baluchistan as well as in northern Sind. Status FREQUENT to COMMON.

HABITS

It is perhaps more gregarious than other Sandgrouse, congregating sometimes in flocks of over 100 when on migration or when flighting in to suitable drinking places. They come to drink later than other species, slightly later than *P. coronatus* at Manjand (author obs.), arriving well after sunrise around 9 a.m. in summer and 10.30 a.m. in the cold weather. They flight in very directly, some birds actually landing on the water in their eagerness to drink (observed Lal Sohanran, February 8.30 a.m.). In the hot weather, birds are also reliably reported as coming to drink again in the

Plate 18:
(1) Orange-breasted Green Pigeon *Treron bicincta*, male.
(2) Red-collared or Red Turtle Dove *Streptopelia tranquebarica*, both sexes.
(3) Eurasian Collared Dove or Indian Ring Dove *Streptopelia decaocto*, both sexes.
(4) Laughing, or Little Brown Dove *Streptopelia senegalensis*, both sexes.
(5) Eastern Rufous or Oriental Turtle Dove *Streptopelia orientalis*, both sexes.
(6) Spotted or Chinese Dove *Streptopelia chinensis*, both sexes.
(7) Western Turtle Dove *Streptopelia turtur*, both sexes.
(8) Common Wood-pigeon *Columba palumbus casiotis* Eastern sub-species, both sexes.
(9) Hill Pigeon or Eastern Rock Dove *Columba rupestris*, both sexes.
(10) Snow Pigeon *Columba leuconota*, both sexes.
(11) Wedge-tailed Green Pigeon *Treron sphenura*, male.
(12) Yellow-footed Green Pigeon *Treron phoenicoptera*, male.
(13) Eastern or yellow-eyed Stock Dove *Columba rupestris*, both sexes.
(14) Rock Pigeon or Dove *Columba livia*, both sexes.

Map 199: Spotted Sandgrouse *Pterocles senegallus*

in 1933 and found it breeding in four localities, in northern Sind, at Boh-jo-Pat near Reti in Sukkur district, near Khenju in Bugti territory, near Sui in Bugti (formerly Jacobabad) district, and near Kotri in Hyderabad district. In all these locations, nests were on clay flats or barren dry sand and clay soil areas and they avoided stony or gravelly regions. Nests were found in June and July. The full clutch is 3 eggs and unincubated eggs were found on 10 June and 11 July also 2 downy young just hatched on 10 June. No nest is made, a hollow being scratched in the bare soil. The eggs were described as similar to those of *Pterocles coronatus*, having a pale creamy or buff background, blotched with reddish brown and grey-brown under-markings (K. Eates, MS notes unpublished). Ticehurst (*Ibis*, 1923) records oviduct eggs taken from females shot on 24 February near Karachi, and another shot on 20 March near Shikarpur (formerly Jacobabad district), and eggs said to be of this species collected from near Kotri, Hyderabad district on 15 May. Both sexes share incubation duties the female sitting during daylight hours and the male at night. Both parents are said to bring water to their chicks which is given by regurgitation and also the male brings water to his incubating mate, which is fed to her in the same way (Meinertzhagen, 1954). Injury feigning by both sexes when disturbed off the nest has been noted in Iraq (Meinertzhagen, *op. cit.*)

late evening (Capt. Pitman in Stuart Baker, *Game Birds*, Vol. 2, 1921 and Meinertzhagen, 1954). Meinertz-hagen noted that birds coming to drink in the morning already had their crops full and that small weed seeds rather than grain from cultivated crops, constituted their entire stomach contents. Stuart Baker notes that though seeds, particularly *Polygonum* form the bulk of stomach contents examined, insect remains were also found in some stomachs (Stuart Baker, 1921).

Hume noted that males and females came to drink in separate flocks. They are noisy and call frequently when flighting.

BREEDING BIOLOGY

Kenneth Eates (MS notes unpublished) discovered the first authentic nest of this species on the subcontinent

*VOCALIZATIONS

They have louder calls than some sandgrouse and are noisy when flocking together and flying. Their bi-syllabic flight calls appear to be more distinctly enunciated and higher pitched than the calls of *P. coronatus* and their call has been rendered as 'quiddle-cuddle-cuddle' (Meinertzhagen, 1954). In mixed flocks of Coronetted and Spotted Sandgrouse, tape recorded while flying round their waterhole at Manjand in Dadu district, the slower louder calls of *senegallus* were easily recognized and sounded like 'wittoo-wittoo'.

Chestnut-bellied Sandgrouse *Pterocles exustus* Temminck
Common Indian Sandgrouse (Ali and Ripley's 'Handbook')

Plate **17**
Map **200**

DESCRIPTION

Body length 28-32 cms.
Wingspan 65-70 cms.
Tail 9 cms. females, 10-14 cms. males,

This is an elegant little Sandgrouse, slightly smaller than the Spotted (*P. senegallus*) and considerably smaller than the Imperial Sandgrouse (*P. orientalis*).

Hardly any larger than the Collared Dove (*Streptopelia decaocto*) when seen together, it has an even smaller head than *Streptopelia,* with a deep rounded breast and long narrow pintail. In flight the wings are narrow and pointed with rapid wing beats, showing dark chocolate on the under-wing coverts with blackish flight feathers in contrast to the whitish under-wing coverts of *Pterocles coronatus* and *Pterocles alchata*.

The central tail feathers are black tipped in males and finely cross-barred in females. In both sexes the lower belly shades to deep sepia brown, becoming paler chocolate brown in the region of the under-tail coverts. The crown and mantle is sandy brown, the back being marbled or blotched darker grey-brown and paler yellow-buff, whilst the wing shoulder is yellow-buff circled by five narrow parallel curving black lines formed to the tips of the wing coverts. In males the cheeks and throat are rich orange-buff or ochraceous whilst the upper breast is more pinkish clay coloured, circled in mid-breast region by a narrow black line. Females besides having shorter pintails are spotted on the crown and closely cross-barred on the mantle and rump. There is a double circlet of black in the mid-breast region with parallel rows of dark spots on the throat, and the lower belly is pale chestnut closely barred with black. In both sexes the very short but powerful down-curved bill is leaden grey and the iris is dark brown. The orbital skin is greenish yellow and the legs and feet are dark brown with the forepart of the tarsi densely feathered with pinkish buff. Females, like males have the cheeks and throat ochraceous yellow.

HABITAT, DISTRIBUTION AND STATUS

This is the commonest and most widespread sandgrouse in Pakistan, being largely resident throughout the major desert tracts of Thal, Cholistan and Thar, as well as in the dryer broken foothill country of the Punjab Salt Range, Indus Kohistan and the plains around Dera Ismail Khan and Kohat in the NWFP in

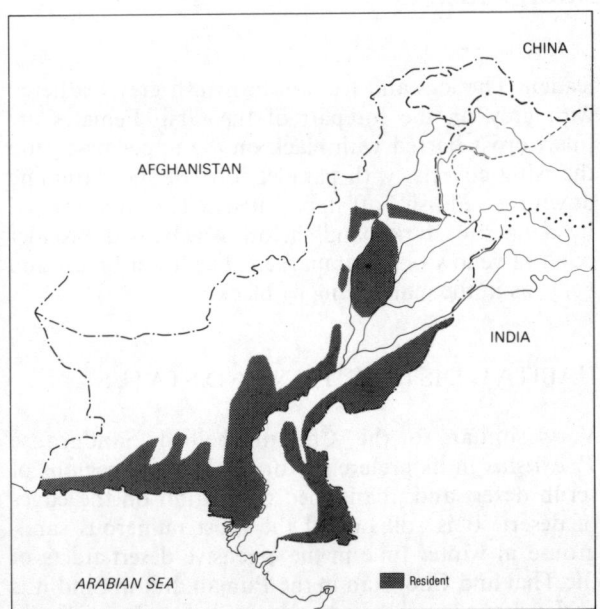

Map 200: Chestnut-bellied or Indian Sandgrouse *Pterocles exustus*

which latter place it is uncommon (Whitehead, 1911). In Baluchistan it occurs in lowland areas along the Makran coast and also on the Sibi desert plain. It is absent from the high plateau regions of northern and central Baluchistan. It is subject to erratic local migratory movements according to food supply and rainfall patterns, but has been found breeding in every district of Sind (Eates, MS notes) as well as in Cholistan. It also breeds at the foot of the Salt Range in Mianwali and Jhelum districts (Waite, 1948). It occurs in the more arid dry farming ('Barani' cultivation) tracts as well as in pure sand dune desert and is more adapted to cultivation, fallow fields on the edge of desert and a thorn scrub biotope than either *P. senegallus* or *P. coronatus*, both of which are confined to more desolate and arid desert regions. Occasionally small flocks can still be seen flying over the periphery of Karachi air field and Malir Cantonment. Status COMMON to ABUNDANT.

HABITS

They are gregarious throughout the year except during the breeding season, usually feeding in small flocks of 5 or 6 up to 20 or 30 scattered birds. They fly in to traditional drinking places, each morning often travelling over 80 kilometres (50 miles) from their feeding grounds. Deep or extensive bodies of water are usually avoided, their preference being for shallow seepages around springs. One such locality on a tributary of the Baran near Allah Dad in the hilly tracts of Sind, has been visited since people have lived in the region. Birds come in from all directions of the compass starting to arrive some 1½ to 2 hours after sunrise, at around 8 a.m. and continuing to fly in up to 10.45 a.m. with a total of some 400 to 500 birds coming to drink (1984) at the same spot. Usually they settle some distance from the water and are both extremely nervous and reluctant to approach until numbers have built up. When they finally fly in with a loud clatter of wings, they wade into the water, cock up their tails and ruffle their breast feathers to wet them to the maximum besides dipping down their heads to drink.

Their food consists of cultivated grains gleaned from stubbles such as *Phaseolus* and *Pennisetum*, young shoots of mustard *Brassica campestris,* as well as seeds of grasses such as *Eleusine aristata, Panicum* also *Tephrosia tennuis, Indigofera cardifolia* and *Pseudanthispteria hispida* (Ticehurst, 1923 and Faruqi, Bump *et al., JBNHS,* 1960).

BREEDING BIOLOGY

The breeding season can be very extended though most clutches are laid during April in Sind. H. Waite in the Punjab found nests from March to May, usually on stony gravel fans at the foot of the Salt Range, but he also found eggs in June and one clutch of 3 eggs on 15 October. (Waite, *JBNHS*, 1948). Tony Gaston found a downy chick estimated to be 2 or 3 days old at the end of January 1976. This was in the low stony hills just west of Sehwan in Dadu district. The parents which gave a distraction display confirmed the identification of the chick (A.J. Gaston, pers. comm., 1977). K. Eates found most nests in February and October (MS notes unpublished). No nest is built, just a hollow is scraped in the bare soil or sand. Both sexes share incubation duties and the normal clutch is 3 eggs. These are rather long cylindrical shaped eggs, glossy in texture and of dark greyish yellow varying to limestone yellow in ground colour, spotted and blotched with reddish brown and with under-markings of pale violet (Eates, MS notes). The incubation period is given as 20 days (Ali and Ripley Handbook, 1969), or 23 days in captivity (G.L. McLean in Campbell and Lack, 1984), and the brooding parent will give a wing injury distraction display when incubation is well advanced as well as when the chicks are young. The downy chick is light reddish brown above, broken with white patterns and lines picked out in black. The under parts are creamy white, darker on the throat and the pattern on the head and back look like a figure of eight (Lowther, 1949). The chicks are nidifugous and able to walk about and feed themselves from the time of hatching. Both parents brood them and shelter them from the hot sun. Practically nothing has been recorded about courtship in this family, but Dharmakumarsinhji notes that the male frequently makes nest scrapes and calls to his mate for several days before the final site is chosen. He bobs his head up and down apparently to entice his mate to the nest scrape, somewhat like a Rock Pigeon (*Columba livia*) (Dharmakumarsinhji, 1972). He also observed the cock bird of one pair fly in with breast and belly feathers wet and after he walked up to the chicks and covered them, they appeared to be sucking water from his breast (Dharmakumarsinhji, op. cit.). In fact the chicks depend on such drinking water brought to them daily in the parent's soaked belly feathers until they are about 4 weeks old and can fly themselves to water.

*VOCALIZATIONS

They call both on the ground and in flight. It is a far carrying rather liquid 'kwit-kwit-kwituroh-kwituroh-kwituroh'. These calls are very short and sharp sometimes with quite a guttural tone initially, then rising to a higher pitched twittering. The females have a more grating lower pitched call and when a flock is calling in flight these blend with the higher pitched twittering calls of the males, producing a characteristic sound impossible to describe but very distinctive for this species.

Imperial Sandgrouse *Pterocles orientalis* (Linnaeus)
Black-bellied Sandgrouse

Plate 17
Map 201

DESCRIPTION

Body length 34-39 cms.
Wingspan 77-84 cms.
Tail length 10-13 cms.

One of the larger Sandgrouse, distinguished by its lack of elongated pintail and black lower belly in both sexes. In flight the under-wing coverts are white contrasting sharply with the black flight feathers and lower belly. Males have a rather blue-grey head and upper breast divided by an orange-ochraceous collar below the cheeks and broadening on the side of the neck, with a narrow black collar on the forepart of the throat extending from the base of the lower mandible and half way around the neck. The blue-grey breast is bounded below by a narrow black line, below which is a broad pinkish mushroom coloured band followed by the all-black belly. The wing coverts and mantle are mottled with dark blue-grey and ochre buff and the narrow pointed wedge shaped tail is cross-barred black. in both sexes the iris is brown and the bill dark leaden. The legs and feet are brownish grey feathered with grey on the forepart of the tarsi. Females are finely cross-barred with black on the upper parts and the wing coverts, with parallel rows of spots running down the reddish buff upper breast. This area framed by a narrow dark band, below which, is a broader band of yellow-ochre unmarked. The lower breast and belly, as in the male, being jet black.

HABITAT, DISTRIBUTION AND STATUS

Very similar to the Chestnut-bellied Sandgrouse *P. exustus* in its preference for the less arid regions of scrub desert and abandoned cultivation on the edges of desert. It is still one of the most numerous sandgrouse in winter time in the extensive desert tracts of the Thal and Cholistan in the Punjab, but in Sind it is curiously uncommon, occurring only very sporadically in the Thar desert, mainly on the borders of the former Jaisalmir state and in lower Sind. K. Eates always

Map 201: Black-bellied or Imperial Sandgrouse *Pterocles orientalis*

HABITS

Rather a shy sandgrouse and not easy to approach. Parties were noted, flying in to drink mid-March at Lal Sohanran when the weather was already quite warm, they circled around nervously and several flocks flew away without settling at all. On other occasions at the same place, however, this species has been observed hurtling into the lake with some birds landing right on the water in their eagerness to drink (Lal Sohanran at 8.30 a.m. late February). Landing directly on the water is characteristic of this very shy species and has been observed in Iraq on the Tigris (Meinertzhagen, 1954). Like its congeners it feeds mostly upon hard seeds gleaned from the ground and considerable quantities of grit were noted in the crops of specimens shot near Rahim Yar Khan (author obs.). Among identified seeds in stomachs were *Indigofera, Tephrosia, Heliotropium, Melilotus* and *Astragalus* (Ticehurst *et. al., JBNHS,* 1922c). E.M. Nicholson noted that when coming to drink they bobbed their heads to and fro like a Pigeon, walking with a rather waddling gait.

BREEDING BIOLOGY

Both sexes share incubation duties, the male by night and the female during the day. The nest consists of a hollow scraped in the bare sand or earth, usually in quite open exposed situations. 2 to 3 eggs are the normal clutch and these vary from pale creamy buff to greenish grey in ground colour with brown smudges and spots and lavendar grey under-marking. Incubation in captive birds was found to take 23-28 days (G.L. McLean in Campbell and Lack, 1984). E.M. Nicholson found small numbers coming to drink 17.6 kms. (11 miles) south of Kalat town on the Surab road near Samandar throughout July and thought that they were nesting on the Dasht-Goron plain, but failed to locate any nests (MS notes, 1946). The egg collected by F. Good from Chaman was partly incubated and found in May on the Khwajah Amran foothills near Chaman.

VOCALIZATIONS

Flight call is rather gruff, being a double 'katarr-katarr' (Stuart Baker, 1928, op. cit.), also rendered as 'tchourou-tchourou' (Heinzel, Fitter and Parslow, 1972).

encountered a few of this species each winter around Karachi and found it common around Virahwah on the eastern borders of Tharparkar district (MS notes). They occur mainly as winter visitors to Pakistan. Whitehead found them on stony plains around Lachi, Dhano and Doaba in Kohat district (*JBNHS,* 1911) and they are common on migration in the upper Kurram valley in November (Col. Dastagir, pers. comm., 1968). Major Biddulph shot a female, presumably on migration, on 19 December near Gilgit (*Stray Feathers,* 1881) and one was shot on 15 March, 1914 near Chitral town (H. Stirling, *JBNHS,* 1914, p.159). In Baluchistan they are mainly spring and autumn passage migrants during September, October and again in March and April, but a few stay to breed in southern Kalat state and in the Chagai (E.M. Nicholson, MS notes unpublished 1946, and Christison, *Ibis,* 1941). It is the common sandgrouse in winter in southern warmer parts of Baluchistan. C.H. Williams was sure it bred on the Samangli plain near Quetta where he saw pairs in the summer months (*JBNHS,* 1929) but this area is now the civil airport. He had eggs brought to him from Chaman further to the north-west by a Mr. F. Good and there was a half-fledged chick in the Quetta Museum obtained from near Quetta on 12 September (Ticehurst, 1927). Status COMMON in winter in Punjab.

Large Pin-tailed Sandgrouse *Pterocles alchata* (Linnaeus)

Plate **17**
Map **202**

DESCRIPTION

Body length 37-38 cms.
Wingspan 78-85 cms.
Central tail feathers 14-19 cms.

This is the only sandgrouse in the region in which the lower breast and belly is pure white with a broad pinkish chestnut breast band narrowly bordered by black in both sexes. They are slightly larger than the Spotted Sandgrouse (*P. senegallus*) and share with that species and the much smaller *P. exustus* a long pintail. In flight the white under-wing coverts and lower belly contrast with the black remiges and the rufous tawny breast. In the breeding season males have a black patch under the chin, the cheeks and throat being ochraceous orange. The mantle and wing coverts are an oily greenish yellow with grey tips to the feathers giving a scaled pattern. The wing coverts are chocolate chestnut on the shoulder, each rounded feather margined with white in a conspicuous pattern, like the scales of a carp. Posteriorly they are yellow-buff with black tips forming concentric black bars over the closed wing. The rump and tail is sandy yellow cross-barred black and the pintails are black tipped. The feathered forepart of the tarsus is white and the bare toes and hind part of the tarsus is olive green. The bill is dark horny brown varying to slaty blue and the iris is brown. The orbital ring of bare skin is blue. In winter plumage males are barred above like the females. Females have the upper parts yellow-buff,

finely cross-barred with black and the cheeks are rufous ochraceous with darker spots, as also the lower throat. The chin is white. The crown is barred, and the pinkish chestnut breast band is narrowly bordered by black as in the male, with the lower belly white. In winter males also lose their black chin and acquire a white throat.

HABITAT, DISTRIBUTION AND STATUS

This is a rather locally restricted sandgrouse having well defined wintering areas and migration routes outside of which it is seldom encountered. It is probably the least numerous of the species which visit Pakistan. Most of the population comprises winter migrants with a small number probably remaining to breed in the extreme south-western corner of Baluchistan and on the Afghan border around Chaman.

They are quite numerous on migration in central Baluchistan, both in the autumn during November and in the spring when they again pass through, in the first half of March. At this time they can be encountered on the plains of Said Hamid and around the Lora river (Meinertzhagen, 1920). In the Chagai in south-western Baluchistan they are common in winter especially around Zangi Nawar lake and also some pass through on spring and autumn passage (Christison, 1941). In Sind, small numbers occur throughout the winter on the Sibi desert plain north-west of Jacobabad, and there is a strong spring passage of migrants through Jacobabad in the spring. They are practically unknown elsewhere, though a very few birds may winter in the border regions of Cholistan around Yazman and Derawar (General E.H. Marden, pers. comm., 1970). It was not recorded around Kohat or Bannu by Whitehead or Magrath (*JBNHS*, Vol. 20, 1910 and *JBNHS*, Vol. 18, 1908), but at the turn of the century it was a common wintering bird on the plains to the west of Mardan (Hume *circa* 1873, quoted in Stuart Baker, 'Game Birds', Vol. 2, 1921). Most of this region is now too heavily cultivated to provide suitable habitat for any Sandgrouse. It may breed in southern Baluchistan though there are no authentic records. A Mr. Laubmann records sighting this species at Panjgur in south-western Baluchistan on 13 July 1911 and this is cited as possible evidence of breeding (Ticehurst, *JBNHS*, Vol. 31, 1928). J.S. Bogle in the N.W.F.P. had eggs, said to be of this species, brought to him by a local hunter, and on 10 June 1900 found 5 or 6 pairs of this Sandgrouse in the same spot at a place 12 miles from Peshawar where he shot a female with an oviduct egg fully formed (Bogle, *JBNHS*, Vol. 13, p. 539). Status SCARCE.

Map 202: Pin-tailed Sandgrouse *Pterocles alchata*

HABITS

This sandgrouse inhabits both arid desert tracts and bushy scrub desert and in this respect resembles *P. exustus* and *P. orientalis.* It is highly gregarious in its wintering grounds and on migration, with some evidence that nesting is also loosely colonial. It is believed to feed mainly upon grass seeds as well as small seeds of pulses and cereal grains and a small quantity of green leaves (Meinertzhagen, 1954). Specimens collected in Mesopotamia had their crops full of *Polygonum argyrocoleum* which appeared to be their favourite food (Pitman in Stuart Baker, op. cit., 1921).

When flighting into drinking places, they are said to travel in enormous flocks and to be more noisy than other sandgrouse species, coming to drink 2 to 3 hours after sunrise in the morning and in very hot weather at mid-day and in late afternoon. Birds have been noted in Iraq pitching right into the open water where they swam buoyantly like gulls (Magrath, *JBNHS*, Vol. 25, 1917).

BREEDING BIOLOGY

In captivity the incubation period is 21 to 23 days, and given as nearly 4 weeks in the wild (Dement'ev *et al.,* Vol. 2, 1951). Both sexes share incubation duties, the female doing most of the sitting during the day and the male at night, (Stuart Baker, op. cit., 1928 and Dement'ev *et al.,* 1951).

No nest is made, the eggs being laid in a bare hollow scraped in the sand. The normal clutch is 2 to 3 eggs, which are variable in ground colour and pattern, being usually creamy buff with red-brown and violet grey blotches. Some eggs are salmon pink or bright warm buff in ground colour (Meinertzhagen, 1954 and Stuart Baker, 1928). Pitman describes nests in Iraq within a few metres of each other and commonly built on patches of vegetation rather than bare earth (cited in Stuart Baker, 1921). Incubating birds when disturbed have been described as injury feigning. Also as deliberately wetting their belly feathers before returning to the nest (Meinertzhagen, 1954 and Stuart Baker, 1928). The newly hatched chicks are fully down covered and precocial.

VOCALIZATIONS

Their flight call is a rapid bi-syllabic 'catarr' or 'quettarr' (Stuart Baker, 'Game Birds', Vol. 2, 1921) also described as 'ka-kra' (Meinertzhagen 1954) and is lower pitched and more guttural than similar calls of other sandgrouse species (author obs. birds encountered on the Crau southern France, October 1984). Their alarm note, given by both sexes, has been described as 'twoi-twoi-twoi' (Stuart Baker, op. cit., 1928).

ORDER COLUMBIFORMES

Pigeons and Doves

FAMILY COLUMBIDAE

Rock Dove, Blue Rock Pigeon *Columba livia* Gmelin

Plate **18**
Map **203**

Columba livia neglecta Northern race

Columba livia intermedia Plains race

DESCRIPTION

Body length 33 cms.
Wingspan 78-84 cms.
Wing length 22.6-23.9 cms. (12 males).
    21.5-23.2 cms. (17 females) (Paludan)
Tail 11-12.7 cms.

As the ancestor of most domesticated pigeons, and a cliff dwelling species which has readily adapted to man-made structures such as city buildings, this is one of the most familiar birds to towns people. They are relatively small headed and plump bodied birds with a broad square tail and a rather weak bill having a swollen fleshy cere at the base. This species interbreeds so readily with feral domestic pigeons that it is possible to see considerable variation in wild populations in the more settled areas. The true wild Rock Dove is slaty blue-grey all over with a broad ashy terminal tail bar and two conspicuous blackish bars encircling the wing coverts. The rump is generally paler grey than the tail and mantle and the sides of the neck are iridescent green, merging to purple-bronze lower down the neck. The legs and feet are crimson to cyclamen pink and the swollen cere is white. The iris is orange. The hill population, found in all the more arid inner Himalayan valleys, *C. livia neglecta* is distinguished by having a much paler rump contrasting with the rest of the body colour. Both sexes are alike in plumage.

HABITAT, DISTRIBUTION AND STATUS

This pigeon is adapted to a very wide variety of habitats and being a strong flyer can travel long distances daily between its roosting and feeding areas.

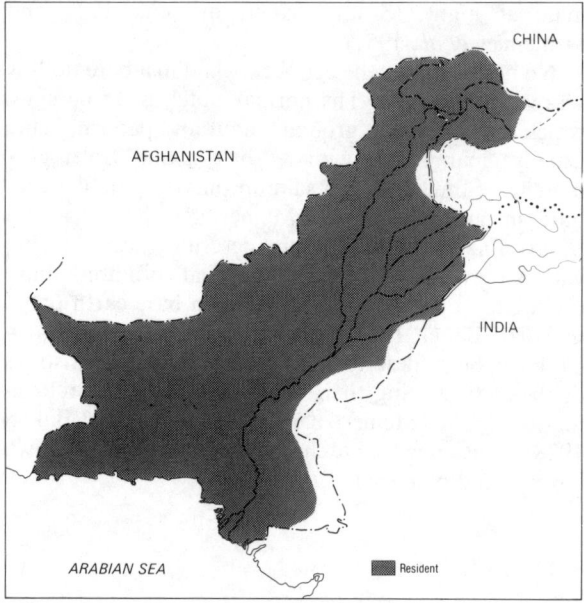

Map 203: Rock Pigeon or Dove *Columba livia*

It is quite specialized in its roosting requirements being strongly attracted to bare rock cliffs, or as a substitute to this, the iron girders of bridges and the ledges of tall concrete and brick or stone buildings. In the dry rocky hills of Baluchistan, far from human habitation it occurs commonly, as well as around the towns and cities of the densely human settled irrigated canal colonies of the Punjab. In Baluchistan it roosts on the ledges provided by the vertical tunnels giving access to 'kharezes' (subterranean irrigation channels), as well as down open wells, as in Malir. In the northern areas the bare rocky cliffs and gorges of Gilgit, Chitral and Baltistan attract considerable numbers and it is one of the more conspicuous birds in these remote northern valleys. Thus it is common from the sea cliffs along the Makran coast up to 3,353 metres (11,000 feet) elevation in the northern regions of Baltistan and Gilgit. Part of the northern population is believed to be partially migratory, descending to the Potohar plateau of northern Punjab as well as to the plains of the Indus valley and its tributaries. Certainly, flocks of migratory pigeons are common in Multan district from late December, but are usually gone again by mid-February. Status ABUNDANT.

HABITS

It tends to be gregarious in roosting and feeding, with flocks often roosting in towns and villages on such structures as warehouses, overhead concrete water towers and grain storage silos. Flocks in day time settle on stubbles where they glean grains from recently harvested crops. They are voracious feeders, taking the seeds of all cultivated crops including cereals, brassica crops (mustard and rapeseed), also leguminous crops such as lentils, chick peas and Mung beans. They also eat green shoots and any weed or grass seeds they encounter and quantities of grit. They are aggressive birds with conspecifics, quarrelling constantly over favoured roosting perches, with males attempting to court females almost every week of the year in the sedentary plains populations. The migratory flocks from the hill regions are less tolerant of human disturbance and may roost in trees in their winter quarters, which the resident population seldom does.

BREEDING BIOLOGY

Display flights around the nesting territory are characteristic with wings held upwards rather stiffly as the birds swoop along the cliff-side or around the steel trestles of a bridge. These glides are interspersed with wing clapping, this noise believed to be created as the carpal areas are struck together over their backs. On the ground males court by bobbing their head up and down with neck and breast puffed out and keeping up a continuous crooning call. The female is approached with the bill pointing downwards. Nesting of the hill populations takes place throughout the summer and early autumn months, an untidy platform of twigs being placed on some ledge or in a hole on which the two glossy broad oval eggs are laid. The incubation and fledging periods are quite short (17 days and 34-36 days respectively), enabling several broods to be reared annually. The young are fed by regurgitated 'crop milk', (made from the shed epithelial cells of the crop and gullet) and this device enables this largely granivorous bird to provide a higher protein and more easily digestible diet to the young squabs. Nesting can often be semi-colonial on suitable cliff sites, e.g. the Gaj-i-Nai in the Kirthar hills of Sind, as well as suitable cliffs in Indus Kohistan, Gilgit and the Hunza valley. Wind fretted holes in these cliffs are favoured sites for nest building in which both sexes participate. Both sexes share incubation duties. The newly hatched chick is very helpless (altricial) and sparsely covered with coarse yellow down, hence the preference of this species for nesting in cliffs or building holes when available, rather than open ledges. The nesting season is presumably less extended in the northern mountain regions, starting from May rather than late February.

*VOCALIZATIONS

On their nest cliff and roosting sites, advertising type calls frequently given, which often echo strangely in the bare rocky gorges inhabited by them in the remoter mountain regions of Pakistan. It is a low moaning 'oorh-oorh' or 'ohoor-ohoo-oor'. During display and courtship males utter the well known more rapid crooning 'goo-roo-goo-coo' or 'coo-roo-coo'.

Hill Pigeon, Eastern Rock Dove or
Turkestan Hill Pigeon *Columba rupestris* Pallas

Plate **18**
Map **204**

Crag Pigeon (in Dement'ev and Gladkov)

DESCRIPTION

Body length 33-35 cms.
Wingspan 78-85 cms.
Wing length 22.2-22.3 cms. (3 females) (Paludan)
Tail 11-13 cms.

A stout bodied pigeon very similar in size and general appearance to the Blue Rock (*Columba livia*) but chiefly differentiated by its tail pattern which consists of a dark ashy band at the tip and base separated by a broad almost white mid-portion, similar in flight to the pattern of the Snow Pigeon (*Columba leuconota*). Also its rump and lower belly are very pale grey almost white. It has the same prominent ashy black double wingbar and blue-grey dense plumage of *Columba livia* with the sides of the neck glossed green and purple, the individual feathers being bifurcated. In flight it can be distinguished from *Columba leuconota* by having only 2 not 3 black wingbars and being blue-grey on the body lacking the dark head and white neck of the latter. Its white banded tail and very white lower belly and rump distinguish it from *Columba livia*. The iris is orange and legs and feet coral red to crimson. The bill is dark slaty black with a whitish cere. Both sexes are alike. Usually the iridescent display feathers on the sides of the neck are less extensive in area than in *Columba livia*.

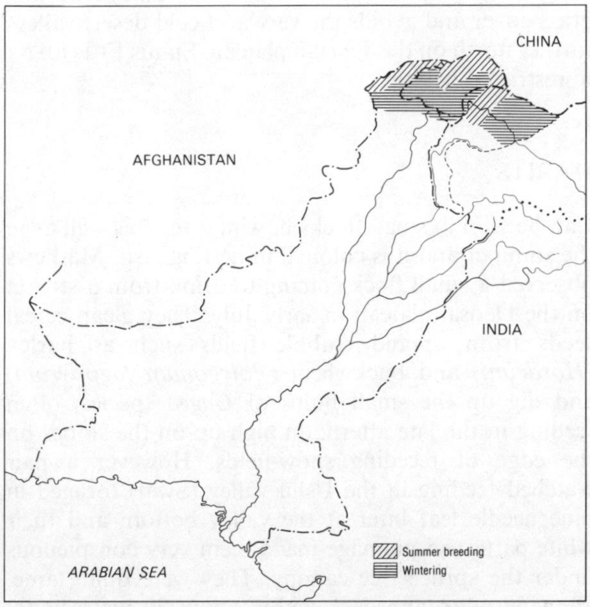

HABITAT, DISTRIBUTION AND STATUS

This pigeon is comparatively restricted in range in Pakistan to the furthest northern inner valleys of the Karakoram, Hindu Kush and Pamirs. It is a montane species, with largely sedentary habits, much more widespread and plentiful further east in central Asia and northern China. In Pakistan it occurs in northern Chitral particularly in the western part bordering Nuristan in Afghanistan, further east in valleys of Gilgit in Yasin and Hunza and the Karakoram ranges in Baltistan from about 2,000 metres (6,500 feet) in winter up to 5,500 metres (17,500 feet) during the summer months. It is often sympatric with *Columba leuconota* in these regions and in winter when it descends to the valleys it may flock with and feed alongside *Columba livia*. It appears to be ecologically allopatric in the summer breeding season from *Columba livia* being only met with in the higher alpine valleys or slopes. Status in Pakistan SCARCE.

HABITS

A gregarious species throughout the year, feeding in flocks in the terraced cultivated fields in winter and nesting colonially in suitable cliffs in the summer. Their feeding habits are similar to *Columba livia*, being mainly granivorous, supplementing their diet with green shoots and leaves and occasionally small mollusca such as land snails. Specimens shot in Nuristan, Afghanistan had crop and stomach crammed with small seeds (Paludan, 1959). They are reported to be quite tame in the vicinity of hill villages like *C. livia*. It is largely a high altitude, non-migratory bird like *C. leuconota* but more commensally adapted than the latter species.

BREEDING BIOLOGY

Salim Ali records that males have a bowing display similar to that of the Blue Rock Pigeon (*C. livia*) ('Handbook,' Vol. 3, 1969) and nothing distinctive seems to have been recorded elsewhere about the display of this pigeon which suggests that display and courtship is similar to *C. livia* (D. Goodwin, 1967). It is an early breeder, nesting in small colonies on cliffs and

Map 204: Hill Pigeon or Eastern Rock Dove *Columba rupestris*

crags (Dement'ev and Gladkov, Vol. 2, 1951). It has an extended breeding season in central Asia and China, where birds often utilize holes or niches in man-made buildings, however, unlike *C. livia* such building sites chosen are usually ruins or unoccupied by man. In the USSR it has been recorded as starting to nest as early as February (Dement'ev and Gladkov, op. cit.) with many young just fledging as late as September in north-eastern Tibet (Goodwin, op. cit.). A platform of twigs or plant stems is placed in the nesting fissure or on a suitable sheltered ledge and the clutch of 2 white

smooth shelled eggs differs in no way from *C. livia*. Both parents share incubation duties and feeding of the nestlings by bill to bill regurgitation.

VOCALIZATIONS

The voice has been variously described as similar to *Columba livia* (Hartert, 1920, and Grummt (1961) in Goodwin) and as a high pitched quickly repeated 'gut-gut-gut'.

Snow Pigeon *Columba leuconota* Vigors

Plate **18**
Map **205**

DESCRIPTION

Body length 34 cms.
Wingspan 70-80 cms.
Wing length 23.8-24.9 cms. (4 males), 23.5 cms. (1 female) (Paludan)
Tail length 12-13.7 cms.

In general body build it is like the Common Rock Dove (*C. livia*) with broad square tail, stout body and small head, but it averages slightly larger in size than *C. livia*. The whole of the neck and breast is pure white with the head contrasting sharply, being a dark blue-grey. The mantle and scapulars are brownish grey with the mid-back showing prominently white between the scapulars. The tail is dark ashy black on the base and tip, separated by a broad crescentic white bar in the middle. The upper-tail coverts and lower rump are black, merging with the black base of the tail feathers. The wings are pale slaty grey with 3 (not 2 as in *C. livia*) curving black wingbars. The outer webs of the primaries and the secondaries are blackish. The legs are rather an orange-red colour and the bill is black with a dark purplish brown cere. The iris is pale lemon yellow. The sexes are alike. In flight the white tail band and white neck and breast are conspicuous, making this pigeon unmistakable. The presence of 3 wingbars in contrast to the 2 of *Columba rupestris* and *Columba livia* is also diagnostic.

HABITAT, DISTRIBUTION AND STATUS

A resident high alpine and mountain dwelling pigeon, which in our area is more widespread in distribution and found in more southerly latitudes than *Columba rupestris*. In winter time it occasionally descends to the broader lower valleys, as low as 1,500 metres (5,000 feet), but can commonly be encountered even in mid-winter, on the snow covered slopes around 1,000 to 1,400 metres (3,050 to 4,600 feet). In Chitral it was seen in small flocks several times between 2 and 4 January

in the main valley (C. Winkler, pers. comm., 1968). In the Kaghan valley it can be seen in the spring from late April till the end of May on the edge of the tree-line at about 3,200 metres (10,500 feet) at Lalazar above Battakundi, at Burawai and on the flanks of Malika Parbat and above Saif-ul-Maluk lake (author obs.). It also occurs in Swat Kohistan with sightings on the slopes of Mankial in the Jabbah valley and in the Ushu valley, north-west of Falakser. In Baltistan it is uncommon but occurs in summer above 3,600 metres (12,000 feet) in Tseri valley and on the Deosai plateau (W.H. Mathews, 1941). In Gilgit also it is not very plentiful but is widespread in the valleys above 3,050 metres (10,000 feet) (Biddulph, 1881), and on the Khunjerab (Hunza) (author obs.). It appears to be absent from the Safed Koh range. It is adapted to fairly mesic alpine slopes with some *Juniperus* and grass cover and avoids the very arid cold desert valleys further north on the Tibetan plateau. Status FREQUENT in restricted habitat.

HABITS

Can be seen in small flocks in winter time as well as in the summer and it is colonial in nesting also. Mathews observed a small flock coming to drink from a stream on the Deosai plateau in early July. They glean cereal seeds from upland stubble fields such as barley (*Hordeum*) and buckwheat (*Polygonum fagopyrum*) and dig up the small bulbs of *Gagea* species, often feeding in the late afternoon high up on the slopes on the edge of receding snow-fields. However a pair watched feeding in the Ushu valley (Swat) foraged in pine needle leaf litter at the valley bottom and their white patterned plumage made them very conspicuous under the spruce tree canopy. They were quite tame, allowing us to approach within about 30 metres (100 feet) and appeared to be picking up seeds (obs. 20 May 2,500 metres/8,200 feet elevation). Many observers have noted how their piebald plumage blends on the rocky screes and stone strewn slopes where they settle to feed, so that they are quite invisible until flushed. If a Golden Eagle (*Aquila chrysaetos*) flies overhead they rely upon freezing to escape detection.

BREEDING BIOLOGY

They nest in small colonies in cliffs or steep rock faces, either in river gorges or on mountain slopes. In Kashmir one such colony, often visited by British ornithologists, was located on cliffs close to Sonamarg village at about 2,800 metres (9,000 feet) elevation, but many nesting sites are typically at higher elevations than this. The same site is used year after year and the nests made of stick platforms, become very fouled with droppings. The eggs are plain white and normally two in number, though Colonel Rattray found a nest containing 3 eggs. In captivity the incubation period is 17 to 19 days (T.H. Newman, 1911) and both parents share incubation duties. Captive birds were noted to have a bowing display with the hinder part of the body and tail being jerked upwards as the head bobbed forward and downward (Goodwin, op. cit.). In the wild they have a display flight similar to the Rock Pigeon (*C. livia*) with wing clapping and sailing on stiffly spread and slightly raised wings in the vicinity of the nesting colony.

VOCALIZATIONS

C. H. Mathews noted that they uttered a prolonged high pitched tremulous 'coo-ooo-oo' whenever taking

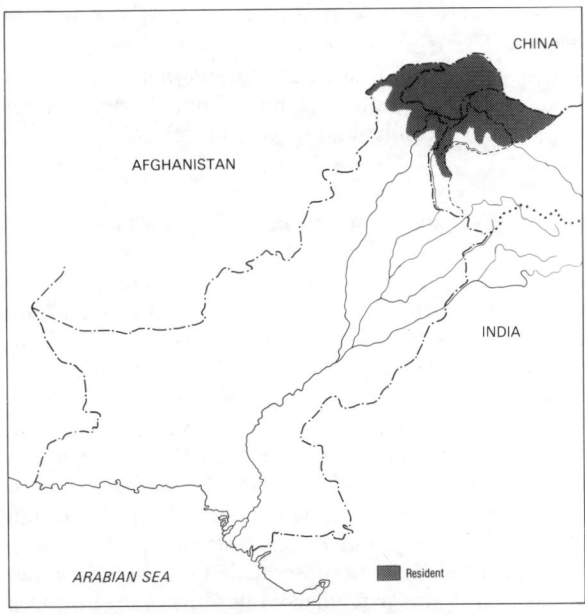

Map 205: Snow Pigeon *Columba leuconota*

flight or alighting, rather reminiscent of the calls of *Acridotheres tristis* (Mathews, *JBNHS*, 1941). This is probably a contact or flight call. During courtship they utter a repeated short croaking call 'not unlike a hiccup' (Finn in Stuart Baker, Vol. 5, 1928). Newman also describes a hiccup-like note twice repeated followed by a double noted 'kuck-kuck' sometimes followed by another 'hiccup', these being interpreted by him as advertising and nest calls (Newman, op. cit.).

Pale-backed Eastern Stock Dove or Yellow-eyed Stock Dove
Eversmann's Stock Dove *Columba eversmanni* Bonaparte

Plate **18**
Map **206**

DESCRIPTION

Body length 30 cms.
Wingspan 72-76 cms.
Wing length 19.4-20.5 cms. (7 males), 19.4 cms. (1 female) (Paludan)
Tail length 9.6-10 cms.
Weight 185-234 gms. (7 males), 183 gms. (1 female) (Paludan)

A very compact pigeon, slightly shorter tailed and smaller in overall size than the Rock Pigeon (*Columba livia*). It is similar in general body shape and colouration to *C. livia* but if examined closely will be found to have only two, or a third partly abbreviated, black bar or rather spots on the wing coverts and to have the mid-back region much paler than in the plains population of *C. livia*. Also its iris is yellow or yellow-brown, surrounded by quite a broad naked orbital ring of pale yellow skin. The head and upper breast is a distinctly mauve-pink tone and the sides of the neck are metallic emerald green. The tail has a dark blackish terminal band, and the under-wing coverts in flight are very pale grey to white, matching the mid-back region. It is smaller than the extra-limital Stock Dove (*Columba oenas*) of western Europe and its mushroom pink head contrasts with the grey head of *C. oenas* as well as the whitish mid-back region which is similar to the northern populations of *C. livia* whereas *C. oenas* has a uniformly grey mantle, back and rump. The bill is greenish slaty with the cere greenish yellow. Sometimes the eye-ring is fleshy yellow and the bill tip brown. The legs and feet are pink to pale carmine. When seen perched alongside *Streptopelia decaocto* it is about equal in body size but with a shorter squarer tail and a noticeably high crowned forehead.

HABITAT, DISTRIBUTION AND STATUS

A winter migrant only to Pakistan, being found mostly in the plains west of the Indus river or in the immediate riverain areas where it is highly gregarious in both roosting and feeding. A small population breeds in northern Afghanistan but probably most of the birds wintering in Pakistan breed in Asiatic USSR, in Turkmenia, Tashkent and the Tien Shan. It has not been recorded in Baluchistan, but Whitehead noted a definite migration in the latter part of April through the Kurram valley and Kohat district (op. cit., 1911). Dr. Ticehurst did not observe it in lower Sind but noted that it was a regular winter visitor to northern Sind (Ticehurst, 1923). The position today is much the same, large flocks occurring only in traditional localities such as the riverain tracts around the Bhegari protection embankment on the west bank of the Indus in Jacobabad district where up to 300 birds have been counted in a small area between November and mid-April over several years. During 10 years observations around southern Sind a small flock (33 birds) was only once encountered roosting in Acacia trees alongside Haleji lake(6 December 1979). Nigel Tucker saw a flock of 12 on 24 February 1986, while motoring to Khinjar lake (Thatta district) (Tucker *in litt.*, 1986.) H. Waite never encountered it in the Salt Range, nor Hugh Whistler in Rawalpindi district, but it does occur occasionally in flocks in the riverain tracts of Punjab around Jhang (Whistler (Rawalpindi district) *Ibis*, 1930 and (Jhang district) *Ibis*, 1922). The author never encountered it in Multan district during 28 years' residence, so it appears rather local and uncommon outside of northern Sind. Status SCARCE but gregarious where it occurs.

HABITS

They roost in trees and when disturbed commonly fly up into trees in sharp contrast to *Columba livia* which prefers buildings or earth cliffs and rock ledges for perching. In Jacobabad district they feed in fairly compact flocks gleaning grains and weed seeds from stubbles especially of rice fields. Flocks may number 20 to 50 birds. When disturbed they are often joined by other feeding flocks and wheel around in tight groups, swerving and twisting in unison. On Bhegari bund (embankment) a roost of over 300 was still congregating in a row of *Acacia arabica* trees in early December 1979 and on 11 April 1980. Their flight is strong and

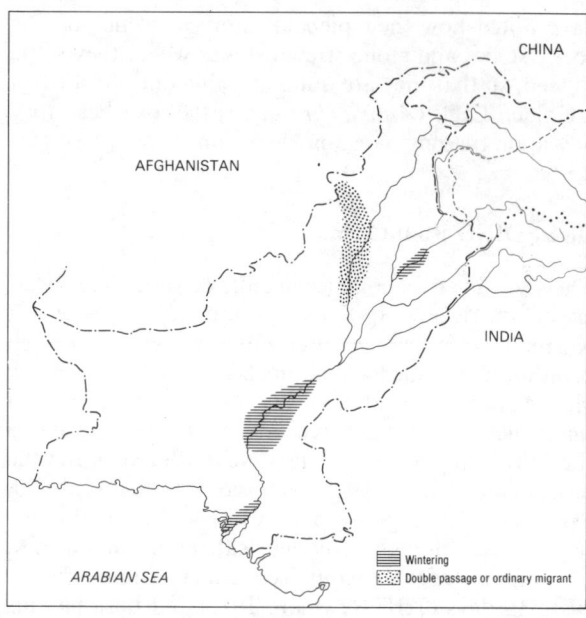

Map 206: Eastern or Yellow-eyed Stock Dove *Columba eversmanni*

rapid and when a flock takes off their wings make a high whirring noise. They appear remarkably silent outside the breeding season, flocks at roost in mid-April not uttering a single call despite many hours of patient waiting and listening. They feed on berries such as *Ziziphus* and mulberries (*Morus alba*) plucked from trees (noted by Whitehead on spring migration in Kohat), as well as grass seeds and seeds of cultivated cereals. However they feed mostly on the ground in their winter quarters.

BREEDING BIOLOGY

Like *C. oenas* it is a cavity nester. Trees with suitable nest holes are scarce in the arid steppic hill country where it breeds but Meinertzhagen found nests in hollow Willow trees (*Salix* spp.) in northern Afghanistan and in Turkmenia it nests in holes in Poplar trees (*Populus alba*) (Dement'ev and Gladkov, Vol. 2, 1951). However most breeding is in earth cliffs both in the USSR and Afghanistan. Paludan found a small colony of 20 pairs in a clay cliff and a 'rather large colony' in a gravel cliff some 10 to 15 metres high on the edge of a river (Paludan, 1959). In such cliff regions they nest colonially. Courtship involves much cooing from the vicinity of the nest site (Dement'ev and Gladkov, op. cit.). Nesting is in full swing in June, July and August. Young 4 to 6 weeks old have been taken in August, September and early October (D. Goodwin, op. cit). The eggs are plain white and presumably two is the normal clutch. Nothing more seems to have been recorded about their courtship display or voice.

A recording made by B.N. Veprintsev on the banks of the Syr-darya in Kazhakstan shows that the breeding season calls of this dove are distinctive from *Columba* *oenas*. Whereas the latter has a rather sonorous single-noted call, that of the Yellow-eyed Stock Dove consists of a few single calls followed by a series of repeated bi-syllabic calls, which can be rendered as 'quooh, quooh, quooh-cuw-gooh-cuw-gooh-cu-gooh'.

Wood Pigeon, Eastern Ring Dove or
Cushat *Columba palumbus casiotis* (Bonaparte)

Plate **18**
Map **207**

DESCRIPTION

Body length 43 cms.
Wingspan 80-86 cms.
Wing length 25.8-27 cms. (2 males),
    25.2-26.5 cms. (2 females) (Paludan)
Tail length 14-15 cms.

A large rather longer tailed and heavier bodied pigeon than *Columba livia*. It is predominantly grey in body colour, but the lower breast and belly are mushroom pink or vinaceous. The crown and nape is pale bluish grey as also the back, wings and tail. The square ended tail has a broad blackish terminal band. At rest the wings show a clear white margin around the wing shoulder which is very conspicuous in flight as a crescentic wingbar. The primaries are blackish grey edged with buff. The fleshy swollen cere is white and the base of the bill is reddish purple or orange. The naked orbital ring is magenta coloured. The legs and feet are vermillion. On the sides of the neck is a conspicuous triangular patch of creamy yellow feathers framed above and below with crimped feathers which are iridescent green and sapphire pink. In the European population this neck patch is pure white. In the hand, the neck patch of *casiotis* is quite ochraceous with pinkish tints. Females are the same in appearance but slightly smaller in size. Sub-adult birds lack the white neck patch (plumage descriptions based on live specimens from Chinji, Punjab Salt Range).

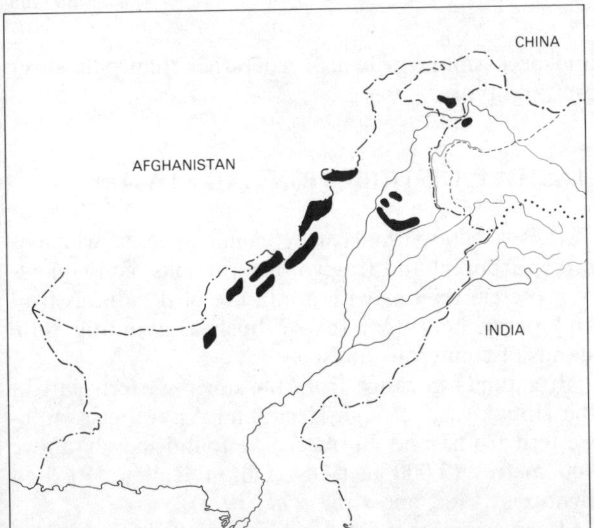

HABITAT, DISTRIBUTION AND STATUS

A resident bird, very local in distribution and uncommon, being confined to lower altitudes and hill tracts with an *Olea cuspidata* and *Acacia modesta* scrub forest cover such as in parts of the Punjab Salt Range and parts of Waziristan. It occurs sparsely in Baluchistan on the lower slopes of major mountain ranges or upland valleys where there are pockets of *Pistacia cabulica* scrub forest and *Juniperus macropoda*. It occurs in Gilgit in the higher valleys where it is distinctly uncommon. Also in the upper Kurram valley in association with *Quercus ilex* scrub forest and *Pistacia integerrima*. Four were noted bove Shalozan on 8 April (author obs.) west of Parachinar. In Baluchistan it is very sparsely distributed but has been noted on the slopes of the Chiltan range and near Wam and Philgar in the hill ranges north-east of Quetta valley. Perhaps, its greatest stronghold is the olive scrub forest of the Kala Chitta hills in Attock (formerly Campbellpur) district, northern Punjab. A flock of 80 was seen returning from the crops in the nearby lowlands, to roost at night in the hills which are clothed with *Olea cuspidata*. Status SCARCE.

HABITS

So unlike the rather bold birds which can be seen frequenting London's parks, the Wood Pigeon in Pakistan is a very shy and wary bird. Typically they perch on the tops of trees at sunrise in their hilly

Map 207: Common Wood-pigeon *Columba palumbus*

retreats to warm themselves in the first rays of the sun, flying with very swift and twisting flight down to the adjacent valleys or cultivated plains to feed on crop stubbles in the early morning and late afternoon, retreating in the middle of the day and again in the evening to the higher hills. Males start calling territorially and in advertisement from early spring. They feed on all kinds of seeds and gleaned grains from cropland, also tender shoots of mustard (*Brassica campestris*) in early winter. In the Hazar Ganji park near Quetta, I have seen them feeding upon the berries of *Pistacia integerrima* and they must also depend partly on the wild fruits of the olive *Olea cuspidata*. In both cases these weak billed birds swallow the fruit whole, and cannot break open the hard shells or get at the kernel which is the main food of some of the larger finches in the same regions. They are also reported to pick ripe mulberries and acorns when available, these two latter items being abundantly available on the lower slopes of the Safed Koh, where this pigeon can be regularly encountered.

BREEDING BIOLOGY

In the Salt Range they build their nest in *Dalbergia sissoo* or *Ziziphus jujuba* trees, and in Baluchistan in *Juniperus macropoda*. The site selected is on a horizontal branch or fork ,where the flimsy platform of twigs and stems is well supported. H. Waite (*JBNHS,*

1948) found nests with young in them on 12 April and 4 June near Choa Saidan Shah in the Salt Range and 3 nests near Sakesar further west in Mianwali district on 13 June and 9 July, so that breeding seems to be spread over most of the spring and summer months. These nests were about 6 metres above ground. Like its European cousin, both sexes share in nest building and in incubation and the normal clutch is 2 eggs which are plain white and glossy. Incubation takes 16 to 17 days and the altricial young are very helpless and weak upon hatching but develop rapidly. In display flights this Wood Pigeon usually flies upward steeply, clapping its wings loudly over its back and then planes down at a shallow angle with wings held horizontally not raised to an angle above the body as in the similar display flight of *C. livia*.

*VOCALIZATIONS

The advertising call is a loud musical 'croo-croo-ku-kuioooh' most commonly heard in the early morning and the evening, or 'krah-kwa-ka-ku-kooh', sometimes rather hoarse sounding but varying in pitch from one individual to another. This calling may be repeated continuously for 2 or 3 minutes on end. Birds have been heard calling in January, March and April and probably do so intermittently throughout the late winter and summer months.

Speckled Wood Pigeon *Columba hodgsonii* Vigors

Illus. **62**
Map **208**

DESCRIPTION

Body length 38 cms.
Wing length 70-80 cms.
Tail length 14-15 cms.

Slightly larger than the Blue Rock Pigeon (*Columba livia*) and very distinctive in colour being a dark plum colour over most of the body with pale pinkish grey neck. The head is pale grey and the feathers of the nape and hind neck are lanceolate, dark purple-black basally with silvery pink tips. The wing coverts and rump are more bluish grey with the wing quills and tail being blackish. The breast is dark reddish purple streaked with pale pink. The wing shoulders are speckled with white and the neck hackles have silvery pink margins and dark purple-brown centres. The irides are greyish white with an orbital ring of grey skin. The bill is black tipped, purple at the base. The legs and feet are usually dull green or olive brown with the claws noticeably bright yellow. Females are duller than the males with their heads more brownish grey

and necks more brownish red rather than pale silver grey and pink.

HABITAT, DISTRIBUTION AND STATUS

This is a Sino-Himalayan endemic species, which is rather arboreal and frugivorous in habits, consequently it prefers regions with a mixture of deciduous fruit and berry bearing trees or bushes, avoiding pure stands of coniferous forest.

It extends in range from the north-western part of the Himalayas, through Nepal and Assam into southwestern China and Burma being found mainly above 900 metres (3,000 feet) (S. Dillon Ripley, 'Revised Synopsis', 1982, and King *et al.*, 1975).

In our region it is very rare and highly nomadic, probably wandering widely in search of suitable fruit. In the early 1880s both Biddulph and Scully encountered them in Gilgit where they suddenly appeared in July in a forested valley at 2,400 metres (8,000 feet) elevation (Biddulph, *Stray Feathers*, Vol. 9, 1881, p. 356, and Scully, *Stray Feathers*, Vol. 10, 1881, p. 136) There have been no records since then from Gilgit, though the author was always on the lookout for them.

In Kashmir (west of our region) it is also decidedly uncommon as Bates and Lowther in 16 years never came across it (*Breeding Birds of Kashmir*, 1952). However it was recorded by other observers, including once at Sonamarg at the head of the Sind valley and again in the Wangat valley, a side tributary of the Sind valley.

In 1983, during pheasant surveys in Azad Kashmir, Kamal Islam came across small parties of this pigeon in September near Salkalla at 2,600 metres (8,600 feet) elevation. Later he again saw one or two across the other side of the Neelum valley above Khuttan (Kamal Islam, pers. comm., 1983). While carrying out pheasant surveys in the Kaghan valley a year later, Richard Grimmett and Craig Robson watched 2 birds feeding in unidentified bushes at about 3,000 metres (10,000 feet) just below the summit ridge of Kadir Gali on 18 April at a time when much of the ground was still carpeted with snow (Grimmett and Robson, pers. comm., May 1984). During an early June visit to the Machiara valley, Azad Kashmir in the same year, the author had a frustratingly short glimpse of a single bird which flew up the valley actually below the point where I was standing. Duke and Eames while surveying for Tragopaus, in June 1989, came across several on both sides of the Indus in Kandia and Palas valleys (W.P.A. Report, 1989). Status RARE seasonal visitor.

HABITS

This species is largely frugivorous and aboreal in habits and rather nomadic according to available food supply. They were observed in September clambering

Illus. 62: Speckled Wood-pigeon *Columba hodgsonii*, male.

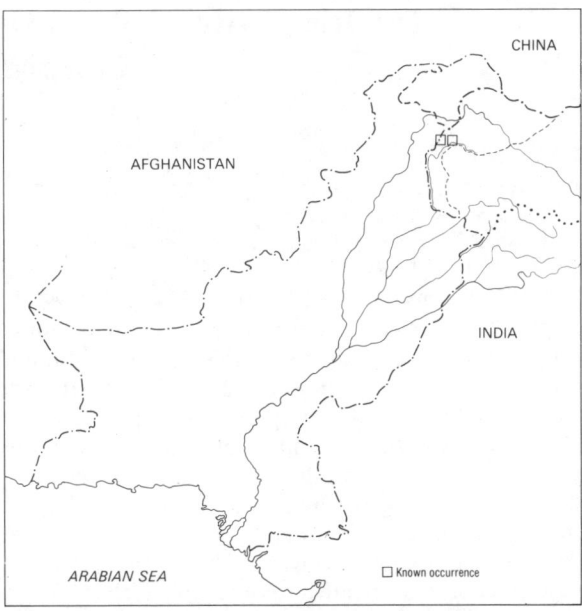

Map 208: Speckled Wood-pigeon *Columba hodgsonii*

over *Lonicera quinquelocularis* bushes feeding on the ripe black berries (Kamal Islam pers. comm., 1983). They were also seen in late September in the Machiara valley eating the acorns from the Brown Oak (*Quercus semicarpifolia*). They have been described as keeping to higher elevations often at the upper limit of the tree-line, but a flock of 4 was seen by the author in Nepal at 1,829 metres (6,000 feet) elevation on 18 March in mixed oak forest and were feeding in a *Ficus* tree. They usually forage in small groups and have the habit of freezing in the trees, where they are clambering after food, when danger approaches, relying upon escaping attention in the thick foliage.

BREEDING BIOLOGY

Said to lay only one egg and to build a typical flimsy stick nest on a horizontal fork or branch of a tree, nesting taking place mainly in May and June. Nothing has been recorded about courtship display or incubation period. However one writer refers to its noisy wing clatters which may indicate wing clapping display flights as in others of the genus.

VOCALIZATIONS

Stuart Baker decribes their call as a very deep 'whick-whrroo-whrroo', the third note more prolonged than the second and of a rather rolling or burry intonation (*Indian Pigeons and Doves*, 1913).

Indian Ring Dove, Collared Dove *Streptopelia decaocto* (Frivaldszky)
Collared Turtle Dove

Plate **18**
Map **209**

DESCRIPTION

Body length 30-32 cms.
Wing length 17.6-18 cms. (2 males),
      16.2-17.7 cms. (4 females) (Paludan)
Tail length 11.5-13.5 cms.
Weight 166-176 gms. (2 males), 156-175 gms. (4 females) (Paludan)

A pale grey-brown dove with a pale lilac pink tint to the head and neck only discernible in good light, and more earthy brownish grey back and wing coverts. The flight feathers are darker grey-brown and also the upper-tail coverts. The tail is blue-grey. The breast pinkish buff, whiter on the lower belly and under-tail coverts. Encircling the hind neck in a half collar is a narrow black band delicately edged with clear white. The legs and feet are crimson-pink and the iris is dark brown. The bill is rather more slender than typical *Columba* species, blackish in colour without any prominent paler cere. The sexes are alike. Juvenile birds lack the black neck collar and are paler grey, less pinkish on head and neck. The blue-grey tail is long and rounded at the tip with a prominent white terminal band (excluding central rectrices), visible only when the bird fans its tail upon alighting or during flight. It is a bigger plumper bird than *Streptopelia senegalensis* with which it is often sympatric.

HABITAT, DISTRIBUTION AND STATUS

This dove is essentially a plains species, avoiding rocky foothill country favoured by *Streptopelia senegalensis*. There is however a summer migration into the broader cultivated valleys of Baluchistan and the NWFP where it breeds. It is the commonest dove throughout the Punjab. It will colonize extensive desert tracts such as the Thal and Cholistan deserts and Thar desert regions where *S. senegalensis* is seldom seen. Whitehead noted a few over-wintering around Kohat and the Kurram valley with an influx of birds in the spring (Whitehead, 1911). An occasional individual can be seen in winter in Baluchistan but almost the entire population arrives in March and leaves again after breeding in October (Meinertzhagen, 1920). It has been noted (author) at Wam at 2,285 metres (7,500 feet) elevation. It has not been recorded in Gilgit and generally avoids the northern Himalayan regions. An occasional individual can be seen with *Streptopelia chinensis* in the Murree foothills in the sub-tropical pine zone. Status ABUNDANT to very ABUNDANT.

HABITS

A largely ground feeding granivorous dove which sometimes congregates in large flocks in cultivated areas, picking up seeds from threshing floors and freshly harvested crops whilst still lying in the field. It thus inflicts quite serious crop losses in parts of Sind, e.g. Sanghar district, on favoured crops such as mustard or rape-seed (*Brassica campestris*) at harvest time. In Multan they have also been noted digging up and eating newly germinated maize crops (*Zea mays*) though this habit is not widespread. Food habit studies at Faisalabad, Punjab (formerly Lyallpur) noted their fondness for mustard seed at the time of harvest and threshing (Afzal Husain and Bhalla, 1937). In Bihar, at the Pusa Agricultural Research Institute their food consisted mainly of cultivated grains including wheat, barley, rice, mustard, and linseed, with a lesser proportion of weeds (Mason and Lefroy, 1912). They roost by night and at intervals during the day in trees and bushes, usually congregating in small flocks. Generally they are more dispersed in feeding and roosting than *Columba* species, but in Sanghar district in late February flocks of over 200 have been seen. Around farmsteads they become quite tame in the presence of livestock and people.

BREEDING BIOLOGY

The display flight of this species can be seen for most of the spring and summer months. The male rises steeply

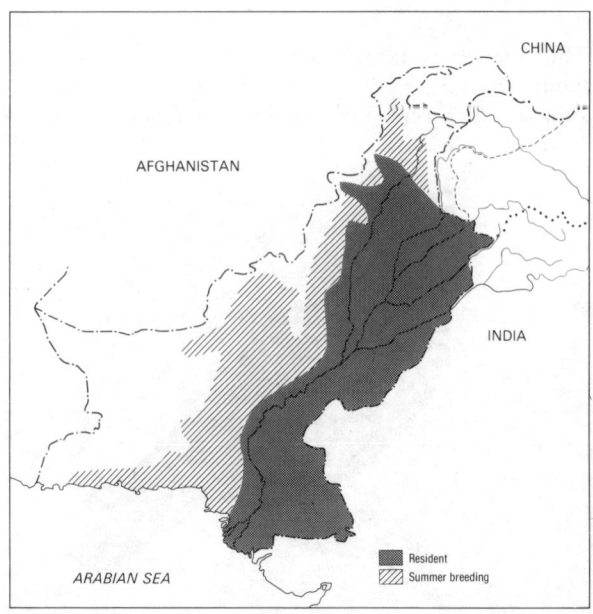

Map 209: Eurasian Collared Dove or Indian Ring Dove
Streptopelia decaocto

ith wings clapped audibly over its back and then with
il spread and wings spread horizontally, planes
own often in a sweeping semi-circle, to the accompani-
ent of display calls 'coo-coo-cuw', the last phrase
uch shorter. On the ground they court by pursuing
males, neck puffed out and head bowing with bill
ointing vertically to the ground. Nesting has been
ell studied in Rajasthan (north-west India). Nesting
from May to September with probably two peak
reeding periods in April and again post-monsoon in
ugust/September. Autopsies of males showed increas-
d spermatogenesis during March/April and again in
ugust/September and commencing from February,
ying away in October. Average weights of breeding
irds (both sexes 130 grams), and seasonal weight
easurements, indicating that some doves born in the
pring are big enough to be able to breed in the
ollowing monsoon season when 6 months old (B. D.
ana, *The Auk*, 1975). The nest is a typical dove's
imsy platform of inter-lacing twigs placed generally
ell inside a thorny bush or taller trees at heights
arying from 1.5 to 2.7 metres. The male aids in
ollecting material and nest building. The majority of
lutches comprise two eggs, plain white and smooth
lossy shelled. 62 per cent of eggs laid in the Rajasthan
tudy were during August and September and 78 per
ent of the clutches were of 2 eggs, only 2 clutches were

of 3 eggs. In the spring the average incubation period
was 18 to 19 days and during the monsoon season
incubation varied from as little as 10 to 17 days. The
nesting period to full fledging being 14 to 18 days in the
spring and 12 to 17 days during the monsoon. Only 31
per cent of eggs laid in the spring hatch, due to heavy
predation by natural predators, including *Corvus
splendens* and *Coracias bengalensis*. In the Rajasthan
study 68.7 per cent of eggs under observation were
destroyed by avian predators (Rana, 1975, op. cit.). In
Baluchistan most nesting starts from late April continuing to end of June.

*VOCALIZATIONS

When excited during agonistic encounters or display
flight it utters a plaintive rather high pitched 'quweeer-
kweeer'. Well described by Bates and Lowther as a
'wheezy peevish were-were', (*Breeding Birds of Kashmir*, 1952). The advertising or territorial song uttered
throughout the day during the summer and often in
the middle of the night, consists of a tri-syllabic 'coo-
cooo-cuh' the last syllable being shorter and lower and
the second or middle syllable being accented and
slightly higher pitched than the first, the whole phrase
lasting barely more than 1 second.

Red Turtle Dove or
Red Collared Dove *Streptopelia tranquebarica* (Hermann)

Plate **18**
Map **210**

DESCRIPTION

Body length 22-23 cms.
Wing length 13-14.4 cms. (Stuart Baker)
Tail length 8.4-9.2 cms.

ntermediate in size between the larger Collared Dove
Streptopelia decaocto) and the smaller Little Brown
Dove (*S. senegalensis*) with a relatively shorter tail and
plumper body than *S. senegalensis*. It is very distinctive
n plumage when viewed perching or on the ground,
with a blue-grey head and neck sharply divided from
he back and wing coverts by a short black collar
encircling the hind neck which is finely edged with
white. The back and wings are a reddish maroon or
purplish chestnut from which its name is derived. The
ail when spread has the outer feathers broadly tipped
white and is dark blue-grey on the central rectrices.
The flight feathers are also blackish grey and the
under-wing coverts and basal half of the tail look
conspicuously white in flight with the base of the tail
darker. The bill is black and the legs and feet are dull
red to brownish red. The iris is dark brown. Females

have slightly browner grey heads and are paler duller
more sandy brown on the upper part of the body. The
breast and flanks are pale pinkish white with the lower
belly and vent white. Their flight is generally more
swift and direct than *S. decaocto*.

HABITAT, DISTRIBUTION AND STATUS

This is an Oriental species extending to Taiwan and
the Philippines but uncommon on the Malaysian
archipelago. It is a summer migrant visitor to Pakistan
from India where it is more or less resident. A bird
ringed on 13 September 1961 in Bhavnagar, Rann of
Kutch was recovered near Karachi on 27 March 1967.
It is abundant in the Punjab plains and a few birds
straggle into the lower valleys of the NWFP but it has
not been recorded in Baluchistan. In northern Sind it
is also plentiful, but is comparatively rare in lower
Sind. In the Punjab first arrivals may be noted about
the first week of March and most have gone again by
early October but further south in Rahim Yar Khan

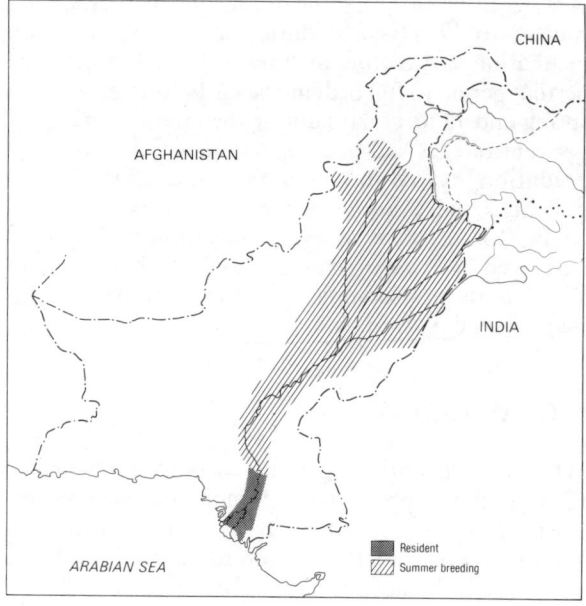

Map 210: Red Collared Dove or Turtle Dove *Streptopelia tranquebarica*

district and northern Sind stragglers may be seen in November and December. It is never seen in the environs of Karachi but a few birds may over-winter in the better wooded parts of lower Sind. In different years one was seen on 11 January near Sujawal (Thatta district) and another on 27 January near Khinjar lake. An occasional bird strays into the main Himalayan valleys and Biddulph collected specimens in Gilgit on 23 June (*Stray Feathers*, 1881). H. Waite noted first arrivals in Jhelum and Mianwali districts from the first week of April. They prefer better wooded tracts such as canal or roadside tree plant-ations and avoid extensive desert regions where *S. decaocto* occurs. When they first arrive they are often in small flocks, but they soon split up and start pair formation and breeding. Status ABUNDANT but mainly in Punjab.

HABITS

A less commensal species than *S. decaocto*, not so dependent for its food on cultivated grains. Weed and grass seeds are included in greater proportion in the diet, though it gleans cultivated grains from stubble fields and more often can be seen feeding on the ground in company with *S. decaocto*. Two specimen examined in Bihar, India had only grass and weed seeds in their crops and stomachs (Mason and Lefroy, 1912). They feed almost exclusively on the ground and will sometimes supplement their diet with buds and green shoots. Ticehurst records them as eating the seeds of maize (*Zea mays*) as well as such weeds as *Amaranthus gangeticus* (op. cit., 1923). In Pakistan they are as tame as the other doves of the Indus plain and not particularly wild or shy.

BREEDING BIOLOGY

The advertising calls of this little dove fill the air during the hot days of March and April, as birds call from the depths of leafy trees. They have display flights similar to *S. decaocto* with wing clapping and soaring followed by down gliding and on the ground or on a horizontal branch the male courts by bobbing its head rapidly up and down rather like *S. chinensis*. They nest quite high up in trees (average height 6.5 up to 8 metres above ground), making the typical flimsy platform of slender twigs. In the northern Punjab nests may be found from late April to mid-May. K. Eates in Nawabshah district of Sind found as many as 9 nests within a stretch of 400 metres between 5 and 10 May and found nests with eggs as late as 23 June and in Thatta district he took a nest on 28 August near Gujjo. The normal clutch is 2 eggs, plain white and both sexes share in nest building and incubation. The young are fed bill to bill by regurgitation and when bigger put their heads right inside their parents' throats (observed Pirawala Forest Plantation, Khanewal).

*VOCALIZATIONS

Their advertising call is a very distinctive one, unlike any other *Streptopelia* doves. It is a very rapid low pitched almost guttural song of four syllables 'cru croo-cu-coo' each lasting less than a second and rapidly followed by another four noted phrase. The first phrase is accentd and the third is very abbreviated with emphasis upon the last 'coo'. These calls are usually repeated in a run lasting half a minute or so before the bird falls silent. From a distance the calls sound like a rapid 'kurr-kurr-kurr'. Other flight or contact calls have not been noted.

Western Turtle Dove *Streptopelia turtur* (Linnaeus)

Plate **18**

DESCRIPTION

Body length 27-30 cms.
Wing length 16.9-17.3 cms. (5 males),
    16.6 cms. (2 females) (Paludan, Afghanistan)
Tail length 11.5-13 cms.
Weight 111-140 gms. (5 males),
    126-131 gms. (2 females) (Paludan, Afghanistan)

This is a rather slender bodied, long tailed dove slightly smaller than *Streptopelia orientalis* as also *S. decaocto*. It is closely similar to *S. orientalis* in the orange-rufous edged feathers with black centres, forming a delicate scale-like pattern on the mantle and wing shoulders. However it is generally a paler pink and grey colour on the rest of its body than *S. orientalis,* also the display area on the side of its neck comprises black feathers tipped almost white forming diagonal parallel broken lines. In *S. orientalis* this neck patch is distinctly blue-grey and black, and it is generally larger and more stoutly built than *S. turtur,* also *S. turtur* has a more brownish rump not blue-grey as in *S. orientalis.* When fanned the tail is blue-grey with the outer tail feathers broadly tipped white. The irides are orange and the feet crimson-red. The bill is purplish black. The naked orbital ring is reddish purple. The throat and upper breast are pinkish white shading to pinkish brown on the breast and the crown and nape are blue-grey, paler on the forecrown. The lower belly and under-tail coverts are pure white and the flight feathers are dull grey to blackish grey. Both sexes are alike in size and plumage but usually the female is a little duller.

HABITAT, DISTRIBUTION AND STATUS

This bird is a rare straggler into the dryer mountain valleys on the northern Himalayas and north-western Baluchistan. Meinertzhagen considered it a summer straggler, a specimen having been shot by Swinhoe on 7 May near Quetta (Meinertzhagen, 1920), also it has been recorded at Chaman and on the Khojak pass on 7 April and 5 May by Murray (Ticehurst, 1927). Major Biddulph considered it a very rare summer visitor to Gilgit appearing at the beginning of May (S. Biddulph, 1881). Ali and Ripley consider it only a winter visitor to Pakistan ('Handbook', 1969). Because of the difficulty of separating it in the field from *S. orientalis* it may be overlooked but it must be very rare. In several winter field trips to Gilgit and one into Baltistan the only species seen (author) was *S. orientalis.* It favours arid thorn bush and scrub dotted areas, avoiding tall tree forest. Status a straggling passage migrant or OCCASIONAL VAGRANT in winter.

HABITS, BREEDING BIOLOGY

Well studied in western Europe. Breeding is extra-limital in Pakistan. It is largely a ground feeding species gleaning grains from cultivated crops as well as weed and grass seeds and young leaves of fodder crops. It has been recorded in Europe eating pine cone seeds and caterpillars also. Nesting is in trees or shrubs, laying the usual two white eggs, and it has a short incubation period (13-14 days) and even shorter fledging period (11-12 days), so that it is often double brooded.

VOCALIZATIONS

Its advertising call is a purring 'turr-turrr' from which it gets its name.

Eastern Rufous Turtle Dove or
Oriental Turtle Dove *Streptopelia orientalis* (Latham)

Plate **18**
Map **211**

DESCRIPTION

Body length 33 cms.
Wingspan 52-60 cms. (Scully and Davison)
Wing length 19.0 cms-20.7 cms. (7 males),
    18.5 cms (1 female) (Paludan)
Tail length 13-14.5 cms.
Weight 165-221 gms. (4 males), 204 gms. (1 female) (Paludan)

This is a larger and darker, more richly coloured version of *Streptopelia turtur*. The back and wing shoulders are beautifully patterned with dark blackish centres and broad chestnut red margins meeting in a point. The crown and rump are blue-grey with the lower neck and upper mantle gradually becoming vinaceous or purple brown. The display patch on the sides of the neck comprises blue-grey and black feathers in a chequer-board pattern in contrast to the paler silvery white and black of the same neck pattern in *S. turtur*. The flanks are pale blue-grey, upper breast is pale pinky buff and under-tail coverts pure white. The mid-breast is darker vinaceous brown. The iris is orange with a purple orbital ring of bare skin. The cere is purple-pink, horny brown on the bill tip and the legs and feet are carmine. In flight it shows more white on the belly than *S. decaocto*. The sexes are alike, but juveniles are somewhat drabber and paler without the chequer-board neck patch.

HABITAT, DISTRIBUTION AND STATUS

Partly migratory in the winter down to the foothill regions of the NWFP and the main vale of Peshawar, around Rawalpindi, Sialkot and eastwards to Lahore. Small numbers regularly winter in Changa Manga forest plantation 67 kilometres (42 miles) south-west of Lahore (a flock of 30 on 14 March 1980). In the northern Himalayan areas it is a very widespread summer breeding visitor with a small population over-wintering in the lower valleys. Thus it has been seen in early December around Gilgit town and in the lower Kaghan valley in late February. On autumn migration it gathers into small flocks and is keenly sought after by sportsmen for shooting, around Mardan in September when it first arrives. In the Murree hill range flocks pass through on passage, lingering through most of September in the forest. The return passage can again be noted from late April to early May. A few stay to nest in the Margalla hills and the Murree hills where they stick to forest and are very shy and secretive. In Gilgit they nest up to 3,350 metres (11,000 feet) even in *Betula utilis* forest and scrub Blue Pine (*Pinus walli-chiana*) and have been found nesting at 3,048 metres (10,000 feet) in *Picea morinda* spruce forest at the base of the Babusar pass in Chilas district. They are particularly common in *Quercus ilex* (Holly Oak) in lower Chitral, Dir and Swat Kohistan where they nest from 2,134 metres (7,000 feet) up to 3,200 metres (10,500 feet). In Baltistan they inhabit all the wider valleys, particularly in the vicinity of orchards and cultivation as far as the Shyok valley. It is a passage migrant through Kohat and breeds on the Safed Koh range and on the Samana hill range (C.H. Whitehead, *JBNHS*, 1907 and 1911). In the lower warmer parts of its summer range it is sympatric with *Streptopelia chinensis*. It has not been recorded in Baluchistan or Waziristan. Status COMMON.

HABITS

Gregarious in winter and on passage, largely solitary or in breeding pairs in its summer haunts. It is a rather shy and secretive bird flying up into a tree where it remains concealed in thick foliage whenever it is disturbed. In the outer Himalayan ranges it is typically associated with mixed deciduous and coniferous forest, whereas in the dryer inner mountain ranges of Gilgit and Baltistan it is associated more with the outskirts of villages and orchard groves. It feeds largely on the ground, gleaning pine cone seeds, acorns as well as the flower buds and seeds of various herbs, the berries of *Viburnum nervosum*, *Berberis* and *Lonicera*. In its winter quarters it feeds more in cropland, gleaning seeds and grains from stubbles.

BREEDING BIOLOGY

They make quite a substantial pad for a nest, often lined with wiry roots, and the normal clutch is two eggs which are plain white in colour. Both sexes assist in nest building, incubation and feeding the young. Nests are usually well concealed in forests, where there is a thicket of younger conifer trees growing up and they are often located quite low down between two and three metres from the ground. The normal season is from late spring to early summer. In Kashmir, Bates and Lowther found the earliest nest on 18 April and the latest in the latter part of June (op. cit. 1952). Prior to nesting, males frequently make display flights which involve soaring over the mountain tree tops with much wing clapping followed by a graceful curving descent on stiffly held wings. Females are courted by running forward with neck puffed out and bobbing the head forward and downward. When the head is bobbed downwards, the neck patches are everted. The head bobbing is rather more rapid than in *S. decaocto* or *S. senegalensis*.

*VOCALIZATIONS

The voice is very reminiscent of the Wood Pigeon (*Columba palumbus*) in cadence and timbre, but rather deeper toned and perhaps hoarser. It is pleasant sounding to the human ear, comprising four rather quickly repeated notes 'kruu-ku-kruu-kooh' with stress upon the first and last syllables and the second and third notes being quite brief. It can also be rendered 'coo-coo-cuk-rooh'. Territorial calling starts from early April, often when there is still snow upon the ground.

Map 211: Eastern Rufous or Oriental Turtle Dove
Streptopelia orientalis

Little Brown Dove *Streptopelia senegalensis* (Linnaeus)
Laughing or Palm Dove of Africa

Plate **18**
Map **212**

DESCRIPTION

Body length 26-27 cms.
Wing length 12.9-13.5 cms. (2 males),
    12.3-13.2 cms. (3 females) (Paludan)
Tail length 10.4-12 cms.
Weight 75-76 gms. (2 males), 75-78 gms. (3 females) (Paludan)

The smallest of the doves and with a distinctive brick orange patch on the side of the neck with a pattern of indistinct small black spots in this display area. Despite its comparatively longer tail and overall longer body length it is slimmer and smaller bodied than *Streptopelia tranquebarica*. The head, neck and breast are a rather dark mauve-pink, much darker than *S. decaocto*, the lower breast gradually shades to creamy white and the under-tail coverts are white. The mantle and inner wing coverts are earthy brown, with the outer wing coverts (lower wing shoulder) grey-blue. The lower rump is slaty blue and the central tail feathers are greyish brown, the outer tail feathers being largely white. In flight the under-wing coverts are grey-blue, darker than the almost white under-wing coverts of *S. decaocto* and the white outer tail tip is particularly conspicuous when the tail is fanned. The iris is dark brown and the legs and feet are purplish red or pink. The bill is blackish. The sexes are alike with the females being slightly paler, less reddish brown on the upper parts and juveniles being altogether duller brown with some pale fringes to the mantle and wing covert feathers.

HABITAT, DISTRIBUTION AND STATUS

In the hot arid stony hills and gullies of the outer foothills of Baluchistan and the NWFP, along the west bank of the Indus, this is a very characteristic species. It is the common breeding dove throughout the Punjab Salt Range and Sind Kohistan and over most of lowland Baluchistan. In the Indus plains it is widespread and sympatric with *S. decaocto* but not so numerous and largely absent from extensive sand dune desert to which *decaocto* has adapted. A small population remains resident in the outskirts of Quetta in winter. It is the resident dove in Lasbela and along the Makran. In the Murree foothills it ascends to about 914 metres (3,000 feet) where it is sympatric with *S. chinensis* in the summer. In summer in Baluchistan it ascends as high as 2,100 metres (7,000 feet) during the breeding season (Mana valley, 19 May) and up to 2,400 metres (8,000 feet) in the Kurram valley (Whitehead, 1911). In many parts of Sind it is commoner than *S. decaocto*. In Gilgit it occurs very rarely in the

summer in the main Indus and Gilgit valleys. There are some local migrations northwards in the spring. A specimen ringed on 15 March 1961 near Bhuj in Kutch was recovered on 27 February 1964 near Hyderabad. Status ABUNDANT.

HABITS

A ground feeder, and not very gregarious, usually found feeding in twos or threes and roosting in small numbers. It is largely granivorous in diet, taking grass and wild herb seeds as well as cultivated cereals and pulse crops. It will also consume small quantities of green leaves, flower buds and in Dadu district of Sind was observed trying to perch on the bending stalks of 'Jowar' (*Sorghum sudanense*) and picking at the ripe seeds (El Sadiq Bashir, pers. comm., 1979). It can run quite quickly over smooth ground and is active in foraging. It is not particularly shy in the presence of humans.

BREEDING BIOLOGY

The courtship of this dove is very similar to other *Streptopelia* species with advertising or display flights in which the male towers into the air, clapping his

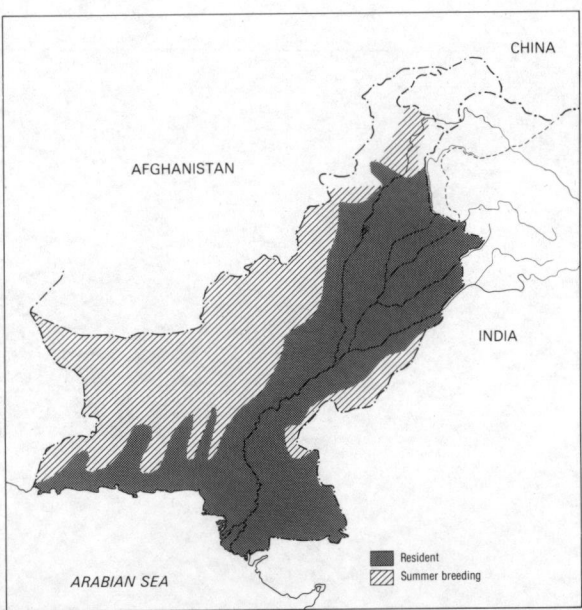

Map 212: Laughing Dove, Little brown Dove or Senegal Dove *Streptopelia senegalensis*

wings over his back and then volplaning down on stiffly out-stretched wings often in a wide sweeping arc. At this stage, however, they never utter the high pitched plaintive toned calls characteristic of *S. deca-octo*. There are minor differences in the male's courtship of the female also when the male advances towards her and bows his head and neck, but the bill is not so depressed, often remaining extended horizontally. At this time the neck feathers are erected displaying the red and black pattern on each side.

They build the usual type of nest platform on a horizontal branch often in quite low thorny bushes (*Acacia senegal*) in Lasbela and often nest in *Euphorbia caducifolia* 'cactus' clumps. The usual clutch is 2 eggs, plain white and the nesting season is prolonged, extending from as early as late January till as late as the end of October. Both sexes take part in nest construction, incubation and feeding of the young by regurgitation. The young chicks are clothed in hairy yellow down and develop very quickly, a pair in my bungalow garden leaving the nest on the thirteenth day after hatching. E.N. Lowther comments that *S. deca-octo* though equally common and perhaps more commensal never nests in gardens (*A Bird Photographer in India*, 1949). He described parent birds feeding fledged ten day old 'squabs' with beaks interlocked, the parent bird pumping its head and neck up and down vigorously as the chick fed.

*VOCALIZATIONS

The advertising call is a more high pitched and rapid sequence comprising six syllables. 'coo-roo-coo-cu-cu-coo'. The latest three uttered in rapid succession. The first three syllables are almost interrogative 'kwoi-koi-koi' with a fractional pause between the first and second syllable and then the final 3 uttered in rapid descending scale. It does not really evoke the sound of laughter as its trivial name suggests.

Spotted Dove or Chinese Dove *Streptopelia chinensis* (Scopoli)   Plate **18**
Map **213**

DESCRIPTION

Body length 28-30 cms.
Wing length 13-15 cms. (Stuart Baker)
Tail length 11.8-14 cms.

About the same size as *S. decaocto* with a similar rather long graduated tail and rather small head and

Map 213: Spotted or Chinese Dove *Streptopelia chinensis*

slender weak bill. The head and neck are bluish grey, the forecrown and cheeks sometimes being almost white. The breast is pinkish brown. The mantle and scapulars are quite dark brown with pale pinkish spots, and the inner wing coverts are drab brown with darker shaft streaks, also giving a spotted pattern. The outer wing coverts are bluish grey and the flight feathers darker brownish black. The tail is greyish black with the outer tail feathers tipped white and the central pair paler brownish grey. The vent and undertail coverts are white and the throat is pinkish white. On the sides of the neck extending broadly around the top of the mantle, is a black and white spotted area formed by the feathers which are bifurcated with black bases and white tips. The irides are orange-yellow, with purple or bluey mauve naked orbital eye-ring. The legs and feet are pinkish red and the bill is blackish. The sexes are alike. Immature birds lack the black and white spotted 'chess-board' display patch on the sides of the neck and are generally paler and browner, less vinaceous on the breast.

HABITAT, DISTRIBUTION AND STATUS

This dove is largely a summer visitor to Pakistan confined to the sub-tropical pine zone (*Pinus roxburghii*) and the lower or outer foot-hill regions of the Himalayas in a relatively narrow belt from lower Chitral, Dir, Swat and the Murree foothills. It is rarely found above about 1,829 metres (6,000 feet) elevation and does not descend onto the Potohar plateau beyond the plains immediately abutting upon the outlying hill spurs. It spreads up the Indus valley through Hazara district and Indus Kohistan as far as Besham. It is the common dove on the Mansehra plain and around Abbottabad and is associated with *Acacia modesta* scrub on the low hills surrounding Tarbela dam and Amb state. A few over-winter in these regions. One seen 10 January Margalla hills (D. Corfield, pers. comm., 1981). A few were also recorded in December and January in Gilgit main valley (Scully, 1880).

The bulk of the population are summer breeding visitors having an east-west migration pattern along the outer foothills of the Himalayas. In winter a few birds are occasionally observed in the plains around Sialkot and Lahore, relatively close to the foothills. In summer on the dryer south and south-west facing slopes of the Murree hills it ascends to 2,000 metres (6,500 feet) and has been collected at Bhurban (H. Waite). There is a noticeable influx of birds in early May around Islamabad and the Margalla hills. It has not been recorded in Sind or Baluchistan and is absent from the southern parts of the NWFP and Punjab. Status COMMON but restricted in habitat.

HABITS

They often associate with village and farm settlements, perching readily upon buildings and nesting in orchards. They feed mainly on the ground picking up grains and seeds, but also occasionally berries (*Berberis ceratophylla*) and green shoots and buds. They are quite tame in the presence of humans but' when disturbed burst upwards with noisy clapping flight, often towering before gliding down to settle in a tree or on telephone wires nearby. They are not particularly gregarious.

BREEDING BIOLOGY

A male watched at Samli forest rest house (Murree foothills) walked rapidly along a horizontal pine branch where a female was sitting. He kept up a continuous crooning with his neck inflated and feathers puffed out and as he approached bobbed his head and neck rapidly up and down to the accompaniment of each 'coo'. The female crouched and copulation ensued, 5 May. Nests in the foothill regions of Pakistan are found from early May to July and consist of the typical pigeon's flimsy platform of twigs built on a horizontal bough or over a horizontal fork. The normal clutch is two plain white eggs. Both sexes take part in nest construction and incubation. The young are fed by regurgitation by both parents and as incubation takes only about 13 days they are probably double brooded. Males have also been noted giving the typical display flights for the family with a wing clapping ascent followed by glides with wings held rigid and the tail partly fanned. The dark blue-grey outer tail feathers with white tips are particularly conspicuous during such displays.

*VOCALIZATIONS

The advertising song is a very soft almost querulous crooning, very evocative of a hot summer's day, each call is rather similar to the previous in pitch and duration and may be syllabized 'krooo-krooo-krooo' sometimes preceded by two shorter more emphatic 'kuk-kuk-krroo-kuku-krooo'. They call throughout the day during the summer months.

Orange-breasted Green Pigeon *Treron bicincta* (Jerdon)   Plate **18**

DESCRIPTION

Body length 29 cms.
Wing length 15.4-17.0 cms. (Stuart Baker).
Tail length 9-11 cms.

This is the smallest of the green fruit pigeons in our region. Its body colouration is predominantly olive or yellow-green and it has a relatively stout bill, high crowned forehead and a broad blue-grey tail rounded at the tip with the outer rectrices bearing a black band, sub-terminally. Males are yellow-green on the head and neck, darkening to olive green on the mantle. The upper breast is narrowly banded with lilac-mauve shading to coppery orange in the mid-breast. The lower belly is yellow-green. The wing coverts are blackish green with creamy yellow tips forming a narrow wingbar. The under-tail coverts are pale cinnamon chestnut with creamy yellow margins. The female has a plain green breast lacking the orange and lilac bands of the male. The naked eye-ring is violet-blue and the iris crimson. The bill varies from pale blue to pale green and the legs and feet are deep crimson.

HABITAT, DISTRIBTUION AND STATUS

This is a forest dwelling fruit pigeon typically associated with evergreen rain forest or moist deciduous forest in both southern India and Sri Lanka and the outer ranges of the north-eastern Himalayas to Bangla Desh and Assam. On 3 January 1938 a boy shot 2 green pigeons found feeding in a Pipal tree (*Ficus religiosa*) at Keamari, Karachi. These were identified by K. Eates as *T. bicincta* (*JBNHS,* Vol. 40, p. 330). It cannot be considered more than an accidental vagrant to Pakistan as there have been no subsequent records.

HABITS AND BREEDING etc.

Gregarious in feeding when large trees are fruiting. Noisy on the wing but very difficult to detect when feeding in a tree as they freeze at the approach of an intruder. *Their calls are fluting whistes which vary in pitch as though running up and down a scale.

Common Green Pigeon or Yellow-footed Green Pigeon
Treron phoenicoptera (Latham)

Plate **18**
Map **214**

DESCRIPTION

Body length 33 cms.
Tail length 11-11.8 cms.

This is a stout bodied fruit pigeon, bigger than the previous two species, with a broad rounded tipped tail which is yellow-green basally, dark grey on the lower part with blackish tips. The forecrown is greenish yellow turning grey on the nape and ear coverts. The throat, upper breast and upper mantle are bright olive yellow and the wings and back are darker olive green with a narrow lilac mauve patch on the wing shoulder and a narrow blue-grey band across the mantle framing the yellow neck collar. The flight feathers are blackish grey with creamy yellow tips to the secondaries and outer wing coverts forming two wingbars. The under-tail coverts are cinnamon chestnut with creamy tips and there are some darker green bars on the lower flanks. The legs and feet are orange or chrome yellow and the stout bill is grey with the cere greenish. The irides are lilac varying to blue. The female is a duller version of the male but not markedly different and she has a more restricted mauve shoulder patch. The yellow feet, contra crimson of all other fruit pigeons, and the grey head serve to distinguish this species in the field.

HABITAT, DISTRIBUTION AND STATUS

This pigeon is a regular winter visitor to the Punjab plains wherever there are Pipal and Banyan fig trees (*Ficus religiosa* and *F. bengalensis*). In the spring they invade the irrigated forest plantations to feed on ripe mulberries. They are locally nomadic and seldom stay to breed in the Punjab though there have been occasional authentic records. In Sind they are uncommon. They are often associated with towns where there are roadside avenues of Pipal trees or in canal housing colonies where similar trees are planted. They have not been recorded in Baluchistan or the NWFP. Earliest arrivals noted 4 October and latest 7 April. Due to tree plantations following the development of

irrigation in the Punjab, this arboreal pigeon has probably extended its range and is certainly more common than 50 years ago. Status COMMON in the Punjab plains only, but erratic in occurrence.

HABITS

They are gregarious, usually being seen in small flocks of 5 to 20 flying swiftly with whirring wings from one feeding place to another. Like all the *Treron* pigeons they are very agile in clambering around drooping branches, being able to reach fruit from the farthest extremities and when feeding in a tree they keep well concealed in the foliage and are often extremely difficult to locate once they have landed in a tree. They have also been recorded feeding on ripe mulberries (*Morus alba*) in Changa Manga plantation (Z.B.

Map 214: Yellow-footed Green Pigeon *Treron phoenicoptera*

Mirza, pers. comm.) and upon Jaman fruits (*Eugenia jambolana*) in Lahore in June (R. Holmes, pers. comm.). Out of over 100 stomach and crops of this species examined in Bihar, India no insect remains were found, and *Ficus* fruit predominated. (Mason and Lefroy, 1912).

BREEDING BIOLOGY

An attempt at nesting was recorded in Daphar irrigated forest plantation in Gujranwala district. The nest was built in an *Acacia arabica* tree in April, 1972. Only the female attempted to incubate the clutch of 2 white eggs. Unfortunately House Crows (*Corvus splendens*) robbed the nest (Akhlaq Ahmad Khan, Conservator of Forests, pers. comm., 1972). It has also bred in Changa Manga forest plantation with nests being

found in April (Z.B. Mirza, pers. comm.) but in parts of India it can be found nesting up to June. The incubation period is said to be 14 days (Goodwin, op. cit.). The eggs are plain white, glossy and broadly oval in shape, and the nest is a slight platform of inter-laced twigs. Stuart Baker describes the courtship of the male during which he sidles up and down a branch with feathers fluffed out and especially the throat and breast puffed up. The wings are lowered and he continuously bows his head as he emits soft whistles (Baker, *Indian Pigeons and Doves*, 1913).

VOCALIZATIONS

A melodious whistling, lower in pitch than *T. bicincta* and often given from February onwards until breeding is finished.

Wedge-tailed Green Pigeon or
Kokla Green Pigeon *Treron sphenura* (Vigors)

Plate **18**
Map **215**

DESCRIPTION

Body length 33 cms.
Wing length 17.2-18.5 cms. (Stuart Baker)
Tail length 11.5-14 cms.

Like all the fruit pigeons this is a very plump rounded bird in body outline with a high crowned forehead and relatively stout bill and short legs. The scapulars and

Map 215: Wedge-tailed Green Pigeon *Treron sphenura*

lower back, the wing coverts and tail are dark olive green. The distinctive features of the male are the maroon-chestnut wing, shoulders and lower mantle and blue-grey 'cape' around the upper part of the mantle. The forecrown is coppery orange as also the upper breast, gradually merging to olive yellow on the flanks. The tail is comparatively long and graduated or pointed as implied by its name, rather than square tipped and the outer tail feathers are dark grey. The under-tail coverts are pale orange-chestnut and extend almost to the tail tip with some dark olive green crescentic bars on the lower flanks. The secondaries and tertials are blackish with creamy yellow narrow margins to the outer webs and tips forming pale yellow wingbars when the bird is at rest. The legs and feet are coral red or crimson and there is a naked orbital ring of blue skin. The irides are crimson-pink, becoming blue in the centre. The bill is blue, soft and swollen at the base. The female lacks the orange crown and upper breast and has no maroon-red on the wing shoulders, being wholly green on the body. Her under-tail coverts are buffy yellow with olive green centres. This pigeon is distinguished from the Pintailed Green Pigeon (*Treron apicauda*) which also has a blue-grey mantle, by lacking the pintail elongation of the central rectrices. *T. apicauda* also lacks the wine red wing shoulder.

HABITAT, DISTRIBUTION AND STATUS

This is a locally nomadic but essentially resident Oriental pigeon, adapted to live in the outer Himalayan forested slopes and foothills. It used to occur regularly in the Murree hills as a summer visitor and there is a specimen in the British Museum collected by H. Waite from Murree on 16 June 1947 at 1,829 metres (6,000 feet) elevation. At the turn of the century it was considered a summer visitor only, around the southern or eastern part of the Murree hills, breeding commonly between 1,800 to 2,100 metres (6-7,000 feet) (Whistler, *Ibis*, 1930). It is apparently very rare or absent now from the Murree hill range as there have been no recent sightings reported and the author over about 16 years of observations has only one record of a solitary bird at Khanspur (now Ayubia), feeding on ripe *Viburnum nervosum* berries in late July 1966. However, in late April 1984 several birds were observed haunting walnut groves (*Juglans regia*) in Malkandi forest, Kaghan valley (R. Grimmett and C. Robson pers. comm., 1984). During a three day stay in the same area in May 1985 the author watched 2 females and a week later flushed a male and it seems probable that a small population is resident in this region which, though of limited area, has a variety of fruit bearing deciduous trees. Status RARE in restricted habitat.

HABITS

A purely arboreal pigeon adept at clambering through and along narrow branches, with its strong feet capable of hanging upside-down to reach ripe berries or fruit. This is an entirely frugivorous bird, moving about when various trees are bearing ripe fruit, feeding in late summer on the berries of *Lonicera, Berberis ceratophylla* and *Viburnum nervosum*, and earlier in the summer consuming the drupes of *Prunus, Ficus* and *Myrica sapuida* (Stuart Baker, 1913). It is most

active in foraging in the early morning and evening, and inclined to be rather sluggish, sitting without movement for long periods during the middle of the day inside a thick foliaged tree. It is also said to be not very gregarious as compared with other fruit pigeons, generally being encountered only in parties of 2 or 3 birds (Stuart Baker, op.cit.).

BREEDING BIOLOGY

Both parents share nest building and incubation duties. The nest, is built on a horizontal bough often quite high up in a tree, up to 15 metres from the ground. The full clutch is 2 eggs which are glossy, plain white and the nest is a typical pigeon's flimsy flat platform of inter-woven twigs. The incubation period of captive birds is 14 days (Goodwin, op. cit.). The young are fully fledged at 12 days. Most nests in the north-west of the Himalayas are built between May and early July (Whistler, op. cit.).

VOCALIZATIONS

Its advertising calls are a rich and musical series of whistling or fluting notes, sometimes a series of over 20 notes are uttered in sequence (Osmaston in Ali and Ripley, Vol. 3, 1969). Whistler describes it as a long melodious but slightly grating whistle sounding like 'why, we what cheer' (*Popular Handbook*, 1949), whilst Magrath syllabized the call as 'ko-kla-koi-oi-oi-oi-oilli, illio-kla' (*JBNHS*, 1909a) During courtship the male fans its tail and puffs out the throat and bows its head while calling (Stuart Baker, 1913). At this time instead of the whistles it utters a low 'coo-coo'.

ORDER PSITTACIFORMES
FAMILY PSITTACIDAE
Parrots, Cockatoos, Macaws and Lory's

Alexandrine or Large Indian Parakeet *Psittacula eupatria* (Linnaeus)   Map **216**

DESCRIPTION

Body length 50-53 cms. including tail 22-36 cms.
Wing length 20.5-23 cms. (males),
   19.4-21.5 cms. (females) (Stuart Baker)
Tail length 29.8-32.8 cms. (males),
   28.6-37.6 cms. (females) (Cooper and Forshaw)

All the parrots of this genus are predominantly green in colour with extra long pointed tails. This is the largest of the Indian parakeets, with the most prom-

inent maroon-red shoulder patch. The powerful bill, forming a semi-circular arc along the culmen of the upper mandible, being deep crimson with a paler orange-red tip. Its body is grass green but a slightly darker greyer toned green than typical specimens of *P. krameri*. The tail is very long and the central rectrices are slender and pointed becoming verditer blue distally, paling at the extreme tip to creamy yellow. Males have a black collar on the forepart of the neck which extends around below the ear coverts to meet under the chin. They also have a broad lilac pink nuchal collar across the nape which becomes narrow on the sides of the neck where it meets the black fore-collar. The legs and feet are a dirty buff colour and the feet are zygodactyl with two toes in front and two behind. The iris is lemon yellow. Females lack the rose pink nuchal collar or the black fore-collar but have small maroon-red shoulder patch, unlike the females of *P. himalayana* or *P. cyanocephala* which lack any red shoulder patch. The under-wing coverts are dark grey-green whereas in *P. krameri* the under-wing coverts are bright olive yellow. Females are slightly smaller than males.

HABITAT, DISTRIBUTION AND STATUS

Largely local and sedentary in the better wooded parts of Punjab, extending up to the sub-tropical pine zone (*Pinus roxburghii*) in the Murree foothills, where it is resident. The population tends to be disjunct with a small number resident in the Margalla and Murree hills, Kahuta and Lehtrar, but practically absent from, the Potohar plateau and around Rawalpindi. It is widespread in the irrigated canal colonies of the Punjab but absent from the more open desert regions such as the Thal and Dera Ghazi Khan. There is a small thriving colony in Kohat, NWFP but it is rare in Peshawar. In Sind it is practically absent except for a few colonies around Sukkur and Ubaro in the extreme north and it is absent from Baluchistan or the northern mountain areas. Curiously a small breeding population persists inside ravines of the Margalla hills at Saidpur but it is still comparatively uncommon in and around Islamabad. Old bungalow and rest house gardens in Multan district, Sahiwal and around Lahore are favourite haunts of this parakeet. They are much less numerous than *P. krameri* and by tradition are more highly prized as cage birds than the latter species. Despite heavy human predation of fledged nestlings for the aviary trade, the species is probably more than holding its own in the Punjab with the development of tree plantations which provide more potential nest sites. Status COMMON.

HABITS

Rather territorial, roosting each night in the same tree groves and keeping in small flocks or family parties throughout the year. They have loud far carrying calls and indulge in much calling when flying around both between feeding places and before settling down in their night time roosts. Their strongly hooked upper mandible and short curved lower mandible have a powerful pincer action enabling them to tear open flower buds and seed pods which are inaccessible as food sources to many other granivorous birds. A captive female when given a pencil reduced this to

splinters within seconds by mouthing it between her mandibles. In the Murree foothills they feed on ripening maize, tearing off the tough spadith to expose the cobs underneath. They are very fond of the flowers and flower buds of such nectar bearing trees as *Salmalia*, *Bauhinia* and *Erythrina*, eating the petals partly as well as the nectaries. They eat a variety of seeds, fruits and vegetables including ripe *Citrus* fruits, guavas (*Psidium guajava*) and the young leaves of vegetables. A study in Faisalabad, based on stomach analysis, showed 52 per cent of the diet was seeds of cultivated crops and 2.7 per cent was weed seeds, with 4.8 per cent vegetables, 19.3 per cent cultivated fruits, 9.8 per cent wild fruits and 11.4 per cent seeds which were neither beneficial nor harmful to agriculture (Husain and Bhalla, *JBNHS*, 1937). Their zygodactyl feet enable them to hang upside down to reach otherwise inaccessible fruit or flowers and they also use their feet to hold ears of grain which they can consume from a tree perch after having hovered over a standing crop to nip off the inflorescence between their mandibles. They are mostly active in feeding in the early morning and late afternoon.

BREEDING BIOLOGY

This rather aggressive bird has a prolonged courtship period involving a lot of appeasement gestures. They nest in tree cavities which they enlarge and deepen with their bills, but usually need the rotten wood of a dead knot, or an old woodpecker's hole to initiate nest

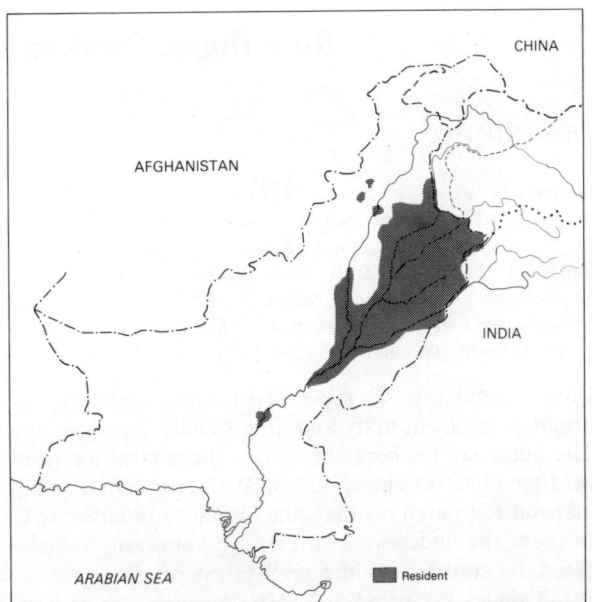

Map 216: Alexandrine or Large Indian Parakeet *Psittacula eupatria*

excavation. A suitable nest hole will be used year after year, possibly by the same pair, as observed in my Khanewal garden. The pair hangs around the nest hole and guards the nest site for many weeks before egg laying commences. This is usually between early February and March in the southern Punjab. Both parents help to excavate the nest hole which is a vertical shaft with an enlarged unlined chamber at the bottom. Only the female incubates and while doing so she is fed by her mate by regurgitation (observation Khanewal garden). The incubation period is 21 days or longer and the normal clutch is 3 to 4 eggs, rather round in shape, plain white and glossless. The newly hatched chicks are altricial and have to be fed by regurgitation which is preceded by the parent bird bobbing its head up and down. As they grow older the fledglings solicit feeding with loud penetrating whistles audible near the nesting tree. They are usually independent after about 8 weeks from hatching. K. Eates in the 1940s found a clutch of 4 eggs on 2 February in a mango tree at Malir (MS notes unpublished). This colony now seems to have died out. During ten years' residence in Malir, the author only 3 times encountered single individuals of this species. Eates also found fledglings barely able to fly in a nest on 15 March at Sukkur in northern Sind (Eates, 1937). In Khanewal the young usually left their nest hole by the end of April after which their parents continued to feed them (by regurgitation) for about 10 days into May. After the pair bond is established, both sexes spend a large part of the day perched close together indulging in mutual billing and preening or nibbling of the nape

area. Nest predation is high. In the Punjab the Monitor Lizard (*Varanus griseus*) often takes the young chicks despite the loud protests of the parents. They are much esteemed as cage birds and in many villages the effort of cutting open the nest hole is considered well worth the sale value of the fledged young.

*VOCALIZATIONS

Like all the *Psittacula* genus, flocks indulge in excited chorusing with one bird initiating responding calls from the whole flock. As they fly swiftly through the air swooping between the tree tops their flight or contact call is a very loud strident scream repeated 2 or 3 times sounding like 'keeak-keeak'. When danger threatens they have harsher more grating 'kraah-kraah' calls also characteristic of the genus. Males have a sort of territorial song consisting of a long sequence of calls of varying pitch and intonation 'keeh-keeh-kah-keah-kah-keeh' etc. Their voice is louder and deeper in pitch than *P. krameri*, each of their calls being distinctive for the species once recognized.

MISCELLANEOUS

Alexander the Great brought caged specimens of this species back to Greece after his invasion of the region that is now Pakistan, in about 324 B.C.

Rose-ringed Parakeet *Psittacula krameri* (Scopoli)

Illus. **63**
Map **217**

DESCRIPTION

Body length 42 cms. including tail 19-28 cms.
Wing length 15.6-17.1 cms. (males),
    15.1-16.1 cms. (females) (Stuart Baker)
Tail length 22.5-26.3 (males) (H. Waite),
    19-24 cms. (females) (Stuart Baker)
Weight (average of 10 wild caught birds) 116 gms.
    (G.A. Smith, *Aviculture Magazine*, 1972)

Closely similar to *P. eupatria* in colouration with a bright grass green body and long slender tapering tail. The outer tail feathers are yellow, the central rectrices verditer blue, becoming yellow at the tip. There is no maroon-red patch on the wing shoulder in either sex. In flight the under-wing coverts are noticeably yellow-green in contrast to the grey-green of *P. eupatria*. Adult males by their third year acquire a pale rose pink nuchal collar and underneath this a narrow black collar widening under the cheeks to meet at the base of

the lower mandible. The female entirely lacks this collar around the neck. In both sexes the plumage of the breast is often very yellow-green. Above the pink nuchal collar in the male is a narrow band of paler aquamarine. The legs and feet are greenish slaty. The bill is smaller and less powerful looking than that of *P. eupatria* and both mandibles are blood red, sometimes the hooked tip of the upper being paler orange. The lower mandible is often suffused blackish. The iris is whitish yellow and when excited the male contracts his pupils and 'blazes' his widened eye.

HABITAT, DISTRIBUTION AND STATUS

A very adaptable species which is common and widespread throughout the Indus basin. Wherever it occurs it is largely sedentary. In the Murree foothills it ascends in summer to 914 metres (3,000 feet) where it breeds sympatrically with *Psittacula cyanocephala* (Manga valley in *Pinus roxburghii*). It comes into the main valleys of Baluchistan around Quetta in the summer months and a small hardy band over-winters in the gardens of Quetta cantonment (observed 16 January, 1984 in sub-zero weather). It is uncommon but occurs in the more arid foothill regions around Kohat (Whitehead, 1911), and it is comparatively uncommon in wild olive forest, preferring the vicinity of villages and cultivated crops. It is absent from extensive desert tracts and also the northern Himalayan regions. It is well adapted to human settlements, roosting and nesting in buildings and feeding on garden crop plants. In winter time birds often congregate in large flocks at attractive feeding sites. A total of over 3,000 was counted in one afternoon on one farm at Umarkot (western Sind) feeding upon ripe maize in November in scattered flocks, each numbering between 200 and 300 birds. Outside of the breeding season they roost colonially, often in tall tree groves on the outskirts of towns. These colonies may number 200 or 300 to over 1,000 and are extremely noisy when settling for the night and also at dawn before they disperse to different feeding grounds. Status ABUNDANT, and a serious agricultural pest where citrus orchards, sunflower and maize are important cash crops.

HABITS

They are gregarious outside of the nesting season, forming large noisy communal roosts in the winter season, in tall tree groves. They disperse in all directions in the early morning from these communal roosts to feed on crops, and again in the late afternoon, roosting in nearby trees, in scattered parties during the middle of the day. Their flight calls, frequently uttered, are loud and far carrying and attract others of the same species. They are principally granivorous, but will also eat all kinds of ripe fruit and feed on the nectar of flowering trees such as *Salmalia malabarica*. They usually cut off a seed pod in their bills and carry this to

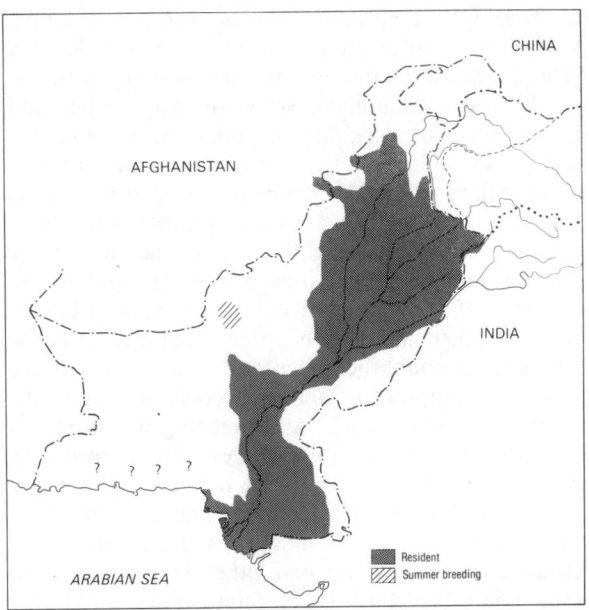

Map 217: Rose-ringed Parakeet *Psittacula krameri*

a convenient perch, holding the pod in their foot which is raised to their bills. They are especially fond of the huge seed pods of the siris tree (*Albizia lebbeck*) as well as Shisham tree pods (*Dalbergia sissoo*). Hybrid maize crops (*Zea mays*) have been damaged up to 30 and 40 per cent by flocks of this parakeet (S. A. Bashir, V.P.C.C. Internal Report, 1980) and Sunflower crops (*Helianthus annuus*) have been destroyed up to 80 per cent (observed Rahim Yar Khan, 1976). They are very destructive to ripening mustard crops (*Brassica campestris*) cutting off the whole inflorescence and flying to a tree perch where only one or two pods are consumed before the branched fruiting parts are dropped and the bird again flies to the crop to cut off another. They also cut ripening ears of wheat in a similar manner, hovering over the standing ears. The scattered green glumes of wheat beneath nearby trees are sure evidence of parakeet activity. They will feed on the ground where they waddle with tail partly raised in rather a ludicrous manner. They can cut into the slippery skin of oranges (*Citrus sinensis*) on the tree and are especially fond of ripe citrus fruit. They also attack Red Pepper crops (*Capsicum minimum*) when the pods are ripening. It is customary for both orchard owners and maize growers to hire people as full time bird scarers to protect these crops during the critical ripening season, from this parakeet.

BREEDING BIOLOGY

They nest in tree cavities and also in holes in man-made buildings. If in a tree, both sexes excavate the

Illus. 63: Rose-ringed Parakeets *Psittacula krameri*.

nest hole, using their bills to tear out splinters of wood. A male was observed excavating a nest hole in a *Salmalia malabaricum* tree in late January with the female close by watching. When she approached and tried to assist by pecking and pulling at the hole, the male pecked at her and she jumped backwards (observations Miani forest, Hyderabad). Courtship involves the male walking towards the female with neck upstretched and head slightly everted and then lifting of one foot. He then jumps backwards and makes similar foot lifting and neck stretching approaches from the other side. Courtship also includes, after the pair bond is established, feeding of the female by the male by regurgitation and this continues after the female starts to incubate. Also preening was noted on 1 December by a pair which successively nibbled each other on the nape. Pairs often perch for prolonged periods in close physical contact. These observations made of wild birds on numerous occasions are at variance with courtship and other behaviour based upon aviary bred birds (G.A. Smith, 1972, *Aviculture Magazine*, Vol. 78 (4)). The nest hole is jealously guarded as this is probably the limiting factor in more frequent and successful breeding and females especially have been noted hanging around nest holes during December and January, even though egg laying had not commenced.

A large communal roost in *Terminalia* trees at Khanewal, gradually broke up each year from early March when pairs dispersed for breeding. In the Punjab the main nesting season is February to April. The eggs are rather glossless, plain white and round in shape and clutches of up to 5 have been recorded though 4 is more normal. Incubation takes 25 to 28 days and is carried out entirely by the female who is fed by the male by regurgitation from his crop. The male also roosts in the nest hole at night with the female but does not share incubation duties. The altricial chicks take about eight full weeks before they fledge and are still partly fed by their parents for a couple of weeks after leaving the nest hole. Newly fledged young just out of the nest hole have been seen being fed by their parents on 1 April in Khanewal and in early May near Sukkur by K. Eates. A breeding pair closely watched in Delhi by Malcolm MacDonald started copulation on 1 February and this continued for six weeks, often at hourly intervals, until the clutch of eggs was completed on about 15 March but the chicks did not leave the nest hole until 27 May, having been first observed at the nest hole on 7 May when about 7 weeks old. In this particular instance the female initially selected the nest hole and enlarged it to her convenience. The male did not assist in nest excavation (MacDonald, 1960).

*VOCALIZATIONS

They have a similar repertoire of calls as the Alexandrine Parakeet, including loud ringing flight or contact calls 'keeah keeah' or 'kee kee'. Advertising or territorial songs are believed to be mainly given by young males and comprise a rather extensive sequence or soliloquy of softer notes in a rising and falling cadence 'kiah kiah ki kiah ki ki'. When danger threatens such as the appearance of a snake, cat or mongoose, their loud scolding screeches are very characteristic, being rather lower pitched and grating calls. Young birds have a similar high pitched whistle to the Alexandrine Parakeet when in the nest and soliciting food.

Plum-headed or
Blossom-headed Parakeet *Psittacula cyanocephala* (Linnaeus)

Illus. **64**
Map **218**

DESCRIPTION

Body length 33-36 cms.
Wing length 13.6-14.6 cms. (males),
   13.3-14.1 cms. (females) (Stuart Baker)
Tail length 20.8-24 cms. (males),
   15.5-19.2 cms. (females) (Stuart Baker)

This is the smallest of the long tailed parakeets inhabiting Pakistan. With a bright grass green body, they are easily recognized in flight by the darkened head of both sexes and the white tips to the tail. The tail is long, slender and graduated, the central rectrices being slightly spatulate at the tip and turning more blue than aquamarine towards the tip with the final portion being clear white. The outer tail feathers are more yellow green. In males the nuchal collar is a very thin black line broadening under the chin, and the whole of the head is a rich purplish red washed with lilac and mauve, particularly in the region of the nape and lower cheeks, with a deeper red-purple around the eyes. The bill, which is less powerfully developed than in *P. krameri*, is golden orange on the upper mandible and black on the lower mandible. There is a small patch of crimson-maroon on the wing shoulder in males. Females have a paler yellow bill, less orange and lack the maroon shoulder patch entirely. They have a narrow yellow green ring around the nape and below the ear coverts, and above this their heads and cheeks are a dull grey or purplish grey. First year males are like females with yellow upper mandibles and dull purplish grey heads. The narrow yellow nuchal collar is quite noticeable and in both sexes the white tail tip is conspicuous. In both sexes the legs and feet are greyish green and the irides are creamy or pale straw. In flight the under-wing coverts are bluish green contra the yellow green of *P. krameri* which is probably the only way that immature birds can be separated with certainty. Immatures of the Slaty-headed Parakeet (*P. himalayana*) also have bluish green under-wing coverts, but look larger and have prominent yellow tail tips.

HABITAT, DISTRIBUTION AND STATUS

It is largely a resident breeding species in the lower Murree foothills with some part of the population dispersing in winter down to the immediate plains areas particularly around Jhelum and Sialkot districts. It is very restricted in distribution in Pakistan and has not been noted above the sub-tropical pine zone (*Pinus roxburghii*) nor west of Murree in similar sub-tropical pine zone which stretches across Hazara district. Thus it does not occur in Swat or the Kaghan valley. This is a locally nomadic parrot outside of the breeding season, because it is less granivorous than *P. krameri* and more specialized in its feeding habits. Small flocks have been noted 5-7 December in Sialkot city, at Chilianwala on 16 January at the foot of the

Map 218: Plum-headed Parakeet *cyanocephala*

Salt Range and there is usually an influx along the Lahore canal from late December to early January around the Punjab University campus (Z.B. Mirza, pers. comm., 1972). They can frequently be encountered in winter around Islamabad (400 metres) and breed up to 1,500 metres (5,000 feet) in the Karor, Lehtrar, and Manga valleys. Status FREQUENT but restricted in distribution.

HABITS

They are more frugivorous parakeets than the other species and they also eat nectar bearing flowers. They prefer small seeds of tall herbs and weeds, to large grains. When *Adhatoda vasica* the dominant ground shrub in the outer foothills is in flower they can be seen in small parties greedily eating the whole of their white flowers (March, Nurpur Shahan). On 7 June they were observed eating the flowers of wild pomegranate *Punica granatum* near Samli Forest Rest House. They will eat red peppers (*Capsicum minimum*) and hover over ripe thistle heads (*Echinops* sp. and *Cnicus* sp.) to peck at the seeds (Sialkot, 5 December). They are destructive of apricots (*Prunus armeniaca*) in the orchards around Tret, and eat wild fruits of *Ficus* as well as *Ziziphus* in season. In winter time they are generally encountered in small parties, rarely forming large flocks. They are often sympatric with *Psittacula krameri* in the breeding season, and in feeding and foraging with *Psittacula himalayana*. Generally this species sticks to well wooded areas and forest and does not feed on the ground as often as *Psittacula krameri*.

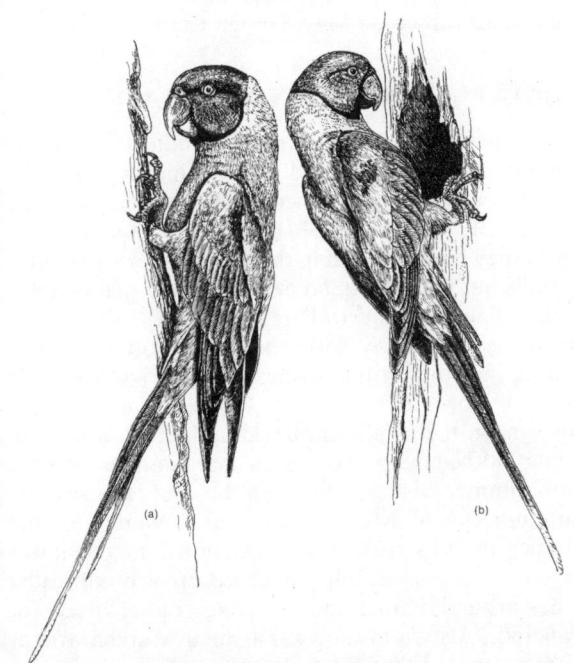

Illus. 64: (a) Slaty-headed Parakeet *Psittacula himalayana*. Male with small maroon red patch on wing shoulder and yellow lower mandible.
(b) Plum-headed Parakeet *Psittacula cyanocephala*. Male showing small maroon shoulder patch and black lower mandible. Both drawn to same scale.

BREEDING BIOLOGY

They nest in tree cavities and in Pakistan favoured sites are tall *Pinus roxburghii* trees which are dead or partly diseased. In such trees, with old woodpecker holes in dead wood, they are easily able to widen and extend such cavities for their own nesting requirements. They will also appropriate the unoccupied nest holes of the Blue Throated Barbet (*Megalaima asiatica*). Suitable nest holes are reoccupied year after year and are guarded and defended against the Rose-ringed parakeet (*P. krameri*) or Mynas (*Acridotheres tristis*) for some weeks if not months, before actual egg laying commences. In Manga valley on 8 May pairs were already guarding nest holes and on 22 May in Lehtrar valley a female was seen being fed by her mate at a nest hole. He bobbed his head and then grabbed the top of her bill to pass regurgitated food. Young were seen on 4 June outside the nest hole, soliciting food by bobbing their heads vigorously up and down and emitting a special high pitched squeaking whistle. The normal clutch is 4 or 5 and the eggs are said to be more spherical than those of other *Psittacula* species. They are plain white in colour. Only the female incubates and she is fed during this period by the male, by regurgitation. Apparently it is mainly the female which feeds the young but she probably passes on some of the food previously fed to her by the male.

*VOCALIZATIONS

Their flight call is a quite distinctive, more inflected, musical 'toowinck - toowinck' than *P. krameri*. Males at rest indulge in long complicated song patterns, perhaps more so than the other long tailed parakeets. These songs consist of a long sequence of rising and falling notes 'queah-queeah', 'kwik-kwink-queeah' etc.

Slaty-headed Parakeet *Psittacula himalayana* (Lesson)
Sometimes considered conspecific with *Psittacula finschii*

Illus. **64**
Map **219**

DESCRIPTION

Body length 39-41 cms.
Wing length 16.2-17.4 cms. (males),
    15.4-16.1 cms. (females) (Stuart Baker)
Tail length 20.5-24.6 cms. (males),
    15-16.1 cms. (females) (Stuart Baker)

It is slightly larger than the common Rose-ringed Parakeet (*P. krameri*). Like others of the genus it has a bright grass green body, yellower on the breast and males have a crimson-maroon patch on the wing shoulder. The tail is long and slender, the central rectrices becoming blue to verdigris with the distal part bright chrome yellow. Often when hurtling across a ravine, the yellow tail tip is visible and serves to separate the species from the sympatric Blossom Headed Parakeet (*P. cyanocephala*) with a white tail tip. The male has the head greyish slaty bordered by a narrow black collar becoming slightly wider under the chin. The grey of the head ends sharply in the mid-crown region and does not extend over the nape. The bill is orange-red, darker than the orange of *P. cyanocephala* but the lower mandible is paler more yellow. The feet and legs are greenish yellow and the irides are creamy white. Males have a distinctly brighter area on the nape, of pale aquamarine or emerald, unlike the more olive green of the rest of the mantle and side of the neck.

Females are slightly smaller with their heads washed greyish green not as intense as that of males and juveniles in their first summer have dull brownish green heads (observed May, Dunga Gali).

HABITAT, DISTRIBUTION AND STATUS

As its scientific name implies this parakeet is adapted to mountainous regions and inhabits higher elevations than *P. cyanocephala*, normally preferring to breed around 2,134 up to 2,438 metres (7-8,000 feet). In winter they move around in the lower valleys and outer foothills but do not extend down into the plains as far as the Bloosom-headed Parakeet often does. In the late winter and early spring they may often be seen feeding on the same bushes as *P. cyanocephala* especially on the nectar of *Woodfordia fruticosa* blossom. In their Pakistan breeding haunts, which are deeply clothed with snow until late March, they are only summer visitors, but can be seen around the summit ridge of Murree from early March. In the summer months they have been noted in the lower Kaghan valley especially on open terrace cultivated slopes around Paras, and a strong colony breeds at Malkandi. Also in lower Swat around Murghazarin at 1,500 metres (5,000 feet) elevation. They have also been noted in small numbers in Chitral main valley in June. A small colony nests at Zeran at 1,750 metres (5,000 feet) in the Safed Koh hills and flocks visit Kohat briefly in the spring (Whitehead, 1911).

In January they can occasionally be seen as low as 460 metres (1,500 feet) in the Margalla hill ravines adjacent to Islamabad. They are essentially associated with Himalayan moist forest, of mixed coniferous and deciduous composition. Status COMMON in restricted localities.

HABITS

Like others of their genus they are swift and agile on the wing and commonly the visitor to the hills will have his first encounter when a group rockets overhead swooping down over the tree tops in noisy screeching flight. They are gregarious throughout the year, preferring to nest in colonies, often with 3 or 4 nests in the same tree.

Their diet comprises the wild fruits and seeds of various trees in season, as well as flowers of trees and shrubs in the spring and in the autumn they will inflict damage to cultivated orchard fruit especially apples and pears (*Malus pumila* and *Pyrus communis*) as well as ripe maize (*Zea mays*). They have been seen feeding in flocks on the downy seeds of *Populus ciliata* (early July), and the flowering racemes of *Pistacia integerrima* in mid-April (Kao forest, Dunga Gali). also eating the rind off green walnuts (*Juglans regia*) (Paras, July). At Malkandi in mid-May they were noted as being particularly fond of the small unripe seeds of Sumac (*Rhus cotinus*), and in November the acorns of *Quercus dilatata*. They are hardy birds being able to survive in forest where the ground is carpeted with snow.

BREEDING BIOLOGY

Not much has been recorded to differentiate this species from the other long tailed parakeets, except for its stronger predilection for nesting in colonies. Observations over several summers in Dunga Gali and Kao forest indicate that colonies spend many weeks around the potential nest site before egg laying actually commences. In the initial period females are actively inspecting or searching for suitable nest holes, and once approved will remain guarding these against intra-specific usurpers for many weeks. Males also show interest in nest sites and it seems likely that these are as yet unmated and that it is mainly the female which selects the nest hole. In a *Quercus dilatata* grove along the water pipe line at Dunga Gali, which attracts nesting colonies of this parakeet each year, there were an estimated fourteen breeding pairs with at least one nest in a *Pinus wallichiana,* but the majority of nests were excavated in the rotten stumps of the higher up lateral branches of the Hill Oak (*Quercus dilatata*). One nest was at least 20 metres above ground and in one tree there were 3 nests quite close together. In Kashmir, Bates and Lowther found five occupied nest holes in one tree (Bates and Lowther, 1952). Most nest

holes are high up and very inaccessible because of the steep ground on which these magnificent oaks grow. Egg laying may start as early as April or late as June. Bates and Lowther found a nest on 10 June in the Kaz-i-Nag range in which incubation was still going on (op. cit., 1952). Fully fledged but flightless young were taken from one nest hole in Dunga Gali by a village boy on 23 July and in another year on 3 July, young just out of the nest, were watched still being fed by their parents by regurgitation. The normal clutch is 3 to 5 eggs, usually 4 and these are plain white spherical ovals. The nest hole is a vertical shaft and unlined. Presumably only the female incubates during which her mate feeds her, and the eggs take about 25 to 26 days to hatch and the chicks about 7 or 8 weeks to fledge. By mid-August in the Murree hills these parakeets disappear completely, and they usually leave the vicinity of their nesting colonies by mid to late July.

At the beginning of the breeding season, what appear to be unpaired females have a very loud distinctive squeaking call which they constantly give from the vicinity of their selected nesting hole. Parties indulge in a lot of excited flight pursuit, dashing in between the trees with amazing speed and dexterity at the beginning of the nesting season, but once pair bonds are established this is less common. Yellow-throated Martens (*Martes flavigula*) are the main potential predator of eggs and chicks, as these keen scented mammals leave no hole uninvestigated and can run with equal facility up or down a vertical tree trunk.

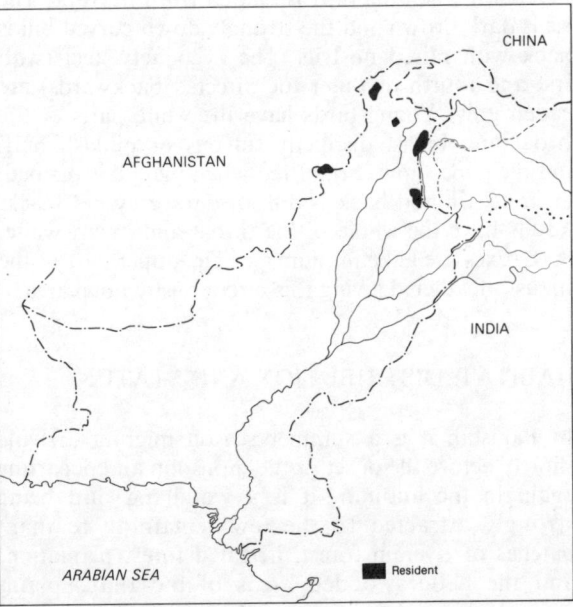

Map 219: Slaty-headed Parakeet *Psittacula himalayana*

*VOCALIZATIONS

The advertising call of the female, a rather hard piercing but tuneless squeak is often heard at the beginning of the mating season. In flight they call noisily like all members of the genus and the call is higher pitched than that of *P. krameri*, and equally strident but more rolling or liquid in timbre, and can be rendered as 'trree-trree'. While roosting or feeding, flocks will break into a noisy chorus of calling when one individual starts and these include more complicated sequences of notes in rising and falling pitch including 'toi-trree-treet-ti-toi-tree' etc.

ORDER CUCULIFORMES

FAMILY CUCULIDAE

Cuckoos, Koels, Coucals and Malkohas

Plate **19**
Map **220**

Pied Crested Cuckoo or Jacobin Cuckoo *Clamator jacobinus* (Boddaert)

DESCRIPTION

Body length 33-36 cms.
Wingspan 45-50 cms.
Wing length 14.6-16.1 cms. (northern population) (Stuart Baker)
Tail length 15.7-15.9 cms. (northern population) (Stuart Baker)

This is a medium sized black and white cuckoo, with a long graduated tail and a bushy hair-like crest on the nape. It is smaller than the Eurasian Cuckoo (*Cuculus canorus*) and rather slim in build. The black plumage extends from the base of the upper mandible through the eye and over the nape, back, wings and tail. There is a broad white patch on the wings at the base of the primaries, and the tail feathers are white tipped. In good light the black plumage is glossed with green. The crest is erectible when the bird is excited and tends to form two separate 'fingers' parted in the middle of the crown when the bird is viewed from in front. The iris is dark brown and the strongly down-curved bill is black with raised nostrils. The zygodactyl feet (with first and fourth or outer toe directed backwards) are leaden grey. Young birds have the white parts of the throat and breast distinctly fulvous or pinkish buff, and the gape shows bright red when the bill is opened. Their crown and back is dull or dark grey not black. Adults have the whole of the throat and breast white. Both sexes are alike in plumage. The upper part of the tarsus is feathered giving this bird a bushy tibial area.

HABITAT, DISTRIBUTION AND STATUS

In Pakistan it is a summer season migrant arriving shortly before the onset of the monsoon and departing again in the autumn. It is a woodland bird being strongly attracted to the few remaining remnant patches of riverain forest, irrigated forest plantations and the better wooded areas of the Indus plains around old gardens and canal housing colonies. It is absent from Baluchistan and from most of the NWFP, and does not normally penetrate beyond the outer foothills of the Himalayas. The origin of these summer visitors has yet to be substantiated by ringing, but they are thought to come from the population which winters in east Africa, now considered the race *Clamator jacobinus pica* (see Becking, *JBNHS,* Vol. 78, No. 2, 1981). Many of these birds 'winter' in the southern hemisphere when the rainy season associated with the summer season of November to January provides an abundance of insect food (Whistler, *JBNHS,* 1928*b* and R. Simmons, *JBNHS* Vol. 34, 1930). The earliest arrivals noted in southern Punjab (Rahim Yar Khan) on 31 May and in lower Sind 16 June (author). H. Waite (1948) noted first arrivals in the Punjab Salt Range on 9 June. A flock of 8 was seen

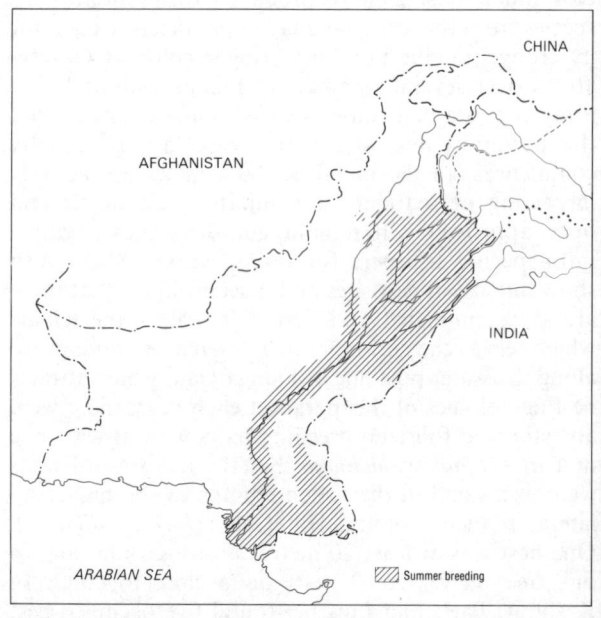

Map 220: Pied or Jacobin Cuckoo *Clamator jacobinus*

on 20 June in Pir Goth riverain forest in Thatta district, suggesting recent migrant arrival, as this bird is normally solitary in behaviour throughout the breeding season except during actual courtship. The latest dates adults were noted 23 October, though juveniles still being fed by their foster parents have frequently been encountered up to the end of October and first week of November. Stray birds are occasionally heard and seen in the month of June and July in the Murree hills, often arriving after dark and attracting attention by their incessant calling. Biddulph collected one as far north as Gilgit on 15 June (op. cit. 1881). Whitehead noted occasional birds around Kohat in September (op. cit. 1911). They are particularly plentiful around Islamabad and the lower Margalla hills. There is a resident population of this cuckoo in peninsular India and Sri Lanka and it is possible that some of this population moves northwards to Pakistan in the monsoon season, at which time they have been observed to disperse widely from their south Indian wintering haunts. However, the majority of birds collected from Pakistan, being larger is size than typical south Indian birds, are attributed to the east African migrant population. Status COMMON.

HABITS

They are insectivorous preferring soft insects especially caterpillars of *Lepidoptera*. In August at Malir a bird was watched with 2 Drongos (*Dicrurus adsimilis*), swooping over a lucerne field and capturing both Gryllidae and caterpillars. Stomach contents of 4 birds from Bihar, India contained Geometrid moth caterpillars and also *Chrotogonus* grasshoppers, Mole Crickets (Gryllotalpidae), Lady bugs (*Coccinella punctata*) and a weevil (Mason and Lefroy, 1912). They are capable of fluttering into the air to seize flying or jumping insects and obtain most of their food from the leaves and twigs of bushes and trees. They are usually solitary in foraging though several observers have noted that upon first arrival, they are often in small groups or flocks and that they indulge in much excited calling and aerial chases.

BREEDING BIOLOGY

It is thought that the pair bond is of very brief duration. Shortly after arrival in their summer grounds, males start advertising by calling. This they do intermittently throughout the day and even during dark moonless nights. These advertising calls are known to attract females in all the cuckoos. In the presence of a female, males have been observed fanning their tails to expose the white terminal spots conspicuously and also raising their head crests. Females when they fly away are hotly pursued by the male.

In the riverain forests and irrigated forest plantations, this species is brood parasitic exclusively upon the Jungle Babbler (*Turdoides striatus*). This same foster parent has been recorded around Delhi. In the Margalla foothills and dryer tracts, they have been noted parasitizing the Common Babbler (*Turdoides caudatus*). In Pakistan, Laughing Thrushes (*Garrulax* species) which have also been recorded as hosts (Ali and Riply, 'Handbook', 1969) do not occur sympatrically with the Pied Crested Cuckoo, which remains largely at lower altitudes. The eggs of this cuckoo are plain pale blue, very similar to the eggs of both *T. caudatus* and *T. striatus* but distinguishable in being larger, more rounded at the poles and glossless. In a year with good monsoon rains, I estimate that over 75 per cent of the Jungle Babblers' nests in the riverain forests of Thatta district are parasitized by this cuckoo. A. J. Gaston in a limited forest area outside Delhi noted 71 per cent of the Jungle Babbler's nests and 38.7 per cent of the Common Babbler's nests parasitized by this species (Gaston, *Journ. Animal Ecology*, Vol. 45, 1976). K. Eates found 9 eggs of this species in Sind. The earliest on 23 June and others throughout July and August. One nest of a Jungle Babbler in Malir in an orchard contained 2 Cuckoo's eggs and 2 Babbler's eggs, the latter smaller and glossier (unpublished MS notes). This was in a mango tree. Usually the cuckoo's egg hatches before the babbler's. Malcolm MacDonald noted in his Delhi garden that the cuckoo's egg hatched 2 days before the babbler's (MacDonald, 1960). Incubation period for Pied Crested Cuckoo's eggs is estimated at 11 days (Gaston, 1976). In such cases the babbler's nestlings usually perish from starvation, the larger cuckoo monopolizing the food supply. The cuckoo nestlings definitely do not evict their nest mates as has been observed for *Cuculus canorus* (J.H. Becking, op. cit.). In one brood reared in the author's garden at Malir, a young fledged cuckoo and a young Jungle Babbler were both watched being fed daily in the first week of October 1983. 3 adult babblers were involved in this feeding. MacDonald also noted no less than 4 adult Jungle Babblers feeding a Jacobin cuckoo fledgling in his garden (MacDonald, 1960). This feeding by nest-helpers, is common with *Turdoides* species (see relevant species accounts) and would be a big advantage to the young cuckoo, which despite being not much larger than its foster parents, develops very quickly. Macdonald noted that within 10 days of hatching, the nestling though only partly fledged, would leave the nest when disturbed by a human, but that it returned to the nest to be fed. At 11 to 12 days old they leave the nest for good (Gaston, 1976). Fledgling cuckoos keep up incessant food begging including a soft inflected whistle accompanied by quivering of the partly spread wings. Their red gape is also prominent. The fledgling is often fed on the ground or in a tree and follows the foraging babbler family group continuing to be fed for

2½ weeks after it was first noted out of the nest (observations Malir in author's garden). Undoubtedly in some instances more than one cuckoo's egg is laid in the same babbler's nest. At Khinjar in the riverain forest on 4 November two newly fledged cuckoos were watched flying together from bush to bush and soliciting food from the accompanying band of Jungle Babblers which appeared to feed both of them indiscriminately. It appeared as though both cuckoos had been reared from the same nest. Late hatched Pied Crested Cuckoos must have to undergo their autumn migration after the adults have departed as no adults have been seen in lower Sind after the third week of October whereas young birds still dependent for food upon their foster parents can frequently be seen up to the end of October and the first week of November

*VOCALIZATIONS

At the beginning of the season males give a rather weak single call, a slightly inflected whistle, monotonously repeated 5 to 12 times and this can be rendered as 'peearr-peearr-peearr' or 'piu-piu-piu'. As the monsoon progresses they call more loudly and intersperse these drawn-out whistles with a 3 syllable phrase, the second and third notes being uttered quickly and rolled together, 'kee-kwi-kwi' or 'piu-pipiarr'. This is preceded and followed by long sequences of 'piu-piu-piu' calls. They are most active in calling at dusk and during the early morning and will call during the night also. By late September they become completely silent.

Large Hawk Cuckoo *Hierococcyx sparverioides* (Vigors)

Synonym *Cuculus sparverioides* Vigors of Ali and Ripley's 'Handbook' (1969)

and S. Dillon Ripley's 'Revised Synopsis' (1982)

Plate **19**
Map **221**

DESCRIPTION

Body length 38-41 cms.
Tail length 17.5-22 cms.

As its name suggests this is rather a hawk-like looking cuckoo with long narrow pointed wings and a long broad tail rounded at the tip. This is cross-barred with 3 narrow creamy bars and bordered above these by darker sepia brown bars, merging gradually into paler grey-brown. There is a broad dark brown sub-terminal bar and narrow creamy white tip. The crown and nape is greyer brown and the throat and breast is white broadly cross-barred with dark blackish brown on the belly with vertical brown streaks on the throat and upper breast. An indistinct band of rufous chestnut encircles the upper breast below the white throat and this area is also overlaid with vertical brown streaks. The bill is down-curved, horny brown on the upper mandible and greenish slaty on the lower with a yellow gape. The iris is yellow as also is a naked eye-ring. The zygodactyl feet are greenish yellow and the tibia is well covered with bushy feathers. Both sexes are alike. Young birds are heavily streaked all over the breast with vertical bars, another character which enhances their likeness to *Accipiter* hawks.

HABITAT, DISTRIBUTION AND STATUS

This is a forest dwelling cuckoo largely confined in distribution to the eastern Himalayan valleys of lower well wooded hills extending eastwards to Malaysia where it is found only in the mountains, as well as

south-west China, Taiwan and the Philippines. The Indian population migrates to peninsular India during the winter, but not as far as Sri Lanka. Its status in Pakistan is unknown. Hugh Whistler did not himself encounter it, but records Colonel Rattray's statement (*JBNHS*, 1905) that it was rare at Murree but fairly common around Dunga Gali. H. Waite did collect a specimen on 2 June at Ghora Gali, below Murree and one in Murree itself on 3 June, both in 1947 (diary

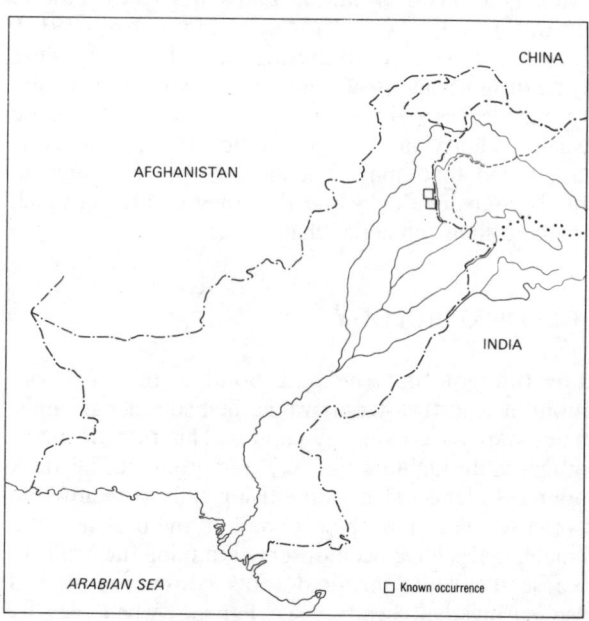

Map 221: Large Hawk Cuckoo *Hierococcyx sparverioides*

notes in manuscript). There are no sightings or records of this bird in recent decades and the author has never encountered it despite regular summer visits to Dunga Gali over 24 years. Paul Goriup thinks he saw one on 14 December 1980 at Nagar Parkar in the extreme south-eastern corner of Sind, which might indicate autumn passage of birds from the north-western part of the Himalayas, but he was not certain of this identification. Status UNKNOWN, possibly VAGRANT.

HABITS AND BREEDING BIOLOGY

This hawk cuckoo lays uniformly olive brown eggs in the nests of two foster species. The Nepal or Lesser Shortwing (*Brachypteryx leucophrys*) and the Streaked Spider Hunter (*Arachnothera magna*). Both these species lay eggs which closely match the Hawk Cuckoos eggs in colour. Turquoise blue eggs in the Stuart Baker collection housed at the British Museum at Tring, have been wrongly attributed to this cuckoo and probably belong to *Cuculus canorus* (See J.H. Becking, 'Notes on the Breeding of Indian Cuckoos', *JBNHS*, Vol. 78, No. 2, 1981). The basis for recent statements that the Large Hawk Cuckoo breeds in the Murree hills (Ali and Ripley, 1969 and Ripley, 'Revised Synopsis', 1982) apparently all derive from Colonel

Rattray's statement that it was common in Dunga Gali (Rattray, *JBNHS*, 1905). He relied heavily upon local native collectors in obtaining eggs and nests and consequently made some errors in identification since there was greater financial reward to such collectors when cuckoo's eggs were found! One such example is *Surniculus lugubris* the Drongo Cuckoo which he claimed to have found in the nest of a Grey Drongo (*Dicrurus leucophaeus*), a species upon which this cuckoo is never brood parasitic. Similarly in the Murree hills neither of the host species for the Large Hawk Cuckoo occur, and *Sparverioides* itself, in its overall distribution, is generally confined to higher rainfall and more sub-tropical moist deciduous or evergreen montane forest in the eastern Himalayas.

In the eastern Himalayas it has been reported egg laying from April to June. The adults feed upon caterpillars, including many hairy species, also spiders, beetles and Hemiptera bugs.

VOCALIZATIONS

Apparently a louder version of the Common Hawk Cuckoo (*Cuculus varius*), described as 'pipeeah-pipeeah' (Salim Ali, 'Handbook', 1969) or 'beer-frever breer-fever' (Fleming, *Birds of Nepal*, 1976).

Common Hawk Cuckoo *Hierococcyx varius* (Vahl)

Brainfever Bird *Cuculus varius* Vahl of S. D. Ripley's 'Revised Synopsis'

Plate **19**
Map **222**

DESCRIPTION

Body length 33-34 cms.
Wing length 17.7-19.3 cms. (Stuart Baker)
Tail length 15.5-19 cms.
Weight (1 female) 104 gms.

Very similar to the Large Hawk Cuckoo (*Hierococcyx sparverioides*). It is smoky grey-brown on the head, ear coverts, mantle and wing coverts with a white throat and belly, cross-barred with dark grey-brown on the lower breast and washed with orange-chestnut on the upper breast. The tail is long, quite broad and rounded at the tip and not graduated as in *Cuculus* species. It bears three narrow dark brown cross-bars, framed distally by narrow creamy bands, and with a broader sub-terminal brown bar. It can be distinguished in the field from the Large Hawk Cuckoo by the absence of bold vertical streaking on the throat and upper breast in adults, as well as by its more greyish, less brown back and white throat and chin. The chin of *sparverioides* bears two blackish streaks and is always darker. The bill is down-curved, with raised nostrils and a greenish colour becoming yellow around the gape. The legs and feet are bright yellow as also the

naked eye-ring. The iris is straw yellow. Juveniles are browner on the back, less grey and heavily streaked with dark brown tear-drop marks all over the breast and belly, with finer streaks on the throat. They lack the cross-bars of the adult's breast. The tail is not graduated as in *Cuculus canorus* and the rufous instead of grey upper breast serve to dintinguish this Hawk Cuckoo in the field from *C. canorus* or *C. saturatus*.

HABITAT, DISTRIBUTION AND STATUS

A summer breeding visitor to the Punjab, which prefers better wooded tracts, especially irrigated forest plantations, riverain forest and older bungalow gardens, orchards and groves. It appears to be totally absent from Baluchistan and Sind. A small number straggle westwards into the vale of Peshawar, the Malakand agency and into Swat and also into the Murree foothills up to Tret, 900 metres (3,000 feet) elevation.

Earliest arrivals heard calling 25 February at Changa Manga forest plantation and 27 March in Pirawala forest plantation near Khanewal (over 20 years' observations). It is plentiful around Sargodha and Mianwali and especially around Rawal lake and the Margalla hills where it arrives at least two months before *Clamator jacobinus*.

It has been seen (and heard) around Saidu Sharif in Swat in July. Latest date seen on 22 October on the banks of the Ravi river, Lahore. Nigel Hacking saw a single bird near Khaur in the Salt Range in mid-February 1979 and the author saw single birds at Balloki headworks and Changa Manga on 25 and 26 February and it is presumed that these were very early migrant arrivals. The increase in tree plantations in the Punjab with the spread of irrigation has undoubtedly favoured the spread of this cuckoo north-westwards. Whistler curiously did not record it in the Rawalpindi district though he noted Rattray's record for *sparverioides* (Whistler, *Ibis,* 1930). Nowadays in Rawalpindi even a non-interested person cannot help but be conscious of the incessant calling of this cuckoo from the end of March onwards, especially as it calls loudly through the hours of darkness. It is common in the olive scrub forest (*Olea cuspidata*) of the Salt Range, as well as on the plains and in the Margalla hills. Its status in winter is not known but in the absence of sight records during that season, it is believed that the population drifts south-eastwards into the warmer latitudes of peninsular India. Status COMMON in Punjab only.

HABITS

A largely arboreal cuckoo which hunts among bushes and in trees for insects' pupae and larvae, feeding mainly upon caterpillars. It will also eat soft fruits such as Banyan figs (*Ficus bengalensis*) and out of 17 birds examined in Bihar, 4 were found with figs in their stomachs though the majority had fed on caterpillars including Noctuid moths, hairy larvae of Lymantriid moths, also ants, *Camponotus compressus* and *Oecophylla smaragdina*, crickets *Gryllodes* and mole-crickets *Gryllotalpa* species, cut-worms (*Agrotis* spp.), *Chrotogonus* grasshoppers and the large nocturnal cricket *Schizodactylus monstrosus* (Mason and Lefroy, 1912). They fly low and swiftly through the trees, rising up abruptly to perch in a tree, just like *Accipiter* hawks. The usual hunting technique is to perch on a tree or telephone wire watching surrounding ground. A bird watched at Changa Manga fluttered down to the grass every half minute or so to seize orthopterous insects.

BREEDING BIOLOGY

It is a brood parasitic species, laying its eggs in the nest of another host species. Accounts of its egg laying habits in India, have never been fully authenticated (J.H. Becking, *JBNHS,* 1981) but in Sri Lanka

Map 222: Common Hawk Cuckoo or Brainfever Birds *hierococcyx varius*

fledglings have been reliably observed being fed by Jungle Babbler (*Turdoides striatus*) foster parents. It is presumed that this species also serves as host in Pakistan as it occurs wherever the Hawk Cuckoo does during the summer season. Stuart Baker's collection of glossy turquoise blue eggs attributed to this species, have upon analysis by electrophoresis of egg white protein and by shell morphology (electron microscope), been proved to be wrongly attributed in many instances, such blue eggs in the collection belonging to both *Cuculus canorus*, and *Clamator jacobinus* (Becking, op. cit.).

*VOCALIZATIONS

The main advertising song is a stereotyped three or four-noted phrase 'whee-whee h'yar ho', the last two notes short and rapidly repeated. It does indeed sound like the words 'brain fever' the first two notes being accented. This can be repeated 5 to 20 times in a rising pitch and tempo, sometimes interspersed by single notes 'wheer-wheer-wheer' each rather short sharp and rising in scale until it again breaks into the 3 noted 'wheeh-pi-whit' or 'brain fever'. Often this calling continues throughout the night with several birds chorusing in close proximity. Another quite different call, recorded in the Margalla hills at dusk, was a repeated loud ringing call without inflection like an electric bell 'trerr-treerrr-trerrr'. These calls are given in quick sequence and in a slightly rising scale in groups of three and have a 'burring' or reverberating quality.

Plaintive Cuckoo or
Grey-bellied Plaintive Cuckoo *Cacomantis passerinus* (Vahl)

Plate **19**
Map **223**

Cacomantis merulinus (Scopoli) of K.H. Voous's 'Recent List'

Indian Plaintive Cuckoo *Cacomantis merulinus passerinus* (Vahl)

of Ali and Ripley's 'Handbook' and Stuart Baker's 'Fauna Series'

TAXONOMY

There appears to be some lack of agreement amongst taxonomists in the treatment of the Grey-bellied and Rufous-bellied forms of the Plaintive Cuckoo. Since the Rufous-bellied form is more truly Oriental and tropical in distribution, it is presumed that K.H. de Voous in his 'List of Recent Holarctic Bird Species' (*Ibis*, 1977) is referring to the Grey-bellied form when he lists only *Cacomantis passerinus* as the Plaintive Cuckoo, since it is this form which extends into the Himalayas. We prefer to follow the treatment given by S. Dillon Ripley in his *Synopsis of the Birds of India and Pakistan* (Revised ed. 1982). He splits them up into two species *Cacomantis passerinus* (Vahl), the Indian Plaintive Cuckoo (or Grey-bellied form), and *Cacomantis merulinus* (Scopoli), the Rufous-bellied Plaintive Cuckoo. This is because B. Biswas (*Ibis*, Vol. 93, pp. 596-8, 1951) has shown that both *passerinus* and *merulinus* overlap in range in the eastern Himalayan foothills, where there is no hybridization or intergrades. This treatment is also followed by King et al., (*Birds of Southeast Asia Field Guide*, 1975), by the *Pictorial Guide to the Birds of the Indian Subcontinent* (Ali and Ripley, BNHS, 1983) and in the *Guide to the Birds of Nepal* (Inskipp and Inskipp, 1985).

DESCRIPTION

Body length 22-23 cms.
Wing length 11.4-12.2 cms. (Stuart Baker),
   11.3-12.0 cms. (Ali and Ripley).
Tail length 9.9-11.9 cms. (Stuart Baker),
   10.5-11.5 cms. (Ali and Ripley).

This is a small slim cuckoo about the size of the White-cheeked Bulbul (*Pycnonotus leucogenys*) with which it associates. It however has a long tail with the outer rectrices shorter and white tipped, so that it is graduated or wedge shaped. The male is confusingly like the Dark Grey Cuckoo Shrike (*Coracina melaschistos*) in appearance, the head, neck and upper breast being dark steely blue-grey. The throat and upper breast being not quite such a dark grey and gradually merging to white on the lower breast. The wing coverts and flight feathers are more brownish with long slender curved wingtips. The tail is darker than the rest of the body, and is a gun metal blue, each tail feather

narrowly white tipped. In good light the tail shows indistinct darker cross rays of fine bars. The iris is brown and the legs and feet are brownish yellow. Females are dimorphic in plumage. They can be similar with a slightly more uniform and paler ashy plumage or they can be bright rufous chestnut on crown, nape and back, with fine black cross-barring all over, the breast and throat being white, also cross-barred dark grey. In this hepatic colour morph, females strongly resemble the Bay Banded Cuckoo (*Cacomantis sonneratii*). They can only be separated by examination of the tail tips which narrow to a point in *sonneratii* whilst those of female 'hepatic' *merulinus* remain uniformly broad down to their rounded tips.

In flight a small white patch shows conspicuously on the under surface of the wings at the base of the primaries. Juvenile birds are always rufous chestnut and cross-barred all over their upper parts.

HABITAT, DISTRIBUTION AND STATUS

This cuckoo is extremely limited in distribution and occurrence. It is somewhat similar to *Streptopelia chinensis* in its habitat preference in Pakistan, entering the outer foothills of the Murree hills and the sub-tropical pine zone, *Pinus roxburghii* but preferring the rather open scrub covered hills rather than dense pine forest, which also comprise this zone. Also it is only found during the summer or monsoon months and does not occur much below 600 metres (2,000 feet) nor above 1,800 metres (6,000 feet) elevation. It has not been noted earlier than 20 May (in the Margalla hills) nor later than the end of September. The author has also encountered it in Swat below the Shangla pass in late May 1985, and this is presumed to be the extreme western limit of its range. It occurs sporadically in Hazara ditrict, in the summer in the Mansehra plain around the side valleys and better wooded ravines, as well as around Abbotabad and eastwards to the Lehtrar and Kahuta valleys. H. Waite collected it at Bansra Gali at 1,500 metres (5,000 feet) 22 June and at Rewat south-east of Murree at 1,500 metres (5,000 feet) on 6 August, both in the Murree hill range. In the monsoon it can be heard in the Daman-i-Koh ravine just to the west of Islamabad down to 450 metres (1,500 feet) elevation and even in the woodland around the shores of Rawal lake. Its arrival often coincides

Map 223: Grey-bellied Plaintive Cuckoo *Cacomantis passerinus*

with that of the Pitta (*Pitta brachyura*) into the Margalla hills, which is in the last week of May. In 1965 a male frequented the forest rest house gardens at Ghora Gali at 1,800 metres (6,000 feet), but it has not been seen or heard at that elevation subsequently. Status LOCAL but FREQUENT.

HABITS

They call mostly at dusk, during the early part of the night and again around dawn. A rather secretive bird, males will ascend to the topmost branch of a bush or tree to call. They feed principally upon caterpillars gleaned from leaves of trees and bushes, being rather arboreal in habits. They like the thick tangled undergrowth of ravines and are rarely seen flying in the open. One was watched with a large hairy caterpillar in its bill which it shook vigorously before swallowing. Dr. Jerdon noted one endeavouring to catch a butterfly with its feet (Mason and Lefroy, 1912), and the hairy larvae of the moth *Nepita conferta*, as well as the bug *Dysdercus cingulatus* have been identified among its prey (Ali and Ripley, 'Handbook', 1969).

BREEDING BIOLOGY

This cuckoo is brood parasitic upon the Tailor Bird (*Orthotomus sutorius*) as well as *Prinia* and *Cisticola* fantail warblers. Where it occurs in Pakistan it is sympatric with *Orthotomus sutorius* and *Prinia hodgsonii* and it is not known which of these two possible host species is favoured. Around Rawal lake *Cisticola juncidis* breeds, but this cuckoo has been heard in this comparatively open low elevation habitat only occasionally. In all instances these host species construct deep purse shaped nests, often with small side entrances which would make it difficult for the cuckoo to directly deposit its egg into the nest chamber, an act which has never been observed. However their eggs are thick shelled and it is believed that they are deposited directly from the bird's cloaca and not transferred by its bill. Several types or colour morphs of eggs from this species have authentically been recorded, these colour types being known as 'gentes'. Eggs in Tailor Bird nests or Fantail Warbler's (*Cisticola*), are light pinkish or bluish white in ground colour marked with reddish brown blotches and speckles, like their host's. Another type of egg adapted to the nests of the Ashy Wren-warbler (*Prinia socialis*) is deep chestnut red in colour, again matching the host's, but this type of egg is mainly found in south India. The eggs are usually rather longer and more oval in shape than the host's and are quite glossy. Egg laying can occur any time between May to September being synchronized with that of the host species' breeding. In the 'Handbook' (Salim Ali, Vol. 3) a remarkable case is recorded in which a chick of this cuckoo, found at Poona in Maharashtra state, India, was being fed both by the host *Nectarinia asiatica* and by a pair of Tailor Birds (*Orthotomus sutorius*). One likely explanation for the Tailor Birds is that the large chick protruding from the Sunbird's nest, with its constant begging had elicited a feeding response from the Tailor Birds because they had recently lost their offspring (Becking, 1981, op. cit.). But the choice of a Sunbird as host is the only known record for this cuckoo and indeed unusual.

*VOCALIZATIONS

The most common advertising call comprises 3 repeated rather drawn-out whistling calls in a rising scale and in a rather minor key which can be rendered as 'keveeear-keveear-keveear'. Another variation involves a more complicated sequence of unevenly stressed plaintive whistles 'wheeeh-whooh-pe-ti-weear pe-ti-weear' with the first two syllables louder and emphasized and the second phrase 'pe-ti-weear', more rapidly repeated. Also 'wheeh-whooh' rising and falling followed by 'pe-ti-weear, pe-ti-weear'. As in most of these Oriental cuckoo species, calling can occur intermittently throughout the hours of darkness.

Indian Cuckoo or Short-winged Cuckoo *Cuculus micropterus* Gould
Plate **19**
Map **224**

DESCRIPTION

Body length 33-34 cms.
Wing length 19-20.7 cms. (both sexes) (Salim Ali)
Tail length 14-16 cms.

This cuckoo looks very similar to both the Common and Himalayan or Oriental Cuckoos (*Cuculus canorus* and *Cuculus saturatus*) both in size and shape and plumage details. The calls of all three species are very distinctive, but identification in the field is otherwise difficult. The Indian Cuckoo is usually browner, less blue-grey on the mantle and wing coverts with bolder more widely spaced cross-barring on its breast, if a large sample of birds can be compared with the European Cuckoo. Also there are often rufous margins to the mantle and upper breast feathers, lacking in *C. canorus*. The best field point, if it can be observed from behind or above, is the rather pale brownish tail with prominent broad darker slaty sub-terminal band. Also its iris is not pale yellow but rather reddish brown. Whilst *C. canorus* has pale golden irides, *C. saturatus* can also have dark irides. The feet and toes are bright yellow and zygodactyl as in all cuckoos and there is a bright yellow orbital ring. The bill is arched on the culmen, of a greenish brown tinge with the nostrils raised. The head, cheeks, throat and upper breast are uniformly pale grey as in all these three cuckoos. In flight the wings are long, and slender which, with short secondaries and long tail, give the bird a very hawk-like appearance. Females are usually browner on the throat and upper breast, less blue-grey and immature birds are very distinctive having pale creamy or white feathers on the crown, nape and breast, each feather tipped blackish in a scale-like pattern. The tail is rufous brown with darker cross-barring, the extreme tips of all tail feathers being white in both sexes and juvenile birds, also there are white spots on the tail quills.

HABITAT, DISTRIBUTION AND STATUS

Col. Rattray claimed that it was the most common cuckoo in Murree (Rattray in Stuart Baker, *JBNHS*, Vol. 17, p. 356) but Hugh Whistler never came across it and doubted his identification. (See discussion below under Breeding Biology). Bates and Lowther never came across it in Kashmir (*Breeding Birds of Kashmir*, 1952) but A.E. Jones collected it in the Simla hills (specimens in BNHS collection at Bombay). In Pakistan it is probably an erratic and only sporadic rainy season visitor to the outer foothill regions of north-west Punjab, rarely penetrating to the higher ranges of the Murree hills. E. Fernando (pers. com., 1968) recorded it in the summer of 1967 and 1968

around Marala barrage in Sialkot district. From the first week of May and through early June a single male was heard calling around Rawal lake near Islamabad in 1983 and its calls were recorded by David Corfield. Comparison of its song with recordings made by the author in Malaysia were identical. The bird was also seen and Corfield noted clearly its rather spaced, bold breast barring. There are no other recent records. Status RARE VAGRANT to north-eastern foothills only.

HABITS AND BREEDING BIOLOGY

Largely an arboreal cuckoo preferring better wooded regions. Its food includes all kinds of soft bodied insects and comprises principally larvae of *Lepidoptera*, but the Mole Cricket *Gryllotalpa africana* and *Sphex lobatus*, *Hypsa alcifron*, and a Melonthid grub *Oxycetonia albopunctata*, have been identified in stomach contents (Mason and Lefroy, 1912).

In breeding the normal hosts in the subcontinent are 'drongos', including *Dicrurus leucophaeus* in the north-west Himalayas (A.E. Jones, 1941), *Dicrurus adsimilis* in the Siwaliks and Duars (foothill ranges) and further east *Dicrurus paradiseus* the Large Racket-tailed Drongo. Colonel Rattray's eggs (7 in number)

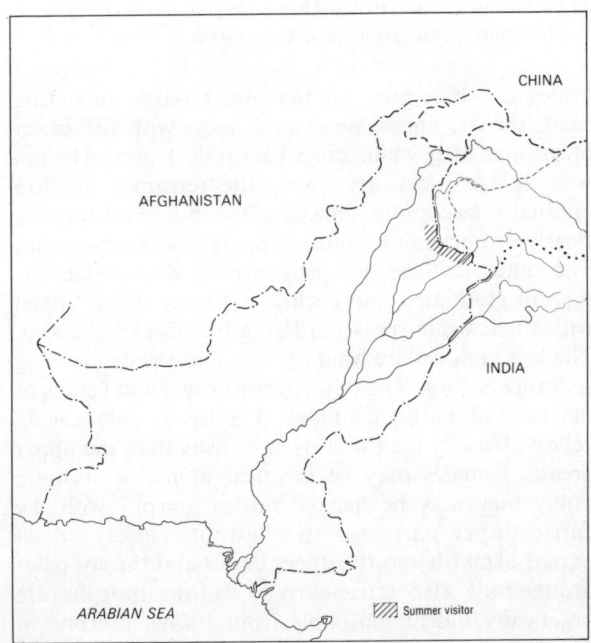

Map 224: Indian or Short-winged Cuckoo *Cuculus micropterus*

collected from Murree and now in the Stuart Baker Collection (British Museum, Tring) are sky blue and recent electron microscopic examination of their shell structure by J.H. Becking, has revealed that most of these are not eggs of cuckoos at all, but of 'babblers' *Timaliinae* (Becking, *JBNHS*, op. cit., 1981). Eggs of this cuckoo appear to be pink or whitish pink in ground colour with violet or carmine pink, ill defined blotches and spots and greyish underlying markings, so that they mimic drongo eggs very closely. In the USSR in Amur land, this cuckoo parasitizes the Brown Shrike (*Lanius cristatus*), where it also lays similar type eggs which match those of its host. The most reliably documented account of the breeding of this cuckoo is described by Dr. Irene Neufeldt 'Life History of the Indian Cuckoo (*Cuculus micropterus* Gould) in the Soviet Union' (*JBNHS*, 1966). The incubation period as described by Neufeldt is 12 days, 2 or 3 days shorter than that of the host (shrike) eggs.

*VOCALIZATIONS

Males call at all times of the day and night during the breeding season, but most incessantly in the evening and after dusk and again pre-dawn. It is a four-noted song on a descending scale, comprising four loud clear fluting calls in two couplets, each pair being repeated on a lower scale 'kwer-kwah-kwah-kurh'. Other authors have rendered its call as 'kyphal-pakka' or 'orange pekoe'. These phrases are cited by Dr. Salim Ali in many of his writings (*The Book of Indian Birds*, 1972, *Field Guide to the Birds of the Eastern Himalayas*, 1977, and 'Handbook', Vol. 3, 1969). Its call comprises simple single toned notes, easy to imitate by whistling and do not suggest words to this writer. An individual male was timed, giving 23 calls per minute (Ali and Ripley, 'Handbook', 1969). Tape recordings (author's collection) of birds from Malaysia and north-west India sound identical and the call is stereotyped.

Eurasian Cuckoo *Cuculus canorus* Linnaeus

Plate **19**
Map **225**

DESCRIPTION

Body length 33 cms.
Wingspan 55-60 cms.
Wing length 20.9-24.3 cms. (males),
   21.3-23 cms. (females) (Ali and Ripley)
Tail length 15.2-18.3 cms. (males),
   15.7-17 cms. (females) (Ali and Ripley)
Weight 90 gms. (1 male) (Salim Ali), 81-128 gms. (males),
   81-91 gms. (females) (Paludan, Afghanistan)

Males are blue-grey on the upper parts, including head, throat, upper breast and back, with the lower breast and belly white, cross-barred dark grey. The tail is a darker slaty grey with the terminal portion gradually becoming darker black but without any clearly distinct sub-terminal band as in *C. micropterus*. The outer feathers are progressively shorter making the tail graduated and each is narrowly white tipped with small white spots bordering the outer webs also. The legs and feet are bright yellow and the orbital ring is orange-yellow. The bill is horny brown and yellow at the base of both mandibles. The iris is pale golden yellow. Usually the throat is paler ashy than the upper breast. Females may be identical to males or more rarely they may be hepatic rufous morphs with the entire upper parts rufous chestnut, closely cross-barred blackish and the upper breast and throat, paler orange-buff also cross-barred. Rufous morphs are practically indistinguishable from rufous morphs of *Cuculus saturatus* but lack prominent dark cross-barring on the rump and the barring on their crown and nape is also less distinct. Hepatic phase females

have been collected in Gilgit (specimen British Museum). The male of this cuckoo is distinguished with difficulty in the field, from *C. saturatus* by the primary wing coverts on the underside of the wing being cross-barred whereas they are white unmarked in *saturatus*. Also the tail tip in *saturatus* is more noticeably black at the base, than that of the Eurasian Cuckoo. *C. canorus* is easier to separate when compared with *C. micropterus*, because the latter also has a conspicuous darker terminal tail bar and paler browner tail in contrast to the overall much darker tail of *C. canorus*. Juveniles of *C. canorus* have a conspicuous white nuchal spot also which *C. saturatus* sometimes lacks.

HABITAT, DISTRIBUTION AND STATUS

Like the European Cuckoo, this bird winters in east Africa and is a summer breeding visitor to the mountainous regions of Pakistan. It may occasionally be seen on passage in the autumn in Sind, but most of the spring migration takes place further westwards through Baluchistan. It prefers higher altitude valleys above 1,800 metres (6,000 feet) and juniper forest-clad slopes in Baluchistan and the higher inner ranges of the NWFP, and is not a bird of the plains. Small numbers haunt the Himalayan coniferous forest zone from about 2,100 metres (7,000 feet) elevation and upwards, where it can be sympatric with *Cuculus saturatus*, but its preferred habitat is the alpine slopes of the outer hill ranges and upland valleys in the inner

Himalayan mountains where it can be sympatric with *Cuculus poliocephalus*. It is common in summer in the Deosai plateau and Baltistan as well as Gilgit and Hunza and in the Kaghan valley, Azad Kashmir and Murree hills. It has been recorded in July up to 3,650 metres (12,000 feet) in alpine meadows and dwarf juniper scrub in Gilgit and Hazara district. It can be first heard calling in Baluchistan from the end of March and in the Himalayas from mid to late April. Birds on passage have been seen (author) at Khanewal, southern Punjab on 17 April, and in Rawalpindi on 21 April, and Islamabad as late as 2 June. However, numbers can be seen and heard in the Murree hills from mid-April, though most of these are still on passage to the inner more northern mountain ranges. They return on autumn migration in September. Seen Rawalpindi 19 September, (H. Whistler, *Ibis*, 1930). H. Waite collected specimens (Brit Mus.) on return migration from Sakesar in the Salt Range on 28 July and 2 August. In Karachi district one was seen (author) feeding in a ploughed field on 23 September. Ticehurst encountered it in lower Sind on 16 September and 1 October, and also saw 2 birds on spring passage on 5 and 20 May (op. cit., 1923). Status COMMON.

HABITS

Like all the cuckoos they are normally solitary in habits and except when calling in the breeding season, inclined to be secretive and not easy to see. They are almost wholly insectivorous, eating a variety of insects, especially larvae of Lepidoptera. Their method of

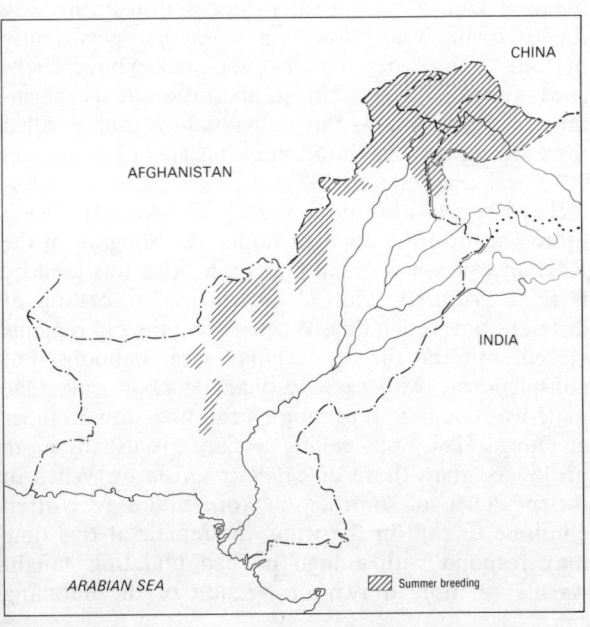

Map 225: Common Eurasian Cuckoo *Cuculus canorus*

foraging is to perch on a leafy branch or close to the ground and watch for insect movement, caterpillars being gleaned from bushes, amongst the leaves of trees and less commonly from the ground. Besides hairy caterpillars, they have been noted catching Hemiptera bugs and Hymenoptera (Paper wasps and bees). They have also been noted in Europe taking the eggs of smaller passerine birds. An individual watched at evening time on the summit ridge of Kadir Gali perched on low rocks and darted to the nearby ground to seize caterpillars. In a space of about 12 minutes it occupied 3 different perches and successfully collected 9 hairy caterpillars (all the same kind with green and brown longitudinal stripes). Each caterpillar was jiggled around in the bill until the head was crushed whereupon it was swallowed whole. There is evidence from the author's observations that both females and males may be territorial in the breeding season though the pair bond is quite transient and females are polyandrous. In the Kao valley catchment, Dunga Gali, a male cuckoo with an aberrant call 'cuck-cuck-kooh' has been seen and tape recorded in the same patch of hillside over a period of 3 successive summers. Though it is possible that more than one individual had this type of call, it is unlikely. Females drive off rival females and are responsible for finding and selecting potential hosts for their brood parasitism.

BREEDING BIOLOGY

Compared with the other *Cuculus* species in the region, it is evident that the Eurasian Cuckoo is more adaptable in its choice of brood parasite hosts, with greater plasticity in egg 'gentes'. Probably populations which lay a particular pattern or colour of egg are more host specific or restricted in their choice, but in Pakistan, *Emberiza* buntings and *Anthus* Pipits with creamy or white eggs speckled with brown, and Desert Finches (*Rhodospiza*) and *Saxicola* chats with pale blue and brown speckled eggs, can all be hosts. In the subcontinent, particularly in the Himalayas, a number of cuckoos' eggs of other species have been incorrectly attributed to this species in the Stuart Baker collection (British Museum at Tring) and probably an even greater number of the plain blue type of *C. canorus* eggs have been wrongly attributed to other species (See J.H. Becking, *JBNHS*, 1981). In Baluchistan where other cuckoo species do not occur, confusion is less likely. Two types of eggs have been collected from that province. One, yellowish white in ground colour, thickly spotted with brownish red and a few purple dots, has been found in the nest of the Pied Bush Chat (*Saxicola caprata*). Another newly hatched cuckoo chick has also been found in the nest of *Anthus similis*, the Persian Rock Pipit on 9 June (Meinertz-hagen, *Ibis*, 1920). C.H. Williams found three cuckoo eggs, all in the nests of the Streaked Scrub Warbler

(*Scotocerca inquieta*) and these were spotted with more grey and dark brown than the egg described by Meinertzhagen. He also found a fledgling cuckoo in the nest of the Desert Finch (*Rhodospiza obsoleta*) which lays a pale blue egg sparsely speckled with dark brown. (Williams and Williams, *JBNHS*, 1929). In the Murree hills a plain blue egg often occurs and is laid in the nests of the Streaked Laughing Thrush (*Garrulax lineatus*) which also has a blue egg, and in the nests of the Dark Grey Bush Chat (*Saxicola ferrea*) which lays a pale blue egg speckled with red brown (Magrath, *JBNHS*, 1908a). Also a plain blue egg has twice been found in the nest of a Himalayan Rubythroat in the Kaghan valley (Whitehead, *JBNHS*, Vol. 23, 1914). In higher alpine pastures such as in Hazara district they chose as hosts Rosy Pipits (*Anthus pelopus*), Stonechats (*Saxicola torquata*) and Blue-fronted Redstarts (*Phoenicurus frontalis*), the two latter species with pale blue eggs.

The breeding biology of this cuckoo has been well studied in Europe (Wyllie, 1981) and is reasonably well known in the subcontinent. Females are polyandrous or promiscuous, mating with a succession of males when they are successful in finding nests of suitable hosts in the right stages. Incubation period varies from 11 to 13 days (average 12.4 days), being invariably 2 to 3 days less than that of the host species (Cramp *et al.*, Vol. 4, 1985). The female lays her egg directly into the host's nest, clinging to the side of side-entrance, globular nests and ejecting the egg against the tiny entrance hole. It has been well documented that the altricial cuckoo chick for the first 5 or 6 days of its life possesses a peculiarly sensitive patch on its mid-back region, any contact with which produces an impulse to heave upwards. Thus all unhatched eggs or nestlings which come into contact with this area are heaved over the nest rim leaving the young cuckoo the sole recipient of the foster parents' food supply. Period from hatching to fledging varies from 17 to 21 days (average 19 days) (Cramp *et al.*, op. cit.). As observed in Kaghan valley and at Dunga Gali male cuckoos have definite calling territories with favoured trees, often on prominent ridges of the mountainside, which they visit irregularly over a period of weeks to call during May and June. Normally, they become less vocal towards the end of June and silent by July, but in high altitudes in parts of Baltistan they continue to call in July. Their calls are strongly attractive to females. Synchronization of the host's breeding cycle is crucial for successful parasitism and it appears that the female cuckoo is successful in searching for and locating suitable nests by watching the nest building activity of its preferred host species. During two consecutive evenings while walking up the Saif-ul-Maluk side valley (of the Kaghan valley), two cuckoos were observed flying into and landing on a hill slope where a pair of Collared Bushchats or Stonechats had built a nest in the stone wall of a terraced field embankment.

On both occasions the alarm rattles of both Stonechats and the bubbling calls of the cuckoos attracted attention. From the bubbling calls of both birds, it is probable that they were both females and presumably territorial rivals, as their behaviour in chasing each other, suggested aggression and competition rather than pair bond behaviour. The Stonechat's nest was complete but contained no eggs and it might be surmised that both female cuckoos had recognized the nest building behaviour of the Stonechat pair, as potential host species and were checking to see if egg laying had commenced. It was in open treeless alpine country so that our presence probably disturbed the cuckoos from making closer investigation, but their purposeful landing on the open ground and visiting the site on 2 consecutive afternoons, indicated almost certainly that they had localized the Stonechat's nest.

*VOCALIZATIONS

For European readers the advertising call of the male cuckoo is too familiar to require description, being a much loved sound woven into the folk music and poetry of many nations. In the Murree hills several individuals have been recorded, but usually only at the beginning of the summer, which persistently reversed the two-noted call and called 'koo-kuk' instead of 'kuk-koo'. When excited or in visual contact with a female, the male utters a series of rather low pitched rapid chuckling calls 'gak-gak-gagagak'. Females when approaching a male utter rather resonant 'bubbling' calls, well rendered as 'quick-quick-quick' by Dr. Salim Ali. This call sounds closely similar to those made by female *Cuculus saturatus*. Another recording made in Dunga Gali in early June and again in May the following year, was of a male that persistently uttered a three-noted call 'cuck-cuck-whoo, cuck-cuck-whoo'. This was not an aberration at the beginning of the season, as this individual invariably called thus throughout the duration of our stay (2 months in 1983 and 2 months in 1984). *C. canorus* is much less inclined to call after dusk than is *C. saturatus*, being more vocal during daylight hours. At Shogran in the Kaghan valley a very tall pine on a ridge was used by both *C. canorus* and *C. saturatus* for calling at different times each day. When calling the bill remains closed, but the throat or gular area 'balloons' out conspicuously with each resonant 'cuckoo' call. One male was counted, repeating its call non-stop 36 times at Dunga Gali. Such calling sessions are usually more prolonged than those of *Cuculus saturatus*. When in the presence of another cuckoo, males will often continue to call on the wing. the female at this time may respond with a high pitched ululating 'laugh-warble', a more drawn-out version of the bubbling call.

Himalayan or Oriental Cuckoo *Cuculus saturatus* Blyth

Cuculus saturatus horsfieldi Moore

Plate **19**
Map **226**

DESCRIPTION

Body length 30-33 cms.
Wingspan 51-57 cms. (Cramp *et al.*)
Wing length (sub-sp. *horsfieldi*) 18.3-19.3 cms. (188 specimens)
(Vaurie) 18.5-19.5 cms. (28 specimens from western Himalayas and USSR) (D.R. Wells)
Tail length 14.5-16 cms.
Bill length 16-17.5 mm. (D.R. Wells)

Though it averages slightly smaller than the Eurasian Cuckoo (*C. canorus*) the difference is not discernible in the field and these two cuckoo species look almost identical. *C. saturatus* is however slightly darker, more slaty blue-grey on the upper parts than *C. canorus* and the cross-barring on its breast is often rather bolder, being broader and more widely spaced. It also has a wider and more prominent sub-terminal dark bar on the tail tip. However, it must be re-emphasized that such differences are not constant. The terminal tail bar also merges gradually with the rest of the tail and is difficult to see if the human observer is on the ground and the bird is calling from a tree, may be little better than a silhouette against the sunlit sky. The inner primary wing coverts on the underside of the wing in *C. saturatus* are plain white whereas in *C. canorus* they are cross-barred, and as they sometimes droop their wings when calling, this field character can occasionally be discerned. Both sexes have orange irides and a chrome yellow or orange ring of naked skin around the eye. The bill is blackish on the upper mandible and greenish on the lower, and comparatively weak looking. The zygodactyl feet and legs, partly covered by the bushy tibia feathering, are yellow. One male observed closely in July in the Murree hills had rufous chestnut margins to the feathers on the crown, and the base of the tail feathers, characters typical of some *Cuculus micropterus* specimens, but this rufescence has not been noted on any of the skins in the Bombay Natural History Society collection. Females are sometimes encountered which are hepatic morphs as in *C. canorus*, being entirely chestnut red on the upper parts, cross-barred with black. One such was observed at Malkandi (Kaghan valley), recognizable because of the prominent broad dark cross-barring on its crown and hind neck, and all across its upper tail coverts. These areas are much less distinctly cross-barred in rufous morphs of *C. canorus*.

HABITAT, DISTRIBUTION AND STATUS

This cuckoo is a summer breeding visitor to the northern mountain regions of Pakistan, more or less confined to mixed deciduous and coniferous temperate forest above 1,800 metres (6,000 feet), in areas which are influenced by the summer monsoon rains, i.e. the lower Kaghan valley, Azad Kashmir (Neelum and Jhelum valleys) and the Murree hill range. It is never encountered in the dryer or colder alpine slopes above the tree line, so favoured by *C. canorus* in the Himalayas, nor at the elevations of the sub-tropical Pine zone (*Pinus roxburghii*) where *Cacomantis merulinus,* or *Cuculus micropterus* are encountered. Where it occurs in such forested areas it is sympatric with *C. canorus*. C.H. Whitehead (*JBNHS*, 1911) did not encounter it in the Safed Koh range, nor has it been recorded from Swat or Gilgit (Scully, *Stray Feathers,* Vol. 10, 1881). Biddulph's specimens were a case of mis-identification (Biddulph, *Stray Feathers,* Vol. 9, 1880, p. 315). In Ali and Ripley's 'Handbook' (Vol. 3, 1969) the population breeding in the north-west Himayalas is ascribed to the nominate sub-species *saturatus,* and these authors consider it is probably resident in the Himalayas.

Recent morphometric analysis of extensive museum collection material by Dr. Wells has shown that the larger sub-species *C. saturatus horsfieldi* is represented in specimens from north-western Himalayan regions, including the Murree hills (D.R. Wells, 'The Genus Cuculus: two amendments to the Handbook' etc., *JBNHS*, Vol. 69, 1972, p. 179). This sub-species

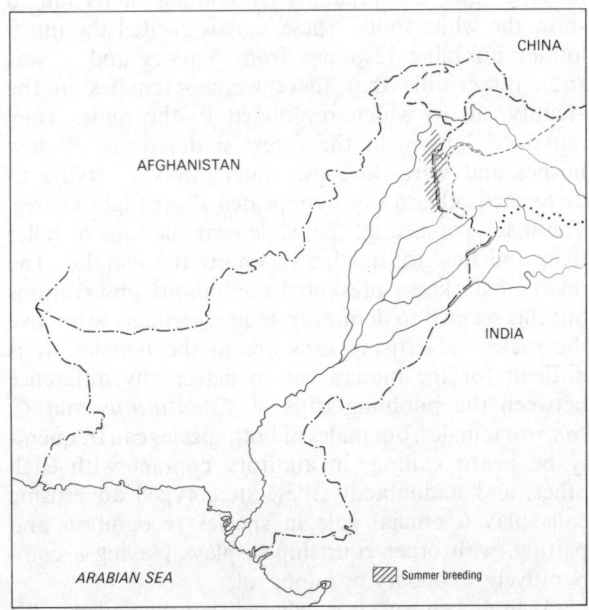

Map 226: Oriental or Himalayan Cuckoo *Cuculus saturatus*

winters in Borneo, Indonesia and the Philippines (C. Vaurie, Vol. 1, 1965, p. 573). Part of this population migrates northwards across the Himalayas to breed in central and eastern USSR but some of these migrants undoubtedly breed in the Pakistan Himalayas. Status FREQUENT but of limited distribution.

HABITS

A strong and swift flyer very reminiscent of a Sparrow-hawk (*Accipiter* spp.). When in flight its long narrow wingtips enable it to turn and twist adroitly among the tree tops. It feeds mostly upon larvae of Lepidoptera, but also takes all soft-bodied insects found in the upper tree canopy including *Tettagonia* bush crickets, Hemiptera bugs, Praying Mantises (Mantoidea) and Cicadas (Cicadidae.) In August in the Murree hills when thousands of cicadas are emerging it has been seen capturing newly hatched adults. It is largely arboreal in feeding, but has been observed fluttering down to Viburnum bushes to pick off caterpillars. It is more crepuscular in calling and general activity than *C. canorus* in the Murree hills, but will call throughout the daytime also. It arrives in the Murree hills in the first few days of May and can be heard calling up to about mid-July, though much less frequently after the end of June.

BREEDING BIOLOGY

On 19 June in Kao forest, Dunga Gali at dusk a male which had been calling earlier, attracted attention because it was uttering excited short rasping and hacking calls, accompanied by fanning of its tail to show the white spots. These signals elicited the much louder bubbling response from females and it was soon discernible that there were 3 females in the vicinity, all of which responded to the male. They remained largely in the forest understorey, in low bushes and were obviously intent only in trying to drive each other away by repeated short flight chases. The male, perched all the while near the tops of taller trees and flew to and fro to follow the females. The onset of darkness prevented continuous observations but this seemed to demonstrate how strongly attractive the males' advertising calls are to the females. It is difficult for the human ear to detect any difference between the bubbling calls of *C. saturatus* and *C. canorus* females, but males of both species can frequently be heard calling, in auditory contact with each other, and undoubtedly these stereotyped advertising calls play a crucial role in species recognition and pairing, with other courtship displays playing a comparatively secondary or minor role.

As in *C. canorus*, it is believed that the pair bond is transient, females copulating with several different males, and that territories of the males are rather loosely overlapping whilst females will drive off rival females near a potential nesting site.

This species in the north-west Himalayas is believed to be brood parasitic only on warblers of the genus *Phylloscopus*, but in Darjeeling Osmaston found two nests of the Rufous-bellied Niltava, *Niltava sundara* which contained cuckoo's eggs. These when sent to Stuart Baker were identified as those of *Cuculus saturatus* and correspond in description exactly to typical eggs of this species (*JBNHS*, Vol. 15, p. 514). In the USSR their eggs have been found in nests of *Anthus* spp. as well as *Phylloscopus* warblers (Dement'ev and Gladkov, Vol. 2, 1952). Further east in Malaysia and Indonesia it is brood parasitic on *Seicercus* Flycatcher Warblers. In the Murree hills *Phyllosocopus occipitalis* is the usual host and the author has twice been lucky to encounter newly fledged cuckoos being fed by the Large Crowned Leaf Warbler. The first on 10 July at Dunga Gali in a thicket of *Viburnum nervosum*, was still unable to fly strongly and had barely sprouted tail feathers. A second on 28 July on the north ridge of Murree, was almost full grown with well developed tail, and was so large that the harassed parents actually had to perch on a twig above its head in order to reach its bright red and seemingly cavernous gape. The mantle feathers of these juveniles were broadly tipped white in a scale-like pattern and there were no clear white patches in the nuchal area which characterize *C. canorus* in the same fledgling stage.

In the past, many nestlings and eggs of this cuckoo from the eastern part of the Himalayas were wrongly attributed to the Small Cuckoo (*Culculus poliocephalus*), thus leading to some confusion in breeding habits. *C. poliocephalus* also breeds in the Himalayas and north-eastwards into China. *C. saturatus*, because of its choice of host, lays comparatively small eggs which are quite variable in size but still considerably bigger than those of its host. They are usually rather long, elliptical and bluish white or buff in ground colour, very sparsely speckled with small spots and tiny lines of reddish brown or purple grey, often with a zone of spots round the broader end. Sometimes these markings are almost absent. Their eggs fairly well mimic those of some *Phylloscopus* Leaf Warblers. It is known that nestlings of this cuckoo have the same sensitive areas in the mid-dorsal region which induces them to push up and outwards any objects coming into contact with their backs. Thus the host species' eggs, or newly hatched nestlings are very soon heaved out of the nest. *P. occipitalis* builds its nest in quite deep cavities (sometimes under tree roots or in holes in walls and trees) and, since the Himalayan Cuckoo is certainly too large to squeeze through such entrance holes, this leads to speculation that eggs might be laid first on the open ground and then transferred by the female in her bill into the nest. Bates and Lowther (*Breeding Birds* of Kashmir, 1952) refer to a fully developed oviduct egg found by Brooks in a female collected on 17 June and give May and June as the principal breeding months. They describe a nest found on 6 June in a cavity in the slim hole of a Witch Hazel (*Parrotia jacquemontiana*) shrub, in which one of the *Phylloscopus* host's eggs had been cracked by the fall of the cuckoo's egg, but they declined to infer whether the egg had been ejected directly from the cuckoo's oviduct while clinging to the side of the *Parrotia* tree trunk, or had been transferred by its bill, but observations on *C. canorus* would suggest that in this species also, the egg is ejected direct from the cloaca into the nest entrance.

*VOCALIZATIONS

Males give their advertising calls from early May to early July and are most vocal at dusk and just before dawn. They tend to have regular territories or calling places, though they will occasionally visit quite new areas and start calling, in their search for a mate. Some birds fluff out their breast feathers, droop their wings and cock up their tails while calling, others depress their tails but also droop their wings. Close inspection of calling birds shows that, as in *C. canorus*, the bill remains closed while calling but the gular area 'balloons' out in a regular manner with each call, whilst the tail is depressed slightly and the head bent forward simultaneously with each 'boom'. The call is four-noted, resonant and low pitched, very similar in tone to that of the Hoopoe (*Upupa epops*). However, the latter gives 3 'whoop-whoop-whoop' calls whereas *saturatus* calls in series of four, rarely of five and when heard close up, the first call is always slightly higher in tone and briefer than the three following. The whole phrase takes barely much more than one second and is repeated at intervals of less than one second. One bird called 48 times before pausing for ten minutes and recommencing from the same tree top. After an interval of silence, the first call is a series of four subdued croaks and during this time the gular area becomes inflated and the next series of calls are all resonant 'hoops'. Females have a loud ringing call not unlike *Picus* Woodpeckers, which can also be likened to a rapid quavering or bubbling noise.

Little Cuckoo or Lesser Cuckoo *Cuculus poliocephalus* Latham
Small Cuckoo of Ali and Ripley

Plate **19**
Map **227**

DESCRIPTION

Body length 25 cms. (Fleming) 27.9 cms.
    (King *et al.*) 26 cms. (Ali and Ripley) 27.2 cms. (Dement'ev)
Wing length 14.2-16.2 cms. (Ali and Ripley) 14.3-16.1 cms.
    (East African and Seychelles) (D.R. Wells)
Tail length 12.6-13.7 cms. (Ali and Ripley)
Bill length 13-14.6 mm. (East African and Seychelles specimens)
    (D.R. Wells)

This cuckoo is a small edition of *C. canorus* and *C. saturatus*, being difficult to distinguish in the field unless size can be accurately gauged, as it is not very much smaller than *C. saturatus* and indeed overlaps in size with the races *C. saturatus lepidus* which breeds in Sabah and Sumatra, and *C. saturatus insulindae* which breeds in Borneo (see D. R. Wells, and J.H. Becking 1975, *Ibis*, Vol. 117). The wing lengths given in Stuart Baker's 'Fauna Series' (Vol. 4, p. 142) are too high and actually refer to specimens of *Cuculus saturatus lepidus* from Sikkim (D.R. Wells, *JBHNS*, Vol. 69, 1972, p. 184).

In general appearance it is slaty blue-grey above with long narrow wings and dark slaty graduated tail. The throat is pale ashy and the lower breast and belly, white boldly cross-barred with dark grey. The outer edge of the webs of each tail feather and the tips are spotted with white. The legs and feet are yellow as also the fleshy eye-ring. The bill is blackish with the base of both mandibles yellow and the nostril slightly raised as in all the *Cuculus* species. The iris is brown not pale yellow. Helpful field points are the rather uniformly dark tail without a very noticeable blacker tip (as in *saturatus*) and the inner lining of the wing shoulder is grey not clear white as in *C. saturatus* or cross-barred grey-brown as in *C. canorus*. Sometimes the rump is less brownish than in *C. canorus* and more concolorous with the back, but it is often paler grey contrasting with the tail. Ben King (*Field Guide to the Birds of South-east Asia*, Collins, 1975) describes the rump as dark and concolorous with the tail, this character serving to separate it in the field, but it is not evident in specimens from Pakistan.

HABITAT, DISTRIBUTION AND STATUS

There is accumulating evidence that this cuckoo spends the northern hemisphere winter in east and south Africa as specimens have been collected from Tanzania, Kenya, Zimbabwe and south Africa from November to April (J.H. Becking, *JBNHS*, 1981, p. 220). It has not been recorded on passage through the Middle East (un-like *C. canorus* which also winters in east Africa), so presumably flies across the Indian

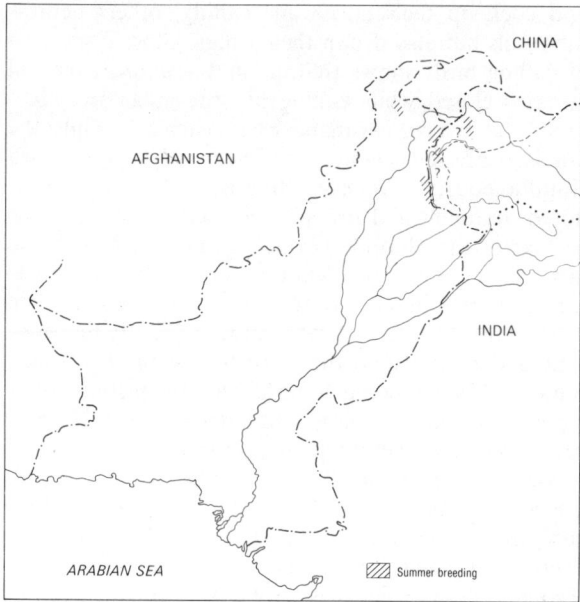

Map 227: Small or Little Cuckoo *Cuculus poliocephalus*

Ocean. It has been recorded in Sri Lanka and south India, presumably birds on passage. During the summer breeding season it spreads across the Himalayas and north-eastwards into China, the Korean peninsula and Japan. It occurs widely but sparsely in the north-western part of the Himalayas as a summer breeding visitor. First arrivals in the Murree hills heard between 26 May and 30 May in 3 consecutive years. It has been regularly heard calling in the lower Kaghan valley at Shahran and in Machiara valley, Azad Kashmir and also once at Miandam (1,800 metres/6,000 feet elevation) in Swat on 28 May (author obs.). It can still be heard calling occasionally up to early August and appears to arrive in its mountain breeding haunts later than the other cuckoos. Characteristically it is a bird of Himalayan mixed coniferous and deciduous forest extending to the upper limit of the tree-line and penetrating farther north-wards into the mountain ranges than does *C. saturatus*. A bird collected by H. Waite from Nurpur Shahan at 610 metres (2,000 feet) in the Margalla hills on 5 September was presumably already on autumn passage. Status FREQUENT

HABITS

Much more vocal at night than *C. canorus* and inclined to be more restless, frequenting a valley or ridge for a day or two and then moving elsewhere. On first arrival it may often be encountered around the bungalows and gardens in Murree and it is less shy and retiring than *C. saturatus* which sticks to uninhabited forest. It has a very distinctive advertising call or song

and can be heard intermittently throughout the day during June on the north side of the Murree ridge whereas further north in the Galis it is seldom encountered. One seen in late July at Dunga Gali was very tame and allowed close approach whilst it perched on a low bush. Stuart Baker describes two specimens, one having a stomach full of small blue Coleoptera (beetles) and the second as having a mass of tiny bees (Baker, 'Fauna Series', Vol. 4, 1927). The bird seen in Dunga Gali had a green unidentified caterpillar in its bill. In Kashmir, Magrath noted it 'coming down to the ground in the evening to search for caterpillars in the undergrowth'.

BREEDING BIOLOGY

There are conflicting statements about the choice of brood parasitic host made by this cuckoo in the north-western Himalayas as Leaf Warblers (*Phylloscopus*) and Wren-babblers (*Pnoepyga*) are cited as foster parents (Ali and Ripley, 'Handbook', 1969). J.H. Becking in his researches believes that it specializes only on one type of host, the small warblers of the genus *Cettia*. Certainly it frequents those parts of the Murree hill range where *Cettia fortipes* occurs most regularly in summer. In Japan where it has been well studied, it regularly lays its eggs in the nest of the Chinese Bush Warbler (*Cettia diphone*). The egg is invariably quite a broad oval, comparatively large for the size of the bird and hence closely approximating in dimensions the comparatively small eggs of *Cuculus*

Plate 19:
(1) Common Hawk Cuckoo or Brain Fever Bird *Hierococcyx varius*, both sexes.
(2) Large Hawk Cuckoo *Hierococcyx sparverioides*, both sexes.
(3) Oriental or Himalayan Cuckoo *Cuculus saturatus*, both sexes.
(4) Common Cuckoo *Cuculus canorus*, both sexes.
(5) Little or Small Cuckoo *Cuculus poliocephalus*, both sexes.
(6) Indian or Short-winged Cuckoo *Cuculus micropterus*, both sexes.
(7) Pied or Jacobin Cuckoo *Clamator jacobinus*, both sexes.
(8a) Grey-bellied Plaintive Cuckoo *Cacomantis passerinus*, female.
(8b) Grey-bellied Plaintive Cuckoo *Cacomantis passerinus*, male.
(9a) Common Koel *Eudynamys scolopacea*, male.
(9b) Common Koel *Eudynamys scolopacea*, female.
(10) Greater Coucal or Common Crow Pheasant *Centropus sinensis*, both sexes.
(11) Sirkeer Malkoha or Cuckoo *Taccocua leschenaultii*, both sexes.

Plate 19

(1)

(2)

(3)

(4)

(5)

(6)

(7)

(8a)

(8b)

(9a)

(9b)

(10)

(11)

Plate 20

facing p. 457

saturatus. When parasitic on *Cettia fortipes* and *C. pallidipes* its egg is terra-cotta red or deep chocolate brown in colour so it closely resembles the host's eggs. B. Osmaston also obtained a terra-cotta coloured egg from the oviduct of a cuckoo of this species which he shot near Darjeeling (*JBNHS.*, Vol. 15, p. 514). Stuart Baker's collectors obtained plain white unmarked eggs, said to be of this species, which would well match the eggs of some *Phylloscopus* leaf warblers, but there is no conclusive evidence as yet that this cuckoo parasitizes Leaf Warblers. In parts of Hazara and Gilgit where I have heard it calling well above the coniferous tree line, the most widespread breeding bird and potential host species is Tickell's Leaf Warbler (*Phylloscopus affinis*), but as in so many parasitic cuckoos a great deal more investigation into the breeding biology is required before concluding definitely that plain white eggs are laid by this species and whether Leaf Warblers are selected as hosts for egg laying.

*VOCALIZATIONS

Magrath notes the predilection of males to call, not at dawn but starting some two hours after sunrise and then again in late afternoon and intermittently during the hours of darkness (cited in Bates and Lowther, 1952). The call is a 6 or 7 noted phrase, the first three notes rising in scale, the remainder 'accelerando' and declining in scale. It is a bit reminiscent of the chattering song of the Yellow-billed Magpie (*Cissa flavirostris*) and has been rendered as 'pretty-peel-lay-ka-beet' (Fleming, *Birds of Nepal*) or 'pot-pol-chip-to-you' (Ben King, *Birds of South-east Asia*). It is strikingly different from the two-noted resonant calls of *C. canorus* and *C. saturatus* and more warbling in timbre and might not be associated with a cuckoo were it not for its persistent stereotyped song given during the hours of darkness. Females have a similar though slightly slower in tempo 'bubbling' or quavering call to *C. canorus*.

Drongo Cuckoo *Surniculus lugubris* (Horsfield)

DESCRIPTION

Body length 25 cms. including tail 13-15 cms.

It has a straight tail forked at the tip and is a glossy metallic black all over so that it strongly resembles the Drongo (*Dicrurus adsimilis*). However, the under-tail coverts are barred white and there is white barring on the base of the outer tail feathers which *Dicrurus* lacks. Its bill is longer and more slender than that of the Common Drongo.

HABITAT, DISTRIBUTION AND STATUS

It is an Oriental species associated with the outer foothills and lower Himalayan tracts, typically found in evergreen rain forest and moist deciduous hill forest and plantations. The *Synopsis of the Birds of India and Pakistan* (1961) and the 'Revised Synopsis' (1982) by Dillon Ripley include the Murree hills within the distributional range of this largely residential cuckoo. This is based only upon a record by Col. Rattray of an egg supposed to be of this species taken from a nest of the Ashy Drongo (*Dicrurus leucophaeus*) on 28 May 1899 at 1,524 metres (5,000 feet) in Murree. He also claimed to have shot a Drongo Cuckoo in the same vicinity (Rattray, *JBNHS*, 1905). More recent studies indicate that this cuckoo does not lay its eggs in the nests of any Drongo species (as cited in Ali and Ripley, Vol. 3, 1969) but that it is brood parasitic on babblers, *Alcippe* species and *Rhopocichla* species. In Indonesia *Stachyris* and *Trichastoma* babblers also serve as hosts (J.H. Becking, 1981, op. cit.). Examination of the series of skins in the British Museum and Bombay Natural History Society collections indicates that Kamaon is about the most western extremity of the range of this species in the Himalayas. It should therefore be excluded from the **Checklist** of Pakistan birds until fresh evidence comes to light.

Plate 20:
(1) Striated or Pallid Scops Owl *Otus brucei*, sexes alike.
(2) Mountain or Spotted Scops Owl *Otus spilocephalus*, sexes alike.
(3) Asian or Collared Scops Owl *Otus bakkamoena*, sexes alike.
(4) Oriental or Indian Scops Owl *Otus sunia*, sexes alike.
(5) Eurasian Scops Owl *Otus scops*, sexes alike.
(6) Asian Barred Owlet *Glaucidium cuculoides*, sexes alike.
(7) Spotted Owlet or Spotted Little Owl *Athene brama*, sexes alike.
(8) Little Owl *Athene noctua bactriana*, sexes alike.
(9) Barn Owl *Tyto alba*, sexes alike.
(10) Collared Owlet or Collared Pygmy Owl *Glaucidium brodiei*, sexes alike.
(11) Dusky Eagle Owl or Dusky Horned Owl *Bubo coromandus*, sexes alike.
(12) Himalayan Tawny Owl *Strix aluco himalayensis*, sexes alike.
(13) Northern Eagle Owl or Great Horned Owl *Bubo bubo*, sexes alike.
(14) Brown Fish Owl *Ketupa zeylonensis*, sexes alike.

Koel *Eudynamys scolopacea* (Linnaeus)

Plate **19**
Map **228**

DESCRIPTION

Body length 40-44 cms.
Wing length 18.2-20.5 cms. (males),
   17.9-20.3 cms. (females) (Ali and Ripley)
Tail length 18-20 cms.
Weight 229 gms. (1 female) (Salim Ali)

Unlike the previous genera of cuckoos described, the Koel is markedly dimorphic in the plumage of the sexes. Males are dark black with metallic blue gloss in good light and have a long slender round tipped tail. The bill is pale greenish and rather strongly built with a down-curved culmen and the iris is ruby red. The zygodactyl feet and legs are blue-grey. Females are dark brown above, profusely spotted with white on crown, throat and mantle and with creamy spots and bars on the flight feathers and tail. The breast is scaled with dark brown tips to the white feathers and the rest of the lower breast and belly is creamy white, with dark brown cross-bars. The bill is more yellowish horny. Newly fledged chicks of both sexes are predominantly black on the head and mantle with their bills blue-black, thus mimicking their foster parents, the House Crow (*Corvus splendens*). The female juvenile is however barred on her under parts and tail, and slightly browner on the crown and mantle. The young males have some chestnut buff tips to wing coverts giving their shoulders a spotted pattern. Their gapes are horny yellowish (observations of young in Malir garden).

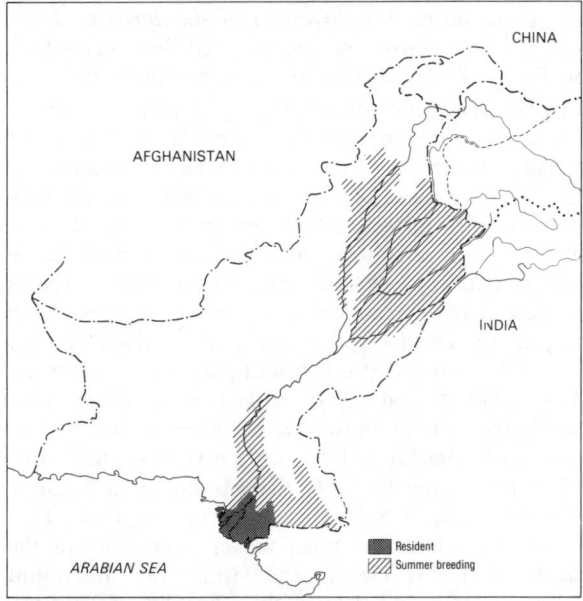

Map 228: Common Koel *Eudynamys scolopacea*

HABITAT, DISTRIBUTION AND STATUS

This is a plains dwelling species which enters the Punjab largely as a summer season breeding visitor migrating south-eastwards into India with the approach of winter. In lower Sind from Hyderabad district southwards there is a resident year round population supplemented in late March by an influx of migrants from India. A few penetrate to the better wooded regions of the NWFP such as around Bannu and Kohat towns. First arrivals have been noted around Islamabad in the north and Multan in the south in early April though most of the population arrives in May. The spread of tree plantation and orchard plantations following irrigation developments in the Punjab have led to a rapid increase in the summering population of this cuckoo since the early part of the century. H. Whistler encountered it very occasionally in Jhang district (*Ibis*, 1922) whereas today it can be heard calling throughout the irrigated tracts of the Punjab in summer. Status COMMON.

HABITS

It is an arboreal cuckoo and largely frugivorous specializing on the fruits of *Ficus* trees in Pakistan. In the summer season they are very noisy, calling day and night but they are less vocal in winter and more inclined to congregate in loose bands when trees bearing ripe fruit are available. They have been observed feeding on the ripe fruit of the Pipal tree (*Ficus religiosa*), Banyan (*Ficus bengalensis*) and *Ficus glomerata,* also upon the flowers of Papaya (*Carica papaya*). They also occasionally take insects, flower nectar, eggs of small birds such as Red-vented Bulbuls *Pycnonotus cafer,* and the berries of palms such as *Areca* and *Caryota urens* and the fruits of the Persian Lilac (*Melia azedarach*). Despite being so noisy, they are furtive when approached and tend to freeze, concealing themselves amongst the foliage. They have never been noted drinking or descending to the ground though fledglings will sometimes hop around on the ground.

BREEDING BIOLOGY

Very little of actual courtship has been recorded or observed. Males can frequently be seen flight chasing females, both calling excitedly. Males when perched in a tree fan their tails and droop their wings when in the vicinity of a female and make short staccato calls unlike their loud advertising whistles. They are brood parasitic on the House Crow (*Corvus splendens*) in Pakistan, which breeds largely during the monsoon season. In India they also lay eggs in the nests of Jungle

Crows (*Corvus macrorhynchos*) but this species is confined to the Himalayan regions in Pakistan where the Koel does not penetrate. They also occasionally adopt the Black Drongo (*Dicrurus adsimilis*) as a host (T.E.H. Smith, *JBNHS*, 1950). K. Eates on two occasions saw female Koels drive incubating crows off their nests with much wing flapping and believes that they then laid eggs in their nests. Immediate subsequent inspection on both occasion revealed two crow's eggs unbroken and one Koel's egg. The female was observed to fly from the nest without carrying any eggs in her bill (unpublished MS notes). The eggs are mostly laid in June and July and are smaller in size than House Crows' eggs but similarly patterned, being greenish grey in ground colour, blotched and speckled all over quite densely with reddish brown. The relationship between Koels and Crows is generally antagonistic with pairs of House Crows frequently observed success-fully harrying and driving Koels away from their territory. Yet Koels on occasion will threaten Crows, opening their bills to expose their crimson red gape which seems to frighten the Crow (observed Malir garden). Also very frequently both species can be seen feeding or perching in the same tree within a few metres of each other, and apparently tolerant of each other's presence. In the author's garden at Malir in the summer at least 3 females and 4 males appeared to be in permanent residence within an area of little more than 0.2 hectares, a surprisingly high density. In the same region generally 2 if not 3 pairs of House Crows also bred each summer and it was possible to obtain a number of observations. Over a ten year period eight crows' nests were under observation when the young were fully fledged and about to leave the nest. In seven instances these contained young Koels. In one nest 4 Koels were reared, 2 males and 2 females (easily separated by plumage differences) and there were no crow chicks surviving. In 4 other instances only one crow chick and one Koel were both observed being fed by the same pair of crows and in another nest there were 2 crow fledglings and one male Koel. In another nest 2 Koels emerged and one young crow, all leaving the nest on the same date and being fed by the same crow parents. In the eighth nest which was seen to be visited more than once by female Koels at the start of egg laying, and from which I was confident that 2 Koel

chicks would emerge, there were instead four crow chicks successfully fledged and no Koels. These observ-ations suggest that more than one female may success-fully parasitize the same Crow's nest, and that the newly hatched Koels make no attempt to evict their sibling crow chicks if they hatch and obtain enough food. The four cases in which only 2 chicks fledged, one being a Koel does however indicate that the shorter incubation period of the Koels, 13 to 14 days, versus 16 to 17 for *Corvus splendens* and their more persistent begging gives them a competitive advantage over their crow siblings some of which probably die of starvation or get accidentally jostled out of the nest, as it is rare for House Crows to lay less than a clutch of 3 or 4 eggs. Dates at which the Koel chicks in this Malir garden were strong enough to leave the nest permanent-ly, varied between 28 June and 22 September. Parents were observed feeding the young Koels for well over 14 days after leaving the nest. One well grown young of unknown age was still being fed by crow foster parents on 15 October. The young Koel was very noisy at this stage, begging whenever crows perched in its vicinity with very hoarse rasping calls and usually fluttering only one of its wings. Its open mouth revealed a bright red throat which seemed to induce the crow parents to regurgitate from their crops.

*VOCALIZATIONS

The advertising call of the male which is so typically associated with the hot weather, is a drawn-out upward inflected whistle or shriek, 'kooweeah, kooweeah', the last part being more emphatic. This call is often repeated 7 or 8 times, immediately eliciting response calls from other males in the vicinity. Often they start calling before dawn and continue inter-mittently throughout the day but tending to call more regularly and vociferously at dusk. Males also call during flight chases and when excited, usually in the presence of conspecifics and this is a different call, comprising a long yodelling continuous shrieking call 'kwa-kwa-kwa-keow-keow' each phrase quite short and more melodious than the drawn-out advertising calls. Females give rapid repeated 'kik-kik-kik-kik' calls when excited, both in flight and when perched in a tree.

Sirkeer Malkoha or Cuckoo *Taccocua leschenaultii* Lesson

Synonym *Phaenicophaeus leschenaultii*

Plate **19**
Map **229**

DESCRIPTION

Body length 42-44 cms.
Wing length 15.3-16.8 cms. (both sexes) (H. Whistler)
Tail length 21-23 cms.
Bill length 30-34 mm. (H. Waite)

This is a rather large cuckoo, about the size of the House Crow (*Corvus splendens*), with a conspicuously long graduated tail, very short rounded wings and a deep powerfully down-curved bill. Its crown, nape and back are a drab earthy brown with rather loose untidy

Map 229: Sirkeer Malkoha or Cuckoo *Taccocua leschenaultii*

plumage. The feathers on the crown and nape are filamentous, giving its head a loose hair-like outline, and both crown, nape and mantle feathers bear narrow darker brown shaft streaks. The throat and upper breast are paler creamy chestnut, becoming more rufescent on the lower breast and some of the breast feathers, particularly on the sides of the neck, having dark shaft streaks. The legs are strongly developed with relatively long tarsi and the feet are zygodactyl. They are slaty grey coloured and the iris is very dark brown, with noticeable long eye-lashes. The most striking features are the cherry red bill strongly down-curved and horny yellow on the tip, and the long graduated tail with broad white tips to the outer tail feathers. These white tips become very conspicuous when the bird fans its tail during display. The outer tail feathers are darker blackish brown but the central tail feathers are the same drab earthy brown as the rest of the upper parts. When at rest, the tertials completely cover the primary wingtips. Both sexes are alike.

HABITAT, DISTRIBUTION AND STATUS

This is an Oriental species which is sedentary in habits and in Pakistan is very thinly and only locally distributed in uncultivated tracts, well covered with dry deciduous thorn scrub along the foothills of the Himalayas or along the coastal belt of Sind where the monsoon influence is stronger. It is practically absent from all but the extreme north-eastern regions of Punjab and the south-eastern border regions of Sind. In the former province a reasonable population sur-

vives along the west bank of the Chenab river in *Acacia modesta* and *Adhatoda vasica* scrub from Marala barrage up to the frontier with India. It also occurs in the northern part of Sialkot district. Around Islamabad in the foothills and especially those ravines flanking the Margalla hills, it also survives in a reasonably strong widespread population. There are no recent records from Lahore or Kasur districts of Punjab.

In Sind it survives very precariously in the Hab valley in nullahs (seasonally dry stream beds) with some *Tecomella undulata* scrub jungle, also around Umarkot in the pockets between high sand dunes, where the thorny Acacia (*Prosopis spicigera*) and *Ziziphus* bushes provide adequate ground cover. It must always have been rare in these Sind localities as C.B. Ticehurst never came across it ('Birds of Sind', *Ibis*, 1923) and K. Eates in over 30 years of residence in Sind only encountered it three times during the 1930s and 1940s, once at Umarkot, once near Ghulamjo Ghot in Sukkur district and the third time in the Pabb hills in Lasbela just west of the Hab valley (unpublished MS notes). Menesse (*JBNHS*, 1939) records seeing occasional birds flushed out of cotton fields while shooting partridges in Hyderabad division both north up to Hala and south of Hyderabad city to Katiar during the late 1930s. The author saw one in December 1979 near Jangshahi in Thatta district, and one in Miani forest plantation just north of Hyderabad in mid-March 1985. It has been reliably reported several times from the Hab valley during the 1970s (J.A.W. Anderson and Jack Coles, pers. comm.). There are many recent sight records from Sialkot and Islamabad regions. Status RARE.

HABITS

They are shy and secretive birds, hard to see because of their habit of running swiftly through the thick underbush and disappearing behind grass clumps when disturbed, rarely offering a good view. They are weak fliers, rarely travelling far on the wing, proceeding by rather laboured flaps and frequent glides, but will perch readily in trees especially when roosting at night. If they think they are observed, they prefer to skulk inside dense clumps of thorn bush. They run with tail cocked up and body crouched low, taking big strides and looking more like a mammal than a bird with their earthy brown plumage. Their food includes berries of *Ziziphus* and also a variety of insects and mollusca which they can dig out of the ground, a feeding method for which their powerful bills are well adapted. Dharmakumarsinhji (1972) records lizards as a favourite food item, as well as mice and the occasional fledgling bird, but noted that in Gujarat State (India), large insects (Orthoptera) comprised the bulk of their diet. One observed (author) in June in the Margalla hills flew overhead with a half grown *Calotes versicolor* lizard in its bill. They also eat locusts (*Schistocerca gregaria*) and caterpillars and *Chrotogonus* grasshoppers. They often associate in pairs and the pair bond may be of fairly long duration. The one observed in Miani forest, foraged on the ground in a ripe mustard crop and appeared to be pecking at the seed pods.

BREEDING BIOLOGY

These cuckoos are non-parasitic, forming stable monogamous pair bonds and building their own nest. E.H. Gill (*JBNHS*, 1923) and Salim Ali (Ali and Ripley, 1974) describe their courtship which involves both sexes bobbing their heads and necks up and down, with the plumage fluffed out and the tail raised up and fanned fully, so that the white tips show up in a conspicuous band. The bills are held open at this time and the birds first thrust their heads into the air often holding this position for a minute (Gill, op. cit.).

Their nests are usually cleverly concealed inside a *Euphorbia cauducifolia* clump or in thickets formed by a pollarded *Prosopis spicigera* tree, and the nest itself consists of a platform of twigs lined in the middle with green leaves. The normal clutch is 2 to 3 egg which are chalky white marked. Both sexes share incubation duties. This is such a shy bird, that E.N. Lowther, an experienced bird photographer, gave up all attempts to erect hides near nests, as even after the chicks had hatched the parent birds deserted the nest immediately after it was discovered (*A Bird Photographer in India*, 1949). The nesting season extends from May to September. A bird seen carrying a lizard in its bill (10 June) was quite undisturbed and appeared to be taking food to its young, though I failed to locate the nest due to the impenetrable nature of the vine festooned thorny bushes into which it had disappeared.

*VOCALIZATIONS

A group of 3 birds watched at Marala uttered a soft 'kokh-kokh-kokh' call rather reminiscent of that of the Coucal (*Centropus sinensis*). E.N. Lowther describes a similar call which he attributes to an alarm call (Lowther, 1949). It has another chattering call similar in tone to the Rose-ringed Parakeet (*Psittacula krameri*), rendered as 'wit-wit-wit-tit-titirrr' (Fleming and Fleming, 1976) whereas Dharmakumarsinhji (1972) likens its call to the Indian Nightjar (*Caprimulgus asiaticus*). It is much less vocal than most cuckoo species outside of the breeding season. The individual encountered at Miani forest uttered short high pitched 'kwit' calls, at intervals of 3 to 5 seconds, after going to roost for the night and until well after darkness had fallen.

Greater Coucal or
Common Crow-pheasant *Centropus sinensis* (Stephens)

Plate **19**
Map **230**

DESCRIPTION

Head and body length 48-53 cms.
Wing length 20.5-23.2 cms. (both sexes), (Stuart Baker)
    22.5-23.9 cms. (Sind - Ticehurst)
Tail length 22-26 cms.
Weight 362 gms. (sex unknown) (Salim Ali)

This is a large, ground dwelling cuckoo, with a broad graduated tail relatively shorter than the tail of the Sirkeer, whereas in body size and bulk it is considerably bigger than the Sirkeer. It has a glistening metallic blue-black plumage over the head, body and tail, offset by bright chestnut wings. The bill is very strong and deep with the upper mandible hooked downwards at the tip. The legs are robust with long tarsi and toes. They are leaden grey to blackish with the hind toe shorter than the outer or third toe which also points backwards in this zygodactyl bird. The claw on the hind toe is straight and very elongated. In good light the tail feathers show light and dark cross rays and there are narrow pale shaft streaks to the feathers of the neck, mantle and breast. The irides are crimson. Juveniles have browner tails with conspicuous cross rays on the central rectrices.

HABITAT, DISTRIBUTION AND STATUS

Very widespread throughout the Punjab and Indus plains and extending into the perennially irrigated tracts of the NWFP. It is confined to areas near major canals or rivers where thick cover prevails, generally dominated by *Saccharum munja* and *Saccharum spontaneum* clumps or dense bushes of *Capparis aphylla* and *Salvadora*. It avoids open desert terrain, hilly and rocky country or intensively cultivated tracts and its stronghold is around the margins of seasonal swamps, lakes and canal headworks where seepage creates ideal conditions. It is a sedentary species. It is absent from Baluchistan and the northern mountain areas. Status COMMON.

HABITS

Though so widespread and common they are generally rather skulking and secretive birds which run into cover or fly laboriously to the nearest clump of thick vegetation, into which they disappear immediately they think they have been detected. They are also rather solitary in habits. Their powerful bills enable them to be omnivorous in diet. They will eat ripe berries and all kinds of insects. I have watched one pecking away at the carcass of a bullock, and they have frequently been observed catching lizards and amphibia, mainly frogs. They are also predators on all small passerine bird nests, eating eggs and nestlings with equal gusto. They have also been recorded successfully catching snakes (*Natrix piscator*), crustaceans, mollusca and a variety of insects. In the study at Pusa Agricultural Research Institute, items identified from 5 birds' stomachs, included *Chrotogonus* grasshoppers, larvae of *Agrotis* species (cut-worms), shells of the watersnails Vivipera and Ampoullaria, Hemiptera bugs, ants (*Camponotus compressus*), small frogs, lizards, spiders, centipedes and beetles (*Copris orientalis* and Carabidae spp.) (Mason and LeFroy, 1912). Their zygodactyl feet enable them to clamber up vertical reed stems with facility and they are well adapted to shelter and roost in reed beds where tree cover is not available.

BREEDING BIOLOGY

The pair bond is long lasting and both sexes take part in nest building and incubation. Courtship involves chasing by the male with both sexes having their breast

Map 230: Greater Coucal or Common Crow Pheasant *Centropus sinensis*

and flank feathers puffed out and their tails cocked up and expanded, to the accompaniment of wing drooping and in the case of the soliciting female sometimes fluttering of the half opened wings. The nest is usually extremely difficult to find, being constructed inside one of those dense thickets so typical of their habitat, where tall grasses grow up through a large spreading clump of thorny bushes. The nest itself is a large globular dome shaped affair with a side entrance and the inner chamber is often lined with green leaves. Usually 3 eggs are laid, less commonly four, which are chalky white unmarked. Most nests are built entirely of coarse grass, leaves and reeds but occasionally twigs are used. They normally start nest building at the onset of the monsoon, nests being found in Multan district during July and August. In lower Sind, however, Eates found nests at Gujjo in May and near Mirpur Sakro on 7 April. An adult was observed in July carrying a small frog in its bill, presumably to feed to its nestlings.

In a study at Ludhiana Agricultural University (east Punjab, India) a nest was constructed in 3 days in a *Saccharum* grass thicket, being made entirely of leaves of this grass. Both sexes took part in building the elongated sphere which was 1.4 metres (4.5 feet) from the ground. A total clutch of 5 eggs was laid at 1, 2 and 3 days' intervals and the incubation period was 17 to 18 days, the chicks hatching within 12 hours. They were black skinned and blind at birth with hair-like down. They are nidicolous and altricial. Two toes pointed backwards (zygodactyl) even at birth. Their eyes opened on the fourth day, but the nest was destroyed by predators before the chicks could fledge (M.S. Dhindsa and H.S. Toor, *JBNHS*, Vol. 78, December 1981).

*VOCALIZATIONS

They are most vocal in the late afternoon and at dusk and again in the early morning. Males give territorial calls which consist of a prolonged series of deep resonant notes sounding like 'hoop-hoop-hoop-hoop'. Usually the bird climbs into a bush or into tall reeds before giving these calls which are repeated 15 to 20 times before the bird falls silent. Whilst calling the bill remains closed and the throat swells out with each 'hoop' and at the same time its head is bowed forward or depressed with bill pointing downwards. A second type of call is a more rapid staccato 'kwok-kwok-kwok' not so resonant and repeated 20 or 30 times. Females have been observed uttering a rasping choked sort of 'meeaow' only audible at close range and this in the presence of a male bird which had earlier been calling territorially (noted Haleji lake, early June at dusk).

ORDER STRIGIFORMES
Owls
FAMILY TYTONIDAE
Barn and Grass Owls

Barn Owl *Tyto alba* (Scopoli)

Plate **20**
Map **231**

DESCRIPTION

Body length 34-36 cms.
Wingspan 85-93 cms.
Wing length 27.5-32.3 cms. (both sexes) (Stuart Baker)
Tail length 11-12.75 cms.

The large rounded skull of this bird with forward directed eye sockets, gives it a completely different appearance from other bird families. The fine hair-like feathers surrounding the two large dark eyes, radiate outwards to form two white facial discs framed in a heart shape by short brown tipped stiff crimped feathers. The bill is pinkish horn coloured, rather long and downward pointing with the upper mandible hooked at the tip. The crown, neck and back are ochraceous buff and grey finely stippled with white and black spots. The breast is creamy white with ochraceous tinges on the lower breast and parallel longitudinal lines of small dark spots. The tarsus is covered with short bristly white feathers and also the upper surface of the toes which are powerfully developed with long curving talons. The flight feathers are pale chestnut buff with obsolescent darker cross-bars and also the tail which is cross-barred darker brown. The irides are very dark brown. Both sexes are alike. The bill of this owl looks comparatively weak but its long tibia and tarsus enable it to stretch out a considerable distance to seize its food prey. Wild and apparently healthy specimens encountered around Karachi were invariably sullied by industrial grime or dirt, their plumage being noticeably dirty.

HABITAT, DISTRIBUTION AND STATUS

This is one of the classical examples of a Cosmopolitan species occurring widely around both hemispheres and adapted to temperate and sub-tropical climates. In Pakistan it is confined to the Indus plains and is rather locally and erratically distributed, being found mainly in the vicinity of older and larger towns where suitable man-made building structures provide daytime roosts and where the commensal rodents which form the bulk of its diet are most easily obtained. It is a sedentary resident species. There is quite a large population on the outskirts of Karachi, particularly in the orchard and vegetable farms of Malir and Landhi where deep open wells provide undisturbed daytime shelter. In recent years it has been noted in old tombs at Mirpurkhas, in a deserted palace at Dera Nawab Sahib in Bahawalpur, on the outskirts of Lahore, and in Thatta town. It appears to be rare in northern Punjab and has not been recorded in the Salt Range or Rawalpindi district (Waite, 1948 and Whistler, 1930). During over 30 years' residence in Sind, Eates only encountered the Barn Owl 3 or 4 times and considered it rare. He saw it once at Mirpurkhas, once at Larkana town and once in Tharparkar district (unpublished MS notes). It avoids forest or desert areas and has not been recorded in Baluchistan or in the northern mountain areas. Status SCARCE.

HABITS

In Malir region outside Karachi, Barn Owls occasionally roost in old neglected buildings but owing to human population pressure, unused buildings are rare and they can often be encountered by daytime roosting in the thick fringe of bushes growing out of the vertical sides of open well shafts or less commonly in tree groves above ground. Though they have been observed hunting by day light in Europe, they seem to have relatively poor vision in bright sunlight, because specimens encountered can often be easily captured by hand. Normally they emerge to hunt well after darkness and are more nocturnal in activity than many owl species. Their hunting technique includes watching from an exposed prominent perch as well as flying low and systematically alongside the edge of poultry houses, hedge rows and ditches where rodents are most likely to be encountered. Their diet around Karachi (based on regurgitated casts found in a poultry shed at Korangi), consists exclusively of small mammals, including *Rattus rattus* the most common food prey, *Mus musculus* second, and *Suncus murinus* the Musk Shrew third in importance. The Bush Rat (*Golunda ellioti*) and Indian Gebril (*Tatera indica*) occasionally occurring also. Elsewhere in India, they have been recorded as occasionally taking bats (Chiroptera) and House Sparrows (Mason and Lefroy, 1912 and Ali and Ripley, 1969). Based upon observations of a pair resident near the author's home, they can occasionally be heard calling to each other in the early part of the night or just before dawn but they are relatively silent most of the year. Their flight is very buoyant with a high aspect wing loading ratio and they can control their flight in and around obstacles,

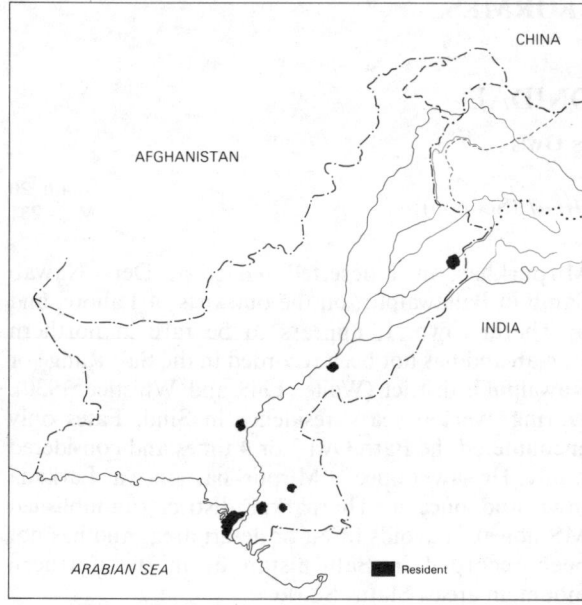

Map 231: Barn Owl *Tyto alba*

descending swiftly and noiselessly to the ground to seize even large and fierce rodents.

BREEDING BIOLOGY

Breeding can apparently take place at any time of the year. Four fledged young taken near Karachi were being hawked for sale, having been taken from a 'nest' on 7 January when about 3 weeks old (F. Koning, pers. comm., 1975). On 26 April 1980 two recently fledged juveniles were also found (author) being hawked for sale by a boy at Korangi creek, which had been found and captured in a disused building. K. Eates records

four chicks found in a nest in the belfry tower of Holy Trinity Church in Karachi in early February and a nest found a by a Mr. Culbertson in a hole in an earth cliff on the banks of the dry Malir river bed in June (K. Eates, MS notes). H. Whistler found a nest in a tree hole in Jhang but the date is not recorded. Nesting takes place when food supplies are favourable, which can occur in winter or summer. The pair bond is stable with the female carrying out all the incubation duties but both parents tend the young. The male is presumed to do most of the hunting and bringing of food prey to his incubating mate and later to bring food which the female tears up and presents to the young when they are very small. The eggs are white roundish ovals and a full clutch varies from 3 to 7 according to available food supply. No nest is built, the eggs being laid on a bare ledge or at the bottom of a natural hole or crevice. Eggs are laid often at 2 day intervals and as in all the owls, hatching is asynchronous so that the first born and largest young are most likely to survive whilst the youngest and weakest starve, if food prey becomes difficult to secure. Incubation in Europe has been recorded as taking from 30 to 31 days (Cramp *et al.*, Vol. 4, 1985), and is by the female alone and the young fledge in 50-5 days but depend upon their parents for food for some 4 or 5 additional weeks (Cramp *et al.*, 1985).

*VOCALIZATIONS

Their calls comprise a rather varied repertoire of very hoarse squawks and shrieks, which are uttered in flight, especially when they first emerge in the evening, but can also be given from some perch such as a telephone pole. Birds when threatened or disturbed in their diurnal roost will give hissing calls, and also like most of the Strigidae, clatter their bills under extreme threat. Their most characteristic flight or contact calls have a distinctly rasping or wheezing tone.

FAMILY STRIGIDAE

TAXONOMY of the *Genus OTUS* Pennant

There are five owls of this genus in Pakistan, all of which are very similar in size and cryptic plumage patterns. Their striking similarity has led to much divergence of opinion in the past as to how these owls should be classified. Studies by the author reinforce the belief that advertising calls play a crucial role in pair formation and intra-specific recognition. Their calls are quite distinctive and recognizable even to the

human ear and enable sympatric or overlapping members to maintain separate breeding populations. All these owls can be distinguished and identified on the basis of wing formula which is presented hereunder. Since this can only be determined from specimens in the hand, the importance of accepting song pattern as species specific in this genus is long overdue (See T.J. Roberts, and Ben King, 'Vocalizations of the Owls of the Genus Otus in Pakistan', *Ornis. Scandinavica*, 1986).

WING FORMULAE of the OWLS of the *Genus OTUS* in PAKISTAN

Primaries are counted from the carpal joint (i.e. in descending order)

Species	Longest Primary	Length of 10th Primary	Length of 9th Primary
Otus bakkamoena	6th or 7th	Shorter than 4th	Between 3rd and 5th
Otus spilocephalus	6th or 7th	Shorter than 1st	Between 2nd and 4th
Otus sunia	7th	Equal to or shorter than 4th	Between 4th and 6th
Otus scops	8th or 9th	Between 5th and 7th	Roughly equal to 8th
Otus brucei	7th or 8th	Between 4th and 6th	Roughly equal to 7th

Indian Scops Owl *Otus bakkamoena* Pennant
Collared Scops Owl of Ali and Ripley

Plate **20**
Map **232**

DESCRIPTION

Body length 23-25 cms.
Wingspan 60-61 cms.
Wing length 13.5-15.2 cms. (Stuart Baker)
Tail length 6.4-7.4 cms. (Stuart Baker)

This beautifully camouflaged little owl is rather larger in size than the other Scops owls of the region and is slightly slimmer and longer bodied than the Spotted Owl (*Athene brama*). Its plumage is mottled with rufous brown, creamy buff and darker brown in a complicated pattern of darker spots, striations and vermiculations with an indistinct creamy buff collar on the hind neck (whence the name Collared Scops) and also along the scapulars or inner wing coverts, forming a diagonal line with creamy and dark brown tipped feathers. The large eyes are framed by facial discs with blackish hair-like feathers radiating up to each corner of the broad flat crown and terminating in short feather 'ear' tufts. The breast is creamy buff with prominent dark brown cross-rayed shaft streaks like a multi-barbed fish harpoon. The tarsi are feathered and also cross-barred but the toes are bare and greenish yellow in colour. The tail is of medium length, chestnut brown with darker cross-barring. The irides of this species are variable but numerous specimens seen in the Punjab (author), all had dark brown eyes. This population has been assigned to the sub-species *Otus bakkamoena plumipes* on the basis of larger size (wing over 162 cms.) and the toes being feathered to the sub-terminal phalanx. This latter character is not constant in specimens examined from the BNHS,

AMNH and BM skin collections but iris colour seems to be a constant character (see Roberts and King, 1986). Specimens collected from Sind are paler in plumage and larger than Punjab specimens and have been assigned to the sub-species *deserticolor* by Dr. C.B. Ticehurst (*Bulletin Brit. Ornith. Club*, No. 42,

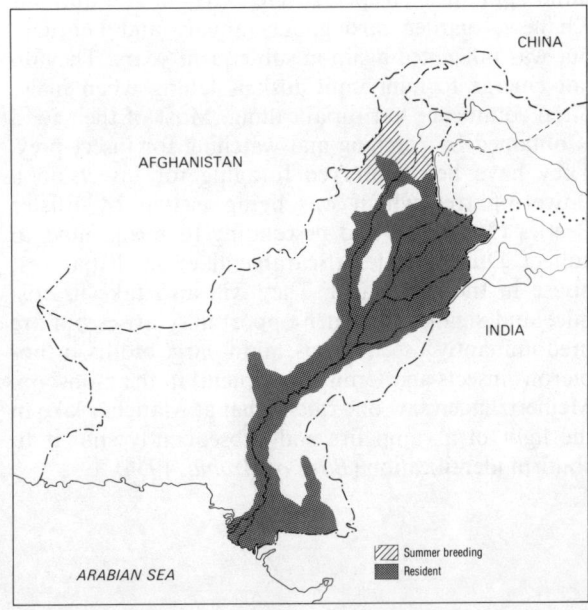

Map 232: Indian or Collared or Asian Scops Owl *Otus bakkamoena*

1922). Furthermore their irides are generally golden brown or hazel colour. One specimen collected near Lahore (Punjab University Museum) was rufous or hepatic in the upper parts of the body plumage.

HABITAT, DISTRIBUTION AND STATUS

They are rather solitary owls, keeping to a fixed territory and using the same roost for varying periods, yet inclined to local migratory movements according to food supply, since they are heavily dependent upon insects. In the summer months small numbers migrate into the Murree foothills and even up to the summit ridges, where they are sympatric with *Otus sunia* and *Otus spilocephalus*. H. Waite (Brit. Mus. coll.) collected 2 specimens at 1,829 metres (6,000 feet) just below Murree on 13 May and another in Murree itself at 2,134 metres (7,000 feet) on 28 June. It was also collected in June from Murree by G.H.T. Marshall. It is a species which requires good tree cover and is therefore uneven in distribution in the Indus plains, being found mainly in irrigated forest plantations or patches of riverain forest or in the shady gardens of old bungalows. It is entirely absent from desert or open treeless country but odd pairs will turn up in every district of Punjab and Sind provinces. Status COMMON.

HABITS

A very secretive bird which roosts close to a tree bole well hidden in foliage and relies upon freezing and its cryptic plumage to escape notice. One roosted in the same spot in a creeper covered tree in the author's Khanewal garden throughout January and February but was not noted again in subsequent years. They do not emerge to hunt until dusk is falling when males often commence territorial calling. Most of their food is obtained by perching and watching for insect prey. They have been observed foraging for insects in a flowering tree which was being visited by Blister Beetles (Meloidae) and descending to the ground to collect Dung Beetles (Scarabaeidae) at Tajjal rest house in the east Narra. They will also take lizards, mice and small birds when opportunity arises but are predominantly insectivorous, taking large moths, orthopterous insects and termites (Isoptera) in the monsoon. Meinertzhagen saw one catch a bat at Manchar lake in the light of a camp fire and subsequently shot it to confirm identification (*Birds of Arabia*, 1954).

BREEDING BIOLOGY

The pair bond is monogamous and stable. They normally nest in natural tree cavities, often quite high up from the ground. Bell found a nest on 13 March in the Sukkur district with two yellow downy young in a cavity formed underneath the foundation of an old Neophron's nest in a tall *Ficus* tree. Another nest was found on 10 April near Hyderabad with downy young (Ticehurst, *Ibis*, 1923). Kenneth Eates (MS notes unpublished) found eggs in Karachi, Hyderabad and Sukkur districts between February and April, all of them in tree holes. The usual clutch is 3 eggs and these are rather spherical and glossy. They are slightly larger than the eggs of *Athene brama*. Occasionally four eggs are laid. It is known that hatching is asynchronous with young showing considerable size disparity in the same nest hole. Both parents have been seen going into the nest hole but it is presumed that, as in most Strigiformes, the female does all the incubation which probably takes 25 to 27 days, though this has not been definitely recorded. Newly hatched young are blind and nidicolous.

*VOCALIZATIONS

The call of this owl is much softer and subdued than that of other Scops species and is difficult to hear from distances much greater than 150 metres. It is a well spaced interrogative 'whut' or 'whuk' sounding rather frog-like, usually repeated after 4 to 6 second intervals. Vocalizations are thought to be very important in pair bond establishment in this species and during the breeding season, the female often responds with a lower pitched short call and both duet rapidly with alternate rapid calls of the same difference in pitch 'di-dahh di-dahh di-dah'. A pair with young in their nest hole were heard duetting on two successive evenings at about 8.30 p.m. at Changa Manga forest plantation in late March. Studies in Gharko riverain forest, Thatta district, where these owls are resident, indicate that they call from early March to December but that during January and early February they appear to lose their voice as suggested by playing back taped territorial calls, which immediately elicit aggressive response including 'dive bombing' and from March to December, normal vocal responses. During January, birds perched nearby, in response to taped playback, emitted only a very faint and spaced sighing noise.

Oriental Scops Owl or Asian Scops Owl *Otus sunia* (Hodgson)

Synonym *Otus pennatus* in *Stray Feathers*

Plate **20**
Map **233**

TAXONOMY

This is treated as a sub-species of *Otus scops*, the European Scops Owl by Ali and Ripley ('Handbook', Vol. 3) and in the 'Revised Synopsis' by Dillon Ripley (1982) and by Vaurie (Vol. 2, 1965). However, its call is very distinctive and stereotyped throughout its known range and in this genus is sufficient basis for specific separation (see Roberts and King, *Ornis Scandinavica*, Vol. 17, 1986, p. 299-306). In 1923 C.B. Ticehurst in discussing the taxonomy of *Otus pennatus* (syn. *Otus sunia*) considered it sufficiently distinctive as to be regarded as a separate species on the basis of wing formula alone (*Ibis*, 1923). Hugh Whistler in the first edition of the *Popular Handbook* (Oliver and Boyd, 1930) also considered it a separate species (p. 346). It is also treated as a separate species by Gallaher (*Birds of Oman*, 1980).

DESCRIPTION

Body length 19 cms.
Wingspan 50-52.4 cms. (Dement 'ev)
Wing length 13.7-15.4 cms. (Stuart Baker)
Tail length 6.1-7.1 cms. (Stuart Baker)

It is indistinguishable in the field from *Otus scops* though if a series of skins are put together they are more richly coloured with more rufescent tones about the neck and wing coverts. It is cryptically plumaged like all the Scops with each feather vermiculated and barred, forming a complicated but disruptive pattern of speckled browns and greys. The facial discs are paler grey-brown and the breast is whitish with scattered dark brown streaks on the feather shafts, having small barbs or cross rays like a harpoon. The wing coverts and scapular feathers are tipped creamy white and dark brown, forming conspicuous diagonal bars. It can only be separated from *O. scops*, when in the hand, by the wing formula, in which the first primary is shorter than the seventh and the fourth is the longest (counted in descending order from the carpal joint, see page 465). This owl has lemon yellow irides and fairly prominent ear tufts or aigrettes. The tarsus is feathered but the toes are naked and dingy yellow in colour. The sexes are alike. There is often a rufous or hepatic morph of this species.

HABITAT, DISTRIBUTION AND STATUS

This is a forest dwelling species which occurs across the Himalayas and also in the forested slopes of the Western Ghats, Nilgiri hills and in the highlands of Sri Lanka. In Pakistan it occurs only in the Himalayas where it is on the westernmost extremity of its range. There is no evidence that it extends westwards beyond the Murree hills though Ali and Ripley include Afghanistan in its range ('Handbook', Vol. 3, 1969). Paludan did not encounter any Scops species in Afghanistan (Paludan, 1959). The sub-species *Otus scops rufipennis* (Sharpe) from the Nilgiri hills and Kerala and *O. scops leggei* from Sri Lanka (Ticehurst 1923 and Ali and Ripley, 'Handbook', 1969) with identical stereotyped calls are considered to belong to *Otus sunia* group. In Pakistan it is confined to the southern part of the Murree hills and extends to the sub-tropical pine zone (*Pinus roxburghii*) at 1,500 metres (5,000 feet). It occurs very sparsely in the northern part of the range, (known as the Galis), perhaps one *sunia* to six *spilocephalus*, and visiting Dunga Gali region in the northern part of the range in some years only, and then not ascending above 2,000 metres (6,500 feet). On 7 June at Samli forest rest house south of Murree, 4 birds could be heard calling territorially within auditory range of each other in Long-needle Pine forest (*Pinus roxburghii*). One was heard calling in a side ravine of the Margalla hills from May to July at about 600 metres (2,000 feet) elevation in 1983 so that they may breed in the better wooded ravines, at this elevation also. Status SCARCE.

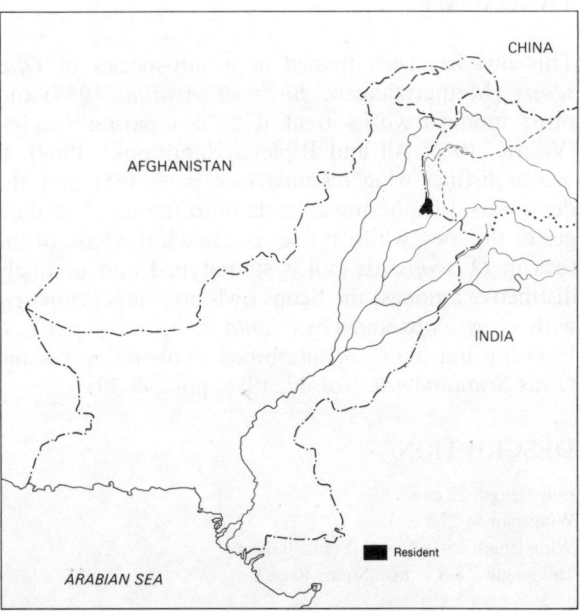

Map 233: Oriental or Indian Scops Owl *Otus sunia*

HABITS

Nothing much is known except for the evidence of its calling. It is a very secretive bird, remaining hidden in dense forest throughout the day, and only emerging to hunt well after darkness. During the breeding season males have definite territories and can be heard calling throughout successive nights from a particular patch of forest. They do not start to call until an hour or less just after dusk but during May and June they call intermittently throughout the night. Their food probably consists largely of insects including cicadas, mantises, beetles and Acrididae, but they will also capture birds, mice and rodents when available (Stuart Baker, Vol. 4, 1927). Their normal flight consists of a rapid flutter and then a dipping glide and they fly swiftly and buoyantly.

BREEDING BIOLOGY

Around Murree, Buchanan collected eggs of this species as early as 12 February. (Stuart Baker, op. cit, p. 436). H. Waite found them breeding at Chariban in the Murree hills at 1,800 metres (6,000 feet) in early May. They nest in holes in trees, occasionally in holes in revetment walls or disused buildings. They lay 3 to 4 eggs, plain white, rather spherical in shape and of a smooth glossy texture. Nothing else is specifically recorded about their breeding behaviour but Stuart Baker describes a pair during the breeding season circling around together on stiffly outspread wings, squawking softly at the same time (op. cit., 1927).

*VOCALIZATIONS

The advertising or territorial call of this owl is very distinctive being a stereotyped short four-noted song, monotonously repeated with such regularity and evenness of tone and tempo as to suggest a mechanical source, e.g. the chugging noise of an old 'donkey pump' as so aptly described by Hugh Whistler in the 'Popular Handbook' (1943). Specimens tape recorded in Peryar Sanctuary, Kerala in December by the author, and in Sri Lanka by Ben King sounded identical to the calls of Murree hill birds. The call can be rendered as 'wut-chu-chraaii', the first note musical and heavily accented, the last two notes more burry and rasping but very rhythmical and even in tempo. This song is usually uttered once every 3 or 4 seconds, sometimes in continuous sequences of over 100 calls. It can be heard from up to 0.8 km. (half mile) distance. It has been heard calling in the Murree hills from early March until the end of September but was not heard during October or December during brief visits to Murree.

Pallid Scops or Striated Scops Owl *Otus brucei* (Hume)

Plate **20**
Map **234**

TAXONOMY

This owl has been treated as a sub-species of *Otus scops* (Meinertzhagen, *Birds of Arabia*, 1954) but most modern works treat it as a separate species, (Vaurie, 1965, Ali and Ripley, 'Handbook', 1969). It has a distinct wing formula (see page 465) and the dense tarsal feathering extends onto the basal phalanges of the toes, whilst it does not reach the base of the toes in *O. scops*. Its call is stereotyped and uniquely distinctive amongst the Scops owls, and it is sympatric with *Otus scops* and *Otus sunia* in different parts of Pakistan but does not interbreed (Roberts and King, *Ornis Scandinavica*, Vol. 17, 1986, pp. 299-305).

DESCRIPTION

Body length 22 cms.
Wingspan 54-57.5 cms.
Wing length 15-16.1 cms. (Stuart Baker)
Tail length 7.8-8.2 cms. (Stuart Baker)

It is slightly larger than *Otus scops* but smaller than *O. bakkamoena* and generally more grey-brown in plumage, lacking any rufous tones when compared with *O. bakkamoena* or *O. sunia*. The crown, nape and mantle are finely vermiculated with dark brown striations and with scattered narrow dark brown shaft streaks. Its facial discs are greyish without any noticeable concentric rings and its ear tufts are not very prominent. Its wings are comparatively long, almost extending to the tips of its short tail which is cross-barred brown and buff with darker speckling. The iris is yellow, the toes are grey and the bill is horny brown, darker at the tip. Both sexes are alike.

HABITAT, DISTRIBUTION AND STATUS

This Scops Owl is adapted to semi-desert arid foothill and rocky country, except in winter, avoiding both flat plains and heavy bush or forest cover. It extends from the eastern Mediterranean through Syria, Iraq and Iran and also down into the Arabian peninsula. In Pakistan it is a summer migrant visitor to the lower valleys and gorges of Gilgit, and central Baluchistan. It occurs on the steeper cliffs in the Margalla hills, the southern bluffs of the Salt Range and also in the

Kirthar Range in western Sind. It has been heard calling in the Kirthar in February, but there is some winter migration down to the plains of Sind and India, as one was collected in the rocky outcrops east of Umarkot in January (Blanford, *Stray Feathers,* 1877 p. 245) and another near Hyderabad on 16 December (S. Doig, *Stray Feathers,* 1878, p. 505). The orginal type specimen was collected in rather untypical habitat, in Maharashtra near Ahmednagar (A.O. Hume, *Stray Feathers,* 1873, p. 8). A specimen collected (seen by author) in December from Ambala in east Punjab, India, was sent to the Bombay Natural History Society in January 1982. In the Margalla hills it is sympatric with *Otus sunia* and around Quetta, Baluchistan it is sympatric with *Otus scops* though generally keeping to lower elevations than the latter. H. Waite collected it at Sakesar in the Salt Range in March and July (Brit. Mus. coll.). The author has recorded its territorial calls from the Margalla hills, Hazar Ganji in the Chiltan range south-west of Quetta, at Karchat in the Kirthar Range, Dadu district and at the foot of the Salt Range north of Kushab. Status SCARCE.

References in J. Burton (1973) *The Owls of the World* to *Otus brucei* to the effect that it interbreeds 'in the upper Indus region with the eastern population of the Oriental Scops Owl *Otus sunia* plainly reveal the dangers of reaching conclusions without first hand field experience of their calls and distribution. *Otus sunia* and *Otus brucei* are sympatric in the Margalla hills where they do not interbreed, whilst *O. sunia* never penetrates into the dryer colder northern Himalayan regions where *O. scops* comes in summer, e.g. in Gilgit and where both *O. scops* and *O. brucei* are

sympatric. (Roberts and King, *Ornis Scandinavica,* 1986). As in the Caprimulgidae the scops owls are strictly nocturnal in hunting and evolutionary pressures have placed great emphasis upon crypsis in their plumage and the ability to escape detection during daytime by remaining immobile. Intra-specific recognition and pair formation, it can be logically inferred, thus depends heavily upon correct vocal signals.

HABITS

It often roosts by day on the ground in a rock crevice or under an overhang, usually on a precipitous cliff where it will not be disturbed by grazing domestic stock. It is typically associated with stunted bushes of *Pistacia* and *Acacia* in these places. It only emerges to hunt well after darkness and is a swift and strong flyer being recorded as catching insects on the wing in Arabia (Cheeseman in Stuart Baker, 1927). It has the typical undulating flight of the genus with strong rapid wing beats alternating with a dipping glide. It is mainly insectivorous, catching beetles, moths, locusts and possibly small bats and birds when opportunity presents itself. (Meinertzhagen 1954 and Ticehurst, 1926). It appears to be solitary outside of the breeding season and rather local and erratic in wandering, being entirely absent some years from previous regular haunts (observation Hazar Ganji over six years and Karchat during four years).

BREEDING BIOLOGY

A cavity nester, utilizing holes in tree boles, holes in walls, earth cliffs and in Iraq apparently utilizing old Magpie nests (*Pica pica*). The only authentic nest from Pakistan was one found on 16 April 1925 in a tree hole on the slopes of Takhatu range. There was no nest lining and only one egg but the female which perched nearby was also shot and contained a fully developed oviduct egg (Williams and Williams, *JBNHS,* Vol. 33, 1929). Christison also found newly fledged young in the hills east of Nushki but gives no date (Christison, *Ibis,* 1941). The eggs are plain white rather spherical in shape. Nothing has been recorded about the normal clutch size, incubation period etc. but presumably the female does the incubating and the male brings food both to his incubating mate and subsequently for the young.

*VOCALIZATIONS

This little known desert dwelling owl was probably first tape-recorded, calling by this author (see *Birds of Western Palearctic,* Vol. 4, 1985,). Many recent authors have given conflicting accounts of its calls, e.g. Mei-

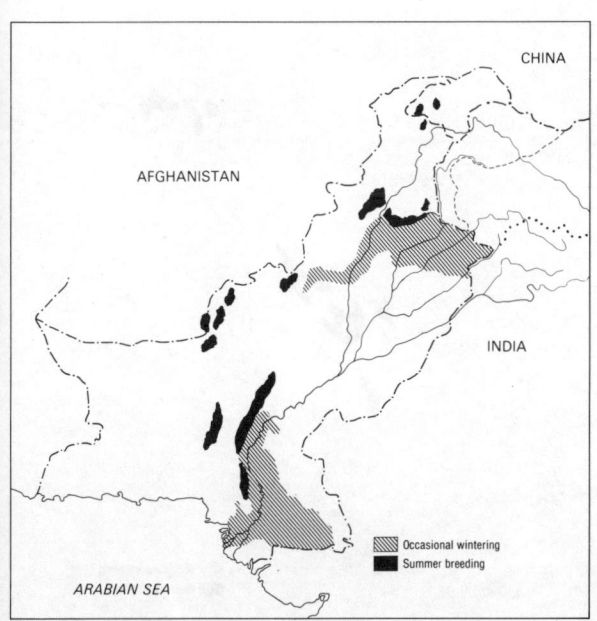

Map 234: Striated or Pallid Scops Owl *Otus brucei*

nertzhagen (1954), who said its call was indistinguishable from the Common Scops (*Otus scops*). Heinzil, Parslow and Fitter (1972) give its call as 'ukh ukh'. The main advertising or territorial call of the male is an extraordinarily 'un-birdlike' hollow resonant low pitched 'whoop' note repeated at precise intervals like a metronome (about 8 times in 5 seconds) and sometimes given non-stop for 3 or 4 minutes on end. It has a rather ventriloquial quality due perhaps to echoing of sound waves from the steep cliffs which it haunts, plus its ability to twist its head from side to side while calling. Besides the 'whoop-whoop-whoop' calls, it gives a longer and irregular call usually uttered once

and then repeated after 5 or 6 seconds. This 'whoo' is also rather hollow and deep sounding. One may speculate as to whether this second type of call is given by the female (see Sonagrams of both calls, Roberts and King, *Ornis Scandinavica,* 1986). It is, as repeated above, a very difficult species to see in the field, being active only after dark and relying upon crypsis and an inaccessible roosting place during its diurnal roosting. A period of almost five years elapsed between the first hearing of this call and the author's success in correctly identifying the source! It has been heard calling in the Kirthar from mid-February and in the Margalla hills from April to mid-October.

Eurasian Scops Owl *Otus scops* (Linnaeus)

Plate **20**
Map **235**

DESCRIPTION

Body length 19 cms.
Wingspan 49-54.5 cms. (sub-species pulchellus)
Wing length 15-15.8 cms. (sub-species *pulchellus*) (Stuart Baker)
Tail length 6.6-7.1 cms. (sub-species *pulchellus*) (Stuart Baker)

This is a somewhat variable owl over its wide Palearctic range. C. Vaurie recognizes 6 sub-species excluding the *sunia* group (Vol. 1, 1965). Three of the sub-species listed in the *Handbook of the Birds of India and Pakistan* (Ali and Ripley, 1969) are preferably treated as a separate species *Otus sunia* (see above account). The sub-species occurring in Pakistan is listed as *pulchellus* by Ali and Ripley and as *turanicus* by Vaurie. Both races are distinguished by being paler and greyer in plumage than the nominate race. Thus they are convergent in plumage to *Otus brucei*. They average slightly smaller than *O. brucei* and the breast is more boldly streaked with dark brown and 'barbed' shaft streaks. There are fine horizontal vermiculations on the greyish white breast and the upper parts are delicately patterned in rufous brown, dark brown and grey with spots, streaks and barring. The facial discs are pale brown with some concentric darker rings and the ear tufts are quite conspicuous. The flight feathers and tail are cross-barred and spotted with cream and darker brown and there is a conspicuous diagonal row of white spots formed by the tips of the scapulars and a thinner shorter bar formed by the pale tips to the inner wing coverts. The tarsi are feathered but this does not extend onto the base of the toes. It has yellow irides, and the bill is horny black. The sexes are alike. Compared with closely similar (and sometimes sympatric) *Otus brucei* the scapular feather tips are white not creamy and the streaking on the crown is thicker and more regular. However, the most reliable distinction is in the wing formula, the eighth or ninth primaries (descending order) being longest (with the eighth and

ninth roughly sub-equal), whilst the seventh and eighth are longer in *O.brucei* and the ninth and seventh primaries are roughly sub-equal.

HABITAT, DISTRIBUTION AND STATUS

In the summer months this little owl extends from the Mediterranean through the Middle East and to central Asia. In Pakistan it is a summer breeding visitor to the higher mountain ranges between 1,500 and 3,000 metres (5,000 and 10,000 feet) in Baluchistan and it also occurs in the lower valleys of Gilgit and Chitral, and in Swat Kohistan between 1,500-2,450 metres

Map 235: European or Eurasian Scops Owl *Otus scops*

(5,000-8,000 feet). In these regions it is characteristically associated with dryer rocky hills and valleys where *Juniperus* scrub forest or Holly Oak (*Quercus ilex*) grow. In winter a few straggle down to the plains of Sind and two have been collected as far south as Bombay (Ali and Ripley, 1969). Specimens were collected on 13 November at Hyderabad and another in Karachi on 16 November, both in the late 1870s (Ticehurst, *Ibis*, 1923). However, it is possible that some of the wintering population travels as far as east Africa where this species is a regular winter visitor and evidence of this was obtained by Cumming who collected a specimen in September from Ormara on the Makran coast (Ticehurst, 1927). There are no more recent winter time records. In Baluchistan it is quite numerous in the juniper forests of Murdar, Zarghun, Wam and Kaliphat and birds can be heard calling from mid-March, often when snow is still on the ground. Specimens from these regions are in the Punjab University Museum (coll. Z.B. Mirza, June 1969) and in the British Museum (coll. H. Waite, March 1937). It has been heard (author) within auditory range of a calling *Otus brucei* in the Hazar Ganji reserve, west of Quetta, in April and May of 1978 and 1979 at 1,500 metres (5,000 feet) elevation. In Gilgit it was collected by Dr. Scully from Punial on 21 May (*Stray Feathers*, 1881) and the author has heard them calling in the Kargah nullah in mid-June 1981. Scops Owls (*Scops giu* synonym *Otus scops*) have been recorded in Chitral (Perreau, *JBNHS*, 1910). Whitehead did not record them in the Safed Koh range or upper Kurram valley. Status summer visitor, COMMON.

HABITS

Shy and secretive by day and only coming out to hunt after darkness. They are largely insectivorous in diet, including various Coleoptera and nocturnal moths, grasshoppers and locusts with an occasional small bird or lizard (Meinertzhagen, 1954). Many writers have commented on their habit, if observed in the daytime, of sleeking down their feathers and stretching their body upright with eyes closed, to take maximum advantage of their disruptive plumage pattern. They hunt insects both on the wing and by watching and pouncing from a perch. They have also been observed hunting on foot, seizing small items with their bill (Cramp *et al.*, Vol. 4, 1985). In the summer and spring months, males call territorially for an hour or so, starting immediately after dusk has fallen and again becoming vocal an hour before sunrise.

BREEDING BIOLOGY

No nest has been authentically described from Pakistan, but a pair was watched going in and out of a small hole in a tree fork of *Juniperus macropoda* on 20 March 1981 at Ziarat (2,400 metres 8,000 feet) by Ben King (pers. comm.) and by their behaviour, appeared to have selected the site for breeding. In Iran they have been recorded breeding in holes in old wells and buildings as well as tree holes. The normal clutch is 3 to 4 eggs, which are plain white roundish ovals. Egg laying has been recorded from April to June. In Ziarat region observations throughout the spring and summer indicate that males call territorially throughout March and April with the majority falling silent or calling only occasionally in May and June, presumably when they are busy collecting food for their incubating mates and later on for their young. In the first week of June only one male was heard calling in a region where at least 5 different individuals were heard calling in April and March in the same and previous years. Only the female incubates, which takes 24-25 days (Cramp *et al.*, Vol. 4, 1985). The eggs hatch asynchronously and the young take 21-29 days to fledge (Cramp *et al.*, 1985).

*VOCALIZATIONS

The territorial or advertising song is a single short flute or gong-like call 'piu', or 'tonk'. It varies much in pitch between individuals, probably an important factor in spacing out territories and in individual mate selection. The calls are uttered at irregular intervals, usually 3 to 5 seconds apart and may be repeated 20 to 30 times. It is a resonant and far carrying call and the author well remembers the excitement of his first visit to the juniper forests of Ziarat in early March when 5 or 6 of these little owls started to call as dusk fell, their irregular timing and variation sounding indeed 'like an erratic peal of small bells', as so aptly described by Cheeseman in Iran (quoted in Stuart Baker, Vol. 4, 1927). Scops owls heard calling in Italy by the author and recorded in France (Roche) are virtually indistinguishable, confirming that this is a very stereotyped call, despite the wide distribution of this owl and the number of sub-species which have been described.

Spotted or Mountain Scops Owl *Otus spilocephalus* (Blyth)

Plate **20**
Map **236**

DESCRIPTION

Body length 18-20 cms.
Wing length 13.7-15.1 cms. (Stuart Baker)
Tail length 7.7-9.0 cms. (Stuart Baker)

Smaller than the Collared Scops Owl and generally rather darker in plumage with less prominent 'aigrettes' or ear tufts. The facial disc is generally rather pale brown and the crown and nape have noticeable black spots in scattered lines, formed by the feather tips. The hind neck and breast is pale greyish buff finely vermiculated with darker lines and with scattered darker spots. The dorsal plumage tends to be a richer more chestnut brown with rufous tinges in the tail and flight feathers, which are cross-barred in lighter and darker brown. Compared to the other five Scops owls the breast is less heavily streaked. The wing scapulars are white tipped and also the inner wing coverts forming prominent white diagonal lines on the shoulder. The tarsus is buff and well feathered down to the base of the toes. The bill is dull yellow and the toes are fleshy brown. The iris is golden yellow. The sexes are exactly alike. In the eastern part of its range a rufous or hepatic morph is more common. In the field its call notes are the only certain means of identification, though the absence of prominent shaft streaks on the breast is helpful.

HABITAT, DISTRIBUTION AND STATUS

This is an Oriental species occurring in mountain forest through most of south-east Asia, from Indonesia, Thailand, Cambodia to Malaysia. In the subcontinent it is confined to Himalayan forest and occurs in Blue Pine (*Pinus wallichiana*) and mixed deciduous and coniferous forest from 1,800 to 3,000 metres (6,000 to 10,000 feet) elevation. In Pakistan it occurs mainly in the Murree hill range, not descending in summer below about 1,800 metres (6,000 feet). It also occurs in the forested side valleys of the Neelum river in Azad Kashmir (Salkalla by Kamal Islam, pers. comm., and Machiara as high as 3,000 meters, author's obs.). It is sympatric with *Otus sunia* in the southern parts of the Murree hill range where both species can often be heard calling together. In 1983 a single bird was heard calling below Shogran in the lower Kaghan valley, Hazara district (D. Corfield, pers. comm., 1982), but it appears to be no more than an erratic visitor to the lower Kaghan valley (observations based on 12 years of early summer visits to the region). It has not been heard in areas of similar habitat in other parts of Hazara district. It is not known whether this owl migrates eastwards into warmer Himalayan climates during the winter or whether there is only north south altitudinal migration, as these are very secretive owls, difficult to locate unless they call. In view of their insectivorous diet and the heavy winter snowfall experienced in the Murree hills it is assumed that they are locally migratory in winter. Status FREQUENT in Murree hills only.

HABITS

They call loudly from April to the end of October and these calls indicate a territorialism outside of the breeding season, as well as revealing something about their density. In the Kao valley below Dunga Gali in the summer months, normally about 6 calling males can be heard over an area of about 800 hectares of forest in an area which rarely harbours more than 1 pair of *Otus sunia*. They are extremely difficult to see, being very secretive and relying upon natural concealment during the day, roosting in the shelter of Ivy (*Hedera*) covered tree boles or dense leafy areas in pollarded oaks (*Quercus dilatata*). Like all Scops they stretch themselves upright and freeze with eyes almost closed when disturbed, thus making themselves almost invisible against the mottled bark and lichen covered background of their perches. They are largely insectivorous in diet. Swarms of cockchafers (Melolonthinae) emerge in early May in the Murree hills and it

Map 236: Mountain or Spotted Scops Owl *Otus spilocephalus*

is presumed that this owl was the species dimly seen by the author in early May in Dunga Gali, swooping low over a forest pathway to snatch at these insects as they noisily drone on 'take-off' from their daytime burrows. They also feed on large moths, and during July and August upon cicadas and *Longicornis* (Cerambycidae) beetles. They probably take small birds and rodents when opportunity is offered.

BREEDING BIOLOGY

Very little is known about the breeding habits of this species. They nest in tree holes, mainly in June, in the Murree hills and the normal clutch is 2 to 3 eggs, which are pure white roundish ovals of smooth glossy texture. They often appropriate disused woodpecker's holes or barbet's holes and the nest cavity is not lined in any way, the eggs being laid on the bottom. H.W. Wells, while on a mammal survey for the Bombay Natural History Society, shot a female from a nest containing chicks on 4 July 1923 near Bhurban in the Murree hills at 1,900 metres (6,075 feet) (Whistler, 1930), and Capt. Marshall obtained 2 unincubated eggs and shot the male bird (skin in Brit. Mus. coll.) from a nest hole only 4.5 metres up in the trunk of a dead tree on 1 June 1872 near Murree at 1,800 metres (6,000 feet) (Cock and Marshall, *Stray Feathers*, 1873). Buchanan took a clutch of 3 eggs, incubated, on 28 June 1901 at Changla Gali also in the Murree hills (Whistler, *Ibis*, 1930). It is believed that only the female incubates and that the male brings her food to the nest hole during this time and that both parents feed the young.

*VOCALIZATIONS

The advertising or territorial calls of these little owls are very resonant and far carrying and can be heard up to 1.6 kms. (1 mile) distance over open valleys. They comprise a flute-like double whistle which varies in pitch from one individual to another. No doubt this variation helps in spacing out territories and individual mate selection. The whistle can be rendered as 'toot-too' repeated every 5 or 7 seconds, with about a half second interval between the first and second 'toot' which is slightly more emphasized than the first, but of the same pitch. Each individual starts calling at dusk; they never call during daytime. On moonlight nights in May and June they can be heard throughout the night. It is however a very stereotyped unvarying call and recordings of birds calling in Thailand made by Ben King could not be differentiated from Murree hills' recordings (See Roberts and King, 1986).

Northern Eagle Owl *Bubo bubo* (Linnaeus)
Rock Eagle Owl of Whistler, Indian Great Horned Owl of Ali and Ripley

Plate **20**
Map **237**

DESCRIPTION

Body length 56-66 cms.
Wingspan 160-188 cms.
Wing length 37-43.3 cms.
Tail length 19-31 cms.
Weight 2,100-2,700 gms. (males),
   3,075-3,260 gms. (females) (Dement'ev *et al.*)

The northern population (*turcomanus*) of this owl is much larger than that which breeds in the plains. They are smaller than a Greater Spotted Eagle (*Aquila clanga*) (author obs. wild birds in proximity) but with their loose soft plumage and short tails, look almost as bulky. This owl is similar in colouration and plumage pattern to the Long-eared Owl (*Asio otus*) but more tawny buff on the breast and under-tail coverts. Its throat area is almost whitish and upper breast is boldly streaked with vertical rows of dark brown tear drops. These streaks become narrowed on the lower breast which is sparsely vermiculated with wavy horizontal cross-barring. The wings and tail are rufous chestnut with darker brown cross-barring and the crown mantle

and wing coverts are barred and mottled with chestnut and darker brown with some paler whitish buff on the median wing coverts. Its ear tufts or aigrettes are set well apart and are long and conspicuous. The most striking feature of this handsome owl is its huge orange irides. The bill is blackish horny and the tarsi are short and powerful being densely clothed in rufous buff feathers down to the tip of the toes. The claws are long and strongly recurved. Females are noticeably larger with less distinct dark streaking on the upper breast (observations 2 different breeding pairs, Thatta district).

HABITAT, DISTRIBUTION AND STATUS

This owl is very widely but thinly distributed in the hilly tracts of Pakistan including the more arid steppic montane regions of Baluchistan and the NWFP. It is absent from the well cultivated and settled plains regions but will occasionally be found along old disused irrigation channels where the high embankments are covered with bushes, and also spreading into Tamarisk scrub in the more thinly settled regions of Sind. It occurs throughout the lower valleys of Gilgit (a specimen collected 4 January by Biddulph) and Whitehead considered it the common owl around Kohat (Whitehead, 1911). It has been noted in recent years (author) in low hills north of Quetta at Beleli and at Gupis in Gilgit at 3,200 metres (10,500 feet) elevation. It is resident and breeding in the Punjab Salt Range and Sind Kohistan. H. Waite collected specimens (Brit. Mus. coll.) from Attock district and Sethi in the Jhelum district in the Salt Range. Christison considered it resident in the mountain ranges of the Chagai (*Ibis*, 1941). Though subject to local wandering at times, it is largely sedentary and resident where it occurs. It is absent from the coniferous forest areas of the outer Himalayan ranges, but has been recorded in lower Swat, Dir and Chitral. Status FREQUENT.

HABITS

The pair bond is monogamous and life-long (Cramp *et al.*, 1985). Also two pairs have been under regular observations over 5 years (R. Passburg, N. van Zalinge and author) in different localities in Thatta district, where they can be encountered within a few hundred metres of each other in all months of the year. They roost by day on the ground under a bush or in the shelter of a rock overhang, but in Quetta they have been noted roosting in a plantation of Ash trees high up in the branches (*Fraxinus excelsior*). In the winter breeding season, males start calling territorially from late afternoon, at intervals of 15 minutes to half an hour and as dusk approaches these calls become more frequent (every 2 to 3 minutes) and they usually fly out and perch on a prominent mound or rock just before darkness fully descends, after which they commence hunting. They are occasionally sympatric with the Dusky Horned Owl (*Bubo coromandus*) in lower Sind. Their food comprises mostly nocturnal rodents. In Gilgit the Migratory Hamster (*Cricetulus migratorius*) which does not hibernate in winter, is an important food source and in Baluchistan remains of the Afghan Hedgehog (*Hemiechinus megalotis*) and Persian Jirds (*Meriones persicus*) have been found in a nest of this species (J.A.W. Anderson, pers. comm.). Analysis of 35 pellets of this owl from a single roost near Manchar lake (Dadu district) revealed that the very fierce fossorial rat *Nesokia indica* comprised 31.6 per cent of their composition, with *Tatera indica* 49.1 per cent and *Millardia meltada* 14 per cent. There were remains of single *Rattus rattus, Gerbillus nanus* and the Shrew *Suncus stoliczkanus*, also a few fragments of insects Coleoptera and Orthoptera (G. Fulk and A.R. Khokar, *Pakistan Journal of Zoology*, 1976). In the Kohat district Whitehead (*JBNHS* 1911) observed them feeding on Tiger Frogs (*Rana tigrina*) and Rosy Starlings (*Sturnus roseus*). It is obviously a resourceful and opportunistic hunter, capable of taking a variety of mammalian, avian, reptilian and even insect and amphibian prey. Salim Ali records it as preying upon the Indian Roller (*Coracias bengalensis*), lizards and frogs ('Handbook', 1969). A captive Eagle Owl is known to have lived 68 years (Landsborough Thomson, ed., 1964).

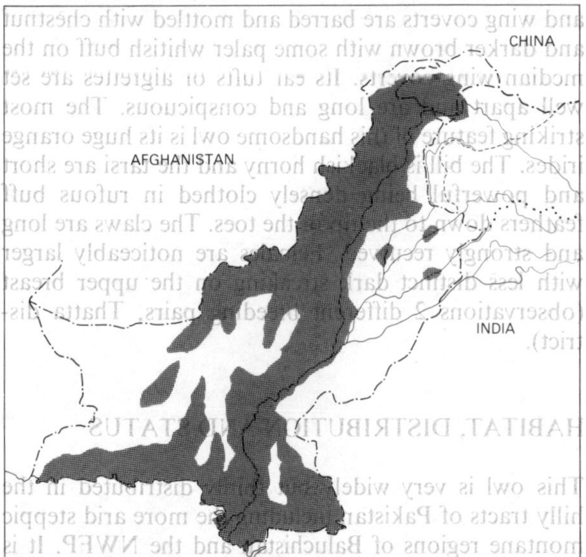

BREEDING BIOLOGY

The two pairs observed in Thatta district at Kalankot and Shah Hussain both commenced breeding in November in 3 years, and one started in early December in the fourth year. As indicated the same general territory is used year after year but the nest location varies. The pair bond appears to be permanent. In the warmer southern latitudes, no nest is normally made, the eggs being laid in a bare scrape on the earth. One at Shah Hussain was in a sheltered hollow between a bush and a rock and one at Kalankot was on the top of a large isolated limestone rock but sheltered under a bush (*Commiphora mukul*). Usha Ganguli (*Guide to the Birds of Delhi*, 1975) records fresh eggs and full clutches being found in the Delhi area in January. K. Eates (unpublished MS) found 4 eggs in December and 2 unincubated eggs on 2 March in the limestone hill ridges at Rohri (Sukkur district). He also found well incubated eggs on 18 March in an earth cliff near Rehri in Korangi creek, Karachi and on 26 February a nest in a deep hole on the banks of a water course at Jang Shahi. On 26 March in Tharparkar district near Umarkot he found two young, with their wing quills

Map 237: Northern Eagle Owl *Bubo bubo*

sprouting, in an unusual location. This was a hole about 45 cms. in diameter and 60 cms. deep in the side of a sand hill, which from claw marks appeared to have been partly excavated by the owls themselves. At Shah Hussain the pair laid five eggs by 20 November 1984 and these were noticeably variable in size. They are pure white and rather round in shape. Only the female incubates, starting with the first egg laid, and this takes 34 to 36 days. The eggs are laid at 2 or 3 day intervals and the eggs hatch asynchronously. At birth the young are blind and nidicolous, being covered with white down. At about 7 weeks of age the oldest from a brood of three at Shah Hussain could fly a little, and was the size of a domestic hen but the youngest of its two siblings still had a wholly down covered head and was much smaller and could not fly (observed 7 January 1982). On 4 February all three young could fly but were still being fed by both parents. The following years the same pair hatched 3 young which was estimated to have occurred between 5 and 10 December. The usual clutch is 2 to 3 eggs, and the clutch of 5 was exceptional. In the far northern areas such as Gilgit and Baltistan breeding commences in May or even later and sometimes the disused cliff nest of an

eagle or raven is used (Meinertzhagen, 1927). When approached the young owls will clatter their mandibles together and hiss, threat gestures common to all the owl family. Both parents look after their young, including bringing food and guarding the nest area, and a female with downy young less than 2 weeks old resorted to injury feigning to divert attention from the nest (Dharmakumarsinhji, *JBNHS*, Vol. 41, p. 177).

*VOCALIZATIONS

When warning their young against danger, the adult owls utter a 'kueck-kueck' quacking type of call, rather like some of the calls made by *Strix aluco*. At dusk both sexes call to each other during the breeding season. The males territorial call is a double resonant hoot, but not very loud considering the size of the bird. The first note is short and accented, the second more drawn-out and rising in tone 'tu-whooh'. A female when flying low over the male on the ground, elicited a rapid series of calls 'tur-tur-tur-tur' like the first phrase of their double hooting call. Fully grown young when approached issued hissing threat calls.

Dusky Horned Owl *Bubo coromandus* (Latham)

Plate **20**
Map **238**

DESCRIPTION

Body length 58 cms.
Wing length 38-41.5 cms. (Stuart Baker),
   40.3-41.2 cms. (females) (Salim Ali)
Tail length 19.5-21 cms.

Much the same size as the Eagle Owl though averaging slightly larger than the plains' population of *Bubo bubo bengalensis*. In overall body colouration it is rather a drab umber brown with a pale grey-brown breast and lacks the warm chestnut tones of *Bubo bubo*, or the whitish area around the throat. Its flight and tail feathers are brown with pale mottled cross bands and tips and the breast is marked all over with narrow dark shaft streaks. The back, wing coverts and crown being more heavily mottled and streaked with dark brown. The ear tufts or aigrettes are large and when the bird is alert, stand up close together on its crown, round tipped and very like a cat's ears in silhouette. There are some white and creamy spots on the wing coverts and the facial discs are framed by narrow black lines. The powerful bill is horny whitish and the irides are golden yellow, not orange as in *Bubo bubo*. The tarsi and toes are well feathered with pale grey-brown feathers. Females are similarly plumaged but slightly larger. When heard close up, in flight its wings make an audible swishing sound.

HABITAT, DISTRIBUTION AND STATUS

This is an Oriental species, confined to well watered tracts with tree cover, in lowland regions. It occurs in

Map 238: Dusky Eagle Owl or Dusky Horned Owl *Bubo coromandus*

Thailand, Malaysia and south-east China and all over the Indian subcontinent except in the Himalayas. In Pakistan it is absent from Baluchistan and the NWFP being more or less confined to the riverain tracts, especially where remnants of forest remain. It has adapted to irrigated forest plantation in the Punjab and in the east Narra and Sanghar in *Prosopis spicigera* groves alongside seepage lakes. Recent sightings include Pirawala and Changa Manga forest plantations in Multan and Kasur districts of Punjab and at Lal Sohanran in Bahawalpur. A pair has been noted resident in the same grove of *Terminalia* trees in Gharko forest, Thatta district over a seven year period. They have been observed on the Ravi river near Abdul Hakim in *Populus euphratica* scrub and in scrub jungle and tall *Saccharum* grass between two canals at Marala in Sialkot district. At Ghauspur jheel in Jacobabad district a pair or two haunts the tamarisk scrub bordering the eastern shore of the lake. They are sedentary wherever they occur. Status FREQUENT.

HABITS

Pair bonds are probably long lasting as they are often encountered in the autumn and spring months in pairs. They roost by day in leafy trees or occasionally near the ground on the lower branches of Tamarisk thickets. They start to call from early afternoon but only do so rather intermittently until dusk is falling when they often fly to a prominent bare branch to survey their surroundings while the last rays of light are fading. They are bold and aggressive hunters taking birds rather than mammals more frequently than does the Eagle Owl *Bubo bubo*. T.R. Bell, of the forest service recorded them in the riverain forest at Sukkur feeding mainly upon House Crows (*Corvus splendens*) and Coucals (*Centropus sinensis*) with the remains of two young porcupines (*Hystrix indica*) in one nest with young (Ticehurst, 1923). Eates examining casts below a roost in Nawabshah found plentiful remains of Collared Doves (*Streptopelia decaocto*) and Crows (*Corvus splendens*), as well as the Indian Roller (*Coracias bengalensis*), a Paddy Bird (*Ardeola grayii*), a Palm Squirrel (*Funnambulus pennatus*), rats (*Rattus rattus*) and a frog (MS notes unpublished). Salim Ali notes their habit of decapitating larger prey before taking it to their feeding perch or nest and notes Water Beetles (*Dytiscus* sp.) besides a variety of birds and mammals in their regurgitated pellets ('Handbook', 1969).

BREEDING BIOLOGY

Nesting takes place mostly in December and January as in the case with *Bubo bubo* and they normally appropriate the nest of a raptor, high up in a leafy tree, in contrast to the ground nesting Eagle Owl. At Gharko forest the old nests of Brahminy Kites very high up in *Terminalia* trees are used. Bell found old vulture, and Pallas's Fish Eagle nests being appropriated. The earliest nest with eggs found by K. Eates was in the beginning of December in an old Kite's nest and he took 2 fresh eggs from this nest on 14 December. The normal clutch is 2 eggs, occasionally only one well incubated egg is found. The eggs are plain white and spherical. A nest with 2 eggs robbed by Eates on 14 December was again visited on 17 January and the same pair had laid 2 more eggs which were fresh, unincubated (K. Eates, MS notes). Fledged young, sprouting wing quills can be found from early February. One nest seen by the author in the pollarded crown of a *Prosopis spicigera* tree near Hatungo Goth in Khipro taluka of Sind appeared from the vantage point of camel back, to have only one young, with wing quills just sprouting on 22 February, and the flat platform type of nest looked similar to that built by *Neophron* Vultures. On a visit to Bharatpur sanctuary (India) a female was seen brooding two downy young on 6 February in an *Acacia arabica* tree on an island in what looked like the old nest of a Cormorant.

*VOCALIZATIONS

During the breeding season pairs frequently indulge in responsive calling and are quite noisy. The territorial call is very characteristic, involving a series of rapid throaty notes, the first two emphasized and distinct and the remaining 4 or 5 accelerando and also dying away. The tempo is very reminiscent of an old steam railway engine when shunting but the notes are not hissing but deep and throaty 'wruck-whruck-wruk-kuk-uk-uk-yk-k-k'. In the east Narra four widely scattered owls were heard calling within auditory range of each other in late February. They can be heard calling from October to March and at the beginning of the breeding season when pairs call responsively to each other, the tones become deeper, more throaty and more rapidly repeated, without the initial emphasized notes, becoming rattling arpeggios.

Brown Fish Owl *Ketupa zeylonensis* (Gmelin)

Bubo zeylonensis of Ali and Ripley

Plate **20**
Map **239**

DESCRIPTION

Body length 53-56 cms.
Wingspan 145-150 cms.
Wing length 35.5-40.6 cms. (Stuart Baker)
Tail length 18.5-21 cms.
Weight 1,105 gms. (female - March - Nepal).(Diessel Horst)

This large owl averages slightly bigger than *Bubo bubo*, the Eagle Owl and can easily be recognized in the field because it has the same warm ochraceous or chestnut buff breast colour of the Eagle Owl but without any heavy dark streaking. The upper breast and throat region is white and the breast has fine dark shaft streaks with horizontal delicate cross rays. The crown and back are pale chestnut with dark brown streaking and mottling and the wing coverts and flight feathers are dark brown with whitish or cream mottling and indistinct bars. The ear tufts or aigrettes are well developed but are characteristically held rather horizontally protruding sideways rather than upwards. The facial discs are creamy with radiating narrow darker hair-like lines and the irides are yellow. The tarsi and toes are bare of feathers in this owl, an adaptation to its aquatic food prey but when squatting on a branch or even on the ground this feature is difficult to observe. The bill is greenish horny, darker on its tip and the legs and toes are dirty yellow in colour. The tail is boldly cross-barred brown and creamy. The sexes are alike though females average slightly larger in size.

HABITAT, DISTRIBUTION AND STATUS

This is a sedentary owl confined to the vicinity of perennial streams or rivers which contain fish and preferring those streams which have good tree and bush cover along their banks. In Pakistan it is extremely rare and there are few recent records. Whitehead shot two specimens along streams in Kohat district (*JBNHS*, Vol. 20, 1911). Whistler encountered both a juvenile bird and an adult in the eroded earth cliffs along the Sohan river in the Potohar (*Ibis*, 1930). Ticehurst did not encounter it in Sind but Scrope Doig did in the east Narra (*Stray Feathers*, 1879b). K. Eates only encountered it once in over 30 years in Sind (*JBNHS*, Vol. 40, No. 4, 1939). This owl extends westwards to the Middle Eastern countries of the Persian Gulf (Hue and Etchecopar, 1970) and may occur in the Hingol and Hinglaj rivers in the Makran. It evidently still survives on the Hab river as an adult was seen near the mouth of the Hab river perched on the sandy beach at tide's edge one evening in October

1980 by Mr. and Mrs. Vollum. (pers. comm., 1980). It was apparently hunting Ghost Crabs (*Ocypoda rotundata*) which are plentiful towards sunset along these beaches. Status RARE.

HABITS

They sleep by day in some leafy tree or a hollow in an earth cliff sheltered by a growing bush. They do not usually emerge to hunt until darkness falls. Crabs, both fresh water and marine, have been noted as an important item in their diet (Whistler, 1949, and Mason and Lefroy, 1912). One shot by Whitehead near Kohat had its stomach crammed full of fragments of fresh water crabs plus a few fish bones (Whitehead, *JBNHS*, 1911). Salim Ali records one as consuming a Monitor Lizard (*Varanus* sp.) 28 cms. long and another one feeding upon the putrefying carcass of a crocodile ('Handbook', 1969). K. Eates found remains of Lesser Whistling Teal (*Dendrocygna javanica*) and fish bones in a nest near Hyderabad (K. Eates, *JBNHS*, 1939, p. 751). They will catch partly stranded fish and feed on frogs and a variety of insects, small rodents and birds according to opportunity. The soles of their feet are covered by prickly scales, like those of the Osprey (*Pandion haliaetus*) which would aid in gripping slippery fish.

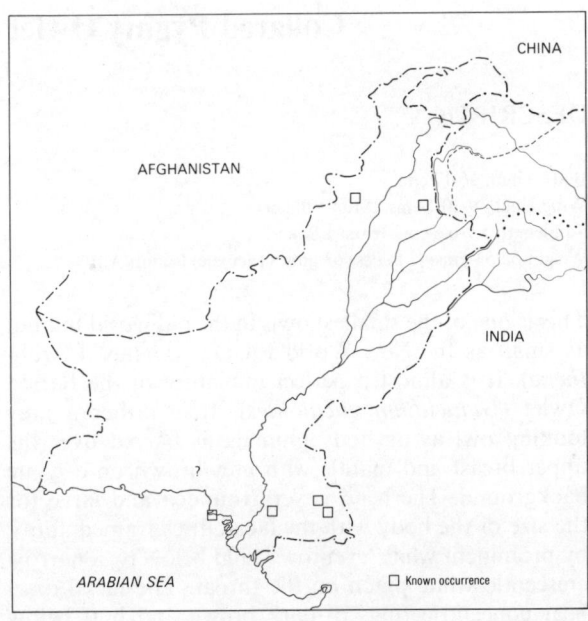

Map 239: Brown Fish Owl *Ketupa zeylonensis*

BREEDING BIOLOGY

They probably pair for life as they are generally encountered throughout the year in pairs. They build no nest, laying their eggs on a cliff or ledge on the steep bank of a stream. The normal clutch is 2 eggs which are plain white, smooth and slightly glossy in texture. H. Whistler (*Ibis*, 1930) found a 'full grown nestling' (presumably about 2 months old) on 11 May in Rawalpindi district. K. Eates found a nest in the junction of massive bifurcating boughs in a *Ficus bengalensis* tree containing 1 egg on 29 December 1922 by a small lake in Hyderabad district (Eates, *JBNHS*, Vol. 40, 1939). E.N. Lowther in central India found most eggs in January and once an egg on 25 December. (*Bird Photographer in India*, 1949). Stuart Baker gives the incubation period as about 5 weeks and it probably is the same as that of *Bubo bubo* (35 to 36 days). Only the female incubates.

*VOCALIZATIONS

A pair tape-recorded in late February at Wilpattu National Park, Sri Lanka indulged in duetting each night, usually several hours after dark and they did not call at sunset before emerging to hunt as *Bubo bubo* characteristically does.

Their calls are low pitched, dolorous and almost the same pitch as a man's voice. Usually one bird starts single spaced 'whu' calls and then the other bird in the pair joins in, but at a much lower pitch, so that the calls sound like 'why' (pause) 'why' (pause) 'wh-o-o-h-hoah' (lower) 'woh-o-oh-hoah' or 'boo' (short) 'bo-o-oh' (longer) 'boh' (lower in pitch) 'boo-bo-o-o-h, boh' (dropping in tone). Whistler describing the same duetting calls syllabizes this as a deep hollow sounding 'hu-who-ho' (Whistler, 'Popular Handbook', 1949).

Snowy Owl *Nyctea scandiaca* (Linnaeus)

STATUS

On 3 March 1876 a local hunter encountered several of these huge white owls on the Mardan plains adjacent to the Swat river and one specimen was secured and later preserved (Stuart Baker, Vol. 4, p. 421). There have been no subsequent records since then from any part of the subcontinent until an adult female was

observed on 26 December 1989 in late afternoon by a party of British Embassy staff in the Margalla hills. Though apparently asleep in a deciduous tree, it was being mobbed by Scimitar Babblers and Lanceolated Jays. Its pure white face and breast with dark grey feather margins in a chevron pattern precluded confusion with any other species (N. Churchill, *in litt.* to author). Status RARE STRAGGLER.

Collared Pygmy Owlet *Glaucidium brodiei* (Burton)

Plate **20**
Map **240**

DESCRIPTION.

Body length 16-17 cms.
Wing length 9-10.1 cms. (Stuart Baker)
Tail length 5.7-6.6 cms. (Stuart Baker)
Weight 52-53 gms. (2 males) 63 gms. (1 female) (Salim Ali)

This is one of the smallest owls in the old world but not as small as the New World Elf Owl (Genus *Micranthene*). It is almost a perfect miniature of the Barred Owlet (*Glaucidium cuculoides*). It is rather a grey looking owl as its body plumage is barred over the upper breast and mantle with grey-brown on a white background. The head is very rounded and large for the size of the body with the facial discs framed above by prominent white 'eyebrows', and below by a narrow crescentic white patch on the throat. The facial discs bear concentric rings of dark brown and buff below the eyes. The crown is spotted with creamy buff not

cross-barred as in *G. cuculoides*. There is a striking and curious pattern on its nape and hind neck, comprising two outwardly curving dark lines like eyebrows with white spots below and a white crescentic nape patch beneath the white spots. This pattern looks remarkably like the face of an owl and may serve to deter mobbing birds when the owlet turns its head around. The legs are densely feathered white but the toes are bare and greenish yellow in colour. The tail and flight feathers are cross-barred dark brown and white and the scapulars are tipped white forming interrupted lines down each side of its mantle. The bill is greenish horny and the large wide set eyes are golden yellow. The sexes are alike. Juvenile birds have vertical bold tear drop marks on the upper breast, in contrast to the cross-barring of adults. It is so well proportioned with powerful feet and large head that it looks quite a fearsome hunter, but in size it is no bigger than a Crimson-breasted Barbet (*Megalaima haemacephala*).

HABITAT, DISTRIBUTION AND STATUS

A Himalayan owl which extends westwards through Chitral, Swat and Hazara district in the lower Kaghan valley and in the Neelum valley of Kashmir. Several males were heard calling territorially throughout the day in early May in the Machiara forest, Azad Kashmir (author, 1984 and 1985). It is common in the Murree hills from 1,800 to 2,750 metres (6,000 to 9,000 feet) and numbers can be encountered around Shogran and Manshi forest on both sides of the Kaghan valley. In winter there is some altitudinal migration to the foothills and they can be seen and heard on the outskirts of Islamabad, even calling territorially in the Margalla ravines up to late March when they return to the coniferous forest of the higher hills where they breed. They are comparatively uncommon in Dir, Chitral and Swat but Perreau (*JBNHS*, Vol. 19, 1910) collected a specimen at 1,500 metres (5,000 feet) in February in Chitral main valley. They have been heard in the Safed Koh range by Whitehead (op. cit., 1911), and are apparently absent from Afghanistan (Paludan, 1959).

This is an Oriental species, which occurs in Cambodia, Thailand, Malaysia, Borneo, China and Taiwan and is widespread wherever steep hillsides are clothed with forest. Status COMMON.

HABITS

Usually solitary in hunting and roosting, they become active by late afternoon, not waiting for darkness.

Individuals can also be seen hunting at all hours of the day. In the breeding season territorial males call a lot throughout the day as well as during the night. Despite their diminutive size, they are fierce hunters and will successfully attack or kill birds as large as the Streaked Laughing Thrush (*Garrulax lineatus*) (observed Dunga Gali in good light at 7 p.m.) Their main hunting technique is to select a prominent perch and to watch and listen. They have been observed at dusk on a fence post listening to the drone of Cockchafers (Melolonthinae) and swooping down from their perch to seize the beetle in flight. In Dunga Gali they catch cicadas in a similar manner during the monsoon season and have been observed tearing to pieces and eating a Tailor Bird (*Orthotomus*), near Rawalpindi (H. Whistler, op. cit., 1930) and a Crested Black Tit (*Parus melanolophus*) (author) in Dunga Gali.

They can often be located in daytime because of the scolding cries of mixed parties of Paridae when they encounter this owl. While attempting to study variation in colour and prominence of the face mask on the back of its head, the author has frequently circled right around this owl perched in a tree, and it is amusing to see its unwinking yellow eyes full face, watching one from whatever angle, as they really seem capable of turning their heads through a 270° arc. Population densities in the Murree hills are somewhat less than *Otus spilocephalus*, with an average of 3 males calling in a typical region (Kao valley, Dunga Gali) over a 4 year period compared with 5 or 6 territorial males of the Spotted Scops (*O. spilocephalus*).

BREEDING BIOLOGY

The breeding season can be quite extended as occupied nest holes have been found between May and July, and Whistler gives July as the main breeding month for the Murree hills (*Ibis*, 1930). They are tree cavity nesters, their small size enabling them to take over Hill Barbet (*Megalaima virens*) nest holes and those of the Himalayan Pied Woodpecker (*Dendrocopos himalayensis*) without having to enlarge the entrance hole. The only nest found by the author was in a much lopped *Prunus cornuta* tree in a rather open stretch of hillside at about 2,650 metres (8,500 feet) elevation and this was too small a hole to have been made by the Scaly-bellied Woodpecker (*Picus squamatus*). It was discovered because one bird was uttering soft squeaking calls nearby which were followed by the presumed female flying out of the hole. This was at about 4.30 p.m. in the last week of May. The normal clutch is 3 to 4 eggs which are almost spherical and pure white. Nothing is known about the incubation period or role of the sexes in incubation but the former is presumed to be about 25 days. The eggs are known to hatch asynchronously and both parents bring food to their young. On 27 June two young were encountered which had apparent-

Map 240: Collared Owlet or Collared Pygmy Owl
Glaucidium brodiei

ly just left the nest. One still had tufts of down on its head and the other fully feathered and more ready to fly on being approached, had vertical streaks on its upper breast indicating it also, was a juvenile. The adult birds could be heard giving warning calls in the distance but did not aproach closer despite my watching the young till dusk fell.

*VOCALIZATIONS

Bates and Lowther (op. cit., 1952) record that when inspecting a nest hole occupied by this owlet, the incubating adult inside hissed and snapped its mandibles with loud clicks. The advertising or territorial

calls are given from March till November and possibly occasionally throughout the winter in its lower altitude wintering quarters. It is a stereotyped song consisting of four clear flute-like notes all about the same in pitch, the first and last more spaced and drawn-out and the two middle notes uttered rapidly and close together but with equal emphasis 'poop-pupu-poop' or 'took-tuk-tuk-took'. Each series of calls lasts about 1.75 seconds and is repeated after intervals of about 1 second, sometimes only 6 or 7 times, but at the beginning of the breeding season, often in continuous runs of 20 or 30 calls and even during the middle of the day. A pair at their nest hole were heard giving shorter rather softer squeaking calls.

Asian or Himalayan Barred Owlet *Glaucidium cuculoides* (Vigors)   Plate **20**
Map **241**

DESCRIPTION

Body length 22-23 cms.
Wing length 14.5-16.2 cms. (Stuart Baker)
Tail length 7.5-9 cms.

This owl is slightly larger than the Spotted Owlet (*Athene brama*) and considerably bigger than its near relative *Glaucidium brodiei*. It is a rather grey looking owl being pale creamy on the upper parts with narrow horizontal cross-bars of dark grey-brown extending from the forecrown to the upper-tail coverts. The tail is more broadly cross-barred dark brown and the scapulars are creamy white tipped, forming a conspicuous row of spots down each side of the back. It has relatively small round facial discs and no ear tufts or aigrettes and there are concentric dark rings around the eyes. The irides are golden yellow and the bill is yellowish green, darker grey on the cere. There is a white patch on the throat and the breast is white with the pectoral and flank regions barred horizontally. The white area dividing the middle of the barred breast is a good field point, also the rows of white spots on the scapulars and wing coverts. There is also a broad white 'eyebrow' line over the facial discs. It lacks the 'eyes' and dark eyebrow lines on the back of its head, conspicuous in *G. brodiei*. The feathered tarsus is cross-barred with the toes dull greyish green. The sexes are alike. Juveniles have the breast streaked, not cross-barred, and are more rufous about the crown and mantle with creamy spots rather than cross-bars.

HABITAT, DISTRIBUTION AND STATUS

This is an Oriental species occurring from south China, Assam, Bangla Desh and across the Himalayas to Pakistan. It also occurs in Indonesia and 'Indo-China'- but not in Malaysia. It is associated in the

Himalayas with lower elevations and foothill country, being most prevalent between 460 metres (1,500 feet) and 1,800 metres (6,000 feet), preferring scrub jungle and terraced cultivation in the outer valleys to the coniferous forests of the higher hills. In Pakistan it has been seen in Hazara district at Balakot and in the Mansehra plain at Battagram but its main stronghold is in the Murree foothills including the Margalla hills, Lehtrar valley, Tret, Kahuta and around Mangla dam catchment. In the Neelum valley, Azad Kashmir, it has been seen (author) around Pateka and Muzaffarabad. It has not been noted further west in Swat or Chitral. Status FREQUENT.

Map 241: Asian or Himalayan Barred Owlet *Glaucidium cuculoides*

HABITS

This owl becomes active in the late afternoon, occasionally throughout the day,and is not averse to hunting before the sun has set. Like most of the smaller owls it seeks much of its prey by choosing a prominent perch and keeping a watchful eye. It has an undulating flight consisting of rapid wing beats alternating with a dipping glide with the wings practically closed. When mobbed by smaller passerines or agitated by close human observation it has the curious habit of wagging its tail from side to side (not up and down like *Elanus caeruleus*). One was observed at Lehtrar on a mud threshing floor where it had successfully killed a sparrow (*Passer domesticus*) on the ground in bright daylight (about 4.30 p.m.), but a large part of its diet comprises insects such as grasshoppers, beetles and cicadas. It has also been recorded as taking lizards, frogs, a Quail (*Coturnix coturnix*) and white ants (Isoptera) (Mason and Lefroy, 1912).

BREEDING BIOLOGY

Pairs have been noted together from March to August and the breeding season may be rather extended. A pair was observed copulating on 21 March at Tret in an *Acacia modesta* tree (760 metres or 2,500 feet elevation). The same pair was noted in a stream fed ravine at this spot early in April but their nest could not be located. On 3 July above another narrow ravine containing a stream near Balakot (Hazara district) 3 fledged young were observed coming to the mouth of

their nest hole and being fed by both parents though it was mid-afternoon and bright sunlight. This nest was in the hole left by a rotten branch in a pollarded White Mulberry (*Morus alba*) about 6 metres from the ground and the location was at about 1,200 metres (4,000 feet) elevation on a rather bare treeless slope of terraced fields. The normal clutch is 3 to 4 eggs which are white and rather spherical in shape and a tree hollow is the usual nest site described by other writers (Ali and Ripley, 'Handbook', 1969). They sometimes appropriate barbets' nest holes and are sympatric with *Megalaima asiatica* in the Murree foothills. It is believed that the male shares incubation duties.

*VOCALIZATIONS

During the breeding season, the male has a special call which is uttered at frequent intervals from late afternoon often until well into the night and again in the early morning. Each call lasts from 8-10 seconds and consists of a continuous rapid ululating or rippling warble often starting at a higher pitch and gradually descending in scale, whilst increasing in volume and then dying away again. This song is quite high pitched and melodious in sound. A female at Tret (Murree hills) was heard giving a rather brief bubbling call just before copulation took place. Fleming also describes a single noted call as 'keek' which may be a warning or contact call (Fleming and Fleming, 1976). Another individual when disturbed in mid-afternoon at Pateka (Azad Kashmir) uttered a variety of chuckling calls of varying pitch.

Brown Hawk Owl or Bobook Owl *Ninox scutulata* (Raffles)

STATUS

The 'Handbook' (Ali and Ripley, 1969) and 'Revised Synopsis' (Ripley, 1982) include the Murree hills within the range of this owl. However no evidence has been obtainable of its occurrence even in former times. Hugh Whistler noted that Stuart Baker in his 'Fauna Series' (Vol. 4, 1927) stated that it occurred at Murree but doubted this record as he did not know of its occurrence even farther east in Dharamsala (Whistler,

'Birds of Rawalpindi District', *Ibis*, 1930). H. Waite who collected extensively in the Murree hills and Dharamsala in Kangra district, did not encounter it. It is typically an owl of evergreen moist rain forest or moist deciduous forest and the author has seen and tape-recorded its call both at Chitawan, Nepal and in Sri Lanka but never heard it anywhere in the Murree hills. It is therefore excluded from the **Checklist** of Pakistan birds, as its normal range is well east of Pakistan.

Little Owl *Athene noctua* (Scopoli)

Plate **20**
Map **242**

DESCRIPTION

Body length 23 cms.
Wingspan 54-58 cms. (Cramp *et al.*)
Wing length 15.6-16.8 cms. (sub-species *bactriana*) (Stuart Baker)
Tail 8.4-8.7 cms.
Weight 118-172 gms. (males),
165-260 gms. (females) (sub-species *bactriana*) (Cramp *et al.*)

These are small owls with short tails and large rounded heads, lacking ear tufts. The tarsi are quite long and well feathered right down to the base of the toes. The upper parts are grey-brown, the feathers tipped white on the crown and mantle giving a spotted effect. The wing coverts and flight feathers are mottled and barred with white. The breast is white, with paler brown vertical streaks forming a distinct zone around the upper vertical breast but longer and more spaced on the lower flanks. The facial discs are white and there is a broad white crescentic collar around the throat. The outer edges of the facial discs are framed with dark grey-brown with prominent white 'eye brows' above the eyes. The irides are golden yellow and the eyes are rather widely spaced. The bill is greenish yellow and the toes are a dirty greenish colour. The sexes are alike. This owl is with difficulty separable from *Athene brama*, the similarities being much greater than the differences. The upper breast is spotted without a suggestion of cross-barring as in *A. brama* and the crown tends to have white spots arranged in parallel longitudinal stripes rather than evenly and at random

as in *A. brama*. Their calls are however distinctive. A specimen observed at Ziarat, Baluchistan appeared darker brown in upper body plumage than *A. brama*.

HABITAT, DISTRIBUTION AND STATUS

A small population of this owl breeds in earth cliffs up in Ladakh just to the east of Pakistan territory (film of nesting pair taken by Brigadier Moti Dar in 1983). It is not known if any penetrate into Baltistan.

Its status in Baluchistan is somewhat unclear as it overlaps in distribution with *Athene brama* the Spotted Owlet and these two very similar species could easily be confused. It certainly occurs throughout the western border regions, as specimens have been collected from Chaman (Barnes, *Stray Feathers*, 1881) and Surab in southern Kalat (by Col. Hotson, skin in British Museum, Tring). It has also been collected from Ziarat (Ticehurst, 1927). Both Meinertzhagen (*Ibis*, 1920) and Williams (*JBNHS*, 1929) considered *Athene brama* to be the common owl around Quetta and central Baluchistan and did not encounter *A. noctua*. Similarly Brigadier Christison makes no mention of either of these two owls from the Chagai or Zhob district (*Ibis*, 1941 and *JBNHS*, 1942). The author has encountered in daylight (1985) a pair in a dry earth gully near Mastung, 48 kilometres (30 miles) south-west of Quetta, and a solitary bird perched on a telegraph pole in bright sunlight near Zhob (formerly Fort Sandeman) in 1984. Also in June 1971 one was seen on low rocky hills near Barshore in Pishin district and another in a narrow rocky canyon known locally as a 'tangi' near Ziarat in 1978. It could therefore appear to be as widespread in distribution as *Athene brama*. Its preferred biotope in Baluchistan is dry open country, in the vicinity of erosion earth gulleys or rock cliffs in narrow gorges between 600 and 2,400 metres (2,000 to 8,000 feet). Status locally FREQUENT.

HABITS

They are quite territorial and can be encountered in the same locality throughout the year. They do not seem to be particularly bothered by bright sunlight and are often active in flying abroad from late afternoon. However their hunting technique is usually from the ground rather than in flight and they characteristically select a prominent perch from which to watch and wait for prey. The specimens watched by the author at Mastung and Ziarat displayed unusual agility in clambering over the cliffs and walking along the ground when compared with other owl species and it seems probable that a lot of their prey is captured

CHINA

AFGHANISTAN

INDIA

ARABIAN SEA

■ Occasional breeding

Map 242: Little Owl *Athene noctua*

while the birds are on the ground. Specimens in Ziarat were roosting in natural cavities in the side of a narrow canyon in similar hollows to adjacent ones occupied by Choughs (*Pyrrhocorax pyrrhocorax*) and Whistling Thrushes (*Myiophoneus caeruleus*). No food habit studies have been conducted in Asia but this species in Europe feeds largely upon beetles and moths and even earthworms. In Baluchistan lizards must constitute an important component of their diet and no doubt small birds are captured when opportunity allows.

BREEDING BIOLOGY

Has been well studied in Europe but the only authentic nest found in Pakistan is the record by H.E. Barnes of two young in a tree hole on 3 June near Chaman. The hole was about 3 metres above ground and A.O. Hume confirmed identification of one of the parent birds shot (*Stray Feathers*, Vol. 9, 1880). The normal clutch in Europe is 3 to 4 eggs and they are usually laid on consecutive days (Glue and Scott, *British Birds*, Vol. 73, No. 4, 1980). They are plain white and rather spherical in shape and are incubated only by the female which takes between 28 to 33 days. The young fledge after 30 to 35 days, most of the food in the first week or two being brought by the male to the nest (Glue and Scott, op. cit.). Males are said to perform display flights over their nesting site which could be a hole in an earth cliff or a tree, but in Europe the site is most commonly a hole in a deciduous tree.

VOCALIZATIONS

The territorial call or advertising call of the male is a rather plaintive sounding 'quew' or 'piu' repeated every 2 or 3 seconds. They also have a sharp barking type of call 'werro-werro'. During courtship the male utters a loud 'hooo-oo hooo-oo', and the female shrieks and yelps in answer (Glue and Scott, op. cit.).

Spotted Owlet *Athene brama* (Temminck)
Spotted Little Owl of K.H. Voous's 'Recent List'

Plate **20**
Map **243**

DESCRIPTION

Body length 21-23 cms.
Wing length 15.3-16.9 cms. (males),
    15.9-17.1 cms. (females) (Stuart Baker)
Tail length 7.5-8.5 cms.
Weight 114 gms. (1 male) (Salim Ali)

It averages slightly smaller in size than *Athene noctua* but the population in the north-western part of the subcontinent is closely similar in general colour tones. The crown, mantle and wings being grey-brown, with scattered white spots and white mottling or interrupted barring on the flight feathers and wing coverts. The tail looks short and stumpy and is the same grey-brown tone cross-barred white. The breast is creamy white with the feathers tipped dark grey-brown forming short transverse bars in contra-distinction to the longitudinal streaks on the breast of *Athene noctua*. It also has a broad white collar around the front of the neck and the facial discs are white, with dark brown borders only around the sides of the face. The irides are golden yellow and the eyes are framed above by white eyebrows. The tarsus is comparatively long and well feathered to the base of the toes. The bill is greenish horny, more yellow on the culmen and the feet are yellowish green. Like the Little Owl it has a comparatively large rounded head lacking any ear tufts and with rather a broad flat crown. The sexes are alike.

HABITAT, DISTRIBUTION AND STATUS

This owl is resident and sedentary and very widespread throughout the Indus plains, extending into extensive desert tracts and into the broader valleys and foothill regions of Baluchistan and the NWFP. It is absent from the forested hills of Murree or the northern Himalayan regions. Whistler noted that it did not ascend the Murree foothills above 600 metres (2,000 feet) (*Ibis*, 1930) but Meinertzhagen collected it at 1,600 metres (5,200 feet) in a dryer habitat near Quetta, (*Ibis*, 1920). Whitehead noted that it avoided villages in Kohat district being encountered in the most desolate country (*JBNHS*, 1911). Its familiar shrieking cries can be heard in every large bungalow garden in the Punjab, where the spread of canal and roadside tree plantations has favoured its increase. It will however be encountered in treeless regions, wherever there are suitable earth cliffs in the hilly or sand dune uncultivated tracts of Cholistan, Lasbela, the Thar desert and the Punjab Salt Range, and is as well adapted to desert conditions as to the better wooded and intensively cultivated regions. Status COMMON.

HABITS

They are mainly nocturnal in activity but if disturbed in their diurnal roost, show no discomfiture, flying often to perch on some open exposed site from whence they can watch the intruder. Normally they roost by day in a thick shady tree, preferably in a tree hole or cavity, and in open desert country in a crevice or hole in a cliff. Their flight is characteristically dipping, with rapid silent wing beats followed by dipping swoops with wings semi-closed. If discovered in their diurnal roost, they bob their heads forward in threat gesture as they stare down upon the observer. On the ground they can hop and run with surprising agility. Their food comprises small roosting birds, ghekos, beetles, orthopterous insects, large moths, mice and shrews. Stomachs of eight specimens examined in Bihar (India) contained mostly beetles, Coprids, Carabids and *Catharsius saboeus*, with lesser numbers of *Chrotogonus* and *Gryllotalpa* grasshoppers and mole crickets and no remains of birds or mammals (Mason and Lefroy, 1912). One individual near Thal Jute Mills (Muzaffargarh) in May perched near a road floodlight, fluttering down to the ground to pick up beetles and what appeared to be Giant Water bugs (*Belostomatidae*) attracted to the arc of light thrown by the lamp, but returning each time to its perch in a young tree to eat its prey, which was raised to its bill in one foot, usually the left side. In a laboratory experiment, specimens of this owl captured from near Karachi could successfully capture juvenile Libyan Jirds (*Meriones libycus*) but adults could ward off their attacks (Lay, *Journal Mammalogy*, 1974). Adult males of this sand rat weigh up to 90 gms. (Roberts, 1977) and it is not surprising that this little owl would normally select smaller prey.

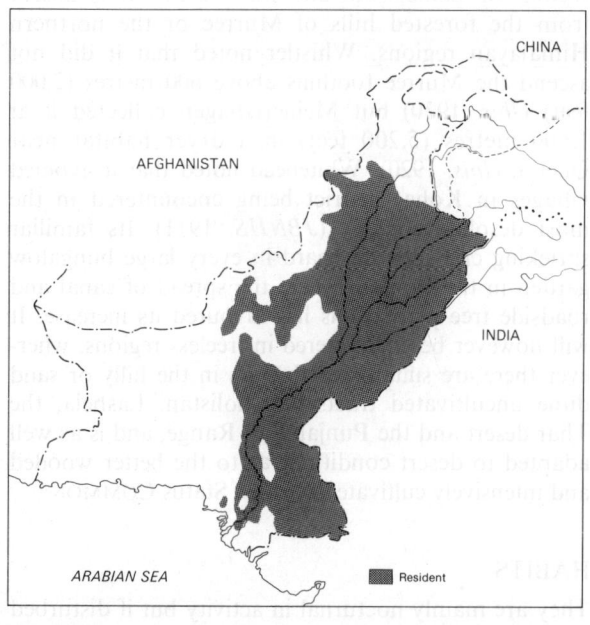

Map 243: Spotted Owlet or Spotted Little Owl *Athene brama*

BREEDING BIOLOGY

A pair has frequented the same site near Gharo village in Thatta district for 3 years since they were discovered. It is in open flat treeless country and they breed in a hole in the mud cliff of a deep drainage channel and can be predictably located within a few hundred metres of this site at all times of the year. Whether this is typical of the species under all conditions, it certainly indicates a remarkable territorial fidelity to a suitable nesting site and the probability of long duration pair bonds.

Most nests are not lined, the 3 to 4 round white eggs being laid on the detritus in the bottom of a tree hollow or in a hole excavated in a vertical cliff or occasionally in the wall of a building. Malcolm MacDonald (op. cit., 1960), recounts a pair in his Delhi garden which used a hole in the wall for both breeding and as a year round shelter. In one year they reared only one young which first emerged from its wall cavity nest on 29 May and was subsequently observed with its parents for a full 3 weeks before becoming independent. K. Eates found nests in limestone cliffs in the Hab valley and once in an old Palm Squirrel's nest (*Funambulus pennanti*) and another in an old crow's nest. He found most eggs during the month of April and noted that the young hatched asynchronously, indicating that incubation starts when the first egg is laid (K. Eates, MS notes). Whistler (*Ibis*, 1930) also noted most eggs laid in April in Rawalpindi district. The period of incubation is about 28 to 30 days and the period to fledging usually slightly less than 5 weeks.

It is probable that, as in its close relative, *Athene noctua*, only the female incubates despite statements to the contrary (Ali and Ripley, 'Handbook' Vol. 3, 1969). A nest site was under observation particularly during the evening, for 6 days in the first week of April. The inaccessible nest was in a brick wall cavity under the roof of an occupied dwelling in Akramabad Ginning Factory (Rahim Yar Khan district). The author was told of the nest by the human occupants who stated that the same site had been used the previous year as the young chicks could be heard giving a wheezing cry through the wall whenever they were hungry, particularly from the early evening onwards. It was observed that throughout each day one owl (presumed to be the male) roosted in a tree nearby. Each evening he became active, calling and was then joined by the second owl observed several times flying out of the nest hole. They sat and called together for some minutes each evening, and on one occasion copulation was observed. No daytime nest changeover was ever observed nor feeding of the brooding bird by its mate during daylight. It was presumed from the early dates (31 March to 5 April) and observation of copulation, that egg laying or incubation was on-going. As known in other related owl species, the male normally hunts and brings food to his incubating mate and also food for the female to feed the nestlings when they are very young. Later, both sexes share in feeding the young.

*VOCALIZATIONS

A great variety of calls from a sort of warbling or babbling call to chattering shrieks, often given by both pairs in duet during March and April. The basic call in these territorial songs or duets is a short strident 'kueek', harsh and not melodious, sometimes repeated rapidly and with increasing speed and loudness. When duetting one bird often utters a shorter quicker 'kuerk-kuerk' and a two-noted 'kuer-veek kuer-veek' also uttered rapidly and in sequences of 4 up to 7 or 8 calls uttered within 1½ to 3 second intervals.

Brown Wood Owl *Strix leptogrammica* Temminck

STATUS

Said to be resident in Pakistan in the Himalayan part of Punjab (Ali and Riply, 1969 and Ripley, 'Revised Synopsis', 1982). No evidence has come to light for this statement except perhaps Stuart Baker's writing that it occurred 'throughout the Himalayas from the extreme west' ('Fauna Series', Vol. 4, 1927). Hugh Whistler's manuscript notes (lodged in the BNHS library) confirm that there are no old records for this owl from the Murree hills or Kashmir region. It is a noisy owl, fairly common in Sri Lanka where the author has encountered it. It would not have easily escaped the notice of ornithologists if it did still occur in the Murree hills. Moreover, its biotope in Pakistan is probably occupied by *Strix aluco*. It is therefore excluded from the **Checklist** of Pakistan until new evidence comes to light.

Tawny Owl *Strix aluco* Linnaeus
Himalayan Wood Owl *Strix aluco nivicola*
Scully's Wood Owl *Strix aluco biddulphi*

Plate **20**
Map **244**

DESCRIPTION

Body length 45-48 cms.
Wingspan 94-104 cms.
Wing length 28.2-31.2 cms. (Stuart Baker)
Tail length 19-21 cms.
Weight 375-392 gms. (males) 380 gms. (1 female)
   (sub-species *nivicola*) (Salim Ali)

After almost daily encounters with a Tawny Owl which roosted in a laurel thicket in the author's home in Wales, when this owl was first seen well in daylight at Shahran in the Kaghan valley it looked strikingly larger and greyer or paler than the nominate sub-species. The race *biddulphi* is supposed to be slightly larger and paler than *nivicola* but the entire Pakistan population is considerably bigger than the European population. The main difference however is in the absence of warmer ochre and russet tones in the plumage, the Pakistan population being mainly dark sienna brown and white. It is a rather squat bulky looking owl with a large rounded head lacking ear tufts and a broad rounded tail extending well beyond the broad wingtips. The facial discs are well formed and show traces of radial darker rings and are narrowly framed by dark brown borders of crimped feathers. The irides are dark brown making the area above the eyes look conspicuously white. A specimen watched in daylight constantly flicked a bluish nictating membrane across its eye. The breast is buffy grey or pale coffee coloured with parallel rows of broad dark brown barbed shaft streaks on the feathers. The crown, mantle and back are mottled and streaked with the feathers being tipped buffy white and bearing darker shaft streaks or obsolescent barring. The plumage is thus quite cryptic. The shaft streaks on the crown and nape are narrower and tend to form parallel lines and the tips of the scapulars are white, forming a prominent diagonal line on the shoulders. The flight feathers and tail are cross-barred buff and darker brown and the tarsus is rather short and well feathered extending down to the toes. There is also some white spotting on the wing coverts. The sexes are alike, though detailed studies have revealed that in Europe females average 26 per cent greater body weight and 4 per cent longer wing length.

HABITAT, DISTRIBUTION AND STATUS

This is a forest dwelling owl confined to the higher colder mountain ranges of Pakistan where it is largely sedentary and resident with limited altitudinal migration in winter. It is, however, quite adaptable to any coniferous forest in the more northerly Himalayan valleys as well as moist coniferous and deciduous mixed forest in the outer ranges which receive heavy monsoon rains. It has been collected in Blue Pine *(Pinus wallichiana)* and spruce forest *(Picea morinda)* in Gilgit, in Deodar forest *(Cedrus deodara)* at Drosh in lower Chitral and in Blue Pine and Chilghoza Pine forest *(Pinus wallichiana and Pinus gerardiana)* in the Shinghar and Takht-i-Suleiman ranges in northern Baluchistan bordering on Waziristan (Biddulph, *Stray Feathers* , 1881, Perreau, *JBNHS*, 1910 and Christison, *JBNHS*, 1942). In the Murree hills and lower Kaghan valley it is not uncommon, birds drifting down to the foothills and especially the Margalla hills in winter when the higher ridges are blanketed with snow. It has been heard calling in Deodar forest near Kalam in Swat. Whitehead also recorded it as a winter visitor to the orchards and wild olive *(Olea cuspidata)* groves in Kohat district (Whitehead, *JBNHS*, 1911). Status FREQUENT.

HABITS

They are known to be very territorial in Europe, living, nesting and hunting throughout the year within a fixed territory. On the flanks of Mukshpuri mountain in Dunga Gali, there is generally a resident pair each summer which has been seen up to the end of October but it is presumed that in the months of January and February they drift down to lower snow-free altitudes. They are more nocturnal than many owl species, rarely being seen moving around or heard calling until after dusk. They hunt largely by ear, perched near the ground on the top of a young sapling or the stump of a pine branch, listening for the rustle of rodents. *Alticola* voles and Long-tailed Field Mice *(Apodemus sylvaticus)*

Map 244: Tawny Owl (including Himalayan & Scully's Wood Owl) *Strix aluco*

share the same forest habitat and are believed to be an important food source, also possibly *Rattus turkestanicus* and a variety of small roosting birds and any large insects which are available such as Cidadas and Cockchafers *(Melolonthiae)*. Their wings are rather broad and their flight is somewhat fluttering, but very controlled and they are well adapted for flitting in and out of the tree boles in the forest.

BREEDING BIOLOGY

In the Shinghar in northern Zhob district, this owl is reported to nest on the ground under boulders (Christison, *JBNHS*, 1942). The typical breeding site is however a hollow in a tree and they have a preference for vertical openings formed by cavities in the forks of branches or at the tip of rotten stumps rather than holes in the side of vertical tree boles. In Kashmir, nests were located also on the ground under rocks on steep ground and also in tree boles in about equal proportion with evidence of one particular tree cavity having been regularly used by this species for many years (Bates and Lowther, 1952). In Dunga Gali pairs have been noted circling around each other at considerable heights and in apparent display flights at dusk. They uttered soft quacking sounds but no wing clapping was heard as has been described for European birds. The normal clutch is 2 eggs, plain white and the female does all the incubation. This takes about 28 days. The male is believed to bring some food to his incubating mate and also to do most of the hunting to feed the young. They take about 5 weeks to fledge but still remain dependent upon their parents for 4 weeks or more after they leave the nest hole. Nesting may start as early as late March, but most eggs are laid in early May. Two newly fledged young hardly able to fly were observed for several days on 31 May to 2 June at Shahran forest rest house (Kaghan valley) and the adults could be heard calling to them each evening.

*VOCALIZATIONS

A quite varied repertoire of calls is made by this owl and calling is often heard in the winter months as well as in the breeding season. A pair watched in May uttered two types of calls. One, an almost quacking sort of low pitched 'ku-wack-ku-wack' uttered as a single call at short intervals. The second owl responded with the typical three-syllable calls 'too-tu-whoo' rather spaced and tremulous. This latter is the territorial or advertising call of males and is given regularly each evening one-half hour after sunset and for upwards of 20 minutes by the male in the breeding season. Sometimes it is a more sonorous call 'kook-ur-kooh' but the stereotyped three-noted phrase is characteristic though it can be shortened to a two-noted 'turr-whooh' when the male is excited. It is presumed that the shorter quacking type calls are made by the female but some authors describe a hunting call which is a variation of this and comprises harsh repeated 'kee-vit-keevit' calls.

Hume's Wood Owl or Hume's Tawny Owl *Strix butleri* (Hume)   Map **245**

DESCRIPTION

Body length 36-38 cms.
Wingspan 95-98 cms.
Wing length 23.8-25.2 cms. (3 males)
   25.5 cms. (1 female) (Cramp *et al.*)
Tail length 13.4-13.7 cms. (3 males)
   14 cms. (1 female) (Cramp *et al.*)
Weight 214-220 gms. (females) (Cramp *et al.*)

Very similar to the Tawny Owl *(Strix aluco)* in general body proportions and plumage pattern, but it is a very pale yellowish plumaged bird, wholly adapted to desert habitats. This owl is stout bodied with broad wings and large rounded head lacking any ear tufts. Whereas the Himalayan Tawny Owl is grey-brown, the predominant colour of this owl on its upper parts is isabelline brown or light yellow dotted with darker brown spots. Its tail and flight feathers are cross-barred isabelline and darker dull brown. The scapulars are pale dirty buff or white tipped and also the wing coverts. The facial discs are white and the irides are orange. The breast, belly and legs are creamy white with only indistinct darker mottling and this lack of dark brown shaft streaks on the breast is the most striking difference between this owl and *Strix aluco*. The sexes are alike.

HABITAT, DISTRIBUTION AND STATUS

The original type was collected from near Ormara in the Makran by Nash (now in the British Museum) and was described by Alan Hume *(Stray Feathers,* Vol. 7, 1878*a*). It remained practically unknown thereafter except for a specimen collected from Saudi Arabia in 1950 and several earlier specimens from Wadi Feiran in the Sinai peninsula (Meinertzhagen, 1954). With the great increase in ornithological surveys and observers in Israel and other parts of the Middle East during the 1960s and 1970s 'oil boom', this owl has recently been rediscovered, captive specimens photographed, and it is considered now to be quite common in the desert parts of Israel (J. Leshem in *Sandgrouse,* Issue No. 2, 1981).

Its range seems to extend only from the Sinai peninsula, the Arabian peninsula (M. C. Jennings, 1981) and the Makran coast. It is apparently very sparsely distributed wherever it occurs but these are areas which are also not well surveyed by ornithologists. It has not been recorded in Pakistan since the original type specimen was described but there have been no resident experienced ornithologists in that region since Cummings was at Ormara in the 1920s and it can be presumed still to be a sparse resident in the extreme south-western corner of Makran.

HABITS

In the Sinai and Saudi Arabia it lives in rocky gorges or canyons in which some water occurs. It shelters by day in rock crevices or occasionally in ruined buildings in Saudi Arabia where Ben King has recently observed it (pers. comm., 1980). In the Makran and around Ormara there is a good rodent population of *Acomys* Spiny Mice, *Meriones crassus* Jirds and *Gerbillus* desert Gerbils, so it can be presumed that this owl is able to prey upon these rodents, as well as small reptiles and insects such as Tenebrionidae and Scarabaeidae (Dung Beetles). Studies in Israel based upon analysis of 3 pellets in 1973 and 52 pellets in 1978 revealed the principal food was *Gerbillus dasyurus*, followed by *Meriones crassus* with *Acomys russatus* and *Crocidura suaveolens* as minor items. Remains of 2 small lizards (*Ptyodactylus* sp.) and 3 small passerine birds (one *Ammomanes deserti*) were also found and 6 scorpions *Nebo hierochonticus* and 29 Acrididae grasshoppers (Lesham, op. cit.).

BREEDING BIOLOGY

Nothing has been recorded.

VOCALIZATIONS

In Israel they can be heard calling from January to April and their call is a clear continuous 'hoo-hoo-hoo', sometimes broken by a short throaty cough (J. Leshem, op. cit.).

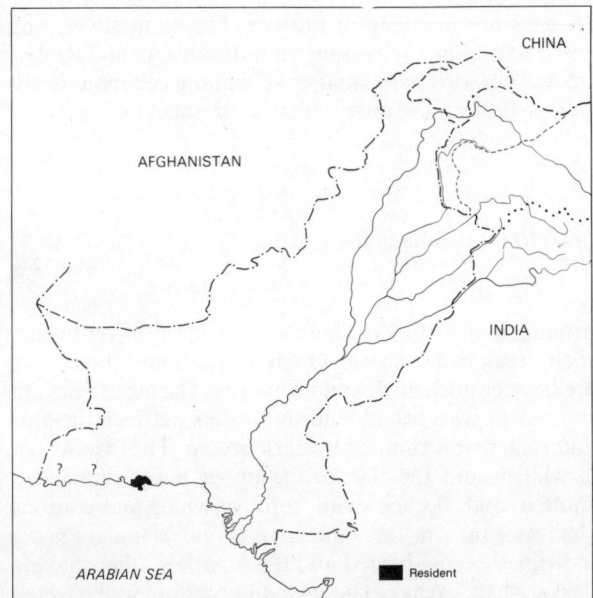

Map 245: Hume's Tawny Owl *Strix butleri*

Mottled Wood Owl *Strix ocellata* (Lesson)

Synonym *Syrnium ocellatum* in Blanford and Oates

DESCRIPTION

Body length 48 cms.
Wing length 33.8-34.6 cms. (Koelz)

This is a large handsomely marked owl, bigger than the Tawny Owl *(Strix aluco)* but lighter in build than the Eagle Owl *(Bubo bubo)*. It has a large round head with prominent round facial discs which are mottled with rufous orange against a white background and marked by darker concentric rings. The irides are dark brown. The breast is mottled white and russet chestnut with narrow blackish cross-bars all over and this is a clearly distinguishing feature enabling it to be separated from other owls with which it is likely to be sympatric (except the much smaller *Glaucidium cuculoides)*. The flight feathers and rounded tail are barred grey-brown and blackish brown whilst the crown, mantle and wing coverts are beautifully marbled and vermiculated with white, black, dark brown and tawny buff. It has a prominent white collar around the front of the neck. The bill is horny black and the feet are brownish flesh coloured. The sexes are alike.

HABITAT, DISTRIBUTION AND STATUS

This Oriental species is confined to the better wooded and higher rainfall regions of the Indian subcontinent in the plains. It has probably diminished in its overall range distribution with increased human settlement and more intensive agriculture in the less well wooded regions of the north-west. Writing in 1916, about the birds around Lahore, A. J. Currie noted that a pair of these owls was residing in the grounds of Government House and another was noted by him 29 kilometres (18 miles) east of Lahore on the old Grand Trunk Road *(JBNHS*, Vol. 24,

June 1916), but he considered it rare even then. There are no records of this owl subsequently from anywhere in Pakistan. During visits to Bombay and the Konkans the author was able to make tape recordings of their calls and subsequent to this, over a three year period, visited several likely localities around Lahore (including Changa Manga) in March and April at sunset when this owl is known to be most vocal. No evidence has been obtainable of its continued survival in Pakistan. H. Waite in his extensive bird collection obtained specimens in August 1935 at Gurgaon in India (east Punjab) but nowhere west of this or within Pakistan territory. R.S. Holmes, an experienced ornithologist also commented on the absence of this owl during his residence in Lahore from 1965 to 1968 (pers. comm.). At the present time therefore its status in Pakistan must be considered as a VAGRANT.

HABITS, BREEDING AND VOCALIZATIONS

Quite a nocturnal owl, remaining well hidden by day in thick evergreen tree groves emerging only towards sunset when it habitually calls for a few minutes before starting to hunt. It is strong enough to kill birds up to the size of a Rock Dove *(Columba livia)* and will eat insects, rodents, crabs, lizards and scorpions (Ali and Ripley, 'Handbook', 1969). The pair observed by Currie in Lahore Government House nested in March and produced 2 eggs, the normal sized clutch for this species. It normally selects a tree hollow and lays its eggs at the bottom of such a cavity without attempting any lining. Its calls are uniquely distinctive. The main advertising call is a resonant quavering 'chuhuaaah' (Ali and Ripley, op. cit.). It also has a hoarser screaming call reminiscent of *Tyto alba* and some ululating or chuckling cries.

Long-eared Owl *Asio otus* (Linnaeus)

Plate **21**
Map **246**

DESCRIPTION

Body length 35-37 cms.
Wingspan 90-100 cms.
Wing length 28.5-30.5 cms. (both sexes) (Ali and Ripley)
Tail length 14-15.4 cms. (both sexes) Ali and Ripley)
Weight 250 gms. (1 female) (Biddulph)

This is a medium sized rather slim bodied owl, smaller than *Strix aluco* with conspicuous feather ear tufts. In general plumage pattern it is a miniature version of the Eagle Owl *(Bubo bubo)* with the same very striking deep orange irides. Its breast is pale russet buff and white with

prominent dark brown shaft streaks on the lower breast, each streak being barbed. On the upper breast the streaks are heavier unbarbed, and form lines. The facial discs are prominent with broad whitish borders between the eyes and each disc is rimmed by dark brown. The throat area is whitish and the rest of the upper body plumage is spotted and flecked with rufous, white and various shades of brown, the wing coverts and scapulars being indistinctly cross-barred and the mantle feathers having darker shaft streaks. The tarsi and toes are well covered with buff feathers down to the tips. The bill is blackish at the tip. The sexes are alike.

HABITAT, DISTRIBUTION AND STATUS

This is essentially a Palearctic owl adapted to well wooded coniferous areas or tree groves in its breeding range which just extends into parts of Pakistan's western and northern border regions. There is some migration in winter down into the plains and it has been recorded throughout Punjab and Sind as a sparse and uncommon winter visitor. Recent records include one roosting in semi-flooded Tamarisks at Sidhnai, Multan district (collected by the author), in mid-November and one collected December 1970 from Changa Manga irrigated forest plantation (specimen in Punjab University Museum) 72 kilometres (45 miles) south of Lahore (collected by Z. B. Mirza) and a dead one picked up by Manchar lake on 10 February by Fred Koning. It was also collected by E. Fernando at Marala in 1966 and by H. Waite on 9 January 1937 from Shahpur (Sargodha district). Numbers have also been seen in Lal Sohanran irrigated forest plantation at Bahawalpur (David Stirling Wyllie, *in litt.*, 1973 and 74). In its breeding haunts it has been recorded breeding in the Neelum/Kishenganga valley in Kashmir but is evidently rare as Bates and Lowther (op. cit., 1952) did not encounter it during 2 visits to that area. In early October 1982, Tim Hurrell photographed one in a pine tree near Murree (identifiable from his photographs) and a specimen was captured in the Murree hills two years earlier which was exhibited in the Islamabad Zoo. H. Whistler saw one in Ayub park on 13 April, 1926 at Rawalpindi. Biddulph considered it a not uncommon summer visitor to Gilgit, recorded in March and April (*Stray Feathers*, 1881). These Gilgit and Murree hills' records may be of birds on passage and its breeding in the Neelum valley (in the vicinity of the Kashmir 'Cease-fire' line) is apparently only sporadic. Captain Perreau obtained one at Drosh, lower Chitral on 2 May (*JBNHS*, Vol. 1910). In Baluchistan, a specimen collected on 2 December was in the Quetta Museum. Major C. H. Williams encountered a family of 5 in Galbraith Spinney, towards the end of March and presumed that they had bred in the vicinity (*JBNHS*, 1929). This would be unusually early for this species to have fully grown and flying young, and it is possible that this was a flock on migration. In the 'Handbook', (Ali and Ripley, Vol. 3, 1969). Christison is cited as having found it breeding in 1941 but there is a possibility of confusion with Williams' record as Christison does not record even sighting this owl in his two published accounts of Baluchistan birds (*Ibis*, 1941 and *JBNHS*, 1942). Status mainly SCARCE winter visitor.

HABITS

On their wintering grounds, in favourable habitat, they will occasionally form colonial roosts and two such instances have been observed in Pakistan. E. Fernando encountered a party of 15 in dense *Saccharum* grass and *Prosopis* thickets at Marala barrage, Sialkot district in November 1966 (pers. comm.), and Stirling Wyllie at Lal Sohanran forest plantation in Bahawalpur district, came across a group of about 12 of these Owls in December 1973, and found them using the same site over a period of at least 2 weeks (pers. comm.).

The one encountered by the author at Sidhnai spill channel was roosting on the ground in *Tamarix dioica* thickets and flew readily for some distance though it was the middle of the day. The ones encountered by Mirza and Stirling Wyllie were roosting in dense leafy trees and both observers commented on their habit of sleeking down their feathers and attenuating their body in a sort of upright stretch when observed, with eyes closed to slits but still watching the observer.

They hunt a variety of small mammals and birds as well as taking insects and reptiles and are much more nocturnal in hunting activity than *Asio flammeus*, the Short eared Owl.

BREEDING BIOLOGY

The one authentic breeding record was published by a Mr. Shelley (*JBNHS*, Vol. 10, 1895). At Gurais near the Neelum (Kishenganga) river he found four well incubated eggs of this owl in a Sycamore tree (*Acer* sp.) on 4 June. It was in the dilapidated remains of a Jungle Crow's nest *(Corvus macrorhynchos)* and he shot one of the parent birds for identification. In Europe where their breeding has been well studied, they do nest by preference in old stick nests in trees, often Carrion Crow's nests in coniferous trees and the males have been described as performing display flights over the nesting territory at dusk, including wing clapping. The normal clutch is 3 to 5 eggs, incubated by the female alone and

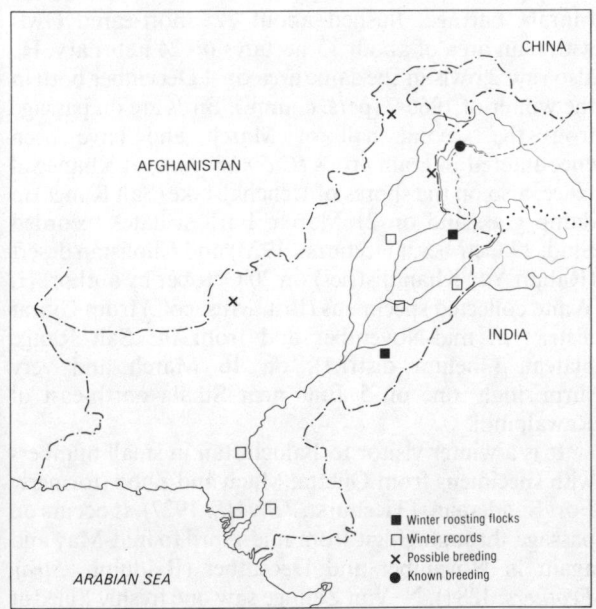

Map 246: Long-eared Owl *Asio otus*

incubation period is about 27 to 28 days, the eggs hatching asynchronously. The young are sufficiently developed to leave the nest 3½ weeks after hatching and family parties of 5 to 8 birds commonly stay together for the early part of the winter.

VOCALIZATIONS

The males territorial call has been described as a 'cooing moan' syllabized as a mournful 'oo-oo-oo'. It is also reported to make several more staccato yelping noises.

Short-eared Owl *Asio flammeus* (Pontoppidan)

Plate **21**
Map **247**

DESCRIPTION

Body length 38 cms.
Wingspan 95-110 cms.
Wing length 29-32.5 cms. (Stuart Baker)
Tail length 13.9-15 cms. (Salim Ali)

Bulkier than the Long-eared Owl, this owl has a rather tawny appearance, its body being golden or rufous buff, heavily streaked all over with darker brown. There is mottled barring and spotting on the wing coverts, flight feathers and tail and the crown and nape bear narrow closer streaked lines. The lower belly is whitish, unstreaked, and the tarsus and toes are feathered whitish. The facial discs are more rounded than in *Asio otus* and the area around the eyes is noticeably dark brown. There are two small ears rather close set and not normally visible in the field. The irides are yellow not orange. In flight it shows a prominent blackish carpal patch above and below the wing and its wings look long for the body. The tail is broadly cross-barred dark brown. Its flight is very buoyant and it glides a lot, low over the ground so that in daylight it is reminiscent of a Pallid Harrier *(Cirus macrourus)*. It is unlikely to be confused with the smaller darker Long-eared Owl *(Asio otus)* which has barbs or cross rays on its breast streaks and darker orange irides. In flight and from the dorsal view it is like *Bubo bubo* but its smaller size with relatively longer narrower wings, and tendency to glide, are helpful pointers.

HABITAT, DISTRIBUTION AND STATUS

This owl undergoes long distance migrations and spreads widely over the subcontinent in winter from its breeding haunts in the Caucasus and Turkestan. It is very characteristically a bird of open country rather than forest and in Pakistan is most likely to be encountered in those limited areas where grass grows, such as on the margins of lakes, in military dairy livestock breeding farms, in the irrigated canal colonies, or in sand dune desert tracts. It has been encountered much more frequently by the author than the Long-eared Owl but both must be considered uncommon and sparsely distributed in Pakistan. It has the habit of roosting colonially in its wintering grounds and may be gregarious on passage. E. Fernando while beating for wild pig in tall *Saccharum* grass between two canals near Marala barrage, flushed about 22 Short-eared Owls within an area of about 15 hectares on 24 February. He also saw 2 owls in the same area on 9 December both in the winter of 1966-7 (pers. comm.). Birds are on passage from the second half of March and have been encountered in gram crops *(Cicer arietum)* at Khanewal twice, also on the shores of Uchchali lake (Salt Range) in damp grassland on 21 March. Earliest dates recorded Sind, 29 October. (Ticehurst, 1923) and Cholistan desert (Rahim Yar Khan district) on 20 October by author. H. Waite collected specimens (Brit. Mus. coll.) from Gujrat district in mid-November and from the Salt Range plateau (Jhelum district), on 16 March and very surprisingly one on 5 June near Sihala north-east of Rawalpindi.

It is a winter visitor to Baluchistan in small numbers with specimens from Quetta, Mach and Zhob (formerly Fort Sandeman) (Ticehurst, *JBNHS*. 1927). It occurs on passage through Gilgit from mid-April to mid-May and again in November and December (Biddulph, *Stray Feathers,* 1881). N. Van Zalinge saw one freshly killed at Pasni (Makran) in late December, 1985 *(in litt.* to author, February 1986). Whitehead also noted it on passage each

Map 247: Short-eared Owl *Asio flammeus*

March at Kohat and collected a specimen on the Samana (Whitehead, *JBNHS,* 1911). Status SCARCE.

HABITS

Can occasionally be watched hunting in bright sunshine. Its flight is very controlled due to a favourable wing loading aspect. It can moreover, rise suddenly, sail slowly on upstretched wings and jinx according to need, as it constantly traverses over open grassland (observations Uchchali lake with light breeze blowing). It was noted in handling specimens collected by Fernando that this owl has a huge external auditory meatus, crescentic in shape and almost as wide as its eye sockets. Its hearing is assumed to be very acute and to be an important factor in locating small ground moving prey. One of the specimens collected by Fernando had 5 *Pipistrellus* bats in its stomach. It is known to be an efficient hunter of rodents and also to take grasshoppers, beetles and small birds. D. Stirling Wyllie during January and February 1975, encountered a flock of these owls roosting by day in a grove of trees alongside the Desert Feeder Canal at Lal Sohanran. The maximum number he encountered was 30 birds and at the end of February there were still 6 or 7 *(in litt.).* This species has been noted in the desert near Lal Sohanran by other observers and on other occasions. They do most of their hunting at dusk and before dawn. The House Mouse *(Mus musculus)* and Hairy-footed Gerbil *(Gerbillus gleadowi)* were both found to be common in the sand dune deserts adjacent to Lal Sohanran (see Roberts, 1977) and being nocturnal rodents, are probably an important item in the diet of this owl.

BREEDING BIOLOGY

Breeding is extra-limital. They are ground nesters laying their eggs in a bare scrape in the open. Studies in Britain indicate that before egg laying commences courtship feeding by the male is an important element in pair bonding. Only the female incubates.

VOCALIZATIONS

During display flight males indulge in wing clapping and a territorial or display song, a deep repeated 'boo-boo-boo'. Its contact call is a high sneezing bark 'kee-aw'. However in its winter quarters it is very silent. It is interesting that the clap noise is produced by the carpal joints striking together below the owl's body.

Tengmalm's Owl or Boreal Owl *Aegolius funereus* (Linnaeus)

DESCRIPTION

Body length 25 cms. (Salim Ali),
    20.7 cms. (1 male), 24.6-27 cms. (5 females) (Dement'ev *et al.)*
    24.-26 cms. (Cramp *et al.)*
Wingspan 54-62 cms. (Cramp *et al.)*
Wing length sub-species *palleus* 15.5-17 cms. (3 males),
    16.4-17.6 cms. (3 females) (Dement'ev *et al.)*
Weight 98-121 gms. (males), 134-215 gms. (females, breeding)
    (Cramp *et al.)*

This is a slightly larger owl than the Spotted Owlet *(Athene brama)* but with a comparatively larger head and longer tail. The facial discs are dull white, but very conspicuous being slightly raised at their upper outer edges, with a broad black border giving the face a perpetually 'astonished' expression. Its upper parts are greyish brown with copious small white spots on the crown merging to larger more scattered spots on the mantle and white feather margins or tips on the wing coverts with larger white spots on the scapulars almost coalescing into a line, as in *Otus scops.* Its breast is white with some pale brown streaking along the flanks and broader blotches across the upper breast.

In flight it has a more direct, less dipping flight than *Athene* owls, and if seen perched, its comparatively large face with 'raised' eyebrows and black bordered facial discs, make it very distinctive and easily recognizable.

The legs and toes are very thickly feathered to the extent that its popular Russian name is 'Shaggy-legged Owl'. The irides are yellow. Females are similar in appearance to males but average slightly larger in size.

HABITAT, DISTRIBUTION AND STATUS

Tengmalm's Owl is widely but thinly distributed across the northern boreal forest zone of the Holarctic region from north America, through Scandinavia and the USSR. In the southern part of its range it is exclusively an alpine dweller, occurring mostly in isolated populations in the Alps, French and Swiss Jura and mountainous regions of central Germany and the Carpathians. A central Asian population *A. funereus palleus* occurs in Turkestan in the Tian Shan and Tarbagatai mountains (Dement'ev *et al.,* 1951).

Up until recently, the only record for the whole subcontinent was that of an adult female with a feathered juvenile collected together by Walter Koelz in Kyelang, Lahul district of Himachal Pradesh, India *(Proceedings Biological Society,* Washingston, Vol. 52, 1939). These birds were collected in a habitat of sub-alpine juniper forest. In 1986, Rudei Hess while making prolonged winter time observatios on Himalayan Ibex *(Capra ibex sibirica)* in the Barpu Glacier valley (N. 36° 17' E. 74° 52') on the Nagar Hunza border, came across a calling male which he was able to examine for about 15 minutes at close range. Hess was familiar with this species from his home in the Swiss Alps (pers. comm., 1988) and on a bright moonlit night, was attracted by the sound of a male calling. By imitating its call, he was able to attract the bird which eventually flew and perched a few metres away from his tent (25 February, 1986, elevation 3,360 metres/11,000 feet, time 1900 hours) (pers. comm., 1988, Dr. Hess, Schonwart, CH-6314, Unterageri).

The Barpu is a tributary valley of the larger Hispar valley and glacier and runs in a north-south axis, and the owl was in an area of sparse *Juniperus* scrub or dwarf trees in a broader part of the valley, with scattered bushes of *Rosa, Hippophae,* and in the gullies *Tamarix* and *Salix* bushes. This appears to be similar habitat to the collection locality in Lahul, India, mentioned above (see Ali and Ripley, *Pictorial Guide,* 1983).

No doubt the bird is a rare resident in the inner dryer Himalayan ranges, and as these regions are seldom explored by ornithologists, it has largely escaped notice. It has also been collected in north-east Tibet at Rangta Gol gorge (Vaurie, 1972).

HABITS

Feeds largely upon small birds in winter with more rodents in the diet during the summer months according to studies in the European Alps (Cramp *et al.,* Vol. 4, 1985). Russian birds have been noted feeding upon Tree Creepers *(Certhia* sp.) and *Parus* Tits as well as rodents (Dement'ev *et al.,* 1951). In the Barpu Glacier valley Migratory Hamsters *(Cricetulus migratorius)* are numerous and probably provide a suitable food prey for this owl, as also Bar-tailed Tree Creepers *(Certhia himalayensis)* and Black-crested Tits *(Parus rufonuchalis).* This owl is largely nocturnal in hunting activity and its usual technique is to search the ground from a suitable elevated tree or boulder perch, flying down to seize its prey on the ground. In winter time when temperatures remain well below freezing even in the day time, this owl is reported (Dement'ev *et al.,* 1951 and Cramp *et al.,* 1985) to store surplus food prey in suitable tree hollows.

VOCALIZATIONS

The call heard by Hess (pers. comm., 1988) was a short spaced whistle 'pu-pu-pu-pu', each call 1 or 2 seconds apart and quite distinctive in timbre and inflection from the single 'toink-toink' whistle of *Otus scops.* The above call is the typical advertising call described by other authors as 'po-po-po' and will carry, on a windless night, for distances of up to 2 kms. (Cramp *et al.,* 1985).

ORDER CAPRIMULGIFORMES
FAMILY CARPRIMULGIDAE

Savanna or Allied or Franklin's Nightjar *Caprimulgus affinis* Horsfield

Plate **21**
Illus. **65**
Map **248**

DESCRIPTION

Body length 25-26 cms.
Wing length 18.1-20.5 cms. (males),
    17.9-20.2 cms. (females) (Ali and Ripley)
Tail length 9.6-12 cms.

This is a smaller nightjar than *C. indicus* but bigger than *C. asiaticus* and slightly bigger than *C. mahrattensis.* It is darkly patterned on the back like *C. indicus* and is more richly pigmented and darker than *C. asiaticus* or *C. macrurus.* Its most noticeable and distinguishing characters are the extensive white areas in the two outer tail feathers of males, which are almost wholly white except for the extreme tips. Females have no white on their tail feathers. The first four primaries have white spots, in the males near their tips and in females these spots are pale tawny. It has two prominent white throat spots and a 'V' of buff formed by the feather tips running down either side of the back. The under parts are rufous with dark brown cross-barring, the back and wing coverts are speckled and vermiculated with very dark brown. There are dark brown spots on the crown, and cross-barring and marbling on the tail. The tarsus is naked and the legs and feet are fleshy brown. Superficially it looks very like *C. indicus* with heavily pigmented rich rufous and brown tones to its plumage. The habitat and calls in Pakistan are quite distinctive from *C. indicus.* The iris is dark brown and the eye is large as in all *Caprimulgus* species, with prominent rictal bristles framing the wide gape.

HABITAT, DISTRIBUTION AND STATUS

This nightjar as far as available evidence goes, is definitely not resident but is an abundant spring migrant visitor to the north-western dry foothill areas, staying to breed and not leaving until October or November. Some birds can be encountered on passage through the Punjab plains, thus Hugh Whistler encountered them in Jhang *(Ibis,* 1922) in late March only and the author saw and heard one calling in two different years around Balloki headworks on the Ravi river on 27 March and 2 April. In Sialkot district adjacent to the foothills E. Fernando encountered flocks of up to 30 birds (and collected 2 specimens) from 30 March up to 5 April in ripening wheat crops alongside the Chenab river but did not hear them calling at all after the end of April (pers. comm., 1967). H. Waite (*JBNHS* 1948) noted their regular arrival in the Salt Range from early May. The author has heard them calling along the foot of the Salt Range near Kushab on 1 April. They breed in bare stony broken country in ravines and hill slopes characteristically covered with scattered bushes of *Dodonaea viscosa* and *Reptonia buxifolia* in Kohat district, Thal and the Salt Range and amongst *Adhatoda vasica* and *Acacia modesta* on the dryer stony ridges of the Potohar plateau where it abuts on to the foothills. They are extremely plentiful around the outskirts of Islamabad and Shakar Parian park. They have been seen and heard around Baran dam, Bannu district from 29 March. C. H. Whitehead (*JBNHS*, 1911) considered it resident around the Thal, north of Kohat district but no skins of this species collected by him during the winter months are in the British Museum collection which received all his bird skins. It is sympatric with *C. europaeus* in many of these regions and around Islamabad also with *C. macrurus*, though normally preferring the lower dryer ridges and not penetrating into the damper ravines typical of *C. macrurus*. Status COMMON.

HABITS

This species is quite gregarious during migration into Pakistan and in its breeding haunts also 3 or 4 birds can be encountered hawking insects at dusk in one spot. They are the only nightjar in the region which habitually call while on the wing. They do not start calling or become active until after sunset. Their flight is often more direct and less fluttering than the forest species, sailing up and down along stream beds or the grass margins of lakes, quite high in the sky. A specimen collected near Delhi had fed upon Click Beetles (Elateridae) and Stag Beetles (Lucanidae) (Donahue, *JBNHS*, 1967). Salim Ali describes their habit of dipping down over water to drink regularly in the evening, like swallows ('Handbook', Vol. 4, 1970).

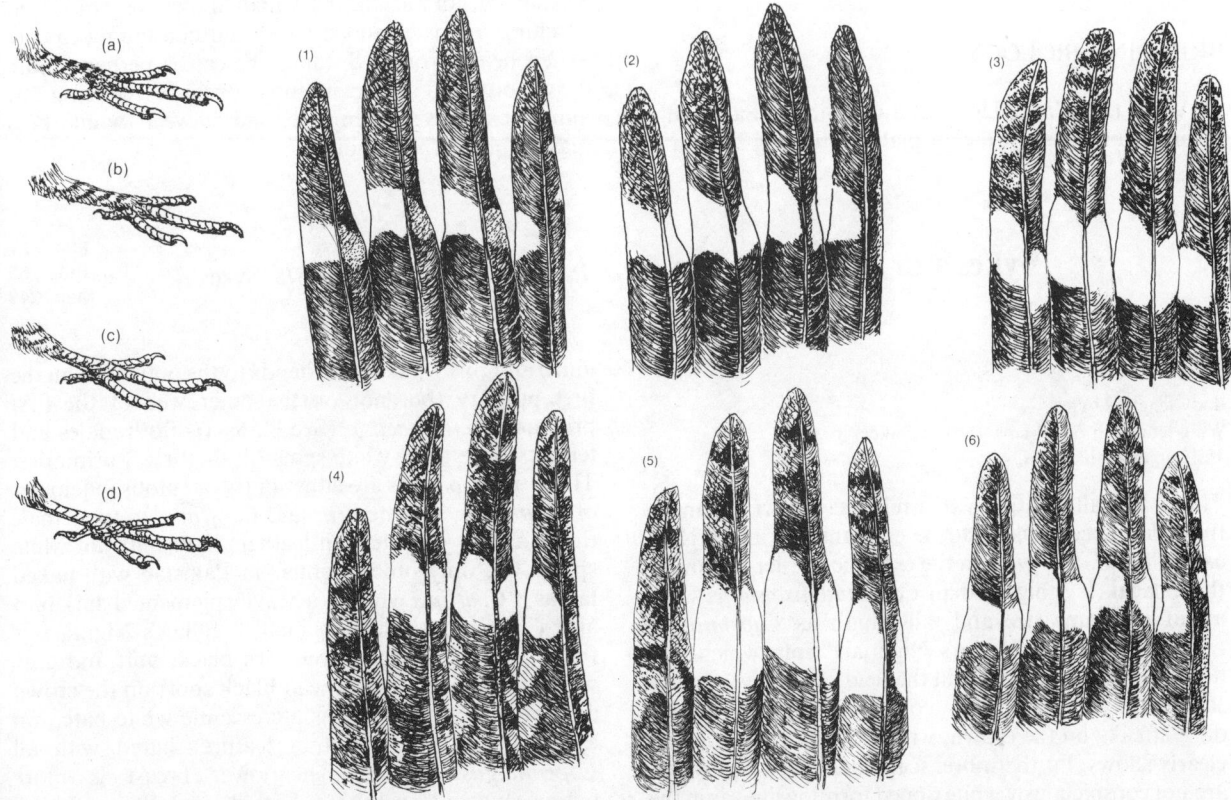

Illus. 65: Nightjars showing wingtips (1st. 4 primaries) of males and amount of feathering on tarsus.
(1) *Caprimulgus indicus hazarae*. (2) *Caprimulgus macrurus*. (3) *Caprimulgus affinis*.
(4) *Caprimulgus europaeus unwini*. (5) *Caprimulgus mahrattensis*. (6) *Caprimulgus asiaticus*.
(a) Left foot of *Caprimulgus indicus*. (b) Left foot of *Caprimulgus europaeus*. (c) Left foot of *Caprimulgus mahrattensis*. (d) Left foot of *Caprimulgus asiaticus*. Note pectinated middle-toe claw.

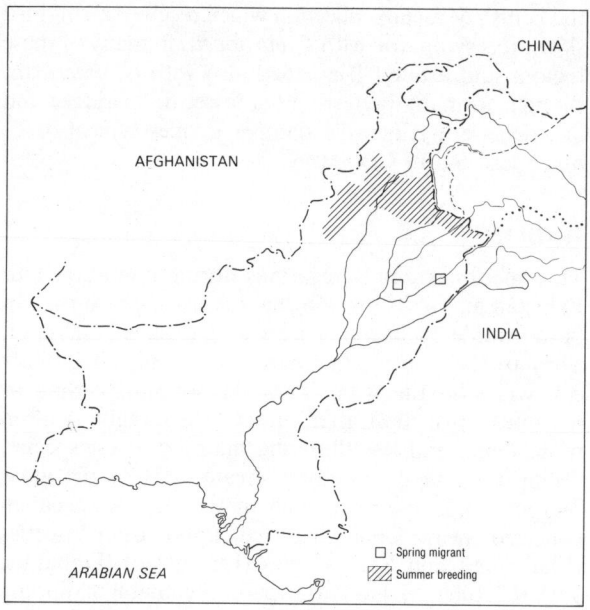

Map 248: Savanna, Allied or Franklin's Nightjar *Caprimulgus affinis*

BREEDING BIOLOGY

H. Waite (*JBNHS,* 1926) found nests in the Salt Range during June and July, the normal clutch being 2 eggs and laid on the bare ground, occasionally in a slight depression but never with any lining. The eggs are usually laid in the lea of a small bush or plant (Waite *JBNHS,* 1926 and 1948). The eggs are rather distinctive being salmon pink to pale fulvous in ground colour, longish ovals, spotted and blotched with deep red and red-brown. Both sexes share incubation duties. If the nest site is disturbed, they have been known to move their eggs some distance to a new location possibly in their mouths (H. Waite, pers. comm., 1965). The chicks are nidicolous and rely upon natural concealment, squatting flat on the ground motionless when the parents make alarm calls. In the Salt Range nesting of this nightjar is in exactly similar habitat and sympatric with *Caprimulgus europaeus.*

*VOCALIZATIONS

The call is unlike any other nightjar's in the subcontinent. It consists of a rather hard strident and shrill 'dheet' or 'chait' uttered with vehemence as the bird flies to and fro. When 3 or 4 are calling in the same vicinity each calls at a slightly different pitch. These calls are usually repeated 2 or 3 times but at intervals of about one per second. At Shakar Parian park near Islamabad they were observed perching upon telephone wires and continuing to call. When heard from a distance the call is perhaps more drawn-out and rising in tone, distance reducing the almost explosive vehemence, and 'chweep' would be a reasonable 'verbal rendition'.

Sykes's or Sind Nightjar *Caprimulgus mahrattensis* Sykes

Plate **21**
Illus. **65**
Map **249**

DESCRIPTION

Body length 23 cms.
Wing length 15.7-17.3 cms. (Stuart Baker)
Tail length 10-10.4 cms.

This is a small nightjar with a relatively short tail and is the palest species likely to be encountered in Pakistan except for *C. aegyptius* in the extreme border regions of the Chagai. It is bigger than *Caprimulgus asiaticus* but about the same size and tail length as *Caprimulgus europaeus unwini* (Hume's Nightjar) from which it is doubtfully distinguishable in the field except by its voice. It is overall a pale sandy grey colour without conspicuous dark streaks on the crown, which *C. europaeus unwini* clearly shows. Furthermore, scapulars and wing coverts are not conspicuously white tipped forming lines as in the case of *C. europaeus* though there are a few buff tipped feathers in this region. The tarsus is almost bare of feathers and dark reddish brown. It is partly feathered in *C. europaeus.* The male has extensive white spots on the first 3 primaries (in descending order from the carpal joint) and this white area extends to the outer web on the first primary (but not on the outer web of the first primary in *europaeus)* (See **ill. 65-4).** Both males and females have clear white spots on the first 3 primaries. These wingtip spots are either buffy or rufous in females of *C. indicus, C. macrurus* and *C. affinis.* In the female the two pairs of outer tail feathers have buff not white spots. The only other nightjar in Pakistan with naked tarsus is *C. affinis* which is a heavily pigmented dark bird and *C. asiaticus* which is smaller. Sykes's Nightjar, is finely vermiculated all over with black, buff and grey with scattered tiny arrowhead black spots on the crown and wing coverts. There is a crescentic white patch on each side of the throat, a feature shared with all *Caprimulgus* species. The lower breast is more ochraceous and finely cross-barred on the lower flanks, darker and more spotted on the upper breast. The iris is dark brown with the eye large and there are prominent down-curved stiff white rictal bristles over the gape. The bill is brown with a very short down-curved culmen but a wide gape extending well below the eye.

HABITAT, DISTRIBUTION AND STATUS

Like all the nightjars of the region this species is locally migratory and can turn up in a wide variety of habitats in winter. It is, however, mainly adapted to semi-desert open tracts with scattered dry tropical thorn scrub such as occurs in the remnant uncultivated patches of the Indus plains and in the sand dune and clay flat extensive deserts of Sibi, Cholistan and the Thar. It is more adapted to such sand dune desert than *C. europaeus* and less commonly encountered in a stony or rocky region such as is favoured by that species, but it does occur in dry hilly areas with rocky outcrops in Lasbela. It is a summer breeding visitor to the foot of the Salt Range (Waite, 1948) and around the low bare hills of Thal and the lower Kurram valley, but absent from the rocky and stony ranges around Kohat (Whitehead, 1911). In Baluchistan it is probably resident on the Sibi desert plain and throughout coastal Makran. Hotson collected it from Buleda and Turbat (specimens in British Museum) in March and December. Its status in the Chagai in western Baluchistan is uncertain. Paludan did not encounter it across the border in adjacent Seistan territory where he collected *C. aegyptius,* nor did Sarudny in his various expeditions to Seistan but he encountered it in southern Fars province of Iran (Paludan, 1959). Christison considered it resident only in the extreme western tip of the Chagai around Robat and Kucha where it was sympatric with *C. aegyptius* (Christison, *Ibis,* 1941). It is fairly widespread throughout Sind and H. Waite collected it from Muzaffargarh and Dera Ghazi Khan districts in southern Punjab (specimens in Brit. Mus. coll.). The author has found it common in winter in Cholistan, particularly at night along the Jue-sher minor in Rahim Yar Khan district and the Desert Feeder Canal in Bahawalpur district. It avoids well cultivated or irrigated tracts and is absent from the northern Punjab. It is sympatric with *C. europaeus* in parts of Sind and the Punjab. Status COMMON.

HABITS

Roosts by day out in the open, preferring the shade of a tamarisk bush or euphorbia clump but often settling in the open when flushed by grazing livestock. It habitually keeps the eyes almost closed in daylight even when alert, probably to increase the natural camouflage of its plumage. At night it often settles on canal side roads or open embankments between bouts of hunting. At Sanghar over a dozen were flushed along a 3.2 km. stretch of embankment in late February. They hunt insects at night, mainly moths and beetles including cockchafers and other night flying insects. Five Melolonthid beetles undigested, each nearly 2 cms. in body length, were found in the stomach of a specimen collected at Sukkur by Mr. Bell (Ticehurst, *Ibis,* 1923). They often fly near the ground, turning and twisting and suddenly rising vertically to seize an insect with perfect balance and timing. They walk with difficulty often raising their wings over their back as they shuffle forward on their tarsi. They settle more readily on the ground and not on bushes as do other Nightjars, but one was observed (author) for a considerable period roosting in a tall mango tree *(Mangifera indica)* at Matheran in Maharashtra, 24 January 1982 in forest in a very a-typical habitat.

BREEDING BIOLOGY

Breeding can be almost semi-colonial. Two or three males will sing territorially in suitable tracts of scrub jungle often within 3 or 400 metres of each other (observed Bhegari-bund, Sukkur). They often call while squatting on the ground in quite open terrain in contrast to *C. europaeus* which usually perches on a bush to call. Besides this territorial advertising, display flights are made when a female is sighted and this involves a very graceful controlled flight up and down with wing clapping and a series of soft clucking calls. No nest whatsoever is made, the eggs being laid on the bare ground often without shelter even of a grass tuft or clod of earth. The clutch is invariably two eggs and these are slightly elongated ovals, and distinctive, being white in ground colour, thickly smeared and blotched with grey but entirely lacking any brown or blackish over-spots as are found in *C. europaeus.* K. Eates in Sind found full clutches in mid-April in Thatta district (MS notes) and C. Whitehead found eggs from mid-March to the end of April *(JBNHS,* 1911). H. Waite found nests on the clay-cum-gravel plains at the foot of the Salt Range amongst

Map 249: Sykes's or Sind Nightjar *Caprimulgus mahrattensis*

Prosopis spicigera scrub forest (op. cit, 1926 and 1948) but the author searching the same area in 1980 at the end of March and early April, while encountering *C. affinis* and *C. europaeus,* failed to encounter this species. The incubation period is about 17 to 18 days and is carried out mainly by the female. The chicks are nidicolous, being hatched already covered with down, cryptically speckled grey and black.

*VOCALIZATIONS

The territorial or advertising songs of the male are very similar to those of *C. europaeus* comprising bouts of long continuous purring and churring. But these calls are only audible at close range, carrying perhaps for up to 200 or 300 metres whereas the much louder churring of *europaeus* carries for distances up to 500 or 600 metres on a still night. Usually this continues unbroken for 3 or 4 minutes, rather like the distant throb of a small motor cycle engine. It does not vary in pitch and lacks the two tone variations of *C. europaeus.* In flight the male utters a low soft 'chuck-chuck' call quite different from the higher pitched 'quoit' calls of *C. europaeus* during display flights.

Little or Indian Nightjar *Caprimulgus asiaticus* Latham

Plate **21**
Illus. **65**
Map **250**

DESCRIPTION

Body length 24 cms.
Wing length 14.1-15.8 cms (Stuart Baker)
Tail length 9.6-11.9 cms.
Weight 46 gms. (1 male) (Salim Ali)

This is the smallest of the six nightjar species occurring in Pakistan with a relatively large head and shorter tail than the others. The upper parts are more richly coloured than in either Syke's or the European Nightjar, with the crown vermiculated with pale grey-brown and having bold dark blackish brown longitudinal streaks. There is a rather rufous chestnut collar around the base of the hind neck and the mantle is darker grey with the scapular feathers broadly margined pale chestnut with dark brown arrow shaped centres forming two conspicuous 'back straps'. The tail is paler grey-brown speckled and mottled with darker cross-bars and the lower breast and belly rufous buff with grey-brown cross-barring. The tarsus is fleshy brown and naked as in *C. mahrattensis.* Both males and females have white spots on the first four primaries (counted in descending order from the carpal joint), but in the female these spots are much smaller and less conspicuous. The two outer pairs of tail feathers are white tipped in both sexes and there are two white patches on either side of the throat. The sexes are alike in this nightjar. Its bill is small and weak, with a large eye and dark brown irides. Six or seven stiff rictal bristles protrude down on either side of the gape from in front of its eyes.

HABITAT, DISTRIBUTION AND STATUS

It is locally migratory and in the winter when silent, difficult to locate. Its an inhabitant of low hill country with thin scattered sub-tropical thorn scrub as well as flat saline tracts, studded with tamarisk bushes. It is absent from Baluchistan, the NWFP and apparently also the Punjab, occurring only in southern Sind from Tharparkar district through Thatta district and to the edge of the hills bordering Lasbela. Holmes and Wright (*JBNHS*, Vol. 65 and 66, 1968-9) heard it calling in Badin district at Sirani, Jati, and Ladiun and Muradani, and encountered it in the cold weather as well as the summer. On the Makli hills near Kalankot Fort, a few can always be encountered but are silent in December and January. K. Eates (MS notes unpublished) considered it rare in Sind, encountering it very irregularly around Sehwan in Dadu district and around Sakrand in Nawabshah district. It has been heard calling at Mirpur Sakro and a road-kill specimen was picked up near that town in November (Rolf Passburg), also

Map 250: Indian or Little Nightjar *Caprimulgus asiaticus* and Large-tailed or Long-tailed Nightjar *Caprimulgus macrurus*

calling on the low stony hills on the northern borders of Haleji lake in mid-March. Apparently neither Waite or Whistler (Brit. Mus. coll.) encountered it in the Punjab nor E. Fernando in Sialkot district. Status resident and partly migratory, COMMON only in Thatta, Badin and Thar Parkar districts.

HABITS

They start to call from February and in the early spring only do so for a brief period each night just after dusk. Like all the nightjars they hunt their insect prey on the wing and only become active after dark. A specimen collected near Delhi had large coleoptera including two Elateridae in the stomach, one large grasshopper (Acrididae) and some crickets (Donahue, *JBNHS,* Vol. 64, 1967), Heterocera moths, Hemiptera bugs, dung beetles *(Onthophagus)* and water beetles (Dysticidae) have been recorded amongst stomach contents in India (Ali and Ripley, Vol. 4, 1970). They have the same habit of perching on the ground on embankment roadways as do other nightjars (observed Pir Goth village, Thatta district) in between bouts of hunting.

BREEDING BIOLOGY

Eates found only one 'nest' near Sujawal, Thatta district in *Salvadora persica* scrub. One egg had been laid on 7 April (Eates, MS notes). In central India, breeding has been recorded between February and September. Two eggs are laid on the bare ground with no attempt at a nest and these are rather cylindrical ovals and cream or pale salmon pink in ground colour, spotted and smeared with reddish brown and inky purple. E.N. Lowther describes the incubating bird as sitting with eye half closed and allowing a human to approach very closely without leaving the eggs (op. cit., 1949). Males near Kalankot were observed performing what appeared to be display flights with wing clapping and the call note was rather short and distinctive. It is not known whether more than one brood is reared, but both sexes are believed to share incubation (Salim Ali, 'Handbook', 1970).

*VOCALIZATIONS

The advertising calls of this Nightjar have been aptly described as reminiscent of a ping pong ball bouncing to a rest on a hard surface. It is not a very loud call but quite far carrying 'chuk-chuk-chuk-chuk-tukarroo'. Its song is remarkably similar in timbre and composition to that of the Lesser Sand Plover *(Charadrius mongolus)* though usually more drawn-out. The whole sequence lasts barely 3 seconds. Sometimes there are 4 or 5 'tuk-tuk-tuk-tuk' calls before the 'chk-uk-uk-urrrh'. Derek Holmes heard them calling in December near Jati but in their haunts well known to this author they have not been heard before late February. A flying male uttered a short sharp 'qwit-qwit' call.

Large-tailed or
Long-tailed Nightjar *Caprimulgus macrurus* Horsfield

Plate **21**
Map **250**
III. **65**

DESCRIPTION

Body length 28-33 cms.
Wing length 20.5-23.5 cms. (Stuart Baker)
Tail length 13-18 cms.

This is slightly larger than *Caprimulgus indicus* and has a proportionately longer and slightly broader tipped tail than other species (as its name suggests). It is a dark pigmented nightjar very beautifully marked with rufous, creamy buff and dark umber brown streaks, mottling and vermiculations. The throat patch is fairly prominent and forms a continuous crescent around the fore throat, unlike *C. affinis* or *C. indicus* where it comprises two separate patches on either side of the neck. There are conspicuously patterned feathers on the edge of the scapulars forming two stripes over its back, margined creamy buff with black centres. The tarsus is fully feathered in this species. The male has four white spots on the first four primary feathers and in the female these spots are pale tawny rufous. There are broad white tips also to the two outer pairs of tail feathers in the male.

Again in the female these outer spots on the tail are rufous buff. There is often an indistinct dull rusty nuchal collar and quite broad blackish brown mesial bands along either side of the crown. The under-tail coverts and lower belly are rufous chestnut, cross-barred blackish. The wing coverts are tipped creamy producing spotted lines on the wing shoulder. The central tail feathers are speckled and cross-barred darker brown and the flight feathers are barred more tawny and dark brown. The iris is dark and the naked toes are fleshy brown. The rictal bristles are whitish at their base. It is more rufous in general body colouration than other species of nightjar occurring in Pakistan.

HABITAT, DISTRIBUTION AND STATUS

It does not seem to have been previously recorded in Pakistan (Ali and Ripley, 'Handbook', 1970). Hugh Whistler did not come across any records for the Murree hills, neither did H. Waite collect it, though he 'worked' the Margalla hills quite systematically. It is only known from the Murree foothills around Margalla hills, lower Lehtrar valley and Kahuta where it is a regular summer visitor in small numbers, confined to lower elevations (460 to 900 metres) in association with sub-tropical dry deciduous forest, particularly *Acacia modesta* stunted trees, with *Adhatoda vasica* and *Carissa opaca* under shrubs. It arrives from mid to late March and can be heard calling at dusk and again in the early morning before dawn, right through until September. In autumn it probably drifts eastwards through to the Siwaliks, into latitudes with a more temperate winter in common with several other Oriental faunal species which come into the Murree foothills only for breeding. It is very local in distribution in Pakistan. Status SCARCE.

HABITS

They seem to be quite gregarious as 3 or 4 can be encountered in one small ravine in the Margalla hills, whilst adjacent similar areas will be devoid of birds. They remain inactive until dusk is well advanced and then often commence rapid low gruff short calls which can only be heard close up and are given from the ground. All kinds of insects are hunted on the wing. Stomach contents of 3 birds collected from Bihar contained remains of 92 large insects, mainly Hemiptera bugs, Blister Beetles (Cantharidae), Dung Beetles *(Onthophagus)* etc. (Mason and Lefroy, 1912).

BREEDING BIOLOGY

Courtship flights have been seen just after dusk with the male, his white spotted wingtips glowing, following the female up and down a narrow ravine, wingtip to wingtip almost touching in graceful buoyant flight as they dive then surge upwards, twisting low over the tree tops (observed early April, Margalla hills). In parts of India, several nests have often been found in the same vicinity and breeding may be loosely colonial. The eggs are a creamy or salmon buff ground colour with grey mottling and dull reddish brown spots. Incubation period is thought to be about 16 to 17 days and in the day time carried out mainly by the female. The young are hatched fully down covered but they are speckled and mottled cryptically with buff and black.

*VOCALIZATIONS

These nightjars, associated with relatively thick jungle and ground cover, have very loud resonant calls. The call is stereotyped and birds heard in Bangla Desh (Cox's Bazaar) and Nepal (Chitawan) could not be differentiated from recordings made in the Margalla hills. The main call of the male is a well spaced but repeated resonant 'chaunk-chaunk'. This may be repeated 4 or 5 times or over 20 times, slowly, deliberately and with slightly varying intervals. Typically it utters 7 or 8 'chaunks' over a period of 5 seconds. The sound has been likened to the noise produced by striking a hollow log with an axe. At the beginning of the evening, before these loud calls commence, birds also give a series of much softer gruffer short calls as though 'tuning up' for the louder 'chaunks'. These low pitched calls are similar to calls uttered by *C. indicus*. Birds have been seen calling while perched on a bare footpath and also while perched on the branches of a *Ficus glomerata* tree.

Jungle, Grey or Japanese Nightjar *Caprimulgus indicus* Latham
Himalayan Jungle Nightjar *Caprimulgus indicus hazarae* Whistler and Kinnear

Plate **21**
Illus. **65**
Map **251**

DESCRIPTION

Body length 28-32 cms.
Wing length 20-21.5 cms. (males), 18.7-20.3 cms. (females),
   (Ali and Ripley)
Tail length 12.5-14.6 cms.

A rather large sized relatively long-tailed nightjar (larger than *C. affinis* but smaller than *C. macrurus*), with rich dark plumage patterns. Like all the family it has a large flat crowned head with huge dark brown eye and seemingly tiny weak bill, though the large gape and strongly hooked tip enables it to crush the hard elytra of beetles and to seize large flying insects over 2 cms. wide. In this species the tarsus is well covered over its upper two-thirds with feathers and there are 4 white spots on the first four primary feathers of the male (counted in descending order from the carpal joint) and 4 white spots sub-terminally on the outer four pairs of tail feathers. The female entirely lacks these white tail spots and her wings have pale rufous spots not white on the primaries.

The tail and flight feathers are spotted and marbled with rufous and dark brown cross-bars and the lower belly is quite rufous with blackish cross-bars. The white throat spot is relatively small in this species, but in the Pakistan population, is white whereas further east in the Himalayas this throat spot becomes pale tawny orange. The crown is speckled and vermiculated without prominent dark streaks as in *C. europaeus* and *C. macrurus*.

There are dark brown streaks on the mantle and chestnut spots on the wing coverts giving a disruptive camouflaged pattern particularly effective amongst dried leaves and when it crouches on the forest floor.

HABITAT, DISTRIBUTION AND STATUS

Though this Oriental species is widely distributed in south-east Asia from the Philippines, south-east China, Thailand and throughout peninsular India and Sri Lanka, it is very restricted in occurrence in the dryer north-west. It is essentially a bird of sub-tropical deciduous or evergreen forest. In Pakistan it is a summer visitor to the Murree hills with a few straggling westwards into the forested slopes of the Neelum valley. There, it is associated with temperate moist Himalayan forest of mixed deciduous and coniferous species at higher altitudes, being common between 1,800 and 2,400 metres (6,000 and 8,000 feet) elevation in the Murree hills. On first arrival in May it can be heard in the lower or outer foothills where it is then temporarily sympatric with *C. macrurus* and *C. affinis*, but in its breeding haunts it is usually the only nightjar species present. It has not been heard or encountered in the Kaghan valley despite the scientific name *hazarae* attached to this Himalayan sub-species, but the northern part of the Murree hills do extend into Hazara district and it has also been collected from the Agrore valley draining into the Mansehra plain. It does not occur anywhere in the plains of Pakistan. Status summer breeding visitor, locally COMMON.

HABITS

This nightjar often hunts high over the trees in the higher air space and is a skilful and buoyant flyer. They frequently call both on the wing and while perched quite high up in Deodar *(Cedrus deodara)* or pine trees. They become active only at dusk when males start calling territorially and hunting their food in the air, turning and twisting with amazing dexterity. Their food includes all kinds of larger insects especially moths, Melolonthinae cockchafers, and other nocturnal beetles, Hemiptera and cicadas in the monsoon months.

BREEDING BIOLOGY

Well incubated eggs of this species were taken by Rattray on 7 June near Dunga Gali *(JBNHS,* 1905) and by Col. Buchanan from Changla Gali; fresh eggs on 28 April, on 4 June slightly incubated, and a clutch of 2 eggs unincubated as late as 18 June (H. Whistler, *Ibis,* 1930). Breeding is probably dependent upon weather conditions becoming settled and is delayed till June if May is a stormy month. The eggs are creamy or white in ground colour with grey and brown smudges and

Map 251: Jungle Nightjar or Grey Nightjar *Caprimulgus indicus*

splotches. As in others of the genus, the normal clutch is 2 eggs which are laid on the bare ground with no attempt at nest lining. Incubation is thought to take 16 to 17 days (Stuart Baker, Vol. 4) and the female does all the day time incubation being relieved by her mate for short periods in the late evening and early morning. The young are fledged in about 17 days and are nidicolous being down covered at birth but unable to move around for the first few days.

*VOCALIZATIONS

The calls of this nightjar indicate a close relationship with *Caprimulgus macrurus.* The males', normal advertising call is a fairly rapidly repeated series of loud ringing 'chunk-chunk-chunk' which can be uttered in flight but is mainly given from perches when such calling is prolonged. These 'chunks' are uttered at the rate of about 3 or 4 per second and generally in strophes of 7 or 8 before a brief pause. They may continue uninterrupted for 6 or 7 minutes. The 'chunks' of *C. macrurus* are much more spaced and resonant. Another call heard is a more rapid, hoarser deeper pitched and less resonant 'quor-quor-quor-quor', repeated many times usually from the ground or a low perch. This may be the females' call or when one bird is in close proximity to another. Courtship and display flight of this Nightjar also includes some rapid low soft calls and occasional wing clapping (observed Dunga Gali Pipeline first week of May).

European Nightjar *Caprimulgus europaeus* Linnaeus
Unwin's Nightjar or
Hume's European Nightjar *Caprimulgus europaeus unwini* Hume

Plate **21**
Illus. **65**
Map **252**

DESCRIPTION

Body length 25-27 cms.
Wingspan 57-64 cms. (Cramp *et al.*)
Wing length 17.2-19 cms. (Stuart Baker)
Tail length 12.5-14.4 cms. (Stuart Baker)

The sub-species *unwini* is a paler and greyer bird than the European breeding population. Slightly longer tailed than *C. mahrattensis* and darker with greyer tones on the back, it is smaller than *C. indicus*, but about the same size as *C. affinis*. Like all the Nightjars of the genus *Caprimulgus*, this is a slender bird with long pointed wings and hawk-like outline in flight. The wing loading ratio is very favourable so that they can become airborne with the minimum of forward thrust and have a peculiarly buoyant flight. Their legs are comparatively weak with short tibia and they shuffle along the ground awkwardly, usually with wings raised over the back to assist in balance. The inner or middle toe is much longer than the first, and its claw is serrated or pectinated as in all the nightjars. In *C. europaeus* the tarsus is completely feathered which distinguishes it from the smaller more sandy buff *C. mahrattensis*. The first three primary feathers of the male have round white spots (counted in descending order from the carpal joint) and the two outer pairs of tail feathers are narrowly white tipped. In the female these white areas on the wing are replaced by rufous buff and the outer tail feathers are entirely barred. The legs and feet are orange-brown. The eye is very large with a dark brown iris. The body plumage is delicately patterned with grey, dark umber brown and rufous in a cryptic pattern. The vermiculated greyish buff crown bears conspicuous dark brown streaks (absent on the crown of *C. mahrattensis*). The flight feathers are barred rufous and dark brown and the long tail is marbled and speckled with darker brown cross-bars. There are two white crescentic patches on either side of the throat and the ear coverts are speckled dark brown framed below by a paler creamy white line as in all the nightjars. The lower belly is creamy buff with blackish cross-bars. It also shows a prominent row of buffy white spots along the mantle and wing coverts.

HABITAT, DISTRIBUTION AND STATUS

This nightjar is very widely distributed in Pakistan in regions of hilly country with stony slopes and rocky ridges but usually rather sparse vegetative cover. It is a summer breeding visitor to the dry inner mountain ranges of Chitral and Gilgit and breeds throughout the hills ranges of Baluchistan. In the Indus plains it is a double passage migrant only, extending from the Karachi coastline and particularly in the sand dune desert tracts of Thar and Cholistan. Probably a large part of this population winters in east Africa and migrates through lower Sind, where it is noticeable from August to mid-October over much the same period and habitat as *Coracias garrulus* which also winters in east Africa. It does breed in Sind in the low stony hills just north of Karachi and in Lasbela and in the hilly tracts of Thatta and Dadu districts. Ticehurst wrongly considered it only a passage migrant and winter visitor to Sind (*Ibis*, 1923) and this has been cited by Stuart Baker ('Fauna Series', 1927, and in Ali and Ripley's 'Handbook', Vol. 4, 1970). In the breeding season specimens have been collected from Baltistan, the Deosai plateau, Gilgit and the hill ranges around Kaliphat and Chiltan in central Baluchistan (Brit. Mus. coll.), Whitehead (*JBNHS*, 1911) considered it the commonest nightjar in the Thal area of Kohat district. It breeds in the Salt Range. The author has heard it calling and seen them in summer on the lower outer slopes of the Margalla hills where it is sometimes sympatric with *C. affinis*, also at Baran dam near Bannu, in Hazar Ganji park near Quetta, and at Ziarat up to 2,900 metres (9,500 feet) elevation in early April when the temperature at night was below freezing. Status COMMON.

HABITS

They are exclusively nocturnal in activity and insectivorous, hunting large insects on the wing, quartering to and fro with buoyant twisting flight alternating glides with twists and swoops as they pounce on some food prey. They start to call as dusk is gathering and in lower Sind the favourite perch is on the top of a *Euphorbia caducifolia* 'cactus' clump. By day they squat on the ground often in the shelter of a bush and will usually not fly more than 2 or 300 metres, low over the ground, before settling again. Moths, orthopterous insects, Gryllidae, *Chrotogonus* grasshoppers and beetles (Scarabidae) are amongst items in their diet.

BREEDING BIOLOGY

On the low stony ridge of hills which run north-west from Chaukandi Tombs, this nightjar regularly arrived from early May and remained to breed in the latter part of that month, within 24 to 32 kilometres (15 to 20 miles) of Karachi city, up to the late 1970s. Increased human settlement has now pushed them further afield. K. Eates found nests near Korangi creek at Rehri village on 18 May and a clutch of 2 eggs as late as 27 July near Mangho Pir hills on the other side of Karachi, west of North Nazimabad. The author found 2 eggs on 21 May near Chaukandi ridge and another clutch in the same area 2 years later, and downy chicks on 22 June. On all occasions the nest was located because of the injury feigning and distraction display of the female which was brooding the two chicks just after sunset. The eggs are long ovals of a white ground colour with grey under-markings and sparsely blotched with dark brown in contrast to the all-grey marking on the eggs of *C. mahrattensis*. The newborn chicks are nidicolous and those seen on 22 June had black bills with close woolly grey down and a few black spots on the back and crown. As is typical of the genus no nest whatsoever is made, not even a scrape, the eggs being laid on the bare ground in both cases perhaps intentionally in the lea of one or two larger stones. In the Punjab Salt Range, H. Waite found most nests in June and July, the earliest date eggs being found was 22 May and the latest 11 August. They were nearly always in the shelter of a bush (Waite, *JBNHS*, 1948). In Baluchistan near Chaman, H.E. Barnes found a fledged but flightless young at the end of May *(Stray Feathers*, 1881). Studies in Europe indicate that the male takes over the rearing and feeding of the first brood and that the female then frequently lays a second clutch. If disturbed at the nest, besides a convincing distraction display, the parent birds have been recorded as moving their chicks (in their bills) and also occasionally their eggs, the latter being indirectly observed by H. Waite in the Salt Range (pers. comm., 1965). Incubation is normally 17 or 18 days and period from hatching to fledging 16-17 days (Cramp *et al.,* Vol. 4, 1985).

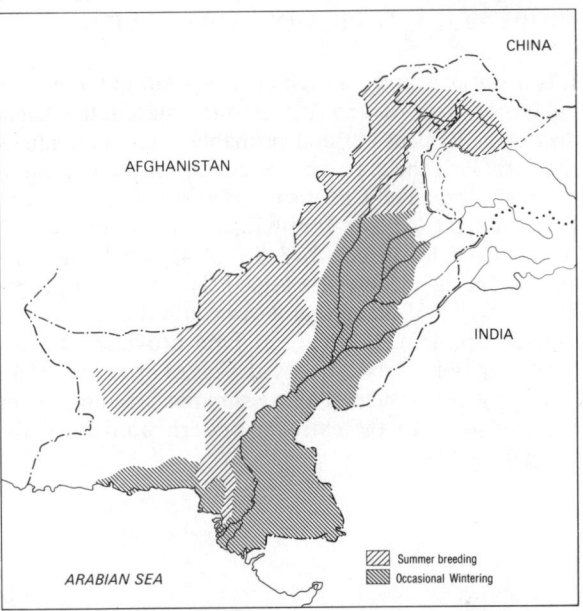

Map 252: European or Unwin's Nightjar *Caprimulgus europaeus unwini*

*VOCALIZATIONS

The advertising song of the male Nightjar is a peculiar steady churring or purring noise, uttered mostly in the early evening and again before dawn, either from the top of a bush, or on the ground. These bouts of calling may cease suddenly after only 15 or 20 seconds, or continue uninterrupted for 2 or even 3 minutes. Usually the churring occasionally varies in pitch, rising and falling, and is quite a loud throbbing noise audible up to 500 metres distance. It is remarkably similar to the call of *C. mahrattensis* indicating in all probability a close phylo-genetic relationship between these two rather desert adapted species. However the call of *C. mahrattensis* is not nearly so loud and does not vary in tone, being rather deeper in pitch.

Males have been watched giving display flights over their territory in Hazar Ganji, Baluchistan and in Karachi district. Wing clapping combined with soft liquid 'quoit-quoit-quoit' calls are combined with rather slow gliding flight.

Egyptian Nightjar *Caprimulgus aegyptius* Lichtenstein

DESCRIPTION

Body length 25 cms.
Wingspan 58-68 cms.
Wing length 20.8 cms. (Robat) (Christison)
Tail length 10.5-11 cms.

A rather small pale sandy coloured nightjar, paler than Sykes's *(Caprimulgus mahrattensis)* and only slightly larger than the latter species. Distinguished from *C. mahrattensis* by the complete absence of any white spots on the wing tips or the outer tail feathers in both the male and female. It has a white crescentic throat patch which extends right across the forethroat instead of two separate spots on either side as in *C. mahrattensis*. It has fewer and smaller black spots on its mantle and wing coverts and only faint cross-barring on its lower flanks and belly. There are no prominent white or creamy spots on the scapulars or wing coverts as in *C. europaeus*. The iris is dark brown. The sexes are alike.

HABITAT, DISTRIBUTION AND STATUS

It is a summer breeding visitor to Seistan in the border regions of south-eastern Afghanistan adjacent to Chagai district of Baluchistan and probably a few individuals regularly visit the extreme western tip of the Chagai in summer. Brigadier Christison collected a specimen at Robat on the Afghan-Iran-Baluchistan frontier on 12 April 1939 (Christison, *JBNHS*, Vol. 43) and he also saw another in the vicinity. He added that the local people knew it well and stated that it came into the area to nest (op. cit., p. 483). It occurs in Fars province of Iran, bordering Baluchistan so may also occur around Turbat.

It must be considered a rare summer breeding visitor confined only to the extreme western borders of the Chagai.

HABITS, BREEDING etc.

Nothing specifically recorded from our region. Like others of the genus no nest is made, the normal clutch of 2 eggs being laid on the bare earth and the nidicolous chicks hatching after 15 or 16 days incubation period. Meinertzhagen describes the eggs as white in ground colour heavily blotched with lilac and pale yellowish brown (op. cit., 1954). Descriptions of their calls appear to be rather variable, being described as 'tukl-tukl' and 'kre-kre-kre' (Heinzel, Fitter and Parslow, 1972) and to 'churr' like the European Nightjar (Bruun and Singer, 1970 and Peterson, Mountfort and Hollom, 1966 ed.). P.J. Sellar (in Cramp *et al*, Vol. 4, 1985) describes the advertising call of the male as a spaced (3 to 4 per second) rapidly repeated series of glottal or chugging 'kowrr' or 'powrr' sounds winding down towards the end of the calling bout.

ORDER APODIFORMES

FAMILY APODIDAE

Swifts

White-throated Needle-tail or
White-throated Spinetail Swift *Hirundapus caudacutus* (Latham)

Plate **21**
Map **253**

Chaetura caudacuta (Latham) of Ali and Ripley (1970)

DESCRIPTION

Body length 20-21 cms.
Wingspan 50-53 cms.
Wing length 19.5-21 cms.
Tail length 5-5.7 cms.

This is a stout bodied swift, larger than the Alpine Swift *(Apus melba)* with a fusiform streamlined body, deep in the chest to accommodate the well developed pectoral muscles and tapering smoothly to a point at the rear with both upper and under-tail coverts dense and extending to the tip of the central rectrices. It has narrow back-swept, sickle shaped wings, a very short bill but wide gape. The eyes are heavily hooded or recessed and the tail in flight, looks short and square. There are 8 to 10 tail feathers (based on examination of USSR specimen in BNHS Museum), the quills of each extending 3 to 5mm. beyond the veins, in stiff black needle-like tips. Its wings and tail are steely blue-black, highly glossed, with the body smoky brown unglossed. The mid region of the back, gradually pales into a large oval straw buff or smoky white patch. From underneath there is a prominent circular white throat patch, also visible when the bird is in side view. The underside of the tail coverts are also white extending up each side of the lower flanks in a prominent 'V'. The short oval tertial feathers are white on their outer halves. The hind toe is directed posteriorly and the legs are very weak and short. In flight its wings make an audible 'zizzing' sound which can frequently be heard when the bird is still out of sight. Close-up the noise produced by their wings is like sawing. The bill is black and the iris is dark brown. The toes are leaden coloured sometimes with a purple tinge. The sexes are alike.

HABITAT, DISTRIBUTION AND STATUS

Widely distributed in summer from eastern Siberia and the Soviet Far East, south to the Himalayas, Indo-Chinese region and Taiwan. Some birds in winter migrate as far as Australia. This is a rather uncommon swift in Pakistan, confined to the outer forest clad mountain slopes and particularly around their summits and shoulder ridges. It is a summer migrant visitor, staying only to breed, and the population is believed to winter in the South Pacific region from as far away as Australia.

They have been noted around Makrah peak and Shogran in the Kaghan valley, the slopes of Jamgarh peak further to the north, also on the west bank of the Kunhar river on the slopes of Musa-ka-Masala at Shahran. A few pairs frequent the northern end of the Murree hills each summer. It was recorded during a pheasant survey in June 1989 in the Palas valley of Indus Kohistan (J. Eames and G. Duke *in litt.*). Bates and Lowther only encountered them in the Kishenganga valley in Kashmir (op. cit. 1952) and they still occur in the Neelum valley above Machiara (Kamal Islam, pers. comm., October 1983). Status SCARCE.

HABITS

First sightings in the Murree hills are usually in the second or third week of May. Weather conditions are often disturbed at this time of year with tall cumulus clouds building up by midday and these aerial hunters probably ascend beyond human vision to the upper air space during such conditions rather than retreating to the plains and lower altitudes where *Apus melba* and *Apus apus* from adjacent mountain areas have been frequently noted in bad weather.

These large swifts are justly claimed to be the fastest flying birds in the avian kingdom. This is when measured over a transverse or horizontal flight path rather than during a power dive. In the USSR they have been estimated to fly at speeds of 170 kms. (106 miles) per hour (Dement'ev and Gladkov, *Birds of the Soviet Union,* 1951). Birds observed at Dunga Gali during foraging, spent a high proportion of their time actually beating their wings, though these are rapid shallow movements. Actual gliding was estimated to be only about one-third of the flying time. Their aerial manoeuvrability, speed and dexterity are quite marvellous to behold. They feed on larger insects including Hymenoptera, Diptera, Hemiptera and Coleoptera. In Pakistan they are almost invariably encountered in pairs from the time of arrival but also occasionally 2 or 3 pairs can be seen hunting together, indicating a gregariousness which seems to characterize most of the Apodiformes. Though no Needle-tailed Swifts have been located breeding in the vicinity of Dunga Gali, the local population (usually 2 or 3 pairs) assembles each evening, an hour or so before sunset, to drink from the twin iron water reservoir tanks at Dunga Gali. This they do by banking steeply and then skimming over the water surface with shallow wing beats and short glides, their lower mandible furrowing the water.

BREEDING BIOLOGY

No nests have been found or described on the Indian subcontinent. In the USSR, where they only occur in mountainous tall tree forest in a similar habitat, they usually nest in cavities in the 'chimneys' formed by tall hollow coniferous trees. They also occasionally nest in cliff crevices, (Dement'ev and Gladkov op. cit. 1951). The same habits have been recorded from their breeding in Japan (Yamashina *Birds of Japan,* 1962), and usually 2 or 3 nests are found in the same tree chimney. The nest is a small bracket or half-cup glued to the vertical side of a hollow tree or on a vertical rock face and made from thin twigs and straws or stems, cemented together by the bird's saliva. The author was very excited to discover 2 pairs one evening flying up to, and clinging vertically to the stem of a tall lopped *Acer* tree at about 2,500 metres (8,000 feet) in Dunga Gali catchment on 22 May 1981. Their excited chattering was accompanied by one bird actually entering a cavity formed by an old Woodpecker hole in the dried and dead tree stem. The next evening one pair was also observed settling on the same tree but they were never seen there subsequently and evidentally the location was not favoured and this behaviour may have been part of early nest site exploration. It can be presumed that these swifts never settle except on a vertical surface and at their nest site. In the USSR and Japan the normal clutch is 2 eggs, laid from late May to early June. These are pure white and rather elliptical ovals in shape. Incubation is believed to be by both sexes and to take about 21 days. The altricial chicks are blind and helpless at birth and devoid even of down. They take about 6 weeks to fledge and develop their wings sufficiently before they leave the nest.

*VOCALIZATIONS

A pair at Dunga Gali chasing each other in what might have been courtship pursuit, called excitedly. It was a high pitched and rapid chitter, 'chit-tit-tit-tit' not like the screams of *Apus apus*. Salim Ali describes loud and lively 'screams' being given in flight ('Handbook', Vol. 4, 1970), but only the rapid chitter has been heard from birds in the Murree hills.

Map 253: White-throated Needletail or White-throated Spinetail Swift *Hirundapus caudactus*

Common Swift *Apus apus* (Linnaeus)
Eastern Swift *Apus apus pekinensis* (Swinhoe)

Plate 21
Map 254

DESCRIPTION

Body length 17 cms.
Wingspan 40-42 cms. (Sub. species *pekinensis* Dement'ev *et al.*)
Wing length 16-18 cms.
Tail length 6.6-7.7 cms.

The population which inhabits the subcontinent averages larger in size than the west European population and is paler sooty brown in body colouration with more extensive pale areas on throat and forecrown. This makes separation in the field from the Pallid or Pale Brown swift *(Apus pallidus)* particularly difficult. In this genus all four toes point forward in contrast to *Chaetura* and *Hirundapus* swifts. It is a slightly slimmer, smaller bird than the White-throated Spinetail but with the same long slender backward curving wing of very shallow camber, a deeply hooded eye and a small weak bill with a large wide gape. The tail is forked and the whole of the body is sooty brown with the throat and forecrown paler buffish and clearly visible in flight, especially the chin which is almost white. The iris is dark brown and the tiny weak feet are purplish brown, equipped with long curved claws to enable them to cling to vertical surfaces. The tarsus is fully feathered extending to the foot. The sexes are indistinguishable.

HABITAT, DISTRIBUTION AND STATUS

A Palearctic species occurring from western Europe and the southern part of the USSR as far as Transbaikal. Winter migrants occur throughout Africa. A summer migrant visitor to the dryer mountain regions of Pakistan extending from north-central Baluchistan through the Himalayas up to the far north. They are quite plentiful in Chitral, Gilgit, Hunza and Baltistan in summer. C. H. Whitehead (1911) noted small numbers only, breeding on the cliffs of the Safed Koh range whilst the author saw a wheeling flock of about 30 over Parachinar on 8 April. W.H. Mathews noted swifts in Baltistan as plentiful throughout July and August, hawking over ridges at 4,575 metres (15,000 feet) (Mathews, 1941). They are rather sparsely distributed in Baluchistan and can be seen in summer around Zarghun, Wam and Baba Kharwari from 2,400 to 2,700 metres (8,000-9,000 feet). There are no resident breeding colonies in the Murree hills but they are plentiful in Hazara district in the Kaghan valley up to the Babusar pass and in Swat around Kalam. They are common around Sost in Northern Hunza in July. Nothing is definitely known of their winter quarters but they probably disperse widely to southern hemisphere regions such as the highlands of east Africa and south Africa (specimens collected Transvaal). They have been noted on passage in the Andaman and Maldive islands in the southern Indian Ocean (specimens in British Museum, Tring).

Despite its scientific name the sub-species *pekinensis* hardly extends into Kansu and south Manchuria provinces of China. Time of first arrival has not been noted, but flocks of over 90 birds have been observed around Hannah lake near Quetta 24-25 March and around Ziarat (Baluchistan) on 15 April. Swifts were noted hawking around the summit of the Lowari pass (Dir) on 7 May in low cloud and the ground carpeted with snow (author) and Biddulph noted their first arrival in Gilgit from early May *(Stray Feathers,* 1881). The striking thing about this species in Pakistan is its close association with remote mountain regions, often far from any human habitation, in contrast to the close urban association of the European population. Status ABUNDANT.

HABITS

This swift is markedly gregarious both in breeding and foraging. Their adaptations to an aerial environment are remarkable and are only gradually beginning to be fully understood. They are apparently sensitive to changes in barometric pressure and possibly magnetic fields, being able to anticipate bad weather and capable of travelling 160 kilometres (100 miles) within hours to pass through advancing storms. Relying for food upon 'aerial plankton', i.e. all sorts of airborne insects available in the upper air layers, they are heavily dependent upon relatively calm clear weather for food and thus must be highly mobile in foraging. Over 23 years of intermittent observations in the Murree hills, the author has several times encountered a flock of 50 to 60 swifts hawking around the summit of Miranjani 3,050 metres (10,000 feet), sometimes over a period of 1 or 2 weeks but there are no known breeding areas within 112 to 128 kilometres (70 to 80 miles) of this peak, and sometimes they are not observed in this region for seven or eight consecutive weeks. Studies of breeding swifts in England have shown that the food brought to the nestlings in the gular pouch of the adults consists of the larger airborne items, with a great variation according to the time of day and seasonal abundance of various insects. Many species of smaller Coleoptera, especially weevils (Curculionidae), also hover flies (Syrphidae), Hymenoptera, Diptera and such soft bodied insects as aphids and plant bugs (Hemiptera). Also a surprising proportion of such flightless creatures as spiders are included in their diet. During many summer visits to Shahran on the flanks of Musa-ke-Masala peak in the Hazara district, it has been noted that each evening around 5 p.m. flocks descend to the amphitheatre around the forest rest house at 2,400 metres (8,000 feet) where they excitedly call together, whereas during the morning and early afternoon they hunt around the ridges, ascending the flanks of this peak hurtling close to the ground at 3,900 metres (13,000 feet) as they cross the ridges and within seconds wheeling out into dizzying space above the valleys often more than 1,000 metres below them.

Swifts sleep in the air, and even collect nesting material in the air, but will rest on vertical surfaces when exhausted by prolonged spells of bad weather.

BREEDING BIOLOGY

Ali and Ripley ('Handbook', 1970 Vol. 4, p. 44) state that no nesting sights in the subcontinent have been described or actual proof of breeding on the subcontinent. This is largely because these highly mobile birds nest colonially on inaccessible vertical cliff faces or precipices and often forage more than a day's march for earth bound humans, from their breeding sites. The author has seen two breeding sites used by this swift. One in Chitral in the Lutko nullah and the other estimated to comprise about 45 pairs in a side valley of the Kaghan valley (Hazara district). Both sites were huge vertical cliff faces into which the birds disappeared inside narrow cracks in the cliff face. Such natural sites are abundant in the rugged Himalayan ranges and this swift does not use man-made artefacts for breeding as does the European population.

The patient and detailed studies carried out upon breeding colonies in Oxford by the Edward Grey Institute are now well known. These studies have revealed that the plain white eggs which are glossless elongated ovals are laid at 2 to 3 day intervals according to food availability and that copulation usually takes place on the nest or near it, rather than aerially as was formerly supposed. Both sexes share incubation duties which last 18 to 19 days and the naked blind altricial chicks take a comparatively long time to fledge and become independent, remaining in the nest for 6 weeks after hatching. When they finally launch themselves into the air they are totally independent and may in all probability remain continuously airborne for 2 or 3 years before reaching sexual maturity and starting to breed

Map 254: Swift or Eastern Swift *Apus apus pekinensis*

themselves (Bromhall, *Devil Birds,* 1980). Usually clutch sizes are 2 to 3 eggs with 3 being most usual (Ali and Ripley, 'Handbook', 1970 and Bromhall, op. cit., 1980). When bad weather makes food unobtainable, the chicks are capable of surviving starvation for short periods of 2 to 3 days without ill effects and also the incubating eggs can withstand chilling for 1 to 2 days without the developing embryo being killed.

*VOCALIZATIONS

As might be expected in such a gregarious species, they frequently call to each other while wheeling around in the sky. Courtship obviously includes high speed chases and unison acrobatics, all the while accompanied by high pitched screams. These screams carry a considerable distance. But they call to each other even long after egg laying is completed and the chicks are ready to fly so that such screaming is not confined to the courtship season.

Pallid Swift or Pale Brown Swift *Apus pallidus* (Shelley)   Map **255**

DESCRIPTION

Body length 17 cms.
Wingspan 42-46 cms.
Wing length 16.4-17.1 cms.
Tail length 6.5-7 cms.

It is hardly separable in the field from the pallid sub-species of the Common Swift *(A. apus pekinensis).* It

does however, have a slightly more extensive white chin and throat area and pale face, and its body is overall a paler more milky brown, less black looking. The primaries constrast dark brown with paler brown secondaries and wing coverts, also the shadowed area around the eyes appears more contrastingly dark than in *Apus apus.* A good description of this species and comparison with the Common Swift is given in *British*

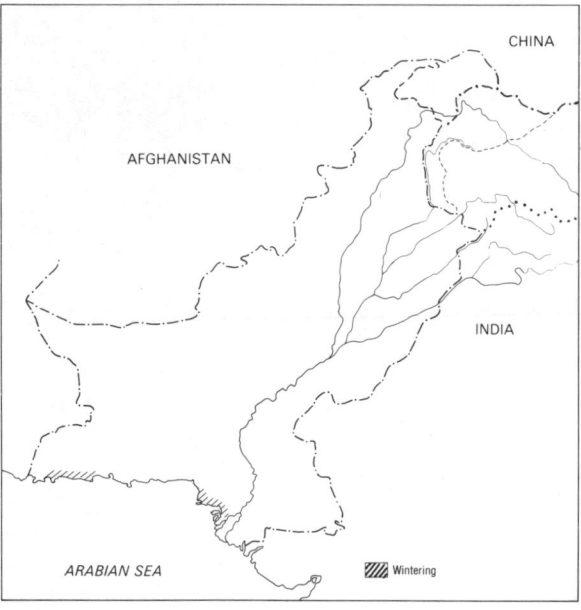

Map 255: Pallid Swift *Apus pallidus*

Museum. N. Van Zalinge saw several in late December 1985 around Gwadar during a Marine Fisheries survey (*in litt.* to author, February 1986). In late October the author watched a pale faced swift hawking around Cape Monze and presumed it was this species. Because of close similarities between *A. apus pekinensis* and *A. pallidus* definite identification was not possible and one cannot rule out the occurrence of stray migrant specimens of *Apus apus* though there are no known records for this species from Sind. Status UNKNOWN but probably not uncommon along extreme western coastal regions of the Makran.

HABITS

Presumed to be closely similar to those of *Apus apus* but generally considered to be slightly more sluggish in foraging with noticeaby slower wing beats and a tendency to forage in the locality of the nest site for a more prolonged period than *Apus apus*

BREEDING BIOLOGY

Breeding is presumed to be extra-limital. They nest colonially and utilize crevices in buildings as well as natural cavities or fissures in rock cliffs. Nesting habits are said to be similar to those of *Apus apus*. A clutch of 2 to 3 eggs etc. is normally laid on a rough pad made of whatever feathers and bits of straw or leaves the birds can catch from the air, and both sexes share incubation duties.

VOCALIZATIONS

Vocalizations have been described as being slightly different from those of *Apus apus,* and as being deeper and not as shrill (Cramp *et al.,* 1985).

Birds (Harvey, 1981). The sexes are alike. It is also said to have shallower wing beats and to indulge in more frequent gliding and not to be such an acrobatic flyer as *Apus apus.*

HABITAT, DISTRIBUTION AND STATUS

This is a largely sedentary species which breeds in coastal areas from north Africa and across the Middle East particularly through the eastern Mediterranean littoral and the Persian Gulf, also Morocco and Algeria. It extends in winter time along the Makran coast up to Karachi. A specimen collected off Manora head (Karachi) by Hume on 7 February 1872 is in the British

---

Pacific Swift, Asian White-rumped Swift or
Fork-tailed Swift *Apus pacificus* (Latham)

Plate **21**
Map **256**

DESCRIPTION

Body length 14-15 cms.
Wingspan 42-42.8 cms.
Wing length 14.7-16 cms.
Outer tail feathers 6.5-7 cms.

This is a relatively small swift of blackish body colour with a narrow white rump band and forked tail. The subspecies *leuconyx* is closely similar to *Apus affinis,* when

seen in the field, being only slightly larger in size than the latter and with similar paler areas around the chin and over the eye. The deeply forked tail is the best field character: that of the House Swift *A. affinis* being square tipped and even when closed, appearing shorter than the pointed tail of *Apus pacificus.* When viewed close-up the breast and belly feathers are narrowly squamated or scaled with whitish narrow margins. The iris is dark brown and the legs and feet are purplish black. The sexes are alike.

HABITAT, DISTRIBUTION AND STATUS

Occurs from eastern Asia south to the Himalayas, northern Indo-Chinese regions and southern Japan, with some winter migrants reaching as far as Australia. It is a summer visitor to the north-western Himalayas, associated with forest clad slopes and higher rainfall regions, and recent evidence of its breeding within Pakistan limits is lacking. Major Magrath (*JBNHS,* 1909*b*) discovered a small colony in limestone cliffs below Changla Gali and this 75 year old record is the basis for inclusion of Murree within its current breeding range (Ali and Ripley, 1970). Numerous searches around Changla Gali and particularly in the vicinity of suitable cliff regions, spanning 23 years by the author had failed to result in any sightings of this species which definitely has no surviving colonies in the Murree hill range. However, on 16 May 1982 (Roberts, *JBNHS,* 1984) a mixed flock of swifts was observed hawking around the summit of Mukhshpuri peak 2,800 metres (9,300 feet) which is on the wetern flank of the Jhelum river and at the northern end of the Murree hills. There were about 20 White-rumped Swifts in the flock, at least 5 or 6 House Swifts *(Apus affinis)* and about 30 Common Swifts *(Apus apus).* Ability to compare and contrast these three species as they flew together greatly facilitated positive identification. May was a late and cold spring in the Murree hills in 1982 and it seemed probable that these birds had not yet settled down to breed and that they might have come from further to the north in Azad Kashmir. Mark Malallieu also had two sightings of a pair of birds around Dunga Gali in mid-April 1987 (*in litt.* to author, 1987). In the summer of 1983, Kamal Islam regularly saw small numbers of this swift from May to August above the Neelum valley and around Salkalla, often in mixed flocks with *Apus apus* and felt certain that they bred somewhere in the valley (pers. comm., October 1983). It is probable that a small summer breeding population visits the upper Neelum valley in Azad Kashmir but in Pakistan as a whole its Status is RARE.

HABITS

Gregarious in foraging and in breeding. They are strong flyers, acrobatic on the wing and like others of the family particularly like to hunt around the relatively open treeless ridges or hill tops in the Himalayas. In India flying termites (Isoptera) have been noted in their stomach contents.

Map 256: Pacific Swift, Asian White-rumped or Fork-tailed Swift *Apus pacificus*

BREEDING BIOLOGY

They nest colonially and normally inside narrow crevices or cracks on vertical cliff faces. They construct small bracket nests, shaped like a half-cup, comprising bits of aerial flotsam such as grass stems, glued together with saliva. The normal clutch is 2 to 3 eggs, unglossy, pure white elongated ovals. Meinertzhagen reported finding nests of this species in Somalia in clusters, two or three nests joining each other. In the Himalayas, breeding has been reported from April to July. Probably both sexes share incubation duties which take 18 to 19 days. Magrath reporting his discovery at Changla Gali (1909*b*) could hear the young squeaking inside the inaccessible 'horizontal rifts in the rock' on 13 July at 2,300 metres (7,600 feet).

VOCALIZATIONS

Individuals in the feeding flock observed on 18 May frequently called. It was a rapid chirrup call, similar to that of *Apus affinis* but shorter and less of a prolonged chitter, and deeper pitched than that of the House Swift.

Alpine Swift *Apus melba* (Linnaeus)

Plate 21
Map 257

DESCRIPTION

Body length 22 cms.
Wingspan 45-60 cms.
Wing length 21-22.8 cms.
Outer tail feathers 7.75-9 cms.

This swift is more stout bodied and larger than *Apus apus*. The wings look very long, narrow and pointed and the upper parts are entirely dark sooty brown, in some lights appearing paler milky brown, with some squamation due to paler margins of the feathers, particularly on the leading edge of the wings and the under-tail coverts. The chin, throat and breast are white in this species with a broad dark brown band encircling the upper breast and some darker scaling in the flank region. The white under parts at once separate this swift from *Apus apus* with which it is sometimes associated and its larger size and more brown, less black upper parts at once separate it from *Apus pacificus* or *A. affinis*. The tail is not so deeply forked as in *Apus apus*. The iris is dark brown and the legs and feet are blackish purple. The sexes are alike.

HABITAT, DISTRIBUTION AND STATUS

It occurs widely from eastern and southern Europe, Africa, Madagascar east to south-western Asia. These swifts are far ranging and powerful flyers, which makes any accurate determination of their seasonal movements, particularly difficult. In Pakistan they are summer breeding visitors to warmer lower altitudes than those nesting sites favoured by *Apus apus*. It seems highly probable (see below) that the northern Himalayan population migrates in winter to the African continent as well as peninsular India. It is a summer breeding visitor widespread over Baluchistan but very local because of its dependence upon suitable cliff precipice nesting sites. Meinertzhagen observed what he presumed to be the start of autumn migration over Quetta from late August and their first arrival in central Baluchistan from late April (Meinertzhagen, 1920). The author has seen a few birds in the Hannah gorge north of Quetta as early as 14 April. In the northern areas it occurs in Chitral, Swat, Indus Kohistan and the Kaghan valley and Murree hill range. In the 1940s there were a few pairs breeding each summer around the summit cliffs of Sakesar in the Punjab Salt Range and H. Waite's specimens collected thence from July to September, are in the British Museum collection.

Some evidence of their migration is available from sighting of birds each spring around Kohat, though there is no evidence of breeding in the Safed Koh range (Whitehead, 1911). In the Punjab plains Hugh Whistler (1922) saw a small group on 25 August near Jhang. The author has sightings in Jabba valley, Mianwali district

31 October (see Mountfort, *Vanishing Jungle,* 1969). Also a party of 3 seen in southern Sind on 30 September 1983 near Var, just east of Mirpur Sakro in Thatta district. These birds were flying in a south-westerly direction and were at that time about 65 kilometres (40 miles) from the seacoast. Again in late September 1984, N. Van Zalinge saw a party of about 20 Alpine Swifts flying out to sea in the Cape Monze area (*in litt.* to author). Several small parties, probably on passage, were seen 1 to 3 March in the Kalla Chitta hills also (author). There are no records for northern Gilgit or Baltistan regions. Being highly gregarious they seem plentiful in the vicinity of suitable breeding colonies. Status COMMON but very local.

HABITS

Nesting is colonial and they forage gregariously outside of the breeding season also. Anyone who has watched this bird, cannot fail to be impressed by the vigor and power of its flight, whether hurtling close to the ground across some exposed mountain ridge, or sweeping past the vertical face of a cliff. Sometimes they wheel around, hunting insects very high in the air, and at other times they dive down between narrow gorges and canyons following the contours of the hill slopes. Their insect food comprises a variety of beetles, Hemiptera bugs, Diptera and Hymenoptera. Specimens shot in India had their stomachs full of the plant bug *Agnoscelis nubila* (Ali and Ripley, 1970). Due to their need for anticipating

Map 257: Alpine Swift *Apus melba*

unfavourable weather and the thunderstorms so prevalent in the spring and summer months, these birds seem to suddenly appear from nowhere and perhaps for one day or several can be seen in a pack, hawking around a certain locality. The largest concentration noted was over 100 birds on 16 August observed at Dunga Gali and in Indus Kohistan about 60 birds at a nesting colony site.

BREEDING BIOLOGY

Traditional sites are reoccupied year after year. The author knows of 2 on the Karakoram highway, the largest one near Komila village and on a vertical cliff face at about 1,000 metres (3,500 feet) where the Indus river runs through a narrow steep sided gorge. The other, closer to Chilas is on a huge cliff face on the left bank (descending) and about one-and-a-half kilometres (1 mile) from the river in an area where the valley widens out. The Komila colony is estimated to consist of just over 30 pairs. In lower Swat near Murghazar at about 1,500 metres (5,000 feet) elevation there is another colony nesting in a cliff gorge. On the Abbotabad side of the Murree hills another small colony comprising between 11 to 15 pairs has been under observation for about 15 years. This is located below Bhagnotar village at 1,700 metres (5,500 feet), the site being in a narrow vertical sided rock gorge with a stream at the bottom. The proximity of the road bridge crossing this gorge has enabled fairly close observations to be made. In most years Alpine Swifts do not arrive at the nest site until the third week of May or even later. At this stage they indulge in flight chasing and spectacular aerial courtship but rarely enter any of the nest crevices. In most years first arrivals at the colony site are noted between 22 and 26 May. In one year birds were watched entering nest crevices on 29 May. Young can be heard squeaking audibly inside the inaccessible crevices from mid-June (Indus river gorge) and as late as 17 July (Abbotabad hills road). Nests are small brackets of aerial flotsam held together by saliva and similarly glued to the cliff face or occasionally pads of straw and feathers, placed on a horizontal ledge.

The fissures used by birds are usually sedimentary faults which run parallel to the cliff face but with their external opening running vertically down the rock slope. Cracks of only 4 or 5 cms. width appear to be quite acceptable so one can only surmise that the cavities behind them are wider. A group of birds will hurtle down into the gorge seemingly from nowhere and skim along the cliff side without any apparent deceleration, when suddenly one individual swoops up to the crevice and clings momentarily to the vertical rock face before disappearing inside. Grease marks on the smooth rock face from the birds' plumage persisting from one year to the next, clearly indicate much used nesting entrance sites. The normal clutch is 3 eggs sometimes only 2 or as many as 4. They are thin shelled rather long ovals and plain white. Both sexes share incubation which takes 19 to 20 days. Nest material is gathered while on the wing and from the air and consequently vegetable matter, down and feathers are important components.

Many writers report that copulation takes place in the air (Ali and Ripley, Vol. 4, 1970 and H. Arn in Cramp *et al.,* Vol. 4, 1985) and no doubt there is some physical contact and attempted copulation on the wing but studies of *Apus apus* at Oxford confirm that copulation normally takes place at the nest site when both birds are roosting side by side (Bromhall, op cit.) As in all the swifts, the young take a very long time to develop and to be able to fly, varying from perhaps 6 up to 8 weeks depending on the food supply and the weather.

*VOCALIZATIONS

At the nesting colony, they indulge in a loud screaming and rapid chittering or twittering, each burst lasting just over one second. 'skri-skri-ki-ki-ki'.

Both sexes appear to use this call and birds can also be heard twittering from inside rock crevices when birds outside fly past calling. This is the only type of call heard by the author although others describe a screaming 'cheeh-cheeh' (Fleming *et al.,* 1976).

Little Swift, House Swift or
Indian House Swift *Apus affinis* (J.E. Gray)

Plate **21**
Map **258**

DESCRIPTION

Body length 12-15 cms.
Wingspan 34-35 cms.
Wing length 12.2-13.8 cms.
Tail length 3.7-4.5 cms.

The smallest swift in our area with a noticeably short tail, only slightly forked in appearance. It has a sooty black body with long narrow back-swept wings of shallow camber typical of the genus. The rump is white and more extensive in area than the white rump patch of *A. pacificus.* The throat is also whitish and the forecrown is pale creamy white. The iris is dark brown and well recessed. The claws are long and sharp for clinging to vertical surfaces and the legs are very short and weak. The under-tail coverts as in all swifts are particularly dense and well contoured, extending down to the tip of the rectrices, an important streamlining adaptation, for this is an area of turbulence during flight. The sexes are alike. In the field it generally looks blacker than either *Apus apus pekinensis* or *Apus melba* and is closely similar to *Apus pacificus* except for the unforked tail.

HABITAT, DISTRIBUTION, AND STATUS

A very widely distributed swift across Africa, the Middle East, India, Malaysia, the Philippines, southern China and Taiwan. Partly residential and partly locally migratory, occurring throughout the Indus plains and in many mountain areas as well. It is a summer migrant to the more accessible northern areas such as the main vale of Swat, the Murree hills and around Abbotabad. In Sind and Punjab it is largely resident. Man-made structures are widely used for nesting colonies particularly crevices under the rafters of older houses with timber construction in the roof. However a variety of sites can be used for nest building such as under bridge culverts, on top of horizontal roof beams and natural fissures in rock faces. There are often large noisy breeding colonies in the older parts of congested busy cities such as Frere Road in Karachi, the old bazaar in Peshawar, and Anarkali in Lahore. In Baluchistan it is a summer visitor and can be seen in small colonies around some of the higher peaks where it nests in caves or rock fissures, e.g. on Murdar peak and Kaliphat. Meinertzhagen (*Ibis,* 1920) noted first arrival around Quetta from 22 April. Whitehead considered it only a summer visitor around Kohat (Whitehead, 1911) but the author has observed a colony in Bannu town in March.

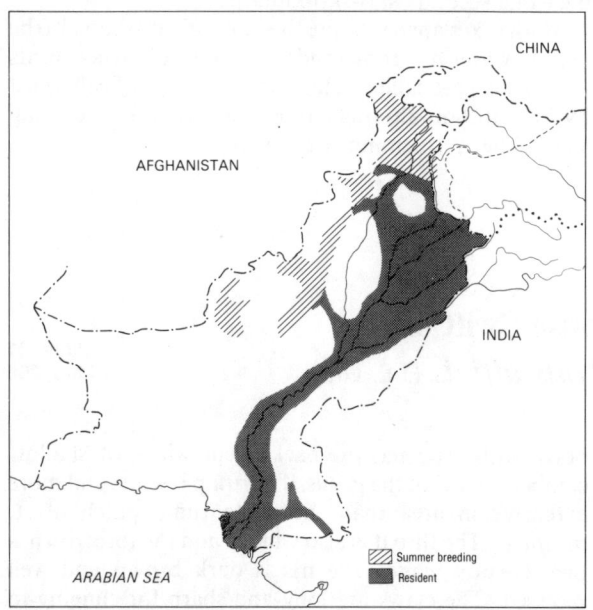

Map 258: Little Swift, House Swift or Indian House Swift
Apus affinis

A colony in Peshawar cantonment was observed using their old nests as roosting places each night during January. Similarly a nesting colony in the Suleiman hills on the summit at Fort Munro is used year round for roosting and shelter. A colony under the Malakand pass tunnel, on the other hand seems to be used only in the summer season. A small colony was noted in early May under a road bridge culvert in Indus Kohistan in the heart of steep mountainous country below the Shangla pass. They have not been recorded in Gilgit or Chitral. A few nest in rock crevices at Ayubia and Khanspur in the Murree hill range. Status ABUNDANT.

HABITS

Highly gregarious in nesting, roosting and foraging. Their nests are untidy accumulations of straw, feathers and other airborne rubbish, more substantial than those of the much larger Spinetail *(Hirundapus)* swifts. Usually two or three nests are placed one against the other on some horizontal ledge, and these are used year around for shelter and night time roosting. They become very verminous and insect infested. Local migrations are however performed by most colonies in the Punjab which seem to disappear for a few weeks during the coldest part of January and February.

Their food consists of aerial plankton obtained at great heights as well as low flying insects when skimming over canals and over roof tops. Arachnida, Chironomid midges, Diptera and small Coleoptera have been found in their stomachs. (Ali and Ripley, Vol. 4, 1970). In Madurai, south India, during a plague of 'Chilli' White Aphis *(Myzus persicae),* these insects swarmed in huge vertical columns and flocks of up to 50 House Swifts were observed feeding on these swarms (Thirumurthi and Krishna Doss, *JBNHS,* 1981).

BREEDING BIOLOGY

They are quite adaptable and flexible as nesting activity has been noted from early February (Karachi) up to October (young in nest, Khadeji falls near Karachi). They sometimes breed in large mono-specific colonies as well as only one or two pairs amidst colonies of other species such as Cliff Swallows *(Hirundo fluvicola)*. Often in cases where isolated pairs breed, they expropriate a Cliff Swallow's nest adding their own lining of feathers and straw inside the mud cup (two such occupied nests under a canal bridge at Sidhnai near Abdul Hakim). Both sexes take part in nest building and incubation. In undisturbed colonies, the same nests may be repaired and reoccupied for several successive decades or even longer. A colony in part of the old Peshawar Club has been in continuous occupation for over 80 years. The normal clutch is 3 eggs, less commonly 2 or 4 and they are thin shelled, plain white rather pointed ovals. Nests vary in structure. Vegetable down and bits of straw are cemented together with saliva and may be flimsy brackets barely able to contain 3 developing young as seen in a vertical sandstone cliff at Khadeji falls (56 km. north of Karachi). Some nests comprise almost 90 per cent of feathers glued together and are huge misshaped balls (Fort Munro nests). K. Eates, (MS notes unpublished) describes finding nests in a colony at Umarkot in Sind, each containing 2 or 3 fresh eggs on 28 February and quite a few of these nests had short 10 cms. (4 ins.) long entrance tunnels leading to a rounded egg chamber. No mud is used in nest construction as in Hirundines, only material which can be gleaned from the air in flight. Incubation is prolonged and varies according to weather conditions extending from 18 up to

26 days (Razack 1968). The young take 37 to 43 days to become independent and leave the nest (Razack, op. cit.).

*VOCALIZATIONS

A high pitched chittering rapidly repeated, 'chik-chik-chik-chik-chik'. Birds call in flight, particularly around the nest colony, and also while roosting in the nest in response to calls of birds flying past. They also make more subdued 'chirring' calls while on the nest particularly in the early morning before leaving the colony to start foraging.

ORDER CORACIIFORMES
FAMILY ALCEDINIDAE
Kingfishers

White-throated Kingfisher, White-breasted Kingfisher or
Symrna Kingfisher *Halcyon smyrnensis* (Linnaeus)

Map **259**

DESCRIPTION

Body length 25-28 cms.
Wing length 11.8-12.8 cms. (Stuart Baker),
      11.9-13.1 cms. (Forshaw and Cooper, 1983)
Bill length 6-6.7 cms. (Stuart Baker),
      5.9-6.9 cms. (Forshaw and Cooper, 1983)
Weight 75-108 gms. (males),
      79-101 gms. (females), (Forshaw and Cooper)

This is a slightly smaller Kingfisher than *Ceryle rudis* with an even heavier looking bill which is broader and more flattened at the base. It is a dark chocolate or maroon-brown colour all over its head and body except for a large round white patch extending from the base of the lower mandible down to the centre of the breast. The tail, back, rump and wings are iridescent blue, occasionally reflecting an almost verdigris green colour at certain angles. The bill is dull dark red and also the legs and feet. In flight the wings show a conspicuous white patch at the base of the primaries visible from above and below. The under-wing coverts are brown with the median coverts on the upper surface showing black.

HABITAT, DISTRIBUTION AND STATUS

It occurs from the Middle East, across India, southern China, the Indo-Chinese region, Taiwan and the Philippines. A very widespread and ubiquitous kingfisher throughout the Punjab and Indus plains. It is largely absent from the dry mountainous tracts of Baluchistan and the NWFP but occurs up to the lower part of the Bolan pass in Baluchistan (author obs.) and

sparsely in the Miranzai and Kurram valleys of the NWFP (Whitehead, 1911). It occasionally penetrates the larger rivers valleys in the lower mountain regions of the north. It occurs (author obs.) on the Kunhar river up to Balakot in Hazara district and the vale of Swat as far as Miandam, also along the Jhelum river up to

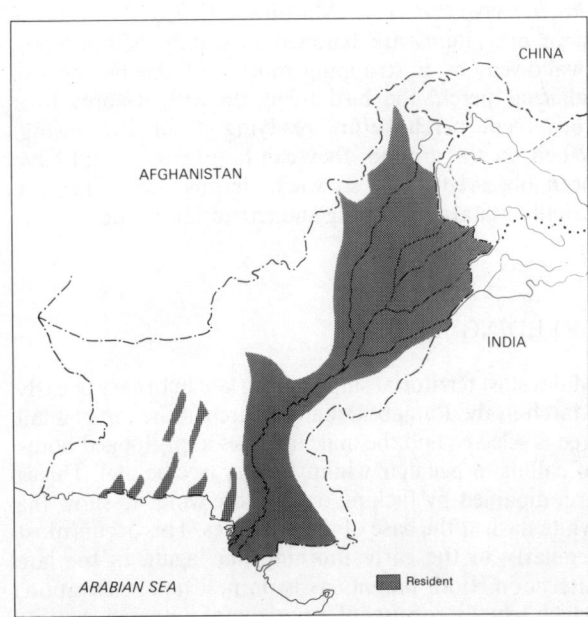

Map 259: White-throated or Smyrna Kingfisher *Halcyon smyrnensis*

Muzaffarabad in Azad Kashmir. Being adaptable to hunt a wide range of animal food, it is not tied to major wetlands or rivers as *Ceryle rudis,* and occurs widely along canals, around village 'tanks' and in irrigated forest plantations. It also occurs very sparsely in the mangrove creeks of the Indus delta. It is largely sedentary where it occurs. Status COMMON.

HABITS

They never hover in the air over the water like *Ceryle rudis* and *Alcedo atthis,* invariably waiting on some perch, often over dry land, as well as near water from whence they dive down at an angle to seize prey. They freely perch on telephone lines along flooded roadside borrow pits. They are quite frequently encountered far from any permanent water, though they prefer the vicinity of trees and irrigation channels even if these are temporarily dry. There have been many observations on the feeding habits of this Kingfisher published in the *Journal of the Bombay Natural History Society.* They have been recorded diving upon and capturing *Calotes versicolor* lizards on dry land (*JBNHS,* Vol. 6, p. 758), capturing small birds (*JBNHS,* Vol. 13, p. 184) and a White throated Munia *(Evodice malabarica),* (Roberts, *JBNHS,* 1965a). Also land crabs *(Paratelphusa* spp.) mice, grasshoppers (Acrididae and Crotogonus spp.) and other insects (*JBNHS,* Vol. 12, p. 562). K. Eates saw one take a newly fledged Sand Martin *(Riparia paludicola)* and beat it on a branch like a fish before swallowing it whole (K. Eates, MS notes). Insects normally comprise the bulk of the diet during the greater part of the year, and orthopterous insects are most commonly taken, e.g. locusts and the cricket *Brachytrypes achatinus* (Mason and Lefroy, 1912). Most large prey items are battered to a pulp before being swallowed, by a stropping motion of the bill on the adjacent perch, the bird flying up with its prey to a convenient perch before readying it for swallowing. When on the ground, they can hop forward and have been observed to do so when chasing prey. They are usually solitary in hunting and territorial throughout the year.

BREEDING BIOLOGY

Males start territorial singing from late February to early March in the Punjab. Usually a perch at the top of a tall tree is selected and the male indulges in prolonged bouts of calling, a peculiar whinnying cry (see below). This is accompanied by flicking open of the wings to show the white flash at the base of the primaries. This occurs most regularly in the early morning and again in the late afternoon. Both parents assist in nest hole excavation, which usually consists of a horizontal tunnel in the side of an open well or silt bank of the spoil from an irrigation channel. Less typical nest sites regularly used in the

authors garden at Khanewal included holes excavated in mud brick walls of stables and once in a partly excavated natural hollow in a Eucalyptus tree. In lower Sind, digging of nest holes in a gravel bank has been seen on 28 March and at Khanewal on 27 March and young in the nest, with wing quills still ensheathed, seen on 12 April. The normal clutch is from 5 to 7 eggs, pure white, glossy in texture and rather spherical in shape. In Malir a pair nested in the side of a well over a period of three years using a hole excavated by Bank Mynas (*Acridotheres ginginianus*) in one instance. The young are altricial (born naked with eyes closed). As the inside of the nest chamber becomes very foul with regurgitated bones, it is perhaps a valuable ecological adaptation that the flight feathers remain protected by a waxy sheath during the first half of the nestling period. Both parents assist in incubation and feeding young. Flight Capt. Mervyn Sequira, a member of the Bombay Natural History Society, succeeded in getting a pair to breed in captivity and made a number of fascinating observations (pers. comm., 1981). He constructed a large open air aviary in his garden and an artificial nest tunnel from wet mud wrapped around a bamboo which was subsequently withdrawn. The young hatched after 16 to 18 days and eventually left the man-made nest chamber at 50 days of age when they were well able to fly. The parent birds were noted to be active in hunting throughout the night, being adept at catching Gheko lizards (*Hemidactylus flaviviridis*) and mice (*Mus musculus*) which ventured into their enclosure. When the chicks were very small, they were fed only on insects and this was done bill to bill, not by regurgitation. Cockroaches (*Periplaneta americana*) were a much sought food item. When the adults caught mice, an especially favoured prey, they

Plate 21: Sexes are alike for all the following species.
(1) Long-eared Owl *Asio otus.*
(2) Short-eared Owl *Asio flammeus.*
(3) Indian or Little Nightjar *Caprimulgus asiaticus.*
(4) Jungle or Grey Nightjar *Caprimulgus indicus hazarae.*
(5) European or Unwin's Nightjar *Caprimulgus europaeus unwini.*
(6) Sykes's or Sind Nightjar *Caprimulgus mahrattensis.*
(7) Large-tailed or Long-tailed Nightjar *Caprimulgus macrurus.*
(8) Savanna or Allied or Franklin's Nightjar *Caprimulgus affinis.*
(9) White-throated Needletail or White-throated Spinetail Swift *Hirundapus caudactus.*
(10) Pacific Swift, Asian White-rumped or Fork-tailed Swift *Apus pacificus.*
(11) Alpine Swift *Apus melba.*
(12) Common Swift or Eastern Swift *Apus apus pekinensis.*
(13) Little Swift, House Swift or Indian House Swift *Apus affinis.*

Plate 21

(1)

(2)

(3)

(4)

(5)

(6)

(7)

(8)

(9)

(10)

(11)

(12)

(13)

Plate 22

(1)

(2)

(3)

(4)

(5)

(6)

(7)

(8)

(9)

(10)

gave a special type of food call, discernibly different from their calls when other prey was captured.

In Malir, fresh eggs have been taken from early March to April (K. Eates, MS notes). Newly fledged young with their parents were seen at Malir on 9 June. These juvenile birds had dull brown, not red bills. Also the male parent had a noticeably heavier and brighter red bill than his mate (author's observation of a pair breeding in Malir garden). This same pair which reared three young in Malir, produced a second brood, the male recommencing territorial calling from the third week of July. A. E. Butler also found a nest with fresh eggs in Karachi on 20 July (Ticehurst, *Ibis,* 1923), so it is probable that a second brood is often reared during the monsoon season when there is an abundant insect food supply.

*VOCALIZATIONS

The advertising or breeding season call comprises a series of rapidly repeated rather tremulous calls falling in pitch at the end of each strophe, 'kli-kli-kli-kli-kli-kli'. It is a loud shrill call, sounding almost like a 'whinny' and is repeated every 2 or 3 seconds, often for several minutes on end. Outside of the breeding season when disturbed or flying off across part of their territory they utter a loud raucous cackle 'pit-pit-pit-pit'.

Common Eurasian Kingfisher *Alcedo atthis* (Linnaeus) Map **260**
Small Blue Kingfisher

DESCRIPTION

Body length 18 cms.
Wing length 7.1-7.7 cms. (Stuart Baker),
   6.9-7.6 cms. (Forshaw and Cooper)
Tail length 2.9-3.7 cms. (Forshaw and Cooper)
Bill length 3.9-4.8 cms.
Weight 30-31 gms. (males), 25 gms. (1 female)
   sub-species *bengalensis* (Forshaw and Cooper, 1983)

This is rather a tiny kingfisher with a short stubby tail and comparatively large head set on a short neck, typical of the family. It is orange-chestnut on the breast, under-tail coverts and ear coverts, with a white throat and small white patch on the lower cheeks. The head and a loral patch are iridescent cobalt blue with paler horizontal striations. The back and rump are iridescent turquoise blue and also the wings with paler tips on the wing coverts forming a spotted pattern. The straight sharp pointed bill is black on the upper mandible and shaped like a dagger. The base of the lower mandible shows pink or reddish. The legs and feet are very small and weak with the second and third toes partly fused. The iris is brown and both sexes are alike.

HABITAT, DISTRIBUTION AND STATUS

It occurs throughout the Palearctic except the Arctic zone, and south through Malaysia. Indonesia to Australia and as far as the Solomon Islands. This kingfisher is much more closely tied to an aquatic environment than *H. smyrnensis* and in the Punjab plains, will rarely be encountered away from natural rivers or larger lakes. It is commoner in Sind province and more likely to be seen around smaller swamps or alongside irrigation canals. It penetrates into the mountain areas of central Baluchistan in the summer months, being encountered on Surkhab stream near Pishin in late June and the Baleli nulla north of Quetta in April and May, and at Wam (2,150 metres/7,000 feet) near Ziarat in June. It has also been seen (author) on Zangi Nawar lake in the Chagai desert on 4 May. Meinertzhagen recorded it in all winter months around Quetta (Meinertzhagen, 1920) and considered it a winter visitor, but a bird on the Surkhab was seen going into a nest hole in a gravel bank on 9 May and a few pairs probably breed in Baluchistan. It has been recorded as a year round resident in Chitral (Perreau, 1910) and up to 1,067 metres (3,500 feet) in hill streams of Kohat (Whitehead, 1911) and it penetrates all the northern areas including around Skardu in Baltistan (R. Orr, *in litt.,* 1984). It is sympatric with the Pied Kingfisher and White-breasted throughout the plains but in lowland areas it is the rarest of the three species inhabiting

Plate 22: (1) Blue-tailed Bee-eater *Merops philippinus,* sexes alike.
(2) Green, Little Green or Small Green Bee-eater *Merops orientalis,* sexes alike.
(3) European or Golden Bee-eater *Merops apiaster,* sexes alike.
(4) Blue-cheeked Bee-eater *Merops superciliosus,* sexes alike.
(5) Indian Grey Hornbill *Tockus birostris,* male.
(6) European or Kashmir Roller *Coracias garrulus,* sexes alike.
(7) Indian Roller or Blue Jay *Coracias benghalensis,* sexes alike.
(8) Coppersmith or Crimson-breasted Barbet *Megalaima haemacephala,* sexes alike.
(9) Blue-throated Barbet *Megalaima asiatica,* sexes alike.
(10) Great Barbet or Great Himalayan Barbet *Megalaima virens.*

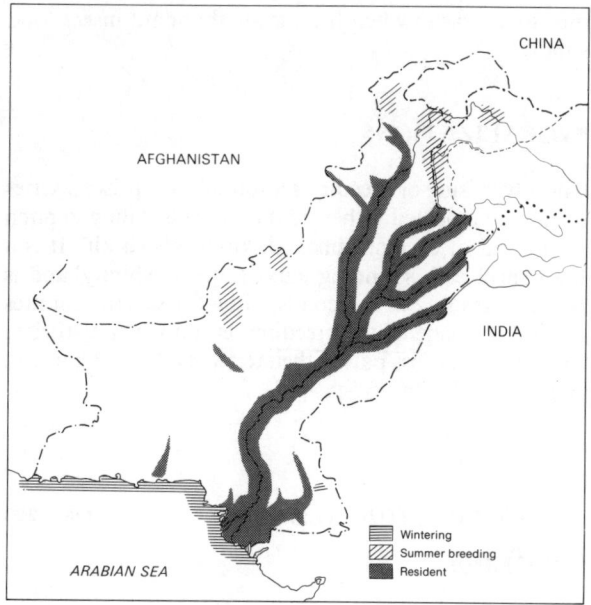

Map 260: Common Kingfisher, Eurasian or Small Blue Kingfisher *Alcedo atthis*

(Whitehead, 1911). One watched at Khadeji falls, a fresh water stream 51 kms. north of Karachi, successfully caught a fish about 6 cms. long which was carried crosswise in its bill to a nearby boulder where it was beaten on the rock before being eaten head first. The fish from this pool were later identified as *Ophiocephalus striatus* (Dr. Farooq Ahmed, Zoological Survey). In the Indus delta, another Kingfisher with a very large fish (estimated nearly 10 cms. in length) managed to swallow this whole after much juggling in its bill. This was accomplished while perched on a mud bank in a situation where there was no hard perch or rock on which the fish could be macerated. On the feeder canal which supplies Haleji reservoir, one was seen repeatedly diving into small shoals of the fish *Puntius sophore* with success in about 1 out of 3 or 4 dives. They have also been recorded carrying small frogs to their nest hole (Ali and Ripley, 'Handbook', Vol. 4, 1970). Their flight comprises of rapid wing beats often skimming low over the water surface.

BREEDING BIOLOGY

Very well studied in Europe where the monogamous pair bond is long lasting and often two broods are reared. In Pakistan they have been found nesting from March to June and both parents excavate the nest hole, which consists of a slightly upward sloping horizontal tunnel with a widened end chamber. The eggs are glossy plain white and normally number 5 to 7 in a complete clutch. Both sexes share incubation duties which take up to 20 days. The chicks are altricial and develop slowly being fledged by 3½ weeks of age, but still having their flight feathers ensheathed in waxy cylinders giving them a rather reptilian appearance up to about 20 days after hatching. Hugh Whistler describes nest tunnels only 5 cms. in diameter and up to 0.9m. in depth. The floor of the nest chamber always becomes littered with a smelly debris of regurgitated bones and fish remains and the protection of the tail and wing feathers of the developing young in quill sheaths over most of the fledgling period, no doubt compensates for the lack of nest hygiene. Courtship involves a lot of flight chasing with the male sometimes uttering a piping song while on the wing.

Pakistan. In winter they are abundant throughout the Indus mangrove creeks as well as along the Makran seacoast and at Sonmiani lagoon. It is believed that these are winter migrant visitors and have been described as belonging to a larger paler sub-species *pallasii* (Ticehurst, 1923).

During a two day voyage on 10-11 December through the Indus delta creeks 42 *Alcedo atthis* were counted, usually as solitary individuals. It has also been seen fishing in the rock pools on Cape Monze headland on the open seacoast in November and December. In Swat, it has been noted up to 1,200 metres at Murghazar and Miandam. Status partly resident, partly local migrant. COMMON only in lower Sind, elsewhere FREQUENT.

HABITS

They are territorial in winter, spending the day hunting from a series of favourite perches and vantage points, sometimes from the prow of a deserted fisherman's boat or an erosion exposed mangrove root in the Indus delta creeks: usually perched on the reed stem of a submerged *Phragmites* in the seepage zones of the larger rivers. When flying from one hunting point to another, they utter very high pitched penetrating squeaks. Occasionally they will also hover over open water like a *Ceryle rudis*, plunging bodily into the water after fish. Their diet consists mainly of small fish but they will also catch tadpoles, water beetles and dragonflies (Odonata), which they have been observed catching on the wing

*VOCALIZATIONS

Throughout the year they utter a flight call which may act as a territorial or contact call. It consists of short sharp and very shrill squeaks uttered usually as they fly over the water. Their song during the breeding season consists of a shrill piping two-noted call, repeated 'chichee-chichee' much weaker in tone and volume than the other kingfishers of the region.

Pied Kingfisher *Ceryle rudis* (Linnaeus)

Small Pied Kingfisher

DESCRIPTION

Body length 28-32 cms.
Wing length 13.3-14.2 cms. (Stuart Baker)
Tail length 6.6-7.4 cms. (Forshaw and Cooper)
Bill length 6-7 cms.
Weight 70-100 gms. (Forshaw and Cooper)

This is a typical kingfisher in body proportions, with a comparatively massive head and short neck, stumpy tail with small weak feet. There is a shaggy spiked crest on the back of the crown. The bill is comparatively heavier than in *Ceryle lugubris* and jet black, as also the legs and feet. The outer and centre toes are joined together basally (syndactyl). The face and breast are white with a broad black band encircling the upper breast and a second broad black band running through the eye to the nape where it extends down to the mantle. Males have a second narrow breast band and some spotting in the flank region and throat. The crown is black, including the crest, with a few narrow white streaks. Females have only a single pectoral black band hardly meeting in the middle of the breast. They are otherwise similar in size and plumage to the males. The flight feathers are black and wing coverts boldly patterned with black and white, as is the mantle which is predominantly black with white scaling formed by the feather tips. The tail is white with a broad black terminal band. In general colouration, this kingfisher looks chequered black and white and is not so finely speckled or greyish as *Ceryle lugubris*.

HABITAT, DISTRIBUTION AND STATUS

It occurs in east Africa, the Middle East, across India, southern China and the Indo-Chinese region. A very

widely distributed kingfisher throughout the Punjab and Sind plains, but closely tied to large fresh water bodies or flowing canals and rivers. It usually avoids mangroves or coastal creeks, and except in major rice growing tracts is absent from stagnant pools or roadside borrow pits so much favoured by *Halcyon smyrnensis*. In the northern mountain areas, it penetrates a few of the lower broader river valleys. Thus it ascends the Jhelum river up to the main vale of Kashmir and occurs on the Swat river up to 16 kms. (10 miles) north of Miandam (900 metres/3,000 feet elevation). Whitehead also recorded it up to 1,400 metres (4,500 feet) in the Kurram valley (Whitehead, 1911). It is largely absent from Baluchistan and the far northern regions such as Chitral and Gilgit. In Thatta district it is very numerous and can be seen perched on telephone lines alongside every road above flooded borrow pits. For example, 47 were counted along a 300 metre (1,000 feet) stretch of such telephone wires above a small swamp near Mirpur Sakro in early February. It is largely a sedentary species. Status COMMON in Punjab, ABUNDANT in lower Sind.

HABITS

It will regularly perch near water to watch for fish moving near the surface, but it also has the very characteristic habit of hovering over water, its body held almost vertically with tail depressed and wing beating over its back in quite a laboured manner. If it sees a fish, it at once goes into a steep vertical dive with wings closed, plunging into the river below but sometimes it checks its dive and again hovers closer to the water surface. Its diet is predominantly fish and it is less insectivorous than *Halcyon smyrnensis*. Stomachs of 5 specimens collected in Bihar, India during July and August, contained entirely, remains of fish (Mason and Lefroy, 1912). Mukherjee, who collected 300 adults in the mangrove creeks of the Sunderbans (west Bengal, India) found 57 per cent of the diet by weight was fish of which three-quarters numerically were *Puntius* sp., *Mystus* sp., *Mugil parsia* and *Ambassis* sp., 17 per cent were Crustaceans and 26 per cent were large aquatic insects (Mukherjee, 1975). Ali and Ripley state that the mainly fish diet may be supplemented with aquatic insects and tadpoles ('Handbook', Vol. 4, 1970). In Thatta district of Sind, it has been seen successfully catching small specimens of *Ophiocephalus striatus* and the Gar fish *(Xenentodon cancila)* (Identification by Dr. Farooq Ahmed, Zoological Survey). Large fish are held crosswise in the bill and battered on a perch before being swallowed head first. They regurgitate pellets containing indigestible fish scales and bones which often accumulate below favourite feeding perches (Forshaw and Cooper,

Illus. 66: Pied Kingfisher or Small Pied Kingfisher *Ceryle rudis*. Male showing double breast band.

Map 261: Pied Kingfisher or Small Pied Kingfisher *Ceryle rudis*

1983). In areas where they are plentiful, often 4 or 5 individuals can be seen hunting in the same vicinity and in this respect it is slightly more gregarious than the equally common White-breasted Kingfisher.

BREEDING BIOLOY

This has been extensively and intensively investigated in south Africa (Douthwaite, 1978 and Fry, 1980*b*). Nesting in the African continent is often semi-colonial with the mated pair of birds often being helped by 1 to 5 unmated males who bring food to the nestlings and guard the nest hole against predators (Reyer, 1980). In the Indian sub-continent such colonial nesting and cooperative breeding with nest helpers, has not been noted except for parts of Kashmir where Col. Phillips noted traditional sites where they nested in colonies of up to a dozen pairs (B. Phillips, *JBNHS*, Vol. 46, p. 94). Bates and Lowther however, in commenting upon Phillip's observation considered that colonial nesting was not usual in Kashmir. In Reyer's study, conducted in Kenya, helper males were initially driven away from the nest hole but after 2 to 4 days were tolerated and allowed to bring food to the young. Such behaviour was not tolerated however by mated pairs nesting in relatively isolated territories with a plentiful food supply (Reyer, op. cit.). Perhaps there are significant ecological differences between breeding conditions for *Ceryle rudis* in Africa and the Indian subcontinent. Certainly more

detailed studies in Pakistan on their breeding biology would be of great interest.

In a normal situation the pair bond is monogamous and both sexes share in digging the nest hole. This is often excavated in the comparatively low silt bank of a canal and within 20 cms. of the water level (author obs. Thatta district), but a preferred site is in higher sand or earth cliffs of river banks and some metres above water level. The nest hole is usually 7 to 8 cms. in diameter and averages greater in depth than typical nest burrows of *Halcyon smyrnensis* (Eates, MS notes). The tunnel is sometimes up to 150 cms. in depth. The nest chamber is unlined and roughly double the width of the entrance tunnel. The nesting season in Pakistan is mostly during March and April. The earliest date nest hole excavation was observed in Sind by K. Eates was on 10 February (Eates, MS notes). In Kashmir, eggs have been found as early as 1 March and up to early June (Bates and Lowther, 1952).

Five to six eggs are normally laid (Eates, MS notes based on over 100 eggs collected). They are plain white, glossy in texture and on average are slightly larger than the eggs of the White-breasted Kingfisher. At Hab dam reservoir both sexes were observed entering a nest hole in a gravel cliff on 22 February but from their behaviour incubation had not started nor probably egg laying. On 27 March a pair was watched bringing fish to their nest hole in another creek along the shores of the Hab dam reservoir. A nest hole near Bahawalpur was excavated on 25 April and found to be only 35 cms. deep. It contained 3 young with wing quills still ensheathed (D. Stirling-Wyllie, *in litt.*, 1976). The female does most of the incubation, and this takes about 18 days (Cooper and Forshaw, 1983). The nestlings leave the nest hole 24 to 25 days after hatching (Douthwaite, 1978).

As in all the kingfishers, the actual nest chamber becomes littered with regurgitated fish scales and bones that are strong smelling. Phillips noted adult birds using nest chambers as a nightly roost (B. Phillips, *JBNHS*, 1946)

*VOCALIZATIONS

A comparatively noisy kingfisher with more social interactions outside of the breeding season than other species in the region, and they are particularly noisy in the early stages of courtship when groups of birds tend to display together, either perched on the ground or on reed stems (obs. Haleji lake). Their calls are softer and more querulous in tone than those of the White-breasted Kingfisher. The basic call is a rapid high pitched 'chitter-chitter' or 'chirruk-chirruk-chirruk'. They call both on the wing and when perched. Courtship involves group displays, with from 3 to 8 birds flight chasing and sitting on the ground calling to each other.

Crested Kingfisher *Ceryle lugubris* (Temminck)

Large Pied or

Himalayan Pied Kingfisher (of Ali and Ripley) *Megaceryle lugubris* (Temminck)

Greater Pied Kingfisher of Cooper and Forshaw

DESCRIPTION

Body length 35-41 cms.
Wing length 17.7-19 cms. (Stuart Baker)
Tail length 10.3-11.4 cms. (Forshaw and Cooper)
Bill length 6.8-8 cms.

Bigger than either the Pied Kingfisher *(Ceryle rudis)* or the White-breasted Kingfisher *(Halcyon smyrnensis).* This large and handsome kingfisher looks more grey from a distance than *C. rudis,* being speckled all over the mantle and the wing coverts with much smaller black and white spots. The large head is framed by a rather bushy upstanding crest from forecrown to nape, which is speckled black and white in front, being blacker around the hind crown and occiput with two white bands in between. A narrow black line encircles the lower cheek and the black nape area extends forward through the eye to the back of the bill. The breast and the belly are white with a spotted band around the upper breast. The stumpy tail and flight feathers are cross-barred with white. The black bill is long and sharp pointed at the extreme tip and looks comparatively slender in this species and slightly down-curved. It is white at the extreme tip and the basal one-third is pale grey except along the commissure. The legs and feet are greenish grey and the iris is almost black. As in all the Alcedinidae the three middle toes are fused together in their basal one-third portion.

The female is separable in having the pectoral region spotted with dark chestnut brown not black and also the flank feathers and under-wing coverts are often rusty cinnamon. In a pair seen at Rawal lake (see below), the breast of the male was entirely chestnut brown whereas that of the female appeared much less distinct and more blackish. Some males also show traces of chestnut brown in the pectoral regions.

HABITAT, DISTRIBUTION AND STATUS

It extends across the Himalayas, through Burma, southern China, northern Indo-Chinese region to Japan. A rather sparsely distributed kingfisher throughout the lower river valleys of the Himalayas, extending up the Neelum valley in Azad Kashmir which is close to the westernmost extremity of its range. However, in 1969 during a visit to Nuristan in north-eastern Afghanistan, F. Koning and J. Vieillard obtained clear views of this species on the Kamdesh stream (F. Koning, pers. comm., 1985). This is also cited in Forshaw and Cooper (1983). It is not included in Paludan's *Afghanistan Checklist*

(1959). Its habitat preference is for fast flowing, even turbulent, mountain streams along the banks of which are over-hanging bushes or tree groves. It will not usually be encountered over lakes or open water and rarely ascends rivers above 1,500 metres (5,000 feet) and often descends in winter as low as 460 metres (1,500 feet). It has not been recorded in Hazara district or further westwards, but a few pairs still survive on the Neelum river around Salkalla in Azad Kashmir and lower down between Ghori (730 metres/2,250 feet) and up to Mirpur (1,000 metres/3,250 feet). There is some local or latitudinal migration in winter. Hugh Whistler (*Ibis,* 1930) encountered one on the Leh stream in Rawalpindi on 8 January 1926 (this area is now a heavily polluted sewer). In 1982 a single bird was seen haunting a shallow rocky stream draining into Rawal lake at Islamabad by David Corfield between 16 February and 28 November. The following year, Corfield and the author, located a pair on the same stream and it was hoped that they might attempt breeding but regrettably one was apparently shot. It was again sighted in the Sohan stream near Tret in May 1987 (R.F. Nana, pers. comm.) and another on the northern fringes of Rawal lake on 6-7 August 1987 (Mallalieu, *in litt.*). Status RARE RESIDENT.

HABITS

A somewhat secretive and shy species and hence easily overlooked. This kingfisher has very distinctive habits

Illus. 67: Crested Kingfisher or Large Pied or Himalayan Pied Kingfisher *Ceryle lugubris.* Male drawn from specimen at Rawal lake.

being very territorial. They stick to the small stretches of river and spend the greater part of the day perched in a leafy tree or bush usually close to the water but often unnoticed by passing humans. They feed, as far as is known, exclusively on fish which they obtain by plunging into the river from their perch. Sometimes they will perch on top of a large boulder in the middle of the river and often in stretches of the river where the current is swift and the water foam tossed, but they instinctively know which eddies and pockets of slack water are likely to harbour fish. They do not hunt by hovering in the air over open water as is characteristic of the smaller Pied Kingfisher. When disturbed or flying to a new 'beat', they characteristically fly very swiftly and low over the water but with comparatively slow wing beats. Salim Ali reports that they will catch and consume fish up to 17 cms. in length (Ali and Ripley, 1970). In the vale of Kashmir, they have earned a reputation for being harmful to trout hatcheries (Bates and Lowther, 1952). The shallow stream frequented by a pair at Rawal lake contained only small developing carp, *Labeo rohita* (identified by Dr. Ataur Rahim, Islamabad). They were 6 to 10 cms. in length. This kingfisher has a curious habit of wagging its tail up and down while perched.

BREEDING BIOLOGY

Bates and Lowther noted both sexes taking part in nest hole excavation and watched one such nest tunnel being excavated in a sandy cliff on 27 April near Salkalla (Azad Kashmir) (Bates and Lowther, 1952). The normal clutch is 4 to 5 eggs which are pure white, almost spherical in shape. An enlarged rounded nest chamber is excavated at the end of a horizontal tunnel which might extend to a depth of 1.5 m. (Bates and Lowther, 1952 and Yamashina, 1961). It is believed that the female does most or all of the incubation duties but both parents have been observed feeding their young (Austin and Kuroda, 1953). The eggs take about 20 days to hatch and the young leave the nest hole about 40 days after hatching (Cooper and Forshaw, 1983). The new born chicks are

Map 262: Crested Kingfisher or Large Pied or Himalayan Pied Kingfisher *Ceryle lugubris*

altricial (devoid of down and blind). They take about 28 days to fledge and when nearly fully grown, will come to the mouth of the nest tunnel to be fed. The nest chamber itself is not lined but soon becomes rather foul with a litter of regurgitated fish bones and scales.

VOCALIZATIONS

On the whole a more silent species than other kingfishers of the region. Fleming and Fleming describe their flight call as a nasal 'klik' (op. cit. 1976) and Yamashina writes that it is similar in tone to the call of the Greater Spotted Woodpecker and renders it as 'cket-cket'. Whistler also describes a loud 'ping' call similar to one of the calls of the Red-wattled Lapwing *(Lobatus indicus)* (Cited in Ali and Ripley's 'Handbook' 1970).

FAMILY MEROPIDAE

Bee-eaters

Little Green Bee-eater *Merops orientalis* Latham

Plate **22**
Map **263**

DESCRIPTION

Body length 21-23 cms. including elongated central rectrices, which extend 6-7 cms. (Sind specimens-Ticehurst) beyond outer tail feathers.
Wing length 8.9-9.9 cms.
Bill length 2.3-3.3 cms (Ticehurst)
Weight 15-18 gms. (India), 17 gms. (Sri Lanka) (C.H. Fry, 1984)

A bright green bird with a long slender down-curved bill of scimitar shape and elongated stiletto shaped central pair of tail feathers. The forecrown is bright golden yellow deepening to chestnut brown or bronzy orange on the hind crown (description of specimen collected February in Multan district). The wings are a mossy green-brown and the lower belly is olive brownish being less bright grass green than the rest of the head and body. The bill is black and also a broad loral streak from the base of the bill through the eye. The chin and throat are verditer blue (or copper green), narrowly framed below by a black line. The primaries are black tipped. In flight the shallow cambered wings look triangular, giving the bird a kite shaped silhouette and they are coppery bronze in colour. The irides are crimson and the legs and feet, which are weak and small, are yellowish brown. The sexes are alike. The outer or fourth toe is united with the longer middle toe and the inner or second toe is also joined at the base to the middle toe.

HABITAT, DISTRIBUTION AND STATUS

It occurs from northern and central Africa through Arabia, to Iran, south-western Asia and the Indo-Chinese region. Equally adapted to arid and open country or well wooded areas, it occurs throughout the plains of Punjab and Sind penetrating even the uncultivated desert tracts of the Thal, Cholistan and Thar deserts as well as the low hills of Sind Kohistan and Lasbela. It avoids extensive mountain or forested areas and occurs sparingly in the outer Murree foothills in summer, up to the sub-tropical pine zone (Kahuta at 1,200 metres/4,000 feet) (author obs.) Except for coastal Makran, it is no more than a straggler to Baluchistan, though it spreads in summer throughout the foothill and valley regions of Kohat, Bannu and the Peshawar vale. In the northern Punjab and the NWFP, most of the population is locally migratory, dispersing in winter, southwards to warmer regions where insect food is more readily available. Small parties can however, still be seen in January and February around Peshawar, and Kohat town in N.W.F.P. and Lahore in Punjab but there is a marked influx of birds into such regions from the beginning of March. They mostly move southwards again from early October (observations Multan district over 28 years). It is entirely absent from the northern Himalayan regions. Status ABUNDANT.

HABITS

They are often gregarious in winter, assembling in the evening into foraging flocks of 30 to 40 birds and roosting together in trees as darkness falls. They are, however, not very colonial in breeding, nest burrows often being excavated in isolation. Most of their food is obtained by aerial sallies returning to some high up and prominent perch in a tree or on a telegraph wire from which the surrounding air space is surveyed. Their flight consists of rapid wing beats followed by graceful twisting glides, their kite shaped wings enabling them to swoop upwards with great speed and agility or to bank steeply when in pursuit of larger insects. They capture all kinds of flying insects including Odonata and small beetles (*Myllocerus* spp). But their principal food prey consists of Hymenoptera including bees both *Apis indica, Apis florea,* Paper wasps *Polistes hebraeus* and Dipteran flies, the occasional beetle (Carabids, Weevils) and Lady birds (*Coccinella* 7 *punctata*) (Mason and Lefroy, 1912), based on examination of over 30 stomachs of birds collected in Bihar during July and August. In Malir they have been seen feeding on a small red dragonfly, often swooping low over a pool to do so.

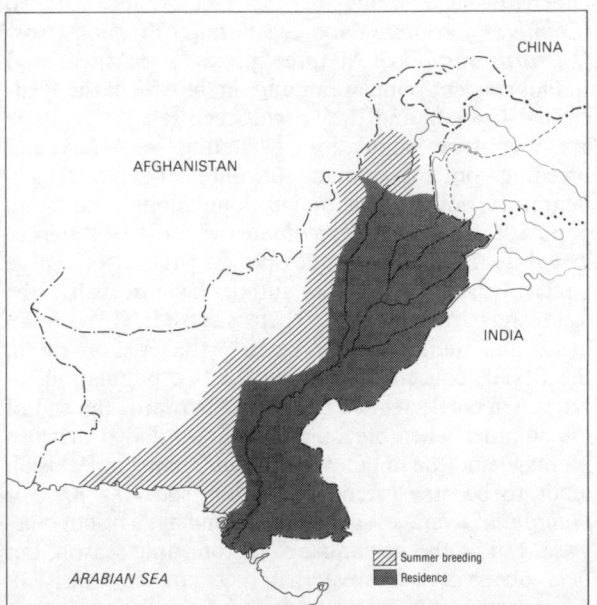

Map 263: Little Green or Small Green Bee-eater *Merops orientalis*

BREEDING BIOLOGY

Unlike other Bee-eaters of the region, this species is not very gregarious or colonial in its nesting habits. Both sexes share in nest excavation and they make a horizontal burrow, sometimes in a low lying earthern embankment of a small irrigation channel or even on relatively flat ground. More commonly, nests are excavated in the sand bank of some borrow pit. Besides digging with their bills, they kick the sand out vigorously with both feet as the tunnel progresses, often disappearing entirely inside and reversing to do this. These nest excavations have been watched on 27 March in Thatta district and 9 April in Rahim Yar Khan district. Kenneth Eates found most nests in Sind during May but also found eggs between March and June. The nest holes excavated by him, varied from 45 cms. in depth in hard gravelly soil to over 1.8 m. in silt banks (MS notes). The normal clutch is 5 to 6 eggs which are thin shelled and pure white and are roundish ovals in shape. The entrance hole of occupied nests reveal tell-tale parallel grooves or 'railway lines' made by the birds wing shoulders (carpal joint) as they shuffle into the nest hole. The young hatch asynchronously and presumably egg laying extends over a period of 10 days or more. Both sexes share in incubation duties and the young are naked upon hatching acquiring a plumage identical to that of adults before leaving the nest. The incubation period has not been recorded. The chicks are fed by both parents, wholly on insects and the unlined widened nest chamber soon becomes littered with the chitinous undigested parts of their insect diet. As in the kingfishers, no nest hygiene is practised, but their growing body feathers remain encased in quill sheaths until shortly before they are ready to leave the nest, which may be about 28 days after hatching.

*VOCALIZATIONS

They call both on the wing and when perched and like all the genus, their calls are rather pleasant liquid sounding notes, 'trri-trri'. Another variation is a slightly more rapid and shorter note 'tin-tin-tin' repeated four or five times.

Large Green Bee-eaters

TAXONOMIC NOTE

Two large green bee-eaters, of very similar appearance and habits, overlap in their breeding range mainly within, Pakistan. These are the Blue-cheeked or Persian Bee-eater *(Merops superciliosus persicus)* and the Blue-tailed Bee-eater *(Merops philippinus javanicus)*. The concept of a super species seems particularly helpful in defining the relationship between the two species, as their shared characters are more significant than their adaptive differences.

Originally in definitive works, they were treated as two separate species *Merops persicus* and *Merops philippinus,* e.g. in Blanford and Oates *(The Fauna of British India,* 4 Vols., 1889-98). Subsequently on the basis of close morphological similarities they were combined as sub-species *Merops superciliosus persicus* and *Merops superciliosus javanicus* (Stuart Baker, 'Fauna Series', Vol. 4, 1927). This treatment was followed by Peters in his 'Checklist of the Birds of the World' (Vol. 4, pp. 234-5, 1945). In 1950, Daniel Marien studying the Walter Koelz collection of *Merops* skins in the American Museum, opined that they were 'quite distinct species, both on the basis of distinct ecological separation, morphological differences and thirdly due to differences in moult patterns' (Marien, 1950, *JBNHS,* Vol. 49). Subsequent works (Ripley's 'Synopsis', 1961 and Revised ed. 1982, Ali and Ripley's 'Handbook' Vol. 4, 1970) followed Mariens revision, separating them as *Merops superciliosus* (Blue-cheeked) and *Merops* *philippinus* (Blue-tailed). Professor H. Fry considers that the Blue-cheeked Bee-eater should be considered a separate species, *Merops persicus,* regarding them as a distinct group within a single inter-continental super species (Fry, 1984 and *in litt.* to author 1987). There are close parallels between the population of these Large Green Bee-eaters and of the Pied-crested Cuckoo *(Clamator jacobinus)* and the Drongo or King Crow *(Dicrurus adsimilis).* All three species have African and Indian resident populations and, in the case of the Pied-crested Cuckoo and the Large Green Bee-eaters, there are migratory populations wintering in Africa and breeding on the Indian subcontinent with locally migratory but resident Indian populations. The entire population of *Clamator jacobinus* despite the differences in habits, is treated as one species. Studies on Large Green Bee-eaters by the author have revealed the following perplexing facts. There is considerable overlap in size and plumage colour (based on the series of skins in the BNHS collection) between the two populations of Large Green Bee-eaters particularly towards the end of the summer when breeding birds are in faded or worn plumage and the tail of *Merops philippinus* in Pakistan tends to become greener. Generally speaking *Merops philippinus* averages smaller in size and has a bright blue-green tail at the beginning of the breeding season, but field observations reveal that there are no consistent differences in rump or crown colour nor the amount of pale blue or white surrounding the black loral eye streak.

The ecological preferences of both species during breeding appears to be identical i.e., proximity to water and a good supply of Odonata (dragonflies) and suitable earth cliffs in open country for colonial breeding. A nesting colony near Sidhnai spill channel (Multan district) of *Merops superciliosus* was observed in two years to include 3 or 4 birds having all the plumage characteristics of *Merops philippinus*. Similarly a colony of Blue-tailed Bee-eaters near Rawal lake was noted to include at least 3 individuals having plumage characteristics similar to *Merops superciliosus*. There is a good deal of evidence that the two species overlap in the Punjab and that there is some admixture of breeding pairs, though actual hybridization has not been proved. Tape recordings of their calls (made by the author) have not been analyzed spectro-graphically but to the human ear sound closely similar.

Merops philippinus may be largely non-migratory in the eastern part of its range (south China, Thailand and Indo-China countries). In India it performs a relatively short migration, mainly from Sri Lanka and south India to the northern plains area where it breeds. It is suggested that climatic factors over its distribution may have resulted in a greater concentration of Carotenoides and Melanins in plumage pigments associated with relatively higher humidity, whilst the short migration distance enables the moult to be completed in one phase after breeding. In the case of *Merops superciliosus* the long migration distance, in some instances from south of the African equator to Soviet central Asia, has resulted in a two-phased moult pattern, with eclipse plumage being assumed in the breeding grounds and the flight feathers being renewed on their wintering grounds. This is a similar moult pattern to *Merops apiaster,* another long distance migrant. Differences in moult patterns within the same species according to geographical races have been noted in other species (Landsborough Thomson, 1984. *Dictionary of Birds* p. 488) and seem to be the only non-controversial basis for specific separation of these two Bee-eaters in Mariens paper (op. cit., *JBNHS,* 1950). As will be seen in the detailed species accounts, the breeding range of these two species is largely allopatric but the question of hybridization where overlap occurs certainly deserves further study.

Blue-cheeked Bee-eater *Merops superciliosus* Linnaeus
Persian Bee-eater *Merops superciliosus persicus* Pallas

Plate **22**
Map **264**

Merops persicus C.H. Fry, 1984

DESCRIPTION

Body length 28-31 cms.
Wing length 14.5-16.4 cms.
Bill length 3.9-4.5 cms. (Ticehurst), 3.8-4 cms. (females)
    (C.H. Fry. 1984)
Central tail feathers 13.5-14.8 cms.
Outer tail feathers 8.8-9 cms.
Weight 41.7-57 gms. (Lake Chad, Africa),
    46-53.7 gms. (Spring, Kazakhstan) (C.H. Fry, 1984).

It is a slender graceful bird with elongated central tail feathers, comparatively large head borne on a very short neck and sharply down-curving slender black bill. Bigger than the Common Green Bee-eater *(Merops orientalis)* it is much slimmer in build than the European Bee-eater *(Merops apiaster)*. It also has a comparatively longer, more slender bill than *Merops apiaster*. Overall its body colour is bright green but with coppery hues on the flight feathers. The chin and throat are butter yellow turning gradually brick red or chestnut on the lower throat. A broad black line through the eye is framed above and below by pale aquamarine blue, becoming almost white on the forecrown. The central rectrices extend for 6 to 7 cms. beyond the outer tail feathers and the tail is green not blue with the rump and upper-tail coverts rather more blue-green. The irides are red and the feet are fleshy to plumbeous with the middle toes fused to the second and fourth outer toes in the basal part. Compared with *Merops philippinus,* the Blue-tailed Bee-eater, the chestnut throat patch is usually more restricted in area, whereas the copper sulphate blue area around the eye is more extensive and conspicuous becoming more whitish on the forecrown. In flight the under-surface of the wings is coppery with the flight feathers tipped blackish. The sexes are alike but females average slightly smaller in size.

HABITAT, DISTRIBUTION AND STATUS

This species winters over a wide range of countries in east Africa, both north and south of the Equator. It breeds in north Africa, Israel, Iran, Iraq and south-western USSR from the Caucasus to Kazakhstan. They cross the Arabian Sea arriving on the Karachi and Makran coasts in about the third week of April up to the end of the first week of May. Observations over a nine year period from Malir (about 4-5 miles inland from the seacoast) indicate that the majority of birds arrive in small groups or parties, flying during the hours of darkness. They habitually call to each other to maintain contact during migration. Over 25 such records are of birds heard calling between 1930 hours and 0330 hours. Only 3 observations are of birds arriving over the coast before

1800 hours or after 0800 hours in the morning. These noted arrival times corroborate those made by K. Eates, an ornithologist who was stationed in Karachi most of the 1940s. He wrote that he had regularly heard first migrant arrivals during the hours of darkness in early summer. They spread through most of the Indus plains for breeding from central and southern Sind up to lower and central Punjab. They are closely attached to swamp and wetland areas for nesting because of their food preferences (see below) and consequently generally avoid dry mountainous areas. Some of the migrant population passes through the northern Himalayan regions to breed in central Asian regions of the USSR. (specimens collected in Gilgit by. J. Scully and flocks seen by him in late autumn. *Stray Feathers*, 1887). They also pass through northern Baluchistan, colonies staying to breed around Sheikh Wasil and Saranan in Quetta district and Kushdil Khan lake (small numbers 5 to 10 pairs) and Zangi Nawar lake (estimated 200 birds) in the Chagai. (Ticehurst, 1927 and author obs. 1974, 1984 and 1985). In the Potohar and Attock district of the Punjab and Peshawar district of the NWFP they are largely replaced by the Blue-tailed Bee-eater *(Merops philippinus)*. Whitehead found it fairly common around Bannu in summer *(JBNHS,* 1911) whereas Whistler did not record it from Rawalpindi district *(Ibis,* 1930). H. Waite found breeding colonies around Nammal lake in the Salt Range but noted only *M. philippinus* further east near Choa Saidan Shah in Jhelum district *(JBNHS,* 1948). In southern Punjab the species predominates. They do not normally stay to breed in lower Sind but from mid-August there is a return influx of birds into the rice growing tracts of Thatta and Badin districts and a few of

these birds linger until late October. Latest date autumn migrants 1 December near Mirpur Sakro. Three pairs were still noted on 30 October at Band Kushdil Khan, Baluchistan. Migrating flocks have been seen passing over the Hab valley at a great height on 7 November at a time when the local population had long since departed. Status summer breeding visitor. Status COMMON to ABUNDANT.

HABITS

These bee-eaters are purely insectivorous and in Pakistan are specialized feeders upon Odonata (dragonflies). They have also been noted feeding upon wasps *(Polistes hebraeus)* and butterflies *(Danais chrysippus)* particularly upon first arrival overland (Karachi) after spring migration (observations at Malir). No doubt they catch and eat Hymenoptera and Diptera whenever available but the dominant food prey is dragonflies. Breeding colonies along the Omani coastline (N. E. Africa) were found to subsist entirely upon grasshoppers and cicadas in equal proportions (C.H. Fry *in litt.*, 1987). They are colonial or gregarious in foraging and roosting, and frequently utilize tree tops, telegraph wires and fence posts near their nest colony or feeding areas, as look-out and feeding points, hunting by sallying out into the air, rising with rapid wing beats and then gliding until an insect is spotted, whereupon they twist or dive abruptly. The captured insect is beaten or rubbed on a perch, but if small, may be eaten while still in flight. They have been recorded preying exclusively upon Honey Bees *(Apis mellifera)* when this insect is plentiful. Studies in Africa showed that 65.5 per cent of the diet comprised Odonata, 17 per cent Hymenoptera and 17 per cent other insects (cicadas, assassin bugs, beetles, mantids and grasshoppers) (C.H. Fry, *The Bee-eaters,* 1984).

Generally they show strong attachment to lakes and swamps but occasional small breeding colonies have been found in dry wadis near the coast in Lasbela (author) and in Zhob district of northern Baluchistan. Similar situations were noted by Holmes and Wright in Jacobabad district *(JBNHS,* 1968). In stormy weather and generally at night time, they roost inside their nest burrows. Outside of the breeding season they roost in trees often huddling in bodily contact.

BREEDING BIOLOGY

Pair bonding is monogamous and nesting is colonial, with burrows being dug into low sand or silt banks and occasionally into fairly flat ground. In various parts of Sind province K. Eates excavated over one hundred nest holes varying from 0.9 metres to 1.5 metres in depth and found the maximum clutch to be 5 eggs but clutches of 4 well incubated were not uncommon. Most egg laying in northern Sind started from late June with a peak in early

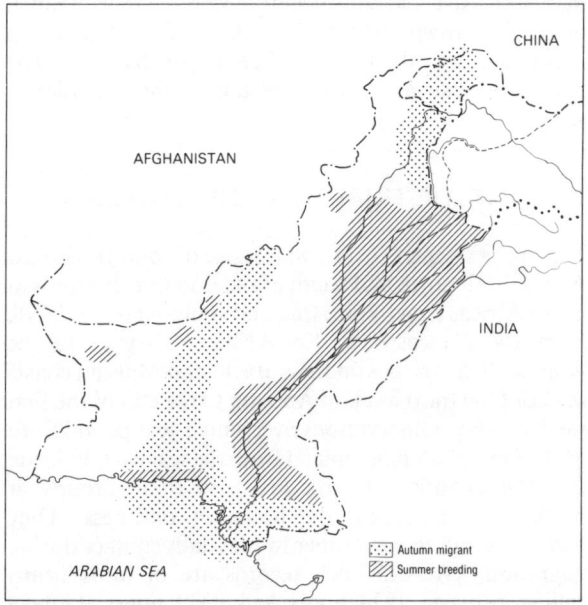

Map 264: Blue-cheeked Bee-eater *Merops superciliosus*

July (K. Eates, *JBNHS,* Vol. 40, 1939). He found most colonies comprising 20 pairs or fewer. In one colony near Rohri (Sukkur district) all nests excavated had fresh eggs on 11 July, suggesting fairly synchronous nest building activity. In Baluchistan, Meinertzhagen found clutches of 3 and 4 eggs on 2 June and many colonies were noted in central and coastal Makran (Ticehurst, *JBNHS,* Vol. 31, 1927). Williams and Williams found a small colony at Yaru in Pishin on 6 July with each nest containing young (*JBNHS,* 1927). Copulation was observed by Derek Holmes on 11 May at Jamrao headworks, east Narra (pers. comm., 1967). The female does most of the incubation which starts with the first egg laid. In the day time the female is relieved after 3 hours spells of incubation by her mate for shorter half-hour spells (C.H. Fry, op. cit., 1984). Incubation period is 24 to 26 days. The chicks hatch naked and with eyes closed. They have flesh coloured bills and legs at this stage (K. Eates, MS notes). The enlarged nest chamber has plenty of room for the adults to turn around, i.e., may be 9 cms. wide and 12 cms. high and 15-20 cms. long (K. Eates. MS notes). No nest lining is made and no nest hygiene is practised, the chamber becoming littered with regurgitated insect remains, and the sides smeared with faeces. The chicks have their growing feathers protected by sheaths which protect their feathers for about the first half of their fledging period, i.e., for about two weeks. Period from hatching to full fledging is believed to be near 30 days.

*VOCALIZATIONS

The flight and contact call frequently uttered while on the wing and also from a perch is a loud and liquid 'quirrip-quirrip' lower in pitch than the calls of *Merops orientalis.* It is a far carrying call, audible in calm weather for up to 1.6 kms. (1 mile) distance.

Blue-tailed Bee-eater *Merops philippinus* Linnaeus

Plate **22**
Map **265**

Merops superciliosus javanicus Horsfield in Stuart Baker, 'Fauna Series'

DESCRIPTION

Body length 28-31 cms. including tail streamers (Stuart Baker)
Wing length 13-14 cms. (males), 12.7-13.5 cms. (females) (C. H. Fry, 1984) 12.1-13.9 cms. (Stuart Baker)
Bill length 3.6-4.8 cms. (Stuart Baker)
Central tail feathers 12-14.5 cms.
Outer tail feathers 8.4-8.7 cms.
Weight 32-45 gms. (Malaysia, both sexes) (C. H. Fry, 1984).

They are shorter winged and average slightly smaller in size than the Blue-cheeked Bee-eater, otherwise they are closely similar in body colouration, except for the tail which is blue. The throat is butter yellow, changing gradually in the lower throat and upper breast to brick red or chestnut which is generally more extensive and more conspicuous than in the Blue-cheeked Bee-eater. The wing coverts, mantle, crown and lower breast are green. A pale superciliary line and stripe below the black eye band, is less whitish and generally less conspicuous than in *Merops supercilious.* The bill is black and the iris crimson. The small weak legs and feet are dark brown and the middle toe is fused in its basal portion with the two other toes. Birds seen in Malaysia (author) were more brightly coloured with deeper blues and yellows, than specimens which visit Pakistan in summer.

In flight their under-wings are coppery bronze with each flight feather narrowly black tipped. The sexes are alike.

HABITAT, DISTRIBUTION AND STATUS

This is an Oriental species occurring widely in south-east Asia, from India, south-eastern China and Indo-China.

Wintering in the Philippines and Celebes it is a summer migrant visitor to the north-eastern parts of Pakistan bordering the foothills. Noted 24 April at Balloki headworks, 72 kms. (45 miles) south of Lahore. Whilst the earliest date recorded in the vale of Peshawar is 2 April (Briggs and Osmaston, 1928). It breeds colonially in Rawalpindi, Jhelum and Sialkot districts of Punjab, and in Peshawar and Mardan districts of the NWFP,

Map 265: Blue-tailed Bee-eater *Merops philippinus·*

and probably also in parts of Lahore and Kasur districts. Whitehead recorded it as very common around Rawalpindi and not uncommon around Bannu but only saw *M. superciliosus* around Kohat (op. cit., 1911). Hugh Whistler only recorded *M. superciliosus* around Jhang (op. cit., 1922). H. Waite collected specimens around Dandot and Choa Saidan Shah in Jhelum district of the Salt Range, but only found *superciliosus* in Mianwali district. The author noted foraging flocks of this bee-eater at Taunsa barrage on the Indus (Dera Ghazi Khan district) on 4 November but these may have been on southward passage. It does not extend southwards into Sind and in Multan district the common breeding bee-eater is *M. superciliosus*. Shortly after breeding this bee-eater migrates in a south-easterly direction across northern India. Latest date seen in Pakistan, 4 November. Part of the population in Malaysia and Thailand is resident and in northern India it is largely a summer migrant, wintering in southern peninsular India, Sri Lanka and possibly Burma and Assam. Status COMMON only in northern Punjab.

HABITS

Gregarious in foraging and colonial in breeding. Observations at a nest colony near Rawal lake indicated that more than 80 per cent of food prey brought back to the colony comprised Odonata with 6 different dragonfly species clearly discernible but not specifically identified. Other prey included the Yellow wasp (*Polistes hebraeus*) and Hornets (*Vespa orientalis*). In Nepal they have been noted catching Pierid butterflies over a chickpea (*Cicer arietum*) crop, (Fleming and Fleming 1976). Examination of 13 stomachs of specimens collected between July and September in Bihar, India, included a high proportion of bees (*Apis indica* and *Apis florea*) as well as dragonflies (*Crocothemis servillia*), and Leaf Cutter Bees, *Megachile carbonaria* (Mason and Lefroy, 1912). The author saw one at Yalla in Sri Lanka battering a huge Carpenter Bee (*Xylocopa dissimilis*) on its wire perch. They catch most of their prey by aerial sallies and are accomplished gliders and flyers. Prey is

taken to a convenient post and if venomous, often rubbed or wiped before swallowing. Dietary habits vary as Odonata predominate in Pakistan but in Malaysia Hymenoptera predominate (C.H. Fry, *The Bee-eaters*, 1984). They roost at night in their nest burrows during the breeding season but while on passage congregate at night in trees.

BREEDING BIOLOGY

Occasionally found nesting solitarily. H. Waite found one such nest in the bank of a stream at Choa Saidan Shah in the Salt Range. The burrow contained 5 incubated eggs on 18 May (*JBNHS*, 1984). Usually they nest colonially. Near Rawal lake, a colony of about 20 pairs had tunnels in a 15 metre high loess earth cliff. The burrow entrances tended to be rectangular in shape and all were on the western aspect of the cliff, none on its southern face. Whistler excavated nests in Rawalpindi district and found the burrows to vary from 1.5 up to 2 m. in depth. (Whistler 'Popular Handbook', 1949). Both sexes take part in nest excavation and incubation of the eggs which are plain white, rather rounded ovals. A clutch of 6 eggs is normal, sometimes 7 (Stuart Baker, Vol. 4, 1927). David Corfield observed copulation by several pairs between 23 and 25 May (pers. comm.) near Rawal lake and most egg laying starts mid-May. Incubation period is thought to be 24 to 26 days and the male sometimes feeds the female during this period as she does most of the incubating. The young hatch naked and blind and develop comparatively slowly taking about 25 to 30 days to fledge.

VOCALIZATIONS

Very similar to the calls of *Merops superciliosus*. Comprising liquid trilling 'treep-treep-treep' calls but these are slightly higher pitched than those of the Blue-cheeked Bee-eater. When excited the notes are elided into a short warble.

European Bee-eater *Merops apiaster* Linnaeus

Golden Bee-eater of USSR

Plate **22**
Map **266**

DESCRIPTION

Body length 27 cms.
Wing length 14.5-16 cms. (Stuart Baker)
Bill length 3.5-4.4 cms.
Central tail feathers 13.5-14.8 cms.
Outer tail feathers 8.8-9 cms.
Weight 51-78 gms. (both sexes)
   (Kazakhstan and Western Europe) (C.H. Fry, 1984)

It is a heavier bodied, slightly larger bee-eater than *Merops superciliosus* and its central tail feathers are less pin pointed and comparatively shorter. In fresh breeding plumage upon arrival in Pakistan, it is a truly gorgeous bird with heavy black eye stripe, butter yellow throat encircled by a narrow black line joining the eye. The immediate forecrown is greenish white and rest of the crown, nape and mantle is bright russet chestnut and the breast is greenish blue or turquoise. The scapulars are creamy tan or russet yellow and the tail is greenish blue. The legs and feet are small and weak and brown in colour. The iris is crimson and the sharp pointed and down-curved bill is black and comparatively shorter and stouter than in *Merops superciliosus* or *Merops philippinus*. The sexes are alike.

HABITAT, DISTRIBUTION AND STATUS

It breeds from Spain across southern Europe to Afghanistan and southern USSR from Transcarpathia to the Altai. Summer migrant visitors are confined to the dryer mountainous regions of Pakistan arriving over the Makran coast from their wintering grounds in central and southern Africa in mid-April. They stay to breed from central Baluchistan (Kalat, Chagai, Khuzdar) through the NWFP, Chitral northwards to lower Swat and Indus Kohistan. Some of the migrant population also continues northwards into central Asiatic Russia to breed. The author noted them as extremely abundant and widespread throughout Zimbabwe in Africa up to late March. In northern Baluchistan where they breed, during a 322 km. road journey on 29 April, none were noted but on the return journey on 2 May, 4 or 5 small groups were conspicuous, perched on telegraph wires indicating recent arrival. Return passage starts from late July, flocks being seen over Islamabad on 25 July and over Sakesar in the Salt Range on 23 to 28 August in four different years (H. Waite, specimens in the British Museum). Autumn migration was also noted to start in late July from around Quetta (Nurse, 1902). They have also been heard and seen on spring passage as early as 30 March at Chinji in the Salt Range (author) and over Rawal lake as late as 15 May. (D. Corfield, *in litt.*, 1982). Surprisingly neither Scully nor Biddulph found evidence of breeding in Gilgit, but in 1982 the author found a huge breeding colony (estimated at over 100 nests) at Chilas on the Indus (Gilgit) and another in the northern part of Indus Kohistan about 64 kms. downstream in the Indus valley. They have not been noted in Dir, but a few were noted on the outskirts of Mingora in Swat in late May (author obs.).

A few colonies breed in Chitral (Perreau, *JBNHS*, Vol. 19, 1910). Whitehead noted that they breed freely in the Kurram valley from about 900-2,100 metres. In Baluchistan it is only a passage migrant in the Makran but Christison noted it breeding from around Nushki up to Shingar in the Zhob district (Christison, 1941 and 1942). The author noted small colonies, usually 3 to 4 pairs, in earth banks of erosion gullies in Pishin, Loralai and Muslim Bagh in different years. Status COMMON.

HABITS

Largely gregarious in foraging and breeding though not in such large gatherings which are typical for *Merops superciliosus*. They are mainly dependent upon hymenopterous insects for their food and are strong flyers, hunting both in the upper air space and occasionally low over the ground, sailing around in graceful glides, calling to each other frequently as they hunt. They often catch insects with an audible bill snapping and generally return to a low bush or rock perch to de-sting and consume their prey. Around Quetta, Meinertzhagen noticed their food as comprising

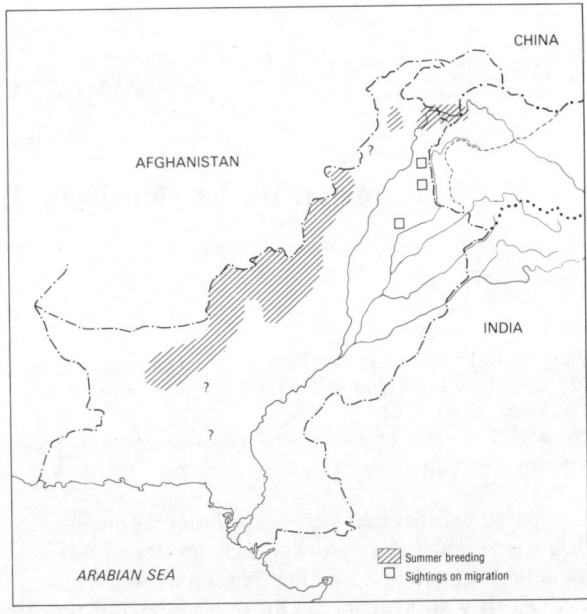

Map 266: Golden or European Bee-eater *Merops apiaster*

mostly wasps, hornets (*Vespa orientalis*) and bees (*Apis mellifera* and *Apis indica*) (*Ibis*, 1920). Insect prey is usually seized from underneath by an upward swoop and clasped in the tip of the mandibles. Venomous wasps and bees are jiggled around in the bill until held head first and then wiped on the perch to de-venom them before being swallowed (C. H. Fry, pers. comm., 1982). If the bee escapes during such jiggling, the bee-eater has no difficulty in darting after and recapturing it. Ticehurst recorded that they are a great nuisance to bee-keepers around Mastung and Kalat as reported by Colonel Hotson. He also noted that they catch the Yellow Wasp (*Polistes hebraeus*) and Hornets *(Vespa orientalis) (JBNHS*, 1927). Major Nurse also recorded this wasp as being captured by birds around Quetta (Nurse, op. cit., 1902). In the USSR, they have been reported as serious predators on honey bees (Dement'ev and Gladkov, 1951) and a very full account of their food and feeding habits, is presented in *Birds of the Western Palearctic* (Vol. 4, 1985), showing that wasps and bees usually form a major component of their diet. A colony nesting on the Omani coast was feeding exclusively upon cicadas and grasshoppers (Acrididae) (C. H. Fry, *in litt.*, 1987).

BREEDING BIOLOGY

They breed colonially, excavating horizontal tunnels in earth cliffs or sloping tunnels if on flat ground. Burrows average 1.5 metres in length and may take more than 14 days to excavate. Birds were watched excavating nest holes on 26 April near Kach, Baluchistan. Soil is loosened with the bill and then kicked out backwards with the feet. Breeding in Europe has been well studied (L. Koenig, 1960, Priklonskiy and Lavrovskiy, 1974 and

Von Blotzheim and Bauer, 1980). Meinertzhagen, around Quetta, noted nest hole excavation from 15 April and found the first eggs on 9 May (*Ibis*, 1920). Males start courtship feeding during nest excavation. The normal clutch is 4 to 7 eggs laid at intervals of 1 to 2 days. Incubation continues 26 days from completion of the clutch. The spherical eggs are plain white and incubation starts with the first egg laid and is mainly carried out by the female. The young are altricial and hatch at 1 to 2 day intervals. The nest chamber becomes foul as the sides get plastered with excreta and the floor littered with chitinous remains of insects. Adults do not remove faecal sacs and the nest acquires a powerful stench of ammonia (Cramp *et al.*, Vol. 4, 1985). The chicks open their eyes at 8 days of age and fledge at 30 days. In Kashmir, Bates and Lowther recorded the normal clutch as 5 or 6 eggs with 8 the highest number found and noted that nest tunnels generally sloped slightly upwards. The nest hole tended to be rectangular and between 0.75 and 1.25 metres long (op. cit., 1952). In the USSR, Dement'ev noted that the breeding pair excavated up to 12 kgs. of earth and took up to 20 days to complete nest excavation (Dement'ev and Gladkov, 1951). The nestlings have their growing feathers protected by sheaths for most of their time in the nest burrow so that they look spiny like porcupines.

*VOCALIZATIONS

The call of this bee-eater is noticeably lower in pitch, more melodious to the human ear, and with a more rolled component than the Large Green Bee-eaters. It is a double 'quilp-quilp' or tre-tre', slightly evocative of the calls of *Pterocles exustus*.

FAMILY CORACIIDAE

Rollers, Blue-jays

Eurasian Roller (Kashmir Roller) *Coracias garrulus* Linnaeus

Plate **22**
Map **267**

DESCRIPTION

Body length 31 cms.
Wing length 18-21- cms. (Stuart Baker), 19.6-21 cms. (males), 19.6-20.5 cms. (females) (Cramp *et al.*, 1985)
Tail length 12.2-13.5 cms.
Weight 127-135 gms. (males), 113-190 gms. (females) Sub-species *semenowi* (Cramp *et al.*, 1985)

Compared to other members of the order Coraciiformes they have comparatively strong well developed feet, but the outer toes are fused in their basal part to the central toe, i.e., they are syndactyl. This roller is a stout-bodied bird with large head and short neck and a powerful,

slightly down-curved bill, hooked at the tip. The bill is black and the iris pale brown. The sexes are alike in plumage and in the breeding season, they have turquoise blue-green throats and breasts extending around the neck and ear coverts, nape and hind crown. The forecrown is usually rather 'frowzled' in outline and of a pale whitish chestnut whilst its back and wing shoulders are pale reddish fawn or sandy without any streaking. The tail and flight feathers are also pale aquamarine or turquoise with the outer tail feathers slightly longer and pointed (only visible at close range) and tipped dark cobalt blue. Also the upper-wing shoulder (formed by the lesser wing coverts) showing dark blue when the bird

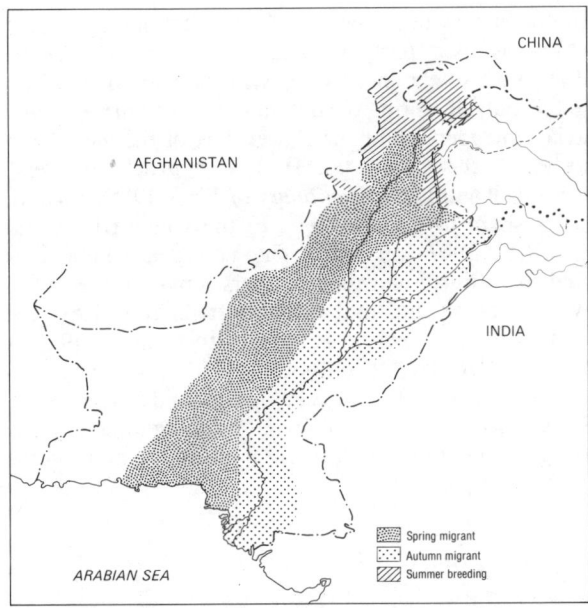

CHINA

AFGHANISTAN

INDIA

ARABIAN SEA

▨ Spring migrant
▨ Autumn migrant
▨ Summer breeding

Map 267: European Roller or Kashmir Roller *Coracias garrulus*

is at rest and as a dark blue line on the leading edge of the wing in flight. The central rectrices are blackish green and the tips of the primaries and secondaries are dark blue-black. The throat and upper breast show pale blue shaft streaks and the under-tail coverts tend to be whitish. Immature birds are more sandy brown on throat and upper breast and generally duller. There is also a partial moult into eclipse plumage after breeding with the autumn migrant population, as it passes through the Indus plains, showing similar dull brown plumage to the juveniles. The legs and feet are dull greyish yellow. In flight, it can be distinguished from the Indian Roller *C. benghalensis* by the lack of continuous dark blue terminal tail band and dark blue band through the middle of the wings, both conspicuous features in *C. benghalensis*. Also by its blue throat and upper breast (lilac or chestnut in *C. benghalensis*).

HABITAT, DISTRIBUTION AND STATUS

This roller is a summer breeding visitor to the dryer mountain valleys of the north and north-western regions of Pakistan with part of the population continuing on migration further north to breed in central Asia. They also breed in southern Europe, northern Africa, the Middle East and Iran. They winter both in east Africa and in southern Africa and most of the Pakistan population begins to arrive from early May, travelling along a comparatively western flight path through the Arabian peninsula and gulf, but returning in autumn by a more eastern route which largely follows the Indus basin. In Baluchistan, Meinertzhagen noted first arrivals on 7 April (*Ibis,* 1920) and a loose aggregation of 9 birds

was seen (author) near Pishin on 8 May and scattered individuals in Zhob district, northern Baluchistan on 1 and 2 May still on migration (author). In Mansehra district of the NWFP in the low eroded hills between Balakot, the Batrasi pass and Mansehra, where they breed, numbers have been noted building up during the first and second weeks of May. They do not stay to breed in Baluchistan but do breed in small numbers around Kohat and Bannu in the NWFP as well as the Malakand Agency, Dir, main vale of Swat, in Chitral and also in Gilgit in the lower valleys. It has been noted sparingly in Chilas and all along the Karakorum highway in summer. Surprisingly it has not been noted in Baltistan (W.H. Mathews, 1941). H. Whistler noted a few pairs staying to breed in the loess earth cliffs of the Potohar in Rawalpindi district (*Ibis,* 1930). Whitehead noted it breeding up to 900 metres in the Kurram valley and as common around Thal (op. cit., 1911). Biddulph (1881) noted first arrivals in Gilgit on 30 April. In its breeding grounds, it rarely occurs below 900 metres (3,000 feet) or above 1,300 metres (4,500 feet). The return autumn passage starts very early with a few individuals noted passing through the Murree hills (Tret-Gora Gali) from the first week of August. Whistler (*Ibis,* 1922) noted a strong autumn passage through Jhang (Punjab) between 25 July and 25 September and the author noted a similar strong passage around Khanewal from the first week of August with the latest date seen, 10 September (28 years of observation). In lower Sind, they can be seen from 30 August and they are in good numbers throughout the first half of September with the latest date seen 7 November near Karachi University. Occasionally birds have been noted on spring passage in Sind. Two were seen near Khairpur, Sukkur district, on 10 April and two were noted on 26 April at Malir. Status FREQUENT.

HABITS

They are often sympatric with the Indian roller in the lower outer foothill regions around northern Punjab and NWFP. They are territorial with a long lasting pair bond and during breeding, can be seen aggressively chasing off such hole nesting species as Mynas *(Acridotheres tristis)* and even the equally aggressive tree nesting Drongo *(Dicrurus adsimilis).* They are rather omnivorous in diet and spend most of the day on some prominent perch near open ground where they can watch for prey. Small lizards (Lacertids), frogs and orthopterous insects are major items in their diet. They will also take beetles, locusts and crickets. Stuart Baker also records them as catching fish fry from drying out pools and also field mice (*Apodemus* spp.) ('Fauna Series' Vol. 4, 1927). Bates and Lowther recorded a pair feeding their young mainly on small frogs (op. cit., 1952). Most food is captured on the ground after a graceful swoop from a nearby perch. On the ground they progress by hopping. Large prey is taken back to the perch, usually on an overhead telegraph wire, and battered on the wire before being swallowed. Their flight consists of rather slow stiff wing beats but is strong and they occasionally take insects on the wing. On passage, they are silent but during breeding and territorial displays, they are quite noisy.

BREEDING BIOLOGY

Like all the genus, males perform spectacular aerial displays around their nesting territory. These include swooping, soaring and stalling accompanied by loud raucous calling. Bates and Lowther once noted a bird perform a complete aerial loop (op. cit., 1952). They are cavity nesters and in Kashmir have been found utilizing natural holes in trees as well as holes in earthern banks, particularly favouring eroded, enlarged nest holes of bee-eaters or kingfishers on river banks (Bates and Lowther, op. cit.). In Pakistan along the Karakorum highway and Kaghan valley approach road, a favoured site for nesting is in the 'weep' holes built in the stone work bridges or culvert revetments. Here they vie with mynas for possession of choice holes. Both parents take part in nest excavation where there is suitable soft soil. Most nesting starts in mid-May and eggs may be laid in the latter part of May. No nest lining is brought and the normal clutch is 4 to 6 eggs, (most commonly 4 in Kashmir,) (Bates and Lowther, 1952). These are white, glossy and rather spherical in shape. As is characteristic of the family, no nest hygiene is practised, the nest cavity soon becoming

foul with excreta on the walls and a litter of regurgitated chitinous insect parts or bachtracian bones. The female does most of the incubating with the male perched on guard outside during the daytime, ready to drive off any avian intruder. The male shares part of the incubation duties in this family (C. H. Fry in article, 'Rollers', Campbell and Lack, *Dictionary of Birds*, 1985) and both sexes share in bringing food to the young (Bates and Lowther, 1952, op. cit.). The young are nidicolous (naked) at birth and their feathers remain protected in waxy sheaths until they are almost ready to leave the nest. Incubation period is thought to be about 19 days (Ali and Ripley, 'Handbook', 1970) and period to fledging between 3 or 4 weeks (Austin and Singer, 1961). Both parent birds were observed bringing food to newly fledged young (4 birds), recently out of the nest, on the Malakand pass on 20 July.

VOCALIZATIONS

It utters rather harsh grating cries somewhat similar to the calls of *Coracias benghalensis*. During nuptial displays, these calls are quickly repeated 4 or 5 times rising to a crescendo 'kee-kee-keeah'. Dr. Scully writing of Gilgit town said that this roller 'makes the day hideous with its constantly uttered harsh grating cry' (*Stray Feathers*, 1887). They will also call from a perch. Their calls are rather corvid-like in tone.

Indian Roller or Blue Jay *Coracias benghalensis* (Linnaeus)

Plate **22**
Map **268**

DESCRIPTION

Body length 31-35 cms.
Wing length 17.7-20 cms. 18.4-20 cms. (males), 17.5-19 cms.
(females), (nominate sub-species) (Cramp *et al.*, 1985)
Tail length 11.7-14 cms.
Weight 166 gms. (males),
166-176 gms. (females) (Ali and Ripley, 1970)

Like the Eurasian Roller this is a rather stoutly built crow-like bird with comparatively large head and short neck. Its bill is broad at the base and slightly down-curved at the tip, being brownish black on the upper mandible. The legs and feet are brownish yellow with the outer toe partly free, but like the inner one fused to the middle toe (syndactyl). The ear coverts, chin and throat are vinaceous or violet grey with pale shaft streaks. The forecrown is whitish buff and midcrown turquoise blue. The nape, mantle and wing coverts are sandy brown, less chestnut in tone than the same region in *Coracias garrulus*. The breast is paler rufous with only the lower flanks turquoise. The square tail is banded basally and terminally with dark cobalt blue and in the middle pale green-blue. The central rectrices are greyish black. The lesser wing coverts (top of wing shoulder) are dark blue

and also the tops of the primaries and a wing bar extending from the base of the primaries through to the secondaries. In flight, the green-blue wing shoulders and contrasting pale and dark blue flight feathers make a striking and beautiful pattern. It can be distinguished from the Eurasian Roller generally by always having a rufous chestnut nape and streaked lilac and chestnut throat, also by its terminal dark blue tail bar and the dark blue bar in mid-wingtip. The sexes are alike. The irides are brown.

HABITAT, DISTRIBUTION AND STATUS

It occurs from Oman in the west, across southern Iran and east to the Indo-Chinese region. A very widespread and somewhat sedentary bird throughout the Indus plains, extending into the lower valleys of the NWFP and westwards along the Makran coast in Baluchistan. Absent from the inner Himalayan valleys, there is some local migration northwards of breeding birds into the subtropical pine zone of the Murree foothills and into Hazara district and Swat where it is often sympatric with the Eurasian Roller in the summer months. They are summer breeding visitors around Kohat and the lower Kurram valley as well as around Kahuta (Rawalpindi district). They are absent from extensive desert tracts as well as from the higher hill tracts of central and northern Baluchistan. Status COMMON.

HABITS

They are territorial both in winter and during the breeding season. Their preference is for rather open country including intensively cultivated tracts, canal colonies or semi-desert and they avoid extensive desert tracts. Their hunting technique involves mostly sitting on a conspicuous over-head perch (telegraph wires are strongly favoured) and watching for ground moving prey which they seize after gliding gracefully to the ground. They can hop along the ground and their powerful hook tipped bills enable them to seize a variety of larger Coleoptera and Orthoptera besides lizards and amphibians. In a study by Pusa Agriculture Institute, stomachs of 17 birds were examined in all months between January and June and again in October. Most of the food prey was found to be insects especially crickets and grasshoppers *(Acridium aeruginosum, Gryllotalpa africana* and *Chrotogonus* spp. and also *Trox* and *Myrmecocystus* beetles), (Mason and Lefroy, 1912). They are of great benefit to agriculture in the high proportion of injurous insects in their diet, including caterpillars of Noctuid moths and larvae of

Melolonthidae beetles (Mason and Lefroy, op. cit.). They also catch frogs and lizards when opportunity is offered and will rob other birds' nests of both eggs and young. Near Khabbaki lake in the Salt Range, the author watched one battering on a wire and swallowing with apparent difficulty a large frog. At Lehtrar, a pair with a tree hole nest in a *Pinus roxburghii* was watched hunting cicadas but generally they do not hunt insects on the wing, preferring to catch them on the ground. They are usually not gregarious and are generally solitary in the winter months. They are more often active hunting in the early morning and evening, catching insects right up to dusk.

BREEDING BIOLOGY

The males start territorial displays from mid-February in the Punjab. This consists of climbing flights with steep swoops and upward stalls, occasionally with complete aerial loops. These displays are accompanied by loud croaking calls. The pair bond is monogamous and unlike *Coracias garrulus,* a collection of grass, bits of paper and rag are carried by both sexes to line the nest which is usually in a natural tree cavity. Occasionally holes in clay or sand cliffs are utilized and Eates found nests at Sukkur in the walls of deserted buildings (MS notes). Scrope Doig (*Stray Feathers,* 1879) found nests in April and May in the east Narra. At Khanewal, a pair started carrying nest material to a hole in a Siris tree *(Albizzia lebbek)* in mid-March. The normal clutch is 4, rarely 5 eggs which are plain white and glossy roundish ovals in shape. There is apparently considerable interval between egg laying, and the chicks hatch asynchronously. Incubation is mainly by the female with the male perching on guard near the nest site. His aggressive chasing away of birds as big as Black Kites *(Milvus migrans)* and House Crows *(Corvus splendens)* observed at Khanewal, often reveals the nest site. Incubation takes 17 to 19 days (B. S. Lamba in Ali and Ripley, 'Handbook', 1970). The chicks are naked and blind at birth and take about 4 weeks to fledge and leave the nest. Both sexes bring food to the young and lizards and frogs have been noted as prominent items at this time.

*VOCALIZATIONS

More vocal in the spring but males will also call territorially in the autumn months. The typical call is a repeated rasping or grating 'kraaah-kraaah-kraaah' often rising to a crescendo as the bird performs its aerial evolutions. When very excited, these calls vary in pitch rising to high screeching tones. They will also call from a perch and when disturbed, often utter a single croaking flight or alarm call.

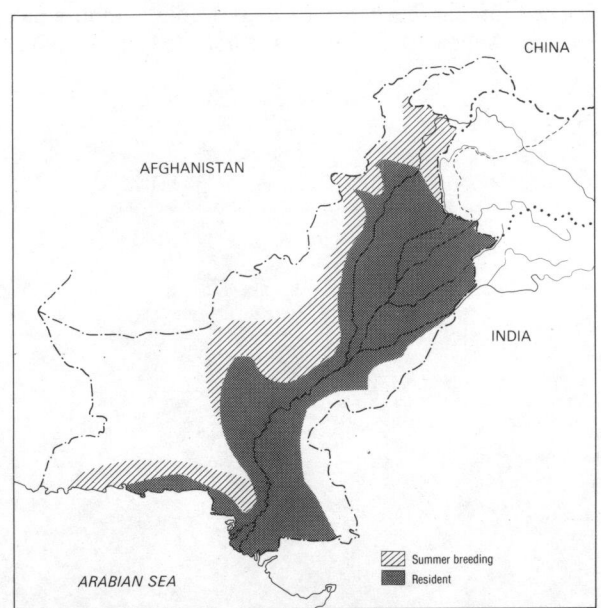

Map 268: Indian Roller or Blue-jay *Coracias benghalensis*

FAMILY UPUPIDAE

Hoopoes

Hoopoe *Upupa epops* Linnaeus

Map **269**

DESCRIPTION

Body length 29-31 cms.

Wing length 13.6-15.3 cms.

12.8-16 cms. (subspecies *orientalis*), 13.8-15.2 cms. (nominate subspecies *epops* (Stuart Baker)

Tail length 9.8-10.9 cms.

Bill length 5-6.3 cms.

Weight 64-77 gms. (males),

57-69 gms. (females) (Afghanistan, Paludan).

A smaller bird more slender in build than the Blue Jay (*Coracias benghalensis*) with a long thin down-curved black bill adapted for probing in the ground. It has a distinct and narrow neck with rather broad wings and square tipped tail. Its most conspicuous feature is the erectile crest of long narrow feathers down the centre of the crown like that of a cockatoo. Its head and body are orange-brown or salmon pink, with black wings and tail prominently barred with white on the wing coverts and the base of the black tipped flight feathers. The scapulars are pinkish buff, cross-barred black and in flight the whole of the upper surface of the wings bear transverse black and white bars. The tail has one broad white band in an inverted 'V' pattern, surrounded by black. The crest feathers are black tipped showing varying amounts of narrow white bars sub-terminally. The iris is brown and the legs and feet are slaty brown. The sexes are alike. Often upon alighting, the crest is erected in a fan shape and whenever the bird is excited or alarmed. When at rest the crest curves downwards in a narrow tail from the nape. The lower belly is whitish and the nape and mantle often are more ashy fawn, less salmon pink than the head and breast. The sexes are alike and juveniles are duller in plumage being more buffy grey, less pink on the back and breast.

HABITAT, DISTRIBUTION AND STATUS

Breeds in Europe, Africa, the Middle East, China, the Indo-Chinese region, Malaysia and Indonesia (Sumatra). The Hoopoe inhabits areas which are lightly wooded and preferably where there is some grass covering the ground. They generally avoid extensive desert as well as true forest.

There are no ringing data for this species on the subcontinent (Ali and Ripley, 1970), but in the north-west including Pakistan, there is a migrant population wintering in Africa, as well as a locally migrant but largely resident population, some of which winters but does not breed in Sind, and also a population which migrates right across the Himalayas to breed extra-limitally in central Asia.

In the northern Himalayas it is a summer breeding visitor penetrating into the furthermost northern valleys including Chitral, Gilgit, Hunza and Baltistan. In Sind it is mainly a winter visitor with a marked autumn passage noted around Karachi from mid-August to mid-September, of birds which probably winter in east Africa. In the Punjab it is partly resident and breeding throughout the Indus plains. In Baluchistan it is a summer visitor breeding in the higher mountain valleys. A few pairs breed in summer in the Murree hills up to 2,438 metres (8,000 feet) elevation, whilst they have been noted in the Kaghan valley around Battakundi and Burawai breeding at 3,048 metres (10,000 feet) beyond the tree-line. It is partly resident, partly a passage migrant and partly locally migratory. Status COMMON.

HABITS

A rather specialized feeder, being insectivorous and obtaining most of its food from beneath the ground surface by probing with its bill. It generally seeks areas with some grass or herb ground cover, probably because the plant roots provide food for various insect larvae. They probe rapidly with jabbing movements into the ground as they walk along, sometimes with the bill partly opened. It has been recorded as feeding upon small worms (Annelida), larvae of Ant-lions (Myrmeleonidae

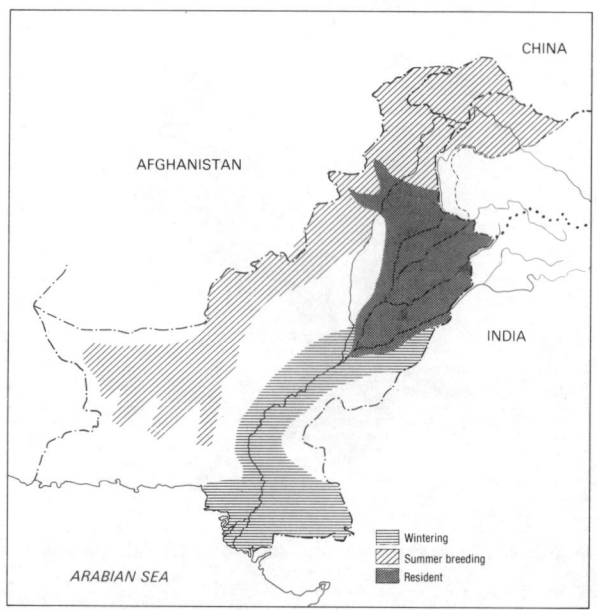

Map 269: Hoopoe *Upupa epops*

spp.) (observed Zangi Nawar, Chagai probing into holes of Ant-lion larvae, author, May 1984). It is generally very beneficial to agriculture because of the high proportion of injurious insects in its diet. Studies in Bihar, India based upon examination of 24 stomachs, collected between February and July and September and November indicated that a large percentage of prey comprised subterranean feeding larvae of Elaterid Beetles, *Agrotis* larvae (Cutworms), larvae of Hemiptera bugs, Melolonthid Beetles, *Anomala, viridis* (*Chrotogonus* grasshoppers, and caterpillars of Noctuid and Geometrid moths, as well as occasional ants *(Camponotus compressus), Anomala* and Mole Crickets *(Gryllotalpa africana)* (Mason and Lefroy, 1912). Occasionally it will be seen fluttering into the air to catch insects or foraging by turning over leaf litter and rubbish to search for insects, but most of its food is obtained by probing. It is not very gregarious in habits and generally will not tolerate a con-specific feeding in the near vicinity, driving the newcomer away aggressively. However in autumn passage in lower Sind loose aggregations of up to 12 or 15 birds have been noted roosting and feeding together and these are thought to be birds which winter in east Africa.

BREEDING BIOLOGY

The pair bond is monogamous and they are territorial during nesting, rival males indulging in fierce chases, both in flight when they utter grating or harsh rattling calls, also on the ground with bills locked together, as each tries to drive away its rival. They are cavity nesters, preferring natural holes or cracks in old trees, but will also utilize holes in gravel banks (observed Surkhab valley, Baluchistan) and holes in occupied mud-brick dwellings (observed in Khanewal). Around Rawalpindi it was noted as egg laying in April (Whistler, op. cit., 1930). The nest itself comprises a pad of straw, animal hair and feathers inside the cavity and this ultimately becomes very matted and foul with the excreta of the chicks as the parents carry out no nest hygiene. The normal clutch is 4 to 7 eggs with 5 being common. These are skim-milk blue, rapidly becoming stained brown (Bates and Lowther, 1952). Only the female incubates and she is a close sitter being brought food during this period by her mate. The normal incubation period is 15 to 17 days (Cramp *et al.*, 1985). The chicks are blind but covered with fluffy white down at birth and for the first 15 days after hatching, their sprouting feathers remain protected by waxy sheaths. In Sind the first record of breeding for that province, was a nest discovered by J.A.W. Anderson in 1949 in Landhi, under the eaves of a water tank. It contained half fledged young in late April (pers. comm., 1965). Normally in the Punjab, males start calling territorially from mid-March and egg laying starts early April. Both parents bring food to their young

but in the early stages the male brings most food, which is presented to the nestlings by the female (Skead, *Ibis,* 1950). Both parents were watched making frequent visits to a nest under the eaves of a brick building from 15 May in Khanewal and in a subsequent year in the same building from 22 May. The young can be heard calling, with rasping hisses and croaks inside the nest hole, for about 3 weeks after hatching (author obs. nest in own garden). Fledging period is 26 to 29 days (Cramp *et al.,* 1985).

The young are fed almost entirely upon caterpillars and Lefroy observed one pair making 286 visits to the nest hole with food over a 6 hour period. Feeding started from about 6 a.m. (Mason and Lefroy, 1912). In the first hour after daylight this pair made 56 visits, each time with large insects (crickets or caterpillars). Both parents continue to feed their young for at least 6 days after leaving the nest (author obs. Khanewal). In the Murree hills on 22 May presumed courtship behaviour was observed when a pair pursued each other from branch to branch in an oak tree *(Quercus dilatata).* They siddled up to each other and jabbed with their bills in a sort of fencing or bill stropping manner with their tails being wagged sideways all the while. No threat calls were uttered during this time which usually accompanies bill jabbing when the encounter is purely aggressive. At Burawai (3,000 metres/10,000 feet elevation) in the Kaghan valley a pair was still feeding their young in a cavity under the rest house roof on 8 July.

*VOCALIZATIONS

The territorial or advertising call of the male is repeated monotonously throughout the day in the plains from early March until the male becomes busy with food gathering as incubation progresses. It is a resonant low pitched 'whoop-whoop-whoop' call (whence its name) closely similar in timbre to the call of the Himalayan Cuckoo *(Cuculus saturatus).* The latter always sings in strophes of 4 calls whereas the Hoopoe utters calls in pairs or series of three 'hoop-hoop-hoop' and this is continuously and regularly repeated at about 3 second intervals often for 5 or 10 minutes at a stretch. A male watched calling had its crest lowered and bill held horizontally and its whole head shook as it pumped out the 3 'hoops'. Rival males or highly excited birds, utter grating or hissing calls which are almost Jay-like in tone. Nestlings when they get bigger utter hissing and rasping noises when begging to be fed. Rival males when pursuing each other from tree to tree were heard uttering long drawn-out rasping calls (early March, Balloki, Punjab) and a courting male and female, the latter being pursued from tree to tree, uttered cat-like mewing calls as well as the rasping hissing noises (author obs., Dunga Gali).

FAMILY BUCEROTIDAE

Hornbills

Grey Hornbill *Tockus nasutus* (Linnaeus) K.H Voous's 'Recent List'

Plate **22**
Illus. **68**
Map **270**

Tockus birostris (Scopoli) in Ali and Ripley's 'Handbook'

DESCRIPTION

Body length 45-61 cms.
Wing length 19.6-22.8 cms.
Tail length 26.4-30.2 cms.
Bill length 8-10.5 cms.

Ali and Ripley ('Handbook', 1970 and 'Synopsis', 1982) consider the Indian Grey Hornbill as a separate species *birostris,* distinct from the African Grey Hornbill, and to this author there do appear several small differences. This is the only representative in Pakistan of this fascinating old world family inhabiting the tropical regions of Asia and Africa. It is a large grey-brown bird with a noticeably long narrow graduated tail and comparatively huge head and bill. The tail is grey-brown with dark blackish brown sub-terminal bands and narrow white tips forming a conspicuous pattern when it fans its tail. The head is generally a darker grey colour almost brown and the flight feathers are also more brownish and darker with white tips to the primaries, whilst the under parts are light grey becoming white

Illus. 68: Indian Grey Hornbill *Tockus birostris*. Showing reduced casque development and slightly shorter bills in females (two upper heads). More extensive casque and blacker bills in males (two lower heads). Drawn from museum specimens.

ventrally and on the well feathered tibia. The eye is large and the iris orange to reddish brown in colour and framed by prominent eye lashes. There is also a small area of naked leaden coloured skin around the eye. The bill is laterally compressed and very deep basally and curves downwards to a sharp point with the cutting edge of the upper mandible serrated. In both sexes there is a prominent casque or 'horn' on the culmen, but this is more developed in the male (See **ill. 68**). The bill is generally black especially on the casque and proximally, but it becomes horny white on the culmen and towards the tip. The legs and feet are slaty grey and the outer toe is fused to the middle or central toe. The wings are rather short and broad and as the under-wing surface practically lacks softer covert feathers, their wings make quite a buzzing noise during flight. Juvenile birds lack the casque on the upper mandible and have paler yellowish bills, also their primary wingtips lack the white tips present in adults. The dark sub-terminal band on the tail feathers has a greenish gloss in some lights. The scapular feathers in adults are pale grey-brown with the secondaries blackish brown. The sexes are alike in plumage.

HABITAT, DISTRIBUTION AND STATUS

It occurs throughout India from the Himalayan foothills south to Kerala. This is a hornbill which occurs in relatively open country, avoiding deep forest, and it favours roadside tree avenues, old gardens and orchard plantations. In Pakistan it occurs only in the extreme north-eastern regions of the Punjab plains from Sialkot district through Lahore and Kasur districts spreading down as far south as Sahiwal district. It can be seen all year round in Lahore and Kasur area, but in Sahiwal it is not usually seen in winter, being a summer breeding visitor arriving in early March each year (Mohammad Mohsin, pers. comm.). Similarly pairs arrived in Sialkot at Marala barrage canal colony in the first week of March each year (E. Fernando, pers. comm.). It has not been recorded in the Rawalpindi area. During 28 year's residence in the Punjab, the author observed an occasional young bird straggling as far south as Multan district in April, but these individuals did not stay. Status SCARCE.

HABITS

Their flight consists of a few rapid wing beats with the primaries bent upwards at the tips, followed by glides. In winter time they are somewhat gregarious and go around in family parties of 4 to 8 birds. They are largely arboreal and depend upon fruit for their diet but occasionally descend to the ground to pick up large insects when they hop rather clumsily. They feed principally on berries and especially upon figs. Examination of stomach contents of 6 birds collected in Bihar revealed only *Ficus* seeds and pulp in their stomachs, no insect remains (Mason and Lefroy, 1912). However, Dr. Jerdon found one which had eaten a large Praying Mantid (Mantodea). Pipal Figs *(Ficus religiosa)* and Banyan Figs *(Ficus bengalensis)* are staple items, also *Ixora* berries and occasionally *Calotes versicolor* garden lizards, Odonata dragonflies and locusts *(Schistcocerca gregaria),* (E.N. Lowther, 1949). It has also been recorded bringing nidicolous nestling to feed to its mate during incubation (Lowther, op. cit.).

BREEDING BIOLOGY

It is thought that the pair bond is life-long. At Renala Khurd on Mitchell's Fruit Farms, the same *Salmalia malabarica* tree cavity was used by breeding Grey Hornbills every year for a period of at least 7 years though there is no proof as to whether this was the same pair. (Mohammad Mohsin, pers. comm., 1975). The details of their unique nesting habits have been well studied. A natural tree hollow or cavity is used and this may be deepened or enlarged somewhat by both birds. There is no lining placed in the nest hole. The female enters the cavity and seals herself in before starting egg laying. The male brings mud as well as insect remains to the nest hole and this is mixed with the female's own faecal matter and trowelled into position with much bill stropping. It eventually dries into an almost concrete consistency, with only a narrow more or less vertical opening slit, through which the female can insert her open bill. This slit is about as wide as a man's finger. The normal clutch is 2 to 3 eggs which are dull, unglossy and plain white, rapidly becoming stained brown. Normally they are laid over a considerable time interval, with excavated nests revealing young 7 or 8 days old together with one egg still chipping (E.N. Lowther, 1949). In the African Grey Hornbill *(Tockus nasutus)* (considered by some authorities to be con-specific) eggs are laid from 1 to 7 days' intervals (Alan Kemp in *Birds of Africa,* Vol. 3, 1988). The female invariably spends the first 2 or 3 days in the nest hole, completing the sealing of the entrance, and egg laying may not commence until 5 to 11 days from sealing the nest hole (Kemp, op. cit., 1988). Incubation is about 24-26 days and after the young are about 19 to 34 days old, the female breaks open the sealed nest entrance and joins her mate to assist in bringing food to the young (Kemp, op. cit., 1988). The nest hole is re-plastered and sealed up at this time by the young themselves. During the female's period of voluntary incarceration, she is fed entirely by her mate, who brings food to the nest about once every hour. This

is regurgitated up from its crop till it reaches the bill tip when it is delicately passed to the open bill tip of the female, who squeals (appreciatively ?) upon receiving each food item. The male has been observed tapping with his bill on the tree hole to attract his brooding mate. E. N. Lowther watched one male successively regurgitate and feed 24 'Pipal' figs to his mate during one such visit and at another nest during a 14 hour watch, the male made 12 visits to feed his incubating mate, with no visits during the hottest hours of mid-day (E. N. Lowther, 1949).

Most egg laying starts from late April and the female breaks out of the nest in late June or early July. Lowther reports of two nests under observation in India that one female broke out of on 4 June and another on 6 July. In the latter instance the young also were fully fledged and left the nest on 13 July (op. cit., 1949). In early July 1972, at Renala Khurd, Mohsin watched the parent hornbills feeding 3 fully fledged young after they had come out of the nest. The young are fully fledged at 30 days but do not leave the nest hole until 43-49 days after hatching (Alan Kemp, op. cit.).

While in the nest and incubating, the female also moults all her flight feathers (rectrices and remiges) but is reported to be in good condition and often quite fat upon emerging from the nest hole. Apparently all the rectrices are shed simultaneously, followed by the remiges. Nest hygiene is much better in this family due to certain adaptive behaviour. Both the incubating female and the chicks practice high velocity defecation through the nest slit. Furthermore undigested fruit seeds and drupes are cast up or regurgitated through the mouth in an enclosed capsule formed by sloughing of epithelial lining of the gizzard. Thirdly a symbiotic fauna of scavenging insects

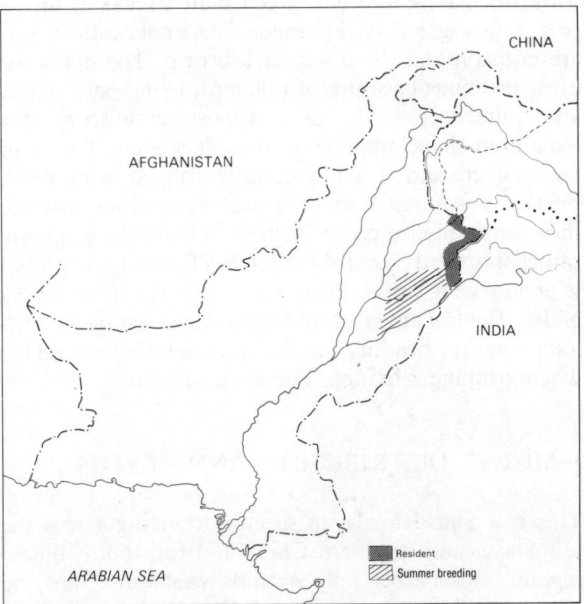

Map 270: Indian Grey Hornbill *Tockus birostris*

develops in the nest cavity which feed upon food residues.

*VOCALIZATIONS

They utter a variety of squealing and cackling cries. One call is reminiscent of the quavering squeal of the Black Kite *(Milvus migrans)* rendered at the time as 'pyee-pyee-pyee'. Other calls are hoarser and reminiscent of a White-backed Vulture *(Gyps bengalensis)*. They are more noisy in the early morning and will call during flight as well as when perched. They also utter more staccato, short repeated cackling and squealing type calls. When displaying the pair point their bills skywards and utter short rapid piping notes 'pi-pi-pi-pi-pipipieu-pipipieu-pipipea' (author obs. and tape recording Bharatpur, India).

ORDER PICIFORMES
FAMILY CAPITONIDAE
Barbets

Great Barbet or Great Hill Barbet *Megalaima virens* Boddaert

Plate **22**
Map **271**

DESCRIPTION

Body length 32-33 cms.
Wing length 14-15.2 cms.
Tail length 9.2-10.9 cms.
Bill length 4.2-4.8 cms.

This is one of the largest representatives of this family inhabiting either south-east Asia or Africa. It is a rather long bodied stout looking bird with a short neck, large head and large pale yellow bill. The tail is short and square and bright emerald green as also the rump and flight feathers. The head and neck is steely blue-black, sometimes with violet tinges and there is a streaked collar of greenish yellow around the base of the hind neck. The mid-breast region, lower back and wing coverts are uniform dark olive brown. The lower belly is streaked with broad olive and blue-green shaft streaks against a pale yellow-green background. The under-tail coverts are crimson and the iris is dark brown. The bill is the most prominent feature of this bird, being pale yellow and quite deep at the gape, almost reminiscent of a banana in shape and colour. It is blackish on the tip of the culmen and is conspicuously fringed with black bristles, a feature common to the whole family, whence their old fashioned name 'Barbet'. In flight the wings are rather short and rounded so that they fly with noisy rapid whirring wing beats, alternating with short swooping glides. The feet are greenish grey and zygodactyl with the long outer toe bending backwards alongside the hind toe when gripping a branch. The sexes are alike.

HABITAT, DISTRIBUTION, AND STATUS

This is a Sino-Himalayan species occurring across the Himalayas into southern China and the Indo-Chinese region. This barbet only extends westwards into the Murree hill range and the Jhelum valley of Azad Kashmir, where it is resident and breeding. It is a bird of mixed coniferous and deciduous forest in the temperate climate zone occurring between 600 metres (2,000 feet) and 2,600 metres (8,500 feet). In winter there is some altitudinal migration or drift into warmer snow free zones.

Curiously it has never extended its range westwards into Hazara district and the Kaghan valley or even as far north as Thandiani, the last group of hills in the Murree hill range. Also it is absent from the Neelum valley in Azad Kashmir (Kamal Islam, pes. comm., 1983). In the sub-tropical dry deciduous zone of the Margalla hills and Murree foothills it also is a sparse resident with numbers augmented in winter. They occasionally straggle as far

Map 271: Great Barbet or Great Himalayan Barbet *Megalaima virens*

from the foothills as the Salt Range. H. Whistler recorded one in Jhelum district on 1 April (Whistler, *JBNHS*, 1914*b*) and H. Waite found a pair breeding in the fruit gardens of Choa Saidan Shah (600 metres/2,000 feet) in the Jhelum district portion of the Salt Range on 5 June 1926, when he excavated the nest hole and took out 3 young almost ready to fly (specimens in British Museum, and Waite, *JBNHS*, Vol. 31, 1926, p. 826). A few birds may be encountered throughout the year in the Margalla hills descending to the outskirts of Islamabad. H. Waite collected specimens from Nurpur Shahan in mid-April and the author has seen them on the Margalla ridge top in July and August as well as heard them calling in April and May. During a two day visit to Choa Saidan Shah in early December 1973 the author did not come across any evidence of this species. Status Locally COMMON.

HABITS

Largely arboreal and poor fliers. They feed principally upon berries and fruits but will take large insects when available. In Dunga Gali they have been noted feeding upon the orange berries of climbing ivy *(Hedera nepalensis)* and the black berries of *Prunus cornuta* a large deciduous tree. Also when the berries of *Viburnum nervosum* are ripe they descend to the forest under-storey. During the monsoon individuals have also been seen feeding upon cicadas as they emerge, and one was seen with a praying mantids in its bill and another with a hornet *(Vespa orientalis)* (author obs.). At Tret they have been seen in early April feeding on the nectar and petals of *Salmalia malabarica* the Silk Cotton tree. In early May one was observed in Dunga Gali, systematically eating the catkin flowers and buds of *Ulmus wallichiana*. On 10 March when snow still carpeted the northern slopes they were seen on the north ridge of Murree. Generally they associate in pairs during the summer but are solitary in feeding, rather secretive and though frequently heard, not easy to see, their green and olive plumage blending well with tree foliage.

BREEDING BIOLOGY

They are very vocal during the spring and summer, and call incessantly during the day, with both sexes often indulging in anti-phonal duetting. The female is apparently attracted to the males calling and will fly into the tree often perching in close proximity, where they engage in much tail wagging from side to side, not up and down, and head bobbing as part of their courtship repertoire. At this time with each call of the male, the female responds with a much shorter lower pitched squawk. The pair bond is thought to be long lasting perhaps being renewed each year if a suitable nest site remains available. Both sexes take part in excavation of the nest which is a hole in a tree, usually on the underside of a sloping branch and not in an exposed vertical tree bole as is often chosen by woodpeckers. They usually bore into rotting or dead wood and the nest cavity comprises a surprisingly small orifice (7 cms. in diameter), extending vertically for a few centimetres before bending at right angles and descending parallel with the branch. A nest hole re-used over several years at Dunga Gali was about 45 cms. deep (probed with a stick), with a widened nest chamber unlined and in an *Aesculus indica* tree. The eggs are plain white, longish ovals (less round than woodpecker eggs) and both sexes share incubation duties. Both sexes will use the nest hole regularly for roosting, particularly during thunderstorms and at night time. The normal clutch is 3 eggs and the incubation period is quite short, believed to be between 13 and 15 days. In a study on *Megalaima viridis* in south India, period to fledging was 38 days. (Yahya, 1988), and may be slightly longer in this larger species. Both parents bring food to their young which are believed to be largely insectivorous when they are very small, supplemented after a few days with berries which ultimately constitute their main diet.

*VOCALIZATIONS

Perhaps the most familiar sound in the Murree hills is the loud plaintive ringing cry of this barbet, repeated from the topmost branch of a tree 'curr-liew curr-liew curr-liew' often repeated several hundred times, the sound carrying across the valleys for up to 1.5 km. distance. One individual timed with a stop-watch called 62 times in a minute (R. Orr, *in litt.,* 1976). When calling thus, the bill remains closed but the throat balloons out and deflates in a pumping action with each call. They call most persistently in the morning and evening and one male will elicit similar territorial calling from several others in the vicinity. The plaintive call is reminiscent in tone to the wailing of a Herring Gull *(Larus argentatus)*. In June it starts at 4.20 a.m. with first light. The approach of rain or a thunderstorm also seems to trigger off their calling.

When two birds are in close proximity, they engage in a variety of different softer calls including a harsh 'miaowing', (possibly the female?) and a querulous ululating soft 'querrhrh-rh' also repeated several times. This is usually when they are in visual contact and is accompanied by head bobbing and is believed to be an important element in strengthening the pair bond.

Blue-throated Barbet *Megalaima asiatica* (Latham)

Plate **22**
Map **272**

DESCRIPTION

Body length 22-23 cms.
Wing length 10-11.25 cms.
Tail length 6-6.8 cms.
Bill length 2.6-3 cms.

A large headed dumpy bird with short neck and tail and heavy swollen bill. Its whole body, including breast and wings is bright moss green except for the ear coverts, lores, chin and throat which are verditer blue. The crown and a small patch on either side of the blue throat area, is crimson with some of the feathers showing black bases. The mid-crown area is divided by a black band running across from eye to eye and the red area of the hind crown and nape is also narrowly divided from the blue by a narrow black line. The bill is coloured horn or pale greenish white with a small crimson spot at the gape. It is generally blackish on the culmen and tip, with prominent black bristles surrounding the base of the bill. The iris is brown with an orange orbital ring of naked skin. The legs and feet are greenish grey with the fourth or outer toe zygodactyl. The sexes are alike. Sometimes a small yellowish band shows in front on the black band across the top of the mid-crown region.

HABITAT, DISTRIBUTION AND STATUS

It occurs across the Himalayan foothill region to southern China, the Indo-Chinese region and extending

Map 272: Blue-throated Barbet *Megalaima asiatica*

into north Borneo. A resident species, narrowly confined to the outer Murree foothill zone especially the Margalla hills, Lehtrar, Manga and Kahuta valleys. It occurs between 450 metres (1,500 feet) and rarely ascends over 1,675 metres (5,500 feet). It has not been recorded as more than an occasional straggler into orchards in the adjoining plains areas of Rawalpindi district. It has not spread westwards beyond the Margalla hills and judging from Whistler's account of this barbet's occurrence in the Murree hills in the first two decades of this century (Whistler, 1930) when he only had one sight record, it appears to have recently spread and colonized the area. In May 1985 two males were heard (author) calling around Muzaffarabad (Azad Kashmir). In the Manga valley where they breed the author has heard, during April, May and June as many as 5 or 6 males calling simultaneously. It is not normally sympatric with the Great Himalayan Barbet *(Megalaima virens)*, the latter living at higher altitudes, except in winter time. Status locally COMMON.

HABITS

They are very vocal like all their genus and can be heard calling in all months of the year, but in May and June while breeding they are particularly noisy. They are not strong flyers having the characteristic whirring noisy flight of the family, alternating with dipping glides. They are strictly arboreal and feed on a variety of wild fruits and berries including the large figs of *Ficus glomerata* observed on 13 April in the Margalla hills. Salim Ali also records one as catching a Mantid which it battered against the perch before swallowing. Another individual was seen above Tret, pecking at and chewing the flowering racemes of the wild Pistachio tree *(Pistacia integerrima)*. They are not particularly gregarious being usually encountered solitarily.

BREEDING BIOLOGY

Nothing has been specifically recorded as distinct from the rest of the genus. The author has encountered two occupied nest holes, one at Lehtrar in the dead bole of a *Pinus roxburghii* which appeared to be an old Woodpecker's hole, and the other in a *Ficus infectoria* tree at Tret, both estimated at about 900 metres elevation and in shady ravines in which a stream flowed. Both sexes are known to take part in nest hole excavation and this is usually a short hole on the underside of a sloping branch and thus well protected from the weather. The normal clutch is 3 to 4 eggs, rather elongated blunt ovals in shape, thin shelled and white. There is no nest lining. Both parents share incubation duties and nest hygiene duties, removing faecal sacs from their nestlings when they visit the hole to feed them. This is in sharp contrast to the hole nesting Coraciiformes. The incubation period is believed to be about 2 weeks, and both parents bring food to their young. The pair bond is monogamous and probably of long duration with much mutual duetting and displaying in close visual contact with tail wagging and head bobbing and twisting during April and May. Courtship feeding by the male is also important. Most eggs are laid during the month of May.

*VOCALIZATIONS

The territorial or advertising calls of this barbet are similar to those of the Brown-headed Barbet *Megalaima zeylanica*, which does not occur in Pakistan. It consists of a series of rippling calls at the rate of about seven in five seconds (Fleming *et al.*, 1976) sounding like 'chakro-chakarroo-chakarro'. This is sometimes varied with a more rapid stuttering 'chukururr-chukururr' and occasional rather subdued 'pork' calls.

Coppersmith or
Crimson-breasted Barbet *Megalaima haemacephala* (P.L.S. Muller)

Plate **22**
Map **273**

DESCRIPTION

Body length 14-17 cms.
Wingspan 3-3.8 cms.
Bill length 1.6-2 cms.

The smallest of the barbets in our region, being about the size of a House Sparrow *(Passer domesticus)* but with a chunkier body and larger head. It has a rather swollen bill, curved on the culmen and horny black in colour. The base of the bill is whitish and is prominently rimmed by forward curving black bristles. The cheeks, chin and throat are bright lemon yellow, each patch framed by a black border and with a restricted crimson gorget on its breast below the yellow throat patch. The forecrown is also crimson, framed behind by a broad black band passing over the crown. There is a smaller yellow patch above the eye also and a red naked orbital ring. The zygodactyl feet are coral red and the iris is brown. The rest of the upper parts are green with the breast pale greenish white streaked with olive. The sexes are alike.

HABITAT, DISTRIBUTION AND STATUS

An Oriental species occurring from India and Pakistan across to the Indo-Chinese region, Malaysia, the Philippines and southern Yunnan. It also occurs in Sri Lanka. It is confined to the Indus plains in the Punjab, being very rare in northern Sind and absent from lower Sind, Baluchistan or the NWFP. In summer there is some westward extension or local migration of its range for breeding, but the bulk of the Punjab population is resident and sedentary throughout the year. It becomes silent in the winter months and because of its unobtrusive habits it largely escapes notice. It becomes noisy and conspicuous in Rawalpindi and Islamabad area each summer around late April when males can be heard calling incessantly and in the Margalla foothills it is then sympatric with the Blue-throated Barbet *(Megalaima asiatica)*. At Khanewal, over about 15 years that records were kept, it was first heard calling most years from the beginning of March. However it has been sighted in

November in Lahore (author) and Rawalpindi (Whistler, 1930), and occasionally can be heard calling in December and January (e.g. 25 January in Khanewal). It has also been seen on 30 January at Nurpur Shahan in the Margalla hills. The spread of irrigated cultivation and tree plantations has favoured this species in the Punjab, particularly in villages where *Ficus* trees are planted for shade. Status COMMON.

HABITS

Largely frugivorous and because of their arboreal habits, and green colouration rather difficult to see though their incessant calling in summer betrays their ubiquitous presence. They normally roost and call from the topmost canopy and branches of tall trees especially *Ficus*

Map 273: Coppersmith or Crimson Breasted Barbet *Megalaima haemacephala*

religiosa, the Pipal and *Salmalia malabarica,* the Silk Cotton. In the early morning they are fond of perching on a dead snag and sunning themselves before commencing to feed. Their diet consists almost wholly of berries and fruits especially figs from *F. bengalensis* and *F. religiosa* trees. They will occasionally take Mantids and Bush Crickets (*Tettagonia* spp.) disturbed from the foliage while foraging. Their flight consists of rapid wing beats and appears quite strong and direct. In winter time, sometimes 4 or 5 will congregate in larger fig trees.

BREEDING BIOLOGY

They excavate their own nest hole, sometime low down at others quite high up. One nest on the underside of a lateral branch but close to the main stem of a Persian Lilac *(Melia azedarach),* was only 7 metres from the ground whilst another had been excavated directly into the underside of a horizontal side branch of a Silk Cotton tree, and was fully 15 metres from the ground. They are quite territorial in defending their nest site particularly in driving off hole nesting birds such as Mynas *(Acridotheres tristis)* and Golden-backed Woodpeckers *(Dinopium benghalense),* though House Sparrows may be tolerated in the nest tree (observed Lahore). Both sexes may take part in nest hole excavation which occupies them for many weeks (Malcolm Macdonald, 1960). The tunnel ends in an enlarged nest chamber which is unlined and may extend from 20 to 40 cms. down into the tree branch.

Studies in Delhi by Malcolm Macdonald indicated that the female does most of the nest hole excavation and she often starts this in the autumn, completing a hole deep enough to act as a winter roost and recommencing excavation the following February or March. The interval from the start of first egg laying till the last youngster to leave the nest, appeared to be fully eight weeks with evidence that the young hatch asynchronously (Macdonald, op. cit.).

The normal clutch is 3, with up to 4 occasionally laid, and both sexes share incubation duties and bringing food to the young. The breeding season is extended;

February to September in the Delhi area (Macdonald, op. cit.). Often two broods are reared, and the same nest hole may be reoccupied year after year. In Mohalandar mango orchard, outside Lahore, a pair was observed bringing food to a second brood of young (visible at the nest hole) on 30 August whilst another pair appeared to be incubating eggs on 10 April in another Lahore garden. In Islamabad two fully fledged but flightless young appeared to have fallen out of their nest hole on 25 June. When put in a cage and hung on a shady tree nearby, both parents continued to bring food which they presented through the cage bars. When they demonstrated an ability to fly on about 5 July the cage door was left open and they left after 2 days (David Corfield, pers. comm., 1983). These young were fed exclusively upon fruit and gave incessant begging calls, much higher in pitch than their parents'. Incubation period is believed to be about 14 days while the period to fledging is about 35 days (Malcolm Macdonald, op. cit.).

*VOCALIZATIONS

The short 'pook-pook-pook' or 'tuk-tuk-tuk' calls, monotonously repeated at regular intervals are one of the familiar sounds in the northern Punjab and the hotter the day, the more incessantly the male seems to call. Salim Ali recalls one individual giving 108 'pooks' per minute and repeating its call 204 times non-stop ('Handbook', Vol. 4, 1970). The noise has been likened to the regular hammering of a coppersmith on his vessels, but in Pakistan it is more akin to the high pitched exhaust of single-stroke flour mill or tubewell engines, a very typical noise in the irrigated canal colonies. When calling, the bill remains closed, but the sides of the throat inflate like two small air sacs and often the head is rotated from side to side with each 'tuk-tuk', lending a ventriloquial effect to the sound.

A Barbet when defending its nest hole and driving off a Golden-backed Woodpecker *(Dinopium benghalense)* was heard (author) to give a churring call or buzzing call, not described in the 'Handbook' (Ali and Ripley, 1970).

FAMILY INDICATORIDAE

Honeyguides

Orange-rumped Honeyguide *Indicator xanthonotus* Blyth
Himalayan Honeyguide in Fleming and Fleming, 1976
Yellow-rumped Honeyguide (in King and Dickinson, 1975)

DESCRIPTION

Body length 15 cms.
Wing length 8.2-9.6 cms.
Tail length 5.5-6.1 cms.
Bill length 8-9 mm.

A small sparrow sized bird of a drab olive grey colour except for a bright orange-yellow patch on the forecrown and a similar glistening yellow patch extending from the rump to the upper back region. The upper parts are dark grey, washed with a tinge of olive and the breast and belly are pale grey with darker grey longitudinal streaks. The wings with only 9 primaries, are comparatively long and pointed and the tail is of medium length and square tipped. The flight feathers are dark grey with the primary and secondary feathers margined with olive yellow. The chin and throat region are tinged with yellow.

Females have the yellow areas reduced and less prominent, the upper back region being almost iridescent white or pale lemon rather than the deep sulphur yellow of the male.

Both sexes have short deep almost finch-like bills, of a horny yellow colour and framed by stiff black rictal bristles. The legs are pale horny green and the irides dark brown, with a pale green orbital ring (Stoliczka, 1873).

HABITAT, DISTRIBUTION AND STATUS

This is an endemic Himalayan species, confined to the outer and lower ranges of the mountains from Burma in the east, to Garhwal Himalaya (Uttar Pradesh) in the west. Formerly the range extended westwards into Pakistan where it was recorded in the Murree hills. Colonel Delme Radcliffe collected a specimen (now in the British Museum) in June 1867 from Kalabagh (Hazara district in the Galis). This record has been wrongly transcribed as Kalabagh on the Indus river in Mianwali district in Dillon Ripley's 1961 'Synopsis', also wrongly quoted as occurring in Bannu and the Afghan frontier (Ripley's 1982 'Revised Synopsis' and in Ali and Ripley's 'Handbook', Vol. 4, 1970). Stoliczka also collected a specimen from Dunga Gali (Murree hills, Hazara district) on 4 July 1873 (*Stray Feathers*, Vol. 1, 1873). Subsequently Colonel Buchanan reported seeing it at the turn of the century at Changla Gali (Whistler, *Ibis*, 1930) and Major Magrath reported the last sighting also in Dunga Gali (which is only 4.8 kms. from

Kalabagh) in April 1908 (*JBNHS*, 1909*a*). Neither Buchanan nor Magrath collected specimens and there are no subsequent records from Pakistan and Whistler writing in 1929 (*Ibis*, 1930), considered that it must have been very rare even at that time.

More recently Hugh Whistler saw it on 2 April 1929 at Truin near Dharmsala (Himachal Pradesh) but there have subsequently been no sightings that far west (Ali and Ripley, 'Handbook', Vol. 4, 1970) and its range today probably does not extend west of Garhwal (Uttar Pradesh) (S.A. Hussain, *JBNHS*, Vol. 75, 1978). It is however commoner in the eastern Himalayas from Nepal and particularly in Bhutan (Hussain and Salim Ali, *JBNHS*, Vol. 80, 1984). It must be PRESUMED EXTINCT in Pakistan since the early part of this century.

BIOLOGY

Recent studies on this relatively unknown species (Hussain and Ali, 1984), have revealed its almost total dependence upon bees wax obtained from the abandoned combs of the Giant Rock Bee *(Apis dorsata)*. It is significant that in the cooler western Himalayas, the migratory Rock Bee does not ascend into the Himalayan foothills or outer ranges, and it nowhere occurs in the Murree hill range, being essentially, a winter visitor to the plains of Pakistan. In the summer months there is some dispersal of honeyguides (which then feed upon non-Hymenopterus insects), at a time when bee combs are a scarce resource, and it must be presumed that the population recorded at the turn of the century in the Murree hills represented such seasonally nomadic birds which would not be breeding. The males defend traditional bee-nesting rock cliffs all the year round and breeding only occurs when females visit such sites to feed on the bees wax during the summer months (Hussain and Ali, 1984). In the warmer and more humid central and eastern parts of the Himalayas this colonial bee builds new combs and breeds from 1,500-3,500 metres (5,000-11,500 feet) which coincides with the altitudinal range of the Honeyguide. In the whole of its range, the eggs or young of the Himalayan Honeyguide have never been discovered, and as yet no details are known of the breeding biology of this presumed brood-parasitic species, which has a resource based, non-harem polygynous system of breeding.

FAMILY PICIDAE

Woodpeckers, Piculets and Wrynecks

Eurasian Wryneck *Jynx torquilla* Linnaeus

Plate **23**
Map **274**

DESCRIPTION

Body length 17.5-19 cms.
Wing length 8.3-9 cms.
Tail length 6-7 cms.

With their rather straight stout bill and zygodactyl feet (fourth or outer toe pointing backwards) their relationship with the true woodpeckers is apparent, but they are not so highly adapted to exploit vertical tree trunks. The sexes are alike and they are cryptically plumaged grey-brown all over, the feathers being finely speckled and vermiculated in chestnut buff and dark brown. The under parts are yellowish grey finely cross-barred darker and the lower cheeks and throat are a warm ochraceous colour with similar horizontal darker striations. There is a conspicuous dark brown stripe down the middle of the crown and darker brown loral streak and a paler supercilium. Also there are three parallel dark chestnut brown stripes down the nape and mantle, the outer lines extending down to the scapulars. The eye is somewhat small for the size of the head, giving it a rather reptilian appearance. The bill is short and thick at the base and dark brown. The flight feathers are cross-barred chestnut and dark brown and the tail is indistinctly cross-barred with brown bands bordered by blackish lines and with darker spots and streaks in a disruptive pattern.

HABITAT, DISTRIBUTION AND STATUS

A Palearctic breeding species from Scandinavia east through Europe and the USSR to Japan, which winters in Africa, the Indian subcontinent and the Indo-Chinese region. The Wryneck is a widespread but sparsely distributed winter visitor to the plains of Punjab and Sind, with quite a marked spring and autumn passage which is more conspicuous in lower Sind and Baluchistan. It over-winters in the Punjab, particularly favouring the better wooded areas around Lahore and Sialkot, and irrigated forest plantations further south. Migrant birds may be encountered from early April up to early May throughout Baluchistan wherever there are a few scattered trees. Latest date seen 2 May at Shinghar in Zhob district and temporary visits to the author's garden in Multan district noted in three consecutive years between 20 and 22 April. Its status as a breeding bird is rather poorly known. Biddulph and Scully both recorded it as a summer breeding visitor to Gilgit and not uncommon (Biddulph, 1881 and Scully, 1887). In a July visit to Naltar valley (Gilgit) in 1986 Mallalieu saw this bird which was probably breeding in the vicinity (pers.

comm., Janurary 1987). The author also found one breeding pair at Naran in the Kaghan valley in 1969 but in subsequent years could locate none. Perreau records it as breeding in Chitral (*JBNHS*, 1910) and Bates and Lowther also found it breeding at low elevations around the main vale of Kashmir (op. cit., 1952) and up to the lower slopes of side valleys. Status FREQUENT.

HABITS

It is a solitary bird outside of the breeding season and with its rather sluggish habits and cryptic plumage easily overlooked. With a very long thin protrusible tongue they are adapted to feed principally upon ants (Formicidae) especially tree-bark dwelling ants (*Crematogaster* spp.) and less commonly the large ground dwelling *Camponotus* species. They will also take small beetles and other insects when encountered. A specimen in Khanewal, Punjab was watched on a footpath licking up *Camponotus* ants with rapid flicks of its tongue which must be coated with a sticky substance. Lacking the stiff tail feathers of woodpeckers they generally forage by working along horizontal branches or hopping diagonally around sloping trunks. They also frequently forage on the ground, hopping forward with facility, doing so more commonly than most

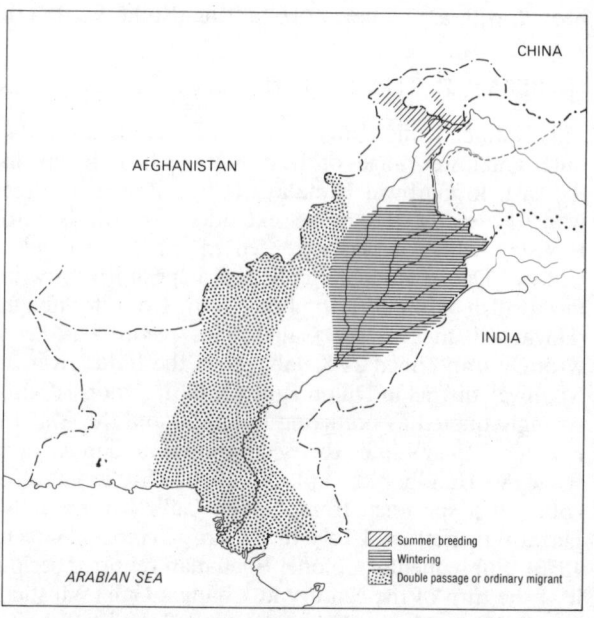

Map 274: Eurasian Wryneck *Jynx torquilla*

woodpecker species. They can cling to vertical trunks quite well and perch on branches; both crosswise like Passerines or horizontally along the branch. They are not particularly shy of man, often allowing close approach and also have the habit of sitting motionless on a branch for prolonged periods. When truly alarmed they twist their heads to and fro in rather a grotesque fashion, a habit which gave rise to their trivial name. The migrant population which passes through Baluchistan and the Indus plains, is thought to winter mostly in peninsular India.

BREEDING BIOLOGY

They form stable pair bonds and at the onset of the breeding season males call continuously during daylight hours and thus become very noticeable. At the onset of nesting, however, they are more secretive and cease calling. The plain white unglossy eggs are deposited in tree holes and clutch sizes are large, varying from 6 to 8. A favoured site is in the end of a rotting side branch or split stump usually quite high from the ground. Occasionally an old woodpecker's hole is appropriated. They are believed to be unable to bore a completely new tree hole, but can deepen and shape existing tree cavities to suit their purpose. No nest lining is made and both sexes share incubation duties, though most of the sitting

is done by the female and this takes 12 to 14 days. The young are born blind and naked and take a further 21 to 24 days to fledge and leave the nest hole. Food is brought to them by both parents. Bates and Lowther (op. cit., 1952) in Kashmir found most clutches of eggs laid during May, rarely up to early June, whilst a nest hole discovered in Naran (Kaghan valley) was in an old Mulberry tree *(Morus alba)* close to the river Kunhar at 2,400 metres (8,000 feet) elevation and on 17 July the parents were watched (author) carrying insect food and disappearing inside the nest hole. Courtship behaviour has not been described in detail but the advertising calls of the males presumably act partly to attract females and a pair was watched on autumn passage (first week September) at Malir outside Karachi excitedly chasing each other up and down a tree trunk and uttering rapid chittering squeaks, behaviour similar to courting *Dendrocopos* woodpeckers.

VOCALIZATIONS

Silent in winter but on their nesting grounds and occasionally in the autumn, males utter a series of 4 or 5 squeaking calls not dissimilar to some of the smaller woodpecker species 'cheet-cheet-cheet-cheet' or 'quee-quee-quee'. From about mid-May with the onset of egg laying they cease to call.

Speckled Piculet *Picumnus innominatus* Burton
Spotted Piculet in Fleming and Fleming

Plate **23**
Map **275**

DESCRIPTION

Body length 10 cms.
Wing length 5.4-6 cms.
Tail length 3-3.5 cms.

This tiny woodpecker has a stumpy tail and short neck so that in overall body length it is smaller than a House Sparrow *(Passer domesticus)* though bulkier in body size than the larger Leaf Warblers *(Phylloscopus* sp.) which are of equal body length. Its tail lacks stiffened quills as in the larger woodpeckers but it has fully developed zygodactyl feet, grasping thin twigs with two toes around the back of its perch. It is olive green or golden brown on the mantle and wing coverts with darker brown tail and flight feathers. The outer tail feathers are white and the central pair of rectrices are white on the outer webs. This white band down each side of the tail shows conspicuously when it flutters to another feeding point. The forecrown on males is pale orange darkening to olive green on mid-crown and a narrow white superciliary line extends from behind the eye. There is also a darker cheek band from behind the eye bordered by a second narrow white line below the ear coverts. The breast is dull white

or pale greenish white with conspicuous black spots or scale-like marks radiating in parallel longitudinal lines from the throat. The feet are grey-green and the short, pointed bill is horny black, becoming quite stout basally. Females are closely similar, only lacking the orange on the forecrown. The irides are red-brown in both sexes.

HABITAT, DISTRIBUTION AND STATUS

A largely Sino-Himalayan species found in the foothill or lower hill zone from Afghanistan across through India, Bangla Desh, Arunachal Pradesh, Malaysia and south-western China. It is extremely rare and local in Pakistan being a more or less sedentary species confined to predominantly deciduous forest at lower elevations, across the Himalayas.

Ali and Ripley in the 'Handbook Series', (1970) give the distribution as the Himalayas from Abbotabad and Murree. The summit ridge of Murree itself (2,100 metres/7,000 feet) is well above its preferred habitat, but the Rev. Storrs Fox saw one in Murree in November 1934 (Storrs Fox, 1937). In the dry deciduous scrub forest of the Margalla hills (600 metres/2,000 feet) this

Map 275: Speckled Piculet *Picumnus innominatus*

piculet is an occasional visitor. One sighting in mid-July 1977 (Kamal Islam, pers comm.) and a single female in the same ravine April 1982 (D. Corfield, pers. comm.). In the past two decades the Margalla hills have been intensively surveyed by keen amateur ornithologists based in Islamabad (including an estimated several hundred hours of observations by the author), and these are the only sightings so far obtained.

Small breeding colonies survive in the Neelum valley of Azad Kashmir at Khuttan and Salkalla (Kamal Islam, pers. comm., 1983). In the Kaghan valley at Malkandi reserve forest at 1,000 metres (3,500 feet) elevation a small colony estimated at 7 or 8 pairs (based upon calling males) was located in 1984. In Afghanistan it has been collected by Dr. El Kulmann (16 December 1963) and by J. Niethammer (11 July 1965) and seen by several others in forest habitat at 2,000 metres (6,500 feet) in Pechtal, Nuristan (J. Niethammer, *Journal fur Ornithologic*, Vol. 108, 1967). This was a new record as it was not observed by Paludan (1959) nor by Hue and Etchecopar (*Les Oiseaux du Proche et du Moyen Orient*, 1970). Status RARE and local.

HABITS

Observations of the small population at Malkandi indicate that outside of the breeding season, they tend to forage in mixed species flocks, in association with White Eyes *(Zosterops)*, Leaf Warblers (*Phylloscopus* spp.) and Tits (*Aegithalos* and *Parus* spp.). Whilst foraging they are remarkably active and agile, frequently hanging upside down from pendulous twigs and creepers, perching sideways like a passerine with equal facility, as well as clinging upside down to vertical branch surfaces like a tit. Because of their diminutive size, they can exploit the slender outermost twigs and also forage in quite small bushes close to the ground surface. Individuals were noted hopping diagonally upwards on the stem of a *Lonicera quinquelocularis* bush like a *Dendrocopus* woodpecker, and also fluttering from twig to twig in *Spirea vaccinifolia* bushes, like a White Eye. They dig and probe vigorously with their bills whilst hunting for food and if a small moth is disturbed will flutter out in pursuit and seize it in mid-air. They seem to have a preference for insects and their larvae, which inhabit tree and leaf buds at the extremities of twigs and growth points, and both Geometrid moths and their looper caterpillars were observed being taken in mid-May. In winter they probably rely more heavily upon wood boring larvae. Spiders and their egg cases are eaten, and particularly bark or wood boring beetles such as Curculionidae and Cerambycidae (Weevils and Longhorn beetles). Mason and Lefroy, (op. cit., 1912) examined stomachs containing the eggs and larvae of wood boring beetles. Ali and Ripley ('Handbook', Vol. 4., 1970) report that they feed mainly upon ants, their eggs and pupae. Tree dwelling ants (such as *Crematogaster* spp.) are not well represented in Pakistan. Apparently they never descend to the ground to feed on ants as Wrynecks *(Jynx)* so frequently do.

Usually they associate in pairs or in groups of 3 or 4, and in the early morning at Malkandi, have been watched sunning themselves, perched motionless for long periods on the topmost branches of tall trees (*Fraxinus* spp.) (author obs.). They do hammer into the bark and crevices of tree branches, like their congeners the larger woodpeckers, with surprising vigour and audible tapping noises.

BREEDING BIOLOGY

Not much has been recorded since publication of the *Nidification of Birds of the Indian Empire* (Stuart Baker, 1935, 4 volumes), but the pair bond is stable and both sexes share in nest hole excavation, selecting decaying wood or dead branches to drill their circular entrance hole, barely 2.5 cms. in diameter. This can be located low or high in a tree and is often bored into quite a small trunk or branch. Both sexes share incubation duties, thought to last 11 days (Stuart Baker, op. cit) and the usual clutch is 3 to 4 eggs, pure white and rather rounded ovals. Both parents also bring food to the young. Egg laying in the western Himalayas appears to start from early to mid-May (Craig Robson, pers. comm., 1984) and males calling territorially in Malkandi seemed to favour a zone between 1,300 and 1,375 metres (4,000 and 4,300 feet). The record of its occurrence at 3,000 metres (9,800 feet) in the north-west Himalayas (Ali and Ripley, 'Handbook', 1970) would appear untypical as far as the Pakistan population is concerned. Like all the woodpeckers the young are nidicolous and altricial at birth, but develop very rapidly, being able to climb about outside the nest hole, about 11 days after hatching.

*VOCALIZATIONS

Like most woodpeckers, they often call while feeding and this is presumably a contact call. It is a weak high pitched squeak 'sik-sik-sik' uttered at regular intervals. Advertising or territorial calls by males consist of longer more rapid sequences, rather more vehement but still weak and 'tinny' compared with woodpeckers. These calls can be rendered as 'ti-ti-ti-ti-ti' (tape recorded May 1984). They will also drum with rapid hammer blows of their bill on a hard dry wood surface, as an advertising call, producing a surprisingly loud 'tatoo' noise 'brrrrh br-r-rr-h', considering the size of the bird. One individual on 12 May 1985 at Malkandi, repeatedly drummed on one of the higher vertical branches of an *Acer pictum* tree, and the taps were not so rapid as is typical of *Dendrocopos himalayensis*.

Small Yellow-naped Woodpecker *Picus chlorolophus* Vieillot
Lesser Yellow-naped Woodpecker (in King *et al.*)

DESCRIPTION

Body length 26.5-28 cms.
Wing length 16.3-17.7 cms.
Tail length 12.7-13.5 cms.

This is a medium sized woodpecker predominantly olive green on the back, but with a noticeable spiky tufted crest on its nape of bright yellow shiny feathers. Unlike the Greater Yellow-naped Woodpecker *(Picus flavinucha)* which has a yellow chin and throat and green forecrown, in *P. chlorolophus* the forecrown and small malar spot at the base of the lower mandible are crimson-red, whilst the throat is barred olive brown and white and the upper breast is grey-green. The lower breast and vent is again barred like the throat in brown and dull greenish white. The flight feathers are chestnut brown or maroon with a greenish wash and the tail is darker blackish olive with bronze sheen to the central rectrices. The female also has a golden yellow occipital crest but lacks the red on the forepart of the face, having only a short crimson superciliary streak behind the eye. In both sexes the bill is horny brown becoming pale green at the base of the lower mandible. The irides are reddish brown and the legs and feet, a greyish to greenish black colour. The nuchal crest is erectile when the birds are excited.

HABITAT, DISTRIBUTION AND STATUS

An Oriental species found in the Himalayan foothills and in moist deciduous or evergreen forest in south India and Sri Lanka. It extends through Indo-China, southern China, Malaysia and Sumatra. This woodpecker is confined to predominantly deciduous forest in the lower or outer Himalayan regions, and further south and east in peninsular India it is adapted to evergreen tropical rain forest and moist deciduous lowland forest. Volume 4 of the 'Handbook' (Ali and Ripley, 1970) gives the western limit of its range in the Himalayas at Dharmsala in India. In Pakistan it has not been recorded in former times and must be considered a straggler or rare vagrant. On 4 June 1982, the author with David Corfield watched a single male individual foraging in a pure stand of Silver Oak *(Quercus incana)* at 1,980 metres (6,500 feet) elevation at the lower end of Kao forest in the Murree hills).

Captain Marshall in 1882 recorded it as occurring at Chamba (slightly to the west of Dharmsala) in the headwaters of the Ravi river (India) at N32° 35', E76° 07' and Hugh Whistler in writing of the birds of the Murree hills cited Marshall's specimens as the westernmost authentic record *(Ibis,* 1930). Status VAGRANT

HABITS AND BREEDING

Specimens watched by the author in the Nepal Terai (Chitawan) and in Sri Lanka (Sinharaja) were encountered with mixed foraging parties in company with Minivets, Drongos and Laughing Thrushes and this is apparently a common habit of the species. Besides hopping diagonally up vertical tree stems in conventional woodpecker fashion, they are quite agile in clambering down to the tips of pendant branches to peck at tree-ant nests *(Crematogaster* spp.) or reach nectar bearing flowers. Mason and Lefroy mention Coleoptera and Formicoidae as principal insect prey and describe them as breaking dried dung pats to search for insects. They generally forage in pairs or singly, not uncommonly with other woodpecker species. They nest in April and May, drilling a typical woodpecker's hole preferably into a decaying part of a tree trunk. The 3 to 4 plain white eggs are incubated by both sexes and the nidicolous chicks are also fed by regurgitation, bill tip to bill tip by both parent birds.

VOCALIZATIONS

Apparently has a rather plaintive toned single note call 'cheenk' which appears to be an advertising call (Ali and Ripley, 'Handbook', 1970) and which is repeated from the top of a bare branch every 15 seconds or so sounding reminiscent of some raptors (Fleming *et al.,* 1976). They also give trilling calls of 5 or 6 rapid ascending notes (Desiree Proud, cited in Ali and Ripley, 1970).

Black-naped Green Woodpecker *Picus canus sanguiniceps* Baker
Grey-headed Woodpecker *Picus canus* Gmelin (in K.H. Voous, Ben King etc.)

Plate **23**
Map **276**

DESCRIPTION

Body length 31-33 cms.
Wing length 14.5-16.5 cms.
Tail length 9.8-11.6 cms.

Superficially similar to the Scaly-bellied Green
Woodpecker it is slightly smaller and lighter in build and
in the field the bill shows noticeably black or dark blue-
grey contra the yellowish bill of *Picus squamatus*. Also
the breast and belly is unmarked without any scale-like
darker border to the feathers, being uniformly dull
greenish grey. The back, rump and wing coverts are olive
green with the rump more yellowish and the wing coverts
more brownish olive. The flight feathers and tail are
barred blackish and white with a greenish wash. In the
male only the forecrown is crimson-red, the nape and
hind crown being black (sub-species *sanguiniceps*).
There is a short broad black malar line and the throat
and cheeks are blue-grey. Females lack any red on the
forecrown, being black flecked with grey extending in a
'V' down the back of the nape. They can be separated
from female Scaly-bellied Woodpeckers (*P. squamatus*)
by their uniformly green breast and lack of any
conspicuous paler creamy superciliary stripe running
from behind the eye. Both sexes have brownish crimson
irides and bluish grey or olive grey legs and feet with long
curved claws and long toes. The feet are zygodactyl.

HABITAT, DISTRIBUTION AND STATUS

It occurs in central Europe, across the USSR in the
deciduous forest zone of Transcaspia, southern Siberia
and the Soviet Far East as well as China, Japan and
Malaysia, extending down to Sumatra and east to
Taiwan. In the Himalayas it extends across India, Nepal,
Arunachal Pradesh and Bangla Desh. It inhabits mixed
coniferous and deciduous forest at lower elevations, in
the western Himalayan part of its range, and is relatively
local and restricted in distribution within Pakistan,
having been recorded only in the southern part of the
Murree hill range. H. Waite collected it at Bhurban 18
May, 1939 and below Murree on 9 June, 1947 (Brit.
Mus. coll.). Major Magrath believed that it bred on the
lower slopes of Thandiani (Magrath, 1907). The Rev.
Storrs Fox during two years' continuous residence at
Murree only saw two pairs once, in October 1935 (E.A.
Storrs Fox, 1937). Recent records are perplexingly
lacking. The author twice encountered this species in the
forest which clothes the northern slopes below Murree
town at about 1,980 metres (6,500 feet) on 22 March
1969 and at Tret at 914 metres (3,000 feet) on 17 June
1967 but has failed to obtain any subsequent sightings
despite many visits to the same area. It apparently never
extended its range eastwards into Azad Kashmir or
Hazara district and may have disappeared from some of

Map 276: Black-naped or Grey-headed Green Woodpecker
Picus canus

Plate 23:
(1) Scaly-bellied Woodpecker *Picus squamatus*, female on left, male on right.
(2) Grey-headed Woodpecker *Picus canus*, female on left, male on right.
(3) Lesser Golden-backed Woodpecker *Dinopium benghalense*, female above male.
(4) Himalayan Woodpecker *Dendrocopus himalayensis*, male above, female below.
(5) Sind Woodpecker *Dendrocopus assimilis*, female above, male below.
(6) Fulvous-breasted Woodpecker *Dendrocopus macei*, female above, male below.
(7) Brown-fronted Woodpecker *Dendrocopus auriceps*, female above, male below.
(8) Wryneck *Jynx torquilla*, both sexes.
(9) Grey-capped Pygmy Woodpecker or Grey-headed Pied Woodpecker *Dendrocopus canicapillus*, female above male below.
(10) Yellow-fronted Woodpecker, Mahratta Pied or Yellow-crowned Pied Woodpecker *Dendrocopos mahrattensis*, female above, male below.
(11) Rufous-bellied Woodpecker *Dendrocopus hyperythrus*, female above, male below.
(12) Speckled Piculet *Picumnus innominatus*, male.

Plate 23

(1)

(2)

(3)

(4)

(5)

(6)

(7)

(8)

(9)

(10)

(11)

(12)

its former haunts due to destruction of suitable forest habitat on the lower hill slopes. Status SCARCE to RARE.

HABITS

Similar to those of the Scaly-bellied Green Woodpecker foraging generally solitarily except during the breeding season. It will also readily descend to the ground where it hops along rather jerkily, in search of ants and termites. It feeds mainly by working its way upwards on the trunks of trees, energetically stabbing, probing and tapping with its bill into the tree surface. It has the same method of feeding using its long thin protusible tongue to pick up ants or extract wood boring larvae from their galleries. It is reported to supplement its diet with flower nectar and in winter time with berries.

BREEDING BIOLOGY

Typical of the genus, both sexes share the task of boring a circular nest tunnel into the vertical side of a tree bole.

The entrance hole is circular and the tunnel itself bends down at right angles to the entrance, being widened at the bottom to form a nest chamber. Both sexes share incubation duties as well as feeding the young. The usual clutch is 4 to 5 eggs which are rounded ovals, white and glossy. Incubation period is believed to be not more than 17 days. In common with other woodpeckers, the altricial and helpless hatchlings develop fast and are probably able to fly out of the nest hole after about 20 days.

VOCALIZATIONS

Its contact and flight calls are similar to those of *Picus squamatus* but slightly more melodious and less strident. Major Magrath describing birds at Thandiani, likened the call to a 'loud repeated whistle' (op. cit., 1908a). It also has a more rapid repeated 'ke-ke-ke-ke-ke' call (Fleming *et al.,* 1976). During the breeding season males drum repeatedly, especially in the early morning hours, often using the same tree branch each day.

Scaly-bellied Green Woodpecker *Picus squamatus* Vigors

Plate **23**
Map **277**

DESCRIPTION

Body length 35 cms.
Wing length 15.5-17.2 cms.
Tail length 12.7-13.6 cms.

It is slightly larger and bulkier than the European Green Woodpecker *(Picus viridis)* but in overall colouration shows close affinities. The head is comparatively large with a stout wedge shaped and straight bill and a medium length wedge shaped tail. The bill is horny yellow in colour (a useful field point in separating it from *Picus canus*) and the nostrils are covered by short stiff bristles. The back and wing coverts are olive green becoming more yellow olive on the rump. The flight feathers and tail are cross-barred blackish green and pale whitish green. The cheeks, sides of neck and upper breast are a dingy greyish buff or pale greyish brown whilst the lower belly and breast is conspicuously patterned pale white or greenish white, each feather margined dark grey in a pattern of pointed overlapping scales. Males have the whole of the crown extending in a 'V' down the nape, crimson-red, flecked with black and there is a prominent black and grey speckled malar streak framing the lower cheeks with a narrower blackish loral line separated from the crown by a dull white superciliary streak extending from behind the eye. The iris is pink. Females are similar except that their entire crown and nape is black, usually flecked with grey. The feet are stout with relatively long toes, the third or outer toe pointing backwards

(zygodactyl). The tarsus is also quite long for a woodpecker and coloured like the toes, a greenish grey. In flight the olive yellow rump area shows conspicuously. The scale-like pattern of the breast feathers extends onto the under-tail coverts.

HABITAT, DISTRIBUTION AND STATUS

It occurs from Transcaspia (southern Turkmenia) through north-eastern Iran, south-western Afghanistan and again in north-western Afghanistan across the Himalayas to central Nepal. A resident species preferring Himalayan coniferous forest but adaptable to exploit a wide altitudinal range, even extending sparsely down to sub-montane or foothill regions. Small resident populations survive in the Juniper scrub forests of the higher ranges in Baluchistan (seen Shingar 1984 in *Pinus gerardiana* and it is common in Ziarat in *Juniperus macropoda*). It has also been recorded in bush studded desert stream beds as far south as Kalat state and the Chagai in southern Baluchistan (Christison, 1941). In the Punjab Salt Range a small resident population still survives around Choah Saidan Shah (author obs. 1975), and has been collected from the Sakesar range (H. Waite *JBNHS,* 1948). A small resident population occurs in the Margalla hills in sub-tropical dry deciduous forest as low as 600 metres (2,000 feet) elevation. In the extreme northern areas it is a sparse resident inhabiting those arid valleys wherever groves of willows, poplars or orchard plantations provide minimum suitable habitat. The author noted them in February in snow covered areas in Baltistan around Skardu, Gul and Bara villages (the latter on the Shyok river). Mathews also found them sparingly in July throughout the Indus valley in Baltistan (*JBNHS,* Vol. 42, 1941). They occur in Gilgit and Swat Kohistan up to the limit of the tree line but their main stronghold and the habitat in which they are most plentiful is mixed Himalayan coniferous forest and temperate deciduous trees from 1,500 to 3,050 metres (5,000 to 10,000 feet). Status COMMON.

HABITS

They have noisy whirring flight and a typically dipping or undulating trajectory in common with other woodpeckers and are easily detected from their frequent loud ringing contact calls. They feed principally upon ants (Formicidae) and forage by chiselling and hammering into bark crevices and the outer wood of vertical tree boles and branches. They will readily descend to the ground to forage and do so more frequently than the sympatric Himalayan Pied Woodpecker *(D. himalayensis).* Their tongues are extremely long and protrusible, the tip being horny and armed with barbs and also supplied with viscid mucous. This enables them to hook out wood boring larvae of beetles such as Curculionidae and Cerambycidae as well as rapidly catching ants without having to capture them first between the mandible tips. In winter time they will supplement their diet with berries and at lower warmer elevations, termites (Isoptera) are important in the diet.

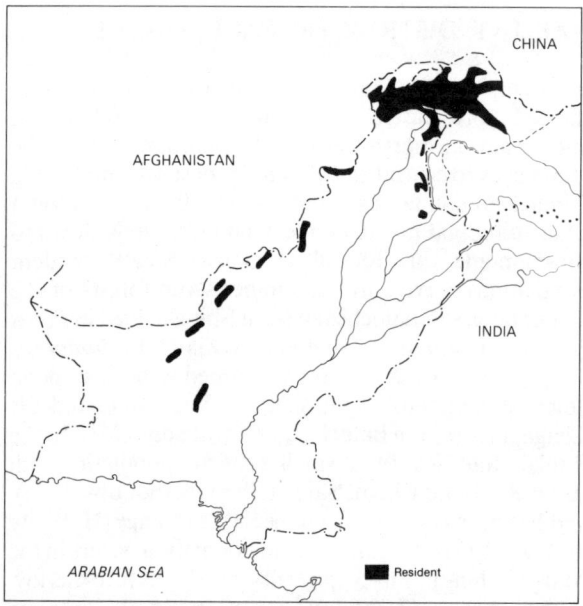

Map 277: Scaly-bellied Green Woodpecker *Picus squamatus*

The pair bond is long lasting and they often forage in pairs throughout the summer and autumn whilst juveniles, after becoming independent, tend to associate together while feeding.

BREEDING BIOLOGY

Both sexes share in nest hole excavation and examination of trees in regions inhabited by this woodpecker, indicates that many incompleted hole borings are made before a pair finally selects its nest site. In Dunga Gali, Murree hills however, the same tree hole has been used for four successive years by a (?) pair of these woodpeckers. Deciduous trees appear to be preferred over coniferous trees according to observations in the Murree hills. The hole entrance is neatly made and circular, being just over 6 cms. in diameter. The cavity angles down into the tree hole and may descend another 20 to 30 cms. with quite a large nest chamber being excavated especially where dead and decaying timber has been selected. No nest lining is made, the female laying 4 to 6 round white and glossy eggs which rest upon the rotting fragments of wood at the bottom of the nest hole. Incubation takes about 17 days (obs. Dunga Gali) and is shared by both sexes. The chicks are born blind and naked and are fed by both parents on insect food which is regurgitated and passed from bill tip to bill tip. Most clutches are laid from early to mid-May in the Murree hills, though a pair was seen entering a nest hole in Ziarat, Baluchistan on 12 April and young were being fed in the nest hole on 15 June at Naltar in Gilgit, which from their rather weak cries when hungry appeared to be still less than 2 weeks old. After about 14 days of age the young are fully feathered and eagerly come to the nest hole entrance to be fed and when hungry emit a surprisingly loud wheezing 'chuff-chuff' call, audible from over 100 metres distance. At Malkandi forest after a warm dry summer with early breeding, a family party of 5 young which could actively forage for themselves, was noted in late May persistently pursuing one parent bird (the male) and begging and being frequently fed, bill to bill.

At the onset of breeding males give a loud ringing territorial or advertising call and indulge in much drumming (i.e., rapid hammering on hollow or resonant branches) with their bills. Pairs also indulge in much *sotto voce* squeaking and chittering as they pursue each other up and down tree trunks in close proximity.

*VOCALIZATIONS

The calls of this woodpecker, though characteristic and easily recognizable once familiarity is gained, are very variable. They can be upwards or downwards inflected and vary in pitch. The usual flight call or contact call is 'kuik-kuik-kuik' repeated quite rapidly 3 or 4 times or occasionally uttered only as a single call. The advertising call is a more melodious inflected or two-noted quavering call 'klee-guh-kleeguh' rapidly repeated 7 or 8 times interspersed with bouts of 'drumming'. Birds in close visual contact often utter subdued but rapid 'chissuh-chissuh' squeaking calls and fully fledged young before leaving the nest have loud wheezy calls like the hissing of an old steam engine.

Lesser Golden-backed Woodpecker *Dinopium benghalense* (Linnaeus)

Golden-backed Woodpecker of Stuart Baker and of Ali and Ripley

Black-rumped Goldenback of King *et al.*

Plate **23**
Map **278**

Synonym *Brachypternus aurantius* of Blanford and Oates

DESCRIPTION

Body length 25-29 cms.
Wing length 13.6-14.7 cms.
Tail length 8.2-9.3 cms.
Bill length 2.8-3.7 cms.

This medium sized woodpecker is conspicuously patterned with a black and white striped face, black bill, all-black wedge shaped tail and flight feathers, offset by olive golden mantle and scapulars and in the male a spiky crimson crest of shiny feathers. The face and sides of throat are white with black streaks through the eye and in a collar around the hind neck, leaving a narrow white superciliary line extending down to the nape. The throat is black, streaked with white and the breast and belly is spotted black and white in a scale-like pattern formed by the white feather centres margined with black. The female has the forepart of the crown black speckled with white in parallel lines and only the hind part of the crown is crimson on the occiput. Both sexes can erect their crests when excited, into a bushy upstanding crest. The neck of this species always looks narrow and sinuous giving a distinctive silhouette when compared with the shorter thicker neck of *Picus* species. The legs and feet are grey-green with long toes. The feet are zygodactyl and there are 4 toes. The irides are reddish brown. From the back view it is a broad shouldered but short bodied woodpecker, less cylindrical in body shape than either the Green *(Picus)* or Pied *(Dendrocopos)* woodpeckers.

HABITAT, DISTRIBUTION AND STATUS

Endemic to the Indian subcontinent, it occurs in Sri Lanka, Banga Desh, Nepal and India but not extending into Burma. This woodpecker is largely sedentary and resident throughout the Indus plains and the main vale of Peshawar. It is rare but occurs around Bannu and Dera Ismail Khan. In the irrigated canal colonies of the Punjab it is common but is encountered much less frequently in lower Sind. It avoids extensive forest plantation to which *Dendrocopos assimilis* is better adapted, seeming to prefer old gardens, roadside and canalside tree plantations. Likewise it avoids extensive desert tracts to which *P. mahrattensis* seems better adapted. It has not been recorded in the hilly tracts or in any part of the northern Himalayan areas. Status COMMON.

HABITS

Their characteristic dipping flight as they go from tree to tree is typical of the family. They are usually tame in the presence of humans, often foraging busily in a tree below which a group of people may be talking loudly. They tend to maintain long lasting pair bonds so that pairs can be seen feeding together in all months of the year. They hop jerkily up tree boles and around lateral branches, being able to cling underneath horizontal branches with ease. Their principal food is ants, particularly the large

Map 278: Lesser Golden-backed Woodpecker *Dinopium benghalense*

black *Camponotus* ants. Studies in Bihar farming areas, based on examination of 16 stomach contents collected between January, March to June and again in October and December, revealed a very occasional spider, centipede or Geometrid moth larva and also tree figs *Ficus* fruit being taken, whilst the bulk of their food comprised ants *(Camponotus compressus* and *Oecophylla smaragdina* and *Meranoplus bicolor)*, Weevils, *(Myllocerus discolor, Astycus lateralis)*, and occasional Tenebrionid beetles *(Mesomorpha* and *Derosphaerus* spp.) (Mason and Lefroy, 1912). Occasionally they descend to feed on the ground particularly in the vicinity of ants nests.

BREEDING BIOLOGY

In Sind, nests with eggs were found on 22 April near Landhi in a mango tree and in Pirawala (Multan district) in a tamarisk tree with young being fed on 20 April. K. Eates took nests from Sirin trees *(Albizzia lebek)* most frequently, but also found a few nests in mango *(Mangifera indica)*, Shisham *(Dalbergia sissoo)* and Kikar *(Acacia arabica)* trees. He found three partly fledged young in a nest at Nawabshah on 24 April. Both sexes excavate the nest hole, which is neat and circular with an entrance hole typically about 8 cms. in diameter. The entrance hole bends vertically downwards into a widened nest chamber in which the plain white rounded and glossy eggs are laid without any nest lining. The usual clutch in Sind is quite small, with 2 eggs often being

incubated and rarely more than 3 young reared (Eates, MS notes). Incubation period is believed to be between 17 and 19 days, with the young taking another 3 weeks before they leave the nest hole. Both parents feed the young by regurgitation of a semi-digested paste of insects. The young are blind and naked at birth. When they are fully fledged but still in the nest chamber they will produce a quite ominous sounding hissing noise if their nest tree is accidentally struck. Ali and Ripley (Vol. 4, 'Handbook', 1970) report that sometimes a second brood is reared. Apparent courtship behaviour was observed (author) in Miani forest (Hyderabad, Sind) as early as 10 December during which the female crouched on a sloping branch of a Sirin tree *(Albizzia)* whilst the male with crest erected, jerkily moved backwards and forwards along the same branch on either side of her and both birds uttered high pitched squeaks.

*VOCALIZATIONS

The contact call is usually a single rather strident note 'kierk'. In flight and while clinging to a tree they also emit whinnying calls, consisting of rapidly repeated 'kyi-kyi-kyi' notes, which are reminiscent of the calls of the White-breasted Kingfisher *(Halcyon smyrnensis)*. A male heard and watched drumming on 25 March made a rather feeble noise in bouts of 2 to 3 seconds duration but the tree being used as a sounding board was a healthy Shisham *(Dalbergia sissoo)* and a harder surface would presumably have produced more resonance.

Sind Pied Woodpecker *Dendrocopos assimilis* (Blyth)

Plate **23**
Map **279**

Dryobates scindianus of Stuart Baker

Synonyms *Picoides assimilis* of Ali and Ripley

DESCRIPTION

Body length 22 cms.
Wing length 11.1-12.3 cms.
Tail length 6.7-7.3 cms.
Bill length 2.4-2.5 cms.

This is a slightly smaller woodpecker than the very similar Himalayan Pied Woodpecker *(D. himalayensis)*. In the male, the immediate forecrown is white, this region and the nostrils being covered by short bristly feathers. The crown is crimson streaked with black and extending to the hind crown region, whilst in females the whole of the crown and nape are black. The hind neck, mantle, and upper-tail coverts are black as also the sharp pointed central pairs of rectrices. The outer two pairs of tail feathers are barred with white and also the primaries and secondaries are checkered with white spots forming comparatively broad bars when the wing is closed. The

throat and breast is rather a dingy white with some obsolete darker streaking on the lower flanks and the under-tail coverts are a paler crimson pink than the crown. The zygodactyl feet, in both sexes are blue-grey with the irides varying from brown to crimson.

The chief distinguishing features between *assimilis* and *himalayensis* are as follows. In *assimilis* the bill is bluish slaty without the black tip of *himalayensis*. The bristly white forecrown and the breast are dingy white, whereas these regions are fulvous buffy in *himalayensis* though in the subspecies *D. himalayensis albescens* which occurs in the inner forested mountain ranges of Pakistan, the forecrown and breast is greyish white rather than fulvous. The scapulars are white in *assimilis*, all-black in *himalayensis* and the white wing bars are broader in *assimilis*. The sides of the head are all-white except for the down-curving black malar stripe in *assimilis*, whereas this black line extends upwards behind the ear coverts in *himalayensis*.

HABITAT, DISTRIBUTION AND STATUS

This is a resident endemic woodpecker occurring from extreme south-eastern Iran, through Baluchistan and parts of Sind and Punjab. It does not occur in India. It is characteristic of the Indus riverain forest in the plains and adapted also to dry sub-tropical thorn forest in the sub-montane or foothill tracts. It is nowhere common but is widely distributed throughout Sind and Punjab, being especially adapted to desert regions where stunted and scattered thorn trees such as *Prosopis spicigera* and *Acacia modesta* or *Acacia senegal* line the dry seasonal gullies or 'wadis'. It is rather rare in lower Sind, more common in northern Sind and the Punjab, frequenting canal and roadside tree plantations and more recently has adapted well to irrigated forest plantations. It is resident in the Wild Olive (*Olea cuspidata*) and 'Phalai' (*Acacia modesta*) scrub forests of the Punjab Salt Range including the Kalla Chitta hills and occurs in *Acacia senegal* thorn scrub in Lasbela and as far north-west as Kalat and Chaman (Barnes, 1880) in Baluchistan. Christison encountered it in Kalat (*JBNHS*, Vol. 43, 1942), and Meinertzhagen also reported it from Chaman (*Ibis*, 1920). In north-western Zhob district, the author encountered it in 1984 at about 1,500 metres (5,000 feet) elevation in the foothills and ravines below Shingar, which were sparsely dotted with *Pistacia integerrima* and *Fraxinus excelsior* and *Olea cuspidata* scrub forest; it was not noted in the Chilghoza Pine *(Pinus gerardiana)* forest higher up. Whitehead noted it around Kohat in Acacia thorn scrub up to 900 metres (3,000 feet) (*JBNHS*, 1911). Status FREQUENT but local.

HABITS

The author has never observed it feeding on the ground but it is an active restless little woodpecker which creeps up and around the relatively slender branches of the *Acacia* and *Pistacia* trees of the habitat it favours, industriously hammering, poking and twisting its bill into every crack and crevice for insect larvae, pupae and especially ants. Nothing has been specifically recorded about its food habits but *Camponotus* ants and the larvae of wood boring beetles are certainly important items in its diet. It is frequently encountered solitarily but like all the family will reveal its presence by occasional contact calls.

BREEDING BIOLOGY

Whistler (*Ibis*, 1930) records them in Rawalpindi district and the Potohar as breeding early, with the young being hatched by the end of April and H.E. Barnes wrote that they bred during April and May in Chaman district of Baluchistan (*Stray Feathers*, 1880). In Sind, Ticehurst (*Ibis*, 1923) noted one boring its nest hole at the end of February and wrote that egg laying was usually in

Map 279: Sind Pied Woodpecker *Dendrocopos assimilis*

March. In Pirawala forest plantation, Multan district, there is a good population and the author noted that the majority of nest holes were drilled in the dead or drying out branches of fairly stout *Tamarix aphylla* trees, whilst at Chinji in the Salt Range a nest hole was noted only 2 metres from the ground in an *Acacia modesta* tree. Christison also noted its propensity to bore nest holes less than 2 metres from the ground (*JBNHS*, 1942). As far as is known both sexes share in nest hole excavation and in incubation duties. The normal clutch is 3 to 4 plain white eggs, which are small glossy ovals (K. Eates. MS notes). Incubation lasts 15 or 16 days and is mostly by the female, but the male is very active in bringing food to the young after they are hatched. The young are naked and blind at birth and as in all the woodpeckers become very noisy in the nest hole as they get older, poking their heads frequently out of the hole when they are hungry.

*VOCALIZATIONS

A rather weak high pitched series of calls are typical of this species 'chir-rir-rirrh-rirrh'. Drumming is confined to the early part of the breeding season and may be done by both sexes (obs. Hugh Whistler). The noise produced varies according to the substrate, being quite surprisingly loud on hard dry wood (e.g. on a tamarisk tree on 19 March at Pirawala). Another male bird in the Salt Range drumming on the stem of an *Acacia modesta* made a relatively weak 'br-r-r-rh' sound. Whistler rendered their advertising calls as 'toi-whit toi-whit toi-whit', rapidly repeated (cited in Ali and Ripley, 'Handbook', 1970).

Himalayan Pied Woodpecker *Dendrocopos himalayensis* (Jardine and Selby)

Picoides himalayensis in Dillon Ripley's 'Revised Synopsis'

Synonym *Dryobates himalayensis* in Stuart Baker's 'Fauna Series'

Plate **23**
Map **280**

DESCRIPTION

Body length 23-35 cms.
Wing length 12.3-13.6 cms.
Tail length 7.7-8.5 cms.
Bill length 2.8-3.3 cms.

This pied woodpecker is slightly larger in size than the Sind Pied *(D. assimilis)* and is a higher elevation species unlikely to occur alongside *assimilis*. Males have the whole of the crown to the nape crimson-red, flecked with black and also narrowly bordered by black. The immediate forecrown and nostrils are however covered in bristly fulvous buff feathers. The back and tail are also black with the central rectrices sharply pointed and having very stiff quills (rachis). The outer tail feathers are barred with white at their tips. The flight feathers are checkered with black and white forming narrow white bars when the wing is closed and the wing shoulder forms a solid white patch on either side of the mantle, but the scapulars are black in contrast to *assimilis* which has white scapulars. The face, throat and sides of breast are dingy white, with the lower breast and belly usually more dingy and in the sub-species *D. himalayanesis* showing a fulvous wash. Its breast is always less white than in the Sind Pied Woodpecker. However the population inhabiting the inner forested hills of Pakistan *(D. himalayensis albescens)*, has the breast more greyish

white than fulvous, so this is not always a reliable field character for distinguishing it from the Sind Pied Woodpecker. Females have the whole of the crown and nape black. Both sexes have a prominent black moustachial streak extending from the base of the mandible and curving down the side of the neck as well as up around the rear of the ear coverts (this black area lacking in *assimilis*). The bill is darker than that of *assimilis,* and is blackish at the tip becoming blue-grey at the base of the lower mandible. The legs and feet are more greenish grey and the irides are red-brown in both sexes. Juveniles have the whole of the crown crimson-red like the male.

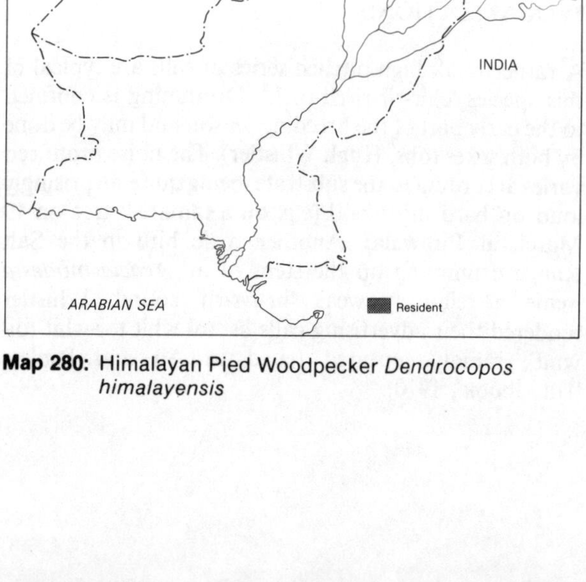

HABITAT, DISTRIBUTION AND STATUS

It is a Himalayan species extending westwards into extreme north-eastern border regions of Afghanistan, and across as far as western Nepal. This is the commonest woodpecker throughout the forest areas of the Himalayas and northern mountain regions. It is a sedentary species, remaining at higher altitudes even during winter snows and generally preferring to inhabit a zone between 1,975 and 3,200 metres (6,500 to 10,500 feet) elevation, and characteristically associated with coniferous forest. It is sympatric with the Scaly-bellied Green Woodpecker *(Picus squamatus),* but more or less allopatric with the Brown-fronted Pied Woodpecker *(Picoides auriceps)* which favours lower altitudes. In the Murree hills and Galis it is very numerous, but also occurs less plentifully in Chitral, Swat, Indus Kohistan and the Hazara district. It is rare in Gilgit but occurs where there are small patches of mixed Blue Pine *(Pinus wallichiana)* and Juniper forest. In Naltar valley it is common in the spruce forest but rare in other parts of Gilgit and in this respect is less adaptable to open terrain with few trees, than is *Picus squamatus*. Whitehead (1911) reported it as common in the Safed Koh range up to the limit of the tree-line, descending in winter down to the Samana hills as low as 1,370 metres (4,500 feet). Status COMMON.

HABITS

More arboreal than the Scaly-bellied Green Woodpecker, only rarely descending to the ground and more confined to extensive forested areas. They are often seen foraging solitarily in the winter months though pairs will keep together in the summer, even after the breeding season is over. Like all woodpeckers, they frequently

Map 280: Himalayan Pied Woodpecker *Dendrocopos himalayensis*

keep in contact with loud ringing cries, and spend most of their day actively foraging for insects and their larvae on the vertical surface of trees or the main side branches, being able to hop upwards or downwards with equal facility. They are more adaptable in diet than the Green Woodpeckers (*Picus* species) being less dependent upon ants as food. In fact, in winter they supplement their diet with acorns *(Quercus dilatata)* and the seeds of the Blue Pine *(Pinus wallichiana)* extracted from the pendant cones. Their main diet however, comprises beetles, especially the larvae of wood boring beetles and weevils and even spiders and their egg cases, also larvae of lepidoptera when available. They are less inclined to descend to lower valleys in winter than the Scaly-bellied Woodpecker which may share the same forest areas in summer.

BREEDING BIOLOGY

Both sexes take part in nest hole excavation and they have a distinct preference in the Murree hills for making their nest hole in deciduous trees and in particular the Bird Cherry *(Prunus cornuta)*. The hole is often located under the shelter of a lateral branch and is drilled into healthy green wood as often as dried decaying areas. Observations by Bates and Lowther (op. cit., 1952) indicate that the male does most of the nest hole excavation. The entrance hole is usually a perfect circle and of about 5 cms. diameter, with the widened nest chamber at the bottom, sometimes no wider than 10 cms. 3 to 5 eggs are laid and are white with a high gloss and the female is thought to do most of the incubating. The young become quite noisy and vocal as they get bigger and can be heard inside the nest hole keeping up an incessant buzzing squeak. Both parents bring food to their young and they take about 3 weeks from hatching before they are ready to leave the nest. Incubation is thought to take 2½ weeks or slightly less and the young are altricial and nidicolous on hatching. Most clutches are laid during the second half of April or early May. Fully fledged young have been seen coming to the nest hole entrance to be fed on 5 June in Sharan (Kaghan valley) and as early as 21 May in the upper catchment region of Dunga Gali. In years with a cold or wet spring breeding may be delayed. Copulation was observed on 7 May in Dunga Gali, the female squatting along a horizontal bough with head lowered and neck extended, after both birds had been perched close together in visual contact for some time, on the same branch. The male remained nearly 30 seconds on her back. Absence of any elaborate preliminary courtship suggested that this was a well established pair bond in which part of the clutch had already been laid. At the beginning of the breeding season pairs in visual contact often pursue each other around a tree bole, both sexes keeping up a rapid high pitched excited 'chissik-chissik' call.

*VOCALIZATIONS

The contact call is generally less loud and softer in tone than that of the Scaly-bellied Green Woodpecker, often being uttered as a single note repeated at considerable intervals of half to 1½ minutes. It has also a flight or alarm call comprising a more rapid repeated sequence of cries 'tri-tri-tri-tri'. They also engage in drumming at the beginning of the breeding season and both sexes have been observed doing this on some hard dried tip of a decayed branch, producing a rapid 'br-r-r-r-h' noise, often as loud as that from the larger Scaly-bellied Green Woodpecker.

Rufous-bellied Pied Woodpecker *Dendrocopos hyperythrus* (Vigors)

Rufous-bellied Sapsucker of Fleming and Fleming

Plate **23**
Map **281**

Hypopicus hyperythrus of Dillon Ripley's 'Revised Synopsis'

DESCRIPTION

Body length 20-23 cms.
Wing length 12-13.6 cms.
Tail length 6.9-8.7 cms.

This is a medium sized woodpecker, noticeably smaller than the Himalayan Pied *(D. himalayensis)* but usually a bit larger than the Brown-fronted *(D. auriceps)*, with which two species it is sympatric in Pakistan. This is a beautiful woodpecker when seen closely with the whole of the throat, sides of neck and breast a rich chestnut red, offset by crimson under-tail coverts and black and white chequered back and wings with clearly defined horizontal white bars across the mantle and wings. The wedge shaped tail is nearly all-black and in male birds the whole of the crown extending down the nape and curving around the sides of the neck, is shiny crimson-red. Females have the crown black with parallel lines of clear white spots. The strongly formed wedge shaped bill usually shows yellowish at the base of the lower mandible, blacker on the culmen and tip. The irides are reddish brown and the zygodactyl feet and legs are dark olive grey. There is no black border to the male's red crown as in *D. himalayensis,* and the ear coverts are usually paler creamy white. The feathers on the tibia are also whitish with black cross-barring. There is a very indistinct darker malar streak in both sexes. Juveniles also have black crowns, spotted with white.

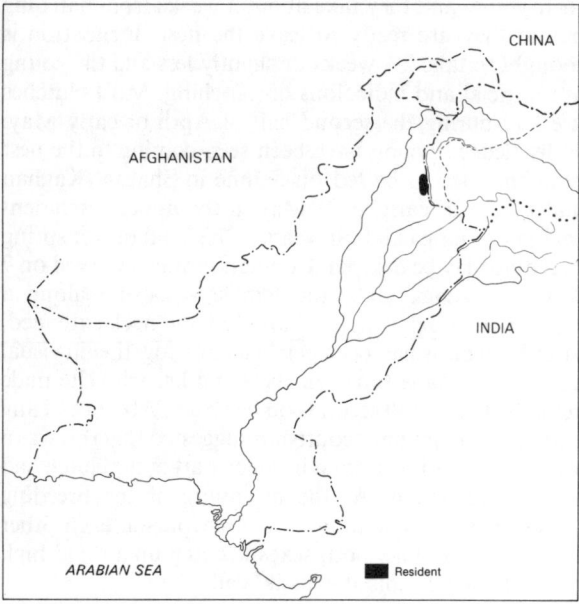

Map 281: Rufous-bellied Woodpecker or Rufous-bellied Sapsucker *Dendrocopos hyperythrus*

HABITAT, DISTRIBUTION AND STATUS

It occurs across the Himalayas at lower altitudes, extending to Bangla Desh, Burma, northern Thailand, Indo-China, China and Korea. This is a forest dwelling woodpecker confined to the Himalayan regions where deciduous trees are plentiful or dominant. In Pakistan it has only been recorded in the Murree hills and usually at elevations between 1,520 and 1,980 metres (5,000 and 6,500 feet). It is characteristically associated with such broad leaved trees as the Himalayan Poplar *(Populus ciliata)* and the False Dogwood *(Cornus macrophylla)*. It has been recorded in mixed coniferous and broad leaved forest below Murree at about 1,800 metres (6,000 feet) elevation (Roger Holmes, pers. comm., 1968 and Clyde Priddy, specimen collected Ghora Gali, 1965). In the Galis it is found in Kao forest below 1,900 metres (6,500 feet). It does not appear to have been recorded from Hazara district or Kashmir, but Magrath (1908*a*) encountered it on the slopes below Thandiani, the northern-most group of hills in the Murree and Gali range. Status SCARCE.

HABITS

Usually encountered solitarily or in pairs and preferring shady ravines and to forage on the main stem of trees. In mid-October in Kao forest a family party of two adults and three juveniles was observed feeding in the same vicinity over a week long period. Besides puncturing the cambium layer of deciduous trees for sap-sucking, they dig and probe in bark in typical woodpecker fashion for the larvae of wood boring insects and will 'lick-up' ants and termites *(Isoptera)*. Whistler observed one flying out from a Horse Chestnut tree *(Aesculus indica)* and successfully catching *Pierid* butterflies *(Ibis,* 1930), and in Kao forest they have been most frequently observed working along the branches in the upper canopy of trees.

BREEDING BIOLOGY

Whistler (op. cit., 1930) noted that fresh eggs could be found in the latter half of April and during May, in the Murree hills. They bore a typical woodpecker's nest hole, with circular entrance in a vertical tree trunk usually 5 or 6 metres from the ground and the 4 or 5 plain white eggs being laid at the bottom of a widened nest chamber without any lining added to the excavation. As far as is known both sexes share incubation duties and feeding of the young, which are born blind and naked. No other details appear to have been recorded.

VOCALIZATIONS

A rather silent species, with a very distinctive high pitched call which can be given in a sequence of rapidly repeated calls 'chit-chit-chit-r-r-r-r-h', less strident than the 'chek-chek' calls of *D. himalayensis* and more commonly uttered in trills. Magrath in describing birds at Thandiani recorded their calls as the 'running down' noise of a small clock or the running out of a large fishing reel *(JBNHS,* Vol. 19, 1909*a*). They will also occasionally drum with their bills on the upper dried out branches of a tall tree and this drumming may be performed by both sexes.

Yellow-fronted Woodpecker *Dendrocopos mahrattensis* (Latham)

Plate **23**
Map **282**

Mahratta Pied Woodpecker or

Yellow Crowned Pied Woodpecker *Picoides mahrattensis* of D. Ripley's 'Synopsis', 1982

Synonym *Leiopicus mahrattensis* of Stuart Baker

DESCRIPTION

Body length 17-18 cms.
Wing length 9.4-11 cms.
Tail length 5.4-6.4 cms.
Bill length 2-2.6 cms.

This is a small woodpecker even less in size than the Brown-fronted *(D. auriceps)* which it closely resembles. The whole of the wedge shaped tail including central rectrices is barred black and white, unlike *D. auriceps* and the mantle, scapulars and wing coverts are speckled black and white with a less distinct pattern of horizontal barring than in *D. macei* or *D. auriceps*. The white spots are somewhat larger and closer together on the back in this species. The male has the whole of the crown golden yellow turning scarlet and crimson on the occiput with a slight crest, whilst the female has the whole of the crown a more olive yellow lacking any crimson on the hind part of the crown. Neither sex has any dark malar streak, the lower part of the cheeks and sides of neck being streaked darker grey-brown. The throat and breast is dull creamy or dingy white streaked brownish grey, especially along the flanks, and the middle of the belly extending well up between the thighs, is crimson-pink varying to orange-red. In both sexes the straight pointed bill is plumbeous or slaty grey and the irides are dark red. The zygodactyl feet are blue-grey. The under-tail coverts in this species are not crimson as in most other *Dendrocopos* woodpeckers, but are white, streaked or cross-barred with grey-brown. The forecrown tends to be iridescent straw yellow in both sexes and the whole crest is erectile when the birds are excited. The sides of the neck in some specimens are quite chestnut brown and the dorsal pattern is more suggestive of marbling than a ladder pattern.

HABITAT, DISTRIBUTION AND STATUS

It occurs throughout India, Sri Lanka, extending eastwards into Burma and Laos. An inhabitant of lowland or plains areas it is virtually absent from Baluchistan or any part of the northern foothill areas. It is well adapted to open country with quite sparse tree cover, frequenting canal and roadside tree plantations but is sometimes rather rare or local in distribution. It occurs around Peshawar but Whitehead did not encounter any around Kohat or the lower Kurram valley though he did find it common in the vale of Peshawar (Whitehead, 1911). It can be seen in Islamabad and in both Jhelum and Mianwali districts and is quite plentiful in the well established canal colonies of the Punjab especially around Faisalabad and Sahiwal. In Sind it occurs very sparingly in Thatta district, and in most of that province is less frequently encountered than in the Punjab, but curiously it is resident in the sparse *Prosopis spicigera* thorn scrub of the Thar desert. Specimens seen near Chhor and Umarkot (author obs. 1975-6). Whistler considered it a common resident in the Potohar region of northern Punjab (*Ibis*, 1930). H. Waite collected one specimen (in British Museum) in the Murree foothills at Kotli 1,500 metres (5,000 feet), but there are no recent sightings in this area. It is often sympatric with the Golden-backed Woodpecker and the Sind Pied Woodpecker within extensive riverain forest and wild olive forest in the Salt Range where *Dendrocopos assimilis* occurs. Status FREQUENT.

HABITS

Usually seen singly or in pairs and almost exclusively arboreal, rarely descending to the ground. They are quite tame in the presence of humans, and hunt in typical woodpecker fashion, hopping jerkily up the trunks of trees and along lateral branches, clinging upside down

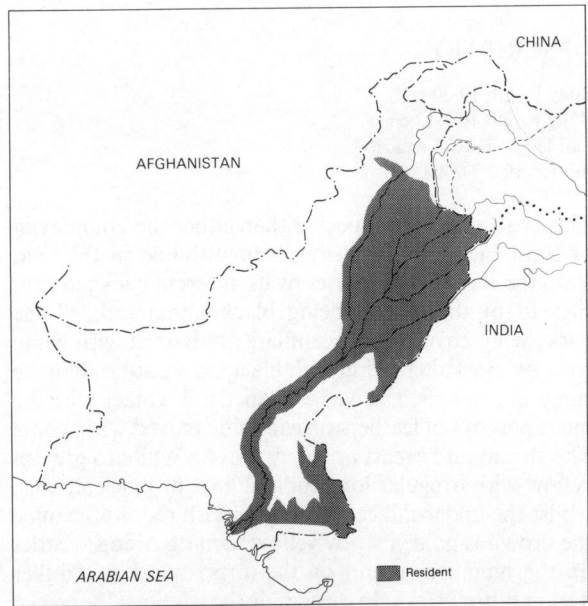

Map 282: Yellow-fronted Woodpecker, Mahratta Pied or Yellow-crowned Pied Woodpecker *Dendrocopos mahrattensis*

with equal facility. They are quite adaptable in exploiting flower nectar, fruit and berries when available, as well as bark dwelling insects and their larvae. Stomach contents of 3 birds collected in May, July and September from cropland in Bihar (India) showed a high proportion (90 per cent) of injurious insects in their diet, including Geometrid moth larvae, Elaterid beetle larvae (Click Beetles), *Myllocerus* weevils and Buprestid larvae (wood boring beetles). (Mason and Lefroy, 1912). In the author's garden at Khanewal (Punjab) they were observed in January taking the nectar from Silk Cotton flowers *(Salmalia malabarica).* Adults were observed bringing dragonflies (Odonata) to their nestlings (Ali and Ripley, 'Handbook', 1970).

BREEDING BIOLOGY

Whistler noted most eggs laid during April for this species in Rawalpindi district *(Ibis,* 1930) and K. Eates (unpublished MS notes) found most eggs in March in Sind. He noted that the nest hole is a neat circle about 2.8 cms. in diameter, usually drilled into a dead or decaying branch and quite commonly very high up in a tree. In Nepal, Fleming and Fleming (op. cit., 1976) noted nest holes frequently near the ground. Ali and Ripley (op. cit.

'Handbook', 1970) describe the nest hole as up to 4 cms. in diameter and with the vertical tunnel descending 15 to 40 cms. In Sind the normal clutch is 3 eggs and fresh eggs with incomplete clutches were found on 22 February and 23 March by K. Eates whilst another nest with 3 young was found on 27 March. It is believed that both sexes assist in nest hole excavation and incubation of the eggs. These are plain white glossy and rather elongated ovals. Both sexes feed the young by regurgitation. Period of incubation and fledging has not been recorded but is believed to be around 14 to 15 days for the former.

*VOCALIZATIONS

Rather a silent species outside of the early spring and onset of breeding. Their contact or advertising calls are relatively weak short calls, shriller in tone than that of *D. assimilis.* A pair calling together on 23 March in Multan uttered single notes 'peek-peek' at irregular intervals. A male near Shahpur in early March was presumed giving advertising calls, a rapidly repeated 'kik-kik-kik-r-r-r-r-h'. This individual was also heard drumming on an *Acacia arabica* tree, making a relatively weak or muffled sound.

Brown-fronted Woodpecker *Dendrocopos auriceps* (Vigors)

Plate 23
Map 283

Picoides auriceps of Dillon Ripley's 'Revised Synopsis'

Dryobates auriceps of Stuart Baker's 'Fauna Series'

DESCRIPTION

Body length 19-20 cms.
Wing length 10.5-13 cms.
Tail length 6.8-7.3 cms.
Bill length 2.1-2.4 cms.

This is a smaller woodpecker than either the Himalayan or Sind Pied, and is easily distinguishable in the field from the former two species by its different back pattern. Instead of the mantle being black unmarked, all the back, wing coverts and scapulars are barred with white rows of spots alternating with black, in a pattern like the rungs of a ladder. The wedge shaped tail is black with the outer pair of tail feathers on each side barred with white. The throat and breast are dirty greyish white to greyish yellow with irregular longitudinal lines of black streaks, whilst the under-tail coverts are pinkish red. In the male the crown is golden straw yellow turning orange-scarlet on the hind crown and on the forecrown it is a duller darker buffy brown. In the female the whole of the crown is a greenish yellow. Both sexes have a darker greyish buff loral streak below the ear coverts becoming blacker on the side of the neck. The legs and feet (which are

zygodactyl) are greyish green and the straight dagger shaped bill is bluish grey, darker on the tip. The irides are red-brown, being duller brown in females and redder in males. Juveniles have buffy brown crown rather like the female.

HABITAT, DISTRIBUTION AND STATUS

A Himalayan species occurring as far east as central Nepal and in the west as far as Nuristan in Afghanistan. This woodpecker is confined to a relatively narrow altitudinal range being typically associated with the sub-tropical pine zone *(Pinus roxburghii)* between 900 and 1,800 metres (3,000-6,000 feet) elevation. Occasionally it descends to the lower limit of the 'Chir' Pine *(P. roxburghii)* zone and has been noted in the Margalla hills in January up to early April, and will also occasionally straggle up to the summit ridge of the Murree hills about 2,438 metres (8,000 feet). In Pakistan it is more or less restricted to the outer Murree foothills, extending eastwards into Lehtrar, Kahuta and Poonch and occurring also in the lower side valleys of Mansehra district on the Batrasi pass and in Hazara tribal division, always in these zones in *Pinus roxburghii.* Further north Fulton found it fairly common in lower Chitral around 1,200 metres (4,000 feet) (Fulton, *JBNHS,* 1904). Whitehead (*JBNHS,* 1911) noted it as a common winter visitor to Kohat and Kurram districts, and there is a small resident population still surviving in the Takht-e-Suleiman range on the Shingar along the border between south Waziristan and Zhob district of Baluchistan in Chilghoza *(Pinus gerardiana)* forest. Birds noted on the Shingar at the beginning of May 1984 by the author were only observed searching the trunks of *Pinus gerardiana* and no courtship activity was noted. Status locally COMMON.

HABITS

They forage singly or in pairs and often follow or associate with mixed feeding parties of tits and minivets. They appear to descend very rarely to the ground and hunt food in typical woodpecker fashion, propped upright on their stiff tail feathers, and to hop rapidly sideways, vertically upwards and even downwards on the trunks and lateral branches of trees. Besides tapping the bark and probing cracks and knot holes in their search for insect larvae, they are able to supplement their diet with berries and the seeds of *Pinus roxburghii.* In Mussoorie (India) they have been reported damaging ripe fruit such as pears, pecking into the ripening fruit on the tree (Mason and Lefroy, 1912). They will also forage for insect larvae and pupae on bushes in the forest understorey.

BREEDING BIOLOGY

Christison found several nests on the Shingar, always fairly high up, from 4.5 to 7.5 metres from the ground, the nest holes being bored into the underside of a branch of a Chilghoza Pine *(P. gerardiana)* (Christison, *JBNHS,* 1942). The full clutch is usually 4 eggs which are plain white with hard glossy shells. Christison excavated a nest hole on the Shingar which contained 4 eggs on 5 June (*JBNHS,* 1942). A nest hole in Dunga Gali was utilized for two successive years by a pair and was located at 12 metres up in a dying Himalayan Elm *(Ulmus wallichiana).* On 5 June both parents were seen carrying

Map 283: Brown-fronted Woodpecker *Dendrocopos auriceps*

food into the nest hole whereas the following year on 25 May the young were further advanced and could be seen looking out from the nest hole. Little has been specifically recorded about the breeding of this species, but presumably both sexes assist in nest hole excavation, as well as incubation of the eggs, though it is likely that the female performs most of the incubation and that the nidicolous chicks, develop rapidly being able to leave the nest in under three weeks.

VOCALIZATIONS

Its calls are more rapid and squeaky than those of the Himalayan Pied (*P. himalayensis*) and the contact or advertising call is a series of rapidly repeated cries 'chitter-chitter-chitter-r-r-rh'. They will also give single short sharp cries 'chik-chik' when feeding in the vicinity of their mate. Fleming (*et al.,* 1976) describes a variety of calls including a subdued 'tu-whit' contact call and a high 'cheek-cheek-cheek-rrrr' presumably similar to the 'chitter-chitter-r-r-rh' calls described above.

Fulvous-breasted Woodpecker *Dendrocopos macei* (Vieillot)
Fulvous-breasted Pied Woodpecker *Picoides macei* of Dillon Ripley's 'Revised Synopsis'
Dryobates macei of Stuart Bakers's 'Fauna Series'

Plate **23**
Map **284**

DESCRIPTION

Body length 19 cms.
Wing length 11-12 cms.
Tail length 6.5-7 cms.

This is a small pied woodpecker, averaging about the same size as the Brown-fronted Pied *(Dendrocopos auriceps)* in the western part of the Himalayas. It has a ladder pattern on its back, of parallel black and white bars, extending over the scapulars and secondaries. The wedge shaped tail is black with only the outer tail feathers spotted with white. In the male the whole of the crown and nape is crimson-red whilst in the female this area is black. The rest of the face, throat and upper breast is dingy yellow-grey with a brownish buff cast especially noticeable in the pectoral region. One female observed closely (author) had the breast olive grey in colour. A black moustachial streak extends from the base of the bill curving down the side of the neck, in both sexes, and there are faint greyish streaks in the region of the lower breast and belly. The under-tail coverts are crimson. Both sexes have red-brown irides and greenish grey legs and feet. The bill is blackish on the culmen and paler blue-grey at the base of the lower mandible. The nostrils and immediate forecrown are covered with dull white bristly feathers. In the field it is closely similar to *Dendrocopos auriceps* but the crown pattern is distinctive being solidly black in females, and bright crimson in males, in contrast to the yellow-buff or orange-yellow crown of *auriceps*. The two species are only likely to occur on the same ground during the winter months and the heavier streaking on the breast of *auriceps* is also an easily seen field character.

HABITAT, DISTRIBUTION AND STATUS

An Oriental species occurring in the foothill region of the Himalayas and the Terai, eastwards through Bangla Desh, Burma, Thailand, southern Indo-China and Indonesia as far east as Bali. This is a sedentary woodpecker, associated typically with deciduous forest at low elevations along the Himalayan foothills. The only places where the author has encountered this species are in the side ravines of the Margalla hills and at elevations between 600 and 900 metres (2,000 and 3,000 feet) in a tropical dry deciduous scrub forest biotope. Records by Captain Marshall (1873) and Colonel Rattray for the Murree hills (1905) do not give details of elevation but it is unlikely that it would occur in the vicinity of Murree which lies above 1,800 metres (6,000 feet) with a predominantly coniferous forst cover. It does occur however, above this altitude in the eastern Himalayas (Ali and Ripley 'Handbook', 1970). Status local and RARE.

HABITS

When encountered this woodpecker appears relatively tame in the presence of humans. In the Margalla hills, it is well adapted to search the stems and boles of small trees as well as bushes, usually flying first to the very bottom of a tree near ground level and then working its way methodically upwards. It also frequently descends to the ground to hop about in search of ants (observed Nurpur Shahan). It behaves in typical woodpecker fashion, tapping vigorously on the branch or trunk in likely places, digging and probing with its bill for insect larvae. It is reported to supplement its insect diet with berries and seeds (Ali and Ripley, 'Handbook', 1970). Only single individuals have been observed in Pakistan.

Map 284: Fulvous-breasted Woodpecker *Dendrocopos macei*

BREEDING BIOLOGY

In the Murree hills, Marshall (in Whistler, *Ibis,* 1930) found a clutch of 3 unincubated eggs on 2 June at 1,980 metres (6,500 feet). They excavate a typical woodpecker's rounded tunnel in a tree and April to May is the usual breeding season (Ali and Ripley, op. cit.).

Both sexes have been recorded, taking part in nest hole excavation, incubation duties and feeding of the young. The plain white eggs are glossy rounded ovals and clutches of 4 to 5 are uncommon, 3 being the normal.

VOCALIZATIONS

Individual birds heard calling in the Margalla hills uttered single noted calls at irregular intervals 'pik-(pause)-pik', not very loud or ringing. A male bird on 24 May was watched drumming on a *Ficus bengalensis* tree growing out of a rock crevice. Short bursts of 1 to 2 seconds' duration were produced over a period of about 8 minutes by this individual and they were not very loud, possibly due to the relatively soft nature of the substrate. It did not call during the drumming and subsequently flew away out of sight. Fleming describes longer call sequences 'peek-peek-trrrrrr' (Fleming *et al.*, 1976).

Grey-capped Pygmy Woodpecker *Dendrocopos canicapillus* (Blyth)
Grey-headed Pied Woodpecker *Picoides canicapillus* in Ripley's 'Revised Synopsis'

Yungipicus hardwickii mitchellii in Stuart Baker's 'Fauna Series'   Plate **23**
Map **285**

DESCRIPTION

Body length 13.75-15 cms.
Wing length 8.4-9.4 cms.
Tail length 4.0-4.7 cms.
Bill length 1.7 cms.

This diminutive woodpecker looks like a smaller edition of the larger ladder-backed *Dendrocopos* woodpeckers, having a relatively long wedge shaped tail with stiffened central tail feathers and a relatively long neck and bill, features which separate it from the piculets. The back and wings are black with white spots forming cross-bars which also extend over the scapulars. The throat and breast are dirty white, streaked all over with dark grey and the lower belly is more fulvous in tone. The rump in this species is conspicuously black, a helpful field point in distinguishing it from the slightly smaller Brown-capped Pygmy Woodpecker *(D. nanus)* which has a white rump. It also lacks any crimson under the tail which is dull fulvous, streaked black, serving to distinguish this species from either *D. macei* or *D. auriceps.* Both sexes have the crown ashy grey extending down to the nape, with a prominent whitish supercilium extending from above the eye and down the sides of the neck, the ear coverts being streaked blackish grey. The grey crown is broadly edged with black, particularly in the nape region, and in males a narrow crimson streak shows beneath the black border on either side of the nape. This red patch is entirely lacking in females and is almost impossible to see in the field, particularly in the population inhabiting the western Himalayas (sub-species *mitchellii*). The bill is dusky black and also the legs and feet in both sexes. Sometimes the toes have an olive tinge. The irides are red-brown or hazel brown. The iris of the Brown-capped Pygmy Woodpecker *D. nanus* is conspicuously pale lemon yellow. *D. canicapillus* averages slightly larger and has a heavier bill than *D. nanus.*

HABITAT, DISTRIBUTION AND STATUS

An Oriental species found across the lower regions of the Himalayas, through Indo-China, Malaysia, China, Korea, Taiwan and the extreme south-eastern corner of the USSR (Primorye Krai) and Japan. It is a resident species confined to the lower or outer foothill regions of the Himalayas extending to the adjacent 'Terai' or 'duars' and typically associated with deciduous tree forest. In Pakistan it is mainly a sporadic visitor with possibly a small resident population, confined to elevations between 450 and 914 metres (1,500 and 3,000 feet) in the Margalla hills and lower foothills of Kahuta, Lehtrar

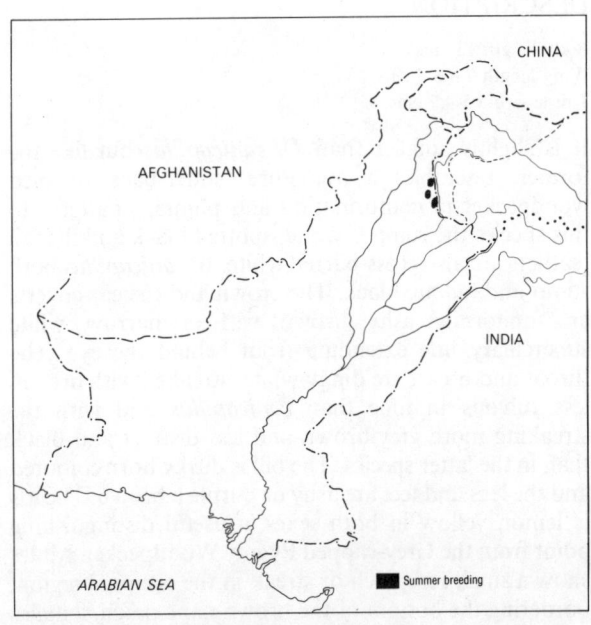

Map 285: Grey-capped Pygmy Woodpecker or Grey-headed Pied Woodpecker *Dendrocopos canicapillus*

and Poonch. The first authentic record was of family parties observed at Lehtrar at 1,060 metres (3,500 feet) in 1910 by Hugh Whistler (*JBNHS*, Vol. 22, p. 626 and specimens in the British Museum). These were originally identified as *Yungipicus pygmaeus*. On 27 November 1966, a male was watched for a considerable period in an Islamabad garden (Guy Roberts, pers. comm.) and the author has one record from the Margalla hills on 5 April 1974, an individual watched drumming from one of the upper branches of a stunted *Salmalia* tree. On 23 May 1986, a pair was watched feeding on a *Salmalia* tree in the Manga valley (900 metres/3,000 feet) (author and M. Mallalieu). There are no other authentic records and it must be considered RARE and very local.

HABITS AND BREEDING

Their small size enables them to exploit the outermost thinner branches and twigs of trees so that they tend to forage in the top of the tree canopy. They are extremely active and agile, clinging sideways as well as underneath slender twigs and jabbing vigorously and rapidly at bud axils and crevices in the bark. Birds watched by the author were tame and allowed close approach. The male

and female seen on 23 May appeared by their behaviour to have already established a pair bond. Both sexes share in nest hole excavation, a shaft burrowed 10 to 20 cms. deep down through a thin branch of a tree, having an entrance hole usually less than 3 cms. in diameter and often located on the underside of sloping branches. The eggs are glossy, plain white with a normal clutch of 4, and are laid mostly in April and May. Both sexes share in incubation and feeding of the young. Stuart Baker in his 'Nidification' (4 Vols. 1932-5) believed the incubation period to be 12-13 days.

VOCALIZATIONS

Its calls are relatively high pitched and weak as might be expected from its size and tend to taper off in a slur 'tit-tit-erh-r-r-r-r-h'. The individual observed in Margalla hills attracted attention by calling in a rapid squeaky 'chitter-r-r-r-h'. Their drumming calls, produced by rapid blows of the bill tip are also rather muted in tone but the noise carries a considerable distance. They also emit single cries as a contact call rendered by Fleming as 'tzit' (op. cit. 1976).

Brown-capped Pygmy Woodpecker *Dendrocopos nanus* (Vigors)
or Indian Pygmy Woodpecker

Synonym *Dendrocopos moluccensis*

DESCRIPTION

Body length 13 cms.
Wing length 7.4-8.3 cms.
Tail length 3.5-4.2 cms.

It is slightly smaller than *D. canicapillus* but like the former, resembles a miniature ladder-back or pied woodpecker in conformation and plumage pattern. In this species the rump is white, spotted black and the tail feathers are also cross-barred white. In *canicapillus* both rump and tail are black. The crown and the ear coverts are uniformly ashy brown with a narrow white superciliary line extending from behind the eye. The throat and breast are dingy white, streaked with brown, less fulvous in tone than *canicapillus* and with the streaking more grey-brown and less distinct and black than in the latter species. The bill is dusky horn coloured and the legs and feet are ashy or purplish brown. The iris is lemon yellow in both sexes, a useful distinguishing point from the Grey-capped Pygmy Woodpecker. Males show a small red patch or streak in the occipital region, bordering the bottom of the brown nape patch, females

lack this red spot. In the field this red spot is very difficult to discern.

HABITAT, DISTRIBUTION AND STATUS

Ali and Ripley in volume 4 of the 'Handbook' give Kahuta in the Murree foothills as the western limit of the distribution of this species, referring to Whistler's specimens of Pygmy Woodpeckers in the British Museum, which are in fact, *D. canicapillus*. *Yungipicus pygmaeus* was synonymized by Stuart Baker with *yungipicus hardwickii mitchellii* which is the western sub-species of the Grey-capped Pygmy Woodpecker, whereas *Y. hardwickii brunniceps* is now considered a seperate species, the Brown-capped, and is confined to sub-tropical zones and allopatric with the Grey capped. Until further evidence is forthcoming, this species is excluded from the **Checklist** of birds of Pakistan. H. Waite collected specimens of this woodpecker (now in British Museum) in the plains of east Punjab in Ambala and Hoshiarpur (India).

Glossary of Terms

A. Terms Associated with Irrigation

Barrage A concrete or masonry structure built across a river to hold a series of lock gates through which water-flow can be regulated or entirely held up in order to feed major irrigation canals.

Borrow pit Excavations alongside the protective embankment of canals, drainage channels, rivers or railway tracks, which frequently become filled with water, sometimes creating permanent swamps.

Headpond The reservoir created upstream of an irrigation barrage, usually much broader than the natural river channel but not wholly contained by artificial embankments.

Headworks An irrigation barrage with attendant protective embankments to control the river and prevent it by-passing the barrage, which together with the start of major canals often creates an extensive marshy area.

Minor A smaller irrigation channel taking off from a canal.

Seepage zone In a country with very low rainfall and high surface water evaporation such as Pakistan, most swamps and wetlands are created by subterranean seepage not surface drainage or run-off. Irrigation canals, deliberately constructed above the surrounding land, so that irrigation can take place by gravity water flow without the aid of pumping, has led to seepage alongside such canals wherever they are unlined. Seepage also occurs in low lying areas adjacent to the main river systems because of the gradual buildup of the river beds from siltation and the generally very flat topography of the surrounding land.

Spoil banks All the major rivers of Pakistan from which irrigation water is drawn, carry heavy silt loads and some of this settles out as the water velocity decreases in smaller man-made channels. This necessitates regular silt clearance or digging out of such canals, the resultant spoil of this, forming steep-pitched banks alongside the channel.

Water logged area A region formerly dry and agriculturally productive which has developed into a semi-swamp due to rising ground water. Such areas may be dry on the surface but with a very high ground water table and will only support plants adapted to swampy conditions. Due to concentration of soluble salts on the ground surface as a result of evaporation, most waterlogged areas are very saline.

Weep hole A deliberately constructed hole in a stone or concrete supporting wall or revetment to allow for natural drainage of water through the wall. A common feature of the bridge abutments where road bridges cross ravines in hilly areas.

B. Vernacular Terms

Bajra Usually grown as a green fodder crop—*Pennisetum typhoides*.

Band An earthern man-made embankment often built to control flooding upstream of an irrigation barrage headworks.

Banyan *Ficus bengalensis*. A huge evergreen wild fig tree often planted by village ponds or tanks in the Salt Range. Its roots grow down from horizontal branches and take root.

Barani Tracts of cultivation dependent upon natural rainfall and without artificial irrigation.

Bazaar A commercial and retail vending region of a town or city.

Ber Thorny trees or shrubs of the *Ziziphus* genus, which bear small plum-like drupes or fruit, often several times in the course of the summer.

Berseem *Trifolium alexandrinum*. Egyptian Clover.

Chappati A flat round pancake of oven-baked, unleavened whole wheat flour.

Chilghoza *Pinus gerardiana*, the Edible Seed Pine. A very xerophytic pine tree with beautiful silvery green bark and not normally growing below 2,100 metres (7,000 feet) elevation.

Chinar *Platanus orientalis*. A tall tree often planted around villages in the dryer inner Himalayan valleys.

Chir	*Pinus roxburghii,* the Long Needled Pine or Chir is confined to the sub-tropical or warmer outer foothills between 900-1,800 metres (3,000-6,000 feet) elevation.
Dhand	Term used in Sind to describe a perennial or seasonal lake or large pond or area of swampland.
Doab	Term used in Punjab for the tract of land between two major rivers thus the Bari Doab lies between the Ravi river on the north and the Sutlej on the south.
Gali	A term generally applied to mountain ridges or saddles in Hazara district. Plural 'Galian' probably derived from Hindkoh language. Not to be confused with the English term 'gully' meaning ravine.
Gol	A term widely used in Chitral for a valley or short amphitheatre surrounded by mountain slopes such as a 'col'.
Gujar	A nomadic tribe of sheep and cattle herders who migrate each summer up to traditional alpine grazing pastures in Hazara district and Azad Kashmir.
Jaman	*Eugenia jambolana.* An evergreen tropical tree which bears small black fruits in June.
Jheel	Generally applied to a perennial lake, and more commonly used in the Punjab.
Jhogi/Jogi	A tribe of professional snake-catchers or snake charmers, who are often semi-nomadic.
Jowar	*Sorghum sudanese.* Usually grown as a green fodder crop and sown in the summer season.
Khareze	A term used in Baluchistan to describe skilfully excavated subterranean tunnels which run from the foot of a mountain range down to the lower slopes of valleys. These tunnels are designed to collect ground water by seepage for crop irrigation lower down the valley floor. They are excavated by hand by means of regularly spaced vertical access shafts.
Kikar	*Acacia arabica.* A thorny tree highly adapted to both periodic inundation and saline soils and therefore the dominant species in the riverain forest and regions close to the main rivers.
Killi	A term used in Baluchistan to describe a fortified or walled village but often now part of the name of quite modern towns.
Kohistan	A mountainous tract such as that of Sind province west of the Indus river.
Kutchas	Tracts alongside main river channels, subject to periodic inundation but regularly cultivated by local farmers after floods subside.
Lora	A term used in Baluchistan to describe a small

	stream usually flowing through a very barren arid tract and eventually drying up or disappearing below ground without becoming part of any major river system.
Maidan	Originally confined to wide open grassy areas in mountainoous regions often equivalent to an 'alp' in Europe, and also applied to sports' fields or open grassy areas even in towns in the plains.
Marg	Term used in Kashmir for an open grassy slope or 'alp'.
Mirbar	Professional fishing tribes or castes in Sind, especially in Dadu district.
Mohanna	Professional fishing tribes or castes, widespread in Punjab as well as Sind and confined to exploiting fresh water fish.
Neem	*Azedarachta indica.* A tropical deciduous tree, bearing drupe-like fruits in June and July and widely grown in the plains.
Nullah	A ravine or valley.
Pat/Patts	Smooth flat clay beds between sand dunes, generally characterized by saline conditions and very sparse vegetation.
Pipal	*Ficus religiosa.* A huge spreading fig tree widely planted in towns for shade. It grows wild in the foothills especially in the Margalla hills. It is often a strangling fig on other trees.
Potohar	The high plateau region between the Salt Range escarpments and the Himalayan foothills, consisting of loess soil deposits. Characteristically this area is scarred by very deep erosion gullies.
Powindah	A Pushtu word for nomadic tribes of herders.
Shikari	Any hunter of game, but more particularly applied to a local villager who makes this his profession and who acts as a guide to amateur sportsmen.
Taluka	Administrative sub-district, a term generally used in Sind and equivalent to the Tehsil in Punjab.
Tangi	A term used in Baluchistan, applied to narrow canyons or ravines especially in the Juniper forest zone.
Tehsil	An administrative sub-district roughly equivalent to an English local council district or parish and a sub-division of a larger district.
Wadi	A dry ravine or stream bed which may only carry water occasionally during flash floods.
Zikari	Professional fishing tribes or castes confined to the Makran and Lasbela seacoast with rather unorthodox religious beliefs and practice, according to Muslim orthodoxy.

C. Technical Terms

Agonistic	Behaviour between two animals or birds relating to threatened or potential combat.
Allopatric	Mutually exclusive geographically. Species which may occur so close together that their geographic ranges may be contiguous, but they still do not overlap.

Allopreening	The preening of one bird by another, usually between mates or closely related birds.
Altricial	Helpless when hatched.
Arthropods	Animals with jointed feet including spiders, insects, crustacea and myriapoda.
Boreal	Coniferous or needle-leaf forest zone pre-

Commissure dominating between the Arctic and the temperate deciduous forest zone.
The meeting line between the two mandibles when viewed laterally.

Congeneric (congener) Two or more species usually placed in the same genus, or at least closely related.

Cosmopolitan A species of very wide geographical distribution that is able to live in a variety of climates and geographical regions.

Culmen The dorsal ridge of the upper mandible.

Cursorial Adapted for running over the ground.

Dimorphic Occurring in two different forms or appearances, generally applied to males and females of the same species with consistently different size or plumage pattern.

Disjunct In an ecological context, not occurring continuously over a region, but localized into widely separated populations.

Eclectic In an ecological context, very broad in choice of habitat or able to survive successfully over a wide range of different habitats.

Endemic A group or kind (Taxon) of animal, bird or plant which is local and of limited distribution.

Filamentous Hair-like, but used especially to describe ornamental plumes, in which the barbs are widely separated because the whole feather is elongated and developed for the purposes of display.

Fluvial Of or pertaining to rivers.

Forb A broad-leaved herb.

Frugivorous Fruit eating.

Gonys The ridge or protrusion formed by the junction of the two lower (mandible) jaw bones or rami.

Granivorous Feeding on grains or seeds.

Gular The throat region.

Hallux The hind toe; actually the first toe anatomically and usually directed backwards in a bird's foot.

Halophytic Relating to plants able to withstand a high concentration of salt in the soil.

Holarctic In the distribution of birds and animals, pertaining to the whole northern and Arctic region.

Insectivorous Feeding on insects, but usually used in the wider sense to include all arthropods.

Insolation Exposure to the sun's rays, usually pertaining to temperature increase due to refraction of the sun's rays from different ground surfaces having different absorptive properties.

Inter-specific Reactions between two or more birds of different species.

Intra-specific Reactions between two or more birds of the same species, within the same species.

Lamellate Composed of thin plates or scales often used to describe covering of a bird's tarsus (leg).

Mesic A site that provides good water availability.

Monoga A mating system in which a long lasting pair bond is maintained.

Neritic The part of the pelagic (sea) zone which generally lies above the lowest tide and particularly relating to coastal inlets, lagoons and river estuaries.

Nictating membrane A third eye-lid composed of a transparent fold of skin.

Nidicolous Young birds which remain in the nest after hatching.

Nidifugous Young birds that leave the nest immediately or soon after hatching.

Nuchal Pertaining to the nape.

Nuptial Pertaining to the breeding season particularly in reference to display plumage.

Occipital Pertaining to the back of the head.

Oriental One of the main geographic regions with a relatively homogenous and continuous distribution of birds, plants and animals indigenous to the region and covering south-east Asia and the western Pacific.

Palearctic Another of the main geographic regions with a relatively homogenous plant and animal fauna indigenous to the area, and covering Europe, north Africa and the Arctic, boreal and temperate zones of Asia.

Pectinate Provided with a serrated or comb-like edge.

Pelagic Zone formed by the open waters of the ocean.

Polyandry A mating system in which females regularly mate with a succession of different males.

Polygyny A mating system in which males regularly mate with two or more females in the course of the breeding season.

Polymorphic Used in the context of birds to describe a species occurring in two or more differently appearing forms e.g. with varying plumage colour pattern.

Precocial Active immediately after hatching.

Ptilopaedic A young bird covered with down when hatched.

Pullus A young bird until it is fledged.

Pyriform Pear-shaped especially used in describing an egg.

Rachis The shaft of a feather, also called the quill, which bears the vane.

Rectrices Main tail feathers.

Relict Isolated and usually discontinuously occurring populations of a bird or animal which formerly occurred in a much wider distribution.

Remiges The main flight feathers in the wing that is primaries and secondaries.

Rictal Pertaining to the gape area, usually applied to bristles in that area.

Scree A precipitous stony slope.

Squamated Looking like scales, usually referring to a pattern produced by small overlapping feathers which are bordered by a contrasting colour e.g. a white feather with narrow black edges.

Steppic A region characterized by low rainfall insufficient for extensive tree growth and more typically characterized by a discontinuous cover of grass (bunch grass) with scattered bushes.

Sympatric Taxa (or different kinds) of animals, birds and plants which overlap geographically.

Syndactyl Having the third and fourth toes coalescent or fused for part of their length.

Talus	A sloping mass of broken rock or stones, usually lying at the base of a cliff.
Tarsus	Strictly speaking the region of the 'ankle' but used in these volumes to cover the leg formed by the fusion of the tarsal and metatarsal bones.
Taxon (plural taxa)	General term for a sub-division or category used in classification of birds, plants and animals.
Tibia	The thigh, or in reality that part of the leg above the ankle joint comprising the fused tibia and proximal tarsal bones.

Toti-palmate	Having all four toes (including the hind toe) joined by webs.
Vane	The web of a feather.
Web	The more or less coherent series of barbs which knit together on each side of the rachis or quill of a feather.
Xeric	Adapted to dry conditions or a site which provides dry conditions.
Xerophytic	A plant with capacity to withstand drought.
Zygodactyl	Having two toes directed forwards and two backwards.

APPENDIX II
Bibliography

A

Abdulali, Humayun, **1940:** 'Swifts and Terns at Vengurla Rocks', *JBNHS*, Vol. 41, No. 3, pp. 661-5; **1942:** 'The Terns and Edible-nest Swifts at Vengurla, West Coast, India', *JBNHS*, Vol. 43, No. 3, pp. 446-51; **1947:** 'The Movements of the Rosy Pastor in India', *JBNHS*, Vol. 46, No. 4, pp. 704-8; **1950:** 'Occurrence of the Chestnut-bellied Nuthatch (*Sitta castaneiventris*) in Sind—A Correction'. *JBNHS*, Vol. 49, No. 2, pp. 303-4; **1953:** 'The Pied Myna and Bank Myna as Birds of Bombay and Salsette', *JBNHS*, Vol. 51, No. 3, p. 737; **1964:** 'The Food and other Habits of the Greater Flamingo (*Phoenicopterus roseus*) in India', *JBNHS*, Vol. 61, No. 1, pp. 60-8; **1965:** 'Notes on Indian Birds 6—The Occurrence of the Pygmy Cormorant (*Halietor [Phalacrocorax] pygmeus*) in Baluchistan', *JBNHS*, Vo. 62, No. 3, p. 553; **1968:** 'A Catalogue of the Birds in the Collection of the Bombay Natural History Society-2. Anseriformes', *JBNHS*, Vol. 65, No. 2, pp. 418-30.

Ahmad, Ashiq, **1985:** 'Sighting of Whooper Swans (*Cygnus cygnus*) in Baluchistan', *JBNHS*, Vol. 82, No. 1, p. 192.

Ahmad, Mushtaq and Ali, Akhtar, **1979:** 'Evaluation of Damage caused by House Sparrows *Passer domesticus*, to different Varieties of Lentil (*Lens esculentus*) in Faisalabad, Pakistan', *Pakistan Journ. Zoology*, Vol. 1, No. 1, pp. 177-9.

Akhtar, S.A., **1947:** 'Ab-Istadeh, a breeding place of the Flamingo [*Phoenicopterus ruber roseus* (Pallas)] in Afghanistan', *JBNHS*, Vol. 47, No. 2, pp. 308-14.

Alam, Mohammed, **1982:** 'The Flamingos of Sambhar Lake', *JBNHS*, Vol. 79, No. 1, pp. 194-5.

Alekseev, A.F., **1980:** Houbara Bustard (*Chlamydotis undulata macqueeni*) in the North-west Kyzyl Kum (in Russian) *Zoology Zhurnal*, Vol. 59, No. 8, pp. 1263-6.

Alexander, Horace G., **1950:** 'Some Notes on the Genus *Phylloscopus* in Kashmir', *JBNHS*, Vol. 49, No. 1, pp. 9-13; **1951:** 'Some Notes on Birds in Lahul', *JBNHS*, Vol. 49, No. 4, pp. 608-13; **1969:** 'Some Notes on Asian Leaf-Warblers (Genus *Phylloscopus*)', privately printed—Truex Press, Oxford.

Ali and Ambedkar, V.C., **1956:** 'Notes on the Baya Weaver *Ploceus philippinus*' *JBNHS*, Vol. 53, pp. 381-9; **1957:** 'Further Notes on the Baya Weaver Bird *Ploceus philippinus*', *JBNHS*, Vol. 54, pp. 491-502.

Ali, Salim, **1927:** 'A Sind Lake', *JBNHS*, Vol. 32, No. 3, pp. 460-4; **1931a:** 'The Nesting Habits of the Baya *Ploceus philippinus*', **1931b:** *JBNHS*, Vol. 34, No. 4. pp. 947-64; 'Casualties among the Eggs and Young of Small Birds', *JBNHS*, Vol. 34, No. 4, pp. 162-7; **1931c:** 'The Role of Sunbirds and Flower Peckers in the Propagation and Distribution of the Tree Parasite *Loranthus longiflorus* in the Konkan (W. India)', *JBNHS*, Vol. 35, No. 1, pp. 144-9; **1941:** 'The Birds of Bahawalpur (Punjab)', *JBNHS*, Vol. 42, No. 4, pp. 704-47; **1945a:** 'More about the Flamingo (*Phoenicopterus ruber roseus*, [Pallas]) in Kutch', *JBNHS*, Vol. 45, No. 4, pp. 586-93; **1945b:** *The Birds of Kutch*, Oxford University Press, Bombay, 175 pp; **1945c:** 'The Avocet (*Recurvirostra avosetta*, Linn.) Breeding in India', *JBNHS*, Vol. 45, No. 3, pp. 420-1; **1946:** 'An Ornithological Pilgrimage, to Lake Manasarowar and Mount Kailas', *JBNHS*, Vol. 46, No. 2, pp. 286-308; **1949:** *Indian Hill Birds*, Oxford University Press, 188 pp; **1960:** 'Flamingo City' Revisited: Nesting of the Rosy Pelican, (*Pelecanus onocrotalus*) in the Rann of Kutch', *JBNHS*, Vol. 57, pp. 412-5; **1962:** 'WHO Bird Migration Study', *JBNHS*, Vol. 59, No. 1, p. 130 and No. 3, pp. 922-9; **1963:** 'A Note on the Eastern Spanish Sparrow, *Passer hispaniolensis transcaspicus*, Tschusi, in India', *JBNHS*, Vol. 60, pp. 318-21; **1969:** *Birds of Kerala*, 2nd edition, Oxford University Press, Calcutta and Madras; **1974:** 'Breeding of the Lesser Flamingo, *Phoeniconaias minor* (Geoffroy) in Kutch', *JBNHS*, Vol. 71, No. 1, pp. 141-4; **1977a:** *The Book of Indian Birds*, 10th edition (Revised), Bombay Natural History Society, Bombay; **1977b:** *Field Guide to the Birds of the Eastern Himalayas*, Oxford University Press, New Delhi, 265 pp.

Ali, Salim and Ripley, S. Dillon, **1968-74:** *Handbook of the Birds of India and Pakistan*, Vol. 1 1968, Vol. 2 1969, Vol. 3 1969, Vol. 4 1970, Vol. 5 1973, Vol. 6 1971, Vol. 7 1972, Vol. 8 1973, Vol. 9 1973, Vol. 10 1974, Oxford University Press, Bombay.

Ali, Salim and Ripley, S. Dillon with Dick, John Henry, **1983:** *A Pictorial Guide to the Birds of the Indian Sub-continent*, Oxford University Press, New Delhi.

Amadon, Dean, **1966:** 'The Superspecies Concept', *Syst. Zool.*, Vol. 15, pp. 245-9.

Amadon, Dean and Brown, Leslie, **1968:** See: Brown and Amadon, *Eagles, Hawks and Falcons of the World*.

Ambedkar, V.C., **1956, 1957:** See: Ali, Salim and Ambedkar, 'Notes on the Baya Weaver'; **1985:** 'Occurrence of the Sandwich Tern (*Sterna sandvicensis*) in India—A Ring Recovery', *JBNHS*, Vol. 82, No. 2, p. 410; **1987:** 'Recovery of an Indian Golden Oriole (*Oriolus oriolus kundoo*) in the USSR', *JBNHS*, Vol. 83, Centenary Supplement, pp. 211-2.

Amstutz, Bruce and Amstutz Mark C., **1977:** *A Checklist of the Birds of Islamabad and Nathia Gali*, published Asian Culture Study Group.

Attenborough, David, **1979:** *Life on Earth—A Natural History*, William Collins Sons and Co., London.

Austin, Oliver L., **1961:** *Birds of the World*. Paul Hamlyn, London, 317 pp.

Austin, Oliver L. and Kuroda, N., **1953:** 'The Birds of Japan: Their Status and Distribution', *Bull. Museum of Comparative Zoology*, Harvard, Vol. 109, pp. 278-637.

B

Baker, E.C. Stuart, **1906:** 'The Oology of Indian Parasitic Cuckoos',

Part 2, *JBNHS*, Vol. 17, pp. 352-74; **1913:** Game Bird Series, '*Indian Pigeons and Doves*', Witherby and Co., London, 260 pp; **1921:** *Indian Game Birds, Snipe, Bustards and Sandgrouse*, Vol. 2, published Bombay Natural History Society, London, 328 pp; **1930:** *Indian Game Birds, Pheasants, Bustard, Quail*', Vol. 3, published Bombay Natural History Society, London 341 pp; *Fauna of British India, Birds*, 2nd. ed., Vol. 1 1922, Vol. 2 1924, Vol. 3 1926. Vol. 4 1927, Vol. 5 1928, Vol. 6 1929, Taylor and Francis, London; **1932-5:** *The Nidification of Birds of the Indian Empire*, 4 Vols., Taylor and Francis, London.

Baker, Robin (Dr.), **1980:** (ed.) *The Mystery of Migration*, Macdonald and Jane's, London, 256 pp.; **1910:** 'Proceedings of the Baluchistan Natural History Society', *JBNHS*, Vol. 20, No. 2, pp. 541-2.

Barnes, Alice D., **1934:** 'The Longtail Duck (*Clangula hyemalis*), First Record for India', *JBNHS*, Vol. 37, p. 549; **1937:** 'Breeding of the Little Indian Nightjar (*Caprimulgus asiaticus*), in the Chinglepur District', *JBNHS*, Vol. 39, No. 4, pp. 865-6.

Barnes, H.E. (Capt.), **1880:** 'Notes on the Nidification of Certain Species in the neighbourhood of Chaman, S. Afghanistan', *Stray Feathers*, Vol. 9, pp. 212-20; **1881:** 'A List of Birds observed in the Neighbourhood of Chaman, S. Afghanistan', *Stray Feathers*, Vol. 9, pp. 449-60.

Barton, F.J.H. (Major), **1902:** 'The Painted Sand-grouse (*Pterocles fasciatus*) and the Wood-snipe (*Gallinago nemoricola*) in the Peshawar Valley', *JBNHS*, Vol. No. 3, pp. 606-7.

Bashir, El Sadig A., **1979:** 'Notes on Parakeet damage to Maize, Sunflower and Fruit Crops in Pakistan and suggested methods of control', Vertebrate Pest Control Centre Report, Pakistan Agricultural Research Council, Islamabad.

Basil-Edwardes, S., **1923:** 'Nidification of the Himalayan Tree-creeper (*Certhia himalayana*)', *JBNHS*, Vol. 29. No. 2, pp. 557-9; **1926:** 'A Contribution of the Ornithology of Delhi', *JBNHS*, Vol. 31, No. 2, Part 1, pp. 261-73 and *JBNHS*, Vol. 31, No. 3, Part 2, pp. 567-78.

Bates, R.S.P., **1936:** 'On the Birds of Kishenganga Valley, Kashmir', *JBNHS*, Vol. 38, pp. 520-39; **1938:** 'Rosefinches and other Birds of the Wardwan Valley', *JBNHS*, Vol. 40, No. 2, pp. 183-90; **1939:** 'Bird Photography in India', *JBNHS*, Vol. 40, No. 4, pp. 666-80; **1942:** 'A Month in the Kazinag Range', *JBNHS*, Vol. 43, pp. 60-72; **1943:** 'Note on the Feeding Habits of the Little Bittern (*Ixobrychus minutus*)', *JBNHS*, Vol. 44, No. 2, pp. 179-81; **1948:** 'Astanmarg', *JBNHS*, Vol. 48, No. 1, pp. 38-46; **1949:** 'The Merbal Glen and some Birds of the Pir Panjal', *JBNHS*, Vol. 48, No. 3, pp. 399-411; **1950:** 'The Lower Sind Valley, and some further Observations on Bird Photography', *JBNHS*, Vol. 49, No. 2, pp. 178-87.

Bates, R.S.P. and Lowther, E.N.H., **1952:** *Breeding Birds of Kashmir*, Oxford University Press, Bombay, 355pp.

Becking, J.H., **1981:** 'Notes on the Breeding of Indian Cuckoos', *JBNHS*, Vol. 78, No. 2, pp. 201-31.

Beg, A.R., **1975:** 'Wildlife Habitats of Pakistan', Bulletin No. 5 (Botany Branch), Pakistan Forest Institute, Peshawar.

Beg, Mirza Azhar, and Qureshi, Junaid Iqbal, **1972:** 'Birds and their Habitats in the Cultivated Areas of Lyallpur District and Vicinity', *Pakistan Journ. Agric. Science*, Vol. 9, pp. 161-6; **1975:** See Qureshi, J.I. and Beg, M.A.; **1976:** 'Birds of Scrublands of the Lyallpur Region', *Pak. Journ. Agric. Science*, Vol. 13, No. 4, pp. 17-22; **1980:** Birds of the University Campus (Faisalabad) and their Economic Significance', *Pak. Journ. Agric. Science*, Vol. 17, Nos. 1 and 2, pp. 7-12.

Bell, H.L., **1981:** 'Information on New Guinean Kingfishers, Alcedinidae', *Ibis*, Vol. 123, No. 1, pp. 51-61.

Bell, J. and Bruning D., **1974:** 'Handrearing Hemipodes at the New York Zoological Park', *International Zoo Yearbook*, No. 14, pp. 196-8.

Bell, T.R., **1906:** 'Occurrence of *Aegithaliscus coronatus*, Severtz, in Sind, *JBNHS*, Vol. 17, No. 1, pp. 244-5.

Berger, Andrew, See Van Tyne and Berger *Fundamentals of Ornithology*.

Betham, R.M., **1905:** 'Notes on Birds Nesting Round Quetta', *JBNHS*, Vol. 16, pp. 747-50; **1907:** 'Further Notes on Birds Nesting round Quetta', *JBNHS*, Vol. 17, pp. 828-32; **1916:** 'Birds Nesting Round Ferozepore', *JBNHS*, Vol. 24, pp. 829-33; **1920:** 'The Desert Lark', *JBNHS*, Vol. 27, pp. 400-1.

Bianchi, Gabriella, **1985:** 'Field Guide to the Commercial Marine and Brackish-water Species of Pakistan', Food and Agriculture Organisation, Rome, 200 pp.

Biddulph, J., **1881:** 'The Birds of Gilgit', *Stray Feathers*, Vol. 9, pp. 301-66; **1882:** 'Further Notes on the Birds of Gilgit', *Stray Feathers*, Vol. 10, pp. 157-78.

Birch, George, **1914:** 'Egret Farming in Sind', *JBNHS*, Vol. 23, No. 1, pp. 161-3; **1921:** 'Egret Farming in Sind, *JBNHS*, Vol. 27, No. 4, pp. 944-7.

Biswas, Biswamoy, **1951:** 'Notes on the Taxonomic Status of the Indian Plaintive Cuckoo *Cuculus passerinus*, Vahl', *Ibis*, Vol. 93, pp. 596-8; **1960a:** 'The Birds of Nepal', *JBNHS*, Vol. 57, No. 2 Part 1, pp. 278-308; **1960b:** 'The Birds of Nepal', *JBNHS*, Vol. 57, No. 3, Part 2, pp. 516-46; **1961:** 'The Birds of Nepal', *JBNHS*, Vol. 58, No. 1, Part 3, pp. 100-34, Vol. 58, No. 2, Part 4, pp. 444-74, Vol. 58, No. 3, Part 5, pp. 653-77; **1962:** 'The Birds of Nepal', *JBNHS*, Vol. 59, No. 1, Part 6, pp. 200-27, Vol. 59, No. 2, Part 7, pp. 405-29, Vol. 59, No. 3, Part 8, pp. 807-21; **1963:** 'The Birds of Nepal', *JBNHS*, Vol. 60, No. 1, Part 9, pp. 173-200, Vol. 60, No. 2, Part 10, pp. 388-99, Vol. 60, No. 3, Part 11, pp. 638-54; **1966:** 'Birds of Nepal', *JBNHS*, Vol. 63, No. 2, Part 12, pp. 365-77.

Blanford, W.T., **1878:** 'Wild Swans in Sind', *Stray Feathers*, Vol. 7, pp. 99-101. See Oates, E.W. and Blanford, W.T., *The Fauna of British India—Birds*.

Boehme, R.L., **1984:** See Flint *et al., A Field Guide to the Birds of the USSR*.

Bogle, J.S., **1901:** 'The Eastern Pintailed Sandgrouse (*Pterocles alchatus*), Breeding in India', *JBNHS*, Vol. 13, p. 539-40.

Bombay Natural History Society, Editors, **1971:** 'Recovery Ringed Birds', *JBNHS*, Vol. 68, No. 1, p. 249-73.

Briggs, F.S. and Osmaston, B.B., **1928:** 'A Note on the Birds of Peshawar District', *JBNHS*, Vol. 32, No. 4, pp. 744-61.

Bromhall, Derek, **1980:** *Devil Birds—the Life of the Swift*, Hutchinson and Co., (publishers) Ltd., London.

Brooke, M. de L., **1979:** 'Differences in the Quality of Territories held by Wheatears, (*Oenanthe oenanthe*)', *Journ. of Animal Ecology*, Blackwell Scientific Public'n., Vol. 48, pp. 21-3.

Brooks, W. Edwin, **1874:** 'On the Indian White-throats', *Stray Feathers* Vol. 2, p. 332; **1875:** 'Notes upon a collection of Birds made between Mussoori and Gangaotri in May 1874', *Stray Feathers*, Vol. 3, pp. 244-57; **1879a:** 'Description of another new *Reguloides*', *Stray Feathers*, Vol. 8, pp. 389-93; **1879b:** 'Ornithological Observations in Sikkim, the Punjab and Sind', *Stray Feathers*, Vol. 8, pp. 464-89.

Brown, Leslie, **1959:** 'The Mystery of the Flamingoes', *Country Life Ltd.*, London.

Brown, L. and Amadon, Dean, **1968:** *Eagles, Hawks and Falcons of the World*, Vol. 1 and 2, Country Life Books (Hamlyn Publishing Group), 945 pp.

Brown, L. Urban, E.K. and Newman, K., **1982:** *The Birds of Africa*, Vol. 1, Academic Press, London, pp. 521.

Bruning, D., See Bell, J.

Bruun, Bertel and Singer, Arthur, **1970:** *The Hamlyn Guide to Birds of Britain and Europe*, Hamlyn, London.

Buchanan, K., **1903:** 'Nesting Notes from Kashmir', *JBNHS*, Vol. 15, pp. 131-3.

Bundy, Graham, **1985:** 'Grey Hypocolius', *British Birds*, Vol. 78, (February), pp. 93-5.

Burkill, I.H., **1909:** 'A Working List of the Flowering Plants of Baluchistan', reprinted 1969, Govt. of Pakistan Press, Karachi.

Burton, J., (Ed.), **1973:** *The Owls of the World,* Peter Lowe, Netherlands pp. 94-115.

Butler, E.A. (Capt.), **1877a:** 'Astola, a Summer Cruise in the Gulf of Oman', Vol. 5, pp. 283-304; **1877b:** 'Additional Notes on the Birds of Sindh', *Stray Feathers,* Vol. 5, pp. 322-8; **1879:** 'Further Additions to the Sindh Avifauna', *Stray Feathers,* Vol. 8, pp. 386-9.

C

Campbell, Bruce and Lack, Elizabeth (eds.), **1985:** *A Dictionary of Birds,* Calton (Poyser) and Vermillion (Buteo).

Carruthers, Douglas, **1949:** *Beyond the Caspian, a Naturalist in Central Asia,* Oliver and Boyd, Edinburgh 289 pp.

Champion, Sir Harry G., Seth, S.K. and Khattak, G.M., **1965:** *Forest Types of Pakistan,* Pakistan Forest Institute, Peshawar.

Christison, A.F.P., **1939a:** 'Notes on Birds Nesting in the Kushdil Khan Lake, Quetta', *JBNHS,* Vol. 41, p. 420-2; **1939b:** 'On the Occurrence of the European Redstart (*Phoenicurus phoenicurus*) in British Baluchistan, *JBNHS,* Vol. 41, p. 434-5; **1940:** *Handbook of the Birds of Northern Baluchistan,* D.H.Q. Press, Quetta. 170 pp; **1941:** 'Notes on the Birds of Chagai, *Ibis,* pp. 531-6; **1942:** 'Some Additional Notes on the Distribution, of the Avifauna of Northern Baluchistan', *JBNHS,* Vol. 43, pp. 478-87.

Chugtai, Mohammad Idrees, **1987:** 'Survey of Zangi Nawar Lake', Pakistan Forest Institute, Handwritten Report.

Clark, Richard-J., **1975:** 'A Field Study of the Short-eared Owl *Asio flammeus* (Pontopiddan) in North America', *Wildlife Monographs,* No. 47, Wildlife Society Inc., USA.

Cloudsley-Thompson, J.L., **1965:** *Desert Life,* Pergamon Press, Oxford.

Cobb, E.H., **1938:** 'An Extension of the Range of the Western Horned Tragopan *Tragopan melanocephalus',* *JBNHS,* Vol. 40, No. 3, p. 569.

Cock (Capt.) and Marshall, C.H.T., **1873:** 'Notes on a Collection of Eggs made at Murree', *Stray Feathers,* Vol. 1, pp. 348-58.

Cole, F.H., **1931:** 'Occurrence of Gold-fronted Finch *Metaponia pusilla* at Sukkur, Sind', *JBNHS,* Vol. 35, No. 1, p. 207.

Collias, Nicholas and Collias, Elsie C., **1984:** *Nest Building and Bird Behaviour,* Princeton University Press, New Jersey, 336 pp.

Comber, E., **1907:** 'The Birds of Chitral', *JBNHS,* Vol. 18, No. 1, p. 186.

Cooke, Fred, **1978:** 'Early learning and its effect on population structure. Studies of a wild population of Snow Geese', *Zeit. Tierpsychol,* Vol. 46, pp. 344-50; **1985:** Article on 'Polymorphism', in Campbell B. and Lack, E. (eds.), *Dictionary of Birds,* Calton and Vermillion.

Corfield, David M., **1983:** *Birds of Islamabad, Pakistan and the Murree Hills,* Asian Study Group, Islamabad.

Cotes, E.C., **1884:** 'The Food of the Rosy Pastor or Jowari Bird (*Pastor roseus*)', *JBNHS,* Vol. 9, No. 1, pp. 66-8.

Cramp, S. and Simmons, K.E.L. (eds.), **1977:** *The Birds of the Western Palearctic,* Oxford University Press, Vol. 1, Ostrich to Ducks; **1979:** Vol. 2, Hawks to Bustards; **1982:** Vol. 3, Waders to Gulls.

Cramp, S. (ed.), **1985:** Vol. 4, Terns to Woodpeckers.

Cumming, J.W.N., **1905:** 'Birds of Seistan, being a list of birds shot or seen in Seistan by members of the Seistan Arbitration Mission 1903-5'. *JBNHS,* Vol. 16, Part 4, pp. 686-99; **1908:** 'Report of the Baluchistan Natural History Society', *JBNHS,* Vol. 18, pp. 941-5.

Cumming, W.D., **1899:** 'Occurrence of the Green-billed Shearwater *Puffinus chlororhyncus* on the Mekran Coast', *JBNHS,* Vol. 12, p. 766; **1903:** Note on *Hieraaetus fasciatus',* *JBNHS,* Vol. 15, No. 1, pp. 145-6.

Cunningham, G., **1933:** 'Occurrence of the European Bustard (*Otis tarda tarda*) in the North West Frontier Province', *JBNHS,* Vol. 36, No. 3, p. 752.

Currie, A.J., **1909:** 'Local Bird Migration', *JBNHS,* Vol. 19, No. 1, p. 265; **1916a:** The Birds of Lahore and the Vicinity'. *JBNHS,*

Vol. 24, No. 3, pp. 561-77; **1916b:** 'Occurrence of the Bristled Grass Warbler (*Chaetornis locustelloides*) at Lahore', *JBNHS,* Vol. 24, No. 3, p. 593; **1916c:** 'Notes on the Punjab Cuckoos (Cuculinae)', *JBNHS,* Vol. 24, No. 4, p. 594.

Curry-Lindahl, Kai (Professor), **1958:** 'Internal timer and Spring Migration in an Equatorial Migrant, the Yellow Wagtail (*Motacilla flava*), *Arkiv for Zoologi,* Serie 2, Vol. 11, pp. 541-57; **1963:** 'Vara Faglar i Norden' (*The Birds of Scandinavia and Finland*), Vol. 4, 2nd. ed., Stockholm, pp. 1535-2294; **1981:** *Bird Migration in Africa: movements between six continents,* Vol. 1 and 2, Academic Press, London.

D

Daniel, J.C., **1983:** *The Book of Indian Reptiles,* Bombay Natural History Society, 141 pp.

Dansereau, Pierre, **1957:** *Biogeography—an Ecological Perspective,* Ronald Press Co., New York, 394 pp.

Davidar, Priya, **1985:** 'Ecological Interactions between Mistletoes and their Avian Pollinators in South India', *JBNHS,* Vol. 82, No. 1, pp. 45-60.

Davidson, J., **1898:** 'A Short Trip to Kashmir', *Ibis,* Vol. 13, pp. 1-42.

Delacour, Jean, **1951:** *The Pheasants of the World,* Country Life Ltd., London, 351 pp.

Delacour, Jean and Peter Scott, **1954:** *The Waterfowl of the World,* Country Life Ltd., London, Vol. 1, 284 pp; **1956:** Vol. 2, 232 pp; **1959:** Vol. 3, 270 pp.

Dement'ev, G.P., Gladkov N.A., Ptushenko, E.S., Spangenberg, E.P. and Sudilouskaya, A.M., **1951a:** *Birds of the Soviet Union,* Vol. 1, 645 pp; **1951b:** Vol. 2, 475 pp; **1951c:** Vol. 3, 676 pp.

Dement'ev, G.P., Gladkov, N.A., Isakov, Yu. A., and Kartashev, N.N., 1952. Vol. 4. 683 pp.

Dement'ev, G.P., Meklenburtsev, R.N., Sudilouskaya, A.M., and Spangenberg, E.P., **1951:** Vol. 2, 475 pp; Vol. 3, 676 pp.

Dement'ev, G.P., Gladkov, N.A., Sudilouskaya, A.M., Spangenberg, E.P., Boehme, L.V., Volchanetskii, I.B., Voinstvenskii, M.A., Gorchakovskaya, N.N., Korelov, M.N. and Austamov, A.K., **1954a:** Vol. 5, 797 pp; **1954b:** Vol. 6, 751 pp. Translated from Russian by Israel Program for Scientific Translations, Jerusalem, 1966-70.

Desai, J.H. and Malhotra A.K., **1979:** 'Breeding biology of the pariah kite *Milvus migrans* at Delhi Zoological Park', *Ibis.* Vol. 121 No. 3, pp. 320-5; **1986:** 'Breeding Biology of the Bay-backed Shrike *Lanius vittatus* at National Zoological Park, New Delhi, *JBNHS,* Vol. 83, No. 1, pp. 200-2.

Dewar, Douglas, See Wright and Dewar, *The Ducks of India.*

De Zylva, T.S.U. (Dr.), **1984:** *Birds of Sri Lanka,* Trumpet publishers, Colombo, 133 pp.

Dharmakumarsinhji, R.S., **1939:** The indian Great Horned Owl *Bubo bubo bengalensis, JBNHS,* Vol. 41, No. 1, pp. 174-7; **1950:** 'The Lesser Florican *Sypheotides indica,* its Courtship, Display, Behaviour and Habits', *JBNHS,* Vol. 49, No. 2, pp. 201-16; **1954:** *Birds of Saurashtra, India,* Times of India Press, Bombay, 561 pp; **1962:** 'The Great Indian Bustard *Choriotis nigriceps* at Nest', *JBNHS,* Vol. 59, No. 1, pp. 173-84.

Dharmakumarsinhji, R.S. and Lavkumar, K.S., **1972:** 'Sixty Indian Birds', Publications Div'n. Ministry of Information and Broadcasting Govt. of India, New Delhi, 100 pp.

Dhindsa, Manjit Singh, and Toor, H.S., **1981:** 'Some Observations on a nest of the Common Crow Pheasant *Centropus sinensis', JBNHS,* Vol. 78, No. 3, pp. 600-2.

Dickinson, E.C., **1966:** 'Notes on some Birds seen in Kashmir', *JBNHS,* Vol. 63, No. 1, p. 203-4.

Dodsworth, P.T.L., **1911:** 'Notes on the Nidification of *Microcichla scouleri,* the Little Forktail', *JBNHS,* Vol. 21, No. 1, pp. 257-61; **1912:** 'Habits, Food and Nesting of the Great Himalayan Barbet *Megalaima marshallorum', JBNHS,* Vol. 21, No. 2, pp. 681-4.

Doig, Scrope B., **1878a:** Letter to the Editor, *Stray Feathers* Vol. 7, pp. 466-7; **1878b:** Letters of the Editor, Breeding of *Phalacrocorax carbo, Stray Feathers,* Vol. 8, No. 2, pp. 466 and 468; **1879a:** 'Some Notes on Sindh Birds', *Stray Feathers,* Vol. 7, pp. 502-6; **1879b:** 'Birds Nesting on the Eastern Narra (Sind)', *Stray Feathers,* Vol. 8, No. 2, pp. 369-79; **1880:** 'Birds Nesting on the Eastern Narra, Sind Additions and Alterations', *Stray Feathers* Vol. 9, pp. 277-82.

Donahue, Julian P., **1967:** 'Notes on a Collection of Indian Birds, mostly from Delhi', *JBNHS,* Vol. 64, No. 3, pp. 410-29.

Donald, C.H., **1917:** 'The Raptores of the Punjab', *JBNHS,* Vol. 25, pp. 231-48; **1918:** 'The Birds of Prey of the Punjab', *JBNHS,* Vol. 26, No. 1, Part 1, pp. 247-65, *JBNHS,* Vol. 26, No. 2, Part 2, pp. 629-55; **1919:** 'The Birds of Prey of the Punjab', *JBNHS,* Vol. 26, Part 3, pp. 826-35; **1920a:** 'Birds of Prey of the Punjab', *JBNHS,* Vol. 26, Part 4, pp. 1000-20; **1920b:** 'Birds of Prey of the Punjab', *JBNHS,* Vol. 27, Part 5, pp. 128-40; Vol. 27, Part 6, pp. 280-300; **1921;** Vol. 27, Part 7, pp. 606-15.

Douthwaite, R.J., **1978:** 'Breeding Biology of the Pied Kingfisher *Ceryle rudis* on Lake Victoria', *Journal E. African Natural History Society,* Vol. 31, No. 166, pp. 1-12.

Dyrcz, Andrzej, **1977:** 'Nest Helpers in the Alpine Accentor *Prunella collaris',* Ibis, Vol. 119, No. 2, p. 215.

E

Eates, Kenneth R., **1926:** 'A note on the Nidification of the Western Reef Egret (*Lepterodius asha*) in Karachi city, Sind', *JBNHS,* Vol. 31, No. 3, pp. 823-5; **1937a:** 'The Distribution and Nidification of the Greater Spotted Eagle (*Aquila clanga*) in Sind', *JBNHS,* Vol. 39, No. 2, pp. 403-5; **1937b:** 'The Status of the Koel (*Eudynamis scolopaceus*) in Sind', *JBNHS,* Vol. 39, No. 2, pp. 406-13; **1937c:** 'The Distribution and Nidification of the Large Indian Paroquet (*Psittacula eupatria nepalensis*) in Sind', *JBNHS,* Vol. 39, No. 2, pp. 414-8; **1937d:** 'A Note on the Distribution and Nidification of the Northern Yellow-fronted Pied Woodpecker (*Leiopicus mahrattensis blanfordi*) in Sind', *JBNHS,* Vol. 39, No. 3, pp. 628-30; **1937e:** 'The Distribution and Nidification of the Rock Horned Owl (*Bubo bubo bengalensis*) in Sind', *JBNHS,* Vol. 39, No. 3, pp. 631-3; **1937f:** 'The Behaviour of Jerdon's Little Ringed Plover (*Charadrius dubius jerdoni*) with Young', *JBNHS,* Vol. 39, No. 3, pp. 636-8; **1938:** 'Occurrence of the Lesser Orange-breasted Green Pigeon (*Dendrophasa bicincta*) at Keamari, Sind', *JBNHS,* Vol. 40, No. 2, pp. 330-1; 'The Status of the Koel *Eudynamis scolopaceus* in Sind', *JBNHS,* Vol. 40, No. 2, p. 328; **1939a:** 'A Note on the Resident Owls of Sind', *JBNHS,* Vol. 40, No. 4, pp. 750-5; **1939b:** 'The Status and Nidification of the Persian, Bee-eater (*Merops persicus*) in Sind', *JBNHS,* Vol. 40, No. 4, pp. 756-9; **1940-1950:** Manuscript Notes on Breeding Habits of Various Sind Birds, (Untyped and pages not numbered), unpublished, approx. 500 pages. Now lodged with British Museum of Natural History sub-department of Ornithology, Tring; **1943:** 'The Arabian Large-crested Sea Tern (*Sterna bergii velox*) breeding off the Sind coast', *JBNHS,* Vol. 44, No. 2, pp. 302-3; **1968:** *'Birds of Sind'*, Chapter 3, Zoology Section pp. 53-63 and 'List of the Birds of Former Sind and Khairpur', (Revised in 1934), Appendix 4, pp. 108-27 in *Gazetteer of West Pakistan, the Former Province of Sind,* ed. Sorley, H.T., Published by Govt. of Pakistan.

Endler, John A., **1985:** Article 'Speciation', in Campbell, B. and Lack, E., (eds.) *Dictionary of Birds,* Calton and Vermillion.

Etchecopar, R.D., **1970:** See Hue and Etchecopar, *Les Oiseaux du Proche et du Moyen Orient.*

F

Fairbrother, Lieut., **1880:** 'Himalayan Snowcock Chicks', *Stray Feathers,* Vol. 9, p. 207.

Faruqi, S.A. and Bump, G., **1957:** A Study of the Seasonal Foods of three Pakistan Game Birds', Prog. Reports U.S. Fish and Wildlife Service.

Fitter, Richard, See Heinzel, Fitter and Parslow, *The Birds of Britain and Europe.*

Fleming, R.L., **1955:** 'The Bone Dropping Habit of the Lammergeier', *JBNHS,* Vol. 32, No. 4, pp. 933-5.

Fleming, R.L. (Sr.), and Fleming, R.L. (Jr.), **1976:** *Birds of Nepal* Published by Robert Fleming in Kathmandu, Nepal, 349 pp.

Flint, V.E., Boehme, R.L., Kostin, Y.U. and Zuznetson, A.A., **1984:** *A Field Guide to Birds of the USSR,* Translated from the Russian by Natalia Bourso-Leland, Princeton University Press, 353 pp.

Fogden, M.P.L., **1964:** 'The reproductive behaviour and taxonomy of Hemprich's Gull *Larus hemprichii',* Ibis, Vol. 106, pp. 299-320.

Ford, E.B., **1945:** 'Polymorphism', *Biological Review,* Vol. 20, pp. 73-88.

Foreshaw, Joseph and Cooper, William T., **1973:** *Parrots of the World,* Lansdowne Press, Melbourne, 583 pp; **1983:** *Kingfishers and related birds—Alcedinidae,* Lansdowne editions, Melbourne, Vol. 1 *Ceryle* to *Cittura,* 282 pp; **1985:** Vol. 2, *Halcyon* to *Tanysiptera,* 559 pp.

Frith, H.J. (ed.), **1979:** (Revised) *The Readers Digest Complete Book of Australian Birds,* Readers Digest Pty. Ltd., Sydney.

Fry, C. Hilary, **1964:** 'Red-necked Kestrel *Falco chiquera* hunting bats', *Bull. Nigerian Ormithological Society,* Vol. 1, No. 4, p. 19; **1969:** 'The Evolution and Systematics of Bee-eaters (Meropidae)', *Ibis,* Vol. 3, pp. 557-92; **1980a:** 'The Biology of African Bee-eaters', *The Living Bird,* Vol. 2, pp. 75-112; **1980b:** 'The Evolutionary Biology of Kingfishers (Alcedinidae)', *The Living Bird,* Cornell, New York, pp. 113-60; **1983:** 'Honey-bee Predation by Bee-eaters, with Economic Considerations', *Bee World,* Vol. 64, No. 2, published by International Bee Research Assn., pp. 65-78; **1984:** *The Bee-eaters,* T. and A.D. Poyser Ltd. 304 pp.

Fulk, George W. and Kokar, Abdul Rauf, **1976:** 'Short-tailed Mole Rat (*Nesokia indica*), Prey of the Great Horned Owl (*Bubo bubo*)', *Pakistan Journal of Zoology,* Vol. 8, No. 1, pp. 97.

Fulton, H.T. (Capt.), **1904:** 'Notes on the Birds of Chitral', *JBNHS,* Vol. 16, pp. 44-64.

G

Gallaher, M.D., **1977:** 'Birds of Jabal Akhdar, in Scientific Results of the Oman Flora and Fauna Survey, 1975', *Jour. Oman Studies,* Special Report (No. 1) pp. 27-58; **1979:** 'The Sooty Falcon in Oman', *Journ. of Oman Studies.*

Gallaher, M.D. and Rogers, T.D., **1980:** 'On Some Birds of Dhofar and other parts of Oman', Special Report, No. 2, the Oman Flora and Fauna Survey 1977 (Dhofar), *Journal of Oman Studies,* pp. 347-85.

Gallaher, M.D. and Woodcock, Martin, **1980:** *Birds of Oman,* Oxford Univ. Press, 312 pp.

Gallaher, M.D., Scott, D.A. Ormond, R.F.G., Connor, R.J. and Jennings, M.C., **1984:** 'The Distribution and Conservation of Seabirds breeding on the Coasts and Islands of Iran and Arabia', *ICBP Technical Bulletin,* No. 2.

Ganguli, Usha, **1975:** 'A Guide to the Birds of Delhi Area', Indian Council of Agricultural Research, New Delhi.

Gaston, A.J., **1976:** 'Brood Parasitism by the Pied-crested Cuckoo *Clamator jacobinus',* Journal Animal Ecology, Vol. 45, pp. 331-48; **1977:** 'Social Behaviour with Groups of Jungle Babblers (*Turdoides striatus*)', *Animal Behaviour,* Vol. 25, pp. 828-48; **1978:** 'Ecology of the Common Babbler *Turdoides caudatus',* Ibis, Vol. 120, No. 4, pp. 415-32; **1980:** 'Census Techniques for Himalayan Pheasants including Notes on Individual Species', *World Pheasant Assoc. Journal* No. 5, pp. 40-53; **1981:** 'Present Distribution and Status of Pheasants in Himachal Pradesh, Western Himalayas', *World Pheasant Assoc. Journal,* No. 6, pp. 10-30; **1984:** 'Is Habitat destruction in India and Pakistan beginning to affect the Status of Endemic Passerine birds?', *JBNHS*, Vol. 81, No. 3, pp. 636-41.

Gibson-Hill, C.A., **1948**: 'The Storm Petrels occurring in the northern Indian Ocean and adjacent seas', *JBNHS*, Vol. 47, No. 3, pp. 443-9.

Gibson, R.E., **1918**: 'Comb Duck (*Sarcidiornis melanonotus*) in Sind', *JBNHS*, Vol. 25, No. 4, pp. 747-8.

Gill, E.H., **1924**: 'Plumage display by the Sirkeer Cuckoo (*T. leschenaulti*)', JBNHS, Vol. 29, p. 299.

Gladkov, N.A., See Dement'ev and Gladkov, *Birds of the Soviet Union.*

Glue, David and Scott, Derek, **1980**: 'Breeding Biology of the Little Owl', *British Birds*, Vol. 73, No. 4, pp. 167-80.

Glutz von Blotzheim, J.N., Bauer, K.M. and Bezzel, E., **1971**: *Handbuch der Vogel Mitteleuropas*, Vol. 1-4, Frankfurt am Main; **1980**: Vol. 9, *Columbiformes, Piciformes;* Weisbaden.

Goethe, F., **1957**: *Behaviour*, Vol. 2, pp. 310-7.

Goodwin, Derek, **1967**: *Pigeons and Doves of the World*, British Museum (Natural History), 446 pp.; **1976**: *Crows of the World*, British Museum (Natural History), 354 pp.; **1982**: *Estrildid Finches of the World*, Oxford University Press, 328 pp.

Goriup, P.D., **1981**: 'The Houbara Bustard, Houbara Conservation. and Research in Pakistan', *Western Tanager*, Los Angeles Audubon Soc., Vol. 48 (4) pp. 3-6; **1982**: 'Behaviour of Black-winged Stilts, *British Birds*, Vol. 75, pp. 12-24.

Grimmett, Richard, See Redman, Lambert and Grimmett, *Observations of Scarce Birds in Nepal.*

Grimmett, Richard and Robson, Craig, **1984**: 'Report on a Survey of the Western Tragopan in Pakistan', submitted to National Council for the Conservation of Wildlife (unpublished).

Grummt, W., **1961**: 'Ornithologische Beobachtungen in der Mongolei', *Beitrage zur Vogelkunde*, Vol. 7, pp. 349-60.

Gupta, P.D., **1974**: 'Stomach contents of Great Indian Bustard *Choriotis nigriceps*', *JBNHS*, Vol. 71, No. 2, pp. 303-4.

H

Hancock, James and Elliot, Hugh, **1978**: *The Herons of the World*, London Editions 304 pp.

Harmata, A.R., **1982**: 'What is the Function of Undulating Flight Display in Golden Eagles?' *Raptor Research* (Florida) Vol. 16, No. 4, pp. 103-9.

Harris, M.P., **1970**: 'Abnormal Migration and Hybridization of *Larus argentatus* and *L. fuscus* after interspecies fostering experiments', *Ibis*, Vol. 112, No. 4, pp. 488-98.

Harrison, C.J.O., **1969**: 'The Identification of the Eggs of the smaller Indian Cuckoos', *JBNHS*, Vol. 66, No. 3, pp. 478-88.

Harrison, Colin, **1982**: *An Atlas of the Birds of the Western Palearctic.*, Wm. Collins and Sons, London.

Harrison, Peter, **1983**: *Seabirds—An Identification Guide*, Croom Helm Ltd., London, 448 pp.

Hartert Ernst, **1910-22**: *Die Vogel der Palaarktischen Fauna*, 3 vols, Friedlander u. Sohn. Berlin.

Hartshorne, Charles, **1973**: *Born to Sing—an Interpretation and World Survey of Bird Song*, Indiana Univ. Press, 304 pp.

Harvey, W.G., **1981**: 'Pallid Swift: New to Britain and Ireland', *British Birds*, Vol. 74, No. 4, pp. 170-8.

Hasan, S. Azhar, **1964**: 'Birds of Manchar Lake', *Agriculture Pakistan*, Vol. 15, No. 3, pp. 259-83.

Hayman, Peter, Marchant, John and Prater, Tony, **1986**: *Shorebirds: An Identification Guide to the Waders of the World*, Croom Helm Ltd., 412 pp.

Heinzel, H., Fitter, R. and Parslow, J., **1972**: *The Birds of Britain and Europe with North Africa and the Middle East*, pub. Collins, London.

Henry, G.M., **1971**: *A Guide to the Birds of Ceylon*, De Silva and Sons, Kandy, Sri Lanka, 457 pp.

Hirons, G., **1980**: 'The Significance of Roding by Woodcock *Scolopax rusticola*: an Alternative Explanation based on Observations of Marked Birds', *Ibis*, Vol. 122, No. 3, pp. 350-4.

Hirschfeld, Eric, Kjellen, Nils and Ullman, Magnus, **1988**: *Birdwatching in Pakistan*, 14 Feb.-6 Mar., 1988, privately published by Hirschfeld, S-211-43. Malmo, Sweden.

Hogan-Warburg, A.J., **1966**: 'Social Behaviour of the Ruff, *Philomachus pugnax'*, *Ardea*, Vol. 54, pp. 102-229.

Holmes, Derek A. and Wright, John O., **1968**: 'The Birds of Sind: a Review', *JBNHS*, Vol. 65, No. 3, pp. 533-56; **1969**: *JBNHS*, Vol. 66, No. 1, Part 2, pp. 8-30.

Holmes, J.R.S., Roberts, T.J. and Savage, C.D.W., **1968**: 'Red-necked Grebe, *Podiceps griseigena* sighted in West Pakistan', *JBNHS*, Vol. 64, No. 3, pp. 555-7.

Holmes, Peter R., Holmes, H.J. and Parr, A.J., **1983**: Report of the Oxford University Expedition to Kashmir 1983.

Hollom, P.A.D., see Peterson, Mountfort and Hollom, *Field Guide to the Birds of Britain and Europe.*

Holloway, Colin W., and Khan, Khan Mohammad, **1974**: 'Management Plan for Kirthar National Park, Sind, Pakistan, July 1973-78', published by Sind Government Press, for Sind Wildlife Management Board.

Houston, D.C., **1980**: 'A possible function of sunning behaviour by Griffon Vultures, *Gyps* spp., and other large soaring birds', *Ibis*, Vol. 122, No. 3, pp. 366-9.

Hudson, Corrie, **1920**: 'Some Birds observed in South Waziristan', *JBNHS*, Vol. 27, No. 3, pp. 402-3.

Hue, Francois and Etchecopar, R.D., **1970**: *Les Oiseaux du Proche et du Moyen Orient*, N. Boubee et Cie, Paris, 948 pp.

Hughes-Buller, R., **1908**: *Baluchistan: Imperial Gazetteer of India*, Reprinted 1976 by Sheikh Mubarak Ali, Lahore, 216 pp.

Hughes, F.L., **1917**: 'Note on the Great Brown Vulture (*Vultur monachus*) in Captivity', *JBNHS*, Vol. 25, No. 2, pp. 298-300.

Hume, Alan Octavius, **1873**: 'Contributions to the Ornithology of India, Sindh No. 1', *Stray Feathers*, Vol. 1, pp. 44-50. Sindh No. 2, pp. 91-289; **1878a**: '*Asio butleri, sp. nov.' Stray Feathers*, Vol. 7 pp. 316-8.

Hume, Alan Octavius and Marshall, C.H.T., **1878-80**: *The Game Birds of India, Burma and Ceylon*, 3 Vols. Published by the authors, Calcutta.

Hume, A.O. and Oates, E.W., **1889-90**: *Nests and Eggs of Indian Birds*, 3 Vols., 2nd. ed., R.H. Porter, London.

Husain, M. Afzal and Bhalla, Hem Raj, **1937**: 'Some Birds of Lyallpur and their Food', *JBNHS*, Vol. 39, No. 4, pp. 831-42.

Hussain, S.A., **1978**: 'Orange-rumped Honey Guide (*Indicator xanthonotus*) in the Garhwal Himalayas', *JBNHS*, Vol. 75 (2), pp. 487-8.

Hussain, S.A. and Ali, Salim, **1984**: 'Some Notes on the Ecology and Status of the Orange-rumped Honey Guide *Indicator xanthonotus* in the Himalayas', *JBNHS*, Vol. 80 (3), pp. 564-74.

Hutchinson, R.G., **1943**: 'The Distribution of the Grey Hornbill (*Tockus birostris*) and Tickell's Flower-Pecker (*Piprisoma agile?*)', *JBNHS*, Vol. 44, No. 2, pp. 296-7.

Hutson, H.P.W., **1954**: *The Birds about Delhi*, Published by Delhi Bird Watching Society, 210 pp.

I

Immelmann, Klaus, **1985**: Article—'Estrildid Finches' in Campbell, B. and Lack, E. (eds.), *Dictionary of Birds*, Calton and Vermillion.

Inglis, C.M., **1903**: 'The Birds of the Madhubani sub-division of the Darbhanga district, Tirhut, with notes on species noticed elsewhere in the District', Part 5, *JBNHS*, Vol. 14, No. 4, pp. 70-7.

Ingram, Collingwood, **1978**: 'Carriage of the Young and related adaptations in the Anatomy of the Woodcock *Scolopax rusticola'*, *Ibis*, Vol. 120, No. 1, p. 67.

Inskipp, Carol and Inskipp, Tim, **1985**: *A Guide to the Birds of Nepal* (1985), Croom Helm, London.

Islam, K., **1983:** In Gaston, Islam and Crawford, 'The Current Status of the Western Tragopan *Tragopan melanocephalus*', *WPA Journal,* No. 8.

Islam, K. and Crawford, J.A., **1985:** 'Brood Habitat and Roost Sites of Western Tragopan in northeastern Pakistan', *WPA Journal,* No. 10, pp. 7-14.

J

Jalal, Hamid, Merriam, Aimee, Qayyum, Abdul and Stacey, Tom (eds.), **1977:** *Pakistan, Past and Present,* Stacey International, London, 288 pp.

Jarridge, Jean-Francois and Meadow, R.H., **1980:** 'The Antecedents of Civilization in the Indus Valley', *Scientific American,* Vol. 243, No. 2, pp. 122-33.

Jayakar, S.D. and Spurway, H., **1965:** 'Incubatory adaptations of the Yellow Wattled Lapwing *Vanellus malabaricus*', *Zool. Zahrb, Abt. allgemeine Zool. u. Physiol.,* pp. 53-72.

Jennings, Michael C., **1981:** *The Birds of Saudi Arabia:* A checklist, published by the author, Whittlesford, Cambridge.

Jerdon, T.C., **1862-4:** *The Birds of India,* 2 Vols. (3 Parts) published by the author, Calcutta.

Johnsingh, A.J.T. and Anandham, K. Param, **1982:** 'Group care of White-headed Babbler *Turdoides affinis* for a Pied-crested Cuckoo *Clamator jacobinus* chick', *Ibis,* Vol. 124, No. 2, pp. 179-83.

Jones, A.E., **1912:** 'Notes on Birds from Lahore', *JBNHS,* Vol. 21, No. 3, pp. 1073-4; **1915:** 'Nesting of the Hobby (*Falco subbuteo*) near Simla, N.W. Himalayas', *JBNHS,* Vol. 23, No. 3, pp. 579-81; **1919:** 'Further Notes on the Birds of Ambala District, Punjab', *JBNHS,* Vol. 26, No. 2, pp. 675-6; **1921:** 'Bird Notes from the Campbellpur-Attock District, Western Punjab', *JBNHS,* Vol. 27, No. 4, pp. 794-802; **1927:** 'Further Notes on the Birds of the Ambala District', *JBNHS,* Vol. 31, No. 4, pp. 1000-8; **1932:** 'The nesting of the Besra Sparrow Hawk *Accipiter virgatus affinis* in Simla', *JBNHS,* Vol. 35, No. 2, p. 208; **1938:** 'Nesting of the Booted Eagle (*Hieraeetus pennatus*) in the Simla Hills', *JBNHS,* Vol. 40, No. 3, p. 568; **1941:** 'Presumptive Evidence of the Nidification of the Indian Cuckoo, *Cuculus micropterus*', *JBNHS,* Vol. 42, No. 4, pp. 931-3; **1947:** 'The Birds of Simla and Adjacent Hills', Part 1, *JBNHS,* Vol. 47, No. 1, pp. 117-25, Part 2, *JBNHS,* Vol. 47, No. 2, pp. 219-49; **1948:** Part 3, *JBNHS,* Vol. 47, No. 3, pp. 409-32.

Jourdain, F.C.R., See Witherby and Jourdain, *The Handbook of British Birds.*

K

Kahl, M. Philip, **1970:** 'Observations on the Breeding of Storks in India and Ceylon', *JBNHS,* Vol. 67, No. 3, pp. 453-61; **1972:** 'The Stork. A Taste for Survival', in *Marvels of Animal Behaviour,* Allen, T. (ed.) National Geographic Society, Washington.

Karim, Syed Imtiaz, **1985:** 'Breeding Evidence of Flamingos (*Phoenicopterus roseus*) in Mudflats beyond Gularchi, District Badin (Sind)', Records Zoological Survey of Pakistan, Vol. 10, No. 1 and 2, p. 121.

Kefford, H. Kingsley, **1945:** 'Peculiar Behaviour of the Bronze-winged Jacana *Metopidius indicus*', *JBNHS,* Vol. 45, No. 2, p. 238.

Khan, Wazir Mohammad and Shah, Iqmail Hussain, **1982:** 'Population Dynamics of the Koklass Pheasant *Pucrasia macrolopha* in Malkandi Forests, Pakistan', *Pheasants in Asia,* ed. C.D.W. Savage and M.W. Ridley, Published by WPA.

Khanum, Zakia and Ahmed, Farooq Mohammad and Ahmed, Manzoor, **1980:** 'A Checklist of Birds of Pakistan with Illustrated Keys to their Identification', Records Zoological Survey of Pakistan, Vol. 9, No. 's 1 and 2, Govt. of Pakistan Press, Karachi, 138 pp.

Khanum, Zakia and Qadri, M.A.H., **1972:** 'Fresh Records of some

Birds in West Pakistan', *Pakistan Journ. of Zoology,* Vol. 4, No. 2, pp. 219-21.

King, Ben, Woodcock, Martin and Dickinson, E.C., **1975:** *A Field Guide to the Birds of Southeast Asia,* Collins, London.

Kinnear, Norman B., See Ludlow and Kinnear, *A Contribution to the Ornithology of Chinese Turkestan.* See H. Whistler, *Popular Handbook of Indian Birds,* 4th. ed. revised by Kinnear.

Knights, C.R., **1985:** 'Great Crested Grebe *Podiceps cristatus* passing breast feathers to its one-day old young', *British Birds,* Vol. 78, No. 5, p. 213.

Knystautas, A., **1987:** *The Natural History of the USSR,* Century Hutchison Ltd., London.

Knystautas, A. and Liutkaus, A., **1984:** *In the World of Birds,* Vilnius, Mokslas, USSR, 280 pp.

Kobayashi, Keisuke, **1958:** *Birds of Japan in Natural Colours* (Text in Japanese), Hoikusha, Osaka.

Koelz, W., **1939:** 'Occurrence of Tengmalm's Owl in Lahul, India', Proceedings Biological Society, Washington, No. 52, p. 80.

Koenig, L., **1960:** '*Merops apiaster* (Meropidae) Jugendentwicklung, *Encyclopaedia Cinematographica,* Instit. Wiss. Film. Gottingen.

Koning, F.J., **1971:** 'Notes on the Winter Distribution of the Stifftail *Oxyura leucocephala* in Turkey', *Ardea,* Vol. 59, pp. 53-5; **1972:** 'Notes on the Birds of Prey in the Indus Valley', *JBNHS,* Vol. 73, No. 3, pp. 448-55.

Koning, F.J. and Walmsley, J.G., **1973a:** 'Quantitative Angaben uber die in der Turkei uberwinternden anatiden', *Bonn Sool. Beitr,* Vol. 24, pp. 219-26; **1973b:** IWRB Mission to West Pakistan (February 1973) *IWRB Bulletin* No. 35, pp. 64-73.

Kostin, Y.V., See Flint *et al., A Field Guide to the Birds of the USSR.*

Krebs, J.R. and Davies, N.B., **1981:** *An Introduction to Behavioural Ecology,* Blackwell Scientific Publications, Oxford, 292 pp.

L

Lack, David, **1940:** *Pair Formation in Birds, Condor,* Vol. 42, pp. 269-86; **1954:** *The Natural Regulation of Animal Numbers,* Oxford University Press, London; **1968:** *Ecological Adaptations for Breeding in Birds,* Methuen and Co. Ltd., London, 409 pp.

Lamba, B.S., **1969:** 'The Nidification of some Common Indian Birds, Part 12, the Koel (*Eudynamys scolopacea*)', *JBNHS,* Vol. 66, No. 1, pp. 72-80.

Landsborough Thomson, (Sir A.) (ed.) **1964:** *A New Dictionary of Birds,* Thomson Nelson and Sons Ltd., London, 906 pp.

Lapersonne, V.S., **1928:** *A Collecting Trip to Ladak,* Part 1, Vol. 32, No. 3, pp. 505-17, Part 2, Vol. 32, No. 4, pp. 650-9; **1931:** 'Description of and Notes on the Female Chestnut-mantled Koklass (*Pucrasia m. castanea*), from Chitral', *JBNHS,* Vol. 34, No. 4, p. 1062.

Lavee, Daphna, **1983:** *The Influence of Grazing and Intensive Cultivation on the Population Size of Houbaras in the Northern Negev in Israel,* Tel-Aviv University, Israel.

Lavkumar, K.S., **1972:** See Dharmakumarsinhji and Lavkumar.

Lay, Douglas M., **1974:** 'Differential Predation on Gerbils, (*Meriones*) by the Little Owl *Athene brahma*', *Journal of Mammalogy,* Vol. 55, No. 3, pp. 608-14.

Leshem, Josi, **1981:** 'The Occurrence of Hume's Tawny Owl in Israel and Sinai', *Sandgrouse 2, Journ. of Orn. Soc. for the Middle East,* pp. 100-2.

Leslie, N.A., **1956:** 'Red-breasted Merganser *Mergus serrator* in Sind', *JBNHS,* Vol. 53, No. 4, p. 708.

Lester, C.D., **1894:** 'The Flamingo Breeding in India', *JBNHS,* Vol. 8, No. 4, pp. 553-4.

Lindsay-Smith, J., **1913:** 'Common Grey Quail (*Coturnix communis*) Breeding in the Lyallpur district', *JBNHS,* Vol. 22, No. 1, pp. 200-1; **1914a:** 'Notes on Doves in the Punjab', *JBNHS,* Vol. 23, No. 2, p. 364; **1914b:** 'A Note on the Nesting of some Birds found in the Multan District', *JBNHS,* Vol. 23, No. 2, pp. 365-7; See also Smith, J. Lindsay.

Lister, M.D., **1954**: 'A Contribution to the Ornithology of the Darjeeling Area', *JBNHS*, Vol. 52, No. 1, pp. 20-68.

Liutkus, A., **1984**: See: Knystautas and Liutkus, *In the World of Birds*.

Livesey, T.R., **1921**: 'Nest of Nakta or Comb Duck *S. melanonotus*', *JBNHS*, Vol. 27, No. 3, p. 637.

Lofts, B. and Murton, R.K., **1968**: 'Photoperiodic and Physiological adaptations regulating Avian breeding cycles and their Ecological Significance', *Journ. Zoology*, Vol. 155, pp. 337-94.

Loke, Wan Tho., **1946**: 'A Bird Photographer in Kashmir', *JBNHS*, Vol. 46, No. 3, pp. 431-6; **1952**: 'Kashmir Revisited', *JBNHS*, Vol. 51, No. 1, pp. 121-7.

Lowther, E.H.N., **1949**: *A Bird Photographer in India*, Oxford University Press, 150 pp. See Bates and Lowther, *Breeding Birds of Kashmir*.

Ludlow, Frank, **1920**: 'Notes on the Nidification of Certain Birds in Ladak', *JBNHS*, Vol. 27, No. 1, pp. 141-6; **1927-28**: 'Birds of the Gyantse Neighbourhood - Southern Tibet', *Ibis*, Part I, pp. 644-59 Part 2, pp. 51-73, Part 3, pp. 211-32; **1928**: 'Dongtse, or Stray Bird Notes from Tibet', *JBNHS*, Vol. 33, pp. 78-83; **1950**: 'The Birds of Lhasa', *Ibis*, Vol. 92, No. 1, pp. 34-45.

Luthin, Charles S., **1987**: 'Status of and Conservation Priorities for the World's Stork Species', *Journ. of Colonial Water Birds Society*, USA.

M

MacDonald, D.W. and Henderson, D.G., **1977**: 'Aspects of the Behaviour and Ecology of Mixed species Bird Flocks in Kashmir', *Ibis*, Vol. 119, No. 4, pp. 481-93.

MacDonald, Malcolm, **1960**: *Birds in My Indian Garden*, Jonathan Cape, London, 192 pp.

Madge, S.G., **1987**: 'Display of Thick-billed Flowerpecker *Dicaeum agile*', *JBNHS*, Vol. 83, Centenary Supplement, p. 213.

Madoc, G.C., **1956**: 'An Introduction to Malayan Birds', Revised ed., The Malayan Nature Society, Kuala Lumpur, 234 pp.

Magrath, H.A.F., **1908a**: 'Notes on the Birds of Thandiani', *JBNHS*, Vol. 18, No. 2, pp. 284-99; **1908b**: 'Notes on the Birds Found at Bannu, NWFP', *JBNHS*, Vol. 18, No. 3, pp. 684-5; **1909a**: 'Bird Notes from Murree and the Galis', *JBNHS*, Vol. 19, No. 1, pp. 142-56; **1909b**: 'Bird Notes from Dunga Gali', *JBNHS*, Vol. 19, pp. 753-5; **1912a**: 'Bird Notes by the Way in Kashmir', *JBNHS*, Vol. 21, No. 2, pp. 545-52; **1912b**: 'More Bird Notes by the Way in Kashmir', *JBNHS*, Vol. 21, No. 4, p. 1304-14; **1912c**: 'The Notes of Pallas's Eagle (Haliaeetus leucoryphus) and some Water Birds', *JBNHS*, Vol. 21, No. 2, p. 662; **1916**: 'Additions to the Birds of Kohat and Kurram', *JBNHS*, Vol. 24, No. 3, p. 601; **1917**: 'Large Pintailed Sandgrouse (*P.a. caudata*) settling on Water', *JBNHS*, Vol. 25, No. 1, p. 149.

Marchant, John H., **1986**: 'Identification, habits and status of Great Knot', *British Birds*, Vol. 79, No. 3, pp. 123-35.

Marien, Daniel, **1950**: 'Notes on some Asiatic Meropidae (Birds)', *JBNHS*, Vol. 49, No. 2, pp. 151-64.

Marler, Peter R. ed. with Allen, Thomas B. (General Editor), **1972**: *The Marvels of Animal Behaviour*, National Geographic Society, 422 pp.

Marshall, C.H.T., See Cock and Marshall, 'Notes on a Collection of eggs made in and about Murree'.

Marshall, J.T., **1978**: 'Systematics of smaller Asian Night Birds based on Voice', *American Ornith'l Union Monographs*, No. 25.

Marshall, T.E. (Capt.), **1902**: 'Notes on Birds near Quetta', *JBNHS*, Vol. 14, pp. 601-6; **1903**: 'Notes on Birds near Quetta', *JBNHS*, Vol. 15, pp. 351-5.

Mason, C.W. and Maxwell-Lefroy H., **1912**: *The Food of Birds in India*, Memoirs of the Department of Agriculture in India, published Dept. of Agric. in India. Calcutta, 371 pp.

Mathew, D.N., **1971**: 'A Review of the Recovery Data obtained by the Bombay Natural History Society's Bird Migration Study Project', *JBNHS*, Vol. 68, No. 1, pp. 65-85.

Mathews, W.H., **1918**: 'Birds nesting in the Bhillung Valley, Tehri Garhwal', *JBNHS*, Vol. 25, No. 3, pp. 495-7.

Mathews, W.H., **1941**: 'Bird Notes from Baltistan', *JBNHS*, Vol. 42, No. 3, pp. 658-63.

Maxwell-Lefroy, H., See: Mason and Lefroy, *The Food of Birds in India*.

Mayers, James, **1984**: 'Studies on the Ecology of Himalayan Snowcock *Tetraogallus himalyensis* in Hunza during the summer of 1984', in Cambridge Karakoram Expedition Report 1984, ed. Bamber J., Cambridge.

Mayr, Ernest, **1963**: *Animal Species and Evolution*, Cambridge Massachusets; **1969**: *Principles of Systematic Zoology*, New York.

McCann, Charles, **1939**: 'The Flamingo (*Phoenicopterus ruber antiquorum*)', *JBNHS*, Vol. 41, No. 1, pp. 12-38.

McKelvie, Colin Laurie, **1986**: *The Book of the Woodcock*, published by De Brett Carriage Ltd., London.

McLean, G.L. (Dr.), **1985**: Article: 'Sandgrouse', in Campbell, B. and Lack, E. (eds.), *Dictionary of Birds*, Carlton and Vermillion.

Meinertzhagen, R., **1900**: 'Reported occurrence of the Greater Black-backed Gull (*Larus marinus*) in Rajputana', *JBNHS*, Vol. 13, No. 2, p. 374; **1914**: 'Birds Nesting at Quetta', *JBNHS*, Vol. 23, No. 2, pp. 362-3; **1920**: 'The Birds of Quetta', *Ibis*, pp. 132-95; **1927**: 'Systematic results of Birds collected at High Altitudes in Ladak and Sikkim', Part 1, *Ibis*, Vol. 3, No. 3, pp. 363-422, Part 2, *Ibis*, Vol. 3, No. 4, pp. 571-633; **1928**: 'Some Biological Problems connected with the Himalaya', *Ibis*, Vol. 4, pp. 480-533; **1954**: *The Birds of Arabia*, published Oliver and Boyd, London, 624 pp; **1959**: *Pirates and Predators*, Oliver Boyd, Edinburgh, 230 pp.

Menesse, N.H., **1939**: 'The Punjab Sirkeer Cuckoo (*Taccocua leschenaultii*) in Sind', *JBNHS*, Vol. 41, No. 1, pp. 172-3; **1942**: 'The Distribution of the Nukta or Comb Duck in Sind', *JBNHS*, Vol. 43, No. 1, p. 106; **1943**: 'Occurrence of the Golden Oriole and Common Cuckoo in Sind', *JBNHS*, Vol. 44, No. 2, p. 296.

Merriam, C. Hart., **1894**: 'Laws of Temperature Control of the Geographic Distribution of Terrestial Animals and Plants', *National Geographic Magazine*, Vol. 6, pp. 229-38.

Mian, Afsar, **1984a**: 'A contribution to the Biology of Houbara, 1982-3 Wintering Population in Baluchistan', *JBNHS*, Vol. 81, No. 3, pp. 537-45; **1984b**: (August) 'Survey of the Houbara Bustard (*Chlamydotis undulata macqueeni*) in the Baluchistan: 1983-4 Population Studies', Report to World Wildlife Fund, Pakistan on sponsored Research Project; **1985**: 'Biology of the Houbara Bustard (*Chlamydotis undulata macqueeni*) in Baluchistan: 1984-5 Wintering Population', Report to World Wildlife Fund, Pakistan.

Mian, A. and Surahio, M.I., **1983**: 'Biology of Houbara Bustard (*Chlamydotis undulata macqueeni*) with reference to Western Baluchistan', *JBNHS*, Vol. 80, No. 1, pp. 111-8.

Miller, Keith, **1982**: *Continents in Collision*, published George Philip, Royal Geographic Society, 212 pp.

Minton, Sherman A. (Jr.) **1966**: 'A Contribution to the Herpetology of West Pakistan', *Bulletin of the American Museum of Natural History*, Vol. 134, Article 2, 184 pp.

Mirza, Zahid Beg, **1965**: 'Addition to the Recorded Birds of Lahore', *Pakistan Jour. Science*, Vol. 17, p. 215; **1973**: 'Note on the Distribution of some Birds in West Pakistan', *Pakistan Jour. of Zoology*, Vol. 5, No. 2, pp. 203-5; **1978**: 'Pheasant Surveys in Pakistan', *American Pheasant and Waterfowl Society Magazine*, Vol. 78, No. 1.

Mohan, R.S. Lal, **1986**: 'Recovery of a Ringed Sandwich Tern *Sterna sandvicensis* from Rameswaram Island, Tamil Nadu', *JBNHS*, Vol. 83, p. 664.

Moreau, R.E., **1972:** *The Palearctic-African Bird Migration Systems,* Academic Press, London, 384 pp.

Morse, Pamela S., **1979:** 'Pairing and Courtship in the North American Dipper', *Bird Banding,* Vol. 50, No. 1.

Morzer Bruyns, W.F.J. and Voous, K.H., **1965:** 'Food of Bridled Tern' *Ardea* Vol. 53, No. 1 and 2, p. 79.

Mountfort, Guy, **1969:** *The Vanishing Jungle—the Story of the World Wildlife Fund Expeditions to Pakistan,* Collins, London, 286 pp; **1971:** 'Occurrence of the Chaffinch *Fringilla coelebs* in Gilgit', *Ibis,* Vol. 113, p. 109.

Moynihan, M., **1953:** 'Some displacement activities of the Black-headed Gull', *Behaviour,* Vol. 5; **1955:** *Behaviour,* Supplement 4.

Mukherjee, A.K., **1963:** 'An Analysis of the Food of the Grey Quail, *Coturnix coturnix* (Linnaeus) in Western Rajasthan', *Pavo* Vol. 1, No. 1, pp. 32-4; 'Food Habits of Water Birds of the Sunderbans 24 Parganas District, West Bengal, India-4; **1969:** Part 1, *JBNHS,* Vol. 66, No. 2, pp. 345-60; **1971a:** Part 2, *JBNHS,* Vol. 68, No. 1, pp. 37-64; **1971b:** Part 3, *JBNHS,* Vol. 68, No. 3, pp. 691-716; **1974:** Part 4, *JBNHS,* Vol. 71, No. 2, pp. 180-200; **1975:** Part 5, *JBNHS,* Vol. 72, No. 2, pp. 422-447.

Murray, J.A., **1878:** 'Further Additions to the Sindh Avifauna', *Stray Feathers,* Vol. 7, pp. 108-23; **1884:** *The Vertebrate Zoology of Sind,* Richards London.

N

Naik, R.M., **1984:** See Parashârya and Naik, 'Juvenile Plumage of the Little Egret Compared with that of the White-phase Indian Reef Heron', *JBNHS,* Vol. 81, No. 3, pp. 693-5.

Naik, R.M., Para Sharrya, B.M., Patel, B.H. and Mansuri, A.P., **1981:** 'The Timing of Breeding Season and Inter-breeding between the Colour Phases of the Indian Reef Heron *Egretta gularis',* *JBNHS,* Vol. 78, No. 3, pp. 494-7.

Nasir, E. and Ali, S.I., (eds.), **1972-89:** *Flora of Pakistan,* published by National Herbarium, Pakistan Agricultural Research Council, Islamabad, 191 Fascicles, No. 1—Flacourtiaceae, No. 2—Hamamelidaceae etc. etc., No. 191—Boraginaceae.

Neelakantan, K.K., **1956:** 'Some Observations on the Breeding Behaviour of the Chestnut Bittern *Ixobrychus cinnamomeus* and the Black Bittern *Dupetor flavicollis',* *JBNHS,* Vol. 53, No. 4, pp. 704-8.

Neginhal, S.G., **1982:** 'The Birds of Ranga Nathittu', *JBNHS,* Vol. 79, No. 3, pp. 581-93.

Nethersole-Thompson, D., **1979:** *Greenshanks,* T. and A.D., Poyser, Berkhamsted, 275 pp.

Neufeldt, I., **1966:** 'Life History of the Indian Cuckoo *Cuculus micropterus micropterus* in the Soviet Union', *JBNHS,* Vol. 63, No. 2, pp. 399-419.

Newman, T.H., **1911:** 'The Snow Pigeon', *Aviculture Magazine,* (Third Series) No. 2, pp. 173-8.

Niethammer, Jochen, **1970:** 'Die Flamingos am Ab-i-Istada in Afghanistan', *Natur und Museum,* Vol. 100, No. 3, Frankfurt a M., pp. 201-10.

Niethammer, Von G., **1967:** 'On the breeding biology of *Montifringilla theresae',* *Ibis,* Vol. 109, No. 1, pp. 117-8.

Niethammer, Von G, and Niethammer, J., **1967:** 'New Records of Afghanistan Bird Fauna' (in German), *Jour. fur Ornithologie,* Vol. 108, No. 1.

Noble, G.K. and Schmidt, A., **1938:** Social Behaviour of the Black-crowned Night Heron', *Auk,* Vol. 55, pp. 7-40.

Nurse, C.G., **1902a:** 'Occurrence of the Red-breasted Merganser *Merganser serrator,* near Quetta', *JBNHS,* Vol. 14, No. 2, pp. 400-1; **1902b:** '*Merops apiaster* Breeding in Baluchistan', *JBNHS,* Vol. 14, No. 3, p. 627; **1903:** 'The Enemies of Butterflies', *JBNHS,* Vol. 15, No. 2, pp. 349-50.

O

Oates, E.W. and Blanford, W.T., **1889-98:** *Fauna of British India—*

Birds, Ist. ed., (Vol. 1 and 2 by E.W. Oates, 1889-90), (Vol. 3 and 4 by T. Blanford 1895-8), Taylor and Francis, London.

Olsen, Klaus M. and Christensen, Steen, **1984:** 'Identification of Juvenile Skuas', *British Birds.* Vol. 77, No. 9, pp. 443-50.

Osborn, W. (Lieut. General), **1902:** 'Notes on the Himalayan Nutcracker *Nucifraga hemispica',* *JBNHS,* Vol. 14, pp. 628-9.

Osborne, Patrick, Collar, Nigel and Goriup, Paul, **1984:** *Bustards,* Dubai Wildlife Research Centre with International Council for Bird Preservation.

Osmaston, B.B., **1897:** 'Birds Nesting in the Tons Valley', Part 1, *JBNHS,* Vol. 11, No. 1, pp. 64-72; **1898:** 'Birds nesting in the Tons Valley', Part 2, *JBNHS,* Vol. 11, No. 3, pp. 468-73; **1901:** 'Birds Nesting in the Tons Valley', *JBNHS,* Vol. 13, No. 3, p. 542; **1918:** 'Further Notes on Birds Nesting in the Tons Valley', *JBNHS,* Vol. 25, pp. 493-5; **1923a:** 'Note on the Nidification and Habits of some Birds in British Garhwal', *JBNHS,* Vol. 28, No. 1, pp. 140-60; **1923b:** 'Birdlife in Gulmarg', *JBNHS,* Vol. 29, No. 2, pp. 493-502; **1925:** 'The Birds of Ladakh', *Ibis,* Vol. 1, No. 3, pp. 663-719; **1926:** 'Birds Nesting in the Dras and Suru Valleys', *JBNHS,* Vol. 31, No. 1, pp. 186-97; **1927a:** 'Notes on the Birds of Kashmir', *JBNHS,* Vol. 31, No. 4, Part 1 pp 975-99; **1927b:** 'Notes on the Birds of Kashmir', *JBNHS,* Vol. 32, No. 1, Part 2, pp. 134-53.

P

Pakistan, **1977:** *Pakistan—Past and Present:* Ed. Jalal, Hamid, Merriam, Aimee, Qayyum, Abdul and Stacey, Tom, published Stacey International, London.

Pakistan, Govt. of, **1965:** 'Census Report of Tribal Agencies, Malakand, Mohmand, Khyber, Kurram, North Waziristan and South Waziristan', Ministry of Home and Kashmir Affairs, Govt. Press, Karachi.

Palmes, P. and Briggs, C., **1986:** 'Crab Plovers *Dromas ardeola* in the Gulf of Kutch', *Forktail,* No. 1, published Oriental Bird Club, Sandy, Beds.

Paludan, Knud, **1940:** 'Contributions to the Ornithology of Iran', Danish Scientific Investigations in Iran, No. 2, pp. 11-54; **1951:** *Vinensk. Medd. Dark Natur Foren,* Vol. 114, pp. 1-128; **1959:** 'On the Birds of Afghanistan', *Vidensk Medd Dansk Naturh:* Vol. 122.

Parashârya, B.M. and Naik, R.M., **1984:** The Juvenile Plumage of the Little Egret compared with that of the White-phase Indian Reef Heron', *JBNHS,* Vol. 81, No. 3, pp. 693-5.

Parker, R.N., **1956:** *A Forest Flora for the Punjab with Hazara and Delhi,* 3rd. ed., Govt. Printing Press, Lahore, 584 pp.

Payne, R.B. and Risley, C.J., **1976:** *Systematics and Evolutionary Relationships among the Herons (Ardeidae),* Miscellaneous publications, Museum of Zoology, University of Michigan, No. 150.

Perreau, G.A. (Capt.), **1910:** 'Notes on the Birds of Chitral', *JBNHS,* Vol. 19, pp. 901-22.

Perrins, C.M. and Birkhead, T.R., **1983:** *Avian Ecology,* Blackie, Glasgow, 221 pp.

Peters, J.L., **1960:** (E. Mayr and J.C. Greenway, eds.) *Checklist of Birds of the World* 9, Museum Comparative Zool., Cambridge, Mass.

Peterson, Roger, Mountfort, Guy and Hollom, P.A.D., **1966:** *A Field Guide to the Birds of Britain and Europe,* Collins, London.

Phillips, B.T., **1946:** 'A Bird Photographer's Musings from Kashmir', *JBNHS,* Vol. 46, No. 1, Part 1, pp. 89-103; *JBNHS,* Vol. 46, No. 3, Part 2, pp. 487-500; *JBNHS,* Vol. 47, No. 1, Part 3, pp. 84-102.

Phillips, T.J., **1941:** 'On the Occurrence of the Chaffinch *Fringilla coelebs* in Waziristan', *JBNHS,* Vol. 42, p. 439; **1947:** 'Occurrence of the Waxwing (*Bombycilla garrulus*) in Baluchistan', *JBNHS,* Vol. 47, No. 1, p. 160.

Phillips, W.W.A., **1939:** 'Nests and Eggs of Ceylon Birds', 3 parts,

Ceylon Journal of Science, Vol. 21; **1956:** 'Bird Migration in Relation to Ceylon', *Journal of Royal Asiatic Society* (Ceylon Branch, New Series, No. 5(I), pp. 25-41.

Pierce, R.J., **1986:** 'Observations of Behaviour and Foraging of the Ibisbill *Ibidoryncha struthersii* in Nepal', *Ibis,* Vol. 128, No. 1, pp. 37-47.

Pilbeam, David R., Behrensmeyer, A.K., Barry, John C., and Ibrahim Shah, S.M., **1979:** 'Miocene Sediments and Faunas of Pakistan, *Postilla',* No. 179, Peabody Museum, Yale University, Connecticut.

Polunin, Oleg and Stainton, Adam, **1984:** *Flowers of the Himalayas,* Oxford University Press, Delhi, India, 580 pp.

Ponomareva, T., **1979:** 'The Houbara Bustard, Present Status and Conservation Prospects', in Russian [translated for ICBP by M.G. Wilson], *Okhota I Okhotnich'e Khoxyaistud,* Vol. 2, pp. 26-7.

Porter, R.F., Willis, Ian, Christensen, Steen and Nielson, Bent Pors, **1974:** *Flight Identification of European Raptors,* T. and A.D. Poyser Ltd., Berkhamsted.

Prater, S.H., **1936:** 'The Longtailed Duck (*Clangula hyemalis*) in Sind', *JBNHS,* Vol. 38, No. 4, p. 831; **1965:** *The Book of Indian Animals,* 2nd ed. Revised, Bombay Natural History Society and Prince of Wales Museum, Bombay, 323 pp.

Priklonskiy, S.G. and Lavrovskiy, V.V., **1974:** 'On the Ecology of Golden Bee-eaters and perspectives on their protection in central reaches of the Oka River', (in Russian) Proceed: 6 All-union Ornith. Congress, No. 2, pp. 106-8.

Proud, Desiree, **1949:** 'Some Notes of the Birds of the Nepal Valley', *JBNHS,* Vol. 48, No. 4, pp. 695-719; **1951a:** 'More Bird Notes from Nepal Valley', *JBNHS,* Vol. 49, No. 4, pp. 784-5; **1951b:** 'Some Birds seen on the Gandak-Kosi Watershed in March 1951', *JBNHS,* Vol. 50, No. 2, pp. 355-66; **1955:** 'More Notes on the Birds of the Nepal Valley', *JBNHS,* Vol. 53, No. 1, pp. 57-78.

Pym, A., **1982:** 'Identification of Lesser Golden Plover and Status in Britain and Ireland', *British Birds,* Vol. 75, No. 3, pp. 112-24.

Q

Qureshi, Junaid Iqbal, **1972:** 'Notes on the Useful feeding Activities of Birds in Field Areas of Lyallpur', *Pakistan Journal of Agricultural Science:* Vol. 9, pp. 151-66.

Qureshi, J.I. and Beg, Mirza Azhar, **1975:** 'Birds of Wetlands of Lyallpur Region', *Pakistan Journal of Agricultural Science:* Vol. 12(1-2), pp. 103-8.

R

Radcliffe, H. Delme, **1915a:** 'List of the Birds of Baluchistan', Part 1, *JBNHS,* Vol. 23, pp. 745-57; **1915b:** 'List of the Birds of Baluchistan', Part 2, *JBNHS,* Vol. 24, pp. 156-69.

Rafi, Mohammad, **1973:** Vegetation Types of Quetta-Kalat Region', Punjab Forest Dept., Govt. Press, Lahore.

Ramzan, M. and Toor, H.S., **1973:** 'Damage to Maize Crop by Rose-ringed Parakeet (*Psittacula krameri*) in the Punjab', *JBNHS,* Vol. 70, No. 1, pp. 201-3.

Rana, B.D., **1975:** 'Breeding Biology of the Indian Ring Dove (*Streptopelia decaocto*) in the Rajasthan Desert', *The Auk,* Vol. 92, No. 2, pp. 322-32.

Randall, J.E., Allen, G.R. and Smith-Vaniz, W.F., **1978:** *Illustrated Identification Guide to Commercial Fishes,* Food and Agricultural Organisation, Rome, Regional Fishery Survey and Development Project, Oman, Qatar, Saudi Arabia, and United Arab Emirates.

Rands, M.R.W., Ridley, M.W. and Lelliott, A.D., **1984:** 'The Social Organization of Feral Peafowl', *Animal Behaviour,* Vol. 32, pp. 830-5.

Rane, Ulhas, **1984:** 'Unusual Communal Nest Feeding in Southern Small Minivet *Pericrocotus cinnamomeus cinnamomeus',* *JBNHS,* Vol. 81, No. 2, pp. 473-4.

Rashid, Haroon Er, **1967:** *Systematic List of the Birds of East Pakistan,* The Asiatic Society of Pakistan, Dacca, 144 pp.

Rattray, R.H., **1897a:** 'Notes on Nests taken from March to June at Kohat and Mussooree, Northwestern Provinces', *JBNHS,* Vol. 10, No. 4, p. 628; **1897b:** 'Nesting of the Little Forktail (*Microcichla scouleri*)', *JBNHS,* Vol. 2, p. 334; **1898:** 'The Nesting of the Red-tailed Chat *Saxicola chrysopygia',* *JBNHS,* Vol. 12, No. 2, p. 225-6; **1899a:** 'Birds collected and observed at Thull during five months in 1898, and Notes on their Nidification', *JBNHS,* Vol. 12, No. 2, p. 337-47; **1899b:** 'Occurrence of the Rufous-bellied Niltava (*Niltava sundara*) at Murree', *JBNHS,* Vol. 12, No. 3, p. 579; **1899c:** 'The Red-tailed Chat *Saxicola chrysopygia* a Correction', *JBNHS,* Vol. 12, No. 3, pp. 579-80; **1905:** 'Birds Nesting in the Murree Hills and Gullies', Part 1, *JBNHS,* Vol. 16, No. 3, pp. 421-8; Part 2, *JBNHS,* Vol. 16, No. 4, pp. 657-63.

Razack, K.M.A., **1968:** 'Some Observation on the Biology of the House Swift *Apus affinis',* Ph. D. Thesis, M.S. University of Baroda, India.

Redman, N.J., Lambert, F.L. and Grimmett, R., **1984:** 'Some Observations of Scarce Birds in Nepal', *JBNHS,* Vol. 81, No. 1, pp. 48-53.

Reeve, M.B.P., **1938:** 'Occurrence of the Long-tailed Duck *Clangula hyemalis* near Quetta', *JBNHS,* Vol. 40, No. 2, pp. 333-4.

Reyer, Heinz-Ulrich, **1980:** 'Flexible helper structure as an Ecological adaptation in the Pied Kingfisher (*Ceryle rudis rudis*)', *Behavioural Ecology and Sociology,* Vol. 6, published Springer-Verlag, pp. 219-27.

Richmond, C.W., **1895:** 'Catalogue of a Collection of Birds made by Doctor W.L. Abbott in Kashmir, Baltistan and Ladak, with notes on some of the species, and a description of a new species of *Cyanecula',* *Proceedings US National Museum,* Vol. 18, No. 1078, pp. 451-591.

Ridley, M.W., **1983:** 'A Review of the Ecology and Behaviour of Button-quails', *The World Pheasant Assn., Journal* No. 8.

Ridgway, **1912:** *Colour Standards and Color Nomenclature,* Washington, D.C.

Ripley, S. Dillon, **1961:** *A Synopsis of the Birds of India and Pakistan, together with those of Nepal, Sikkim, Bhutan and Ceylon,* Bombay Natural History Society, 702 pp; **1982:** 2nd. ed. Revised, Bombay Natural History Society, 652 pp; **1968-74:** See Ali, S. and Ripley, S.D., *Handbook of the Birds of India and Pakistan.*

Roberts, T.J., **1965:** Food of the White-breasted Kingfisher *Halcyon smyrnensis',* *JBNHS,* Vol. 62, No. 1, pp. 152-3; **1967:** 'Epilogue on a Sind Lake', *JBNHS,* Vol. 64, No. 1, pp. 13-21; **1968:** See Holmes, J.R.S., Roberts and Savage, 'Red-necked Grebe *Podiceps griseigena* Sighted in West Pakistan'; **1970:** 'A Note on the Pheasants of West Pakistan', *The Pakistan Jour. of Forestry,* Vol 20, No. 4, pp. 319-26; **1972:** 'A Brief Examination of Ecological changes in the Province of Sind and their consequence on the Wildlife Resources of the Region', *Pakistan Journal of Forestry,* Vol. 22, No. 2, pp. 89-96; **1973:** 'Conservation Problems in Baluchistan with particular Reference to Wildlife Preservation', *Pakistan Jour. of Forestry,* Vol. 23, No. 2, pp. 117-27; **1974:** 'Interesting Distributional Records for Pakistan', *JBNHS,* Vol. 70, No. 3, pp. 552-4; **1975:** 'Ornithological Records for Pakistan', *JBNHS,* Vol. 72, No. 1, pp. 201-4; **1977:** *The Mammals of Pakistan,* Ernest Benn Ltd., London, 384 pp; **1978:** 'Unusual Ornithological Records for Pakistan', *JBNHS,* Vol. 75, No. 1, pp. 216-9; **1980:** 'Bird Notes from Baluchistan Province, Pakistan', *JBNHS,* Vol. 77, No. 1, pp. 12-20; **1981a:** 'Ornithological Notes from Pakistan', *JBNHS,* Vol. 78, No. 1, pp. 73-6; (ed.) **1981a:** *Handbook of Vertebrate Pest Control in Pakistan,* Pakistan Agricultural Research Council in collaboration with FAO; **1982:** *A Synopsis of the Birds of India and Pakistan— Review;* **1984:** 'Recent Ornithological Records from Pakistan', *JBNHS,* Vol. 81, No. 2, pp. 399-405; **1985:** 'Zangi Nawar—Portrait of a Unique Lake in the Desert', *JBNHS,* Vol. Vol. 82, No. 3, pp. 540-7.

Roberts, T.J., Grimmett, Richard and Robson, Craig, **1985:** 'Commentary on a Pictorial Guide to the Birds of the Indian Sub-continent', *JBNHS,* Vol. 82. No. 3, pp. 567-72.

Roberts, T.J. and King, Ben, **1986:** 'Vocalizations of the Owls of the Genus *Otus* in Pakistan', *Ornis Scandinavica,* Vol. 17, No. 4, pp. 299-305.

Roberts, T.J. and Landfried, Steven (Dr.), **1983:** 'Hunting Pressures on Crane Migration through Pakistan', Proceedings of International Crane Conference, Bharatpur, India, Buraboo Wisconsin, unpublished.

T.J. Roberts and Savage, C.D.W., **1970:** 'On the Occurrence of *Haliaeetus albicilla* in West Pakistan', *JBNHS,* Vol. 66, No. 3, pp. 619-22.

Roos-Keppel, G., **1918:** 'Occurrence of the European Great Bustard *Otis tarda,* near Peshawar', *JBNHS,* Vol. 25, p. 745.

S

Sankaran, Ravi and Rahmani, Asad R., **1985-6:** 'The Lesser Florican', *Annual Report* 2, Bombay Natural History Society.

Savage, C.D.W., **1965:** 'Wildfowl Survey in Southwest Asia', *Wildfowl Trust 16th. Annual Report,* Slimbridge, pp. 123-5; 31968: 'Red-necked Grebe *Podiceps griseigena* again sighted in West Pakistan', *JBNHS,* Vol. 65, No. 3, p. 773.

Schafer, Ernst, **1938:** 'Ornithologische Ergebnisse zweier Forschung-reisen nach Tibet', *Journ. fur Ornithologie Sonder,* Heft 1, p. 349.

Schauensee, Rodolphe Meyer de, **1984:** *The Birds of China,* 602 pp.

Schuz, E., **1957:** 'Ringing Recovery of Spoonbill', *Vogelwarte,* Vol. 19, pp. 41-44.

Sclater, Philip, **1858:** 'On the General Geographical distribution of the Members of the Class Aves', *Journ. Proceedings Linnean Society,* London (Zoology), Vol. 2, pp. 130-45.

Scott, Derek A., **1978:** 'The Birds of the Seistan Basin, Iran' (unpublished Checklist submitted to Iranian Govt.).

Scott, Derek A., Hamadani, H.M. and Hosseyni, A.A. Mir, **1975:** *A Working Checklist of the Birds of Iran with notes on Status and Distribution* (in Farsi), published Iran Dept. of the Environment, Tehran.

Scully, J., **1876:** 'A contribution to the Ornithology of Eastern Turkestan', *Stray Feathers,* Vol. 4, No. 1, pp. 41-205; **1887:** 'Contribution to the Ornithology of Gilgit', *Stray Feathers,* Vol. 10, No. 1, pp. 88-146.

Seth-Smith, D., **1903:** 'On the breeding in captivity of *Turnix tanki* with some notes on the habits of the species', *Avicultural Mag.* No. 1, pp. 317-24.

Severinghaus, Dr. Sheldon R., **1979:** 'Observations on the Ecology and Behaviour of the Koklass Pheasant in Pakistan', *The World Pheasant Assn. Journal,* No. 4, 1978-79, pp. 52-69.

Sharma, Indira Kumar, **1976a:** 'Pestilence and feeding habits of the Peafowl (*Pavo cristatus*)', *Agricultural Research Newsletter,* Vol. 4, Nos. 10-12, published Karnal, India, pp. 1-3; **1976b:** 'Ecological studies of pestilence of grain crops by birds', *Mysore Journ. Agricultural Science,* Vol. 10, pp. 471-8.

Sharma, Satish Kumar, **1987:** 'Colour Selection by the Black-throated Weaver Bird *Ploceus benghalensis*', *JBNHS,* Vol. 83, Centenary Supplement, pp. 214-6.

Shelley, B.A.G. (Lieut.), **1895:** 'The Nesting of the Long-eared Owl (*Asio otus*) in India', *JBNHS,* Vol. 10, No. 1, p. 149.

Shirt, D.B., **1983:** 'The Avifauna of Fuerteventura and Lanzarote, in 'Bustard Studies', *Journ of ICBP Bustard Group,* No. 1, January 1986.

Shivrajkumar, Yuvraj, **1962:** 'Recoveries of ringed migratory birds at Hingolgadh, Jadan, Saurashtra', *JBNHS,* Vol. 59, No. 3, p. 963.

Simmons, K.E.L., **1955:** 'Studies on Great Crested Grebes', *Aviculture Magazine,* Vol. 61, London.

Simmons, R.M., **1930:** 'Migration of the Pied-crested Cuckoo (*Coccystes jacobinus*)', *JBNHS,* Vol. 34, No. 1, pp. 252-3.

Sinclair, J.C. and Nicholls, G.H., **1980:** 'Winter Identification of Greater and Lesser Sand Plovers', *British Birds,* Vol. 73, No. 5, pp. 206-13.

Skead, C.K., **1950:** 'A Study of the African Hoopoe *Upupa epops africana' Ibis,* Vol. 92, pp. 434-63.

Smith, G.A., **1972:** 'Some observations on Ring-necked Parakeets (*Psittacula krameri*)', *Aviculture Magazine,* Vol. 78, No. 4, pp. 120-37.

Smith, J. Lindsay, **1913:** 'Common Grey Quail (*Coturnix communis*) breeding in the Lyallpur District', *JBNHS,* Vol. 22, No. 1, p. 200; **1915:** 'Egret Farming in Sind', *JBNHS,* Vol. 23, No. 3, pp. 582-3.

Smith, T.E.H., **1950:** 'Black Drongos fostering a Koel', *JBNHS,* Vol. 49, No. 2, p. 304.

Smythies, Bertram E., **1953:** *The Birds of Burma,* Revised ed., Oliver and Boyd, Edinburgh, 668 pp.

Snow, David W., **1978:** *An Atlas of speciation in African Non-passerine Birds,* London.

Sorley, H.T. (Dr.) ed., **1968:** *The Former Province of Sind—The Gazetteer of West Pakistan,* Govt. Printing Press, Karachi, 811 pp.

Stegmann, B., **1934:** *Journ. Ornith.,* Vol. 82, pp. 340-80.

Stevens, H., **1914:** 'Notes on the Birds of Upper Assam', *JBNHS,* Vol. 23, No. 2, pp. 234-68; **1915:** *JBNHS,* Vol. 23, No. 4, pp. 547-70; *JBNHS,* Vol. 23, No. 4, pp. 721-36.

Stewart, Ralph R., **1957-1958:** 'The Flora of Rawalpindi District, West Pakistan', *Pakistan Journal of Forestry,* Rawalpindi, 163 pp; **1967:** 'Checklist of the Plants of Swat State, Northwest Pakistan', *Pakistan Journal of Forestry,* Vol. 17, No. 4, pp. 457-528; **1972:** *An annotated Catalogue of the Vascular Plants of West Pakistan and Kashmir, Flora of Pakistan,* ed. Nasir, E. and Ali, S.I., Pakistan Agricultural Research Council; **1982:** *History of Exploration of Plants in Pakistan's adjoining areas,* in Flora of Pakistan series, ed. Nasir and Ali, Islamabad.

Stirling, H.D., **1914:** 'Imperial Sandgrouse *Pterocles arenarius* in Chitral', *JBNHS,* Vol. 23, p. 159.

Stockley, C.H., **1930:** 'Notes on Birds of Baluchistan', *JBNHS,* Vol. 34, No. 2, p. 575.

Stoker, R.M., **1881:** 'Female Scaup, *Fuligula marila,* near Attock', *Stray Feathers,* Vol. 10, p. 158; **1883:** 'Golden-eye Duck *Clangula glaucium,* near Attock', *Stray Feathers,* Vol. 10, p. 424.

Stoliczka, F., **1873:** '*Indicator xanthonotus* the Honeyguide from Murree Hills', *Stray Feathers,* Vol. 1, p. 425-7.

Stoney, R.F., **1938:** 'Woodcock, Wood Snipe, Pintail Snipe and Jack Snipe in one day!', *JBNHS,* Vol. 40, No. 2, pp. 331-2.

Storor, C.R., **1948:** 'Fishing with the Indian Darter *Anhinga melanogaster* in Assam', *JBNHS,* Vol. 47, No. 4, pp. 746-7.

Storrs-Fox, E.A. (Rev.), **1933:** 'Occurrence of the Blue-throated Barbet (*Cyanops asiatica*) at Murree', *JBNHS,* Vol. 36, No. 3, p. 750; **1937:** 'Notes on Murree Birds', *JBNHS,* Vol. 39, No. 2, pp. 354-7.

Stuart Baker, E.C., See Baker, Stuart E.C.

Sudilovskaya, A.M., See Dement'ev *et al, Birds of the Soviet Union,* 5 Vols.

Surahio, M.I., **1982:** 'Houbara Bustard in Pakistan, Conservation and Research', WWF/IUCN Project No. 855, Annual Report (unpublished).

Sushkin, P.P., **1938:** *Birds of the Soviet Altai and Northwest Mongolia,* 2 Vols. (in Russian), Moscow.

Svensson, Lars, **1984:** *Identification Guide to European Passerines,* 3rd. ed. Revised, British Trust for Ornithology, Tring.

T

Taylor, Neale W., **1985:** 'Houbara Bustard Conservation and Management in Pakistan (1983)' *ICBP Project Report PK-IUCN/WWF 855.*

Tewary, P.D., Kumar, Vinod, and Prasad, B.N., **1983:** 'Influence of Photoperiod in a Sub-tropical Migratory Finch, the Common Rosefinch, *Carpodacus erythrinus', Ibis,* Vol. 125, pp. 115-20.

Thirumurthi, S. and Doss, D. Krishna, **1981:** 'A note on the Feeding Habits of Swifts (Apodidae, Apodiformes)', *JBNHS,* Vol. 78, No. 2, pp. 378-9.

Ticehurst, Claude B., **1922a:** 'The Birds of Sind', *Ibis,* July, Part 1, pp. 526-72; **1922b:** Oct. Part 2, pp. 605-62; **1922c:** 'Determination of Races of *Otus bakkamoena* in India', *Bull. of British Ornith. Club,* Vol. 42, pp. 57-122; **1923a:** Jan., Part 3, pp. 1-43; **1923b:** April, Part 4, pp. 235-75; **1923c:** July, Part 5, pp. 438-74; **1923d:** Oct., Part 6, pp. 645-66; **1924a:** Jan., Part 7, pp. 110-46; **1924b:** July, Part 8, pp. 495-518; **1926a:** 'On the Down Plumages of some Indian Birds', *JBNHS,* Vol. 31, No. 2, pp. 368-78; **1926b:** 'Some Notes on the Second Edition of the Fauna of British India—Birds', Vols. 1 and 2, *JBNHS,* Vol. 31, No. 2, pp. 490-9; **1926c:** 'The Birds of British Baluchistan', Part 1, *JBNHS,* Vol. 31, No. 3, pp. 687-711; **1927a:** Part 2, *JBNHS,* Vol. 31, No. 4, pp. 862-81; **1927b:** Part 3, *JBNHS,* Vol. 32, No. 1, pp. 64-97; **1930:** 'Notes on the Fauna of British India—Birds', Volumes 4, 5 and 6 (New ed.), *JBNHS,* Vol. 34, No. 2, pp. 468-90; **1938:** *A Systematic Review of the Genus Phylloscopus',* published by Trustees British Museum (Natural History), London, 193 pp.

Ticehurst, C.B., Buxton P.A. Cheeseman, R.E., **1922c:** 'The Birds of Mesopotamia', Part 4, *JBNHS,* Vol. 28, No. 4, pp. 937-56.

Turcek, F.J. and Kelso, L., **1968:** 'Ecological aspects of food transportation and storage in the Corvidae', *Communications in Behavioural Biology,* Part A, No. 1, pp. 277-97.

U

Urban, Emil K., Fry, Hilary C. and Keith, Stuart, **1986:** *The Birds of Africa,* Vol. 2, Academic Press Inc. (London) Ltd. 532 pp.

V

Vanderwall, Stephen B. and Balda, Russell P., **1977:** 'Co-adaptations of the Clark's Nutcracker and Pinon Pine for efficient seed harvest and dispersal', *Ecological Monographs,* Vol. 47, No. 1, pp. 89-111.

Van Rhijn, J.G., **1973:** 'Behavioural Dimorphism in male Ruffs *Philomachus pugnax',* *Behaviour,* Vol. 47, pp. 153-229.

Van Tyne, Josselyn and Berger, Andrew J., **1965:** 'Fundamentals of Ornithology', John Wiley and Sons, London, 624 pp.

Vaurie, Charles, **1949:** 'Revision of the Family Dicruridae', *Bulletin American Museum Natural History* No. 93, pp. 199-342; **1959:** *The Birds of the Palearctic Fauna—a Systematic reference, Passeriformes,* published, H.F. and G. Witherby Ltd., London, 762 pp; **1965:** *Non-Passeriformes,* 763 pp; **1972:** *Tibet and its Birds,* H.F. and G. Witherby Ltd., London, 407 pp.

Venour, Walter, **1907:** 'Occurrence of the Cheer Pheasant *Catreus wallichi* in the North West Frontier Province', *JBNHS,* Vol. 17, No. 3, p. 812.

Vertebrate Pest Control Centre, See *Handbook of Vertebrate Pest Control in Pakistan,* ed. Roberts, T.J.

Viney, Clive and Phillips, Karen, **1979:** *A Colour guide to Hong Kong Birds,* 2nd. ed., Govt. Printer, Hong Kong.

Von Blotzheim, See Glutz Von Blotzheim.

Voous, K.H. de, **1973:** 'List of Recent Holarctic Bird Species', *Ibis,* Vol. 115, No. 4, pp. 612-38; **1977:** 'List of Recent Holarctic Bird Species, Passeriformes', *Ibis,* Vol. 119, No. 2, pp. 223-50, Vol. 119, No. 3, pp. 376-406.

Vuilleumier, Francois, **1985:** Article 'Zoogeography', in Campbell, B. and Lack, E. (eds.) *Dictionary of Birds,* Calton and Vermillion.

W

Waite, H.W., **1917:** The Breeding of the Gull-billed Tern (*Sterna anglica*), *JBNHS,* Vol. 25, No. 2, pp. 300-1; **1924:** 'The Nidification of the Common or Grey Quail (*Coturnix coturnix*)', *JBNHS,* Vol. 30, No. 1, p. 226; **1925:** 'Breeding of the Painted Sand-grouse (*Pterocles indicus*) in the Punjab Salt Range', *JBNHS,* Vol. 30, No. 4, p. 917; **1926:** 'Note on the breeding of the Genus *Caprimulgus* (Nightjars) in the Punjab Salt Range', *JBNHS,* Vol. 31, No. 3, pp. 821-2; **1934:** 'Birds observed at Fort Munro, Sulaiman Hills', *JBNHS,* Vol. 37, No. 3, pp. 688-93; **1935:** 'Occurrence of the Black-headed Cuckoo Shrike (*Lalage sykesii*), in Hoshiarpur District of the Punjab', *JBNHS,* Vol. 35, No. 1, Misc. Notes; **1937:** 'Some interesting records of birds in the Punjab', *JBNHS,* Vol. 39, No. 4, pp. 861-2; **1938:** 'Some interesting records of birds in the Punjab—a correction', *JBNHS,* Vol. 40, No. 2, pp. 328-9; **1948:** 'The Birds of the Punjab Salt Range (Pakistan)', *JBNHS,* Vol. 48, No. 1, pp. 97-117; **1962:** 'Notes on the range of certain birds as given in S.D. Ripley II (1961): A Synopsis of the Birds of India and Pakistan', *JBNHS,* Vol. 59, No. 3, pp. 959-63; **1963:** 'The Common Hawk Cuckoo (*Cuculus varius varius*) in the Punjab', *JBNHS,* Vol. 60, No. 1, p. 260.

Wall, F., **1912:** 'Rambling Notes on Natural History in Chitral Aves', *JBNHS,* Vol. 21, No. 2, pp. 617-9.

Ward, A.E., **1906:** 'Birds of the Provinces of Kashmir and Jammu and Adjacent Districts' *JBNHS,* Vol. 17, No. 1, Part 1, pp. 108-13, *JBNHS,* Vol. 17, No. 2, Part 2, pp. 479-85, *JBNHS,* Vol. 17, No. 3, Part 3, pp. 723-9, *JBNHS,* Vol. 17, No. 4, Part 4, pp. 943-9; **1908:** 'Further Notes on Birds of the Provinces of Kashmir and Jammu and Adjacent districts', *JBNHS,* Vol. 18, pp. 461-4.

Wathen, M.L. (Mrs.), **1923:** 'Ornithological notes from a trip in Ladak', *JBNHS,* Vol. 29, Part 3, pp. 694-702.

Watson, H.W.A., **1914:** 'Note on the habits of the Kalij Pheasant', *JBNHS,* Vol. 23, No. 1, p. 159.

Watson, J.W., **1903:** 'Notes on birds near Quetta', *JBNHS,* Vol. 15, Part 1, pp. 144-5.

Wayre, Philip, **1969:** 'A Guide to the Pheasants of the World', *Country Life,* London, 175 pp.

Wayre, Philip and Harrison, C.J.O., **1972:** 'Display of the Koklass', *The Outdoorman,* Vol. 2 No. 7 and 8, Paragon Publishers, Karachi.

Wells, D.R., **1972:** 'The Genus *Cuculus:* two amendments to the Handbook of the Birds of India and Pakistan', *JBNHS,* Vol. 69, No. 1, pp. 179-85.

Wells, D.R. and Becking, J.H., **1975:** 'Vocalisations and Status of Little and Himalayan Cuckoos, *Cuculus poliocephalus* and *C. saturatus,* in Southeast Asia, *Ibis,* Vol. 117, pp. 366-71.

Whistler, Hugh, **1911:** 'Some winter visitors to Rawalpindi', *JBNHS,* Vol. 21, No. 1, p. 261-2; **1912a:** 'Immature plumages of Lammergayer (*Gypaetus barbatus*)', *JBNHS,* Vol. 21, No. 2, p. 663-5; **1912b:** 'Nestling Plumage of the Great Stone Plover (*Esacus recurvirostris*)', *JBNHS,* Vol. 21, No. 3, p. 1074; **1913a:** 'The Pale Sand Martin *Riparia riparia diluta* (Sharpe and Wyatt)', *JBNHS,* Vol. 22, No. 2, p. 393; **1913b:** 'Occurrence of the Himalayan Pigmy Woodpecker *Jungipicus pygmaeus* in Rawalpindi district', *JBNHS,* Vol. 22, No. 3, p. 626; **1914a:** 'The Small Indian Pratincole *Glareola lactea',* *JBNHS,* Vol. 22, No. 4, p. 804; **1914b:** 'Interesting Birds from Jhelum District, Punjab', *JBNHS,* Vol. 23, No. 1, p. 153; **1914c:** 'Notes on Doves in the Punjab', *JBNHS,* Vol. 23, No. 1, p. 157; **1915:** 'Some birds in Hissar District, Punjab', *JBNHS,* Vol. 24, No. 1, p. 190; **1916:** 'Notes on some Birds of the Gujranwala District, Punjab', *JBNHS,* Vol. 24, No. 4, p. 689; **1918a:** 'Notes on the Birds of Ambala District, Punjab', Part 1, *JBNHS,* Vol. 25, No. 4, pp. 665-81; Part 2, *JBNHS,* Vol. 26, No. 1, pp. 172-91; **1918b:** 'The White-necked Stork in the Punjab', *JBNHS,* Vol. 25, p. 746-7; **1919:** 'Some birds of Ludhiana District, Punjab', *JBNHS,* Vol. 26, No. 2, pp. 585-98; **1920:** 'Some notes on the Genus *Caprimulgus* in the Punjab', *JBNHS,* Vol. 27, No. 2, p. 363-70; **1922:** 'Birds of the Jhang District', *Ibis,* Vol. 4, pp. 259-309; **1927:** '*Gyps indicus jonesi sub-sp. nov.',* *Bulletin of the British Ornithologists Club,* Vol. 47, p. 74; **1928a:** 'Further

Notes on birds about Simla', *JBNHS*, Vol. 32, No. 4, pp. 726-32; **1928b**: 'The migration of the Pied-crested cuckoo (*Clamator jacobinus*)', *JBNHS*, Vol. 33, No. 1, pp. 136-44; **1930**: 'The Birds of the Rawalpindi District, N.W. India' *Ibis*, Part 1 pp. 67-119, Part 2 pp. 247-79. Materials for the Ornithology of Afghanistan; **1944a**: Part 1, *JBNHS*, Vol. 44, No. 4, pp. 505-19; **1944b**: Part 2, *JBNHS*, Vol. 45, No. 2, pp. 61-72; **1945a**: Part 3, *JBNHS*, Vol. 45, No. 2, pp. 106-22; **1945b**: Part 4, *JBNHS*, Vol. 45, No. 3, pp. 280-332; **1945c**: Part 5, *JBNHS*, Vol. 45, No. 4, pp. 462-85; **1949**: *Popular Handbook of Indian Birds*, 4th. ed., Revised by Kinnear, N.B., Oliver and Boyd, Edinburgh, 560 pp.

Whitaker, J.I.S., **1905**: *Birds of Tunisia*, London.

Whitehead, C.H.T., **1906**: 'Notes on the Occurrence of certain Birds in the Plains of Northwest India', *JBNHS*, Vol. 17, No. 1, pp. 243-4; **1907a**: 'Some additions to the Birds of India', *JBNHS*, Vol. 18, No. 1, pp. 190-1; **1907b**: 'Nesting notes from the North West Frontier', *JBNHS*, Vol. 18, No. 1, pp. 191-3; 'On the Birds of Kohat and the Kurram Valley, northern India'; **1910**: Part 1 *JBNHS*, Vol. 20, No. 1, pp. 169-97; **1911**: Part 2, *JBNHS*, Vol. 20, No. 3, pp. 776-99; Part 3, *JBNHS*, Vol. 20, No. 4, pp. 954-80; **1914**: 'Some notes on the birds of the Kaghan Valley, Hazara, North West Frontier Province', *JBNHS*, Vol. 23, No. 1, pp. 104-9.

Whitehead, W.A. (Capt.), **1931**: 'Notes on the White-headed Duck or Stiff Tail (*Erismatura leucocephala*)', *JBNHS*, Vol. 35, pp. 211-2.

Whymper, S.L., **1902**: 'Birds nesting in Kumaon, *JBNHS*, Vol. 14, pp. 624-6; **1910**: 'Birds nesting in Garhwal', *JBNHS*, Vol. 19, No. 4, pp. 990-1; **1911**: 'Birds nesting in the Nila Valley, Garhwal', *JBNHS*, Vol. 20, No. 4, pp. 1157-60.

Williams, C.H. (Major) and Williams, C.E., **1929**: 'Some notes on the Birds breeding round Quetta', *JBNHS*, Vol. 33, pp. 598-613.

Wintle, C.C., **1975**: 'Notes on the breeding habits of the Kurrichane Button Quail', *Honey Guide*, No. 82, pp. 27-30.

Witherby, H.F., Jourdain, F.C.R., Ticehurst, Norman F. and Tucker, Bernard W., **1943-4**: *The Handbook of British Birds*, 5 Volumes, Revised ed., H.F. and G. Witherby Ltd., London.

Wright, John, See Holmes, D.A. and Wright, *The Birds of Sind: A Review*.

Wright, R.G. and Dewar, Douglas, **1925**: *The Ducks of India*, H.F. and G. Witherby, London, 231 pp.

Wyllie, I., **1981**: *The Cuckoo*, Pub. B.T. Batsford, London

Y

Yahya, H.S.A., **1988**: 'Breeding Biology of Barbets, *Megalaima* spp. (CAPITONIDAE: PICIFORMES) at Periyar Tiger Reserve, Kerala', *JBNHS*, Vol. 85, No. 3, pp. 493-511.

Yamashina, Yoshimaro, **1961**: *Birds of Japan—a Field Guide* (in English), Tokyo News Service Ltd., 233 pp.

Young, Llewellyn, Garson, P.J. and Kaul, Rahul., **1987**: 'Calling Behaviour and Social Organization in the Cheer Pheasant: Implications for Survey Technique', *World Pheasant Assoc. Journ.* No. 12, pp. 30-43.

Z

Zacharias, V.J. and Gaston, A.J., **1983**: 'Breeding Seasons of Birds at Calicut, Southwest India', *Ibis*, Vol. 125, No. 3, pp. 407-11.

Zink, Robert M. and Eldridge, Jan L., **1980**: 'Why does Wilson's Petrel have yellow on the webs of its feet? *British Birds*, Vol. 73, No. 9, pp. 385-7.

APPENDIX III

Gazetteer of Locations in Pakistan

In this Gazetteer localities mentioned in India, Afghanistan or elsewhere outside Pakistan are not included, nor are the coordinates given for every village and town located in Pakistan as this would be an onerous task beyond the specialized needs of this book. The aim has been to give map references for every place mentioned in the text, including besides villages, some forest rest houses, streams, hill ranges and peaks, as well as lakes or reservoirs.

In this attempt, about 5 per cent of those places mentioned have not been located on any map. For example, the location of a few places written on collector's labels in museums and in the diary notes (unpublished) of E.M. Nicholson and K. Eates are such instances, as also some localities mentioned by Captain Fulton who worked in Chitral at the turn of the century. Despite this shortcoming, an estimated 15 per cent of the places actually listed hereunder, are not shown on any available maps, and for this reason it has been considered essential to include such a gazetteer.

Besides the above problem there is one of transliteration. Places in Pakistan are known only by their original Persian, Arabic, Saraiki, Sindhi or other names and the English spelling has been variable, particularly in older maps and publications. Since these older spellings are likely to be encountered by anyone studying the ornithology of the region, they have in most cases been included. Thus 'q' and 'k' are often interchangeable, 'oo' and 'u', as well as 'a' and 'e'. In a few instances the same place name occurs in widely different localities. For example there are two Jabbah's in the Punjab Salt Range and Swat, two Ziarats in southern Chitral and in Baluchistan and two Dubers, one a tributary of the Indus in Kohistan and one a tract of riverain forest in Sind.

For security reasons detailed maps of most regions of Pakistan have been unavailable for many years and this has made the compilation of a gazetteer particularly difficult. Heavy reliance has been placed upon old World War II aeronautical charts of the 1:1,000,000 scale which are still obtainable in Britain.

Name	Approximate Location	Coordinates Lat.(N) Lon.(E)	Name	Approximate Location	Coordinates Lat.(N) Lon.(E)
Abbottabad	Hazara District	34° 08′ 73° 12′	Atharan Hazari	Jhang District	31° 11′ 72° 07′
Abdul Hakim	Multan District	30° 33′ 72° 07′	Athmaqan	Azad Kashmir	34° 15′ 73° 40′
Agram Pass	Chitral, Afghan border	36° 18′ 71° 31′	Attock	Campbellpur District	33° 54′ 72° 14′
Ahmadpur East	Bahawalpur Division	29° 08′ 71° 14′		(now Attock District)	
Akara Dam	Makran, West of Gwadar	25° 17′ 62° 15′	Ayub National		
Alipurai			Park	Rawalpindi	33° 34′ 73° 05′
or Alpurai	Indus Kohistan	34° 54′ 72° 37′	Ayubia,		
Alipur	Muzaffargarh	29° 23′ 70° 54′	see Khanspur	Murree hills	
Allahdad	South of Karchat				
	and Kirthar Range	25° 38′ 67° 34′	Baba Kharwari	South of Ziarat,	
Amandara	Malakand Agency	34° 33′ 71° 57′		Sibi District	30° 18′ 67° 43′
Amb	Town, Amb State,		Babusar Pass	Between Hazara District	
	west of Hazara District	34° 18′ 72° 51′		and Chilas	35° 09′ 74° 06′
Anam Bostan	Nushki District,		Babusar Village	Village, Chilas District	35° 12′ 74° 03′
	Afghan border	29° 42′ 65° 50′	Bach Mission		
Ara	Salt Range, not located		Hospital	Mansehra plain	34° 22′ 73° 13′
	on map		Badin	Formerly Tharparkar	
Arandu	South-western Chitral	35° 19′ 71° 34′		District now H.Q'rs	
Arifwala	Sahiwal District	30° 17′ 73° 04′		Badin District	24° 39′ 68° 51′
Arkari Nullah	North-western Chitral	36° 05′ 71° 47′	Bahaudin	See Mandi Bahauddin,	
Astola Island	Makran coast	25° 17′ 63° 50′		Gujrat District	32° 34′ 73° 29′
Astor river			Bahawalnagar	Town and headquarters	
and town	Gilgit Agency, south-			northern district of	
	eastern district	35° 22′ 74° 52′		Bahawalpur Division	29° 59′ 73° 16′

Name	Approximate Location	Coordinates Lat.(N) Lon.(E)
Bahawalpur	Town, Bahawalpur Division	29° 23′ 71° 39′
Bahrein or Bahrain	Swat State	35° 14′ 72° 32′
Bajaur	North Malakand Agency	34° 45′ 71° 20′
Bajwat	Sialkot District, not located on map	
Bakar Lake	Sanghar District	26° 05′ 69° 10′
Bakkar	Mianwali District, Thal	31° 37′ 71° 03′
Balakot town	Hazara, Kaghan valley	34° 33′ 73° 19′
Balgatar	North central Makran	26° 10′ 63° 45′
Balloki headworks	Lahore District on Ravi river	31° 13′ 73° 52′
Baltit	Hunza, Gilgit Agency	36° 19′ 74° 41′
Band Kushdil Khan Lake	See Kushdil Khan	
Band Murad Khan	Hab valley	25° 07′ 67° 02′
Bannu	Town, Bannu District, NWFP	33° 00′ 70° 37′
Bara Gali	Murree hills	34° 06′ 73° 22′
Bara Village	Village on Shyok river, Baltistan	35° 13′ 76° 18′
Baran Dam	Kurram river, north-west of Bannu	33° 03′ 70° 32′
Baran River	Dadu District, Sind Kohistan	25° 25′ 68° 05′
Baratpur	Lahore District on Indian border	31° 41′ 74° 28′
Bargo or Bargu Bala	On Gilgit river	36° 03′ 74° 08′
Barian	Murree hills	33° 58′ 73° 23′
Baroghil	See Brughal, N. Chitral	36° 51′ 73° 22′
Barshore	Central Pishin District	30° 50′ 67° 17′
Basal	Hazara District, Kaghan valley	35° 04′ 73° 56′
Basal	Attock District	33° 33′ 72° 16′
Batapur	Near Lahore	31° 34′ 74° 30′
Batrasi Pass	Hazara District, near Mansehra	34° 24′ 73° 21′
Battagram	Hazara Tribal Territory	34° 40′ 72° 59′
Battakundi	Hazara District, Kaghan valley	34° 56′ 73° 46′
Battal	Hazara Tribal Territory	34° 34′ 73° 20′
Bela	Town, west of Karachi	26° 13′ 66° 19′
Beleli	North of Quetta	31° 16′ 66° 54′
Beori Nullah	South-eastern Chitral	35° 29′ 71° 48′
Besal	Village, northern Kaghan valley	35° 05′ 74° 03′
Besham	Karakoram Highway	34° 56′ 72° 53′
Besima or Bisemah	South-eastern Kharan District	24° 59′ 64° 27′
Besti Nullah	North-western Chitral	36° 10′ 71° 35′
Bhagnotar	Northern Murree foothills	34° 07′ 73° 18′
Bhakkar	Muzaffargarh, Thal, see Bakkar	31° 37′ 71° 03′
Bhalwal	Sargodha District	32° 16′ 72° 53′
Bharakao	Murree foothills	33° 44′ 73° 09′
Bhaun	See Bhoun	
Bhegari Bund or Flood Protection Embankment	North of Sukkur city	27° 43′ 68° 54′
		28° 00′
Bhimbal or Bhimbel	Kaghan valley	34° 53′ 73° 36′
Bhogamarg		
Bhogar Marg		
Bhog Marg	Alternative spellings: Hazara District	34° 35′ 73° 15′
Bhong	North-west of Sadiqabad	28° 24′ 69° 57′
Bhoun	Salt Range	32° 52′ 72° 46′
Bhurban	Murree hills	33° 55′ 73° 27′
Bichla Valley	Kaghan valley, left bank	34° 42′ 73° 37′
Bidook Pass	Lasbela, near coast	25° 04′ 66° 50′
Birir Nullah	Chitral	35° 32′ 71° 41′
Bolan	Pass, Sibi District	29° 53′ 67° 14′
Brughal or Baroghal	Northern Chitral	36° 51′ 73° 22′
Bubak	Dadu District	26° 25′ 67° 44′
Buleji	Fishing village, near Karachi	24° 51′ 66° 45′
Buner or Bunir	District in south-eastern Swat	34° 25′ 72° 15′
		34° 40′ 72° 55′
Buni	Village, north-western Chitral	36° 12′ 72° 04′
Bunja Valley	East bank of Kaghan valley	34° 42′ 73° 36′
Bunji	Gilgit on Indus river	35° 41′ 74° 37′
Bumboret Valley	Chitral	35° 41′ 71° 38′
Burewala	Sahiwal District	30° 09′ 72° 42′
Burzil Pass	South-east Astor or Azad Kashmir	34° 51′ 75° 07′
Campbellpur	North-west Punjab	33° 47′ 72° 22′
Cape Monze	Karachi coast	24° 52′ 66° 39′
Chabar or Chahbar	Iranian coast just west of Pakistan coast	25° 20′ 60° 35′
Chacharan		
Chachran	Western Bahawalpur Division on Indus	28° 52′ 70° 27′
Chachro	Tharparkar District	25° 06′ 70° 16′
Chagai	Town south-west Baluchistan, Chagai District	29° 18′ 64° 44′
Chak or Chack Dhand	South of Umarkot, Tharparkar District	25° 10′ 69° 41′
Chakdarra	Malakand Agency	34° 39′ 72° 01′
Chak Jabbi	Kala Chitta hills, Campbellpur	33° 42′ 72° 17′
Chaklala	Rawalpindi	33° 37′ 73° 06′
Chakri	Salt Range	32° 47′ 73° 28′
Chak Shahana	Multan District	30° 12′ 71° 59′
Chakwal	Salt Range	32° 56′ 72° 52′
Chaman	Pishin District on Afghan border	30° 55′ 66° 25′
Changa Manga	Lahore District	31° 05′ 73° 59′
Changla Gali	Murree hills	33° 59′ 73° 23′
Charsadda	Peshawar District	34° 09′ 71° 49′
Charwa	Sialkot District: not located on map	
Chashi	North-western Gilgit in Ghizar valley	36° 13′ 72° 41′
Chashma Barrage	Dera Ismail Khan District	32° 27′ 71° 19′
Chashma Spring		32° 24′ 71° 18′
Chattar	Murree foothills	33° 47′ 73° 09′
Chatteji Lake	Just north of Haleji lake	24° 52′ 67° 48′
Chauhar Jamali	See Chuhar Jamali	
Chaukandi Ridge (Tombs)	East of Karachi	24° 52′ 67° 17′
Chautar or Chauter	Loralai District	30° 22′ 67° 55′
Cherat (Hills)	Southern Peshawar District	33° 49′ 71° 52′
Chib	South-west Makran	26° 17′ 63° 07′
Chichawatni	Sahiwal District	30° 32′ 72° 42′
Chilas	Southern Gilgit Agency	35° 25′ 74° 06′
Chilianwala	Gujrat District	32° 39′ 73° 37′

Name	Approximate Location	Coordinates Lat.(N)	Lon.(E)
Chiltan	Near Quetta on Northern Kalat border	30° 02′	66° 52′
Chiniot	Sargodha District	31° 43′	72° 59′
Chinji Forest Resthouse	Salt Range	32° 42′	72° 22′
Chinna Creek	Near Rann of Kutch	24° 02′	67° 27′
Chistian	Bahawalnagar District	29° 48′	72° 52′
Chitral	Town, Chitral State	35° 50′	71° 47′
Chitral Gol Reserve	West of Chitral town	35° 53′	71° 45′
Choa Saidan Shah	Salt Range	32° 42′	72° 59′
Chor or Chhor (old maps)	Tharparkar District	25° 31′	69° 46′
Chuhar Jamali	Thatta District	24° 24′	67° 59′
Chutran Village	Shigar Valley, Baltistan	35° 43′	75° 24′
Clifton	Karachi	24° 48′	67° 02′
Dabeji or Dhabeji	Thatta District, east of Pipri	24° 45′	67° 32′
Dadu town	Dadu District, north-western Sind	26° 44′	67° 47′
Dalbandin	Chagai District	28° 53′	64° 26′
Daman Ghar Range	Hills in west central Baluchistan	30° 57′	67° 55′
Daman-i-Koh	Margalla hills	33° 44′	73° 06′
Dambh Village	Sonmiani lagoon	25° 27′	66° 34′
Dandar Marshes	Near Thal, Kohat District	33° 21′	70° 32′
Dandikot	Buner District in Swat	34° 38′	72° 33′
Daphar Forest Plantation	Sargodha District	32° 24′	73° 08′
Daraban	Dera Ismail Khan District	31° 43′	70° 21′
Darel Range	Gilgit Agency, northern Chilas	35° 46′	74° 06′
Dargai	Border of Mardan District and Malakand	34° 29′	71° 54′
Darkot River	Yasin, N. Gilgit	36° 14′	73° 23′
		36° 39′	
Darya Khan	Mianwali District	31° 47′	71° 06′
Darzi Chach	Nushki, Baluchistan	29° 41′	65° 37′
Dasht Valley	North-west of Takhatu	30° 25′	67° 00′
		30° 37′	67° 25′
Dasu	Karokoram Highway, Indus Kohistan	35° 19′	73° 16′
Daudkhel	Salt Range, Mianwali District	32° 52′	71° 34′
Deosai Plateau		34° 45′	75° 15′
		35° 10′	75° 50′
Deg Nullah	Shekhupura District	31° 40′	74° 10′
Dera Ghazi Khan	Punjab, Trans-Indus District	30° 03′	70° 38′
Dera Ismail Khan	N.W.F.P., southern part	31° 48′	70° 54′
Dera Nawab Sahib	Bahawalpur Division (also known as Dera Nawab)	29° 07′	71° 17′
Derawar or Darawar	Cholistan desert (also known as Dharawar)	28° 46′	71° 20′
Dhanial	Forest, Hazara District	34° 36′	73° 38′
Dherki	Northern border of Sukkur District	28° 03′	69° 42′
Dho Lake	Thatta District, near Ladiun	24° 21′	68° 04′
Dhoro Naro	East Nara, east of Mirpurkhas	25° 31′	69° 34′
Digri	Western Tharparkar District	25° 09′	69° 07′
Dina	Jhelum District	33° 02′	73° 37′
Dipalpur	Bahawalnagar District	30° 40′	73° 39′
Diplo	Tharparkar District, near Rann of Kutch	24° 28′	69° 34′
Dir	Town, Dir State, north of Malakand Agency	35° 12′	71° 52′
Doaba	Western border of Kohat	33° 24′	70° 45′
Dodhanwala Toba	Cholistan	29° 08′	72° 51′
Dokri	Larkana District	27° 22′	68° 05′
Dorah Pass	Chitral, Afghan border	36° 07′	71° 14′
Dok Dusra	Northern Dir State	35° 32′	72° 13′
Domel	Azad Kashmir	34° 21′	73° 28′
Dormushkh Nullah	Gilgit	35° 53′	74° 11′
Dost Ali Lake or Dost Allee	Larkana District	27° 39′	67° 53′
Drazinda	Dera Ismail Khan District	31° 42′	70° 08′
Drosh	South part of Chitral State	35° 33′	71° 47′
Duber	Forest reserve, Sukkur District	27° 32′	68° 23′
Duber or Dubair Valley	Indus Kohistan	35° 09′	72° 52′
Dullewala	Thal	31° 49′	71° 26′
Dundikot	Swat State	34° 17′	72° 30′
Dunyapur	Multan District	29° 47′	71° 43′
Dura Pass	Chitral State	36° 07′	71° 15′
Dunga Gali	Murree hills	34° 03′	73° 22′
Faisalabad formerly Lyallpur		31° 25′	73° 07′
Falakser	Peak, Swat Kohistan	35° 42′	72° 48′
Fateh Jhang	Attock District, Salt Range	33° 34′	72° 39′
Feroza	North of Khanpur, Bahawalpur District	28° 45′	70° 49
Fort Abbas	Cholistan, Bahawalnagar District	29° 12′	72° 52′
Fort Dharawar	See Derawar		
Fort Munro	Dera Ghazi Khan District	29° 54′	69° 59′
Fort Sandeman (Now Zhob)	Northern Baluchistan	31° 21′	69° 28′
Gabral	Indus Kohistan	35° 37′	72° 57′
Gabrial or Gabral	Swat Kohistan	35° 36′	72° 31′
Gadap Plain	Karachi District	25° 03′	67° 23′
Gadra	North-east border of Tharparkar District	25° 38′	70° 36′
Gakuch	Gilgit river near junction of Karumbar river	36° 11′	73° 45′
Gambat	Khairpur State	27° 19′	68° 32′
Gandhala	Salt Range, not shown on map	32° 48′	73° 20′
Garhi Habibullah	Azad Kashmir	34° 23′	73° 24′
Garho (see also Gharo)	Thatta District	24° 18′	67° 36′
Garm Chashma	Chitral	36° 02′	71° 44′
Ghabi Dero or Ghabi Dhero	Indus delta (ancient ruins)	24° 37′	67° 16′
Ghadabar Garh	Hills in Loralai District	30° 21′	69° 09′
Gharial	Murree hills	33° 55′	73° 27′
Gharibwal	Salt Range	32° 41′	73° 11′
Gharko Forest	Thatta District	24° 32′	67° 53′
Gharo (see also Garho)	Thatta District	24° 44′	67° 36′
Ghauspur Lake	Jacobabad District	28° 09′	69° 05′
Ghazi Ghat	Muzaffargarh District on Indus river	30° 06′	70° 56′
Ghizar	River/valley in north-western Gilgit	36° 12′	72° 40′
			73° 40′

Name	Approximate Location	Coordinates Lat.(N) Lon.(E)	Name	Approximate Location	Coordinates Lat.(N) Lon.(E)
Ghizri Creek	Karachi	24°47' 67°05'	Hasan Abdal	Attock District	33°48' 72°42'
Gholum Ullah	Thatta District	24°36' 67°48'	Hasilpur	Northern Bahawalpur District	29°43' 72°33'
Ghora Dhaka	Murree hills	34°02' 73°26'	Hathungo		
Ghora Gali	Murree hills	33°52' 73°20'	Village	Sanghar District on border of desert	25°47' 69°29'
Ghotki	Sukkur District	27°59' 69°19'	Havelian	South-east Hazara District	34°03' 73°09'
Ghulam Mohammed Barrage	Hyderabad District	25°22' 68°18'	Hazar Ganji		
			Park	Chiltan hills, Quetta District	30°04' 66°52'
Ghulam (Ullah)	Town, Thatta District	24°36' 67°48'	Hindu Bagh, renamed		
Gilgit	Town, Gilgit Agency	35°54' 74°18'	Muslim Bagh	North central Baluchistan	30°49' 67°47'
Gishk	Mountain range, northern Kalat	29°12' 66°52'	Hinglaj	South-east Makran	25°45' 65°35'
			Hingol	River, eastern Makran	25°45' 65°33'
Gittidas or Gitidas	Northern Hazara District	35°07' 73°59'	Hoshab	South central Makran	26°01' 63°55'
Gizri Village	Near Karachi	24°46' 67°06'	Hub or		
Gojra	Faisalabad District	31°09' 72°42'	Hab River	Lasbela/Sind border	25° 66°50'
Gol Nullah	Chitral State (also called Chitral Gol)	35°53' 71°45'	Hushe	Baltistan	35°27' 76°23'
Gol Village	Baltistan on Indus river	35°14' 75°51'	Hyderabad	City, southern Sind	25°24' 68°22'
Gora Dhaka	See Ghora Dhaka		Ibrahim Hydari		
Gora Gali (see Ghora Gali)	Murree hills	33°52' 73°20'	Fishing Village	Mouth of Korangi creek	24°48' 67°08'
Guddu Barrage	Sind, Bahawalpur border	28°24' 69°45'	Isa Khel	Bannu District, N.W.F.P.	32°41' 71°17'
Gujar Khan	Rawalpindi District	33°15' 73°18'	Ishkoman		
Gujjo	Thatta District	24°44' 67°47'	Ishkuman	Northern Gilgit	36°32' 73°49'
Gujranwala	Gujranwala District, north-west Punjab	32°09' 74°12'	Islamabad	Rawalpindi District	33°43' 73°05'
Gujrat	Gujrat District, north-west Punjab	32°34' 74°04'	Islam Headworks	On Sutlej river, formerly Palla Headworks, Multan District	29°49' 72°33'
Gulistan	Pishin District	30°37' 66°35'	Islamkot	Thar desert	24°41' 70°10'
Gulmarg	Rahimyar Khan District on Cholistan border	28°19' 70°26'	Jabba Valley	Swat Kohistan	35°20' 72°40'
Gungri Lake	South of Ladiun, Thatta District	24°16' 68°02'	Jabbi or Chak Jabbi	Mianwali District, Salt Range	32°55' 71°42'
Gupis	North-western Gilgit	36°13' 73°27'	Jabho Lake	Badin District on borders of Rann of Kutch	24°18' 68°37'
Gurais	Neelum valley, Azad Kashmir	34°37' 74°51'	Jacobabad	Town, north-west Sind	28°17' 68°26'
Gurchani Hills	North-west of Bugti Territory, Baluchistan	29°38' 69°50'	Jajjian Abbas		
			Jajja Abbasian	Bahawalpur Division	28°47' 70°34'
Gwadur or Gwadar	Makran coast	25°07' 62°20'	Jallo Park	East of Lahore	31°34' 74°30'
Gwaldai	Northern Dir (also known as Gwaldri)	35°25' 72°04'	Jamal Dinwali	Bahawalpur Division	28°29' 70°03'
Gwambuk Kaul	Central Makran	26°09' 64°15'	Jamgarh Peak	Kaghan valley, north of Kadirgali	34°42' 73°43'
Hab River	See Hab		Jamroa Head	Nawabshah District	26°26' 68°52'
Hab Chowki	Border between Lasbela and Karachi District	25°02' 66°54'	Jamrud	West of Peshawar	34°00' 71°19'
Hab Dam	Karachi District, Lasbela boundary	25°05' 67°00'	Jandola	South Waziristan	32°19' 70°07'
Hadeiro Lake			Jangri	Near Baran river, Sind Kohistan	25°24' 67°59'
Haidero Lake	Thatta District	24°50' 67°53'	Jang Shahi	Thatta District north of Haleji lake	24°52' 67°47'
Hafizabad	Gujranwala District	32°04' 73°41'	Jared	Kaghan valley	34°40' 73°33'
Hala	Northern Hyderabad District	25°48' 68°25'	Jati or Jatti	Southern Thatta District	24°22' 68°17'
Haleji Lake	Thatta District	24°49' 67°44'	Jebri	Eastern Kalat	27°30' 66°38'
Hangu	Kohat District	33°32' 71°04'	Jhal Jhao	Eastern Makran or Hingol river	26°18' 65°34'
Hannah Lake	Near Quetta	30°05' 67°57'	Jhang or Jang		
Haramosh Range	Baltistan	35°45' 74°50' ext to 75°35'	Maghiana	Jhang District	31°16' 72°19'
			Jhelum	Town, Jhelum District	32°57' 73°44'
Harappa	Sahiwal District	30°37' 72°52'	Jherruck or		
Harboi Hills	Kalat Division	28°55' 66°42'	Jerruk	Indus river, southern Hyderabad District	25°04' 68°25'
Haripur	Southern Hazara District	33°59' 72°59'	Jhika Gali or		
Harnai	Sibi District	30°05' 67°57'	Jhikka Gali	Near Murree	33°57' 73°23'
Haro Stream	From Dunga Gali to Indus river, South of Campbellpur	33°45' 72°15' 34°04' 73°23'	Jhimpir	Thatta District, right bank of Indus	25°02' 68°01'
Haroonabad	Bahawalpur District	29°36' 73°08'	Jhol	South-west of Hyderabad city	25°22' 68°21'

Name	Approximate Location	Coordinates Lat.(N) Lon.(E)
Jhuddo or Jhudo	Tharparkar District	24° 58' 69° 17'
Jinnah Barrage	Indus river near Kalabag	32° 55' 71° 29'
Jiwani	South-west Makran coast	25° 02' 61° 45'
Jora Valley	Left bank north end of Kaghan valley	34° 49' 73° 53'
		34° 55' 73° 57'
Kabirwala	Multan District	30° 24' 71° 52'
Kach	West of Takhatu Range	30° 26' 67° 17'
Kacha Range	Extreme western Chagai	29° 28' 61° 15'
Kachlagh	North of Quetta	30° 21' 66° 54'
Kadir Gali	East bank, Kaghan valley	34° 41' 73° 42'
Kaghan or Kagan	Village from which valley is named (in Hazara District)	34° 47' 73° 32'
Kahuta	Murree foothills	33° 36' 73° 23'
Kajar Island and Kajar Creek	Indus delta near Indian border	23° 49' 68° 07'
Kala Bagh	On Indus, north of Mianwali	34° 04' 71° 36'
Kala Bagh	Murree hills, Hazara District	34° 04' 73° 22'
Kala Chitta or Kalla Chitta	Attock District	33° 40' 72° 20'
Kalam	Swat Kohistan	35° 31' 72° 35'
Kalankot Lake	Thatta District	24° 42' 67° 53'
Kala Nullah	Kaghan valley, a tributary of Bunja	34° 45' 73° 35'
Kalat	Town, northern Kalat Division	29° 02' 66° 34'
Kaldi	Tharparkar	24° 30' 69° 17'
Kalian	Sialkot District, not located on map	
Kalinjhar Hills	Nagar Parkar, extreme south-east	24° 17' 70° 38'
	Tharparkar District	24° 27' 70° 50'
Kaliphat or Khalifat	Peak in Northern Sibi District	30° 18' 67° 42'
Kallar Kahar Lake	Salt Range	32° 47' 72° 43'
Kalri Lake, see Keenjhar Lake	Thatta District	24° 56' 68° 03'
Kalurkot	Thal	32° 09' 71° 15'
Kamalia	Southern Faisalabad District	30° 43' 72° 43'
Kambar Lake	Larkana District. old name for Drigh lake, (see Drigh)	
Kamila or Kumila	Karakoram Highway	35° 14' 73° 14'
Kand Forest Rest House	Hazara District	34° 09' 73° 56'
Kandhkot	Jacobabad District	28° 13' 69° 11'
Kangar Kot	Sind, not marked on map	24° 17' 69° 22'
Kanori Hills	Nowshera District	33° 54' 72° 01'
Kanshian Valley	Left bank, Kaghan valley	34° 32' 73° 26'
Kanti	Southern Chitral	35° 35' 71° 41'
Kao Forest	Below Dunga Gali, Murree hills	34° 03' 73° 25'
Karak Lora	West of Pishin town	30° 05' 66° 45'
		30° 33'
Karchah Chashma	West of Manjand, Dadu District	25° 55' 68° 10'
Karchat	Southern Dadu District	25° 46' 67° 44'
Kargah Nullah	Gilgit Agency	35° 56' 74° 13'
Karimabad	Hunza	36° 21' 74° 41'

Name	Approximate Location	Coordinates Lat.(N) Lon.(E)
Karo Nai	Sukkur District at head of East Narra	27° 35' 69° 02'
Karochi Dak	South-west Makran	25° 51' 63° 46'
Karrarkar	Buner District, Swat	34° 26' 72° 13'
Kasbo or Kasba	Rock outcrop in Tharparkar	24° 17' 70° 48'
Kashmor or Kashmore	Jacobabad District	28° 25' 69° 35'
Kasur	Kasur District	31° 08' 74° 27'
Katalpur Lake	Near Ravi river, Khanewal District	30° 36' 71° 57'
Kaurang or Karang	Indus Kohistan	35° 30' 73° 01'
Kaur Bridge	NWFP., not located on map	
Kawai or Kuwai	Lower Kaghan valley, Hazara	34° 38' 73° 26'
Kelat	See Kalat Town	
Keti Bandar	Indus mouth	24° 08' 67° 27'
Keti Shahu	Indus river, Sukkur District	27° 54' 68° 55'
Khabbaki Lake	Salt Range	32° 37' 72° 14'
Khadeji Falls	Karachi District	25° 06' 67° 32'
Khai Creek	Indus delta	24° 31' 67° 13'
Khair-i-Murat	Hills in Salt Range	33° 27' 72° 47'
Khairpur	Town, northern Sind	27° 32' 68° 47'
Khairpur Nathan Shah	Khairpur District	27° 06' 68° 44'
Khalifat	See above, Kaliphat, northern Sibi District	30° 18' 67° 42'
Khanewal	Khanewal District	30° 18' 71° 56'
Khanki Headworks	Gujrat District on Chenab	32° 24' 73° 58'
Khanna	Rawalpindi District	33° 32' 73° 07'
Khanori Hills	Malakand Agency	34° 27' 72° 14'
Khanpur	Bahawalpur Division	28° 38' 70° 38'
Khanspur, now Ayubia	Murree hills (Hazara District)	34° 02' 73° 24'
Khapalu	Shyok valley, Baltistan	35° 11' 76° 22'
Khar	Hab valley	25° 18' 67° 08'
Kharan	South-east Kalat	28° 34' 65° 26'
Kharrar Jheel	Sahiwal District	30° 52' 73° 31'
Kharunjhar hills	South-east Tharparkar	24° 20' 70° 42'
Khaur	Attock District	33° 14' 72° 27'
Khawas or Khowas	North-east of Ziarat	30° 28' 67° 34'
Khenju	Bugti Territory	28° 32' 68° 53'
Kheti Bandar or Keti Bandar	See above, Keti Bandar, Indus mouth	24° 08' 67° 27'
Khewra	Salt Range	32° 38' 73° 01'
Khinjar Lake	Thatta District, alternative name for Kalri lake	24° 56' 68° 03'
Khipro Lake	Sanghar District, eastern Sind	25° 49' 69° 21'
Khirgi or Kirgi	South Waziristan	32° 18' 70° 11'
Khirsar	Forest Reserve, Indus river, Southern Sind	24° 36' 68° 02'
Khojak Pass	Pishin District, near Chaman	30° 52' 66° 28'
Khokush Gul	Valley in north-western Gilgit	35° 54' 72° 33'
		36° 10' 39'
Khudi Creek	Indus Delta	24° 35' 67° 12'
Khudo or Khude	North-east Lasbela	26° 50' 66° 55'
Khujoorag Hills	Southern Swat, Buner	34° 37' 72° 20'
Khunjerab Pass	Northern border of Hunza	36° 52' 75° 27'
Khushab	Sargodha District	32° 17' 72° 21'
Khuttan	Neelum valley, right bank	34° 35' 73° 45'
Khuzdar	South-east Kalat Division	27° 53' 66° 36'
Khwaja Amran Range	South of Chaman	30° 20' 66° 17'
		30° 45' 66° 31'

Name	Approximate Location	Coordinates Lat.(N) Lon.(E)	Name	Approximate Location	Coordinates Lat.(N) Lon.(E)
Khyber Agency	North-west of Peshawar	34°05' 71°05'	Lodhran	Multan District	29°32' 71°37'
Kila Saifullah, see Qila Saifullah	Zhob, Baluchistan		Lora	Nullah or river, Chagai	29°30' 65°15'
Killick Pass	North-west boundary of Hunza, the most northerly point in Pakistan	37°04' 74°42'	Lora	River, Pishin District	30°22' 66°53'
			Loralai or Lorelai	Town, Loralai District, north-east Baluchistan	30°23' 68°38'
Kingri	Eastern Loralai District	30°26' 69°48'	Lowarai Pass, or Lawarai Pass	Northern Dir/Chitral border	35°22' 71°47'
Khinjar Lake	Muzaffargarh District	29°54' 70°57'	Lowari Sharif	Badin District	24°31' 68°53'
Kirani	Chiltan hills, Quetta	30°11' 66°57'	Lower Topa	Near Murree	33°55' 73°23'
Kirthar Range	Sind hill tracts, right bank of Indus	25°44' 67°10' / 27°15'	Luddan	Multan District	29°54' 72°34'
			Ludak Sar	Swat Kohistan	35°41' 72°46'
Kohat	Town, Kohat District	33°34' 71°26'	Lulusar	Lake, northern Hazara District	35°05' 73°56'
Koh-i-Maran Range	North-east Kalat	29°26' 66°52'	Lung Lake	See Lang	
Koh-i-Safed Range	Upper Kurram valley	34°00' 70°00'	Lutko Nullah	Northern Chitral	36°05' 71°20' / 71°45'
Koh-i-Sultan Range	Western Chagai	29°05' 62°40' / 29°15' 62°55'	Lyallpur, renamed Faisalabad	City, Lyallpur District	31°25' 73°07'
Kolwa	Baluchistan	26°2' 64°39'	Mach	Western Sibi District	29°52' 67°21'
Kolwahkupp Hills	Northern Makran	26°10' 64°50'	Machiara Sanctuary	Azad Kashmir, west bank Neelum river	34°31' 73°37'
Kot Addu	Muzaffargarh District	30°28' 70°59'	Madyan	Swat	35°08' 72°32'
Kot Diji	Khairpur District	27°21' 68°42'	Maggowal	Gujrat District	32°29' 73°54'
Kotli	Southern Azad Kashmir in Poonch	33°31' 73°55'	Mahboob Shah Lake	Thatta District	24°25' 67°59'
Kotrato Island	Korangi creek	24°43' 67°07'	Mahendri	Swat, not located on map	
Kotri	Hyderabad District	25°21' 68°19'	Mahodan or Mohidan Lake	Swat Kohistan	35°38' 72°40'
Kulachi	Dera Ismail Khan District	31°55' 70°28'	Mailsi	Multan Division	29°42' 72°12'
Kuldan	South-east Makran	26°09' 64°15'	Makhi Dhand	East Nara	26°00' 69°00' / 26°12' 69°12'
Kullanch	Southern Makran	25°35' 63°12'	Makhra or Makra Peak	Hazara District, east of Shogran	34°33' 73°28'
Kululai	Rest House in southern Swat Kohistan	35°18' 72°35'	Makrash or Makrach Lake	Salt Range, Jhelum District	32°39' 72°49'
Kund Lake	Lasbela, near Hab chowki, Lasbela	25°01' 66°50'	Makli Hills	Thatta District	24°46' 67°57'
Kundai	Muzaffargarh District north of Panjnad	29°12' 70°42'	Malach	Murree hills, below Dunga Gali	34°03' 73°21'
Kundian	Mianwali District	32°27' 71°29'	Malakand	Pass, Malakand Agency	34°34' 71°57'
Kushakgarh	Eastern Kohat District	33°29' 71°53'	Malezai Stream	South of Gulistan, Chaman District	30°32' 66°42'
Kutabpur	See Qutabpur		Malika Parbat Peak	Hazara District, Kaghan valley	34°48' 73°43'
Lachi in Koi	South of Kohat	33°23' 71°18'	Malir	Karachi	24°59' 67°13'
Ladam Sar	Bahawalpur District	29°22' 71°59'	Malkal	Azad Kashmir, not located on map	
Ladiun Lake	Thatta District	24°18' 68°03'	Malkandi Forest	Kaghan valley	34°41' 73°35'
Lahore	North-east Punjab	31°33' 74°19'	Mall Creek	Indus delta	23°52' 67°40'
Lak Bidok	Lasbela	25°12' 66°45'	Mana Valley	Loralai District	30°28' 67°42'
Lakhat	Forest, Nawabshah District	26°36' 67°53'	Manchar	Dadu District	26°24' 67°38'
Lakhi	Dadu District, Sind Kohistan	26°16' 67°53'	Mandi Bahauddin	Gujrat District	32°36' 73°28'
Lalamusa	Gujrat District	32°42' 73°57'	Manga Valley	Murree foothills	33°47' 73°22'
Lalazar	Hazara District, Kaghan valley	34°54' 73°46'	Mangla Dam	Jhelum District	33°08' 73°40'
Lal Suhanran or Sohanra	Bahawalpur District	29°21' 71°58'	Manjand	Dadu District	25°54' 68°12'
Lalusar	See Lulusar		Mankial Peak	Swat Kohistan	35°19' 72°44'
Landi Kotal	Khyber Agency	34°07' 71°20'	Manshera	Hazara District	34°19' 73°12'
Landi or Landhi	North of Karachi	24°52' 67°13'	Manshi Forest	Kaghan valley, west bank	34°41' 73°25'
Lang or Lung Lake	Larkana District	27°30' 68°03'	Manur Khatta	Kaghan valley	34°49' 73°38'
Larkana	Town, north-western Sind	27°32' 68°12'	Marachak Reserve	Takhatu Range	30°19' 67°07'
Laspur Nullah	Chitral	36°12' 72°28'			
Lawarai	See Lowarai				
Leh Stream	East of Rawalpindi	33°33' 73°04' / 33°51' 73°24'			
Lehtrar	Murree foothills	33°42' 73°26'			
Leiah	Muzaffargarh District, Thal	30°57' 70°57'			

Name	Approximate Location	Coordinates Lat.(N) Lon.(E)		Name	Approximate Location	Coordinates Lat.(N) Lon.(E)	
Marala Barrage	Sialkot District on Chenab river	32° 40′	74° 29′	Muridke	Shekhupura District	31° 47′	74° 15′
Marchar	Below Shangla Pass, Indus Kohistan	34° 50′	72° 33′	Murree	Town, Murree hills	33° 54′	73° 22′
Mardan	Town, Mardan District	34° 19′	71° 56′	Musa-ke-Masala	Peak, Hazara District	34° 52′	73° 31′
Margalla Hills or Marghala Hills	Murree foothills	33° 48′	73° 10′	Muslim Bagh, Formerly Hindu Bagh		30° 49′	67° 47′
Marri Mungkhtar Hills	North of Karachi	25° 17′	67° 17′	Muzaffar Garh	Town, Muzaffargarh District	30° 04′	71° 12′
Mashelakh Reserve	West of Quetta	29° 55′	66° 30′	Muzaffarabad	Town, Azad Kashmir	34° 22′	73° 28′
		30° 17′	66° 50′	Nagar or Naghr	Southern Chitral	35° 29′	71° 43′
Mashkai Stream	Makran, tributary of Hingol	26° 22′	65° 12′	Nagar or Nagir	Gilgit Agency	36° 16′	74° 44′
		27° 30′	65° 47′	Nagar Parkar	South-east Tharparkar	24° 20′	70° 46′
Massan Valley	Mianwali District, Salt Range	32° 52′	71° 43′	Naltar and Naltar Valley	Gilgit	36° 07′	74° 14′
Mastuj	Northern Chitral	36° 17′	72° 31′	Nalwah	Rahimyar Khan District on edge of desert	28° 14′	70° 26′
Mastung	Northern Kalat	29° 48′	66° 52′	Nammal	Lake, Salt Range	32° 40′	71° 49′
Matiltan	Swat Kohistan	35° 34′	72° 10′	Nanga Parbat	Gilgit, Astor District	35° 13′	74° 35′
Mauripur	West of Karachi	24° 52′	66° 56′	Nankana Sahib	Lahore District	31° 27′	73° 43′
McLeod Ganj	Bahawalnagar District	30° 09′	73° 44′	Naran	Hazara District, Kaghan valley	34° 53′	73° 39′
Mekhtar	West of Loralai	30° 28′	69° 22′	Narowal	Sialkot District	32° 07′	74° 52′
Metla	Bahawalpur District	28° 51′	70° 54′	Nathan Shah	Dadu District	27° 06′	67° 43′
Mian Channu	Multan District	30° 27′	72° 22′	Nathia Gali	Murree hills	34° 04′	73° 24′
Miandam	Swat	35° 09′	72° 30′	Naukot	Tharparkar District	24° 52′	69° 27′
Miani Hor	Alternative name for Sonmiani lagoon			Naundero or Naudero	Larkana District	27° 40′	68° 21′
Miani Forest	Hyderabad District	25° 27′	68° 23′	Nawabshah	Town, Nawabshah District	26° 15′	68° 24′
Mianwali	Town, Mianwali District	32° 13′	71° 33′	Nawan	Azad Kashmir, Jhelum valley	33° 32′	73° 40′
Minapin Glacier	Hunza valley left bank	36° 14′	74° 32′	Nihing Valley	West central Makran	26° 07′	62° 12′
Mingora	Swat	34° 47′	72° 22′				62° 44′
Miran Shah	North Waziristan	33° 01′	70° 04′	Noa Kandi or Nokundi	Western Chagai	28° 49′	62° 46′
Miran Jani Peak	Murree hills	34° 06′	73° 25′	Nowshera	Peshawar District	34° 00′	71° 59′
Miranzai Valley	North of Thal, Kohat/Kurram border	33° 20′	70° 33′	Nuri Forest	Kaghan valley, left bank	34° 38′	73° 36′
		33° 35′		Nurpur	Salt Range	32° 39′	72° 31
Mirpur	Jacobabad District	28° 12′	68° 48′	Nurpur Shahan	Margalla hills, Rawalpindi District	33° 44′	73° 07′
Mirpurkhas	Town, Tharparkar District	25° 32′	69° 01′	Nushki	Chagai or Nushki District	29° 33′	66° 02′
Mirpur Sakro	Thatta District	24° 32′	67° 38′	Okara	Sahiwal District	30° 48′	73° 27′
Misgar	Northern Hunza	36° 46′	74° 46′	Ormara	Makran coast	25° 13′	64° 40′
Mithankot	Dera Gazi Khan District	28° 57′	70° 22′	Pab range	Eastern Lasbela From	25° 30′	66° 25′
Mithi Village	Tharparkar desert	24° 43′	69° 48′			27° 35′	67° 05′
Moach Plain	North-west of Karachi	24° 55′	67° 16′	Pabbi	West of Peshawar	34° 00′	71° 48′
		24° 60′	67° 35′	Pail	Salt Range	32° 38′	72° 28′
Mochko	On Hab river	25° 00′	66° 52′	Padak or Padag	Chagai desert	29° 01′	65° 15′
Moen Jodaro	Larkana District (see Mohenjo Dero)	27° 18′	68° 08′	Panjar	Extreme southern foothills of Murree hills	33° 38′	73° 33′
Mogi Wala Toba	Bahawalnagar District, Cholistan	28° 50′	73° 02′	Panjpur	Northern Makran	26° 56′	64° 06′
Mogli	Salt Range, not located on map			Panjkora River	Dir	35° 08′	71° 55′
Mohenjo Dero	Larkana District	27° 18′	68° 08′	Panjnad headworks	On Chenab river on Muzaffargarh and Bahawalpur border	29° 21′	71° 02′
Mohib Shah	Muzaffargarh District	29° 37′	70° 44′	Parachinar	Upper Kurram Agency	33° 53′	70° 07′
Mohidan or Mahodan Lake	Ushu valley, Swat	35° 38′	72° 40′	Paras	Hazara District	34° 39′	73° 31′
Momin	Peak, South Waziristan, not located			Parkuta	Baltistan	35° 07′	75° 57′
Montgomery, renamed Sahiwal		30° 39′	73° 06′	Parom	See Prom		
Mughalpura	Lahore	31° 33′	74° 20′	Pasni	Makran coast	25° 15′	63° 28′
Mukshpuri Mountain	Murree hills	34° 04′	73° 26′	Pasrur	Sialkot District	32° 15′	74° 40′
Multan	City, Multan District	30° 11′	71° 26′	Pasu	Hunza	36° 28′	74° 53′
Munian Wala	Bahawalnagar, Cholistan	30° 10′	73° 43′	Pateka	Neelum valley, Azad Kashmir	34° 27′	73° 34′
Murdar Peak	Quetta, Pishin District	30° 07′	67° 07′	Patriata Forest Rest House	Murree foothills	33° 50′	69° 56′
Murghazar	Swat	34° 39′	72° 19′	Peshawar	City, Peshawar District	34° 01′	71° 33′

Name	Approximate Location	Coordinates Lat.(N) Lon.(E)	Name	Approximate Location	Coordinates Lat.(N) Lon.(E)
Petaro	Near Hyderabad	25° 31′ 68° 18′	Renala Khurd	Sahiwal District	30° 53′ 73° 34′
Phandar or			Robat	Extreme western tip of	
Phander	North-west Gilgit	36° 10′ 73° 05′		Chagai	29° 48′ 60° 55′
Philgar	Mountain ridge, central		Rohri	Sukkur District	27° 41′ 68° 54′
	Baluchistan	30° 26′ 67° 26′	Rohtas	Jhelum District, Salt Range	32° 58′ 73° 36′
Phoosani Lake	Near Tando Bago, Badin		Rondu	Baltistan	35° 36′ 75° 09′
	District	24° 44′ 68° 55′	Rondu	Chitral, see Arandu	
Phulra	Cholistan	29° 10′ 72° 50′	Rupal	Astor, Gilgit Agency	35° 11′ 74° 43′
Pind Dadan					
Khan	Salt Range, Jhelum District	32° 36′ 72° 57′	Sadikabad	Town, southern part of	
Pindi Bhattian	Shekhupura District	31° 55′ 73° 17′		Bahawalpur	28° 18′ 70° 02′
Pindi Gheb	Attock District	33° 13′ 72° 16′	Sadori Lake	Sanghar District	26° 12′ 69° 07′
Piplan	Northern Thal, Mianwali		Safed Koh		
	District	32° 17′ 71° 22′	Range	Upper Kurram Agency	34° 00′ 70° 00′
Pipri	On old Thatta road, 20 miles		Sahiwal	Formerly Montgomery town,	
	west of Karachi	24° 52′ 67° 22′		Sahiwal District	30° 39′ 73° 06′
Pir Chela Hills	Hazara District, north of		Sahiwal	Sargodha District	31° 58′ 72° 19′
	Garhi Habibullah	34° 27′ 73° 23′	Saidpur	Margalla hills	33° 44′ 73° 07′
Pir Chinasi	East bank, Jhelum river,		Saidpur Valley	Azad Kashmir, west bank	
	Azad Kashmir	33° 55′ 73° 38′		Neelum river	34° 32′ 73° 28′
Pir Hasimar	Kotli District, Azad Kashmir	33° 42′ 73° 48′	Saidu Sharif	Swat	34° 44′ 72° 21′
Pirawala or			Saif-ul-Maluk		
Pirowala	Khanewal District, forest		Lake	Upper Kaghan valley	34° 52′ 73° 41′
	plantation	30° 21′ 72° 02′	Sai Nullah	Gilgit Agency, Chilas	35° 45′ 74° 20′
Pir Ghal	Waziristan	32° 47′ 69° 46′	Sakesar	Salt Range, Mianwali District	32° 33′ 71° 57′
Pir Patho	Thatta District near Indus	24° 33′ 67° 52′	Sakhi Sarwar	Dera Ghazi Khan District	29° 58′ 70° 17′
Pishin	Town, Pishin District	30° 39′ 67° 00′	Sakra	Mardan and Swat boundary	34° 27′ 72° 17′
Pithoro Lakes	Tharparkar District	25° 28′ 69° 22′	Saligran or		
Pitiani Creek	Indus delta	24° 23′ 67° 17′	Salgran	Jhelum valley, east of	
Pitti Creek	Indus delta	24° 42′ 67° 09′		Rawalpindi	33° 31′ 73° 34′
Popalzai Forest/			Salkallah	Neelum valley, Azad	
Plain	Western Pishin District	30° 44′ 66° 47′		Kashmir	34° 33′ 73° 54′
Potowar or			Samana Hills	North-western Kohat District	33° 37′ 70° 30′
Potohar	Plateau in Northern Punjab	33° 38′ 73° 00′			70° 50′
Prom	North central Makran	26° 38′ 64° 21′	Samandar	Kalat on Dasht plain	28° 50′ 66° 27′
Punial	North-west of Gilgit town	36° 07′ 74° 02′	Samli Forest		
			Rest House	Murree foothills	33° 43′ 73° 24′
Qadirabad Link			Samilzai Valley	North of Samana hills, near	
Canal	Gujrat District	32° 18′ 73° 29′		Kohat	33° 30′ 70° 37′
Qaidabad	Mianwali District, Thal	32° 18′ 71° 54′			71° 20′
Qalandarabad	Hazara District, near		Samundri	Faisalabad District	31° 04′ 72° 58′
	Mansehra	34° 22′ 73° 13′	Sandeman Tangi	North-east Baluchistan	30° 23′ 67° 41′
Qila Saifullah	Zhob District, Baluchistan	30° 43′ 68° 22′	Sandho Lake	Badin District on borders of	
Quetta	Central Baluchistan	28° 14′ 66° 56′		Rann of Kutch	24° 10′ 68° 40′
Qutabpur	Multan District	29° 54′ 71° 47′	Sanghar Town	Sanghar District, western	
				Sind	26° 03′ 68° 57′
Raghai River	Kharan District, Baluchistan	27° 20′ 65° 20′	Sangriaro	Lake, Sanghar District	26° 03′ 69° 12′
Rahimyar Khan	Town, Rahimyar Khan		Sarai Alamgir	Gujrat District near Jhelum	32° 53′ 73° 47′
	District	28° 24′ 70° 18′	Saranan	Pishin District	30° 35′ 66° 52′
Rahwali	Gujranwala District	32° 14′ 74° 20′	Sargodha	Town, Sargodha District	32° 04′ 72° 41′
Rajanpur	Town, Dera Ghazi Khan		Sarhad	Northern Sukkur District	28° 03′ 69° 25′
	District	29° 06° 70° 17′	Sari Forest		
Rajari Forest	Hyderabad District	25° 37′ 68° 23′	Rest House	Hazara District	34° 34′ 73° 32′
Raiwind	Lahore District	31° 15′ 74° 12′	Sarkhand or		
Rakaposhi	Peak, Gilgit	36° 09′ 74° 30′	Sakrand	Nawabshah District	26° 08′ 68° 16′
Rakhshani	Plateau, south-western Kalat	27° 05′ 64° 50′	Sehwan	Dadu District	26° 26′ 67° 51′
Rama Lake	Astor, Gilgit Agency	35° 18′ 74° 46′	Shah Bandar	Indus mouth	24° 08′ 67° 53′
Ranikot Fort	Dadu District	25° 55′ 67° 52′	Shahdadpur	Hyderabad District	25° 56′ 68° 38′
Ras Koh Range	Southern Chagai	28° 50′ 65° 06′	Shahdara	Lahore ourskirts on Ravi	31° 37′ 74° 18′
Rasul barrage	Gujrat District	32° 42′ 73° 33′	Shahid Pani	Hazara District	34° 36′ 73° 21′
Ratto or Rattoo	Astor, Gilgit	35° 19′ 74° 49′	Shah Nurani	Peak, Pab hills, Lasbela	27° 30′ 66° 35′
Rawal Lake	Near Islamabad	33° 40′ 73° 96′	Shahi Mahar	Ghizar valley, north-western	
Rawalpindi	City, northern Punjab	33° 36′ 73° 03′		Gilgit	36° 12′ 72° 43′
Razmak	South Waziristan	32° 42′ 69° 52′	Shahpur	Sargodha District	32° 17′ 72° 28′
Rehri Village	On Korangi creek, east of		Shahpur	Nawabshah District, Sind	26° 35′ 67° 58′
	Karachi	24° 49′ 67° 13′	Shah Yaki	Thatta District	24° 22′ 68° 02′
Rej Khan Peak	Above Manshi Forest,		Shaikh Badin or		
	Kaghan valley	34° 43′ 73° 29′	Shaikh Budin	Peak, Dera Ismail Khan District	32° 18′ 70° 48′

Name	Approximate Location	Coordinates Lat.(N) Lon.(E)	Name	Approximate Location	Coordinates Lat.(N) Lon.(E)
Shalozan	Upper Kurram valley	33° 57′ 70° 01′	Sukh Beas	Old river bed, east of Lahore	31° 00′ 74° 00′
Sham	Bugti Territory, south-eastern Baluchistan	29° 20′ 69° 40′			31° 15′ 75° 00′
Shandur Pass	North-western Gilgit	36° 08′ 72° 30′	Sukkur	Town, Sukkur District	27° 42′ 68° 52′
Shangla Pass	Swat/Kohistan border	34° 54′ 72° 36′	Suleimanki	Barrage on Sutlej river	30° 24′ 73° 52′
Shankar Garh	Astor, Gilgit Agency	35° 02′ 74° 56′	Sumbak Range	South-east of Kirthar Range, Dadu District	25° 25′ 67° 55′
Sharan Forest Rest House	Hazara District	34° 43′ 73° 20′			26° 10′
			Sumbal Pani	Kala Chitta	33° 37′ 72° 23′
Sharig	Sibi District	30° 11′ 67° 41′	Sunari	Lake in Sanghar District	26° 02′ 69° 07′
Shastun	Iranian Baluchistan	27° 22′ 62° 20′	Suntsar	South-west Makran	25° 30′ 62° 00′
Shekhupura	Shekhupura District	31° 43′ 73° 59′	Surab or Sorab	Southern Kalat	28° 29′ 66° 15′
Shenkar Garh	See Shankar Garh		Sui	Bugti, Dera Ghazi Khan border	28° 37′ 69° 17′
Shigar	Baltistan	35° 25′ 75° 44′			
Shikarpur	Formerly Jacobabad District, now in Shikarpur District	27° 57′ 68° 38′	Surjana Hills See Soorjana	Sind Kohistan	
			Surkhab Lora or River	Pishin District	30° 34′ 67° 20′
Shimshal	Hunza, Gilgit Agency	36° 27′ 75° 18′			
Shingar or Shinghar Range	Hills in Zhob District	31° 20′ 69° 45′ 31° 40′	Tajjal Rest House	East Nara	26° 52′ 68° 59′
			Takatu or Takhatu	Mountain, Quetta District	30° 23′ 67° 07′
Shinghai Gah Nullah	Gilgit	35° 51′ 74° 17′	Takht-i-Suleiman		
Shinwari	On Afghan, Khyber border	33° 50′ 70° 50′	Takht-i-Sulaiman	Boundary, Baluchistan and Dera Ismail Khan	31° 36′ 69° 59′
Shishpar Nullah	Hunza, Gilgit Agency	36° 50′ 74° 55′			
Shogran	Hazara District	34° 37′ 73° 28′	Takkar	Limestone outcrop in Khairpur State	27° 15′ 68° 49′
Shojh or Shoj	Hazara District, Kaghan valley	34° 53′ 73° 38′	Talaganj	Salt Range, Attock District	32° 56′ 72° 25′
Shorab	Central Kalat, see Sorab				
Shorkot	South of Jhang	30° 50′ 72° 04′	Tando Masti Khan	South of Khairpur	27° 26′ 68° 39′
Shost	North-east Chitral on the Yarkhun river	36° 45′ 72° 57′	Tando Mohammed Khan	Hyderabad District	25° 07′ 68° 31′
Sialkot	Town, Sialkot District	32° 30′ 74° 32′			
Sibi	Town, Sibi District	29° 33′ 67° 54′	Tank or Tonk	Dera Ismail Khan, South Waziristan	32° 13′ 70° 22′
Sidhnai	On Ravi river, Multan District	30° 35′ 72° 04′	Tank Zam Valley	South Waziristan	32° 19′ 70° 13′
Sikaram or Sitaram	Peak, Safed Koh Range, Kurram	34° 01′ 69° 58′	Tarbat	See Turbat	25° 59′ 63° 05′
Singal or Singhal	On Gilgit river, Gilgit	36° 07′ 73° 54′	Tarbela Dam	Mardan and Hazara boundary	34° 07′ 72° 49′
Siran Nullah	Hazara District	34° 06′ 72° 54′ 34° 37′ 73° 15′	Tarinda Mohammed Puna	Muzaffargarh District, not located on map	29° 30′ 71° 02′
Siranda Lake	Lasbela District	25° 32′ 66° 38′	Tarki or Traki	Salt Range, Jhelum District	33° 03′ 73° 26′
Sir Creek		24° 15′ 68° 16′	Tarli Serai	Northern Kaghan valley	34° 58′ 73° 56′
Sitaram Peak	Safed Koh Range	34° 01′ 69° 58′	Taror Dhand (Lake)	South of Rohri	27° 33′ 69° 02′
Skardu	Baltistan	35° 17′ 75° 38′	Tatepur	Multan District	30° 13′ 71° 39′
Soan Valley	Salt Range	32° 37′ 72° 12′	Tatta or Thatta	Town, Thatta District	24° 45′ 67° 56′
Sodhi	Salt Range, Sargodha District	32° 35′ 72° 17′	Taunsa	Barrage, Muzaffargarh and Dera Ghazi Khan Boundary	30° 42′ 70° 46′
Sohan Stream	Salt Range and Potohar Punjab	33° 02′ 71° 43′ 33° 35′ 73° 07′	Taxila	Rawalpindi District	33° 43′ 72° 49′
Sohtagan	Western Chagai	28° 03′ 62° 50′	Teep Valley	North-east Swat	35° 17′ 72° 43′
Sonmiani	Lasbela	25° 25′ 66° 35′	Teru	Pass, Chitral/Gilgit boundary	36° 13′ 72° 44′
Sonmiani Lagoon	Lasbela coast	25° 20′ 66° 30′	Thal (Thall in older books)	Western border of Kohat District	33° 22′ 70° 33′
Soonari Lake	See Sunari				
Soorjana Hills See Surjana Hills	Sind Kohistan	25° 10′ 67° 45′ 25° 24′ 67° 55′	Thano Bula Khan or Thana Bula Khan	Dadu District	25° 21′ 67° 51′
Sorab	See Surab		Thandiani	Hazara District	34° 14′ 73° 22′
Sost	Village in northern Hunza	36° 42′ 74° 49′	Thatta	See Tatta	
Spezand	Northern Kalat	29° 58′ 66° 55′	Tirich Mir	peak Chitral	36° 14′ 71° 49′
Sujawal or Sojawal Town	Thatta District	24° 36′ 68° 05′			
Sujawal Bridge	On Indus	24° 40′ 67° 58′			

Name	Approximate Location	Coordinates Lat.(N) Lon.(E)		Name	Approximate Location	Coordinates Lat.(N) Lon.(E)	
Toba or				Wah	Attock District	33°48′	72°43′
Toba Kakar	Range of hills, north-western			Walhar	Southern Rahimyar Khan		
	Baluchistan	30°55′	67°00′		District	28°11′	69°59′
		31°15′	69°00′	Wali Creek	Indus delta, west of Haideri		
Toba Tek Singh	Faisalabad District	30°57′	72°28′		creek	24°03′	67°30′
Torghar Range	East of Zhob (formerly Fort			Wam	Northern Sibi District	30°27′	67°43′
	Sandeman)	31°08′	69°40′	Wana or Wano	South Waziristan	32°18′	69°34′
		31°50′	69°55′	Warah	Larkana District	27°26′	67°48′
Tori Rest House	Jacobabad District	28°09′	69°05′	Warsak dam	Mohmand Agency	34°09′	71°25′
Trakama Pass	Hazara District Azad			Wassowari Lake	West of Jamrao Head	26°36′	68°50′
	Kashmir border	34°21′	73°26′	Wazirabad	Gujranwala District	32°24′	74°08′
Traki	See Takri, Salt Range,						
	Jhelum District	33°03′	73°26′	Yarkhun	Nullah and Yarkhun village,		
Tret	Murree foothills	33°51′	73°19′		northern Chitral	36°47′	73°02′
Trimmu	Jhang District, barrage on			Yakhmach	Western Chagai, Baluchistan	28°45′	63°52′
	Chenab river	31°09′	72°08′	Yakh Tangai	Swat	34°55′	72°38′
Tseri Village	Baltistan	35°32′	75°28′	Yasin	North-western Gilgit	36°22′	73°20′
Turbat	See Tarbat, south-western			Yazman	Cholistan	29°08′	71°46′
	Makran	25°59′	63°05′	Yusuf Khel	Mohmand Agency	34°24′	71°22′
Turikho Gol or							
Valley	Northern Chitral	36°24′	72°23′	Zahirpir	Rahimyar Khan District	28°48′	70°32′
		36°33′	72°28′	Zambaza Range	Hills, Zhob District	31°03′	69°24′
Uch or Uch Sharif	Bahawalpur District	29°14′	71°04′	Zangi Nawar			
Uchchali	Lake, Mianwali District,			Lake	Chagai, West of Nushki	29°25′	65°47′
	Salt Range	32°33′	72°01′	Zarghun	Peak, Quetta and Sibi		
Umarkot or					boundary	30°13′	67°18′
Umerkot	Tharparkar	25°22′	69°44′	Zhob River	North-east Baluchistan	31°32′	69°10′
Urak or Uruk	Valley, Quetta District	30°16′	67°11′				69°50′
Ushu	Swat Kohistan	35°32′	72°39′	Zhob Town			
Uthal	Lasbela	25°47′	66°37′	(Formerly			
Utrot or Utror	Swat Kohistan	35°29′	72°27′	Fort			
Utzu	Southern Chitral	35°30′	71°40′	Sandeman)	North-east Baluchistan	31°21′	69°28′
				Ziarat	Northern Sibi, Baluchistan	30°22′	67°44′
Vihari or Vehari	Multan District	30°03′	72°22′	Ziarat	South-eastern Chitral	35°23′	71°47′
Virahwah	Extreme south-east			Zizri	Sibi District, Baluchistan	30°17′	67°42′
	Tharparkar District	24°31′	70°47′				

Index